Springer Textbooks in Earth Sciences, Geography and Environment

The Springer Textbooks series publishes a broad portfolio of textbooks on Earth Sciences, Geography and Environmental Science. Springer textbooks provide comprehensive introductions as well as in-depth knowledge for advanced studies. A clear, reader-friendly layout and features such as end-of-chapter summaries, work examples, exercises, and glossaries help the reader to access the subject. Springer textbooks are essential for students, researchers and applied scientists.

More information about this series at http://www.springer.com/series/15201

Francisco Castro Rego • Penelope Morgan
Paulo Fernandes • Chad Hoffman

Fire Science

From Chemistry to Landscape Management

Springer

Francisco Castro Rego
Instituto Superior de Agronomia
Centro de Ecologia Aplicada Prof. Baeta Neves
Universidade de Lisboa
Lisbon, Portugal

Penelope Morgan
Department of Forest, Rangeland, and Fire Sciences
University of Idaho
Moscow, ID, USA

Paulo Fernandes
Departamento de Ciências Florestais e Arquitetura Paisagista
Universidade de Trás-os-Montes e Alto Douro
Vila Real, Portugal

Chad Hoffman
Department of Forestry and Rangeland Stewardship
Colorado State University
Fort Collins, CO, USA

ISSN 2510-1307 ISSN 2510-1315 (electronic)
Springer Textbooks in Earth Sciences, Geography and Environment
ISBN 978-3-030-69817-1 ISBN 978-3-030-69815-7 (eBook)
https://doi.org/10.1007/978-3-030-69815-7

This Springer imprint is published by the registered company Springer Nature Switzerland AG
The registered company address is: Gewerbestrasse 11, 6330 Cham, Switzerland

To Will

Foreword I

The field of wildland fire science is more complex than one may think. Not only must one know the intimate details of combustion science and fire behavior, but that knowledge must then be interpreted in the context of fire ecology and wildland fuel dynamics. Many believe that this extremely wide continuum of topics needed to understand wildland fire sciences only spans chemistry, combustion physics, heat transfer, fuel moisture dynamics, smoke transport, emissions, climate, weather, and wind for fire behavior. But, the really important topics critical to wildland fire science are those that cover the full breadth of ecology, such as phenology, eco-physiology, and morphology. It is fire ecology that gives us the deep understanding needed for solving the myriad problems in fire behavior and subsequently fire management. To further complicate matters, each of these diverse wildland fire disciplines, whether it be fire behavior or ecology, includes intrinsic hierarchically nested time and space scales that overlap and interact. Fuel moisture, for example, depends on both knowledge of the water diffusion process (dead biomass) and plant phenology (live biomass) at the plant, community, stand, and landscape scales.

I have studied fire and its ecology for over 30 years at the Missoula Fire Sciences Laboratory in Montana, USA, and I have found that this "soup to nuts" coverage of wildland fire science is rare in the diverse books that have been written on the subject. Too often, the complexity of fire science precludes comprehensive coverage of all relevant topics; most authors tend to cover only those areas in which they have the greatest knowledge or the topics that are the most studied. Some books, for example, are quite detailed in their description of fire physics but totally over-generalize wildland fuel dynamics. Others focus on fire ecology without a comprehensive coverage of the elements of fire behavior. Some fire behavior books tend to focus only on the physics of combustion and ignore fuel moisture dynamics. This is important because over-generalizing some wildland fire science topics breeds bias in scientific study and fire management approaches. Wildland fuels, for example, were historically described using fire behavior fuel models—an abstraction supposed to represent expected fire behavior. Today, we use more comprehensive classifications of fuels to describe and communicate, and to use as inputs to models.

This textbook, in three parts, appears one of the few that provides the broad spectrum of coverage across all subjects that are needed to fully understand the dynamics of wildland fire science. In the introduction, the authors provide a detailed overview of wildland fire science that introduces the reader to the book and its subjects. Then in Part I, we learn of fine-scale dynamics of fire chemistry and the reactions that lead to ignition processes, especially the concept of flammability (Chap. 1). I especially liked the addition of moisture dynamics related to ignition processes. Combustion processes are covered from the breaking of chemical bonds to the heat of combustion along various pyrolysis pathways. In the next chapter (Chap. 2) we learn about how this chemistry and pyrolysis eventually result in smoke, and in Chap. 3 we learn how the breaking of chemical bonds generates heat, specifically the heat of combustion, along the pyrolysis highways. I especially was grateful for the coverage of char creation in combustion because char is now a major carbon pathway in fire ecology. Then in Chap. 4 we learn of the pre-ignition processes, requirements for ignition, and the breakdown of fuels to release the volatile gases using a wonderful example. And last, in Chap. 5, we move out of the sphere of chemistry and on to physics and heat transfer, where the four primary modes are introduced—convection, radiation, conduction, and mass transport—replete with equations, interpretations, and examples. Readers should have a wonderful grasp on the mechanistic processes that govern combustion science.

In Part II, we move up in scale from the fine-scale dynamics of combustion to fuels, fire behavior, and fire effects. The descriptors of fuels and fire behavior are covered in Chap. 6 including the components of a fuelbed, physical fuel characteristics, and fuel classifications. A mélange of fuel characteristics are presented but most are used for fire behavior prediction. Missing are the important ecological processes that control these characteristics, but they are described in later chapters. The five phases of fire spread are then discussed in Chap. 7 from fire geometry to acceleration. Fire propagation has always been important to fire management because of its influence on fire safety, but it is also one of the most difficult to predict of all fire behavior processes. The chapter also includes fire growth and two-dimensional spread. Then, unlike many other wildland fire books, there is a chapter on extreme fire behavior (Chap. 8), which is becoming incredibly important in fire management because it results in our greatest social and biological impacts. There is a wonderful treatment of large fires and their consequences. Crown fires are covered in this chapter in great detail with great reverence to Charlie van Wagner, the pope of crown fire behavior and effects. I especially like the inclusion of canopy moisture dynamics in crown fire spread and the thorough treatment of firebrands and embers and their lofting and dispersal, along with other major topics in this exhaustive chapter. Since I am a fire ecologist, I was finally rewarded with a delightful chapter (Chap. 9) on fire effects on plants, soils, and animals. A boatload of information was synthesized in this chapter from fire effects on plant stems, seeds, and roots; a comprehensive review of fire and soil heating; and animal responses to fire. Another novel addition in this book is the coverage of contemporary society and its role in fire science (Chap. 10). Protecting people and homes from heat and smoke is the primary thrust of this chapter, including such diverse topics of safe zones, smoke inhalation, the wildland-urban interface, and impacts on ecosystem services. I appreciated the juxtaposition of the costs and effects of fire suppression vs. benefits gained from society by suppression. Societal and ecosystem resilience to fire impacts is also presented in this chapter.

In Part III, we learn about wildland fire science at landscape scales and how to manage fuels and fire across all relevant scales. In Chap. 11, we find out what drives fuel dynamics and how these fuels can be appropriately managed over time. It is in Chap. 12 where this book separates itself from most of the others. Fire regimes and their measurements are covered first; fire regimes are spatio-temporal expressions of fire and many feel that this is a very complicated subject because fire regimes are created by the complex interactions of vegetation, fuels, climate, topography, and ignitions. The authors delved into ways that fire regime information can be integrated into management using concepts of historical ranges of variability, refugia, and self-organized criticality. This is very useful information for planning and implementing various treatments, which are then covered in the next chapters. Integrated landscape fire management is discussed in detail in Chap. 13 using various examples and case studies. Objectives for integrated fire management are covered (e.g., biodiversity, fuel reduction) along with the tools needed for implementing actions (e.g., prescribed fires).

The book ends with a look to the future. Chapter 14 covers climate change, global change, development, exotics, and a variety of other issues facing fire management. Included in this chapter is a very timely treatment of the Australian fire season of 2019–2020 and its importance to global fire. Also covered in this chapter are nascent technologies that will serve to advance fire science across the globe including new remote sensing platforms, LiDAR, wireless networks, big data, and simulation models. New education and integrated approaches to fire science are also discussed.

There are four underlying themes embedded in this book that, in my opinion, set it apart from many other fire books. First is the concept of integration—everything in fire science must be integrated to reach the fullest potential from designing fuel treatments to building models, to managing fire. This book does a great job of emphasizing integration in fire sciences. Next is the concept of scale. Scale issues are uniquely highlighted across all pages of this book from chemical bonds to heat transfer to fuel ignition to fire spread to stand conditions to landscape processes. Third, the authors have integrated the anthropologic influences on wildland fire science including human ignitions, health concerns, community resilience, and historical peoples and their use of fire. And last is the concept of "learn by doing," where there are useful examples, case studies, and, most importantly, spreadsheets to understand the content of this book. I thoroughly enjoyed this book and feel it is destined to be a solid reference for wildland fire scientists and managers and a textbook for all students studying fire.

Missoula, MT, USA Bob Keane
April 2021

Foreword II

Fire management practice has been dominated by developing country reactions to fires, with a focus on damaging wildfires. This has stimulated and established a long and deep body of research and operational practice that middle income and wealthy nations have benefitted from. Bringing that knowledge together in one place is an enormous and sustained effort, undertaken by four highly experienced and esteemed members of the fire community. It is timely and very very welcome.

As reflected in this book, fires have been used by humans for millennia and play a critical role in many ecosystems. People are the cause of ~90% of fires globally through a combination of limited access to alternative approaches to fire, poor practice, accidents, weak understanding of fire risk, machinery, negligence, and carelessness. In developing countries the dominant factor in fires starting is the need to use fire as a tool where there are no viable alternatives to the use of fire for hunting, favoring preferred plants for food or fiber, clearing for agriculture and grazing, easing travel, and controlling pests, all of which continue today.

For that context, and also for developed countries, where fires are damaging, they are a landscape problem. They are not a problem resulting only from insufficient or inadequate means of suppression but also from the situation of fuel continuity and accumulation of fuels from vegetation, human activities, and sources of ignition. The solution is resilient landscapes that balance the hazards, reduce fire risks, and can be sustained. This is Integrated Fire Management, a key chapter in this book. Underpinning that is the need to understand fire, fuels, and landscapes starting from the fundamentals; something this book provides well.

Key to successfully integrating ecology, society, and fire management with methods and technologies is an effective analysis of the situation. What is the ecological role and impact of fire; the social, cultural, and economic context; who is starting fires and why; what are the characteristics of the fuels in the area and how does fire behave in them under different burning conditions; and what other factors or threats are exacerbating the fire problem, such as land tenure, illegal logging, invasive species, or climate change?

In working and collaborating with colleagues, agencies, students, and interested fire and land management staff in developing countries on all continents, there has always been a problem when they request texts, information, and guidance to improve their understanding of fire management.

Sound practice in fire management has been well documented but mainly focused on descriptions of needs, requirements, and approaches for readiness to fight fires and fire suppression that date from 1953 by FAO. There are reference books on fire management that present the thinking and approach to a suite of fire management topics. The material that underpins a sound understanding of fire management tended to be scattered among key texts, and studying it required access to a series of books, most of which pre-date the digital era and nearly all of which are out of print. Not all topics were set down well or completely; where the physical was well covered the ecology may not be and where ecology was covered people were not.

Obtaining the knowledge of science, process, and systems that provide the basis for this sort of analysis is not simple or easy and requires access and time to a multitude of reference materials and the time to process them. This book brings this together, first fire as a chemical and physical process; then fuels, fire behavior, and effects followed by managing fuels, fires, and landscapes. That the chapters set out learning outcomes and in many cases are accompanied by interactive spreadsheets reinforces the concepts and deepens understanding of processes, inputs, and outcomes.

Having these four experienced fire sector actors create that all in one place is a wonderful contribution. I very much look forward to directing interested parties to a single volume that covers chemistry, physics, fuels, fire behavior, managing fuels, and landscapes, rather than a pile of sections in multiple texts.

My congratulations to Francisco Castro Rego, Penelope Morgan, Paulo Fernandes, and Chad Hoffman on the book and deep thanks for its preparation. It will have an important place in informing and educating fire managers, students, and interested individuals with particular value for those in developing countries.

Rome, Italy
April 2021

Peter Moore

Acknowledgments

This book has been in development since 2008, when Francisco Castro Rego visited the University of Idaho and taught a course in wildland fire behavior. It is impossible for us to thank everyone who has provided inspiration, support, and encouragement. However, we give a very special thank you to two groups of people. To our colleagues and mentors, we thank you for the numerous discussions over the years, many of which have, in one form or another, found their way into these pages. We thank the students, postdoctoral researchers, and fire managers we have worked with. Your endless enthusiasm and interest in wildland fire science and management over the years have truly inspired us. May you all thrive as you continue to lead and shape how people think about, manage, and understand fire.

We owe an enormous debt of gratitude to Sarah Bunting, who tirelessly worked on obtaining and tracking copyright permissions, formatting figures, and references, reorganizing tables, and many other book details. Most importantly, she brought her lightness of being and immense kindness to our team.

We are especially grateful to the 22 authors of the 12 case studies for their willingness and efforts. The case studies greatly enrich the ideas in this book.

We also thank Kari Greer for allowing us to use her photographs. Kari is often photographing people and land as large fires burn in the western USA, and in so doing, she communicates the awe, hard work, and professionalism fire people bring to their work. Her photographs bring beauty and insights to our pages. We thank all the researchers and journals who granted us permission to use their photographs and figures. Both Heather Heward and Catarina Sequeira drew or adapted figures for us. We thank the Instituto Superior de Agronomia and the University of Trás-os-Montes e Alto Douro for supporting some expenses associated with the production of the book. We are also grateful to Bob Keane and Peter Moore, who gave their time and insights to help us clarify concepts and generally improve this book. We are humbled by what they wrote in the forewords. Moreira da Silva, Harold Biswell, Jan van Wagtendonk, Stephen Bunting, Leon Neuenschwander, and other early proponents of applied fire science, prescribed burning, and fire ecology have inspired our work. We honor the many practitioners and keepers of fire and fire culture, past, present, and future.

We have learned that it is difficult to write a book. It is much harder than any of us anticipated. Thus, we confirmed the hypothesis put forth by Joseph Epstein that "it is better to have written a book than to actually be writing one." With this in mind, we cannot thank our families enough for their support and encouragement during the writing of our book, even on nights and weekends.

Introduction

Photograph by Kari Greer

Why We Wrote This Book

We are fascinated by fires, their power, and their beauty, and by peoples' attitudes about fires. We love to teach about fires and to learn about fires from the observing fires themselves, and the many conversations, science, and stories about fires. Like all humans, we are fire people.

Globally, almost every place has a fire history, reflecting the fire environment in which we live. Globally, many fires have and will occur. Some will be large, and both flames and smoke will affect people. Fires shaped and will continue to shape ecosystems, with substantial implications for people and nature. Science can help inform the challenging complexities of fire-related issues today, including global change, escalating firefighting costs, threats to people and property, ecological values, and impacts of fire. As we adapt, we will shape future fires and smoke, as well as the ecosystems and ecosystem services that are influenced by fires. We must grapple with the ecological imperative of fire and manage ways to live with and use fire.

Fire is a good servant and a bad master (one of our favorite proverbs). Because fire has an essential and pervasive influence in forests, woodlands, shrublands, and grasslands, many plants and animals have evolved with fire. Many species are dependent on fire or similar disturbances to survive and thrive. Fires have so shaped vegetation that Bond and Keeley (2005) described fires as global herbivores in many ecosystems, for the vegetation biomass present is far less than the biomass expected based on climate and site productivity alone. Fires have shaped the structure, composition, and diversity of vegetation (Fig. 1). The ecological roles of fire include rejuvenating habitats by consuming live and dead vegetation, releasing nutrients and space for new growth, and favoring some plants and other life over others, adding to landscape composition. Because fires burn differently from place to place, fires can foster biodiversity at multiple scales. Thus, fire is central to life on Earth (Fig. 2). However, fires can also threaten people, their property, and ecosystem services they value. Ever since humans first used fire, humans have used fire to manage vegetation and for many other cultural purposes. Humans can use and manage fires to get more of the positive benefits and fewer negative outcomes when areas burn.

Fire is many things. For us, fire is the manifestation of coupled human-natural systems. For example, where there is fire there is smoke. The particulates in smoke pose a health hazard to people while also affecting visibility. Concerns about protecting people and property from fires are often disincentives to reintroducing fire and yet our path forward needs to embrace fire to create resilience and lower fuel loads and potential future smoke impacts. It is a paradox that the more we suppress fires, the more intensely the next fire will likely burn as fuels accumulate, and the subsequent fires will likely produce more particulates. Another example is how fire is often an agent of climate change. Climate influences many aspects of where and how fires burn, and then how vegetation responds, and in some areas that have recently become too hot and dry, forests are not regenerating after fires. Without fire, the transition from forest to shrublands, woodlands, or grasslands might happen more slowly. However, many trees and other plants are more likely to survive fires if the area burned in a prior fire that thinned vegetation, consumed biomass and fuels, and stimulated plants and nutrient cycles.

Fire science has made major strides forward in the last few decades—just in time to face the many fire challenges and opportunities ahead. We draw upon the many different ideas from global fire science. We have learned much about fire behavior,

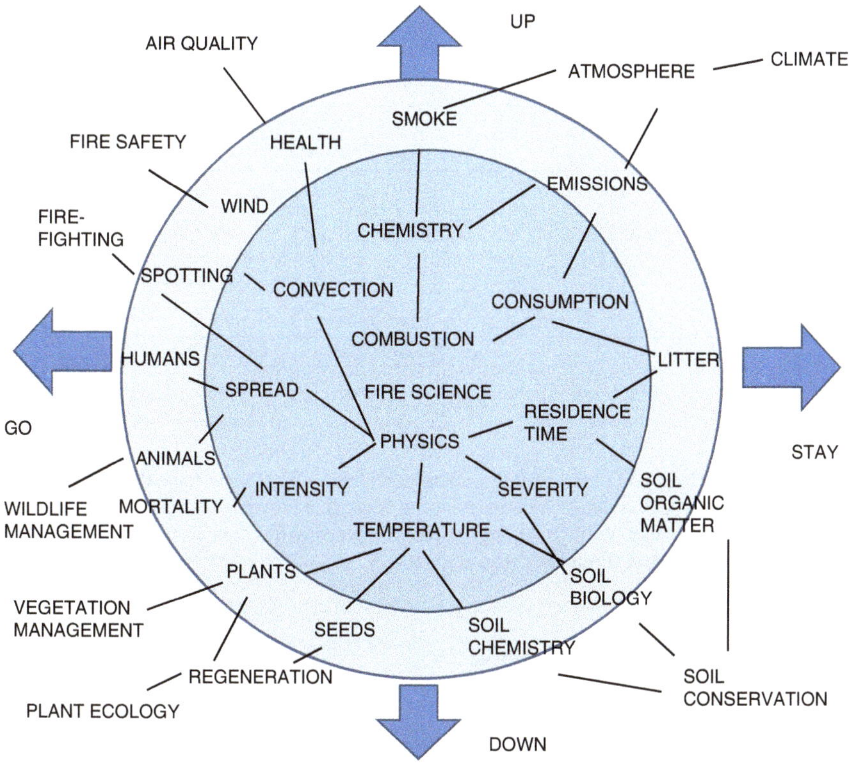

Fig. 1 Fire has implications across multiple scales in time and space, for the heat of fire travels up into the air, down into the earth, and across space and time to influence almost all terrestrial landscapes. Certainly fire is central to vegetation management and to many of the ecosystem services people care about. Fire is part of social-ecological systems

more recently about the ecological effects of fire, and even more recently how fire and smoke affect people and respond to social, political, and economic conditions. Still, we struggle to clearly connect fire behavior to fire effects. Part of the challenge is fire metrology (Kremens et al. 2010), as we seldom have spatially and temporally coincident measurements of before, during, and after fires. Without that, fire ecology is severely limited because we often don't know what was present before fires and how they burned and how that legacy influences the vegetation response that fire ecologists are trying to understand and predict.

Humans are fire species. We have used fire almost as long as humans have been able to walk (Fig. 2). Humans need fire. At the same time, people have changed both fires and the environment greatly, enough so that we are now in the Anthropocene. Many fires are burning in novel environmental conditions.

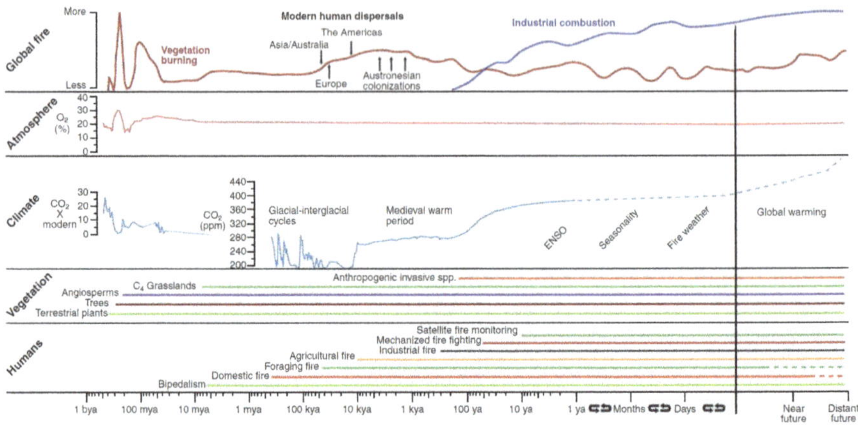

Fig. 2 Fires have been part of the Earth's system for millennia. Fires have shaped and responded to the atmospheric oxygen essential to combustion, and vegetation provides the fuels when the climate is conducive to fire spread. While lightning ignites fires, globally it is humans who shape when and where and how fires burn. (From Bowman et al. 2009)

How This Book Is Organized

This book has three parts. Each of these can be read independently of the others, but they are designed to build logically from one to the other. Thus, the book is organized around the fire triangles at different scales (Fig. 3).

- Part I focuses on combustion and heat transfer processes. The first two chapters are chemistry-based, as they cover the chemical conditions required for ignition and combustion as a chemical reaction with implications for smoke. The following three chapters focus on the production of heat through combustion, the heat required for pre-ignition and flames, and the physical processes by which that heat transfers away from fires.
- Part II addresses fuels, fire behavior, and effects. We start with fuel and fire behavior descriptors and then address how fires propagate. What makes some fires extreme? We address fire effects on plants, soils, and animals in ecosystems from all over the Earth. The ways fire affects people and people affect fires are covered in one chapter. By necessity, the chapters in Part II are longer and have fewer equations. We also emphasize more applications and implications.
- In Part III, we address managing fuels, fires, and landscapes. This part begins with a chapter on fuels dynamics and management. Our discussion of fire regimes and landscape dynamics leads us to examples of landscape management. We cover integrated fire management with examples from all around the world. We use eight global success stories as case studies, and then we discuss lessons learned. In our last chapter, we discuss the trends globally that are influencing fires and

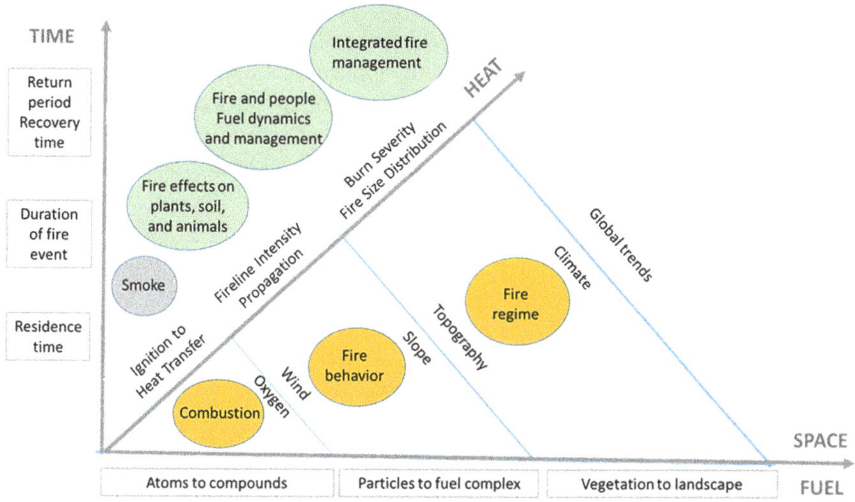

Fig. 3 Fire triangles illustrate how this book is organized. Different temporal and spatial scales are associated with different fire and heat variables. These triangles are nested within each other showing that all fire characteristics are related and interdependent

their effects now and in the future. We especially focus on social trends, fire science trends, and the implications for education and training about fires. These trends all pose both challenges and opportunities for the future.

Every chapter includes learning outcomes to help guide you. All of the references cited in the text are listed at the end of each chapter. In lieu of a glossary, terms are defined in the text, often with examples that provide valuable context for understanding. The index will guide readers to where key terms are discussed in detail and used in examples.

Our book is about both fire behavior and fire effects AND how they are linked. We hope it will inform integrated fire management. This is not a book about firefighting and suppression and their related activities such as ignition control, fire detection, or the organization of fire management. We include prescribed burning and other fire management strategies, as well as relevant social science. The straightforward learning path incorporates multiple fire models and links them to historical fires as case studies, and practical, current applications. Our approach is strongly process-based, comprehensive, and quantitative. We provide interactive spreadsheets that our students have found valuable for learning and understanding without requiring that students program the models. Change the inputs and immediately see how the graphical display of the outputs changes. This problem-solving approach with practical applications makes the science approachable. Case studies and many examples enable learners at multiple levels. We envision this book as useful to students in undergraduate courses at the middle to upper levels, in graduate classes, and for learners who are practitioners. In short, this book is for anyone who wants to learn more about wildland fires. Parts can be readily incorporated into professional training courses. Indeed, we draw upon all we have learned from and with practitioners of fire. Throughout this book, we emphasize the logical progression of ideas. We quickly build from concepts to engaging applications from around the world to ground the concepts in reality. We are honest about the many challenges and heartened by the many different solutions people have developed. The text is designed to be engaging, highly relevant to students and practitioners of wildland fire science and applications, and readable by any student of fire.

Why We Need to Live With and Use Fire Now

Society faces many fire challenges today and in the future, including many large fires that threaten people and property, smoke that poses a health hazard, and fire seasons around the globe that are almost 20% longer than they were three decades ago. At the same time, fire is used to suppress fires, and in the conservation of wildlife habitat, livestock grazing, fuels management, and ecological restoration. To live with fire as a good servant, rather than a bad master, we must understand it. Understanding is critical to forecasting the implications of global change for fires and their effects. Fire science has burgeoned and technology is rapidly changing, and we will need to adapt as the world itself is rapidly changing.

Addressing the fire challenges of today and tomorrow will require engaging with fire. This is more than living with or coexisting with fire, for we must also embrace fire in all the complex ways in which fire, people, and environments interact. Throughout our book, we emphasize concepts and encourage critical thinking central to being innovative and effective.

As we were writing this book, wildfires were burning worldwide with consequences that should concern all people. Examples follow. With 20 million ha burned in Siberia (Alberts 2020), climate change is favoring fires that in turn advance changing climate in this and other far northern regions. Some of the 18.6 million ha burned in Australia in 2020 (See section 14.2.3) burned in ecosystems not well adapted to fires, suggesting that fires will burn in novel ways in novel places in the future. In the Amazon and elsewhere, land use and people have contributed to extensive and numerous fires that have threatened and displaced many Indigenous people and animals. Many people were evacuated and came back to burned homes when more than 2.3 million ha burned in 2020 in California, Oregon, and Washington (Whang and Maggiocomo 2020) with more elsewhere in the western USA. Many millions of hectares burned in Indonesia with smoke visible from space. The smoke from these and other fires blanketed cities and threatened peoples' health. Globally, the fire season has lengthened by almost 20% in the last three decades (Jolly et al. 2015) with warmer droughts. Societal costs are high, including both the direct costs of suppression and the ecosystem services that are affected. The fires highlight and make more imperative the need for society to understand and find ways to live well with fire. Understanding fire behavior and effects well can inform innovative, integrated fire management to increase the positive effects of fire (fuel

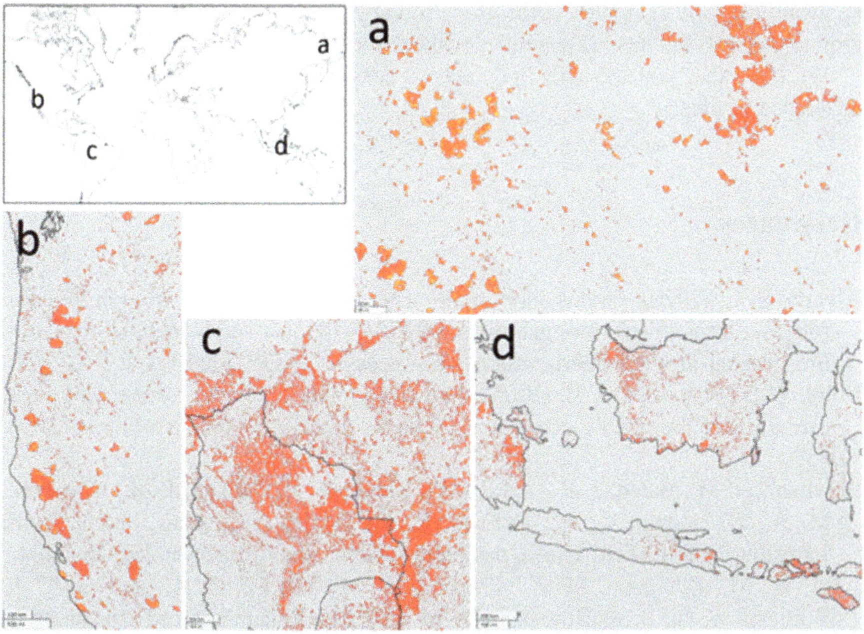

Fig. 4 Burned areas in (**a**) Siberia, (**b**) California, Washington, and Oregon in the USA, (**c**) the Amazonian region of Bolivia, Brazil, and Paraguay, and (**d**) Indonesia between October 2019 and October 2020. Maps of burned areas (red) interpreted from MODIS and VIIRS NRT satellite imagery as part of the Global Wildfire Information System, https://gwis.jrc.ec.europa.eu/static/gwis_current_situation/public/index.html

reduction, enhanced habitat for many plants and animals, key ecosystem services increased) while reducing the negative impacts of fire (loss of human lives, smoke emissions that threaten human health, carbon emissions that threaten global health, etc.). We hope readers of our book will learn where and how fire can be a good servant rather than a bad master. The ideas in our book become more relevant with every fire and with every discussion about fire and fuels policy.

We wrote this for all of you who are students of fire. Some will seek the basic concepts, while others will read for the linkages between concepts and applications. All of the chapters include examples, with colorful and meaningful figures. We have taught fire science in leading international fire education programs, and we have communicated with fire professionals and other people affected by fires. The core content for this book, including case studies and interactive spreadsheets for illustrating and comparing tools, has been used in teaching applied fire science courses at the University of Idaho and Colorado State University in the USA, and Instituto Superior de Agronomia and Universidade de Trás-os-Montes e Alto Douro, in Lisbon and Vila Real in Portugal, as well as in many training programs for professionals. We draw from experience successfully teaching these key concepts and applications. Our examples come from forests, grasslands, and woodlands from around the world—these case studies make the science relevant and meaningful.

We hope that you too will find beauty, possibility, and the means to make a difference in fires.

References

Alberts, E. C. (2020). Photos show scale of massive fires tearing through Siberian forests. https://news.mongabay.com/2020/07/photos-show-scale-of-massive-fires-tearing-through-siberian-forests/n. Accessed 29 Oct 2020.

Bond, W. J., & Keeley, J. E. (2005). Fire as a global 'herbivore': the ecology and evolution of flammable ecosystems. *Trends in Ecology and Evolution, 20*(7), 387–394.

Bowman, D. M., Balch, J. K., Artaxo, P., Bond, W. J., Carlson, J. M., Cochrane, M. A., D'Antonio, C. M., DeFries, R. S., Doyle, J. C., Harrison, S. P., & Johnston, F. H. (2009). Fire in the Earth system. *Science, 324*(5926), 481–484.

Jolly, W. M., Cochrane, M. A., Freeborn, P. H., Holden, Z. A., Brown, T. J., Williamson, G. J., & Bowman, D. M. (2015). Climate-induced variations in global wildfire danger from 1979 to 2013. *Nature Communications, 6*(1), 1–1.

Kremens, R. L., Smith, A. M., & Dickinson, M. B. (2010). Fire metrology: Current and future directions in physics-based measurements. *Fire Ecology, 6*(1), 13–35.

Whang, O., Maggiocomo, T. (2020). Western wildfires have now burned an area bigger than New Jersey. National Geographic. https://www.nationalgeographic.com/science/2020/09/western-wildfires-have-now-burned-area-bigger-than-new-jersey/. Accessed 29 Oct 2020

Contents

About the Authors

Francisco Castro Rego is Professor and Former Director of Centro de Ecologia Aplicada Prof. Baeta Neves, Instituto Superior de Agronomia in Lisbon, Portugal. Dr. Rego was lead of the Fire Paradox project (http://www.fireparadox.org/) funded by the European Union with 36 partners from 16 countries that advanced fire science and fire policy. Dr. Rego's research spans the ecological effects, behavior, and management of wildfires: http://comparatistas.academia.edu/FranciscoRego. He is currently serving on the commission to investigate and recommend policy changes to address recent large fires in Portugal. He is the lead author of *Applied Landscape Ecology* published in 2019 by Wiley.

Penelope Morgan is Emeritus Professor in the Department of Forest, Rangeland, and Fire Sciences, University of Idaho, Moscow, Idaho, USA. Dr. Morgan's extensive teaching and research are in the ecological effects of wildland fires, with implications for fire management. She is certified as Senior Fire Ecologist and was recently recognized with Lifetime Achievement Award by the Association for Fire Ecology. In 2021, the International Association for Wildland Fire gave her the Ember Award for Fire Science in recognition of sustained excellence. She was deeply honored to contribute thanks to learning from and with many students, collaborators and professionals in fire. May those embers ignite the world!

Paulo Fernandes is Associate Professor in Departamento de Ciências Florestais e Arquitetura Paisagista at Universidade de Trás-os-Montes e Alto Douro, Vila Real, Portugal. Dr. Fernandes is an internationally recognized expert in fuels and fire science. His research focuses on both basic fire science and applications in fire ecology and fire management, especially for wildland fire in forests: https://publons.com/researcher/453867/paulo-m-fernandes/.

Chad Hoffman is Associate Professor in the Department of Forest, Rangeland and Watershed Stewardship, and Co-director of the Western Forest Fire Research Center at Colorado State University, Fort Collins, Colorado, USA. Dr. Hoffman teaches and conducts research related to wildland fire and fuel dynamics. His research lab works on a variety of questions related to how interactions among fire, fuels, topography, and the atmosphere influence fire behavior and effects across spatial and temporal scales using a combination of field, laboratory, and modeling approaches: https://scholar.google.com/citations?user=EEOXuBQAAAAJ&hl.

PENELOPE MORGAN, FRANCISCO REGO, PAULO FERNANDES and CHAD HOFFMAN (LEFT TO RIGHT)

List of Symbols

Roman Symbols

a, a_1, a_2	Empirical coefficients (Coefficients)
A_f	Flame angle (Degrees °, radians)
A_T	Flame tilt angle (Degrees °, radians)
A_p	Fire plume angle (Degrees °, radians)
A	Area, sectional area, projected area (cm^2, m^2, hectare)
AC	Atmospheric stability component for the Haines index (Dimensionless)
a_i	Area of the element i (single fire or single patch) (km^2, hectares, acres)
A_{land}	Area of a landscape (km^2, hectares, acres)
b, b_1, b_2	Empirical coefficients (Coefficients)
b_{fc}	Minimum buoyancy to generate the pyrocumulonimbus (m)
BC	Atmospheric moisture component for the Haines index (Dimensionless)
BA	Stand basal area ($m^2\ ha^{-1}$)
c	Speed of light ($3.00 \times 108\ m\ s^{-1}$)
C	Costs of suppression and pre-suppression (Euros, dollars)
CBD	Canopy bulk density ($kg\ m^{-3}$)
CBH	Canopy base height (m)
CFB	Crown fraction burned (Proportion)
C_d	Drag coefficient (Dimensionless)
C-Haines	C-Haines index (Dimensionless)
C_k	Fraction of the crown killed (Dimensionless, ratio)
C_p	Specific heat capacity at constant pressure ($J\ K^{-1}\ g^{-1}$, $kJ\ K^{-1}\ kg^{-1}$)
C_{pa}	Specific heat capacity of air at constant pressure ($J\ K^{-1}\ g^{-1}$, $kJ\ K^{-1}\ kg^{-1}$)
C_{pd}	Specific heat capacity of the dry fuel ($J\ K^{-1}\ g^{-1}$, $kJ\ K^{-1}\ kg^{-1}$)
C_{pi}	Specific heat capacity of gas i ($J\ K^{-1}\ g^{-1}$, $kJ\ K^{-1}\ kg^{-1}$)

C_{pwl}	Average specific heat capacity of liquid water (J K^{-1} g^{-1}, kJ K^{-1} kg^{-1})
C_{pwv}	Average specific heat capacity of water vapor (J K^{-1} g^{-1}, kJ K^{-1} kg^{-1})
d	Diameter of a cylinder or a fuel particle (m)
D	Flame depth (m)
DBH	Diameter at breast height (1.27 m above ground) (cm)
DC	Drought code (%)
DMC	Duff moisture code (%)
dNBR	differenced Normalized Burn Ratio (Ratio, dimensionless)
DT_{850}	Dewpoint temperature at the atmospheric height of 850 hPa (°C, K)
erf	The complement of the Gauss error function (Function)
E(NVC)	Expected net value change (Euros, dollars)
EFFM	Estimated fine dead surface fuel moisture (%)
EMC	Equilibrium moisture content (%)
EOFR	Equivalent oxygen to fuel ratio (Ratio)
F_{ab}	View factor between surface a and b (Ratio, dimensionless)
FFMC	Fine fuel moisture code (%)
FMC	Fuel moisture content (equivalent to M but in %) (%)
FR	Fire rotation, a measure of fire frequency (Years)
FRP	Fire radiative power (kW, MW)
FSG	Fuel strata gap (m)
g	Gravity acceleration constant (9.8 m s^{-2})
G_i	Mass of gas i per unit mass of fuel (Ratio, dimensionless)
h	Convection heat transfer coefficient (W m^{-2} K^{-1})
h_c	Heat of combustion, the same as low heat of combustion ΔHL (kJ kg^{-1})
h_P	Planck's constant (6.626×10^{-34} J s)
H	Flame height (m)
H_A	Heat release (kJ m^{-2})
Haines	Haines index (Dimensionless)
HDW	Hot-Dry-Windy Index (Dimensionless)
HT	Tree height (m)
I_B	Fireline intensity, Byram's intensity (kW m^{-1})
I'_s	Critical surface fireline intensity (kW m^{-1})
I_{min}	Critical minimum horizontal heat flux (kW m^{-2})
I_p	Propagating heat flux (kW m^{-2}, kJ m^{-2} s^{-1})
I_{p0}	Propagating heat flux for no-wind no-slope condition (kW m^{-2}, kJ m^{-2} s^{-1})
I_R	Reaction intensity, combustion rate, area-fire intensity (W m^{-2}, kW m^{-2})
k	Thermal conductivity of a given material (W m^{-1} K^{-1})
k_B	Boltzmann constant (1.381×10^{-23} J K^{-1})
L	Flame length (m)

LFL	Lower flammability limit (g m^{-3}, %)
LFL_T	Lower flammability limit at temperature T (g m^{-3}, %)
L_v	Latent heat of vaporization (J g^{-1})
m	Mass (g, kg, t or ton)
m/A	Surface density (kg m^{-2}, g cm^{-2})
M	Fuel moisture in relation to fuel dry weight (Ratio [the same as FMC if expressed as %])
M_x	Moisture of extinction (Ratio, %)
MCE	Modified combustion efficiency (Ratio)
MFR	Mass flow rate (kg m^{-2} s^{-1})
MFR_{min}	Critical minimum value for the mass flow rate (kg m^{-2} s^{-1})
MIR	Mid-infrared band reflectance (Ratio, dimensionless)
n	Number (of moles, or fires) (Count)
NBR	Normalized burn ratio (Ratio, dimensionless)
NBR_{offset}	Average NBR in unchanged areas outside the fire perimeter (Ratio, dimensionless)
$NBR_{postfire}$	NBR after fire (Ratio, dimensionless)
$NBR_{prefire}$	NBR before fire (Ratio, dimensionless)
N_c	Convective number (Dimensionless)
NIR	Near Infrared band reflectance (Ratio, dimensionless)
NT	Number of trees (count)
NVC	Net value change (Euros, dollars)
p_i	Probability of fire of intensity i (Dimensionless, probability)
P	Pressure (atm)
P_b	Total radiative power of a black body (kW m^{-2})
P_g	Total radiative power of a grey body (kW m^{-2})
P_{char}	Proportion of char from combustion (Proportion in fuel dry weight)
P_i	Proportion of constituent i in the fuel (Proportion in fuel dry weight)
PFT	Pyrocumulonimbus firepower threshold (GW)
PM	Particulate matter (emission factor) (Grams per 1000 g of fuel)
P_m	Probability of tree mortality (Dimensionless, probability)
PWR	Power of the fire (MW)
q_{cond}	Heat flux by conduction between two surfaces (W m^{-2})
q_{conv}	Convective heat flux (W m^{-2})
q_{rad}	Radiative heat flux received at a surface (W m^{-2}, kW m^{-2})
Q	Heat supply (J g^{-1})
Q_b	Heat for the separation of bound water from the fuel (J g^{-1})
Q_{dig}	Heat to increase the temperature of dry fuel to ignition (J g^{-1})
Q_{ig}	Heat for pre-ignition (J g^{-1})
Q_m	Heat to change fuel moisture to water vapor at ignition temperature (J g^{-1})
Q_{wl}	Heat to change liquid water in fuel to water vapor (J g^{-1})
Q_{wv}	Heat to increase temperature of water vapor from fuel to ignition (J g^{-1})

R	Rate of spread (m s^{-1}, m min^{-1}, m h^{-1}, km h^{-1})
$R(t)$	Rate of spread at time t (m s^{-1}, m min^{-1}, m h^{-1}, km h^{-1})
$R(W)$	Rate of spread when the width of the fire front is W (m s^{-1}, m min^{-1}, m h^{-1}, km h^{-1})
R_{crown}	Crown fire rate of spread (m s^{-1}, m min^{-1}, m h^{-1}, km h^{-1})
$R_{surface}$	Rate of spread of the surface fire (m s^{-1}, m min^{-1}, m h^{-1}, km h^{-1})
R_{global}	Global rate of spread in crown fires (m s^{-1}, m min^{-1}, m h^{-1}, km h^{-1})
R_{min}	Minimum threshold for rate of spread (m s^{-1}, m min^{-1}, m h^{-1}, km h^{-1})
R_s	Steady-state rate of spread (m s^{-1}, m min^{-1}, m h^{-1}, km h^{-1})
$R_s(0)$	Steady-state rate of spread of reference (no-wind, no-slope, or no moisture) (m s^{-1}, m min^{-1}, m h^{-1}, km h^{-1})
$R_s(M)$	Steady-state rate of spread with fuel moisture M (m s^{-1}, m min^{-1}, m h^{-1}, km h^{-1})
$R_s(S)$	Steady-state rate of spread with slope S (m s^{-1}, m min^{-1}, m h^{-1}, km h^{-1})
$R_s(U,S)$	Steady-state rate of spread with wind speed U and slope S (m s^{-1}, m min^{-1}, m h^{-1}, km h^{-1})
$R_s(U_z)$	Steady-state rate of spread with wind speed U_z at height z (m s^{-1}, m min^{-1}, m h^{-1}, km h^{-1})
$R_s(U_{zm})$	Steady-state rate of spread with midflame wind speed U_{zm} (m s^{-1}, m min^{-1}, m h^{-1}, km h^{-1})
RBR	Relativized burn ratio (Ratio, dimensionless)
RdNBR	Relativized differenced normalized burn ratio (Ratio, dimensionless)
RF_{ij}	Response function of resource j to a fire of intensity j (Euros, dollars)
RH	Air relative humidity (%)
S	Slope angle (Degrees °, radians)
S_c	Critical mass flow rate (kg m^{-2} s^{-1})
S_e	Proportion of silica-free minerals (Proportion in fuel dry weight)
S_m	Surface area-to-mass ratio (m^2 kg^{-1})
SD	Spotting distance (m, km)
SL	Firebrand travel during lofting (m, km)
SPI	Standard Precipitation Index (Dimensionless)
t	Time (seconds, minutes, hours, years)
t_B	Burn-out time, reaction time (s, min)
t_c	Critical time for cambium kill (s, min)
t_{ig}	Time for ignition (s, min)
t_l	Duration of lethal heat (s, min)
t_R	Flame residence time (s, min)
T	Temperature (°C, K)
T(x)	Temperature at a distance x (°C, K)
T(x,t)	Temperature at distance x at time t (°C, K)
T_{700}	Temperature at the atmospheric height of 700 hPa (°C, K)
T_{850}	Temperature at the atmospheric height of 850 hPa (°C, K)

T_a	Ambient temperature (°C, K)
T_{ad}	Adiabatic flame temperature (°C, K)
T_i	Temperature inside (°C, K)
T_{ig}	Temperature for pilot-flame Ignition (°C, K)
T_L	Temperature for cell death (°C, K)
T_o	Temperature outside bark (°C, K)
T_v	Temperature of vaporization or boiling point (°C, K)
u (λ, T)	Spectral radiant emittance (W m^{-2} μm^{-1})
UFL	Upper flammability limit (g m^{-3}, %)
UFL_T	Upper flammability limit at temperature T (g m^{-3}, %)
U	Wind velocity (km h^{-1})
U_{10}	Wind velocity at 10 m height (km h^{-1})
U_6	Wind velocity at 6.1 m height (km h^{-1})
U_z	Wind velocity at height z (km h^{-1})
U_{zm}	Midflame wind velocity (km h^{-1})
v_t	Terminal velocity (m s^{-1}, m min^{-1}, m h^{-1}, km h^{-1})
vk	Von Karman constant (0.41)
V	Volume (m^3)
V_e	Entrainment velocity (m s^{-1}, m min^{-1}, m h^{-1}, km h^{-1})
w	Load, amount of fuel available for flaming combustion (kg m^{-2}, t ha^{-1})
w_a	Weight of a cylinder of ambient air (kg m s^{-2}, N)
w_f	Weight of a cylinder inside the flame volume (kg m s^{-2}, N)
w_L	Fuel load of litter (kg m^{-2}, t ha^{-1})
w_{L0}	Initial fuel load (t ha^{-1})
w_{LS}	Maximum (or steady-state) fuel load (t ha^{-1})
w_t	Total fuel consumption (kg m^{-2}, t ha^{-1})
w_z	Net buoyance force (kg m s^{-2}, N)
W	Width (of the fire front) (m)
W_{AF}	Wind adjustment factor (Ratio, dimensionless)
W_z	Vertical velocity, updraft (m s^{-1}, m min^{-1}, m h^{-1}, km h^{-2})
x	Distance (μm, m, km)
z	Height (m)
z_0	Surface roughness (m)
z_d	Zero plane displacement (m)
z_{fc}	Minimum height the plume must rise to generate a thunderstorm (m)
zm	Midflame height (m)
z_{max}	Maximum firebrand height (m)
z_v	Vegetation height (m)

Greek Symbols

α	Thermal diffusivity (m^2 s^{-1}, mm^2 s^{-1})
β	Packing ratio (Dimensionless)
δ	Fuel depth, height (m)

ε	Emissivity (Ratio, dimensionless)
γ	universal gas constant (8.21 10^{-5} m^3 atm K^{-1} mol^{-1})
λ	Wavelength (μm)
ξ	Ratio of propagating heat flux to reaction intensity (Ratio, dimensionless)
ρ	Density of a given material (kg m^{-3})
ρ_a	Density of air (kg m^{-3})
ρ_b	Bulk density (kg m^{-3})
ρ_{be}	Effective bulk density (kg m^{-3})
ρ_i	Density of gases inside the volume of the flame (kg m^{-3})
ρ_p	Particle density (kg m^{-3})
σ	Surface area-to-volume ratio (m^{-1})
σ_{SB}	Stefan-Boltzmann constant (5.67×10^{-11} kW m^{-2} K^{-4})
ΔD	Moisture indicator, a dewpoint depression term (°C, K)
ΔH	Net energy release (enthalpy change) (kJ mol^{-1}, kJ g^{-1})
$\Delta H'_P$	Enthalpy of formation of products (kJ mol^{-1}, kJ g^{-1})
$\Delta H'_R$	Enthalpy of formation of reactants (kJ mol^{-1}, kJ g^{-}1)
ΔH_{char}	Energy available in the char (kJ mol^{-1}, kJ g^{-1})
ΔH_{fuel}	Total energy content of the fuel (heat of combustion) (kJ mol^{-1}, kJ g^{-1})
ΔHH	Higher heat of combustion (kJ mol^{-1}, kJ g^{-1})
ΔHH_i	Higher heat of combustion of substance i (kJ mol^{-1}, kJ g^{-1})
ΔH_i	Net energy release of substance i (kJ mol^{-1}, kJ g^{-1})
ΔHL	Lower heat of combustion (kJ mol^{-1}, kJ g^{-1})
$\Delta H_{volatiles}$	Energy available for volatiles (kJ mol^{-1}, kJ g^{-1})
ΔHY	Heat yield (kJ mol^{-1}, kJ g^{-1})
ΔT	Temperature difference (°C, K)
Δx	Distance between two points (m)

Part I
Combustion and Heat Transfer Processes

Photograph by Kari Greer

Fire is a rapid exothermic chemical reaction, called combustion, between a fuel and an oxidant that results in the release of energy and a variety of chemical products termed emissions. Because this reaction results in the release of energy, we call it an exothermic reaction. Chemical reactions that require heat energy are termed endothermic. When there is an ignition, three elements in the right proportions are necessary for combustion: fuel, heat, and oxygen. For combustion to occur, the

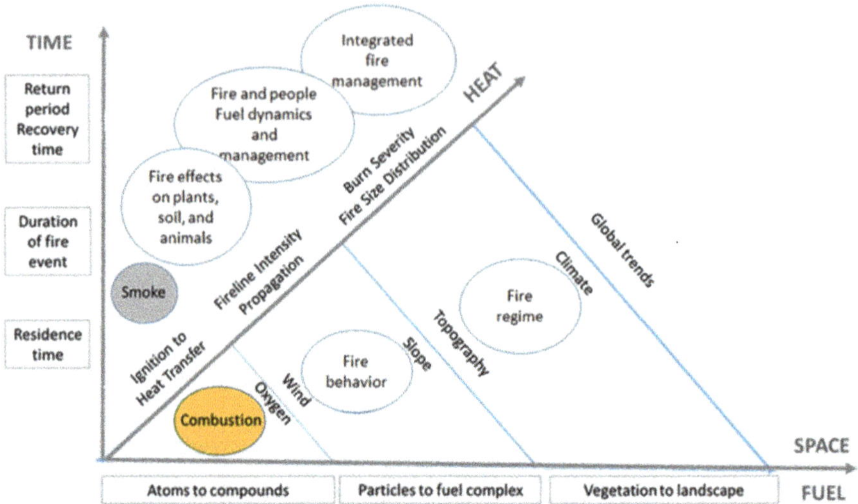

Fig. I.1 The combustion triangle with fuel (here as atoms and compounds), oxygen, and heat, from ignition to heat transfer, which are all necessary elements for fires to burn. This triangle is the first in the nested triangles for other temporal and spatial scales

fuel must be heated to its ignition temperature and mixed with an oxidant at a sufficient concentration. The need for all three elements to occur is commonly represented in the fire triangle (Fig. I.1). The heat released during combustion sustains the combustion reaction.

Michael Faraday presented one of the earliest scientific explanations of the various processes involved in combustion (Fig. I.2). In 1861, he published his course of six lectures on the "Chemical History of a Candle" delivered at the Royal Institution of Great Britain. As Faraday indicated, "there is no better, there is no more open door by which you can enter into the study of natural philosophy than by considering the physical phenomena of a candle. There is not a law under which any part of this universe is governed, which does not come into play, and is not touched upon, in these phenomena."

The main processes occurring during fires are similar to those of a burning candle (Fig. I.3). In a candle's flame, fire is sustained by continuous inputs of air and flammable gases that are produced from the volatilization of solid fuel itself continuously heated by radiation from the flame. During the process, part of the heat generated in combustion is lost, and combustion products go upwards to the atmosphere in convection currents.

In vegetation fires, the combustion processes are similar. Before vegetation fuels can ignite, they must first be dried and be heated enough that thermal degradation occurs (this is called pyrolysis) to release volatile gases. As long as there is enough

Fig. I.2 The British scientist Michael Faraday (1791–1867) author of the book on "The Chemical History of a Candle". (Photograph by John Watkins)

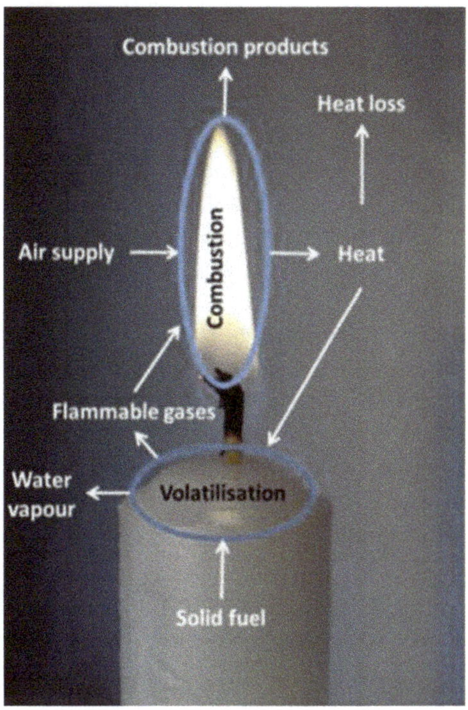

Fig. I.3 Every time you light a candle, you can see the processes of combustion. All of the same processes apply when vegetation burns

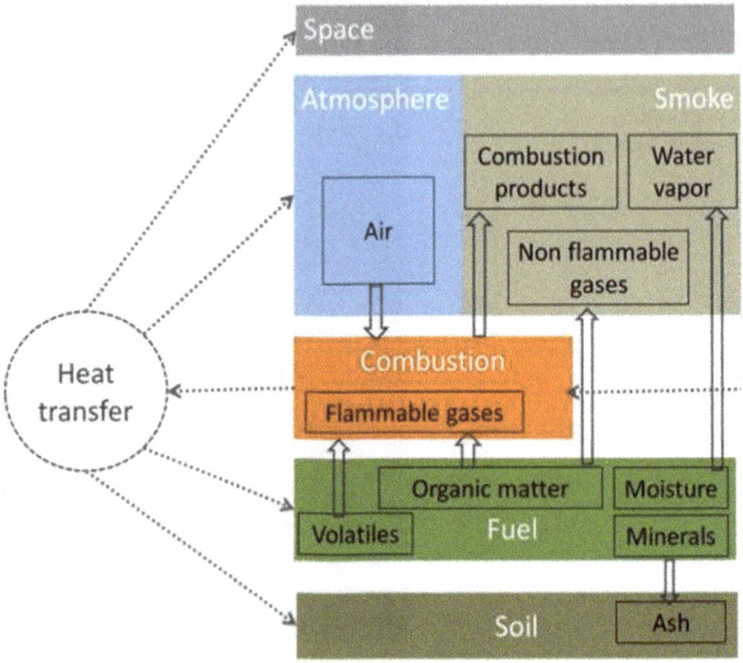

Fig. I.4 Combustion is central to sustained burning. Combustion is initiated by heating fuels until they release flammable gases that can then ignite and burn to release combustion products and heat. Part of the heat generated in the combustion process is used in the preignition of unburned fuels. Heat is also transferred into soil, air, and space

heat, flaming combustion can occur, then smoldering combustion as the oxidation of fuels slows, and eventually extinction. During flaming and smoldering combustion, fuel and oxygen combine in the combustion reaction to generate combustion products, including heat. The main elements and processes occurring in fires are shown in Fig. I.4.

In Chap. 1–5 of our book, *Fire science from chemistry to landscape management,* the subsequent chapters, we use this model (Fig. I.4) to understand the functioning of the system based on the principles of chemistry, conservation of matter, and conservation of energy. We will show:

- the mixtures of flammable gases and air that allow conditions for ignition
- how the chemical composition of fuels and air supply determine the products of combustion, emissions, and smoke
- how heat is produced from the combustion reaction
- what are the heat requirements for pre-ignition and flames
- how heat is transferred from the combustion zone

In each of the following chapters, different processes within this general model are shown and illustrated. All of them are at play in fires, whether small or large.

References

Faraday, M. (1861). In W. Crookes (Ed.), *A course of six lectures on the chemical history of a candle*. London: Griffin, Bohn & Co.

Watkins, J. (1922). *Photograph of Michael Faraday*. Wikimedia Commons. Retrieved May 15, 2020, from https://commons.wikimedia.org/w/index.php?curid=2525521.

Chapter 1
Chemical Conditions for Ignition

Learning Outcomes
Upon completion of this chapter, you will be able to:

1. Understand how pyrolysis results in the production of volatile gases from solid fuels as they are heated,
2. Describe the relative importance of the factors influencing ignitability, including the lower and upper limits of flammability and ignition temperatures, and
3. Understand the difference between ignitability and flammability of wildland fuels.

1.1 What Conditions Are Required for Ignition?

In this chapter, we discuss the ignition of wildland fuels and describe the factors that influence when and how fuels ignite (Fig. 1.1). For a fire to occur, the organic matter must first be converted to flammable gases which then ignite. During this process, fuels are heated enough to drive off moisture so the fuels are dry enough to burn. As fuel is heated, volatile compounds in the fuel become flammable gases, and it is those gases that burn when there is enough oxygen. Thus, we need heat, fuel, and oxygen, the three elements of the Fire Triangle for combustion (See Part II overview). In this chapter, we focus on how fuel chemistry influences ignition, given there is a heat source. This initial heat source can be lightning, matches, embers, flames, or another source of pilot ignition. In subsequent chapters, we will describe processes of fire, including pyrolysis, and then both flaming and smoldering combustion.

F. Castro Rego et al., *Fire Science*, Springer Textbooks in Earth Sciences, Geography and Environment, https://doi.org/10.1007/978-3-030-69815-7_1

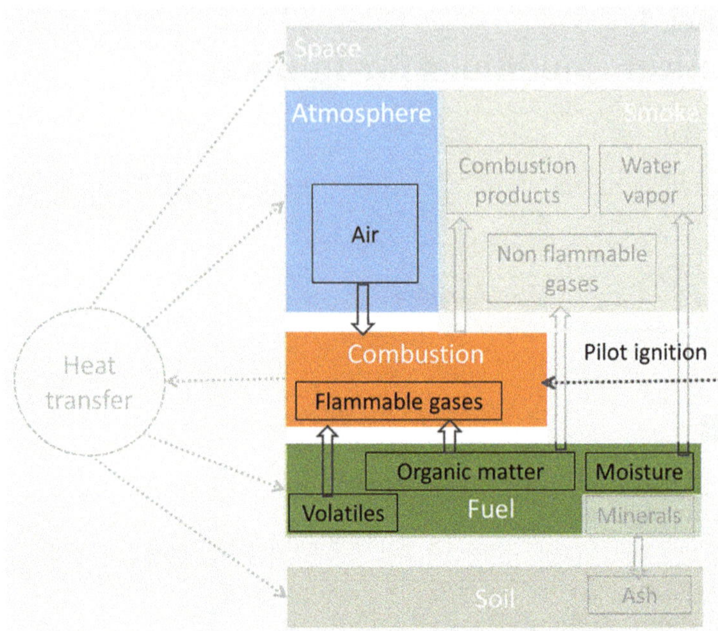

Fig. 1.1 The components of the process involved in the ignition and the start of the combustion process within our global conceptual model. Before combustion can occur, fuels have to release flammable gases (through pyrolysis) and moisture (through dehydration)

Why are some fuels more flammable than others? Mutch (1970) hypothesized that some plants evolved to be flammable in order to be more competitive. The results of scientific evaluations of that hypothesis are mixed, but it makes for great discussions. The volatile oils that make some plants more flammable also play an essential role in mediating plant function and ecological interactions, including influencing drought and herbivory resistance.

1.2 Ignitability and Flammability

For a fire to start, we need heat from an ignition source. Ignition sources could be lightning, embers, matches, downed electrical powerlines, and many other sources associated with human-related causes. People commonly ignite fires, sometimes purposefully and sometimes accidentally. Powerlines can ignite fires when trees fall on them, or when highly charged wires touch one another, or the lines fall onto the ground as the result of a storm or other disturbance. Sometimes, fires are ignited by embers from another fire. Most of the ignition sources responsible for fires result in piloted ignition. That is when the energy required for the ignition of the flammable gases is supplied by an external source such as a flame. The spatial and temporal variability in ignition sources, along with climate and vegetation, are the major factors controlling the pattern of fires over space and time. See the discussion of fire regimes in Chap. 12.

Before fuels can ignite, they go through a pre-ignition phase that removes water from the fuel through dehydration and converts the solid fuel to flammable gases through a process called pyrolysis (Fig. 1.1). Pyrolysis, a word that originates from the Greek words "pyro" (fire) and "lysis" (separation), can be simply defined as the thermal degradation of solid fuel. A flame, by definition, is a combustion reaction where both the fuel and oxidant are in a gaseous phase. To burn in either smoldering or flaming combustion, solid fuels must be exposed to enough heat such that the water is dehydrated and the solid fuel goes through chemical decomposition by thermal degradation (pyrolysis) to release volatile gases that are then combusted when mixed with oxygen at the correct proportion. See Chap. 2 for more on pyrolysis and the chemistry of combustion.

The concepts of flammability and ignitability are often used interchangeably to describe the ease with which fuel is combusted. However, special attention should be given to their definitions. Although both terms are used to refer to the general ability of that substance to burn (Anderson 1970), flammability consists of multiple metrics that quantify not just the ease of ignition but also the behavior of the fire once ignited. For example, some definitions of flammability also consider sustainability (how well combustion will continue), and combustibility (velocity or intensity of combustion) (Anderson 1970). We define ignitability as the ability of a fuel to ignite and produce flames. Ignitability can be estimated based on the time needed for ignition or the probability of ignition given a heat source.

1.3 Ignitability Limits

Combustion is a rapid exothermic chemical reaction between a fuel, such as gases produced during pyrolysis of wood, and an oxidant, such as the oxygen in the atmosphere. However, the chemical properties of the gases involved in combustion can significantly influence their ease of ignition and flame properties.

Two concepts are central to the ignitability of different substances. Auto-ignition temperature which is the temperature at which a substance may ignite without an external ignition source and pilot-ignition temperature which is the temperature at which a substance or mixture may ignite when exposed to an ignition source. Reference values for the auto and pilot ignition temperatures of various fuels are shown in Table 1.1.

The lower the ignition temperature, the easier a fuel can ignite. However, ignition not only depends upon temperature but also upon the relative concentration of fuel and oxygen. The upper and lower bounds of these concentrations are referred to as the flammability limits (or explosive limits). The lower flammability limit (LFL) of a gas is the concentration threshold below which the mixture of fuel and air is too lean (lacks sufficient fuel) to burn. The upper flammability limit (UFL) of a fuel is the

Table 1.1 Lower and upper flammability limits at the standard temperature of 25 °C and at 1 atm pressure and temperatures of auto-ignition and boiling point for compounds commonly involved in the flaming process in wildland fires. (Compiled from The PubChem Project: https://pubchem.ncbi.nlm.nih.gov/, The NIST Chemistry Webbook: https://webbook.nist.gov/chemistry/, Chemspider: http://www.chemspider.com/, Barboni et al. 2011 and Della Rocca et al. 2017)

Class	Class/ compound	Formula	Molar weight ($g\ mol^{-1}$)	Lower flammability limit LFL (% vol. and $g\ m^{-3}$ in parentheses)	Upper flammability limit UFL (% vol. and $g\ m^{-3}$ in parentheses)	Autoignition (°C)	Compound	Boiling point (°C)
Combustion products	Carbon dioxide	CO_2	44	Non flammable				−78
	Carbon monoxide	CO	28	12.5 (143)	74.0 (846)	605		−191
	Methane	CH_4	16	5.0 (33)	15.0 (98)	580		−161
Terpenoids	Isoprene	C_5H_8	68	1.0 (28)	9.7 (269)	220		34
	Monoterpenes	$C_{10}H_{16}$	136	0.7 (39)	6.1 (339)	255	α-pinene	155
							Camphene	161
							Sabinene	164
							β-pinene	166
							Myrcene	167
							Limonene	176
							Terpinolene	187

	Sesquiterpenes	$C_{15}H_{24}$	204	0.5 (42)	5.2 (433)		β-caryophyllene	263
							α-humulene	276
	Oxygenated terpenoids	$C_5H_{18}O$	86	1.5 (5.3)	9.4 (330)	380	Prenol	97
		$C_{10}H_{18}O$	154	0.6 (38)	4.1 (258)	269	Eucalyptol	176
							Linalool	198
							4ol terpinen	209
							α-terpineol	220
		$C_{10}H_{20}O_2$	196	0.7 (56)	4.2 (336)	287	Linalyl acetate	220
							Geranylacetate	242
		$C_{20}H_{34}O$	290	0.4 (47)	4.9 (580)		Manoyl oxide	338
Other volatile organic compounds		C_2H_4O	44	40 (718)	60 (1078)	185	Acetaldehyde	20
		C_3H_6O	58	2.5 (59)	12.8 (303)	465	Acetone	56
		CH_4O	32	6.0 (78)	36.0 (470)	440	Methanol	65
		$C_2H_4O_2$	60	4.0 (98)	19.9 (487)	485	Acetic acid	118

concentration threshold above which the mixture is too rich in fuel (deficient in oxygen) to burn. The lower and upper flammability limits (LFL and UFL) for common compounds are shown in Table 1.1.

The upper and lower flammability limits can be reported as either a percentage (%) or as a mass per unit volume (g m^{-3}). The lower flammability limit as a percent can be converted to mass per unit volume through Eq. (1.1).

$$LFL\left(g\,m^{-3}\right) = \left(\frac{LFL(\%)}{100\%}\right) \times \left(\frac{Molar\ weight\ \left(g\,mol^{-1}\right)}{Molar\ volume\ \left(m^3\,mol^{-1}\right)}\right) \tag{1.1}$$

Recall that the molar weight of an ideal gas is a known constant (8.206×10^{-5} m^3 atm mol^{-1} K^{-1}), but the molar volume of a gas (m^3 mol^{-1}) is dependent on its absolute temperature (K) and pressure (atm) according to Eq. (1.2):

$$Molar\ volume = \frac{Gas\ Constant\ \times Temperature}{Pressure} \tag{1.2}$$

For example, we can use Eqs. (1.1) and (1.2) to calculate the value of the lower flammability limit in concentration LFL(g m^{-3}) for α-pinene, a common monoterpene in *Pinus* species around the world. Experimental studies indicate that α-pinene has a lower flammability limit expressed as percentage volume LFL(%) of 0.7% (Table 1.1). The molar weight of α-pinene ($C_{10}H_{16}$) is 136 g mol^{-1}. From Eq. (1.2), we can compute the molar volume of the gas at a pressure of 1 atm and a reference temperature of 25 °C (298 K) as 0.0245 m^3 mol^{-1}. Using this value in Eq. (1.1) we get the value of LFL of 39 g m^{-3} for α-pinene that is shown in Table 1.1.

The lower and upper flammability limits (Table 1.1) are established for standard pressure and a reference temperature of 25 °C (Gharagheizi 2008, 2009), indicated as LFL_{25} and UFL_{25}, respectively. However, flammability limits are a function of temperature (T), with the lower flammability limit decreasing with temperature and the upper flammability limit increasing with temperature. The equations for the adjustment of the lower and upper flammability limits (LFL_T and UFL_T) as a function of temperature (T in °C) are of the form:

$$LFL_T = LFL_{25}[1 - \alpha_1(T - 25°C)] \tag{1.3}$$

$$UFL_T = UFL_{25}[1 + \alpha_2(T - 25°C)] \tag{1.4}$$

where α_1 and α_2 are empirical constants determined experimentally as $\alpha_1 = 7.80 \times 10^{-4}$, and $\alpha_2 = 7.21 \times 10^{-4}$ (Arnaldos et al. 2001; Chetehouna et al. 2014; Zabetakis 1965). These equations indicate, for example, that for temperatures around 200 °C, the lower flammability limit decreases by around 13% and the upper flammability limit increases by a similar percentage when compared with the values shown as LFL and UFL in Table 1.1 that correspond to the reference values LFL_{25} and UFL_{25}. More precise procedures to estimate the flammability limits of mixtures of different compounds may be found in Courty et al. (2010).

1.4 Mixing Between Fuel Gases and Air

The mixing between fuel gases and oxygen can occur before ignition (called a premixed flame) or they can mix during combustion (called a diffusion flame). A Bunson burner is a common device that produces a premixed flame. The gasoline internal combustion engine with spark-ignition or the diesel engine with autoignition are other common examples of premixed flames (Quintiere 1998).

In wildland fires, the combustible fuel and oxygen are not premixed and come together from the two sides of a reaction zone through molecular and turbulent diffusion. This results in a diffusion flame, in which the burning rate is determined by the rate at which fuel and oxygen are transported (diffused), brought together, and mixed in proper proportions (within flammable limits) for reaction (Glassman and Yetter 2008).

If the production rate of flammable gases and oxygen supply is constant, we have a quasi-steady combustion rate. This is the same as the flame of a candle where we can observe different colors, from the darker interior, where we have the fuel gases in concentrations above the flammability limits, to the bright flame surface where combustion occurs as the mixture between air and fuel is between the flammable limits.

In order to demonstrate that "there are clearly two different kinds of action—one the *production* of the vapor, and the other the *combustion* of it" Faraday (1861) in his lectures presented an experiment placing a tube in the flame to get "the vapor from the middle of the candle produced by its own heat" and pass it through the tube to the other extremity where he lit it, obtaining "absolutely the flame of the candle at a place distant from it" (Fig. 1.2).

There are also situations where the production of flammable gases does not coincide in time with combustion in flames. This is the case, for example, of compartment fires where the accumulation of flammable gases may accumulate through time and sudden exposure to air or a spark may dramatically increase fire growth to the full involvement of a room. These flashover events cause significant problems in fire safety in houses motivating the increasing use of various fire safety technologies, including residential smoke detectors and smoke control systems (Quintiere 1998). See Chap. 8, for a discussion of similar types of behavior during extreme fires.

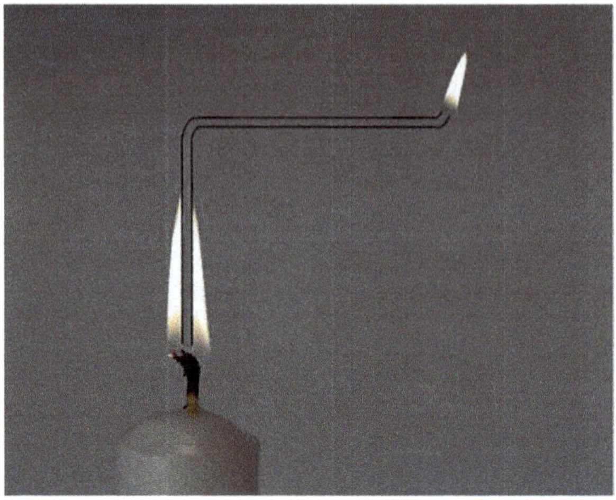

Fig. 1.2 Faraday (1861) showed the generation of flames in two different places from the vapor produced from the heated wax in a candle. One of the flames was next to the candle. The other flame was distant from the candle. The fuel passing through the tube with flammable gases from the interior of the flame only mixes with the air outside the tube where the mixture felt within the flammability limits and, provided a heat source by a match, ignited and produced flames

1.5 Ignitability of Wildland Fuels

In typical wildland fuels, both solid and liquid components have to be considered when evaluating ignitability. Wildland fuels at ambient temperature are composed of both liquids (e.g., fuel moisture and volatile compounds) and solids (e.g., cellulose, hemicelluloses, lignin, minerals, and some volatile compounds). All these materials have to go through a pre-ignition phase in order to release flammable gases before ignition can occur, as will be discussed in Chaps. 2 and 4.

The moisture present in wildland fuel is vaporized through a process called dehydration before combustion occurs. Dehydration occurs when enough heat is supplied to the fuel such that the temperature reaches the boiling point (or vaporization point). The boiling point is the temperature at which a liquid changes into a gas.

The same process of volatilization occurs for Volatile Organic Compounds (VOC) particularly for a class of compounds known as terpenoids, that are present in leaves and other plant parts. A class of terpenoids is terpenes, which consist of different numbers of C_5H_8 units, from a single unit (isoprene) to two units (monoterpenes) or three units (sesquiterpenes). Another class of terpenoids is oxygenated terpenoids, which are oxygen-containing derivatives from terpenes of a single unit (as prenol), of two units (as eucalyptol), or of more units (as manoyl oxide). Other VOCs that are not terpenoids have varying compositions. All of these compounds are easily volatilized when heated as they have low boiling points and are easily transformed into flammable gases. The boiling points for these various compounds are shown in Table 1.1.

For the solid component of wildland fuels, the processes have some complexity. Most plant tissues are composed of cellulose, hemicelluloses, and lignin, which have to go through the process of pyrolysis to produce flammable gases. Depending on the oxygen supply, the combustion can be complete or incomplete, and different phases might be defined. Details of pyrolysis are explained in Chap. 2. The complete combustion of typical wildland fuels produces carbon dioxide (which is not burnable), but the first phases of combustion or incomplete combustion also produce carbon monoxide and methane (flammable gases) that will burn if and when appropriate conditions occur. For the products of incomplete combustion, as for all other gases, we can define the lower and upper flammability limits and the temperatures of auto-ignition.

Ignitability is often measured in fuel samples in laboratory experiments, using cone calorimeter or epi-radiators as heat sources (Fig. 1.3). A sample of fuel is exposed to constant heat flux, and the time until ignition is recorded (Valette and Moro 1990). Alternatively, ignitability is expressed as the number of successful ignitions of a given fuel sample subject to a specific heat flux for a given time. These experimental approaches inform our understanding of some of the main factors involved in flammability (e.g., Weise et al. 2005; Madrigal et al. 2009; Ubysz and Valette 2010).

From laboratory results, it is clear that the presence of fuel moisture influences ignitability by increasing the time required to vaporize the moisture and produce and ignite flammable gases (Fletcher et al. 2007; Davies and Legg 2011). Figure 1.4 shows the relationship between fuel moisture content (%) and ignition time for

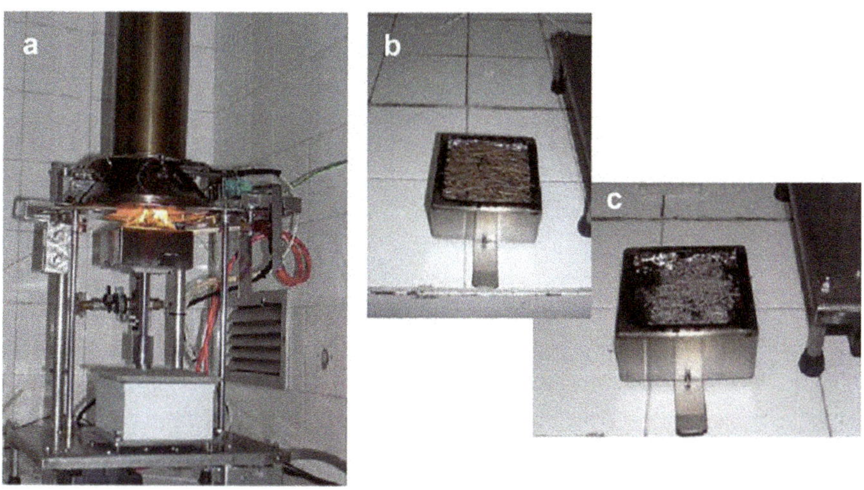

Fig. 1.3 (**a**) Experimental device at INIA-CIFOR laboratory (Madrid) showing the cone calorimeter (**b**) sample of *Pinus pinaster* needles before the experiment (**c**) sample of *Pinus pinaster* needles after the experiment. (Photographs by Mercedes Guijarro)

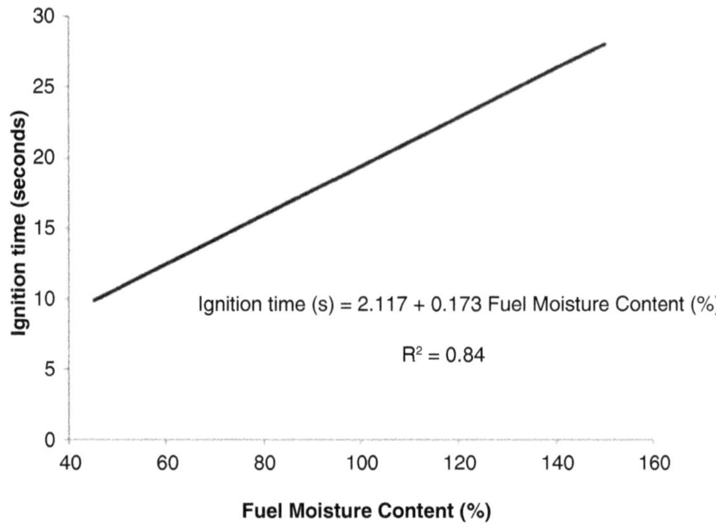

Fig. 1.4 Relationship between ignition time (in seconds) and fuel moisture content (%) for *Erica arborea* leaves and twigs using the equations presented. (Data from Moro 2006 as cited by Ubysz and Valette 2010)

leaves and twigs of *Erica arborea.* For similar values of fuel moisture, ignitability also depends on the physical characteristics, chemical composition, and the proportion of flammable VOC in the fuel (e.g., Alessio et al. 2008; Pausas et al. 2016) (Fig. 1.5).

The results from laboratory experiments on the ignitability of fuel samples provide valuable insights that help to interpret field observations. However, these results should not be simply used as a measure of the ignitability of a fuel complex. At the scale of the fuel complex (including vegetation components and litter), many other factors have to be accounted for, such as the fuel arrangement, or the mixture of fuels. Also, direct associations of these results with fire behavior are difficult to ascertain, mostly because laboratory studies seldom replicate the heat fluxes of wildland fires (Fernandes and Cruz 2012).

1.6 Implications

Both the types of fuel (do the plants contain many volatile organic compounds?) and the environmental conditions (how dry is the fuel?) influence how readily fuels ignite in vegetation fires. We focus upon the factors influencing when, how, and whether fuels will ignite whether that ignition is from lightning, matches, embers, flames, or another source. Without ignition, there will be no fire. In the next chapters, we will further explore the products of combustion, including heat.

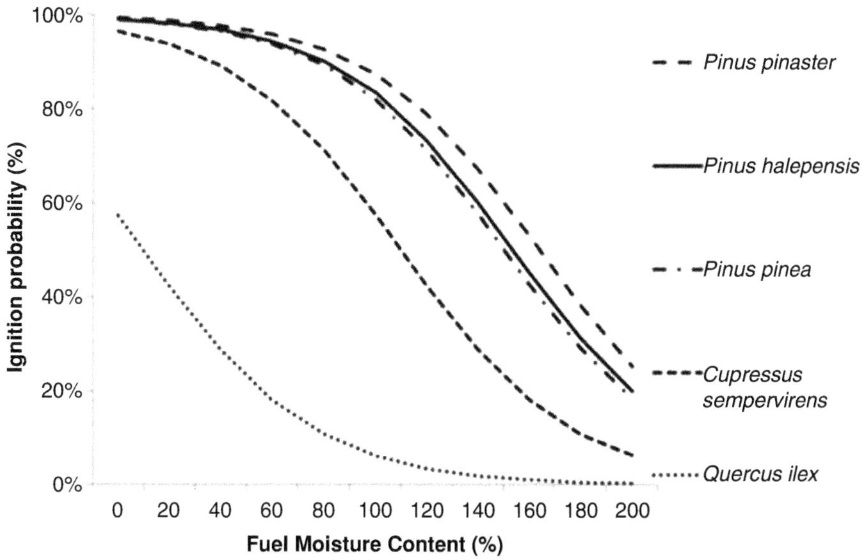

Fig. 1.5 Ignition probability of samples of leaves and litter of various species exposed to a heat flux of 25 kW m^{-2}. The main factor associated with the probability of ignition was fuel moisture content. Differences between species have been attributed to differences in physical characteristics of fuel samples but also to differences in stored terpenoids. *Quercus ilex* is a species that does not store terpenoids in the leaves. (Adapted from Della Rocca et al. 2017)

References

Alessio, G. A., Penuelas, J., Llusia, J., Ogaya, R., Estiarte, M., & De Lillis, M. (2008). Influence of water and terpenes on flammability in some dominant Mediterranean species. *International Journal of Wildland Fire, 17*, 274–286.

Anderson, H. E. (1970). Forest fuel ignitability. *Fire Technology, 6*, 312–319.

Arnaldos, J., Casal, J., & Planas-Cuchi, E. (2001). Prediction of flammability limits at reduced pressures. *Chemical Engineering Science, 56*, 3829–3843.

Barboni, T., Cannac, M., Leoni, E., & Chiaramonti, N. (2011). Emission of biogenic volatile organic compounds involved in eruptive fire: Implications for the safety of firefighters. *International Journal of Wildland Fire, 20*, 152–161.

Chetehouna, K., Courty, L., Garo, J. P., Viegas, D. X., & Fernandez-Pello, C. (2014). Flammability limits of biogenic volatile organic compounds emitted by fire-heated vegetation (*Rosmarinus officinalis*) and their potential link with accelerating forest fires in canyons: A Froude-scaling approach. *Journal of Fire Sciences, 32*(4), 316–327. https://doi.org/10.1177/0734904113514810.

Courty, L., Chetehouna, K., Goulier, J., Catoire, L., Garo, J. P., & Chaumeix, N. (2010). On the emission, flammability and thermodynamic properties of Volatile Organic Compounds involved in accelerating forest fires. In VI International Conference on Forest Fire Research, Portugal.

Davies, G. M., & Legg, C. J. (2011). Fuel moisture thresholds in the flammability of *Calluna vulgaris*. *Fire Technology, 47*, 421–436.

Della Rocca, G., Madrigal, J., Marchi, E., Michelozzi, M., Moya, B., & Danti, R. (2017). Relevance of terpenoids on flammability of Mediterranean species: An experimental approach at a low

radiant heat flux. *iForest – Biogeoscience and Forestry, 10*(5), 766–775. https://doi.org/10.3832/ifor2327-010.
Faraday, M. I. (1861). In W. Crookes (Ed.), *A course of six lectures on the chemical history of a candle*. London: Griffin, Bohn.
Fernandes, P., & Cruz, M. (2012). Plant flammability experiments offer limited insight into vegetation–fire dynamics interactions. *The New Phytologist, 194*, 606–609.
Fletcher, T. H., Pickett, B. M., Smith, S. G., Spittle, G. S., Woodhouse, M. M., & Haake, E. (2007). Effect of moisture on ignition behaviour of moist California chaparral and Utah leaves. *Combustion Science and Technology, 179*, 1183–1203.
Gharagheizi, F. (2008). Quantitative structure-property relationship for prediction of the lower flammability limit of pure compounds. *Energy & Fuels, 22*, 3037–3039.
Gharagheizi, F. (2009). Prediction of upper flammability limit percent of pure compounds from their molecular structures. *Journal of Hazardous Materials, 167*, 507–510.
Glassman, I., & Yetter, R. A. (2008). *Combustion* (4th ed.). San Diego: Academic.
Madrigal, J., Hernando, C., Guijarro, M., Diez, C., Marino, E., & de Castro, A. J. (2009). Evaluation of forest fuel flammability and combustion properties with an adapted mass loss calorimeter device. *Journal of Fire Sciences, 27*, 323–342.
Moro, C. (2006). Méthode de mesure de l'inflammabilité du combustible forestier Méditerranéen. Internal technical report. (p. 14).
Mutch, R. W. (1970). Wildland fires and ecosystems–A hypothesis. *Ecology 51*(6), 1046–1051.
Pausas, J. G., Alessio, G. A., Moreira, B., & Segarra-Moragues, J. G. (2016). Secondary compounds enhance flammability in a Mediterranean plant. *Oecologia, 180*(1), 103–110.
Quintiere, J. G. (1998). *Principles of fire behavior* (p. 258). Boston: Delmar.
Ubysz, B., & Valette, J. C. (2010). Flammability: Influence of fuel on fire initiation. In J. S. Silva, F. Rego, P. Fernandes, & E. Rigolot (Eds.), *Towards integrated fire management—Outcomes of the European Project Fire Paradox. Research Report 23* (pp. 23–34). Joensuu: European Forest Institute.
Valette, J. C., & Moro, C. (1990). Inflammabilités des espèces forestières méditerranéennes, conséquences sur la combustibilité des formations forestières. *Revue Forestière Française XLII, numéro spécial, 1990*, 76–92.
Weise, D. R., White, R. H., Beal, F. C., & Etlinger, M. (2005). Use of the cone calorimeter to detect seasonal differences in selected combustion characteristics of ornamental vegetation. *International Journal of Wildland Fire, 14*, 321–338.
Zabetakis, M. G. (1965). *Flammability characteristics of combustible gases and vapors. Bulletin 627*. Washington, DC: Bureau of Mines.

Chapter 2
From Fuels to Smoke: Chemical Processes

Learning Outcomes
After reading this chapter and using the interactive spreadsheet, we expect that you will be able to:

1. Explain the chemical equations for both complete and incomplete combustion of fuels. In particular, you should be able to identify the key inputs and how changing their values alters the predicted values,
2. Summarize how flaming and smoldering combustion differ with respect to smoke composition, and
3. Use the interactive spreadsheet to estimate emission factors, particulate matter, and smoke production for wildfires and prescribed fires.

2.1 Introduction

The chemical breakdown of organic compounds in live and dead plants (also called organic matter) provides the gases which ultimately combust and provide the energy that fuels the spread of fires and their effects. In addition to heat, combustion emits several other products that we call smoke (Fig. 2.1). Smoke often travels far from the flames, where the particulates and other compounds can pose health hazards to people and affect travelers' visibility.

Supplementary Information The online version of this chapter (https://doi.org/10.1007/978-3-030-69815-7_2) contains supplementary material, which is available to authorized users.

F. Castro Rego et al., *Fire Science*, Springer Textbooks in Earth Sciences, Geography and Environment, https://doi.org/10.1007/978-3-030-69815-7_2

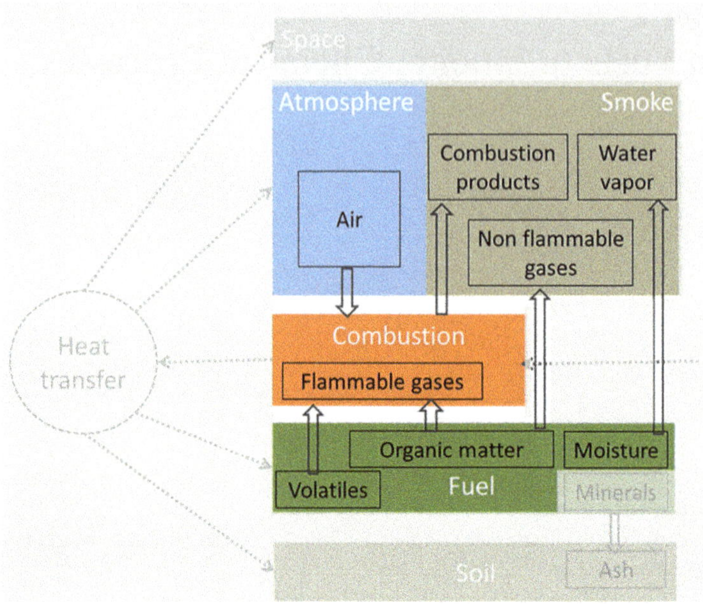

Fig. 2.1 Diagram of the chemical flows included in this chapter, all of which influence the rate of combustion in vegetation fires. When fuel is heated, it dries, and the resulting water vapor is what makes smoke appear white or light gray. When fuel is heated, flammable gases are produced from both volatile oils in the organic matter as well as the organic matter biomass itself. When mixed with oxygen and ignited, combustion occurs. Smoke often contains carbon dioxide, water, and other gases from the combustion of organic matter. These and the particulates of partially consumed organic matter make smoke gray. The particulate matter in smoke is an air pollutant and can be harmful for people to breathe

2.2 Combustion at the Level of Atoms and Molecules

All physical substances, including the fuels that drive combustion during wildland fires, consist of atoms that bond together to form molecules. At its simplest, combustion involves only atoms of three elements, carbon (C), hydrogen (H), and oxygen (O). Atoms of the same element may form electrically neutral groups of bonded atoms such as hydrogen (H_2) and oxygen (O_2) (Fig. 2.2). When two or more atoms of a single element combine, they form a molecule. When atoms of two different elements combine to form a molecule, it is called a binary compound. Combinations of C and H can form different hydrocarbons from methane (CH_4) to octane (C_8H_{18}). Combinations of C and O atoms combine to form carbon monoxide (CO) and carbon dioxide (CO_2), and the combination of H and O forms water (H_2O) (Fig. 2.2). Atoms of the three elements may form organic compounds such as the carbohydrates of the general formula $C_m(H_2O)_n$, including cellulose ($C_6H_{10}O_5$) and glucose ($C_6H_{12}O_6$) (Fig. 2.2).

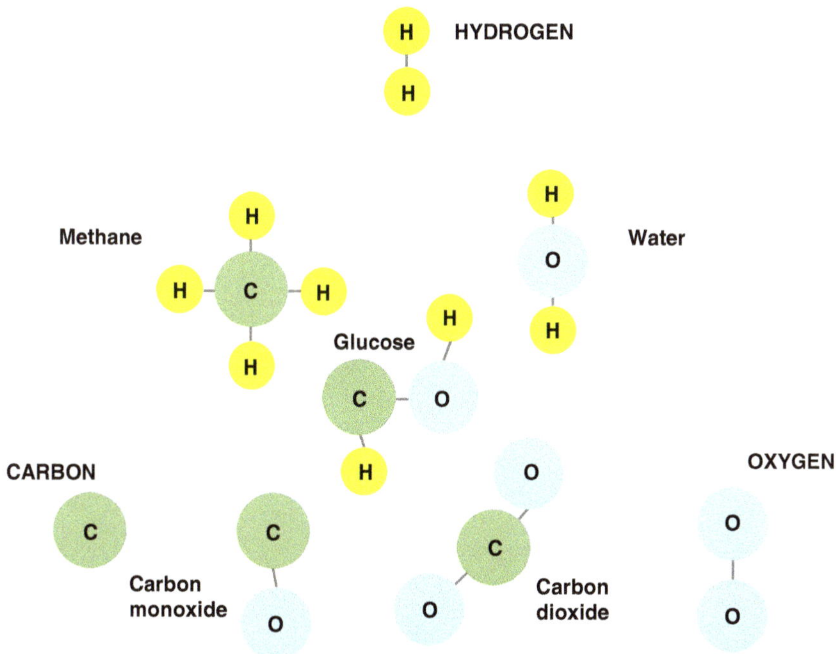

Fig. 2.2 C, H, and O combine into molecules that can be involved in combustion. These are the basic building blocks of organic matter that can burn. For simplification, glucose is represented here as CH_2O

Combustion involves the chemical breakdown of the bonds between atoms. The most straightforward combustion reaction involves the burning of methane, a natural gas, in pure oxygen. The complete combustion of methane occurs when each molecule of methane reacts with two molecules of oxygen to produce one molecule of carbon dioxide and two molecules of water vapor. If oxygen is limited, incomplete combustion occurs, and different products are emitted. For example, if three atoms of oxygen are available per atom of carbon during combustion, carbon monoxide will be produced instead of carbon dioxide. If the supply of oxygen is limited even further, the combustion products also begin to include carbon particles. Understanding the differences between complete and incomplete combustion and the role of oxygen supply is critical to estimate the amount of heat and smoke generated during a fire (Fig. 2.3). Note that heat from combustion is addressed in Chaps. 3, 4, and 5 with ecological implications in Chaps. 9 and 10.

In its simplest form, combustion can be visualized as occurring in a single step reaction whereby fuel is mixed with an oxidizer in the correct proportions and temperatures, resulting in the ignition of the mixture and the release of heat and emissions. For example, a single-step model of the combustion of methane in pure oxygen would result in the production of heat and some combination of carbon dioxide, carbon monoxide, and water (Fig. 2.3). In reality, however, combustion

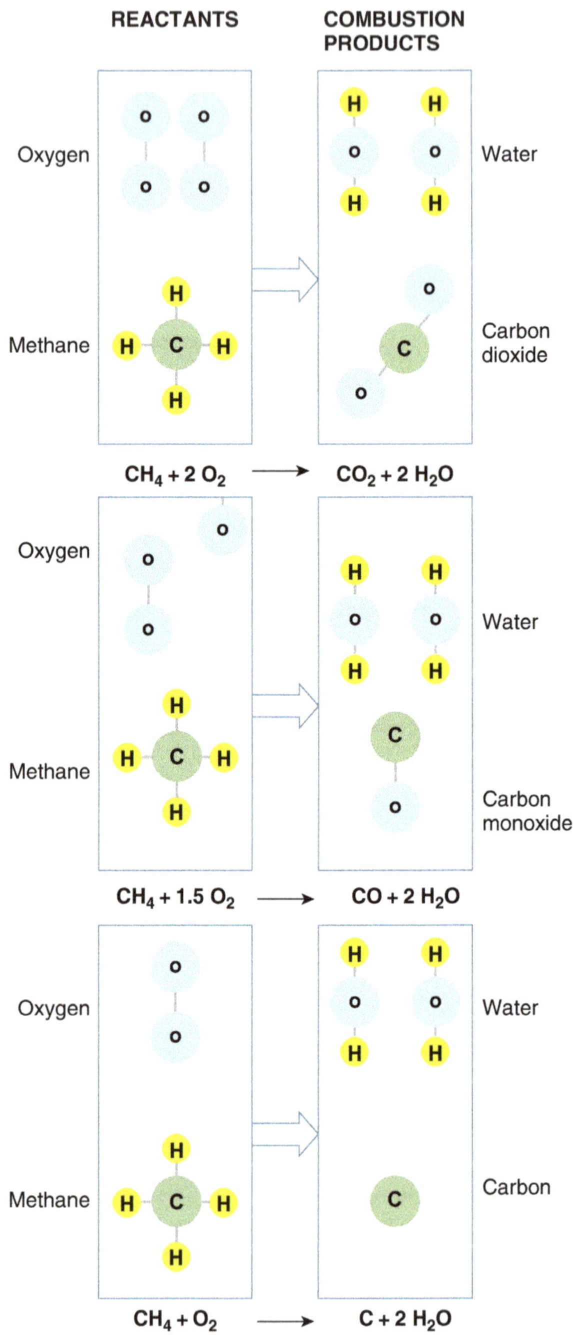

Fig. 2.3 Different possible combustion reactions of burning methane with different levels of oxygen supply, decreasing from top to bottom

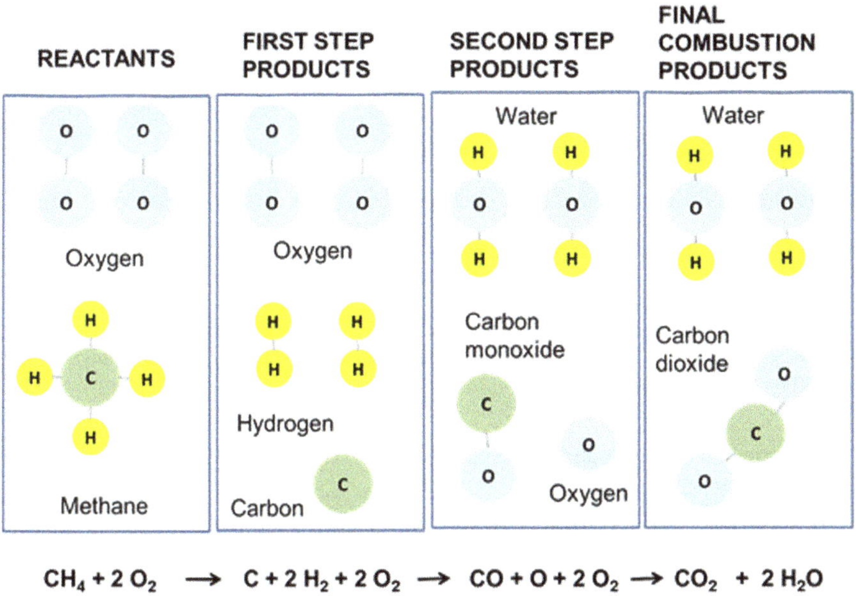

Fig. 2.4 Sequence of reactions of the combustion of methane as a fuel gas showing the intermediate products at each step

reactions are often more complex, involving multiple steps. The complete combustion of methane in pure oxygen, whereby CH_4 is decomposed into carbon and hydrogen molecules, occurs in three steps (Fig. 2.4). The carbon is oxidized to form carbon monoxide, while hydrogen and oxygen combine to form water. Finally, the carbon monoxide is further oxidized to carbon dioxide.

The chain reaction sequence has important consequences as these processes tend to occur at different places and at different moments during combustion. The sequence of reactions is even more complicated when dealing with wildland fuels such as wood or leaves of grass and trees.

2.3 Combustion of Solid Fuels

Managers and scientists generally recognize three types of combustion: flaming, smoldering, and glowing (Fig. 2.5). During a wildfire, flaming, smoldering, and glowing combustion can co-occur, and in many cases, one type of combustion leads to another type. These different types of combustion vary in their rates of spread, heat release rates, and emissions.

Flaming combustion is easily recognized by the bright colors of the hot gases present in flames. Flaming combustion results from a rapid oxidation reaction at high

Fig. 2.5 The three types or phases of combustion. (**a**) During flaming combustion gases from thermal degradation of solid fuel burn. (**b**) During smoldering combustion, smoke is abundant. (**c**) During glowing combustion, oxygen combines with fuel molecules at the surface of the solid fuel. (Photographs by (**a**) Oscar https://en.wikipedia.org/wiki/Flame#/media/File:DancingFlames.jpg, (**b**) Terrie Jain, and (**c**) Jens Buurgaard Nielsen https://en.wikipedia.org/wiki/Smouldering#/media/File:Embers_01.JPG)

temperatures with an abundant oxygen supply resulting in the release of carbon dioxide and water vapor.

Smoldering, or glowing, combustion is a non-flaming form of combustion recognized by an abundant production of smoke with large amounts of particulate matter (mostly carbon) and carbon monoxide (Ward 2001). Smoldering combustion occurs at a slower rate with a limited oxygen supply than flaming combustion, resulting in lower temperatures, rates of fire spread, and intensities. Smoldering combustion is much slower than flaming combustion because the combustion reaction takes place at the surface of solid fuels, and oxygen molecules have to diffuse to the solid surface to combine with fuel molecules. When smoldering combustion heats the solid fuel to a high enough temperature such that it radiates in the visible spectrum, it is often referred to as glowing combustion. Dense organic matter layers, such as duff in forests or the mulch that results from chipping trees and shrubs, usually burn by smoldering combustion.

The combustion of fuels during vegetation fires results in the production of char as well as unburned material. Char is a residual carbon material resulting from incomplete combustion after the removal of water and volatile organic compounds from the fuel by heat. See Chap. 9 for discussion of carbon in soils.

Pyrolysis, a word that originates from the Greek words "pyro" (fire) and "lysis" (separation), is defined as the thermal degradation of solid fuel into gases under the influence of heat. Because solid fuels do not burn directly, they first must go through pyrolysis to produce the volatile gases that combust. The pyrolysis of cellulose has been very well described (Fig. 2.6). The decomposition of lignin and hemicelluloses follows the same general pattern, with lignin being far more resistant to thermal degradation.

We contrast flaming and smoldering combustion using the chemical equations for glucose. For complete combustion, glucose first decomposes into carbon monoxide and hydrogen. In the presence of enough oxygen, these two gases burn in flaming combustion, producing carbon dioxide and water (Fig. 2.7). For smoldering combustion, pyrolysis first results in water vapor leaving behind solid carbon (char). Then if oxygen is present, carbon may undergo smoldering combustion, producing

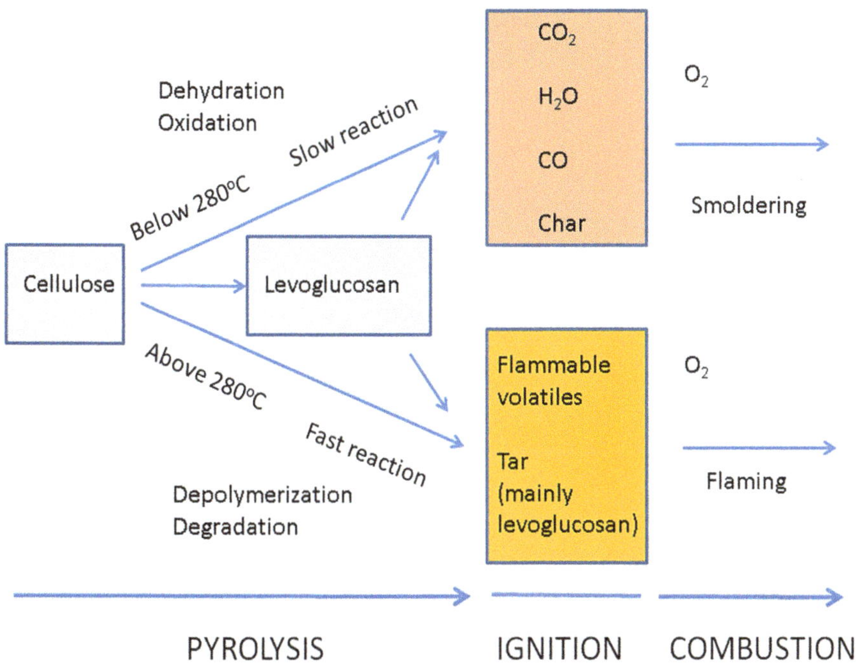

Fig. 2.6 The pyrolysis of cellulose can have two pathways, producing gases and char (top colored box) or flammable volatiles and tar (lower colored box). If there is ignition, these are then the compounds that burn in smoldering (top) and flaming combustion. (Adapted from Philpot 1971; Chandler et al. 1983)

the same final products, carbon dioxide, and water if combustion is complete (Fig. 2.7).

Similarly, wood and cellulose can burn with either flaming or smoldering combustion. The final products are the same for flaming and glowing combustion processes if enough oxygen is available for complete combustion. However, the intermediate products and the different rates of combustion have important implications for managing fires.

2.4 Combustion Completeness and Emission Factors

Because the law of conservation of mass states that matter cannot be created nor destroyed in a chemical reaction such as combustion, we can use a mass balance approach to understand what happens to all of the atoms that make up the fuels that burn. The standard unit of measurement for the amount of a substance is a mole and is defined as having $6.02214076 \times 10^{23}$ particles. The molar mass of a substance is

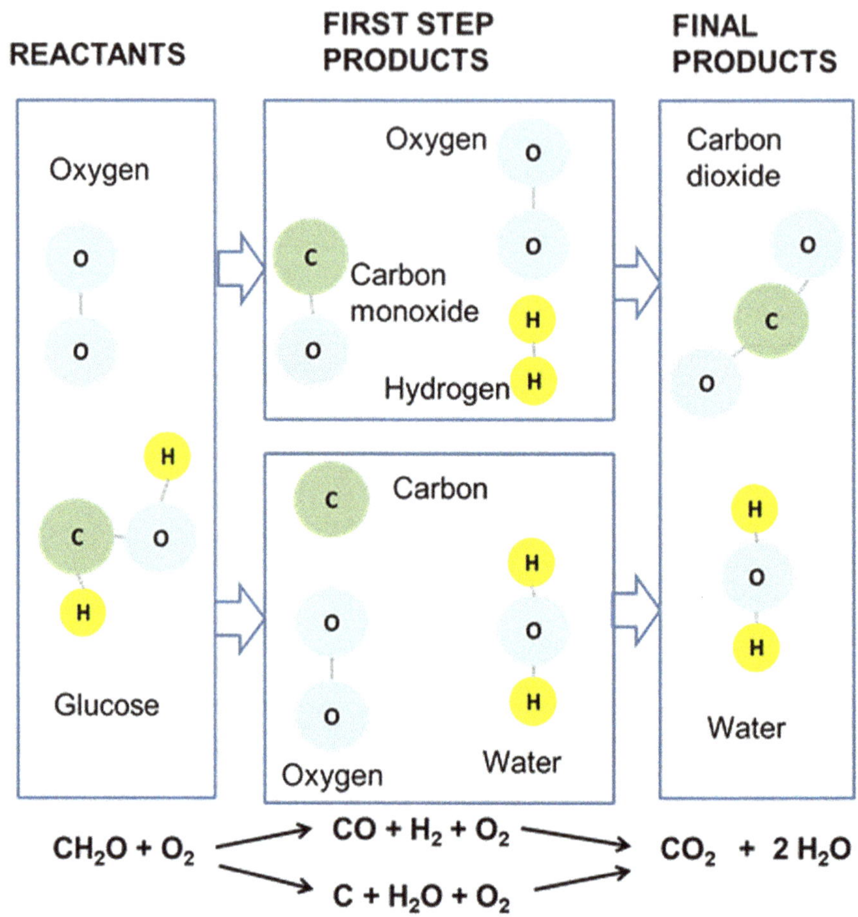

Fig. 2.7 Two possible paths of glucose combustion. For simplification purposes glucose is represented as CH_2O. The upper path corresponds to flaming combustion and the lower path to smoldering combustion. In both processes, if enough oxygen is available for combustion to be complete, the final products are carbon dioxide and water vapor

the mass of a sample in grams for 1 mol. The molar masses of selected molecules are presented in Table 2.1.

In this chapter, we focus on three main elements (C, H, and O), but other elements present in fuels or the atmosphere, such as nitrogen and sulfur, could also be considered. Nitrogen is the fourth most abundant element in Earth's biomass (between 2% and 4%, according to Larcher 1977) and is a minor component of wildland fuels. Whereas, sulfur is an essential macro element in plants, with a concentration ranging from 0.1 to 0.45% on a dry weight basis, and plays a critical role in building amino acids and the formation of chlorophyll in plants. Although elements such as nitrogen and sulfur play important roles in plant function, they are

Table 2.1 Molar masses of important molecules involved in fire processes based on their composition and the atomic weight of the elements. (Compiled from Atkin and Jones 2005; Byram 1959; Perrett 2007)

		Element and atomic weight					
Molecule	Formula	Carbon 12	Hydrogen 1	Oxygen 16	Nitrogen 14	Sulfur 31	Molar mass (g/mol)
Carbon (atom)	C	1					12
Hydrogen	H_2		2				2
Oxygen	O_2			2			32
Nitrogen	N_2				2		28
Water	H_2O		2	1			18
Carbon monoxide	CO	1		1			28
Carbon dioxide	CO_2	1		2			44
Methane	CH_4	1	4				16
Isoprene	C_5H_8	5	8				68
Octane	C_8H_{18}	8	18				114
Monoterpenes	$C_{10}H_{16}$	10	16				136
Eucalyptol	$C_{10}H_{18}O$	10	18	1			154
Wood (equivalent)	$C_6H_9O_4$	6	9	4			145
Cellulose	$C_6H_{10}O_5$	6	10	5			162
Glucose	$C_6H_{12}O_6$	6	12	6			180
Lignin	$C_{10}H_{12}O_3$	10	12	3			180
Proteins (empirical)	$C_{400}H_{620}O_{120}N_{100}S_1$	400	620	120	100	1	8771

often ignored for practical purposes during combustion calculations. For example, the equivalent chemical composition of wood suggested by Byram (1959) and used throughout this book excludes all minor elements in plant materials, including nitrogen and sulfur.

Emission factors are expressed as the ratio of the mass of product yielded divided by the mass of fuel consumed (Eq. 2.1). Emission factors are critical for regulatory agencies and managers as they are used to predict the impact of fires on air quality.

$$Emission\ factor = \frac{Mass\ of\ product\ released}{Mass\ of\ fuel\ consumed} \tag{2.1}$$

Emission factors are often determined experimentally by measuring the mass of fuel consumed and the mass of the product of interest released. However, emission factors can also be estimated based on conservation of mass and knowledge of the chemical composition of the fuel and the products:

$$Emission\ factor = \frac{Number\ of\ molecules\ of\ the\ product\ released \times Molar\ mass\ of\ the\ product}{Number\ of\ molecules\ of\ fuel\ involved \times Molar\ mass\ of\ the\ fuel\ compound} \tag{2.2}$$

For example, the complete combustion of wood can be represented by the simplified chemical equation:

$$C_6H_9O_4 + 6.25\ O_2 \rightarrow 6\ CO_2 + 4.5\ H_2O \tag{2.3}$$

Using the corresponding molar masses the above equation can be represented by an equation of the masses of reactants and products for 145 g of wood as:

$$145g\ wood + 200g\ oxygen \rightarrow 264g\ carbon\ dioxide + 81g\ water\ vapor \tag{2.4}$$

To calculate the emission factor for carbon dioxide from the complete combustion of wood (grams of carbon dioxide per 1 g of wood) we have.

$$Emission\ factor\ for\ carbon\ dioxide = \frac{264g\ carbon\ dioxide}{145g\ wood} = 1.82 \tag{2.5}$$

Likewise, the emission factor for water vapor during the complete combustion of wood is:

(continued)

$$Emission\ factor\ for\ water\ vapor = \frac{81g\ water\ vapor}{145g\ wood} = 0.56 \tag{2.6}$$

If combustion is complete, no carbon monoxide is produced, and thus its emission factor is zero. Similar calculations could be made for typical flaming combustion of wood when the oxygen supply is not maximum and we have only 6 mol of oxygen reacting with 1 mol of wood or for typical smoldering combustion with even less oxygen when we have 5.5 mol of oxygen per mole of wood. These two types of combustion would result in different emission factors for carbon dioxide (1.67 and 1.37, respectively) and for carbon monoxide (0.10 and 0.29, respectively) (Table 2.2).

Smoldering and flaming combustion, or the combustion completeness, significantly influence the emission factors used by fire and air-quality managers to forecast the impacts of wildfires (Table 2.2). For example, smoldering combustion, which is less efficient than flaming combustion, emits greater amounts of CO. Combustion is never 100% complete in vegetation fires. Incomplete combustion products include carbon monoxide (CO), methane (CH_4), other hydrocarbons (CH), and particles as soot (mostly C). Several measures of the completeness or efficiency of combustion can be computed.

The Equivalent Oxygen to Fuel Ratio (EOFR) uses the comparison of the oxygen used in the actual combustion with the oxygen that would be used by the same quantity of fuel during complete combustion. The Equivalent Oxygen to Fuel Ratio is the inverse of the Equivalence Ratio (ϕ) proposed by some authors (e.g., Drysdale 1985):

$$\begin{aligned} EOFR &= \frac{mol\ of\ \ O_2\ consumed\ during\ combustion}{mol\ of\ O_2\ consumed\ in\ complete\ combustion} \\ &= \frac{1}{Equivalence\ ratio\ (\Phi)} \end{aligned} \tag{2.7}$$

Rather than use the amount of oxygen consumed during combustion, as done in the Equivalence Ratio and the Equivalent Oxygen to Fuel Ratio, combustion completeness can also be estimated by evaluating the composition of the products. Ward and others (1996) proposed the use of a Modified Combustion Efficiency (MCE) metric, which is based on the ratio of moles of carbon released as CO_2 to the sum of the moles of carbon released as CO_2 and CO:

$$MCE = \frac{mol\ of\ CO_2}{mol\ of\ CO_2 + mol\ of\ CO} \tag{2.8}$$

Table 2.2 Calculation of the emission factors of carbon dioxide, water vapor, and carbon monoxide for dry wood and for different types of combustion completeness. Calculations are based on the chemical composition of the fuel, on the chemical reaction, and on the atomic weights of the elements involved. See text for the equations used in the calculations

		Number of atoms in one molecule of substance					Number of molecules involved in reaction				Combustion completeness		Consumption factor	Emission factors		
		Elements and atomic weight					Reactants	Products			Equivalent oxygen-to-fuel ratio	Modified combustion efficiency		CO_2	H_2O	CO
		C	H	O	Molar	Type of	O_2	CO_2	CO	H_2O						
Substance	Formula	12	1	16	mass	combustion	32	44	28	18	EOFR	MCE	O_2			
Wood (equivalent)	$C_6H_9O_4$	6	9	4	145	Complete	6.25	6.0	0.0	4.5	1.00	1.00	1.38	1.82	0.56	0.00
						Flaming	6.0	5.5	0.5	4.6	0.96	0.92	1.32	1.67	0.56	0.10
						Smoldering	5.5	4.5	1.5	4.7	0.88	0.75	1.21	1.37	0.56	0.29

In wildland fires, smoldering and flaming combustion generally yield an MCE value between 0.75 and 0.92. Typical values for EOFR, MCE, and emission factors for complete, flaming, and smoldering combustion of wood are shown in Table 2.2.

Incomplete combustion results in a variety of additional products besides CO_2, H_2O, and CO, many of which are regulated as air pollutants. A more complete balanced equation of incomplete combustion of wood as a function of the Equivalent Oxygen to Fuel Ratio (EOFR) is shown in Eq. (2.9) (Ward 2001):

$$\begin{aligned} C_6H_9O_4 + (6.25\,EOFR)O_2 \rightarrow\ & [6 - 8.85(1 - EOFR)]CO_2 \\ & + [4.5 - 1.40(1 - EOFR)]H_2O \\ & + 6.65\,(1 - EOFR)CO \\ & + 0.70\,(1 - EOFR)CH_4 \\ & + 1.50\,(1 - EOFR)\mathrm{C} \end{aligned} \tag{2.9}$$

Furthermore, the fraction of methane computed using Eq. (2.9) can be divided as true methane (59.2%) and other hydrocarbons (40.8%), and the fraction of carbon can be subdivided by sizes, with particulates less than 2.5 μm (2.5 microns) in diameter comprising 69.2% of and particles above 10.0 μm comprising 18.4%.

Examples of the results of emission factors calculated for different values of the Equivalent Oxygen to Fuel Ratio (EOFR) using the Excel spreadsheet that is shown in this chapter are presented in Table 2.3.

Both the amount and composition of particulate emissions differ for flaming and smoldering combustion. As flaming combustion is more complete, the particulates from flaming combustion tend to be higher in ash minerals than those from smoldering combustion with more sodium (Na), potassium (K), chlorine (Cl), or sulfur (S) (Ward 2001). These latter elements, along with N, are part of the fuel that is burned.

The amount of particulate matter and gases emitted during combustion can be predicted using information about the type of fire, combustion efficiency, the amount

Table 2.3 Emission factors for the main products of wood combustion computed from the equations presented in the text with different values of the Equivalent Oxygen to Fuel Ratio (EOFR)

Emission factors (grams per 1000 g of fuel)			
Type of combustion	Complete	Flaming	Smoldering
Equivalent Oxygen to Fuel Ratio (EOFR)	1.00	0.93	0.80
Modified Combustion Efficiency (MCE)	1.00	0.92	0.75
Water (H_2O)	559	546	523
Carbon dioxide (CO_2)	1821	1632	1283
Carbon monoxide (CO)	0	90	257
Methane (CH_4)	0	3	9
Other hydrocarbons (CH)	0	2	6
Particulate matter (mostly C)			
PM < 2.5 μm	0	6	17
PM < 10.0 μm	0	7	20
Total particles	0	9	25

Table 2.4 Emission factors (grams per 1000 g of fuel) for different types of fuels burned in prescribed fires and wildfires. (Data from Urbanski 2014; Peterson et al. 2018)

	Prescribed fires			Wildfires	
Pollutant	Northwestern conifer forests	Western shrubland	Grassland	Northwestern conifer forest	Boreal forest
Carbon dioxide (CO_2)	1598	1647	1705	1600	1641
Carbon monoxide (CO)	105	74	61	135	95
Methane (CH_4)	5	4	2	7	3
Other hydrocarbons (CH)	27	18	17	34	23
Particulate matter (PM <2.5 μm)	18	7	9	23	22
Nitrogen oxides (NO_x)	2.1	2.2	2.2	2.0	1.0
Ammonia (NH_3)	1.6	1.5	1.5	1.5	0.8
Nitrous oxide (N_2O)	0.2	0.3	0	0.2	0.4
Sulfur dioxide (SO_2)	1.1	0.7	0.7	1.1	1.1

of fuel consumed, and empirically derived emission factors (Table 2.4). The Fire Emissions Production Simulator (FEPS) (Anderson et al. 2004), the First Order Fire Effects Model (Lutes 2017), and CONSUME (Prichard et al. 2007) are examples of modeling tools that are commonly used to predict wildfire emissions.

Field observations of emissions have suggested that the emission factors vary across fuel types and for prescribed fires and wildfires (Table 2.4). Prescribed fires in grasslands tend to have greater combustion efficiencies and, therefore, result in greater amounts of CO_2 and lower amounts of CO than prescribed fires and wildfires in coniferous forests. However, it is important to remember that the actual emissions during a wildfire or prescribed fire will vary with the amount of fuel consumed, fuel moisture, ignition pattern, wind, temperature, and other factors. Note that in coniferous forests of the US, the observed emissions of incomplete combustion products (CO, CH_4, other CH, and particulates) are only slightly higher in wildfires than in prescribed fires (Table 2.4; Urbanski 2014; Peterson et al. 2018). Andreae and Merlet (2001) summarized emission factors for more than 90 pyrogenic chemical species emitted from various types of biomass burning, from savannas and grasslands to tropical forests, to burning biofuel, charcoal, or agricultural residues.

2.5 From Emissions to Smoke Composition

Predicting the amount and composition of emissions from fires is important to develop strategies to minimize health impacts on downwind populations, and reduce the potential for hazardous travel conditions associated with impaired visibility

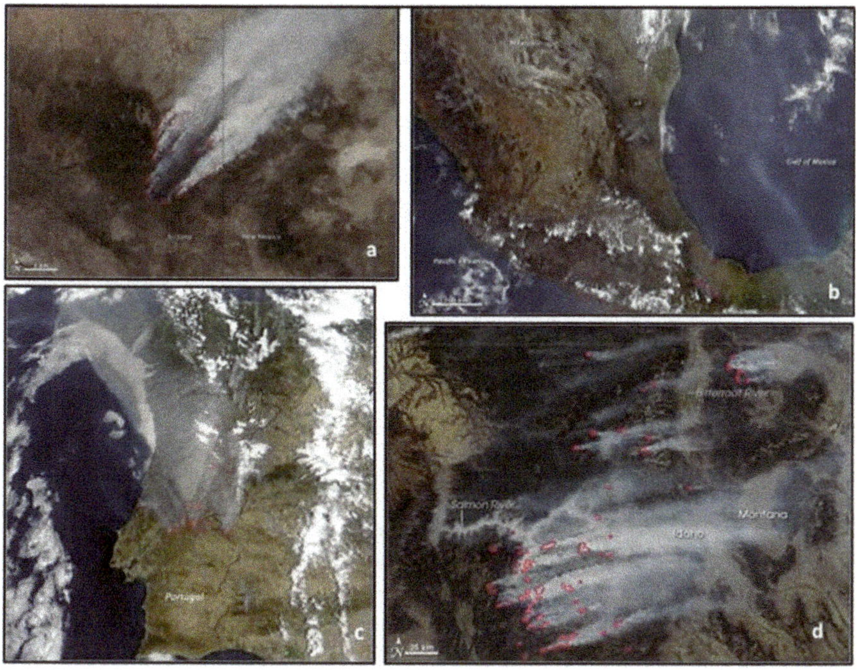

Fig. 2.8 Smoke from wildland fires often spreads far from the source, affecting air quality. When fires are widespread, many people can be affected by reduced visibility and enough particulates in the air to pose a health hazard. Note that the color of smoke can tell us what is burning. Higher prevalence of water vapor produces whiter smoke, while particulates make grey or black smoke. Here are satellite images from (**a**) Arizona in 2011, (**b**) Mexico in 2011, (**c**) Portugal in 2003, and (**d**) Idaho and Montana in 2007. (Images from Peterson et al. 2018)

(Fig. 2.8). However, besides information about the type of fire, combustion efficiency, the amount of fuel consumed, and emission factors, we also need to consider the elements and molecules in the atmosphere, including oxygen and nitrogen, while calculating the amount and composition of smoke (Table 2.5).

The quantity of air available significantly influences the combustion process and the products emitted. As shown in Eq. (2.4), the complete combustion of 145 g of wood requires 200 g of oxygen. Since oxygen comprises only 23.14% of the atmosphere, each gram of oxygen used in the combustion reaction involves 4.32 g of air, of which 75.52% (3.26 g) is nitrogen. In addition to nitrogen, we should also consider water vapor in the atmosphere, commonly reported as relative humidity (RH%). Relative humidity is calculated as the ratio of the current amount of water vapor in the atmosphere over the amount of water vapor in the atmosphere at the saturation point for a given temperature. The mass of water in the atmosphere can be computed based on the relative humidity and temperature (T_a °C) following Eq. (2.10):

Table 2.5 The normal composition of dry air in percentage volume and percentage mass by constituent gas (Atkin and Jones 2005)

	Molar mass	Normal composition of air	
Constituent gas	(g mol^{-1})	% volume	% mass
Nitrogen (N_2)	28	78.09	75.52
Oxygen (O_2)	32	20.95	23.14
Argon (Ar)	40	0.93	1.29
Carbon dioxide (CO_2)	44	0.03	0.05

$$Mass\ of\ H_2O\ in\ air = (Mass\ of\ O_2 + N_2) \times (RH\%/100\%) \times (0.000216)\,exp\,[0.0656(T_a)] \quad (2.10)$$

Fuel moisture is another important variable that significantly influences combustion and emissions. When the fuel moisture content is greater than some level, known as the moisture of extinction, fuels will not burn, while dryer fuels can burn readily. In general, fuel moisture is measured as the ratio between the mass of water and that of dry fuel. This is commonly expressed as a percentage calculated as the (wet weight—the oven-dry weight)/oven-dry weight. Thus, if the fuel moisture content is 10%, the mass of water in 1100 g of wood is 100 g, and this mass of water must be added to both sides of the combustion equations. In the products, this mass is added to the 559 g of water vapor emissions produced during the combustion of dry wood. The resulting N oxides and water vapor need to be considered in the combustion equations. They are essential to the thermodynamics of combustion (See Chap. 4) and are particularly important in the composition of smoke.

2.6 Implications

Understanding the products generated during combustion is critical as they are ultimately responsible for many of the effects on people, plants, animals, soils, and other parts of the ecosystems. The heat generated from combustion is discussed in Chaps. 4 and 5, while the effects of heat and smoke on plants, ecosystems, and people are discussed in Chaps. 9 and 10.

The composition of smoke includes many different gases, liquids, and solid particulates (mostly carbon), all of which are products of combustion and pose considerable risk to human health. Although it is true that "when there is fire, there is smoke," the smoke is often carried far downwind (Fig. 2.8). Smoke management is a growing concern among managers and policymakers due to the potential for long-distance transport and risk to human health of many emissions,

INPUTS:		LOW HEAT OF COMBUSTION	EQUIVALENT AIR-TO-FUEL RATIO
		2716 kJ / mol	0.92 dimensionless
COMPOSITION OF FUEL			
Carbon (C)	6.00	FUEL MOISTURE	AIR RELATIVE HUMIDITY
Hydrogen (H)	9.00	0.70 Mf = moisture / dry fuel	60 (percent %)
Oxygen (O)	4.00		
		FUEL INITIAL TEMPERATURE	AIR TEMPERATURE
Molar mass	145.0 g / mol	298 K	298 K

2. HEAT OF COMBUSTION (DRY FUEL)		3. HEAT OF PREIGNITION	
per gram of fuel	18.7 kJ / gram	Total heat of preignition	2.8 kJ / gram
per gram of oxygen	14.8 kJ / gram	To raise the temperature of wood	0.6 kJ / gram
per gram of air	3.4 kJ / gram	To raise the temperature of liquid water	0.2 kJ / gram
Ratio oxygen / fuel	1.3 (gram / gram)	To vaporize the water	1.6 kJ / gram
		To raise the temperature of water vapour	0.4 kJ / gram

4. HEAT IN COMBUSTIBLE PRODUCTS		4. ADJUSTED HEAT YIELD	
Heat in CO, CH_4 and C	1.68 kJ / gram	per gram of fuel	14.3 kJ / gram
		per mole of fuel	2072 kJ / mole

6. EMISSIONS		5. ESTIMATED FLAME TEMPERATURE	
Emission factors		Estimated temperature of gases in the flame	1706 K
Water vapor (H_2O)	1309 g / kg	Heat capacity of the mixture of products (per mol of fuel)	1472 J/K
Carbon dioxide (CO_2)	1605 g / kg	Estimated increase in temperature	1408 K
Carbon monoxide (CO)	103 g / kg		
Methane (CH_4)	4 g / kg		
Other Hydrocarbons (HC)	3 g / kg		
Particles < 2.5μm	7 g / kg		
Particles < 10μm	8 g / kg		
Total Particles	10 g / kg		

Fig. 2.9 A worked example using the interactive Excel spreadsheet on COMBUSTION_v2.0, including the required inputs that you can readily change, the global balanced equations, and the composition of smoke resulting from the composition of fuels and the completeness of the combustion. Use this to explore the relative influence of input factors on the heat (under what conditions will fires burn?) and emissions (these are components of smoke) products of combustion

an increasing human population, longer fire seasons, and widespread biomass burning (Bowman et al. 2009). Smoke can also affect visibility that puts highway and airport traffic at risk for accidents and affect the air quality over national parks and other areas where people go for recreation and to see beauty (Peterson et al. 2018). Globally, biomass burning and smoke emissions are increasingly targeted for reduction as part of addressing climate change and human health (Bowman et al. 2009; Johnston et al. 2012).

2.7 Interactive Spreadsheet: COMBUSTION

We suggest using the interactive spreadsheet COMBUSTION_V2.0 to explore the relationship of smoke composition to fire and fuel chemistry. See Fig. 2.9 to understand an example worked using the spreadsheet, and then adjust the inputs or outputs to see the implications of different fuel compositions and air supply. What input conditions will result in the most complete combustion? What combustion conditions have the least emissions, and which have the most emissions?

We encourage you to use the interactive spreadsheet (online supplementary material) to explore the relative importance of inputs for predicted emissions. Being able to interpret those results and how they change with inputs will strengthen your ability to explain how and why flaming and smoldering differ in the products and particulates produced.

References

Anderson, G., Sandberg, D. V., & Norheim, R. A. (2004). *Fire emissions production simulator (FEPS) user's guide version 10*. Seattle: USDA Forest Service Pacific Northwest Research Station.

Andreae, M. O., & Merlet, P. (2001). Emission of trace gases and aerosols from biomass burning. *Global Biogeochem Cycles, 15*(4), 955–966.

Atkin, P. W., & Jones, L. (2005). *Chemical principles: The quest for insight* (3rd ed.). New York: WH Freeman.

Bowman, D. M., Balch, J. K., Artaxo, P., Bond, W. J., Carlson, J. M., Cochrane, M. A., D'Antonio, C. M., DeFries, R. S., Doyle, J. C., Harrison, S. P., & Johnston, F. H. (2009). Fire in the earth system. *Science, 324*(5926), 481–484.

Byram, G. M. (1959). Combustion of forest fuels. In K. P. Davis (Ed.), *Forest fire: Control and use* (pp. 61–89). New York: McGraw-Hill.

Chandler, C., Cheney, P., Thomas, P., Trabaud, L., & Williams, D. (1983). *Fire in forestry: Forest fire behaviour and effects* (Vol. 1). New York: Wiley.

Drysdale, D. (1985). *An introduction to fire dynamics*. Chichester: Wiley.

Johnston, F. H., Henderson, S. B., Chen, Y., Randerson, J. T., Marlier, M., DeFries, R. S., Kinney, P., Bowman, D. M., & Brauer, M. (2012). Estimated global mortality attributable to smoke from landscape fires. *Environmental Health Perspectives, 120*(5), 695–701.

Larcher, W. (1977). *Ecofisiologia vegetal*. Barcelona: Omega.

Lutes, D. C. (2017). *FOFEM 6.4: First order fire effects model user guide*. USDA Forest Service Rocky Mountain Research Station, Fire Modeling Institute. Retrieved April 20, 2019, from https://www.firelab.org/sites/default/files/images/downloads/FOFEM6_Help_Aug2017.pdf.

Perrett, D. (2007). From 'protein' to the beginnings of clinical proteomics. *Proteomics: Clinical Applications, 1*(8), 720–738.

Peterson, J., Lahm, P., Fitch, M, George, M., Haddow, D., Melvin, M., Hyde, J., & Eberhardt, E. (Eds.). (2018). *NWCG smoke management guide for prescribed fire* (PMS 420-2 NFES 1279). Retrieved April 20, 2018, from https://www.nwcg.gov/sites/default/files/publications/pms420-2.pdf.

Philpot, C. W. (1971). *The pyrolysis products and thermal characteristics of cottonwood and its components* (Res Paper INT-107). Ogden: USDA Forest Service Intermountain Forest and Range Experiment Station.

Prichard, S. J., Ottmar, R. D., & Anderson, G. K. (2007). *Consume 3.0 user's guide*. USDA Forest Service Pacific Northwest Research Station, Pacific Wildland Fire Sciences Laboratory. Retrieved April 20, 2019, from https://www.fs.fed.us/pnw/fera/research/smoke/consume/consume30_users_guide.pdf.

Urbanski, S. P. (2014). Wildland fire emissions, carbon, and climate: Emission factors. *Forest Ecology and Management, 317*, 51–60.

Ward, D. (2001). Combustion chemistry and smoke. In E. A. Johnson & K. Miyanishi (Eds.), *Forest fires: Behavior and ecological effects* (pp. 55–77). San Diego: Academic.

Ward, D. E., Hao, W.-M., Susott, R. A., Babbitt, R. A., Shea, R. W., Kauffman, J. B., & Justice, C. O. (1996). Effect of fuel composition on combustion efficiency and emission factors for African savanna ecosystems. *Journal of Geophysical Research, 101*(23), 569–576.

Chapter 3
Heat Production

Learning Outcomes
Upon completion of this chapter, we expect you to be able to

1. Explain where the heat in fires comes from in chemistry terms,
2. Differentiate between higher and lower heat of combustion,
3. Understand the concept of heat yield,
4. Clearly explain in your own words how fuel characteristics and combustion completeness influence heat production, and
5. Use the interactive spreadsheet to evaluate how changes in fuel and combustion inputs alter the heat production from fires.

3.1 Heat Production

In the previous chapter, we discussed the effect of fuel and oxygen supply during combustion on smoke production and composition. In this chapter, we investigate the other major product released during combustion: heat. Recall that combustion reactions are always exothermic; that is, they result in the release of energy in the form of heat or light. The primary fuel involved in wildland fires is composed of dead and living plant material produced through photosynthesis. Photosynthesis is a process that uses light energy from the sun, water, and carbon dioxide to create chemical energy that is stored in molecules such as glucose, which are used to make other more complex substances such as cellulose. When plants or plant parts die,

Supplementary Information The online version of this chapter (https://doi.org/10.1007/978-3-030-69815-7_3) contains supplementary material, which is available to authorized users.

F. Castro Rego et al., *Fire Science*, Springer Textbooks in Earth Sciences, Geography and Environment, https://doi.org/10.1007/978-3-030-69815-7_3

decomposition breaks down photosynthesis products into simpler organic compounds and releases the stored energy. Both decomposition and combustion can be thought of as the reverse of photosynthesis in that they break down plant substances into their chemical constituents and release the stored energy. The major difference between combustion and decomposition is the rate at which the reaction occurs (Byram 1959). Thus, photosynthesis is:

$$\begin{aligned} &\textit{Carbon dioxide} + \textit{Water} + \textit{Energy}\ (\textit{solar radiation}) \\ &\quad \rightarrow \ \textit{Plant substances} + \textit{Oxygen} \end{aligned} \tag{3.1}$$

while combustion and decomposition can be described as:

$$\begin{aligned} &\textit{Plant substances} + \textit{Oxygen} + \textit{Energy}\ (\textit{heat}) \\ &\quad \rightarrow \textit{Carbon dioxide} + \textit{Water} + \textit{Energy}\ (\textit{heat}) \end{aligned} \tag{3.2}$$

Decomposition and combustion of wildland fuels are both often incomplete, so not all organic matter is consumed but is often broken into smaller pieces. As explained in Sect. 3.4, some organic compounds are more readily decomposed than others. Rotting wood is higher in lignin than living or recently dead wood, which affects the potential for the remaining organic matter to burn in combustion.

In Chap. 2, we discussed the principles of conservation of mass and species. In this chapter, we analyze the chemical equations of combustion from the conservation of energy. Within a given system, the amount of energy remains constant as energy can not be created nor destroyed. However, energy can be converted from one form to another (e.g., potential energy can be converted to kinetic energy). The amount of heat produced during combustion can be evaluated based on either the strength of the fuel's chemical bonds or chemical constituents. We start this by building upon fire chemistry principles introduced in Chaps. 1 and 2 by exploring how various chemical bonds influence the net energy release. Next, we discuss how the chemical components of wildland fuels (organic matter, minerals, volatiles, and moisture) influence flammable vapor production and heat release during combustion (Fig. 3.1).

3.2 The Net Energy Release in Combustion and the Strength of Chemical Bonds

From the perspective of the conservation of energy, combustion reactions can be understood as a two-step process. First, energy is required to break the chemical bonds of the reactants (endothermic step). Second, energy is released (exothermic step) when the new chemical bonds are formed. These two steps can be visualized as occurring in two different zones within a flame. First, the fuel molecules within the flame are heated such that they split into atoms; these atoms then react with oxygen in the reaction zone to form new molecules (Fig. 3.2). In a combustion reaction, the

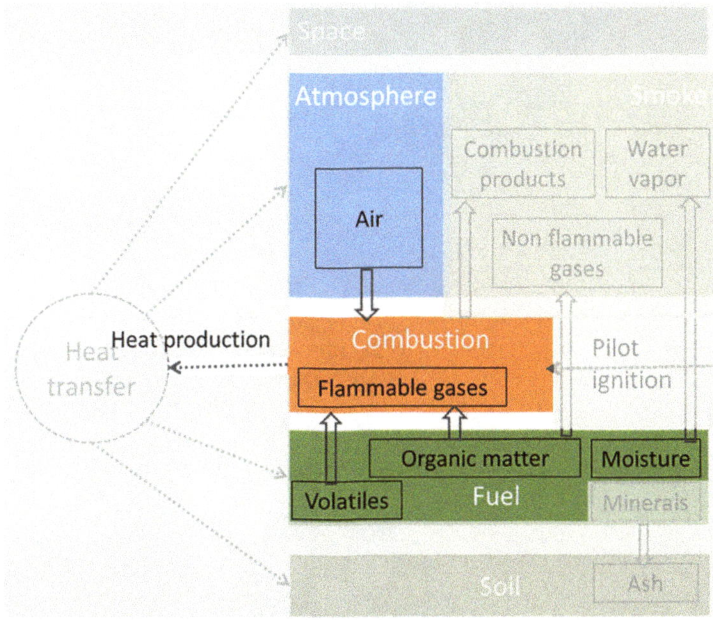

Fig. 3.1 Heat production is influenced by fuel and fuel moisture, by the combustion of flammable gases, and by oxygen in the air. In this chapter, we focus on the colored portions of this diagram of our general conceptual model

energy released in the second step is always greater than the energy absorbed in the first step. Therefore, combustion always results in the release of energy.

The amount of energy absorbed and released during combustion depends upon the strength of the chemical bonds associated with both the fuel and products involved in the combustion reaction. The strength of the chemical bonds increases as the number of electron pairs in the bond increases. Single (–), double (=), or triple (Ξ), bonds share one, two, and three electrons, respectively. The strength of the bonds is commonly measured by the average bond dissociation energy expressed in kilojoules per mole or kilojoules per gram (Table 3.1). The net energy released during combustion can be estimated as the difference between the heat released in forming the bonds of the products and the heat absorbed in breaking the bonds of the reactants (Eq. 3.3):

$$\begin{aligned}\text{Net energy release } (\Delta H) = {} & \text{Energy to form the bonds of products} \\ & - \text{Energy to break bonds of reactants}\end{aligned} \quad (3.3)$$

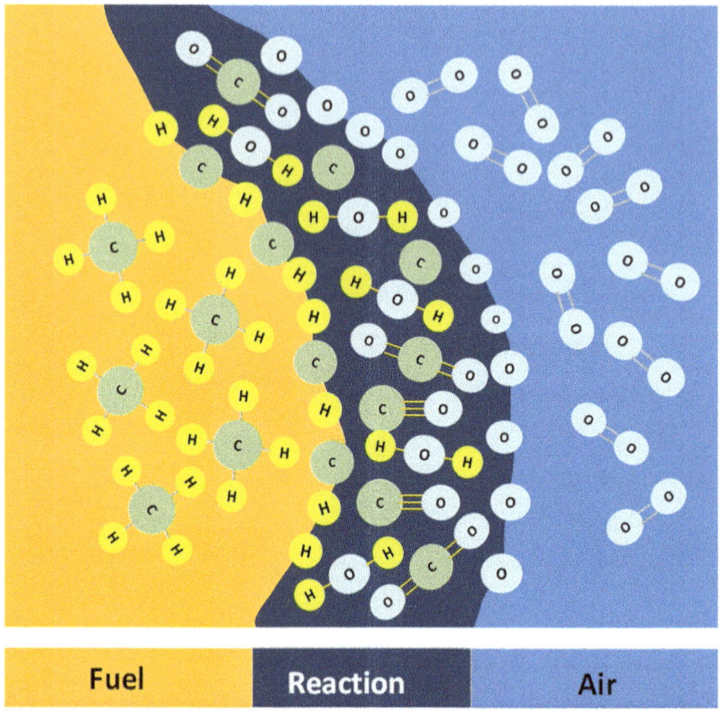

Fig. 3.2 Methane burning in the reaction zone of a flame. The chemical bonds of the methane molecules of the fuel inside the flame *(yellow, left)* break by heat into carbon and hydrogen atoms (this thermal degradation is pyrolysis) that react with oxygen atoms after the break of heated oxygen molecules in the air *(blue, right)*. In the reaction zone *(middle)*, combustion produces molecules of carbon dioxide, water, and carbon monoxide as well as carbon particles

Table 3.1 Typical bond strengths measured by their dissociation energy ($kJ\ mol^{-1}$) between atoms of carbon (C), hydrogen (H), and oxygen (O) of different types (single bond –, double bond =, and triple bond Ξ) (From Atkin and Jones 2005)

Chemical bond	Bond strength ($kJ\ mol^{-1}$)
C–C	348
C=C	612
CΞC	837
C–H	413
C–O	360
C=O	804
CΞO	1062
H–H	424
O–H	460
O=O	497

For example, we can calculate the net energy release for the complete combustion of methane in oxygen (Fig. 3.3), assuming the reactants include one molecule of methane, CH_4, and two molecules of oxygen, $2O_2$. Methane contains four single bonds between the carbon and hydrogen atoms (4 C–H), and oxygen has two double bonds between oxygen atoms (2 O=O) (Fig. 3.3). The products of this reaction include one molecule of carbon dioxide, CO_2, and two molecules of water, 2 H_2O. Carbon dioxide has two double bonds between the atoms of carbon and oxygen (C=O), and water has four single bonds between the oxygen and hydrogen atoms (4 O–H). Using Eq. (3.3) and the bond strengths in Table 3.1, we get:

$$\begin{aligned} \textit{Net energy release}\ (\Delta H) &= [2(804) + 4(460)] - [4(413) + 2(497)] \\ &= 3448 - 2646 = 802\ kJ\ mol^{-1} \end{aligned} \quad (3.4)$$

We can convert the net energy released from kJ mol^{-1} to kJ g^{-1} by dividing by the molar mass of methane, 16 g mol^{-1} as (Eq. 3.5):

$$\begin{aligned} &\textit{Net energy release}\ \left(\Delta H\ in\ kJ\ g^{-1}\right) = 802\ kJ\ mol^{-1}/16\ g\ mol^{-1} \\ &= 50.1\ kJ\ g^{-1} \end{aligned} \quad (3.5)$$

Calculations similar to those presented for methane can be done for all substances for which the chemical composition and the nature and number of bonds are known. The chemical composition and the nature, strength, and number of chemical bonds associated with common molecules composed of C, H, and O that are common in wildland fuels can be seen in Fig. 3.4. The calculations of the net energy release from bond nature, strength, and number are summarized for the same molecules in Table 3.2.

An alternative way to estimate the net energy release (ΔH) of a combustion reaction is by using the concept of enthalpy of formation. The standard enthalpy of formation of a compound is the change in enthalpy when one mole of a substance is formed from its pure elements. Values for the standard enthalpy of formation for many compounds are available in many classical books in chemistry (e.g., Atkin and Jones 2005). Because there is no energy associated with the formation of an element in its standard state (e.g., oxygen or graphite), the standard enthalpy of formation is zero.

We can then compute the net energy release by taking the difference between the enthalpies of formation of reactants and that of products of the reaction:

$$\textit{Net energy release}\ (\Delta H) = \Delta H'_R - \Delta H'_p \quad (3.6)$$

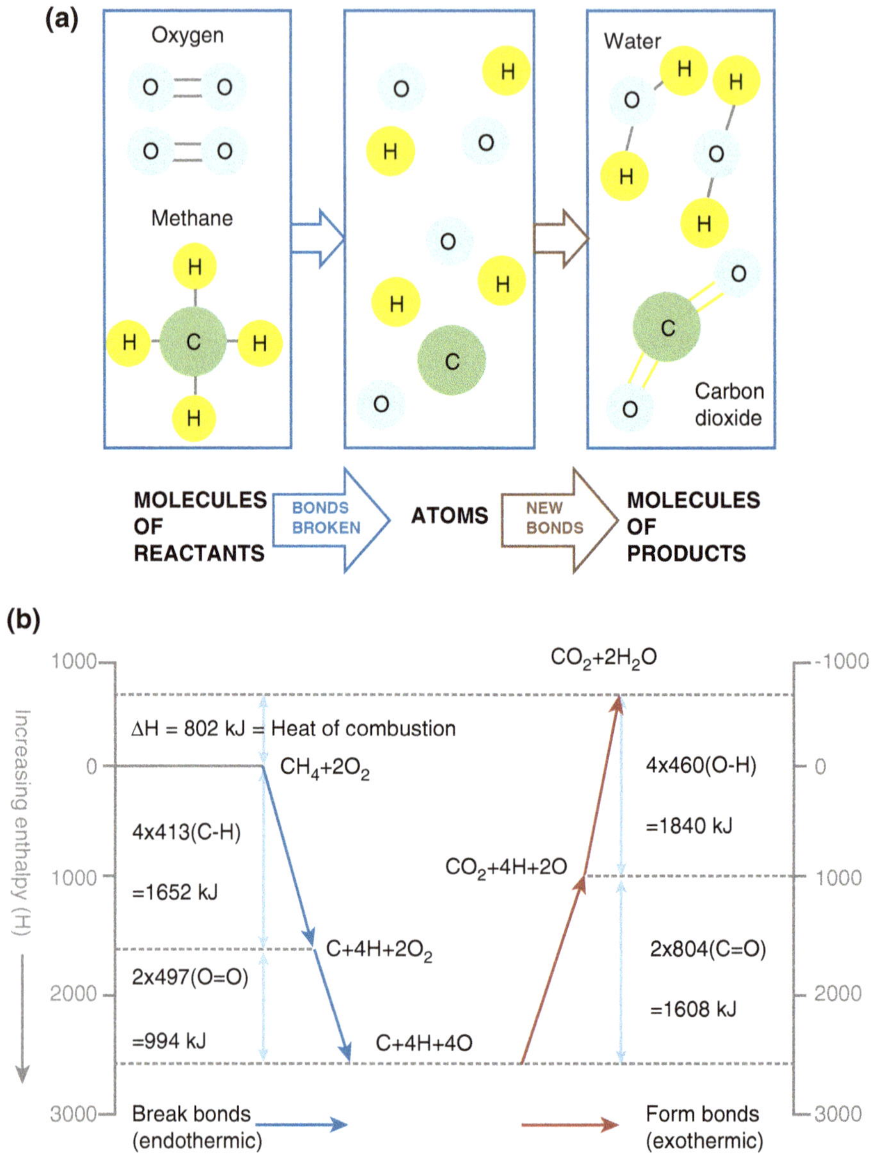

Fig. 3.3 The complete combustion of methane. (**a**) In the first step, four chemical bonds C–H in methane and two double bonds O=O of oxygen are broken absorbing energy resulting in dissociated atoms of C, H, and O. in the second step these atoms are combined to form two double bonds C=O in carbon dioxide and four single bonds O–H in water, releasing energy. (**b**) Applying the values for the strength of the bonds in Table 3.1 and the calculations explained in the text example, we can represent the energy absorbed and released from a mole of methane and conclude that its complete combustion results in a net energy release (heat of combustion) of 802 kJ mol^{-1}

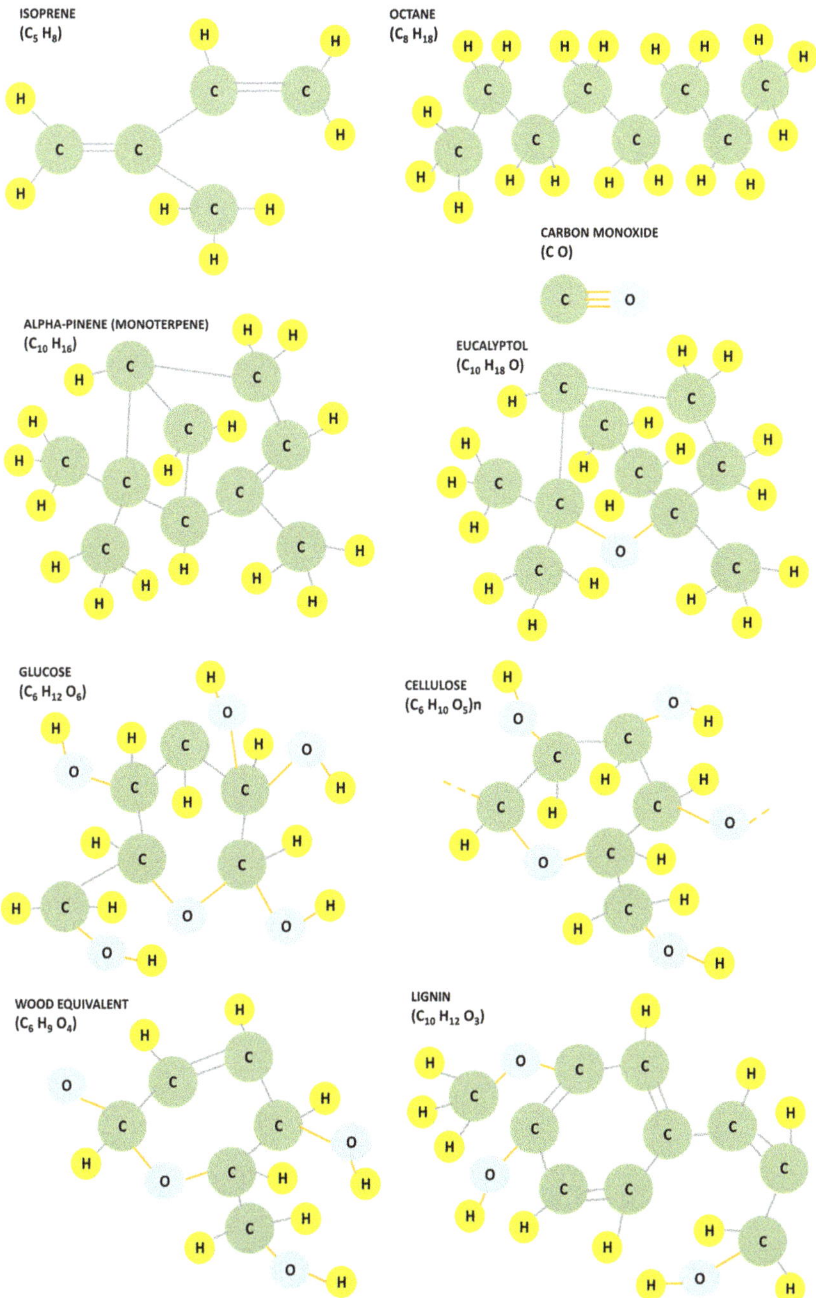

Fig. 3.4 Simplified representations of molecules commonly involved in combustion during vegetation fires. Both the chemical formulas and the bonds are shown

Table 3.2 Examples of calculations of the net energy release of various substances based on their chemical composition and on the nature, strength, and number of the chemical bonds

		Number of atoms and bonds in one molecule of the substance													Number of molecules				Bond energy ($kJ.mol^{-1}$)				Energy balance (kJ/g)		
		Elements and atomic weight				Types of bonds and corresponding strengths									Reactants		Products		Reactants		Products				
		C	H	O	Molar mass	C–C	C=C	C≡C	C–H	C–O	C=O	C≡O	H–O	O=O						O_2	CO_2	H_2O			
Substance	Formula	12	1	16	(g. mol^{-1})	348	612	837	413	360	804	1062	460	497	Fuel	O_2	CO_2	H_2O	Fuel	497	1608	920	Breaking bonds	Forming bonds	Net balance
Oxygen	O_2			2	32									1	1	-1	0	0	497	−497	0	0	0.0	0.0	0.0
Water	H_2O		2	1	18								2		1	0	0	1	920	0	0	920	51.1	51.1	0.0
Carbon dioxide	CO_2	1		2	44						2				1	0	1	0	1608	0	1608	0	36.5	36.5	0.0
Carbon monoxide	CO	1		1	28							1			1	0.5	1	0	1062	249	1608	0	46.8	57.4	10.6
Methane	CH_4	1	4		16				4						1	2	1	2	1652	994	1608	1840	165.4	215.5	50.1
Isoprene	C_5H_8	5	8		68	2	2		8						1	7	5	4	5224	3479	8040	3680	128.0	172.4	44.4
Octane	C_8H_{18}	8	18		114	7			18						1	12.5	8	9	9870	6213	12,864	8280	141.1	185.5	44.4
Monoterpenes	$C_{10}H_{16}$	10	16		136	10	1		16						1	14	10	8	10,700	6958	16,080	7360	129.8	172.4	42.5
Eucalyptol	$C_{10}H_{18}O$	10	18	1	154	9			18	2					1	14	10	9	11,286	6958	16,080	8280	118.5	158.2	39.7
Wood (equivalent)	$C_6H_9O_4$	6	9	4	145	4	1		7	5			2		1	6.25	6	4.5	7615	3016	9648	4140	73.9	95.1	21.2
Cellulose	$C_6H_{10}O_5$	6	10	5	162	5			7	6			3		1	6	6	5	8171	2982	9648	4600	68.8	88.0	19.1
Glucose	$C_6H_{12}O_6$	6	12	6	180	5			7	7			5		1	6	6	6	9451	2982	9648	5520	69.1	84.3	15.2
Lignin	$C_{10}H_{12}O_3$	10	12	3	180	5	4		10	4			2		1	11.5	10	6	10,678	5716	16,080	5520	91.1	120.0	28.9

The strength of the chemical bonds used in the calculations are the same as in Table 3.1

where: ΔH is the net energy release, $\Delta H'_R$ is the enthalpy of formation of reactants, and $\Delta H'_P$ is the enthalpy of formation of the products.

To demonstrate the use of the standard enthalpy of formation to estimate the net energy released, let us revisit the complete combustion of methane in oxygen. As shown in Fig. 3.3, one mole of methane (CH_4) reacts with two moles of oxygen (O_2) to produce one mole of carbon dioxide (CO_2) and two moles of water (H_2O). From calorimeter studies, the following values are given by Atkin and Jones (2005) for the enthalpy of formation: methane (CH_4 gas) as -74.8 kJ mol^{-1}, carbon dioxide (CO_2 gas) as -393.5 kJ mol^{-1}, and water (H_2O gas) as -241.8 kJ mol^{-1}. Because oxygen is in its standard state, its standard enthalpy of formation is 0 kJ mol^{-1}. Substituting these values into Eq. (3.6) results in the following:

$$\begin{aligned} \textit{Net energy release } (\Delta H) &= -74.8 - [-393.5 + 2(-241.8)] \\ &= 802.3 \; kJ\; mol^{-1} \end{aligned} \tag{3.7}$$

This is the same result as obtained in the previous example using Eq. (3.4).

3.3 Energy Release and Heat of Combustion

In classical fire literature, the term heat of combustion is generally defined as the net energy released when a substance undergoes complete combustion under standard conditions (Drysdale 2011). However, the terminology associated with the heat of combustion can be confusing as the terms heating value, energy value, heat content, and calorific value are also occasionally used as synonyms. Furthermore, there are two distinct, but related heat of combustion estimates that are used to describe the net energy released: the higher and the lower heat of combustion.

The difference between higher heat of combustion (ΔHH) and lower heat of combustion (ΔHL) can be explained based on how the water produced during combustion is considered. Recall from Eq. (2.4) that during the combustion of 1 g of dry wood, approximately 0.56 g of water vapor is produced. If we measure the total amount of energy released after all of the combustion products have returned to the initial temperatures and states, we would have estimated a value called the high heat of combustion (or gross heat of combustion or higher heat content). The distinguishing feature of the high heat of combustion is that the water vapor produced during combustion condenses back to a liquid state, thus releasing the energy that was initially used to vaporize it, called the latent heat of vaporization.

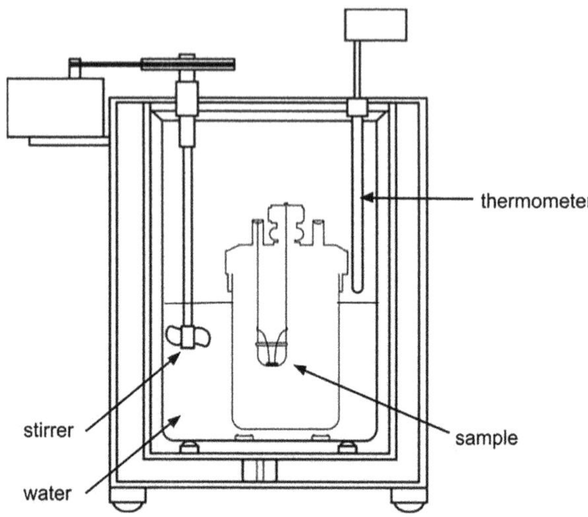

Fig. 3.5 A schematic representation of a bomb calorimeter, generally made from stainless steel, where the combustion reaction occurs at constant volume and without heat flow to the exterior of the bomb. The increase in the temperature of the water is used to estimate the net heat release in the combustion (Adapted from Polik 2000)

The high heat of combustion is useful in industrial settings where the condensation of water is an important consideration, such as estimating the energy from a gas-fired boiler. The estimation of the high heat of combustion of wildland fuels is practically impossible to do from first principles as we seldom know the precise chemical composition of the fuels. Therefore, the high heat of combustion is often estimated using a bomb calorimeter (Fig. 3.5).

In wildland fires, the energy associated with water vapor production does not significantly contribute to fire behavior or effects. Therefore, it should be subtracted from the high heat of combustion, providing an estimate of the lower heat of combustion. The lower heat of combustion is always smaller than the higher heat of combustion, as the water vapor produced during combustion does not condense back to a liquid, and the latent heat associated with the water vapor is not recovered.

It is important to recognize that the calculations of net energy released (ΔH) (Sect. 3.2) correspond to the lower heat of combustion (ΔHL) since they incorporate the energy required to create water vapor. Also, it should be noted that the values presented in Table 3.2 are theoretical and higher than those commonly used in practice for the lower heat of combustion for wildland fuels. Although the low heat of combustion is often considered a constant, differences in the fuel composition among species and throughout a growing season can influence estimates of the lower heat of combustion.

The measurement of the energy released in combustion is often made in dry samples of the fuel of interest using a bomb calorimeter (Fig. 3.5). This can be done for pure substances, but it is particularly important for actual wildland fuels where the estimation of the energy release is practically impossible to be made from first principles using bond strength as we seldom know the precise chemical composition of the fuels. Because the water vapor released in the combustion of a dry sample is allowed to condense, the estimation of the net energy release by bomb calorimetry results in the value of high heat of combustion (ΔHH).

The lower heat of combustion can be estimated after measuring the higher heat of combustion. If we measure the total amount of energy released after all of the combustion products have returned to their initial temperatures and states, we would have estimated the high heat of combustion (or gross heat of combustion or higher heat content). The distinguishing feature of the high heat of combustion is that the water vapor produced during combustion condenses back to a liquid state, releasing the energy that was initially used to vaporize it, called the latent heat of vaporization. The difference between higher and lower heat of combustion can be illustrated with the example of methane.

To demonstrate the differences between the higher and lower heat of combustion, we can continue to use the complete combustion of methane in oxygen. Based on measurements in a bomb calorimeter, it has been determined that the higher heat of combustion (ΔHH) for methane is 890 kJ mol^{-1}. The combustion reaction associated with bomb calorimeter measurement is:

$$\begin{aligned} &CH_4(gas) + 2O_2(gas) \rightarrow CO_2(gas) + 2H_2O(liquid) \\ &\Delta HH = 890\ kJ\ mol^{-1} \end{aligned} \tag{3.8}$$

Notice that in this reaction the products include carbon dioxide and liquid water. However, if we were to measure the energy released while the water was still a gas we would have the following:

$$\begin{aligned} &CH_4(gas) + 2O_2(gas) \rightarrow CO_2(gas) + 2H_2O(gas) \\ &\Delta HL = 802\ kJ\ mol^{-1} \end{aligned} \tag{3.9}$$

The difference between these two values is because for the higher heat of combustion we considered that the water vapor condensed back to a liquid, while for the lower heat of combustion the water is in vapor form. If this water vapor is allowed to condense, an additional 88 kJ is given off as heat. This difference corresponds to the energy required for the vaporization of water (44.0 kJ mol^{-1} for two moles of water). We can also express the higher and lower heat of combustion in terms of energy per unit mass by dividing each by the molar mass of methane (16 g) which results in 55.6 and 50.1 kJ g^{-1}, respectively.

For common wildland fuels, similar calculations can be done. Recall from Eq. (2.4) that during the complete combustion of one gram of common wildland fuels, like dry wood, approximately 0.56 g of water vapor is produced. The energy associated with converting this water to vapor is 1.4 kJ g^{-1}, which can be estimated by multiplying 0.56 g of water vapor by the latent heat of vaporization for water at 25 °C, which is 2.44 kJ g^{-1}. The low heat of combustion can then be estimated by subtracting 1.4 kJ g^{-1} from the higher heat of combustion. The values for the high and low heat of combustion do not vary widely between different types of wildland fuels and thus, for many practical purposes, the values of 20.1 kJ g^{-1} for the high heat of combustion and of 18.7 kJ g^{-1} for the low heat of combustion are often assumed as constant for most wildland fuels (Van Wagner 1972; Alexander 1982). However, some differences exist, which will be discussed in the next sections.

3.4 Estimating Heat Release from Fuel Composition

The heat generated during combustion can be estimated globally or as the sum of the heat from various fuel components involved in the combustion reaction and, in particular, the amount of volatiles and char. Recall that the relative proportions of char and volatile gases produced depend upon combustion completeness (See Chap. 2). In Fig. 3.6 we provide a graphical representation of various fuel components, their mass, and effects on energy associated with wildland fires. Note that both the higher and lower heat of combustion are estimated based on dry fuels. Thus, neither the fuel moisture nor the mass of minerals, which are not combusted during a wildfire, contribute to the calculation of the heat of combustion of fuels. After removing minerals and fuel moisture, the remaining mass of dry fuel results in the production of volatiles and char. The heat generated from volatile compounds contributes to flames and fire spread. In contrast, the heat from char is primarily

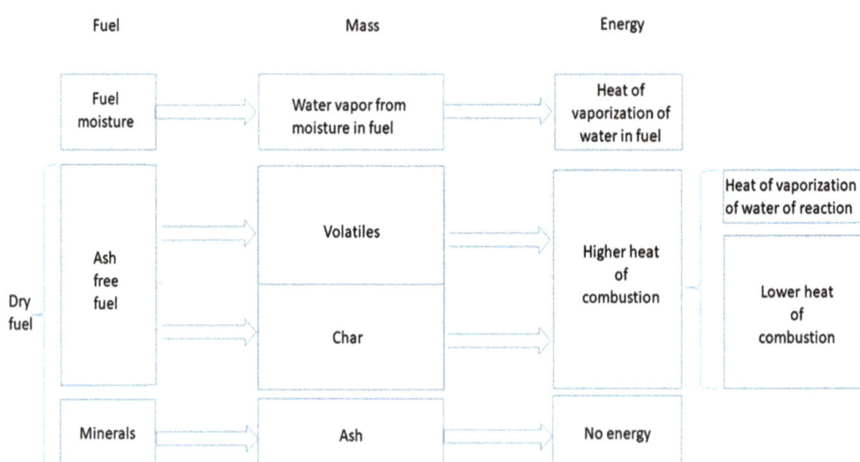

Fig. 3.6 A schematic diagram showing the relationship between fuel components and the corresponding mass and energy components included in the combustion of vegetation fuels

released in the slower process of smoldering combustion and plays an important role in determining fire effects on plants and soils (See Chap. 9).

Typically, estimates of the values for the heat of combustion of a fuel (ΔH_{fuel}) do not distinguish between energy produced through the combustion of volatiles ($\Delta H_{volatiles}$) vs. energy produced through char (ΔH_{char}). However, since the rate of heat release is different between char and volatiles, it can be useful to partition the heat of combustion into various components:

$$\Delta H_{fuel} = \Delta H_{volatiles} \times \rho_{volatiles} + \Delta H_{char} \times \rho_{char} \tag{3.10}$$

For many thermodynamic calculations, the higher heat of combustion of char can be considered a constant for wildland fuels, ΔH_{char} at around 29.2 kJ g^{-1} (Rothermel 1976) or 32.0 kJ g^{-1} (Susott 1982). Because the proportion of char and the higher heat of combustion are both easily measured experimentally, we can calculate the heat of combustion for volatile products (per unit weight of the original fuel) from the conservation of energy equation (Susott et al. 1975):

$$\Delta H_{volatiles} \times \rho_{volatiles} = \Delta H_{fuel} - \Delta H_{char} \times \rho_{char} \tag{3.11}$$

where $\Delta H_{volatiles} \times \rho_{volatiles}$ is the energy produced due to volatile combustion, ΔH_{fuel} is the heat of combustion of the fuel, and ρ_{char} is the proportion of total fuel mass that produces char. $\Delta H_{volatiles}$ is the higher or lower heat of combustion depending on ΔH_{fuel} is the higher or the lower heat of combustion of the fuel.

The proportion of the fuel mass that is converted to char or volatiles depends on the fuel type (Albini 1980). Therefore, to estimate the partitioning of heat between char and volatiles, we need to know more about the composition of the fuels typically involved in wildland fires. Plant tissues' main components can be organized in classes, from organic compounds such as cellulose, hemicellulose, lignin, and extractives, to minerals and water (Fig. 3.7).

The chemical composition of plant materials (Fig. 3.7) differ by plant parts (e.g., wood, stems or foliage), in different conditions (e.g., heartwood vs. rotten wood), and from different species (e.g., western larch (*Larix occidentalis*) vs. ponderosa pine (*Pinus ponderosa*) vs. saltbush (*Atriplex cuneata*)), all of which result in differences in the proportions of char and volatile production and heat of combustion. Rothermel (1976) determined that the heat of combustion was higher for partially decomposed ('punky') wood of Douglas-fir (*Pseudotsuga menziesii*) due to the high percentage of lignin compared to solid wood. In addition, seasonal variations in the chemical composition of vegetation are also important. Philpot and Mutch (1971) found a 4–7% increase in extractive content for ponderosa pine needles during the fire season. Philpot (1971) found an increase of extractives during the fire season from 8 to 12% in aspen (*Populus tremuloides*) leaves.

Cellulose and hemicellulose are the two major substances in plant tissues. They are both polymers derived from glucose (Fig. 3.4) that differ in their chain lengths. However, they have similar combustion properties and are usually combined into a

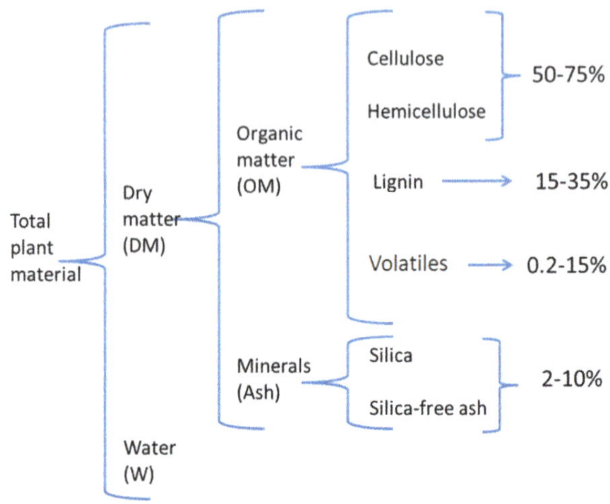

Fig. 3.7 Typical values for the main substances in vegetative material involved in the combustion of wildland fuels. (Adapted from Chandler et al. 1983)

single constituent for practical applications. The low heat of combustion for cellulose and hemicellulose has been estimated to be around 16.1 kJ g^{-1} (Chandler et al. 1983). Both cellulose and hemicellulose tend to produce volatile gases during pyrolysis. However, the presence of inorganic materials, and especially silica-free minerals, particularly phosphates, promotes the formation of char at the expense of flammable volatiles.

Although silica does not affect the combustion rate of cellulose (Philpot 1968), Rothermel (1976) found that the proportion of char formed from cellulose combustion depends upon the proportion of the silica-free mineral content (S_e) of the fuel (Eq. 3.12). In the absence of silica-free minerals, the amount of char produced by cellulose is less than 1% (0.48–0.92%, Susott 1982).

$$\textit{Proportion of char from cellulose} = 0.092 + 0.5\ S_e^{\ 0.462} \tag{3.12}$$

Lignin gives wood its stiffness. It is a heterogeneous polymer that is primarily derived from coniferyl alcohol (see Fig. 3.4 for a simple representation of lignin chemistry). Although the chemical composition of lignin differs in hardwoods and softwoods, their combustion properties are similar. Lignin has a low heat of combustion that is about 50% greater than cellulose (Table 3.3) due to the more complex chemical bonds, which require more energy to break apart, resulting in slower decomposition and combustion. That is why rotten or "punky" wood (i.e., wood that has decayed to the point of being soft) has high percentages of lignin and why it generally burns more by smoldering than by flaming combustion and does not contribute directly to the spread of flames in fires (though it does smolder and can be ignited by embers that then start spot fires (see Chap. 8). The high resistance of lignin to thermal degradation explains Rothermel's estimation (1976) that the

Table 3.3 Indicator values of basic substances in plant material (ash-free) for the proportion of char produced and their corresponding values for the higher heat of combustion ΔHH determined by bomb calorimetry (From Rothermel 1976)

Substance	Proportion of char from combustion	Higher heat of combustion ΔHH (kJ g^{-1})
Cellulose and hemicellulose	0.092	16.1
Lignin	0.624	24.5
Volatiles	0.285	32.3

average proportion of dry weight for lignin used to produce flames was as low as 0.376 and that the remaining proportion (0.624 of dry weight) produced char.

Volatiles are low molecular-weight organic compounds that can greatly influence flammability and fire behavior. There are a number of important volatiles that can influence fire behavior, including resins that have alpha-pinene, essential oils such as eucalyptol, and terpenes such as isoprene and isoprene polymers. Volatile compounds such as these have low boiling points "and can form flammable, volatile mixtures well in advance of the flame front in a forest fire" (Chandler et al. 1983). Differences in flammability between plant species are often attributed to variation in volatile oil content. For example, *Eucalyptus,* which is often considered highly flammable, can contain up to 3% of highly flammable oils such as eucalyptol (or 1,8-cineol) (Sebei et al. 2015). The low heat of combustion for volatile compounds is estimated to be around 32.3 kJ g^{-1} (Chandler et al. 1983), which is twice as much as cellulose. Rothermel (1976) indicated that an average proportion of 0.715 of the initial dry mass of volatiles is used to produce flames, whereas only a proportion, 0.285, of the mass of volatiles produces char.

Minerals, including silica and silica-free ash, also influence the organic matter consumed in fires. Their presence interferes with the combustion process by causing an increase in char production and a consequent decrease in flammable volatiles. Because minerals do not burn, they are often used in fire retardants to suppress flammability. As mentioned earlier in this chapter, silica and silica-free ash have different effects on cellulose combustion.

Char is produced from the combustion of all vegetation fuels. Typical values for the proportion of char from the combustion of the main components of plant material in fuel are shown in Table 3.3.

The higher heat of combustion for any fuel per unit of dry weight (ΔHH_{fuel}) can be estimated using an average of the values of the basic substances (ΔHH_i) as in Table 3.3, weighted by their proportions in the fuel (P_i):

$$\Delta HH_{fuel} = \sum\nolimits_{i=1}^{n} (P_i x \, \Delta HH_i) \quad (3.13)$$

If the proportions are expressed as a fraction of dry matter, the proportion of minerals should be considered with $\Delta HH = 0$, as minerals do not contribute to the heat of combustion.

The consideration of the mineral content of the fuels can be important in some specific situations. In rotten fuels, the decomposition of organic material results in a higher percentage of minerals in the chemical composition. Lower duff, where organic material is mixed with soil material, can also have high mineral content, as soil fauna commonly mix the organic layers on top of the soil with the surface mineral soil. This is one of the reasons that the compact duff layers don't burn readily in flaming combustion. Minerals can determine the high heat of combustion for plants that have high mineral content.

Mineral content influences the ignition probability of organic soils. Frandsen (1987) showed that the limits for smoldering ignition of a mixture of peat moss with water and mineral soil depended on fuel moisture and mineral content. Later Frandsen (1997) extended his studies to ignition tests on organic soil samples from Alaska and the northern and southeastern United States showing that, even for completely dry peat moss, ignition does not occur when the mineral content of the fuel is above 81.5% and this limit is much reduced with increasing fuel moisture.

In this example, we are going to estimate the higher heat of combustion for leaves of valley saltbush (*Atriplex cuneata*) using Eq. (3.14) and the higher heat of combustion values in Table 3.3. Valley saltbush grows in semi-arid parts of the southwestern portion of the USA and is considered to have low flammability. The saltbush leaves' composition, measured as proportions of dry weight, consists of 0.464 cellulose, 0.327 lignin, 0.023 extractives (volatiles), and 0.180 of silica-free ash (Rothermel 1976). Substituting the higher heat of combustion values from Table 3.3 into Eq. (3.14) results in the following:

$$\begin{aligned}\Delta HH_{fuel} &= 0.464 \times 16.1\ kJ\ g^{-1} + 0.327 \times 24.5\ kJ\ g^{-1} + 0.023 \\ &\quad \times 32.3\ kJ\ g^{-1} \\ &= 16.2\ kJ\ g^{-1}\end{aligned} \quad (3.14)$$

We can now estimate the mass of char as a proportion of the initial mass of the fuel using Eq. (3.12). By summing the product of the proportion of each component in the fuel by the corresponding proportion of char formed, we get:

$$\begin{aligned}P_{char} &= 0.464 \times \left(0.0917 \times 0.180^{0.462}\right) + 0.327 \times 0.624 + 0.023 \\ &\quad \times 0.285 \\ &= 0.358\end{aligned} \quad (3.15)$$

Thus, we conclude that 35.8% of valley saltbush foliage will become char during combustion and the remaining 64.2% will produce volatile gases that are likely to burn in flaming combustion.

With a measurement or estimate of the higher heat of combustion of fuel ($\Delta HH_{fuel} = 16.2\ kJ\ g^{-1}$), a constant value for the higher heat of combustion of

(continued)

char ($\Delta HH_{char} = 29.2\ kJ\ g^{-1}$), and the proportion of char produced from the fuel, we can use Eq. (3.10) to calculate the energy available from volatiles:

$$\begin{aligned}\Delta HH_{volatiles} x \rho_{volatiles} &= \Delta HH_{fuel} - \Delta HH_{char} x \rho_{char} \\ &= 16.2\ kJ\ g^{-1} - 29.2\ kJ\ g^{-1} \times 0.358 \\ &= 5.7\ kJ\ g^{-1} \end{aligned} \tag{3.16}$$

The variation in the chemical composition of wildland fuels directly results in different estimates for the lower heat of combustion and its partitioning between volatiles and char (Fig. 3.8). The typical ranges of variation of characteristics of common wildland fuel types and the partitioning of the higher heat of combustion between energy for volatiles and char are shown in Table 3.4.

3.5 Estimating Heat Yield

The higher and lower heats of combustion are best thought of as the maximum heat generated during combustion. However, in real wildland fire scenarios, other factors such as incomplete combustion and fuel moisture will decrease the amount of heat generated relative to the lower heat of combustion. These additional heat losses can be subtracted from the lower heat of combustion (ΔHL) to estimate a parameter known as heat yield (ΔHY). The relations are shown in Fig. 3.9. Byram (1959) suggests that the heat yield can be physically thought of as equivalent to the quantity of heat per unit weight of the fuel burned, "which passes through a cross-section of the convection column, above a fire burning in a neutrally-stable atmosphere."

The heat yield and the heat of combustion are often very different from one another, especially when there is considerable incomplete combustion or fuel moisture. The estimates of the lower and higher heat of combustion assume complete combustion occurs (i.e., that the oxygen in the air is not limiting the combustion process, see Chap. 2). However, in actual wildland fires, there is typically a combination of both flaming and smoldering combustion occurring at the same time. Smoldering combustion releases less heat and at a lower rate than flaming combustion.

The values for the lower heat of combustion (ΔHL) reported in Table 3.2 refer to the energy release for complete combustion of a dry fuel where only CO_2 and H_2O are produced. However, when incomplete combustion occurs, carbon monoxide (CO), methane (CH_4), and carbon (C) are produced. The net energy released from incomplete combustion of dry fuel, or its heat yield (ΔHY), can be estimated as the difference between the energy in reactants, which is the high heat of combustion

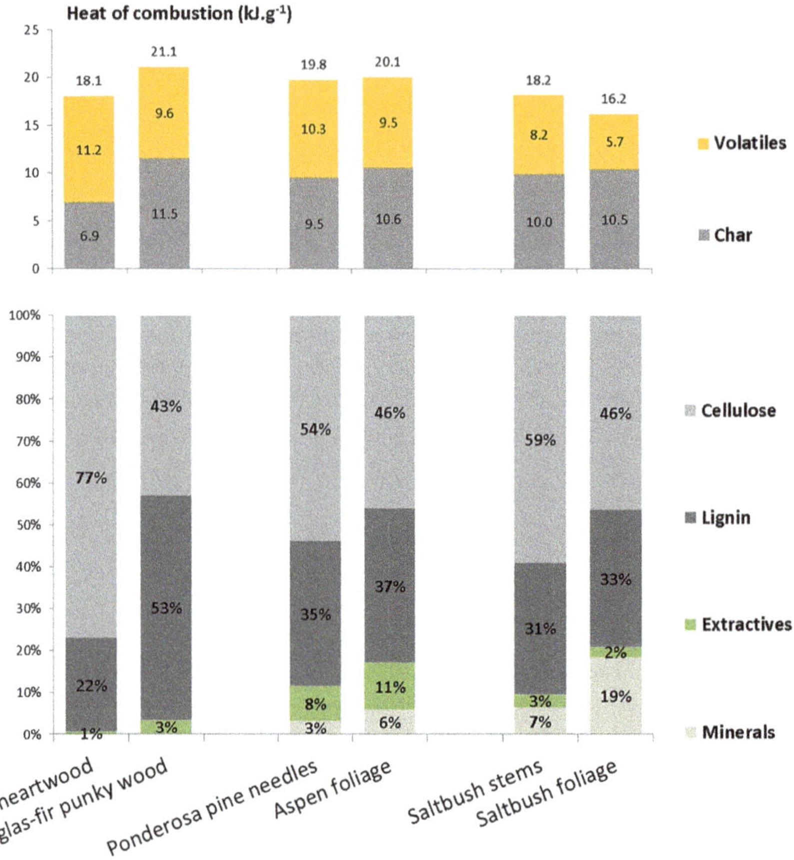

Fig. 3.8 Variation of the chemical composition of fuels and effects on the higher heat of combustion (ΔHH). As cellulose degrades more easily than lignin, decaying fuels have progressively higher lignin content and therefore higher values of heat of combustion due to char. However, the heat of combustion by volatiles is higher when cellulose is higher. Ponderosa pine (*Pinus ponderosa*) needles and aspen (*Populus tremuloides*) foliage show high values of heat of combustion due to volatiles mainly from cellulose and extractives. The lower values of heat of combustion for both stems and foliage of valley saltbush (*Atriplex cuneata*) are due to the high percentages of minerals. (Data from Rothermel 1976)

(ΔHH), the energy for the vaporization of the water of reaction and the energy in the products from incomplete combustion.

The implications of incomplete combustion on the heat yield of a dry fuel can be illustrated using the values of 20.1 kJ g^{-1} for the high heat of combustion of common wildland fuels, the value of 1.4 kJ g^{-1} for the vaporization of water, and the values for the lower heat of combustion of products from Table 3.2. We consider that the products are methane and carbon particles, with a value of 30.0 kJ g^{-1} for C. From the amounts of products generated by the various types

(continued)

Table 3.4 Reference values of important characteristics for typical fuel materials in wildland fires showing ash content (%), the fraction of char from combustion (%), the higher heat of combustion of the fuel (ΔHH), and its partitioning between energy for volatiles and energy for char (Data from Susott 1982)

Fuel type	Ash content in fuel (%)	Fraction of char from combustion (%)	Higher heat of combustion of fuel (kJ g^{-1})	Energy for volatiles (kJ g^{-1})	Energy for char (kJ g^{-1})
Grasses	6.5–9.5	21.7–24.6	19.4–20.2	12.0–12.2	7.1–8.2
Foliage	1.5–7.1	25.2–34.0	20.6–23.3	10.9–15.8	7.5–10.6
Stems	2.2–6.1	22.3–27.9	20.0–22.4	10.9–15.2	7.2–9.1
Wood	0.2–0.6	15.4–23.7	19.6–21.0	12.6–14.6	5.0–7.6
Rotten wood	0.2–0.2	21.3–40.6	20.3–23.1	10.4–13.6	6.8–12.6
Bark	0.5–17.7	27.9–46.9	21.5–24.0	7.7–12.8	8.9–14.3
Duff	31.2–34.1	35.5–38.8	20.3–23.3	8.9–11.1	11.4–12.2

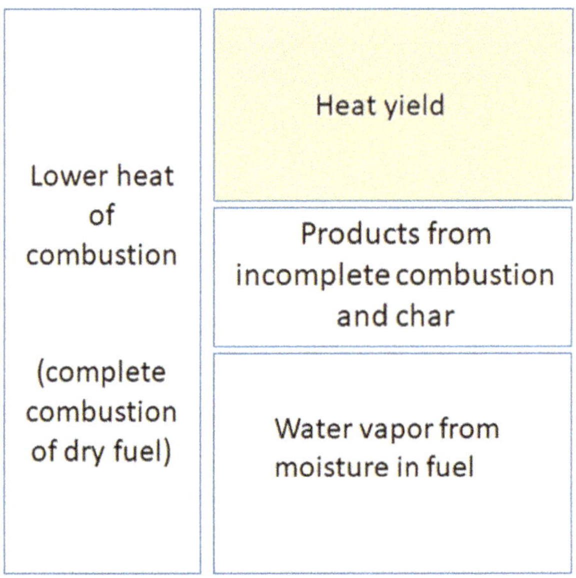

Fig. 3.9 Schematic diagram of the relation between heat yield, lower heat of combustion and heat losses from incomplete combustion and char, and from vaporization of water in fuel moisture

of combustion per unit weight of the fuel (emission factors in Table 2.3) we can calculate the energy in products. By subtraction from the low heat of combustion, we calculate heat yield (Table 3.5).

Besides combustion completeness, fuel moisture is the other critical factor that influences the heat yield. Reductions in the heat yield due to fuel moisture occur as energy is used to heat the water from ambient temperature to boiling point,

Table 3.5 Calculations of the heat yield associated with wood combustion

		Type of combustion					
		Complete		Flaming		Smoldering	
Equivalent oxygen to fuel ratio (EOFR)		1.00		0.93		0.80	
Modified combustion efficiency (MCE)		1.00		0.92		0.76	
Substances/compounds	Heat of combustion	**Mass**	**Energy**	**Mass**	**Energy**	**Mass**	**Energy**
	(kJ g^{-1})	(g kg^{-1} of fuel)	(kJ g^{-1} of fuel)	(g kg^{-1} of fuel)	(kJ g^{-1} of fuel)	(g kg^{-1} of fuel)	(kJ g^{-1} of fuel)
Fuel (ΔHH)	20.1	1000	20.1	1000	20.1	1000	20.1
Water of reaction	–	559	−1.4	546	−1.4	523	−1.3
Oxygen (O_2)	–	1379	–	1283	–	1103	–
Low heat of combustion (ΔHL)			**18.7**		**18.7**		**18.8**
Carbon dioxide (CO_2)	–	1821	–	1632	–	1283	–
Carbon monoxide (CO)	10.6	–	–	90	−1.0	257	−2.7
Methane (CH_4)	50.1	–	–	5	−0.2	15	−0.8
Carbon (C)	32.0	–	–	9	−0.3	25	−0.8
Total in products			**0.0**		**−1.5**		**−4.3**
Heat yield (ΔHY)	(kJ g^{-1} of fuel)		**18.7**		**17.2**		**14.5**

vaporize the water, and heat the water vapor to the ignition temperature. The effect of fuel moisture content on the heat yield can be predicted from measurements or estimates of fuel moisture before or during a fire, unlike combustion completeness. An additional factor that should be considered when estimating heat yield is the effect of mineral content. The effect of mineral content on heat yield is commonly handled by reducing the net fuel load consumed, however it can also be considered as a reduction in heat generation as shown in Fig. 3.10.

3.6 Implications

Heat is one major product of combustion and is central to understanding how fires spread and how they impact soils, plants, people, and ecological processes. The burning characteristics and the heat production associated with different types of fuel depend upon fuel characteristics. We saw how the mineral content and the fuel moisture content of the fuel influence heat yield. These two factors are known to be

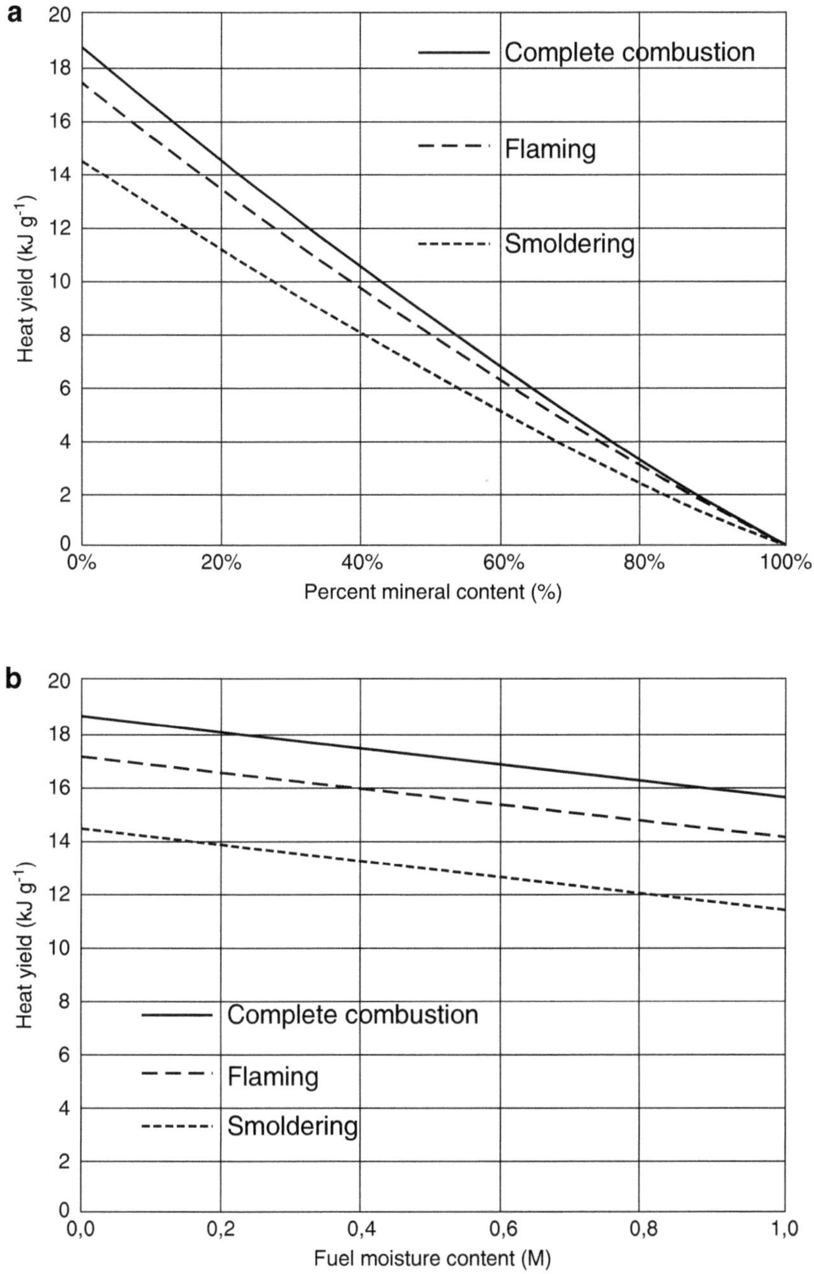

Fig. 3.10 Heat yield as a function of combustion completeness and (**a**) percent mineral content and (**b**) fractional fuel moisture content. The results from Table 3.5 are at the left side of these graphs and the effect of mineral content and fuel moisture are shown by the slope of the lines, considering that minerals do not contribute to heat yield and considering an average value of 3.0 kJ g^{-1} for the energy losses due to fuel moisture

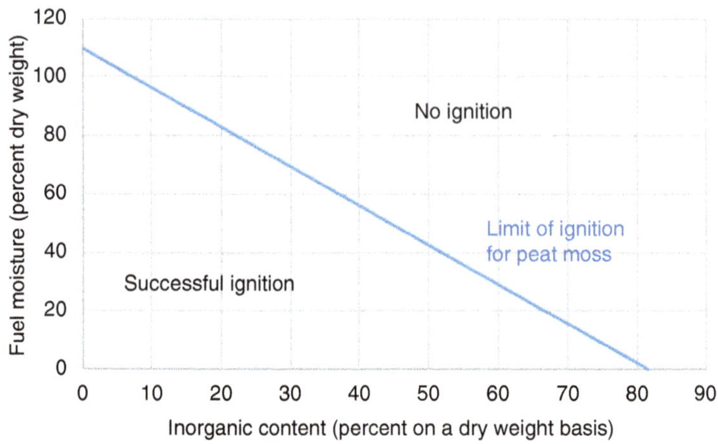

Fig. 3.11 Ignition limits established experimentally for peat moss and extended to various organic soils (Data from Frandsen 1987, 1997)

determinant of the possibility of ignition as studied for peat mosses by Frandsen (1987, 1997) as shown in Fig. 3.11.

Minerals and water are also used by people to extinguish fires. Water is widely used to cool the fire below the point of ignition, and it can be seen as an artificial means for increasing fuel moisture so much that the fire slows or stops burning. Similarly, applying sand, dirt, or mineral soil to mix with fuel can be seen as a way to increase the mineral content of the fuel with all the subsequent reductions in heat yield and therefore in fire behavior.

In Chap. 4, we will investigate the amount of heat energy required to ignite fuels, and in Chap. 5, we will look at the transfer of the heat generated during a fire. Then in Chap. 7, we will link heat production, the heat of pre-ignition, and heat transfer to investigate the spread of wildland fires.

3.7 Interactive Spreadsheet: COMBUSTION

We use the interactive spreadsheet, COMBUSTION_v2.0, to illustrate the energy balance of combustion at the same time that mass calculations are made. Details of the inputs, intermediate results, and outputs of the system are shown in Fig. 3.12. We encourage adjusting the inputs to see how the intermediate results and outputs change. Consider the sensitivity of the outputs to different inputs to the combustion process, and relate that to the potential for fires to continue burning, and perhaps to burn with great intensity. The heat of pre-ignition, heat transfer during fires, and fire spread are each addressed in subsequent chapters in our book. The importance of fuel moisture in the probability of ignition by various sources is illustrated in the spreadsheet "MASS_TRANSFER_SPOTTING_v2.0" and in chapter 8.

INPUTS:		LOW HEAT OF COMBUSTION	EQUIVALENT AIR-TO-FUEL RATIO
		2716 kJ / mol	0.92 dimensionless
COMPOSITION OF FUEL			
Carbon (C)	6.00	FUEL MOISTURE	AIR RELATIVE HUMIDITY
Hydrogen (H)	9.00	0.70 Mf = moisture / dry fuel	60 (percent %)
Oxygen (O)	4.00		
		FUEL INITIAL TEMPERATURE	AIR TEMPERATURE
Molar mass	145.0 g / mol	298 K	298 K

2. HEAT OF COMBUSTION (DRY FUEL)		3. HEAT OF PREIGNITION	
per gram of fuel	18.7 kJ / gram	Total heat of preignition	2.8 kJ / gram
per gram of oxygen	14.8 kJ / gram	To raise the temperature of wood	0.6 kJ / gram
per gram of air	3.4 kJ / gram	To raise the temperature of liquid water	0.2 kJ / gram
Ratio oxygen / fuel	1.3 (gram / gram)	To vaporize the water	1.6 kJ / gram
		To raise the temperature of water vapour	0.4 kJ / gram

4. HEAT IN COMBUSTIBLE PRODUCTS		4. ADJUSTED HEAT YIELD	
Heat in CO, CH_4 and C	1.68 kJ / gram	per gram of fuel	14.3 kJ / gram
		per mole of fuel	2072 kJ / mole

6. EMISSIONS		5. ESTIMATED FLAME TEMPERATURE	
Emission factors		Estimated temperature of gases in the flame	1706 K
Water vapor (H_2O)	1309 g / kg	Heat capacity of the mixture of products (per mol of fuel)	1472 J/K
Carbon dioxide (CO_2)	1605 g / kg	Estimated increase in temperature	1408 K
Carbon monoxide (CO)	103 g / kg		
Methane (CH_4)	4 g / kg		
Other Hydrocarbons (HC)	3 g / kg		
Particles < 2.5μm	7 g / kg		
Particles < 10μm	8 g / kg		
Total Particles	10 g / kg		

Fig. 3.12 Details of inputs, intermediate results, and outputs of the spreadsheet, COMBUSTION_v2.0, showing the relevant parts of the energy balance allowing to derive heat yield from the lower heat of combustion. Note that these calculations are only approximations

References

Albini, F. A. (1980). *Thermochemical properties of flame gases from fine wildland fuels*. Res Paper INT-243. Ogden: USDA Forest Service Intermountain Forest and Range Experiment Station.

Alexander, M. E. (1982). Calculating and interpreting forest fire intensities. *Canadian Journal of Botany, 60*(4), 349–357.

Atkin, P., & Jones, L. (2005). *Chemical principles: The quest for insight* (3rd ed.). New York: WH Freeman.

Byram, G. M. (1959). Combustion of forest fuels. In K. Davis (Ed.), *Forest fire: Control and use*. New York: McGraw-Hill.

Chandler, C., Cheney, P., Thomas, P., Trabaud, L., & Williams, D. (1983). *Fire in forestry* (Vol. 1: Forest fire behaviour and effects). New York: Wiley.

Drysdale, D. (2011). *An introduction to fire dynamics*. New York: Wiley.

Frandsen, W. H. (1987). The influence of moisture and mineral soil on the combustion limits of smoldering forest duff. *Canadian Journal of Forest Research, 17*(12), 1540–1544.

Frandsen, W. H. (1997). Ignition probability of organic soils. *Canadian Journal of Forest Research, 27*(9), 1471–1477.

Philpot, C. W. (1968). *Mineral content and pyrolysis of selected plant materials*. Res Note INT-84. Ogden: USDA Forest Service Intermountain Forest and Range Experiment Station.

Philpot, C. W. (1971). *The pyrolysis products and thermal characteristics of cottonwood and its components*. Res Paper INT-107. Ogden: USDA Forest Service Intermountain Forest and Range Experiment Station.

Philpot, C. W., & Mutch, R. W. (1971). *The seasonal trends in moisture content, ether extractives, and energy of ponderosa pine and Douglas-fir needles*. Res Paper INT-102. Ogden: USDA Forest Service Intermountain Forest and Range Experiment Station.

Polik, W. F. (2000). *Bomb calorimetry*. Holland: Department of Chemistry, Hope College. Retrieved June 18, 2020, from http://www.chem.hope.edu/~polik/Chem345-2000/bombcalorimetry.htm.

Rothermel, R. C. (1976). Forest fires and the chemistry of forest fuels. In *Thermal uses and properties of carbohydrates and lignins*. New York: Academic.

Sebei, K., Sakouhi, F., Herchi, W., Khouja, M. L., & Boukhchina, S. (2015). Chemical composition and antibacterial activities of seven Eucalyptus species essential oils leaves. *Biological Research, 48*(1), 1–5.

Susott, R. A. (1982). Characterization of the thermal properties of forest fuels by combustible gas analysis. *Forest Science, 28*(2), 404–420.

Susott, R. A., DeGroot, W. F., & Shafizadeh, F. (1975). Heat content of natural fuels. *Journal of Fire and Flammability, 6*, 311–325.

Van Wagner, C. E. (1972). *Heat of combustion, heat yield, and fire behavior*. Info Rep PS-X-35. Chalk River: Canadian Forest Service, Petawawa Forest Experiment Station.

Chapter 4
Heat for Pre-ignition and Flames

Learning Outcomes
Upon completion of this chapter, you will be able to

1. Explain the relationship between heat and temperature,
2. Understand how fuel characteristics influence the heat required for pre-ignition,
3. Describe in your own words how the low heat of combustion and excess air influence the estimated flame temperature, and
4. Use the interactive spreadsheet, titled COMBUSTION, to explore the implications of changing inputs for predicted outputs, then interpret those implications and why they are important.

4.1 Introduction

Before fuels can ignite, they go through a pre-ignition phase that removes water and other liquid volatile compounds from the fuel through dehydration and distillation and converts the solid fuel to flammable gases through a process called pyrolysis (See Chap. 1, Fig. 4.1). These gases are then heated up until combustion occurs. The temperature at which ignition occurs is called the ignition temperature (See Sect. 1.3). The energy required to dehydrate fuel, convert the solid fuel to flammable vapors through pyrolysis, and heat the gas mixture up to the ignition temperature is termed heat of pre-ignition. In this chapter, we investigate the various components associated with estimating the heat of pre-ignition. Following our discussion of the

---

Supplementary Information The online version of this chapter (https://doi.org/10.1007/978-3-030-69815-7_4) contains supplementary material, which is available to authorized users.

F. Castro Rego et al., *Fire Science*, Springer Textbooks in Earth Sciences, Geography and Environment, https://doi.org/10.1007/978-3-030-69815-7_4

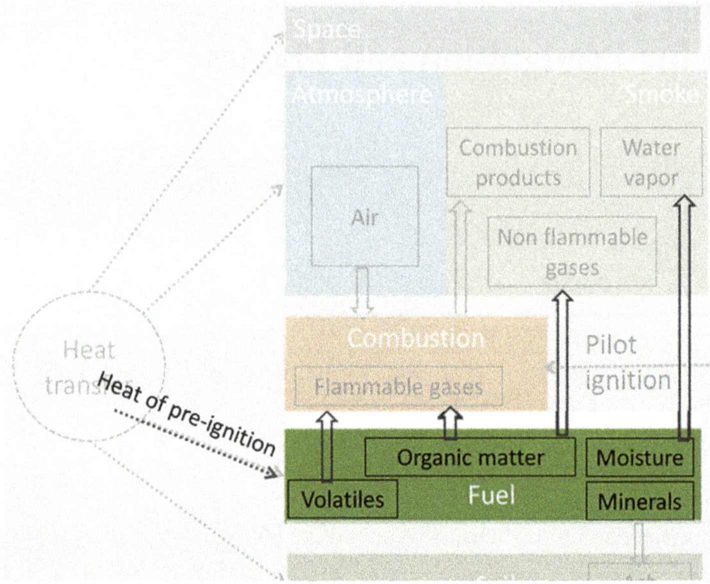

Fig. 4.1 Diagram showing the heat of pre-ignition required to increase the temperature of the fuel (with all its components) to release flammable gases. In this chapter, we focus on the colored portion of this diagram

heat of pre-ignition, we calculate the adiabatic flame temperature based upon the low heat of combustion and heat yield calculations from Chap. 3 and the concepts of heat capacity that we illustrated in the estimation of the heat of pre-ignition. For fires to burn, there must be enough heat from ignition and burning of one piece of fuel to drive off moisture, thermally degrade the solid organic matter in the fuel, and volatilize the flammable gases. The energy required for raising the temperature of the fuel from ambient to the ignition temperature is called the heat of pre-ignition. If less heat is produced than is needed to ignite additional fuel, the fire will go out. Fire suppression actions often exploit this understanding to know when and where particular actions will be effective.

4.2 From Heat Supply to Temperature Rise: Specific Heat Capacity

In Chap. 3 we discussed the heat of combustion with no mention of temperature. Heat is "energy that is transferred as the result of a temperature difference between a system and its surrounding" (Atkin and Jones 2005). In contrast, the temperature is an "intensive property that determines the direction in which heat will flow between two objects in contact". The relation between the two is expressed by the heat capacity, defined as "the ratio of heat supplied to the temperature rise produced"

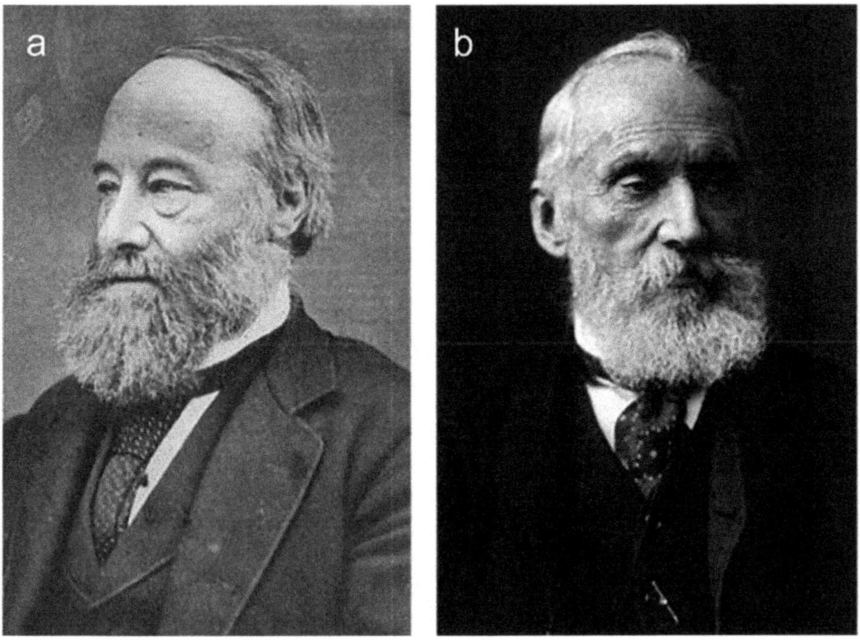

Fig. 4.2 Two pioneering British scientists whose names are now used for the units of energy (Joule) and temperature (Kelvin) in the International System. (**a**) James P. Joule (1818–1889) (Shuster and Shipley 1917) (**b**) William Thomson, Lord Kelvin (1824–1907) (Dickinson n.d.)

(Atkin and Jones 2005). The heat capacity is an extrinsic property since the amount of heat required to raise the temperature of an object depends upon the size of the object. Dividing the heat capacity by the mass of the sample heated results in an estimate of the specific heat capacity. The specific heat capacity can be thought of as the amount of heat energy that is required to raise a unit mass of a substance by one degree of temperature.

The units of the International System for heat (or energy) are Joules (J) and for temperature are degrees Kelvin (K), after the names of two important British scientists (Fig. 4.2). The unit Joule is the heat required to raise the temperature of 1 g of water by 0.24 K. Alternatively, we can say that we need 4.2 J to raise the temperature of 1 g of water by 1 °K.

The relation between heat, temperature, and specific heat capacity is:

$$\Delta T = \frac{Q}{C_p} \tag{4.1}$$

where ΔT is the temperature difference in degrees K, Q is the heat supplied in Joules per unit mass (J g^{-1}), and C_p is the specific heat capacity which has units of energy per unit of temperature per unit mass of the substance (Joules per degree Kelvin per gram, J K^{-1} g^{-1}). Using Eq. (4.1) it is possible to estimate the temperature rise of an object based on its exposure to heat and its heat capacity.

Table 4.1 Reported values of specific heat capacity at constant pressure (C_p) for various materials involved in wildland fires

		Specific heat capacity (C_p) ($J\ K^{-1}\ g^{-1}$)		
		Temperature (K)		
Phase	Material	300 K	600 K	1200 K
Solid	Woody fuels	1.1–2.0		
	Minerals	0.8–1.0		
	Humus	1.8–2.0		
	Soils	1.0–1.8		
Liquid at 300 K	Water	4.18	2.02	2.43
	Isoprene	2.24	2.51	3.51
Gas at 600–1200 K	Monoterpenes	1.84		
	Eucalyptol	1.76		
Gas	Nitrogen	1.04	1.08	1.20
	Oxygen	0.92	1.00	1.12
	Carbon dioxide	0.85	1.08	1.28
	Carbon monoxide	1.04	1.09	1.22
	Dry air	1.01	1.05	1.18

Data from Anderson (1969), Rothermel (1972), Chandler et al. (1983), Jury et al. (1991), Dickinson and Johnson (2001), Atkin and Jones (2005) and Incropera et al. (2006)

Estimates of specific heat capacity (C_p) vary among different materials, with temperature, and for different states of matter (Table 4.1). For example, it takes between two and four times more energy to raise the temperature of liquid water by 1 °K than it does to raise the temperature of woody fuels. Similarly, the amount of energy required to raise the temperature of one gram of liquid water by 1 °K requires about two times more energy than vapor water (Table 4.1). During a phase change, such as from a liquid to a gas, the heat capacity is technically infinite because the energy is used to change the state of matter and not increase the temperature. For these reasons, the temperatures associated with the heating of wildland fuels to the pilot ignition temperature are divided into temperature ranges and by the state of matter and assigned an average value of C_p (Table 4.1).

4.3 From Heat Supply to Phase Changes: Latent Heat of Vaporization

In the pre-ignition process, the total energy required is based on both the fuel's heat capacity and the energy associated with phase changes. The amount of energy absorbed or released during a phase change without changing its temperature is called the latent heat. The amount of energy associated with changing a solid to a liquid is termed the latent heat of fusion and the amount of energy associated with

Table 4.2 Temperatures of vaporization (T_v), latent heat of vaporization (L_v), piloted-ignition temperature, and the auto-ignition temperatures for water and three volatiles: isoprene, monoterpene, and eucalyptol)

Liquid	Temperature of vaporization (T_v in °C)	Latent heat of vaporization (L_v in J g^{-1})	Piloted-ignition temperature (°C)	Auto-ignition temperature (°C)
Water	100	2257	–	–
Isoprene	34	425	−54	427
Monoterpenes	155	263	38	255
Eucalyptol	176	267	269	269

Temperatures are reported in °C as common in many references (Data summarized from NIST 2018, Royal Society of Chemistry 2020, and NCBI n.d.)

changing a liquid to a gas is called the latent heat of vaporization (L_v generally measured in J g^{-1}).

In wildland fuels, the concept of latent heat is especially important due to the formation of water vapor during combustion, any additional water associated with fuel moisture, and the distillation of liquid volatile compounds. The temperature at which a compound changes from a liquid to a solid is called the temperature of vaporization (or boiling point) (T_v generally expressed in °C or K). Table 4.2 shows common values of the temperature of vaporization (T_v) and the latent heat of vaporization (L_v), and the piloted- and auto-ignition temperatures for water and three volatile compounds commonly involved in vegetation fires (isoprene, monoterpenes, and eucalyptol).

From Table 4.2 it can be seen that the latent heat of vaporization of volatiles is much lower than that of water. Also, it shows that isoprene vaporizes at relatively low temperatures, 34 °C, and can ignite readily if given an ignition source, whereas monoterpenes and eucalyptol require higher temperatures to vaporize, but they have lower auto-ignition temperatures.

4.4 Evaluating the Heat of Pre-ignition for Wildland Fuels

The total heat required for pre-ignition (Q_{ig}) expressed per unit mass of fuel can be estimated as the sum of two main components:

1. The heat required to increase the temperature of the dry fuel to the ignition temperature (Q_{dig}) and
2. the heat required to heat liquid water from fuel moisture to water vapor up to the ignition temperature (Q_m).

4.4.1 Estimating the Main Components of the Heat of Pre-ignition

The first component to be considered is the energy required to raise the temperature of the dry fuel to ignition temperature. This can be calculated as:

$$Q_{dig} = C_{pd}\left(T_{ig} - T_a\right) \tag{4.2}$$

where Q_{dig} is the amount of energy in J required to heat a unit mass of a substance from its ambient temperature (T_a) to the ignition temperature (T_{ig}), and C_{pd} is the specific heat (heat capacity) of the substance (J g^{-1} K^{-1}).

The second component required to calculate the heat of pre-ignition is the energy needed to heat the liquid water in the fuel from its ambient temperature to water vapor at ignition temperature (Q_m) and it is expressed per unit mass of fuel. This component is calculated from three main sub-components: the energy required to heat liquid water in the fuel up to the boiling point (Q_{wl}), the latent heat of vaporization (L_v), and the energy needed to raise the temperature of the water vapor up to the ignition temperature (Q_{wv}). These values are expressed per unit of mass of water. To express the total energy required to heat the water from its ambient temperature to water vapor at the ignition temperature per unit mass of dry fuel (Q_m), we multiply the sum of the three components discussed above by fuel moisture (M) represented as a ratio between liquid water and dry matter in the fuel:

$$Q_m = M\ (\ Q_{wl} + L_v + Q_{wv}) \tag{4.3}$$

The amount of energy required per unit mass of water to raise the temperature of the liquid water from the ambient temperature to the boiling point (Q_{wl}) can be estimated by multiplying the difference between ambient temperature (T_a) and vaporization temperature (T_v) by the specific heat capacity for liquid water (C_{pwl}):

$$Q_{wl} = C_{pwl}(T_v - T_a) \tag{4.4}$$

After the liquid water is converted to a vapor, requiring the latent heat of vaporization (L_v), additional energy is needed to increase its temperature from boiling temperature to the temperature of ignition. This is estimated as:

$$Q_{wv} = C_{pwv}\left(T_{ig} - T_v\right) \tag{4.5}$$

where Q_{wv} is the amount of energy per unit mass of water required to raise the temperature of water vapor from the temperature of vaporization (T_v) to the ignition temperature (T_{ig}), and C_{pwv} is the specific heat of water vapor.

Byram (1959) adds another component which is the amount of heat required for the separation of bound water from the fuel. Bound water is water that is chemically bonded to cellulose through hydrogen bonds. More energy is required to remove this water as fuel moisture decreases (Chandler et al. 1983). This heat of separation of bound water was approximated to a maximum value of 120 kJ per gram of dry fuel by Byram (1959), but it is often not considered in Q_{ig} calculations as it is a relatively small amount of energy (Van Wagner 1972).

4.4.2 Combining the Components of the Heat of Pre-ignition of the Fuel

The amount of heat required to raise dry fuel and the fuel moisture to the ignition temperature is the heat of pre-ignition (Q_{ig}), which can be estimated by summing the two main components discussed in Sect. 4.4.1:

$$Q_{ig} = Q_{dig} + Q_m = Q_{dig} + M\ (\ Q_{wl} + L_v + Q_{wv}) \tag{4.6}$$

The calculation of the heat of pre-ignition of wood illustrates the process. First, we compute the energy required to raise the temperature of dry fuel to ignition (Q_{dig}). For this example, let us assume that the initial temperature of the dry wood is 300 K (27 °C) and that the ignition temperature is 600 K (327 °C). Using Table 4.1 we can consider the specific heat capacity of dry wood as 1.30 J K^{-1} g^{-1}. Substituting these values into Eq. (4.2), we see that it takes 390 J of energy to raise 1 g of dry wood to the ignition temperature (Eq. 4.7):

$$Q_{dig} = 1.30(600 - 300) = 390\ \mathrm{J\ g^{-1}} \tag{4.7}$$

Next, we can calculate the amount of energy it takes to raise the temperature of one gram of water from initial (or ambient) fuel temperature to the boiling point (Q_{wl}). With an ambient temperature of 293.15 K (20 °C), the boiling temperature of 373.15 K (100 °C), and the specific heat for liquid water of 4.20 J K^{-1} g^{-1} (Table 4.1) we have:

$$Q_{wl} = (4.20) \times (373.15 - 293.15) = 336\ \mathrm{J\ g^{-1}} \tag{4.8}$$

Once the liquid water is heated up to the boiling point, we calculate the energy required to convert it from a liquid to a vapor. Assuming standard atmospheric pressure, the amount of energy required to convert 1 g of liquid water to vapor water is 2257 J g^{-1}. We treat L_v as a constant.

Using a specific heat for water vapor C_{pwv} of 1.92 J K^{-1} g^{-1}, a temperature of ignition (T_{ig}) of 593.15 K and a boiling temperature (T_v) of 373.15 K, the

(continued)

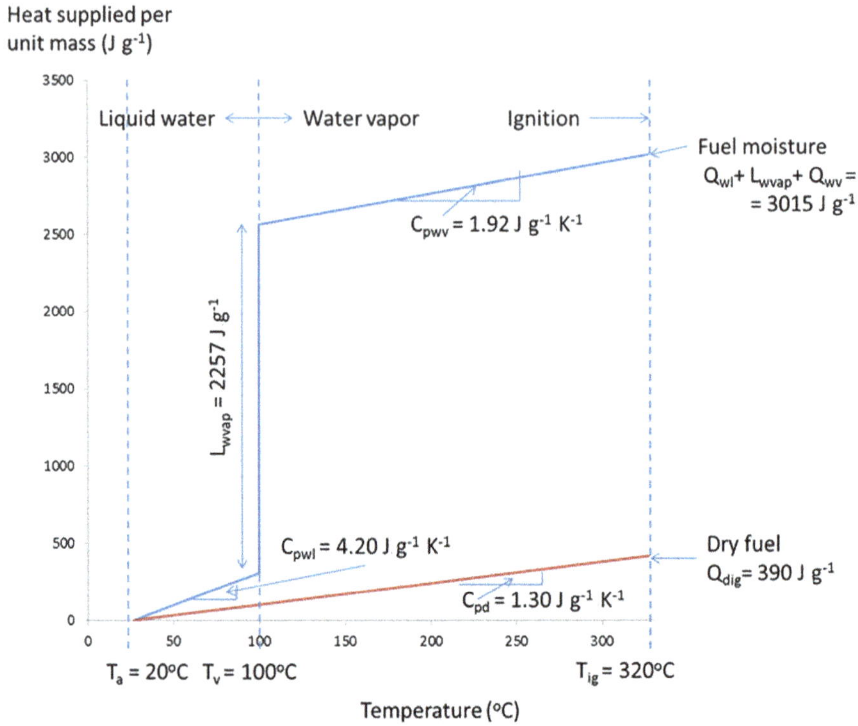

Fig. 4.3 Comparison of heating curves for dry fuel (as wood) and water (fuel moisture), with heat supplied per unit mass (J g^{-1}) and the resulting temperature increase (°C). These values represent heat supplied per unit mass of dry fuel (*red line*) and per unit mass of water (*blue line*), independently. The heat required to raise 1 g of liquid water to water vapor at ignition temperature is greater than that required to raise 1 g of dry fuel to ignition temperature

energy needed to raise the temperature of water vapor to ignition temperature is:

$$Q_{wv} = (1.92) \times (593.15 - 373.15) = 422 \text{ J g}^{-1} \quad (4.9)$$

By combining the values obtained using Eqs. (4.5) and (4.7) and assuming a latent heat of vaporization of 2257 J g^{-1} we calculate that 2986 J of energy is required to heat 1 g of water from an initial temperature of 300 K to an ignition temperature of 593.15 K. Assuming 100% fuel moisture content (M = 1), we get:

(continued)

$$Q_m = 1\,(336 + 2257 + 422) = 3015\ \mathrm{J\,g^{-1}} \tag{4.10}$$

With these results, we represent the relationship between heat supplied per unit mass of water in fuel moisture and the heat supplied per unit mass of dry fuel, and the corresponding temperature rise is shown in Fig. 4.3.

Using the boiling, ignition and ambient temperatures and specific heat capacities, and the heat of vaporization for water from the example above, we can approximate the heat of pre-ignition for wood as:

$$Q_{ig} = 390\ \mathrm{J\,g^{-1}} + (M)\,3015\ \mathrm{J\,g^{-1}} \tag{4.11}$$

Some fuel types may require considering additional components when estimating the heat of pre-ignition. For fuels that have a large proportion of liquid volatile compounds, the heat of pre-ignition can include an additional adjustment for the energy required to heat these compounds from ambient to ignition temperature. The inclusion of liquid volatile compounds would include three parts: (1) the energy required to heat the compound from ambient temperature to its heat of vaporization, (2) the latent heat of vaporization for the compound and, (3) the energy required to heat the compound in gas up to ignition temperature. This addition would mirror the calculations presented for fuel moisture in Sect. 4.5. The effect of a common liquid volatile compound (eucalyptol) on the heat of pre-ignition for a typical fuel with 10% volatiles and a fuel moisture content of 30% is shown in Fig. 4.4. Fuel moisture has the largest effect on the heat of pre-ignition followed by the dry fuel and finally the liquid volatile compounds.

The presented heat of pre-ignition calculations assume all heating during pre-ignition occurs in the solid phase. In other words, it is considered that the heat energy required to dehydrate and bring the fuel and water vapor to ignition temperature is sufficient for assessing fire behavior.

4.5 Flame Temperatures

In Sects. 4.2 through 4.4, we assessed how much energy is required to raise fuel temperature from ambient to ignition temperature, represented in the grey boxes in Fig. 4.5. Following ignition, the energy produced by combustion increases the temperature of combustion products, including water vapor, nitrogen, and excess air, leading to an overall increase in flame temperature, as shown in yellow boxes in Fig. 4.5.

The adiabatic flame temperature is the theoretical maximum temperature of the flaming gases produced during combustion. This theoretical value is estimated by assuming that there are no heat losses due to radiation from the flame. In other words, all the energy produced during combustion is used to heat the products. Therefore,

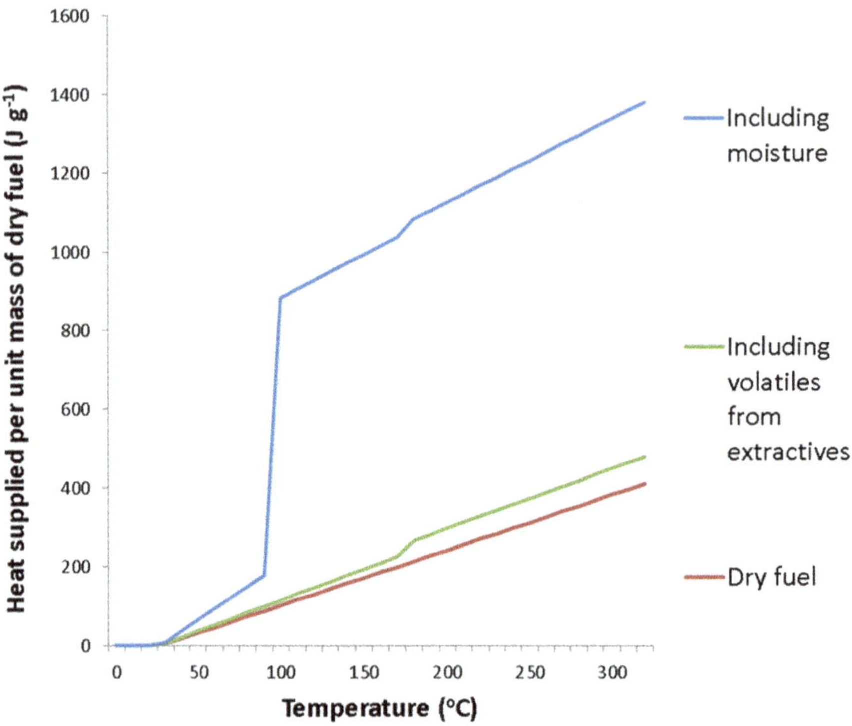

Fig. 4.4 The heat supplied per unit mass of dry fuel (Q_{ig} in J g^{-1}) accumulated for all fuel components as a function of temperature (°C). The example illustrates heating from an initial temperature of $T_a = 27$ °C (300 K) to the ignition temperature of $T_{ig} = 320$ °C (close to 600 K). The example uses an extractive content (mass of extractives per mass of dry fuel) of 0.1 and a 0.3 moisture content (M). The extractive considered here is eucalyptol with boiling temperature T_v = 176 °C, latent heat of vaporization L_v = 276 J g^{-1}, and heat capacities (C_p) of 1.8 J g^{-1} K^{-1} and 1.2 J g^{-1} K^{-1} for the liquid and gas phases, respectively. All values follow Tables 4.1 and 4.2

the adiabatic flame temperature is never attained in the real world because flames emit radiation to their surroundings.

To estimate adiabatic flame temperature, we first compute the amount of energy required to increase the temperature of the products based on their mass (G_i) and specific heat capacity (C_{pi}):

$$\textit{Sum of heat capacities for the products} = \sum\nolimits_{i=1}^{n} \left(G_i C_{pi}\right) \tag{4.12}$$

where G_i is the mass of gas i involved per unit mass of fuel (dimensionless), and C_{pi} is the specific heat of gas i (J g^{-1} K^{-1}). The products include all gases involved in the combustion process, including any excess oxygen, nitrogen in the air, water vapor, carbon dioxide, and carbon monoxide. We then estimate the increase in temperature by dividing the amount of energy generated during combustion by the sum of the

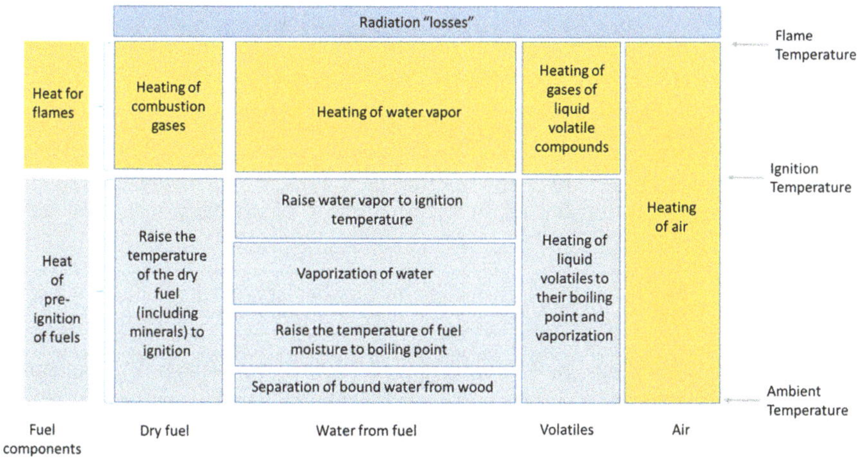

Fig. 4.5 The energy components mentioned in the text. The heat of pre-ignition of fuels is the energy absorbed by dry fuel, fuel moisture, and volatiles to reach ignition temperature. Heat for flames is the energy absorbed by air, gases of volatile compounds, water vapor, and combustion products to reach flame temperature. Radiation "losses" are also represented; this is heat radiated into space

product's heat capacities (Eq. 4.13). The maximum heat potentially generated during combustion is commonly estimated using the low heat of combustion (ΔHL). To increase the temperature of the gases in the flame, we calculate:

$$\Delta T = \frac{\Delta HL}{\sum_{i=1}^{n} (G_1 C_{pi})} \quad (4.13)$$

where ΔHL is the low heat of combustion in J g^{-1}, C_{Pi} is the specific heat of gas i (J g^{-1} K^{-1}), and ΔT is the temperature increase of gases (°C or K). The final adiabatic flame temperature is then estimated by adding the increase in temperature (ΔT) to the initial ambient temperature (T_a):

$$\text{Adiabatic flame temperature} = T_a + \Delta T \quad (4.14)$$

For example, we can calculate the adiabatic flame temperature for the complete combustion of methane in oxygen using Eq. (4.14) and the stoichiometric chemical equation (Eq. 4.15) and it's equivalent in terms of mass and energy per unit mass (grams) of fuel (Eq. 4.16):

(continued)

$$CH_4 + 2O_2 \rightarrow CO_2 + 2H_2O + 802 \text{ kJ mol}^{-1} \tag{4.15}$$

$$1gCH_4 + 4gO_2 \rightarrow 2.75gCO_2 + 2.25gH_2O + 50.1 \text{ kJ} \tag{4.16}$$

From Table 2.5, we know that the ratio of the mass of the proportions of nitrogen in air (75.52%) to oxygen (23.14%) is 3.26. Therefore, each gram of oxygen combustion in dry air involves 13.04 g of nitrogen that must be accounted for:

$$1gCH_4 + 4gO_2 \rightarrow 2.75g\ CO_2 + 2.25g\ H_2O + 13.04g\ N_2 + 50.1 \text{ kJ} \tag{4.17}$$

Using specific heats for a temperature of 1200 K from Table 4.1, we can estimate the amount of energy required to increase the temperature of the gases:

$$\begin{aligned}\sum\nolimits_i (G_i C_{pi}) &= (2.75 \times 1.28) + (2.25 \times 2.43) + (13.04 \times 1.2) \\ &= 24.6 \text{ Jg}^{-1}\text{K}^{-1}\end{aligned} \tag{4.18}$$

Thus, for the complete combustion of 1 g of methane in the air, it will take approximately 24.6 J of energy to raise the temperature of the products by 1 degree K.

Assuming a low heat of combustion of 50.1 kJ per gram of methane (Table 3.2) and an initial temperature of 298 K (25 °C) the adiabatic flame temperature for methane is estimated as:

$$T_{ad} = 298\text{K} + \left(\frac{50100 \text{ J g}^{-1})}{24.6 \text{ J g}^{-1}\text{K}^{-1}}\right) = 2335\text{K} \tag{4.19}$$

When wildland fuels burn, estimating flame temperature is far more complex than we assumed using Eq. (4.14). The temperature of a flame generally varies across its width and height, attaining maximum values near its base and decreasing with height (Wotton et al. 2012). The flame flow is turbulent in wildland fires, so the temperatures at any given point in the flame, especially near the edges, will fluctuate widely. Thus, measurements of flame temperature will record large fluctuations in time with averages of approximately 800–1000 °C (Quintiere 1997). The theoretical higher and lower limits of these fluctuations are the adiabatic flame temperature and the ambient air temperature. In addition, in Eq. (4.13) we used the low heat of combustion, which, as shown in Chap. 3, Sect. 3.4, is typically much greater than the actual heat yield during a fire. In the example calculation of adiabatic flame temperature in Sect. 4.5, we assumed that the combustion reaction occurred in pure oxygen. The adiabatic flame temperature for a given fuel will always be greater when combustion occurs in pure oxygen than when combustion occurs in the air because some portion of the low heat of combustion is used to raise the temperature of the nitrogen in the air. Furthermore, the air entrained into the flame can be up to two times greater than what is required in the stoichiometric combustion reaction (Steward 1964; Van Wagner 1974; Nelson 1980), which would result in further reductions in flame temperature. We can account for the effect of excess air on flame temperature by introducing an entrainment factor ranging from zero (no excess air entrainment) to 2 (excess entrainment twice the value of the stoichiometric amount).

4.6 Implications

For fires to be sustained, there must be enough heat from one burning piece of fuel to ignite the next piece of fuel. Thus, this is a balance between the heat available and the heat needed to heat the next piece of fuel enough to drive off moisture, thermally degrade the solid organic matter in the fuel, and volatilize the flammable gases. This process, therefore, takes energy. The energy required for raising the temperature of the fuel from ambient to the ignition temperature is called the heat of pre-ignition. If there is less heat produced than is needed to ignite additional fuel, the fire will go out. Based on dead fuel moisture content, both prescribed burning and fire suppression actions exploit this understanding to know when and where particular strategic actions will be effective, and perhaps where no action is needed to contain fires.

Estimations of flame temperature are useful to understand the processes involved even if flame temperatures vary more within a flame than they do between flames. In the next chapter, we use this knowledge in the context of heat transfer from fires. Heat transfer from flaming and smoldering combustion is critical to connecting fire behavior to fire effects, though geometry, the variables controlling the flow of air and heat, and material properties also influence flame temperatures and how heating from flaming and smoldering combustion affects people and ecosystems.

In this chapter, we also saw how heat capacity and latent heat are used to calculate the adiabatic flame temperature. The flame temperatures and emissivity are critical for estimating heat fluxes and the effects of fire on the plants and animals and fire safety (Chaps. 9 and 10). The relative importance of fuel composition, fuel moisture, and completeness of combustion to flame temperature can be explored using the interactive spreadsheet included in this chapter.

4.7 Interactive Spreadsheet: COMBUSTION

Please use the interactive spreadsheet, COMBUSTION_V2.0, to explore the effects of fuel composition, fuel moisture, completeness of combustion, excess air, and other factors on the heat of pre-ignition and flame temperature. With the graphical output, you can readily visualize how outputs change with different input conditions. By evaluating how much outputs change in response to small changes in the inputs, you can evaluate how sensitive the estimates could be to environmental conditions,

INPUTS:		LOW HEAT OF COMBUSTION	EQUIVALENT AIR-TO-FUEL RATIO
		2716 kJ / mol	0.92 dimensionless
COMPOSITION OF FUEL			
Carbon (C)	6.00	FUEL MOISTURE	AIR RELATIVE HUMIDITY
Hydrogen (H)	9.00	0.70 Mf = moisture / dry fuel	60 (percent %)
Oxygen (O)	4.00		
		FUEL INITIAL TEMPERATURE	AIR TEMPERATURE
Molar mass	145.0 g / mol	298 K	298 K

2. HEAT OF COMBUSTION (DRY FUEL)		3. HEAT OF PREIGNITION	
per gram of fuel	18.7 kJ / gram	Total heat of preignition	2.8 kJ / gram
per gram of oxygen	14.8 kJ / gram	To raise the temperature of wood	0.6 kJ / gram
per gram of air	3.4 kJ / gram	To raise the temperature of liquid water	0.2 kJ / gram
Ratio oxygen / fuel	1.3 (gram / gram)	To vaporize the water	1.6 kJ / gram
		To raise the temperature of water vapour	0.4 kJ / gram

4. HEAT IN COMBUSTIBLE PRODUCTS		4. ADJUSTED HEAT YIELD	
Heat in CO, CH_4 and C	1.68 kJ / gram	per gram of fuel	14.3 kJ / gram
		per mole of fuel	2072 kJ / mole

6. EMISSIONS		5. ESTIMATED FLAME TEMPERATURE	
Emission factors		Estimated temperature of gases in the flame	1706 K
Water vapor (H_2O)	1309 g / kg	Heat capacity of the mixture of products (per mol of fuel)	1472 J/K
Carbon dioxide (CO_2)	1605 g / kg	Estimated increase in temperature	1408 K
Carbon monoxide (CO)	103 g / kg		
Methane (CH_4)	4 g / kg		
Other Hydrocarbons (HC)	3 g / kg		
Particles < 2.5μm	7 g / kg		
Particles < 10μm	8 g / kg		
Total Particles	10 g / kg		

Fig. 4.6 Details of inputs, intermediate results, and outputs of the spreadsheet, COMBUSTION_v2.0, showing the relevant parts of the energy balance used in deriving the maximum estimated flame temperature from the lower heat of combustion, combustion completeness, and fuel moisture. Note that these calculations are only approximations

errors in measurement (e.g., of fuel moisture) or assumptions. Details of the inputs, intermediate results, and outputs of the system are shown in Fig. 4.6. We encourage adjusting the inputs to see how the intermediate results and outputs change. Consider the sensitivity of the outputs to different inputs to the combustion process and relate that to the potential for fires to continue burning and perhaps to burn with great intensity. Heat transfer during fires and fire spread are each addressed in subsequent chapters in our book.

References

Anderson, H. E. (1969). *Heat transfer and fire spread*. Res Pap INT-69. Ogden, UT: USDA Forest Service Intermountain Forest and Range Experiment Station.

Atkin, P. W., & Jones, L. (2005). *Chemical principles: The quest for insight* (3rd ed.). New York: WH Freeman.

Byram, G. M. (1959). Combustion of forest fuels. In K. P. Davis (Ed.), *Forest fire control and use*. New York: McGraw-Hill.

Chandler, C., Cheney, P., Thomas, P., Trabaud, L., & Williams, D. (1983). *Fire in forestry* (Vol. 1: Forest fire behaviour and effects). New York: Wiley.
Dickinson. (n.d.). *Photograph of William Thomson, Lord Kelvin*. Wikimedia Commons. Retrieved June 13, 2020, from https://commons.wikimedia.org/w/index.php?search=Lord+Kelvin&title=Special%3ASearch&go=Go&ns0=1&ns6=1&ns12=1&ns14=1&ns100=1&ns106=1#/media/File:Lord_Kelvin_photograph.jpg.
Dickinson, M. B., & Johnson, E. A. (2001). Fire effects on trees. In E. A. Johnson & F. Miyanishi (Eds.), *Forest fires, behavior and ecological effects*. San Diego, CA: Academic Press.
Incropera, F. P., DeWitt, D. P., Bergman, T. L., & Lavine, A. S. (2006). *Fundamentals of heat and mass transfer* (6th ed.). New York: Wiley.
Jury, W. A., Gardner, W. R., & Gardner, W. H. (1991). *Soil physics*. New York: Wiley.
National Center for Biotechnology Information (NCBI). (n.d.). *PubChem*. Retrieved April 18, 2019, from https://pubchem.ncbi.nlm.nih.gov/.
National Institute of Standards and Technology (NIST). (2018). *NIST chemistry WebBook, SRD 69*. Retrieved April 18, 2019, from https://webbook.nist.gov/chemistry/.
Nelson, R. M. (1980). *Flame characteristics for fires in southern fuels*. USDA Forest Serv Res Paper SE-205. Asheville, NC: Southeastern Forest Experiment Station.
Quintiere, J. G. (1997). *Principles of fire behavior*. New York: Delmar Publishers.
Rothermel, R. C. (1972). *A mathematical model for predicting fire spread in wildland fuels*. USDA Forest Serv Res Pap INT-115. Ogden, UT: Intermountain Forest and Range Experiment Station.
Royal Society of Chemistry. (2020). *Chemspider*. Retrieved April 18, 2019, from http://www.chemspider.com/Default.aspx.
Shuster, A., & Shipley, A. E. (1917). *James Prescott Joule*. Wikimedia Commons. Retrieved June 13, 2020, from https://commons.wikimedia.org/wiki/File:SS-joule.jpg.
Steward, F. R. (1964). Linear flame heights for various fuels. *Combust Flame, 8*, 171–178.
Van Wagner, C. E. (1972). *Heat of combustion, heat yield and fire behavior*. Canadian Forest Service Information Report PS-X-35. Chalk River, ON: Petawawa Forest Experiment Station.
Van Wagner, C. E. (1974). *A spread índex for crown fires in spring*. Canadian Forest Service Information Report (P. 55). Chalk River, ON: Petawawa Forest Experiment Station.
Wotton, B. M., Gould, J. S., McCaw, W. L., Cheney, N. P., & Taylor, S. W. (2012). Flame temperature and residence time of fires in dry eucalypt forest. *International Journal of Wildland Fire, 21*(3), 270–281.

Chapter 5
Heat Transfer

Learning Outcomes
After reading this chapter and using the interactive spreadsheet, we expect that you will be able to

1. Explain the four different modes of heat transfer, including how they differ and when each is dominant,
2. Describe in your own words the processes of heat transfer in vegetation fires, and
3. Use the interactive spreadsheet to explore the relative importance of the key factors influencing heat transfer from fires, and then discuss the implications of your findings.

5.1 Introduction

When plants photosynthesize, they make chemical bonds to form organic compounds. In vegetation fires, a large amount of heat is released as the chemical bonds in organic compounds are broken, as shown in the previous chapters. In this chapter, you will learn about how the heat generated during a wildfire is transferred to soil, fuel, atmosphere, and space (Fig. 5.1). The implications of heat transfer presented in this chapter will be developed further in our discussion of fire behavior and effects on plants, soils, and animals in Chaps. 7, 8, and 9.

The understanding of the modes of heat transfer has many applications. Consider, for example, the case when a woodland savanna burns. We know that heat from fires

Supplementary Information The online version of this chapter (https://doi.org/10.1007/978-3-030-69815-7_5) contains supplementary material, which is available to authorized users.

F. Castro Rego et al., *Fire Science*, Springer Textbooks in Earth Sciences, Geography and Environment, https://doi.org/10.1007/978-3-030-69815-7_5

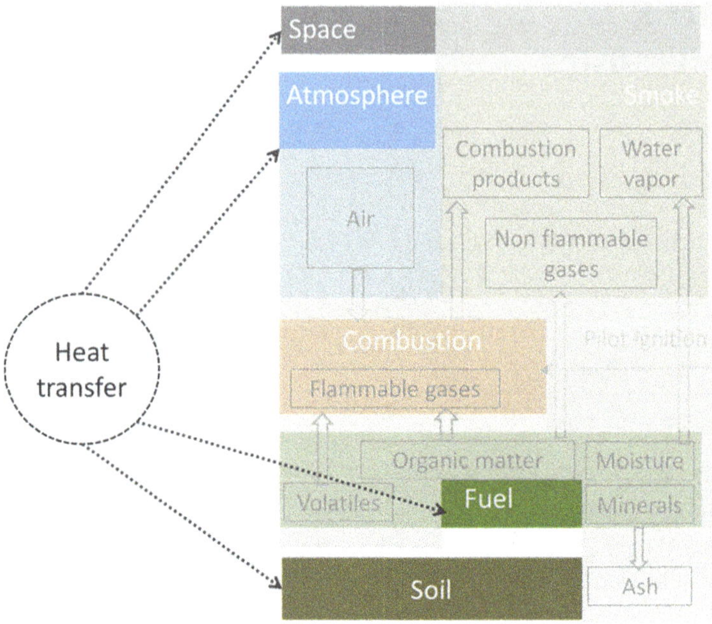

Fig. 5.1 When a fire is burning, the heat produced is transferred through convection, radiation, conduction, and mass transport. Some heat goes into the soil, some into the fuel, and some into atmosphere and space. How much heat goes where has implications for both the effects of and management of fires

can kill above-ground living tissue in vegetation, but we also know that bark can provide insulation protection. Thus, while small trees, grasses, and shrubs are often top-killed, larger trees often survive when the surface fuels burn. With their big, scattered trees in a sea of grasses, Savannas support many plants and animals that are well adapted to frequent fires, but how do they survive if those surface fires burn with very high intensity? When fires burned in Australian forests, how did the heat from fires transfer to where it affected animals such as koalas (Fig. 5.2)? Think about these applications as you learn about heat transfer, for these processes are key to the effects of fires.

5.2 Modes of Heat Transfer

All important processes occurring in fires, from molecular kinetics to large scale fire behavior and fire effects, are associated with the transfer of energy due to temperature differences. For sustained combustion to occur, the heat energy generated (Chap. 3) and transferred to unburned fuel must be sufficiently large to heat the fuel to ignition temperature (Chap. 4). Similarly, understanding the transfer of heat energy from the combustion zone to the soil, plants, and the atmosphere is critical to estimate many important fire effects (Chaps. 9 and 10). In this chapter, we focus on understanding the various modes of heat transfer, which is defined as "thermal energy in transit due to a spatial temperature difference" (Incropera et al. 2007).

Fig. 5.2 A koala bear seeking protection in a tree during 2020 Australian bushfires. Heat may be transferred from the fire to koalas by radiation from flames, by mass transport of firebrands, by convection with hot gases, and by direct consumption of tree leaves and bark, all contributing to temperatures that burn the animal skin. Similar processes govern the safety of fire personnel and buildings (see Chap. 10), or trees (see Chap. 9). (Photograph by Andrea Izzotti)

There are four modes of heat transfer. In addition to radiation, conduction, and convection, the mass transport of solid hot particles, called firebrands, also occurs. Firebrand transport occurs when winds carry solid burning material meters or kilometers ahead of the fire front. This phenomenon is often called spotting and can play an important role in fire spread (See Chap. 7) and is associated with extreme fire behavior (See Chap. 8).

One way to visualize heat transfer is to envision the interactions between a hot and cool molecule (Fig. 5.3). Radiation occurs when energy is transferred from the hot molecule to the cool molecule through space by electromagnetic waves. Conduction occurs when energy is transferred through collisions between neighboring molecules or atoms. Convection is the transport of energy due to the bulk movement of molecules within a fluid by the movement of the energized molecule, and solid mass transport occurs when solid particles are transported through space (Fig. 5.3).

The four modes of heat transfer can also be understood by visualizing a spreading fire (Fig. 5.4). Radiation is the heat energy you feel when standing several meters away from a fire. Radiation in a wildfire occurs mostly in the form of infrared waves and visible light and spreads out in all directions. As you get closer to the fire, the amount of heat you feel increases. Convection is the hot air that is transported away from the flames due to buoyancy. Convective transport is most obvious when you see smoke being lifted up and away from the fire. However, convective heat transfer

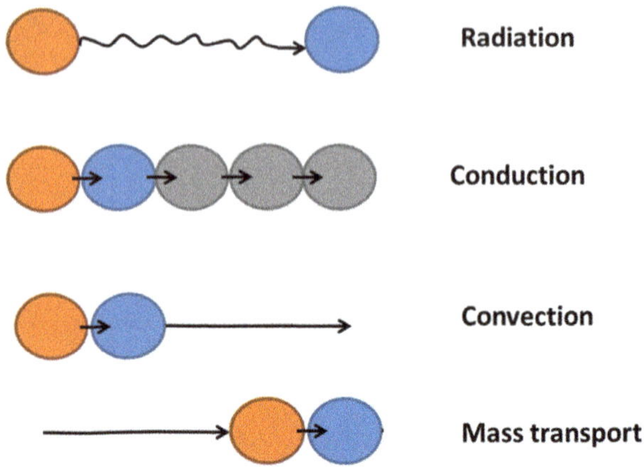

Fig. 5.3 Representation of heat transfer modes from the source molecule (*orange*) to the receiver molecule (*blue*) by radiation with the transport of energy by electromagnetic waves, conduction with energy transfer between adjacent molecules, convection where there is a movement of the energized molecule, and solid mass transport where the source molecule moves

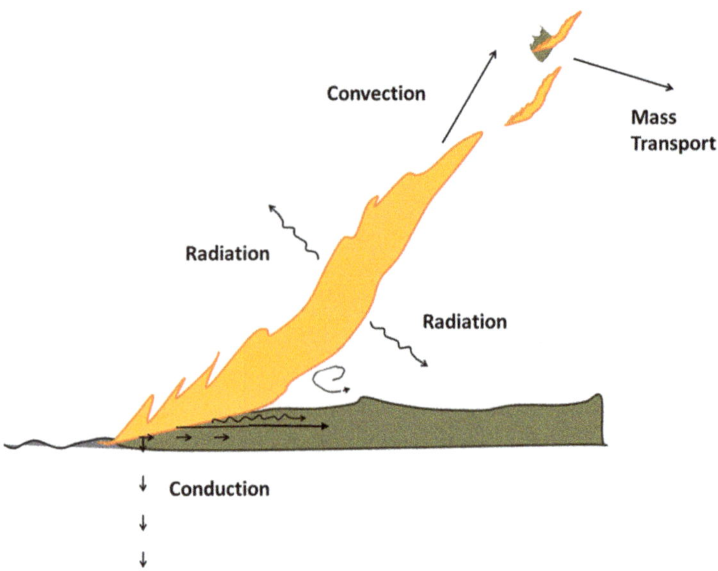

Fig. 5.4 A simplified view of the main heat transfer modes involved in different parts of a spreading fire

also commonly occurs near the base of the flame where buoyancy induced vorticity and instabilities can significantly increase local heat transfer rates (see Chap. 7 for further discussion). Solid mass transport, or spotting, is often dependent upon convective flows that carry the burning material away from the flames. Convection,

Fig. 5.5 James Watt, British mathematician and engineer (1736–1819) (Partridge 1806)

radiation, and solid mass transport are the dominant heat transfer modes associated with fire spread. Because gases such as air and wildland fuel beds are poor conductors, we often ignore the role of conductive heat transfer in fire spread. However, conduction is critical for understanding many fire effects such as heat transfer through the soil or cambium heating through a tree bole.

Heat transfer rate can be measured as the amount of energy transferred per unit time (e.g., Watts or Joules per second, Eq. (5.1)), or as heat flux which is the energy transferred per unit of time and unit area of the corresponding surface (e.g., Watts per square meter, Eq. (5.2)). The unit of heat transfer, Watt, is named in recognition of the work of James Watt (Fig. 5.5). Thus,

$$\text{Heat transfer rate (Watts)} = \text{Energy (Joules)}/\text{Time (seconds)} \tag{5.1}$$

$$\begin{gathered}\text{Heat flux (Watts per square meter)} = \\ \text{Energy (Joules)}/\text{Time (seconds)}/\text{Area (square meters)}\end{gathered} \tag{5.2}$$

5.3 Radiation

All objects with a temperature above absolute zero emit energy via electromagnetic radiation because of vibrational and rotational movement of their molecules and atoms (Fig. 5.3). Unlike conduction and convection, heat transfer by thermal radiation does not necessarily need a material medium for energy transfer. Radiative heat transfer can pass through a vacuum such as in outer space, a liquid, or a gas such as air. The transmission of radiative energy through a vacuum is more efficient than through gases and liquids because the latter contain atoms and molecules which can

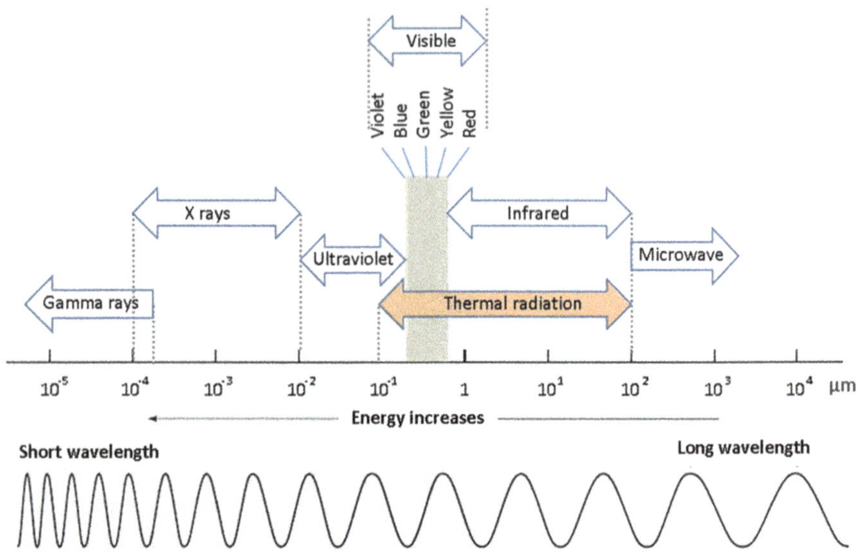

Fig. 5.6 The electromagnetic spectrum with the range of wavelengths including thermal radiation. (Data from Lillesand and Kiefer 1979)

absorb, or reflect, radiative energy. However, the ability of an object to absorb or reflect radiative energy depends on the wavelength emitted, the temperature, and emissivity.

The electromagnetic spectrum includes various types of radiation, including short wavelengths (Gamma rays) to long wavelengths (microwave). The range called "thermal radiation"includes wavelengths between approximately 0.1 and 100 µm, including all the infrared, visible, and some of the ultraviolet portion of the electromagnetic spectrum. The power emitted by an object is directly proportional to its frequency (ν) and inversely proportional to wavelength (λ). The energy associated with radiation increases as the wavelength decreases (Fig. 5.6).

The spectral emittance, the energy emitted per unit wavelength interval, of an ideal "blackbody" as a function of temperature and wavelength can be calculated using Eq. (5.3). A "blackbody" is an idealized object that absorbs all radiation that it receives and emits radiation at a consistent frequency that depends only on its temperature:

$$u(\lambda, T) = 2\pi h_p c^2 \lambda^{-5} \left\{ exp\left[\frac{h_p \times c}{k_B \times T \times \lambda}\right] - 1 \right\}^{-1} \tag{5.3}$$

where u (λ, T) is the spectral radiant emittance (W m^{-2} μm^{-1}), λ is the wavelength (µm), T is the absolute temperature (K), h_P is Planck's constant (6.626×10^{-34} J s), k_B is Boltzmann constant (1.381×10^{-23} J K^{-1}), and c is the speed of light (3.00×10^8 m s^{-1}). This relationship between emittance and wavelength for

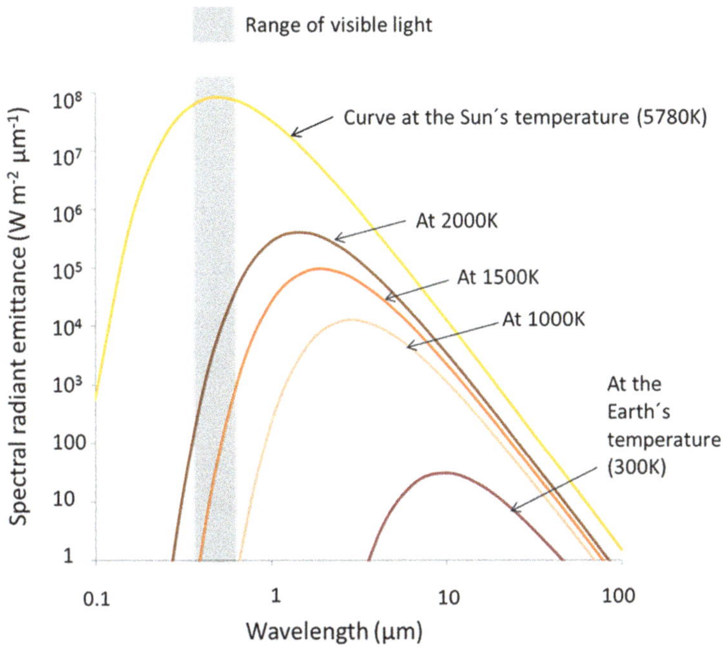

Fig. 5.7 Spectral radiant emittance, or the energy emitted by an ideal "blackbody"per unit wavelength interval (μm) within the thermal radiation range (wavelength between 0.1 and 100 μm) as a function of absolute temperature and wavelength. Note the logarithmic nature of the two axes, allowing for a representation from the temperature of the Sun to the temperature of Earth through values typical of wildland fires

different temperatures is illustrated in Fig. 5.7 using wavelength in microns ($1\ \mu m = 10^{-6}$ m).

The area under a spectral radiant emittance curve provides an estimate of the total radiative power (P_b) of a blackbody. The power radiated from a "blackbody" can be estimated using the Stefan-Boltzmann law (Eq. 5.4), which states that the total radiative power emitted by a "blackbody" per unit surface area for all wavelengths is directly proportional to the fourth power of the object's temperature:

$$P_b = \sigma_{SB} T^4 \tag{5.4}$$

where P_b is the total radiative power (kW m^{-2}), σ_{SB} is the Stefan-Boltzmann constant (5.67×10^{-11} kW m^{-2} K^{-4}), and T is the absolute temperature of the emitting blackbody (K).

Recall that "blackbody" objects represent an ideal emitter in that for a given temperature no real object emits more radiation than a "blackbody". To account for differences in radiative power between a "blackbody" and real-world objects, we can include an additional term into Eq. (5.4) called the emissivity (ε). Objects which deviate from the "blackbody" assumption are commonly referred to as "grey body"

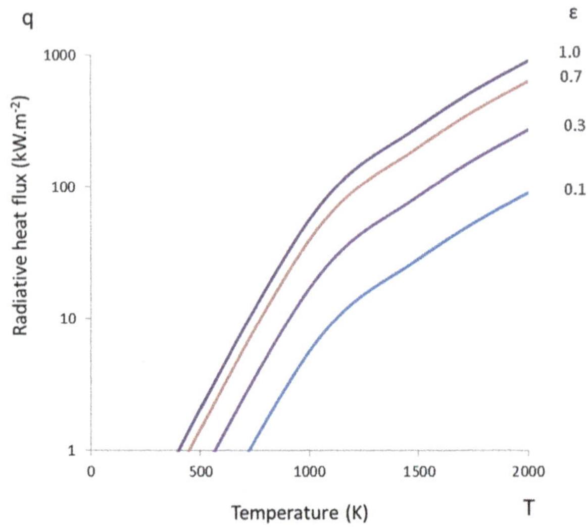

Fig. 5.8 Radiative power of an object with emissivity from 1 to 0.1 across a range of temperatures

objects. The emissivity is the ratio of actual radiation emitted to the maximum radiation emitted by a "blackbody". Total radiative power for a "grey body" (P_g) object can be estimated with Eq. (5.5). The differences in radiative power for various emissivities across a range of temperatures is shown in Fig. 5.8:

$$P_g = \varepsilon \sigma_{SB} T^4 \tag{5.5}$$

Due to the heterogeneous and turbulent nature of flames, it can be difficult to estimate the flame temperature (see Chap. 4) and emissivity for real-world fires. The radiation emitted from vegetation fires is due to hot gases (e.g., carbon dioxide, carbon monoxide, and water) and soot particles present in flames. The intensity of flame radiation depends on both the composition of the flames and the temperature. Flames that have greater amounts of soot tend to have higher emissivities but lower temperatures than flames with less soot. For example, flames with a significant amount of soot might have a mean temperature of 1200 K and an emissivity value of 1.0, resulting in a radiative power of 117 kW m^{-2}. In contrast, flames with little soot might have a mean temperature of 1400 K with an emissivity of 0.50, resulting in a radiative power of 109 kW m^{-2}. Experimental measurements of flame emissivity range from 0.25 to 0.94. For most wildland fire applications, an emissivity value ranging from 0.9 to 1 is assumed.

In Eq. (5.5) we estimated the total radiative power emitted per square meter of an object such as a flame given its temperature. However, many wildland fire applications in fire behavior and effects are concerned about the exchange or transfer of that heat energy between the flame and some other object. A common method for estimating the exchange of radiative energy between two objects is to include the concept of a view factor (sometimes called a shape factor) in Eq. (5.5). The view factor (F_{ab}) is the proportion of the radiative power that leaves object a and is intercepted by object b. The radiative heat flux (q_{rad} in kW m^{-2}) received at surface b from surface a can be estimated as:

$$q_{rad} = F_{ab}P_g = F_{ab}\varepsilon\sigma_{SB}\,T^4 \quad (5.6)$$

View factors are purely a function of the geometry associated with objects a and b. When calculating a view factor, it is common to assume that radiation is emitted in all directions and that the medium between the two objects or surfaces is neutral (i.e., the medium does not absorb, emit or scatter the radiation). The equations to estimate the view factor for several simplified 2- and 3-dimensional scenarios can be found in heat transfer textbooks such as Incropera et al. (2007).

For example, let's use Eq. (5.6) to estimate the radiative heat flux between two surfaces and estimate the total radiative heat flux (q_{rad}) transported from a flame of length (L) with a flame angle of 45° (A), an emissivity of 0.9, and a temperature of 1000 K to the unburned fuel ahead of the flame (b). Figure 5.9 provides a visual representation of this example.

The view factor for the scenario identified in Fig. 5.9 can be estimated using the equation for a long symmetrical wedge:

$$F_{ab} = 1 - sin\,\frac{A}{2} \quad (5.7)$$

Substituting all parameters into Eqs. (5.6) and (5.7) and solving for the radiative heat flux, we see that 31.6 kW per square meter are transferred from the flame to the unburned fuel ahead of the fire:

$$q_{rad} = F_{ab}P_g = F_{ab}\varepsilon\sigma_{SB}T^4 = 0.62 \times 0.9 \times 5.67 \times 10^{-11} \times 1000^4$$
$$= 31.6\,kW\,m^{-2} \quad (5.8)$$

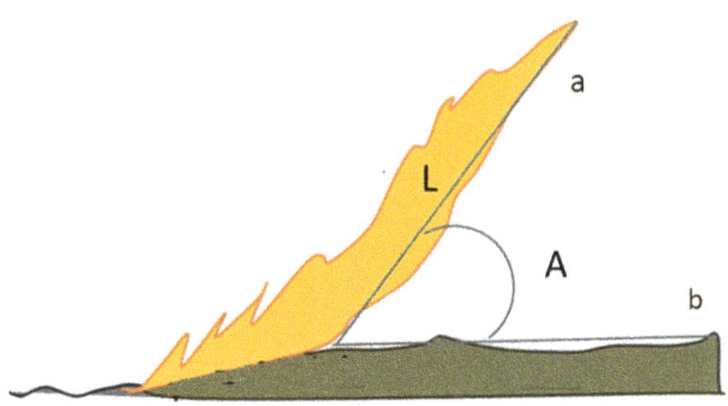

Fig. 5.9 Visual representation of a flame and the unburned fuel ahead showing flame length (*L*) and the flame angle (*A*) between the two surfaces *a* and *b*

Another important application of heat transfer by radiation is in determining safety distances for fire personnel and others. This can be based on the geometry, emissivity, temperature of the flames, and the thresholds for skin injury. See Chap. 10.

5.4 Conduction

Conduction is thermal energy transfer that occurs through collisions between particles and the movement of electrons. Although conduction can occur in all phases of matter (i.e., solid, liquid, and gas), it is most commonly associated with heat transfer within a solid object or between two solid objects that are in contact with each other. Conduction is a primary mode of heat transfer during the pre-ignition phase of large solid fuels and in determining fire effects on trees via injury to the vascular cambium, and roots (see Sect. 9.3) and soil heating that can affect soil organisms and nutrients such as nitrogen (See Sect. 9.6). Conductive heat transfer can also be an important component of large log combustion and fire spread in densely packed fuelbeds such as those that result from mastication when trees, shrubs, and other fuels are chipped or mulched (See Chap. 11).

A theory of conductive heat transfer was first proposed in 1807 by the French scientist Joseph Fourier (Fig. 5.10). Fourier's law indicates that, under steady-state conditions, the rate of heat transfer through a material is directly proportional to the temperature difference and inversely proportional to the distance traveled (Fig. 5.11).

Fig. 5.10 The French mathematician and physicist Jean-Baptiste Joseph Fourier (1768–1830) (Boilly n.d.)

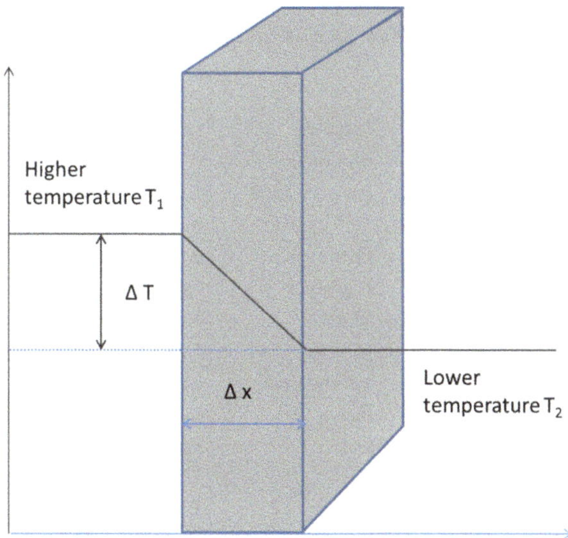

Fig. 5.11 One-dimensional steady-state heat transfer by conduction between two surfaces at different temperatures. The negative slope $-\Delta T/\Delta x$ is equal to q/k where q is the heat flux and k the thermal conductivity of the material. For any given heat flux lower conductivities result in higher temperature differences

This relationship can be represented in two equivalent forms: an integral form and a differential form. The one-dimensional forms of these two equations are shown below:

$$q_{cond} = -k\frac{\partial T}{\partial x} \tag{5.9}$$

$$q_{cond} = -k\ \frac{\Delta T}{\Delta x} \tag{5.10}$$

where q_{cond} is the heat flux between the two surfaces (W m^{-2}), ΔT is the temperature difference between the two surfaces (K), Δx is the distance between the ends (m), and k is the thermal conductivity of the material (W m^{-1} K^{-1}). For steady-state one-dimensional conduction, the heat transfer through a given material depends only on the temperature gradient, the distance between the surfaces, and the material's thermal conductivity. The minus sign in the equation indicates that the heat flow is in the opposite direction to the temperature gradient ($\Delta T/\Delta x$).

For non-steady-state conductive heat transfer conditions, i.e., when the temperature at the boundary of the object changes, as in most situations in wildland fires, we may want to account for the variations in temperature with time. To account for these fluctuations, we need to incorporate not just the rate at which temperatures are changing but also the object's thermal diffusivity. Thermal diffusivity is a measure of a material's ability to conduct energy relative to its ability to store energy. Materials with higher thermal diffusivity transfer energy more rapidly and reach steady-state conduction faster than those with low thermal diffusivity. Metals and gases often have high conductivity and diffusivity coefficients, while bark, wood, and dry soils have low values (Table 5.1). Thermal diffusivity (α) can be estimated

Table 5.1 Typical values of thermally relevant characteristics of some materials, including those related to wildland fires (Data from Geiger 1980; Incropera et al. 2007; Jury et al. 1991; Martin 1963)

Types of materials		Materials		Density $kg\ m^{-3}$	Heat capacity $kJ\ kg^{-1}\ K^{-1}$	Conductivity $W\ m^{-1}\ K^{-1}$	Diffusivity $mm^2\ s^{-1}$
Gases		Carbon dioxide		1.77	0.85	0.017	11
		Air	300 K	1.16	1.01	0.026	23
			1000 K	0.35	1.14	0.067	168
		Hydrogen		0.08	14.31	0.183	158
Liquids		Engine oil		884	1.91	0.15	0.09
		Water		1000	4.18	0.61	0.15
		Mercury		13,529	0.14	8.54	4.53
Solids	Wood	Yellow Pine		640	2.81	0.15	0.08
		Fir		418	2.72	0.13	0.10
		Oak		545	2.39	0.17	0.13
	Bark	Cork		225	1.70	0.04	0.10
		Others		300	1.70	0.06	0.12
	Soils	Peat	Dry	700	1.20	0.10	0.12
			Wet	1000	3.20	0.40	0.13
		Sand	Dry	1500	0.80	0.20	0.17
			Wet	2000	1.20	1.20	0.50
		Clay	Dry	1600	0.70	0.20	0.18
			Wet	2100	1.00	1.20	0.57
	Metals	Iron		7870	0.45	80	23
		Aluminum		2702	0.90	237	97
		Copper		8933	0.39	401	117
		Gold		19,300	0.13	317	127
		Silver		10,500	0.24	429	174

(Eq. 5.11) by dividing the thermal conductivity of a material (k) by the product of the material density (ρ) and its specific heat capacity (C_p):

$$\alpha = \frac{1000k}{\rho C_p} \tag{5.11}$$

where α is thermal diffusivity ($mm^2\ s^{-1}$), k is thermal conductivity ($W\ m^{-1}\ K^{-1}$), ρ is the density of the material ($kg\ m^{-3}$), and C_p is the heat capacity per unit mass of the material ($kJ\ kg^{-1}\ K^{-1}$).

The differential form of Eq. (5.9) for non-steady-state conduction is (Eq. 5.12):

$$\frac{\partial T}{\partial t} = \alpha \frac{\partial^2 T}{\partial x^2} \tag{5.12}$$

where α is the thermal diffusivity of the material, δT/δt is the instantaneous rate of temperature change, and $\delta^2T/\delta x^2$ is the second derivative of temperature change along the x gradient.

Assuming that the surface temperature of a solid infinite slab is equivalent to the flame temperature, Eq. (5.12) can be solved using a Laplace transformation as used by Spalt and Reifsnyder (1962) and used by many authors, including Peterson and Ryan (1986):

$$\frac{T_x - T_b}{T_i - T_b} = erf\left(\frac{x}{2\sqrt{\alpha t}}\right) \tag{5.13}$$

where T_x is the temperature at a distance x from the surface, T_b is the average temperature on the outside of the bark, T_a is the ambient temperature, t is the time of exposure to temperature T_f, α is thermal diffusivity, and erf is the Gauss error function.

Several authors have used Eq. (5.10) to investigate tree mortality due to cambium heating during a fire (Dickinson and Johnson 2001) by rearranging Eq. (5.10) to solve for the residence time (t_R) of bark heating required to heat the cambium to a critical temperature (T_x) given the bark thickness (x) and thermal diffusivity (α). They all concluded, as did Hare (1965), that the duration of a heat pulse required to kill the cambium of a tree is directly proportional to bark thickness squared.

For example, using Eq. (5.13) we can estimate the residence time required to kill the cambium of a tree with a bark thickness of 1.5 cm. Assume an ambient temperature of 30 °C (T_i), an outside bark temperature of 500 °C (T_b), a lethal temperature of 60 °C for vascular cambium (T_x), and a thermal diffusivity (α) of 1.35×10^{-7} m^2 s^{-1} for bark. The first step is to substitute the temperatures into the left-hand side of equation:

$$\frac{T_x - T_b}{T_i - T_b} = \frac{60 - 500}{30 - 500} = 0.936 \tag{5.14}$$

Using an inverse error function table and substituting this value into Eq. (5.13), we get:

$$1.31 = \left(\frac{x}{2\sqrt{\alpha t}}\right) \tag{5.15}$$

Rearranging Eq. (5.15) to solve for t (s) and x (mm) results in:

$$t = 1.08 \times x^2 \tag{5.16}$$

where t is the critical residence time (s) required to heat the vascular cambium up to a critical temperature of 60 °C given a bark thickness of x in

(continued)

millimeters. Assuming a bark thickness of 15 mm the critical residence time would be 243 s. This implies that cambium mortality is unlikely for trees of this bark thickness, unless downed dead woody fuels or dense duff and litter layers, which take longer to burn, are in close proximity to the tree.

Several authors from Australia (Gill and Ashton 1968; Vines 1968) and North America (Hare 1965) studied the relationships between bark thickness and cambium temperature and found that the cambium temperature data compared fairly well with this simple model (Dickinson and Johnson 2001). However, it is important to recognize that the temperature on the outside of the bark varies as a function of height and around the circumference of a tree bole and through time, and that bark thickness varies along the height of the bole. Cambial heating and tree mortality are discussed further in Chap. 9.

Alternatively, one-dimensional non-steady-state conduction heat transfer problems, for simple geometries, can be approximated by solving Eq. (5.12) using finite difference methods (Rego and Rigolot 1990; Dickinson and Johnson 2004; Mercer and Weber 2007). In this approach, the object of interest, such as soil, is partitioned into a uniform grid, and the derivatives are replaced by finite differences between neighboring points. This approach has been particularly useful in modeling the thermal regime of soils (Jury et al. 1991). The finite-difference equation can be expressed through its discrete equivalent:

$$T(x, t + \Delta t) = \alpha\, T(x - \Delta x, t) + (1 - 2\,\alpha)\, T(x, t) + \alpha\, T(x + \Delta x, t) \tag{5.17}$$

where T(x,t) is the temperature at distance x at time t, Δt and Δx are the increments in time and space, and α is the coefficient for thermal diffusivity. If the distance step (Δx) is 1 mm, and the time step (Δt) is 1 s, it is useful to express diffusivity (α) in $mm^2\ s^{-1}$ which is equivalent to $10^{-6}\ m^2\ s^{-1}$. It can be seen by the equation that for very small diffusivities (α close to zero), the temperature remains practically unchanged. Simulations of heat transfer at different depths from a surface with a given diffusivity can be made with the spreadsheet presented at the end of the chapter.

5.5 Convection and Solid Mass Transport

Convective heat transfer can play an important role in determining fire spread and fire effects on plants and people. Convection is the transport of energy due to the bulk movement of molecules of a fluid (i.e., gases and liquids). The term convection is used to describe the combined heat transport due to diffusion, the movement of particles due to a concentration gradient, and advection, the bulk movement of the flow. In wildland fires, this bulk movement includes not just gases and liquids, but also solid particles. When these particles are burning, their transportation can be an important mechanism driving fire propagation. In general, there are two types of convection, natural and forced convection. If the fluid movement is artificially imposed by something like a pump or fan, it is called forced convection. If, as typically in wildland fires, the flow is due to buoyancy forces, it is called free, or natural, convection.

The convective heat flux (q_{conv}) transferred between an object with a given surface temperature (T_f) and the surrounding fluid at temperature (T_a) is dependent upon the temperature difference and a heat transfer coefficient (h) (Eq. 5.18):

$$q_{conv} = h\left(T_f - T_a\right) \tag{5.18}$$

where h is a convection heat transfer coefficient (W m^{-2} K^{-1}). The convection heat transfer coefficient depends upon a number of factors including the type of fluid, flow properties, the geometry of the surface, and the temperature differences. The heat transfer coefficient increases as a function of the flow velocity and is greater during turbulent than laminar flow. There are numerous correlations that can be used to estimate the convective heat transfer coefficient for various cases including natural and forced convection for a range specific geometry and flow conditions. The heat transfer coefficient for air during natural convection across a range of conditions varies from 5 to 25 W m^{-2} K^{-1} (Incropera et al. 2007). The convective heat transfer flux during a fire can be estimated experimentally by subtracting estimates of the radiative heat flux from total heat flux which can be measured using a variety of sensors. Further discussion on the role of convective heat transfer and solid mass transport in determining fire spread, crown fire ignition, and fire effects on plants and people can be found in Chaps. 7, 8, 9, and 10.

5.6 Implications

The heat from fires will result in ecological and other effects only if it can transfer from the fire to the ecosystem. All fires produce heat that transfers by some combination of radiation, convection, conduction, and mass transport. When you understand these physical processes, combined with the chemical and heat production processes covered in previous chapters, you can then understand fire as an ecosystem process.

In the remainder of the book, we will discuss fuels, fire behavior and fire effects, and then fuel dynamics and landscape management. In all of these aspects of fire, heat transfer processes are involved. We emphasize the application of heat transfer in understanding the ecological effects of fires and how that information can inform effective fire science and integrated fire management.

5.7 Interactive Spreadsheets: RADIATION_Fireline_Safety, CONVECTION, CONDUCTION_Soils_Plants, and MASS_TRANSFER_Spotting

We encourage you to use the interactive spreadsheets to explore the relative importance of the key factors influencing heat transfer. In this chapter, we show in the spreadsheet, CONDUCTION_Soils_Plants_v2.0, that you can readily change the inputs and then graphically display how outputs change (Fig. 5.12). In this way, you can evaluate how sensitive the outputs are to changes in one or more of the inputs. You can see how the temperature and duration of the flaming and smoldering phases affect the maximum temperature attained at different soil depths. The same applies to the heat transfer through tree bark. The lethal temperature is an input and you can see graphically at what distance the lethal temperature is attained. You can also understand how different soil types or moisture conditions affect heat transfer by changing the corresponding diffusivity values. You will learn more about the effects of heat transfer by conduction in soils and bark when you learn more about how heat from fires influences ecological systems (Chap. 9). In Chaps. 8–10, you will have other examples of practical applications of heat transfer, also by using spreadsheets RADIATION_Fireline_Safety_v2.0, CONDUCTION_Soils_Plants_v2.0, CONVECTION_Crown_Scorch_v2.0, and MASS_TRANSFER_Spotting_v2.0.

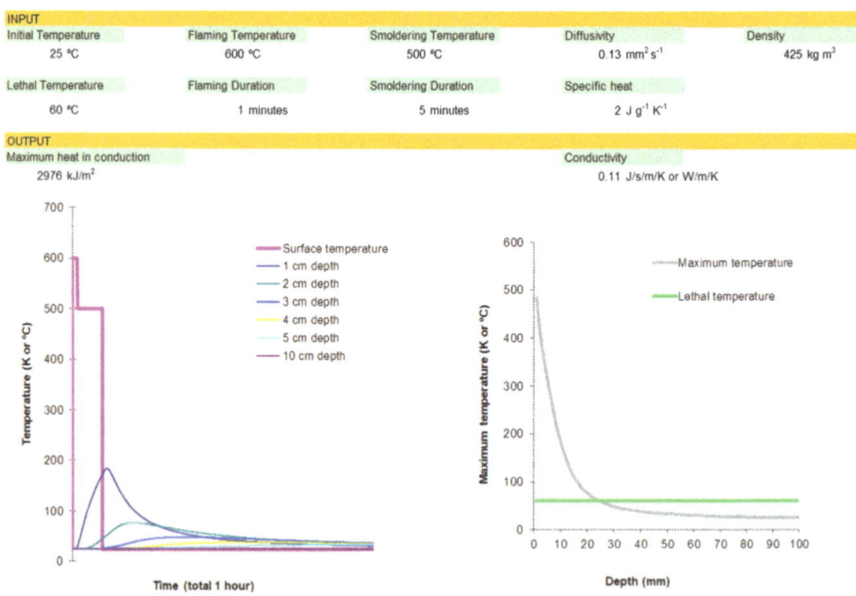

Fig. 5.12 Details of inputs and outputs of the spreadsheet, CONDUCTION_Soils_Plants_v2.0, showing how the duration and temperatures of flaming and smoldering influence the evolution of temperatures for 1 h at various soil depths with a given diffusivity value. The graph at the right shows the maximum temperatures attained in comparison with a given lethal temperature

Using each of these spreadsheets will help you understand one of the four processes of heat transfer, including radiation, convection, conduction, and mass transfer. These spreadsheets are for learning. For prediction, we recommend you use one of the fire behavior prediction systems because they will include multiple modes of heat transfer at the same time depending on the conditions. The equations used in our spreadsheets are also used in the fire behavior prediction systems.

References

Boilly, L.-L. (n.d.). *Portrait of Joseph Fourier*. Wikimedia Commons. Retrieved June 13, 2020, from https://commons.wikimedia.org/wiki/File:Fourier2.jpg.

Dickinson, M. B., & Johnson, E. A. (2001). Fire effects on trees. In E. A. Johnson & K. Miyanishi (Eds.), *Forest fires: Behavior and ecological effects* (Vol. 1–9, pp. 477–526). San Diego: Academic.

Dickinson, M. B., & Johnson, E. A. (2004). Temperature-dependent rate models of vascular cambium cell mortality. *Canadian Journal of Forest Research, 34*(3), 546–559.

Geiger, R. (1980). *Das Klima Der Bodennahen Luftschicht (Manual de microclimatologia: O Clima da Camada de Ar Junto ao Solo)* (I. Gouveia, C. Cabral, A. Lobo de Azevedo, & Fundação Calouste Gulbenkian, Trans.). Lisbon.

Gill, A. M., & Ashton, D. H. (1968). The role of bark type in relative tolerance to fire of three central Victoria eucalypts. *Australian Journal of Botany, 16*, 491–498.

Hare, R. C. (1965). Contribution of bark to fire resistance of southern trees. *Journal of Forestry, 63*, 248–251.

Incropera, F. P., DeWitt, D. P., Bergman, T. L., & Lavine, A. S. (2007). *Fundamentals of heat and mass transfer* (6th ed.). New York: Wiley.

Jury, W., Gardner, W. R., & Gardner, W. H. (1991). *Soil physics* (5th ed.). New York: Wiley.

Lillesand, T. H., & Kiefer, R. W. (1979). *Remote sensing and image interpretation*. New York: Wiley.

Martin, R. E. (1963). Thermal properties of bark. *Forest Products Journal, 13*(10), 419–426.

Mercer, G. N., & Weber, R. O. (2007). Modeling heating effects. In E. A. Johnson & K. Miyanishi (Eds.), *Plant disturbance ecology: The process and the response* (pp. 371–396). New York: Academic.

Partridge, J. (1806). *James Watt*. Wikimedia Commons. Retrieved June 13, 2020, from https://commons.wikimedia.org/wiki/File:Beechey_James_Watt.jpg.

Peterson, D. L., & Ryan, K. C. (1986). Modeling postfire conifer mortality for long-range planning. *Environmental Management, 10*(6), 797–808.

Rego, F., & Rigolot, E. (1990). Heat transfer through bark: A simple predictive model. In J. G. Goldammer & M. J. Jenkins (Eds.), *Fire in ecosystem dynamics: Mediterranean and northern perspectives* (pp. 157–161). The Hague: SPB Academic Publishing.

Spalt, K. W., & Reifsnyder W. E. (1962). Bark characteristics and fire resistance: a literature survey. New Orleans: USDA Forest Service, Southern Forest Experiment Station.

Vines, R. G. (1968). Heat transfer through bark and the resistance of trees to fire. *Australian Journal of Botany, 16*, 499–514.

Part II
Fuels, Fire Behavior and Effects

Photo by Kari Greer photo/USFS

Fuels are the link between fire behavior and effects (Keane 2015, Fig. II.1), so fuel dynamics and management are central influences on how fires burn and their consequences for vegetation, people, and other ecosystem components. Fuels are vegetation biomass, and thus fuels accumulate wherever biomass accumulation exceeds decomposition, and thus in almost all terrestrial ecosystems. Further, that biomass is ecologically important, and how it burns can greatly alter fire effects. In

Fig. II.1 Fuels are the link between fire behavior and effects (Keane 2015). (Photo by Kari Greer/ USFS taken in Idaho, USA, in 2008)

the absence of fires, biomass can accumulate to fuel fires when they ignite when it is hot, dry, and windy enough to carry fire. When those fires burn under extreme conditions, they can threaten lives, homes, be difficult and costly to manage, and have lasting effects for future fires.

These ideas are the central themes to the chapters in Part II of our book, *Fire science from chemistry to landscape management.* In this part of our book, we have five substantial chapters. After a summary of the descriptors used for fuels and fire behavior in Chapter 6, we address how fires spread from point to point and across landscapes in Chapter 7. For extreme fire behavior, we characterize the science behind crown fires and spotting, as well as what makes some fires more extreme than others. In the next chapter, we discuss the effects of fire on plants, soils, water, and animals. In Chapter 10, we address the direct and indirect effects of fire on people and what people can do to protect homes and communities when fires threaten, all within the context of recognizing the risks, costs, and ecosystem services of fires. In all of these chapters, we draw upon examples from around the world. We link to the fire science learned in prior chapters of this book. Throughout this book, we emphasize fire as part of ecosystems, including people, as we make the case that fire is part of a social-biophysical system.

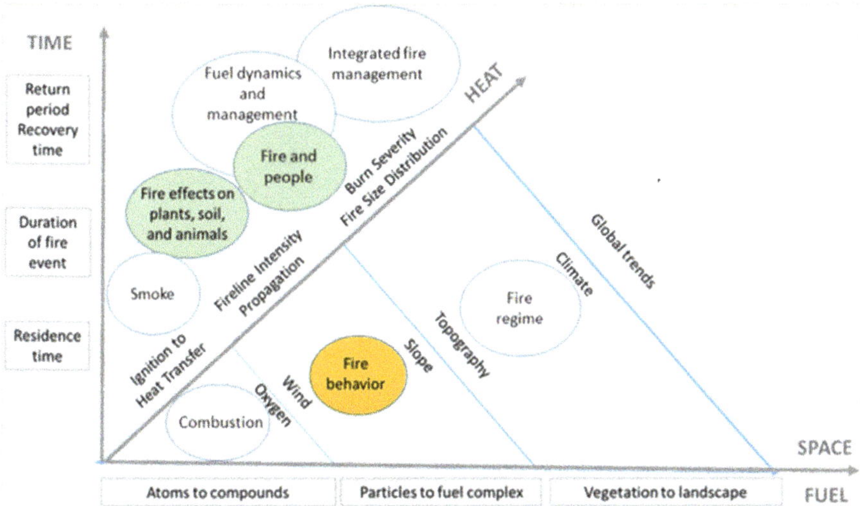

Fig. II.2 The fire behavior triangle represents the influence of the fuel complex, wind, and slope on the behavior of whole fires, including aspects as fireline intensity or propagation. The association between fire behavior and effects on ecosystems and humans is discussed in this part of the book

Hopefully, by better understanding the relations between fuels, fire behavior, and effects, our perceptions of fire are becoming more complete. Perhaps we will come to recognize that the carefully planned "good" fires can often prevent "bad" fires.

Reference

Keane, R. E. (2015). *Wildland fuel fundamentals and application*. Switzerland: Springer.

Chapter 6
Fuel and Fire Behavior Description

Learning Outcomes
Upon completion of this chapter, we expect you to be able to

1. Apply the descriptors of fuels to an area of vegetation with which you are familiar,
2. Distinguish amongst fuel particles, fuelbeds, fuel components, and fuel layers,
3. Diagram a fire burning in a grassland and apply the fire behavior descriptors to the flaming and smoldering combustion, and
4. Explain why the rate of spread and fireline intensity or flame size are crucial metrics in fire management

6.1 Introduction

Vegetation mediates both the effects of fire on ecosystems and human impacts on the fire regime. Independent of its biological nature, characterizing vegetation's ability to burn (i.e., as fuels) is useful for describing and modeling fire behavior and fire effects for multiple fire management applications. Fuel properties are highly variable in space and time. Analysis of fuels is a matter of scale, from the combustion at the scale of individual flames to the fire behavior scale to the landscape scale (Pyne et al. 1996).

This chapter addresses extrinsic fuel properties, those that exert quantifiable influence on fire behavior characteristics. In contrast, fuels' intrinsic properties are fundamental to fire ignition and spread (Pyne et al. 1996) but are unlikely to result in substantial variation in fire behavior at the scale measurable in the field (Cheney 1981). Intrinsic properties of fuels include chemical composition (including mineral

F. Castro Rego et al., *Fire Science*, Springer Textbooks in Earth Sciences, Geography and Environment, https://doi.org/10.1007/978-3-030-69815-7_6

content and readily volatilized constituents), heat content, particle density, thermal conductivity, and diffusivity (Chaps. 1–4). Fuel moisture is highly dynamic (See Chap. 11).

Wildland fires are classified according to the dominant fuel type being burned (e.g., a shrub fire is one burning in a shrubland) and the layer of fuel supporting fire spread (ground, surface, or crown fire). Ground fires burn organic matter in the soil, spreading very slowly. Surface fires can spread either slowly or rapidly as they burn leaf litter, fallen wood, and plants near the soil surface. Crown fires burn the foliage of trees and shrubs with high intensity. Fire behavior is described for various parts of a fire. This chapter introduces the fire behavior descriptors that are commonly measured or estimated by wildland fuel and fire managers and by fire scientists.

Fuels and fire behavior descriptors are generally shown as measures of central tendency, as means or medians of different measurements, but it should be recognized that both fuels and fires are generally quite variable in space and time. Measures of variability, as ranges or standard deviations, should be considered in sampling and analysis, for variability in fuels influences variability in fire behavior and effects. Measurements of central tendency and associated variability derive from sampling and statistical analyses. Many options are possible to optimize sampling to ensure unbiased and precise estimates while minimizing costs. Further discussion about sampling measurements of fuels and fire properties is beyond the scope of this book. In this chapter, we simply present the most commonly used extrinsic fuel properties and fire descriptors. Keane (2015) addressed many fuel-related topics, including fuel properties and inventory.

6.2 The Wildland Fuel Hierarchy

Wildland fuel description can be approached as a top-down hierarchy in terms of spatial resolution. Distinct levels of structural organization correspond to different scales of observation and heterogeneity in varying degrees. Here we will loosely follow Keane (2015), from coarse to fine scales (Fig. 6.1):

1. **Fuelbed is** a generic description of the complex of fuels occupying a given area.
2. **Fuel layer** or stratum results from the vertical stratification of the fuelbed into ground, surface, and canopy fuels (Fig. 6.2) that correspond with different combustion environments, respectively ground (or subterranean, or smoldering), surface, and crown fire. Although the boundary can be subjective, ground fuels typically comprise organic matter that does not contribute to flaming combustion. Ground fuels are in an advanced state of decomposition and overlaying the mineral soil; included are the duff (or humus) layer, peat, roots, and rotten woody fuels. Surface fuels refer to litter plus vegetation (grasses, forbs, and shrubs, including mosses and lichens in boreal ecosystems) within 2 m above ground (Keane 2015), especially for the purpose of using Rothermel's fire spread model (Chap. 7). Litter consists of fallen leaves, woody elements of various sizes,

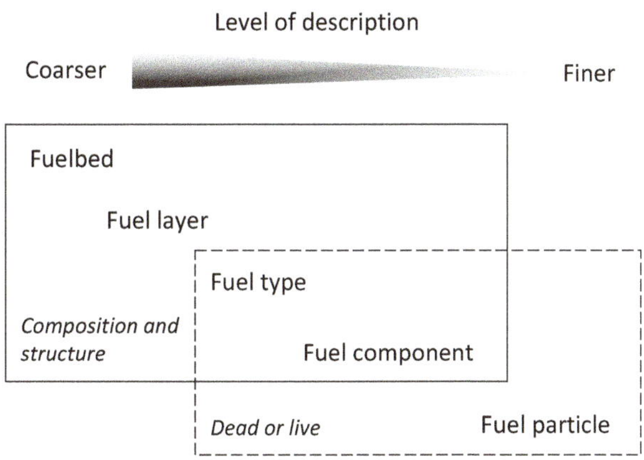

Fig. 6.1 Levels of wildland fuel description based on Keane (2015), from coarser (fuelbed) to finer (fuel particle), and how their properties are defined in broad terms

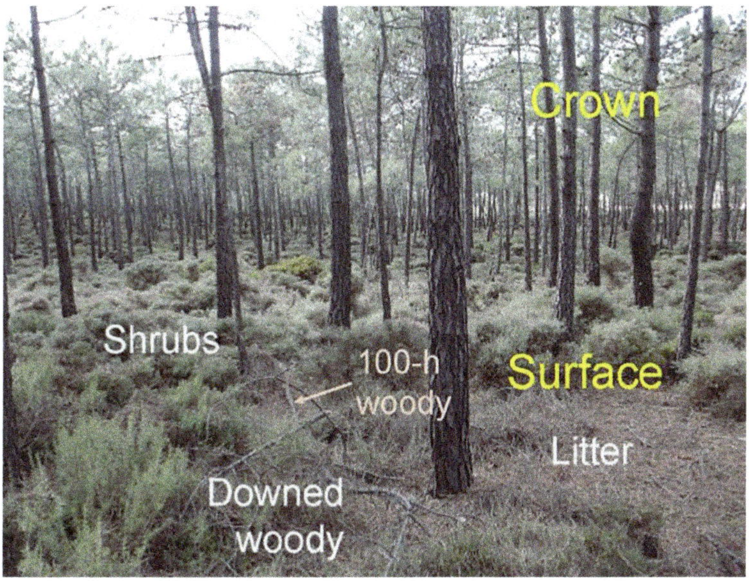

Fig. 6.2 Fuels in a maritime pine (*Pinus pinaster*) forest on the southwestern coast of Portugal consists of three fuel layers: ground (not visible), surface, and crown fuels. Surface fuels include three fuel types (litter, downed woody, and shrubs). One example fuel component is 100-h woody fuels (i.e., 25–75 mm in diameter). As litter and shrubs are codominant, surface fire behavior will develop in a litter-shrub fuel complex. (Photograph by P. Fernandes)

and bark and typically comprises a decomposing (fermentation) layer (F) and the fresh (undecomposed) layer (L) that drives fire spread. Slash (or activity) fuels will add to the previous categories after silvicultural operations, including canopy fuel treatments. The canopy (or aerial) layer encompasses all biomass above 2 m, regardless of its nature, and is often divided into ladder and crown fuels. Note that the same term can differ in meaning among countries or systems. In Australia, surface fuels designate litter only and the following layers are also considered: near-surface fuels composed of grasses and low shrubs and suspended material from the overstory, elevated fuels (tall shrubs), and bark fuels on the bole and branches (Gould et al. 2007).

3. **Fuel type** describes the generic nature of fuels or the dominant fuels in the fuelbed, e.g., "grass". Again, the understanding of "fuel type" varies. In the Canadian Forest Fire Behavior Prediction (FBP) System, fuel type refers to a vegetation type with distinctive fire behavior characteristics.
4. **Fuel component** subdivides the fuel type as a function of size (e.g., "fine woody fuels") or physiological condition (e.g., "live shrub foliage"). The distribution of fuels by size class (usually <6, 6–25, 25–75, and >75 mm diameter) and whether they are dead or live profoundly influences fire behavior and both direct and indirect effects of fires (through fuel moisture content dynamics, Chap. 11).
5. **Fuel particle** refers to the individual fuel elements or units (e.g., twig, leaf, needle) that form the fuel complex at coarser scales and often define the calculation of aggregate fuel properties at those scales (Fig. 6.1).

6.3 Fuel Description

Various extrinsic fuel metrics are used to describe fuels (Table 6.1). These include the size and shape of particles and the amount and structure (including metrics of compactness and continuity) of fuelbeds, fuel layers, and fuel components.

The size and shape of individual particles influence ignitability, heat release rate, and burn duration. The size and shape of fuels are integrated into fire-spread models through surface area-to-volume (or mass) ratio. Size classes as a function of particle diameter or thickness are used to define nominal rates of dead fuel moisture content response (time-lag) to variation in atmospheric conditions (see Chap. 11). Size classes are used in summarizing the fuel loads and moisture contents that are the basis for fire behavior modeling and fuel inventories.

Fuel load, defined as the mass per unit area, plays a central role in fire science and management. Along with fuel moisture content (Chap. 11), fuel load determines fuel consumption and the amount of heat released during combustion. The amount of fuel consumed influences fire effects in vegetation and soils (See Chap. 9) and the effectiveness of fire control operations. Fuel load is needed to calculate fuel structure descriptors such as bulk density (Table 6.1), and fire intensity is an input to fire behavior and effects models. Fuels treatments are often designed to reduce fire hazard (See Chaps. 10 and 11). The degree to which fuel load contributes to fire behavior, especially to the flaming front properties, is termed fuel availability and depends on fuel structure and moisture. Fuel depth, bulk density, and packing ratio are descriptors of fuel structure that affect heat transfer and fire-spread rate. These descriptors are included in fire-spread rate models. Finally, the vertical discontinuity between the surface and canopy fuel layers is used to assess the likelihood of crown fire.

Table 6.1 Extrinsic fuel variables definitions, compiled from various sources

Scale	Variable	Symbol	Definition	Units
Fuel particle	Diameter	d	Diameter of a cylinder (generalized assumption)	m
	Surface area-to-volume ratio	σ	Particle surface area divided by its volume	m^{-1}
	Surface density	m/A	Mass of the particle (m) divided by its projected area (A)	$kg\ m^{-2}$, $g\ cm^{-2}$
	[a]Surface area-to-mass ratio	S_m	Particle surface area divided by its dry weight	$m^2\ kg^{-1}$
Fuel complex	Load	w	Dry weight per unit area	$kg\ m^{-2}$, $t\ ha^{-1}$
	Depth, height	δ	Fuel layer or fuel complex thickness	m
	Bulk density	ρ_b	Dry weight per unit volume	$kg\ m^{-3}$
	[b,c]Packing ratio	β	ρ_b divided by particle density, the fuel bed volume fraction occupied by fuel	Dimensionless
	Canopy base height	CBH	Vertical distance between ground surface and the live canopy base	m
	[d]Fuel strata gap	FSG	Distance from the top of the surface fuel to the lower limit of the canopy constituted by ladder and live fuels	m

[a]Rossa and Fernandes (2018)
[b]Countryman and Philpot (1970) refer to $1/\beta$ as porosity
[c]Rothermel and Anderson (1966) define porosity as the void volume per unit of fuel surface area (m)
[d]Cruz et al. (2004)

The methods and variables used to describe wildland fuels are intrinsically dependent on the input requisites of the adopted fire behavior models. Empirical models for fire-spread rate, developed for specific or generic fuel types, seldom include the effect of more than one descriptor of fuel structure (Cruz et al. 2015). They can also altogether disregard fuel variation within a vegetation type, such as in the Canadian Forest Fire Behavior Prediction (FBP) System (Forestry Canada Fire Danger Group 1992).

Rothermel's (1972) model of surface fire spread relies on sets of numerical fuel characteristics called fuel models that represent the fuels complex for fire behavior predictions. A set of 13 stylized fuel models organized in 4 groups (grass, shrub, litter, slash) depending on the vector of fire spread has been developed (Anderson 1982). Fuel models are widely used to predict fire behavior characteristics with applications that employ Rothermel's model. Each fuel model is described in terms of fuel load by size class, surface area-to-volume ratio, fuel depth, heat content, and moisture of extinction. As the 13 fuel models are insufficient to account for the variability in fuel characteristics found across vegetation types and ecosystems, including when multiple fuel layers are considered, additional fuel models have been developed, including a set of forty for the USA (Scott and Burgan 2005). Of most importance for management applications, the parameters of custom fuel models should be calibrated and optimized such that predicted fire behavior matches observed fire behavior (Hough and Albini 1978; Cruz and Fernandes 2008; Ascoli et al. 2015).

Various methods have been developed to assess and quantify fuels, directly and indirectly, and destructively or not. Catchpole and Wheeler (1992) reviewed the existing types of techniques and distinguished between:

1. **Direct sampling**, based on destructive (and costly) fuel sampling and is usually restricted to research studies that require accurate estimates of fuel load. Fuels are harvested from within quadrats of variable size, are bagged, and then processed in the laboratory, which includes sorting, weighing, and oven drying by size class and dead or live condition. Direct sampling also includes scoring or rating (from nil to extreme) fuel hazard by fuel layer (Gould et al. 2007), and measurements of fuel structure, such as the depth (or height) of the existing fuel layers, and the ground covered by individual fuel layers. In the latter case, linear or planar intercept (or transect) methods are used (Van Wagner 1968; Brown 1971), but also point contact techniques.
2. **Calibrated visual estimation**, where comparison with reference information is used to estimate fuel loads (Keane and Dickinson 2007) or assign fuel models (Anderson 1982). Photo keys or photo series are a common tool for this purpose (Fig. 6.3). Photo series typically associate each photo to a fuel model and quantify fuel loads for the depicted situation, plus fire behavior characteristics and fire control difficulty for a given weather scenario.
3. **Double sampling**, where a fuel parameter is estimated using a two-stage approach. In the first stage, a sample is taken to develop either a ratio or regression that relates a variable of interest to another more easily measured variable. In the second phase, the more easily measured variable is collected, and the variable of interest is estimated using the ratio or regression developed in the first phase. For example, a regression that relates fuel load and a metric of fuel structure such as cover or depth or even time since disturbance (see Chap. 11) can be developed by collecting data on the two variables. The secondary, often larger sample, of the more easily measured variable can be collected and the variable of interest can be estimated by applying the developed regression to the measured variable.

6.4 Fire Description

Wildland fires are often simply classified qualitatively according to the dominant fuel type being burned or by the layer of fuel supporting fire spread. However, quantitative metrics describing fires are available and very useful in fire management.

Ericaceous Shrub 17

Site Information	
Location	Clear Creek State Forest, Jefferson County, Pennsylvania
Overstory metrics	103 trees/acre; basal area 95 feet/acre; stocking 69 percent
Overstory composition	Chestnut oak, red maple, northern red oak, white oak
Understory composition	Mountain laurel
Aspect/elevation/slope	South; 1,800 feet; 5 percent
Stand history	Gypsy moth mortality and salvage harvest in late 1980s

Fuel Characteristics			
Leaf litter	3.4 tons/acre	Litter cover	100 percent
1-hour woody	4.1 tons/acre	Shrub cover	95 percent
10-hour woody	4.5 tons/acre	Ave. fuel height	3.3 feet
100-hour woody	6.6 tons/acre	Ave. fuel moisture	15 percent
1,000-hour woody	8.2 tons/acre	Optimal fuel model	5*
Total fuels	26.8 tons/acre	Min/max fuel model	10*, 4*

Weather and Fire Behavior Data			
Air temperature	64 °F	Burn date/time	01 May 2003, 1200
Relative humidity	33 percent	Firing device	Drip torch
Wind direction	South	Firing pattern	Strip-head fire
Wind speed	1 to 3 miles/hour	Flame lengths	5 to 10 feet
Cloud cover	50 percent	Rates of spread	5 to 10 feet/minute
Days since rain, amt.	3, 0.25 inch		

Fig. 6.3 An extract of the photo guide for estimating fuel loads and fire behavior in mixed-oak forests of the Mid-Atlantic Region in the USA (Brose 2009). This photo guide includes photographs taken before and during burning, the fuel model(s) assigned, site and detailed fuel data, and observed fire behavior and the corresponding weather conditions

From a top view (Fig. 6.4), one fire includes at any point in time different locations with different behavior. Therefore the description of fire behavior is commonly provided for a specific part of a fire. The ignition location is referred to as the origin of the fire. In the absence of wind and slope and in perfectly homogeneous fuels, fires spread with a circular shape following ignition. However, in the presence of wind or slope, fires spread in an elliptical shape. The portion of the fire that is spreading upslope and with the wind is called the headfire and is associated with the most active fire behavior. The section of the fire spreading against the wind and/or slope is called the backfire, and typically has the shortest flames and is the slowest moving portion of the fire. The section of the fire perimeter associated with the backfire is commonly referred to as the back, heel, or rear of the fire. The fire spreading on the sides is moving perpendicular to the wind and is referred to as flank fire. Fire behavior along the flanks is somewhere between the headfire and backfire. The shape of the fire perimeter is tightly linked to variability and interactions among the fire environment, topography, and fuels. This variability can result in a number of unique features along the fire perimeter. Fingers are formed when part of the fire front spreads faster than the surrounding front, resulting in the formation of a long narrow strip of fire. Spot fires are formed when firebrands are transported beyond the fire front and ignite new fires. These present especially hazardous conditions for fire managers. Spotting is discussed further in Chap. 8. In other cases, variability in fire

Fig. 6.4 Fires grow from the point of ignition, spreading faster with the wind and up the slope. The head, flank, and back of fires differ in their fire behavior characteristics. All fires have islands that are unburned or burned with such low severity that they are refugia for plants and animals from the effects of fires. Fire suppression efforts often progress from the back to the head of the fire, as the latter can be too intense for effective and safe attack. (Drawn by Heather Heward, University of Idaho)

behavior and spread can result in patches of unburned fuels within the fire perimeter. These are often referred to as islands or fire refugia; see Chap. 12 for more discussion on fire refugia as they can influence the survival of animals and rates of vegetation recovery post fire.

For fire management purposes, flame front characteristics, and in particular, how fast fires spread, have been the primary focus of fire behavior measurement and modeling. Additional metrics describe the amount, rate, and duration of heat release, as well as flame geometry (Table 6.2). We draw upon Cheney (1981) and Alexander (1982) for the following fire descriptors.

From a side view (Fig. 6.5), the flaming fire front has three dimensions: depth (D), height (H), and length (L). Flame depth (D) increases linearly with the rate of spread (R), and D/R defines the residence time of the flame (t_R). Flame size, either H or L, is a visible and obvious manifestation of energy release. Thus, flame size is a common fire descriptor, despite subjectivity in definition and measurement.

From the perspective of energy, Byram (1959) coined the concept of fireline intensity (I_B) to quantify the heat release rate in the active combustion zone per unit length of the fire front (See Figs. 6.4 and 6.5). Fireline intensity is calculated as:

Table 6.2 Fire behavior metrics definitions. Compiled from Byram (1959), Cheney (1981, 1990), Alexander (1982), and Andrews (2018), unless otherwise stated

Variable	Symbol	Definition	Units
[a]Rate of spread	R	Linear advance of the flaming fire front per unit of time	[b]$m\ s^{-1}$, $m\ min^{-1}$, $m\ h^{-1}$, $km\ h^{-1}$
Residence time	t_R	Flaming combustion duration, or the length of time for the flame front to pass a given point	s, min
Burn-out time, reaction time	t_B	Total combustion duration, or the time for all fuel fractions to burn out	s, min
Flame height	H	Mean extension of the flame front measured vertically from the ground	m
Flame angle	A_f	Angle between the fire front and the unburned fuel bed	°
Flame tilt angle	A_T	Angle between the vertical and the fire front	°
Flame depth	D	Width of the flaming front (the active combustion zone, i.e. with continuous flame)	m
[c]Flame length	L	Mean distance from the flame extremity to the mid-point of the flaming front	m
Reaction intensity, combustion rate, area-fire intensity	I_R	Heat release per unit area per unit of time within the flaming front	$kW\ m^{-2}$
Heat release	H_A	Heat release per unit area within the flaming front	$kJ\ m^{-2}$
[d]Fireline intensity	I_B	Heat release within the flaming front per unit time and unit length of the fire front	$kW\ m^{-1}$
[e]Power of the fire	PWR	Heat release per unit time within the flaming zone, integrated around the fire perimeter	MW

[a]Equivalent metrics can be calculated for the perimeter (e.g., $m\ h^{-1}$) and area ($ha\ h^{-1}$) of the fire as a whole (Byram 1959)
[b]Usually dependent on the scale of application
[c]Calculated from H and A_f or A_T; some authors consider the leading flame edge rather than its central axis (Nelson and Adkins 1986; Catchpole et al. 1993)
[d]Fire intensity (Byram 1959), Byram's fireline intensity, or frontal fire intensity
[e]Harris et al. (2012)

$$I_B = h_c\, w\, R \tag{6.1}$$

in units of $kW\ m^{-1}$ and where h_c is the low heat of combustion ($kJ\ kg^{-1}$), w is the amount of fuel available for flaming combustion ($kg\ m^{-2}$), and R is the rate of fire spread ($m\ s^{-1}$). To estimate h_c, the low heat content ΔHL is generally used but should ideally be adjusted for losses due to water evaporation, radiation, and incomplete combustion to give heat yield ΔHY (see Chap. 3). However, given the

Fig. 6.5 Side view of an experimental surface headfire in a maritime pine stand (*Pinus pinaster*) in Portugal. Flame geometry descriptors and the energy release stages during and after the passage of the flame front are shown. (Adapted from Cheney 1981 and Alexander 1982 with background photograph by P. Fernandes). Note that smoldering is an important component of fire behavior with consequences for ecological effects

inherent difficulty in making these adjustments, and because heat content varies little among different fuels, h_c is often assumed constant at about 18,000–18,700 kJ kg^{-1} (e.g., Albini 1976; Forestry Canada Fire Danger Group 1992). The greatest uncertainty in calculating I_B lies in estimating w, especially in fuel complexes that are highly heterogeneous in compactness, moisture, and particle size. In the end, the impact of these shortcomings is relatively minor because variation in I_B is primarily a function of R.

Multiple empirical relationships have been derived over the years that relate fireline intensity I_B and flame length L for different fuel types (Alexander and Cruz 2012). Byram (1959) relationship is the most commonly used and indicates that L (m) scales approximately at the square root of I_B (kW m^{-1}):

$$L = 0.0775\, {I_B}^{0.46} \tag{6.2}$$

For crown fires the relationship of Thomas (1963) based on convection theory is more commonly preferred, e.g., Rothermel (1991):

$$L = 0.02665\, I_B^{0.667} \tag{6.3}$$

The reciprocals of these equations can be used to estimate I_B from L, with equation coefficients respectively 259.833 and 2.174 for Eq. (6.2), and 229 and 1.5 for Eq. (6.3).

I_B is a simple quantity that cannot be observed or measured but is a single figure synthesis of the most relevant properties of a wildland fire front. Fire managers commonly use fireline intensity to assess fire control difficulty (Hirsch and Martell 1996) and aboveground fire effects such as crown scorch height on trees (Van Wagner 1973; Michaletz and Johnson 2006). However, as discussed at length by Cheney (1990), I_B is of limited value for comparing fires in different fuel structures because the same value can correspond to quite different fire behaviors. For example, a fast-moving fire in grassland can have the same I_B as a slow-moving fire in slash fuels because the latter consumes much more fuel.

Remote sensing technology is allowing substantial advances in the realms of pyrogeography. Fire radiative power (FRP) is the rate of energy emission per unit area (in kW or MW) from fires as derived from thermal remote sensing in the middle infrared region. FRP correlates with reaction intensity and I_B and can be calculated in kW m^{-1} for a fire front by summing individual pixel FRP values and dividing the result by the length of the fire front (Wooster et al. 2003). Thus, FRP can be expressed as the radiant component of I_B.

Many of these fire descriptor variables that describe the energy output of a wildland fire front are interrelated. For example, heat release per unit area (H_A, kJ m^{-2}) is calculated as the product of heat of combustion (h_c, kJ kg^{-1}) by fuel load (w, kg m^{-2}). Reaction intensity (I_R, kW m^{-2}), the heat per unit area released per unit time, is calculated by dividing heat per unit area (H_A, kJ m^{-2}) by the residence time (t_R, s). Also, the rate of spread (R, m s^{-1}) can be calculated by dividing flame depth (D, m) by residence time (t_R, s). These relations combined with Eq. (6.1) are shown in the following equalities:

$$I_R = \frac{H_A}{t_R} = \frac{h_c w}{t_R} = \frac{I_B}{D} = \frac{I_B}{t_R R} \tag{6.4}$$

From Eq. (6.4) we see that I_B can be estimated as:

$$I_B = I_R t_R R \tag{6.5}$$

This is the approach followed by Albini (1976) and used in the fire behavior modeling and effects tools based on Rothermel's model. Using Eq. (6.5) avoids estimating the amount of fuel consumed in the active flaming front but requires a good estimation of t_R, which currently cannot be provided because fuel complex characteristics are ignored in its calculation (Catchpole et al. 1998). Consequently, Eq. (6.5) typically produces I_B values that are 2–3 times lower than those calculated according to Byram's (1959) formulation (Cruz and Alexander 2010). As a result, predictions of crown fire transition based on I_B from Eq. (6.5) are not consistent with the transition model (Van Wagner 1977, see Chap. 7). The reader should be aware of these differences and implications.

6.5 Implications

Fuel chemistry, combustion characteristics, and energy production and transfer were discussed in Chaps. 1–5 without specific considerations of the extrinsic physical properties of fuel particles and complexes that influence fire behavior. In this chapter, we focus on fuel descriptors, beyond their intrinsic thermochemical properties, and on the most commonly used fire behavior descriptors. These descriptors are attempts to simplify and model entities that vary in space and time, and the recognition of this variability is important. However, for understanding the relationships between fuels and fire and for practical implications in fire management, simple descriptors are fundamental to support strategies and decisions.

The limits for wildfire suppression or for the use of prescribed fire are generally established as a function of flame length or fireline intensity. The difficulties of controlling a fire and the threat to the safety of fire fighters and people, in general, are associated with fast-spreading fires. Furthermore, models for the rate of spread require knowledge of the main characteristics of the fuel particles and the fuel complexes involved. These issues associated with fire propagation are discussed in Chap. 7.

For the understanding of the processes of crown fires, the concept of canopy bulk density is fundamental, as are the concepts of crown base height or fuel strata and gaps between them. For spotting, the properties of fuel particles, such as surface density, are relevant to understanding their possible role as firebrands. These aspects are discussed in Chap. 8.

Fire behavior is directly associated with fire effects. For example, whether trees and large shrubs survive fires often depends on the intensity of flaming combustion. The degree of canopy scorch and stem char heights are correlated with H, L, and I_B. Other ecological effects are more closely related to duff consumption and soil heating. These depend on the downward heat flux and are controlled by the moisture, compactness, and composition (e.g., coarse woody fuels with long burn-out times) of the organic matter on and in the soil (Hartford and Frandsen 1992). The use of total fuel consumption w_T instead of w in the calculation of I_B, H_A, and I_R (in this case using burn-out time t_B), embraces all of the energy release phases (See Chap. 2), and is therefore expected to increase the association between fire behavior metrics and some fire effects (See Chap. 9).

The recognition that fires and fire regimes may change with time and that fuels are dynamic entities that are influenced by management are discussed in Chaps. 10 and 11 using the descriptors presented in this chapter. Many other relations between fuel and fire properties are present in the integrated fire management examples of Chap. 13.

References

Albini, F. A. (1976). *Estimating wildfire behavior and effects* (Gen Tech Rep INT-30). Ogden: USDA Forest Service Intermountain For Range Experiment Station.

Alexander, M. E. (1982). Calculating and interpreting forest fire intensities. *Canadian Journal of Botany, 60*, 349–357.

Alexander, M. E., & Cruz, M. G. (2012). Interdependencies between flame length and fireline intensity in predicting crown fire initiation and crown scorch height. *International Journal of Wildland Fire, 21*, 95–113.

Anderson, H. E. (1982). *Aids to determining fuel models for estimating fire behavior* (Gen Tech Rep INT-GTR-122). Ogden: USDA Forest Service, Intermountain Forest and Range Experiment Station.

Andrews, P. L. (2018). *The Rothermel surface fire spread model and associated developments: A comprehensive explanation* (Gen Tech Rep RMRS-GTR-371). Fort Collins, CO: USDA Forest Service, Rocky Mountain Research Station.

Ascoli, D., Vacchiano, G., Mott, R., & Bovio, G. (2015). Building Rothermel fire behaviour fuel models by genetic algorithm optimisation. *International Journal of Wildland Fire, 24*, 317–328.

Brose, P. H. (2009). *Photo guide for estimating fuel loading and fire behavior in mixed-oak forests of the Mid-Atlantic Region* (Gen Tech Rep NRS-45). Newtown Square: USDA Forest Service Northern Research Station.

Brown, J. K. (1971). A planar intersect method for sampling fuel volume and surface area. *Forest Science, 17*, 96–102.

Byram, G. M. (1959). Combustion of forest fuels. In K. Davis (Ed.), *Forest fire: Control and use*. New York: McGraw-Hill.

Catchpole, W. R., & Wheeler, C. J. (1992). Estimating plant biomass: A review of techniques. *Australian Journal of Ecology, 17*, 121–131.

Catchpole, E. A., Catchpole, W. R., & Rothermel, R. C. (1993). Fire behavior experiments in mixed fuel complexes. *International Journal of Wildland Fire, 3*, 45–57.

Catchpole, W. R., Catchpole, E. A., Butler, B. W., Rothermel, R. C., Morris, G. A., & Latham, D. J. (1998). Rate of spread of free-burning fires in woody fuels in a wind tunnel. *Combustion Science and Technology, 131*, 1–37.

Cheney, N. P. (1981). Fire behaviour. In A. M. Gill, R. H. Groves, & I. R. Noble (Eds.), *Fire and the Australian Biota* (pp. 151–175). Canberra: Australian Academy of Science.

Cheney, N. P. (1990). Quantifying bushfires. *Mathematical and Computer Modelling, 13*, 9–15.

Countryman, C. M., & Philpot, C. W. (1970). *Physical characteristics of chamise as a wildland fuel* (Res Pap PSW-66). Berkeley: USDA Forest Service Pacific Southwest Rocky Mountain Research Station.

Cruz, M. G., & Alexander, M. E. (2010). Assessing crown fire potential in coniferous forests of western North America: A critique of current approaches and recent simulation studies. *International Journal of Wildland Fire, 19*, 377–398.

Cruz, M. G., & Fernandes, P. M. (2008). Development of fuel models for fire behaviour prediction in maritime pine (*Pinus pinaster* Ait.) stands. *International Journal of Wildland Fire, 17*, 194–204.

Cruz, M. G., Alexander, M. E., & Wakimoto, R. H. (2004). Modeling the likelihood of crown fire occurrence in conifer forest stands. *Forest Science, 50*, 640–658.

Cruz, M. G., Gould, J. S., Alexander, M. E., Sullivan, A. L., McCaw, W. L., & Matthews, S. (2015). Empirical-based models for predicting head-fire rate of spread in Australian fuel types. *Australian Forestry, 78*, 118–158.

Forestry Canada Fire Danger Group. (1992). *Development and structure of the Canadian Forest Fire Behavior Prediction System* (Inf Rep ST-X-3). Ottawa: For Canada.

Gould, J. S., McCaw, W. L., Cheney, N. P., Ellis, P. E., & Matthews, S. (2007). Field guide-fuel assessment and fire behaviour prediction in dry eucalypt forest. Ensis-CSIRO. In *Canberra*. Perth: ACT and Department of Environment and Conservation.

Harris, S., Anderson, W., Kilinc, M., & Fogarty, L. (2012). The relationship between fire behaviour measures and community loss: An exploratory analysis for developing a bushfire severity scale. *Natural Hazards, 63*, 391–415.

Hartford, R. A., & Frandsen, W. H. (1992). When it's hot, it's hot ... or maybe it's not! (surface flaming may not portend extensive soil heating). *International Journal of Wildland Fire, 2*, 139–144.

Hirsch, K. G., & Martell, D. L. (1996). A review of initial attack fire crew productivity and effectiveness. *International Journal of Wildland Fire, 6*, 199–215.

Hough, W. A., & Albini, F. A. (1978). *Predicting fire behavior in palmetto-gallberry fuel complexes*. Research Paper SE-RP-174. Asheville, NC: USDA Forest Service, Southeastern Forest Experiment Station, 48 p.

Keane, R. (2015). *Wildland fuel fundamentals and applications*. Cham: Springer.

Keane, R. E., & Dickinson, L. J. (2007). *The photoload sampling technique: Estimating surface fuel loadings from downward-looking photographs of synthetic fuelbeds* (Gen Tech Rep RMRS-GTR-190). Fort Collins: USDA Forest Service Rocky Mountain Research Station.

Michaletz, S. T., & Johnson, E. A. (2006). A heat transfer model of crown scorch in forest fires. *Canadian Journal of Forest Research, 36*, 2839–2851.

Nelson, R. M., Jr., & Adkins, C. W. (1986). Flame characteristics of wind-driven surface fires. *Canadian Journal of Forest Research, 16*, 1293–1300.

Pyne, S. J., Andrews, P. L., & Laven, R. D. (1996). *Introduction to wildland fire* (2nd ed.). New York: Wiley.

Rossa, C., & Fernandes, P. M. (2018). Empirical modelling of fire spread rate in no-wind and no-slope conditions. *Forest Science, 64*, 358–370.

Rothermel, R. C. (1972). *A mathematical model for predicting fire spread in wildland fuels* (Res Pap INT-115). Ogden: USDA Forest Service Intermountain Rocky Mountain Research Station.

Rothermel, R. C. (1991). *Predicting behavior and size of crown fires in the Northern Rocky Mountains*. Research Paper INT-438. USDA Forest Service, Intermountain Research Station, Ogden, UT. 46 p.

Rothermel, R. C., & Anderson, H. E. (1966). *Fire spread characteristics determined in the laboratory* (Res Pap INT-30). Ogden: USDA Forest Service Intermountain Rocky Mountain Research Station.

Scott, J. H., & Burgan, R. E. (2005). *Standard fire behavior fuel models: A comprehensive set for use with Rothermel's surface fire spread model* (Gen Tech Rep RMRS-GTR-153). Fort Collins: USDA Forest Service Rocky Mountain Research Station.

Thomas, P. H. (1963). The size of flames from natural fires. *Symposium (International) on Combustion, 9*, 844–859.

Van Wagner, C. E. (1968). The line intersect method in forest fuel sampling. *Forest Science, 14*, 20–26.

Van Wagner, C. E. (1973). Height of crown scorch in forest fires. *Canadian Journal of Forest Research, 3*, 373–378.

Van Wagner, C. V. (1977). Conditions for the start and spread of crown fire. *Canadian Journal of Forest Research, 7*(1), 23–34.

Wooster, M. J., Zhukov, B., & Oertel, D. (2003). Fire radiative energy for quantitative study of biomass burning: Derivation from the BIRD experimental satellite and comparison to MODIS fire products. *Remote Sensing of Environment, 86*, 83–107.

Chapter 7
Fire Propagation

Learning Outcomes
After reading this chapter, we expect you to be able to

1. Identify the locations where fire spread rate and intensity will be highest in an area ignited with each of the fire ignition patterns shown in Fig. 7.7, then explain in your own words how this is likely to change with wind speed and direction,
2. Describe the conditions for fast, steady-state fire spread and why changing each of the legs of the fire triangles can increase or decrease the fire spread rate, and
3. Predict the effect of changing the value of each of multiple inputs into fire behavior models and then use the interactive spreadsheet to test your predictions for a wide range of input values,
4. Identify the heat transfer processes taking place in smoldering and flaming fires and justify their differences in spread rate, and
5. Use the three interactive spreadsheets to explore how the graphs of predicted outputs are influenced by changing the fire growth, fire rate of spread, and wind profiles.

7.1 Introduction

Understanding fire propagation is critical for assessing fire risk, and planning and decision making during fire suppression and prescribed burning operations. The rate of spread is the most common measure of fire propagation, although the area burned

---

Supplementary Information The online version of this chapter (https://doi.org/10.1007/978-3-030-69815-7_7) contains supplementary material, which is available to authorized users.

F. Castro Rego et al., *Fire Science*, Springer Textbooks in Earth Sciences, Geography and Environment, https://doi.org/10.1007/978-3-030-69815-7_7

is also used. The rate of spread is commonly described relative to the direction of the prevailing winds. The forward or head fire rate of spread is the spread rate in the direction of the prevailing winds or upslope, the backfire rate of spread is the spread in the opposing direction from the prevailing winds or slope and the rate of spread perpendicular to the wind is referred to as the flanking rate of spread. Throughout this chapter, we use the term rate of spread to refer to the forward or head fire rate of spread unless otherwise noted.

Following a point or line ignition, the rate of spread of a fire increases until a steady-state state rate of spread is reached. The development phase of fires is commonly referred to as initial fire growth or build-up (Luke and McArthur 1978). During the initial fire growth, the rate of fire spread, and consequently fireline intensity, are accelerating. Fire acceleration is defined as the rate at which the fire rate of spread is increasing. During the initial fire growth, there are four distinct phases. During the initial growth of a fire, fire suppression may be successful. An understanding of initial fire growth is used to manage fireline intensity during prescribed fire operations.

Following the initial growth phase, fires reach a steady-state rate of spread, assuming uniform fuels, weather, and topography. In many cases, the term 'quasi steady-state' rate of spread is used to acknowledge that this is a time-averaged value. The steady-state rate of spread can be assessed using a heat balance approach, whereby the rate of spread depends upon the ratio of heat transferred from the fire to the unburned fuel and the amount of energy required to ignite the fuel ahead of the fire.

Given that fires rarely, if ever, burn under constant fire, weather, and topographic conditions, or with homogeneous fuels, it is critical to recognize that fire spread and behavior is dynamic in time and space. Understanding and predicting fire propagation and growth become increasingly complex due to the spatial and temporal interactions among fuels, weather, topography, and the fire plume on fire spread. A common approach to deal with spatial and temporal heterogeneity is to identify discrete areas and time periods that are relatively homogeneous and invoke the steady-state rate of spread assumption. Predictions of fire propagation through heterogeneous landscapes and variable topography, winds, and fuels based on the assumption of a steady-state fire spread are commonly used by managers around the world to inform decisions for a variety of purposes.

The equations used to predict the spread of fires illustrate the influence of key variables. Many different approaches exist. Most are informed by both empirical observations and theoretical understanding of fire growth. Some very useful models are mostly empirical.

7.2 Initial Fire Growth

Initial fire growth has been studied in laboratory experiments (e.g., Viegas 2004). These experiments have obvious scale limitations but reveal some of the basic aspects of the initial development of a fire from a point source (Fig. 7.1).

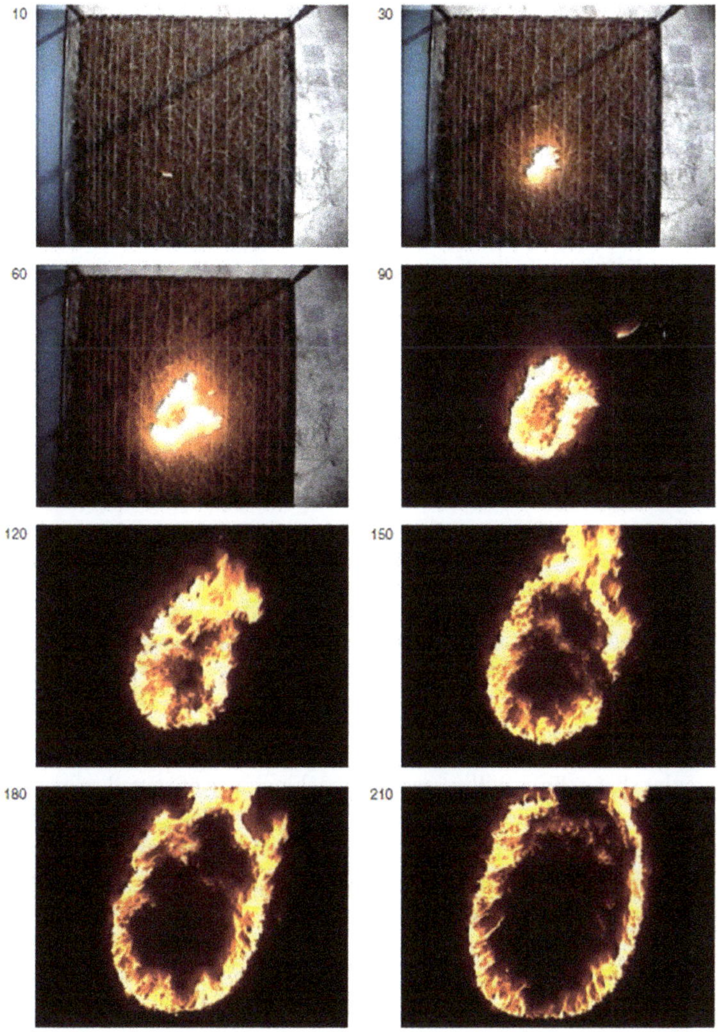

Fig. 7.1 Initial fire growth from a point source in laboratory experiments with fuelbeds of dead needles of *Pinus pinaster* showing initial fire growth. The wind is blowing from left to right and the slope gradient is from bottom to top. Time steps of 30 s between each frame. (From Viegas 2004)

Initial fire growth consists of four distinct phases. The simplest situation is a point source of ignition with no wind and no slope (Fig. 7.2).

In the *first phase*, the hot air above the burning area rises and is replaced by ambient air around the fire perimeter (Chandler et al. 1983). These indrafts cause the flames to tilt inward, and fire spreads outwards slowly, maintaining a circular shape. As described by Cheney and Gould (1997), in this first phase, "the convection above small fires can draw flames towards a single convective center and can maintain fires

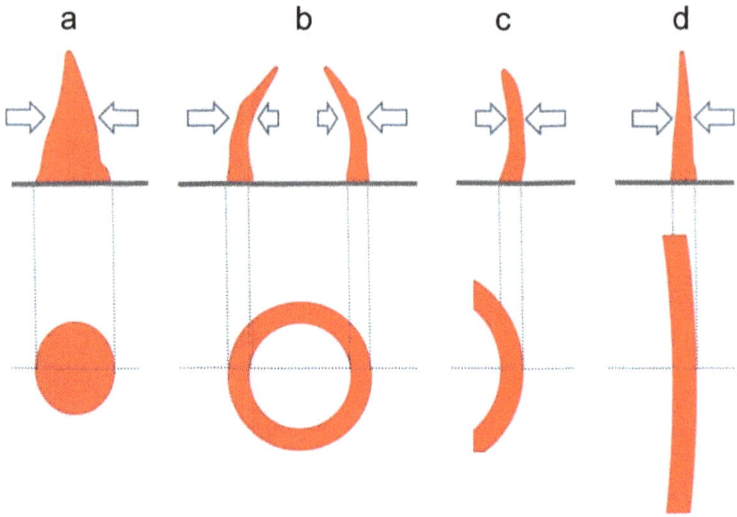

Fig. 7.2 Schematic representation of the four phases of initial growth of a fire starting from a point source in no-wind, no-slope conditions in a homogeneous fuel. (**a**) First phase where ambient air tilts the flame to the center of the flaming area. (**b**) Second phase with some ambient air already available from inside the fire perimeter but flames still leaning inwards. (**c**) Fire spreads with indrafts almost equivalent from both inside and outside of the fire perimeter. (**d**) Fire spreads as a line with equivalent indrafts. The top row is the view from the side, while the bottom row is the view from above

with a circular shape. Fires will spread with backing rates of spread around the entire perimeter". The duration and size of the fire during the first phase depends strongly on flame residence time. In heavy fuels with long residence times, such as brush or slash, fire may grow as large as half a hectare with flames 4 or 5 m high during the first period of acceleration (Chandler et al. 1983).

The *second phase* begins when the fuel that was ignited in phase 1 is consumed. In the second phase, a burned area develops inside the fire perimeter, allowing ambient air to flow into and outwards from the burned-out area towards the fire perimeter. This flow pattern pushes the flames away from the center of the burned area and towards the unburned fuels along the perimeter of the fire. The tilting of the flames towards the unburned fuels increases the net heat transfer from the flames to the unburned fuels, and the rate of fire spread accelerates.

In the *third phase,* the acceleration of fire growth is dependent upon the length and curvilinearity of the fire front. The speed of the fire increases as a function of the fireline length and the interaction of the buoyant plume and ambient air. Short firelines allow for the ambient wind to flow around the flaming front, which limits the convective heat transfer from the flames to the unburned fuel. As firelines get longer, the ambient air is forced through the flame zone rather than around it, increasing the heat transfer from the flames to unburned fuel.

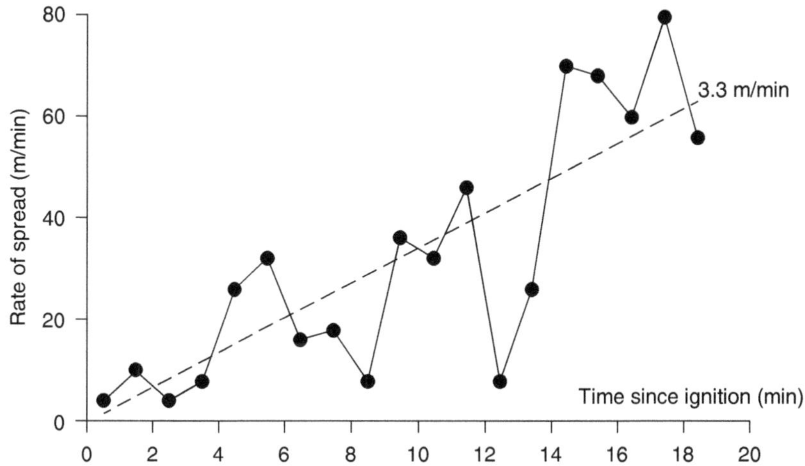

Fig. 7.3 Initial acceleration with fluctuations before a quasi-steady state is established (Cheney and Gould 1997)

The *fourth phase* is based on the assumption of a fireline of infinite length moving at a steady-state spread rate of fire, as discussed further in Sect. 7.3.

During the early phases of fire growth following ignition, the rate of spread, fireline intensity, area, and perimeter of the fire all accelerate. The fire continues to accelerate until it approaches a quasi-steady state rate of spread (Fig. 7.3).

One of the primary variables associated with the initial growth phase of fires is the length of time it takes to reach the steady-state rate of spread. The acceleration of the fire rate of spread during this phase depends upon the properties of the fuels complex, fuel moisture, wind speed, slope, and fire-atmospheric interactions. Weber (1989) related acceleration to the curvature of the fire front. The greater the curvature, the faster the rate of acceleration. As the fire front approaches a straight fireline, the acceleration decreases, and fire spread rate approaches a quasi-steady state.

7.2.1 Models of Acceleration of Fire Fronts

The rate of spread of a fire front during the initial growth phase depends upon both the time since ignition (t) and the steady-state rate of forward spread (R_s). Scientists (e.g., Van Wagner 1985; McAlpine and Wakimoto 1991) have expressed the relationship between the instantaneous rate of spread of a fire front during the initial growth phase and the steady-state rate of spread as:

$$R(t) = R_s\left[1 - e^{(-a_1 t)}\right] \tag{7.1}$$

where R(t) is the head fire rate of spread at time t after ignition (m min^{-1}), R_s is the steady-state equilibrium spread rate (m min^{-1}), t is the time elapsed from the start of the fire (min), and a_1 is an empirical coefficient (min^{-1}) that determines the rate of acceleration. For Canadian forests, Van Wagner (1985) suggested that it is reasonable to assume it takes a fire about 30 min to reach 90% of the equilibrium rate of spread, which is equivalent to using a value of 0.077 min^{-1} for a_1 in Eq. (7.1). However, models in use (FCFDG 1992) indicate that 20 min after ignition from a point source, the fire spread rate reaches 90% of the steady-state spread rate, which corresponds to $a_1 = 0.115$ min^{-1}.

The introduction of the correction for initial acceleration has profound effects on simulated initial fire growth, rate of spread, fire perimeter (represented as an ellipse), and area (Fig. 7.4).

The empirical coefficient, a_1, depends upon fuel type, ignition pattern (i.e., line vs. point ignition), wind speed, and fuel moisture. For example, stronger winds may increase the steady-state rate of spread, but also can increase the time to achieve steady-state conditions. Changes in wind direction perpendicular to spread can increase the fire front width reducing simultaneously the time required to reach steady-state rate of spread (Finney 2019). This can represent major threats for safety and for fire control.

Fires ignited as lines rather than points typically have greater values for a_1, indicating more rapid acceleration and reduced time to steady-state conditions. For example, empirical research in Australian grasslands (Fig. 7.5) indicates a value for a_1 of 0.3 min^{-1}, which corresponds to a fire reaching more than 90% of its steady-state fire spread in 8 min.

Because the width of the fire front at the head of the fire increases with time, an equivalent approach can be developed that directly uses the head fire width rather than time (e.g., Cheney and Gould 1995; Anderson et al. 2015):

$$R(W) = R_s\left[1 - e^{(-b_1 W)}\right] \tag{7.2}$$

where R(W) is the rate of spread of a fire front as a function of its width W (m), and b_1 is an empirical coefficient (m^{-1}).

The direct relationship between the width of the fire front and the rate of spread has been established based on experiments (Fig. 7.5). The values of b_1 were estimated to be 0.04–0.06 m^{-1} for open grasslands and 0.03–0.04 m^{-1} for woodlands (Cheney and Gould 1995). If the ignition line is sufficiently long, approximately 50 m for shrublands (Anderson et al. 2015) and more than 100 m for forests (Cheney et al. 2012), fires may not exhibit any significant acceleration and will spread at its quasi-steady rate of spread immediately following ignition. In practice, these relationships are important in the initial phases of fire fighting or when planning and using prescribed fire with line ignition patterns.

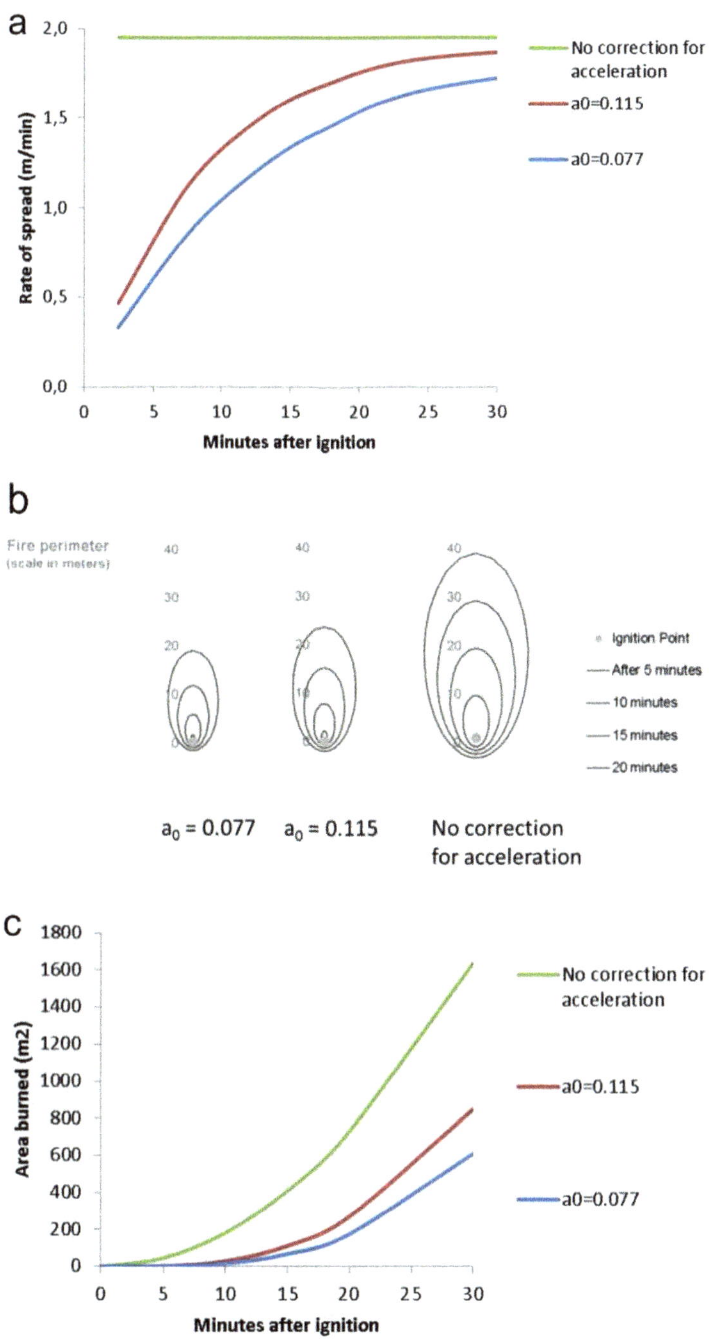

Fig. 7.4 Simulations of initial fire growth considering acceleration at the initial phases (with $a_1 = 0.077$ and $a_1 = 0.115$) and without the correction for initial acceleration assuming that steady-state occurs from the beginning. The steady-state spread rate considered was 2 m min^{-1}, typical of logging slash (McRae 1999), and a wind speed of 3 m s^{-1}. (**a**) Fire rate of spread. (**b**) Fire shapes for the first 20 min after ignition. (**c**) Area burned with the three options (acceleration at different rates and no consideration of acceleration)

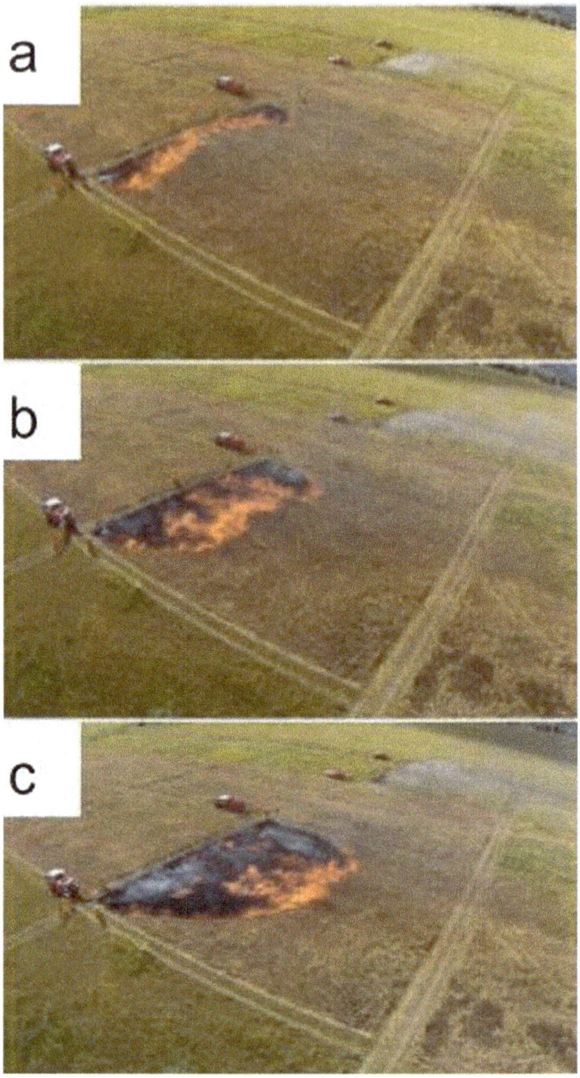

Fig. 7.5 Fire experiments in grassland plots (33 m × 33 m) in Australia with the initial growth of fire from a line ignition at (**a**) 20 s, (**b**) 30 s, and (**c**) 40 s after ignition. In the foreground plot, fires burned with grasses 100% cured. Ignition was simultaneous in the background plot with grasses 48% cured. (From Cruz et al. 2016)

The models for initial fire growth, from point or line sources, are simplifications. Conceptually, fires will eventually reach a quasi steady-state rate of spread. However, fires burning in real landscapes are constantly reacting to temporal and spatial variability in fuels, topography, and fire weather which results in variability in fire rate of spread and complex fire perimeters (Fig. 7.6). Further discussion of the spatial and temporal controls of the fire rate of spread can be found in Sect. 7.4.

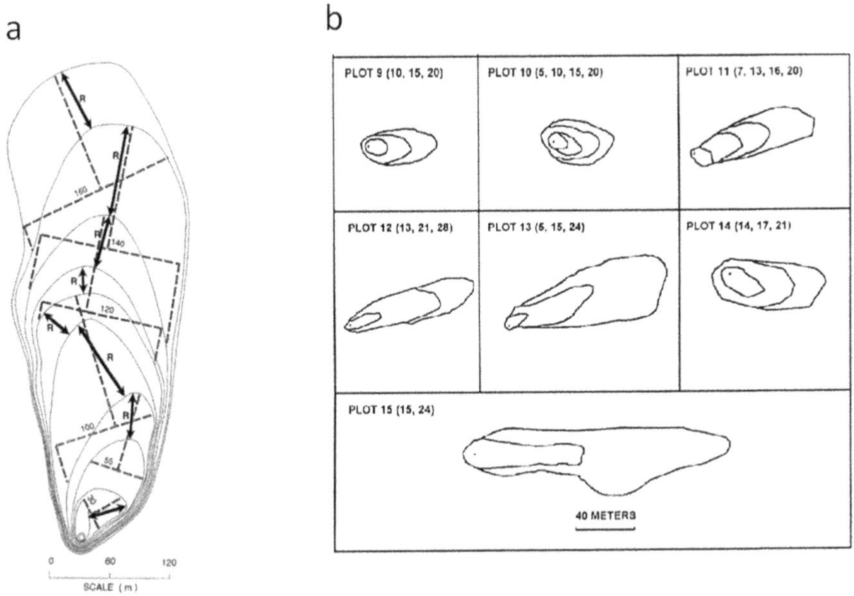

Fig. 7.6 The variable geometry of fire shapes during initial fire growth. (**a**) Changing environmental conditions result in complex fire shapes (Finney 2004). (**b**) Examples of different shapes from initial growth in different plots in jack pine slash. (From McRae 1999)

7.2.2 The Practical Use of Understanding Initial Fire Growth

Successful firefighting operations depend greatly on the understanding of the initial growth of wildfires. Fire fighters have long been aware of the need for rapid intervention for initial attack to succeed before fires attain rates of spread or intensity beyond suppression thresholds. If firefighting starts rapidly, the fireline intensity may still be within control capacity.

Understanding initial fire growth can be important. A good example is the megafire of Pedrógão Grande that burned on June 17, 2017 in Portugal, where 66 people were killed. The inquiry commission created by the Portuguese Parliament (CTI 2017) estimated that, under the observed extreme weather and fuel conditions, the fire would become uncontrollable by any means (fireline intensity >10,000 kW m^{-1}) as soon as 11 min after ignition (Fig. 7.7). This analysis, considering the initial fire acceleration, provides a clear example of the importance of rapid initial attack before direct attack is ineffective or unsafe.

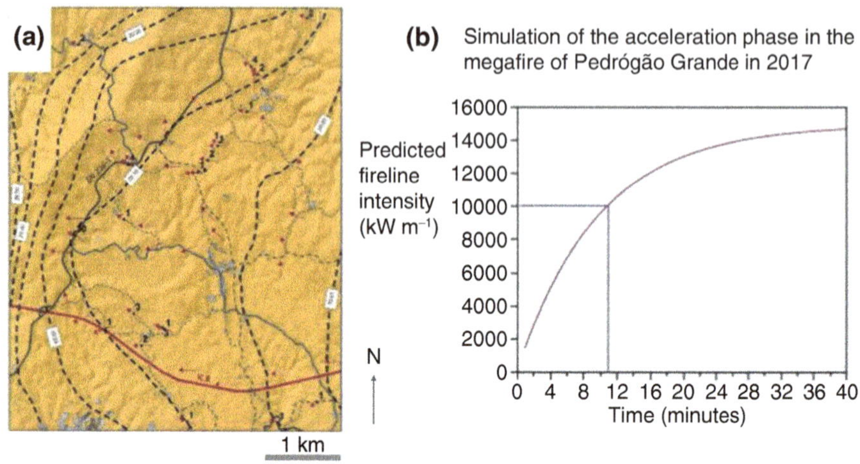

Fig. 7.7 The fast development of the Pedrógão Grande 2017 megafire in Portugal shown in (**a**) map with fire propagation with dashed lines representing fire perimeters at 10 min intervals, and (**b**) in the prediction of rapidly increasing fireline intensity with time after ignition. Fire control capacity was reached at 11 min after ignition and it was not possible to contain the wildfire during its build-up phase under the prevailing fire weather conditions and with the resources available. (From CTI 2017)

Fires with narrow heads may be spreading at rates and with intensities which are well below the potential rate of spread and may be controllable. However, conditions may change. Sudden shifts in wind direction (e.g., wind shifting to blow at right angles can cause a flank fire to immediately reach its steady-state rate of spread) have been associated with disaster fires and fire fighter fatalities (the Dead Man Zone, Cheney et al. 2001).

Prescribed burning also benefits from understanding initial fire acceleration. Fire managers design ignition patterns (Fig. 7.8) to manipulate fire behavior and effects during prescribed burning (Finney 2019).

Prescribed burning managers commonly increase or decrease fireline intensity and ensure that management objectives are met. For example, managers can use backing or heading fires, the firing methods (aerial ignition vs. hand ignition), as well the pattern, rate, and spacing of ignitions. For example, it is common for managers to alter the distance between line ignitions to help maintain the desired fireline intensity when using strip head fires during prescribed fire operations (Fig. 7.8b). Relatively short distances between firelines prevent the fires from reaching a steady-state rate of spread and thus limit the fireline intensity. Similarly, the spacing between point-source ignitions (Fig. 7.8c) can be chosen to maintain the fire spread rate and intensity in the accelerating phase. This has been used to reduce intensity and crown scorch of high-value trees by locating spot ignitions near them (Weatherspoon et al. 1989).

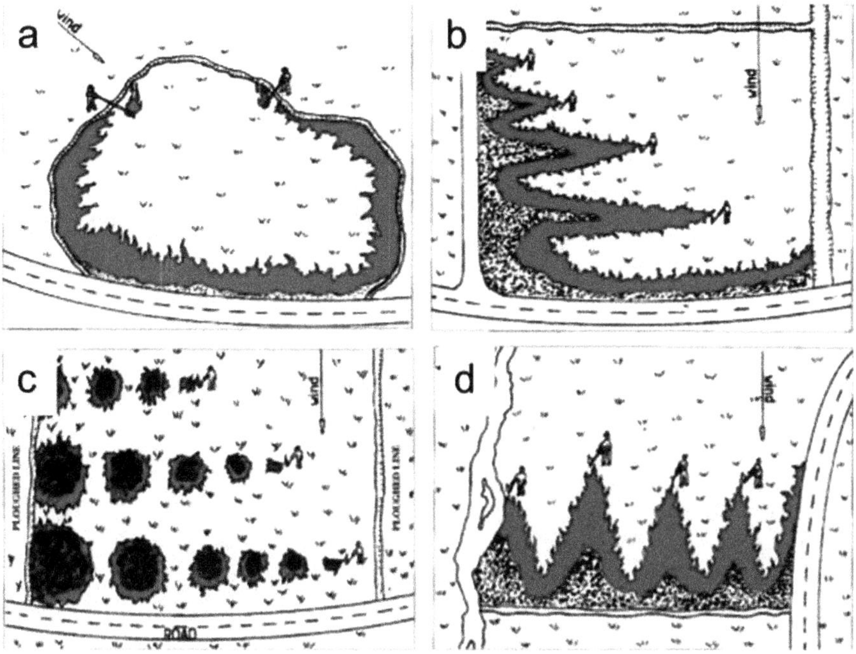

Fig. 7.8 Examples of typical ignition patterns used in prescribed fire. Patterns are named for their dominant fire spread. Managers choose patterns to achieve desired fire intensity, effects, and safety for those igniting and containing the fire. (**a**) Ring pattern: fire is started from the downside to the upwind side from a control line that surrounds the area to burn, and firing continues rapidly on both sides of the starting place. The fire's edge becomes shaped like a horseshoe with all flames moving towards the center due to the suction created by the fire until they meet in the center of the area. (**b**) Strip head fire pattern: fire starts from downwind in staggered lines perpendicular to the wind. When lines are ignited close to one another, the fire intensity is less because the fire spread has not fully accelerated. (**c**) Dot pattern: fire starts from downwind from points along lines perpendicular to the wind. (**d**) Flank ignition pattern: fires start from downwind in lines parallel to the wind. (From Heikkila et al. 2010)

7.3 The Steady-State Spread Rate of a Fireline

After the initial fire growth stage, the rate of fire spread tends to stabilize towards a steady-state rate of spread (R_s). The steady-state rate of spread can be estimated based on a heat balance approach which considers the amount of heat generated and transferred to the unburned fuel, and the amount of energy required to ignite the unburned fuel. Knowledge about the fire rate of spread is important for a variety of fire management applications, including fire suppression operations, fire fighter safety, community fire protection planning, and fuels management.

7.3.1 Heat Balance and Fire Spread

From a conservation of energy perspective, the rate of steady-state fire spread is determined by the ratio between the net heat flux received by the fuel and the heat required for fuel ignition:

$$\text{Rate of spread} = \text{Net heat flux received by the fuel}/\text{Heat required for fuel ignition} \tag{7.3}$$

The net heat flux received by the fuel depends upon the heat produced during combustion (see Chap. 3) and the heat transfer from the flame to the unburned fuel. As discussed in Chap. 5, the relative contributions of heat transfer through radiation, convection, and conduction are greatly influenced by the wind speed and the slope. In the absence of wind and slope, called the no-wind no-slope situation, radiative heat transfer from a flame to the unburned fuel has been considered to be the dominant mode of heat transfer. However, as wind speed, slope, or both, increase, the fire spread rate increases, and the dominant mode of heat transfer changes to convection, including heating by the diffusion of turbulent eddies from the flames, rather than radiation (Fig. 7.9).

As the energy from combustion is absorbed by the unburned fuel, the fuel temperature begins to rise. Fuel moisture evaporates, pyrolysis occurs, and fuels are heated to ignition temperature. The amount of energy required to ignite the unburned fuel depends upon ignition temperature, fuel characteristics, fuel moisture, and the amount of fuel ignited. As discussed in Chap. 4, the amount of energy

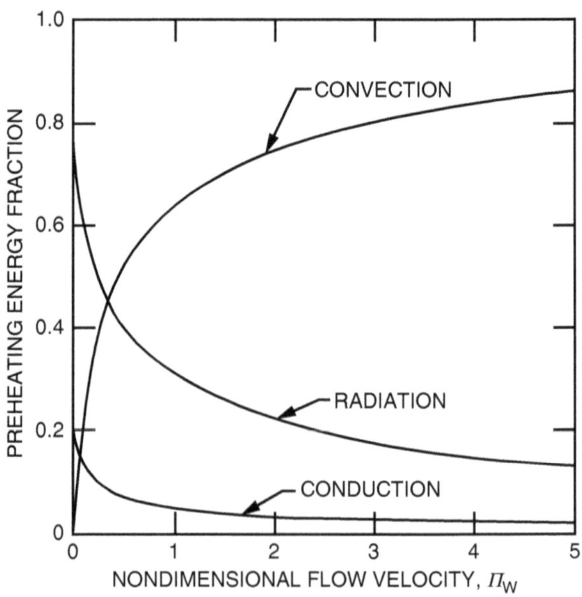

Fig. 7.9 The fractions of pre-heating energy from the various heat transfer processes involved as related to a gradient of wind speed and spread rate. The dominant role of radiation in no-wind conditions decreases and it is progressively replaced by convection. Conduction plays a limited role in flaming combustion, significant only with low wind and slow spread conditions. (From Pagni and Peterson 1973)

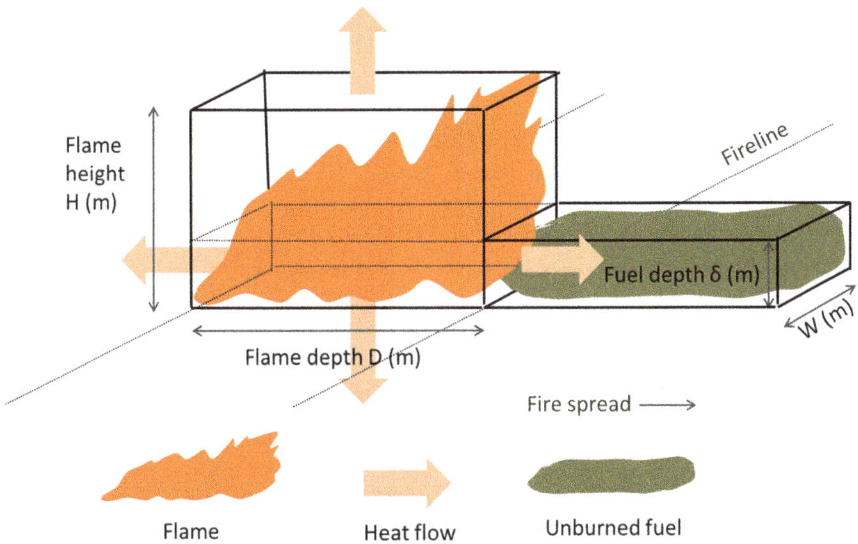

Fig. 7.10 Illustration of the conservation of energy principle applied to the one-dimensional steady-state fire spread. Fire is spreading from left to right. The flame zone has height (H) and depth (D), and the unburned fuels in the fuel bed have a vertical depth δ and a width W. Heat transfer from the combustion zone is represented as arrows, vertically (upward and downward) and horizontally (forward and backward). It is assumed that, within a fireline, there are no net lateral heat flows

required for fuel ignition (Q_{ig}) consists of three components: the amount of energy required to evaporate fuel moisture, the amount of energy to heat up the dry fuel to ignition temperature, and the energy associated with converting liquid water to water vapor. To estimate the total heat required to ignite a fuelbed, the heat of pre-ignition is adjusted because the entire fuelbed doesn't need to be heated to ignition temperature. This adjustment can be achieved by multiplying the heat of pre-ignition (Q_{ig}) by the effective bulk density (ρ_{be}), which represents the amount of fuel that must be ignited for a fire to spread.

A visualization of a one-dimensional steady-state fire spreading through a fuelbed is shown in Fig. 7.10. The fuelbed is represented by a homogeneous porous fuel layer which is typically characterized by its moisture content, the bulk density or fuel load, and the typical surface area-to-volume ratio of the individual fuel particles. The interface between the fire front and the unburned fuel advances when the heat transferred from the flame to the unburned fuel is sufficiently large to increase the temperature of the unburned fuel to the ignition temperature. This process is then repeated again and again as the fire continues to spread.

If we assume that all energy is transferred through the combustion zone with an area Wδ, the steady-state conservation of energy implies that the total energy transferred during the flame residence time (t_R) is equal to the energy used to dehydrate and ignite the unburned fuel with a volume DWδ:

$$I_p \tau_R W \delta = \rho_{be} Q_{ig} D W \delta \quad \text{or} \quad I_p = \frac{D \rho_{be} Q_{ig}}{\tau_R} \tag{7.4}$$

where I_p is the propagating heat flux (kW m^{-2}, or kJ m^{-2} s^{-1}), t_R is the flame residence time (s), δ and W are the vertical and horizontal dimensions of the fuel cell (m), ρ_{be} is the effective fuel bulk density (kg m^{-3}), D is the flame depth, and Q_{ig} is the heat of preignition (kJ kg^{-1}). The fuel cell is defined as the smallest volume that retains the characteristics of the fuelbed. The steady-state rate of spread (R_s) can be estimated by dividing the flame depth (D) by the residence time τ_R:

$$R_s = D/\tau_R \tag{7.5}$$

Substituting Eq. (7.5) into Eq. (7.4) and rearranging to solve for the rate of spread (R_s) results in the following relationship:

$$R_s = \frac{I_p}{\rho_{be} Q_{ig}} \tag{7.6}$$

where the steady-state rate of spread (R_s m s^{-1}) is equal to the propagation heat flux (I_p kW m^{-2}) divided by the product of the effective fuel bulk density (ρ_{be} kg m^{-3}) and the heat required for pre-ignition (Q_{ig} kJ kg^{-1}). This is the fundamental equation for fire propagation.

The rate of fire spread increases with the propagating heat flux and decreases with effective bulk density and heat of pre-ignition. The relative strength of each of these components is influenced by fuel characteristics, fuel moisture, wind speed, and slope (Fig. 7.11).

A heat balance approach similar to Eq. (7.6) has been used by many scientists to represent a one-dimensional steady-state rate of spread (e.g., Thomas et al. 1964; Anderson 1964, 1969; Frandsen 1971; Williams 1977; Van Wagner 1977a). This heat balance approach is the foundation for the Rothermel (1972) model applied in prediction systems developed by USDA Forest Service scientists and used worldwide. Although approaches similar to Eq. (7.6) are widely accepted due to their physical reasoning, it is challenging to estimate the propagating heat flux, the effective fuel bulk density, and the heat of pre-ignition, and thus to apply these equations in the real world. These issues are discussed in the next sections.

7.3.2 Estimating Fire Spread

The propagating heat flux is a complex variable that incorporates both the quantity of heat produced during combustion and the rate of heat transfer from the combustion zone to the unburned fuel. Over the last 75 years, scientists have developed a variety of approaches to model the propagating heat flux, including statistical approaches

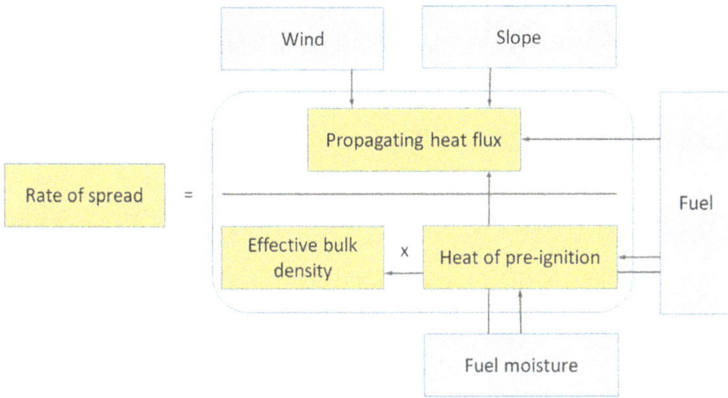

Fig. 7.11 The main components for estimating fire rate of spread from a heat balance approach (yellow boxes) and the main influences (grey boxes). The numerator is the heat source, the propagating heat flux, or the net heat flux transferred to the unburned fuels. The denominator is the heat sink, the heat required for fuel ignition, with its two sub-components, effective bulk density and heat of pre-ignition. Wind and slope influence the propagating heat flux, and fuel moisture determines the heat of pre-ignition. Fuel physical and chemical characteristics are involved in both heat source and heat sink

which do not include any explicit heat transfer mechanisms, empirical approaches which lump all heat transfer mechanisms together, and physical approaches which explicitly distinguish between the various modes of heat transfer. Furthermore, physical approaches can vary in their complexity depending upon their treatment and inclusion of convective, conductive, and radiative heat transfer and cooling. In Sects. 7.3.2.1–7.3.2.3, we discuss three different approaches for estimating the propagating heat flux, including (1) the use of a conduction-only approach applied to smoldering fires (Williams 1977), (2) the use of a radiation-only approach applied to a fire burning under a no-wind no-slope condition (Van Wagner 1967; Telitsyn 1973), and (3) the empirical approach used by Rothermel (1972). We discuss the effects of wind and slope on the propagating heat flux and the rate of fire spread in Sect. 7.3.3.

The Spread of Smoldering Fires Modeled from Conduction

Smoldering combustion is a slow, low-temperature, flameless form of combustion. Smoldering occurs in all fires but is typically associated with the burning of downed dead woody fuels and organic soils, including duff and peat (See Chap. 2). Smoldering combustion can persist for several hours, days, or months following the passing of the flaming front and is characterized by greater burnout times compared to fires that primarily burn fine fuels like grasses. Smoldering combustion is commonly responsible for the bulk of fuel consumption when the forest floor is deep and dry and contributes to smoke production (See Chap. 10) and many fire effects (See Chap. 9).

The propagating heat flux during smoldering fires can be estimated by assuming that conduction is the only significant heat transfer mechanism (Drysdale 1999; Weber 2001), although convection and radiation do occur. Following this approach, the propagating heat flux of a smoldering fire (I_p) is equal to the conductive heat flux from the combustion zone to the unburned fuel that can be approximated as:

$$I_p = q_{cond} = k\,\frac{\left(T_{ig} - T_a\right)}{x} \tag{7.7}$$

where I_p is the propagating heat flux, q_{cond} (W m^{-2}) is the conductive heat flux received by the unburned fuel, k is the thermal conductivity of the material (W m^{-1} K^{-1}), T_{ig} is the temperature of ignition (K), T_a is the ambient temperature, equivalent to the temperature of the unburned fuel, and x is the distance over which heat is being transferred by conduction (m). The distance (x) over which heat is being transferred in smoldering fires is often considered to be on the order of 0.01 m (Palmer 1957; Drysdale 1999; Miyanishi 2001; Weber 2001).

By combining the Eqs. (7.6) and (7.7), we have the rate of spread of smoldering fires:

$$R_s = k\,\frac{\left(T_{ig} - T_a\right)}{x\,\rho_{be}\,Q_{ig}} \tag{7.8}$$

Because conduction is the primary heat transfer mode of the propagating heat flux in smoldering fires, the resulting rate of spread is always small, even when the fuel is completely dry. For dry fuels, the heat of pre-ignition in Eq. (7.8) can be simplified to include only the heat required to raise the temperature of the dry fuel to ignition temperature (Q_{dig}). This can be calculated by multiplying the specific heat capacity of the dry fuel (C_{pd}) and the temperature difference between the ambient temperature (T_a) and ignition temperature (T_{ig}). Since the thermal diffusivity (α) is related to conductivity (k), effective bulk density (ρ_{be}), and specific heat capacity (C_{pd}), we can simplify the equation for the rate of spread of a smoldering fire by conduction in a dry fuel as:

$$R_s = \frac{k\,\left(T_{ig} - T_a\right)}{x\,\rho_{be}\,C_{pd}\,\left(T_{ig} - T_a\right)} = \frac{k}{x\,\left(\rho_{be}\,C_{pd}\right)} = \frac{\alpha}{x} \tag{7.9}$$

For example, we may want to know what is the rate of spread of fire in dry peat, which has a thermal diffusivity (α) of 1.2×10^{-7} m^2 s^{-1} (or 0.12 mm^2 s^{-1}). If we consider an inception distance (x) of 0.01 m (or 10 mm) we can calculate the rate of propagation as:

$$R_s = \frac{\alpha}{x} = \frac{0.12}{10} = 0.012 \text{ mm s}^{-1} \tag{7.10}$$

The rate of spread (R_s) is 0.012 mm s^{-1}, spreading about 43 m per hour. This indicates a very slow rate of spread. As most smoldering fires spread in moist fuels, the rate of spread would be much slower than calculated here for dry fuels.

Equation (7.9) provides a crude estimation (e.g., the correct order of magnitude) of the rate of spread of smoldering fires (Drysdale 1999) for a variety of types of dry fuels. The rate of spread of smoldering fires, even in dry fuels, is very slow compared to typical flaming combustion. The rate of spread of smoldering fires is affected by the moisture and mineral content of fuels, the porosity of the fuelbed, and wind speed. The moisture content, which can range from less than 10 to over 300%, influences the heat of pre-ignition, which acts as a heat sink and is thus critical to the fire rate of spread. The effect of fuel moisture can be modeled using Eq. (7.8). More sophisticated models have been developed that can account for other variables and modes of heat transfer in smoldering fires (e.g., Rein et al. 2008).

The Spread of Flaming Fires Modeled from Radiation

The propagating heat flux associated with the spread of a flaming fire front depends upon a combination of radiative, convective, and conductive heat transfer. However, in the absence of wind and slope, it is commonly assumed that radiative heat transfer dominates (Fig. 7.9). Thus, other forms of heat transfer can be ignored when estimating the propagating heat flux in no-wind, no-slope conditions where fire spreads at the same rate in all directions. Examples of a radiation-only approach were reviewed in Weber (1989) and include those of Van Wagner (1967) and Telitsyn (1973).

Under this approach, the propagating heat flux for a no-wind no-slope condition (I_{p0}) is assumed to be equal to the radiative heat flux (q_{rad}):

$$I_{p0} = q_{rad} = F_{ab}\epsilon\,\sigma_{SB}\,T^4 \tag{7.11}$$

where I_{p0} is the propagating heat flux for a no-wind no-slope condition, q_{rad} (kW m^{-2}) is the heat transferred from the flame to the unburned fuel, F_{ab} is the view factor, ε is the emissivity, σ_{SB} is the Stefan-Boltzmann constant

(5.67×10^{-8} W m^{-2} K^{-4}), and T is the temperature of the flame (K). As discussed in Chap. 4, flame temperature can be highly variable at a single location due to turbulence, with typical mean values ranging from 1000 to 1200 K (Anderson 1969; Beer 1991; Grishin 1996; Telitsyn 1996; Quintiere 1998). View factor and emissivity were discussed in Chap. 5. The view factor (F_{ab}) represents the proportion of total radiative energy leaving the flame received by the unburned fuel and is a function of the geometry of the flame and the unburned fuels. Emissivity (ε) is the relative ability of a flame to emit thermal radiation and depends upon the nature of the flame. Flame emissivity values for fires spreading in pine needle fuelbeds generally range from 0.16 to 0.28 (Anderson 1969) but can vary depending upon the flame depth (D):

$$\varepsilon = 1 - e^{(-a_1 D)} \tag{7.12}$$

where a_1 is an empirical coefficient 0.1–0.3 m^{-1} (Thomas 1965) and D is flame depth (m).

Assuming that radiation is the only mode of heat transfer associated with the propagating heat flux and including Eq. (7.11) in Eq. (7.6), we can estimate the steady-state rate of spread as:

$$R_s = \frac{F_{ab}\, \epsilon\, \sigma_{SB}\, T^4}{\rho_{be} Q_{ig}} \tag{7.13}$$

This simplified approximation for the steady-state rate of spread is based on the assumption that the propagating heat flux can be completely represented by radiative heat transfer. Although a radiation-only approach is commonly used to represent fire spread for no-wind, no-slope conditions, radiation is not the sole mechanism of heat transfer under such conditions. Anderson (1969) indicated that radiation only accounted for 40% of the total heat transfer under no-wind and no-slope conditions. Indicating that the heat transfer mechanisms involved with fire spread, even under no wind and no slope conditions, are complex.

The Spread of Flaming Fires Modeled from Reaction Intensity and Propagating Flux Ratio

A widely used empirical approach for estimating the propagating heat flux ratio was developed by Rothermel (1972). It is based on the use of an empirical ratio between the propagating heat flux and the reaction intensity of the fire. Reaction intensity (I_R) is the rate of energy release per unit area.

If we assume that all of the mass within the fuelbed is consumed during flaming combustion over a given residence time, reaction intensity I_R (kW m^{-2}) can be estimated by multiplying bulk density ρ_b (kg m^{-3}), the low heat of combustion ΔHL (kJ kg^{-1}) and the fuelbed depth δ (m) together and dividing by the residence time t_R (s):

$$I_R = \left(\frac{\rho_b \Delta H L \delta}{t_R}\right) \quad (7.14)$$

Rothermel (1972) proposed that the no-wind no-slope propagating heat flux (I_{p0}) can be related to reaction intensity (I_R) through an empirically-derived value called the propagating flux ratio (ξ):

$$I_{p0} = \xi I_R \quad (7.15)$$

Based upon a series of laboratory experiments, Rothermel (1972) related the propagating heat flux ratio (ξ) to the typical surface area-to-volume ratio (σ in m^{-1}) and the packing ratio (β) of the fuelbed:

$$\xi = \frac{\left[\left(0.792 + 0.36\sigma^{0.5}\right)(\beta + 0.1)\right]}{192 + 0.0791\,\sigma} \quad (7.16)$$

where the packing ratio is estimated as the ratio of bulk density (ρ_b) to particle density (ρ_p):

$$\beta = \frac{\rho_b}{\rho_p} \quad (7.17)$$

The ratio of the propagating flux to the reaction intensity for a variety of fuel types is shown in Fig. 7.12. For the same reaction intensity, the propagating heat flux increases with increasing packing ratio and decreases with fuel particle size.

7.3.3 The Effects of Wind and Slope on Fire Spread

Wind speed and slope are two critical variables, along with fuels, that affect the propagating heat flux and, therefore, the rate of spread of a fire front. In this section, we address the individual effects of wind and slope and then discuss their combined effects.

The Effect of Wind Speed

The effect of wind on the propagating heat flux and the rate of spread of fires has been analyzed using several different approaches. The questions associated with modeling wind speed above and within vegetation canopies will be discussed later. Here we will use z as vertical height and U_z as the wind speed at the z height, with z_m corresponding to midflame height. As the ambient wind speed increases, the flames are pushed or tilted towards the unburned fuel, which increases the effectiveness of

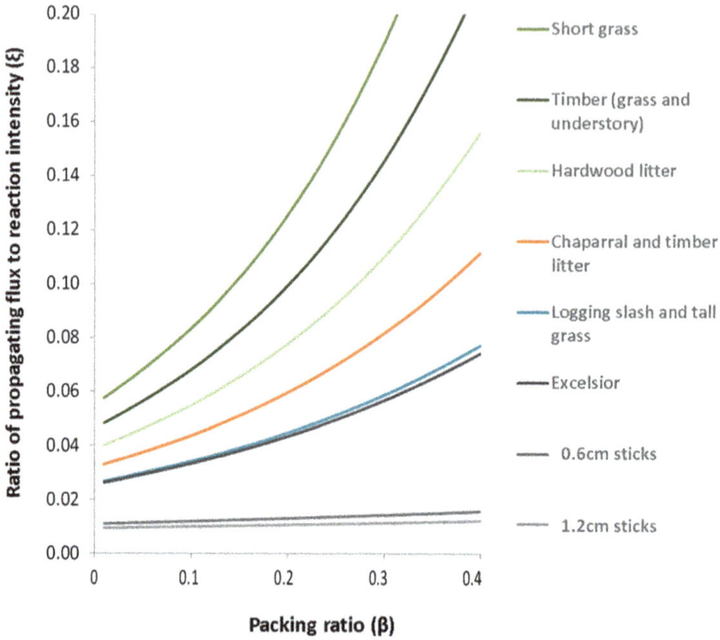

Fig. 7.12 The ratio of propagating flux to reaction intensity (ξ) as a function of the fuel packing ratio (β) and particle size (σ) using equations from Rothermel (1972) and Frandsen (1973) for no wind and no slope. Different particle sizes, expressed as the surface area-to-volume ratio (σ in m^{-1}), were used for the different fuels: short grass), timber grass and understory ($\sigma = 9843\ m^{-1}$), hardwood litter ($\sigma = 8202\ m^{-1}$), chaparral and timber litter ($\sigma = 6562\ m^{-1}$), logging slash and tall grass ($\sigma = 4921\ m^{-1}$), excelsior ($\sigma = 4757\ m^{-1}$), 0.6 cm-diameter sticks ($\sigma = 495\ m^{-1}$), and 1.2 cm-diameter sticks ($\sigma = 247\ m^{-1}$)

radiative and convective heat transfer, the propagating heat flux, and ultimately the fire rate of spread. From a radiation-only model perspective, wind may be seen as tilting the flame angle, increasing the configuration factor F_{ab}, and thereby (Eq. 7.13) contributing to increase the rate of spread of fire. However, as different heat transfer processes operate simultaneously, the spread rate of a wind-driven fire $R_s(U_z)$ can be modeled as a function, f(U), of the no-wind no-slope rate of spread (R_s) and a measure of the wind velocity at a given height (U_z). The height at which wind is measured can vary among studies. Special attention should be given to the idea of "midflame" wind speed (U_{zm}). The concept of midflame wind speed was developed for use with the Rothermel (1972) surface fire spread model, and it is meant to represent the average wind speed over a range of heights that affects surface fire spread rather than the open wind speed, which is commonly collected at 6.1 m (in the US) or 10 m above open ground.

One of the earliest attempts to estimate the effect of wind on fire rate of spread was made by Rothermel and Anderson (1966). Their work was based on experimental laboratory fires burning in pine litter. They used a function f(U) with an exponential form to adjust the no-wind no-slope rate of spread $R_s(0)$ based on the

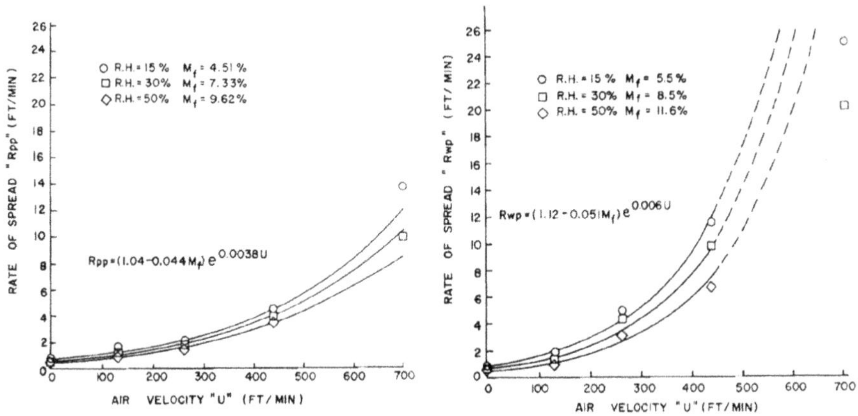

Fig. 7.13 The influence of midflame wind speed and fuel moisture on the rates of spread of fires burning in fuelbeds of pine needles of (**a**) *Pinus ponderosa* and (**b**) *Pinus monticola*. Original results with a rate of spread, R, and midflame wind speed, U, in feet per min and fuel moisture "H_f"as a percentage of dry fuel weight. (From Rothermel and Anderson 1966)

midflame wind speed (U_{zm}) and an empirical coefficient (a_1) that indicates the importance of the effect of wind. Their model for the wind effect on the fire rate of spread is:

$$R_s(U_{zm}) = R_s(0)\ f(U_{zm}) = R_s(0)\ e^{(a_1 U_{zm})} \tag{7.18}$$

where $R_s(U_{zm})$ is the steady-state rate of spread at midflame wind speed U_{zm}. Wind increases the fire rate of spread downwind, especially at higher wind speeds. The response to wind varies with fuel moisture (Fig. 7.13).

McArthur (1967) and Luke and McArthur (1978) also used a graph represented by Eq. (7.18) to describe the relationship between open wind speed 10-m above-ground (U_{10} in km h^{-1}), the no-wind, no-slope rate of spread, and the steady-state rate of fire spread for eucalypt and grass fuels in Australia. Using data from a series of experimental fires, they established relationships that correspond to values of the empirical coefficient, a_1, of 0.145 and 0.0842 km^{-1} h for grasslands and forests, respectively (Noble et al. 1980). These relationships provided a foundation for the Australian fire danger rating systems.

Another common approach is to represent the effect of wind speed on the fire rate of spread as a power function:

$$R(U_z) = R_s(0)\ U_z{}^b \tag{7.19}$$

where the exponent b is derived empirically from experiments with measurements made at various heights. The exponent depends upon the type of fuel and conditions of the experiments (Fendell and Wolff 2001). Estimates for b range from 0.42 to 0.65 for *Calluna vulgaris* and *Ulex europaeus* shrubland in the United Kingdom (Thomas

Table 7.1 The coefficients a and b of Eq. (7.20) for the different fuel sizes (different surface area-to-volume ratios) representing different fuel types. Data from Rothermel (1972)

	Surface area to volume ratio		Coefficients	
Fuel type	ft^{-1}	m^{-1}	a	b
Short grass	3500	11,483	0.061	2.07
Timber (grass and understory)	3000	9843	0.068	1.91
Hardwood litter	2500	8202	0.076	1.73
Chaparral and timber litter	2000	6562	0.086	1.53
Logging slash and tall grass	1500	4921	0.103	1.31

1971), to 0.87–1.83 for shrublands in Spain and Portugal (Vega et al. 1998; Fernandes 2001; Marino et al. 2008), to 2.22–2.67 for *Eucalyptus marginata* litter in Australia (Burrows 1999a, b).

Many modifications of these simple equations have been proposed. The approach of Rothermel (1972) is a modification of the simple power function:

$$R_s(U_2) = R_s(0)\left(1 + a\ U_2{}^b\right) \quad (7.20)$$

The wind speed input to Eq. (7.20), although described as midflame wind speed, is usually assumed to be the wind at a height of ~2 m. The values of the coefficients a and b in Eq. (7.19) (Table 7.1) are related to the characteristic fuel surface area-to-volume ratio.

The effect of wind on the fire rate of spread using Eq. (7.24), the values in Table 7.1, and those reported in Noble et al. (1980) for Australia are shown in Fig. 7.14. The effect of wind on rate of spread is more pronounced in fuel types with a relatively high surface area-to-volume ratio (e.g., short grass) than in fuel types composed of coarse fuel elements and consequently characterized by relatively low surface area-to-volume ratios, i.e., logging slash.

The association of these results with heat transfer processes is not clear. We hypothesize that the effect of the convective upward airflow reduces the effect of wind. The influence of wind on fires will be further discussed for crown fires and extreme fire behavior (Chap. 8).

The curvilinear relationships in Fig. 7.14 are not followed by other empirical fire-spread models, which indicate a near-linear response of the fire rate of spread to wind speed in a variety of fuel types (Cruz et al. 2015). The curvilinear relationships predicted by power and exponential functions are in part an artifact of the lack of response to weak winds (Fernandes et al. 2000) and cannot be extrapolated to strong winds. To overcome this limitation, Cheney et al. (1998, 2012) suggested using different wind functions to describe the relationship between rate of spread and wind speed above and below a threshold wind speed of ~5 km h^{-1}. Similarly, Butler et al. (2020) suggested that when wind speeds are less than 36 km h^{-1}, fire rate of spread increases linearly at a rate of around 3% of the wind speed. With higher wind speeds,

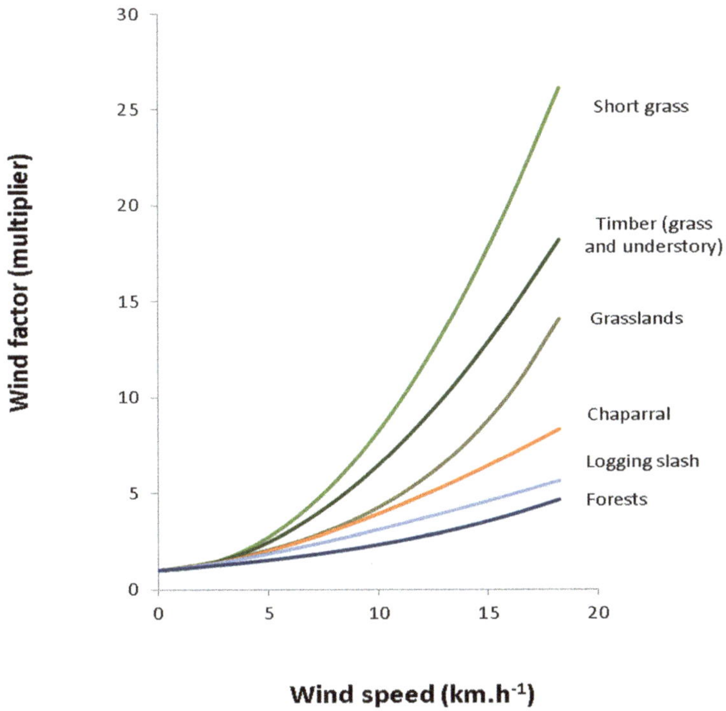

Fig. 7.14 The general form of the multiplying factors of wind from Rothermel (1972) for short grass, timber (grass and understory), chaparral, and logging slash in the USA, and for grasslands and forests in Australia (Noble et al. 1980). Contemporary Australian models in operational use have wind factors of ~1 (Cruz et al. 2015). Note that wind speeds are measured differently in the USA (z = 6.1 m) and Australia (z = 10 m)

the fire rate of spread response to wind remains approximately linear at a rate of about 13% of the wind speed.

The existence of an upper limit for the wind effect has been a matter of discussion. The original work of McArthur (1969) reproduced by Rothermel (1972) indicated the upper limits of wind effects on fire spread (Fig. 7.15). In the Canadian fire behavior prediction system, fire-spread rate levels off at very high values of the Initial Spread Index, a function of wind speed and dead fuel moisture content (FCFDG 1992).

The upper limits of the effect of wind on fire spread are not simple to evaluate. These limits are known to exist as wildfires have been reported to be blown out by very high winds. These thresholds have also been used for fighting by using explosives to blow out wildfires or by using hand-held leaf-blowing devices to fight low-intensity fires. However, the wind limits are difficult to establish in practice. Despite the work of McArthur (1969), other authors also working with grass fires in Australia (Cheney and Sullivan 2008) found no reliable evidence to indicate a sharp decrease in the rate of spread at wind speeds above 50 km h^{-1}, and

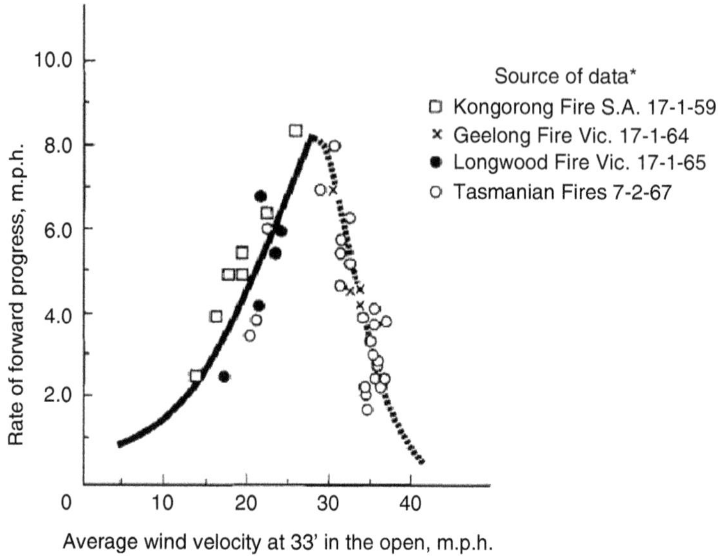

Fig. 7.15 The upper limits of fire spread and the wind effect in Rothermel (1972) using McArthur's (1969) rate of spread data for grass fires. The data are for average wind velocity at 33 feet above the ground in the open; units are miles per hour

indicated that the effects of very high wind speeds, above 80 km h^{-1}, were uncertain. A reanalysis of the original data and more recent grassfire data do not support the wind speed limit. This limit has since been removed from the operational version of the Rothermel model (Andrews et al. 2013). Butler et al. (2020) also did not find a wind limit, at least up to the maximum wind speed of 100 km h^{-1} explored in their laboratory study of fire spread in pine needle beds.

Modeling Wind Speeds Above and Within a Vegetation Layer

One of the practical challenges of modeling the effect of wind on the fire rate of spread is defining where wind should be measured. This is not a problem in laboratory experiments where wind speed is controlled. However, in field experiments, most instrumentation used to measure wind speed cannot survive a fire, and wind measurements are commonly collected at some distance from the fire front. The height of wind measurements is especially important. Models are available to estimate surface (1.5–2 m height) wind speed from measured wind speed at a different height.

Wind speeds and direction are commonly measured at a standard height (e.g., 10 m) above the ground to estimate the surface wind speed used in fire behavior prediction. Given that wind speed generally increases with height above the ground, measurements of wind speed at one height can be converted to the wind speed at another height if the wind profile is known or can be estimated.

A common approach used to estimate wind speed above a vegetation layer is the power law wind profile:

$$U_{z2} = U_{z1} \left(\frac{z2}{z1} \right)^{a} \tag{7.21}$$

where U_{z2} is the unmeasured (and to be estimated) wind speed at a given height above the ground, z2, U_{z1} is the measured wind speed at a given height, z1, and the exponent, a, is an empirical constant dependent on atmospheric stability. It is common to assume that the exponent, a, is a constant ($a = 0.143$ or 1/7). However, in areas where the wind flow is significantly affected by tall or dense vegetation, the use of the constant 1/7 value can lead to significant errors in wind speed estimates, and other approaches are required. One of the major effects on wind speed near the Earth's surface is the presence of vegetation. Here we will distinguish between estimating wind speeds above and within vegetation.

The logarithmic wind profile provides more reliable estimates of the wind profile closer to the Earth's surface than the power law profile (Eq. 7.21). It accounts for the effects of vegetation on the wind flow. For a neutral stable atmosphere, the logarithmic wind profile is:

$$u_z = \frac{u_*}{vk} \left[ln \left(\frac{z - z_d}{z_0} \right) \right] \tag{7.22}$$

where u_z is the wind speed at height z, u_* is the friction velocity (m s^{-1}), vk is the von Karman constant ($\cong$0.41), z_d is the zero plane displacement (m), and z_0 is the surface roughness (m).

By using Eq. (7.22), we can estimate the wind speed at the height of interest U_{z2} as a function of wind speed measured at a different height U_{z1}:

$$U_{z2} = U_{z1} \frac{ln\left((z2 - z_d)/z_0\right)}{ln\left((z1 - z_d)/z_0\right)} \tag{7.23}$$

The characteristics of the vegetation determine the coefficients of the zero plane displacement, z_d, and of the surface roughness z_o. The zero plane displacement z_d can be estimated for dense crops or forests as 60–80% of the average height of the vegetation (Thom 1975). The roughness length (z_o) is an adjustment factor that accounts for objects or roughness near the surface on the wind flow. Roughness depends on the general terrain characteristics (Fig. 7.16). Alternatively, roughness length (z_o) has been estimated as 0.13 times the vegetation height (Albini 1983).

Within vegetation, the wind profile does not follow the logarithmic and power laws indicated above. In general, wind speed decreases with the height from the top of the vegetation canopy towards the surface. The exact nature of this decrease is

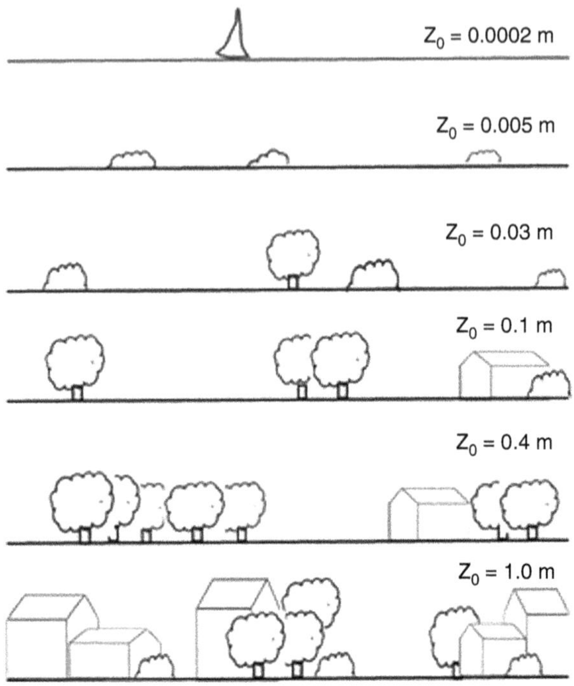

Fig. 7.16 Roughness values (z_0) of different terrain based upon the European Wind Atlas (Troen and Petersen 1989). Values are $z_0 = 0.0002$ m typical of seas and lakes, $z_0 = 0.005$ m for land with negligible vegetation, $z_0 = 0.03$ m for open areas with few trees and bushes, $z_0 = 0.1$ m for large open areas with windbreaks, trees, and buildings, $z_0 = 0.4$ m for forests or urban areas with well-spaced building without tall trees, and $z_0 = 1.0$ m for tall forests or densely built town area with openings. Roughness values can be even larger

directly related to the amount, characteristics, and arrangement of vegetation. The vertical wind profile differs for a dense forest with dense midstory and understory underneath the overstory layer and a forest with a single canopy layer (Fig. 7.17). The latter is characterized by greater wind velocity throughout the canopy profile and increased wind speed below the canopy.

The estimation of wind speed within a canopy is complex where the structure of vegetation is complex. However, estimates of wind speed underneath the tree canopy are required by fire spread models based on surface (1.5–2 m height) wind speed. The effect of vegetation layer on the wind velocity can be estimated using an exponential function (Fritschen 1985):

$$U_z = U_{zv} e^{\left[b\,\left(1-\frac{z}{zv}\right)\right]} \tag{7.24}$$

where zv is vegetation height (m), and b is an empirical parameter called the wind velocity attenuation coefficient, which depends upon the structure and density of the canopy. Typical values for b for forests range from 2 to 5 (Cionco 1965, 1978; Kunkel 2001). Example calculations of the wind profile based on Eqs. (7.21), (7.22), and (7.24) based on wind speed measured at 10 m above the ground are shown in Fig. 7.18.

The above equations try to represent the vertical wind profile above and within vegetation. They are useful for understanding the variables to take into account to

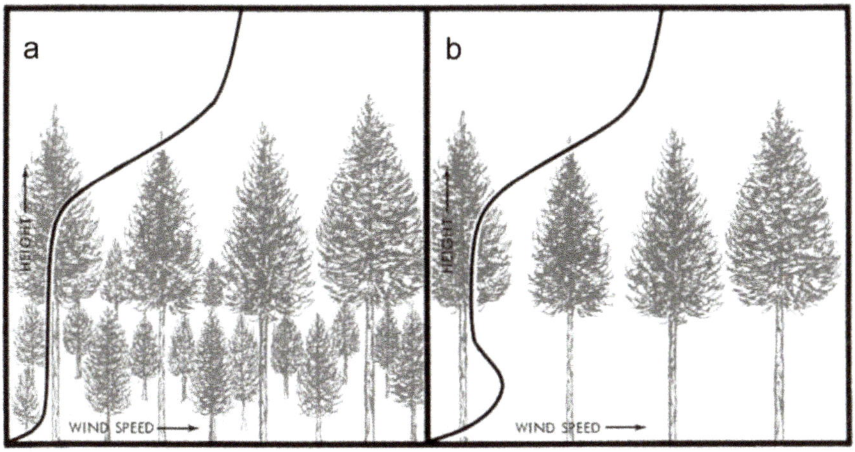

Fig. 7.17 Vertical wind profiles for (**a**) a dense, two-age class forest and (**b**) a single-age class forest (Schroeder and Buck 1970)

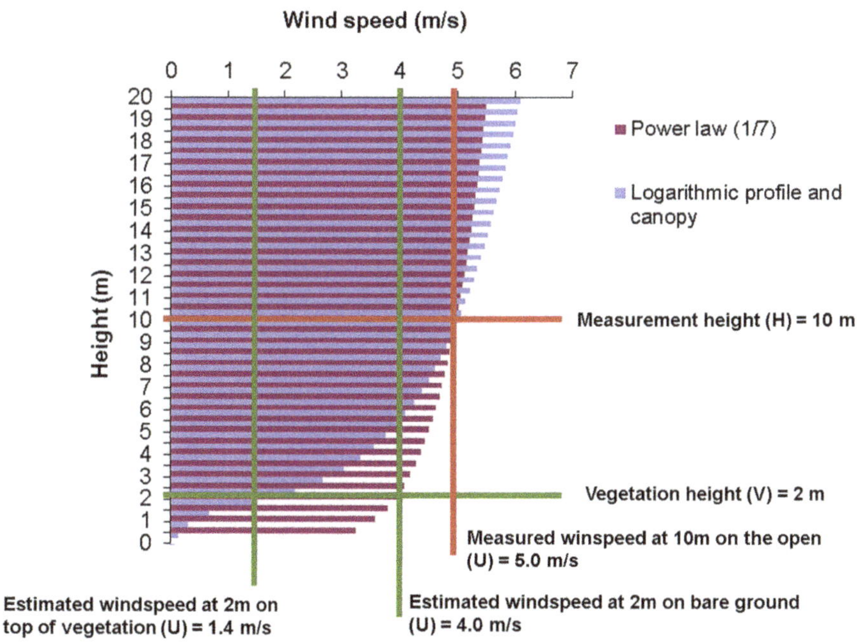

Fig. 7.18 The wind profile modeled from a wind speed of 5 m s^{-1} measured at 10 m height, using the Eq. (7.21) for bare ground (power law), Eq. (7.22) for height above a 2 m tall vegetation layer, and Eq. (7.24) for within the vegetation using a wind velocity attenuation coefficient b equal to 3.0 (see text for details)

reduce wind measurements taken at a fixed height to the height of interest for fire behavior, the midflame or surface (~2 m) height.

The current practice is to adjust wind speed measurements at a given height above the ground (U_z) to the midflame wind speed (U_{zm}) height directly by multiplying the measurements by a wind adjustment factor WAF (Albini and Baughman 1979; Rothermel 1983; Andrews 2012). The WAF is always below unity and assumes that vegetation cover is continuous and on flat ground. Albini and Baughman (1979) defined two separate WAF equations for sheltered and unsheltered surface fuels. The unsheltered WAF is based on estimating the wind speed at a height twice the depth of the fuelbed, and thus depends upon the surface fuel height. The sheltered WAF depends on the fraction of the crown space occupied by vegetation and the assumption that wind speed below the canopy top is fairly constant with height. Sheltered surface fuels tend to have WAF in the 0.1–0.3 range, while unsheltered fuels have WAF of 0.4–0.6 (Rothermel 1983). These systems use a reduction factor of 1/1.15 to convert wind speeds measured at 10 m to wind speed at 6.1 m before using the WAF. In summary:

$$U_{zm} = WAF\ U_{6.1} = WAF\ U_{10}/1.15 \tag{7.25}$$

USDA Forest Service fire behavior prediction tools such as Behave Plus and FlamMap integrate WAF.

The Effect of Slope

The effects of slope on the rate of spread of a fire have been addressed similarly to wind. From a radiation perspective, the slope decreases the flame angle, thereby increasing the configuration factor F_{ab} and, consequently the propagating heat flux ratio and rate of spread (Eq. 7.11). The effect of slope on rate of spread has been represented using a curvilinear function (Curry and Fons 1938; McArthur 1967; Rothermel 1972; Hwang and Xie 1984; Weber and Mestre 1990; FCFDG 1992). A multiplier factor has been applied in Australia since the early work of McArthur (1967), in Russia (Sheshukov 1970), in the USA (Rothermel 1972), or in Canada (Van Wagner 1977b). In all cases, the equations are derived from experimental and field data.

Similar to the effect of wind modeled by Eq. (7.18), one empirical equation for the effect of slope angle (S in degrees) on the steady-state rate of spread of fire (R_s) was fitted by Noble et al. (1980) to the rule of thumb of McArthur (1967) as:

$$R_s(S) = R_s(0)\ e^{(a_1 S)} \tag{7.26}$$

where $R_s(0)$ is the no-wind no-slope rate of spread and a_1 is a parameter indicating the effect of slope on fire spread, with $a_1 = 0.069$ for both grasslands and forests; similarly, Fernandes et al. (2009) obtained $a_1 = 0.062$ for pine forest. This functional form has the advantage of considering the spread of fires downslope. In this case, when estimating fire spread downslope, the value of slope is negative (S < 0), and the slope factor, $e^{(a_1 S)}$, is below unity, indicating a reduction of the propagating heat flux and a slower spread downhill. This is very useful when wind and slope are opposing one another. The combined effects of wind and slope are discussed later in this chapter.

A different expression for the effect of slope on fire spread was proposed by Rothermel (1972):

$$R_s(S) = R_s(0)\left[1 + a_2 \, tan\,(S)^2\right] \tag{7.27}$$

The coefficient, a_2, indicating the importance of slope in fire spread, was considered to be dependent on one of the characteristics of the fuel, the packing ratio (β):

$$a_2 = 5.275\,\beta^{-0.3} \tag{7.28}$$

When the fuelbed is more compact, the proportion of the fuelbed space occupied by fuel (the packing ratio β) is high, the value of a_2 is lower and the slope effect is smaller. The graphical representation of the ratio $R_s(S)/R_s(0)$, the slope factor, as a function of slope S and the packing ratio β, using Eq. (7.26)–(7.28), is presented in Fig. 7.19.

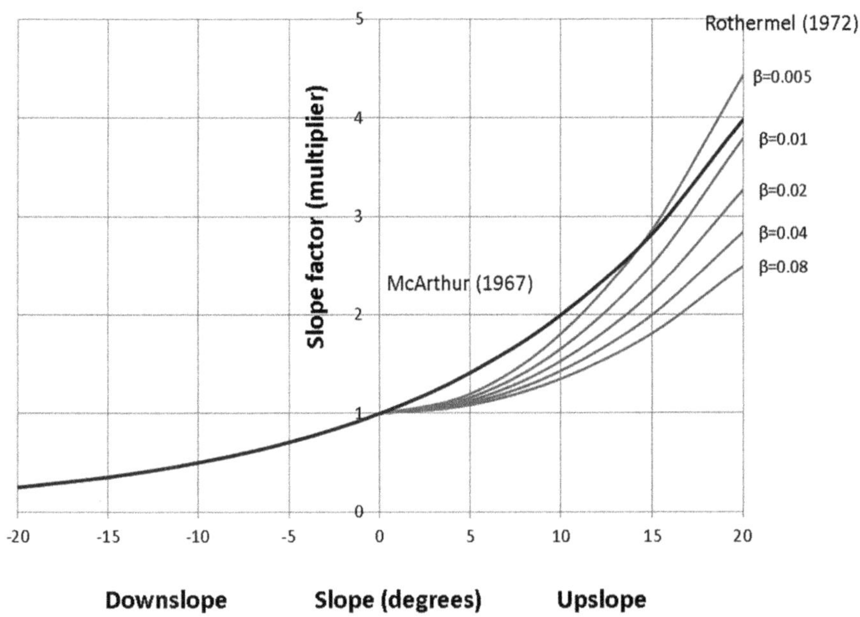

Fig. 7.19 The effect of slope on fire-spread rate according to McArthur (1967) and to Rothermel (1972) for different packing ratios (β)

The slope effect is more pronounced for low packing ratios (β) (Fig. 7.19). As the packing ratio (β) is the ratio of bulk density (ρ_b) to particle density (ρ_p), it follows that the rate of spread of fire in denser fuelbeds, with high bulk density, is less influenced by slope than in lighter fuel beds.

Based on the previous approaches and using the value of $\beta = 0.04$, representative of many litter layers, Van Wagner (1977b) proposed a different equation:

$$R_z(S) = R_s(0)\ e^{\left[\ 3.533\ \tan(S)^{1.2}\right]} \tag{7.29}$$

Notwithstanding the slope effects described above, large fires in undulating landscapes spread at an average rate that approximates what would be observed in no-slope conditions, as a result of slower propagation downslope (Sullivan et al. 2014).

Often the steady-state rate of spread of a fire front is estimated in relation to a specific wind and slope. The combined effects of wind and slope are discussed next.

The Combined Effect of Wind and Slope

Wind and slope together influence the rate of spread of fires. Increased wind velocity and slope affect the efficiency of heat transfer from the flame to the unburned fuel. For a head fire, an increase in wind or slope tilts the flame closer to the unburned fuel, increasing the amount of radiant and convective heat transfer to the unburned fuels. In contrast, during a backing fire, increased wind velocity or slope will cause the flame to tilt away from the unburned fuels, resulting in a reduction of radiative heat transfer.

Together, wind and slope can result in very significant and rapid increases in the fire spread rate (Fig. 7.20), much greater than the potential increases due to changes in fuel or fire type. Slope and wind are synergistic when both wind and slope are aiding the spread of a headfire.

The effects of wind and slope on the rate of spread of a fireline have been assessed independently by deriving empirical equations fitting the data of laboratory or field experiments. The empirical approach is due to the difficulties associated with the adequate modeling of convection, including turbulence. Many laboratory experiments were designed to address each of these two factors independently. Some laboratory experiments were designed to work in no-slope conditions to assess the isolated effect of wind, while others operate in no-wind situations to evaluate the independent effect of the slope. After the evaluation of the independent effects, these are typically combined in a single equation.

Because it is assumed that slope and wind influence fire spread through the same mechanisms, they are often treated as additive. Evaluation of the effect of these factors is generally done independently and later combined in a mathematical equation. Rothermel (1972) used an equation of the form:

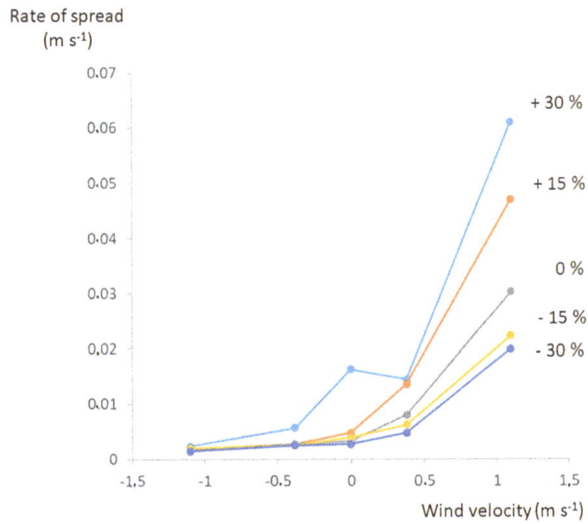

Fig. 7.20 Mean observed rate of spread as a function of wind velocity (m s^{-1}) and slope (%) in experiments by Weise and Biging (1997). Wind and slope interact to affect the rate of fire spread. Negative values of wind speed and slope indicate respectively fire spread against the wind (backfire) and downslope, as opposed to fire spread in the direction of the wind (headfire) and upslope

$$R_s(U,S) = R_s(0)\ (\ 1 + f(U) + f(S)\ ) \tag{7.30}$$

where $R_s(U,S)$ is the head fire rate of spread calculated from the no-wind no-slope rate of spread $R_s(0)$, a wind speed function f(U), and a slope function f(U). The wind and slope functions were those referred to in Eqs. (7.20) and (7.27). This equation includes the three factors of the fire triangle. $R_s(0)$, the no-wind no-slope rate of spread is only dependent on fuel characteristics, including fuel moisture, which is also partially weather-dependent. The wind speed function f(U) represents another weather effect, and the slope function f(S) represents the effect of topography.

Other possibilities of combining wind speed and topography exist. In the Canadian Forest Fire Behavior Prediction System (FCFDG 1992) fire-spread rate is predicted from the Initial Spread Index (ISI), a function of wind speed and fuel moisture. A no-wind ISI is computed to estimate the no-wind rate of spread, which is multiplied by the slope factor in Eq. (7.29) to obtain the slope-adjusted no-wind rate of spread. Then, the latter is converted into a wind speed equivalent to the slope and an overall ISI is calculated to estimate fire-spread rate as determined by the combined effects of wind and slope.

The remaining main approach adopts a multiplication equation of the form:

$$R_s(U,S) = R_s(0)\ f(U)\ f(S) \tag{7.31}$$

In this case, the headfire rate of spread $R_s(U, S)$ is a function of the no-wind no-slope rate of spread $R_s(0)$ times a factor f(U) representing the effect of wind on the fire rate of spread and a second factor, f(S), which represents the effect of slope on the rate of spread. These functions are different from those in Eq. (7.30).

7.3.4 *The Effect of Physical Fuel Properties on Fire Spread*

Different fire-spread modeling philosophies and options consider different fuel variables as inputs. The Rothermel (1972) model and its applications are based on the fuel model concept. A fuel model is a set of quantitative fuel characteristics, including fuel load by size class and condition, fuel depth, and surface-area-to volume ratio, which are direct inputs to the model, or they can be used to calculate other properties such as bulk density or packing ratio (See Chap. 6). Similarly to Rothermel (1972), Catchpole et al. (1998) were able to model in a controlled laboratory setting how fire spread is impacted by variation in a number of physical fuel properties, namely surface area-to-volume ratio (σ), packing ratio (β), particle density (ρp), and low heat of combustion (ΔHL). Their resulting model is based on the results of 357 experimental fires across a range of particle sizes, fuel bed depths, and packing ratios.

Field-based fire behavior models are developed from statistical analyses of data, preferably from field experiments carried out in large plots and covering the widest variation possible in environmental conditions. Robust empirical models have been shown to perform acceptably for operational purposes, including the prediction of fast-spreading wildfires under extreme conditions (Cruz and Alexander 2013; Cruz et al. 2018). However, these models are usually unable to capture and distinguish between the influences of individual fuel characteristics, especially because of the overwhelming dominance of wind and fuel moisture, and natural heterogeneity and correlation between fuel properties. Consequently, the analysis of field data seldom succeeds in identifying more than one descriptor of the fuel complex to add to wind speed and dead fuel moisture in a descriptive equation of fire-spread rate.

The Effect of Bulk Density

Fuels inventory commonly provides measures of the fuel load (kg m^{-2}) and depth (m), which is then used to estimate fuelbed bulk density (ρ_b, kg m^{-3}) (See Chap. 6). However, fuelbed bulk density is not necessarily the most appropriate variable to predict fire spread. Alternatively, a related variable called the effective bulk density (ρ_{be}) can be used. The effective bulk density represents the amount of fuel per unit volume involved in the absorption of heat during the pre-ignition phase. As indicated in Eq. (7.6), the effective bulk density is inversely related to the fire rate of spread.

Rothermel (1972) was one of the first to indicate that fuelbed bulk density is by itself a poor predictor variable of fire spread rate as fuelbeds often include various types of fuels in different size classes. To overcome this limitation, Rothermel (1972) and Frandsen (1973) proposed the use of an effective bulk density (ρ_{be}) as a function of fuel size, measured by the surface-area-to volume ratio (σ in m^{-1}) of the fuel pieces:

$$\rho_{be} = \rho_b e^{\left(\frac{-452.8}{\sigma}\right)} \tag{7.32}$$

The effective bulk density (ρ_{be}) is always smaller than the fuelbed bulk density (ρ_b) as only a fraction of the fuel bulk density needs to be heated to ignition temperature. If the fuelbed is primarily composed of fine fuel particles, such as grasses, with surface area-to volume ratios of around $\sigma = 10{,}000\ m^{-1}$, the difference between ρ_{be} and ρ_b is less than 5%. For fuelbeds primarily composed of thicker fuel particles with $\sigma = 5000\ m^{-1}$, such as litter or shrub particles, the difference between the effective bulk density and bulk density is less than 10%. This correction is important only when coarse woody fuels are a relevant component of the fuel complex. When only fine fuels are considered (high σ), the value of the effective bulk density (ρ_{be}) approaches that of the fuelbed bulk density (ρ_b).

This inverse relationship between effective bulk density and fire rate of spread indicates that for the same fuel load more compact fuels will have slower propagation rates. This justifies that fuel beds of loose long needles propagate fire more rapidly than more compact fuel beds of short needles. Also, fuel treatments that reduce fuel depth maintaining fuel load produce fuel beds with high effective bulk density. This occurs after mastication, in which machines chip or mulch trees and shrubs with the resulting chips spread on the soil surface. Some canopy fuels become compact surface fuels (See Sect. 11.4). Masticated fuels are seldom removed. If the masticated fuels burn, the deep beds of chips are more likely to smolder than to burn with flames (See Sect. 11.4). Thus, mastication can reduce the potential fire intensity and rate of spread but can increase the difficulty of extinction during fire suppression (Kreye et al. 2014).

Other Physical Fuel Properties Used in Fire-Spread Models

Contemporary field-based fire spread models are multiplicative and most commonly take the form of:

$$R_s = b_0\ U^{b1}\ e^{b2\ M}\ F^{b3} \tag{7.33}$$

where b_1 is a positive coefficient denoting an increase in the rate of spread with higher wind speed, b_2 is negative indicating a decrease in the rate of spread with higher dead fuel moisture content, and *F* is a simple descriptor of the fuel complex. An adjustment for the effect of slope through one of the slope factors currently available is subsequently introduced.

The fuel variable F in Eq. (7.33) varies substantially among models. Fuelbed bulk density is not straightforward enough for most end users of these models. Because it is calculated by dividing fuel load by fuel depth (δ), bulk density is inversely related to height for a given fuel load. This results in a positive correlation of fire spread rate and fuel height, as observed in shrublands (Vega et al. 1998; Fernandes et al. 2000; Fernandes 2001; Anderson et al. 2015) and pine forests (Fernandes et al. 2009), and modeled for a variety of fuels by Rossa and Fernandes (2018). For the same fuel load, a decrease in fuel depth or height results in a decrease in the fire-spread rate.

In other fire-spread equations, fuel load takes the place of fuel depth, with which it is highly correlated, unless fuel cover is uneven. Early operational models for prescribed burning or wildfire in eucalypt forest featured increased fire-spread rate with increased fuel load (McArthur 1967; Sneeuwjagt and Peet 1985). This provided a solid rationale for fuel reduction treatments, but subsequent research showed that fuel load should be replaced by fuel structure metrics. The currently recommended model for dry eucalypt forest (Cruz et al. 2015) includes near-surface fuel height and hazard scores for surface (litter) and near-surface fuels as inputs (Cheney et al. 2012). Fuel age can be used in lieu of the fuel structure metrics in this model (Fig. 8.1). Marsden-Smedley and Catchpole (1995) used fuel age to express the fuel complex effect on the fire-spread rate for the buttongrass moorlands of Tasmania.

The prediction of fire-spread rate for fuel types from Australian arid regions, such as spinifex grasslands (Burrows et al. 2018) and mallee-heath shrublands (Cruz et al. 2013), includes fuel discontinuity metrics as inputs. Finally, Australians equations for fire-spread rate in grassland, e.g., Cheney et al. (1998), all include the effect of curing level (%) on the fire-spread rates.

Differently from any other approach, the Canadian Forest Fire Behavior Prediction System (FCFDG 1992) does not consider variation in fuel properties other than fuel type. This is similar to using stylized fuel models as in the USA, except that model inputs are strictly based on fire weather. Also, the effects of wind speed and dead fuel moisture content are accounted for indirectly via the Initial Spread Index (ISI) of the Fire Weather Index System (FWI) (see Chap. 8). For each fuel type, totaling 16 options (e.g., C1—Spruce-Lichen Woodland, C6—Conifer Plantation, or D1—Leafless Aspen), an initial rate of spread is calculated using an S-shaped asymptotic curve. The rate of spread is then adjusted for fuel availability using a buildup effect based on the Build-up Index of the FWI System.

7.3.5 *The Effect of Fuel Moisture on Fire Spread*

Fuel moisture content (M) is calculated as the mass of water per unit dry mass of the fuel. Increasing fuel moisture decreases the propagating heat flux and increases the heat of pre-ignition (Fig. 7.11). The combination of these two effects results in an overall decrease in the fire rate of spread. In general, experiments evaluating the effect of fuel moisture on fire rate of spread do not separate the effect of moisture on the heat of pre-ignition and on the propagating heat flux independently. However, the effect of fuel moisture in decreasing the propagating heat flux has been expressed

Fig. 7.21 Experimental fire spreading through a fuel crib to understand heating ahead of a spreading fire in the Northern Forest Fire Laboratory of the USDA Forest Service (Frandsen 1973)

separately as in the exponential form proposed by Catchpole et al. (1998), as shown later in this chapter. Here we focus on the effect of fuel moisture in the heat of pre-ignition.

The effect of fuel moisture content on the heat of pre-ignition and fire spread rate has been studied experimentally since the 1960s (Thomas et al. 1964). Substantial developments relating fire rate of spread to fuel moisture were based on experiments in the Northern Forest Fire Laboratory in Missoula, Montana using fuelbeds of pine needles or with prepared particles of known dimensions (Anderson 1964, 1969; Rothermel and Anderson 1966; Frandsen 1971, 1973) (Fig. 7.21). The theoretical framework derived from these experiments was based on the conservation of energy principle (Fig. 7.22).

The effect of fuel moisture content on fire spread rate was one of the first problems analyzed in laboratory experiments. Anderson (1969) and then Rothermel (1972) approached the relationship between the fire rate of spread and fuel moisture through empirical polynomial equations (Fig. 7.23).

The steady-state rate of spread, R_s, is inversely related to the heat of pre-ignition, Q_{ig} (Eq. 7.6). As seen in Chap. 4, fuel moisture (M) influences the heat of pre-ignition. Fuel moisture has a pronounced effect on the fire rate of spread. When fuel is completely dry ($M = 0$), the fire rate of spread is maximized. On the other hand, when moisture exceeds the moisture of extinction ($M > M_x$), fires will

Fig. 7.22 Richard Rothermel watching an experimental fire in the wind tunnel at the Missoula fire laboratory in the early 1960s. (From Wells 2008)

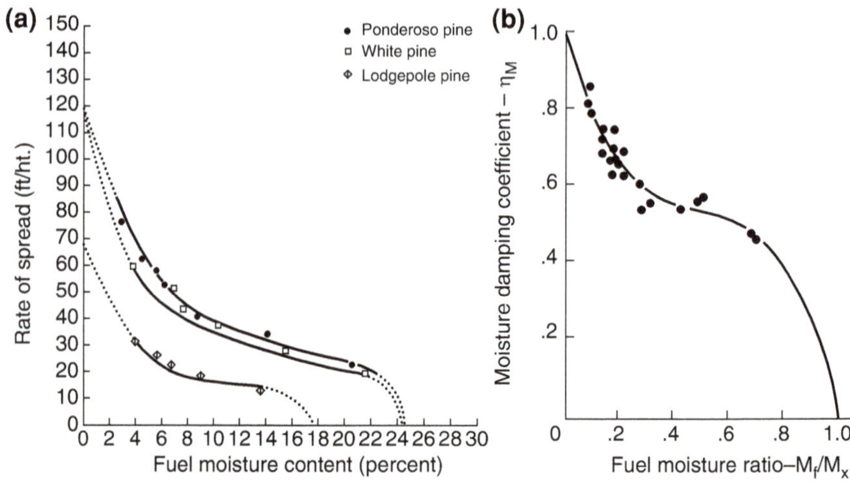

Fig. 7.23 (**a**) Rate of spread as a function of fuel moisture content (Anderson 1969), and (**b**) moisture damping coefficient (estimated rate of spread as a proportion of the maximum rate of spread with no fuel moisture) as a function of the fuel moisture ratio (fuel moisture content as a proportion of moisture of extinction) used by Rothermel (1972). Rate of spread as a function of fuel moisture content shows the artificial adjustment of the curves at the higher end of moisture content to derive the concept of moisture of extinction

not spread. Between these two values, the fire rate of spread responds continuously to changes in fuel moisture.

Maximum Potential Fire Spread Rate at Zero Fuel Moisture Content

We can estimate the maximum theoretical no-wind no-slope rate of spread $R_s(0)$, assuming that fuel moisture content is nil. For completely dry fuel, the heat of pre-ignition (Q_{ig}) consists of only the energy required to heat the dry fuels to the ignition temperature (Q_{dig}). Combining Eqs. (4.2) and (7.6) we can define the steady-state rate of spread for totally dry fuels as:

$$R_s(0) = \frac{I_p}{\rho_{be} Q_{dig}} = \frac{I_p}{\rho_{be}\, C_{pd}\, (T_{ig} - T_a)} \tag{7.34}$$

where the steady-state rate of spread at zero moisture $R_s(0)$ (m s^{-1}) is computed by dividing the propagation heat flux (I_p, kW m^{-2}) by the product of the effective fuel bulk density (ρ_{be}, kg m^{-3}) and the heat required for pre-ignition of dry fuel (Q_{dig}, kJ kg^{-1}). In this case, Q_{dig} is calculated as the product of the specific heat capacity of dry fuel C_{pd} and the difference between ignition temperature (T_{ig}) and ambient temperature (T_a).

For example, we can calculate the maximum potential fire spread rate at zero moisture $R_s(0)$. We draw upon the pioneering works of Anderson (1964, 1969) and Rothermel and Anderson (1966). They illustrated the calculation of a maximum theoretical rate of spread $R_s(0)$ with zero moisture ($M = 0$) in no-wind, no-slope conditions. The authors used fuelbeds with needles of ponderosa pine (*Pinus ponderosa*) and western white pine (*Pinus monticola*). In their experiments, they measured the temperatures at the combustion zone to be 1140 and 1052 K for ponderosa pine and western white pine. Assuming an emissivity of (ε) of 1, they estimated the resulting radiative heat flux (Eq. 5.5) received within the fuelbed to be 95.8 and 69.4 kW m^{-2}, respectively, which they assumed to be equal to the propagating heat flux ratio. The authors also indicated values for the bulk density (ρ_b) of the two fuelbeds to be around 32 kg m^{-3}, which are good approximations of the effective bulk density (ρ_{be}). Using a value of 1.1 J K^{-1} g^{-1} for the heat capacity of pine needles (C_{pd}) and a temperature difference between ambient (T_a) and ignition (T_{ig}) temperatures of 300 K, we can estimate the maximum theoretical rate of spread $R_s(0)$ for $M = 0$. In this case, we would have these equations:

For ponderosa pine needles:

(continued)

$$R_s(0)\left(\text{m min}^{-1}\right) = \frac{60\ \text{s min}^{-1} \times 95.4\ \text{kW m}^{-2}}{32\ \text{kg m}^{-3} \times 1.1\ \text{J K}^{-1}\text{g}^{-1} \times 300\ \text{K}}$$
$$= 0.54\ \text{m min}^{-1} \tag{7.35}$$

For western white pine needles:

$$R_s(0)\left(\text{m min}^{-1}\right) = \frac{60\ \text{s min}^{-1} \times 69.4\ \text{kW m}^{-2}}{32\ \text{kg m}^{-3} \times 1.1\ \text{J K}^{-1}\text{g}^{-1} \times 300\ \text{K}}$$
$$= 0.39\ \text{m min}^{-1} \tag{7.36}$$

These two estimates for the maximum rates of spread with zero fuel moisture can be seen in the graph with the statistical fit presented in Fig. 7.24.

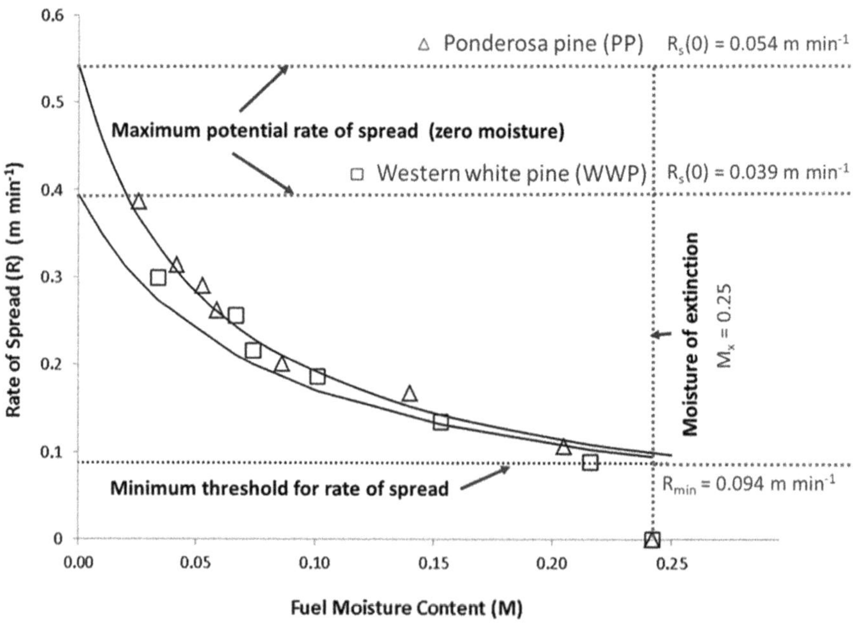

Fig. 7.24 Results of fire spread from laboratory experiments of Anderson (1964, 1969) and Rothermel and Anderson (1966) with needles of ponderosa pine and western white pine (triangles and squares) compared with results from a simple model with only one parameter for the effect of fuel moisture. A minimum threshold for the rate of spread R_{min} is established on the vertical axis equivalent to the moisture of extinction value M_x on the horizontal axis

Fire-Spread Rate Decreases with Increasing Fuel Moisture

The rate of spread of fire decreases with increasing fuel moisture. However, the equations governing this relationship deserve some discussion. The results of the experiments of Anderson (1969) and Rothermel (1972) are used in the fire behavior prediction systems developed by USDA Forest Service scientists. The relationship between the rate of spread and fuel moisture was approached statistically with a third-order polynomial equation (Fig. 7.24). However, this equation artificially forces the curve to reach a value of moisture at which the fire will not spread, which is the Moisture of Extinction (M_x) (Fig. 7.23). Because of its simplicity, the M_x concept has been widely used, despite depending on additional influences, such as wind speed and location on the fire perimeter, back or head fire (e.g., Fernandes et al. 2008).

Combining the fundamental equation for fire propagation (Eq. 7.6) with the full equation of the heat of pre-ignition (Eq. 4.6), we have a heat balance equation that relates the rate of spread R_s with fuel moisture M:

$$R_s(M) = \frac{I_p}{\rho_{be}[Q_{dig} + M\ (\ Q_{wl} + L_v + Q_{wv})]} \tag{7.37}$$

where fuel moisture M in the denominator is multiplied by the sum of three components: the energy required to heat liquid water in the fuel up to the boiling point (Q_{wl}), the latent heat of vaporization (L_v), and the energy needed to raise the temperature of water vapor to ignition temperature (Q_{wv}).

An empirical parameter for the moisture effect (a) may be introduced in Eq. (7.37) to fit the model to observed data. Rate of spread as a function of fuel moisture can be written as:

$$R_s(M) = \frac{I_p}{\rho_{be}[Q_{dig} + a\,M\ (\ Q_{wl} + L_v + Q_{wv})]} \tag{7.38}$$

The experiments by Anderson (1969) and Rothermel (1972) allow us to illustrate the use of Eq. (7.38). The empirical parameter, a, was estimated from the experimental data. In this case, the values of a were estimated to be 1.4 and 2.0 for the western white pine and the ponderosa pine fuelbeds, respectively. Comparison of the historical data of Anderson and Rothermel with models fitted to that same data illustrates how other simple models can also provide a good fit (Fig. 7.24).

This example shows that the use of only one empirical parameter allows fitting the model to the observed values.

Many empirical equations relating fire rate of spread with fuel moisture content have been proposed (e.g., Rossa 2017). A simple empirical model equivalent to Eq. (7.38) for estimating the rate of spread as a simple function of fuel moisture is:

$$R_s(M) = \frac{b_0}{\rho_{be}[b_1 + b\,M]} \quad (7.39)$$

where $R_s(M)$ is the steady-state rate of spread (m s^{-1}), ρ_{be} is the effective bulk density (kg m^{-3}), M is the fuel moisture (dimensionless), b_0 is a parameter that may be interpreted as the propagating heat flux (I_p in mW m^{-2}), b_1 is a parameter that can be interpreted as the product C_p ΔT (kJ g^{-1}), and b_2 is a parameter reflecting the product a (Q_{wl} + L_v + Q_{wv}). This approach was exemplified for laboratory studies, but it can also be applied to field experiments data.

Using data from a comprehensive group of burn experiments in highly diverse fuelbeds, Rossa and Fernandes (2018) developed empirical models of fire spread rate in no-wind no-slope conditions with varying bulk density and fuel moisture content. Here, we use their dataset of 220 experimental fires to illustrate the use of Eq. (7.39) and interpret the parameters b_0, b_1, and b_2 fitted to the data. The effective bulk density (ρ_{be}) was taken as identical to measured bulk density(ρ_b). The fitted equation is:

$$R_s(M) = \frac{b_0}{\rho_b\,(b_1 + b_2 M)} = \frac{48.0}{\rho_b(640 + 4600M)} \quad R^2 = 0.794 \quad (7.40)$$

where $R_s(M)$ is the rate of spread (m s^{-1}). In the numerator, the value of $b_0 = 48.0$ kW m^{-2} may be interpreted as the propagating heat flux, I_p. In the denominator, ρ_b is bulk density (kg m^{-3}), the parameter $b_1 = 640$ kJ kg^{-1} corresponds to Q_{dig}, the heat required to raise 1 kg of dry wood to ignition, and $b_2 = 4600$ kJ kg^{-1} corresponds to the heat requirements for fuel moisture until ignition (Q_{wl} + L_v + Q_{wv}), around 3000 kJ kg^{-1}, multiplied by the empirical coefficient a, around 1.53. M is fuel moisture. The observed rate of spread is similar to that predicted with Eq. 7.39 (Fig. 7.25).

Thresholds for Fire Spread and Moisture of Extinction

The no-wind fire spread in dead fuels such as pine needles rarely occurs when fuel moisture content is above 25% (Fig. 7.24). This upper limit is often referred to as the moisture of extinction. As seen above, at fuel moistures below the moisture of extinction, fire rates of spread increase as dead fuel moisture decreases.

The threshold for fires to spread is often defined in terms of the moisture of extinction (M_x), but it can also be expressed as a minimum rate of spread (R_{min}). Fire

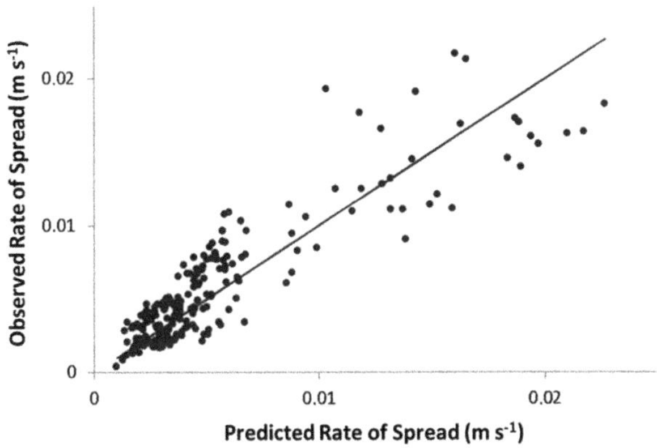

Fig. 7.25 Approximation of the no-wind no-slope rate of spread of fires based on bulk density ρ_b and fuel moisture M using Eq. (7.40) and data from 220 experimental fires compiled by Rossa and Fernandes (2018)

managers have widely used the concept of moisture of extinction (M_x) to represent the limit beyond which surface fires will not spread. However, for crown fires, a minimum threshold for fire spread (R_{min}) is often defined as the limit below which fire will not sustain spread. That is to say, when a crown fire's rate of spread falls below the threshold value, it will no longer propagate as a crown fire, though it may still spread as a surface fire or even exhibit torching.

Based on a combination of experimental data and field observations, Thomas (1967) and Van Wagner (1977a) suggested that, in no-wind no-slope conditions, this critical rate of spread was inversely related to bulk density:

$$R_{min} = \frac{S_c}{\rho_b} \tag{7.41}$$

where R_{min} is the minimum threshold for rate of spread (m s^{-1}), S_c is a constant (kg m^{-2} s^{-1}), and ρ_b is fuel bulk density (kg m^{-3}). The value of the constant, S_c, represents the critical mass flow rate of fuel into the flaming front to sustain fire spread. Thomas (1967) estimated S_c to be between 0.06 and 0.08 kg m^{-2} s^{-1}. Van Wagner (1977a) estimated S_c to be around 0.05 kg m^{-2} s^{-1}. This latter value has been used extensively for crown fire thresholds. The same principle applies to surface fuels.

For example, using a bulk density (ρ_b) of 32.0 kg m^{-3} for pine needle fuelbeds (Anderson 1964, 1969; Rothermel and Anderson 1966) and a critical mass flow rate (S_c) 0.05 kg m^{-2} s^{-1} (Van Wagner 1977a), we can calculate the minimum spread rate R_{min} using Eq. (7.41):

$$R_{min} = \frac{0.05\ \text{kg m}^{-2}\ s^{-1}}{32.0\ \text{kg m}^{-3}} = 0.094\ \text{m min}^{-1} \tag{7.42}$$

The lower limit for fire spread can be set in the fuel moisture axis, using the concept of moisture of extinction, M_x, in this case, around 0.25. It can also be

(continued)

estimated by identifying the minimum threshold for rate of spread evident on the y-axis, in this case, around 0.094 m min^{-1}. Thus, M_x and R_{min} represent two different approaches to estimating the lower limits of fire spread (Fig. 7.24).

7.4 Spatial and Temporal Variability of Fire Spread

Spatial and temporal variability of fire spread drivers, e.g., topography, wind, fuel moisture, and fuel characteristics (fuel load, bulk density) are critical for understanding and predicting fire behavior and effects in real landscapes. Although these factors can be assumed as constant at relatively fine spatial and temporal scales, fire behavior is influenced by how variable the influencing factors are in space and time. The influence of spatial and temporal variability has resulted in the development of landscape fire spread models that explicitly account for variability in the spread, intensity, and growth of fires. We follow this sequence of ideas in this section.

7.4.1 Spatial Variability in Fuels or Topography in the Landscape

Spatial variability in the landscape leads to variability in fire spread. Rapid changes in fire spread are especially important for fire fighter safety and developing fire suppression and community protection strategies. Rapid changes in fire spread are associated with temporal and spatial variability in the fuels complex, fuel moisture, wind speed, or topography. Changes in the physical characteristics and distribution of fuels can be represented by the various fuel types, such as pine litter, shrub, and grass, or transitions in fire type (i.e., transition from a surface fire to a crown fire). However, if fuels change abruptly, the fire-spread rate can rapidly change in response and not at steady-state. This occurs, for example, when a surface fire transitions into a crown fire because the energy flux of the surface fire is great enough to heat the canopy foliage to the ignition point (Weise et al. 2018). Following the initialization of a crown fire, the rate of spread can increase two to four times above that of pine or hardwood litter. Significant changes in fire rate of spread also occur when a fire spreads from pine litter into grasses where the change in fuel type can result in up to a 15-fold increase in the fire rate of spread (Fig. 7.26). See Chap. 8 for more on crown fires.

At forest edges, the fire rate of spread and behavior can change due to differences in fuel type, fuel moisture, and wind speed. Wind flow across a forest edge goes

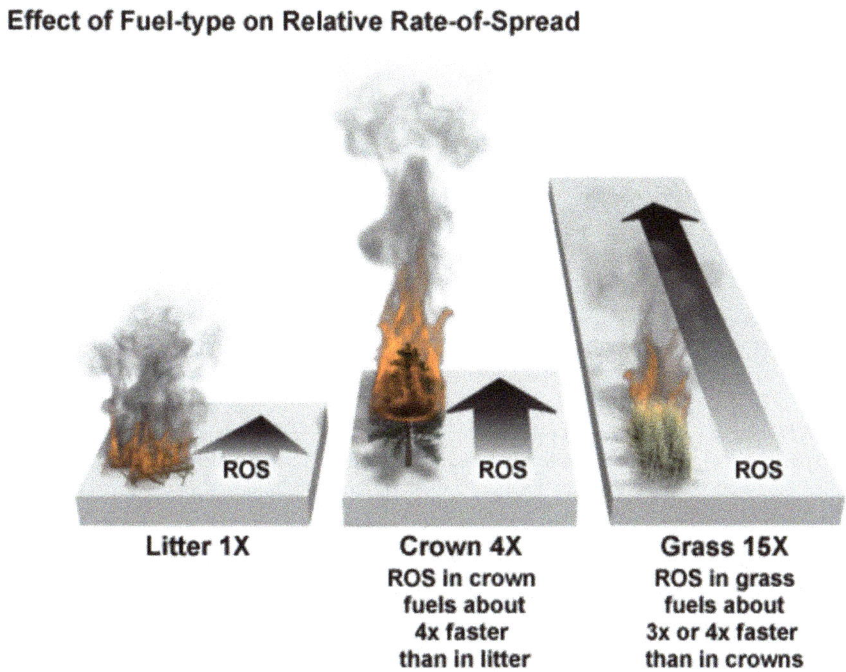

Fig. 7.26 Rapid transitions in fire rate of spread occur when very different fuel types alternate in the landscape, such as litter, crown fuels, and grass (UCAR 2012)

through several distinct regions as it adjusts to changes in vegetation structure (Belcher et al. 2003). In the area just upwind of the forest edge, called the impact region, the wind decelerates slightly. Just downwind of the forest edge is the adjustment region where the streamwise flow slows down within the canopy due to drag, causing an upward motion from the canopy to the atmosphere. This region is sometimes referred to as the enhanced gustiness zone due to the high probability of strong gust events. Further downwind from the edge is the canopy interior. In the canopy interior, the wind flow is fully adjusted to the characteristics of the canopy. Along the top of the forest are the canopy shear layer and the roughness-change region, which are the areas where most of the mass and heat exchange occurs between the vegetation and atmosphere. As wind flow exits a forest, the canopy shear layer detaches from the canopy and reattaches at some distance downwind, typically 2–5 times the height of the vegetation) before reestablishing a new boundary layer in the clearing. The area between the edge and the reattachment area is referred to as the recirculation zone and is characterized by a reversal of the wind direction (Fig. 7.27). Although there has been little research that has evaluated changes in fire behavior across forest edges, it is reasonable to expect the changes in fire behavior due to the changing wind speed, wind direction, and vegetation types.

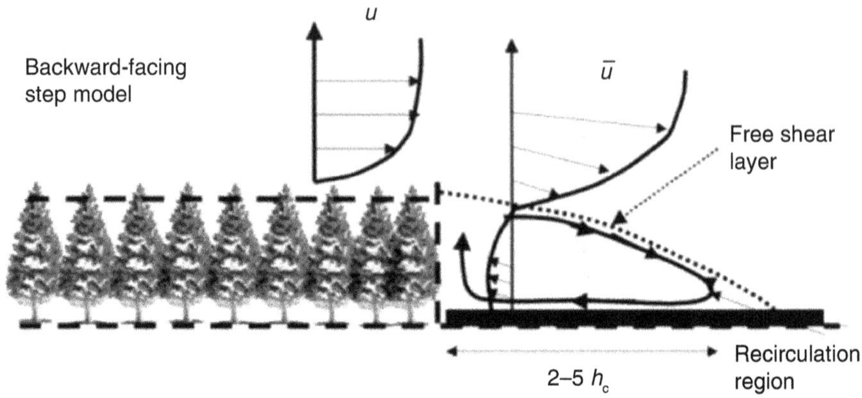

Fig. 7.27 Forest edges alter wind patterns. Arrows indicate direction. U represents wind speed and hc is the height of the canopy. These wind patterns can alter fire behavior from what is experienced within the forest or in the adjacent open vegetation. (From Detto et al. 2008)

Fuel continuity, which is defined as the extent of continuous fuel particles in a fuelbed, can directly and indirectly influence the rate of fire spread and behavior. The indirect effects of fuel continuity are primarily due to the effects of heterogeneous fuels on wind speed and turbulence. For example, forests that are characterized as having tree clumps and openings between the clumps tend to have greater within canopy wind speeds than forests with evenly spaced trees due to wind channeling in the openings between the clumps (Pimont et al. 2011; Hoffman et al. 2015; Parsons et al. 2017). These forests also tend to have greater turbulence below the canopy layer, which can play an important role in driving fire spread when there is also considerable discontinuity in the surface fuel layer (Linn et al. 2013). The theoretical effects of fuel discontinuity on fire spread is illustrated in Fig. 7.28.

At finer scales, spatial variations in fuel physical properties (e.g., fuel loading, depth, bulk density) and continuity can alter the rate of fire spread and surface fireline intensity. The effect of changes in surface fuel loading on the rate of spread is partially due to alterations in mass loss rate and heat transfer to the unburned fuels ahead of the fire front. Although our understanding of the spatial variability in surface fuel properties in many ecosystems is limited, several studies in dry pine forests in the US have indicated that variability in fuel loading occurs at scales of less than 10 m (Hiers et al. 2009; Keane et al. 2012; Vakili et al. 2016) and that this variability influences the spread and intensity of surface fires. Hiers et al. (2009) provided an interesting example of a fire spreading through a 12-m^2 area where a small patch of bare ground caused the fire to split into two separate heads that then merge together resulting in a local increase in fireline intensity (Fig. 7.29).

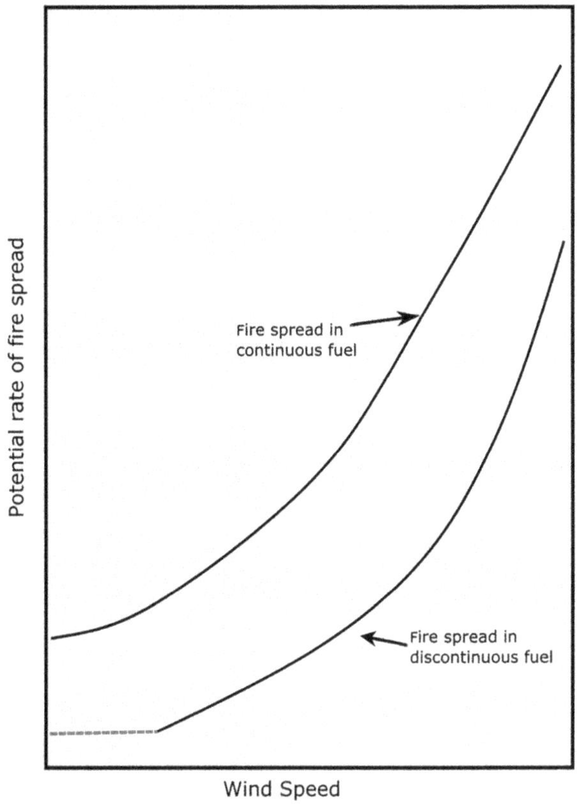

Fig. 7.28 Theoretical rate of fire spread in continuous and discontinuous surface fuels across a range of wind speed. Fire spread in continuous fuels increases exponentially with wind speed. Fires burning in discontinuous fuels do not spread until a certain wind speed is reached (*shown in red*); above that threshold, the rate of spread increases exponentially with wind speed

7.4.2 Integrating the Variability of Weather, Fuel, and Topography in Fire Spread Prediction

We highlighted steady-state fire spread in the previous sections. However, the behavior of wildland fires varies in space and time. Spatial variability in fuels (type, dead or live condition, and physical properties) and topography must be recognized for realistic predictions of fire spread rates. The same is true for the effects of temporal variability of weather on the rate of spread of fires.

Wind is the factor that shows the highest variability during the propagation of a fire. Wind direction and speed may change abruptly, whereas dead fuel moisture (the main fuel characteristic that can vary significantly within the time frame of a wildfire) responds to variations in air relative humidity, ambient temperature, and solar radiation in a more gradual and predictable way (Chap. 11). An example from the wildfire in Pedrógão Grande, Portugal, in June 2017 illustrates these changes well (Fig. 7.30). The fire escalated with a wind change, similarly to other deadly wildfires, and only at night the conditions of higher relative humidity would result in

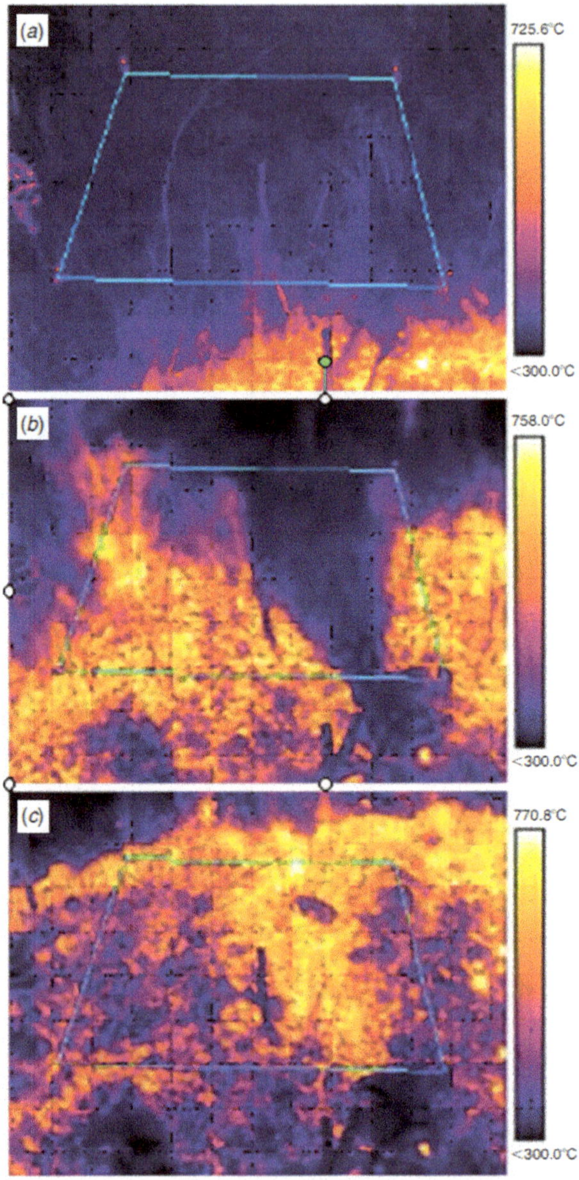

Fig. 7.29 A time sequence of thermal infrared imagery of fire spread through a 12 m^2 area. (**a**) The fire spreads from the lower right through the area. (**b**) A small patch of bare ground is located at the bottom right. As the fire spreads into the plot the patch of bare ground causes the fire to split into two separate flanks. (**c**) As these flanks spread past the area with bare ground, they merged back together resulting in an area of greater fireline intensity and longer residence time than would be expected if the fire was not altered by the area of bare ground. (Images from Hiers et al. 2009)

higher dead fuel moisture which, combined with lower wind speeds, would allow for improved firefighting effectiveness.

Although we can assess the effects of variations in the three components of the fire triangle (Fig. 7.31), fuels, wind, and topography independently, they interact across a wide range of temporal and spatial scales. Topography includes elevation, slope, and aspect, all of which can influence the type, amount, and moisture content

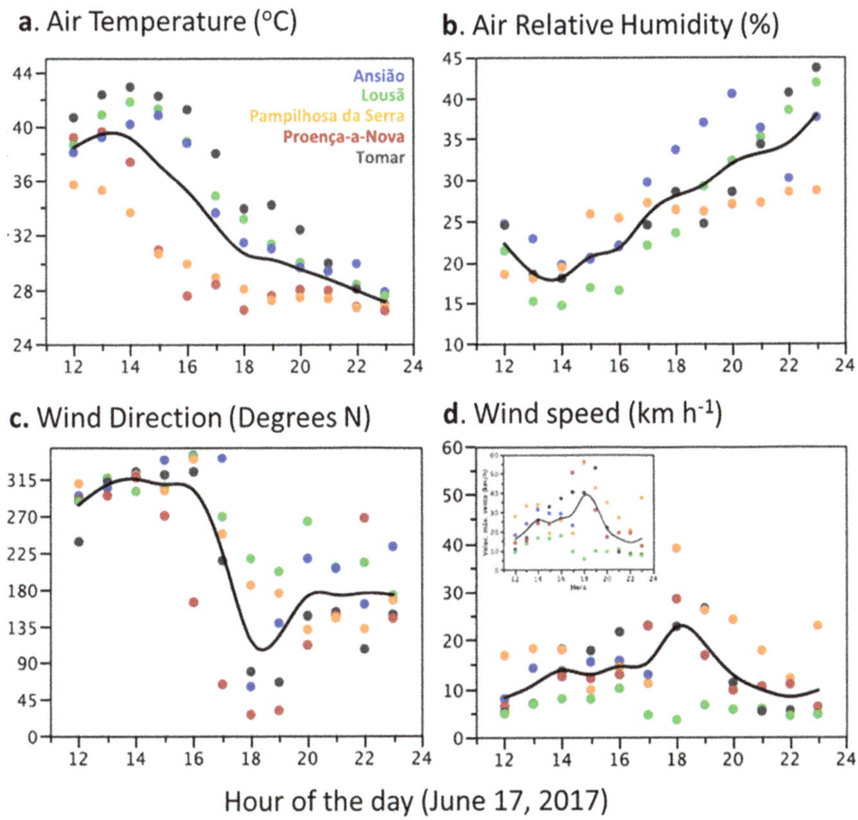

Fig. 7.30 Changes in air temperature (**a**), relative humidity (**b**), and wind direction (**c**) and speed (**d**) in five weather stations (different colors) around Pedrógão Grande, Portugal, during 12 h of the first day of a major wildfire. Air temperature and relative humidity are inversely correlated with values changing gradually from 14:00 to 23:00. The wind changed abruptly in direction and speed around 18:00, which ultimately caused a blow-up and 66 people died. (Adapted from CTI 2017)

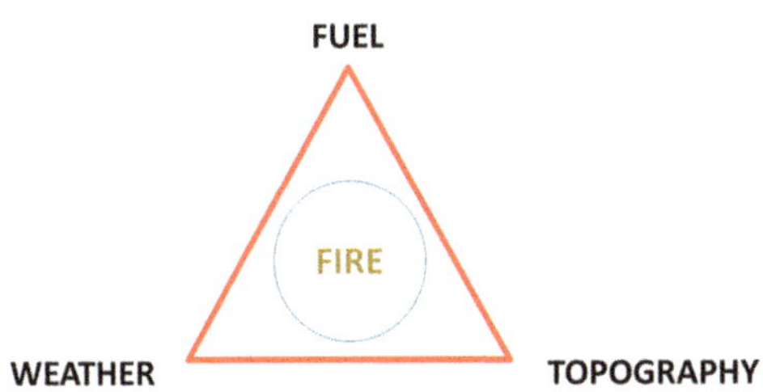

Fig. 7.31 The fire behavior triangle where fire behavior is a function of fuel characteristics, weather (especially wind), and topography (slope). See this fire triangle within the broader spatial and temporal scales of whole fires and recurring fires that are part of the fire regime (Figs. II.2 and III.1). Spatial and temporal variability on each of the corners results in changes in fire behavior and consequent effects

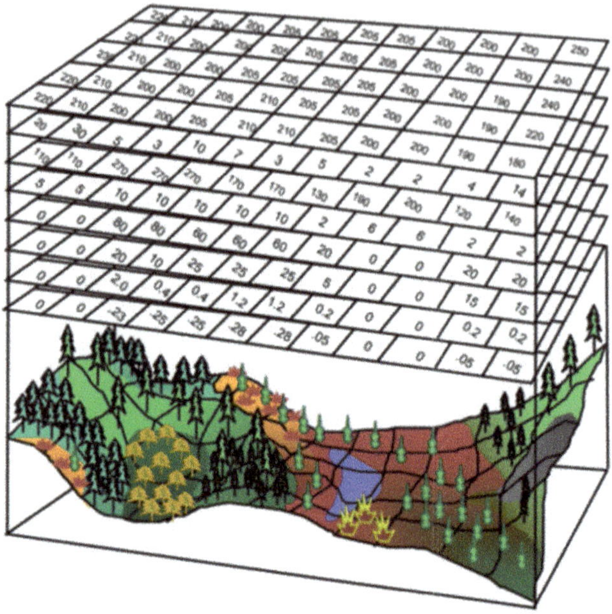

Fig. 7.32 The different spatial data layers required for running FARSITE or FlamMap (Finney 1998, 2006)

of the fuels. The influence of elevation and aspect on fuel type has been well studied. Local wind, air temperature, and relative humidity are dependent on topography and vegetation.

Spatial variability in a landscape affects fire behavior. In general, fire growth systems calculate fire spread for a point (or cell) in the landscape independently, using spatially variable topography (slope), fuel characteristics, and wind as input spatial data. This was the approach taken in FARSITE (Finney 1998) and FlamMap (Finney 2006) that use the same fire models as in Behave Plus, the model of Rothermel (1972) for surface fire spread, and the models of Van Wagner (1977a) and Rothermel (1991) for crown fire initiation and spread. FARSITE has been used first in the USA (Finney and Ryan 1995) as in the Selway-Bitterroot Wilderness Complex (Keane et al. 1998). The input variables need to be available spatially to apply these systems (Fig. 7.32). Equivalent fire growth software has been developed in Canada and Australia based on the fire behavior models adopted in these countries (Parisien et al. 2005; Tolhurst et al. 2008; Tymstra et al. 2010).

The propagation of fire through a heterogeneous landscape with variable wind speeds has been a mathematical challenge with various approaches. Fire spread in the various directions may be estimated at each cell or point but fire growth has to be derived from these points or cell estimates. The first step to incorporate variable wind and slope in the calculations of fire growth is to estimate fire spread at each time step at each vertex of the fire perimeter. One common approach is based on the Huygens'

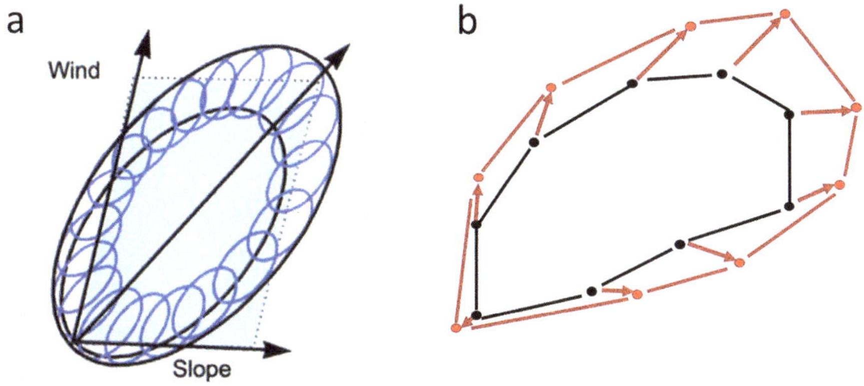

Fig. 7.33 (**a**) The geometric representation of fire progression using the Huygens'principle with elliptical wavelets as in FARSITE (Finney 1998). (**b**) The representation of fire progression through the movement of the nodes of the fire perimeter as in the TIGER system (Giannino et al. 2017)

principle where fire spread calculations determine the shape and direction of an elliptical wavelet at the perimeter of the fire at each timestep and the perimeter at the next time step in the connection of the more extreme points of the ellipses (Fig. 7.33a, Finney 1998). An equivalent alternative is the geometric fire perimeter spread model developed in TIGER where the fire perimeter is represented as a simple closed polygon formed by a dynamic population of linked nodes and the speed and direction of movement is computed for each node (Fig. 7.33b).

Other fire propagation models have since been proposed with a special reference to level set methods (Mallet et al. 2009) where the change of the position of the fire front is calculated through the use of hyperbolic conservation laws on Eulerian grids from given values of rate of spread of infinitesimal segment of fire front estimated as a function of known local parameters relating to topography, vegetation, and meteorology.

One of the interesting features of the approach in the TIGER system (Giannino et al. 2017) is that the spatial fire progression is reversible, as one of its modules allows for a simulation to be run backward in time (Fig. 7.34). This approach has been used by the Italian National Guard (*Carabinieri*) responsible for the investigation of the causes of the origin of fire by locating the approximate starting point of the fire from the final fire perimeter, together with spatial information on wind, topography, and fuels.

The interactions between wind, topography, and vegetation are difficult to model. This is why many fire behavior simulations are done using a single value for wind speed and direction, assuming a homogeneous wind field. This is probably adequate for some applications, but the resulting predictions may be poor. Wind speed and directions are often the most influential factors affecting fire behavior. Wind speed and direction can be predicted using weather forecasts (Wagenbrenner et al. 2016), but topography and vegetation affect wind speed and direction. The spatial variability in the wind due to complex terrain can be accounted for by using the WindNinja program (Forthofer et al. 2009) to calculate wind speed and direction.

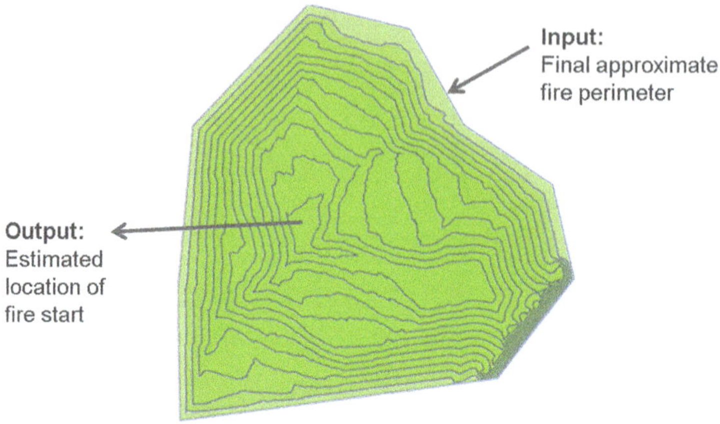

Fig. 7.34 Backward simulation of fire progression from an approximate final perimeter of a fire occurred near Salerno, Italy, in 2015, showing the estimated location of fire start using the TIGER system. The lines inside the final perimeter represent the perimeters at each 1 h step (Giannino et al. 2017)

Wind patterns over heterogeneous landscapes have also been considered in the TIGER software developed under the European Project FIRE PARADOX. In this case, the basis for the wind simulations was the WAsP software produced by the Wind Energy Department (VEA) of the Risø National Laboratory in Denmark, for wind resource assessment, siting, and energy yield calculations for wind turbines and wind farms. WAsP Engineering is used for the calculation of wind conditions for sites located in all kinds of terrain. The European Wind Atlas describes the foundation of the wind atlas methodology and the WAsP models (Troen and Petersen 1989). Under this system, it is necessary to set up a modeled landscape as a map describing the topography and surface aerodynamic roughness length (z_0) based on the terrain and vegetation characteristics (Fig. 7.16). For the simulation of wind conditions near the surface, it is possible to use the wind (speed and direction) observed for geostrophic height, and the WAsP program predicts the speed and direction of the wind at any other point in the landscape (Fig. 7.35).

Some of the variability in fire behavior in space and time may be predicted by tools that integrate fuel, topography, and weather, as these examples suggest. These tools can be applied to understanding fire behavior in the landscape and to the planning of fuels treatments. However, there are still numerous sources of uncertainty in the prediction of fire behavior.

7.5 Limitations, Implications, and Applications

Understanding fire spread is critical for fire suppression, prescribed burning, fuels management in general, civil protection, and other aspects of fire management. Our understanding of wildland fire behavior has resulted in the development of various

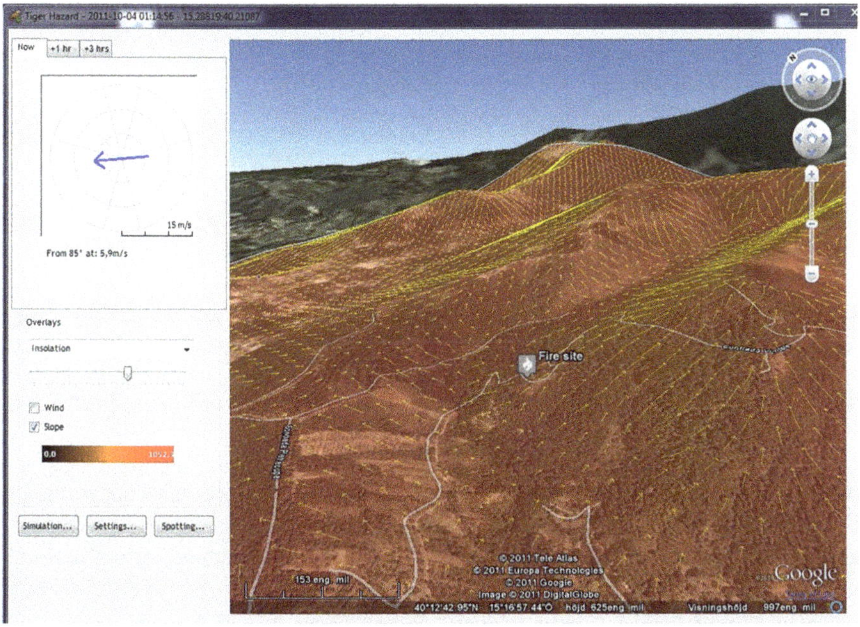

Fig. 7.35 Example of output from TIGER using WAsP calculations of wind speed and direction in complex terrain and heterogeneous vegetation. Example by Mazzoleni and Heathfield (2017)

guides and models that are designed to help predict fire behavior and aid in developing strategies to manage fire and fuels. For example, fire behavior potential is implicit in the fire danger rating systems of the USA (Deeming et al. 1972), Canada (Stocks et al. 1989), and Australia (Noble et al. 1980) (See Chap. 8). Models for predicting fire spread have been used by fire managers worldwide. The models have the same general basis but different ways of combining the effects of wind, slope, and fuel characteristics, especially fuel moisture, on fire spread rate. However, as with any model it is critical that users of fire behavior models understand the conceptual underpinnings and assumptions that went into their development.

A limitation of current approaches is in the "engine" of the systems. For example, US modeling tools rely on the basic equations developed by Rothermel (1972), based on the conservation of energy principle with empirical adjustments. No explicit reference to heat transfer mechanisms is made in these approaches. However, other models exist for fire spread and fireline intensity with different assumptions. No single model is always appropriate. Models vary because models are used for all sorts of applications. Empirical models may run more quickly than more complex process-based models, or the data required for the latter may not be available.

Throughout this chapter, we have looked at how fire spreads, and the effects of fuels, wind, and slope. Although there have been considerable gains in understanding and predicting fire spread over the last three quarters of a century many questions

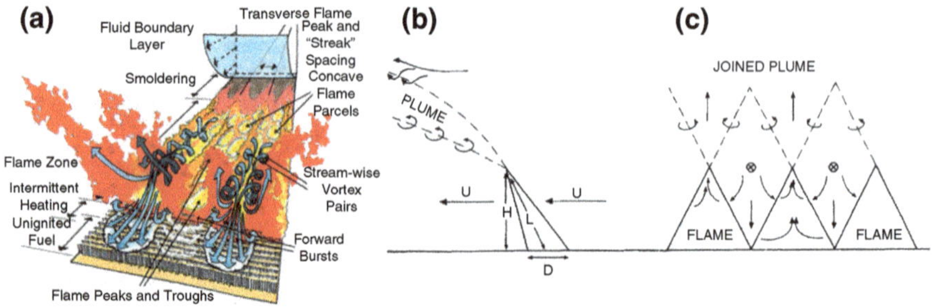

Fig. 7.36 The processes at the fire front are complex. In spite of that complexity, some general patterns are always observed in the interaction of a fire front with the airflow around it. (**a**) The complexity of processes. (**b**) The interaction between the ambient wind and flames at the fire front viewed from the side, and (**c**) looking in the direction that the fire is spreading. (**a**) From Finney et al. (2015), (**b**, **c**) From Beer (1991)

remain unanswered. Recent experimental results and advancements in computational models are playing an important role in advancing our understanding and ability to predict fire spread in a variety of situations. For example, recent research has highlighted the role of convective heat transfer and buoyant-induced diffusion of turbulent eddies from the flames on fire propagation. As the buoyant gases from combustion interact with the ambient wind field, the fire front develops a series of peaks and troughs (Fig. 7.36a). The peaks and troughs are associated with alternating updraft and downdraft regions where cold air comes down to replace the warm ascending air and winds can pass through the fire front. In areas where these trough formations occur, the flames are forced downward and towards the unburnt fuel resulting in increased intermittent convective heat transfer (Fig. 7.36b, c).

Finney et al. (2015) proposed that the intermittent increases in heating due to contact with the flame are an important heat transfer mechanism in surface fire spread. More research is underway to improve our understanding of the processes at the fire front and their effects on fire propagation. Without a better understanding of the air flow induced by the fire front, it is not possible to model the interactions between fire fronts adequately. This limits our expertise in the use of fire in suppression and in prescribed burning operations when interacting ignition points or lines are used.

Another problem in the current use of existing fire behavior prediction systems is that they are often applied to support the suppression of fires burning under conditions quite different than the limited number of experimental fires on which many of the models are based. As models are widely used by managers for predicting fire behavior even for large, intensely burning fires, fire behavior analysts have to combine these models with their experience and expertise. In the cases where fires burn over many days or weeks, the long-term fire behavior analysis must consider the changing topography, fuels, and weather. If those environmental conditions are extreme, then fire behavior is more difficult to predict, as they also have to include spotting and complex fire-atmosphere interactions, as described in Chap. 8.

In the prediction of fire propagation, variability, and scale can be crucial. Variability in the wind, fuels, fuel moisture, and topography over space and time greatly influences fire behavior and fire effects. Scaling will continue to be challenging when dealing with existing spatial and temporal variability. Approaches effective at fine temporal and spatial scales exist, but they do not address the need for approaches at other scales. Addressing variability across spatial and temporal scales is an area of rapidly developing scientific understanding with implications for how variability in pre-fire conditions influences fire behavior and fire effects. Existing models are mostly deterministic and thus may not effectively represent the uncertainty associated with the processes involved. Fires may "skip" areas, especially if they are spreading by spotting (See Chap. 8), leaving some areas unburned. The prevalence and ecological importance of these unburned islands, called refugia, are addressed in Chaps. 9 and 12.

Despite the limitations, models are extremely useful for making predictions. In this book, we emphasize their value for building understanding, such as identifying the driving factors and the ways they interact. We developed the interactive spreadsheets to help you as readers and students of fire to strengthen your understanding of fire behavior and effects.

7.6 Interactive Spreadsheets: FIRE_GROWTH, FIRE_RATE_OF_SPREAD, and WIND_PROFILE

We include three different interactive spreadsheets in this chapter (Figs. 7.37, 7.38, and 7.39). Our spreadsheet, FIRE_GROWTH_v2.0, contains the equations and graphical display of predictions for initial fire growth for different inputs as shown

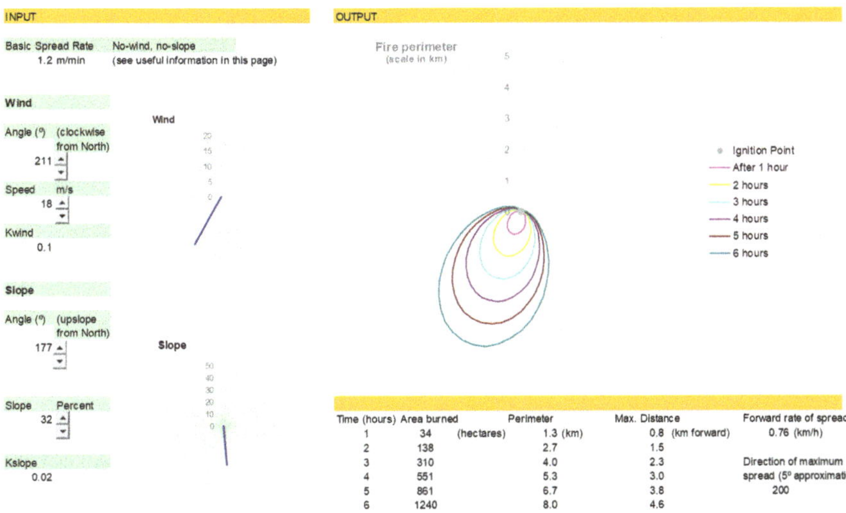

Fig. 7.37 Our interactive spreadsheet, FIRE_GROWTH_v2.0, will help you visualize graphically how changing the equations and graphical display of predictions for initial fire growth for different inputs as shown in Fig. 7.4

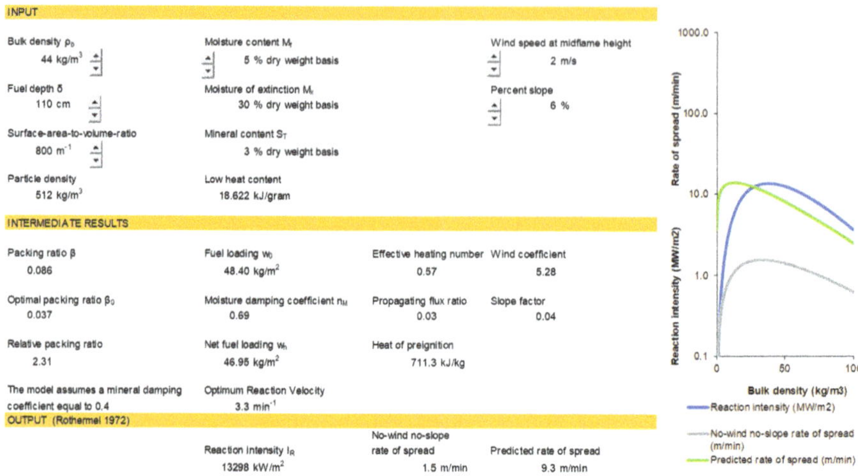

Fig. 7.38 As illustrated here, you can use our interactive spreadsheet, FIRE_RATE_OF_SPREAD_v2.0, to explore how changing the inputs, individually and together, alter the steady-state rate of fire spread predicted using the equations described in this chapter

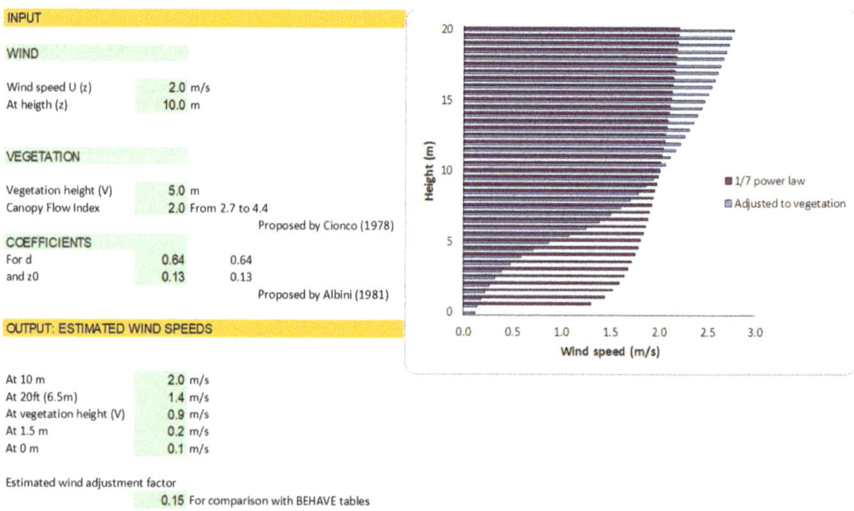

Fig. 7.39 Observed wind speeds vary at different heights above the ground and within and just above vegetation as illustrated here with a screen capture from our interactive spreadsheet, WINDPROFILE_v2.0. The wind speeds can be so altered by vegetation that they substantially change fire spread from what would be predicted from wind high above the ground without the influence of vegetation

in Fig. 7.4. Our spreadsheet, FIRE_RATE_OF_SPREAD_v2.0, is an interactive spreadsheet that uses Rothermel's (1972) equations as in Eq. (7.30) to derive a steady-state fire spread. You can use our spreadsheet, WINDPROFILE_v2.0, to estimate the wind speed at different heights above the ground. The predicted wind speed varies with wind and vegetation characteristics (Fig. 7.18).

We encourage you to deepen your understanding of the factors associated with fire propagation. With these interactive spreadsheets, you can visualize the effects of changing the inputs to the fire growth, fire spread, and vertical wind profile models. We suggest you compare the outputs of these spreadsheets with other predictions from other systems. However, we do not recommend using these spreadsheets for predicting fire behavior for fire operations and other applications. We designed these spreadsheets for learning, not science and management applications.

References

Albini, F. A. (1983). *Potential spotting distance from wind-driven surface fires* (Res Pap INT-309). Ogden: USDA Forest Service Intermountain Forest and Range Experiment Station.

Albini, F. A., & Baughman, R. G. (1979). *Estimating wind speeds for predicting wildland fire behavior* (Res Pap INT-RP-221). Ogden: USDA Forest Service, Intermountain Forest and Range Experiment Station.

Anderson, H. E. (1964). *Mechanisms of fire spread* (Research Progress report No 1. Res Pap INT-8). Ogden: USDA Forest Service Intermountain Forest and Range Experiment Station.

Anderson, H. E. (1969). *Heat transfer and fire spread* (Res Pap INT-69). Ogden: USDA Forest Service Intermountain Forest and Range Experiment Station.

Anderson, W. R., Cruz, M. G., Fernandes, P. M., McCaw, L., Vega, J. A., Bradstock, R., Fogarty, L., Gould, J., McCarthy, G., Marsden-Smedley, J. B., Matthews, S., Mattingley, G., Pearce, G., & van Wilgen, B. (2015). A generic, empirical-based model for predicting rate of fire spread in shrublands. *International Journal of Wildland Fire, 24*, 443–460.

Andrews, P. L. (2012). *Modeling wind adjustment factor and midflame wind speed for Rothermel's surface fire spread model* (Gen Tech Rep RMRS-GTR-266). Fort Collins, CO: USDA Forest Service Rocky Mountain Research Station.

Andrews, P. L., Cruz, M. G., & Rothermel, R. C. (2013). Examination of the wind speed limit function in the Rothermel surface fire spread model. *International Journal of Wildland Fire, 22*, 959–969.

Beer, T. (1991). The interaction of wind and fire. *Boundary-Layer Meteorology, 54*, 287–308.

Belcher, S. E., Jerram, N., & Hunt, C. R. (2003). Adjustment of a turbulent boundary layer to a canopy of roughness elements. *Journal of Fluid Mechanics, 488*, 369–398.

Burrows, N. D. (1999a). Fire behavior in jarrah forest fuels: 1. Laboratory experiments. *CALMScience, 3*, 31–56.

Burrows, N. D. (1999b). Fire behavior in jarrah forest fuels: 2. Field experiments. *CALMScience, 3*, 57–84.

Burrows, N., Gill, M., & Sharples, J. (2018). Development and validation of a model for predicting fire behaviour in spinifex grasslands of arid Australia. *International Journal of Wildland Fire, 27*, 271–279.

Butler, B., Quarles, S., Standohar-Alfano, C. D., Morrison, M., Jiménez, D., Sopko, P., Wold, C., Bradshaw, L. S., Atwood, L., Landon, J., O'Brien, J., Hornsby, B., Wagenbrenner, N., & Page, W. (2020). Exploring fire response to high wind speeds: Fire rate of spread, energy release and flame residence time from fires burned in pine needle beds under winds up to 27 m s^{-1}. *International Journal of Wildland Fire, 29*, 81–92.

Catchpole, W. R., Catchpole, E. A., Butler, B. W., Rothermel, R. C., Morris, G. A., & Latham, D. J. (1998). Rate of spread of free-burning fires in woody fuels in a wind tunnel. *Combustion Science and Technology, 1998*(1), 1–37.

Chandler, C., Cheney, P., Thomas, P., Trabaud, L., & Williams, D. (1983). *Fire in forestry*. New York: Wiley.

Cheney, N. P., & Gould, J. S. (1995). Fire growth in grassland fuels. *International Journal of Wildland Fire, 5*(4), 237–247.

Cheney, N. P., & Gould, J. S. (1997). Fire growth and acceleration (Letter to the editor). *International Journal of Wildland Fire, 7*(1), 1–5.

Cheney, N. P., Gould, J. S., & Catchpole, W. R. (1998). Prediction of fire spread in grasslands. *International Journal of Wildland Fire, 8*, 1–15.

Cheney, P., Gould, J., & McCaw, L. (2001). The dead-man zone—A neglected area of firefighter safety. *Australian Forestry, 64*(1), 45–50.

Cheney, N. P., Gould, J. S., McCaw, W. L., & Anderson, W. R. (2012). Predicting fire behaviour in dry eucalypt forest in southern Australia. *Forest Ecology and Management, 280*, 120–131.

Cheney, P., & Sullivan, A. (2008). Grassfires: fuel, weather and fire behaviour. Collingwood: CSIRO Publishing.

Cionco, R. M. (1965). A mathematical model for airflow in a vegetative canopy. *Journal of Applied Meteorology, 4*, 517–522.

Cionco, R. M. (1978). Analysis of canopy index values for various canopy densities. *Boundary-Layer Meteorology, 15*(1), 81–93.

Comissão Técnica Independente (CTI). (2017). *Análise e apuramento dos factos relativos aos incêndios que ocorreram em Pedrógão Grande, Castanheira de Pera, Ansião, Alvaiázere, Figueiró dos Vinhos, Arganil, Góis, Penela, Pampilhosa da Serra, Oleiros e Sertã, entre 17 e 24 de Junho de 2017*. Relatório preparado para a Assembleia da República. Lisboa, Portugal. Retrieved September 20, 2020, from https://www.parlamento.pt/Documents/2017/Outubro/Relat%C3%B3rioCTI_VF%20.pdf.

Cruz, M. G., & Alexander, M. E. (2013). Uncertainty associated with model predictions of surface and crown fire rates of spread. *Environmental Modelling and Software, 47*, 16–28.

Cruz, M. G., McCaw, W. L., Anderson, W. R., & Gould, J. S. (2013). Fire behaviour modelling in semi-arid mallee-heath shrublands of southern Australia. *Environmental Modelling and Software, 40*, 21–34.

Cruz, M. G., Gould, J. S., Alexander, M. E., Sullivan, A. L., McCaw, W. L., & Matthews, S. (2015). Empirical-based models for predicting head-fire rate of spread in Australian fuel types. *Australian Forestry, 78*, 118–158.

Cruz, M. G., Sullivan, A., Kidnie, S., Hurley, R., & Nichols, D. (2016). *The effect of grass curing and fuel structure on fire behaviour: Final report* (Rep No EP166414). Prepared for the Country Fire Authority of Victoria by Land and Water, CSIRO.

Cruz, M. G., Alexander, M. E., Sullivan, A. L., Gould, J. S., & Kilinc, M. (2018). Assessing improvements in models used to operationally predict wildland fire rate of spread. *Environmental Modelling and Software, 105*, 54–63.

Curry, J. R., & Fons, W. L. (1938). *Transfer of heat and momentum in the lowest layers of the atmosphere. Memo 65, 66*. London: Great Britain Meteorological Office Geophys.

Deeming, J. E., Lancaster, J. W., Fosberg, M. A., Furman, R. W., & Schroeder, P. (1972). *National fire-danger rating system* (Res Pap RM-84). Fort Collins: USDA Forest Service Rocky Mountain Forest and Range Experiment Station.

Detto, M., Katul, G. G., Siqueira, M., Juang, J. Y., & Stoy, P. (2008). The structure of turbulence near a tall forest edge: The backward-facing step flow analogy revisited. *Ecological Applications, 18*(6), 1420–1435.

Drysdale, D. (1999). *An introduction to fire dynamics* (3rd ed.). New York: Wiley.

Fendell, F. E., & Wolff, M. F. (2001). Wind-aided fire spread. Chapter 6. In E. A. Johnson & K. Miyanishi (Eds.), *Forest fires—Behaviour and ecological effects* (pp. 171–223). San Diego: Academic.

Fernandes, P. M. (2001). Fire spread prediction in shrub fuels in Portugal. *Forest Ecology and Management, 144*, 67–74.

Fernandes, P. M., Catchpole, W. R., & Rego, F. C. (2000). Shrubland fire behaviour modelling with microplot data. *Canadian Journal of Forest Research, 30*, 889–899.

Fernandes, P. M., Botelho, H. S., Rego, F. C., & Loureiro, C. (2008). Using fuel and weather variables to predict the sustainability of surface fire spread in maritime pine stands. *Canadian Journal of Forest Research, 38*, 190–201.

Fernandes, P. M., Botelho, H. S., Rego, F. C., & Loureiro, C. (2009). Empirical modelling of surface fire behaviour in maritime pine stands. *International Journal of Wildland Fire, 18*, 698–710.

Finney, M. A. (1998). *FARSITE: Fire area simulator—Model development and evaluation* (Res Pap RMRS-RP-4, revised 2004). Ogden: USDA Forest Service Rocky Mountain Research Station.

Finney, M. A. (2004). Landscape fire simulation and fuel treatment optimization. In J. L. Hayes, A. A. Ager & J. R. Barbour (tech Eds.), *Methods for integrating modeling of landscape change: Interior Northwest Landscape Analysis System* (Gen Tech Rep PNW-GTR-610, pp. 117–131). Portland: USDA Forest Service Pacific Northwest Station.

Finney, M. A. (2006). An overview of FlamMap fire modeling capabilities. In P. L. Andrews, B. W. Butler (comps) *Fuels Management—How to Measure Success: Conference Proceedings* (pp. 28–30) March 2006, Portland. *Proceedings RMRS-P-41* (pp. 213–220). Fort Collins: USDA Forest Service Rocky Mountain Research Station.

Finney, M. A. (2019). Fire acceleration. In S. Manzello (Ed.), *Encyclopedia of wildfires and wildland-urban interface (WUI) fires*. Cham: Springer.

Finney, M. A., & Ryan, K. C. (1995). Use of the FARSITE fire growth model for fire prediction in US National Parks. In *Proceedings of the International Emergency Management and Engineering Conference*, May 1995, Sofia Antipolis (pp. 183–189).

Finney, M. A., Cohen, J. D., Forthofer, J. M., McAllister, S. S., Gollner, M. J., Gorham, D. J., Saito, K., Akafuah, N. K., Adam, B. A., & English, J. D. (2015). Role of buoyant flame dynamics in wildfire spread. *Proceedings of the National Academy of Sciences, 112*(32), 9833–9838.

Forestry Canada Fire Danger Group (FCFDG). (1992). *Development and structure of the Canadian Forest Fire Behavior Prediction System* (Information Report ST-X-3). Ottawa: Forestry Canada.

Forthofer, J., Shannon, K., & Butler, B. (2009). *4.4 simulating diurnally driven slope winds with windninja*. Missoula: USDA Forest Service Rocky Mountain Research Station.

Frandsen, W. H. (1971). Fire spread through porous fuels from the conservation of energy. *Combustion and Flame, 16*, 9–16.

Frandsen, W. H. (1973). *Effective heating ahead of spreading fire* (Res Pap INT-140). Ogden: USDA Forest Service Intermountain Forest and Range Experiment Station.

Fritschen, L. J. (1985). Characterization of boundary conditions affecting forest environmental phenomena. In B. A. Hutchison & B. B. Hicks (Eds.), *The forest-atmosphere interaction* (pp. 3–23). Dordrecht: Springer.

Giannino, F., Ascoli, D., Sirignano, M., Mazzoleni, S., Russo, L., & Rego, F. C. (2017). A combustion model of vegetation burning in "Tiger" fire propagation tool. In *Proceedings of the International Conference of Computational Methods in Sciences and Engineering (ICCMSE-2017)* (vol. 1906(1), p. 100007).

Grishin, A. M. (1996). Mathematical modelling of forest fires. In J. G. Goldammer & V. V. Furyaev (Eds.), *Fire in ecosystems of boreal Eurasia* (Forestry sciences) (Vol. 40, pp. 285–302). Dordrecht: Springer.

Heikkila, T. V., Gronqvist, R., & Jurvélius, M. (2010). *Wildland fire management. Handbook for trainers*. Rome: FAO.

Hiers, J. K., O'Brien, J. J., Mitchell, R. J., Grego, J. M., & Loudermilk, E. L. (2009). The wildland fuel cell concept: An approach to characterize fine-scale variation in fuels and fire in frequently burned longleaf pine forests. *International Journal of Wildland Fire, 18*(3), 315–325.

Hoffman, C. M., Linn, R., Parsons, R., Sieg, C., & Winterkamp, J. (2015). Modeling spatial and temporal dynamics of wind flow and potential fire behavior following a mountain pine beetle outbreak in a lodgepole pine forest. *Agricultural and Forest Meteorology, 204*, 79–93.

Hwang, C. C., & Xie, Y. (1984). Flame propagation along matchstick arrays on inclined base boards. *Combustion Science and Technology, 42*(1–2), 1–12.

Keane, R. E., Garner, J. L., Schmidt, K. M., Long, D. G., Menakis, J. P., & Finney, M. A. (1998). *Development of input data layers for the FARSITE fire growth model for the Selway-Bitterroot Wilderness Complex, USA* (Gen Tech Rep RMRS-GTR-3). Ogden: USDA Forest Service Rocky Mountain Research Station.

Keane, R. E., Gray, K., Bacciu, V., & Leirfallom, S. (2012). Spatial scaling of wildland fuels for six forest and rangeland ecosystems of the northern Rocky Mountains, USA. *Landscape Ecology, 27*(8), 1213–1234.

Kreye, J. K., Brewer, N. W., Morgan, P., Varner, J. M., Smith, A. M. S., Hoffman, C. M., & Ottmar, R. D. (2014). Fire behavior in masticated fuels: A review. *Forest Ecology and Management, 314*, 193–207.

Kunkel, K. E. (2001). Surface energy budget and fuel moisture. Chapter 9. In E. A. Johnson & K. Miyanishi (Eds.), *Forest fires—Behaviour and ecological effects* (pp. 303–350). San Diego: Academic.

Linn, R. R., Sieg, C. H., Hoffman, C. M., Winterkamp, J. L., & McMillin, J. D. (2013). Modeling wind fields and fire propagation following bark beetle outbreaks in spatially-heterogeneous pinyon-juniper woodland fuel complexes. *Agricultural and Forest Meteorology, 173*, 139–153.

Luke, R. H., & McArthur, A. G. (1978). *Bushfires in Australia*. Canberra: Australian Government Publishing Service.

Mallet, V., Keyes, D. E., & Fendell, F. E. (2009). Modeling wildland fire propagation with level set methods. *Computers & Mathematics with Applications, 57*(7), 1089–1101.

Marino, E., Guijarro, M., Madrigal, J., Hernando, C., & Díez, C. (2008). Assessing fire propagation empirical models in shrub fuel complexes using wind tunnel data. In J. De Las Heras, C. A. Brebbia, D. Viegas, & V. Leone (Eds.), *Modelling, monitoring and management of forest fires, WIT transactions on ecology and the environment* (Vol. 119, pp. 121–130). Southampton: WIT Press.

Marsden-Smedley, J. B., & Catchpole, W. R. (1995). Fire behaviour modelling in Tasmanian buttongrass moorlands II. Fire behaviour. *International Journal of Wildland Fire, 5*, 215–228.

Mazzoleni, S., & Heathfield, D. (2017). *The heritage of Fire Paradox project: Tiger propagation models as training and prevention tools*. Università di Napoli Federico II. AF3 Forest Fire Models Workshop, Rome. Retrieved February 10, 2020, from http://www.vigilfuoco.it/aspx/download_file.aspx?id=23228.

McAlpine, R. S., & Wakimoto, R. H. (1991). The acceleration of fire from point source to equilibrium spread. *Forest Science, 37*(5), 1314–1337.

McArthur, A. G. (1967). *Fire behaviour in eucalypt forests* (Leafl. No. 107). Canberra. Australian Capital Territory: Commonwealth of Australia, Forestry and Timber Bureau, Forest Research Institute.

McArthur, A. G. (1969). The Tasmanian bushfires of 7th February, 1967, and associated fire behaviour characteristics. In *The Technical Cooperation Programme. Mass Fire Symposium*, Canberra 1969. Melbourne: Defense Standard Labs.

McRae, D. J. (1999). Point-source fire growth in jack pine slash. *International Journal of Wildland Fire, 9*(1), 65–77.

Miyanishi, K. (2001). Duff consumption. Chapter 13. In E. A. Johnson & K. Miyanishi (Eds.), Forest fires—Behaviour and ecological effects (pp. 437–475). San Diego: Academic.

Noble, I. R., Gill, A. M., & Bary, G. A. V. (1980). McArthur's fire-danger meters expressed as equations. *Australian Journal of Ecology, 5*, 201–203.

Pagni, P. J., & Peterson, T. G. (1973). Flame spread through porous fuels. In *International Symposium on Combustion* (vol. 14(1), pp. 1099–1107). Amsterdam: Elsevier.

Palmer, K. N. (1957). Smoldering combustion in dusts and fibrous materials. *Combustion and Flame, 1*, 129–154.

Parisien, M., Kafka, V., Hirsch, K. G., Todd, J. B., Lavoie, S. G., & Maczek, P. D. (2005). *Mapping wildfire susceptibility with the BURN-P3 simulation model* (Inf Rep NOR-X-405). Edmonton: Nat Res Canada, Can For Serv, Northern For Centre.

Parsons, R., Linn, R., Pimont, F., Hoffman, C., Sauer, J., Winterkamp, J., Sieg, C. H., & Jolly, W. (2017). Numerical investigation of aggregated fuel spatial pattern impacts on fire behavior. *Land, 6*(2), 43.

Pimont, F., Dupuy, J. L., Linn, R. R., & Dupont, S. (2011). Impacts of tree canopy structure on wind flows and fire propagation simulated with FIRETEC. *Annals of Forest Science, 68*(3), 523.

Quintiere, J. G. (1998). *Principles of fire behavior*. Clifton Park: Delmar Publishers.

Rein, G., Cleaver, N., Ashton, C., Pironi, P., & Torero, J. L. (2008). The severity of smouldering peat fires and damage to the forest soil. *Catena, 74*(3), 304–309.

Rossa, C. G. (2017). The effect of fuel moisture content on the spread rate of forest fires in the absence of wind or slope. *International Journal of Wildland Fire, 26*(1), 24–31.

Rossa, C. G., & Fernandes, P. M. (2018). Empirical modeling of fire spread rate in no-wind and no-slope conditions. *Forest Science, 64*(4), 358–370.

Rothermel, R. C. (1972). *A mathematical model for predicting fire spread in wildland fuels* (Res Pap INT-115). Ogden: USDA Forest Service Intermountain Forest and Range Experiment Station.

Rothermel, R. C. (1983). *How to predict the spread and intensity of forest and range fires* (Gen Tech Rep INT-143). Ogden: USDA Forest Service Intermountain Forest and Range Experiment Station.

Rothermel, R. C. (1991). *Predicting behavior and size of crown fires in the Northern Rocky Mountains* (Res Pap INT-438). Ogden: USDA Forest Service Intermountain Forest and Range Experiment Station.

Rothermel, R. C., & Anderson, H. E. (1966). *Fire spread characteristics determined in the laboratory* (Res Pap INT-30). Ogden: USDA Forest Service Intermountain Forest and Range Experiment Station.

Schroeder, M. J., & Buck, C. C. (1970). *Fire Weather. Agricultural Handbook 360*. Washington, DC: USDA Forest Service.

Sheshukov, M. A. (1970). Effect of steepness of slope on the propagation rate of fire. *Lesnoe Khozyaystvo, 1970*(1), 50–54. Translation 185672 Forest Fire Res Inst. Ottawa.

Sneeuwjagt, R. J., & Peet, G. B. (1985). *Forest fire behaviour tables for western Australia* (3rd ed.). Perth: Western Australia Department of Conservation and Land Management.

Stocks, B. J., Lawson, B. D., Alexander, M. E., Van Wagner, C. E., McAlpine, R. S., Lynham, T. J., & Dube, D. E. (1989). The Canadian forest fire danger rating system: An overview. *The Forestry Chronicle, 65*(6), 450–457.

Sullivan, A. L., Sharples, J. J., Matthews, & Plucinski, M. P. (2014). A downslope fire spread correction factor based on landscape-scale fire behaviour. *Environmental Modelling and Software, 62*, 153–163.

Telitsyn, H. P. (1973). Flame radiation as a mechanism of fire spread in forests. In N. H. Afgan & J. M. Beer (Eds.), *Heat transfer in flames* (Vol. 2, pp. 441–449). New York: Wiley.

Telitsyn, H. P. (1996). A mathematical model of spread of high-intensity forest fires. In J. G. Goldammer & V. V. Furyaev (Eds.), *Fire in ecosystems of boreal Eurasia* (Forestry sciences) (Vol. 40, pp. 314–325). Dordrecht: Springer.

Thom, A. S. (1975). Momentum, mass and heat exchange of plant communities. In J. L. Monteith (Ed.), *Vegetation and the atmosphere* (pp. 57–109). London: Academic.

Thomas, P. H. (1965). *The contribution of flame radiation to fire spread in forests. Fire Research Note 594*. Boreham Wood: Joint Fire Research Organisation.

Thomas, P. H. (1967). Some aspects of the growth and spread of fire in the open. *Forestry, 40*, 139–164.

Thomas, P. H. (1971). Rates of spread of some wind-driven fires. *Forestry, 44*, 155–175.

Thomas, P. H., Simms, D. L., & Wraight, H. G. H. (1964). *Fire spread in wooden cribs. Fire Res Note 537*. Boreham Wood: Joint Fire Research Organisation.

Tolhurst, K., Shields, B., & Chong, D. (2008). Phoenix: Development and application of a bushfire risk management tool. *Australian Journal of Emergency Management, 23*, 47.

Troen, I., & Petersen, E. L. (1989). *European wind atlas*. Roskilde: Risø National Laboratory.

Tymstra, C. R., Bryce, W., Wotton, B. M., Taylor, S. W., & Armitage, O. B. (2010). *Development and structure of Prometheus: The Canadian wildland fire growth simulation model* (Rep NOR-X-417). Edmonton: Can For Serv, Nat Res Canada, North For Centre.

University Corporation of Atmospheric Research (UCAR). (2012). *National Wildlife Coordinating Group. Unit 12: Gauging Fire Behavior and Guiding Fireline Decisions*. Retrieved February 10, 2020, from http://stream1.cmatc.cn/pub/comet/FireWeather/S290Unit12GaugingFireBehaviorandGuidingFirelineDecisions/comet/fire/s290/unit12/print_2.htm.

Vakili, E., Hoffman, C. M., Keane, R. E., Tinkham, W. T., & Dickinson, Y. (2016). Spatial variability of surface fuels in treated and untreated ponderosa pine forests of the southern Rocky Mountains. *International Journal of Wildland Fire, 25*(11), 1156–1168.

Van Wagner, C. E. (1967). *Calculation of forest fire spread by flame radiation* (Rep No. 1185). Chalk River: Canadian Department of Forestry and Rural Devel, Petawawa Forest Experiment Station.

Van Wagner, C. E. (1977a). Conditions for the start and spread of crown fire. *Canadian Journal of Forest Research, 7*, 23–34.

Van Wagner, C. E. (1977b). Effect of slope on fire spread rate. Bi-monthly research notes. *Fisheries and Environment Canada Forestry Services, 33*(1), 7–8.

Van Wagner, C. E. (1985). Fire spread from a point source. Memo PI-4-20 dated January 14, 1985 to P.H. Kourtz (unpublished). Chalk River: Canadian Forest Service Petawawa National Forest Institute.

Vega, J. A., Cuinas, P., Fonturbel, T., Pérez-Gorostiaga, P., & Fernandez, C. (1998). Predicting fire behaviour in Galician (NW Spain) shrubland fuel complexes. In D. X. Viegas (Ed.), *Proceedings of the 3rd International Conference on Forest Fire Research & 14th Fire and Forest Meteorology Conference*, Coimbra, November 1998 (pp. 713–728). Coimbra: ADAI, University of Coimbra.

Viegas, D. X. (2004). Slope and wind effects on fire propagation. *International Journal of Wildland Fire, 13*, 143–156.

Wagenbrenner, N. S., Forthofer, J. M., Lamb, B. K., Shannon, K. S., & Butler, B. W. (2016). Downscaling surface wind predictions from numerical weather prediction models in complex terrain with WindNinja. *Atmospheric Chemistry and Physics, 16*, 5229–5241. https://doi.org/10.5194/acp-16-5229-2016.

Weatherspoon, C. P., Almond, G. A., & Skinner, C. N. (1989). Treecentered spot firing—A technique for prescribed burning beneath standing trees. *Western Journal of Applied Forestry, 4*(1), 29–31.

Weber, R. O. (1989). Analytical models for fire spread due to radiation. *Combustion and Flame, 78*, 398–408.

Weber, R. O. (2001). Wildland fire spread models. In E. A. Johnson & K. Miyanishi (Eds.), *Forest fires. Behavior and ecological effects* (pp. 151–170). San Diego: Academic.

Weber, R. O., & Mestre, N. D. (1990). Flame spread measurements on single ponderosa pine needles: Effect of sample orientation and concurrent external flow. *Combustion Science and Technology, 70*(1–3), 17–32.

Weise, D. R., & Biging, G. S. (1997). A qualitative comparison of fire spread models incorporating wind and slope effects. *Forest Science, 43*(2), 170–180.

Weise, D. R., Cobian-Iñiguez, J., & Princevac, M. (2018). Surface to crown transition. In S. L. Manzello (Ed.), *Encyclopedia of wildfires and wildland-urban interface (WUI) fires*. Cham: Springer.

Wells, G. (2008). *The Rothermel fire-spread model: Still running like a champ*. Fire Sci Digest 2. US Joint Fire Science Program, Boise ID. Retrieved September 21, 2020, from https://www.firescience.gov/Digest/FSdigest2.pdf.

Williams, F. (1977) Mechanisms of fire spread. In *16th International Symposium (International) on Combustion* (pp. 1281–1294). Pittsburgh: The Combustion Institute.

Chapter 8
Extreme Fires

Learning Outcomes
After reading this chapter, we expect you to be able to

1. Compare and contrast the definitions of extreme fires, using examples from around the globe to illustrate,
2. Explain using your own words how altering fuel conditions can alter the probability of a surface fire becoming a crown fire,
3. Identify how even within extreme fires, patches of fuels may remain unburned, and
4. Using the interactive spreadsheets, explain which factors are most important in increasing the potential for spotting and rapidly spreading crown fires that qualify as being extreme in their behavior.

8.1 Introduction: Extreme Fires

Extreme fires are generally viewed as those that have great social, economic, and political impacts. Extreme fires often threaten many people, many homes, and many other valued assets at the same time. They are often spectacular events that garner great media attention. Though few extreme fires occur, they burn more area, cost more money, and have more social, economic, and ecological impacts than many other fires combined. In years of widespread fires, many fires burn large areas, and fire suppression resources are stretched thin. Further, where there is one extreme fire, there are commonly many large fires burning at the same time within a region. Many

Supplementary Information The online version of this chapter (https://doi.org/10.1007/978-3-030-69815-7_8) contains supplementary material, which is available to authorized users.

F. Castro Rego et al., *Fire Science*, Springer Textbooks in Earth Sciences, Geography and Environment, https://doi.org/10.1007/978-3-030-69815-7_8

fires may ignite at the same time from lightning associated with dry cold fronts with erratic and changing wind direction as the weather fronts move across large areas. Ignitions by people, whether accidental or arson, are more likely when it is hot, dry, and windy. This magnifies the social and economic impacts. More extreme fires are forecast for the future worldwide as more warm droughts and more lightning are expected with changing climate. Also, more fires will probably have more extreme social and political implications as more people and their homes will be in the path of the fires (See Chap. 10 for more on expanding wildland-urban interface in many areas). Changes in national policies are often triggered by extreme fires. This was the case of the comprehensive legislative reform after the catastrophic fire seasons of 2003 and 2005 in Portugal (Mateus and Fernandes 2014), and has occurred worldwide. Extreme fires are opportunities to implement fully measures that previously were not generally accepted (Fernandes et al. 2017). Because of the relevance of their impacts and consequences we all are aware of the importance of extreme fires. What do we mean exactly by labeling some fires extreme?

The term "extreme fires" may refer to very different aspects, from fires with significant social, economic, or ecological impacts to especially large fires (Lannom et al. 2014) or fires exhibiting unusual behavior (Tedim et al. 2018). Different approaches have been proposed in the definition of extreme fires, based on statistics, on the magnitude of impacts, on the difficulty of control, or on fire behavior characteristics.

Extreme fires have characteristics that "go beyond those exhibited by the majority of fires" (Pyne et al. 1996). Statistically, extreme fires can be defined as those exceeding a given threshold with the 99th percentile or 100-year return period since a previous fire being the most commonly identified (McPhillips et al. 2018). However, this interpretation requires the previous choice of the characteristic (s) considered, which could be fire size, intensity, severity, or rate of spread. Under this interpretation, the definitions of extreme events should not be conflated with their impacts or effects.

Extreme fires can also be defined as those having significant impacts on people, buildings, and other properties and infrastructure. When they greatly affect people, these extreme wildfire events and their impacts are often described as "megafires", "disasters", "firestorms", and "catastrophes". This approach has been used by many authors, with extreme fires being considered as those declared as a disaster by a national government, by having caused human fatalities or burned many homes. Of particular concern is the considerable loss of human life in fires in recent years, including Australia (2009, 2020), Greece (2007, 2018), Portugal (2017), and the USA (2018, 2020). Because of these impacts, extreme fires have been identified by the difficulty of direct fire control methods. The thresholds are then associated with the limits where fire control is considered to be ineffective. Thus, the National Wildfire Coordination Group (NWCG 2020) indicates that "extreme implies a level of fire behavior characteristics that ordinarily precludes methods of direct control action". By these criteria, extreme fires are those exceeding fire suppression thresholds.

Extreme fires may be identified based on fire behavior. In a synthesis of knowledge of extreme fire behavior, Potter and Werth (2016) indicated that extreme fire behavior should be considered as "fire spread other than steady surface spread,

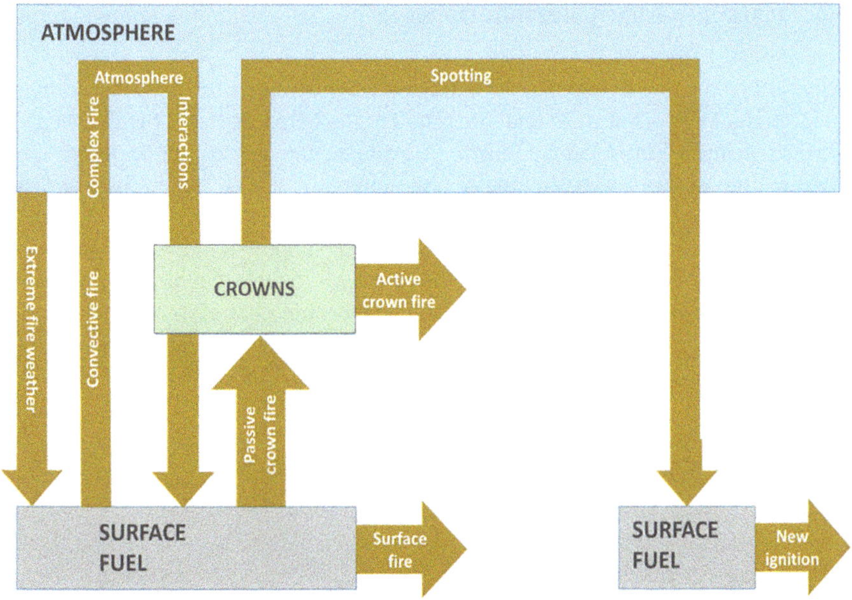

Fig. 8.1 The various processes potentially involved in extreme fires. Extreme fire weather combined with suitable surface fuel may result in extreme surface fire behavior, as characterized by high rates of spread, fireline intensity, and large fire sizes. When fuels are organized in a surface layer and a canopy or crown layer, fires may be sufficiently intense to enable crowning that, under some conditions, will develop as an active crown fire. Spotting may also occur, causing new ignitions ahead of the main fire front. Finally, complex fire-atmosphere interactions might develop, causing unexpected fire behavior

especially when it involves rapid increases". This definition, which is akin to the recent concept of dynamic fire behavior (e.g., Lahaye et al. 2018), indicates that extreme fire behavior is beyond the quasi steady-state surface fire spread that was the object of the preceding chapters of this book. Under this interpretation, extreme fire behavior typically includes crowning, spotting, and complex fire-atmosphere interactions that are not easily predictable, defying fire suppression and imposing safety concerns. According to NWCG (2020), in extreme fires, "one or more of the following is usually involved: high rate of spread, prolific crowning and/or spotting, presence of fire whirls, and strong convection column. Predictability is difficult because such fires often exercise some degree of influence on their environment and behave erratically, sometimes dangerously".

In this chapter, we consider extreme fires in these various interpretations. First, we consider extreme fire characteristics (fire size, fireline intensity, rate of spread) in relation to different thresholds and drivers, including extreme fire weather. Second, we consider separately the various phenomena beyond those involved in surface fires. These phenomena are crown fires (both passive and active), spotting, and complex fire-atmosphere interactions. Although these phenomena do not imply that fires are extreme, extreme fires commonly involve one or more of these phenomena (Fig. 8.1).

8.2 Extreme Fire Characteristics

Extreme fires can be defined as those having extreme characteristics. The simplest characteristic often used is the final fire size. The "extremeness" of a fire event can be assessed using a statistical approach to evaluate the degree of departure from "normal" fire sizes in a given spatial and temporal context. Other characteristics associated with fire behavior, such as fireline intensity and spread rate are of more importance when the focus is the difficulty of fire suppression and socioeconomic damages. When the main interest is in the ecological effects, characteristics such as burn severity might be used. We discuss extreme fire characteristics as fire size, intensity, and spread rate in the next sections.

8.2.1 Extreme Fire Size: The Statistical Approach

From the statistical point of view, extreme events are those in which some characteristics are greater than a "normal" or typical fire. This statistical approach is used in the analysis of extreme events in other hazards, such as droughts or floods. For these hazards, a standard precipitation index (SPI) is often used to determine the extremeness of a precipitation event. The precipitation data are typically fitted to a Gamma or a Pearson Type III distribution and then transformed to a normal distribution. The SPI values can be interpreted as the number of standard deviations by which the observed anomaly deviates from the long-term mean. An extreme event is defined as deviating more than two standard deviations from the mean (extremely dry when $SPI < -2$ or extremely wet when $SPI > 2$) as shown in Fig. 8.2. Equivalent analyses using the statistical approach can also be used to define extreme fires.

In defining extreme fires from a statistical point of view, we are generally interested in large sizes of fires as the basis to look at statistical extremes. Modeling large forest fires as extreme events can be done by looking at the probability distributions of fire size and, in particular, the probability of fires exceeding a given threshold. Different types of distributions have been applied, including Gumbel, Frechet, Weibull, or Pareto distributions, with or without truncation, used by Alvarado et al. (1998) in the statistical analysis of large fires from 1961 to 1981 for the Province of Alberta, Canada. Moritz (1997) used a similar approach, with a generalized cumulative distribution of extremes to characterize the changes of the extremal fire regime of Los Padres National Forest on the central coast of California from 1911 to 1991. The spatial and temporal extremes of wildfires in Portugal from 1984 to 2004 were modeled using extreme value statistics based on the generalized Pareto distribution (de Zea Bermudez et al. 2009). An elegant approach to look at the size distribution of wildfires uses the analogy of a sandpile at a critical state where the adding of new sand grains causes avalanches of different sizes with a distribution

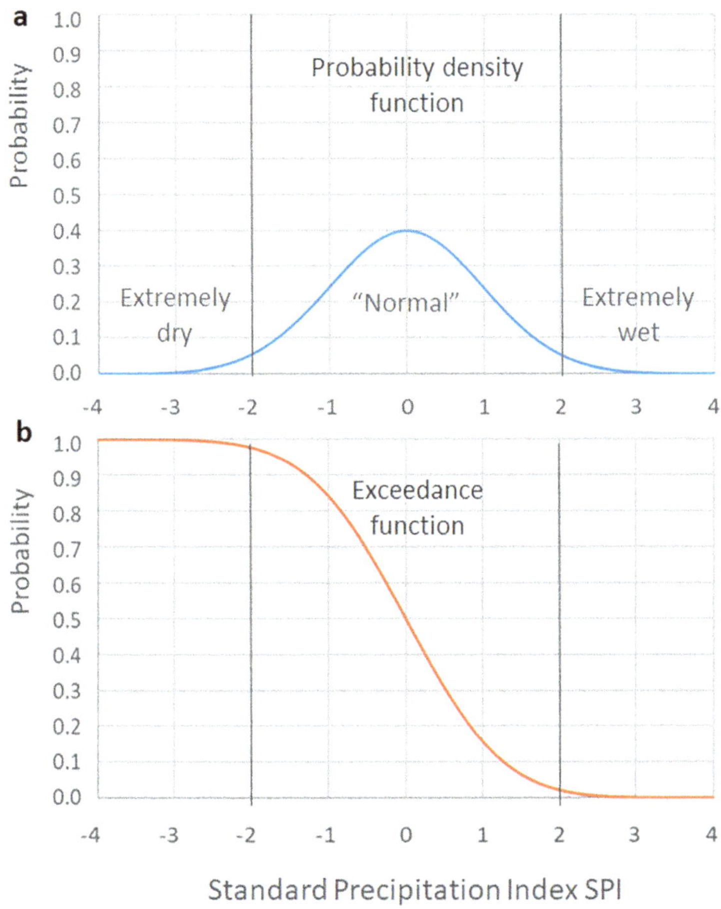

Fig. 8.2 Statistical analysis of the distribution of precipitation using a normal distribution after transformation of the original distribution of precipitation. (**a**) Probability distribution function and (**b**) Exceedance function, the probability that a precipitation event exceeds a given value of the Standard Precipitation Index (SPI). SPI thresholds for extreme events are those with precipitation more than two standard deviations from the mean), indicating a strong deviation from the mean of the adjusted "normal" curve. A value of SPI above 2 represents an extremely wet situation which occurs only in 2.3% of the cases. Similarly, an extremely dry situation, which also occurs in 2.3% of the cases, is defined by a value of SPI below 2

with a characteristic power law. This concept was proposed and used for the fires in Liguria (Italy) from 1986 to 1993 by Ricotta et al. (1999). See the following example.

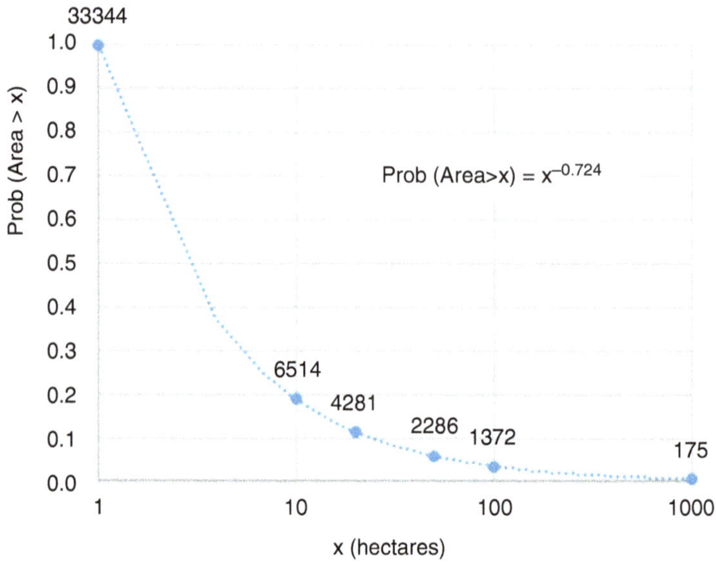

Fig. 8.3 Plot of the size distribution of wildfires (>1 ha) in Portugal (2010–2019) showing the probabilities of exceedance of different size thresholds. Of the 33,344 wildfires recorded, only 175 fires exceeded 1000 ha in area burned

Using the power law suggested by Ricotta et al. (1999) and the data from the Portuguese Forest Services from the 33,344 wildfires (above 1 ha) recorded between 2010 and 2019 we can model the probabilities of fire size exceeding a certain threshold x as:

$$Prob\ (Area > x) = x^{-b} \tag{8.1}$$

where x is the area threshold and b is a scaling coefficient.

The size thresholds used were 10, 20, 50, 100, and 1000 ha. The probabilities were estimated as the number of fires exceeding the threshold as a proportion of the total number of fires (Fig. 8.3).

With the equation fitted, it is possible to compute the probability of exceedance for any fire size. By reversing Eq. (8.1), one can estimate the approximated fire size for any given probability of exceedance.

Fire size is an important characteristic because it is often related to other characteristics more difficult to measure, such as fireline intensity or rate of spread. Larger fires do have high rates of spread (Fig. 8.4), even if the size of increasingly larger fires is less and less correlated with the fire-spread rate. This is partly because most of the area burned results from one or more short-duration fire growth periods, and thus the mean rate of spread may not capture the actual rate of spread associated with

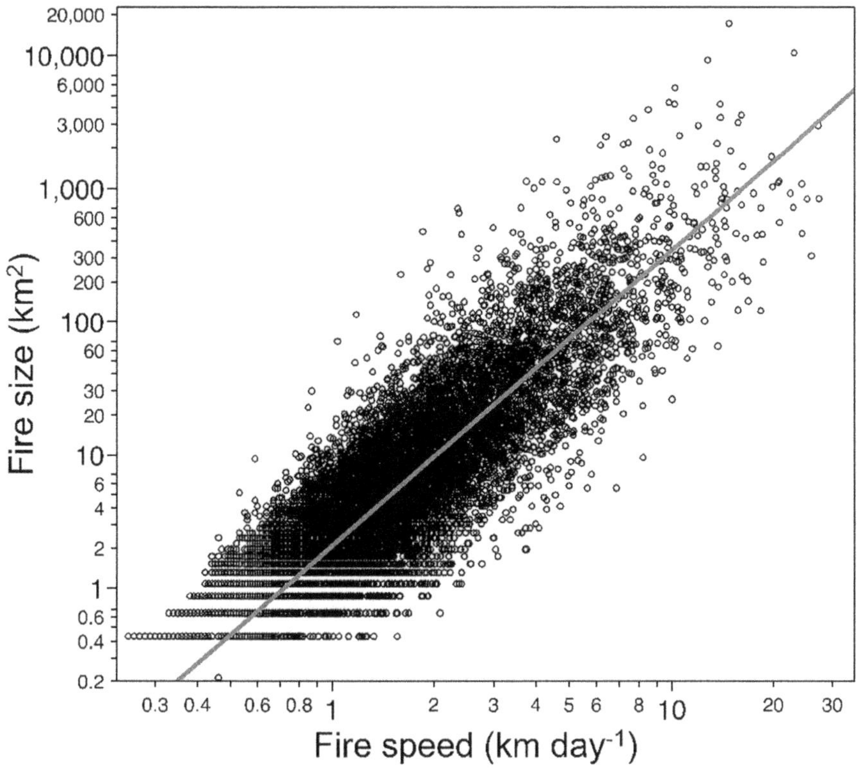

Fig. 8.4 Relationship between log-transformed fire size and fire speed for open shrubland fires (≥21 ha) in the Global Fire Database (Andela et al. 2019) for the year 2017. Data from http://www.globalfiredata.org/fireatlas.html. Despite the strength of the overall relationship ($R^2 = 0.84$, $n = 21{,}712$), variation in fire size is increasingly less explained by fire speed as fires become larger, with R^2 of 0.72, 0.28, 0.15, and 0.03 for size classes of respectively <100, 100–1000, 1000–10,000, and >10,000 km^2

most of the burned area. Laurent et al. (2019) found that for fires from all around the globe, fire size and fire intensity are related. However, they varied proportionally in temperate and boreal forests only, with the relationship saturating at moderate intensity values in the other biomes.

Another fire characteristic that may be used to define the thresholds of extreme fires is fire radiative power (FRP, see Chap. 6), the rate of emitted radiative energy by the fire at the time of the observation. FRP is associated with other fire characteristics such as fireline intensity (Wooster et al. 2005) and it has been used by various authors in their analyses (e.g., Laurent et al. 2019). Bowman et al. (2017) used these measurements to define extreme wildfire events as those reaching the 99.997th percentile of the cumulative FRP in a 10 × 10 km global grid (2002–2013).

The socioeconomic and environmental impacts of wildfire are often implicitly assumed to be a function of fire size rather than the outcome of extreme fire behavior.

However, the impacts of fire are much more related to other fire characteristics (Moreira et al. 2020). From the ecological point of view, the impacts are much more related to severity, duration, and fuel consumption than to size or fireline intensity (see Chap. 9). Also, social and economic impacts are especially related to the difficulty of wildfire control, which is more dependent on spread rate and fireline intensity. On the other hand, while very large fires occur in all Earth's flammable biomes, most of the fires that became disasters are located in the western USA, southern Europe, and southeastern Australia, mostly burning in the interface between urban areas and flammable forests (see Chap. 10). The statistical definition of extreme fires depends completely on the fire characteristics or the impacts of interest in each analysis.

Another obvious limitation of the use of statistical thresholds is that they depend not only on the fire characteristic considered (fire size, rate of spread, fireline intensity, fire radiative power, or other) and/or on the impact of interest (social, economic, ecological) but also on the temporal and spatial context of the analysis. Using the statistical approach, fires of similar characteristics or impacts might be considered as "extreme" in one region but not in another region, depending on the context. Other approaches using different thresholds have been proposed and used in practice.

8.2.2 *Extreme Fire Behavior: The Resistance to Control Approach, Features, and Drivers*

Fire suppression thresholds are dependent on the possibilities of fire control and thus are less dependent on geographical context than statistical thresholds. Schroeder and Chandler (1966) used several thresholds to distinguish four types of fire behavior: fire out, when sustained ignition is not possible; no spread, if ignition occurs but the fire won't spread; actionable, when fire control efforts can succeed; and critical, if effective fire suppression is unlikely.

Fire-spread rate determines the number and type of firefighting resources needed to confine fire growth to a given perimeter or size. However, fireline intensity is ordinarily adopted to define thresholds to categorize and describe how difficult it is to control a spreading fire (Hirsch and Martell 1996). Rules of thumb based on fireline intensity are part of decision-making aids for fire management and evolved from work carried out in the 1960s in the frame of prescribed burning in eucalypt forests (Hodgson 1968). Notwithstanding some variation, the existing tables usually indicate that control by direct suppression methods is precluded at a fireline intensity of ~4000 kW m^{-1}, a tentative lower limit for crowning in conifer forest, although containment lines will be challenged at lower intensities, especially if spotting occurs (Alexander 2000).

Extreme fire behavior comes in various degrees. For example, in jack pine (*Pinus banksiana*) forest in Canada, 2000–4000 kW m^{-1} corresponds to 'very vigorous or extremely intense surface fire', while 'violent physical behavior' is probable at 30,000 kW m^{-1} (Alexander and De Groot 1988). Fireline intensity varies over five orders of magnitude, which, in a given fuel type, will primarily and secondarily be a function of the forward fire rate of spread and fuel consumption, respectively. Consequently, extreme fire behavior is largely dependent on how fast a wildfire can spread. Extreme fire behavior, as described by its rate of spread, flame size, and heat release, is typically observed under surface atmospheric conditions of strong wind, plus high temperature and low relative humidity, which bring the moisture content of fine dead fuels to low levels. Daily weather patterns characterized by the overnight persistence of relatively high temperature and relatively low relative humidity are particularly critical in this regard (Werth et al. 2016). Drought conditions ensuing from extended rainless periods add to the amount of fuel available for combustion, including coarse woody fuels, deep duff, or greener vegetation that otherwise would not burn, and homogenize fuel moisture across the landscape. Then fuels on most sites become available to burn, even sites that otherwise are quite mesic and many that have abundant biomass because they are quite productive.

Extreme fire behavior is always associated with extreme conditions of the fire drivers, those described using the notion of the fire environment triangle that comprises fuels, the atmosphere (weather), and topography (Countryman 1972). This concept allows fires to be perceived and classified as wind-, topography- or fuel-driven, with their respective dominances often shifting over time and space. It also suggests that fire behavior is an outcome not just of the individual influences but also of the interactions between the three types of factors, and even more so in the case of extreme and large fires. For example, those forces can be negatively aligned, as when a fire front spreads downslope and against the wind or, on the contrary, act synergistically, as in the case of an upslope fire run with slope and wind aligned in the same direction.

Wind is the main engine of fire spread, and the drastic changes to fire behavior imposed by approaching cold fronts or thunderstorms are well known. The response of fire-spread rate to wind speed is near-linear, as shown by contemporary empirical models in a variety of vegetation types (Cruz et al. 2015). However, the influence of fine dead fuel moisture content on the fire rate of spread is exponential. In dry eucalypt forests, small decreases in fuel moisture content at its low end ($< 8\%$) can escalate fire spread rate and flame height when combined with high wind or high fuel hazard (Fig. 8.5).

Increasingly steeper terrain will further increase the rate of spread, at least during initial fire development and on a local scale (Sullivan 2017). As a rule of thumb, the rate of spread doubles for every 10° increase in terrain slope. Complex topography is often a significant contributor to extreme fire behavior not just because of direct slope effects but also because topography interacts with wind and other weather elements (Sharples 2009):

1. Spatiotemporal variability in temperature and humidity due to topographically-determined variation in insolation and elevation, including nocturnal temperature inversions and cold-air drainage.

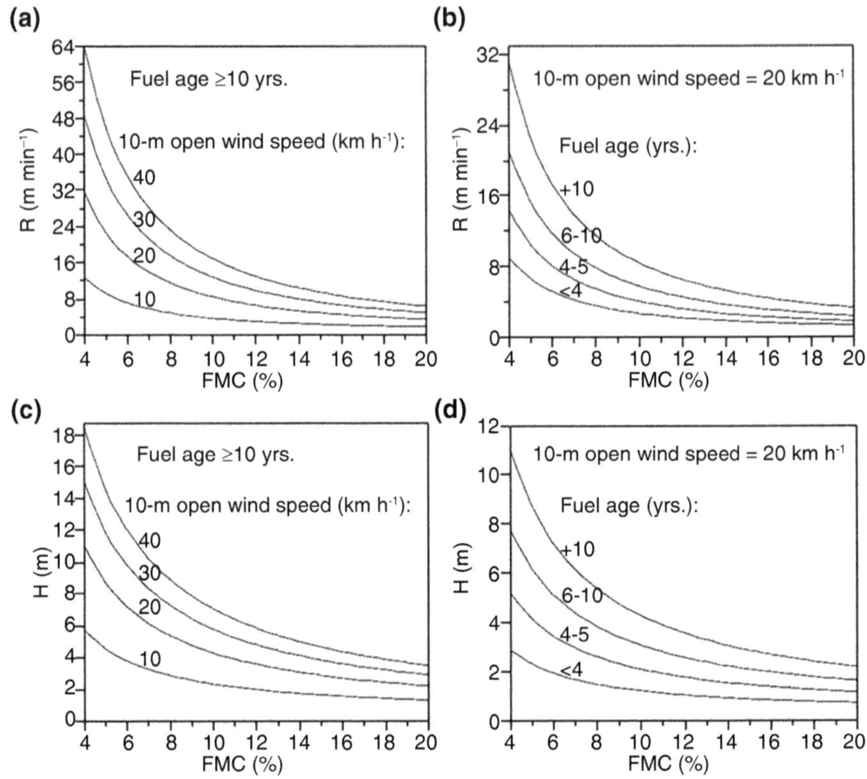

Fig. 8.5 Variation in the forward rate of fire spread (R) and flame height (H) in dry eucalypt forest and flat terrain as a function of dead fine fuel moisture content (FMC) and for constant (**a**, **c**) and variable (**b**, **d**) fuel age and constant (**b**, **d**) and variable (**a**, **c**) 10-m wind speed. Fuel age is a proxy for the effects of fuel hazard scores (surface, near-surface, and elevated layers) on fire behavior, assuming typical values for each age class. (Drawn from models in Cheney et al. 2012 and Gould et al. 2011)

2. The diurnal cycle of upslope or up-valley (anabatic) and downslope or down-valley (katabatic) winds, respectively at daytime and at night, obeying pressure gradients determined by differential heating.
3. Orographic channeling of winds, namely when upper-level winds change in direction and strength when arriving at the surface and flowing through valleys.
4. Foehn winds, which are warm, dry, and strong downslope flows as an outcome of adiabatic compression on mountain lee sides, e.g., the Santa Ana winds in southern California.
5. Disturbance of wind by rough terrain causing turbulence, namely eddies, e.g., reverse flows (rotor winds) forming on the lee slope of mountain ridges.

Fast-spreading wildfires include both wind-driven and plume-driven fires dominated by strong fire-atmosphere interactions under unstable atmospheric conditions (Table 8.1). All exhibit rates of spread of at least 3 km h^{-1}, which, for a typically available fuel load range of 10–30 t ha^{-1} and the assumption of 18,000 kJ kg^{-1} for

Table 8.1 Selected examples of fast-spreading and high-intensity wildfire runs and the prevailing atmospheric and dead fuel moisture conditions

Year	Reference	Country	Fire name or location	Vegetation	Wind speed at 10 m (km h^{-1})	Air relative humidity (%)	Air temp. (°C)	1 h fuel moisture (%)	Rate of spread (km h^{-1})
1930	Stocks and Walker (1973)	Canada	Garden Lake and Dog Lake, Ontario	Conifer forest	48	46	28	10	3.4
1959	USDA Forest Service (1960)	USA	Pungo, North Carolina	Conifer forest	32	67	28	13	3.2
1966	DeCoste et al. (1968)	USA	Gaston, South Carolina	Conifer forest	27	27	26	6	4.8
1967	Anderson (1968)	USA	Sundance, Idaho	Conifer forest	71	23	20	6	9.7
1968	Kiil and Grigel (1969)	Canada	Lesser Slave Lake, Alberta	Conifer forest	46	30	21	8	6.4
1971	Wade and Ward (1973)	USA	Air Force Bomb Range, North Carolina	Conifer forest	32	35	22	7	2.9
1980	Simard et al. (1983)	USA	Mack Lake, Michigan	Conifer forest	33	24	27	7	3.4
1980	Alexander et al. (1983)	Canada	Cold Lake Air Weapons Range, Alberta	Conifer forest	36	21	26	5	3.6
1983	Stocks (1989)	Canada	Red Lake, Ontario	Conifer forest	51	35	23	8	5.9
1983	Keeves and Douglas (1983)	Australia	Belgrave, Victoria	Eucalypt forest	70	5	41	3	12.3
1986	Pearce et al. (1994)	New Zealand	Awarua	Wetland	46	76	12	16	3.8
1987	Noble (1991)	Australia	Boonoke, New South Wales	Grassland	47	7	41	2	23.0
1990	Cheney et al. (1998)	Australia	Junee, New South Wales	Grassland	37	12	40	3	11.4
1994	Butler and Reynolds (1997)	USA	Butte City, Idaho	Shrubland	37	10	35	4	8.9

(continued)

Table 8.1 (continued)

Year	Reference	Country	Fire name or location	Vegetation	Wind speed at 10 m (km h^{-1})	Air relative humidity (%)	Air temp. (°C)	1 h fuel moisture (%)	Rate of spread (km h^{-1})
1994	Butler et al. (1998)	USA	South Canyon, Colorado	Shrubland	55	9	28	3	10.5
2001	Quintilio et al. (2001)	Canada	Chisholm, Alberta	Conifer forest	47	29	25	7	5.4
2003	Donoghue et al. (2003)	USA	Cramer, Idaho	Conifer forest	53	12	34	6	9.1
2009	Cruz et al. (2012)	Australia	Kilmore East, Victoria	Eucalypt forest	63	10	41	4	9.2
2009	Tolhurst (2009)	Australia	Murrindindi, Victoria	Eucalypt forest	54	9	42	3	12.4
2011	Alexander et al. (2013a)	Canada	Lethbridge, Alberta	Crop stubble	69	35	12	7	8.3
2015	Burrows (2015)	Australia	Esperance, Western Australia	Grassland	54	4	42	1	14.3
2017	Coen et al. (2018), Nauslar et al. (2018)	USA	Tubbs, Santa Rosa, California	Miscellaneous	73	7	33	4	6.5

the heat of combustion, corresponds to a fireline intensity range of 15,000–45,000 kW m^{-1}. Mean 10-m height wind speeds in the open lower than 30 km h^{-1} are seldom present. However, fast-spreading fires can happen when relative humidity is relatively high, and air temperature is relatively low, and thus fine dead fuels are relatively moist. The most striking case is the Awarua fire in New Zealand, which spread at 3.8 km h^{-1} in a shrub-grass complex under a dead fuel moisture content of 16% (Pearce et al. 1994). The amount of fuel available for combustion is unknown for many of these fires, but note that the fastest fire in Table 8.1, possibly the fastest ever recorded reliably, attained 23 km h^{-1} in a grassland carrying a fuel load of just 1.6 t ha^{-1} (Noble 1991). Both cases illustrate how cured, fine, aerated grass fuels foster rapid fire spread.

Spotting is a frequent phenomenon and can occur over long distances that are difficult to predict, thus posing significant difficulties in fire control and fire safety. Crown fires are more difficult to control than surface fires due to higher rates of spread and intensities. Larger flames from crown fires dictate larger fire fighter safety zones. Spotting and increased radiation from large flames make structures more difficult to defend from crown fire than from surface fire (Cohen and Butler 1998). Also, complex fire-atmosphere interactions can introduce very significant difficulties in wildfire suppression. Tedim et al. (2018) proposed seven fire categories, of which the upper three are beyond any suppression capacity and accommodate fire behavior variation in extreme wildfire events, those that exceed 10,000 kW m^{-1} and feature those phenomena, which the next sections discuss.

8.3 Crown Fires

Crown fires involve forest and shrub canopies burning. Crown fires are considered dual-layer fires that usually involve both the surface fuel layer and the crown fuel layer (Weise et al. 2018). Extreme fires commonly involve crown fires, but not all crown fires are extreme. Crown fires are historically common in many places, and some plants and animals are well adapted to survive and thrive after crown fires (See Chap. 9). However, the effects of crown fires are more severe and last longer than those of surface fires. Near-total tree mortality often occurs in non-sprouting species, though patches may be unburned or burned with low severity, so trees survive even in extreme fires (See Chap. 12). Smoke production will be great, and foliar nutrients may be lost from the site. Crown fires can occur in a wide variety of forest types. Increasingly, crown fires are taking place in forest types not historically prone to crown fires, such as ponderosa pine forests (Mutch et al. 1993). A significant risk to life and property exists wherever forest stands prone to crown fire burn close to residential or recreational developments (Scott and Reinhardt 2001).

Crown fires were classified in early literature (Kylie et al. 1937; Woods 1944) into two basic types, the "running crown fire", which generates enough heat for crown-to-crown spread, and the "dependent crown fire" which depends upon the heat generated by the surface fire for its spread (Alexander and Cruz 2016). These two types of crown fires are also related to two different steps. After crowning, fire may propagate in the canopy or not, according to the conditions for fire spread. Van Wagner (1977, 1993) distinguished and modeled separately the different forms of crown fire spread: passive, active, and independent. Passive crown fires occur when there is torching of the tree crowns by the surface fire but canopy characteristics and wind do not allow for fire spread through the tree crowns. Active crown fires occur when high winds allow for fast fire spread under appropriate canopy characteristics and fire propagates simultaneously in both surface and tree crowns. A very unusual behavior associated with very high wind speeds and/or steep terrain and closed and dense canopies is that of an independent crown fire when fire propagates in the canopy independently of the surface fire (Fig. 8.6). These different modes or types of crown fires have been used extensively in different applications such as FARSITE and FlamMap (Finney 2006).

Two separate and consecutive steps are required for a crown fire to develop. First, the fire has to move vertically from the understory layer to the canopy. Secondly, fire can spread horizontally in passive or active modes. The Canadian forester, Charles Van Wagner (Fig. 8.7), was the first to recognize and analyze these separate steps. We will approach these steps separately, analyzing first the conditions for crowning and the start of a crown fire and then the conditions for its spread and the prediction of its spread rate.

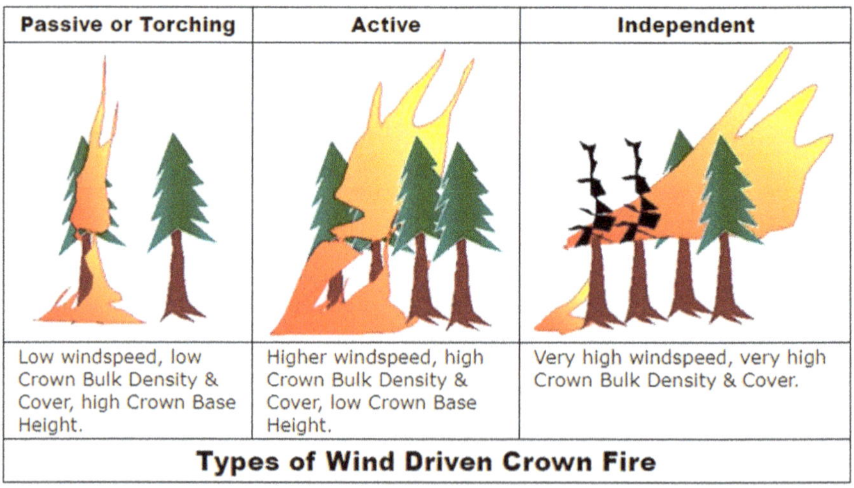

Fig. 8.6 Types of crown fires. From FARSITE technical documentation: (http://www.fire.org/downloads/farsite/WebHelp/technicalreferences/tech_crown_fire.htm)

Fig. 8.7 Charles Van Wagner was a senior scientist at the Canadian Forest Service. He pioneered the study of crown fires in conifer forests (MacIver et al. 1989)

8.3.1 Crown Fire Initiation

There are several approaches to understand crown fire initiation based on different fire behavior and fuel characteristics. Empirical models of the likelihood of crown fire initiation in conifer forests were reviewed by Alexander et al. (2013a, b), Cruz et al. (2004), and Alexander and Cruz (2016). The authors concluded that fire behavior characteristics such as fireline intensity and other associated variables, including wind speed, residence time, fine dead fuel moisture content, or surface fuel consumption, were relevant to estimate the likelihood of crown fire initiation. Also, fuel characteristics such as foliar moisture content and crown base height (or the similar fuel strata gap) are important variables for crown fire initiation (Table 8.2).

Empirical models indicate which fire and fuel variables are important for crown fire initiation. However, they do not show how these variables play a role in the process. To understand how a crown fire is initiated, it is possible to see it as based on a surface fire of sufficient intensity to raise the temperature at the canopy above a certain temperature threshold for the ignition of the crown. If this threshold of temperature is achieved, the crown will ignite.

Under the approach that crowning is expected to occur if a certain temperature threshold is attained at the canopy base from a fire with a given intensity, it is necessary to estimate how the increase of temperature changes with height above the ground. Thomas (1963, 1967) was one of the first to develop equations relating surface fireline intensity (I_B) to the temperature rise above ambient air temperature (ΔT) at a certain height (z):

Table 8.2 Crown fire initiation models with the relevant variables used and their positive (+) or negative (–) effect on crown fire initiation (Table based upon review by Cruz et al. 2004)

Model/variables	Fireline intensity	Foliar moisture content	Wind speed	Residence time	Crown base height (or fuel strata gap)	Estimated fine dead fuel moisture content	Surface fuel consumption
Van Wagner (1977)	+	–			–		
Alexander (1998)	+	–	+	+	–		
Cruz (1999)			+		–	–	+

$$\Delta T = \frac{a\, I_B^{2/3}}{z} \tag{8.2}$$

where ΔT is the change in temperature (°C or K), a is a proportionality constant, I_B is the fireline intensity (kW m^{-1}) as defined by Byram (1959), and z is the height above the ground or above the surface fuel (m). The proportionality constant a in Eq. (8.2) was estimated by Van Wagner (1973) to be a = 4.47 for typical fires. Alexander and Cruz (2012) reviewed several laboratory and field studies in different conditions indicating different values for the proportionality constant a. Recall that Byram's intensity of the surface fire is computed as:

$$I_B = h_c w\, R \tag{8.3}$$

The fireline intensity I_B is in units of kW m^{-1}, h_c is the heat of combustion (typically the low heat of combustion ΔHL in kJ kg^{-1}), w is the amount of fuel available for flaming combustion (kg m^{-2}), and R is the rate of fire spread (m s^{-1}). A similar approach was proposed by Alexander (1998), taking into account that the increase of temperatures above the fireline depends on wind and slope. For the same fireline intensity, calm wind situations in flat terrain will generate higher temperatures above the fireline. The effect of wind is included using its influence in the angle between the fire plume and the ground, the fire plume angle, A_p. In no wind situations, the fire plume angle is 90° and the wind has no effect. The effect of slope is included by using the slope angle S. In flat terrain where S is zero, there is no slope effect. The full empirical equation to be used in predicting the increase in temperature at the crown base height (z = CBH) is:

$$\Delta T = 13.9\, I_B^{2/3}\ sin\left(A_p\right)/[CBH\ sin\,(90 - S)] \tag{8.4}$$

In no-wind, no-slope conditions, Eq. (8.4) is equivalent to Eq. (8.2). In these approaches, a critical temperature is computed and the process of crowning is expected to occur if a certain temperature increase threshold at the base of the crown layer is attained. Alexander (1998) indicates that to initiate combustion of the entire crown fuel layer in pines, the temperature at the base of the crown should attain or exceed the critical threshold of 400 °C (673 K).

We are often interested in knowing what is the height z where a given temperature will be attained in a fire with a given intensity. In this case, we can rearrange Eq. (8.2) to solve for z, and we have:

$$z = \left(\frac{a}{\Delta T}\right) I_B^{2/3} \tag{8.5}$$

where a, ΔT and I_B are the same as in Eq. (8.2). The practical application of this relationship is illustrated in the following example:

The relation between fireline intensity and temperature increase at a certain height has been very useful to predict the height at which crown ignition is expected to occur but also to predict other variables such as scorch height. This example illustrates both calculations and highlights possible comparisons.

Let us start with crown ignition. For the temperature to rise from ambient temperature (20 °C as an example) to the ignition temperature (set as 320 °C) the temperature difference would be ΔT = 300 °C and Eq. (8.5), using a = 4.47, simplifies to:

$$Ignition\ height\ z\ (m) = 0.015\ I_B^{2/3} \tag{8.6}$$

If crown base height (CBH) is less than this estimated ignition height, crowning is expected to occur.

The same Eq. (8.5) can be used to estimate the height at which the lethal temperature for leaves is reached, the crown scorch height (See Chap. 9 for discussion of fire effects on trees including lethal temperature for plant tissues). Necrosis occurs when foliage is heated enough to be killed but not enough to burn, appearing brown rather than black. Assuming 60 °C as the temperature that would cause foliage necrosis, if the ambient temperature is 20 °C, the required rise of temperature for scorch would be ΔT = 40 °C. For ΔT = 40 °C and a = 4.47 the Eq. (8.5) simplifies to:

$$Scorch\ height\ z\ (m) = 0.112\ I_B^{2/3} \tag{8.7}$$

If the crown base height (CBH) is less than the estimated scorch height, the canopy is expected to be scorched to some extent. Crown scorch volume or crown scorch height as percentages of crown volume and crown length, respectively, are indicators of damage to trees; when these are very high, trees may die.

This example shows that ignition height and scorch height can be related using the same type of equation. Estimates of scorch height are very useful for prescriptions of understory burning while estimates of ignition height are generally more useful in the context of attacking fires with potential for crowning (Fig. 8.8). The relationships between the different variables have been used in practice. It is common to observe that crowning is initiated when surface flame height is close to the base of the canopy and it is suggested by the comparisons of Eqs. (8.6) and (8.7) that the relationship between ignition height and scorch height is a simple proportion. This is in line with the commonly used rule of thumb (de Ronde 1988; Luke and McArthur 1978) that "flames associated with prescribed burning are likely to cause scorch within a zone equivalent to six times flame height" (Alexander 1998).

Fig. 8.8 (**a**) Prescribed underburning where crown base height is expected to be at least six times flame height to avoid crown scorch. (**b**) Experimental crown fire (Fernandes et al. 2004), where flames from the surface reach the canopy and crowning occurs. (Photographs in maritime pine (*Pinus pinaster*) stands in northern Portugal by P. Fernandes)

Finally, we can also rearrange Eq. (8.2) to estimate crowning thresholds as a function of fireline intensity:

$$I_B = \left(\frac{z\,\Delta T}{a}\right)^{3/2} \tag{8.8}$$

Van Wagner (1977) suggested the use of the concept of a critical surface fireline intensity (I'_s in kW m^{-1}) as the minimum intensity required to ignite canopy fuels, as a function of the canopy vertical distance from the ground (z or crown base height CBH in meters) and temperature increase (ΔT). Van Wagner (1977) also assumed that the temperature rise (ΔT) required for crown ignition varies with the heat of pre ignition (Q_{ig} in kJ kg^{-1}), and therefore with foliar fuel moisture (Eq. 4.6). This is why crown fires are more likely following long-term droughts that dry soils when foliar moisture of trees and shrub crowns are relatively low. Based on the relations presented in Eqs. (8.5)–(8.8), Van Wagner estimated the critical surface fireline intensity as:

$$I'_s = \left(0.01 \times CBH \times Q_{ig}\right)^{3/2} = \left[0.01 \times CBH \times (460 + 26\,FMC)\right]^{3/2} \tag{8.9}$$

where I'_s is the critical surface fireline intensity (kW m^{-1}) estimated as a function of the canopy vertical distance from the ground (or crown base height CBH in meters) and foliar moisture content is FMC (in %). This formulation has very interesting practical applications. Often, in practical terms, the fireline intensity of the surface fire is estimated from Byram's equation (8.3). Using this criterion, if I_B is higher than

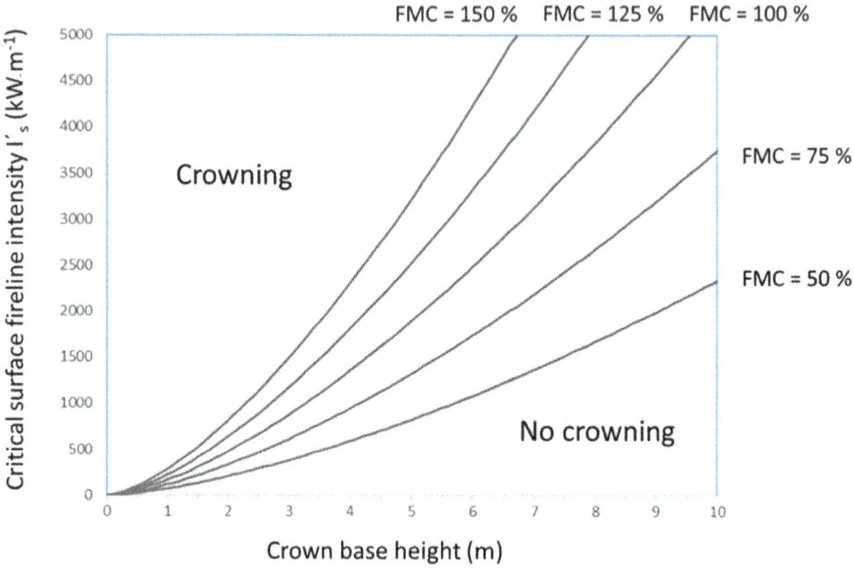

Fig. 8.9 Critical surface fireline intensity for crowning as a function of crown base height (CBH) and foliar moisture content (FMC) based on Eq. (8.8)

the critical surface fireline intensity ($I_B > I'_s$), a transition from surface fire to crown fire is expected (Fig. 8.9).

In the above approaches, we used the heat balance in Eq. (8.2) as the basis for the calculations. A different approach based on residence time was proposed by Xanthopoulos and Wakimoto (1993). These authors estimated the time required for ignition (t_{ig} in seconds) of needles of different conifers based on the temperature at which they were exposed (°C) and their foliar moisture content (FMC in %). The temperature of the needles at crown base height is calculated as the ambient temperature (T_a in °C) plus the temperature difference (ΔT in °C). The equations developed can be used to predict the time for ignition of needles at the base of the crown using the temperature at crown base height as:

$$t_{ig} = 291.9\ exp\left[-0.00664(T_a + \Delta T) + 0.00729\ FMC\right] \qquad \textit{Pinus ponderosa} \tag{8.10}$$

$$t_{ig} = 1408.9\ exp\left[-0.00990(T_a + \Delta T) + 0.00691\ FMC\right] \qquad \textit{Pinus contorta} \tag{8.11}$$

$$t_{ig} = 3607.7\ exp\left[-0.01072(T_a + \Delta T) + 0.00397\ FMC\right] \qquad \textit{Pseudotsuga menziesii} \tag{8.12}$$

Xanthopoulos and Wakimoto (1993) assumed, as others did (Alexander 1998), that the duration of heating received at the crown base during the active flaming stage of combustion at the ground surface could be considered to be equal to the residence time (t_R) of the fire's flaming front. In this approach, if the time required to ignition (t_{ig}) for needles at the crown base height with a given FMC and exposed to a certain temperature ($T_a + \Delta T$) is lower than the time of exposure or residence time (t_R) the needles are expected to ignite and crowning is expected to occur.

8.3.2 The Conditions for Active Crown Fire Spread

After crowning, active crowning is defined by the spread of fire across the crown space between the different crowns. The likelihood of crown fires to spread, or the transition from passive to active crown spread is, among other factors, associated with variables that characterize the tree and shrub canopies and the vertical structure of the fuel complex. Van Wagner (1977) was one of the first to point out that crown fires were common in Canadian coniferous forests but not in deciduous hardwoods. The importance of vertical structure was highlighted by Nunes et al. (2019), who compared the vertical structure of burned and unburned forests dominated by the same tree species in the Iberian Peninsula (Fig. 8.10). Unburned forests always had less cover in the lower strata, indicating the importance of surface fuel loads to burn likelihood. However, differences were high in the more Mediterranean-type forests of the evergreen *Quercus suber,* where the understory vegetation gets drier and more flammable in summer. For *Pinus pinaster* forests, the unburned stands show a little less understory cover and a little more canopy cover. The differences were greatest for burned and unburned temperate deciduous forests of *Quercus robur,* likely because the high foliar moisture and little understory fuel counteract the spread of a crown fire.

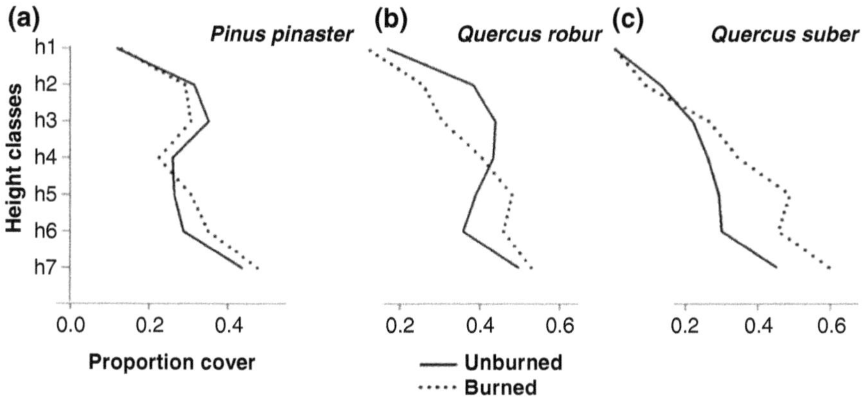

Fig. 8.10 The vertical structure of forest stands of three types dominated, respectively, by (**a**) maritime pine (*Pinus pinaster*), (**b**) a deciduous oak (*Quercus robur*), and (**c**) an evergreen oak (*Quercus suber*) measured in the national forest inventories of Portugal and Spain and that subsequently burned or not. (Adapted from Nunes et al. 2019)

The thresholds for active crown fire spread were proposed by Van Wagner (1977), who indicated that "the simplest general statement on how fire spreads through a fuel layer is the basic heat balance linking". Here we recall the heat balance equation:

$$I_p = R\,\rho_b\,Q_{ig} \tag{8.13}$$

where I_p is the propagating horizontal heat flux (kW m^{-2}), R is the rate of spread (m s^{-1}), ρ_b is the fuel bulk density of the canopy (kg m^{-3}), and Q_{ig} is the heat of preignition (kJ kg^{-1}).

Fuel bulk density (ρ_b) is often applied as an average for relatively homogeneous fuels. A value of ρ_b can be applied for an individual tree crown, for example. However, when referring to the canopy of a stand, the concept of Canopy Bulk Density (CBD) refers to a property of the stand, not the individual crown, and represents the average bulk density for the whole volume of crowns, including empty volumes. This is why the value of CBD can be managed by thinning and why the CBD values show a typical pattern from crown base height to stand height. Scott and Reinhardt (2005) provided photographs and stand data for five Interior West conifer forests that had been thinned to progressively lower values of crown bulk density. Their photographs and data are useful for estimating canopy fuel characteristics and effects of thinning (Fig. 8.11).

LiDAR data have been used by several authors (e.g., Andersen et al. 2004), while others have used other indirect methods to estimate forest canopy bulk density (Keane et al. 2005). In Spain, Ruiz-González and Álvarez-González (2011) used

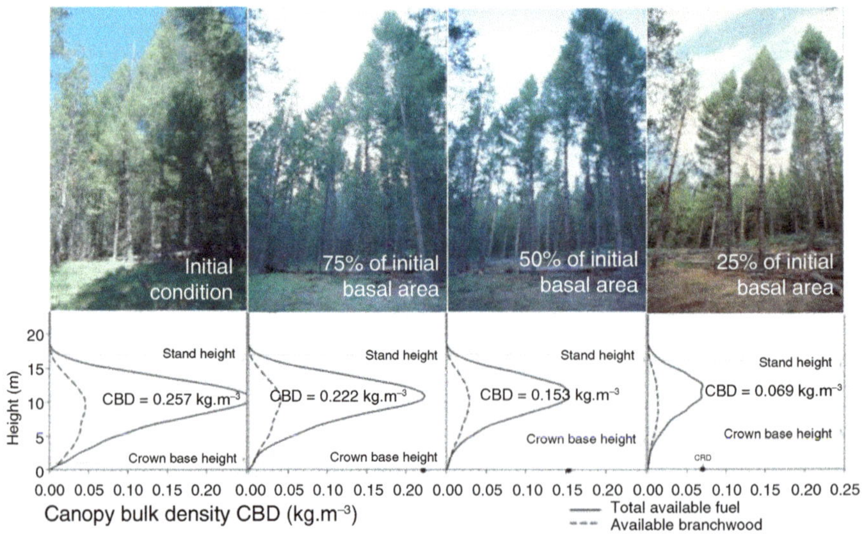

Fig. 8.11 Photographs of Douglas-fir/lodgepole pine forests with different levels of thinning and the resulting vertical distribution of crown bulk density (CBD) and location of the maximum CBD value. (From Scott and Reinhardt 2005)

canopy bulk density and canopy base height equations to assess crown fire hazard in *Pinus radiata* plantations. All these approaches are useful to guide crown fuel treatments to mitigate potential fire behavior (See Chap. 11 for discussion of fuels treatments).

As canopy bulk density is clearly important in determining the conditions for a crown fire to spread, the indication of critical thresholds for CBD can be of practical use. Alexander et al. (2013b) suggested that a minimum value of about 0.1 kg m^{-3} for the bulk density of the canopy represents a critical threshold above which active crowning is likely if crowns are ignited. Canopy bulk density can be reduced by thinning to reduce tree density and basal area, and crown base height can be raised by pruning the lower branches of the crown (See Chap. 11).

Canopy bulk density is an essential component of the approach by Thomas (1967) and Van Wagner (1977) where the threshold for crown fire spread is established by setting a limiting value for a new variable, the mass flow rate (MFR in kg m^{-2} s^{-1}). Mass flow rate represents the quantity of fuel that starts to burn per unit vertical area of the crown space per unit time. If the fire is spreading horizontally through the crowns with a canopy bulk density ρ_b (kg m^{-3}) at a rate R (m s^{-1}), the mass flow rate is the product of the rate of spread and bulk density. Using this relation, we can calculate the mass flow rate MFR as:

$$MFR = R\,\rho_b = \frac{I_p}{Q_{ig}} \qquad (8.14)$$

Under this approach, it was proposed that MFR has a critical minimum value (MFR_{min}) below which a flame cannot form and propagate through the crowns. The predicted MFR, calculated from the rate of spread and canopy bulk density, can be compared with a critical minimum value MFR_{min}. If the predicted MFR is lower than MFR_{min}, the fire is considered to be too slow or the canopy to be insufficiently dense to allow the fire to spread through the crowns. Alternatively, from Eq. (8.14), we can consider that a low value of MFR indicates that fireline intensity is too low or foliar moisture is too high to sustain crown fire spread.

Estimations of the critical mass flow rate (MFR_{min}) have been made by Thomas (1967) in experimental fuelbeds and by Van Wagner (1977) in a pine plantation, resulting in similar values of MFR_{min} = 0.06–0.08 kg m^{-2} s^{-1} (Thomas 1967) or 0.05 kg m^{-2} s^{-1} (Van Wagner 1977).

We recall from Chap. 4 that heat of pre-ignition (Qig in kJ kg^{-1}) is very much dependent upon fuel moisture content (FMC, in %) and can be estimated using approximate equations. Here we use:

$$Q_{ig} = (460 + 26\,FMC) \qquad (8.15)$$

Equations (8.14) and (8.15) can be combined and rewritten to find the propagating horizontal heat flux (I_{min} in kW m^{-2}) threshold for a crown fire to spread as a function of foliar moisture content (FMC in %):

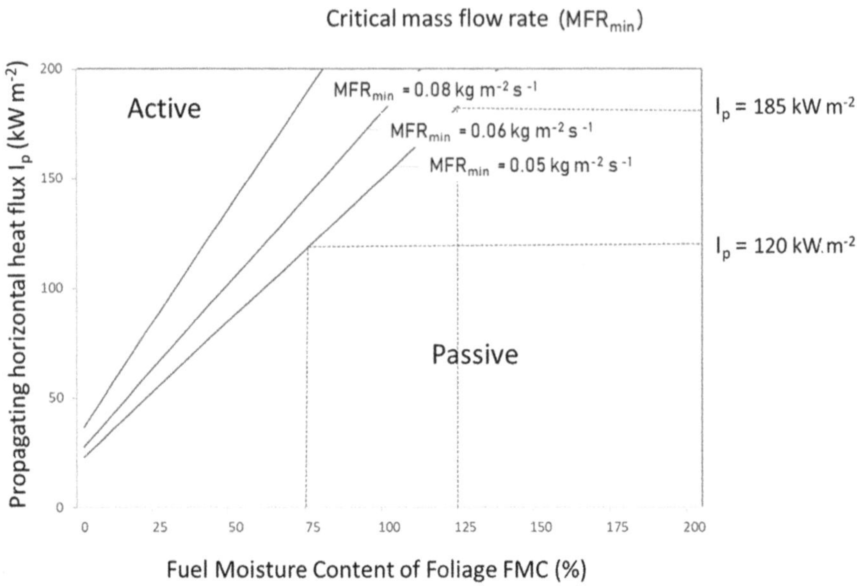

Fig. 8.12 The influence of foliar moisture content (FMC) and the propagating horizontal heat flux (I_p) in distinguishing passive from active crown fires. The lines show the critical horizontal heat flux (I_{min}) using the thresholds for critical mass flow rate of 0.06 and 0.08 (Thomas 1967) and of 0.05 (Van Wagner 1977). The heat flux of 120 kW m^{-2} corresponds, in Van Wagner's model, to a FMC of 75%. A I_p of more than 185 kW m^{-2} would be required for a fire to carry actively through a canopy with FMC of 125% or more

$$I_{min} = MFR_{min}\, Q_{ig} = MFR_{min}(460 + 26\, FMC) \tag{8.16}$$

The graphical representation of this relationship for the various values of the critical mass flow rate (MFR_{min}) is shown in Fig. 8.12.

The prediction of a transition from a passive crown fire to an active crown fire can be made by comparing a known value of the propagating heat flux I_p with the critical value I_{min}, which is a function of foliar moisture content FMC, according to Eq. (8.16) and Fig. 8.12. Van Wagner (1977) indicated that at a temperature around 1000 °C, the crown flame is capable of radiating at most about 125 kW m^{-2} from a vertical front. This indicates that crown fires would be sustainable just by radiation when FMC is below 75%, which can occur in Mediterranean shrubland fires in late summer or fall when they burn intensely. Chandler et al. (1983) indicated as a general rule of thumb that crown fire potential in conifers increases whenever FMC drops below 100%, which is not common. Above a foliar moisture content of 125%, Van Wagner (1967) indicated that radiant intensity would be very small to allow a crown fire to spread. These results indicate that, in the absence of wind, when convection is still not very dominant, crown fires can only spread in some particular formations, as in shrubland in drier periods of the year when FMC is low. We discuss the dynamics of live fuel moisture in Chap. 11.

Van Wagner (1977) pointed to foliar moisture as the main canopy factor affecting the possibility of crown fire to spread but cautions that it should not be considered the only factor to take into account. Variables such as the height of the live crown above ground, the bulk density of foliage within the crown space, the shape and the arrangement of leaves, or the presence of highly flammable volatile waxes, oils, and resins are long known to be important for fire spread.

We saw that only in special situations with very low FMC the radiant heat flux is sufficient to allow for crown fires to spread. This means that active crown fires in forest can only occur as a consequence of convective heat in windy conditions and in fires with high rates of spread. We can now reformulate the problem by rearranging Eq. (8.15) to have the thresholds for rate of spread. Assuming a limiting mass flow rate (MFR_{min}) it follows that for any specified value of canopy bulk density (ρ_b) the fire spread rate must have a lower limit (R_{min}) below which crown spread is not sustained:

$$R_{min} = \frac{MFR_{min}}{\rho_b} \tag{8.17}$$

R_{min} is the critical minimum spread rate to make crown fire spread possible and an active crown fire to occur. The relationship of rate of spread with ρ_b and the thresholds between passive and active crown fires are shown in Fig. 8.13.

Equation (8.17) and (Fig. 8.13) show that under low canopy bulk density, only fires with high rates of spread, typically associated with strong winds, can propagate through the canopies. On the other hand, very dense canopies can sustain crown fire spread even at low rates of spread or low wind speeds.

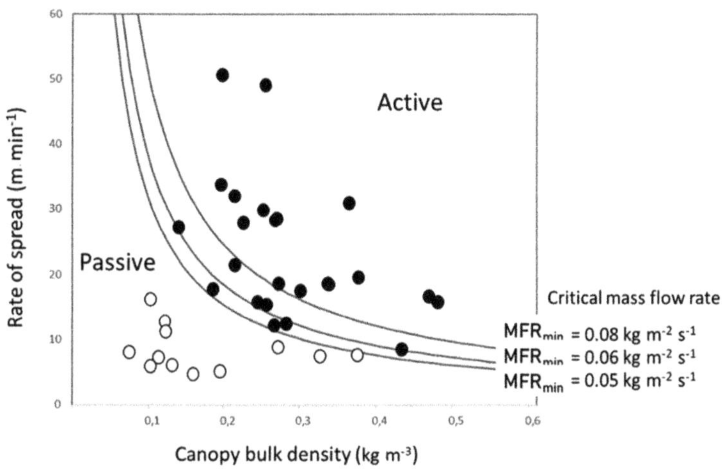

Fig. 8.13 The influence of canopy bulk density and rate of spread in differentiating observed passive fires (empty circles) from active fires (black circles) using data from Cruz et al. (2005) and Van Wagner (1977)'s equations. The lines show the critical rate of spread (R_{min}) using the thresholds for critical mass flow rate MFR_{min} of 0.06 and 0.08 (Thomas 1967) and of 0.05 (Van Wagner 1977)

As discussed, wind speed is a factor of major importance in determining crown fire spread. It is, therefore, relevant to have wind speed thresholds associated with crown fires. Two thresholds have been defined. The Torching Index is the open wind speed at which the predicted surface fire intensity equals the minimum required for crown fire initiation for any given canopy bulk density in conifer forests. Similarly, the Crowning Index is the wind speed at which the predicted crown fire spread equals the minimum needed to maintain a solid flame through the canopy. This is the approach used in many applications as in NEXUS (Scott 1999, 2006). This was also the objective of the empirical model developed by Cruz et al. (2005), indicating the thresholds between passive and active crown fires based on wind speed, canopy bulk density, and the moisture content of fine dead fuel (Fig. 8.14).

The combinations of the main factors that determine the likelihood of a crown fire to transition from passive to active are summarized, in a very practical way, in Fig. 8.14. Canopy bulk density and fine dead fuel moisture content are variables that can be measured or estimated prior to the fire and allow establishing wind speed thresholds for crown fires to initiate and spread. Fernandes et al. (2004) describe how variation in wind speed changed the type of fire propagation in an experimental crown fire. Fire managers know by experience that wind speed thresholds are important for the safe use of prescribed fire and for setting wildfire suppression and safety strategies.

8.3.3 *Crown Fire Rate of Spread*

The likelihood of crown fire initiation is a function of several variables (Table 8.2). It increases with wind speed (by increasing fire spread rate), decreases with increasing

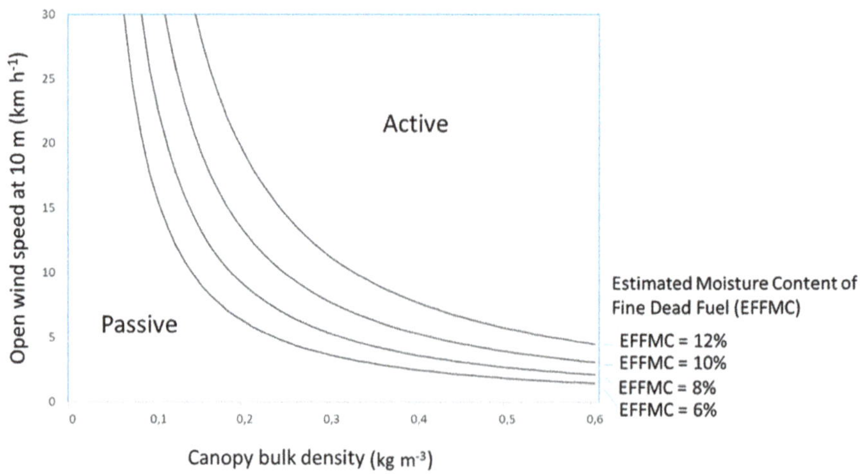

Fig. 8.14 The relative importance of wind speed, canopy bulk density, and moisture content of fine dead surface fuel on the likelihood of an active crown fire using the equation from Cruz et al. (2005)

distance from the surface fuel to the crown base (the fuel strata gap), increases with fuel consumption, and decreases with increased fuel moisture content. For practical applications, it is useful to express this likelihood as a function of a few variables.

After assessing the conditions for crowning and spread it is important to evaluate what are the factors that are associated with the rate of spread of crown fires. Cruz et al. (2005) constructed a statistical model to predict active crown fire spread rate from a database of 25 experimental crown fires in forests of *Pinus banksiana* (jack pine), *Picea mariana* (black spruce), and *Pinus resinosa* (red pine). The best model fit was:

$$R = 11.02\ {U_{10}}^{0.9}\ {\rho_b}^{0.19}\ e^{-0.17\ EFFM} \tag{8.18}$$

where R is the rate of spread of an active crown fire (m min^{-1}), U_{10} is the wind speed at 10 m in open terrain (km h^{-1}), ρ_b is the canopy bulk density (kg m^{-3}), and EFFM is the estimated fine dead fuel moisture content (%). High values for wind speed and canopy bulk density favor higher crown fire spread rates. High dead fuel moisture content decreases the rate of spread.

The simplest way to estimate the rate of spread for passive crown fires is to consider that they are controlled by the surface fire, and therefore they have the same rate of spread. This was the approach initially taken by Finney (2004) in the development of FARSITE.

Another statistical model was developed for the coniferous forests of the northern Rocky Mountains USA. Rothermel (1991) proposed that the average crown fire spread (R) could be estimated as 3.34 times the predictions made with his model for surface fire spread (Rothermel 1972) using Fuel Model 10 (timber litter and understory) and a 0.4 wind adjustment factor (for wind speeds at 6.1 m). The author also proposed that the near-maximum spread rate is 1.7 times faster than the average spread rate. US fire modeling tools adopt the Rothermel (1991) equation.

Some attempts have been made to combine surface and crown fire behavior predictions. Scaling spread rate between predictions of surface and crown fire spread is commonly done using the concept of the "crown fraction burned (CFB)" developed by van Wagner (1989, 1993). CFB represents a transition function between 0 and 1 representing the degree of crowning and it can be calculated according to Van Wagner (1989) as:

$$CFB = 1 - exp\{-0.23(R_s - R_{min})\} \tag{8.19}$$

where R_s and R_{min} are the predicted steady-state rate of spread and the critical spread rates (m min^{-1}), respectively. Alternative ways to calculate CFB have been proposed by Van Wagner (1993). A simple linear transition function was proposed by Scott and Reinhardt (2001).

Alternatively, if we want to estimate a global rate of spread (R_{global}) as a weighted average from the estimated values of the surface fire spread ($R_{surface}$) and of the active crown fire spread (R_{crown}) we can used the Crown Fraction Burned (CFB) as the weighting factor:

$$R_{global} = (1 - CFB) \times R_{surface} + CFB \times R_{crown} \tag{8.20}$$

For a surface fire, the CFB is equal to zero and $R_{global} = R_{surface}$. For an active crown fire (or an independent crown fire), the crown fraction burned is equal to 1 and therefore $R_{global} = R_{crown}$. Many of the current applications such as FlamMap or the Fire and Fuel Extension of the Forest Vegetation Simulator (FFE-FVS) use CFB to estimate a global rate of spread.

Predictions of crown fire behavior have large uncertainties, in part because they are typically based on a small number of observations for which we have measured inputs for the models. This was one of the worths of the International Crown Fire Modeling Experiment (ICFME). The ICFME was a cooperative international experiment carried out between 1995 and 2001 in Canada's Northwest Territories, with more than 100 participants representing 30 organizations from 14 countries (Stocks et al. 2004). The series of 18 experimental high-intensity and highly instrumented crown fires (Fig. 8.15) was fundamental to quantify parameters essential to model the initiation and spread of crown fires.

Crown fires often produce spotting as firebrands are lifted into the air. Crown fires can reflect complex fire-atmosphere interactions. Both spotting and fire-atmosphere interactions are important in extreme fires, as discussed in the next sections.

Fig. 8.15 One of the experimental fires of the International Crown Fire Modeling Experiment, Northwest Territories, Canada (Bunk 2004)

8.4 Spotting

Spot fires are defined as fires "set outside the perimeter of the main fire by flying sparks or embers" (USDA Forest Service 1956). Spotting is probably the most challenging fire behavior characteristic from the perspective of fire suppression, as control difficulty is significantly increased (Byram 1959). Further, spotting can help fires spread through discontinuous fuels and can help to explain the patchiness of fires that greatly influences fire effects and ecosystem recovery (see Chap. 12).

Burning pieces, termed firebrands or embers, are the carriers of fire between the main fire front and other areas ahead of the fire front. Ember showers outside the main fire can ignite burnable fuels on and near houses, or embers can get into roofs and openings in buildings to ignite them despite the surrounding defensible space (see more on protecting homes in Chap. 10). Firebrands can start spot fires across fuel breaks, rivers, or other unburnable areas, producing dangerous conditions for firefighting crews (Potter 2016). Spotting can occur from surface fires, from torching trees, or from slash piles, but spotting is usually associated with crown fire spread as it can generate much greater spotting distances (Fig. 8.16).

The effect of spotting on fire behavior is determined by the density of ignitions from firebrands and the distances the ignitions occur ahead of the main fire. Ignition density from firebrands typically decreases with distance from the advancing flame front (Cheney and Bary 1969; Gould et al. 2007; Alexander and Cruz 2016). As

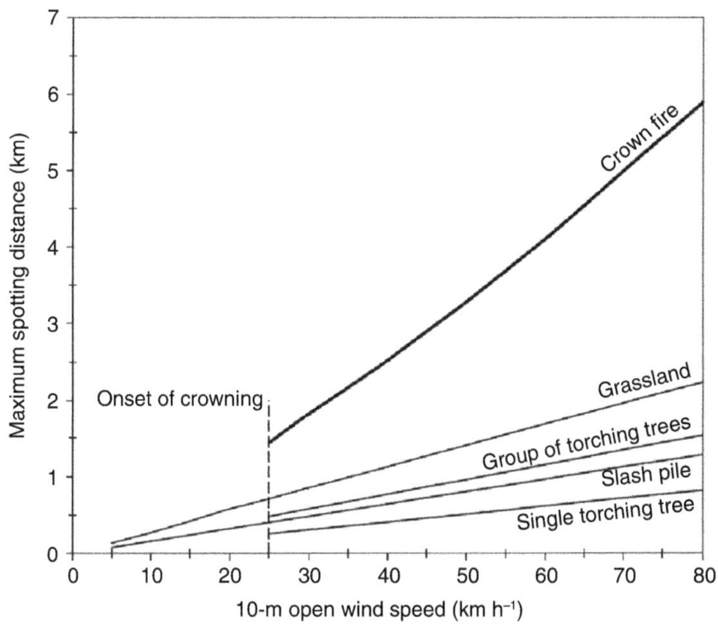

Fig. 8.16 Comparison of maximum potential spotting distance predicted as a function of wind speed for different types of fuels and fire behavior characteristics. (From Albini et al. 2012)

Byram (1959) pointed out, although the spot fires occurring at long distances are spectacular and effective in spreading fire over large areas, the spot fires nearer the main flame front have a much greater effect on fire behavior as showers of burning embers “occasionally produce disastrous firestorm effects by igniting large areas almost simultaneously”. For all these reasons, spot fires are regularly considered a type of extreme fire behavior even though they are a common characteristic of almost all fires burning in many fuel types (Potter 2016).

Records of spotting distances have been gathered for a long time in several regions of the world, from North America and Europe to Australia. In Europe, field observations of spotting distances under the project SALTUS (2001) indicated that spotting distances were not likely to exceed 2 km. In North America, spotting distances up to 5 km are commonly observed on wind-driven crown fires in coniferous forests (Alexander and Cruz 2016). Spotting distances up to 10 km were documented in chaparral fires in California (Countryman 1974) and the forests of the northern Rocky Mountains during the 1910 fires in Idaho and western Montana (Koch 1942). Spotting distances up to 20 km were documented in the 1967 Sundance Fire in northern Idaho (Anderson 1968). Luke and McArthur (1978) indicated that in many eucalypt forests in Australia very long spotting distances “far exceed those recorded in overseas literature”. McArthur (1967) indicated that spotting distances of 10 km were common on many high-intensity eucalypt forest fires. They reported that well-authenticated spot fires 15–20 km in advance of the main fire were recorded in Victoria and New South Wales during the January 1939 fires. In Victoria, one spot fire occurred close to 30 km ahead of the fire source during the March 1965 fires in Gippsland, and spotting up to 33 km of the fire front was documented in the Kilmore East fire that killed 121 people in February 2009 (Cruz et al. 2012).

Eucalypt bark can govern the type of spotting (Luke and McArthur 1978). Some types of eucalypt bark (from trees called stringybarks) are easily torn off tree stems by strong convection currents generated in fires, but the bark pieces are too heavy to travel far (Fig. 8.17). The bark embers generated can produce concentrated short-distance spotting (less than 100 m), allowing for a mass fire or firestorm effect if ember density is enough. Research on short-distance spotting was advanced during the Project Vesta where 104 experimental fires were lit between 1998 and 2001 on summer days in dry south-western Australian eucalypt forests with fireline intensities up to 10,000 kW m^{-1} and flame heights up to 25 m (Gould et al. 2007). In contrast, other types of eucalypt bark (candlebarks) typically break in streamers that may originate long-distance spotting (from 10 km up to more than 30 km) when burning in long unburned forests (Fig. 8.18).

Rothermel and Andrews (1987) suggested that spotting is “one of the most intractable problems” faced by fire scientists. If spotting is split into different stages or components, “these stages are each relatively tractable in terms of the scientific questions they pose” (Potter 2016). Hence, we recognize six successive components in the process of spotting: buoyancy and the development of the fire plume, firebrand generation, the lofting, transport, and fall of the firebrands, and ignition of unburned fuel (Fig. 8.19).

Fig. 8.17 Research on fire behavior in dry eucalypts of Australia showing short distance spotting during Project Vesta. (From Forestry and Forest Products 1998)

8.4.1 Buoyancy and the Fire Plume

All fires involve buoyant air flows, as the characteristics and nature of the flame and plume flow are responsible for the growth of the fire and the transport of heat and smoke up and away from the fires. For vegetation fires, buoyancy is associated with the fire plume. The plume is a column of hot combustion products and flames above a burning fuel source. We use the term fire plume to distinguish flows driven by buoyancy from those associated with forced flow, such as jets.

Buoyancy

Buoyancy is a force that arises due to differences in fluid density created by temperature differences between the column of hot gases and the ambient air. The principle of buoyant forces was introduced around 200 BC by Archimedes of Syracuse, who suggested in his treaty "On Floating Bodies" that any object totally or partially immersed in a fluid is buoyed up by a force equal to the weight of the fluid displaced by the object.

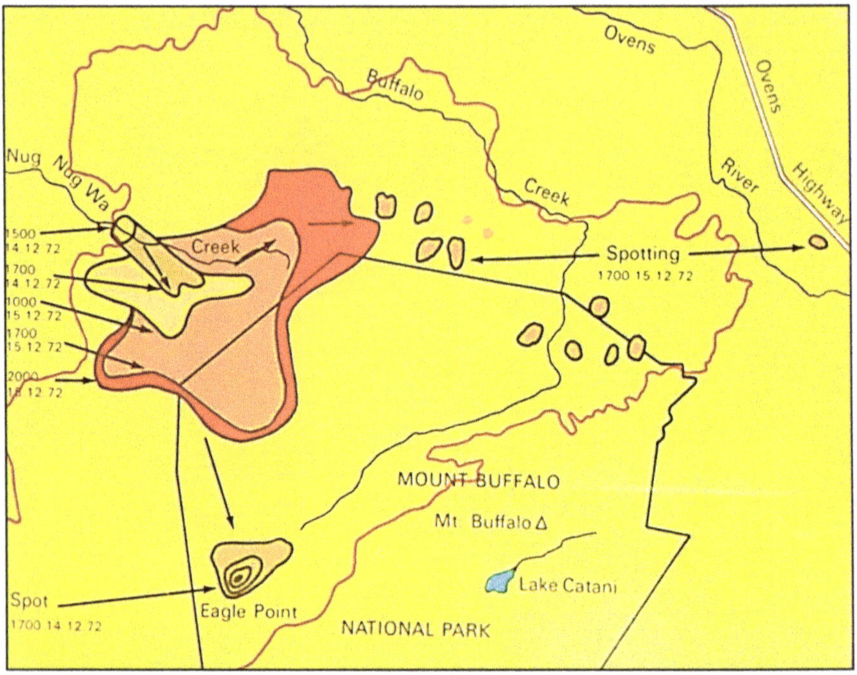

Fig. 8.18 Spotting on the Mount Buffalo fire in Victoria, Australia, on 14 December 1972, reported by the Forests Commission. This fire burned a fuel type that had not burned for at least 35 years during which great masses of candlebark accumulated in eucalypts. The accumulated candlebark was the source for long-distance spotting. (From Luke and McArthur 1978)

The Archimedes principle can be directly applied to the plume of a wildfire where the hot gases have a lower density than the surrounding cool air causing an upward buoyant force. Recall that the ideal gas law states that at constant pressure, the density of a fluid is proportional to the absolute temperature. We can model a flame as a cylinder with a volume (V) with a section area (A) and height (H) with temperature T_i and density ρ_i. In this case, the gases inside the flame have a significantly greater temperature and are less dense than the ambient air (at temperature T_a and density ρ_a) as illustrated in Fig. 8.20.

The driving force that causes the vertical flow is the difference in the weight of the two equivalent columns (Fig. 8.20). This resulting net buoyancy force (in Newtons, or kg m s^{-2}) can be computed as:

$$w_a - w_f = g\,A\,H(\rho_a - \rho_f) \tag{8.21}$$

where w_f is the weight of the column of hot gases inside the volume of the cylindrical flame with density ρ_f (kg m^{-3}), w_a is the weight of an equivalent column of ambient air with density ρ_a (kg m^{-3}), g is the acceleration of gravity (9.807 m s^{-2}), A is the sectional area of the column (m^2), and H is the height of the flame (m).

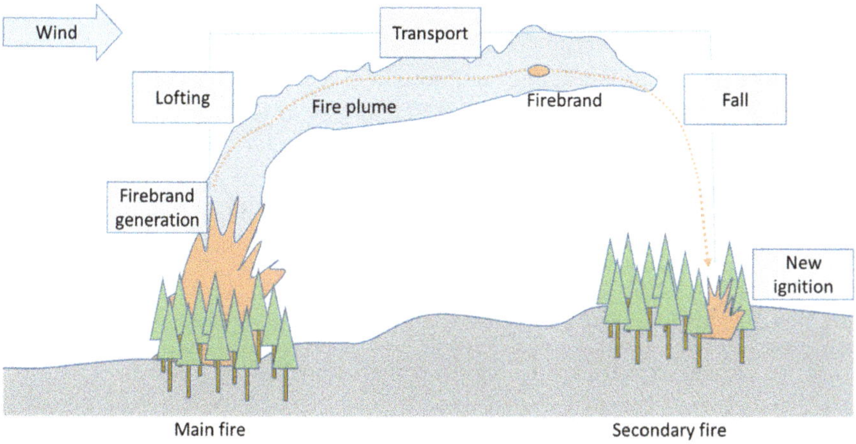

Fig. 8.19 The process of spotting and its idealized components. The fire plume is formed due to buoyancy forces from the fire where firebrands are generated. Firebrands are lofted, transported, and they eventually fall from the plume, possibly originating a new ignition and a secondary fire commonly called a spot fire. When many such firebrands ignite, the resulting fires can spread fire rapidly across a landscape and sometimes contribute to a firestorm

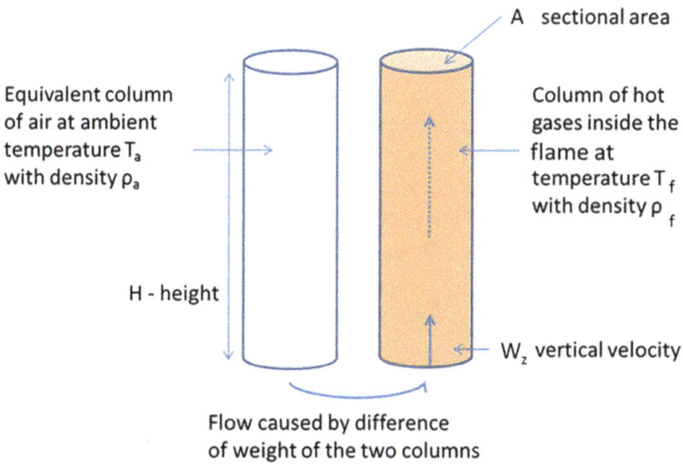

Fig. 8.20 The Archimedes' principle applied to a flame represented as an idealized cylinder. The difference of weight between the volume of air displaced (*left cylinder*) and the volume inside the flame (*right cylinder)* is the cause of the buoyancy flow and the vertical updraft velocity

To understand the lofting process, we need to estimate the buoyancy velocity, which is the updraft flow velocity of air and combustion gases (W_z). In a steady-state combustion reaction with no ambient wind, the forces of the net buoyancy ($w_a - w_f$) and of the upward airflow (w_z) are equivalent:

$$w_a - w_f = w_z \tag{8.22}$$

The force of the airflow applied in a section of area A can be computed as:

$$w_z = 0.5\, \rho_a\, A\, W_z^{\,2} \tag{8.23}$$

where W_z is vertical velocity. As the forces of the net buoyancy and the upward air flow are equal (Eq. 8.22), we can combine Eqs. (8.21) and (8.23):

$$0.5\, \rho_a\, A\, W_z^{\,2} = g\, A\, H\, (\rho_a - \rho_i) \tag{8.24}$$

As we are interested in estimating the updraft velocity W_z, we have:

$$W_z = \left[2\, g\, H\left(1 - \frac{\rho_f}{\rho_a}\right)\right]^{1/2} \tag{8.25}$$

It is often more convenient to estimate updraft velocity from temperatures instead of densities. For that, we have to recall the ideal gas law. The ideal gas law is derived from four simpler gas laws. Charles' law, dating from the 1780s, states that, under constant pressure, the volume of a gas (V) increases as the absolute temperature (T) increases. The Gay-Lussac's Law states that the pressure (P) of a given mass of gas is directly proportional to the absolute temperature (T) of the gas when the volume is kept constant. Avogadro's law (from 1811) states that, under the same conditions of temperature and pressure, a given number of moles of gas molecules (n) occupy the same volume (V) regardless of their chemical identity. Boyle's law states that, for a fixed amount of gas at a constant temperature, the volume of a gas (V) increases as pressure (P) decreases. By combining these four empirical laws, one can write the ideal gas law that explains how the various state variables are related:

$$PV = n\, \gamma_6\, T \tag{8.26}$$

where P is pressure, V is the volume of the gas, n is the number of moles, γ is the universal gas constant, and T is the absolute temperature.

For constant pressure, volume increases with absolute temperature. As density (ρ) is mass (m) over volume (V), density at constant pressure is inversely related to increased absolute temperature (T). We can, therefore, write the equation for velocity as a function of absolute temperatures as:

$$W_z = \left[2\, g\, H\left(1 - \frac{T_a}{T_f}\right)\right]^{1/2} \tag{8.27}$$

This is a very simple model that can be used to estimate the initial vertical velocity of the convective flow from flame height and temperatures of the ambient air and the flame volume. If we assume an ambient temperature (T_a) of 300 K and an average temperature inside the flame (T_f) of 500 K, Eq. (8.29) can be simplified to:

$$W_z = 2.8\, H^{1/2} \tag{8.28}$$

where W_z is the vertical velocity (m s^{-1}), and H is the vertical height of the flame (m).

Telitsyn (1996) measured the updraft flow buoyancy velocity (W_z in m s^{-1}) at the upper level of flames using anemometers in laboratory fires and found a similar result as Eq. (8.28).

$$W_z = 2.5\, H^{1/2} \tag{8.29}$$

The equations above allow for the estimation of the vertical updraft velocity as a function of flame height. However, it is common to predict vertical updraft velocity as a function of fireline intensity. Based on this approach, Raupach (1990) proposed the following equation:

$$W_z = 1.66\, \left[(g\, I_B)/\left(T_a\, \rho_a\, C_{pa}\right)\right]^{1/3} \tag{8.30}$$

where g is the acceleration of gravity (9.8 m s^{-2}), I_B is Byram's fireline intensity (expressed in kW m^{-1} in this equation), T_a is the ambient temperature, ρ_a is the density of air, and C_{pa} is the specific heat capacity of air at constant pressure. For the approximate values of $T_a = 300$ K, $\rho_a = 1.2$ kg m^{-3}, and $C_{pa} = 1$ kJ kg^{-1}, the equation simplifies to:

$$W_z = 0.5\, {I_B}^{1/3} \tag{8.31}$$

The graphical display of Eqs. (8.29) and (8.31) is shown in Fig. 8.21.

Typical values for the vertical velocity derived from Eq. (8.31) have practical applications. Fireline intensities determine updraft velocities, which determine which type of materials can be lofted. Very low-intensity fires with fireline intensity below 70 kW m^{-1}, result in vertical velocities below 2 m s^{-1}. With vertical air velocities below 2 m s^{-1}, only very light burning pieces with large exposed areas, such as broad leaves with a surface density below 0.025 g cm^{-2}, can be lofted.

Fires burning with a fireline intensity of 500 kW m^{-1} have a vertical velocity of 4 m s^{-1}. Needles and bark plates of several pine and eucalypt species with a surface density below 0.1 g cm^{-2} can be carried aloft under these relatively high but not uncommon intensities for prescribed fires. Therefore, the value of $W_z = 4.0$ m s^{-1}, corresponding to a fireline intensity of 500 kW m^{-1}, is considered as the desirable upper limit for a prescribed burn (Ellis 2000, 2010).

High intensity wildfires with fireline intensity of more than 10,000 kW m^{-1} create convective updrafts with a vertical velocity of more than 10 m s^{-1}. This value is associated with the surface density of 0.6 g cm^{-2}, the threshold at which larger

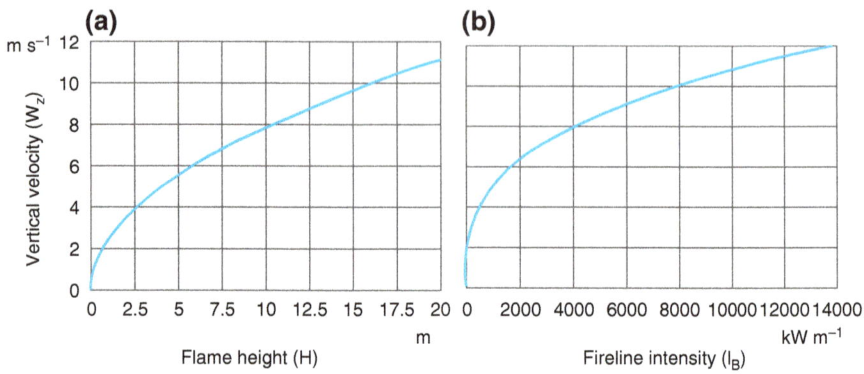

Fig. 8.21 Vertical velocity of the fire plume relative to (**a**) flame height and (**b**) fireline intensity. These are graphs of Eqs. (8.29) and (8.31)

pieces of fuel, such as pine cones or eucalypt capsules, can also be carried in the plume and may cause significant long range spotting.

Under higher fire intensities and stronger convection columns, larger firebrands may be produced. With Inferno fires (Raupach 1990) with fireline intensities of 100,000 kW m^{-1}, the updrafts are estimated to be 23.4 m s^{-1}. Byram (1959) indicated that updraft velocities within the convection column may exceed 30–35 m s^{-1}. In these extreme conditions, firebrand materials may even include logs (Cheney and Bary 1969). Ellis (2000) provided a good discussion on this topic.

Development of the Fire Plume

The airflow within the plumes of most fires is turbulent rather than laminar (Fig. 8.22). In laminar flow, particles within the column of fluid follow a smooth path and do not interfere with each other as they rise. In contrast, turbulent flow is best thought of as irregular flow where the particles in the column of air move in a chaotic manner, resulting in increased mixing and eddy formation (Fig. 8.23). Because the modeling of turbulent flows is difficult, empirical approaches are often taken.

Fire plumes have three unique regions due to the buoyant forces and eddy formation:

1. In the area of persistent flame, heating is continuous, and the gases are accelerating upwards,
2. Where flames are intermittent, combustion occurs irregularly, and
3. Within the buoyant plume, the temperature and velocity of upward movement decrease with height.

As the hot gases generated during combustion rise, cold air entrained into the fire plume increases the plume width (Fig. 8.24). The entrained air also cools the rising

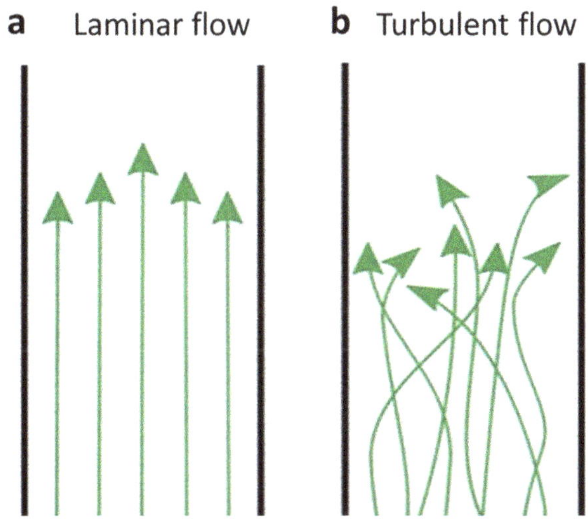

Fig. 8.22 Schematic representation of (**a**) laminar flow regime and (**b**) turbulent flow regime in a smooth tube. (Adapted from Guillom 2008)

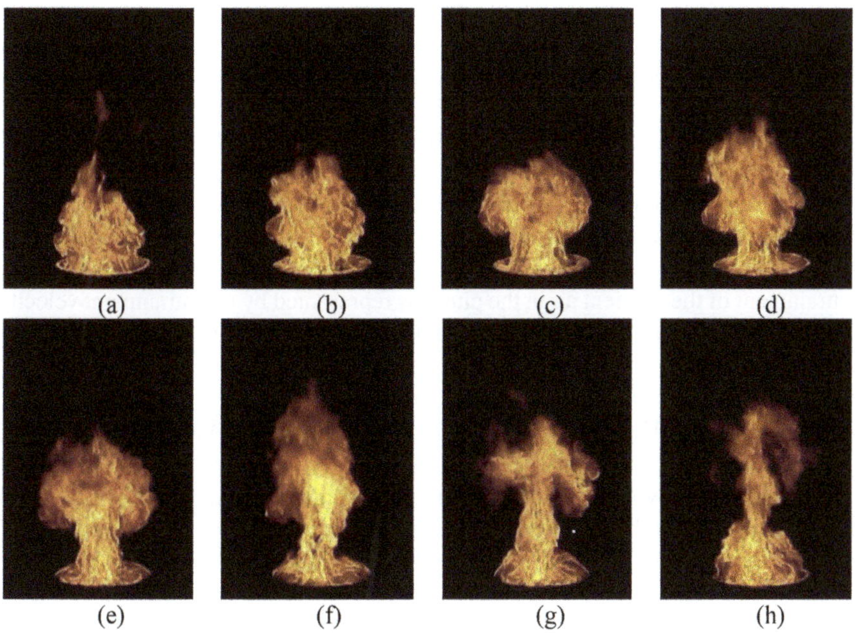

Fig. 8.23 Sequential (**a** to **h**) formation of eddies. (From Raj and Prabhu 2018)

hot gases, which decreases the temperature along the edges of the plume, causing a reduction in the upward velocity of the hot air. This reduction ultimately results in the formation of eddies.

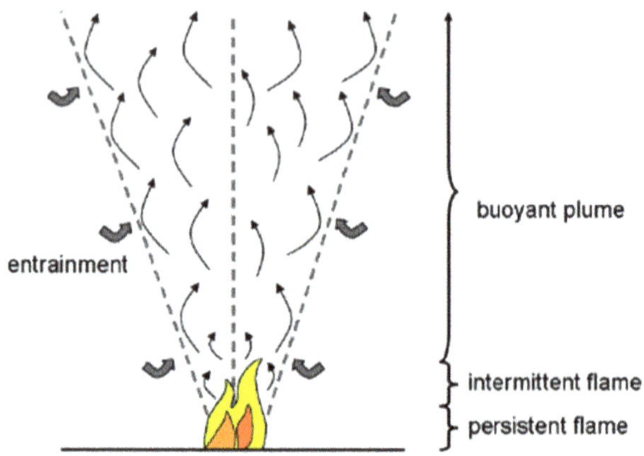

Fig. 8.24 Air entrainment in the buoyant plume. (From Michaletz and Johnson 2007)

In most forest fires the fire plume is observed to expand at a rather constant expansion angle, that can vary from 6° on very intense fires in heavy fuels to 18° on low intensity fires in light fuels, with 12° being a reasonable average (Chandler et al. 1983). The process of expansion of the fire plume is caused by air entrainment, a process that has been studied for a long time. The analysis of the process is based on the fact that when a stream of fluid is in contact with another stream, the eddies which cause transfer of matter between them are characterized by velocities proportional to the relative velocity of the two streams (Lee and Emmons 1961).

For a vertical plume, without wind, the rate at which air is entrained is taken as proportional to the vertical velocity of the fire plume (W_z). However, with wind (U), the fire plume is at an angle (A_p) with the horizontal wind flow, and the mass entrainment of the ambient air in the plume is represented by an entrainment velocity (V_e) which is the sum of the horizontal and vertical components (Mercer and Weber 1994, 2001):

$$V_e = a_1 \left(W_z - U \cos A_p \right) + b_1 U \sin A_p \tag{8.32}$$

For the case of symmetrical plumes around a vertical axis, the coefficients a_1 and b_1 are relatively well documented, with values around $a_1 = 0.1$ and $b_1 = 0.5$ (Krishnamurthy and Hall 1987). For line plumes the values used by Mercer and Weber (1994, 2001) were $a_1 = 0.16$ and $b_1 = 0.5$. In no-wind situations, where $U = 0$, the velocity of air entrainment is only proportional to the vertical air velocity W_z.

The effect of wind in increasing air entrainment justifies the fast cooling of the convection column with height, which is very important in predicting the effect of wind on scorch height in prescribed fire operations (see Chap. 9). Also, the development of the fire plume with entrained air is important to understand complex fire-atmosphere interactions as indicated in Sect. 8.5 in the Chapter.

8.4.2 Firebrand Generation

Firebrand generation has been measured experimentally in a limited number of studies. Firebrand production and characteristics produced from burning small trees of Korean pine (*Pinus koraiensis*) and Douglas-fir (*Pseudotsuga menziesii*) were measured by Manzello et al. (2007a, b). Most firebrands, or embers, collected were small in size (Fig. 8.25) with 70% smaller than 0.3 g mass.

Other scientists have collected data on the nature and density of firebrands and their distance from the main fire. In the experimental fires of the project VESTA in dry eucalypt forests in Australia (*Eucalyptus marginata*), Gould et al. (2007) collected firebrands in polyethylene sheets placed at different distances from the main fire. The density of large firebrands of eucalyptus bark (up to 40 m^{-2}) at close distances decreased exponentially with distance, to less than 1 m^{-2} at 100 m distance from the fire front. In the European project SALTUS, the few firebrands collected up to 2 km from the wildfire front were mostly leaves, needles, and bark fragments from pine trees (*Pinus pinaster*). However, the embers collected far from the fire front are only an unknown fraction of the initial firebrands generated as the firebrands burn and lose mass during lofting, transport, and fall.

Albini (1979) estimated the maximum potential spot fire distance by creating a model for the optimum firebrand, assuming that at least one ideally suited firebrand particle with optimum size exists near the top of a burning tree. Particles smaller than optimum could travel farther, but they would burn up before landing. Particles larger than optimum would still burn when landing but would not travel that far. With this

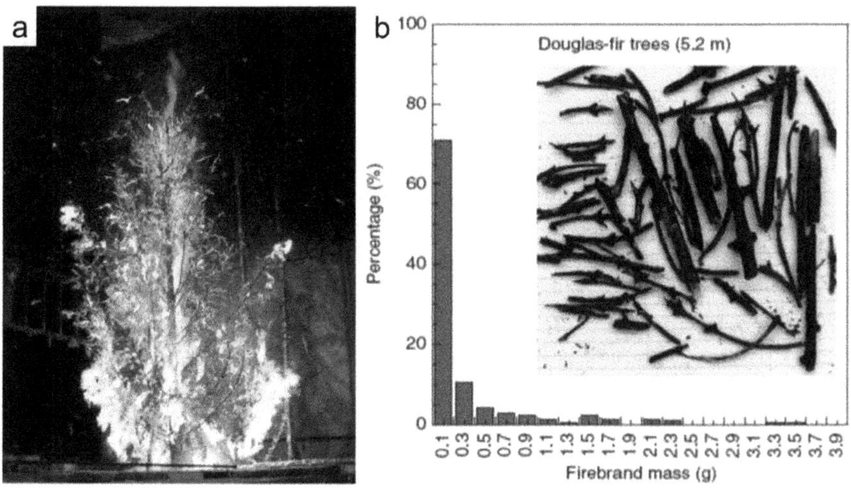

Fig. 8.25 Experimental study of firebrand generation from a small tree (5.2 m tall) of Douglas-fir (Pseudotsuga menziesii) showing (**a**) the burning tree and (**b**) the firebrands generated with their size distribution. (From Manzello et al. 2007b)

assumption, Albini (1979) produced spotting models independent of firebrand size and related only to fuel, fire, and wind variables.

"An individual firebrand starts as a leaf, twig, seed, nut, pine cone, piece of bark, or a small fragment of a larger piece of fuel that was partially consumed" (Potter 2016). All of these fuel components might develop into firebrands with different probabilities due to their different characteristics. The two main parameters to assess to evaluate the potential for spotting of the different fuel components are their aerodynamic characteristics and the burnout time.

8.4.3 Lofting of Firebrands

The Aerodynamic Characteristics of Firebrands

For a firebrand to be lofted, the updraft wind velocity must be greater than the maximum rate of fall that the firebrand would achieve in still air (i.e., the terminal velocity). The terminal velocity is defined as the maximum speed of an object freely falling through a medium such as air. Based on experimental and theoretical work, Tarifa et al. (1965) indicated that "it is an excellent approximation to assume that the firebrands always travel at their terminal velocity".

Studies on terminal velocities of wood samples followed (Muraszew 1974). Clements (1977) was the first to determine terminal velocities of different fuels common in forest fires by measuring the time of their fall from a fire tower. The results showed significant differences between fuel components and species types (Fig. 8.26). The average terminal velocity for all 14 broadleaves tested was 1.8 m s^{-1} (ranging from 1.3 to 2.2 m s^{-1}), while that of the needles of three pine species was 3.7 m s^{-1} (2.9–4.1 m s^{-1}), about twice that of the hardwood leaves. The terminal velocity of pine cones was much higher, averaging 12.1 m s^{-1} (8.6–16.5 m s^{-1}). The only bark studied was that of *Betula papyrifera,* which had an average terminal velocity of 5.6 m s^{-1}.

Terminal velocities of burning firebrands were measured experimentally in a vertical wind tunnel designed by the Bushfire Behaviour and Management Group at CSIRO and built in Canberra specifically for firebrand studies (Knight 2001). This facility was fundamental for the pioneering work of Ellis (2000, 2010), who addressed the aerodynamic characteristics of typical firebrands in Australian forests generated by eucalypt bark (Fig. 8.27). Ganteaume et al. (2011) also evaluated the characteristics of common potential types of firebrands commonly encountered in forest ecosystems in southern Europe.

The terminal velocity of fuel particles derives from their physical characteristics. Tarifa et al. (1965) represented this with a general drag equation for falling objects. They also demonstrated that firebrands in flight orient themselves to achieve maximum drag. At terminal velocity, there is an equilibrium between two forces, a downward force of gravity applied on the particle's mass and a drag force, resistant to the downward movement.

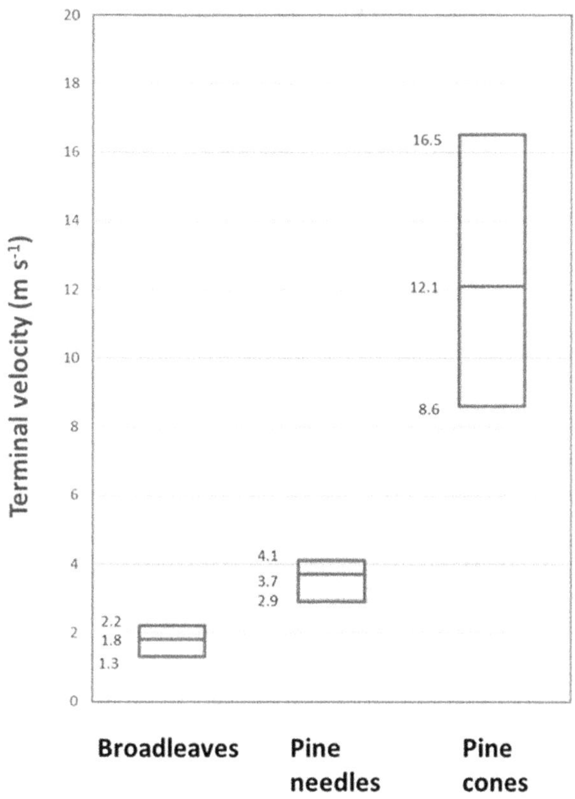

Fig. 8.26 Average terminal velocities for different fuel components. (Data from Clements 1977)

The downward force depends on the mass of the object (m) and gravity (g). The upward (drag) force depends on the density (ρ) of the fluid (air) through which the object is falling, on the falling terminal velocity of the object (v_t), or equivalently the upward velocity of air (W_z), the projected area (A), and the shape of the particle expressed by a dimensionless drag coefficient (C_d). The equilibrium between the two forces can be expressed as:

$$m\,g = 0.5\,\rho\,v_t^2\,A\,C_d \tag{8.33}$$

where m is the mass of the object (kg), g is the acceleration of gravity (m s^{-2}), ρ is the density of the air (kg m^{-3}), v_t is the terminal velocity of the object (m s^{-1}), A is the projected area (m^2), and C_d is a dimensionless drag coefficient. Rearranging Eq. (8.33) to solve for the terminal velocity (v_t) results in:

$$v_t = \left(\frac{2\,m\,g}{\rho\,A\,C_d}\right)^{0.5} = \left(\frac{2\,g}{\rho\,C_d}\right)^{0.5} \times \left(\frac{m}{A}\right)^{0.5} \tag{8.34}$$

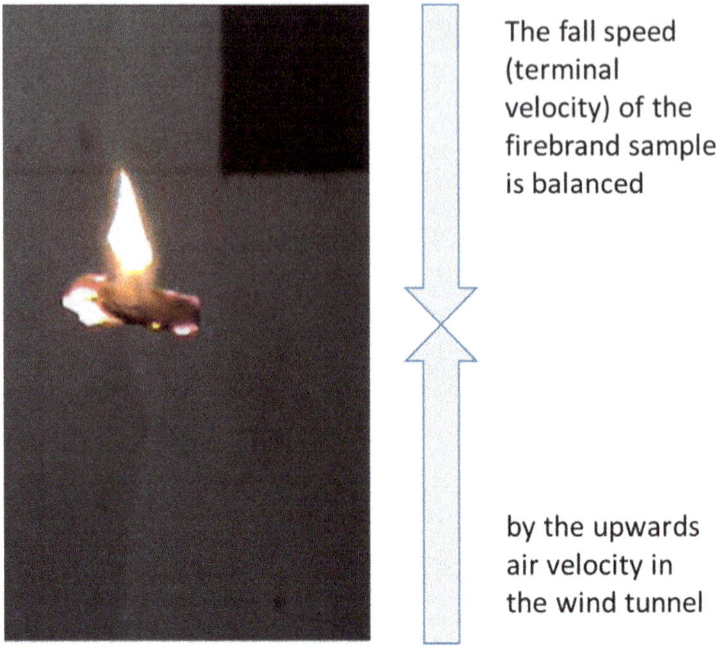

Fig. 8.27 In the vertical wind tunnel of CSIRO, in Canberra, it is possible to determine experimentally the terminal velocity of burning firebrands by finding the balance between the two opposite forces, the gravitational downward force and the upward force generated by vertical air velocity from a blower fan

This equation indicates that the terminal velocity of an object is dependent on the ratio of mass (m) to the area projected (A), which has been termed as surface density (m/A) (Ellis 2010). Assuming a density of air of $\rho = 1.27$ kg m^{-3} and a drag coefficient $C_d = 1$, the standard value for flat plates (Ellis 2010), and expressing the surface density (m/A) in g cm^{-2}, the terminal velocity v_t in m s^{-1} can be estimated as:

$$v_t = 12.4\left(\frac{m}{A}\right)^{0.5} \tag{8.35}$$

The values of the drag coefficient (C_d) have been studied in relation to the shape of the object with values of 0.47–0.50 for spheres and cones to 1.05 for cubes and varying from 0.82 to 1.15 between long and short cylinders. The value of $C_d = 1.17$ was used for wooden disks by Anthenien et al. (2006). Ellis (2010) found that for eucalypt bark pieces, the average values of C_d ranged from 0.77 to 0.85 for samples where no spin occurred to values from 1.13 to 1.39 where spin occurred. In Fig. 8.28, we illustrate the relation between surface density and terminal velocity for different values of the drag coefficient.

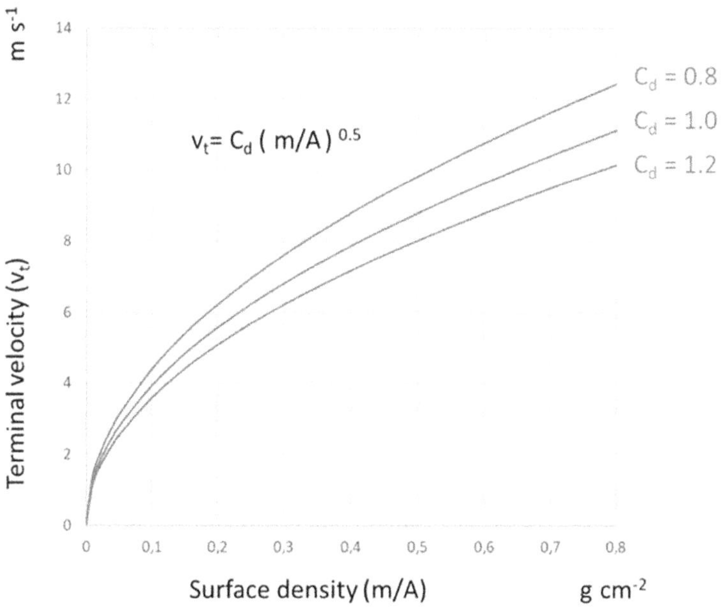

Fig. 8.28 Graphical representation of the relationship between surface density (m/A, the ratio of mass m to projected area A) and terminal velocity (v_t) for typical drag coefficients (C_d) of 0.8, 1.0 and 1.2

The terminal velocity of a fuel particle, or the equivalent upward velocity of air to maintain it floating in the convection plume, is basically a function of two characteristics of the particle, its mass and its projected area. The ratio of mass to projected area, or the surface density, is very different between different types of wildland fuels. Therefore fuel particles with low mass and high projected area, as leaves of broadleaved species, are able to float in plumes of very low-intensity fires, whereas pine cones can only be lifted in fire plumes with high vertical velocity originated by fires burning with high intensity. Typical values of the surface density of the different fuel materials that may be potential firebrands are shown in Fig. 8.29.

Based on Eq. (8.35), broad leaves with a surface density below 0.025 g cm^{-2} can be lifted in a fire plume with updraft velocities as low as 2 m s^{-1}. For most thin bark plates, with surface density values below 0.1 g cm^{-2}, it would take an updraft of about 4 m s^{-1} to loft the bark plates. However, for some bark types, cone scales, and some twigs with a surface density closer to 0.2 g cm^{-2}, a vertical wind velocity of around 6 m s^{-1} is required. For oak acorns, eucalypt capsules, or pine cones, with surface densities above 0.6 g cm^{-2}, transport in the fire plume requires updrafts of around 10 m s^{-1}.

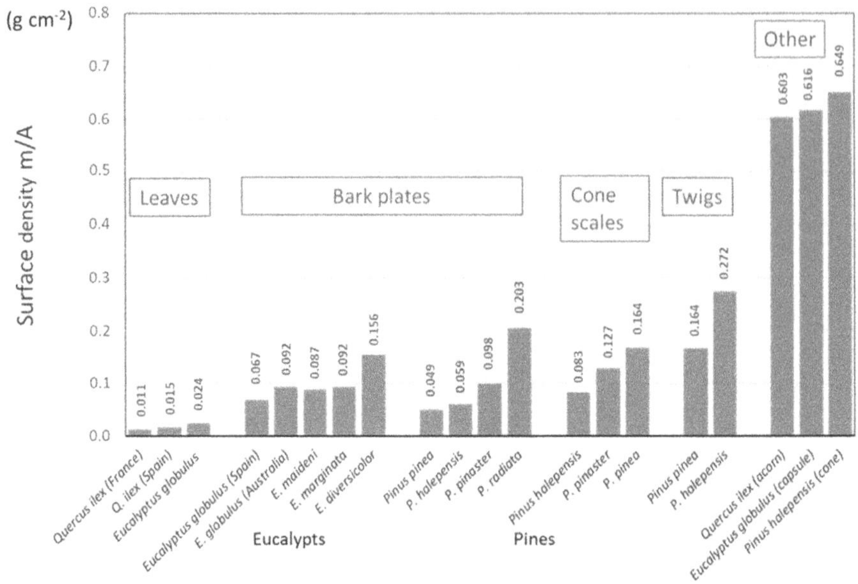

Fig. 8.29 Typical values of surface density (mass per unit surface area projected) based on data from Ellis (2010), Ganteaume et al. (2011), and Rego and Ellis (unpublished data)

The Maximum Potential Firebrand Lofting Height

The lofting of fuels from the ground surface is more difficult than those that are already located at higher places as bark particles in the trunk or leaves or twigs already at the canopy level. The fire's updraft initially accelerates as it rises, causing spotting vertical velocity to be higher at the canopy than at the ground level. Thus, spotting is more prolific in crown fires than in surface fires (Potter 2016). After firebrands are detached from the trees (or from other fuel components), and if their terminal velocities are lower than the updraft velocity created by the fire, the process of lofting starts.

The updraft winds above the flaming front are highly turbulent and, therefore, very difficult to measure in the field. In experimental fires in dry eucalypt forest, Gould et al. (2007) observed alternate updraft and downdraft periods. Because of field measurement difficulties, most initial approaches were based on laboratory experiments and theoretical modeling efforts (Tarifa et al. 1965; Muraszew 1974). More recently, more physically-based fire behavior models using computational fluid dynamics methods, as FIRETEC (Linn et al. 2002) or WFDS (Mell et al. 2005), have been used to model the updrafts, downdrafts, and horizontal movements of air during fires. However, simpler approaches that are more computationally efficient, such as those provided by Albini (1979, 1981, 1983), are widely used and included in operational fire behavior systems such as FARSITE and FlamMap (Finney 2004, 2006).

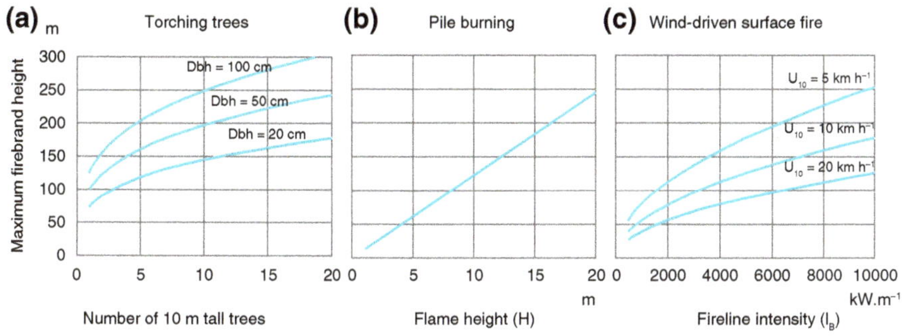

Fig. 8.30 Predictions of maximum firebrand lofting height z_{max} in meters using models for (**a**) torching trees (Eq. 8.36), for (**b**) burning piles of forest debris (Eq. 8.37), and for (**c**) wind-driven surface fires (Eq. 8.38)

The maximum height, z_{max}, at which a firebrand of optimum size would be lofted can be estimated independently of firebrand characteristics (Albini 1979). This approach predicts the maximum firebrand lofting height attainable from one or several torching trees (Albini 1979), by a pile of burning timber debris (Albini 1981), or by a wind-driven surface fire (Albini 1983). Morris (1987) found that variations in fuel parameters in the equations of Albini (1983) for wind-driven fires had little influence on the results. He proposed a simplified model that can be applied to firebrands lofted in any fuel type from torching trees, piles burning, or a surface fire (Fig. 8.30):

The maximum lofting height for firebrands lifted from torching trees is:

$$z_{max} = 25\ NT^{0.303} + 0.5\ HT \tag{8.36}$$

where z_{max} is the maximum lofting height of a firebrand (m), DBH is the tree diameter at breast height (cm), NT is the number of torching trees, and HT is tree height (m).

For firebrands from burning piles, the maximum lifting height is estimated as:

$$z_{max} = 12.2\ H \tag{8.37}$$

where H is the flame height of the burning pile (m).

Finally, the maximum lifting height for a firebrand from a wind-driven surface fire is:

$$z_{max} = 5.63 \left(\frac{I_B}{U_{10}} \right)^{0.5} \tag{8.38}$$

where I_B is Byram's fireline intensity (kW m^{-1}), and U_{10} is the wind speed at 10 m height (km h^{-1}). These equations are represented graphically in Fig. 8.30.

In the next section, we discuss the transport of firebrands, and we will see that it is the wind that mainly governs firebrand transport. As wind speed typically increases with height (Chap. 7), it is necessary to estimate the maximum height that firebrands can attain in order to be able to estimate how far they can go. Only firebrands that are lifted very high into the air can be transported great distances. This topic is discussed in the next section.

8.4.4 *The Transport and Fall of Firebrands: Searching for the Maximum Spotting Distance*

Estimation of the maximum spotting distance that a firebrand can be transported to originate a spot fire is important for fire fighter safety and operational decisions in fire management. The transport of firebrands is governed by the velocity of the air and by the terminal velocity of the firebrand as well as the size and shape of the firebrands and their mass loss during transport (Tarifa et al. 1965; Muraszew 1974). If the primary interest is in predicting maximum spotting distance, one can assume that the maximum velocity of the firebrand within the fire plume is that of the wind-driven fire plume.

In general, wind measurements are made at certain fixed heights, typically at 10 m (here referred to as U_{10}) or 6.1 m (here referred to as U_6). The wind speed at any height (U_z) can be estimated from the wind speed measured at a reference height (for example, U_6) by assuming that the vertical changes in wind velocity follow a power law (see Chap. 7):

$$U_z = U_6 \left(\frac{z}{6.1} \right)^{1/7} \tag{8.39}$$

Conversely, we can make:

$$U_6 = U_z \left(\frac{6.1}{z} \right)^{1/7} \tag{8.40}$$

The prediction of the maximum spotting distance was analyzed by Albini for burning trees (1979), burning piles (1981), and wind-driven surface fires (1983). The maximum spotting distance was approached as having two additive components: the horizontal distance traveled by firebrands during lofting (SL) and the distance traveled after lofting (SD), that is, after attaining the maximum height z_{max}.

For torching trees and burning piles, Albini (1979, 1981) assumed that the firebrands were lofted vertically, and SL was negligible. However, for wind-driven surface fires, Albini (1983) added the predicted horizontal distance that firebrands travel downwind during lofting (SL). Chase (1984) gave the equation based on the measured wind speed (U_6) and the maximum firebrand height z_{max} as:

$$SL = 0.000503\, U_6\, z_{max}{}^{0.643} \tag{8.41}$$

where SL is the downwind drift during lofting (km), U_6 is the measured wind speed at 6.1 m above the ground (km h^{-1}), and z_{max} is the maximum firebrand height (m).

The prediction of spotting distance in flat terrain (SD) after lofting was determined by the estimated maximum firebrand height z_{max}, by the wind speed measured at 6.1 m (U_6) and by the average height of the surrounding vegetation z_v. Chase (1984) gave the equation as:

$$SD = 0.0013\, U_6\, z_v{}^{1/2} \left[0.362 + 0.5 \left(\frac{z_{max}}{z_v} \right)^{1/2} ln \left(\frac{z_{max}}{z_v} \right) \right] \tag{8.42}$$

where SD is the spotting distance in flat terrain after lofting (km), z_v is the average height of vegetation (m), and U_6 and z_{max} have the same meaning and units as before (km h^{-1} and m, respectively).

These models may be seen as overestimating spotting distance as they attempt to predict a maximum spotting distance of an optimal virtual firebrand (Chase 1984). In reality, different types of firebrands have different characteristics that will influence lofting. Furthermore, firebrand consumption during lofting is not considered. Further, as with all empirical models, they may not apply well for conditions outside those for which they were built. This justifies the words of caution by Catchpole (1983), who suggested that Albini's models "may underpredict maximum spotting distance in eucalyptus forests due to the aerodynamic nature of the tree bark".

8.4.5 The New Ignitions from Firebrands

Firebrands, also called embers, are burning fuel particles transported from the main fire. They may ignite spot fires if they land on fuels that can readily burn. The ignition probability from firebrands depends on the state of the firebrands when reaching the surface fuel, their size, and if they are flaming or smoldering. The probability of successful ignition also depends strongly on the characteristics of the fuelbed where the firebrands land, with fuel bulk density and moisture being of particular importance.

Flaming firebrands ignited fuelbeds more readily than smoldering firebrands in Mediterranean fuels (Ganteaume et al. 2009). The type and weight of firebrands were also important, with pine cone scales and eucalypt leaves and bark more likely to ignite fuelbeds. The characteristics of the fuelbeds were also important. Higher bulk density and fuel moisture both increased the time-to-ignition and decreased the probability of successful ignition (Ganteaume et al. 2009). Firebrand characteristics, wind, and the landing fuelbed all interact. Working with small bark samples of

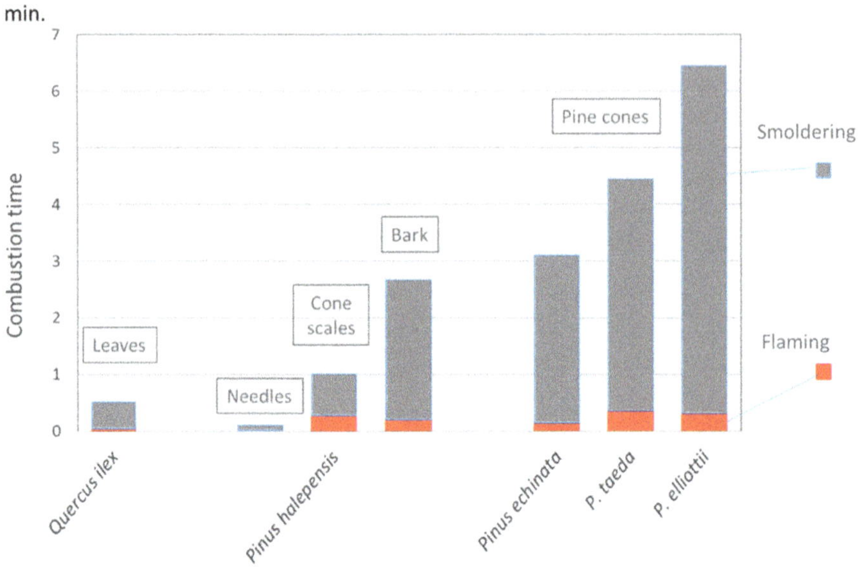

Fig. 8.31 The duration of smoldering combustion is much greater than that of flaming combustion, but both vary with fuel particle type. (Data from Ganteaume et al. 2011 for leaves of *Quercus ilex*, and needles, cone scales and bark of *Pinus halepensis* and from Clements 1977 for pine cones (*Pinus echinata*, *P. taeda* and *P. elliottii*))

Eucalyptus obliqua as firebrands, Ellis (2000) concluded that the probability of ignition was highly dependent on litter moisture content and the surface wind velocity. At 8% dead fine fuel moisture of the receiving fuel and no wind, the probability of ignition by small flaming firebrands was 100%. If those firebrands were smoldering, the probability of ignition was 0%. However, with fine fuel moisture of the receiving fuelbed of 4% and a wind velocity of 1 m s^{-1}, the ignition probability from smoldering firebrands increased to around 60%.

Few firebrands are flaming by the time they land on a fuelbed. Although flaming firebrands are more successful than smoldering firebrands in igniting fuelbeds and starting spot fires, smoldering combustion determines most of the spotting processes, especially long-range spotting as flaming firebrands typically only burn with flames for short periods of time (Tarifa et al. 1965). Ellis (2000) observed reflaming during the flight of firebrands that were smoldering, indicating that this could be significant to the ignition probability by a firebrand.

The relative importance of smoldering and flaming duration of fuel particles of different types can be seen in Fig. 8.31. In all cases, regardless of the total duration of fuel particle combustion, or burnout time (t_B), smoldering is its most important component.

A simple way to estimate maximum spotting distance after lofting (SD, km) is by multiplying the total combustion duration of the fuel particle (burnout time t_B, min) by the wind speed (U_z, km h^{-1}) at height z of transport of the firebrand:

$$SD = U_z \frac{t_B}{60} \tag{8.43}$$

The height of transport can be set at the maximum firebrand height z_{max} mentioned before. An example can illustrate this simple calculation.

Suppose that we have established a maximum firebrand height z_{max} of 200 m for which a wind speed U_z of 30 km h^{-1} is estimated. If our firebrands are dry leaves of *Quercus ilex*, with a burnout time (t_B) of 0.5 min, we could simply estimate the maximum spotting distance using Eq. (8.43) as 0.25 km or 250 m.

This would be a relatively short-distance spotting.

With the exact same wind conditions, $U_z = 30$ km h^{-1}, a pine cone, with a combustion time of $t_B = 6$ min, could travel 12 times more to a maximum SD distance of 3 km.

This illustrates the importance of burnout time in estimating spotting distance.

For any given wind speed, the total duration of combustion sets the limit of the maximum distance that a firebrand can travel and land still burning in the fuel ahead of the main fire. This is a simple way to understand the limits for a successful secondary ignition.

8.4.6 The "Optimal" Firebrand for Long-Range Spotting

Long-range spotting is a critical influence on wildfire suppression effectiveness. Long-range spotting is highly unpredictable as it is impossible to know with precision where and when spot fires will start. If the possible maximum spotting distance is several kilometers, fires could start from firebrands anywhere within that distance. Ecologically, long-range spotting often creates more heterogeneity by creating burned islands (See Chaps. 9 and 12) and spatial variability in crown scorch height. Fire intensities can be higher where multiple fires burn together (See Sect. 8.5.3). Long-range spotting contributes to rapid fire growth and thus to large, extreme fires. This can be especially challenging for fire managers who are often trying to protect people and their homes or other values from multiple fires burning at the same time within a region. If spotting is likely on one of those fires, it is likely on many of those fires.

One approach to estimate long-range spotting is by using the "optimum firebrand material" (Albini 1979) concept. This ideal firebrand for long-range spotting is a

Fig. 8.32 Eucalypt bark types: (**a**) stringybark burning, (**b**) candlebark with ribbons, respectively associated with short and long-distance spotting (Photographs by P. Fernandes), and (**c**) Anthony Mount with Francisco Castro Rego in the forests of Tasmania. (Photograph by Stefano Mazzoleni)

trade-off between combustion time and surface density. Large fuel pieces, like pine cones, tend to burn for a long time, but they generally have large surface density and can only travel within the plume as long as the upward lift of the plume is higher than their terminal velocity and as such are not transported as far as firebrands associated with smaller surface densities. In contrast, small light fuels are easily lifted in the convection column but their combustion time is short, and therefore are not able to spread fires by spotting over long distances.

The optimal situation for long-range spotting comes from eucalypt forests in Australia. Eucalypts have been classified as having either fibrous stringy bark or smooth bark (Fig. 8.32a, b). For those with stringy bark, the bark is held loosely on the stem and is easily torn off by wind during a fire and carried through the air producing abundant short-range spotting. Low-intensity prescribed fire can be used to reduce spotting potential by consuming part of the bark. Eucalypts with smooth bark produce long streamers or ribbons referred to as candlebark which tends to accumulate in the forest floor, especially around trunks, and semi-detached from the trunks up to variable height in the tree. Anthony Mount (Fig. 8.32c) described the process as "as plates or ribbons that form part of the forest floor litter or in other cases the ribbons hang on the trees and in the crotch of the branches and are the main source of fire brands for long distance spotting" (Mount 1969). During fires, "the candlebark pieces burn slowly and have good aerodynamic properties" that "are the source of very long-distance spotting ranging from 10 to 30 km" (Luke and McArthur 1978).

The aerodynamic properties of candlebark were studied by Ellis (2000, 2010). The surface density of the bark is between 0.2 and 0.3 g cm^{-2}, indicating easy transport by a convection plume with an upward air velocity of 5–7 m s^{-1}, which is common in Australian "bushfires". However, these candlebarks burn in a more distinct way. Ellis and Rego (unpublished) found that the cylindrical bark samples burned slowly, floating in the air, in equilibrium with vertical wind velocity, indicating that combustion rate is not influenced by wind velocity. The relative velocity is negligible as the particle and the surrounding air are moving almost at the

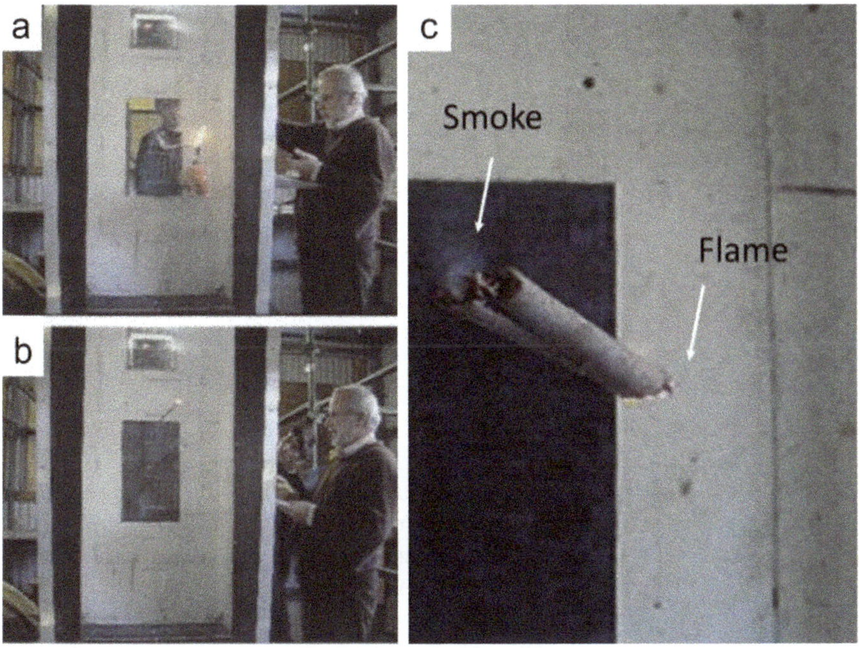

Fig. 8.33 (**a, b**) Experiments in CSIRO's vertical wind tunnel by Peter Ellis (lighting the sample) with the assistance of F.C. Rego (recording). (**c**) A sample of a bark streamer with the typical form of an empty cylinder, as a cigar, floating in the air in equilibrium with vertical air velocity. The bark cylinder burns with flames on one side and smoke on the other side. (Photographs by Francisco Castro Rego)

same speed inside the plume. The cylinder burns from one side to the other, with alternate flaming and smoldering, at a rate of about 1 cm min^{-1}, as a cigar (Fig. 8.33).

The importance of this mechanism can be illustrated with an example. A 30 cm cylinder, very common in these eucalypt forests, could burn for 30 min, which, if the wind speed is 30 km h^{-1}, could land still burning 15 km ahead of the fire front (Eq. 8.43). This process explains the observations of Luke and McArthur (1978) that "a burning piece of candlebark must remain alight for upwards of 30 min or more to produce a spot fire 20–30 km ahead of the main convection centre of the fire".

The duration of combustion or burnout time of these barks is therefore only dependent upon their length allowing for spotting distances several times more than fuel particles of other forest types. It can be recalled that some of the firebrands that burn the longer in pine forests, the pine cones, have a maximum burnout time of no more than 6–7 min (Fig. 8.31).

These results illustrate that fuel characteristics are very important in determining fire behavior and spotting distances.

8.5 Complex Fire-Atmosphere Interactions

Fires and the atmosphere interact across a range of spatial and temporal scales on all fires, but the interplay is especially influential in extreme fire behavior. In extreme fires the convective plume may significantly affect atmospheric patterns that may in turn influence fire behavior. Here we discuss the complex interactions that may occur in extreme fires.

8.5.1 The Relative Strength of Buoyancy and Wind

In large wildland fires, updrafts due to plume rise can reach velocities of 30–60 m s^{-1} (Coen et al. 2004; Filippi et al. 2014; Sullivan 2017). The relative strength of these updrafts depends upon many factors, including the rate of heat release and the velocity of the ambient wind (Fig. 8.34). Fires behave differently when dominated by the upward forces associated with the fire plume, rather than by ambient winds. Plume-dominated fires are less predictable and more likely to result in spotting.

The relative strength of the buoyant plume compared to the ambient wind can be estimated using the non-dimensional Froude or Byram's convective number, N_c

Fig. 8.34 Forces that govern fire behavior. (Background photograph of a northern Portugal wildfire in pine forests by Paulo Fernandes)

(Byram 1959). This is defined as the ratio between the vertical force associated with buoyancy and the inertial force due to horizontal wind flow:

$$N_c = \frac{2\, g\, I_B}{\rho_a\, C_{pa}\, T_a\, (U - R)^3} \tag{8.44}$$

where g is gravity (9.8 m s^{-2}), I_B is Byram's fireline intensity (kW m^{-1}), ρ_a is the density of air (1.2 kg m^{-3}), C_{pa} is the specific heat of dry air at constant pressure (1.005 J kg^{-1} K^{-1}), T_a is the absolute ambient air temperature (K), U is the wind velocity (m s^{-1}), and R is the forward rate of fire spread (m s^{-1}).

When N_c is much greater than 1, fire behavior is dominated by the upward forces associated with the fire plume, with a relatively weak effect of the ambient wind velocities. This results in more vertical flames and a well-defined convective column that is orientated vertically rather than leaning with the wind. Rothermel (1991) suggested that such plume-dominated fires are often associated with relatively low ambient wind velocities. The strong updrafts associated with plume-dominated fires foster rapid fire growth, and the potential for spot fires in part is due to a so-called "reverse wind profile" where the lower level winds have a greater velocity than the upper winds farther above the ground. Plume-dominated fires are often associated with very high fireline intensity, rapid growth, and have contributed to fire fighter fatalities (Rothermel 1991). Where N_C is much less than 1, fire behavior is primarily driven by the ambient wind flow. In these cases, called wind-driven fires, the fire plume is deflected in the direction of the dominant wind flow rather than rising vertically and the growth of the fire is closely related to the wind speed and direction (Fig. 8.35).

Plume-dominated fires often exhibit significant variability in fireline intensity relative to wind-driven fires which can make them more unpredictable. The updrafts and downdrafts within the fire plume can cause erratic winds on the ground that can endanger fire personnel. Furthermore, the fire rate of spread associated with plume-dominated fire is greater than would be expected for a wind-driven fire given the ambient wind speed (Morvan 2014).

8.5.2 Downdrafts Associated with Firestorms

Kerr et al. (1971) identified eight types of large fires as defined by the interactions between wind and the convection column of the fire, providing generic fire behavior features for each type in terms of rate of spread and spotting. The nature of the convection column is an outcome of the heat release rate and temperature and wind speed gradients in the atmosphere.

An unstable atmosphere is one where the temperature decreases rapidly with height. Under these conditions, the rate of upward flow will be greater resulting in increased indrafts into the fire plume. The smoke plume will then grow in height as

Fig. 8.35 Wind-driven (top) and plume-dominated (bottom) fires interact quite differently with the surface winds. This depends on the wind speeds at different heights above the ground relative to the vertical winds within the fire plume. Photographs (**a, e, f**) were taken by Christopher Templeton while serving as a pilot. Photographs (**b, c, d**), were taken by members of elite firefighting crews known as hotshots. All photographs were taken by government employees at work and are not subject to copyright. (Figure made by Heather Heward)

long as there is enough fuel available, fire is burning intensely, and the wind speed decreases with height above the ground. The increased indraft into the fire will amplify fire behavior and phenomena induced by wind gustiness and turbulence such as spotting and fire whirls. Byram (1954) said that a plume-driven fire:

> ", ...bears about the same relation to the ordinary fire that a large railway locomotive bears to a small house furnace. The furnace converts fuel into heat and nothing more. The locomotive on the other hand converts not only fuel into heat but in turn converts a part of its heat energy into the driving or kinetic energy of motion, which is evident in the speed of the locomotive and the cars it pulls."

In very extreme fires, pyrocumulonimbus (pyroCb) clouds (or cumulonimbus flammagenitus) may form (Fig. 8.36). In these situations, an added major concern in wildfire behavior occurs when the convective plume "collapses" resulting in strong outward winds near the ground. This phenomenon is similar to downbursts associated with cumulus clouds and occurs as evaporative cooling causes an air mass to rapidly descend towards the ground and spread out horizontally. The downdraft winds can exceed 100 km h^{-1} and can fan out in all directions when they hit the ground resulting in erratic fire behavior. Lightning may also be produced from the pyrocumulonimbus clouds.

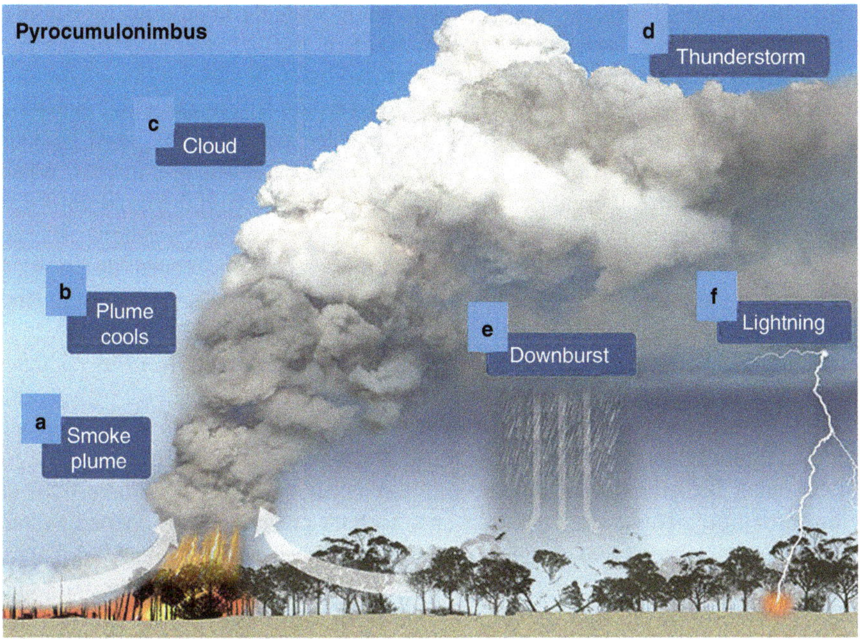

Fig. 8.36 The pyrocumulonimbus clouds that can form over intensely burning fires can influence surrounding winds and the behavior of the fire. Thus, (**a**) upward convection in the smoke plume results in surface winds drawn into the base of the fire from multiple directions. (**b**) As the rising air cools with elevation, the water vapor released during combustion condenses forming (**c**) a cloud. (**d**) A thunderstorm may result from the turbulent circulation of air in updrafts and downdrafts in the pyrocumulonimbus cloud. (**e**) Some of the downdrafts result in intense vertical winds down, enough to cause trees to fall and winds to spread outward along the surface, often resulting in fire spread rapidly increasing and changing directions with the wind. (**f**) Lightning and firebrands may ignite fires far from the area currently burning. Graphic adapted from Australian Bureau of Meteorology http://media.bom.gov.au/social/blog/1618/when-bushfires-make-their-own-weather/. Accessed 21 Nov 2020

Pyrocumulonimbus formation is a particularly severe event that can be decoupled from surface weather, as measured by weather stations in the region. Such decoupling was observed in the deadly wildfires of 2017 in Portugal, with firestorms forming in the evening and at night that coincided with the timing of human fatalities (Castellnou et al. 2018). The processes of downbursts and lightning strikes may further accelerate extreme fire behavior (Fig. 8.36). See Sect. 8.6.2 for an example of a pyroCb event.

8.5.3 Complex Interactions Between the Environment and Fire, and Between Fires

The complex interactions among the fire, fuels, topography, and atmosphere that drive variability in fire behavior are difficult to predict. As we often do not fully understand these influences, we are not able to anticipate the conditions for extreme fire behavior, or we underestimate them. One of the key elements of the fire environment triangle (Part III) is that it recognizes the importance of these complex interactions. These couplings affect numerous aspects of a vegetation fire, including the balance between different modes of heat transfer, turbulent mixing of gaseous reactants, fire spread, and the shape of a fire perimeter and smoke transport. When these aspects interact, their alignment can contribute to blowup fire conditions that appear disproportionally intense in relation to the prevailing conditions (Byram 1954). Such fires take the form of a localized topographically-related 'eruption', a large conflagration with a well-defined moving fire front, or a firestorm producing multiple and coalescing ignitions.

Many different manifestations of extreme fire behavior may occur in wildland fires often as a result of complex interactions, including phenomena such as fire whirls, vortexes, or mass ignition that are not detailed in this book. A good reference for the current understanding of those processes is Werth et al. (2016).

Complex fire-atmospheric interactions can occur when multiple fires burn in close proximity. This topic is of relevance to both prescribed fire operations and suppression fire operations where multiple point or line ignitions occur and merge. The interactions between multiple fires influence flame properties, fuel consumption rates, fireline intensity, and fire rate of spread. In turn, these could influence fire effects.

To understand the interaction between multiple fires, one can first think about a single shrub burning in the absence of wind. In this situation, a vertical plume forms above the shrub with entrainment of cool air into the plume occurring from all sides (Fig. 8.37). However, if a nearby shrub is also ignited, the plume from the second shrub interferes with the entrainment of air into the first plume, resulting in the

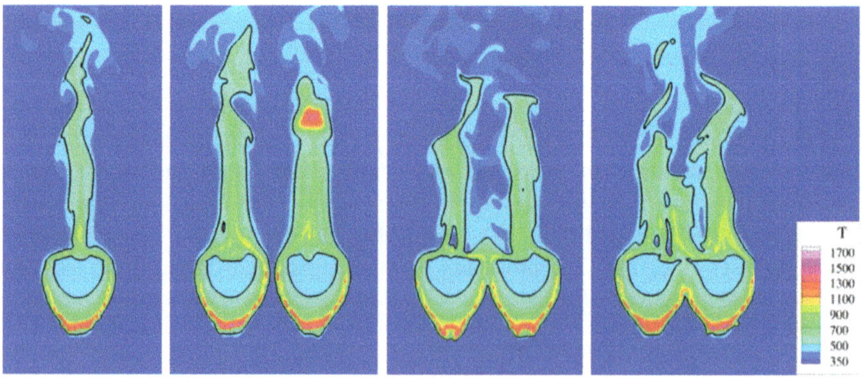

Fig. 8.37 Representations of increasing interactions between flames from burning shrubs from no interaction (a single shrub burning on the left) and flames from two shrubs burning with increasing proximity (from left to right). Flames almost merge when the two shrubs are burning in close proximity (right). The gas-phase temperatures in and above the shrubs are simulated using a computational fluid dynamics model. Absolute temperatures vary from inside the shrubs at around 500 K (light blue) to 1300–1500 K (red to pink). (Adapted from Dahale et al. 2013)

plumes bending towards each other, due to a decrease in the pressure between the two shrubs. As the two plumes merge together, complex feedbacks between the two fires and the atmosphere can result in heat transfer modifications resulting in increased burning rates. The strength of the fire-fire interactions is highly dependent upon the separation distance between the shrubs (d) and the diameter of the shrubs (D). Simulation modeling by Dahale et al. (2013) suggested that at distances of $d/D > 0.1$ the behavior of the burning of two shrubs is essentially the same as a single shrub suggesting that at these distances there are no longer plume interactions (Fig. 8.37).

A common example of fire interactions occurs during prescribed fire operations where multiple lines or points of fire are ignited, or in the use of backfiring operations during a wildfire. In these cases, the fire fronts are purposefully ignited so that their interactions will achieve a given management objective. When fire fronts are close enough to interact and merge, the burning rate of the fire can change dramatically as flame height increases (Werth et al. 2016).

When interactions occur between large fires burning simultaneously, these wildfires are described as mass fires, area fires, or "firestorms" (Countryman 1974). Hundreds or thousands of individual fires may interact over an area and exhibit some "unified" behavior. Such fires are generally described as having very strong indrafts with minimal outward propagation. They have extremely tall convection columns or smoke plumes and burn for long durations until most of the fuel within the perimeter is consumed (Werth et al. 2016). When two fire fronts interact in the vicinity of canyons or ridges, fire fighter fatalities have resulted from the intense fires driven by strong winds. Experiments on the flow and fire spread in canyons (Viegas and Pita 2004) showed strong feedback between fire behavior and winds channeled by topography.

The interaction between fire lines has been studied experimentally in mixed heathland in Galicia, Spain (Vega et al. 2012). The air flow was significantly affected by the fire fronts. In the first phase of their experiments, the backfire and the head fire propagated independently with quasi-steady rates of spread of around 0.03 m s^{-1} (backfire) and 0.25 m s^{-1} (headfire). When the 2 fire fronts were 20 m apart, the head fire accelerated suddenly. Just before their encounter, the progression of the 2 fire fronts accelerated to 0.45 m s^{-1} (backfire) and 0.61 m s^{-1} (headfire). The interaction between fire fronts has also been analyzed in numerical simulations (Morvan et al. 2009, 2011), as shown in Fig. 8.38.

A useful application of the understanding of the interaction between fires is in the use of a counter fire to suppress a wildfire (Fig. 8.39).

Simulations of the interactions between fire lines have provided valuable knowledge about fireline interactions which can assist fire operations. In a study on the wildfire in-draft flows for backfiring (sometimes called counter-fire) operations, Roxburgh and Rein (2008) suggested that the flow field ahead of the main fire can be divided into three zones. Zone 1 occurs relatively close to the main fire front. In this zone, the wind flow is dominated by the local effects of flame dynamics. Zone 2 is characterized as an area downwind of the main fire front. There, the greatest in-draft velocities occur. The length downwind of this zone depends upon both the ambient wind flow and the fireline intensity. Greater ambient winds tend to reduce

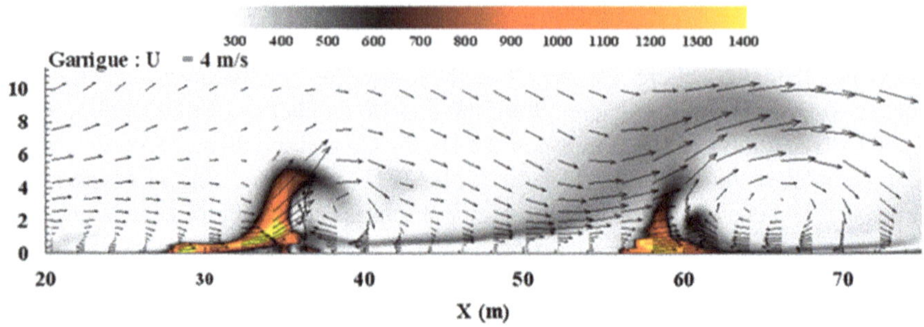

Fig. 8.38 Simulation of interaction between two fire fronts showing the influence in the patterns of wind flows. Arrows indicate wind direction and speed. The color indicates temperature with flames shown in red or orange, while the smoke is gray. (From Morvan et al. 2011)

this zone, while larger fire intensities tend to increase it. This zone has been suggested as the most suitable area for backfiring during operations because it should allow the main fire front to pull in the fire and still result in a significantly wide area of burned fuel. Zone 3 is located the furthest from the main fire front and is an area where fire behavior is dominated by the ambient wind flows.

Using fire as a tool to suppress the advance of a wildfire front has been a subject of many discussions. Fire is an extremely powerful tool in fire suppression and in fuels management, but its use should be based on our understanding of the processes and solid experience in fire management. Also, as wildland fire management strategies increasingly recognize the ecological role of fire and seek to increase the application of managed and prescribed fire there is a need for increased science surrounding the interactions between multiple fires. For example, there is currently a paucity of knowledge about fire-atmospheric interactions associated with complex ignition patterns and how these interactions will impact fire behavior and effects. A research agenda focused on advancing prescribed fire science will inherently rely on advancing our understanding of fire-atmospheric interactions and the interactions among multiple fires. Hiers et al. (2020) suggested that a research agenda focused on prescribed fire is critical for managing resilient ecosystems in the face of global change.

8.5.4 *Other Hypotheses for Unexpected Fire Behavior*

In some special cases, there are reports of unexpected fire behavior that are difficult to explain with current approaches and suggest the exploration of other hypotheses. One of these hypotheses relates to the possible role of flammable volatile gases during a wildfire. Already in 1954, Arnold and Buck, describing blowup fires, indicated that "most fires burn so inefficiently that large quantities of volatile flammable gases are driven off without being burned" and that, "under certain air

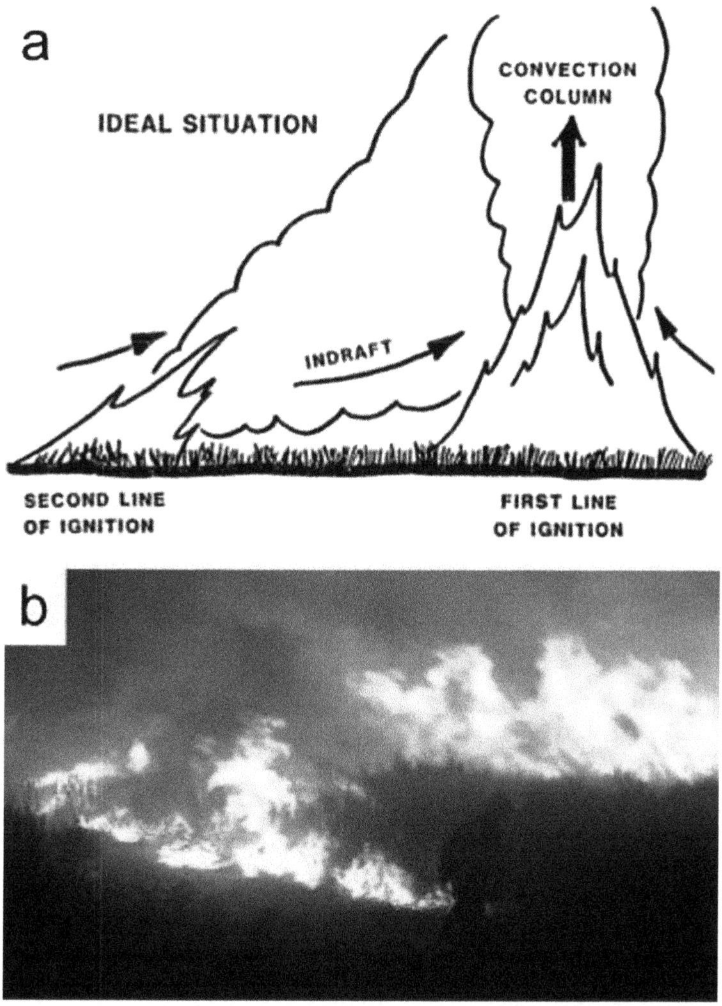

Fig. 8.39 (**a**) Processes of indrafts and convection involved in the interaction between two ignited fire lines. (From Rothermel 1984). (**b**) Cover of the book with the outcomes of the European Project Fire Paradox (Silva et al. 2010) showing the interaction between a wildfire front and counter fire front. (Photograph by Pedro Palheiro)

conditions these gases may be trapped near the ground in low inversions or in poorly ventilated basins or canyons".

By analyzing the concentrations of gases in smoke plumes from wildfires in pine or eucalypt forests in Portugal, various authors (Maleknia et al. 2009; Alves et al. 2011; Evtyugina et al. 2013) confirmed that large quantities of flammable gases are released in wildfires without being burned where they are produced. Researchers have found similar results for *Rosmarinus officinalis* (Chetehouna et al. 2009), *Cistus monspeliensis,* and *Pinus nigra* (Barboni et al. 2011), suggesting that when

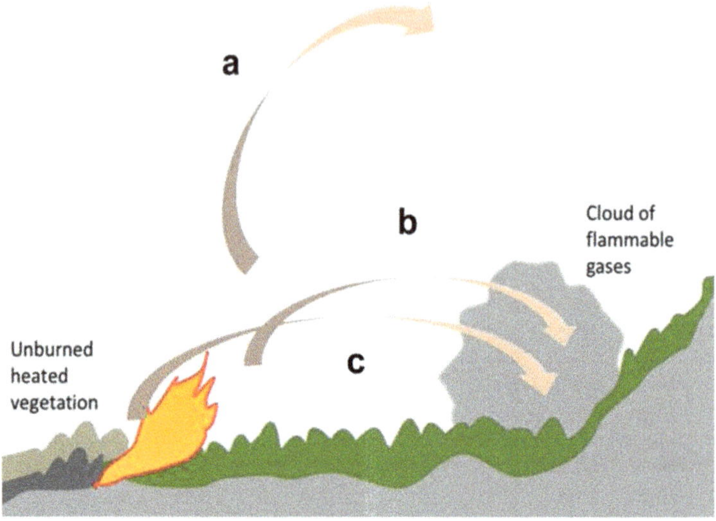

Fig. 8.40 *a* The combustion products typically mix with air by convection with no contribution to fire spread. *b, c* Possible situation where, due to wind and topography, flammable gases originated from incomplete combustion or by heated vegetation may accumulate in a cloud that may burn suddenly and intensely when the ignition source (firebrands or the fire front) arrives

these plants are heated, enough volatile organic compounds (VOCs) could be emitted to support "eruptive" fires to occur or at least to accelerate forest fire spread to endanger fire fighters.

These problems might occur in interaction with topography. Chetehouna et al. (2014) indicated that these processes may accelerate fire spread in canyons. Chatelon et al. (2014) suggested that these situations may arise where and when clouds of flammable gases originating from incomplete combustion or heated vegetation accumulate in a zone where their concentration can reach the Lower Flammability Limit (See Sect. 1.3), triggering the burst of flame when the cloud comes into contact with the fire (Fig. 8.40). These situations have been suggested to have occurred in several extreme wildfires, including the Palasca wildfire in Corsica in 2000 (Dold et al. 2009) and in the very destructive wildfires of Canberra (Australia) in 2003 (Dold et al. 2005; Williams 2007).

This phenomenon is difficult to determine beyond doubt. Sullivan et al. (2007) pointed out that because of the buoyancy of heated gas, these flammable pyrolysis products are probably not found downwind of the fire front, where the turbulent flows typically associated with wildfires are likely to quickly disperse the gases. However, many of these compounds are much heavier than air and may have a tendency to not disperse easily. Recall from Chap. 1 that the molar weight of the main constituents of air are 28 g mol^{-1} for nitrogen and 32 g mol^{-1} for oxygen whereas isoprene (68 g mol^{-1}), monoterpenes (136 g mol^{-1}), and sesquiterpenes (204 g mol^{-1}) are respectively about 2.3, 4.7, and 7.0 times denser than air. The same applies to oxygenated terpenoids such as eucalyptol, with a molar weight of

154 g mol^{-1} that is 5.3 times denser than air. These compounds do not last for long in the atmosphere as they are reactive with air at different degrees. However, most of these volatiles remain active in the air for some hours, and therefore may contribute during that period to the flammability of their mixture in air. There is a great variability between the different volatiles in their lifetime and reactiveness with air.

Our understanding of the phenomena involved in fire behavior at the upper end of its variation is difficult because of their scale and because they are rare by nature and not replicable. We do not understand well the rare phenomena that are not captured by current models, but we know they can threaten the safety of fire fighters and other people by unexpected changes in fire behavior. Generating and testing explanatory hypotheses are the first steps for the necessary scientific development.

8.6 Anticipating and Predicting Extreme Fire Behavior

Extreme wildfire events typically unfold abruptly and often unexpectedly. It is thus critical to forecast the conditions leading up to potentially catastrophic fires, which requires prior understanding and then developing and operationalizing decision-support and communication tools. The upper end of extreme fire behavior, involving complex fire-atmosphere interactions, is difficult to predict. However, the conditions for such phenomena are reasonably understood and so it is important to forecast the atmospheric conditions that may increase the probability of extreme fire events.

Different approaches can be combined to anticipate critical fire behavior situations, including predictions on a daily or multi-day scale, based on fire danger rating, and fire growth projections of ongoing fires. Predictions at short time steps can be made from hourly-resolved fire danger rating or the fire spread prediction of developing fires. Hence, these options vary with the spatiotemporal scope, basically from regional and daily to local and hourly, provided that the data needed are available.

8.6.1 Predictions on a Daily Basis: Fire Danger Rating

The general approach of integrating the various known drivers of fire behavior has been used in fire danger rating systems used by fire management agencies. Fire danger can be defined as the consequence of the "factors affecting the inception, spread, and difficulty of control of fires and the damage they cause" (Chandler et al. 1983). Thus, fire danger incorporates the aspects of individual fire danger elements into calculated and interpretable indices related to fire behavior. The continuum of index values is partitioned into rating classes from Low to Extreme, a scale expected to constitute an objective basis for sound fire management planning and decision-making. Fire danger classes are often defined to indicate resistance to control (or suppression difficulty) of a wildfire. Fire danger rating systems have diverse applications, including:

1. Guidance for land management activities, e.g., use of equipment or prescribed burning;
2. public use restrictions and warnings, namely related to fire bans and access;
3. definition of the preparedness of fire pre-suppression and suppression organizations, e.g., staffing levels committed to fire detection and initial attack;
4. preplan fire-control dispatch levels; and
5. the allocation of fire suppression resources to ongoing fires.

Over the last one hundred years a diversity of fire danger rating systems, ranging from simple equations to complex systems linking multiple equations have been developed. The former are usually straightforward to use and require only a few input variables but are limited in their capabilities, because they are unable to provide a complete picture of fire danger considering its various facets. More complex fire danger rating systems require more input variables and are more versatile and objective.

The input variables and indices are similar for the three better-known fire danger rating systems: the National Fire Danger Rating System (NFDRS) of the USA (Cohen and Deeming 1985), the Canadian Forest Fire Weather Index System (CFFWIS) (Van Wagner 1974; Stocks et al. 1987), and the Forest Fire Danger Index (FFDI) (McArthur 1967), one of the systems in use in Australia (Fig. 8.41). The three systems have in common the reliance on basic weather variables observed over consecutive days as input (wind speed, air temperature, relative humidity, and precipitation), but they differ in how, and to what extent, fuels are described. While the NFDRS distinguishes between fuel models, the Canadian and Australian approaches are for vegetation types.

The CFFWI system is nowadays widely used for both research and management applications at national to regional to global scales. The CFFWIS includes numerical ratings for the moisture content of litter and other surface fine dead fuels (Fine Fuel Moisture Code, FFMC), an indicator of the relative ease of ignition and flammability of fine fuels; the moisture content of duff (Duff Moisture Code, DMC); and the moisture content of deep, compact organic layers (Drought Code, DC). For the relationship of these indexes with fuel moisture dynamics, see Chap. 11. The CFFWIS includes a numerical rating of the expected rate of spread, the Initial Spread Index (ISI), based on wind speed and FFMC. The ISI and BUI are combined to compute the Fire Weather Index (FWI), the overall fire danger index, and an indicator of the expected fireline intensity. From the value of FWI, a Daily Severity Rating (DSR) can be calculated as an indicator of the difficulty of controlling wildfires and the expected effort required for their suppression. In the NFDRS, the fire danger index and analog for fireline intensity is the Burning Index**,** resulting from combining the Spread Component (SC), representing the rate of spread, with the Energy Release Component (ERC), representing fuel consumption. The relative amount of available fuel is also considered in the Australian FFDI through the Drought Factor.

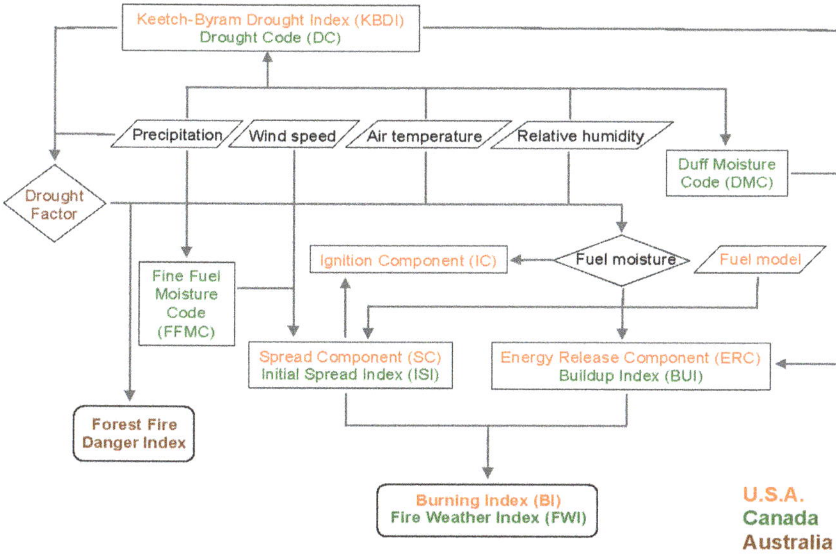

Fig. 8.41 Structure of fire danger rating systems in the USA (National Fire Danger Rating System, NFDRS), Canada (Canadian Forest Fire Weather Index System, CFFWIS), and Australia (Forest Fire Danger Index, FFDI). NDFRS pathways are simplified; unlike other systems, the NFDRS considers the contribution of live fuel moisture content, omitted from the flowchart

Extreme fire behavior is often associated with extended rainless periods and, consequently, significant to extreme drought. Fire danger rating systems integrate the influence of drought on fuel availability, in contrast with simpler fire weather assessment methods, namely those developed in Europe and Russia. The CFFWI uses the Drought Code (DC), commonly used as an indicator of seasonal drought effects on forest fuels and the amount of smoldering in deep duff layers and large dead woody fuels, whereas the NFDRS uses the Keetch-Byram Drought Index (KBDI) (Keetch and Byram 1968). The KBDI is also used in Australia as an indicator of the quantity of rainfall needed to saturate the soil, which then goes into the calculation of the Drought Factor. Alternatively, and because the KBDI was found to underestimate the rate of fuel drying in Tasmania, the Drought Factor can be calculated from the Soil Dryness Index of Mount (1972), derived and tested on the basis that "the long-term drying of the soil has an important effect on fire behavior and a measure of this long-term drying is vital for planning fuel-reduction burning, for the proclamation of fire danger periods and for fire weather forecasts" (Fig. 8.42).

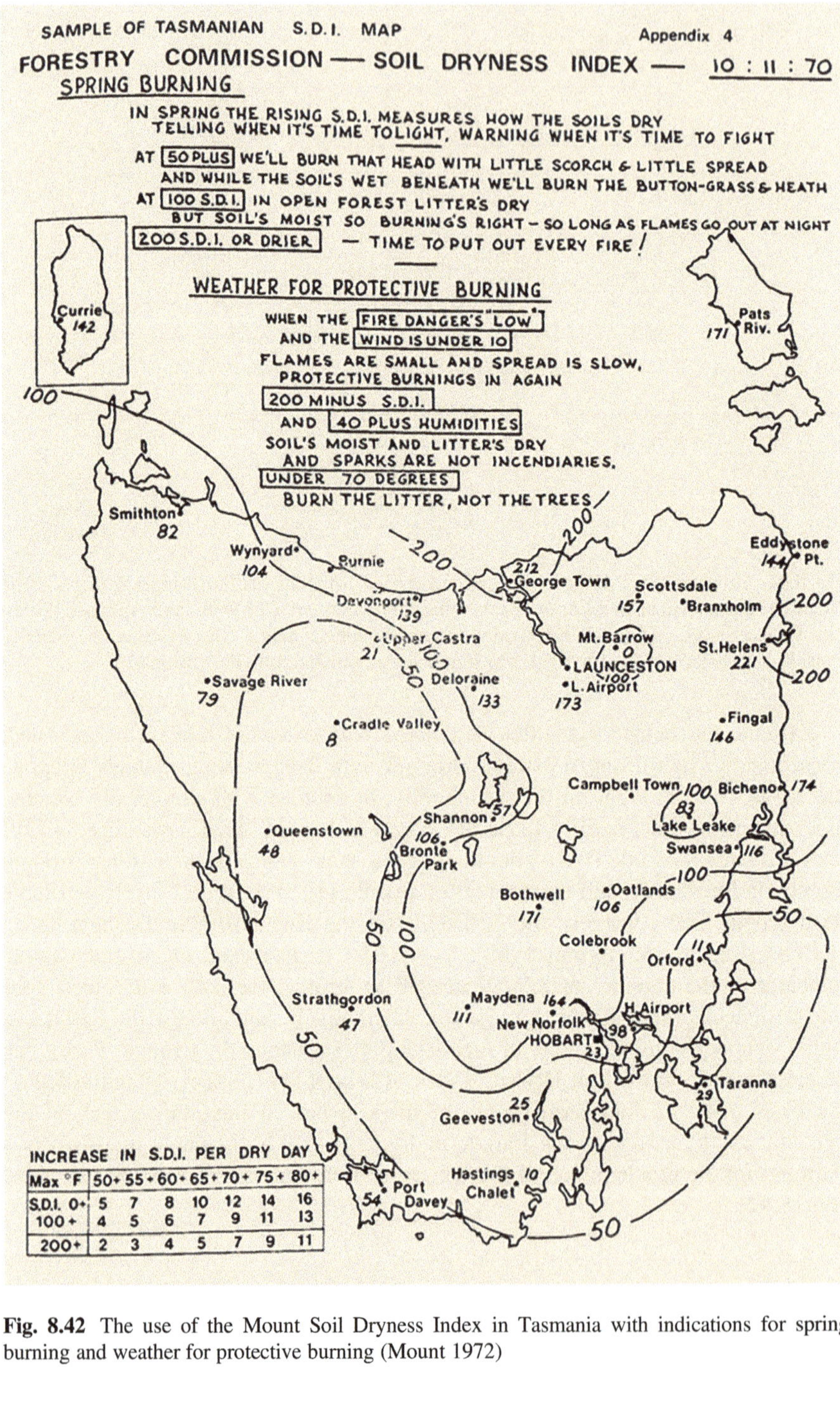

Fig. 8.42 The use of the Mount Soil Dryness Index in Tasmania with indications for spring burning and weather for protective burning (Mount 1972)

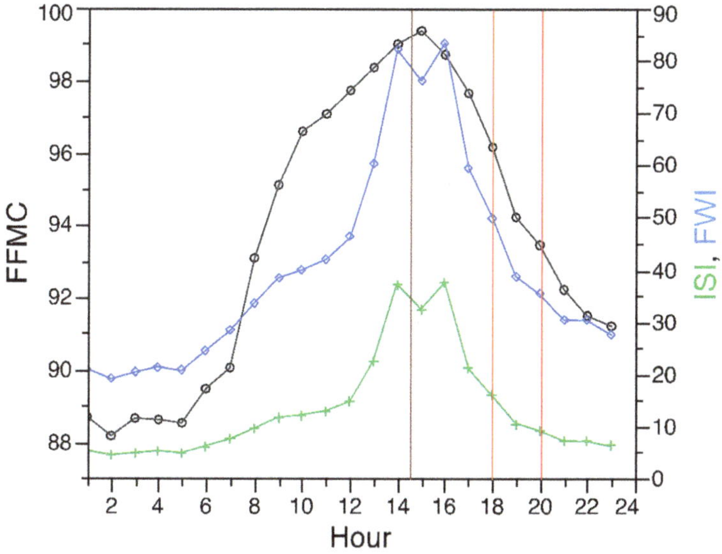

Fig. 8.43 Hourly means for the forecasted indices of the Canadian FWI System for the initial day (June 17, 2017) of the Pedrogão Grande wildfire in Portugal. The FFMC is an indicator of fine dead fuel moisture content. The ISI is an indicator of potential fire spread rate. The FWI is an indicator of potential fireline intensity. From left to right, the vertical lines indicate the timing of ignition, 90° change in wind direction due to a thunderstorm outflow, and downburst occurrence. Built from IPMA data based on the AROME model (Guerreiro et al. 2017)

8.6.2 Predictions on an Hourly Basis

Fire danger rating is typically produced based on weather data forecasted for the early afternoon. However, fire danger resolved at hourly levels can provide more precise and useful information. Understanding the daily variation in fire behavior potential can assist with the planning of the various fire management activities and identify peaks in potential fire activity that could occur outside the early to mid-afternoon period.

As an example, the Pedrogão Grande fire killed 66 people in central Portugal in June 2017 (Guerreiro et al. 2017). Fuel moisture was already relatively low overnight, but then it decreased from 11% to 2% between 7 AM and 3 PM, as estimated from the Canadian FFMC (Fig. 8.43). Combined, the FFMC, ISI, and FWI indices suggest peak fire rate of spread, intensity, and spotting potential at 2 to 4 PM, as usually expected. However, fire development was greatly enhanced during two other time periods. At 6 PM, the fire was hit by strong outflow wind from a nearby thunderstorm that blew perpendicular to the main direction of fire spread and changed the right flank of the fire into a 5 km-long headfire, which then moved at a much higher rate of spread. Similar sudden shifts in wind direction and velocity have been implicated in other catastrophic fires (Sharple et al. 2016). The heat output resulting from this change, combined with a highly unstable atmosphere, resulted in pyrocumulonimbus development and firestorm conditions, with the plume collapsing at 8 PM and the ensuing downburst causing most of the fatalities. In this case, the weather forecast did not predict the wind change. An hourly fire danger rating (Fig. 8.43), if in place, would underestimate the fire growth potential. To increase the capability to predict extreme fire behavior, the Australian national fire danger rating system under development will integrate atmospheric instability and wind changes in the afternoon (Matthews et al. 2019).

Fire growth simulation can assist with fire management planning, and guide fire suppression strategies and tactics and for civil protection. Such predictions can be made in the form of burn probabilities for large fires that are not initially contained or for rapidly developing fires with the potential to impact the human population (Fig. 8.44). Current operational fire spread models were developed with a robust empirical basis allowing for sufficiently accurate projections for many fire management challenges (Cruz and Alexander 2013). A rule-of-thumb that predicts the forward rate of spread in shrubland, eucalypt, and coniferous forests as 10% of mean 10-m open wind speed was found to perform as well as the fire spread models currently used operationally, especially for drier fuels and windier conditions (Cruz and Alexander 2019; Cruz et al. 2020). The rule-of-thumb approach is fast and user-friendly and so is valuable for rapid decision making, especially when homes and people in the WUI are in the path of a fast-approaching fire.

8.6.3 Forecasting Conditions for Blowup Fires

The first significant contribution with operational value regarding the likelihood of plume-driven fires was due to Haines (1988), who developed the Haines Index for describing lower atmosphere stability. The less stable and drier the air mass becomes, the higher the Haines Index and the greater the probability of extreme fire behavior. Stability is derived from the temperature difference between different pressure levels of the atmosphere and dryness is expressed by the moisture content (dew point depression) of the lower atmosphere.

The Haines index can be calculated over three ranges of atmospheric pressure: low elevation (950–850 hPa), mid-elevation (850–700 hPa), and high elevation (700–500 hPa). The Haines index includes a stability component (A) and a moisture component (B) that are weighted equally. The calculation for the mid-elevation (applied for western North America by Winkler et al. 2007) is as follows:

$$\Delta T = T_{850} - T_{700} \tag{8.45}$$

$$\Delta D = T_{850} - DT_{850} \tag{8.46}$$

where ΔT is a stability indicator, the temperature lapse rate, the difference between the temperature at the atmospheric height of 850 hPa (T_{850}) and the temperature at the atmospheric height of 700 hPa (T_{700}), and ΔD is a moisture indicator, a dewpoint

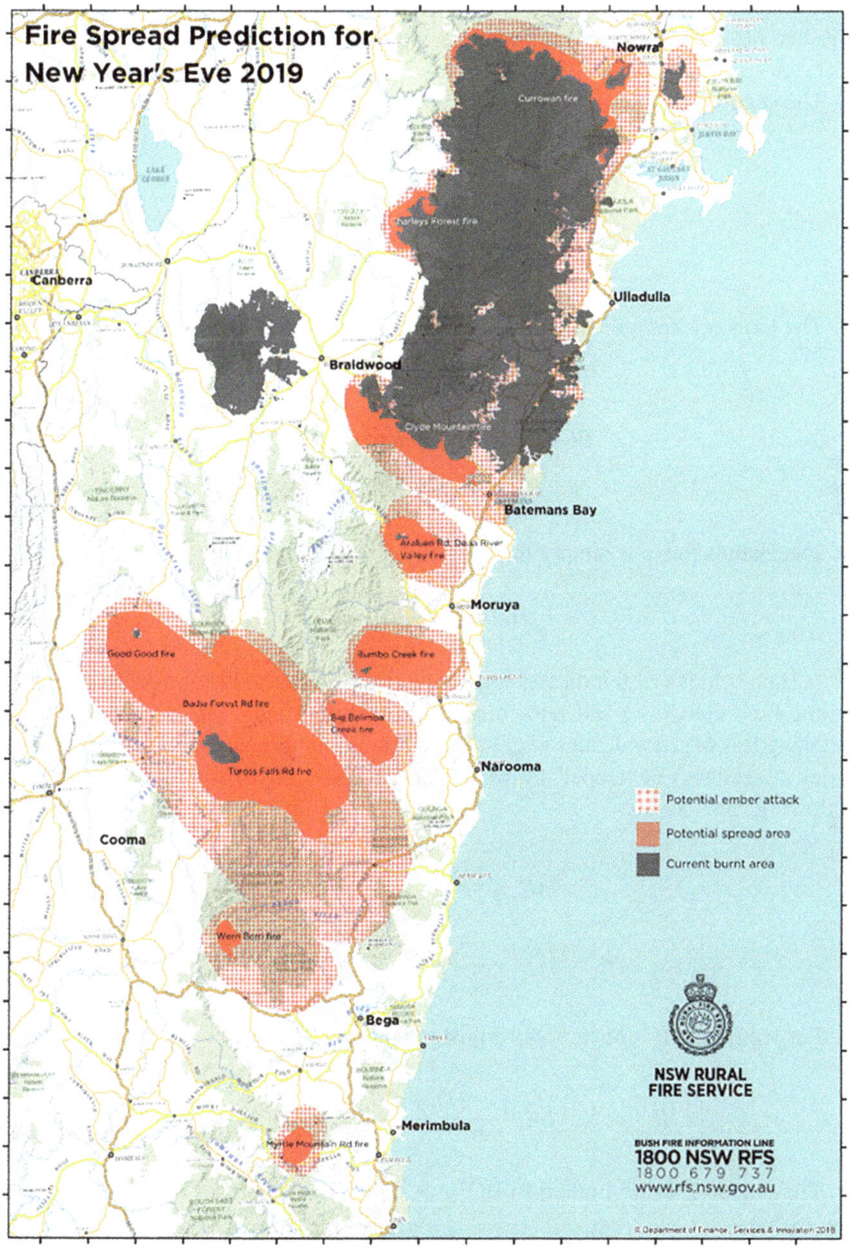

Fig. 8.44 An example of a fire spread forecast distributed to the public for early warning purposes. Produced by the New South Wales Rural Fire Service, Australia, based upon a combination of manual (Cruz et al. 2015) and software (Tolhurst et al. 2008) predictions

depression term, the difference between the temperature at the atmospheric height of 850 hPa (T_{850}), and the dewpoint temperature at the atmospheric height of 850 hPa (DT_{850}).

The stability component (AC) is based on the value of ΔT as follows:

$$AC = 1 \ if \ \Delta T < 6^oC \tag{8.47}$$

$$AC = 2 \ if \ \Delta T \ from \ 6^oC \ to \ 10^oC \tag{8.48}$$

$$AC = 3 \ if \ \Delta T \geqq 11^\circ\mathrm{C} \tag{8.49}$$

The moisture component (BC) is based on the value of ΔD as follows:

$$BC = 1 \ if \ \Delta D < 6^oC \tag{8.50}$$

$$BC = 2 \ if \ \Delta D \ from \ 6^oC \ to \ 12^oC \tag{8.51}$$

$$BC = 3 \ if \ \Delta D \geqq 13^\circ\mathrm{C} \tag{8.52}$$

The Haines Index is simply the sum of the two components:

$$Haines = AC + BC \tag{8.53}$$

A Haines index of 6 indicates high potential for a fire to become large or exhibit erratic fire behavior, 5 indicates medium potential, 4 indicates low potential, and anything less than 4 indicates very low potential. A continuous version of the Haines index (C-Haines) developed by Mills and McCaw (2010) for Australia uses the following calculations:

$$AC = \frac{(T_{850} - T_{700})}{2} - 2 \tag{8.54}$$

$$BC = \frac{(T_{850} - TD_{850})}{3} - 1 \quad \text{for B} < 5 \tag{8.55}$$

For values of BC above 5, the adjusted value of B is:

$$BC = 5 + \frac{(T_{850} - TD_{850})}{6} - 3 \quad \text{for } 5 < \text{B} < 9 \tag{8.56}$$

The value of BC is limited to BC = 9. The C-Haines index is calculated as in Eq. (8.53):

$$C - Haines = AC + BC \tag{8.57}$$

In order to understand if the values of the C-Haines index are extreme, they are then compared with established regional percentiles (Mills and McCaw 2010).

In spite of its wide use in wildland fire management to evaluate the potential for 'large and/or erratic' fire behavior, the Haines index is a subject of debate. There is a need for further developments and some scientists suggest that the Haines index should be revised or replaced (Potter 2018).

Alternatives or refinements of the Haines index are underway, trying to build on the current understanding of the physics involved. This is the case of a new fire weather index called the Hot-Dry-Windy Index (HDW). The HDW uses the basic science of how the atmosphere can affect a fire and takes into account the meteorological conditions both at the Earth's surface and in a 500-m layer just above the surface (Srock et al. 2018). The HDW uses the vapor pressure deficit (VPD) to account for the combined effect of atmospheric temperature and moisture on a fire. Larger VPD implies a faster evaporation rate, and consequently higher fire potential. The HDW is calculated by multiplying wind speed (in m s^{-1}) by VPD (in hPa).

This approach is promising but C-Haines continues to be used. This is the case of the Blow-Up Fire Outlook (BUFO model) developed by McRae et al. (2018) in Australia. The BUFO model consists of a decision-support flowchart where the user combines fire danger, wind speed, and dead fuel moisture content thresholds, whether wind direction will change >45°, C-Haines, and topographic features to establish whether a blowup fire event can occur. In Portugal, 63% of the cumulative area burned by the largest (>2500 ha) fires combine extreme fire danger ratings with high Haines index, and the largest and fastest-growing fires occur when C-Haines is very high (Fernandes et al. 2016). To address this effect of atmospheric instability, Pinto et al. (2020) enhanced the FWI with a C-Haines correction such that the probability of exceedance of energy (FRP) released by fires is better predicted.

A different approach used to predict the potential for extreme fire behavior is to estimate the potential for the formation of a pyrocumulonimbus. The potential for pyroCb cloud formation can be estimated using the Briggs equations to calculate the minimum energy rate required to enter the atmosphere (Fig. 8.45). Following Tory et al. (2018), a pyrocumulonimbus firepower threshold (PFT) could be simply calculated by inverting the Briggs equation as:

$$PFT = a_1 \left(z_{fc}\right)^2 U\, b_{fc} \tag{8.58}$$

where PFT is the minimum amount of energy release (gigawatts) a fire needs to produce for a pyroCb to develop, a_1 is a constant, z_{fc} is the minimum height (km) the plume must rise to before the cloud will form with enough energy to generate a thunderstorm, U is the average wind speed (m s^{-1}) in the layer below that free-convection height, and b_{fc} is the minimum buoyancy the smoke plume must have to generate the pyroCb (related to the temperature difference between the smoke plume and the ambient air needs to be than the air ΔT in °C).

Alternatively, a simpler formulation provided by Kevin Tory can be used (https://phys.org/news/2020-01-bushfires-weather.html):

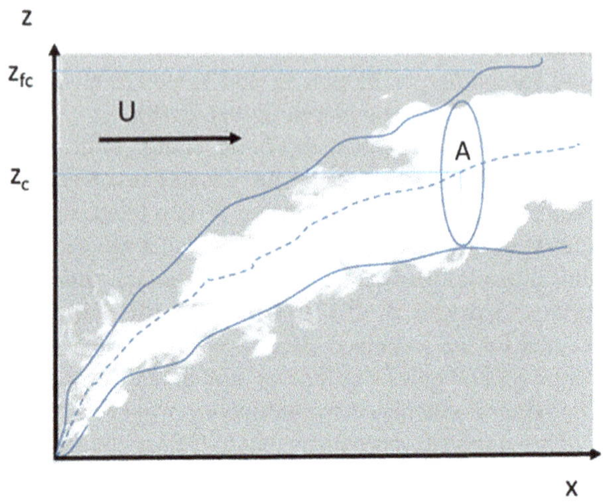

Fig. 8.45 Schematic representation of the variables involved in defining the pyroCb firepower threshold (PFT), z_c is the height of the center of the circular section of the fire plume with area A, z_{fc} is the free-convection height, or the height of the stable layer, and U is the wind speed. (Adapted from Tory and Kepert 2019)

$$PFT = 0.3\left(z_{fc}\right)^2 U\,\Delta T \tag{8.59}$$

This approach has the advantage of showing very interesting properties, as it is a simple and easily understandable formulation requiring inputs that are generally available. The formation of a pyroCb depends on the conditions of the fire and of the atmosphere. The size of the area that is burning and its heat release rate favor the possibility of pyroCb. Light winds that favor pyroCb do not favor intense fires and vice versa. This approach allows for the forecast of pyroCb events with the indication of the predicted heat flux, or fire power.

8.7 Limitations and Implications

Extreme fires burn large areas, are difficult to control, and commonly have significant negative social, economic, or ecological impacts. Given those negative impacts, scientists and managers must understand what factors are related to their occurrence, their effects, and how social-ecological systems recover following an extreme fire. Further discussion on both beneficial and negative ecological effects of fires and ecosystem recovery can be found in Chaps. 9 and 12. Several approaches have been developed to identify and quantify extreme fires, including statistical descriptions and operational approaches that focus on fire control difficulty. Statistical approaches have been based on fire size, burn severity, cost, damage to infrastructure, and human lives lost. Extreme fires that pose significant control issues are often associated with particular types of fire behavior, such as crown fires, spotting, and plume-dominated fires. Extreme fires are often spectacular events that garner great media attention. Changes in national policies are often triggered by extreme fires (Mateus and Fernandes 2014). Though few extreme fires occur, they burn more area, cost more money, and have more social, economic, and ecological impacts than many other fires combined.

Although extreme fire behavior results from the interactions among the topography, fuels, and weather, as discussed in Chap. 7, scientists are still studying the factors that lead to extreme fires and the context around which extreme fires occur. One of the key aspects of identifying and discussing extreme fires is to remember that many of the phenomena we associate with their occurrence commonly occur on fires that are not considered extreme. This is the case of crown fires, spotting, or complex fire-atmosphere interactions.

The implications associated with understanding crown fires are of various nature. From the forestry perspective, management to prevent crown fire spread can be done by understory fuel treatments and silviculture practices (pruning and thinning). Increased foliar moisture of the canopy can be obtained, for example, by mixing conifers with deciduous tree species. The knowledge of fuel and wind conditions can be used to anticipate the likelihood and rate of spread of wildfires threatening fire fighters and people in general. The models presented have been used in the prediction of the probability and speed of active crown fires. However, as cautioned by Van Wagner (1977), other factors might be involved, including the shape and the arrangement of leaves and gaps between tree canopies or tree canopy layers, or the presence of flammable waxes, oils, and resins, factors that are often synergistic in extreme fires.

Spotting can occur under a range of conditions, and does not necessarily result in extreme fire behavior. However, under hot, dry, windy conditions, many spot fires may occur, resulting in rapid increases in the spread rate and mass fire behavior, limiting the ability to control the fire. In other cases, the fire regime of a given ecosystem (See Chap. 12) may be such that the presence of rapid spread or high severity fires might be considered characteristic rather than extreme within the historical context. However, such fires may still be considered extreme due to their resistance to control, costs, or social impacts. The anticipation of the likelihood and potential distance of spotting can be used during wildfires to inform firefighting strategies and enhance safety. Also, the understanding of the processes allows for the recognition of the importance of reducing potentially dangerous long-range spotting firebrands, as eucalypt bark, by fuel treatments. Firebrands can ignite homes if the homes are vulnerable (See Chap. 10 for protecting people and their homes from fires).

The understanding of buoyancy and its interaction with wind is important to anticipate extreme fire behavior. Plume-dominated and wind-dominated fires behave differently, and they require different strategies for their management.

Over the last several decades, extreme fires have emerged as a global phenomenon. Their occurrence has been primarily tied to extreme weather events associated with global climate change and alterations in fuel type and load associated with fire and land management practices. Often the conditions that favor extreme fire behavior are regional and as multiple ignitions are common, managers are often coping with multiple large fires burning at the same time in a region. In many cases, extreme fire behavior has been associated with unique weather events such as droughts or wind systems, which result in low fuel moisture and high wind speeds. In other cases, extreme fire behavior is associated with interactions between the fires' plume and the atmosphere or multiple ignitions associated with spotting. Extreme fire behavior is often erratic and can occur under a variety of weather, fuels, and topographic conditions. Therefore fire managers, scientists, and the public should never assume that extreme fire behavior is impossible (Potter et al. 2016).

Fire managers have developed a number of strategies and predictive tools that can be useful in mitigating extreme fires. Land management strategies focus on modifying the fuels complex to reduce fire intensity and severity and support fire suppression. Fuels management strategies are discussed further in Chap. 11. In addition to treating fuels, fire managers use a variety of predictive tools to anticipate when and where extreme fires are likely. Using these predictions, the public can be alerted to potential dangers and develop appropriate preparation strategies, including staffing levels and evacuation protocols. While many of the current tools available to managers are useful, many were not specifically developed to capture the mechanisms driving extreme fires. Therefore fire behavior analysts often use a mix of predictive models, experience, and training when assessing potential extreme fire behavior. There is a need for more integrated fire science on the topics of extreme fire behavior and effects. See Chap. 14 for future developments.

8.8 Interactive Spreadsheets: CROWNFIRE and MASS_TRANSFER_Spotting

Two different spreadsheets can be used to simulate the processes associated with crown fires and spotting. The corresponding equations were implemented in interactive spreadsheets to simplify calculations and rapidly visualize the effect of the various factors involved. These spreadsheets allow you to explore the implications of the multiple equations without having to solve the equations yourself.

In the spreadsheet CROWNFIRE_v2.0 (Fig. 8.46), you can input different types of data, according to the model used. We suggest that you interpret the implications for changing environmental conditions for the likelihood of crowning and spreading, and to evaluate the effects of changing inputs in the rate of spread of a crown fire.

You can use the spreadsheet MASS_TRANSFER_Spotting_v2.0, to see how far the firebrands from torching trees, burning piles, and wind-driven surface fires can travel (Fig. 8.47). Vary the input variables under different conditions to further test your understanding of spotting potential.

Recall that we do not intend for you to use these spreadsheets as predictive tools. Instead, we expect that they will help you further your understanding of the processes involved and the main factors responsible for crown fire behavior and spotting.

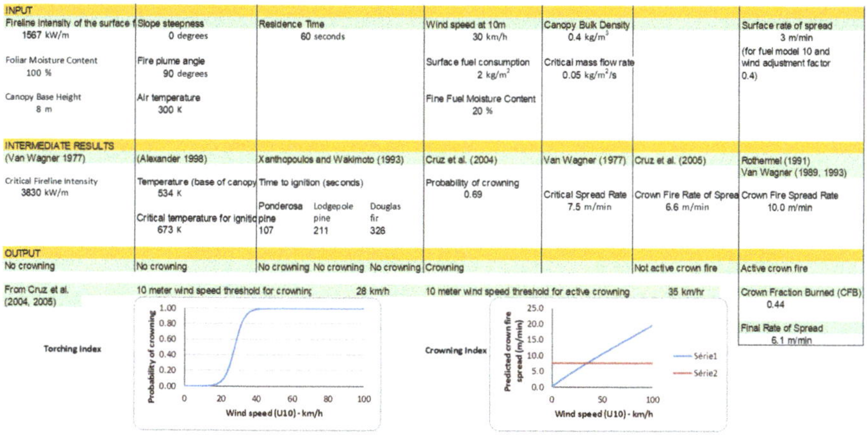

Fig. 8.46 Illustration of the calculations in the interactive spreadsheet, CROWNFIRE_v2.0, for estimating the likelihood and the rate of spread of crown fires. These calculations are based on the equations by Alexander (1998), Cruz et al. (2004, 2005), Rothermel (1991), Van Wagner (1977, 1989, 1993), and Xanthopoulos and Wakimoto (1993). See text for details

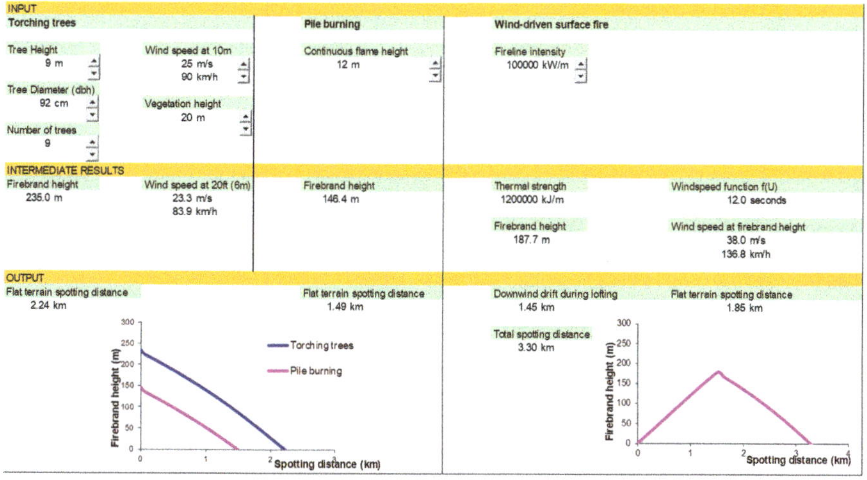

Fig. 8.47 Illustration of spotting calculations based on the models of Albini (1979, 1981, 1983) for torching trees, burning piles, and wind-driven surface fires. This is from the interactive spreadsheet, MASS_TRANSFER_Spotting_v2.0, included in this chapter

References

Albini, F. A. (1979). *Spot fire distance from burning trees—A predictive model* (Gen Tech Rep INT-56). Ogden: USDA Forest Service Intermountain Forest and Range Experiment Station.

Albini, F. A. (1981). *Spot fire distance from isolated sources—Extensions of a predictive model* (Res Note INT-309). Ogden: USDA Forest Service Intermountain Forest and Range Experiment Station.

Albini, F. A. (1983). *Potential spotting distance from wind-driven surface fires* (Gen Tech Rep INT-309). Ogden: USDA Forest Service Intermountain Forest and Range Experiment Station.

Albini, F. A., Alexander, M. E., & Cruz, M. G. (2012). A mathematical model for predicting the maximum potential spotting distance from a crown fire. *International Journal of Wildland Fire, 21*, 609–627.

Alexander, M. E. (1998). *Crown fire thresholds in exotic pine plantations of Australasia*. PhD Thesis. Australian National University, Canberra.

Alexander, M. E. (2000). *Fire behavior as a factor in forest and rural fire suppression* (For Res Bull 28).

Alexander, M. E., & De Groot, W. J. (1988). *Fire behavior in jack pine stands as related to the Canadian Forest Fire Weather Index (FWI) System, poster with text.* Alberta: Canadian Forestry Service & Northern Forestry Centre.

Alexander, M. E., & Cruz, M. G. (2016). Crown fire dynamics in conifer forests. Chapter 9. In P. A. Werth, B. E. Potter, M. E. Alexander, C. B. Clements, M. G. Cruz, M. A. Finney, J. M. Forthofer, S. L. Goodrick, C. Hoffman, W. M. Jolly, S. S. McAllister, R. D. Ottmar, & A. Russell (Eds.), *Synthesis of knowledge of extreme fire behavior: Vol 2 for fire behavior specialists, researchers, and meteorologists* (Gen Tech Rep PNW-GTR-891, pp. 163257). Portland: USDA Forest Service Pacific Northwest Research Station.

Alexander, M. E., Janz, B., & Quintilio, D. (1983). Analysis of extreme wildfire behavior in east-central Alberta: A case study. In *Preprint Volume, Seventh Conference on Fire and Forest Meteorology*; 1983 April 25–29; Fort Collins, CO (pp. 38–46). Boston: American Meteorological Society.

Alexander, M., Heathcott, M., & Schwanke, R. (2013a). *Fire behaviour case study of two early winter grass fires in southern Alberta, 27 November 2011*. Edmonton: Partners in Protection Association.

Alexander, M. E., Cruz, M. G., Vaillant, N. M., & Peterson, D. L. (2013b). *Crown fire behavior characteristics and prediction in conifer forests: A state-of-knowledge synthesis*. Final Report to the Joint Fire Science Program, Boise.

Alvarado, E., Sandberg, D. V., & Pickford, S. G. (1998). Modeling large forest fires as extreme events. *Northwest Science, 72*, 66–75.

Alves, C., Vicente, A., Nunes, T., Gonçalves, C., Fernandes, A. P., Mirante, F., Tarelho, L., de la Campa, A. M. L., Querol, X., Caseiro, A., Monteiro, C., Evtyugina, M., & Pio, C. (2011). Summer 2009 wildfires in Portugal: Emission of trace gases and aerosol composition. *Atmospheric Environment, 45*, 641–649.

Andela, N., Morton, D. C., Giglio, L., Paugam, R., Chen, Y., Hantson, S., van der Werf, G. R., & Randerson, J. T. (2019). The global fire atlas of individual fire size, duration, speed and direction. *Earth System Science Data, 11*, 529–552.

Andersen, H. E., McGaughey, R. J., & Reutebuch, S. E. (2004). Estimating forest canopy fuel parameters using LIDAR data. *Remote Sensing of Environment, 94*, 441–449.

Anderson, H. E. (1968). *Sundance fire: An analysis of fire phenomena* (Res Pap INT-56). Ogden: USDA Forest Service Intermountain Forest and Range Experiment Station.

Anthenien, R. A., Tse, S. D., & Fernandez-Pello, A. C. (2006). On the trajectories of embers initially elevated or lofted by small-scale ground fire plumes in strong winds. *Fire Safety Journal, 41*(5), 349–363.

Barboni, T., Cannac, M., Leoni, E., & Chiaramonti, N. (2011). Emission of biogenic volatile organic compounds involved in eruptive fire: implications for the safety of firefighters. *International Journal of Wild Fire 20*, 152–161.

Bowman, D. M. J. S., Williamson, G. J., Abatzoglou, J. T., Kolden, C. A., Cochrane, M. A., & Smith, A. M. S. (2017). Human exposure and sensitivity to globally extreme wildfire events. *Nature Ecology and Evolution, 1*, 0058.

Bunk, S. (2004). World on fire. *PLoS Biology, 2*(2), e54.

Burrows, N. (2015). *Fuels, weather and behaviour of the Cascade fire (Esperance fire #6) 15-–17 November 2015*. Perth: Science and Conservation Division. Department of Parks and Wildlife.

Butler, B. W., & Reynolds, T. D. (1997). *Wildfire case study: Butte City, southeastern Utah, July 1, 1994* (Gen Tech Rep INT-GTR-351). Ogden: USDA Forest Serv Intermountain Res Stn.

Butler, B. W., Bartlette, R. A., Bradshaw, L. S., Cohen, J. D., Andrews, P. L., Putnam, T., & Mangan, R. J. (1998). *Fire behavior associated with the 1994 South Canyon Fire on Storm King Mountain, Colorado* (Res Pap RMRS–RP–9). Fort Collins: USDA Forest Service Rocky Mountain Research Station.

Byram, G. M. (1954). *Atmospheric conditions related to blowup fires* (Station Paper SE-SP-35). Asheville: USDA Forest Service Southeastern Forest Experiment Station.

Byram, G. M. (1959). Forest fire behavior. In K. P. Davis (Ed.), *Forest fire control and use* (pp. 90–123). New York: McGraw-Hill, 554–555.

Castellnou, M., Guiomar, N., Rego, F., & Fernandes, P. M. (2018). Fire growth patterns in the 2017 mega fire episode of October 15, central Portugal. In D. X. Viegas (Ed.), *Advances in forest fire research*, Chapter 3—Fire management, pp. 447–453.

Catchpole, W. R. (1983). *Letter to Dick Rothermel, October 31* (RWU 2103 files). Missoula: USDA Forest Service Intermountain Forest and Range Experiment Station Northern Forest Fire Laboratory.

Chandler, C., Cheney, P., Thomas, P., Trabaud, L., & Williams, D. (1983). *Fire in forestry. Forest fire behavior and effects* (Vol. I). New York: Wiley.

Chase, C. H. (1984). *Spotting distance from wind-driven surface fires-extensions of equations for pocket calculators* (Res Note INT-RN-346). Ogden: USDA Forest Service Intermountain Forest and Range Experiment Station.

Chatelon, F.-J., Sauvagnargues, S., Dusserre, G., & Balbi, J.-H. (2014). Generalized blaze flash, a "flashover" behavior for forest fires—Analysis from the firefighter's point of view. *Open Journal of Forest, 4*, 547–557.

Cheney, N. P., & Bary, G. A. V. (1969). The propagation of mass conflagrations in a standing eucalypt forest by the spotting process. In *TTCP Mass Fire Symposium*, Canberra.

Cheney, N. P., Gould, J. S., & Catchpole, W. R. (1998). Prediction of fire spread in grasslands. *International Journal of Wildland Fire, 8*, 1–13.

Cheney, N. P., Gould, J. S., McCaw, W. L., & Anderson, W. R. (2012). Predicting fire behaviour in dry eucalypt forest in southern Australia. *Forest Ecology and Management, 280*, 120–131.

Chetehouna, K., Barboni, T., Zarguili, I., Leoni, E., Simeoni, A., & Fernandez-Pello, A. C. (2009). Investigation on the emission of volatile organic compounds from heated vegetation and their potential to cause an accelerating forest fire. *Combustion Science and Technology, 181*, 1273–1288.

Chetehouna, K., Courty, L., Garo, J. P., Viegas, D. X., & Fernandez-Pello, C. (2014). Flammability limits of biogenic volatile organic compounds emitted by fire-heated vegetation (*Rosmarinus officinalis*) and their potential link with accelerating forest fires in canyons: A Froude-scaling approach. *Journal of Fire Sciences, 32*, 316–327.

Clements, H. B. (1977). *Lift-off of forest firebrands* (Res Pap SE-159, p. 11). Asheville: USDA Forest Service Southeastern Forest Experiment Station.

Coen, J., Mahalingam, S., & Daily, J. (2004). Infrared imagery of crown–fire dynamics during FROSTFIRE. *Journal of Applied Meteorology, 43*, 1241–1259.

Coen, J. L., Schroeder, W., & Quayle, B. (2018). The generation and forecast of extreme winds during the origin and progression of the 2017 Tubbs Fire. *Atmosphere, 9*, 462.
Cohen, J. D., & Butler, B. W. (1998). *Modeling potential structure ignitions from flame radiation exposure with implications for wildland/urban interface fire management.* 13th Fire Meterology Conference IAWF, Lorne, pp 81–86.
Cohen, J. D., & Deeming, J. E. (1985). *The national fire-danger rating system: Basic equations* (Gen Tech Rep PSW-GTR-82). Berkeley: USDA Forest Service Pacific Southwest Forest and Range Experiment Station.
Countryman, C. M. (1972). *The fire environment concept*. Berkeley: USDA Forest Service Pacific Southwest Forest Range Experiment Station.
Countryman, C. M. (1974). *Can southern California wildland conflagrations be stopped?* (Gen Tech Rep PSW-7). Berkeley: USDA Forest Service Pacific Southwest Forest Range Experiment Station.
Cruz, M. G. (1999). *Modeling the initiation and spread of crown fires*. MSc thesis, University of Montana, Missoula.
Cruz, M. G., & Alexander, M. E. (2013). Uncertainty associated with model predictions of surface and crown fire rates of spread. *Environmental Modelling & Software, 47*, 16–28.
Cruz, M. G., & Alexander, M. E. (2019). The 10% wind speed rule of thumb for estimating a wildfire's forward rate of spread in forests and shrublands. *Annals of Forest Science, 76*, 44.
Cruz, M. G., Alexander, M. E., & Wakimoto, R. H. (2004). Modeling the likelihood of crown fire occurrence in conifer forest stands. *Forest Science, 50*, 640–658.
Cruz, M. G., Alexander, M. E., & Wakimoto, R. H. (2005). Development and testing of models for predicting crown fire rate of spread in conifer forest stands. *Canadian Journal of Forest Research, 35*, 1626–1639.
Cruz, M. G., Sullivan, A. L., Gould, J. S., Sims, N. C., Bannister, A. J., Hollis, J. J., & Hurley, R. J. (2012). Anatomy of a catastrophic wildfire: The Black Saturday Kilmore East fire in Victoria, Australia. *Forest Ecology and Management, 284*, 269–285.
Cruz, M. G., Gould, J. S., Alexander, M. E., Sullivan, A. L., McCaw, W. L., & Matthews, S. (2015). Empirical-based models for predicting head-fire rate of spread in Australian fuel types. *Australian Forestry, 78*, 118–158.
Cruz, M. G., Alexander, M. E., Fernandes, P. M., Kilinc, M., & Sil, Â. (2020). Evaluating the 10% wind speed rule of thumb for estimating a wildfire's forward rate of spread against an extensive independent set of observations. *Environmental Modelling & Software, 104818*, 133.
Dahale, A., Padhi, S., Shotorban, B., & Mahalingam, S. (2013). *Flame merging in two neighboring shrub fires* (Paper No 070FR-0198). In *Proceedings of the 8th US National Combustion Meeting* (p. 13). Combustion Institute, University of Utah, Salt Lake City, UT.
de Ronde, C. (1988). *Preliminary investigations into the use of fire as a management technique in plantation ecosystems of the Cape Province*. MSc thesis, University of Natal.
de Zea Bermudez, P., Mendes, J., Pereira, J. M. C., Turkman, K. F., & Vasconcelos, M. J. P. (2009). Spatial and temporal extremes of wildfires in Portugal (1984-2004). *International Journal of Wildland Fire, 18*, 983–991.
DeCoste, J. H., Wade, D. D., & Deeming, J. E. (1968). *The Gaston Fire* (Res Pap SE–43). Asheville: USDA Forest Service Southeastern Forest Experiment Station.
Dold, J. W., Weber, R. O., Gill, M., Ellis, P., McRae, R., & Cooper, N. (2005). Unusual phenomena in an extreme bushfire. In G. J. Nathan, B. B. Dally, M. Kalt, et al. (Eds.), *Proceedings of 5th Asia-Pacific Conference on Combustion 2005* (pp. 309–312). Adelaide: University of Adelaide.
Dold, J., Simeoni, A., Zinoviev, A., & Weber, R. (2009). The Palasca fire, September 2000: Eruption or flashover? In D. X. Viegas (Ed.), *Recent forest fire accidents in Europe*. Ispra: JRC-IES, European Commission.
Donoghue, L., Jackson, G., Angel, R., Beebe, G., Bishop, K., Close, K., Moore, R., Newman, E., Schmidt, M., & Whitlock, C. (2003). *Cramer fire fatalities. Accident investigation factual report*. Ogden: USDA For Serv North Fork Ranger District, Salmon-Challis National Forest, Region 4.

Ellis, P. F. (2000). *The aerodynamic and combustion characteristics of eucalypt bark—A firebrand study*. PhD Thesis, Australian National University, Canberra.

Ellis, P. F. (2010). The effect of the aerodynamic behavior of flakes of Jarrah and karri bark on their potential as firebrands. *Journal of Royal Society of Western Australia, 93*, 21–27.

Evtyugina, M., Calvo, A. I., Nunes, T., Alves, C., Fernandes, A. P., Tarelho, L., Vicente, A., & Pio, C. (2013). VOC emissions of smoldering combustion from Mediterranean wildfires in central Portugal. *Atmospheric Environment, 64*, 339–348.

Fernandes, P. M., Loureiro, C., & Botelho, H. S. (2004). Fire behaviour and severity in a maritime pine stand under differing fuel conditions. *Annals of Forest Science, 61*, 537–544.

Fernandes, P. M., Barros, A. G., Pinto, A., & Santos, J. A. (2016). Characteristics and controls of extremely large wildfires in the western Mediterranean Basin. *Journal of Geophysics Research Biogeoscience, 121*, 2141–2157.

Fernandes, P. M., Guiomar, N., Mateus, P., & Oliveira, T. (2017). On the reactive nature of forest fire-related legislation in Portugal: A comment on Mourão and Martinho (2016). *Land Use Policy, 60*, 12–15.

Filippi, J. B., Cruz, M. G., Bosseur, F., & Girard, A. (2014). Investigation of vegetation fire plumes using paragliders tracks and micro-scale meteorological model. In D. X. Viegas (Ed.), *Proceedings of VII International Conference of Forest Fire Research. Advances in Forest Fire Research*. Coimbra: Imprensa da Universidade de Coimbra.

Finney, M. A. (2004) *FARSITE: Fire area simulator—Model development and evaluation* (Res Pap RMRS-RP-4, revised). Fort Collins: USDA Forest Service Rocky Mountain Research Station.

Finney, M. A. (2006). An overview of FlamMap fire modelling capabilities. In P. L. Andrews, B. W. Butler (comps), *Fuels management—How to measure success. Conference Proceedings* (RMRS-P-41, pp. 213–220). Fort Collins: USDA Forest Service Rocky Mountain Research Station.

Forestry and Forest Products. (1998). *Project Vesta Experiment*. CSIRO. Retrieved March 20, 2020, from https://www.scienceimage.csiro.au/image/504/project-vesta-experimental-fire/.

Ganteaume, A., Lampin-Maillet, C., Guijarro, M., Hernando, C., Jappiot, M., Fonturbel, T., Pérez-Gorostiaga, P., & Vega, J. A. (2009). Spot fires: Fuel bed flammability and capability of firebrands to ignite fuel beds. *International Journal of Wildland Fire, 18*, 951–969.

Ganteaume, A., Guijarro, M., Jappiot, M., Hernando, C., Lampin-Maillet, C., Pérez-Gorostiaga, P., & Vega, J. A. (2011). Laboratory characterization of firebrands involved in spot fires. *Annals of Forest Science, 68*, 531–541.

Gould, J. S., McCaw, W. L., Cheney, N. P., Ellis, P. F., Knight, I. K., & Sullivan, A. L. (2007). *Project Vesta—Fire in dry eucalypt forests: Fuel structure, fuel dynamics, and fire behavior*. Canberra: Ensis-CSIRO, Department of Environment and Conservation.

Gould, J. S., McCaw, L., & Cheney, P. (2011). Quantifying fine fuel dynamics and structure in dry eucalypt forest (Eucalyptus marginata) in Western Australia for fire management. *Forest Ecology and Management, 262*, 531–546.

Guerreiro, J., Fonseca, C., Salgueiro, A., Fernandes, P., Lopez, E., de Neufville, R., Mateus, F., Castellnou, M., Silva J.S., Moura, J., Rego, F., & Mateus, P. (2017). Análise e apuramento dos factos relativos aos incêndios que ocorreram em Pedrógão Grande, Castanheira de Pêra, Ansião, Alvaiázere, Figueiró dos Vinhos, Arganil, Góis, Penela, Pampilhosa da Serra, Oleiros e Sertã entre 17 e 24 de junho de 2017. Relatório Final. Lisboa: Comissão Técnica Independente. Assembleia da Républica.

Guillom. (2008). *Laminar and turbulent flows*. Wikimedia commons. Retrieved March 20, 2020, from https://commons.wikimedia.org/wiki/File:Laminar_and_turbulent_flows.svg.

Haines, D. A. (1988). A lower atmospheric severity index for wildland fires. *National Weather Digest, 13*, 23–27.

Hiers, J. K., O'Brien, J. J., Varner, J. M., et al. (2020). Prescribed fire science: The case for a refined research agenda. *Fire Ecology, 16*, 11.

Hirsch, K. G., & Martell, D. L. (1996). A review of initial attack fire crew productivity and effectiveness. *International Journal of Wildland Fire, 6*, 199–215.

Hodgson, A. (1968). Control burning in eucalypt forests in Victoria, Australia. *Journal of Forestry, 66*, 601–605.
Keane, R. E., Reinhardt, E. D., Scott, J. H., Gray, K., & Reardon, J. (2005). Estimating forest canopy bulk density using six indirect methods. *Canadian Journal of Forest Research, 35*, 724–739.
Keetch, J. J., & Byram, G. M. (1968) *A drought index for forest fire control* (Res Pap SE-38). Asheville: USDA Forest Service Research Station.
Keeves, A., & Douglas, D. R. (1983). Forest fires in South Australia on 16 February 1983 and consequent future forest management aims. *Australian Forestry, 46*, 148–162.
Kerr, J. W., Buck, C. C., Cline, W. E., Martin, S., & Nelson, W. D. (1971). *Nuclear weapons effects in a forest environment* (Thermal and Fire, No. DASIAC-SR-112). Santa Barbara: General Electric Co, DNA Information and Analysis Center.
Kiil, A. D., & Grigel, J. E. (1969). *The May 1968 forest conflagrations in central Alberta—A review of fire weather, fuels and fire behavior* (Info Rep A–X–24). Calgary: Canada Department of Fisheries and Forestry, Forest Research Laboratory.
Knight, I. (2001). The design and construction of a vertical wind tunnel for the study of untethered firebrands in flight. *Fire Technology, 37*, 87–100.
Koch, E. (1942). History of the 1910 forest fires in Idaho and western Montana. In *When the mountains roared: Stories of the 1910 fire* (Publ R1–78-30, p. 25). Missoula: USDA Forest Service USDA Forest Service Northern Region.
Krishnamurthy, R., & Hall, J. G. (1987). Numerical and approximate solutions for plume rise. *Atmospheric Environment, 21*, 2083–2089.
Kylie, H. R., Hieronymus, G. H., & Hall, A. G. (1937). *CCC forestry*. Washington, DC: US Govern Printing Office.
Lahaye, S., Sharples, J., Matthews, S., Heemstra, S., Price, O., & Badlan, R. (2018). How do weather and terrain contribute to firefighter entrapments in Australia? *International Journal of Wildland Fire, 27*, 85–98.
Lannom, K. O., Tinkham, W. T., Smith, A. M. S., Abatzoglou, J., Newingham, B. A., Hall, T. E., Morgan, P., Strand, E. K., Paveglio, T. B., Anderson, J. W., & Sparks, A. M. (2014). Defining extreme wildland fires using geospatial and ancillary metrics. *International Journal of Wildland Fire, 23*, 322–337.
Laurent, P., Mouillot, F., Moreno, M. V., Yue, C., & Ciais, P. (2019). Varying relationships between fire radiative power and fire size at a global scale. *Biogeosciences, 16*, 275–288.
Lee, S. L., & Emmons, J. M. (1961). A study of natural convection above a line fire. *Journal of Fluid Mechanics, 11*, 353–369.
Linn, R. R., Reisner, J., Colman, J. J., & Winterkamp, J. (2002). Studying wildfire behavior using FIRETEC. *International Journal of Wildland Fire, 11*, 233–246.
Luke, R. H., & McArthur, A. G. (1978). *Bushfires in Australia*. Canberra: Department of Primary Industry. Forestry and Timber Bureau. CSIRO Division of Forest Research. Australian Government Publishing Service.
MacIver, D. C., Auld, H., & Whitewood, R. (1989). Proceedings of the 10th Conference on Fire and Forest Meteorology. In *Conference on Fire and Forest Meteorology, 17–21 Apr 1989*. Ottawa: Department Forestry Canada.
Maleknia, S. D., Bell, T. L., & Adams, M. A. (2009). Eucalypt smoke and wildfires: Temperature dependent emissions of biogenic volatile organic compounds. *International Journal of Mass Spectrometry, 279*, 126–133.
Manzello, S. M., Maranghides, A., Shields, J. R., Mell, W. E., Hayashi, Y., & Nii, D. (2007a). Mass and size distribution of firebrands generated from burning Korean pine (*Pinus koraiensis*) trees. *Fire and Materials, 33*, 21–31.
Manzello, S. M., Maranghides, A., & Mell, W. E. (2007b). Firebrand generation from burning vegetation. *International Journal of Wildland Fire, 16*, 458–462.

Mateus, P., & Fernandes, P. M. (2014). Forest fires in Portugal: Dynamics, causes and policies. In F. Reboredo (Ed.), *Forest context and policies in Portugal, present and future challenges series: World forests* (Vol. 19, pp. 97–115). Berlin: Springer.

Matthews, S., Sauvage, S., Grootemaat, S., Hollis, J., Kenny, B., & Fox-Hughes, P. (2019). National fire danger rating system: Implementation of models and the forecast system. In *Proceedings for the 6th International Fire Behavior and Fuels Conference April 29–May 3, 2019, Sydney, Australia*. Missoula: International Association of Wildland Fire.

McArthur, A. G. (1967). *Fire behaviour in eucalypt forests* (Leaflet No 107). Forestry and Timber Bureau.

McPhillips, L. E., Chang, H., Chester, M. V., Depietri, Y., Friedman, E., Grimm, N. B., Kominoski, J. S., McPhearson, T., Méndez-Lázaro, P., Rosi, E. J., & Shafiei, S. J. (2018). Defining extreme events: A cross-disciplinary review. *Earth's Future, 6*, 441–455.

McRae, R. H. D., Sharples, J. J., & Badlan, R. (2018). A model for identifying blow-up fire potential. In D. X. Viegas (Ed.), *Advances in forest fire research*, Chapter 1—Fire risk management, pp. 17–22.

Mell, W. E., Charney, J. J., Jenkins, M. A., Cheney, P., & Gould, J. (2005). Numerical simulations of grassland fire behavior from the LANL-FIRETEC and NIST-WFDS models. In *EastFIRE Conference*, May 11–13, George Mason University, Fairfax VI.

Mercer, G. N., & Weber, R. O. (1994). Plumes above line fires in a cross wind. *International Journal of Wildland Fire, 4*(4), 201–207.

Mercer, G. N., & Weber, R. O. (2001). Fire plumes. Chapter 7. In E. A. Johnson & K. Miyanishi (Eds.), *Forest fires. Behavior and ecological effects* (pp. 225–255). San Diego: Academic.

Michaletz, S. T., & Johnson, E. A. (2007). How forest fires kill trees: A review of the fundamental biophysical processes. *Scandinavian Journal of Forest Research, 22*, 500–515.

Mills, G. A., & McCaw, L. (2010). *Atmospheric stability environments and fire weather in Australia—Extending the Haines index* (Tech Rep No 20). Centre for Australian Weather and Climate Research.

Moreira, F., Ascoli, D., Safford, H., Adams, M. A., Moreno, J. M., Pereira, J. M. C., Catry, F. X., Armesto, J., Bond, W., González, M. E., Curt, T., Koutsias, N., McCaw, L., Price, O., Pausas, J. G., Rigolot, E., Stephens, S., Tavsanoglu, C., Vallejo, V. R., Van Wilgen, B. W., Xanthopoulos, G., & Fernandes, P. M. (2020). Wildfire management in Mediterranean-type regions: Paradigm change needed. *Environmental Research Letters, 15*, 011001.

Moritz, M. A. (1997). Analyzing extreme disturbance events: Fire in Los Padres National Forest. *Ecological Applications, 7*, 1252–1262.

Morris, G. A. (1987). *A simple method for computing spotting distances from wind-driven surface fires* (Res Note INT-374). Ogden: USDA Forest Service Intermountain Forest and Range Experiment Station.

Morvan, D. (2014). Wind effects, unsteady behaviors and regimes of propagation of surface fires in open fields. *Combustion Science and Technology, 186*, 869–888.

Morvan, D., Hoffman, C., Rego, F., & Mell, W. (2009). Numerical simulation of the interaction between two fire fronts in the context of suppression fire operations. In *8th Symposium on Fire and Forest Meteorology*, 13–15 October 2009, Kalispell, MT, American Meteorological Society.

Morvan, D., Hoffman, C., Rego, F., & Mell, W. (2011). Numerical simulation of the interaction between two fire fronts in grassland and shrubland. *Fire Safety Journal, 46*, 469–479.

Mount, A. B. (1969). Eucalypt ecology as related to fire. In *Proceedings Tall Timbers Fire Ecology Conference* (Vol. 9, pp. 75–108). Tallahassee: Tall Timbers Research Station.

Mount, A. B. (1972). *The derivation and testing of a soil dryness index using run-off data* (Bull No. 4). Tasmania Forestry Commission.

Muraszew, A. (1974). *Firebrand phenomena*. Aerospace Report ATR-74 (8165–01)-1. El Segundo: The Aerospace Corporation.

Mutch, R. W., Arno, S. F., Brown, J. K., Carlson, C. E., Ottmar, R.D., & Peterson, J. L. (1993). Forest health in the Blue Mountains: A management strategy for fire-adapted ecosystems. USDA Forest Service General Technical Report PNW-GTR-310. 14 pp.

National Wildlife Coordinating Group (NWCG). (2020). Retrieved March 20, 2020, from https://www.nwcg.gov/term/glossary/extreme-fire-behavior.

Nauslar, N. J., Abatzoglou, J. T., & Marsh, P. T. (2018). The 2017 North Bay and Southern California fires: A case study. *Fire, 1*(1), 18.

Noble, I. (1991). Behaviour of a very fast grassland wildfire on the Riverine Plain of south-eastern Australia. *International Journal of Wildland Fire, 1*, 189–196.

Nunes, L., Alvarez-Gonzalez, J., Alberdi, I., Silva, V., Rocha, M., & Rego, F. C. (2019). Analysis of the occurrence of wildfires in the Iberian Peninsula based on harmonized data from national forest inventories. *Annals of Forest Science, 76*, 27.

Pearce, H. G., Morgan, R. F., & Alexander, M. E. (1994). *Wildfire behaviour case study of the 1986 Awarua Wetlands fire* (Fire Tech Transfer Note 5, pp. 1–6). Rotorua: New Zealand Forest Research Institute National Rural Fire Authority.

Pinto, M. M., DaCamara, C. C., Hurduc, A., Trigo, R. M., & Trigo, I. F. (2020). Enhancing the fire weather index with atmospheric instability information. *Environmental Research Letters, 15*, 0940b7.

Potter, B. E. (2016). Chapter 7: Spot fires. In P. A. Werth, B. E. Potter, M. E. Alexander, C. B. Clements, M. G. Cruz, M. A. Finney, J. M. Forthofer, S. L. Goodrick, C. Hoffman, W. M. Jolly, S. S. McAllister, R. D. Ottmar, & R. A. Parsons (Eds.), *Synthesis of knowledge of extreme fire behavior: Vol. 2 for fire behavior specialists, researchers, and meteorologists* (Gen Tech Rep PNW-GTR-891). Portland: USDA Forest Service Pacific Northwest Research Station.

Potter, B. E. (2018). The Haines Index—It's time to revise it or replace it. *International Journal of Wildland Fire, 27*, 437–440.

Pyne, S. J., Andrews, P. L., & Laven, R. D. (1996). *Introduction to wildland fire*. Chichester: Wiley.

Quintilio, D., Lawson, B. D., Walkinshaw, S., & Van Nest, T. (2001). *Final documentation report–Chisholm Fire (LWF-063)*. Edmonton: Alberta Sustainable Resource Development, Forest Protection Division Report I/036.

Raj, V. C., & Prabhu, S. V. (2018). Measurement of geometric and radiative properties of heptane pool fires. *Fire Safety Journal, 1*, 13–26.

Raupach, M. R. (1990). Similarity analysis of the interaction of bushfire plumes with ambiente winds. *Mathematical and Computer Modelling, 13*(12), 113–121.

Ricotta, C., Avena, G., & Marchetti, M. (1999). The flaming sandpile: Self-organized criticality and wildfires. *Ecological Modelling, 119*, 73–77.

Rothermel, R. C. (1972). *A mathematical model for predicting fire spread in wildland fuels* (Res Pap INT-115). Ogden: USDA Forest Service Intermountain Forest and Range Experiment Station.

Rothermel, R. C. (1984). Fire behavior considerations of aerial ignition. In R. W. Mutch (Tech coord), Prescribed fire by aerial ignition. *Proceedings of a Workshop. Intermountain Fire Council* (pp. 143–158).

Rothermel, R. C. (1991). *Predicting behavior and size of crown fires in the northern Rocky Mountains* (Res Pap INT-438). Ogden: USDA Forest Service Intermountain Forest and Range Experiment Station.

Rothermel, R. C., & Andrews, P. L. (1987). *Fire behavior system for the full range of fire management needs* (Gen Tech Rep PSW-101, pp. 145–151). Berkeley: USDA Forest Service Pacific Southwest Forest and Range Experiment Station.

Roxburgh, R., & Rein, G. (2008). Study of wildfire in-draft flows for counter fire operations. *WIT Transactions on Ecology and the Environment, 119*, 13–22.

Ruiz-González, A. D., & Álvarez-González, J. G. (2011). Canopy bulk density and canopy base height equations for assessing crown fire hazard in *Pinus radiata* plantations. *Canadian Journal of Forest Research, 41*, 839–850.

SALTUS. (2001). *Fire spotting: mechanism analysis and modelling. Probabilistic model.* Unpublished final report. EU Project ENV98-CT98-0701.

Schroeder, M. J., & Chandler, C. C. (1966). *Monthly fire behavior patterns. Res Note 112.* Berkeley: USDA Forest Service Pacific Southwest Forest and Range Experiment Station.

Scott, J. H. (1999). NEXUS: A system for assessing crown fire hazard. *Fire Manage Notes, 59*, 20–24.

Scott, J. H. (2006). *Comparison of crown fire modelling systems used in three fire management applications* (Res Pap RMRS-RP-58). Fort Collins: USDA Forest Service Rocky Mountain Research Station.

Scott, J. H., & Reinhardt, E. D. (2001). *Assessing crown fire potential by linking models of surface and crown fire behavior* (Res Pap RMRS-RP-29). Fort Collins: USDA Forest Service Rocky Mountain Research Station.

Scott JH, & Reinhardt. (2005). *Stereo photo guide for estimating canopy fuel characteristics in conifer stands* (Gen Tech Rep RMRS-GTR-145). Fort Collins: USDA Forest Service Rocky Mountain Research Station.

Sharple, J. J., Cary, G. J., Fox-Hughes, P., Mooney, S., Evans, J. P., Fletcher, M. S., Fromm, M., Grierson, P. F., McRae, R., & Baker, P. (2016). Natural hazards in Australia: Extreme bushfire. *Climatic Change, 139*, 85–99.

Sharples, J. J. (2009). An overview of mountain meteorological effects relevant to fire behaviour and bushfire risk. *International Journal of Wildland Fire, 18*, 737–754.

Silva, J. S., Rego, F., Fernandes, P., & Rigolot, E. (Eds.). (2010). *Towards integrated fire management—Outcomes of the European project Fire Paradox* (Res Rep 23). Joensuu: European Forest Institute.

Simard, A. J., Haines, D. A., Blank, R. W., & Frost, J. S. (1983). *The Mack Lake Fire* (Gen Tech Rep NC-83). Saint Paul: USDA Forest Service North Central For Exp Sta

Srock, A., Charney, J., Potter, B., & Goodrick, S. (2018). The hot-dry-windy index: A new fire weather index. *Atmosphere, 9*, 279–289.

Stocks, B. J. (1989). Fire behavior in mature jack pine. *Canadian Journal of Forest Research, 19* (6), 783–790.

Stocks, B. J., & Walker, J. D. (1973). *Climatic conditions before and during four significant forest fire situations in Ontario.* Sault Ste: Canadian Forestry Service, Great Lakes Forestry Research Centre, Information Report O-X-187.

Stocks, B. J., Alexander, M. E., McAlpine, R. S., Lawson, B. D., & Van Wagner, C. E. (1987). *Canadian forest fire danger rating system—User's guide.* Ottawa: Canadian Forestry Service Fire Danger Group.

Stocks, B. J., Alexander, M. E., & Lanoville, R. A. (2004). Overview of the International Crown Fire Modelling Experiment (ICFME). *Canadian Journal of Forest Research, 34*, 1543–1547.

Sullivan, A. L. (2017). Inside the inferno: Fundamental processes of wildland fire behaviour. *Current Forestry Reports, 3*, 150–171.

Sullivan, A., Cruz, M. G., & Cheney, N. P. (2007). Burning bush. *New Scientist, 195*, 27.

Tarifa, C. S., Del Notario, P. P., & Moreno, F. G. (1965). On the flight paths and lifetimes of burning particles of wood. In *10th Symposium (International) on Combustion*, 1 Jan 1965. The Combustion Institute, University of Cambridge, Cambridge (Vol. 10, pp. 1021–1037).

Tedim, F., Leone, V., Amraoui, M., Bouillon, C., Coughlan, M. R., Delogu, G. M., Fernandes, P. M., Ferreira, C., McCaffrey, S., McGee, T. K., Parente, J., Paton, D., Pereira, M. G., Ribeiro, L. M., Viegas, D. X., & Xanthopoulos, G. (2018). Defining extreme wildfire events: Difficulties, challenges, and impacts. *Fire, 1*, 9.

Telitsyn, H. P. (1996). A mathematical model of spread of high-intensity forest fires. In J. G. Goldammer & V. V. Furyaev (Eds.), *Fire in ecosystems of boreal Eurasia* (Forestry sciences) Dordrecht: Springer

Thomas, P. H. (1963). The size of flames from natural fires. In *International Symposium Proceedings Combustion* (Vol. 9, pp. 844–859).

Thomas, P. H. (1967). Some aspects of the growth and spread of fire in the open. *Forest, 40*, 139–164.
Tolhurst, K. (2009). *Report on the physical nature of the Victorian Fires occurring on 7th February 2009*. Parkville: Department of Forest and Ecosystem Science, University of Melbourne.
Tolhurst, K., Shields, B., & Chong, D. (2008). Phoenix: Development and application of a bushfire risk management tool. *Australian Journal of Emergency Management, 23*, 47.
Tory, K. J., & Kepert, J. (2019). *Pyrocumulonimbus firepower threshold: A pyroCb prediction tool.* East Melbourne: Bureau of Meteorology, Bushfire & Natural Hazards.
Tory, K. J., Thurston, W., & Kepert, J. D. (2018). Thermodynamics of pyrocumulus: A conceptual study. *Monthly Weather Review, 146*, 2579–2598.
USDA Forest Service. (1956). *Glossary of terms used in forest fire control. Agric Handbook 104*. Washington, DC: Government Printing Office.
USDA Forest Service. (1960). The "Pungo 1959" fire—A case study. In *Annual report, 1959* (pp. 39–42). Asheville: USDA Forest Service, Southeastern Forest Experiment Station.
Van Wagner. C. E. (1967). *Calculations on forest fire spread by flame radiation. Canadian Department of Forestry and Rural Development Forestry Branch*, Technical Report, Departmental Publication No. 1185. (Ottawa, ON)
Van Wagner, C. E. (1973). Height of crown scorch in forest fires. *Canadian Journal of Forest Research, 3*, 373–378.
Van Wagner, C. E. (1974). *Structure of the Canadian Forest Fire Weather Index. Publ. 1333*. Ottawa: Canadian Forestry Service.
Van Wagner, C. E. (1977). Conditions for the start and spread of crown fire. *Canadian Journal of Forest Research, 7*, 23–34.
Van Wagner, C. E. (1989). Prediction of crown fire behavior in conifer stands. In D. C. MacIver, H. Auld, & R. Whitewood (Eds.), *Proceedings of the 10th Conference on Fire and Forest Meteorology* (pp. 207–212). Ottawa: For Can Environ.
Van Wagner, C. E. (1993). Prediction of crown fire behavior in two stands of jack pine. *Canadian Journal of Forest Research, 23*, 442–449.
Vega, J. A., Jiménez, E., Dupuy, J.-L., & Linn, R. R. (2012). Effects of flame interaction on the rate of spread of heading and suppression fires in shrubland experimental fires. *International Journal of Wildland Fire, 21*, 950–960.
Viegas, D. X., & Pita, L. P. (2004). Fire spread in canyons. *International Journal of Wildland Fire, 13*, 253–274.
Wade, D. D., & Ward, D. E. (1973). *An analysis of the Air Force Bomb Range Fire* (Res Pap SE–105). Asheville: USDA Forest Service Southeastern Forest Experiment Station.
Weise, D. R., Cobian-Iñiguez, J., & Princevac, M. (2018). Surface to crown transition. In S. L. Manzello (Ed.), *Encyclopedia of wildfires and wildland-urban interface (WUI) fires*. Cham: Springer.
Werth, P. A., Potter, B. E., Alexander, M. E., Clements, C. B., Cruz, M. G., Finney, M. A., Forthofer, J. M., Goodrick, S. L., Hoffman, C., Jolly, W. M., McAllister, S. S., Ottmar, R. D., & Parsons, R. A. (2016). *Synthesis of knowledge of extreme fire behavior: Vol 2 for fire behavior specialists, researchers, and meteorologists* (Gen Tech Rep PNW-GTR-891). Portland: USDA For Serv Pacific Northwest Res Station.
Williams, C. (2007). Ignition impossible: When wildfires set the air alight. *New Scientist (1971), 2615*, 38–40.

Winkler, J. A., Potter, B. E., Wilhelm, D. F., Shadbolt, R. P., Piromsopa, K., & Bian, X. (2007). Climatological and statistical characteristics of the Haines Index for North America. *International Journal of Wildland Fire, 16*, 139–152.

Woods, J. B. Jr (1944). *Training manual for smokechasers, lookouts and suppression crews* (Bull 9). Salem: Oregon State Board of Forestry..

Wooster, M. J., Roberts, G., Perry, G. L. W., & Kaufman, Y. J. (2005). Retrieval of biomass combustion rates and totals from fire radiative power observations: FRP derivation and calibration relationships between biomass consumption and fire radiative energy release. *Journal of Geophysical Research Atmospheres, 110*, D24311.

Xanthopoulos, G., & Wakimoto, R. H. (1993). A time to ignition-temperature-moisture relationship for branches of three western conifers. *Canadian Journal of Forest Research, 23*, 253–258.

Chapter 9
Fire Effects on Plants, Soils, and Animals

Learning Outcomes
At the conclusion of this chapter, we expect you will be able to

1. Describe how the fire effects on plants, soils, and animals are linked to fire behavior concepts learned in earlier chapters, including both flaming and smoldering combustion,
2. Describe fire effects on the crowns, stems, and roots of trees and other plants and relate those to fire behavior and post-fire vegetation changes,
3. Discuss how burn severity can be assessed in general, and soil burn severity in particular, as these influence vegetation response and post-fire soil erosion potential,
4. Critically think about how fire can affect nitrogen availability and nutrient dynamics in both the short and long term post fire, and
5. Use the interactive spreadsheets to understand the factors influencing the degree of soil heating at different depths and what this means for ecosystem functions.

9.1 Introduction

Wildfires are a dominant disturbance agent in most terrestrial ecosystems around the world. Fires influence plant mortality, survival, growth, and regeneration, and thus the structure, composition, and dynamics of ecosystems (Payette 1992; Bowman et al. 2009; Pausas and Keeley 2019). Fires may directly affect individual plants

Supplementary Information The online version of this chapter (https://doi.org/10.1007/978-3-030-69815-7_9) contains supplementary material, which is available to authorized users.

F. Castro Rego et al., *Fire Science*, Springer Textbooks in Earth Sciences, Geography and Environment, https://doi.org/10.1007/978-3-030-69815-7_9

through tissue necrosis (i.e., tissue death) due to heat exposure or indirectly by altering physiology and resistance to insect attack or unfavorable environmental conditions. The more visible effects are in the upper parts of the plants but, as recognized already by the Greek philosopher Theophrastus (372 BC–287 BC), plants are composed of two interconnected parts, the aboveground part, more controlled by Air and Fire, and the belowground, mostly dependent on Soil and Water. And if most of the immediately visible effects are due to fire top killing plants, many of the longer-term effects are due to fire influencing soil and the water regime. Also, many other post-fire factors affect fire survival and recovery.

How do plants survive fires? What are the immediate and short-term effects of fires on plants (these are the direct or first-order fire effects)? What are the longer-term effects (indirect or second-order fire effects) of fires on plants, soils, and ecosystems? How do we assess and predict these effects? Why are wildfires so integral to ecosystem function, and how can we use that knowledge to guide management? For instance, forest managers can reduce erosion potential by selecting best management practices during fires and the best treatments after fires to limit soil erosion potential if they understand what influences soil burn severity and vegetation response to fires. In this chapter, we illustrate key concepts with examples from ecosystems around the world.

The ecological effects of fire on plants and habitats (Figs. 9.1 and 9.2) depend on the heating above the ground (typically measured by fireline intensity for which

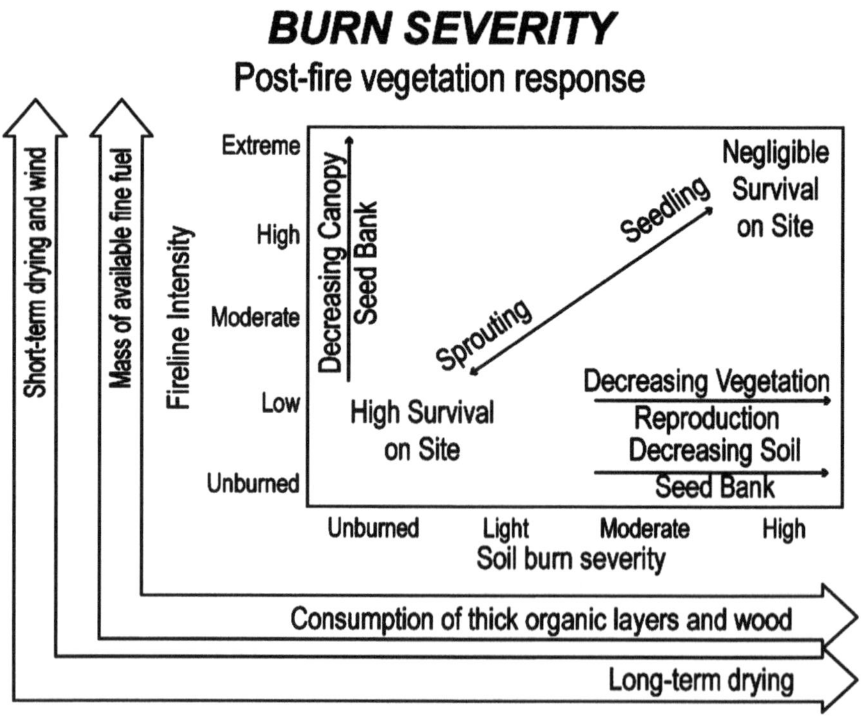

Fig. 9.1 Both fire intensity and soil burn severity influence vegetation and habitat response to fire. The outer arrows illustrate the influence of key environmental variables. (Redrawn from Ryan 2002)

Fig. 9.2 (**a**) Heat during fires directly affects the crowns, stems, and roots of trees. Thick bark can insulate all or part of the cambium underneath it, while the bark on most small trees and shrubs is too thin to effectively protect the cambium. The foliage of small trees, grasses, shrubs, and forbs are consumed or killed even with low-intensity fires, so the degree of soil heating and the degree of exposure of bare mineral soil whether they regrow or establish as new individuals post fire. Note, however, that fire effects are spatially variable, so that small trees and other plants may survive if they are "skipped" by fires. (**b**) Fire effects are commonly assessed visually soon after fires. If meristematic tissue survives in buds, the tree may replace foliage and continue growing. Where twigs and branches are black, the buds have likely been killed, except in particularly fire-resistant species that sprout readily. Where foliage is brown, it is dead but some buds may survive. Indirect fire effects may not be evident for some time; trees that are stressed by drought are more susceptible to insects post fire. (From Hood et al. 2018)

flame length is an indicator or by Fire Radiative Energy (FRE) as described in earlier chapters) and below the ground (here termed soil burn severity). Vegetation survival and recovery through sprouting or seeding depend on whether or not the above-ground portions of plants survived (such as trees that can be a seed source for seedling establishment or that may resprout from the trunk or branches) or below-ground in which the propagules of buds for resprouting or seeds in the soil seedbank are the source of vegetation growth post fire. Clearly, then, what survives the fire above and below ground and what can spread from surviving plants in areas unburned or burned with low severity (these are often called refugia) all influence post-fire vegetation, as do site productivity, climate, land use, and spatial heterogeneity. Here we focus on individual plants and plant communities as well as soils and nutrients. The implications for landscape dynamics and management are addressed in Chap. 12.

Fire ecology is a relatively new science that has steadily expanded in recent decades globally as scientists sought to understand how fires affected plants, animals, and soils, and as managers sought to manage vegetation sustainably. The study of the ecological effects of fire lagged behind the study of fire behavior in the support of fire suppression. However, fire ecology is part of human culture. Fires were one of the few tools that Indigenous people had to manage vegetation, and so they had to observe and understand both fire behavior and fire effects to use fire to accomplish their objectives. Fire ecology is thus part of Traditional Knowledge (see more on this in Chap. 10). In integrated fire management (See Chap. 13), fire ecology helps shape strategic management decisions about fire use, fire suppression, and managing fire effects on ecosystems.

9.2 Heat Transfer Has Implications for Plant Survival and Post-fire Response

Directly and indirectly, fires influence plant survival and response. Immediate and direct fire effects occur as a result of heat-induced damage to plant tissue and function. This heat damage occurs as a direct result of radiative, convective, and conductive heat transfer from the fire to the plant (Figs. 9.2 and 9.3, see Chap. 5). Living plant tissues die when they are exposed to sufficient heat from a fire for long enough. The death of plant tissues can result in either the death of the plant or partial damage to the plant, which can influence plant function. Indirect fire effects result from non-lethal injury to plants and alterations to the immediate post-fire environment that may lead to further stress or the inability to defend against secondary mortality agents such as insects. Plant damage, mortality, and the post-fire response to heating vary as a function of the overall rate or "dose" of heat exposure and, therefore, to variations in fire behavior and heat transfer. For instance, buds and leaves may survive if the temperatures in the buoyant plume above the fire are relatively low, or plants may regrow after the above-ground portions are killed if the seeds and roots are sufficiently deep in the soil to escape soil heating. Both direct and indirect fire effects influence long-term ecosystem response to fire. To understand the potential effects of fire on individual plants and vegetation communities, it is useful to begin by understanding how fires can directly injure plants and what traits are associated with the survival of wildland fires.

Currently, there are two competing hypotheses for how the direct effects of heat from wildfires can damage plants: the cambium necrosis hypothesis and the xylem dysfunction hypothesis (Balfour and Midgeley 2006; Kavanagh et al. 2010; Michaletz et al. 2012; Hood et al. 2018). Although there is growing support for the xylem dysfunction hypothesis, many discussions of fire-induced plant injury and mortality have focused on understanding cambium necrosis. The cambium necrosis hypothesis suggests that the heat transfer from a fire to a plant results in necrosis (death) of phloem, cambium, or buds in the plant roots, stem, and crown (Fig. 9.3)

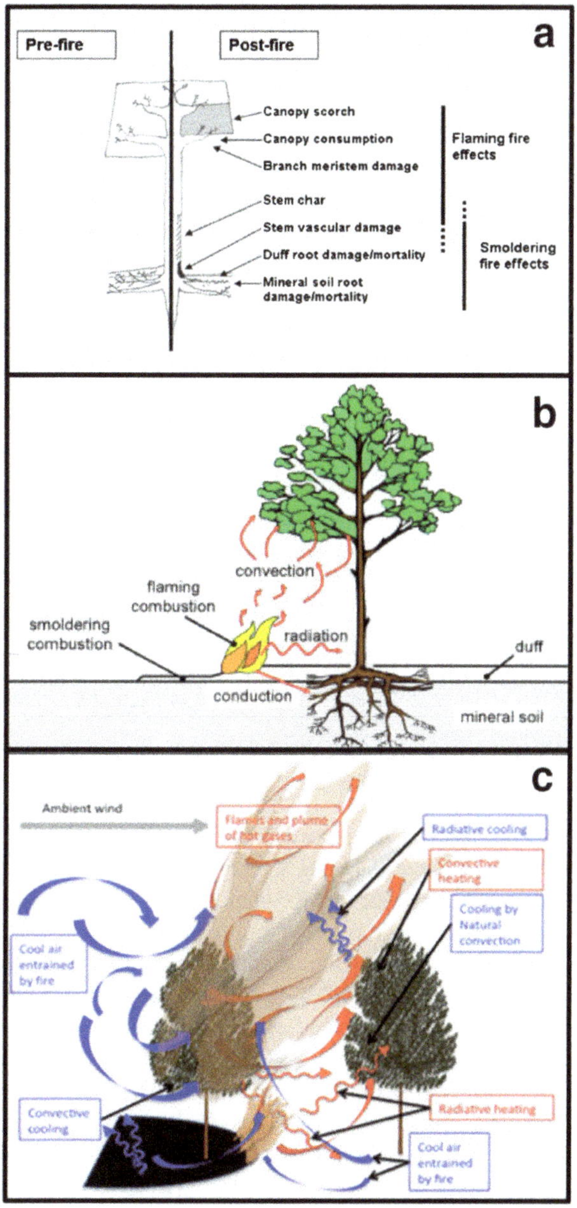

Fig. 9.3 Fires can affect the crown, bole and roots of trees through both flaming and smoldering. Early conceptualizations of the processes evolved from (**a**) one- to (**b**) two-dimensional diagrams of the modes of heat transfer affecting different parts of trees. (**c**) Current approaches reflect interactions between fire, vegetation, and atmosphere in multiple dimensions. (From O'Brien et al. 2018)

which then limits carbon translocation and ultimately results in hydraulic failure and mortality of the individual (McDowell et al. 2008; Hood et al. 2018). Tissue necrosis due to heating is thought to result from the denaturation of proteins and is dependent upon the rate and duration of heat exposure (Dickinson and Johnson 2004). Although many researchers have suggested that the necrosis of tissue can be modeled based on a critical tissue temperature of 60 °C for 1 min (e.g., Brown and DeByle 1987; Steward et al. 1990; Gutsell and Johnson 1996), this threshold is an overly simple generalization. Necrosis can occur when tissues are exposed to elevated temperatures for longer durations or to higher temperatures for shorter duration (Dickinson and Johnson 2004; Pingree and Kobziar 2019).

The xylem dysfunction hypothesis suggests that plant mortality occurs due to damage to the hydrologic function of xylem (Balfour and Midgeley 2006; Michaletz et al. 2012; Thompson et al. 2017). The heat generated during a wildfire can influence plant hydraulic conductivity through two different mechanisms: (1) reduced conductivity due to cavitation, and (2) reduced conductivity due to the deformation of cell walls. Cavitation occurs when the plant stomata are exposed to high vapor pressure deficit within the fire plume. When the stomata cannot close fast enough, and the water tension in the xylem increases beyond a critical threshold, small gaseous bubbles are aspirated and embolisms form (Tyree and Zimmermann 2002; Kavanagh et al. 2010). Ultimately this reduces hydraulic conductivity in the xylem, which can lead to reduced tree productivity, increased vulnerability to insects or diseases, and potentially death (Sperry et al. 1993; Brodribb and Cochard 2009). Hydraulic conductivity can also be impaired due to heat-induced deformations of the cell walls. The deformation of the cell walls results in reduced stomatal conductance, which increases xylem water tensions, increases periods of stomatal closure, and limits carbon assimilation and growth. Reduced growth and fitness results, and in extreme cases, plants die from either hydraulic failure or carbon starvation (McDowell et al. 2008; Sevanto et al. 2014).

9.2.1 *Fire Effects on Plant Crowns*

Thermal injury to tree crowns is frequently evaluated in the field by assessing the portion of foliage and buds that are consumed or scorched during a fire (Peterson and Arbaugh 1986; Ryan and Reinhardt 1988; Fowler and Sieg 2004). Scorching occurs when crown foliage is killed through radiative and convective heat transfer (see heat transfer equations in Chap. 5 and prediction equations in Chap. 8). Scorched foliage looks red or brown in color. In contrast, consumed foliage and twigs are blackened as they are directly involved in the combustion process and contribute to the heat release rate and spread of a fire. In practice, the amount of damage to foliage, twigs, and buds associated with consumption and scorch are lumped into a single estimate of crown damage. However, care should be taken as foliage death doesn't mean that the terminal buds also died. The terminal buds may survive even when surrounding foliage is scorched; then a tree can produce new foliage. When buds are killed, that portion of the tree crown will not recover.

Scorched needles often fall to the ground in the days or weeks after a fire. This needle fall (sometimes called needle cast) can increase surface fuels and, therefore, the potential for the spread of future fires. The fallen needles can contribute to nutrient cycling and can help to protect soils from erosion. When lower branches of trees die, the base of the live crown is higher, which can increase the potential that the tree(s) survive the next surface fire.

The crowns of trees (especially small trees), shrubs, graminoids, and forbs are commonly consumed or scorched by even low to moderate-intensity fires (Fig. 9.2). In general, short trees, with crowns near the heat from the flames, are predicted to experience more crown consumption and scorch for a given fire intensity. However, local variations in the surface fireline intensity and convective cooling can also influence the spatial variability in crown scorch and consumption (Ritter et al. 2020). The terminal buds on branches are better protected in trees with crowns farther above the ground and those with long needles such as ponderosa pine (*Pinus ponderosa*) or longleaf pine (*Pinus palustris*) in the USA, or maritime pine (*Pinus pinaster*) in Europe. Both mechanisms serve to protect the buds from heating, particularly when the duration of heating is short because flame residence time is short.

9.2.2 *Fire Effects on Stems, Especially Vascular Cambium*

The vascular cambium is the primary meristematic tissue located in the stem of trees and shrubs where it produces xylem towards the inside of the stem and phloem and bark to the outside. Vascular cambium is responsible for diameter growth in trees. Damage to the vascular cambium depends upon the rate and magnitude of radiative and convective heat transfer from the flames to the plant stem and the subsequent conduction of heat through the bark to the vascular cambium.

One of the earliest approaches to modeling vascular cambium mortality due to fire is based on modeling heat transfer through bark using a semi-infinite one-dimensional conduction equation (see Chap. 5). In this approach, cambial mortality depends upon two main factors: (1) the temperature and duration of heat on the outside of the bark and, (2) the thickness and thermal properties of the bark. Cambial death can be estimated by comparing the predicted or measured residence time for a fire to the critical residence time estimated with Eq. (5.16). This approach requires estimates of the thermal diffusivity of the bark and the bark thickness. The thermal diffusivity of bark is fairly consistent across species and can be treated as a constant (e.g., 1.35×10^{-7} m^2 s^{-1}). Bark thickness is typically estimated based on species-level empirically-derived regression equations that use the diameter of the tree at breast height (DBH, outside bark) and in some cases the age of the tree. Interestingly, researchers have found that the relationship between the DBH and bark thickness for a given species is generally not influenced by factors such as site productivity or competition, though both factors influence tree diameter. Given that bark thickness varies with bole diameter, necrosis is more likely for small trees than for large trees and in species that tend to have thinner bark. Although regression equations relating bark thickness to tree diameter exist for some tree species (e.g., Hare 1965; Hood and Lutes 2017), there is still a lack of information on many tree species. Spatial variation in vascular cambial mortality occurs across scales. At fine scales, variability in airflow around a tree bole can result in variations in the temperature and duration of bole heating, enough that a tree can survive with only some of the cambium being killed around the circumference of a tree. The greatest temperature and duration of heating are commonly found on the leeward side of tree boles (often the uphill side) where mixing is limited and a significant increase in flame height occurs.

Many plants are readily top-killed by fire. As many shrubs and young trees have very thin bark, and because most plants other than large trees lack bark, the meristematic tissue in their stems is very likely to die in fires. Foliage that is in the flames or close to them is readily killed by heat. When plants are top-killed by fire, they may still survive fires and grow in abundance soon after fires if their seeds or buds on roots or in the root crown of the plants survive on-site.

9.2.3 *Fire Effects on Roots and Buds*

Lethal heating of belowground plant parts is primarily associated with conductive heat transfer through the soil. Soil heating during fires can contribute to the death of large trees in fires, even when those trees live in forests where fires were once frequent. O'Brien et al. (2010) found that more than 40% of longleaf pines (*Pinus palustris*) died post fire in a forest unburned for more than 50 years. During the fire, thick bark protected tree stems and most trees had limited crown scorch, yet many trees died. O'Brien et al. (2010) attributed tree death to loss of fine roots where accumulated needles, cones, and bark chips under tree crowns were consumed, especially if greater than 30% by weight was consumed resulting in soil heating and root damage which led to reduced sap flux post fire so that trees could not supply enough water to crowns and photosynthesis was limited.

In addition to roots, many plants have meristematic tissue (i.e., buds) on underground stems (rhizomes), on above-ground horizontal stems (stolons), in bulbs or corms, or in the root crown (sometimes called a basal caudex or lignotuber) (Fig. 9.4). Such meristematic tissue can also be damaged via conductive heat transfer through the soil. Even low-intensity fires can kill the crowns of low-lying plants, such as shrubs, graminoids, and forbs, so the survival of these plants depends on their ability to resprout following the fire. This means that heat conduction through the soil, and subsequent heating of roots and other belowground plant tissues, is a crucial factor determining if plants will resprout post-fire. Therefore, to predict plants' survival following a fire, it is important to know both the degree and depth of soil heating relative to where the plant propagules are located in the soil.

Basal sprouting can be important for the recovery of many species following fires. Resprouting from epicormic buds is less common but occurs in eucalypts (e.g., *Eucalyptus marginata*) and other Australian trees, savanna trees, oaks such as *Quercus suber* in the Mediterranean Basin or *Q. kelloggii* in California, and even pines such as *Pinus canariensis*, often in association with thick bark (Pausas and Keeley 2017). This allows rapid regrowth of the canopy and forest or woodland resilience to a regime of high-severity crown fires. Sprouting is very prevalent. Most plants have multiple regeneration strategies, so that many plants both resprout and establish from seed if conditions are favorable, and the few initial colonizers may

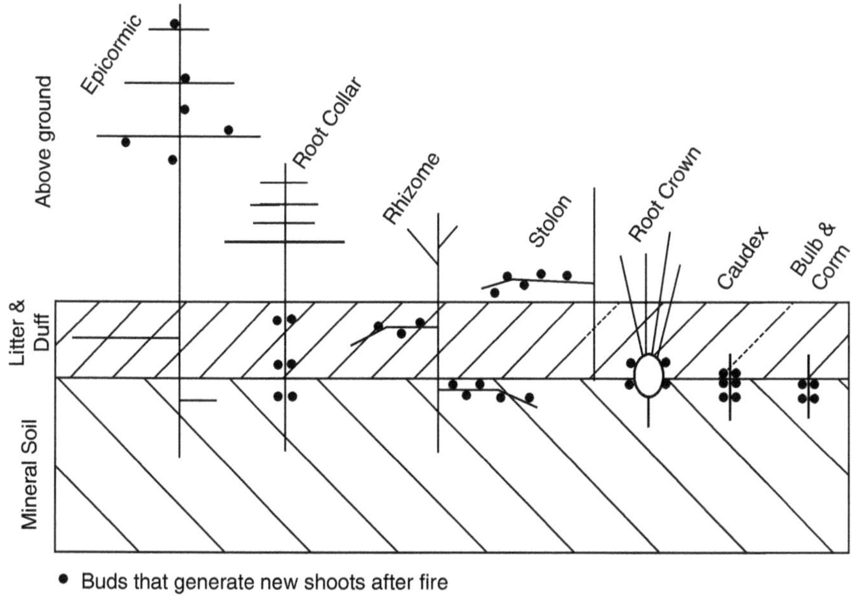

Fig. 9.4 Heat exposure of meristematic tissues (buds) in plants influences their response to fires. If meristems survive the fire, new growth may be stimulated post fire depending on the environmental conditions. (From Brown and Smith 2000)

rapidly produce seed that then establish new individuals. If conditions are not favorable post fire, and we may expect that changing fire frequency, severity, and season and climate change will affect regeneration success.

9.2.4 Heat and Smoke Effects on Seeds, Including Serotiny

Following a fire, many plant species regenerate following a fire by recruiting new individuals from seeds that survived in the canopy or soil layers. The seeds of different species can be more or less tolerant of heat, depending upon various characteristics, including the seed size, shape, mass, and moisture content. In general, seeds with lower moisture content can tolerate more heat than those with greater moisture content. Most plants produce many seeds with different dormancy levels, some of which are stimulated to germinate after fires.

Fire is a double-edged sword for many seeds stored in the soil. Many plants and seeds may be killed or damaged, particularly if the seeds are exposed to heat on or near the soil surface. Many seeds can tolerate extended heating, and the dormancy of some, such as *Ceanothus* in California or *Cistus* in southern Europe, is broken by heat. Germination of the seeds may be stimulated in the post-fire environment. This could be triggered by altered soil chemistry or changes in the post-fire environment, including increased light, temperatures, and moisture at the soil surface. Therefore, the successful germination of soil-stored seeds is a fine balance between mortality caused by excessive heat and receiving sufficient heat to break dormancy or provide the needed environmental conditions for germination. Seeds are numerous, and with the spatial variability in both fuel consumption and soil heating, many seeds will survive and may germinate post fire if other conditions are favorable. Further, additional colonization is likely from seed produced from the plants that survived or established soon after fires.

Serotinous plants delay the opening of cones (or fruits) until favorable conditions for germination occur. For example, serotinous lodgepole (*Pinus contorta*) and Alepo (*P. halepensis*) pine trees in North America and southern Europe produce serotinous cones that are sealed shut with waxes and resins so that cones don't open when seeds are ripe. Instead, the seeds accumulate. When the serotinous cones are exposed to sufficient heat, the waxes and resins melt, the cones open and many seeds are dispersed onto the bare ground, which is favorable for germination. Alternatively, the cones may be consumed in the fire or be exposed to enough heat to cause seed mortality or the cones may be consumed in the fire. Models that predict cone opening based on predicted fire behavior are available for some species (e.g., Alexander and Cruz 2012). The density of serotinous cones and the length of time that plants retain cones or fruit is highly variable across different species and populations of the same species. For example, more of the lodgepole pine cones are serotinous where nonlethal fires and other disturbances are prevalent, and are also more abundant as the time since stand-replacing fire increases.

The recruitment of plants in burned areas occurs when their seeds are carried into burned areas via transport from wind, water, or animals. The recolonization of burned areas depends on several factors, including the size and shape of the burned area, the size of the seeds, and the local topography, weather, and climate. Stevens-Rumann et al. (2018) and Stevens-Rumann and Morgan (2019) highlighted the importance of seed source proximity and climate on whether or not trees can germinate, survive, and grow following fires (see Case Study 12.3).

Many plants are stimulated to germinate post fire, whether by heat exposure, increased light availability, changing soil nutrients, charred wood, smoke, or a combination. For some species, short exposure to smoke enhances germination, while long-duration exposure can inhibit germination, perhaps because of allelopathic compounds in the smoke (Pennacchio et al. 2007). Keeley and Pausas (2018) found that chemical compounds in smoke stimulate the germination of seeds from a diversity of plant species in California chaparral ecosystems. Furthermore, they found that smoke exposure can make the seed coats of long-dormant seeds more permeable to water in soils (Keeley and Fotheringham 1998). In contrast, at least 30 min of smoke exposure reduced the germination of several dwarf mistletoe species (Zimmerman and Laven 1987). Variations in fire-stimulated germination among species within an ecosystem can play an important role in maintaining species diversity and structuring plant communities (Clarke and French 2005).

9.3 Predicting Immediate Fire Effects on Plants

The ability to predict post-fire tree mortality has important management applications. For example, the planning of conifer salvage logging can consider retaining the trees with good prospects of surviving fire. The decision to coppice broadleaf trees can be based on their top-kill likelihood. Prescribed burning can be planned to either avoid tree death or purposefully kill small trees to thin them to accomplish vegetation management goals. Predictive models can help identify environmental conditions for burning that will not likely kill large trees while meeting other objectives such as consuming surface fuel and killing some small trees.

Peterson and Ryan (1986) devised a conceptual approach to model fire-caused conifer mortality. The probability of mortality is computed from the fraction of crown volume killed, the critical time for cambium death, and the duration of lethal stem heating. The crown kill fraction depends on crown kill height (which differs from crown scorch height), tree height, and crown length. Crown kill height is determined from fireline intensity, ambient temperature, wind speed, and the temperature of necrosis for foliar buds. Finally, bark thickness determines the critical time for cambium death, as previously explained. The probability of mortality is given as:

$$P_m = C_k^{\left(\frac{t_c}{t_l} - 0.5\right)} \tag{9.1}$$

where P_m is the probability of tree mortality ($0 \geq P_m \leq 1$), C_k is the fraction of the crown killed ($0 \geq C_k \leq 1$), t_c is the estimated critical time for killing cambium (min), and t_l is the duration of lethal heat (min).

One advantage of Peterson and Ryan (1986) approach is that it estimates the likelihood of tree mortality directly from tree dimensions and fire behavior and can be used to objectively compare fire resistance between species, as Fernandes et al. (2008) did for European pines. However, predictions of tree mortality are mostly based upon empirical data informing logistic regression models (Hood et al. 2018). Models are available worldwide, especially for conifers of the western United States (where there are data). Models are less effective when applied beyond the empirical data on which they were built. Inputs are typically tree species, diameter (as an indicator of bark thickness), height to the base of the live crown, and measures of either crown scorch or flame length.

Some empirical models combine the effects of cambium damage and crown scorch to estimate tree mortality. For instance, Botelho et al. (1998) used a logistic model:

$$P_m = 1/[\ 1 + exp\ (\ b_0 + b_1\ DBH + b_2 C_k\ ] \tag{9.2}$$

where P_m is the probability of mortality, DBH is tree diameter in cm at 1.37 m above ground, C_k is the fraction of the crown killed, while b_0, b_1, and b_2 are regression

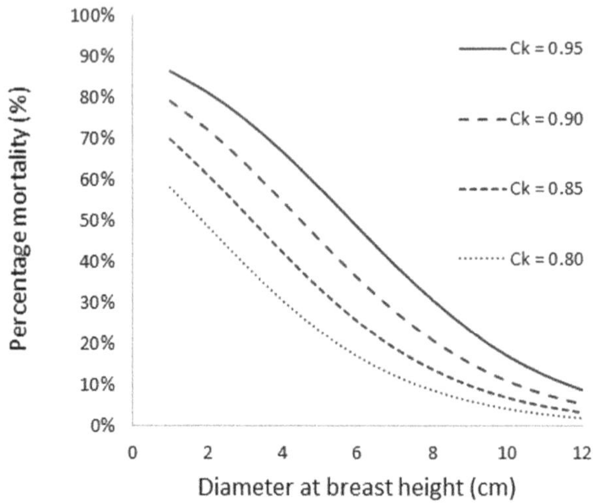

Fig. 9.5 The probability of mortality (P_m) as a function of tree diameter at breast height (DBH in cm) and the fraction of crown killed (C_k) for young *Pinus pinaster* trees after experimental prescribed fires in northern Portugal. (From Botelho et al. 1998)

coefficients. As illustrated in Fig. 9.5, even relatively small trees can survive if crowns are little damaged, and larger trees survive more readily than smaller trees. Others use fireline intensity instead of crown killed or scorch height, and all are affected by air temperature and wind, which influence convective heating effectiveness.

There are many tools useful for predicting plant response to fire. The First Order Fire Effects Model (FOFEM, Hood and Lutes 2017) can be used to predict the degree of crown damage and tree mortality by using fireline intensity in empirical equations for different tree species, along with tree stem diameter and other information. In general, trees that have thicker bark and larger diameter, and trees with crowns high above the flames are more likely to survive fires. The soil heating equations in the FOFEM model have recently been updated to include the soil heating equations developed by Massman et al. (2010) and Massman (2015), see Sect. 10.6 for further information about soil heating and how that can affect plant survival.

Plant functional traits can be very helpful in predicting whether and how plants will resprout or establish from seed (Pausas et al. 2004; Perez-Harguindeguy et al. 2013, 2016). In particular, understanding where the meristematic tissues are relative to heat exposure from fires can be quite useful in predicting plant response to fires, in particular the differences between the growth rates of sprouters and seeders (Fig. 9.6).

The difference between sprouters and seeders is not absolute, as many species show both strategies at different levels or intensities. For example, resprouting intensity, measured by the probability of resprouting, the number of resprouts, and the length of these sprouts, was studied for six Atlantic shrub species by Reyes et al. (2009) showing the diversity of responses.

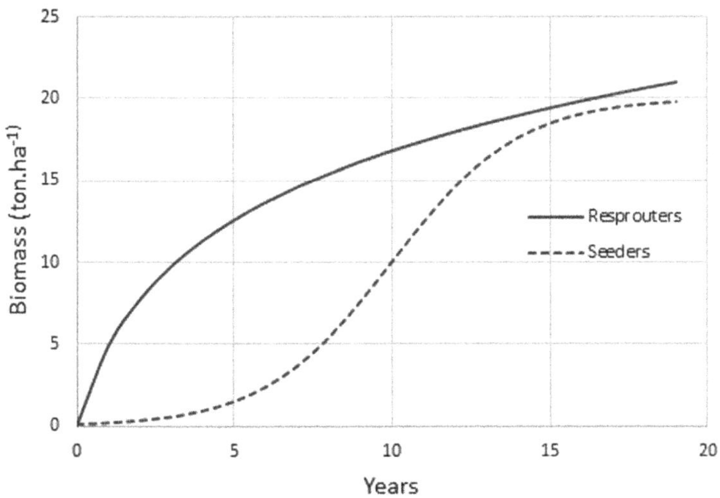

Fig. 9.6 Typical growth curves from resprouters and seeders. Initially, seeders grow more slowly as the plants must form the root system, whereas resprouters take advantage of the surviving root and grow faster initially. This type of curves has been studied for post-fire dynamics of shrub species in France and Spain by Casal (1987) with major differences between sprouters (including many species of the *Papilionaceae* of the genera *Ulex*, *Cytisus*, *Genista*, *Calicotome*, and *Pterospartium*) and many species which recolonize only by seeds (of the *Ericaceae* or the *Cistaceae* families) which become less dominant after fires. The same occurs for herbs and grasses. (Redrawn from Casal 1987)

Plant traits are important in determining which plants are most likely to survive fires and which are most likely to regrow after fires (Perez-Harguindeguy et al. 2013, 2016). Many plants exhibit multiple functional traits that help them survive and thrive through repeated fires. For instance, *Ceanothus* shrubs will often resprout following top removal by fire, then produce flowers and seeds in abundance that then establish new plants in burned areas. Plant functional traits, also called regeneration strategies and life history characteristics, are related to the vital attributes used to simulate the response of plants to fires (See Sect. 12.6).

The Fire Effects Information System (Smith 2010) includes succinct syntheses of published studies of fire effects on individual plant and animal species, some ecosystems, and some fire regimes. While the species are mostly from forests, woodlands, shrublands, and grasslands of the western USA, the information here could be used to support understanding the response of closely related species or those with similar functional traits elsewhere.

Ecosystem process models are increasingly used, including fire effects on tree crowns, stems, and roots (e.g., Keane et al. 2015) as our understanding of the underlying mechanisms improves. This understanding will help as we seek to predict fire effects under warmer and extended droughts or other novel conditions associated with climate change and altered forest composition and structure (Hood et al. 2018).

9.4 Environmental Conditions and Spatial Heterogeneity in Fire Effects Influence Plant Diversity

A plant's ability to survive or recolonize after a fire depends on the functional traits, the fire behavior, and the pre- and post-fire climate (see Fig. 9.3 and Sect. 9.4). The aboveground portions of many plants are readily killed during fires with relatively low fireline intensities. Therefore, it is critical to understand the burn severity, soil heating, and whether bare mineral soil or other favorable microsite conditions are available. The environment prior to, during, and after fires, such as drought, can influence whether seedlings and sprouts can survive. The temporal and spatial variability in plant survival, regeneration and recolonization, and post-fire climate and disturbances play a critical role in determining how plants and plant communities recover from fires.

Fire behavior and effects vary across a range of spatial scales from very fine to broad (Fig. 9.7). For example, some areas can experience stand-replacing fire while other nearby areas may burn at low severity or not at all during the same fire. Locations within wildfire perimeters that are unburned or burned with low severity are commonly referred to as refugia. Refugia allow both fire-resistant and fire-sensitive species to persist when surrounding landscapes burn. The vegetation in refugia is in the "slow" lane with different vegetation trajectories, adding to the biodiversity and resilience of landscapes (Krawchuk et al. 2020). Krawchuk et al. (2016, 2020) and Meddens et al. (2018) described fire refugia characteristics in forests and shrublands of the western USA, while Reside et al. (2014) defined refugia important to biodiversity in Australia.

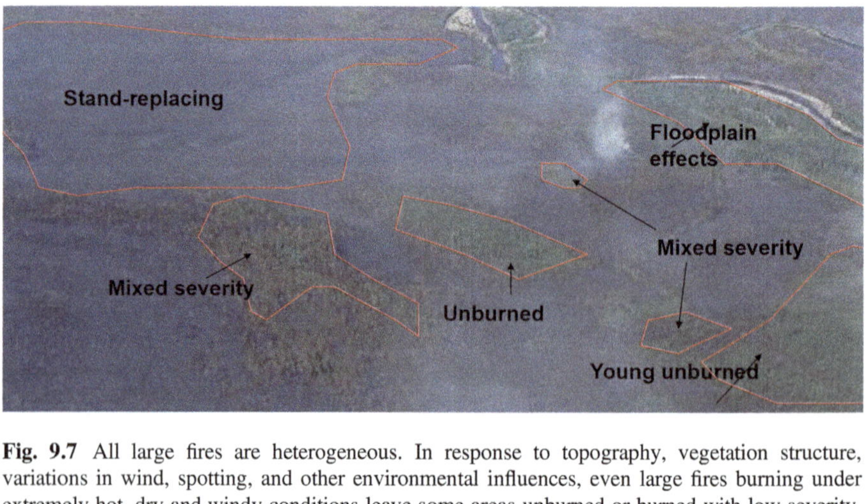

Fig. 9.7 All large fires are heterogeneous. In response to topography, vegetation structure, variations in wind, spotting, and other environmental influences, even large fires burning under extremely hot, dry and windy conditions leave some areas unburned or burned with low severity, while others burn with mixed or high severity (here labeled as stand-replacing). (Photograph by Wendel Hann)

The environmental conditions that occur before and after fires can play an important role in shaping vegetative response following a fire. Plants already stressed by drought are less likely to survive a fire. Further, the same plant species will survive less well when fires burn where it is growing on the dryer sites it can tolerate, such as on the warm, dry end of their geographic distribution. Similarly, plants are more vulnerable when they are burned during extended droughts. Once injured by fires, many plants are more vulnerable to insects and diseases.

Plants and animals may rapidly recolonize burned areas when they survive in nearby areas that were unburned or burned with low severity, and then they can rapidly increase as the plants flourish and flower post fire. Even when only a few plants survive fires, they often grow quickly, flower in abundance, and then colonize surrounding areas. Wagenius et al. (2019) demonstrated that fires can stimulate plants to flower synchronously, thus increasing reproductive success including doubling seed production. Bacteria and fungi may similarly survive on-site or recolonize from fire refugia. Thus, how ecosystems respond to fires depends, in part, on what survives the fire and what can get to and occupy space after fires. It is remarkable how quickly burned areas green up after fires—often new green growth is visible within days or weeks after fires—a source of awe and appreciation for resilience. Some plants and animals are most abundant post fire, and certainly, the diversity of burned landscapes is often high.

9.5 Ecological Implications of Soil Heating

Soil heating during a wildfire can have long-term effects on soil structure, biota, and processes, including soil chemistry, nutrient cycling, water infiltration, and soil erosion potential, all of which can influence vegetation response. Although the organic matter is commonly a source of heat during a fire, it can also act as an insulator if it does not burn completely, thus protecting the soil. After fires, partially burned organic matter often decomposes rapidly, especially when blackened surfaces are warmed by the sun, and there is less shading and evapotranspiration where the fire has reduced living plant biomass. For these reasons, the fate of organic matter is especially important in understanding how fire affects nutrient cycling and soil erosion potential.

9.5.1 Consequences of Soil Heating

When soil is heated, biological, chemical, and physical changes can occur, some of which can have long-term implications for soils and the plants growing in them (Fig. 9.8, Moody and Martin 2009; Pingree and Kobziar 2019; Massman et al. 2010). The heat flux from the soil surface into the soil, the duration of heating, and the properties of the soil are all important to consider when estimating soil

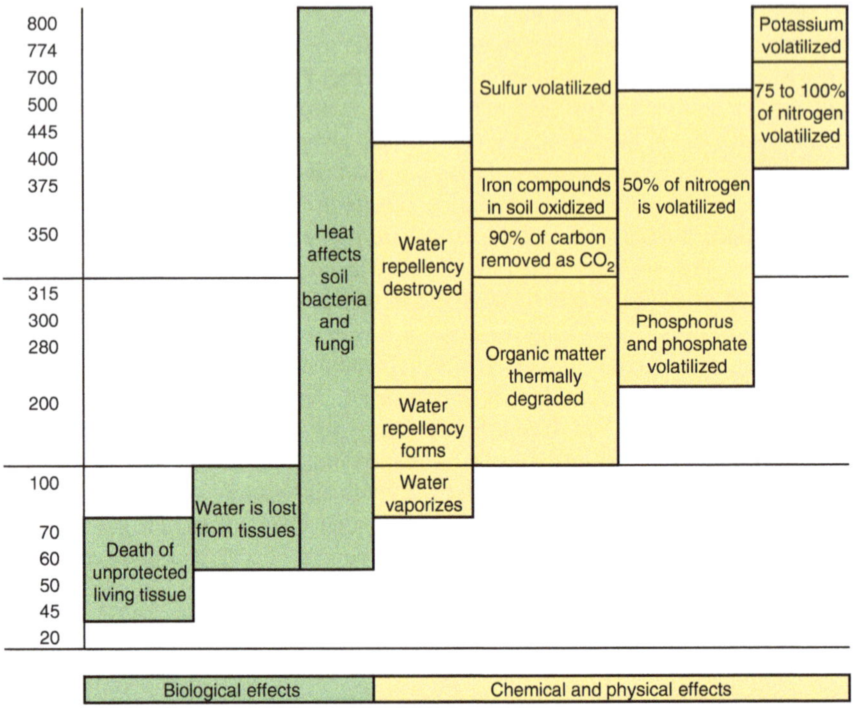

Fig. 9.8 Biological, chemical, and physical changes in soils heated by vegetation fires. Note that the vertical axis is not to scale. Values are combined from multiple sources, including Campbell et al. (1995), Ryan (2002), Moody and Martin (2009), Massman et al. (2010), Pingree and Kobziar (2019), and references therein. The boxes are colored green for biological effects and yellow for chemical and physical effects

effects following fires (Massman 2015). However, in many cases, soil effects are evaluated by comparing the predicted temperature at a given depth with a specific threshold temperature. Biological and chemical changes occur at temperatures far below temperatures observed within flames. Temperatures will be less than 100 °C unless water present in the surface organic layers or in the soil is vaporized; if so, the water vapor can injure living tissue. Water vapor, CO_2, and other gases can move through the soil as heating induces air circulation within soils; these then condense where soil particles are cool enough (Massman et al. 2010).

Soil temperature during a fire decreases rapidly with depth, and peak soil heating lags behind surface soil temperature peaks (Fig. 9.9). The soil temperature, at a given depth, depends on both the duration of heating and the temperature at the soil surface. Therefore although smoldering combustion releases heat more slowly and results in lower temperatures at the surface than flaming combustion, the duration of heating often results in greater soil temperatures. This is especially true if the smoldering material is in contact with the mineral soil. Both the soil and texture influence the rate of conduction through a soil layer. Coarse-textured soil heats more

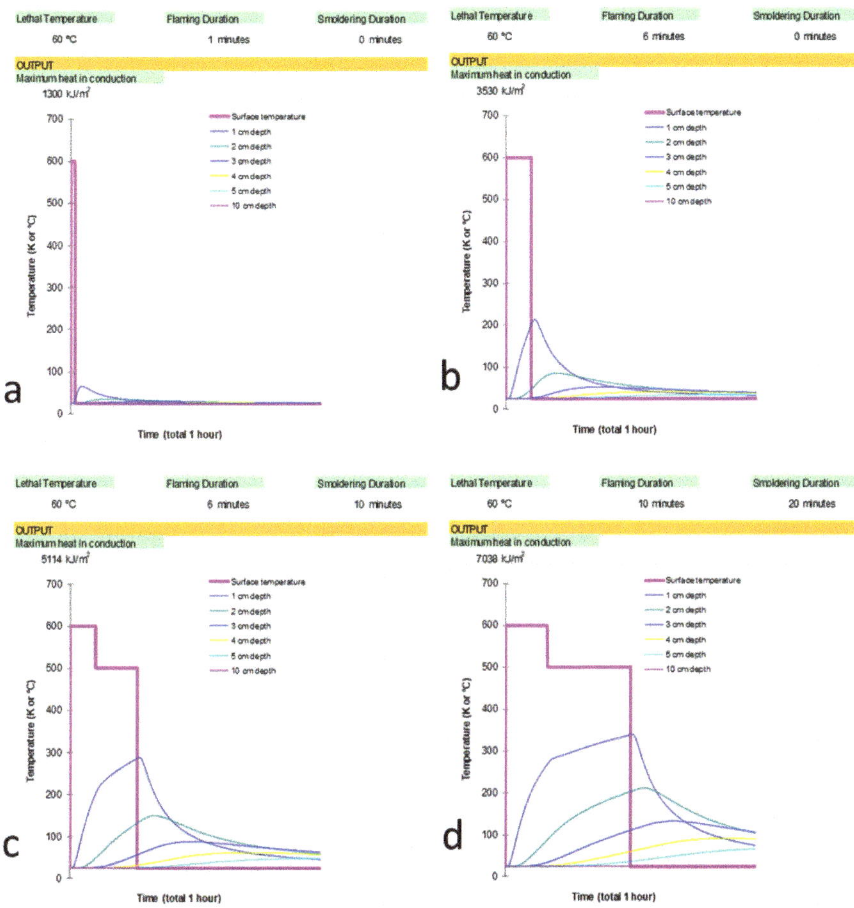

Fig. 9.9 Predicted soil heating under a range of input conditions. (**a**) A grassland with high intensity, short duration surface fire with very little soil heating (30 s flaming with 0 s smoldering), (**b**) pine forest understory with fine fuels (1 min flaming, 3 min smoldering). (**c**) Slash burning long duration and high temperature (5 min flaming, 10+ min of smoldering). (**d**) Peat burning with no flaming, dominated by smoldering combustion (0 flaming, 60 min smoldering). All figures are from our interactive spreadsheet, CONDUCTION_Soils_Plants_v2.0

slowly than those with fine texture. Conductive heat transfer into wet duff and mineral soil can be 20% of that penetrating dry duff and mineral soil and may peak at more than 500 °C where thick duff layers burn and soils are dry (Frandsen and Ryan 1986). Although soil heating is complex given the movement of air, water vapor, and CO_2 within soils (Massman et al. 2010), modeled soil heating is consistent with observations from multiple ecosystems as summarized by Ubeda and Outeiro (2009). Temperatures at the soil surface commonly exceed 200 °C and can be >700 °C during fires. Particularly in grasslands, temperatures at 2 cm into the

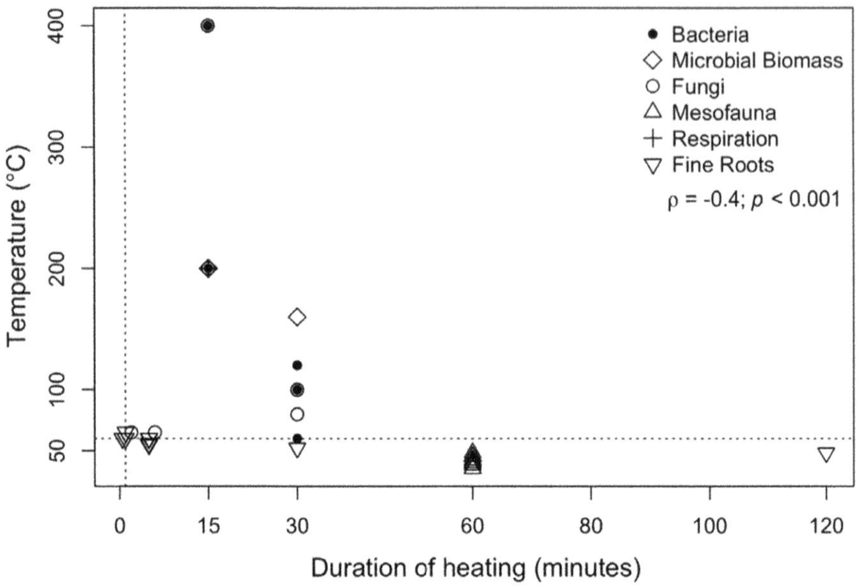

Fig. 9.10 Bacteria, microbes, fungi, soil mesofauna, and fine roots often survive heating to temperatures higher than 60 °C for longer than 1 min. That commonly assumed time-temperature threshold is represented by dotted lines. (From Pingree and Kobziar 2019)

soil seldom exceed 50–100 °C, while temperatures of 66, 177, and 250 °C were measured at 2.5 cm depths in forests with the differences related to surface heating (Ubeda and Outeiro 2009 and citations therein).

Living tissue may not die when heated to a temperature of 60 °C, and thus this common assumption is conservative for several reasons. *First,* soil heating is spatially variable at fine scales and decreases greatly with depth, so soil biota can often survive (Fig. 9.10) and may rapidly recolonize. *Second,* plants and soil biota evolved in a fiery environment where tolerance to soil heating and quick recovery would be an advantage. On the other hand, managers should limit soil heating when they can by not piling debris from logging into deep piles and not creating very deep, dense fuelbeds of chipped, shredded, or masticated fuels as these could burn for a long time at the soil surface. *Third*, soil heating likely has cumulative and additive effects that can best be understood using dose and response ideas (Smith et al. 2016), and by assuming that many plants and microorganisms have adapted to fire in ecosystems that historically burned often.

Unexpectedly severe effects may occur in ecosystems that burn after a long period of fire exclusion with associated accumulation of surface fuels, especially deep layers of organic matter. The high mortality of the old trees that survived many previous fires may be due to injury to fine roots that have grown into the surface organic and soil layers during many years without fire (Varner et al. 2005). Varner et al. (2007) observed up to 60% mortality of large-diameter old longleaf pine (*Pinus palustris*) trees, especially in areas with low duff moisture content when burned. The old, large, relict trees can be very important to wildlife and to people, are often relatively rare, and they potentially have unique genetic qualities, so limiting their mortality is a common challenge for restoring fires to long unburned forests that historically burned frequently. Stephens et al. (2018) and Varner et al. (2007) recommend repeated burns with an incremental reduction of accumulated fuels and other strategies as part of restoration treatments.

Plants and soil biota will likely recover rapidly post fire, particularly if fires burn with low or moderate severity and areas burned with high severity are small. With the high spatial variability in fuels and soil burn severity at fine spatial scales in many different ecosystems, and with the long history of fire in terrestrial ecosystems, there are microsites in most fires where fire-adapted organisms survive to thrive post fire.

Many live bacteria and fungi associated with the organic matter and mineral soil in burned areas can survive and be transported in the convective column associated with wildfires (Kobziar et al. 2018). The transport of bacteria and fungi in fire plumes could help bacteria and fungi rapidly colonize burned areas post fire. The transport of bacteria and fungi in fire plumes likely has important but unknown implications for human and ecosystem health (Kobziar et al. 2018). The viability and distance that microorganisms are transported during a wildfire likely depend upon several factors, including the heat release rate, convective forces, atmospheric stability, the mixing height, and specific traits of the microorganisms.

Duff is the compact layer of partially decomposed leaves, needles, and woody fuels that is found in many forests above the mineral soil and below the litter. When duff is present and is dry enough to smolder, the soil temperature at different depths reflects both radiation and conduction from long-duration heating of burning duff in contact with the soil surface. Burning duff is a source of heat, while unburned duff can insulate the soil from the heat from a fire.

The total energy released at the soil surface during burning reflects the heat energy content of the duff and soil material that is burning, its moisture content, and the duration of smoldering (Frandsen and Ryan 1986). Only some of the heat from fires goes down into the soil, usually less than 25% (DeBano et al. 1976; Packham 1969; Raison et al. 1986). As soils have high heat capacity, temperatures rapidly attenuate with depth. When long-duration heating is present at the soil surface, as when peat layers smolder or when deep piles of logging debris smolder for long periods, soil heating can be enough to greatly affect soils. The greatest soil temperatures usually occur beneath dry heavy slash, particularly with the consumption of large piles of residue or windthrow. Burning large slash piles produces long duration, high-temperature heat pulses that penetrate deep into the soil, potentially altering both physical and biotic characteristics of the soil to significant depths. Due to the high water content of wetland soils, penetration of heat generated by a surface fire can be significantly less than in mineral soils. Organic matter has a lower thermal diffusivity than mineral soil, so penetration of heat is further reduced in organic wetland soils. However, organic soils can become dry enough to burn, producing significant amounts of heat.

9.5.2 The Fate of Organic Matter Influences Soil Processes and Plant Survival

The effects of fire on soil processes depend on how fires affect the organic matter on and in the soil. Organic matter plays multiple roles, for it is a source of heat if it burns, it can insulate the soil from heat during a fire if some or all of it remains unburned, and it acts as a source of nutrients as it decomposes. Before and after fires, organic matter on and in the soil can help hold moisture and nutrients in the soil, and will greatly influence infiltration and runoff, particularly if hydrophobic layers form. The organic matter captures and holds soil moisture and nutrients, and can sometimes cover the bare mineral soil and therefore influence seed germination and establishment, thus potentially affecting what grows after a fire. This is particularly the case for forests where thick layers of duff can develop, but the principles apply to areas without duff, such as grasslands and woodlands with some litter. Sometimes in our management, we create organic layers, such as through mastication of tree and shrub fuels as discussed here and in Chap. 11. Organic matter is crucially important to ecosystems and can be affected by fires in multiple ways.

First, burning organic matter is a source of heat. Burning of downed dead wood, litter, duff, or the very compact fuelbeds that result from mastication or chipping or mulching of trees and shrubs can lead to prolonged heating at the mineral soil surface. Although the intensity of such smoldering fires is low, direct contact with smoldering fuel facilitates soil heating through conduction. If soils are heated for long enough, plant tissues, including seeds and roots, may die (see Sect. 9.3).

Second, unburned or partially burned organic matter can insulate the soil during fires and thus limit soil heating. After the fire, blackened organic matter layers absorb heat from the sun enough that the surface reaches temperatures that may be sufficient to damage or kill plant seedlings.

Third, organic matter on and in the soil is a source of nutrients as it decomposes before or after fires. Post fire, the often greater daytime soil temperatures, moisture, and incoming solar radiation can increase decomposition of any surface organic matter that did not burn in the fire. This unburned organic matter is an important source of nutrients and helps retain nutrients in the soil. Also, soil moisture often increases post fire as there is less transpiration where plants that were actively growing prior to the fire were top-killed by fire. Even if the changes in soil pH and nutrients are short-lived, the immediate post-burn decomposition rates can be high. If much of the organic matter on and in the soil is not consumed by a fire, it will continue to decompose and thus gradually release nutrients needed by plants. Further, the nutrients released during and after fires are less likely to be lost from the site when soils are high in organic matter because the organic matter is important to the cation exchange capacity that loosely "holds" nutrients, thereby making the nutrients less likely to be lost in runoff or leached through the soil.

Fourth, residual organic matter can limit post-fire soil erosion (Robichaud et al. 2013a, b) through the mitigation of raindrop or splash erosion. Raindrop erosion is the first step in the erosion process and occurs when falling raindrops displace soil particles that block soil pores and reduce infiltration. The reduced infiltration associated with raindrop erosion can lead to the formation of rills and gullies. Experimental studies have suggested that soil erosion is reduced when soil cover exceeds 60%. Any organic matter will do, including scorched foliage that falls from tree canopies. Short needles provide better contact with the soil surface, but long pine needles that fall in bunches can create very small barriers to sediment movement. Post-fire soil erosion potential is also higher when intense rain falls soon enough after a fire that soils are dry and mostly bare, on steep slopes, on coarse-textured soils, and in large patches burned with high severity (Scott et al. 2009). Riparian vegetation may help to limit sediment in streams (Luce et al. 2012).

Fifth, the rate of water infiltration into the soil following a fire is influenced by the presence and characteristics of surface organic matter. The presence of soil surface organic matter slows water infiltration into the soil by holding surface water until it can infiltrate. Organic matter is the source of the waxes and resins that form fire-induced hydrophobicity in soils (see Sect. 9.5.5). Hydrophobicity is more likely in coarse-textured soils and where evergreen vegetation is abundant (Doerr et al. 2006).

Sixth, organic layers on the soil surface can be barriers to soil moisture loss and plant establishment. The organic layers may be a barrier to seedling establishment when they limit bare mineral soil exposure, though this effect may not last long as these residual organic matter layers often decompose quickly following fires.

Thus, the fate of organic matter during burning greatly influences soil processes, vegetation response, and soil erosion potential. Combustion is sometimes considered rapid decomposition. Both combustion and decomposition are oxidation processes. Both consume organic matter, result in smaller pieces and simpler compounds, and both are typically incomplete leaving residual organic biomass. Both release nutrients in the biomass. Thus, both are important in making nutrients available to other organisms that are otherwise tied up in the live and dead biomass. We note that fuel consumption is very dependent upon fuel moisture (See Chap. 11).

Fire effects on both surface organic matter and soils are spatially variable. Lentile et al. (2007) found that fire effects on the soil surface vary at spatial scales of 1–10 m. It is important to recognize that the sale of variability identified by Lentile et al. (2007) is smaller than the scale (30 × 30 m) commonly used in fire effects studies that rely on satellite images and aerial photographs. Fire effects on soils tend to be more homogeneous in areas burned at high severity (>70% overstory tree mortality) than areas burned at lower severities (Lentile et al. 2007). Sites burned with moderate severity (20–70% overstory tree mortality) had soil burn severity that varied at very fine spatial scales, as these sites include small unburned areas intermingled with those where rotten logs and stumps had burned to consume all surface organic matter and heat soil sufficiently to change soil color, and many intermediate effects (Fig. 9.11).

Fig. 9.11 Fire effects on soils are often spatially variable at fine scales as evident from these 1-year post-burn photographs from chaparral shrublands in California (CA), mixed conifer forests in Montana (MT), and boreal forests in Alaska (AK). Note the differences from low to moderate and high severity burns in each location. In these sites, burn severity was initially assessed using satellite imagery and then confirmed on the ground based on fire effects on mortality of overstory trees and shrubs. (From Lentile et al. 2007)

9.5.3 Carbon, Pyrogenic Carbon, and Fires

How do soil nutrients fare through repeated fires? How are soil nutrients lost and replenished, and when and where are soil carbon and nutrients most vulnerable? What are the implications for ecosystem function? Soil heating affects all of these during and after fires.

Carbon in forests, both aboveground and in soils, is increasingly important in a rapidly changing world. Page-Dumroese et al. (2003) highlighted the potential of fire and fire suppression to alter carbon pools and fluxes in forest soils and the implications of these alterations on global climate change. Page-Dumroese et al. (2003) emphasized that when surface fuels accumulate in the absence of fire, some of the carbon in those fuels is likely to be lost to the atmosphere when those fuels consume in a future fire. Soil carbon pools will be replenished as vegetation regrows (Page-Dumroese et al. 2003). Limiting deforestation and maintaining forest productivity can help sequester carbon that could be released to the atmosphere.

The influence of fires and repeated fires on global carbon cycles is uncertain, with major implications now and into the future (Santín et al. 2015). Currently, forests

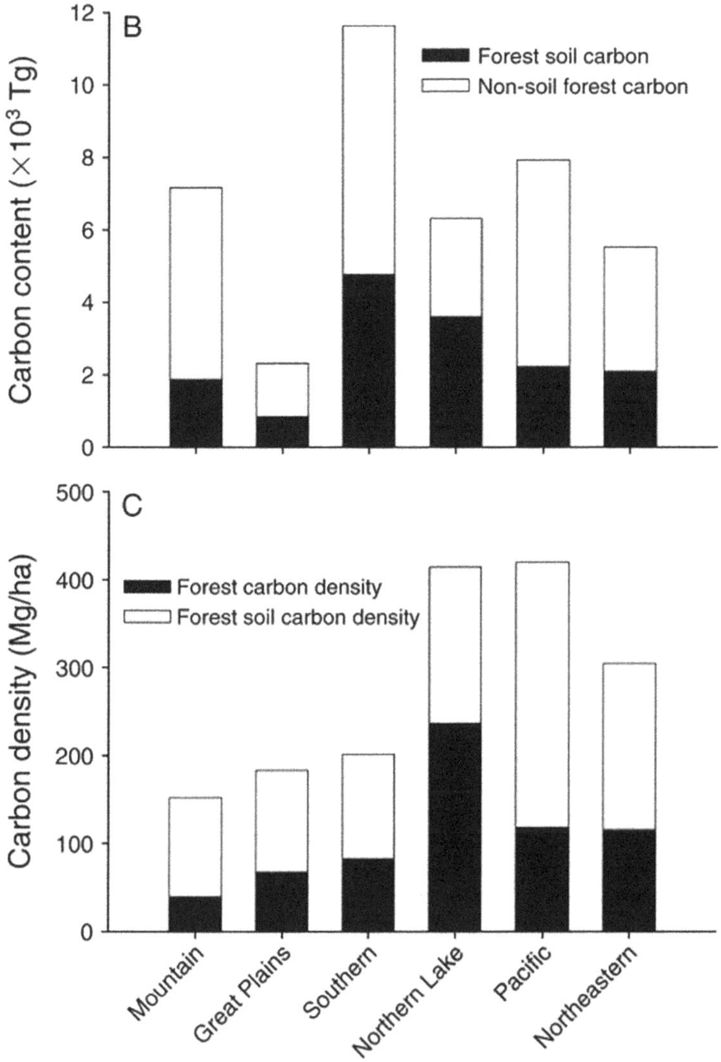

Fig. 9.12 Much of the carbon in forests is in forest soils. These are data from forests in the western USA. (From McKinley et al. 2011)

remove about a quarter of the carbon dioxide that humans add to the atmosphere (McKinley et al. 2011). Carbon in forests accumulates both in the vegetation and soil (Fig. 9.12). Soil carbon is a major carbon sink for two reasons. First, while most carbon in fuels is lost to the atmosphere, only one-third of the carbon from burned biomass may remain on and in the soil of boreal forests (Santín et al. 2015). Second, the charred organic matter on and in the soil (also known as black carbon or pyrogenic carbon) may take decades to many centuries to decay (Santín et al.

2015; Reisser et al. 2016). Pyrogenic carbon makes up an average of 14% (range 12–60%) of the total organic carbon in soils globally, depending on land use, soils, fires, climate, and whether it was leaves or logs, large or small (more of the large logs become pyrogenic carbon compared with needles and other forest floor materials whose high surface-area-to-volume ratio also contributes to more rapid decay) (Santín et al. 2015; Reisser et al. 2016). Although the charred organic matter is less likely to burn than unburned organic matter, repeated fires can reduce the amount of charred organic matter at the soil surface, suggesting that pyrogenic carbon is vulnerable to repeated fires unless it is part of the soil (Tinkham et al. 2016).

Peat fires are often overlooked, yet these highly organic soils sequester carbon on several continents, particularly in northern temperate and boreal zones and in tropical regions (Rein et al. 2008). Fires in peat are globally significant for their emissions both for the atmosphere and for people affected by the smoke (Hu et al. 2018). Peat fires burn the partially decomposed organic matter that *is* the soil, and thus they consume the soil itself as they burn deep and produce emissions that far exceed other fires on a per-area basis (Hu et al. 2018). These highly organic soils burn only after they lose moisture during droughts or through land use, and burning depends on moisture and mineral contents. In the long duration burning typical of peat fires, soil temperatures may exceed 300 °C for more than 1 h (Rein et al. 2008), killing seeds and other plant propagules as well as soil organisms, and post-fire recovery may take many years. The haze that results can greatly impact environmental quality as particulates and volatile compounds pollute the air, especially as the smoke is weakly buoyant and, therefore, near the ground. The impaired visibility associated with peat fires can disrupt transportation, threaten people's health, and have other economic impacts (Hu et al. 2018).

Much forest carbon remains during and after fires. Stenzel et al. (2019) emphasized that much biomass remains after fires, contrary to the assumptions made in many ecosystem process models that all above-ground biomass is consumed when forests and other ecosystems burn (Fig. 9.13). Stenzel et al. (2019) found that only 5% of mature tree biomass is consumed even in intensely burning wildfires with high mortality of trees. Further, the combustion coefficients in models predict emissions much higher (by 59–83%) than actually occurs, even in recent very large wildfires burning under very hot, dry, and windy conditions (Fig. 9.14). Instead, standing dead trees (called "snags") slowly decompose over decades to centuries after a fire while providing wildlife habitat and long-term value to ecosystem processes first as standing trees and then as logs. Grasslands and shrublands commonly have more of their above-ground biomass (and therefore carbon) consumed. In forests, woodlands, shrublands, and grasslands, vegetation rapidly recovers (depending on the climate), and much of the carbon is in the soil where it does not burn even when fires burn severely. Further, fires are often patchy, with many areas unburned or burned with low severity (see Chap. 12). This is what Stenzel et al. (2019) mean by variable severity rather than the assumption that all burned areas are completely toasted.

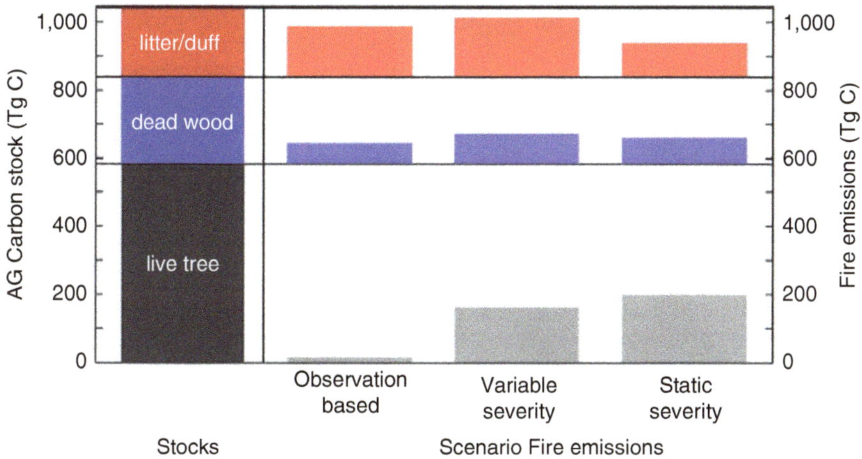

Fig. 9.13 Above-ground (AG) carbon stocks for forests in the western USA showing the amount in live trees, dead wood, and litter/duff in pre-fire carbon pools, and the same pools post fire as a reflection of the area burned 2000–2016 contrasted for what was measured in the forest vs. what was simulated with variable (assuming realistic conditions for fires burning with the range of burn severity) or static (high) burn severity. (From Stenzel et al. 2019)

9.5.4 Nitrogen and Other Soil Nutrients Are Affected by Soil Heating

The size of the soil nutrient pool is often lower post fire, at least temporarily. Paradoxically, the nutrients needed for plant growth, such as nitrogen, are often more available to the flush of new growth of plants post fire. Why? During and immediately post fire, nutrients may be lost if they are volatilized (this is especially likely for nitrogen and phosphorus that are in the organic matter that is consumed), blown away by the wind (via transport of ash and dust) or carried away by water (when nutrients are leached through the soil into the groundwater, or when they are lost through overland flow) (Raison et al. 2009). Post fire, the availability of soil nutrients can increase in multiple ways. *First,* when organic matter burns, the residual ash contains a large proportion of cations that are available for plants. Cations are positively charged ions, including potassium, calcium, sodium, and magnesium; these are readily soluble in water and can be leached into groundwater. *Second,* in the post fire environment, decomposition of residual organic matter often accelerates, as does biological mineralization of nitrogen where the activity of soil fungi and bacteria are favored. Soil pH often increases (this may be short-lived, especially in grasslands), and soils are often warmer and higher in moisture given reduced transpiration when above-ground living biomass is reduced (Wright and Bailey 1982). *Third,* as described below, plants that fix nitrogen often increase post fire, though the nutrient availability is greatly influenced by soil burn severity

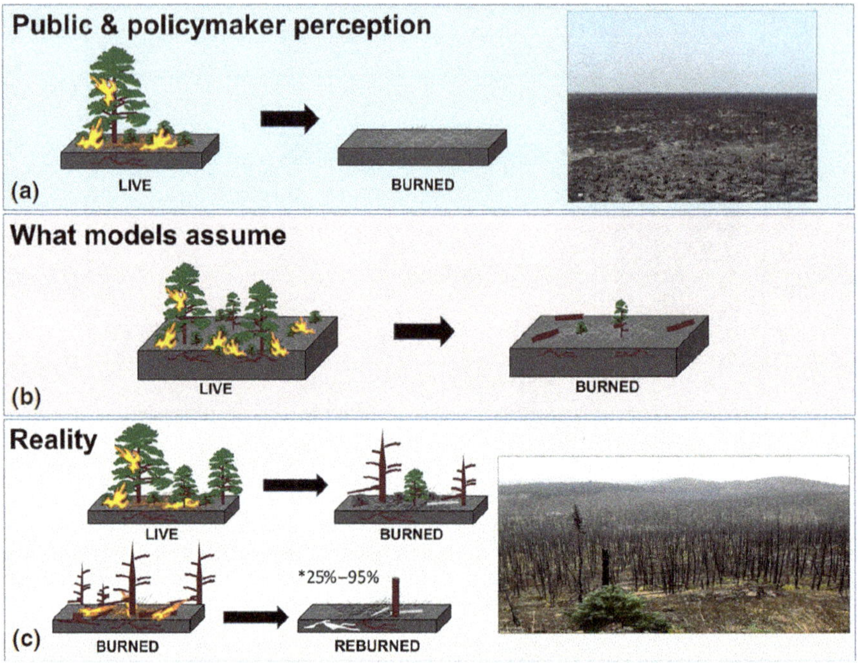

Fig. 9.14 (**a**) People commonly assume that when fires burn intensely all forest biomass is consumed. (**b**) In many simulation models of ecosystem processes the majority of above-ground biomass is assumed to be consumed. (**c**) Field measurements post fire demonstrate that in reality 80–90% of live stem biomass is left after even high-intensity fires to remain as dead, slowly decaying carbon, though when a subsequent fire burns soon after, 25–95% of the carbon in the standing and fallen wood can be consumed and released into the atmosphere as gas or left as ash on the soil surface that may blow or wash away. (Data and figure from Stenzel et al. 2019)

(Raison et al. 2009). *Fourth,* there may be less competition for nutrients post fire if there is less biomass actively growing. However, this is likely to be short term.

Raison et al. (2009) suggested that the negative effects of fires on forest nutrients are likely to be long-lasting especially on sites that were nutrient-limited before the fire, or where wind or water transport of nutrients removes ash or surface soil off-site. They were also concerned about situations where fire frequency is high relative to historical frequency. Changes in nitrogen and other nutrients in fuels are more pronounced where more fuel is consumed. Thus, in savanna fires where mostly grass and litter fuels are consumed, volatilization of nitrogen and loss of other nutrients is proportionately much less than when forests burn with high fuel consumption such as when moist tropical forests are slashed and burned to convert to agricultural uses (Raison et al. 2009).

Fires influence nitrogen availability, both during fires and after fires, across spatial scales, and in both terrestrial and aquatic ecosystems. Nitrogen is a critical nutrient that is often limiting in forests and streams (Vitousek and Howarth 1991; Grimm and Fisher 1986) and can be volatilized during fires, especially if soil burn severity is high. Nitrogen is part of both living and dead organic matter (see Chap. 2) because it is part of proteins and other organic compounds. At moderate temperatures of 300–500 °C, 50% of the nitrogen may be volatilized (Neary et al. 2005) (Fig. 9.8). Nitrogen that is not volatilized remains as part of unburned organic matter or in the soil, or it can be converted through biological action into ammonium. However, the amount of nitrogen available to plants often increases after a fire.

Nitrogen cycling is affected by fires, including both pools and fluxes between pools (Fig. 9.15). Nitrogen (N_2) is abundant in the atmosphere but can't be used by plants until nitrogen-fixing bacteria, and fungi convert it to ammonium (NH_4^+) or nitrate (NO_3^-). Some of the bacteria are symbiotic with nitrogen-fixing plants such as *Ceanothus* or *Lupinus*, but others are free-living in the soil or water. The carbon to nitrogen ratio (C:N, measured by mass) of soil organic matter (Kaye and Hart 1997)

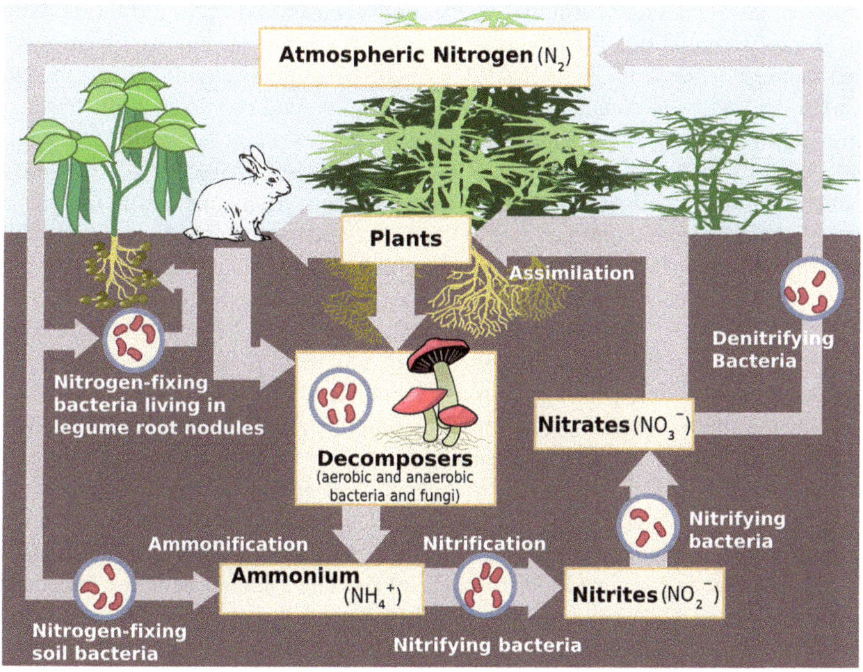

Fig. 9.15 The nitrogen cycle in plants and soils. To be used by plants, nitrogen in the atmosphere must be used by nitrogen-fixing bacteria that can be free-living in the soil or living in symbiosis with plants. When fires occur, much of the nitrogen in above-ground biomass is volatilized unless it is out of reach of the heat from flames (it may be insulated by bark or the crown may not be burned). Plant-available nitrogen may increase post fire as residual organic matter decomposes post-fire, where nitrogen-fixing plants increase, and soil temperature and moisture conditions are favorable to nitrogen-fixing and nitrifying bacteria. We more often measure nitrogen pools (indicated by boxes) than fluxes (indicated by arrows indicating nitrogen movement), but it is the fluxes that greatly influence ecosystem function. There is a similar nitrogen cycle occurring within streams, with Cyanobacteria fixing N, and algae, moss and biofilm using both ammonium and nitrate while producing litter and organic N. (From https://commons.wikimedia.org/wiki/File:Nitrogen_Cycle.svg, WikiCommons, Accessed 18 May 2019)

influences soil microbes and plant growth. Vegetation and fire management can affect soil C:N, particularly if these add organic materials to the soil surface that are high in carbon and low in nitrogen, such as the agricultural straw or other mulch applied to reduce soil erosion potential or deep beds of chipped or shredded organic matter resulting from mastication of tree and shrub fuels. As the added organic matter decomposes, soil microbes may demand increased nitrogen and thereby limit inorganic nitrogen available for plant uptake. Available nitrogen is rapidly assimilated into living organisms. The fine roots and leaves of plants are relatively high in nitrogen, so as they grow and die, there is a rapid turnover of organic nitrogen. Although much of the nitrogen that is released is rapidly absorbed into plants, bacteria, fungi, and other organisms, some nitrogen is lost to the atmosphere through denitrification or by volatilization when organic matter is consumed during a future fire. Some nitrogen can also be lost if the topsoil is eroded by wind or water. In two different meta-analyses of multiple studies, both Johnson and Curtis (2001) and Wan et al. (2001) found that the effects of fire on soil carbon and nitrogen were more pronounced where soils were heated to high temperatures during fires, though carbon and nitrogen abundance returned to pre-fire conditions with time. Further, nitrogen-fixation by plants and bacteria greatly altered the post-fire nitrogen available in soils. Nitrogen is replenished through biological processes that fix atmospheric nitrogen into forms that can be used by plants. Many nitrogen-fixing plants establish post fire, and nitrogen is also fixed by cyanobacteria free-living in the soil and in soil crusts (Fig. 9.16). Raison et al. (2009) reported that nitrogen lost during prescribed low severity fires in subalpine *Eucalyptus pauciflora* forests was replaced, mostly by symbiotic nitrogen fixation, within 9–15 years depending on whether 50% or 75% of the nitrogen was lost as fuels were consumed. Burning frequently could deplete nitrogen through cumulative effects.

Other nutrients are also affected by fires, with the effects on nutrient availability varying with the degree of soil heating, consumption of organic matter on the soil surface, alteration of organic matter in the soil, vegetation and litter cover, and time since fire (Raison et al. 2009) (See Fig. 9.8). Cations are typically deposited in the ash after fires, with their abundance and location post fire often reflecting the amount of organic matter consumed (Neary et al. 2005). The ash may be redistributed by wind and water. The cations are quite soluble in water, and so they may be washed away or carried into and through the soil profile. If soils are high in organic matter and clays, then the cation exchange capacity is high and cations may be held where they are available to plants. The distribution of potassium post fire is important because of its influence on critical plant functions, including transpiration and chemical defenses against insects and diseases. As potassium is most abundant in the leaves, the effects of fire and logging on potassium availability for plants depends greatly on the fate of the organic matter—if the branches and tops of trees were piled and then burned, the potassium and other cations may be highly concentrated where the piles were located, with much less between piles. Phosphorus can be lost through leaching and volatilization. Potassium and phosphorus can be affected by fires when particulate material is carried into smoke columns (Raison et al. 2009). Sulfur can be volatilized at high temperatures, but this is seldom limiting. However, it is important

Fig. 9.16 Soil crusts protect soils and fix atmospheric nitrogen into forms usable by plants. Soil crusts often appear darker in color than surrounding soil as in this picture between the bunches of grasses and cacti. These soil crusts may include cyanobacteria, algae, lichen, mosses, fungi, and bacteria, and though they are harmed by fires, they can reestablish post fire. (Photograph by Hilda Smith, USGS. Public domain, https://www.usgs.gov/centers/sbsc/science/a-field-guide-biological-soil-crusts-western-us-drylands?qt-science_center_objects=0#qt-science_center_objects)

to realize that the effects of fire on forest nutrients are highly variable both on a given burned area and among burned areas, enough that it is difficult to generalize fire effects (Raison et al. 2009).

9.5.5 Hydrophobic Soils

Water repellency in soils can be (but is not always) a major culprit for post-fire soil erosion (Fig. 9.17). Hydrophobic soils are slow to absorb water that can then infiltrate into lower soil layers where plants can use it. Doerr et al. (2006) found that although fire can induce or enhance soil water repellency, fire more often reduces water repellency in surface soils. That is because the waxes and resins that repel water come from evergreen leaves, where they are important in helping plants limit water loss from leaves. These complex compounds decompose slowly and so accumulate in litter and surface soils.

Soils are often water repellent before fires occur (Doerr et al. 2006). When heat from a fire advances into the soil, the waxes and resins causing water repellency are volatilized and transported deeper into the soil layer, where they eventually

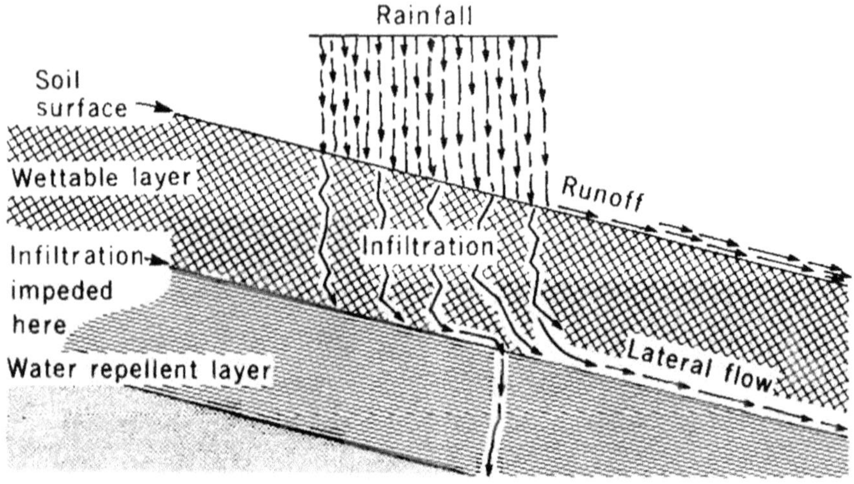

Fig. 9.17 Soil erosion potential is high when rainfall intensity exceeds infiltration, and especially where water repellent layers are below wettable layers on coarse-textured soils on steep slopes. (From DeBano 1981)

condense as they reach a temperature of around 200 °C (Doerr et al. 2006). The condensed waxes and resins coat the soil particles limiting water cohesion and creating a water-repellent layer. The magnitude of fire-induced water repellency is dependent upon the amount and type of organic matter consumed, the soil surface heat flux, the residence time, the temperature gradient within the soil, the texture of the soil, and the water content of the soil (DeBano 1981; Doerr et al. 2006). The potential erosion of the wettable soil layer on top of the hydrophobic layer is greatest when the intensity of rainfall exceeds the infiltration rate in high severity burn areas with little organic matter and on steep slopes (Fig. 9.18) (Parson et al. 2010). Water repellent layers break down with time through decomposition and physical disturbances providing cracks.

9.6 Burn Severity

The immediate and direct ecological effects of wildland fires are commonly referred to as burn severity (Ryan and Noste 1985; Keeley 2009; Morgan et al. 2014). Burn severity is often mapped into distinct categories (i.e., unburned, low, moderate, or high) using either pre- and post-fire satellite imagery or ground-based sampling. Fires that burn with high severity result in pronounced changes in ecosystem conditions. For instance, large patches that burn at high severity have less vegetation, fewer tree seedlings, higher erosion potential, and thus slower post-fire recovery (Morgan et al. 2015; Robichaud et al. 2013a, b; Stevens-Rumann et al. 2018; Stevens-Rumann and Morgan 2019, see Case Study 12.3). In contrast, fires that

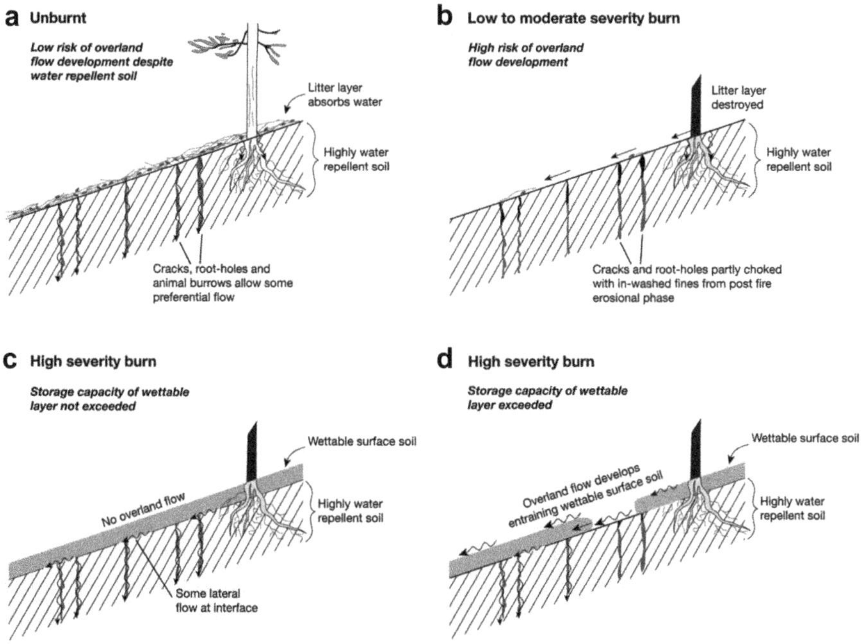

Fig. 9.18 Hydrophobic soils are more likely to contribute to soil erosion potential in areas burned with high severity where there is little soil cover, and where soil heating has resulted in a wettable soil layer on top of a water repellent layer (see text for explanation). If rainfall is of sufficient intensity that the storage capacity of the wettable layer is exceeded, in part because little of the soil water is able to infiltrate into deeper soil, then overland flow develops as water runs over the surface, collecting sediment as it moves. (**a**) Soils are often water repellent before the fire, and post-fire soil erosion potential increases for (**b**) low to moderate burn severity to (**c**) high severity burns, especially if (**d**) there is more water than can infiltrate or be stored in the wettable soil layer. Thus, soil erosion potential depends on burn severity (to consume surface organic matter) combined with high-intensity rainfall and slope and soils susceptible to erosion. (From Doerr et al. 2006)

burn with low severity may have few detectable differences in vegetation cover, structure, or function relative to the prefire condition. Assessing where, why, and how much vegetation changes due to fire is important if we are to understand, predict, and measure fire effects (Morgan et al. 2014). However, this is challenging because ecological change can encompass the response of many different plants, soil erosion, soil processes, and short- or long-term change. As a result, burn severity and related terms are often used inconsistently (Morgan et al. 2014). Therefore, it is crucial to clearly define and identify how burn severity will be measured (Morgan et al. 2014). For more discussion about burn severity as a component of fire regimes and as it influences landscape dynamics and management, see Chap. 12.

Burn severity (both soil burn severity and vegetation burn severity) is often related to but can be quite different from fireline intensity (Fig. 9.19) (Heward et al. 2013). Often, severe fires are also intense fires as when a crown fire burns in a forest when fires occur following extended drought so that the forest floor is very

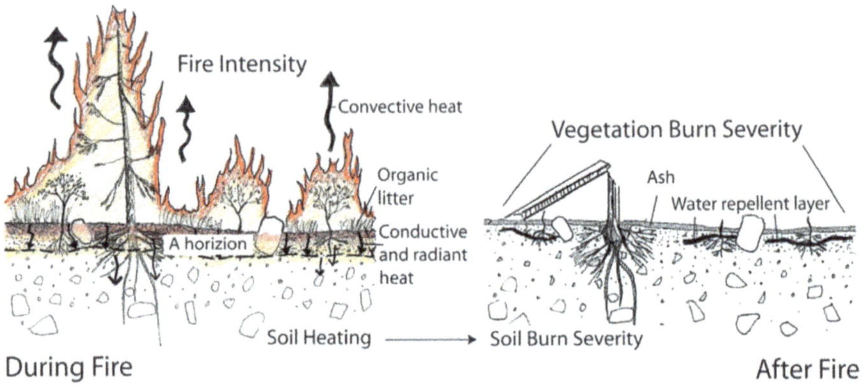

Fig. 9.19 Fireline intensity is related to the rate of heat release during fires and is correlated with flame length. Soil burn severity reflects soil organic matter consumption as well as the heat flux, and thus the temperature and duration of soil heating. (From Parson et al. 2010)

dry and all fuel strata burn. However, burn severity may be poorly correlated with fireline intensity. Consider a fire in a grassland that burns rapidly and with long flames yet very short residence time. There, the intensity is high, yet burn severity is often low because the grasses often grow back rapidly from meristematic tissues at or just below the soil surface, where those tissues mostly escaped soil heating. Similarly, when prescribed burns are conducted in the spring to consume debris left from logging, the duff and litter are often very wet so that even though the fireline intensity could be very high, soil heating and, therefore, soil burn severity are low. In contrast, when the peat burns in a bog, the fireline intensity is very low, but the ecosystem is usually very slow to recover as the organic matter in the soil is consumed in a high severity fire.

Burn severity can be assessed remotely from satellite or airborne imagery in conjunction with field sampling (see further discussion in Chap. 12). This process is aided by the availability of repeated imagery over the same location, and mapping from satellite imagery can be done relatively quickly and consistently over large areas. However, it is very important to be clear about what aspects of burn severity are being assessed. Burn severity assessments are widely used to strategically target post-fire rehabilitation and in long-term planning of vegetation management. Burn severity can be forecast using topography, vegetation, and other variables (Dillon et al. 2011; Parks et al. 2018). Many people have produced maps of high severity fire potential (e.g., Holden et al. 2009; Keyser and Westerling 2017; Parks et al. 2018; Dillon et al. 2011, 2020). Keyser and Westerling (2017) examined high burn severity relative to interannual climate variability at a coarse spatial resolution. Parks et al. (2018) mapped high-severity fire potential, but only for the western USA and only for forest and woodland settings, while Dillon et al. (2011, 2020) mapped it for the USA in all ecosystems at 30-m resolution. In Australia, change in the leaf area index (dLAI) from pre- to post-fire has been used to map burn severity from satellite imagery and from the ground (Boer et al. 2008). Using LAI has the advantage of linking directly to many different ecosystem process models of vegetation dynamics. Other remote sensing approaches to assess fire effects have been used across the globe (Gitas et al. 2012). Around the globe where people want to map fire effects, they are using combinations of sensors and alternative approaches to field data collection to support inferences to overcome the limitations of cloud cover and particulates affecting image quality, scale, and the ecosystem effects they wish to assess (Gitas et al. 2012).

All fires burn with a range of burn severity. Large fires have burned extensive areas around the globe in recent decades, and are expected to increase in many, but not all, areas in the future (Moritz et al. 2012; Dennison et al. 2014). Areas burned with high severity are typically intermixed with unburned islands and areas burned with low or moderate severity in forests (Kolden et al. 2015; Birch et al. 2015; Krawchuk et al. 2016; Meddens et al. 2018). For example, large wildfires in the Rocky Mountains and Southeastern U.S. typically have less than 33% of their area burned with high severity (Birch et al. 2015; Picotte et al. 2016) and 7–10% of the area classified as unburned (Meddens et al. 2018). Birch et al. (2014) found that the proportion of area burned with high severity was not correlated with the area burned. Even when large areas are burned in a single day, there are unburned islands and areas burned with low or moderate severity as fires burn across complex topography with varying wind and fuels. Birch et al. (2015) found that less than 15% of the area burned with high severity except in the most extreme cases (Birch et al. 2015). Spatial variability in fire intensity, weather, and vegetation—influenced by many factors—results in spatial variability in fire effects and long-term ecological benefits or detriments from fire. Understanding the factors influencing burn severity is essential to assessing ecosystems' vulnerability to wildfires now and into the future. For more on burn severity, unburned islands, and how spatial heterogeneity of fire effects influences landscape dynamics, see Chap. 12.

Soil burn severity (Fig. 9.20) is often assessed immediately after fires to determine soil erosion potential. As Parson et al. (2010) explained, in areas classified with a low soil burn severity, some or all of the organic layers on the soil surface remain after the fire so the surface appears brown or black. The organic matter in the soil is largely unaffected because soil heating was minimal. Some areas may even be unburned. In areas classified with moderate soil burn severity, fires have often consumed most (up to 80% of the areal extent) of the organic layers on the soil surface. The soil surface often appears to be blackened (charred) and gray with ash, with much of the vegetation brown or black immediately post fire. The brown needles and leaves that may fall to the ground can help protect the burned soil surface from the impact of rainfall. In areas classified with a high soil burn severity, the organic matter on and in the surface soil is often consumed, leaving behind bare soil with a red hue from oxidized iron compounds, a lack of soil structure, and a powdery feel. Soil burn severity has five elements (Parson et al. 2010, Fig. 9.20) that can be assessed and mapped using satellite imagery complemented with field assessments in the first few days or weeks after fires. These reflect the degree and duration of soil heating:

Fig. 9.20 Five different factors are considered in assessing soil burn severity: ground cover is the amount and quality of the litter or other material covering the soil surface, ash color and depth is the color (black, gray or white) and depth of ash, soil structure is related to the soil aggregates, roots is the number and condition of the fine roots in the surface soil layers, and soil water repellency is an indicator of the infiltration rate of water into the soil. (From Parson et al. 2010)

1. The amount and condition of ground cover. Surface litter and plants help to protect the soil surface from raindrop impact and may keep sediment from moving far in surface water, especially when more than 60% of the surface area is covered (Pannkuk and Robichaud 2003).
2. Ash color and depth. These indicate soil heating and whether the surface organic matter was consumed. Black and gray ash contain organic matter that is charred but not consumed, whereas the organic matter in white ash has been consumed.
3. Soil structure. If the organic matter in the soil is consumed or heated too much, it does not hold the soil aggregates together, will not hold moisture or nutrients, and is not favorable for plant growth post fire.
4. Fine roots. These hold the soil together and can thus limit soil loss to wind and water erosion post-fire. If fine roots are alive, vegetation is more likely to resprout to help protect the soil surface and grow more roots to help hold the soil together.
5. Soil water repellency. Hydrophobic soils are less likely to absorb and hold water, limiting infiltration and increasing surface runoff.

Most assessments of burn severity encompass vegetation as well as soil effects. The visual appearance of the substrate, surface vegetation, and canopy vegetation can be helpful for characterizing burn severity, and therefore the potential for soil erosion and vegetation response (Fig. 9.21). Although measures of burn severity are often summarized by class, continuous measures are preferable to support analysis (they can always be grouped into classes) (Morgan et al. 2014).

The Composite Burn Index (CBI, Key and Benson 2006) was designed for quick assessments of vegetation and soil burn severity and to assist in the interpretation of remotely sensed images of burned areas. CBI values range from 0.0 to 3.0,

Fig. 9.21 (**a**) Low, moderate, and high burn severity in high-density chaparral is judged from the combined fire effects on the substrate, the surface vegetation, and the vegetation canopy. (**b**) Similarly, low, moderate and high burn severity in high-density mixed-conifer forests is judged from the combined fire effects on the substrate, the surface vegetation, and the vegetation canopy. (From Parson et al. 2010)

representing burn severities from unchanged to the highest burn severity possible. The CBI is a composite of ratings for each stratum (substrate, low shrubs, small trees, medium-sized, and large trees) with the observer rating qualitatively the degree to which each was changed by the fire (Key and Benson 2006). The biggest flaw with CBI is that field observers must estimate fire effects without knowing the pre-fire site conditions. A CBI value of 2.25 is commonly used to differentiate moderate and high burn severity inferred using indices of burn severity such as the Normalized Burn Ratio (NBR), differenced Normalized Burn Ratio (dNBR), and Relativized differenced Normalized Burn Ratio (RdNBR) (Dillon et al. 2011, 2020; Key and Benson 2006; Miller and Thode 2007). CBI values of 2.25 or above reflect some combination of consumption of >50% of fuels >7.5 cm in diameter as well as >80% of fuels <7.5 cm in diameter, >95% alteration of foliage of shrubs and small trees, and >85% of foliage on small, medium and large trees made black or brown by the fire. Sikkink et al. (2013) and Kolden et al. (2015) have accumulated CBI data collected from more than 3600 plots and made these readily accessible via the Internet. Although the data are from many sites in the USA, the data are clustered in a few geographic locations with data missing from many ecosystems. Dillon et al. (2020) and Sikkink et al. (2013) improved upon the original CBI and GeoCBI (De Santis and Chuvieco 2009) metrics by incorporating continuous measurements rather than broad classes, and by recording the assessment for each individual substrate and each vegetation strata (rather than a simple average CBI value that had been used previously). This additional information aids interpretation as does including fraction green and fraction charred (Hudak et al. 2007; Lentile et al. 2009) (both can be scaled, meaning they can be readily interpreted on the ground and from remotely sensed imagery with fine to coarse resolution), the size of the smallest remaining twig in shrublands (Keeley et al. 2008), and other indicators of burn severity. Dillon et al. (2020) emphasized that the choice of the index (e.g., RdNBR, dNBR, NBR) and the methods for determining thresholds for severity classes should vary with objective and application.

To advance understanding of how fire effects are related to fire behavior, we need more data that are spatially coincident, physically-based, and readily measured on the ground and remotely in quantitative terms (Kremens et al. 2010). Measurements will be more valuable if they are made consistently, and the resulting data are widely accessible for evaluating and testing models and thus furthering our understanding of fire effects (Kremens et al. 2010). It is a major weakness of many fire ecology studies that we don't know when or how a site burned, yet we know that is important to the fire effects.

Burn severity is related to fuels, topography, weather, and the climatic conditions before and after a fire. The legacy of past disturbances and pre-fire vegetation also influences how severely fires burn and what vegetation develops post fire. Dillon et al. (2011, 2020) modeled high burn severity from satellite imagery for wildlands in the 48 contiguous United States. They found that high burn severity was consistently related to topography (especially elevation), vegetation, and fuel moisture. Based upon western U.S. forests, Parks et al. (2018) argued that burn severity was more strongly influenced by fuels and vegetation than by climate, in contrast to the influence of climate on annual area burned. These results have important implications, for they suggest that managing fuels can alter the effects, the "ecological footprint", of fires. Variability in climate and fire weather does influence the size and severity of large fires (Birch et al. 2015; Dillon et al. 2011; Keyser and Westerling 2017), and it will likely be increasingly important as climate changes (Parks et al. 2016; Moritz et al. 2018). For more on this topic, see Chaps. 12 and 14.

9.7 Fire Effects on Animals

In Australia, three species of raptors, "fire-hawks", purposefully ignite fires by carrying burning sticks into unburned areas, just as indigenous humans have done in many areas (Bonta et al. 2017). Why would this behavior be adaptive?

Fire can greatly alter both the habitat and habitat use by animals. Fire effects on animals vary greatly among species, life history, specific habitat requirements, and the environmental conditions before, during, and after fires, including fire behavior. However, a few generalities can be made. *First,* the indirect effects of fire on animal populations are usually more important than the immediate, direct effects. *Second,* many animals have developed life-history strategies to avoid fires altogether, such as lining in a location that is not likely to burn or to survive fires by fleeing from the burned area or hiding in a safe location within the burn perimeter. Immediate, direct effects of fire include both injury and mortality of individual animals due to heat exposure and inhalation of hot air and smoke (See this chapter). Animal tissues, like plant tissues, are damaged at temperatures well below that observed in flames. Individual animals may die or be injured by smoke inhalation or heat from fires. Unfortunately, data on fire-caused animal injury and mortality are limited, particularly for the many invertebrates that contribute to biodiversity.

Indirect, long-term effects of fire on animal habitat influence long-term population viability. Landscapes change as vegetation responds to the fire effects. Animals may recolonize burned areas immediately or soon after fires. However, the degree to which they do so can depend on the quality of the post-burn habitat, proximity to source populations, and how vulnerable the animals are to predation. Many different species thrive in burned areas. However, changes in fire regimes can often be detrimental to long-term population viability if some aspects of needed habitat are missing, and these are influenced by landscape heterogeneity. The Fire Effects Information System (https://www.feis-crs.org/feis/, accessed 1 May 2019) provides the most complete single source of information on fire effects on animals and their habitat for North America.

Descriptions and maps of vegetation fuels are potentially useful for assessing habitat for plants and animals. The structure of vegetation, including heights of different layers above the ground and the size and arrangement of patches of vegetation, is important to understanding fire hazard. They are also important to animal use. Across the USA, fuels and vegetation are mapped and updated when fires occur by the LANDFIRE program. The related understanding of vegetation

Fig. 9.22 Sandhill pine forests (**a**) before and (**b**) after prescribed burning for restoring sandhill pine forest in Fort Cooper State Park in Florida. (**c**) The Atala hairstreak butterfly and other butterflies, birds and many plants inhabit fire-prone sandhill pine communities in Florida. Wikimedia commons

structure and how fuels and landscapes change through time could be used more in the conservation and management of animal and plant populations.

Here are some examples of fire effects on animals and their habitat. We emphasize here the indirect effects (i.e., the habitat effects). The radiative heat flux from a fire to an animal can be estimated using similar equations to those for humans (see Chap. 10).

Thom et al. (2015) evaluated the implications of forest floor consumption and soil heating for the survival of the Atala hairstreak butterfly (*Eumaeus atala*) pupae in prescribed fires in sandhill pine forests in Florida (Fig. 9.22). Both the butterfly species and the sandhill pine habitat are relatively rare today. This habitat historically burned frequently. In the absence of fire in many recent decades, the host plants used by the butterflies have declined in abundance. Butterfly pupae cannot survive

Fig. 9.23 Grasshoper feeding on sprouts of the shrub, *Pterospartium tridentatum*, 6 months after prescribed fire in a heathland of the Alvão Natural Park, northern Portugal. (Photograph by Paulo Fernandes)

exposure to temperatures over 50 °C, and thus to survive through fires, the pupae must be either in unburned leaf litter or 1–2 cm or more below the soil surface when forests burn. Butterfly eggs and larvae on host plants likely do not survive as they are in the above-ground plants and thus likely exposed to temperatures greater than 350 °C. Based on their findings, Thom et al. (2015) suggested carefully timing prescribed burning to favor host plants and perpetuate the sandhill pine forest communities, and only burning a portion of any area containing butterfly colonies to balance the beneficial effect of fire on the host plants while limiting the lethal effects of fire (Fig. 9.22).

The short and long-term effects of fire on wildlife habitat are important for understanding population response to fire. Many animals will eat the new tender tissue of plants resprouting after fires even when they are unlikely to eat the same plant as the plant matures (Fig. 9.23). Some forbs and shrubs fix nitrogen, and others capture available nitrogen released after the fire as organic matter decomposes. Both of these effects can result in higher protein content in plants as nitrogen is a key component of proteins. Further, newly sprouted plants have less lignin and other cellulose components so they can be readily digested. In many areas around the globe, people burn to improve the quality of forage (grass) and browse (shrubs) for domestic and wild animals. However, people need to burn enough patches at one time, and enough patches over time that the animals and insects do not overuse the freshly burned areas.

For animals, the vertical structure, interspersion of habitat needs (e.g., food next to shelter), the season of fire relative to life history, and size of the habitat patch are often more important than plant species composition except for species that are dependent on particular plants. Fires alter vegetation structure. For instance, fires may consume partially decayed logs on the ground that are often used by many invertebrates, rodents, salamanders, and reptiles. However, fires also may kill or injure trees that become standing snags that support insects and foraging birds, and birds that excavate cavities for nests. Snags and logs provide important habitats for many different birds, small mammals, insects, and other species. Refugia are especially important for unburned or lightly burned areas within large fires, which can be the locations where animals survive fire and recolonize burned areas. Thus, landscape heterogeneity through time is important to population viability, but what size and pattern of burned and unburned patches are needed is not well understood for many species. What is ideal for some (e.g., those needing large areas far from edges) may not be ideal for others (e.g., those needing small openings and edges). In the absence of such information, many managers rely upon the historical range of variability as a clue to what species may need in the future (see related discussion in Chap. 12).

Tingley et al. (2016) found that pyrodiversity supported avian diversity (Fig. 9.24). Using bird observations 1, 5, and 10 years after 97 fires in California, they found the greatest diversity in bird species occurred in areas that burned as a mixed-severity fire. With time, bird species composition became increasingly dissimilar in burned and unburned areas. They suggested that birds respond to the vegetation structural changes in forests post fire, and burn severity altered the vegetation trajectories. Similarly, Ponisio et al. (2016) found more pollinators, more plants, and more plant-pollinator interactions in mixed-conifer forests in Yosemite National Park in California in areas burned with variable fire frequency and burn severity. In contrast, bird diversity was not favored by pyrodiversity in Australia's mallee, a fire-prone ecosystem dominated by *Eucalyptus* spp. trees where bird species diversity did not vary with varying fire frequency (Taylor et al. 2012). There, birds preferred old forests over recently burned areas. Similarly, reptiles did not differ. Taken together these two examples suggest that pyrodiversity can favor biodiversity. However, these effects will vary with the particular animals, habitat preferences, time between fires, burn severity, and the influence of drought or other climatic factors. Davies et al. (2018) found that in an experimental area in northern Australia burned with different fire frequencies, small mammal diversity was highest at intermediate fire frequency. This is consistent with the intermediate disturbance hypothesis (Huston 2014). However, Davies et al. (2018) reported that different species of small mammals had different optimal fire frequencies. Davies et al. (2018) cautioned that simply increasing pyrodiversity will not benefit all species. Similarly, Farnsworth et al. (2014) found that reptile diversity did not differ amongst landscapes with different fire frequencies in semi-arid Australia. Bowman et al. (2016) linked fire regimes and food webs, thus highlighting how herbivory and fire interact with vegetation heterogeneity which in turn affects habitat for animals and subsequent fires (Fig. 9.25). Animal and plant viability in habitats that burn is influenced by metapopulation dynamics which explains how individual subpopulations may be reduced (e.g., in a burned patch that is temporarily unfavorable habitat) and then recolonized from a nearby subpopulation in an area burned differently (perhaps a few years earlier, perhaps unburned) from which individuals may disperse. Luce et al. (2012) described examples where fish and other fauna recovered very quickly in stream reaches severely impacted by fires (stream reaches are segments of streams). Such recovery was only possible, however, where the severely burned

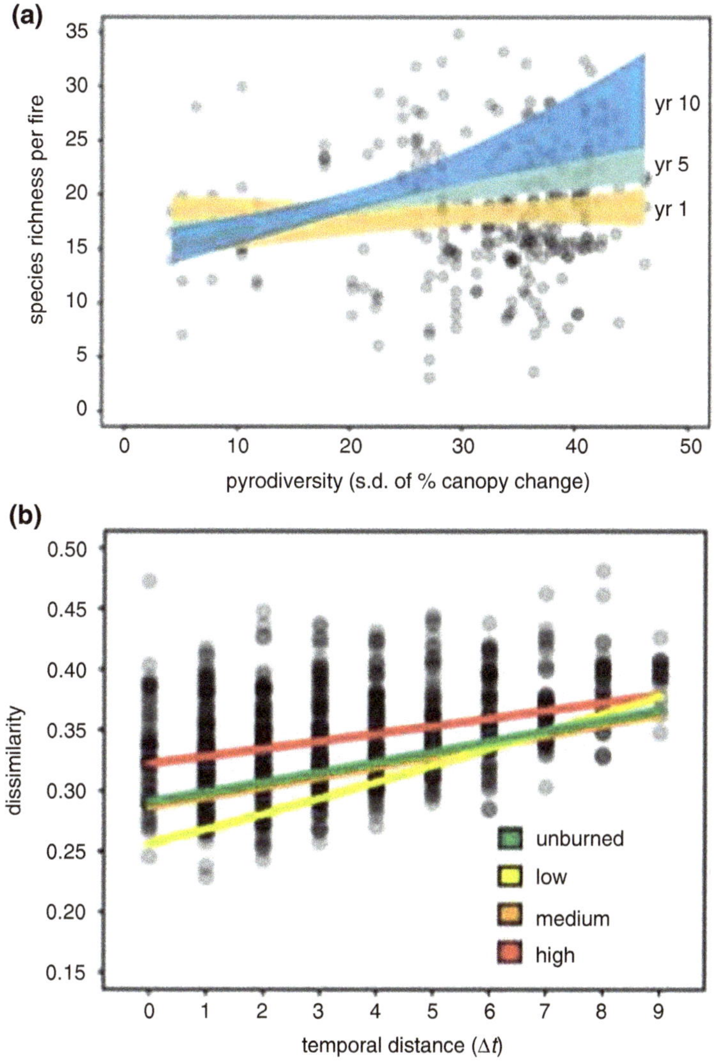

Fig. 9.24 Pyrodiversity begets biodiversity. (**a**) Bird species richness is higher in more diverse landscapes, and (**b**) bird communities differ with burn severity soon after fires but become more similar with time. (From Tingley et al. 2016)

reaches were connected to refugia from severe fire effects. Further, the degree of connectivity depends on how individuals of each fish species move between stream reaches and how the stream and disturbance characteristics influence how suitable different stream segments are as habitat.

In many areas where fires have been less frequent than historically, woody vegetation has increased, and landscapes have become less patchy. Past land uses, such as agriculture, logging, and roads, have fragmented the landscape in other

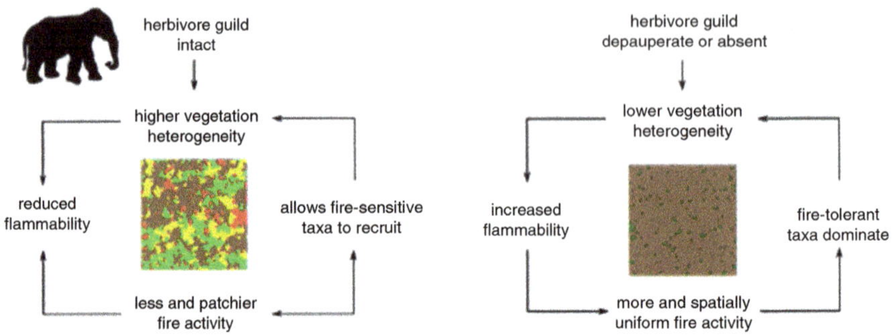

Fig. 9.25 Herbivory can interact with fires to alter vegetation patterns which in turn alters the way fires influence vegetation diversity. Bowman et al. (2016) emphasized the importance of interactions of food webs and fire regimes in influencing vegetation. Note that herbivory can reduce biomass that can fuel fires and can increase fine-scale heterogeneity. (From Bowman et al. 2016)

areas. Fires can be part of management that can increase the habitat values. However, burned areas need to be a large enough proportion of the landscape that the animals don't concentrate on burned areas and use them so heavily that plants are unable to recover from the combined effects of fire and herbivory. Many animal populations and communities suffer in the absence of fire if the habitats they depend upon degrade in quality food and shelter, or the surrounding landscape becomes homogeneous and therefore provides some but not all qualities of habitat a given animal needs. For instance, longleaf pine forests of the southeastern USA support very high biodiversity, but only when frequently burned (Fig. 9.26). Management of longleaf pine forests for red-cockaded woodpeckers, gopher tortoises, and many other wildlife and plant species that are threatened or endangered has led to widespread efforts to restore forest structure and composition (Fig. 9.26). In the absence of fire, habitat for these species degrades, while frequent fire favors open pine forests with diverse understory plants (Van Lear et al. 2005). The plant species richness increases with frequent burning, especially if there are some unburned islands.

9.8 Implications and Management

Given the importance of forests to biodiversity, clean water, carbon sequestration, and other ecosystem services, many have called for increasing the area of forests, managing existing forests well to avoid losses to insects, fire and other disturbances, prompt tree regeneration, and use of wood for biomass and building (McKinley et al. 2011), while others are calling for greatly increasing the pace and scale of forest restoration for resilience and provision of ecosystem services. Forests will burn, so actively managing them for resiliency to future fires is important. Restoring the resilience of forests to future fires through both thinning and burning through managed fires is projected to favor long-term carbon balance over space and time

Fig. 9.26 Longleaf pine forests of the southeastern United States are biodiversity "hotspots" for they can support many plants and animals that are otherwise uncommon. (**a**) Many of these forests are managed to protect and enhance wildlife. (**b**) Many birds and plants benefit from the frequent prescribed burning required to maintain the open stands of pine. (**c**) The holes that red cockaded woodpeckers (*Leuconotopicus borealis*) excavate in live longleaf pine trees provide habitat for them and for many other cavity-nesting birds. (**d**) Gopher tortoises dig deep burrows that are also used by many snakes and other animals as shelter, greatly adding to the biodiversity in these forests. (All photographs by J Morgan Varner at the Tall Timbers Research Station, Florida)

in ponderosa pine and other forests by reducing the severity of future wildfires and, therefore, the mortality of large trees and the variability in forest productivity at landscape scales (Hurteau and Brooks 2011).

Here we discuss three different implications and related post-fire considerations. These include vegetation trajectories (this is succession) and post-fire soil and vegetation treatments, most of which are designed to limit soil erosion potential. We also address the issue of how much high severity fire is desirable or natural.

9.8.1 Vegetation Trajectories

Post-fire vegetation trajectories depend on burn severity, vegetation type, and time since fire (Fig. 9.27, Bright et al. 2019). Areas burned with high severity have, by definition, experienced greater change and may take longer to recover than similar areas burned with low severity (Abella and Fornwalt 2015), especially if high severity patches are large enough to limit seed rain from surviving vegetation. Fires seldom consume all above-ground biomass, and many seeds or other meristematic tissue survive, especially where burn severity is low or moderate. One year after nine forest fires in boreal forests in Alaska, and mixed-conifer forests in Montana, Idaho, and California, Lentile et al. (2007) concluded that understory plant species abundance and diversity were lowest in areas burned with high severity compared to those burned in the same area by moderate or low severity fires. Sampling in many of the same fires in Montana 10 years later, Strand et al. (2019) concluded that burn severity along with climate influenced plant species richness and diversity. The highest plant species diversity both 1 and 10 years after fires were observed in areas burned with moderate severity. Likely this is because areas burned with mixed-severity had diverse microsites suited to survival and establishment of many different plants post fire. One year after fires, areas burned with moderate burn severity included a very fine-scale intermix of areas burned with high soil burn severity (e.g., under rotten logs and old stumps that consumed), others burned with moderate and low soil burn severity, and unburned areas (e.g., skipped by fires). Strand et al. (2019) found few differences in understory plant species richness and diversity 10 years after fires, with the post-fire cover similar to pre-fire cover. Recovery was slower on areas burned with high severity, especially in large patches burned with high severity, and these areas had more off-site and residual colonizers (i.e., plants that resprouted from buds or established from seeds that survived the fire) (Bright et al. 2019). That is consistent with the observation by Lewis et al. (2017) that the cover of green vegetation detected from satellites had increased since immediately post-burn and was similar at about 60% on all burned areas, regardless of burn severity, though it took twice as long to develop on sites burned with high severity compared to sites burned with low severity. These results suggest that burn severity was quite important in the vegetation composition and structure in the early years post-fire. However, the areas burned with different soil burn severity became more similar in the decades post-fire.

Long-term vegetation trajectories post-fire often depend on what plants survive fires or establish post fire. If they are abundant, shrubs and trees often dominate the understory vegetation through time. For instance, in Florida sand pine ecosystems, high severity fires are characteristic, Freeman and Kobziar (2011) found that herbaceous species cover was highest 1 and 2 years after plots burned with moderate severity in a wildfire. More sand pine seedlings established after high severity fire, especially when the pre-fire trees were mature. They expected the differences in oak and sand pine abundance to affect the development of both vertical structure and tree

Fig. 9.27 How severely an area burns, and therefore the degree of ecological change from pre-fire, influences (**a**) post-fire vegetation trajectories, here quantified as the recovery of normalized burn ratio (NBR) relative to the immediate post-fire value. Lines are means with bars showing one standard deviation above and below the mean. (**b**) Fewer of the patches burned with high severity had recovered to pre-burn Normalized Burn Ratio values 9–16 years after fires compared to those burned with moderate or low severity burns. Data are from 8 fires in the western USA that burned during the years 2000–2007. Numbers above the bars are the years post-fire. (From Bright et al. 2019)

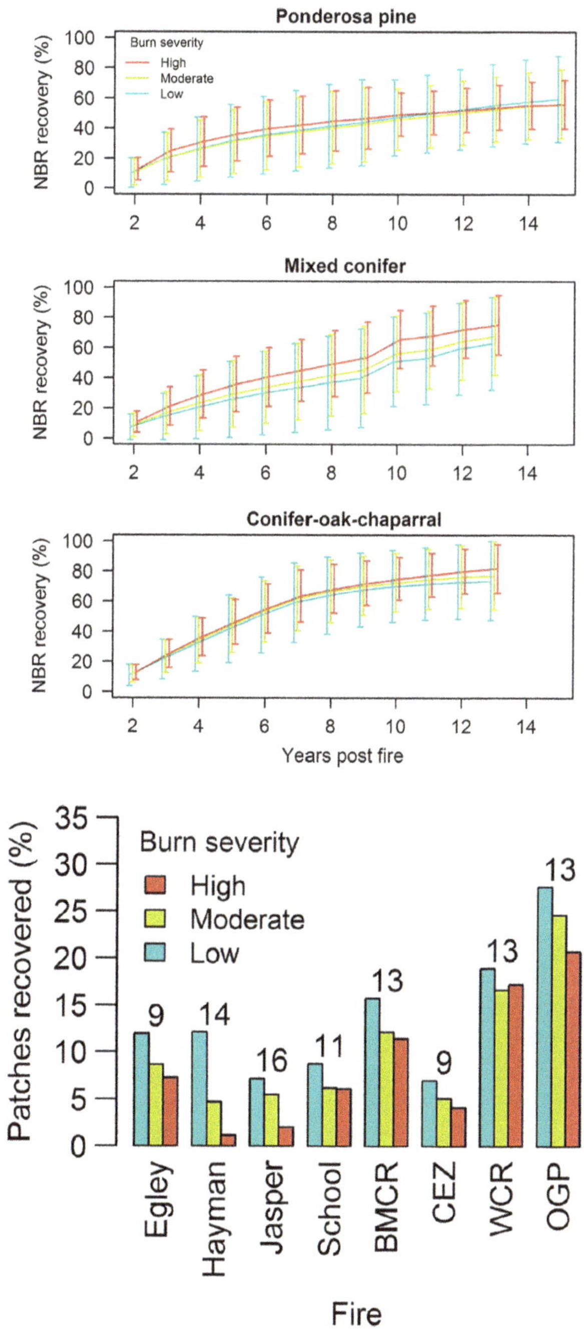

and shrub composition, with these long-term community differences likely to alter the effects of subsequent fires.

Fires may foster the coexistence of many different species if there are areas in a landscape that are unburned along with others that vary in the time since fire and in the effects of those fires. In many ecosystems, the plants that establish post fire are the ones that were present prior to the fire. In Alaska, areas burned with high severity often support aspen (*Populus tremuloides*) while black spruce (*Picea mariana*) forests that burned less severely were more likely to return to dominance by black spruce, likely because of the differences in microsites that resulted from the fires (Johnstone and Kasischke 2005).

Stevens-Rumann et al. (2018) found that post-fire tree regeneration was variable, with few or no tree seedlings on warm, dry sites sampled 3–28 years following 62 large forest fires in the northern Rocky Mountains of the USA. Areas burned with high severity had many burned areas far from seed source trees. Stevens-Rumann and Morgan (2019) found similar patterns in their review of field studies on 152 large forest fires across the western USA, echoing similar observations in the Mediterranean region and elsewhere (Pausas et al. 2009) (see Case Study 12.3 for a related discussion of post-fire tree regeneration (or lack thereof) in recent large forest fires). If trees fail to regenerate in sufficient abundance post fire, forests are replaced by shrublands, woodlands, and grasslands. These ecosystem changes may be accelerated by climate change (Stevens-Rumann et al. 2018; Walker et al. 2018; Davis et al. 2019).

Dodson et al. (2010) found that post-fire mulching treatments applied to reduce soil erosion altered vegetation recovery in forests 2 years after agricultural straw mulch was applied to burned areas. They found higher plant cover and species richness as well as more tree seedlings where mulch covered less than 40% of the ground, but in areas with mulch more than 5 cm deep in the second year after fire, there were few plants. Succession following fires follows multiple pathways (Cattelino et al. 1979), as the vegetation structure and composition at the time of the fire greatly influences what is there after the fire.

The legacy of past fires influences vegetation structure and composition, and therefore how the next fire burns. Burn severity also influences what survives to regrow and what will colonize burned areas, while also influencing the microsites available for plant establishment. This varies with site productivity, burn severity, and time since the previous fire. The legacy of past fires influences future landscape dynamics as they are shaped by fire, people, topography, vegetation, and climate (see related discussion of repeated fires in Chap. 12). Other disturbances before or after fires can also alter post-fire vegetation trajectories, and these vary with site productivity. After all, that is the goal of fuel management (see Chap. 11) and other vegetation management treatments.

9.8.2 *Post-fire Soil and Vegetation Treatments*

Managers conduct many different types of treatments to reduce potential negative impacts of fires on ecosystem goods and services and communities following a fire. Treatments designed to limit soil erosion potential, including seeding and mulching, may be applied within the first few days or weeks following fires. Many burned areas are also planted with tree or shrub seedlings or seeded with grass with the goal of speeding or otherwise managing vegetation post fire (See Case Study 12.3). Salvage logging is done on some sites to recover some of the economic value of logs from trees killed or severely damaged by fire (these are judged to be likely to die soon from injury). However, ecosystems often recover from fires without treatments. Treatments can be expensive and may have unintended consequences, so they should be strategically applied only where they are most needed to protect values at risk.

Mulching can be very effective in stabilizing soil post fire when mulch is spread to cover bare mineral soil to limit soil erosion potential (Bautista et al. 2009; Wagenbrenner et al. 2006; Williams et al. 2014). Mulch can immediately cover bare soil but may not be needed where litter, logs, trees, or other vegetation already provide at least 60% ground cover needed to limit soil sediment movement (Pannkuk and Robichaud 2003). Mulching treatments are expensive, so they are typically strategically targeted where they will help to limit the exposure of values at risk, such as buildings, road infrastructure, and high-quality aquatic habitats (Bautista et al. 2009; Williams et al. 2014), and where high-intensity rainfall events are likely to occur soon after an area burns and before vegetation can establish and grow in abundance (Dodson and Peterson 2010; Robichaud et al. 2013a, b). Mulching treatments are commonly focused on large areas burned with high soil burn severity on steep upper slopes and other locations where soils are highly vulnerable to erosion. Mulching is often done with agricultural straw but also with wood shredded from nearby trees or wood straw derived from waste from lumber manufacturing (Robichaud et al. 2014), but can also be accomplished with wood chips, pine needles, and other materials (Bautista et al. 2009). There have been few studies of the long-term effects of post-fire mulching treatments on vegetation response. Bontrager et al. (2019) sampled six different large forest fires 9–13 years after mulching with agricultural straw in the western USA. They found no differences in understory plant species diversity and richness nor tree seedling density when they compared pairs of otherwise similar sites that were mulched or not immediately post fire. Compared to similar plots within the same burn with little or no mulch, Dodson and Peterson (2010) found less plant cover where mulch cover exceeded 70% of the ground. Dodson and Peterson (2010) found that non-native species were abundant 2 years after mulching on the Tripod fire that burned large areas of mixed-conifer forest in Washington, but Bontrager et al. (2019) found few of the same non-native species 10 years later. Together these results suggest that mulch treatments with agricultural straw can effectively increase ground cover post fire to reduce soil erosion potential while not significantly altering post-fire vegetation trajectories. In contrast, Beyers (2004) and Kruse et al. (2004) found that 2 years after mulching with agricultural straw applied immediately after fires, mulched areas had fewer tree seedlings, less abundant understory vegetation with more non-native species in comparison to similar areas that were not mulched. Although the mechanisms responsible for driving the variability in treatment response remain unknown, local knowledge and judicial monitoring of treatments can be used to support long-term adaptive management following a fire.

Broadcast seeding is one of the most widely applied treatments post-fire to reduce soil erosion potential by increasing post-fire vegetation cover (Peppin et al. 2010; Robichaud et al. 2013a, b). Often, grasses are seeded by hand or from the air as part of post-fire wildfire rehabilitation. Peppin et al. (2010) reviewed 94 studies of post-fire seeding conducted in the western USA in both forests and rangelands. Recent, well-designed studies showed that grass seeding was ineffective at reducing soil erosion (Peppin et al. 2010). Furthermore, 62% of the studies suggested that seeding (mostly done with non-native species) reduced native plant abundance though there are few long-term studies. Seeding done with native, locally adapted seed sources is often recommended as it may be less likely to alter the post-fire recovery of native species. However, native seeds that are certified weed-free can be expensive and may be limited, especially when multiple large fires increase the demand. Morgan et al. (2014) described the Umatilla National Forest native seed program where managers collect seeds of native grasses, then contract with local farmers to plant it to produce more seeds for use on areas burned on the National Forest or to be sold for use elsewhere.

Strategically deciding whether or not to apply treatment post fire, and what to do where, can be informed by careful mapping of burn severity and field reconnaissance (Parson et al. 2010) combined with available predictive tools, local knowledge, and experience (Robichaud et al. 2014). Relying on natural recovery without post-fire treatments is often recommended, but local management objectives and conditions should guide such decisions. Typically, less than 5% of burned areas are treated with emergency stabilization and rehabilitation. This reflects expense and the focus on large patches burned with high severity where soil erosion potential is high (e.g., steep slopes with highly erodible soils and high potential for high-intensity rainfall). With time, with or without treatment, soil erosion potential declines as more vegetation and litter cover the soil, soil water repellency decreases, and reduced sediment is available as organic matter in the soil increases (Scott et al. 2009).

Planting of tree seedlings post-fire is done where policy and management objectives require replacing trees killed by fires. Although managers commonly favor natural regeneration from surviving on-site trees, they may decide to plant to ensure that future forest achieves the desired density or species mix within a specified timeline. As Stevens-Rumann et al. (2018), Stevens-Rumann and Morgan (2019), and North et al. (2019) suggested, planting of tree seedlings needs to focus on those species and sites where planting is more likely to be successful. Successful tree regeneration is less likely (or unlikely) on the hottest and driest sites, sites with competing vegetation, and in areas likely to burn again before tree seedlings can grow to a size to survive fires (See Case Study 12.3). Planting with low and variable density could foster spatial heterogeneity with multiple benefits, such as for wildlife, and increased resilience to future fires (North et al. 2019). Certainly, the likelihood that the plantation will burn again should be considered. Prescribed burning within the young stands may be needed to help manage the fire hazard within these valuable assets, even if there is some mortality. In Portugal, pine plantations previously treated with prescribed burning experienced decreased burn severity when encountered by wildfire (Espinosa et al. 2019). Burning within plantations can increase within-stand heterogeneity, help manage stand density, and reduce fuels (Fernandes and Botelho 2004; Kobziar et al. 2009; North et al. 2019). Mechanical thinning with or without mastication of the resulting fuels is often done in pine plantations in many locations globally. See Case Study 12.3 for discussion of tree seedling regeneration and related post-fire planting strategies informed by science and the experience of managers. See the discussion about mastication as a fuels treatment in Chap. 11.

Salvage logging post fire involves the removal of merchantable trees that are dead, often from fire but also from bark beetles or other disturbances. Salvage logging can be controversial for multiple reasons. The wood does have value and, therefore, can be economically valuable. The timber loses value quickly, so salvage logging is often done soon after the trees die, and sometimes includes trees that were injured and judged likely to die soon. There is little evidence that salvage logging significantly reduces fire hazard if fuels created by logging are treated promptly (Fig. 9.28, Peterson et al. 2009). Compared to areas not salvaged and planted, areas salvaged and planted with trees after the 1987 Silver Fire in Oregon burned some of the same areas more severely in the subsequent 2002 Biscuit Fire (Thompson et al. 2007). Because many woodpeckers and other birds nest and feed in large-diameter snags, salvage logging, which targets the removal of these large dead trees, can greatly reduce their habitat. Salvage logging using existing road networks, on snow, and with low-impact machinery can minimize soil compaction and disturbance that could lead to erosion. Salvage logging is often focused within areas burned with moderate to high severity as that is where there are higher proportions of trees killed. Given the variety of terrain, fire effects, and salvage logging approaches, careful monitoring and more studies are needed to understand both short and long-term effects of post-fire salvage logging.

9.8.3 How Much High Severity Fire Is Natural or Desirable?

Areas burned with high severity, especially in large patches, will always be of concern. By definition, these are areas with a high degree of ecological change due to fire. The ecological changes associated with high severity fire are associated with a variety of negative and positive ecological and social effects, depending on what fire effect is being assessed and what time frame it is being assessed over. Indeed many plants and animals thrive in areas burned with high severity (DellaSala and Hanson 2015; Hutto et al. 2016), and so it is unfortunate that the term “severe” sounds disastrous.

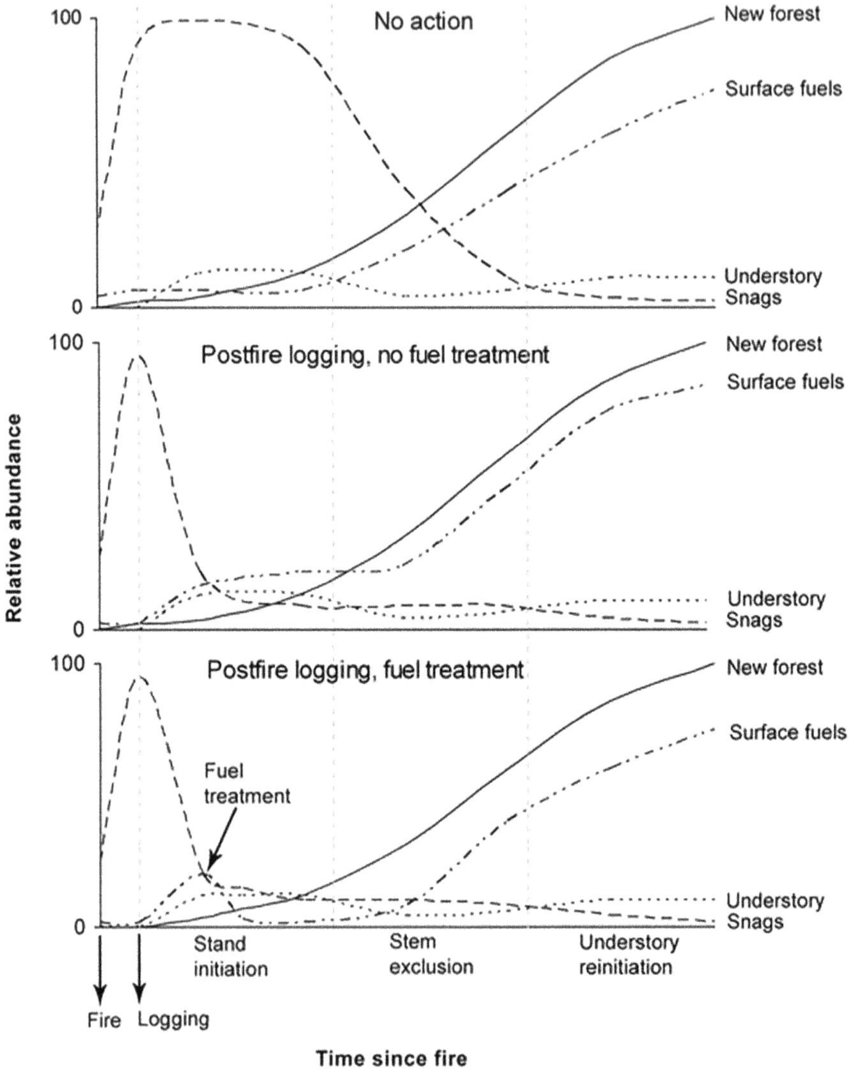

Fig. 9.28 Salvage logging post fire can reduce the number of dead trees (known as snags), but fire hazard is less likely to increase as a result of salvage logging if fuels from logging are treated immediately after logging. (From Peterson et al. 2009)

Certainly, all wildfires include at least some areas burned with high severity, and no doubt fires always had some areas burned with high severity. However, with changed fuels, increasing the likelihood of extended and warmer droughts with changing climate, and with increasingly homogeneous fuel conditions in many areas, we can expect many more and larger areas to burn with high severity. Increasing the area burned under less extreme conditions of weather and fuels could reduce the probability that areas will burn with high severity in the future. How much high severity fire is natural or desirable has and will be debated by both scientists and managers (Moritz et al. 2018). However, there is no shortage of area burned severely as many very large patches have burned with high severity in recent decades. The more relevant questions are about the spatial pattern and post-fire management within these large patches.

Pausas et al. (2009) reviewed fire effects in forests and shrublands of the Mediterranean. They concluded that some areas are sensitive to high severity fires as indicated by soil loss and long-term change in vegetation, but many ecosystems are quite resilient to high severity fires, such as cork oak (*Quercus suber*) woodlands. Ecosystems are less resilient where human actions have greatly altered them. The implications of changing climate and fuels (or both) for fire size, burn severity, and other fire effects are discussed in detail in Chap. 12.

9.9 Conclusions

The effects of fires on plants and soils depend on the fire behavior, the transfer of heat energy from the fire, the properties of organisms and the soil, and the environment. Understanding convective, radiative, and conductive heat transfer is critical for estimating many fire effects, including plant mortality, soil heating, nutrient availability, and post-fire soil erosion potential. Plant functional traits, especially those related to regeneration, are especially useful in predicting post-fire plant response (Sect. 9.3).

There is a pronounced legacy of prior fires that influences what grows and dominates after fires and, therefore, both vegetation trajectories and ecosystem processes. Future fires will interact with topography and heterogeneous vegetation as they burn across landscapes as influenced by the weather.

Fuels management, including the use of prescribed and managed wildfires, can be used to strategically influence fire and fire effects. Prescribed burns often result in less fuel consumption and soil heating than wildfires, and thus reflect ways in which we can use fire and fuels treatments to alter vegetation response and fire effects (see Chap. 11). What we do during and after each fire will be part of what will influence the behavior and effects of the next fire.

9.10 Interactive Spreadsheet: CONDUCTION_Soils_Plants

We encourage readers to use the interactive spreadsheet, CONDUCTION_Soils_Plants_v2.0, to explore how varying intensity and duration of fires can alter the degree of soil heating at different depths (Fig. 9.8). Why does soil heating depend so much on smoldering combustion? The graphs for different conditions can be readily compared, as can the implications for the temperature and duration of soil heating (Fig. 9.7) to draw some inferences about the ecological implications of different burning conditions.

References

Abella, S. R., & Fornwalt, P. J. (2015). Ten years of vegetation assembly after a North American mega fire. *Global Change Biology, 21*(2), 789–802.

Alexander, M. E., & Cruz, M. G. (2012). Modelling the effects of surface and crown fire behaviour on serotinous cone opening in jack pine and lodgepole pine forests. *International Journal of Wildland Fire, 21*(6), 709–721.

Balfour, D. A., & Midgeley, J. J. (2006). Fire induced stem death in an African acacia is not caused by canopy scorching. *Austral Ecology, 31*, 892–896.

Bautista, S., Robichaud, P., & Blade, C. (2009). Post-fire mulching. In A. Cerda & P. R. Robichaud (Eds.), *Fire effects on soils and restoration strategies* (pp. 353–372). Enfield: Science Publishers.

Beyers, J. (2004). Postfire seeding for erosion control: Effectiveness and impacts on native plant communities. *Conservation Biology, 18*(4), 947–956.

Birch, D. S., Morgan, P., Kolden, C. A., Hudak, A. T., & Smith, A. M. S. (2014). Is proportion burned severely related to daily area burned? *Environmental Research Letters, 9*, 064011. https://doi.org/10.1088/1748-9326/9/6/064011\.

Birch, D. S., Morgan, P., Kolden, C. A., Abatzoglou, J. T., Dillon, G. K., Hudak, A. T., & Smith, A. M. S. (2015). Daily weather and other factors influencing burn severity in central Idaho and western Montana, 2005-2007 and 2011. *Ecosphere, 6*(1), 17. https://doi.org/10.1890/ES14-00213.1.

Boer, M. M., Macfarlane, C., Norris, J., Sadler, R. J., Wallace, J., & Grierson, P. F. (2008). Mapping burned areas and burn severity patterns in SW Australian eucalypt forest using remotely-sensed changes in leaf area index. *Remote Sensing of Environment, 112*(12), 4358–4369.

Bonta, M., Gosford, R., Eussen, D., Ferguson, N., Loveless, E., & Witwer, M. (2017). Intentional fire-spreading by “Firehawk” raptors in Northern Australia. *Journal of Ethnobiology, 37*(4), 700–719.

Bontrager, J. D., Morgan, P., Hudak, A. T., & Robichaud, P. R. (2019). Long-term vegetation response following post-fire straw mulching. *Fire Ecology, 15*(1), 22.

Botelho, M., Rego, F. C., & Ryan, K. (1998). Tree mortality models for *Pinus pinaster* of northern Portugal. In *13th Fire and Forest Meteorology Conference*, Lorne, Australia (pp. 235–240).

Bowman, D. M., Balch, J. K., Artaxo, P., Bond, W. J., Carlson, J. M., Cochrane, M. A., D'Antonio, C. M., DeFries, R. S., Doyle, J. C., Harrison, S. P., & Johnston, F. H. (2009). Fire in the Earth system. *Science, 324*(5926), 481–484.

Bowman, D. M., Perry, G. L., Higgins, S. I., Johnson, C. N., Fuhlendorf, S. D., & Murphy, B. P. (2016). biodiversity and fire regimes in food webs. *Philosophical Transactions of the Royal Society, B: Biological Sciences, 371*(1696), 20150169.

Bright, B. C., Hudak, A. T., Kennedy, R. E., Braaten, J. D., & Khalyani, A. H. (2019). Examining post-fire vegetation recovery with Landsat time series analysis in three western North American forest types. *Fire Ecology, 15*(1), 8.

Brodribb, T. J., & Cochard, H. (2009). Hydraulic failure defines the recovery and point of death in water-stressed conifers. *Plant Physiology, 149*(1), 575–584.

Brown, J. K., & DeByle, N. V. (1987). Fire damage, mortality, and suckering in aspen. *Canadian Journal of Forest Research, 17*(9), 1100–1109.

Brown, J. K., & Smith, J. K. (2000). Wildland fire in ecosystems: effects of fire on flora. Gen. Tech. Rep. RMRS-GTR-42-vol. 2. Ogden, UT: US Department of Agriculture, Forest Service, Rocky Mountain Research Station. 257 p. 42.

Campbell, G. S., Jungbauer, J. D., Jr., Bristow, K. L., & Hungerford, R. D. (1995). Soil temperature and water beneath a surface fire. *Soil Science, 159*(6), 363–374.

Casal, M. (1987). Post-fire dynamics of shrubland dominated by *Papilionaceae* plants. *Ecologia Mediterranea, 13*(4), 87–98.

Cattelino, P. J., Noble, I. R., Slatyer, R. O., & Kessell, S. R. (1979). Predicting the multiple pathways of plant succession. *Environmental Management, 3*(1), 41–50.

Clarke, S., & French, K. (2005). Germination response to heat and smoke of 22 Poaceae species from grassy woodlands. *Australian Journal of Botany, 53*(5), 445–454.

Davies, H. F., McCarthy, M. A., Rioli, W., Puruntatameri, J., Roberts, W., Kerinaiua, C., Kerinaiua, V., Womatakimi, K. B., Andersen, A. N., & Murphy, B. P. (2018). An experimental test of whether pyrodiversity promotes mammal diversity in a northern Australian savanna. *Journal of Applied Ecology, 55*(5), 2124–2134.

Davis, K. T., Dobrowski, S. Z., Higuera, P. E., Holden, Z. A., Veblen, T. T., Rother, M. T., Parks, S. A., Sala, A., & Maneta, M. P. (2019). Wildfires and climate change push low-elevation forests across a critical climate threshold for tree regeneration. *PNAS, 116*(13), 6193–6198.

De Santis, A., & Chuvieco, E. (2009). GeoCBI: A modified version of the Composite Burn Index for the initial assessment of the short-term burn severity from remotely sensed data. *Remote Sensing of Environment, 113*(3), 554–562.

DeBano, L. F. (1981). *Water repellent soils: a state-of-the-art* (Gen Tech Rep PSW-46). Berkeley: USDA Forest Service Pacific Southwest Forest and Range Experiment Station.

DeBano, L. F., Savage, S. M., & Hamilton, D. A. (1976). The transfer of heat and hydrophobic substances during burning. *Soil Science Society of American Journal, 40*, 779–782.

DellaSala, D. A., & Hanson, C. T. (2015). Ecological and biodiversity benefits of megafires. In D. A. DellaSala & C. T. Hanson (Eds.), *The ecological importance of mixed-severity fires* (pp. 23–54). Amsterdam: Elsevier.

Dennison, P. E., Brewer, S. C., Arnold, J. D., & Moritz, M. A. (2014). Large wildfire trends in the western United States, 1984–2011. *Geophysical Research Letters, 41*(8), 2014GL059576. https://doi.org/10.1002/2014gl059576.

Dickinson, M. B., & Johnson, E. A. (2004). Temperature-dependent rate models of vascular cambium cell mortality. *Canadian Journal of Forest Research, 34*(3), 546–559.

Dillon, G. K., Holden, Z. A., Morgan, P., Crimmins, M. A., Heyerdahl, E. K., & Luce, C. H. (2011). Both topography and climate affected forest and woodland burn severity in two regions of the western US, 1984 to 2006. *Ecosphere, 2*(12), 130. https://doi.org/10.1890/ES11-00271.1.

Dillon, G. K., Panunto, M. H., Davis, B., Morgan, P., Birch, D., & Jolly, W. M. (2020). *Development of a severe fire potential map for the contiguous United States* (RMRS-GTR-415). Fort Collins: USDA Forest Service Rocky Mountain Research Station.

Dodson, E. K., & Peterson, D. W. (2010). Mulching effects on vegetation recovery following high severity wildfire in north-central Washington State, USA. *Forest Ecology and Management, 260*, 1816–1823. https://doi.org/10.1016/j.foreco.2010.08.026.

Dodson, E. K., Peterson, D. W., & Harrod, R. J. (2010). Impacts of erosion control treatments on native vegetation recovery after severe wildfire in the Eastern Cascades, USA. *International Journal of Wildland Fire, 19*, 490–499. https://doi.org/10.1071/WF08194.

Doerr, S. H., Shakesby, R. A., Blake, W. H., Chafer, C. J., Humphreys, G. S., & Wallbrink, P. J. (2006). Effects of differing wildfire severities on soil wettability and implications for hydrological response. *Journal of Hydrology, 319*(1–4), 295–311.

Espinosa, J., Palheiro, P., Loureiro, C., Ascoli, D., Esposito, A., & Fernandes, P. M. (2019). Fire-severity mitigation by prescribed burning assessed from fire-treatment encounters in maritime pine stands. *Canadian Journal of Forest Research, 49*(2), 205–211.

Farnsworth, L. M., Nimmo, D. G., Kelly, L. T., Bennett, A. F., & Clarke, M. F. (2014). Does pyrodiversity beget alpha, beta or gamma diversity? A case study using reptiles from semi-arid Australia. *Diversity and Distributions, 20*(6), 663–673.

Fernandes, P., & Botelho, H. (2004). Analysis of the prescribed burning practice in the pine forest of northwestern Portugal. *Journal of Environmental Management, 70*(1), 15–26.

Fernandes, P. M., Vega, J. A., Jimenez, E., & Rigolot, E. (2008). Fire resistance of European pines. *Forest Ecology and Management, 256*(3), 246–255.

Fowler, J. F., & Sieg, C. H. (2004). *Postfire mortality of ponderosa pine and Douglas-fir: A review of methods to predict tree death* (Gen Tech Rep RMRS-GTR-132). Fort Collins: USDA Forest Service Rocky Mountain Research Station.

Frandsen, W. H., & Ryan, K. C. (1986). Soil moisture reduces belowground heat flux and soil temperatures under a burning fuel pile. *Canadian Journal of Forest Research, 16*, 244–248.

Freeman, J. E., & Kobziar, L. N. (2011). Tracking postfire successional trajectories in a plant community adapted to high-severity fire. *Ecological Applications, 21*(1), 61–74.

Gitas, I., Mitri, G., Veraverbeke, S., & Polychronaki, A. (2012). Advances in remote sensing of post-fire vegetation recovery monitoring—A review. In L. Fatoyinbo (Ed.), *Remote sensing of biomass-principles and applications, vol. 334* (pp. 143–176). Rijeka: InTech Europe. Retrieved December 16, 2019, from http://www.intechopen.com/books/remote-sensing-of-biomass-principles-and-applications/advances-inremote-sensing-of-post-fire-monitoring-a-review.

Grimm, N. B., & Fisher, S. G. (1986). Nitrogen limitation potential of Arizona streams and rivers. *Journal of the Arizona-Nevada Academy of Science, 1*, 31–44.

Gutsell, S. L., & Johnson, E. A. (1996). How fire scars are formed: Coupling a disturbance process to its ecological effect. *Canadian Journal of Forest Research, 26*(2), 166–174.

Hare, R. C. (1965). Contribution of bark to fire resistance of southern trees. *Journal of Forestry, 63*, 248–251.

Heward, H., Smith, A. M., Roy, D. P., Tinkham, W. T., Hoffman, C. M., Morgan, P., & Lannom, K. O. (2013). Is burn severity related to fire intensity? Observations from landscape scale remote sensing. *International Journal of Wildland Fire, 22*(7), 910–918.

Holden, Z. A., Morgan, P., & Evans, J. S. (2009). A predictive model of burn severity based on 20-year satellite-inferred burn severity data in a large southwestern US wilderness area. *Forest Ecology and Management, 258*(11), 2399–2406. https://doi.org/10.1016/j.foreco.2009.08.017.

Hood, S., & Lutes, D. (2017). Predicting post-fire tree mortality for 12 western US conifers using the First-Order Fire Effects Model (FOFEM). *Fire Ecology, 13*(2), 66–84.

Hood, S. M., Varner, J. M., van Mantgem, P., & Cansler, C. A. (2018). Fire and tree death: Understanding and improving modeling of fire-induced tree mortality. *Environmental Research Letters, 13*, 11304. https://doi.org/10.1088/1748-9326/aae934.

Hu, Y., Fernandez-Anez, N., Smith, T. E., & Rein, G. (2018). Review of emissions from smouldering peat fires and their contribution to regional haze episodes. *International Journal of Wildland Fire, 27*(5), 293–312.

Hudak, A. T., Morgan, P., Bobbitt, M. J., Smith, A. M., Lewis, S. A., Lentile, L. B., Robichaud, P. R., Clark, J. T., & McKinley, R. A. (2007). The relationship of multispectral satellite imagery to immediate fire effects. *Fire Ecology, 3*(1), 64–90. https://doi.org/10.4996/fireecology.0301064.

Hurteau, M. D., & Brooks, M. L. (2011). Short-and long-term effects of fire on carbon in US dry temperate forest systems. *Bioscience, 61*(2), 139–146.

Huston, M. A. (2014). Disturbance, productivity, and species diversity: Empiricism vs. logic in ecological theory. *Ecology, 95*(9), 2382–2396.

Hutto, R. L., Keane, R. E., Sherriff, R. L., Rota, C. T., Eby, L. A., & Saab, V. A. (2016). Toward a more ecologically informed view of severe forest fires. *Ecosphere, 7*(2), e01255.

Johnson, D. W., & Curtis, P. S. (2001). Effects of forest management on soil C and N storage: Meta analysis. *Forest Ecology and Management, 140*(2–3), 227–238.

Johnstone, J. F., & Kasischke, E. S. (2005). Stand-level effects of soil burn severity on postfire regeneration in a recently burned black spruce forest. *Canadian Journal of Forest Research, 35* (9), 2151–2163.

Kavanagh, K. L., Dickinson, M. B., & Bova, A. S. (2010). A way forward for fire-caused tree mortality prediction: Modeling a physiological consequence of fire. *Fire Ecology, 6*, 80–94.

Kaye, J. P., & Hart, S. C. (1997). Competition for nitrogen between plants and soil microorganisms. *Trends in Ecology & Evolution, 12*, 139–143. https://doi.org/10.1016/S0169-5347(97)01001-X.

Keane, R. E., McKenzie, D., Falk, D. A., Smithwick, E. A., Miller, C., & Kellogg, L. K. (2015). Representing climate, disturbance, and vegetation interactions in landscape models. *Ecological Modelling, 309*, 33–47.

Keeley, J. E. (2009). Fire intensity, fire severity and burn severity: A brief review and suggested usage. *International Journal of Wildland Fire, 18*, 116–126.

Keeley, J. E., & Fotheringham, C. J. (1998). Smoke-induced seed germination in California chaparral. *Ecology, 79*(7), 2320–2336.

Keeley, J. E., & Pausas, J. G. (2018). Evolution of 'smoke' induced seed germination in pyroendemic plants. *South African Journal of Botany, 115*, 251–255.

Keeley, J. E., Brennan, T., & Pfaff, A. H. (2008). Fire severity and ecosystem responses following crown fires in California shrublands. *Ecological Applications, 18*(6), 1530–1546. https://doi.org/10.1890/07-0836.1.

Key, C. H., & Benson, N. C. (2006). Landscape assessment: Sampling and analysis methods. In D. C. Lutes, R. E. Keane, J. F. Caratti, C. H. Key, N. C. Benson, S. Sutherland, & L. J. Gangi (Eds.), *FIREMON: Fire Effects Monitoring and Inventory System* (Gen Tech Rep RMRS-GTR-164-CD). Fort Collins: USDA Forest Service Rocky Mountain Research Station.

Keyser, A., & Westerling, A. L. (2017). Climate drives inter-annual variability in probability of high severity fire occurrence in the western United States. *Environmental Research Letters, 12* (6), 065003.

Kobziar, L. N., McBride, J. R., & Stephens, S. L. (2009). The efficacy of fire and fuels reduction treatments in a Sierra Nevada pine plantation. *International Journal of Wildland Fire, 18*(7), 791–801.

Kobziar, L. N., Pingree, M. R. A., Larson, H., Dreaden, T. J., Green, S., & Smith, J. A. (2018). Pyroaerobiology: The aerosolization and transport of viable microbial life by wildland fire. *Ecosphere, 9*(11), e02507. https://doi.org/10.1002/ecs2.2507.

Kolden, C. A., Smith, A. M. S., & Abatzoglou, J. T. (2015). Limitations and utilisation of monitoring trends in burn severity products for assessing wildfire severity in the USA. *International Journal of Wildland Fire, 24*(7), 1023–1028. https://doi.org/10.1071/WF15082.

Krawchuk, M. A., Haire, S. L., Coop, J., Parisien, M. A., Whitman, E., Chong, G., & Miller, C. (2016). Topographic and fire weather controls of fire refugia in forested ecosystems of northwestern North America. *Ecosphere, 7*(12), e01632.

Krawchuk, M. A., Meigs, G. W., Cartwright, J. M., Coop, J. D., Davis, R., Holz, A., Kolden, C., & Meddens, A. J. (2020). Disturbance refugia within mosaics of forest fire, drought, and insect outbreaks. *Frontiers in Ecology and the Environment, 18*(5), 235–244.

Kremens, R. L., Smith, A. M., & Dickinson, M. B. (2010). Fire metrology: Current and future directions in physics-based measurements. *Fire Ecology, 6*(1), 13–35.

Kruse, R., Bend, E., & Bierzychudek, P. (2004). Native plant regeneration and introduction of non-natives following post-fire rehabilitation with straw mulch and barley seeding. *Forest Ecology and Management, 196*(2–3), 299–310.

Lentile, L. B., Morgan, P., Hudak, A. T., Bobbitt, M. J., Lewis, S. A., Smith, A. M. S., & Robichaud, P. R. (2007). Burn severity and vegetation response following eight large wildfires across the western US. *Fire Ecology, 3*(1), 91–108.

Lentile, L. B., Smith, A. M. S., Hudak, A. T., Morgan, P., Bobbitt, M. J., Lewis, S. A., & Robichaud, P. R. (2009). Remote sensing for prediction of 1-year post-fire ecosystem condition. *International Journal of Wildland Fire, 18*(5), 594–608. https://doi.org/10.1071/WF07091.

Lewis, S. A., Hudak, A. T., Robichaud, P. R., Morgan, P., Satterberg, K. L., Strand, E. K., Smith, A. M., Zamudio, J. A., & Lentile, L. B. (2017). Indicators of burn severity at extended temporal scales: A decade of ecosystem response in mixed-conifer forests of western Montana. *International Journal of Wildland Fire, 26*(9), 755–771.

Luce, C., Morgan, P., Dwire, K., Isaak, D., Holden, Z., & Rieman, B. (2012). *Climate change, forests, fire, water, and fish: Building resilient landscapes, streams, and managers* (RMRS-GTR-290). Fort Collins: USDA Forest Service Rocky Mountain Research Station.

Massman, W. (2015). A non-equilibrium model for soil heating and moisture transport during extreme surface heating: the soil (heat-moisture-vapor) HMV-model version. *Geoscientific Model Development, 8*, 659–3680.

Massman, W. J., Frank, J. M., & Mooney, S. J. (2010). Advancing investigation and physical modeling of first-order fire effects on soils. *Fire Ecology, 6*(1), 36. https://doi.org/10.4996/fireecology.0601036.

McDowell, N., Pockman, W. T., Allen, C. D., Breshears, D. D., Cobb, N., Kolb, T., Plaut, J., Sperry, J., West, A., Williams, D. G., & Yepez, E. A. (2008). Mechanisms of plant survival and mortality during drought: Why do some plants survive while others succumb to drought? *The New Phytologist, 178*(4), 719–739.

McKinley, D. C., Ryan, M. G., Birdsey, R. A., Giardina, C. P., Harmon, M. E., Heath, L. S., Houghton, R. A., Jackson, R. B., Morrison, J. F., Murray, B. C., & Pataki, D. E. (2011). A synthesis of current knowledge on forests and carbon storage in the United States. *Ecological Applications, 21*(6), 1902–1924.

Meddens, A. J., Kolden, C. A., Lutz, J. A., Smith, A. M., Cansler, C. A., Abatzoglou, J. T., Meigs, G. W., Downing, W. M., & Krawchuk, M. A. (2018). Fire refugia: What are they, and why do they matter for global change? *Bioscience, 68*(12), 944–954.

Michaletz, S. T., Johnson, E. A., & Tyree, M. T. (2012). Moving beyond the cambium necrosis hypothesis of post-fire tree mortality: Cavitation and deformation of xylem in forest fires. *The New Phytologist, 194*(1), 254–263.

Miller, J. D., & Thode, A. E. (2007). Quantifying burn severity in a heterogeneous landscape with a relative version of the delta Normalized Burn Ratio (dNBR). *Remote Sensing of Environment, 109*(1), 66–80.

Moody, J. A., & Martin, D. A. (2009). Forest fire effects on geomorphic processes. In A. Cerda & P. R. Robichaud (Eds.), *Fire effects on soils and restoration strategies* (pp. 41–79). Enfield: Science Publishers.

Morgan, P., Keane, R. E., Dillon, G. K., Jain, T. B., Hudak, A. T., Karau, E. C., Sikkink, P. G., Holden, Z. A., & Strand, E. K. (2014). Challenges of assessing fire and burn severity using field measures, remote sensing and modeling. *International Journal of Wildland Fire, 23*(8), 1045–1060. https://doi.org/10.1071/WF13058.

Morgan, P., Moy, M., Droske, C. A., Lewis, S. A., Lentile, L. B., Robichaud, P. R., Hudak, A. T., & Williams, C. J. (2015). Vegetation response to burn severity, native grass seeding, and salvage logging. *Fire Ecology, 11*(2), 31–58.

Moritz, M. A., Parisien, M. A., Batllori, E., Krawchuk, M. A., Van Dorn, J., Ganz, D. J., & Hayhoe, K. (2012). Climate change and disruptions to global fire activity. *Ecosphere, 3*(6), 1–22.

Moritz, M. A., Topik, C., Allen, C. D., Hessburg, P. F., Morgan, P., Odion, D. C., Veblen, T. T., & McCullough, I. M. (2018). *A statement of common ground regarding the role of wildfire in forested landscapes of the western United States.* Fire Research Consensus Working Group Final Report. Retrieved January 21, 2019, from https://live-ncea-ucsb-edu-v01.pantheonsite.io/sites/default/files/2020-02/WildfireCommonGround.pdf.

Neary, D. G., Ryan, K. C., & DeBano, L.. F. (2005) *Wildland fire in ecosystems: Effects of fire on soils and water* (RMRS-GTR-42-vol. 4). Fort Collins: USDA Forest Service Rocky Mountain Research Station.

North, M. P., Stevens, J. T., Greene, D. F., Coppoletta, M., Knapp, E. E., Latimer, A. M., Restaino, C. M., Tompkins, R. E., Welch, K. R., York, R. A., & Young, D. J. (2019). Tamm review:

Reforestation for resilience in dry western US forests. *Forest Ecology and Management, 15* (432), 209–224.
O'Brien, J. J., Hiers, J. K., Mitchell, R. J., Varner, J. M., & Mordecai, K. (2010). Acute physiological stress and mortality following fire in a long-unburned longleaf pine ecosystem. *Fire Ecology, 6*(2), 1–12. https://doi.org/10.4996/fireecology.0602001.
O'Brien, J. J., Hiers, J. K., Varner, J. M., Hoffman, C. M., Dickinson, M. B., Michaletz, S. T., Loudermilk, E. L., & Butler, B. W. (2018). Advances in mechanistic approaches to quantifying biophysical fire effects. *Current Forestry Reports, 4*(4), 161–177.
Packham, D. R. (1969). Heat transfer above a small ground fire. *Australian Forest Research, 5*, 19–24.
Page-Dumroese, D., Jurgensen, M. F., & Harvey, A. E. (2003). Fire and fire-suppression impacts on forest-soil carbon [Chapter 13]. In J. M. Kimble, L. S. Heath, R. A. Birdsey, & R. Lal (Eds.), *The potential of U.S. forest soils to sequester carbon and mitigate the greenhouse effect* (pp. 201–210). Boca Raton: CRC Press.
Pannkuk, C. D., & Robichaud, P. R. (2003). Effectiveness of needle cast at reducing erosion after forest fires. *Water Resources Research, 39*(12), 1333. https://doi.org/10.1029/2003WR002318.
Parks, S. A., Miller, C., Holsinger, L. M., Baggett, L. S., & Bird, B. J. (2016). Wildland fire limits subsequent fire occurrence. *International Journal of Wildland Fire, 25*, 182–190. https://doi.org/10.1071/WF15107.
Parks, S. A., Holsinger, L. M., Panunto, M. H., Jolly, W. M., Dobrowski, S. Z., & Dillon, G. K. (2018). High-severity fire: Evaluating its key drivers and mapping its probability across western US forests. *Environmental Research Letters, 13*(4), 044037.
Parson, A, Robichaud, P. R., Lewis, S. A., Napper, C., & Clark, J. T. (2010). *Field guide for mapping post-fire soil burn severity* (RMRS-GTR-243). Fort Collins: USDA Forest Service Rocky Mountain Research Station.
Pausas, J. G., Bradstock, R. A., Keith, D. A., & Keeley, J. E. (2004). Plant functional traits in relation to fire in crown-fire ecosystems. *Ecology, 85*(4), 1085–1100.
Pausas, J. G., & Keeley, J. E. (2017). Epicormic resprouting in fire-prone ecosystems. *Trends in Plant Science, 22*(12), 1008–1015.
Pausas, J. G., & Keeley, J. E. (2019). Wildfires as an ecosystem service. *Frontiers in Ecology and the Environment, 17*(5), 289–295.
Pausas, J. G., Llovet, J., Rodrigo, A., & Vallejo, R. (2009). Are wildfires a disaster in the Mediterranean basin?—A review. *International Journal of Wildland Fire, 17*(6), 713–723.
Payette, S. (1992). Fire as a controlling process in the North American boreal forest. In H. H. Shugart, R. Leemans, & G. B. Bonan (Eds.), *A system analysis of the global boreal forest* (pp. 144–169). Cambridge: Cambridge University Press.
Pennacchio, M., Jefferson, L. V., Havens, K. (2007). Allelopathic effects of plant-derived aerosol smoke on seed germination of Arabidopsis thaliana (L.) Heynh. Hindawi Publishing Corporation. Research Letters in Ecology, Volume 2007: Article ID 65083. https://doi.org/10.1155/2007/65083.
Peppin, D., Fulé, P., Sieg, C. H., Beyers, J. L., & Hunter, M. E. (2010). Post-wildfire seeding in forests of the western United States: An evidence-based review. *Forest Ecology and Management, 260*(5), 573–586.
Perez-Harguindeguy, N., Diaz, S., Garnier, E., Lavorel, S., Poorter, H., Jaureguiberry, P., Bret-Harte, M. S., Cornwell, W. K., Craine, J. M., Gurvich, D. E., Urcelay, C., Veneklaas, E. J., Reich, P. B., Poorter, L., Wright, I. J., Ray, P., Enrico, L., Pausas, J. G., de Vos, A. C., Buchmann, N., Funes, G., Quétier, F., Hodgson, J. G., Thompson, K., Morgan, H. D., ter Steege, H., van der Heijden, M. G. A., Sack, L., Blonder, B., Poschlod, P., Vaieretti, M. V., Conti, G., Staver, A. C., Aquino, S., & Cornelissen, J. H. C. (2013). New handbook for standardised measurement of plant functional traits worldwide. *Australian Journal of Botany, 61*, 167–234. https://doi.org/10.1071/BT12225.
Perez-Harguindeguy, N., Diaz, S., Garnier, E., Lavorel, S., Poorter, H., Jaureguiberry, P., Bret-Harte, M. S., Cornwell, W. K., Craine, J. M., Gurvich, D. E., & Urcelay, C. (2016).

Corrigendum to: New handbook for standardised measurement of plant functional traits worldwide. *Australian Journal of Botany, 64*(8), 715–716. https://doi.org/10.1071/BT12225.

Peterson, D. L., & Arbaugh, M. J. (1986). Postfire survival in Douglas-fir and lodgepole pine: Comparing the effects of crown and bole damage. *Canadian Journal of Forest Research, 16*(6), 1175–1179.

Peterson, D. L., & Ryan, K. C. (1986). Modeling postfire conifer mortality for long-range planning. *Environmental Management, 10*(6), 797–808.

Peterson, D. L., Agee, J. K., Aplet, G. H., Dykstra, D. P., Graham, R. T., Lehmkuhl, J. F., Pilliod, D. S., Potts, D. F., Powers, R. F., & Stuart, J. D. (2009). *Effects of timber harvest following wildfire in western North America* (Gen Tech Rep PNW-GTR-776). Portland: USDA Forest Service Pacific Northwest Res Station.

Picotte, J. J., Peterson, B., Meier, G., & Howard, S. M. (2016). 1984–2010 trends in fire burn severity and area for the conterminous US. *International Journal of Wildland Fire, 25*(4), 413–420. https://doi.org/10.1071/WF15039.

Pingree, M. R. A., & Kobziar, L. N. (2019). The myth of the biological threshold: A review of biological responses to soil heating associated with wildland fire. *Forest Ecology and Management, 432*, 1022–1029. https://doi.org/10.1016/j.foreco.2018.10.032.

Ponisio, L. C., Wilkin, K., M'Gonigle, L. K., Kulhanek, K., Cook, L., Thorp, R., Griswold, T., & Kremen, C. (2016). Pyrodiversity begets plant–pollinator community diversity. *Global Change Biology, 22*(5), 1794–1808.

Raison, R. J., Woods, P. V., Jakobsen, B. F., & Bary, G. A. V. (1986). Soil temperatures during and following low intensity prescribed burning in a Eucalyptus pauciflora forest. *Australian Journal of Soil Research, 24*, 33–47.

Raison, R. J., Khanna, P. K., Jacobsen, K. L. S., Romanyu, J., & Serrasolses, I. (2009). Effects of fire on forest nutrient cycles. In A. Cerda & P. R. Robichaud (Eds.), *Fire effects on soils and restoration strategies* (pp. 225–256). Enfield: Science Publishers.

Rein, G., Cleaver, N., Ashton, C., Pironi, P., & Torero, J. L. (2008). The severity of smouldering peat fires and damage to the forest soil. *Catena, 74*(3), 304–309.

Reisser, M., Purves, R. S., Schmidt, M. W., & Abiven, S. (2016). Pyrogenic carbon in soils: A literature-based inventory and a global estimation of its content in soil organic carbon and stocks. *Frontiers in Earth Science, 4*, 80.

Reside, A. E., Welbergen, J. A., Phillips, B. L., Wardell-Johnson, G. W., Keppel, G., Ferrier, S., Williams, S. E., & VanDerWal, J. (2014). Characteristics of climate change refugia for Australian biodiversity. *Austral Ecology, 39*(8), 887–897.

Reyes, O., Casal, M., & Rego, F. C. (2009). Resprouting ability of six atlantic shrub species. *Folia Geobotanica, 44*, 19–29.

Ritter, S. M., Hoffman, C. M., Battaglia, M. A., Stevens-Rumann, C. S., & Mell, W. E. (2020). Fine-scale fire patterns mediate forest structure in frequent-fire ecosystems. *Ecosphere, 11*(7), e03177.

Robichaud, P. R., Lewis, S. A., Wagenbrenner, J. W., Ashmun, L. E., & Brown, R. E. (2013a). Post-fire mulching for runoff and erosion mitigation: Part I: Effectiveness at reducing hillslope erosion rates. *Catena, 105*, 75–92.

Robichaud, P. R., Wagenbrenner, J. W., Lewis, S. A., Ashmun, L. E., Brown, R. E., & Wohlgemuth, P. M. (2013b). Post-fire mulching for runoff and erosion mitigation; Part II: Effectiveness in reducing runoff and sediment yields from small catchments. *Catena, 105*, 93–111.

Robichaud, P. R., Rhee, H., & Lewis, S. A. (2014). A synthesis of post-fire burned area reports from 1972 to 2009 for western US Forest Service lands: Trends in wildfire characteristics and post-fire stabilization treatments and expenditures. *International Journal of Wildland Fire, 23*(7), 929–944.

Ryan, K. C. (2002). Dynamic interactions between forest structure and fire behavior in boreal ecosystems. *Silva Fennica, 36*(1), 13–39.

Ryan, K. C., & Noste, N. V. (1985). Evaluating prescribed fires. In J. E. Lotan, B. M. Kilgore, W. C. Fischer, & R. W. Mutch (Eds.), *Proceedings, Symposium and Workshop on Wilderness Fire* (Gen Tech Rep INT-182, pp. 230–238). Missoula: USDA For Serv Intermountain Forest and Range Experiment Station.

Ryan, K. C., & Reinhardt, E. D. (1988). Predicting postfire mortality of seven western conifers. *Canadian Journal of Forest Research, 18*(10), 1291–1297.

Santín, C., Doerr, S. H., Preston, C. M., & González-Rodríguez, G. (2015). Pyrogenic organic matter production from wildfires: A missing sink in the global carbon cycle. *Global Change Biology, 21*, 1621–1633. https://doi.org/10.1111/gcb.12800.

Scott, D. F., Curran, M. P., Robichaud, P. R., & Wagenbrenner, J. W. (2009). Soil erosion after forest fire. In A. Cerda & P. R. Robichaud (Eds.), *Fire effects on soils and restoration strategies* (pp. 177–195). Enfield: Science Publishers.

Sevanto, S., Mcdowell, N. G., Dickman, L. T., Pangle, R., & Pockman, W. T. (2014). How do trees die? A test of the hydraulic failure and carbon starvation hypotheses. *Plant, Cell & Environment, 37*(1), 153–161.

Sikkink, P. G., Dillon, G. K., & Keane, R. E. (2013). *Composite Burn Index (CBI) data and field photos collected for the FIRESEV project, western United States*. Fort Collins: USDA Forest Service Research Data Archive. https://doi.org/10.2737/RDS-2013-0017.

Smith, J. K. (2010). Fire effects information system: New engine, remodeled interior, added options. *Fire Management Today, 70*(1), 44–47.

Smith, A. M., Sparks, A. M., Kolden, C. A., Abatzoglou, J. T., Talhelm, A. F., Johnson, D. M., Boschetti, L., Lutz, J. A., Apostol, K. G., Yedinak, K. M., & Tinkham, W. T. (2016). Towards a new paradigm in fire severity research using dose–response experiments. *International Journal of Wildland Fire, 25*(2), 158–166.

Sperry, J. S., Alder, N. N., & Eastlack, S. E. (1993). The effect of reduced hydraulic conductance on stomatal conductance and xylem cavitation. *Journal of Experimental Botany, 44*(6), 1075–1082.

Stenzel, J. E., Bartowitz, K. J., Hartman, M. D., Lutz, J. A., Kolden, C. A., Smith, A. M., Law, B. E., Swanson, M. E., Larson, A. J., Parton, W. J., & Hudiburg, T. W. (2019). Fixing a snag in carbon emissions estimates from wildfires. *Global Change Biology, 25*(11), 3985–3994.

Stephens, S. L., Collins, B. M., Fettig, C. J., Finney, M. A., Hoffman, C. M., Knapp, E. E., North, M. P., Safford, H., & Wayman, R. B. (2018). Drought, tree mortality, and wildfire in forests adapted to frequent fire. *Bioscience, 68*(2), 77–88.

Stevens-Rumann, C., & Morgan, P. (2019). Tree regeneration following wildfires in the western US: A review. *Fire Ecology, 15*, 15. https://doi.org/10.1186/s42408-019-0032-1.

Stevens-Rumann, C. S., Kemp, K. B., Higuera, P. E., Harvey, B. J., Rother, M. T., Donato, D. C., Morgan, P., & Veblen, T. T. (2018). Evidence for declining forest resilience to wildfires under climate change. *Ecology Letters, 21*, 243–252.

Steward, F. R., Peter, S., & Richon, J. B. (1990). A method for predicting the depth of lethal heat penetration into mineral soils exposed to fires of various intensities. *Canadian Journal of Forest Research, 20*(7), 919–926.

Strand, E. K., Satterberg, K. L., Hudak, A. T., Byrne, J., Khalyani, A. H., & Smith, A. M. S. (2019). Does burn severity affect plant community diversity and composition in mixed conifer forests of the United States Intermountain West one decade post fire? *Fire Ecology, 15*(1), 25.

Taylor, R. S., Watson, S. J., Nimmo, D. G., Kelly, L. T., Bennett, A. F., & Clarke, M. F. (2012). Landscape-scale effects of fire on bird assemblages: Does pyrodiversity beget biodiversity? *Diversity and Distributions, 18*(5), 519–529.

Thom, M. D., Daniels, J. C., Kobziar, L. N., & Colburn, J. R. (2015). Can butterflies evade fire? Pupa location and heat tolerance in fire prone habitats of Florida. *PLoS One, 10*(5), e0126755. https://doi.org/10.1371/journal.pone.012675.

Thompson, J. R., Spies, T. A., & Ganio, L. M. (2007). Reburn severity in managed and unmanaged vegetation in a large wildfire. *Proceedings of the National Academy of Sciences, 104*(25), 10743–10748.

Thompson, M. T., Koyama, A., & Kavanagh, K. L. (2017). Wildfire effects on physiological properties in conifers of central Idaho forests, USA. *Trees, 31*(2), 545–555.

Tingley, M. W., Ruiz-Gutiérrez, V., Wilkerson, R. L., Howell, C. A., & Siegel, R. B. (2016). Pyrodiversity promotes avian diversity over the decade following forest fire. *Proceedings of the Royal Society B: Biological Sciences, 283*(1840), 20161703.

Tinkham, W. T., Smith, A. M., Higuera, P. E., Hatten, J. A., Brewer, N. W., & Doerr, S. H. (2016). Replacing time with space: Using laboratory fires to explore the effects of repeated burning on black carbon degradation. *International Journal of Wildland Fire, 25*(2), 242–248.

Tyree, M. T., & Zimmermann, M. H. (2002). Xylem dysfunction: When cohesion breaks down. In H. H. Shugart, R. Leemans, & G. B. Bonan (Eds.), *Xylem structure and the ascent of sap* (pp. 89–141). Berlin: Springer.

Ubeda, X., & Outeiro, L. R. (2009). Physical and chemical effects of fire on soil. In A. Cerda & P. R. Robichaud (Eds.), *Fire effects on soils and restoration strategies* (pp. 105–132). Enfield: Science Publishers.

Van Lear, D. H., Carroll, W. D., Kapeluck, P. R., & Johnson, R. (2005). History and restoration of the longleaf pine-grassland ecosystem: Implications for species at risk. *Forest Ecology and Management, 211*(1–2), 150–165.

Varner, I. I. I. J. M., Gordon, D. R., Putz, F. E., & Hiers, J. K. (2005). Restoring fire to long-unburned *Pinus palustris* ecosystems: Novel fire effects and consequences for long-unburned ecosystems. *Restoration Ecology, 13*(3), 536–544.

Varner, J. M., Hiers, J. K., Ottmar, R. D., Gordon, D. R., Putz, F. E., & Wade, D. D. (2007). Overstory tree mortality resulting from reintroducing fire to long-unburned longleaf pine forests: The importance of duff moisture. *Canadian Journal of Forest Research, 37*(8), 1349–1358.

Vitousek, P. M., & Howarth, R. W. (1991). Nitrogen limitation on land and in the sea: How can it occur? *Biogeochemistry, 13*(2), 87–115.

Wagenbrenner, J., MacDonald, L., & Rough, D. (2006). Effectiveness of three post-fire rehabilitation treatments in the Colorado Front Range. *Hydrological Processes, 20*(14), 2989–3006.

Wagenius, S., Beck, J., & Kiefer, G. (2019). Fire synchronizes flowering and boosts reproduction in a widespread but declining prairie species. *Proceedings of the National Academy of Sciences, 117*(6), 3000–3005. https://doi.org/10.1073/pnas.1907320117.

Walker, R. B., Coop, J. D., Parks, S. A., & Trader, L. (2018). Fire regimes approaching historic norms reduce wildfire-facilitated conversion from forest to non-forest. *Ecosphere, 9*(4), e02182.

Wan, S., Hui, D., & Luo, Y. (2001). Fire effects on nitrogen pools and dynamics in terrestrial ecosystems: A meta-analysis. *Ecological Applications, 11*(5), 1349–1365.

Williams, C., Pierson, F., Robichaud, P., & Boll, J. (2014). Hydrologic and erosion responses to wildfire along the rangeland–xeric forest continuum in the western US: A review and model of hydrologic vulnerability. *International Journal of Wildland Fire, 23*(2), 155–172.

Wright, H. A., & Bailey, A. W. (1982). *Fire ecology United States and Southern Canada*. New York: Wiley.

Zimmerman, G. T., & Laven, R. D. (1987). Effects of forest fuel smoke on dwarf mistletoe seed germination. *Great Basin National Park, 47*(4), 652–659.

Chapter 10
Fire and People

Learning Outcomes
After reading this chapter, you should be able to

1. Build a table of costs and benefits of fires and then describe in your own words how you think those can be balanced to inform people making decisions about fires,
2. Articulate how smoke can compromise human health and name several strategies for reducing the vulnerability of people to smoke from fires,
3. Develop three short statements you can use to inform people about fire, and for one of them how you will adapt them to communicate with people from different perspectives within wildland-urban interface communities, and
4. Explain, based on the fire science you learned in this and previous chapters, one strategy for protecting fire fighters and other people and their homes and communities from fires long-term.

10.1 Introduction

Humans have long used, valued, and feared fire. Fires have been part of the Earth's system for millennia (Fig. 2 in Introduction to this textbook), and people have long influenced how fires burn. People often aggressively suppress fires, usually out of fear of how fire will affect them or those people and resources they care deeply about. People also ignite fires, sometimes accidentally and sometimes on purpose,

Supplementary Information The online version of this chapter (https://doi.org/10.1007/978-3-030-69815-7_10) contains supplementary material, which is available to authorized users.

F. Castro Rego et al., *Fire Science*, Springer Textbooks in Earth Sciences, Geography and Environment, https://doi.org/10.1007/978-3-030-69815-7_10

and people make tremendous efforts to control fire and fire effects. In all human cultures, fire is a symbol of power, warmth, and renewal.

"Fire is a bad master but a good servant". This old saying, and similar sentiments deeply rooted in many different languages, reflect the complex realities of fire. Fires can threaten people and property, yet people can use their understanding of fire to mitigate such threats. People are often affected by fires and smoke, yet people also value the ecosystem services that fires often maintain and sometimes enhance. For instance, fires consume fuels that otherwise accumulate and can lead to future high-intensity fires. Many plants and animals survive and thrive after fires. However, fires can also have negative impacts on the things people value.

Living with fire depends on taking action based upon a sound understanding of fires behavior and effects. Some people suggest that we learn from Indigenous cultures about fires (See Sect. 10.5 and Case Study 13.7). As climate changes, both traditional knowledge and scientific knowledge about fire and ecosystems can help people coexist well with fire (Moritz et al. 2014; Schoennagel et al. 2017).

Fires are what people make of them, and fires will reflect how people perceive them. Thus, although fires are biophysical processes, they are influenced by the social, political, economic, and cultural context in which they occur. The success of fire suppression policies and tactics to protect people and homes can paradoxically lead to the accumulation of fuels that can increase the intensity of subsequent fires. Thus, the fear of negative consequences of fires has led people to make policies and take actions that can lead to fires that adversely affect people. Integrated fire management (see Chap. 13) seeks to increase positive and decrease negative impacts of fires. This often involves balancing the need to protect people and property from fires with the ecological imperative of fires burning.

How do people value the costs and benefits of fire, including ecosystem services? How are people affected by and what will protect fire fighters and other people from heat during fires? How can people reduce the likelihood that their homes will burn in wildfires? The smoke that commonly spreads far from the flames can endanger human health, so how can we manage smoke while using fire proactively and effectively? How might individual people and the communities we live in become fire-adapted so that we can live well with fire? How can we reduce vulnerability and increase the resilience of social-ecological systems to fires? How might we learn together, and how can traditional ecological knowledge complement science to help people? We address these questions in this chapter. We don't review all of the ways fires and smoke affect people. Instead, we highlight important concepts and how they are linked to fire behavior (Parts I and II, including Chaps. 1–5, 7, and 8), and ultimately to fire effects (Chap. 9), fuels management (Chap. 11), changing fire regimes (Chap. 12) and integrated fire management (Chap. 13).

10.2 Different Perspectives About Fire

The relationship between fire and people can be complicated because there are different perspectives. Much active research in social science and environmental economics is devoted to this topic. Here we present different perspectives, starting from those emphasizing the adverse effects of fire, focusing on wildfire damages and other changes due to fire, to those including the beneficial effects of fire under a more comprehensive perspective.

10.2.1 Fire as a Disaster and Change Agent: Vulnerability, and Resilience

This perspective of viewing wildfire only as a hazard is common to the approaches used for other natural hazards, like floods or earthquakes. Many of the concepts and terminology agreed upon internationally provide a common understanding for use by the public, authorities, and practitioners (UNISDR 2009) and allow for the development of indicators to measure global progress in implementing the Sendai Framework for Disaster Risk Reduction 2015–2030. The perspective of fire as a disaster focuses on the negative effects of wildfires. Extreme wildfires represent disasters when they lead to human, material, economic, and environmental impacts. However, this perspective typically does not take into account the possible benefits that fires can also provide. Alternatively, we can view fires as agents of change that can have both positive and negative effects.

Fires change system values, as fires affect people, infrastructures, and ecosystems. The degree to which the system is affected depends upon its exposure and vulnerability. The fraction not affected is the resistance of the system to the event. The accumulated change is a function of the value of the system, its exposure and vulnerability to the event, and the recovery time (Fig. 10.1). Resilience is the ability of a system, community, or society exposed to hazards to resist, absorb, accommodate, adapt to, transform and recover from the effects of a hazardous event in a timely and efficient manner (UNISDR 2009).

The vulnerability of people to fires varies widely. Vulnerability to wildfire and peoples' ability to adapt to fire often varies with race, ethnicity, and economic capacity (Davies et al. 2018). Wildfire vulnerability increases when and where large fires result in large areas burned with high severity. Both social factors and fire behavior and size influence the vulnerability of both individuals and the communities they live in. Older adults are especially vulnerable to both fire and smoke, as is anyone who is not very mobile and has limited financial and social resources.

Resilience depends on the adaptive capacity of people to prepare for, live through, and recover from wildfire (Holling 1985). Resilience will be different for different people and places. Globally, 55% of the world's people live in urban areas, as many people have left many rural areas (UN 2018). North America is mostly urban (82% of all people live in cities), while Africa is mostly rural (43% of people live in urban areas). Abandoning marginally productive agricultural lands increases fuels and fire hazard in many places. Many poor people are so vulnerable that wildfire events can be both devastating and difficult to recover from. Resilience depends on socioeconomic resources, including insurance. Families who rent rather than own their homes may not be eligible for the federal and state funding designed

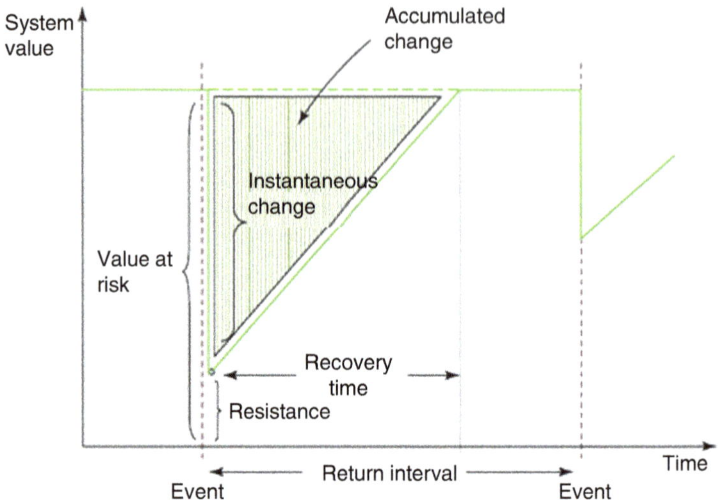

Fig. 10.1 Fires result in changes in system values. The accumulated change (shaded area) depends on the instantaneous change and the recovery time. The recovery rate (instantaneous change/ recovery time) is a fundamental indicator of the capacity of the system's resilience. The fraction changed is an indicator of its vulnerability. Originally developed for fire as a damaging agent, we adapt this perspective to recognize that fires can have both positive and negative effects. Then, vulnerability and resilience are evaluated relative to change due to fires, whether those are positive, negative, or both. (Adapted from Rego and Colaço 2013)

to help people recover from fires (Davies et al. 2018). Sadly, many Native Americans, especially those living on reservations, are vulnerable to fire. Early settlers of central North America learned about fires from Native Americans, and many tribes now are innovative in their use of fires. White people of higher incomes are more likely to live in communities with adaptive capacity for fires. Some rural areas, often described as "amenity communities", are growing fast because they are attractive for recreation and second and third homes.

Recognizing, adapting, and mitigating risks are critical for increasing the resilience of social-ecological systems to fires (Smith et al. 2016). For communities to become more resilient, Schoennagel et al. (2017) emphasized the concept of "adaptive resilience" based on recognizing both the potential and the limitations of fuels management, acknowledging the vital role of wildfire in maintaining many ecosystems and ecosystem services, and embracing new strategies for living with fire. Understanding fire and smoke can help communities develop local strategies to become fire adapted. Outreach advisors working with communities long before and long after fires can aid preparations and recovery and share the messages that fires have benefits as well as costs. Communicating in ways that are meaningful depends on recognizing and appreciating who is listening and when. People vary in their attitudes about fire and protection strategies. Engaging people effectively depends on listening well, understanding, and messaging.

10.2.2 *The Economic Perspective: Costs of Pre-suppression, Suppression, and Net Value Changes*

A second perspective about fires comes from the economic models used to evaluate wildfire management programs. Sparhawk (1925) focused on minimizing costs and losses due to fire. Optimal program levels were based on the trade-offs between pre-suppression costs (fire management costs before fires), suppression costs (during fires), and losses due to fires. According to this model, the optimum pre-suppression budget is the value that minimizes total cost plus losses due to wildfires (Fig. 10.2).

This Least Cost plus Loss model has many drawbacks that limit its practical use. In particular, it is challenging to estimate wildfire damage, or even area burned, only as a function of the pre-suppression budget. The assumptions of the models are difficult to verify and many other factors are involved in the outcome. In the USA, area burned has increased in recent decades with increases paradoxically paralleling investments in fire suppression (Fernandes et al. 2020, Fig. 10.3). Further, while these analyses may indicate how much pre-suppression resources are optimal, the approach does not guide allocating to the many pre-suppression activities that can take place. Although this model has evolved through time (e.g., Gorte and Gorte 1979), in its initial formulation the possible benefits from fires were not considered.

The recognition that some effects of wildfire can be beneficial (e.g., fuel consumption and ecological benefits) led to the development of more comprehensive economic models under the concept of Cost plus Net Value Change (C + NVC) model (Donovan and Rideout 2003). The Net Value Change (NVC) is the difference between losses and benefits to the resource resulting from the fires. The pre-suppression and suppression costs are considered as independent inputs, whereas only pre-suppression was independent in the previous model. The economic analysis of the efficiency of fire management programs is now generally evaluated by this C + NVC model (e.g., Thomas et al. 2017) with resources allocated accordingly (Fig. 10.4). Still, determining the optimal mix of fire-fighting resources for a given fire management program is a necessary condition for identifying the

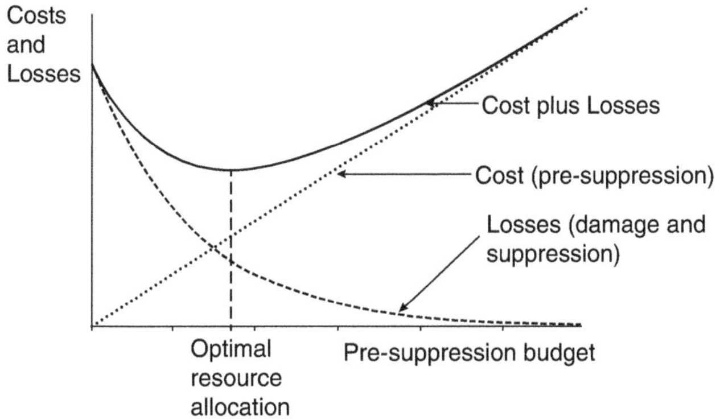

Fig. 10.2 Least Cost plus Loss model for fire management (Sparhawk 1925). The optimal resource allocation for the pre-suppression budget minimizes total Cost plus Losses

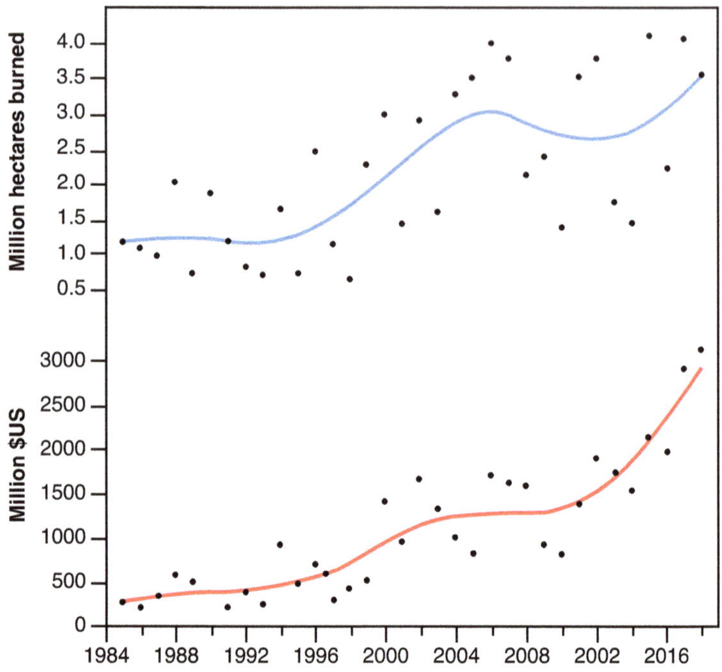

Fig. 10.3 Observed (*dots*) and smoothed (*lines*) area burned in the USA and costs of fire suppression (1985–2018, adjusted for inflation) based on data from the National Interagency Fire Center. (From Fernandes et al. 2020)

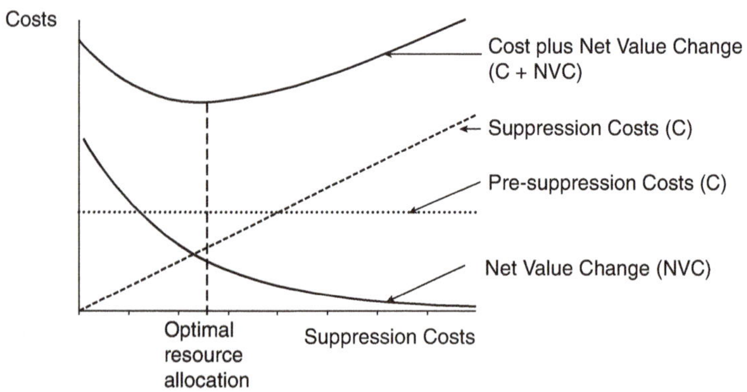

Fig. 10.4 The Cost plus Net Value Change (C + NVC) model for fire management with the indication of the optimal resource allocation for the Suppression Costs, considering Pre-suppression Costs separate from Suppression Costs. The total Cost plus Net Value Change is obtained by adding Pre-suppression and Suppression Costs to the Net Value Change resulting from the wildfires. (Redrawn from Donovan and Rideout 2003)

minimum of the Cost Plus Net Value Change (C + NVC) function (González-Cabán et al. 1986; Mavsar et al. 2010).

The C + NVC model has been widely applied in strategic budgeting in fire management and has integrated benefits from fire and ecological restoration (Rideout et al. 2014, 2017). However, this is challenging as fire management includes multiple objectives. The costs associated with suppression are often easier to estimate than the other costs that may be 2–30 times than the fire suppression (AFE and IAWF 2015). The costs associated with fire management include a diverse array of activities from prevention (including personnel, education, training, detection, enforcement, and equipment), to mitigation (including personnel, fuels management, insurance or disaster assistance), and suppression (including personnel, equipment, training), and post-fire management, as well as legal issues and regulations. This complexity increases when considering the direct and indirect costs of fire management and fire effects in the wildland-urban interface (WUI) fires (Thomas et al. 2017). In WUI fires, because of the people and property values at risk, the tactics and the costs of fire suppression are very different from those in fighting remote fires; WUI fires account for as much as 95% of suppression costs (Schoennagel et al. 2017) and risk to fire personnel.

Fires have sometimes burned electrical power lines and other infrastructure. Sparks from electrical power lines have ignited fires during windy, dry conditions, and the companies distributing electricity have been sued for related fire damages. As a result, Pacific Gas and Electric stopped providing power during high fire danger in California during the summer of 2019 (Abatzoglou et al. 2020). Avista Utilities (2020), a major utility company in the northwestern US, has a comprehensive fire management plan designed to reduce risks to the public, workers, and infrastructure while also limiting the impact of electric system outages due to fires. In addition to hardening the powerline grid by replacing infrastructure such as wooden poles with metal poles in fire-prone areas, the plan calls for managing vegetation to reduce the potential for trees to fall into power lines, improving situational awareness to aid managers, installing automated systems to alter powerline systems in response to fire, and improving operations and emergency personnel. The company works closely with local communities and fire management personnel.

Adverse health impacts of smoke from fires represent a cost to society, but the multiple costs can be difficult to quantify (Kochi et al. 2010; Moeltner et al. 2013). Visits to hospital emergency rooms for respiratory or cardiac complaints due to smoke increased over 3 years in Nevada, USA (Moeltner et al. 2013). More people were exposed to more smoke when fires consumed more fuel close to urban areas in Indonesia, Florida, and elsewhere. Better data on the area burned, the amount of fuels burned each day, and daily medical cost records are needed to inform alternative fire management strategies (Moeltner et al. 2013). Smoke impacts on urban areas are often part of fire suppression decisions. See Sect. 10.4 for information on smoke effects on human health.

Because it can be challenging to value the ecosystem impacts and benefits fiscally, Net Value Change (NVC) is even more difficult to estimate than costs. Quantifying NVC requires information about the direct and indirect effects of fire on the spatial and temporal provision of goods and services, and information about how fire-induced marginal changes in the quality and quantity of goods and services will affect social welfare (Venn and Calkin 2007; Mavsar et al. 2010).

Alternative systems for valuing intangible resources are needed. Rideout et al. (2012) elicited relative values of various natural and cultural resources, from wildlife habitats to archaeological sites. They worked with managers to estimate the relative degree to which the resources would be enhanced or harmed by wildfire in four national parks in the USA.

10.2.3 The Environmental Perspective: Focusing on Ecosystem Services

Ecosystem Services are "the benefits people obtain from ecosystems" (Millennium Ecosystem Assessment 2005). People value many ecosystem services for contributing to health and well-being. Some of these services are provided at the landscape scale. The concept of Landscape Services has also been proposed as a unifying common ground where scientists from various disciplines are encouraged to cooperate in producing a common knowledge base that can be integrated into multifunctional, actor-led landscape development (Termorshuizen and Opdam 2009). In Chap. 9, we discussed both the positive and negative effects of fire on ecosystems. Fire management at the landscape scale will be discussed and exemplified in Chaps. 11–13. Here, the term Ecosystem Services will be used in a broad sense encompassing various scales.

Ecosystem services include (a) provisioning, as ecosystems provide both nutritional and non-nutritional materials, water, and energy; (b) regulation and maintenance in ways that affect human health, safety, and comfort; and (c) cultural values, including how people feel about and see places (Haines-Young and Potschin-Young 2018). People have taken advantage of burned areas and used fire to make openings for grazing, agriculture, and hunting, consume fuels and stimulate the production of desirable biomass, including forage, seeds or fruits, and provide edible, medicinal, or other culturally important plants (Huffman 2013). Fires can consume fuels that would otherwise accumulate to fuel future fires, though fires may also stimulate grass and other surface fuel to grow (See Chap. 11). Fire can be used to regulate carbon (see Case Study 13.1), to create and maintain habitat for plants and animals or to enhance biodiversity, vegetation composition, and to influence pest populations (Pausas and Keeley 2019). Certainly, subsistence hunting and agriculture, and some recreational hunting can be enhanced in burned areas (Huffman 2013; Pausas and Keeley 2019).

Having enough water of sufficient quality is a growing global problem exacerbated by fires (Doerr and Santín 2016). Water quality and quantity are essential ecosystem services as most people depend on streams and other surface water for drinking for people and animals and often for agriculture. Martin (2016) declared fires to be a severe threat to water supply globally, as fire-prone ecosystems, including forests, shrublands, grasslands, and peatlands, provide about 60% of the water for the 100 largest cities in the world. Vegetation fires burn about 4% of the burnable land globally each year. Years of widespread fires are dry years. In droughts, the competing demands for water use for agriculture, industry, drinking water, habitat for fish and other aquatic organisms, and other services often exceed available water. Surface water supplies can be vulnerable to fires if the amount of sediment, debris flows, and wood increases after the lands adjacent to streams, lakes, and reservoirs burn, especially when high-intensity rain falls before vegetation recovers from fires. Areas that burned severely may develop hydrophobic layers in the soil that limit infiltration (See Sect. 9.5). How fires affect vegetation and soils and how quickly vegetation recovers can influence whether surface runoff increases post-fire. Nunes et al. (2018) provide a useful framework for assessing and managing the potential for fires to influence water (Fig. 10.5).

Because of this strong connection between fire and ecosystem services, Pausas and Keeley (2019) consider fire an ecosystem service, summarizing both the evolutionary and socioecological benefits generated by fires. However, while many ecosystem services increase in the short or long-term after fires, many others decrease because of fires. It is, therefore, more appropriate to see fires as a part of a complex network of ecosystem processes whose interactions may translate into services or disservices for society (Sil et al. 2019).

10.2.4 An Integrated Fire Risk Framework

A more comprehensive fire risk framework is needed, one that integrates the definitions from other hazards and the aspects specific to fires. Miller and Ager (2013) proposed a generalized framework for fire risk based upon their review of advances in risk analysis for wildland fire management (Fig. 10.6). They defined risk as the expected loss or gain, similarly to the Net Value Change concept. Risk results from the combination of the likelihood, or probability, of the fires occurring, as well as the intensity of the fire, and the resulting fire effects, that are valued positively or negatively according to the value system (ecological, social, economic) used. This framework acknowledges that both the likelihood of the fire and its intensity are related to fire behavior, ignition, fuels, and weather (Fig. 10.6).

One of the most complex issues in fire economics results from the fact that fires are different in their behavior and, therefore, in their effects. This issue is solved, from a quantitative point of view, by Finney (2005) in his formula to integrate likelihood, intensity, and effects in the calculation of risk as the expected Net Value Change to resource j:

$$E(NVC_i) = \sum_i^n \rho_i \, x \, RF_{ij} \quad (10.1)$$

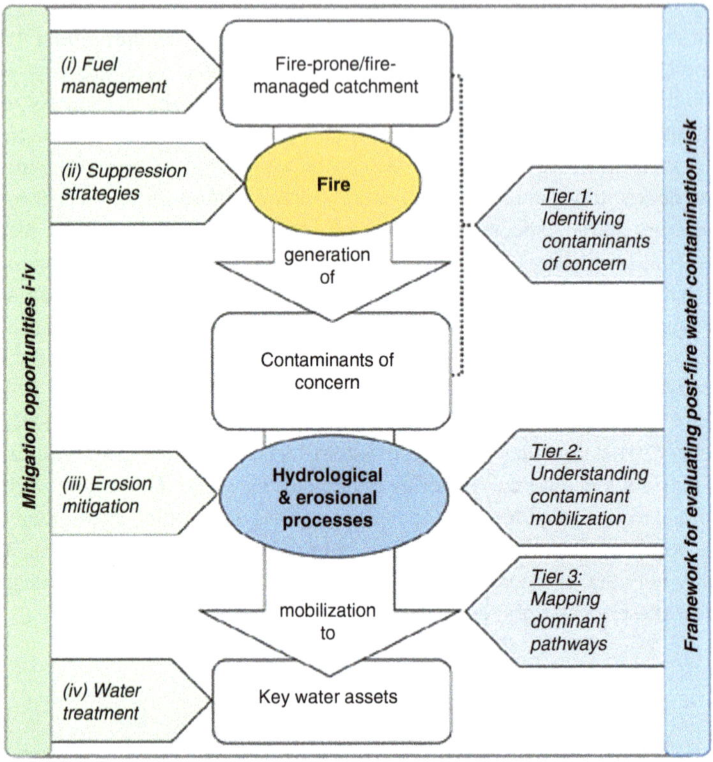

Fig. 10.5 Mitigating the risk of high severity fire is a high priority in key watersheds, such as those that supply drinking water. In some, contamination by sediment and chemicals may also be mobilized during and after fires. This framework is valuable, but it does not reflect the effects of burn severity, size of burned patches and proximity to streams, nor time since fire and degree of vegetation recovery, all of which affect fire effects on water and watersheds. (From Nunes et al. 2018)

where p_i is the probability of fires of intensity i and RF_{ij} is the response function of resource j as a function of a fire at intensity i.

The difficulties in calculating NVC values are the same as before, and the temporal dynamics of risk are not fully integrated. However, in this formulation, fires of different intensity, or severity, will have effects associated with resource values in resource functions. It also allows the integration of different resources in the same analysis.

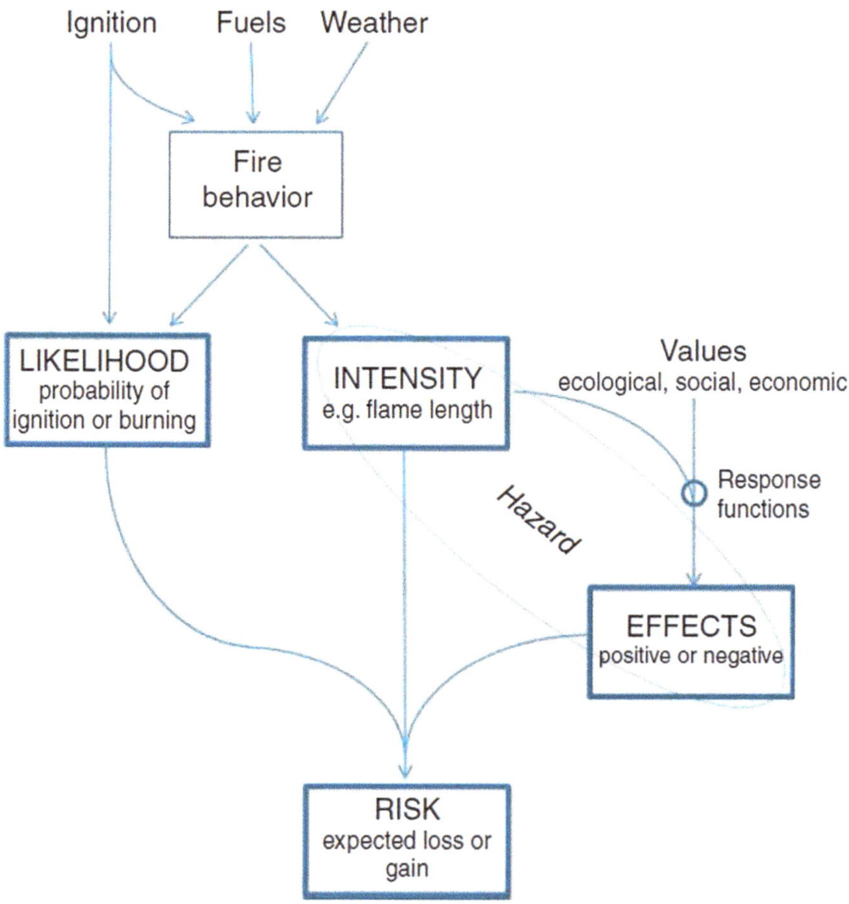

Fig. 10.6 Fire risk can be evaluated as expected loss or gain based upon how likely fires are to occur, their intensity, and both positive and negative effects of fire. (From Miller and Ager 2013)

10.3 Protecting People from Fires

The exposure to heat, embers, and smoke generated during a fire can cause various impacts on human health, property, and infrastructure. People's vulnerability from wildfires depends on several factors, including fire behavior and effects, and people's mobility and health. The safety of people traveling in the area often depends on timely (think early!) advice on escape routes. For residents, shelter in a safe building is usually preferable to trying a last-minute escape. Threats to houses and other infrastructures are often associated with exposure to heat and embers. Here we focus on concepts associated with estimating heat effects on people and buildings. See the discussion about embers and extreme fire behavior in Chap. 8.

The heat from fires can injure or kill people. Fire's effect on people depends upon the level of exposure. Exposure can include both the amount of heat and the duration over which heating occurs. Heat damage to human health, including pain and skin blisters, can be limited by the personal protective clothing (PPE) worn by wildland fire personnel. Similarly, fire fighters develop safety zones to reduce the heat exposure of fire personnel so that they can survive the passing of a fire front without the use of a fire shelter. Drawing upon the concepts in earlier chapters, we first discuss the direct effects of heat from fires on individual people. Then we address strategies for protecting people and their property from fires.

We acknowledge but don't address the mental and physical toll that increasingly long fire seasons in recent decades are having for fire fighters, residents, and politicians.

10.3.1 Fire and Skin

Human skin provides natural protection against radiation, in particular, that from the Sun. The Sun has a temperature of around 5780 K, an emissivity of 1, and it radiates with an average energy flux of 632.8 × 10^2 kW m^{-2}. Taking into account the radius of the Sun (6.96 × 10^8 m) and the distance between Sun and Earth (1.49 × 10^{11} m), the maximum potential radiant heat flux received at Earth's surface (q_{rad} in W m^{-2}) is:

$$q_{rad} = \left(632.8 \times 10^2 \text{ kW m}^{-2}\right)\left(6.96 \times 10^8 \text{ m}\right)^2 \div \left(1.49 \times 10^{11} \text{ m}\right)^2 = 1.38 \text{ kW m}^{-2} \quad (10.2)$$

With an average value of the albedo at around 0.7, the radiant heat flux from the Sun at the Earth's surface is about 1.0 kW m^{-2}. It should be no surprise that 1.0 kW m^{-2} is also the radiant heat threshold to cause pain to a human's bare skin after prolonged exposure (Quintiere 2016).

The effects of radiant heat flux on the human skin have been a subject of many studies (e.g., Wieczorek and Dembsey 2016) that conclude that the human body cannot tolerate elevated temperatures for long periods of time without causing pain, blistering, or other injuries. Humans feel pain when the skin temperature reaches about 43 °C, and exposure to a heat flux of 4 kW m^{-2} for 20 s will cause blisters on bare skin. The relationship between radiant heat flux (q_{rad}) and the exposure time required for a human to feel pain or cause blisters (Fig. 10.7) and can be estimated using Eqs. (10.3) and (10.4) (Stoll and Green 1958, 1959; Quintiere 2016):

$$\textit{Pain threshold} \quad q_{rad} = 30\, t^{-0.75} \quad (10.3)$$

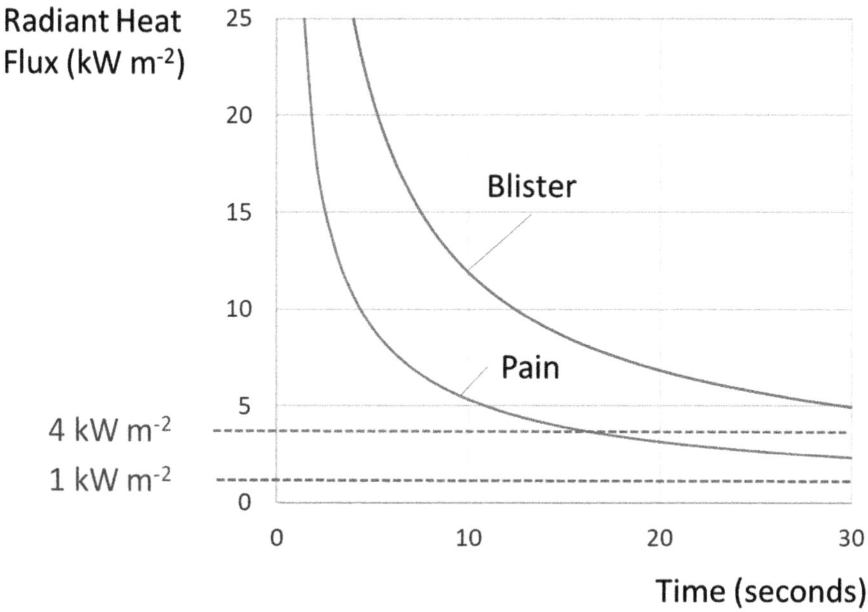

Fig. 10.7 Approximate relations showing the combinations of radiant heat flux (q_{rad}) and exposure time for the thresholds of pain and blister of bare human skin

$$\textit{Blister threshold} \quad q_{rad} = 75\, t^{-0.80} \tag{10.4}$$

where q_{rad} is the radiant heat flux threshold (kW m^{-2}), and t is the time of skin exposure (seconds). The thresholds of exposure to radiant heat for bare skin pain, blisters, or for protected fire fighters have been used to calculate safety distances and safety zones, as discussed next.

10.3.2 Safe Distances from Fires for Fire Personnel and Others

The safety of fire personnel is of concern in all fire operations, whether in fighting wildfires or in prescribed burning. Whether people experience pain or injury from heat exposure from a fire depends on radiant heating (Table 10.1) and the degree to which their skin is protected from heat. There are two strategies for limiting the heat exposure of fire personnel: wearing personal protective equipment and creating safe separation distances between people and flames (Fig. 10.8). This section draws upon the concepts we presented in Chaps. 3 and 5 on heat production and heat transfer from fires.

Table 10.1 Thresholds of pain and injury from radiant heat to unprotected skin, quantified as radiant heat flux (kW m^{-2}). (Adapted from Drysdale 1990; Quintiere 2016; Zárate et al. 2008)

Radiant heat flux (kW m^{-2})	Effect
1.0	Threshold for indefinite skin exposure
2.1	Threshold for pain after 60 s
4.0	Threshold for pain after 20 s, first skin blisters
4.7	Threshold for pain after 15 s, skin blisters after 30 s
6.4	Threshold for pain after 8 s
7.0	Threshold for fire fighters with protective clothes
10.4	Thresholds for pain after 3 s
12.5	Volatiles from wood may be ignited by pilot after prolonged exposure
16.0	Skin blisters after 5 s
29.0	Wood ignites spontaneously after prolonged exposure
52.0	Fibreboard ignites spontaneously in 5 s

Fire fighters use personal protective equipment, including Nomex clothing, to provide protection from the heat and flames and ultimately reduce the risk of injury. Tests of the effectiveness of protective clothing have used different radiant heat flux levels (typically from 1.5 to 10 kW m^{-2}) applied to thermal manikins covered with the test clothing. The time to attain the pain threshold (43 °C) is recorded to evaluate the adequacy of the clothing for the different fire operation activities (e.g., Heus and Denhartog 2017). With a single layer of 210 g m^{-2} Nomex clothing, second-degree burns will occur after 90 s when a fire fighter is subjected to radiant heat fluxes greater than 7.0 kW m^{-2} (Butler and Cohen 1998).

Fire fighters experience heat through a combination of radiation and convection. Historically, wildfire fire safety studies assumed that radiation was the dominant heat transfer mechanism affecting fire personnel. Radiation modeling can be used to estimate the separation distances required between flames and some target, such as a fire fighter or a home, to prevent ignition or injury. Recent work has built upon lessons learned from radiation modeling while incorporating convective heat transfer into the estimation of safety distances.

The radiative power (P_g) from the flame can be calculated as:

$$P_g = \epsilon \sigma_{SB} T^4 \tag{10.5}$$

where ε is flame emissivity, σ_{SB} is the Stefan-Boltzmann constant (5.67×10^{-11} kW m^{-2} K^{-4}), and T is the absolute temperature (K) of the flame (See Chap. 5 for more about radiation). In many studies, authors assume a surface flame temperature of 1200 K and an emissivity of 1 (e.g., Zárate et al. 2008). For high-intensity fires, where there are typically more fire safety concerns, the corresponding radiative powers range from $P_g = 82$ kW m^{-2} to $P_b = 118$ kW m^{-2}.

Fig. 10.8 (**a**) A fire fighter is close enough to feel the heat from a high-intensity experimental fire in Portugal. (**b**) A prescribed burn with much smaller flames in northern Portugal

The transfer of heat between the radiating surface and the object can be estimated using approaches that span a range of detail, accuracy, and applicability. A method that is commonly used in wildland fire safety distance studies is called the solid flame model. In this method, the flame can be represented using a variety of simple geometric shapes, including cylinders and rectangles. The thermal radiation is assumed to be emitted from the surface of the object (Fig. 10.9).

The radiative heat transfer from the flame to an object using the solid flame model can be estimated by multiplying the radiative power by the view factor (F_{ab}) :

$$q_{rad} = F_{ab}P_g = F_{ab}\epsilon\sigma_{SB}\, T^4 \qquad (10.6)$$

The view factor considers the geometry of the flame and the object receiving the radiation, the distance and the angle between the emitter and target, and whether or not the emitter and receiver can "see" each other. The view factor can take on values from 0 to 1. Equations to estimate the view factor for several simplified 2- and

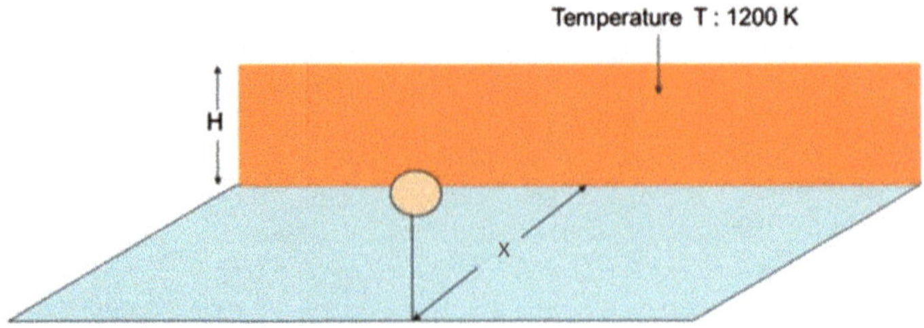

Fig. 10.9 Representation of an object, receiving a radiant heat flux (q_{rad}) from a fireline, represented as a wall of flames of height H, at a temperature of 1200 K (typical of flames), and at a distance x

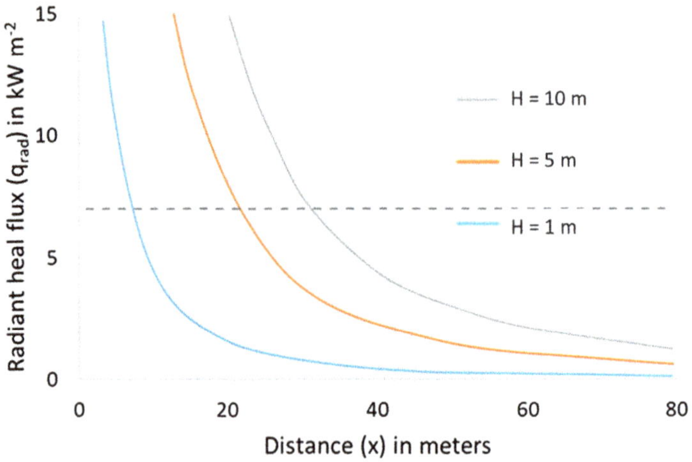

Fig. 10.10 Radiant heat flux (q_{rad}) received as a function of the distance (x) for diverse flame heights (H) from flames with surface temperature T = 1200 K, with an emissivity (ε) of 1.0, $P_g = 118$ kW m^{-2}, and a flame front of 20 m. (From Zárate et al. 2008)

3-dimensional scenarios can be found in heat transfer textbooks such as Incropera et al. (2007). Using the view factor from Zárate et al. (2008) for the scenario shown in Fig. 10.9, the radiative heat flux increases as a function of the flame height and decreases as a function of the distance between the flame and the target (Fig. 10.10). These calculations can be combined with the threshold radiative heat flux from Table 10.1 to identify the safe separation distance.

Safe Separation Distance is defined as "the minimum distance a fire fighter in standard Nomex wildland protective clothing must be separated from flames to prevent radiant heat injury". Using a solid flame model approach, Butler and Cohen (1998) suggested that an appropriate rule of thumb for the safe separation distance is at least four times the maximum flame height. This is the rule of thumb used in the BehavePlus system to calculate fire safety distance (Andrews 2014). There, the flame length is used in place of flame height, as a worst-case estimate. The

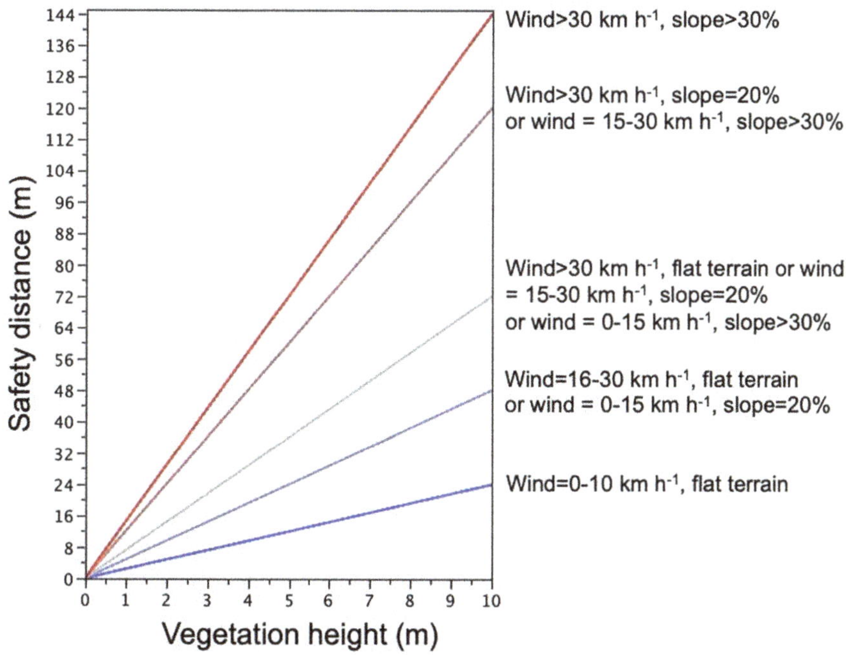

Fig. 10.11 The safe distance and the size of safety zones for fire fighters varies with the height of the surrounding vegetation, wind, and slope. Graph by the authors based upon the rule of thumb in https://wildfiretoday.com/2014/07/11/revised-guidance-for-safety-zones-is-released/, accessed 11 September 2020

idea is for fire personnel to identify both safety zones and escape routes to those safety zones at all times when they are working near fires. Safety zones are sufficiently large that people in their center will be at low risk of injury even if the vegetation surrounding the safety zone burns intensely. A more recent rule of thumb replaces flame height with twice the height of the surrounding vegetation, which eliminates the need to predict flame height. The distance of transport of convective energy ahead of the fire front is at least equal to two or more flame lengths under steep terrain or windy conditions (Butler 2014). To account for convective heat, the previous quantity (8 times vegetation height) is then multiplied by a slope-wind factor that varies from 1 to 6 and increases with wind speed and terrain slope (Fig. 10.11).

10.3.3 Protecting Peoples' Homes

Fires can endanger people due to heat and smoke, and disrupt lives when homes and property burn. Though "no one should ever die to save a house" (Kolden 2013), many fire fighters may risk injury or death to protect people and property. Further, most of the money spent during fire suppression is used to protect homes (Steelman 2016). For more about fire management costs, see Sect. 10.2.

Fires have threatened and burned homes worldwide, including China, Mexico, and southern Europe, not just in Australia and the US, where most of the research has been done on protecting people and property from fires (Mutch et al. 2011). Most of the homes that have been threatened or burned are located in the Wildland-Urban Interface (WUI). Many of the strategies for protecting people and property within the WUI are focused on preparing for fire by managing fuels around homes, reducing the ignitability of the homes themselves, and readying people for early evacuation if needed.

The WUI is commonly defined as an area where buildings meet or intermix with vegetation that can support fires. Sometimes the WUI is divided into two unique areas, the interface, and the intermix, depending upon the density of homes and the amount of vegetation cover. This division effectively distinguishes areas where homes are adjacent to wildland vegetation from areas where homes are interspersed with wildland vegetation. Definitions of WUI can vary from location to location, so it is essential to know what definitions are being used in mapping WUI.

Incorporating fire-resistant building materials and removing fuels from the immediate vicinity around homes can greatly reduce the potential that homes will ignite during a fire. The FIREWISE program addresses the home plus the surrounding Home Ignition Zone up to 30 m from the home (Fig. 10.12). Cohen (2008) developed the Home Ignition Zone concept based upon empirical observations of homes that did or didn't burn in large wildland fires, empirical modeling, and experiments. Recommended treatments are designed to limit the probability that embers will ignite a home or that flammable material on the home will ignite by flame contact. As in Australia (Handmer and Tibbets 2005), many of the homes burned in wildfires are ignited by embers. Cohen (2008), Calkin et al. (2014), and others emphasize that it is the house, the roof, and the fuels within 30 m of the home that are most important. When houses don't ignite from the shower of embers, houses are more likely to survive fires burning surrounding vegetation. If houses have few flammable parts and the flames don't come into contact with them, homes are more likely to survive when even intense fires pass.

The International Wildland-Urban Interface Code (IAWF 2013; International Code Council 2015) is used by local to national governments to guide building and community design to reduce fire risk to homes and the people in them. These codes mainly address home construction based on the science of fire behavior both in and outside of homes. They are designed to limit contact of embers and flames with the house or fuels adjacent to the house—hence the focus on screening, cleaning, and limited fuels in contact with buildings.

Preparation of home is key to avoiding urban disasters. Even the best fire fighters and fire suppression equipment can be overwhelmed when fires threaten many homes at once (Calkin et al. 2014, Fig. 10.13). This is more than creating defensible space, for fire fighters may not be there to defend homes when fires burn near them. Calkin et al. (2014) aptly point out that if homes did not ignite, then they would not burn, and so WUI fires are a home ignition problem rather than a fire control problem. Preparation in advance of fires is key, as is early evacuation.

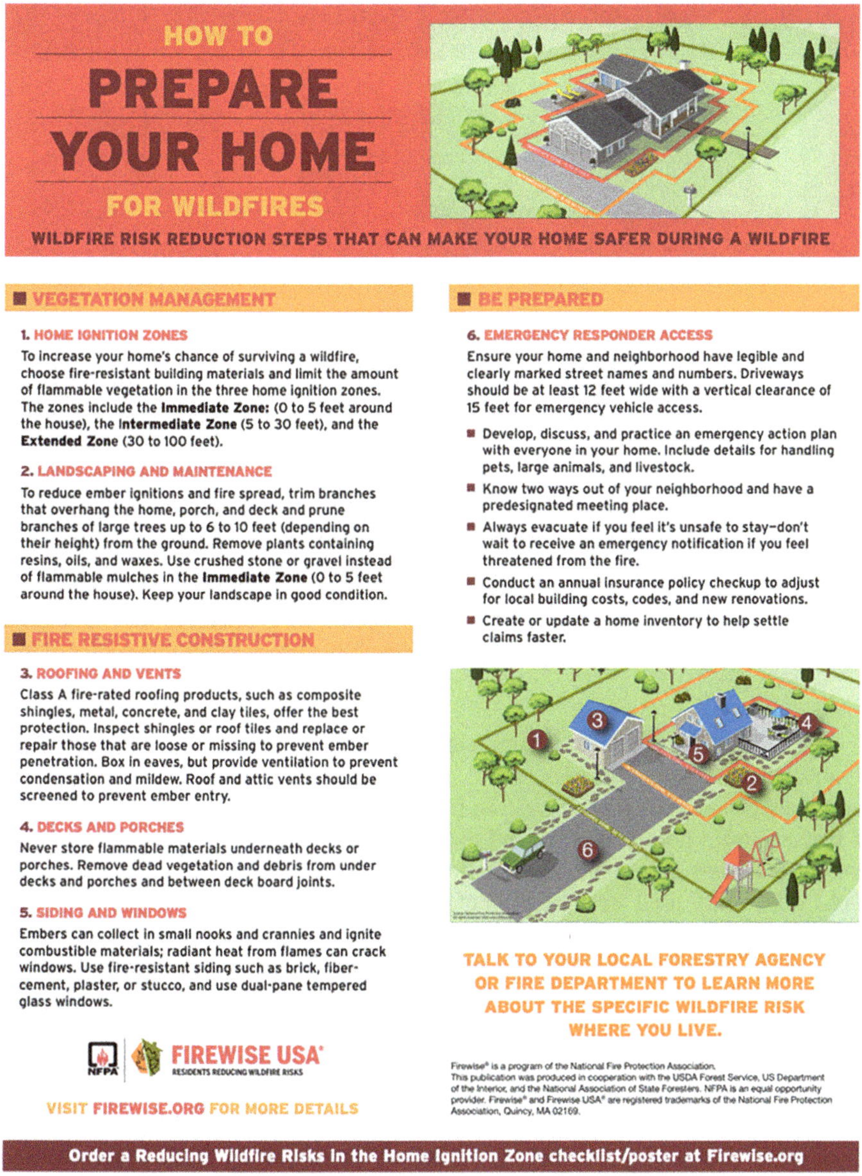

Fig. 10.12 Reducing risk of home ignition during a wildfire involves proactively managing the vegetation around the house, ensuring your home is constructed of fire-resistant materials, and being prepared for evacuating if needed as fires approach. (From NFPA n.d.)

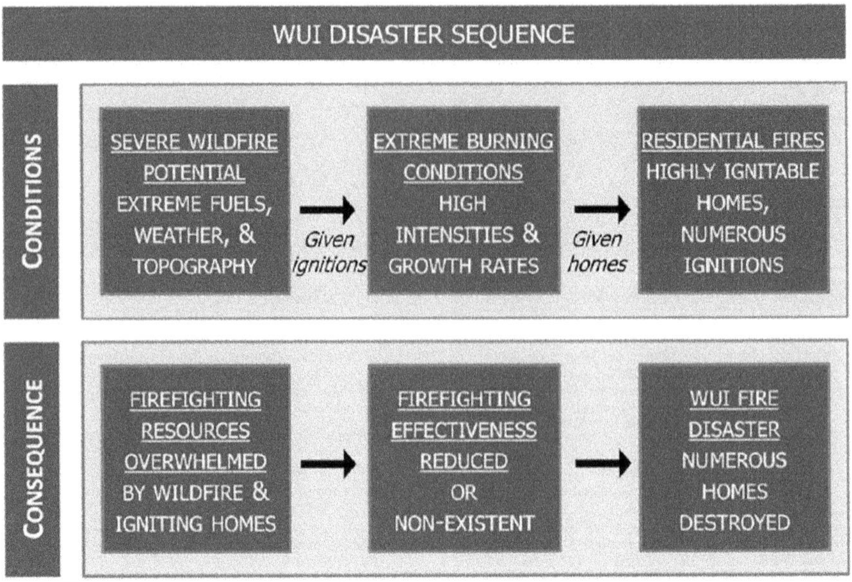

Fig. 10.13 Disastrous losses of homes in the Wildland Urban Interface can be avoided if homes are prepared so they are unlikely to ignite, thus increasing the success of structure protection. (From Calkin et al. 2014)

Fuels management near homes can alter fire behavior, aid fire fighters or homeowners in structure protection, and increase the potential that houses will survive when surrounding vegetation burns (See Chap. 11). Fuels management at a distance from homes well beyond the Home Ignition Zone could alter how fires and their embers approach homes. The effect of fuels treatments far from homes is enhanced when used as part of integrated fire management. Three points are important. *First,* fuel treatments alone, especially if they are limited to public lands, will not fully address the vulnerability of WUI communities to fire, for communities are vulnerable if individual homes are vulnerable. Fuel treatments need to be part of broader fire management strategies that also include prevention to limit ignitions by people and other strategies that help communities become adapted to fire and smoke (see Sect. 10.4.2). *Second*, fuel treatments are less effective as the vegetation regrows. *Third,* only 10% of the total number of fuel treatments completed by the US Forest Service 2004–2013 later burned (2005–2014) (Schoennagel et al. 2017, Fig. 10.14). However, fuels management can help people feel safer and can be part of community-based forest management and landscape management that can contribute to jobs and engage communities in helping themselves thrive. The 2010 WUI in the western United States (Martinuzzi et al. 2015) will grow to cover 40% of the landscape area in some locations (Theobald and Romme 2007). With extensive areas burned in recent decades and projected to increase in many areas, much attention and fire fighting resources are focused on the WUI (Schoennagel et al. 2017). See Chap. 11 for more about fuels

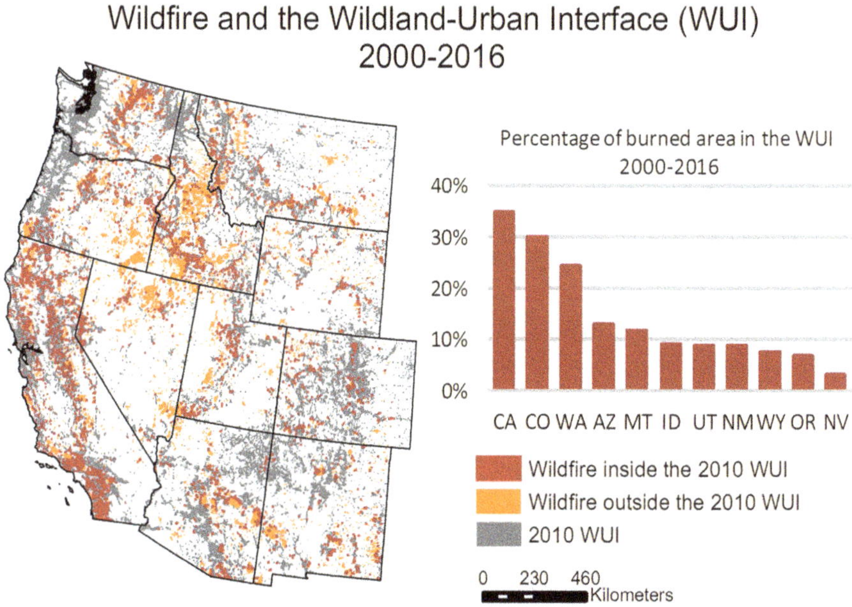

Fig. 10.14 Homes in the Wildland Urban Interface (WUI) are likely to be threatened by fires and smoke when surrounding landscapes burn (Schoennagel et al. 2017)

treatments, including their purpose, effectiveness, and strategic placement in landscapes.

Early evacuation of residents is widely encouraged when fires threaten homes. Most civilian deaths during fires are from heat exposure when fires trapped the people evacuating. Ready, Set, Go! and similar programs are widely used to encourage people to prepare for evacuation in advance in the event of a fire, and then evacuate early. Since the 2009 Black Saturday fire in Australia, in which 173 people died, early evacuation has mostly replaced the shelter in place strategies promoted during the early 2000s (see discussions by Paveglio et al. 2014; McCaffrey et al. 2015). However, some residents (11% in studies cited by McCaffrey et al. 2015) prefer to manage fuels around their homes actively and then stay to protect them in the event of fire despite the challenges and risks of doing so (Paveglio et al. 2014). McCaffrey et al. (2015) found that many emergency responders felt that in the interest of public safety, they needed to provide information to people about how to prepare for fires in case people chose not to evacuate or could not evacuate safely. Also, emergency responders in communities affected by wildfires thought that early evacuation would reduce uncertainty for both residents and emergency responders. Where limited access makes evacuation difficult, early evacuation is especially important. Indeed, if people do evacuate, it is better to do so early rather than at the last minute to avoid the potential for being trapped because of poor roads, smoke limiting visibility, or where trees or power lines and poles have fallen on the road. Traffic snarls when people flee while fire fighters are trying to access key areas for their fire suppression efforts. However, evacuation is emotional, stressful, and costly, particularly when residents don't know whether the homes they left are safe or not (McCaffrey et al. 2015). Planning and practice help people prepare mentally and physically, and both need to fit the people and place. The fire behavior conditions should be considered, including extreme fire weather and the potential for embers and long-term smoke exposure (Mutch et al. 2011).

10.4 Smoke Can Compromise Human Health

Although the heat from fires can pose significant threats to people, inhaling particulates and other components of smoke from burning vegetation is a much more common threat to peoples' health and well-being. Smoke can also affect visibility that can interfere with traffic and therefore contribute to traffic accidents or interfere with views enabled by the exceptionally good air characteristic of many national parks. Smoke is often regulated as air pollution, especially particulate matter. The small airborne particulate matter of various sizes (Fig. 10.15) in smoke can affect visibility and also cause short-term and chronic harm to people. Young children, elderly adults, pregnant women, and people with asthma or other respiratory

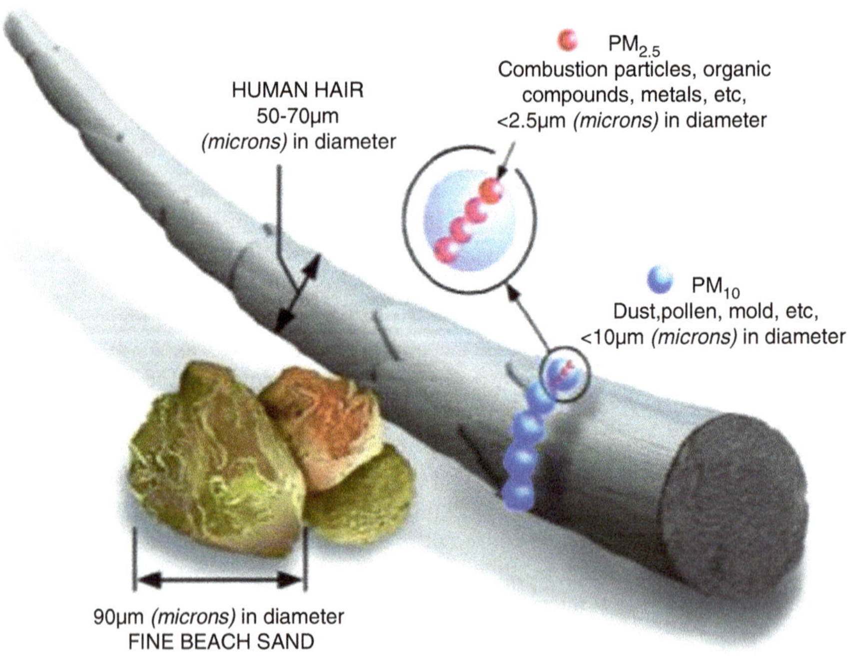

Fig. 10.15 Much of the particulate matter in smoke from vegetation fires is much smaller than a human hair and thus small enough that they can be drawn deep into our lungs. Air quality regulations often limit the concentration of particulates (especially those smaller than 2.5 μm in diameter, PM2.5). (From Peterson et al. 2018)

Table 10.2 National ambient air quality standards for the USA. The air quality index is used in the USA to communicate the health hazards of ambient smoke to the public to encourage people to take care of themselves during smoke exposure from wildland fires. (From the US Environmental Protection Agency (2014) in Peterson et al. (2018)

Air quality	24-h average particulate matter PM < 2.5 μm ($\mu g\ m^{-3}$)
Good	<12
Moderate	12–35
Unhealthy for sensitive groups	35–55
Unhealthy	55–150
Very unhealthy	150–250
Hazardous	>250

ailments are especially sensitive to smoke. Exposure to the smallest particulates, those less than 2.5 μm in diameter, commonly called PM2.5, poses the greatest risk because these fine particulates can be drawn deep into our lungs and can reach our bloodstream. The particulates and the tars and resins that have condensed on them irritate lung tissues. Due to the importance of the particulate matter, the air quality index used in many countries to communicate with the public is focused on particulate matter (Table 2.4 in Chapter 2). In the USA, federal and state regulators set limits based on the Clean Air Act for air pollutants, including particulate matter, carbon monoxide, sulfur dioxide, and nitrogen dioxide. Many air pollution regulations are focused on the concentration and duration of PM2.5 (Table 10.2). Smoke also includes other air pollutants such as carbon monoxide, sulfur dioxide, nitrogen dioxide, volatile organic compounds (VOCs), aldehydes, benzene, as well as metals, soil, pollen, bacteria, and mold spores (Kobziar et al. 2018; Peterson et al. 2018).

Healthy children and adults usually quickly recover from short-term exposure to smoke. However, many people are more sensitive. Chronic exposure is also problematic and may lead to long-term health consequences. Fire fighters exposed to smoke suffer both acute and chronic health hazards. Eye and nose irritation, nausea, and headaches are usually relieved with a brief respite in clean air (Peterson et al. 2018). However, more serious, chronic health effects may result from repeated and long-term exposure to smoke, including that experienced by fire fighters on firelines and in fire camps. These pose occupational safety risks and are being studied (Peterson et al. 2018).

Smoke from wildland fires poses health hazards for people and may cause lung irritation, hospital visits, and in some cases premature death, particularly where biomass burning is widespread (Johnston et al. 2010), such as in the tropics (Fig. 10.16). Historically, at least seven times more area burned in the western USA than currently, and emissions were accordingly high (Leenhouts 1998; Stephens et al. 2007). Fire and smoke are part of most forests, woodlands, shrublands, and grasslands. Although the area burned has increased in recent decades in some regions, the global burned area is decreasing. Expanding intensive farming has resulted in the fragmentation of some tropical savannas and grasslands

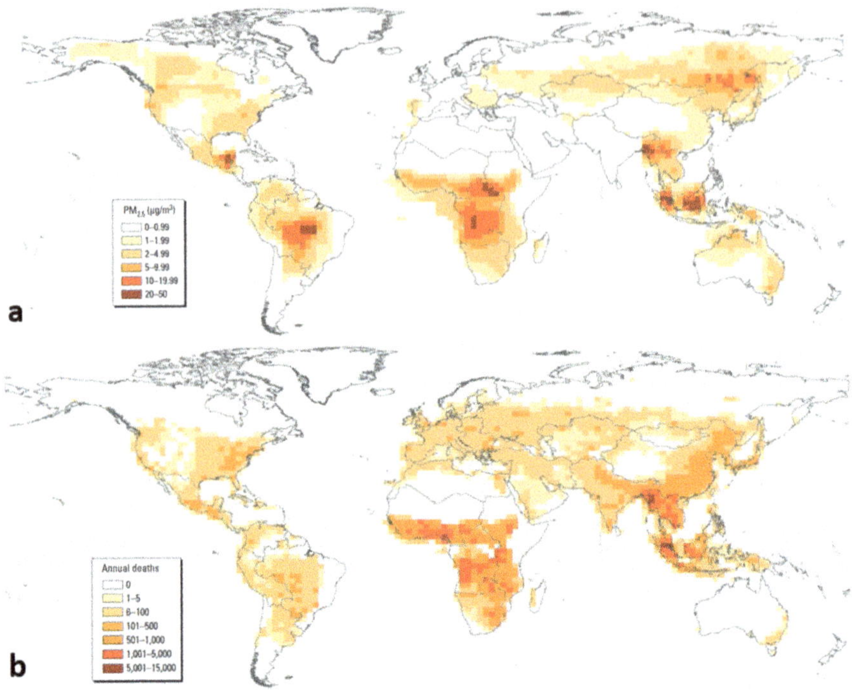

Fig. 10.16 (**a**) Estimated annual average (1997–2006) of fine particulate matter concentrations (PM < 2.5 μm in diameter) from wildland fires in the air people breathe. Estimates are based on a chemical transport model and satellite-based observations. (**b**) Estimated human mortality from smoke from wildland fires. (Both from Johnston et al. 2010)

and so they burn less (Andela et al. 2017). Global change could result in more smoke exposure to more people in many areas (Fig. 10.16, Johnston et al. 2010).

10.4.1 Smoke from Prescribed Fires and Wildfires

In general, prescribed fires produce less smoke than wildfires (Fig. 10.17), though this depends on the fuel type and amount of fuel consumption. There are many reasons for this (Navarro et al. 2018). *First,* prescribed fires are often initially set under conditions that will lead to low-intensity fires that consume less fuel. For example, prescribed fires can be implemented such that there is limited consumption of the duff and large woody fuels to decrease soil heating, potential loss of soil fertility, or carbon emissions. *Second,* prescribed fires often occur over a relatively short time, limiting people's overall exposure to smoke. *Third*, the smoke from prescribed fires is usually more localized than wildfires (Navarro et al. 2018). See Chap. 11 for more discussion about prescribed fires and alternative fuels treatments.

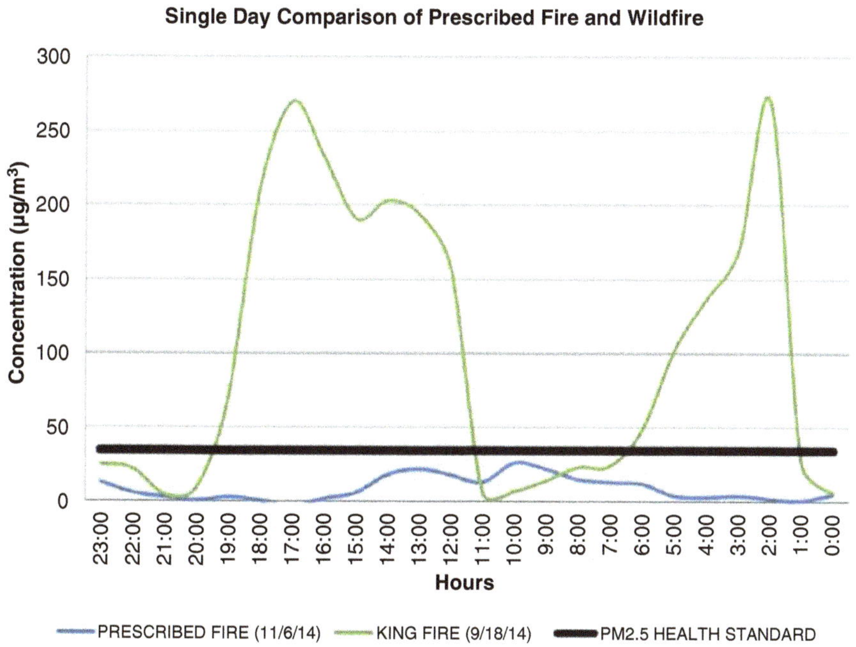

Fig. 10.17 Daily concentration of fine particulates from a distant wildfire (green) and a prescribed fire near Washoe County, Nevada. Currently, in the US, 3.2–3.6 million ha are prescribed burned annually compared to an average of 4 million ha burned in wildfires in 2017. However, in the USA, most of the prescribed burning is in the southeastern USA. (From Peterson et al. 2018)

Generally, air quality regulations are applied to smoke from prescribed fires but not from wildfires. In some countries, air quality regulations are applied to smoke from prescribed fires but not from wildfires because wildfires are considered exceptional events out of our control. Nonetheless, wildfires may contribute significantly to long-term exposure to smoke in some locations (Peterson et al. 2018). Further, future wildfires may burn more intensely and produce more smoke if fuels have accumulated in the absence of prescribed fires or other treatments.

10.4.2 Smoke Management

Smoke management programs are often designed to limit smoke exposure for people, communities, and areas, especially those designated as sensitive (Peterson et al. 2018). Managers utilize weather, smoke, and emissions forecasts (Fig. 10.18) to estimate the potential impacts of smoke and guide management and communication strategies.

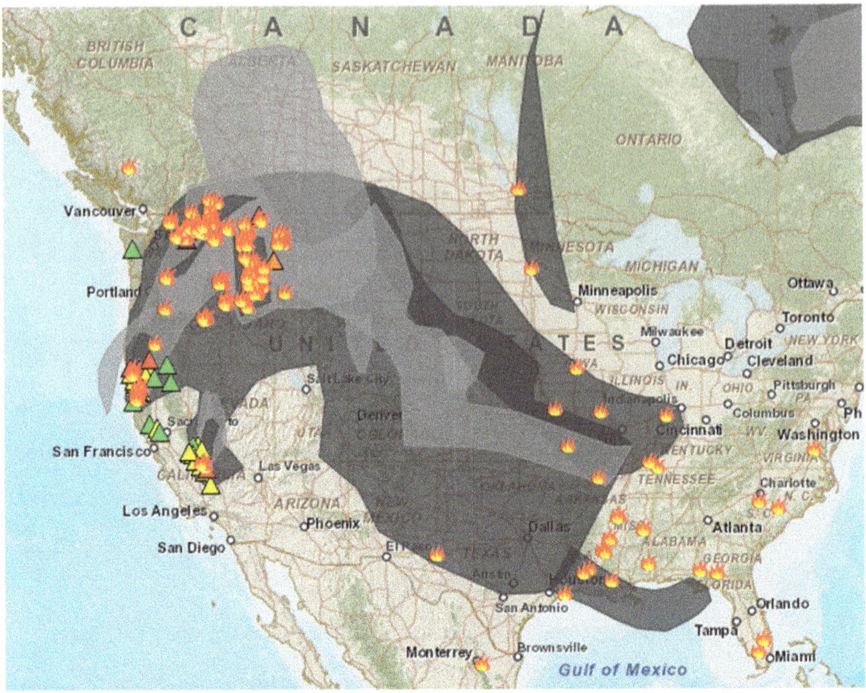

Fig. 10.18 Smoke forecasting tools are used by fire managers and those concerned about smoke impacts on people. On August 24, 2015, many large wildfires burned in the US and Canada. The red symbols indicate actively burning large fires, and the triangles indicate urban areas being subjected to low (green), moderate (yellow), or high (red) health hazards from smoke from fires. Shaded areas indicate smoke in the air, with darker shading indicating more smoke in the air. Clearly, smoke can affect people far from fires. (From EPA 2014)

Smoke forecasting is especially important for prescribed burning operations. Managers can choose to ignite prescribed fires when smoke will carry away rather than into areas with many people. Managers can limit the amount of fuel consumed by burning when the fuel is relatively moist, limiting the area burned, or burning before rains or in the spring before the largest fuels are dry. They may also burn to favor flaming combustion over smoldering combustion (Peterson et al. 2018) when the higher intensity and resulting convection can help carry the particulates up to mix with ambient air. Managers may voluntarily coordinate their burning with others to share the airshed and thus limit total smoke in the air. Despite these efforts, smoke and smoke impacts will happen, particularly during nighttime inversions and for areas close to fires and downwind or downslope from fires. Increasing the area burned for ecological restoration (see Chap. 12) could increase the number of smoky days even if the total particulates are low if relatively little fuels burn in repeated fires. Smoke must be considered in planning prescribed fires and managing ongoing wildfires with less than the most aggressive suppression. In all cases, monitoring smoke and communicating with the public and with air quality regulators are key to success. Providing air filters to schools and childcare centers, or to especially vulnerable people as recommended by their doctors, has helped reduce the negative impacts of smoke. Managing smoke for large burns (whether planned or not) over multiple days is a growing challenge, especially when those fires are managed with limited suppression to reduce costs or to provide natural resource benefits (Schultz et al. 2018, 2019).

10.4.3 Future Opportunities and Challenges

Smoke affects human health and well being. Globally, fires and their smoke have affected 5.8 million people, caused more than 1900 deaths of people, and cost more than US$52 million from 1984 to 2013 (Doerr and Santín 2016). The indirect costs are orders of magnitude more than the direct costs of suppression (Doerr and Santín 2016; AFE 2015). These trends will likely increase as the global human population increases, especially where people move into areas where they could experience fires and smoke. Fire fighters, fire managers, and fire lighters often work in the smoke, but the effects of repeated and extended exposure to smoke are not well understood. Likewise, there has been little study of the long-term implications of the extended exposure of the public to smoke when smoke spreads into towns from fires burning far from the towns. Indeed, many people object to smoke in the air.

Proactively addressing concerns about smoke impacts on air quality will require engagement among fire managers, policymakers, people potentially affected by smoke, and regulators (Peterson et al. 2018). One thing that seems inevitable is that smoke will continue to be part of our landscapes. Many scientists and managers have called for increased prescribing burning (e.g., Schoennagel et al. 2017; Moritz et al. 2018), and for managing for more "good" fires. However, in most regions of the world, the area burned in wildfires far exceeds the area burned in prescribed fires or treated with other methods. Considerations of the smoke effects on human health are central to fire management decisions today and moving forward. Key considerations are: how much smoke is expected, how many people will potentially be exposed to the smoke and for how long, and what alternative strategies can be used to manage smoke exposure.

The reality is, however, that we cannot avoid smoke or fire entirely, and we likely do not wish to avoid fire because of the many ecosystem services that come from burned areas. Prescribed burning provides two benefits as it reduces the fuels available to burn in a future wildfire. It also allows a level of control of how much smoke is produced and where it goes that is not possible with wildfire management (See Chap. 11). Smoke is regulated as a component of air pollution in many places. Even in the absence of regulations, public concerns about smoke can significantly affect public perceptions, enough to limit the use of prescribed fires in some locations. Paradoxically, smoke from prescribed fires is often less acceptable to people if it is perceived as optional while smoke from large vegetation fires can be very unpopular but may be perceived as inevitable. If we can accept that fires will occur and that we cannot fully suppress all wildland fires, then we can have the needed conversations about how we will adapt to more fire and more smoke in the future in some places.

Smoke is just one of multiple barriers for implementing prescribed burning on federal lands in the western USA (Schultz et al. 2018, 2019). Near large populations of people and where air quality is already low, limiting smoke is especially crucial for prescribed fire programs. The greatest challenges to implementing a successful smoke management program are limited funding and availability of trained people and equipment, lack of incentives and internal agency support for prescribed burning, and having enough people with needed expertise available when weather is conducive to prescribed burning (Schultz et al. 2018, 2019). Sharing resources for planning and conducting burns helps, as do programs where people come together with skills and equipment to conduct burns while also documenting the training and experience gained. Prescribed Fire Councils (Coalition of Prescribed Fire Councils n.d.) have also fostered policy, media outreach, partnering, and other ways that local people help each other with their challenges to enable successful prescribed fire programs. For more on these and other approaches, see Chap. 13 for case studies of Integrated Fire Management.

Managing smoke is an essential skill to master if we want to use prescribed fire to foster resilient, fire-adapted communities and landscapes and to have safe, effective, and efficient fire management. These are the goals of the National Cohesive Wildland Fire Strategy that involves all levels of government agencies from federal to state and county as well as non-governmental organizations and the public in the USA (USDA and USDI n.d.). Similar goals guide fire management in other countries, e.g., Canada (Canadian Council of Forest Ministers 2016).

10.5 Communities Becoming Fire-Adapted

Fire Adapted Communities have citizens who work together to coexist and thrive in ecosystems. They work closely with local, state, and federal land management agency personnel and organizations to lessen the need for protection when surrounding ecosystems burn (https://fireadapted.org/, accessed 22 June 2019). Through actions, learning, and communication, communities become more resilient. Fire Adapted Communities can become more so as they gain skills, knowledge, and experience.

Communities differ in their adaptive capacity (Fig. 10.19) for recovery from fires, their experience and acceptance of fire and smoke, and their past exposure to fire (Paveglio et al. 2015, 2018). Adaptive capacity depends on the combination of four different aspects. *First,* interactions and relationships amongst people determine how communities take collective action and the degree to which locals volunteer to reduce risk. *Second,* access to and ability to adapt scientific and technical knowledge affects the degree to which local people and community organizations understand fire suppression responsibilities and accept land use and building standards. *Third,*

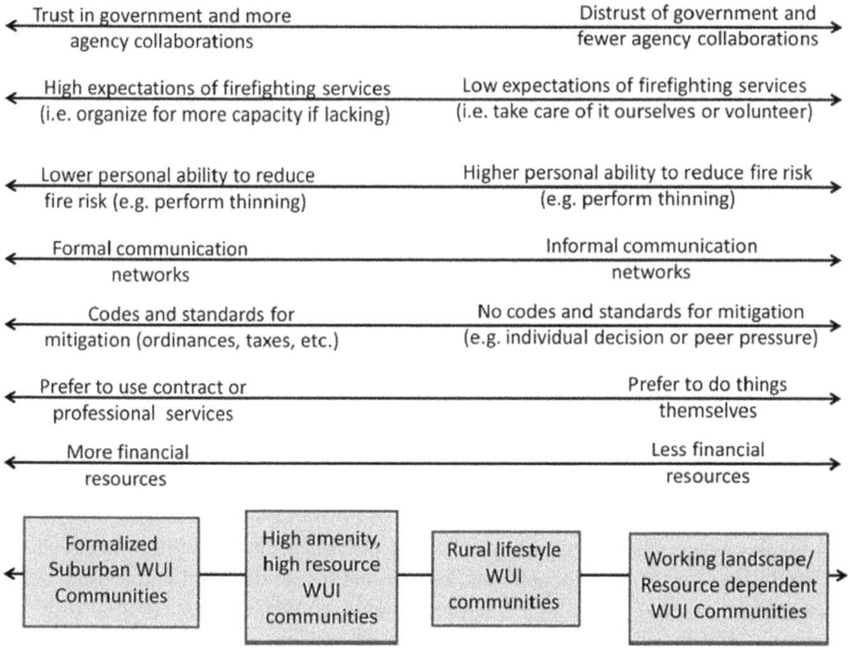

Fig. 10.19 These four archetypes of communities of people who live in Wildland-Urban Interface areas differ in ways that affect what messages and strategies for coping with fire will be useful, and the strategies people are most likely to adopt. The communities are groups of people with similar characteristics, experiences, and ways of functioning. Many towns have a mix of communities of people in them, and some communities may occupy a large geographic area. (From Paveglio et al. 2015)

place-based learning grows with local peoples' experience with wildfire and awareness of wildfire risk. *Fourth,* demographics and structural characteristics include whether there are local wood products operations, patterns of development, and willingness to pay for fire mitigation. Paveglio et al. (2015, Fig. 10.19) described four different archetypes of communities. Each archetype represents groups of people with similar human behavior that will affect what levels of trust exist, what communication strategies will be the most effective, and the strategies communities are most likely to adopt as they adapt to fire. These can inform the pathways for effective action, learning, messaging, and incentives (Carroll and Paveglio 2016; Paveglio et al. 2018). The archetypal communities are not necessarily towns. They are groups of people who identify with each other around common values and perspectives that often reflect their experiences with fire and resource management issues. They are in a place, but communities change as people come and go and as they learn. Understanding who the actors are and the social dynamics are critical for effective community engagement and building adaptive capacity (Paveglio and Edgeley 2017; Paveglio et al. 2018).

Programs such as Fire Adapted Communities Network (https://fireadaptednetwork.org/, accessed 18 June 2019) help people work together to plan for and take actions that will help them prepare for and be resilient to wildfires burning in surrounding landscapes. Some get grants, some share knowledge, and other people work to clear brush, retrofit homes with wildfire-resistant building materials, and develop emergency plans that include evacuation routes, communities can be made safer from fire. Homeowners can design or retrofit their buildings with a fire-resistant roof, screening vents, soffits, and areas under decks to exclude embers. Additionally, they can manage the vegetation and landscaping around their properties. Developers, community planners, and local regulators can insist upon subdivision design and management that decreases rather than increases potential threats from fires to people and their homes (Rasker 2015). The Fire Adapted Communities Learning Network (https://fireadaptednetwork.org/, accessed 21 June 2019) helps share lessons learned elsewhere.

Land-use planning can be an important, proactive part of living with fire. Various strategies can be used during land-use planning to reduce fire risk, incorporate multiple escape routes, to require ignition resistant landscaping, and to incorporate fuel breaks into planned open spaces (Rasker 2015). Zoning, limiting the growth of communities, conservation easements, educational programs, and community assistance are also used in fire-adapted communities to reduce the risk of WUI disasters (Mutch et al. 2011; Rasker 2015; Smith et al. 2016).

10.5.1 Learning Together Through Collaboration

Lack of trust impedes integrated fire management. Sometimes trust can be built with monitoring and stakeholder engagement in land management decision making. Trust of people in leaders and leaders trusting in local people are always important but more so during fires. Gaining and holding trust depends on integrity, transparency, accountability, compassion, and a willingness to listen and try new ideas and approaches.

Fires have brought many communities together (Prior and Eriksen 2013). Many people who might otherwise disagree with one another have collaborated out of both a fear of fire and a sense that people can make their communities safer. Some people have found economic opportunity in community-based forestry around thinning and fuels management. Local approaches to fires change through time (Paveglio and Edgeley 2017).

Increasingly, fire and natural resource managers must work across boundaries between lands managed by different organizations and boundaries within organizations to address barriers to and create opportunities for prescribed burning (Schultz et al. 2018). Resistance to prescribed burning and other fuels management and to smoke is internal to public land management agencies and the public.

Cross-boundary collaboration is not easy (Schultz et al. 2019) but necessary to work at the scales needed to address large fires. Conversations among the many stakeholders and decision-makers involved can be helpful, as can articulating the implications of management alternatives, including no action. Transparency and shared ownership of outcomes are useful. Working with the media is essential, as the media about fires shape people's attitudes about fire and smoke (Paveglio et al. 2011; Paveglio and Edgeley 2017).

10.5.2 *Learning from Traditional Practices and Scientific Knowledge*

Traditional knowledge (TK) can complement scientific knowledge. These different ways of knowing, learning, and teaching (Mason et al. 2012; Lake et al. 2017, Table 10.3) can enrich our understanding of fire from either perspective alone. Both are grounded in observation, learning from trying, and reflecting upon new practices. TK, including Traditional Ecological Knowledge and Indigenous knowledge, is developed by those with long experience in a place and often shared between different generations of people. Many cultural practices developed through millennia through teaching, learning, and adapting from culture to culture and through time. Traditional knowledge about fire draws on this long-term, often anecdotal but immensely deep appreciation for the power of fires to affect plants, consume fuels, and alter landscapes (Lake et al. 2017). Place-based knowledge, including the local expertise from Indigenous peoples or from others (local ranchers and farmers) who have lived and learned in a place for many years, can be immensely valuable as all fires are local. Whatever the source of knowledge, thinking must be broad, flexible, and forward-looking to address the complex challenges fire poses to people in a rapidly changing world. Not all knowledge is wise, and not all ideas are adaptive (Berkes et al. 2000), so users need to be flexible and always willing to question and learn.

Table 10.3 Traditional knowledge and scientific knowledge are complementary. The best managers use both to inform actions with science and learn from observation and local adaptive management and share by example. (Adapted from Berkes et al. 2000, Mason et al. 2012, Huffman 2013)

Traditional knowledge	Scientific knowledge
Qualitative	Quantitative
Intuitive, anecdotal	Intellectual
Place oriented	Short time series and broad generalities
Holistic	Reductionist
Insights shared among practitioners	Researchers share data by publication

People in Indigenous cultures worldwide often used fire skillfully and carefully. Indigenous people ignited fires for clearing land, to fell trees, to provide nutrients to crops, to maintain and improve pasture (against invasion by trees, for instance), to hunt or attract game, to promote medicine and food plants, and in warfare, as well as in many ceremonies (Mason et al. 2012; Huffman 2013). For many Indigenous people, fires were one of the few tools they had for managing vegetation. Fires were essential to life. Those who could ignite, carry, and use fire were often influential and respected within their communities. Indigenous people currently manage or have tenure rights to over 25% of the world's land. Their territories include much of the global biodiversity and forest carbon, so their fire and vegetation management matters to us all.

Management practices around the world often blend scientific with traditional knowledge and local experience. Local wisdom must include humility, recognition of uncertainty, and the need for learning using both traditional ways of seeing and science observations going forward.

10.6 Implications and Management Considerations

"We need a dedicated prescribed fire workforce. Imagine if, for every fire fighter poised and ready to extinguish any start, we also had a fire lighter." Jeremy Bailey, The Nature Conservancy

Imagine a world where fire-wise homes, fire-adapted communities, and fire-resilient landscapes are commonplace rather than exceptional. A world such as this will require people from various backgrounds to work together and take ownership of their collective risk. Fire adapted communities need to expect fires to happen and tolerate smoke. To this end, communities must learn together, whether by biomimicry (Smith et al. 2018) or otherwise thinking "outside the box" or applying practical lessons learned from past fires and other communities. Especially, let's use what social scientists are learning about what shapes understanding and actions by people.

Fire in the WUI is not a public lands issue; it is a private lands issue (Calkin et al. 2014). If homes were less likely to ignite from fires, fires would be less damaging (Calkin et al. 2014). By preparing for fires and managing the fuels within the Home Ignition Zone, homeowners would be less reliant upon fire fighters to protect their homes.

Although we emphasize homes here, both whole communities and the landscapes around them are part of what people consider home. In some cultures, fields of crops or the forest and wildlands are more important than homes. Fires burning far from homes can affect communities through smoke or by changing water supply, altering places special to people, and affecting ecosystem services and long-term sustainability. There is no uniform way to assess the degree to which fires affect both people and the places they love (Smith et al. 2016). Limiting fires and keeping landscapes from burning also has positive and negative effects. Many areas are beautiful

Fig. 10.20 Fire superheroes work effectively with communities in ways that benefit both people and ecosystems in understanding and using fire. (Rick Henion illustration in Brenner et al. 1999)

because they burned in the past, and many burned areas will become lovely given time. The ecosystem services humans depend upon, such as clean water, depend on healthy ecosystems that, in turn, depend upon fires. Yet, people are more likely to vilify fire than celebrate it. Fire fighters are often viewed as heroes. We still need those heroes, but we also need superheroes using prescribed burning and fuels treatments toward future resilience (Fig. 10.20).

Steelman (2016) and Fischer et al. (2016) argued that the current fire management paradigm, which emphasizes fire control and suppression, is financially costly without making significant progress in reducing structure loss and fatalities. The current challenges associated with wildland fires are likely to increase in scale and complexity as climate change continues, the human population grows, and social values about risk and ecosystem services change. Fear of destructive fire and adverse effects on ecosystem goods and services perpetuates increased investment in fire suppression. Incorporating social-ecological perspectives can assist societies in moving from fighting fires to living well with fires (Moritz et al. 2014). Finding solutions to what can seem a "wicked problem" will require embracing the diversity of human attitudes about fire with the biophysical realities of fire as a process (Smith et al. 2016). Working effectively with all of the different people involved depends on us listening. Cultural differences, experience with fire, the trust of government, and appreciation for science and other ways of knowing all influence how we view fire, hear messages, and the sorts of practices we will engage in and support (McCaffrey

2015). In the US, the National Cohesive Strategy (Forest and Rangelands n.d.) is a collaborative effort to create all-lands solutions across the nation that address three goals: to restore and maintain fire resilient landscapes, create fire-adapted communities, and safe, effective fire response. People are central to all of these, and people will be essential to successful integrated fire management.

10.7 Interactive Spreadsheet: RADIATION_Fireline_Safety

We provide an interactive spreadsheet, RADIATION_Fireline_Safety_v2.0, that readers can use to explore the implications of different inputs for the calculation of safe distances and exposure to radiative heating from flames. An example of the output is presented in Fig. 10.21. Note that Chap. 2 also includes an interactive spreadsheet, COMBUSTION_v2.0, that includes the prediction of smoke emissions.

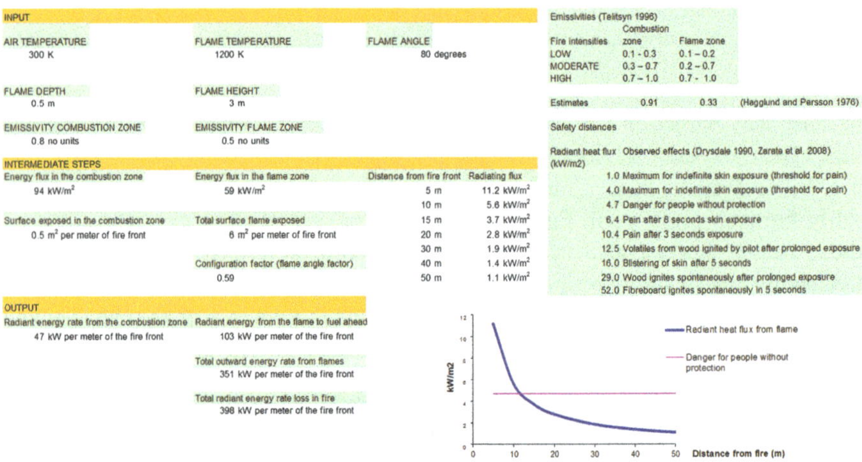

Fig. 10.21 Example of inputs and predictions from the interactive spreadsheet, RADIATION_Fireline_Safety_v2.0, used to evaluate safety distances from fires of different characteristics

References

Abatzoglou, J. T., Smith, C. M., Swain, D. L., Ptak, T., & Kolden, C. A. (2020). Population exposure to pre-emptive de-energization aimed at averting wildfires in Northern California. *Environmental Research Letters, 15*, 094046.

Andela, N., Morton, D. C., Giglio, L., Chen, Y., van der Werf, G. R., Kasibhatla, P. S., DeFries, R. S., Collatz, G. J., Hantson, S., & Kloster, S. (2017). A human-driven decline in global burned area. *Science, 356*, 1356–1362.

Andrews, P. (2014). Current status and future needs of the BehavePlus fire modeling system. *International Journal of Wildland Fire, 23*(1), 21–33.

Association for Fire Ecology (AFE), & International Association of Wildland Fire (IAWF). (2015). *Reduce wildfire risks or we'll continue to pay more for fire disasters*. Retrieved June 21, 2019, from https://fireecology.org/Reduce-Wildfire-Risks-or-Well-Pay-More-for-Fire-Disasters.

Avista Utilities. (2020). *Wildfire resiliency plan*. Spokane: Avista Utilities. Retrieved September 8, 2020, from https://www.myavista.com/safety/were-doing-more-to-protect-against-wildfires.

Berkes, F., Colding, J., & Folke, C. (2000). Rediscovery of traditional ecological knowledge as adaptive management. *Ecological Applications, 10*(5), 1251–1262.

Brenner, J., Peterson, L. E., & Crawford, B. (1999). *Fire in Florida's ecosystems student's workbook*. Leesburg: Florida Department of Agriculture and Consumer Services, Division of Forestry.

Butler, B. W. (2014). Wildland fire fighter safety zones: A review of past science and summary of future needs. *International Journal of Wildland Fire, 23*(3), 295–308.

Butler, B. W., & Cohen, J. D. (1998). Firefighter safety zones: How big is big enough? *Fire Management Notes, 58*(1), 13–16.

Calkin, D. E., Cohen, J. D., Finney, M. A., & Thompson, M. P. (2014). How risk management can prevent future wildfire disasters in the wildland-urban interface. *Proceedings of the National Academy of Sciences, 111*(2), 746–751.

Canadian Council of Forest Ministers. (2016). *Canadian wildland fire strategy: A 10-year review and renewed call to action*. Retrieved April 29¸ 2020, from https://cfs.nrcan.gc.ca/pubwarehouse/pdfs/37108.pdf.

Carroll, M., & Paveglio, T. (2016). Using community archetypes to better understand differential community adaptation to wildfire risk. *Philosophical Transactions of the Royal Society B: Biological Sciences, 371*(1696), 20150344.

Coalition of Prescribed Fire Councils. (n.d.) *Coalition of prescribed fire councils*. Retrieved April 29, 2020, from http://www.prescribedfire.net/.

Cohen, J. (2008). The wildland-urban interface fire problem: A consequence of the fire exclusion paradigm. *Forest History Today. Fall*, 20–26.

Davies, I. P., Haugo, R. D., Robertson, J. C., & Levin, P. S. (2018). The unequal vulnerability of communities of color to wildfire. *PLoS One, 13*(11), e0205825.

Doerr, S. H., & Santín, C. (2016). Global trends in wildfire and its impacts: Perceptions versus realities in a changing world. *Philosophical Transactions of the Royal Society B: Biological Sciences, 371*(1696), 20150345.

Donovan, G. H., & Rideout, D. B. (2003). A reformulation of the cost plus net value change (C+NVC) model of wildfire economics. *Forest Science, 49*(2), 318–323.

Environmental Protection Agency (EPA). (2014). *Air quality index: A guide to air quality and your health*. EPA-456/F-409-002. Retrieved May 3, 2018, from https://www3.epa.gov/airnow/aqi_brochure_02_14.pdf.

Fernandes, P. M., Delogu, G. M., Leone, V., & Ascoli, D. (2020). Wildfire policies contribution to foster extreme wildfires. In F. Tedim, V. Leone, & S. McGee (Eds.), *Extreme wildfire events and disasters* (pp. 187–200). Amsterdam: Elsevier.

Finney, M. A. (2005). The challenge of quantitative risk analysis for wildland fire. *Forest Ecology and Management, 211*(1-2), 97–108.

Fischer, A. P., Spies, T. A., Steelman, T. A., Moseley, C., Johnson, B. R., Bailey, J. D., Ager, A. A., Bourgeron, P., Charnley, S., Collins, B. M., & Kline, J. D. (2016). Wildfire risk as a socioecological pathology. *Frontiers in Ecology and the Environment, 14*(5), 276–284.

Forest and Rangelands. (n.d.). *National cohesive strategy*. Retrieved June 22, 2019, from https://www.forestsandrangelands.gov/strategy/index.shtml.

González-Cabán, A., Shinkle, P. B., & Mills, T. J. (1986). *Developing fire management mixes for fire program planning* (Gen Tech Rep PSW-GTR-88). Berkeley: USDA Forest Service Pacific Southwest Forest and Range Exp Stn.

Gorte, J. K., & Gorte, R. W. (1979). *Application of economic techniques to fire management—A status review and evaluation* (Gen Tech Rep INT-53). Ogden: USDA Forest Service Intermountain Forest and Range Experiment Station.

Haines-Young, R., & Potschin-Young, M. (2018). Revision of the common international classification for ecosystem services (CICES V5. 1): A policy brief. *One Ecosystem, 6*(3), e27108.

Handmer, J., & Tibbets, A. (2005). Is staying at home the safest option during bushfires? Historical evidence for an Australian approach. *Environmental Hazards, 6*, 81–91.

Heus, R., & Denhartog, E. A. (2017). Maximum allowable exposure to different heat radiation levels in three types of heat protective clothing. *Industrial Health, 55*(6), 529–536.

Holling, C. (1985). *Resilience of ecosystems: Local surprise and global change*. New York: Cambridge University Press.

Huffman, M. R. (2013). The many elements of traditional fire knowledge: Synthesis, classification, and aids to cross-cultural problem solving in fire-dependent systems around the world. *Ecology and Society, 18*(4), 3. https://doi.org/10.5751/ES-05843-180403.

Incropera, F. P., DeWitt, D. P., Bergman, T. L., & Lavine, A. S. (2007). *Fundamentals of heat and mass transfer* (6th ed.). New York: Wiley.

International Association of Wildland Fire (IAWF). (2013). *WUI fact sheet*. Retrieved June 18, 2019, from http://sectionb10.org/yahoo_site_admin/assets/docs/WUI_Fact_Sheet_080120131.347133404.pdf.

International Code Council. (2015). *International wildland-urban interface code*. Country Club Hills: International Code Council. Retrieved March 5, 2020, from https://codes.iccsafe.org/content/IWUIC2015/copyright.

Johnston, F. H., Henderson, S. B., Chen, Y., Randerson, J. T., Marlier, M., DeFries, R. S., Kinney, P., Bowman, D. M., & Brauer, M. (2010). Estimated global mortality attributable to smoke from landscape fires. *Environmental Health Perspectives, 120*(5), 695–701.

Kobziar, L. N., Pingree, M. R. A., Larson, H., Dreaden, T. J., Green, S., & Smith, J. A. (2018). Pyroaerobiology: The aerosolization and transport of viable microbial life by wildland fire. *Ecosphere, 9*(11), e02507. https://doi.org/10.1002/ecs2.2507.

Kochi, I., Donovan, G. H., Champ, P. A., & Loomis, J. B. (2010). The economic cost of adverse health effects from wildfire-smoke exposure: A review. *International Journal of Wildland Fire, 19*(7), 803–817.

Kolden, C. (2013). Arizona fire deaths show no one should die for a house. *Washington Post*. Retrieved March 5, 2020, from https://www.washingtonpost.com/opinions/arizona-fire-deaths-show-no-one-should-die-for-a-house/2013/07/05/1c14eaf2-e343-11e2-aef3-339619eab080_story.html?noredirect=on&utm_term=.d33e30c27c73.

Lake, F. K., Wright, V., Morgan, P., McFadzen, M., McWethy, D., & Stevens-Rumann, C. (2017). Returning fire to the land: Celebrating traditional knowledge and fire. *Journal of Forestry, 115*(5), 343–353.

Leenhouts, B. (1998). Assessment of biomass burning in the conterminous United States. *Conservation Ecology, 2*(1), 1. Retrieved May 15, 2018, from https://www.ecologyandsociety.org/vol2/iss1/art1/inline.html.

Martin, D. A. (2016). At the nexus of fire, water and society. *Philosophical Transactions of the Royal Society B: Biological Sciences, 371*(1696), 20150172.

Martinuzzi, S., Stewart, S. I., Helmers, D. P., Mockrin, M. H., Hammer, R. B., & Radeloff, V. C. (2015) *The 2010 wildland-urban interface of the conterminous United States* (Research Map NRS-8). Newton Square: USDA Forest Service Northern Research Station.

Mason, L., White, G., Morishima, G., Alvarado, E., Andrew, L., Clark, F., Durglo, M., Sr., Durglo, J., Eneas, J., Erickson, J., & Friedlander, M. (2012). Listening and learning from traditional knowledge and Western science: A dialogue on contemporary challenges of forest health and wildfire. *Journal of Forestry, 110*(4), 187–193.

Mavsar, R., González-Cabán, A., & Farreras, V. (2010). The importance of economics in fire management programmes analysis. In J. Sande Silva, F. C. Rego, P. Fernandes, & E. Rigolot (Eds.), *Towards integrated fire management—Outcomes of the European Project Fire Paradox. Chapter 3.4. Research report 23* (p. 244). Joensuu: European Forest Institute.

McCaffrey, S. (2015). Community wildfire preparedness: A global state-of-the-knowledge summary of social science research. *Current Forestry Reports, 1*(2), 81–90.

McCaffrey, S., Rhodes, A., & Stidham, M. (2015). Wildfire evacuation and its alternatives: Perspectives from four United States' communities. *International Journal of Wildland Fire, 24*(2), 170–178.

Millennium Ecosystem Assessment. (2005). *Ecosystems and human well-being: Synthesis.* Washington, DC: Island Press.

Miller, C., & Ager, A. A. (2013). A review of recent advances in risk analysis for wildfire management. *International Journal of Wildland Fire, 22*(1), 1–4.

Moeltner, K., Kim, M. K., Zhu, E., & Yang, W. (2013). Wildfire smoke and health impacts: A closer look at fire attributes and their marginal effects. *Journal of Environmental Economics and Management, 66*(3), 476–496.

Moritz, M. A., Batllori, E., Bradstock, R. A., Gill, A. M., Handmer, J., Hessburg, P. F., Leonard, J., McCaffrey, S., Odion, D. C., Schoennagel, T., & Syphard, A. D. (2014). Learning to coexist with wildfire. *Nature, 515*(7525), 58–66.

Moritz, M. A., Topik, C., Allen, C. D., Hessburg, P. F., Morgan, P., Odion, D. C., Veblen, T. T., & McCullough, I. M. (2018). *A statement of common ground regarding the role of wildfire in forested landscapes of the western United States*. Fire Research Consensus Working Group Final Report. National Center for Ecological Analysis and Synthesis. Retrieved April 21, 2019, from https://www.nceas.ucsb.edu/files/research/projects/WildfireCommonGround.pdf.

Mutch, R. W., Rogers, M. J., Stephens, S. L., & Gill, A. M. (2011). Protecting lives and property in the wildland–urban interface: Communities in Montana and Southern California adopt Australian paradigm. *Fire Technology, 47*, 357–377. https://doi.org/10.1007/S10694-010-0171-Z.

National Fire Protection Association (NFPA). (n.d.). *Preparing homes for wildfires*. Retrieved June 18, 2019, from https://www.nfpa.org/Public-Education/By-topic/Wildfire/Preparing-homes-for-wildfire.

Navarro, K. M., Schweizer, D., Balmes, J. R., & Cisneros, R. (2018). A review of community smoke exposure from wildfire compared to prescribed fire in the United States. *Atmosphere, 185* (9), 1–11. https://doi.org/10.3390/atmos905015.

Nunes, J. P., Doerr, S. H., Sheridan, G., Neris, J., Santín Nuño, C., Emelko, M. B., Silins, U., Robichaud, P. R., Elliot, W. J., & Keizer, J. (2018). Assessing water contamination risk from vegetation fires: Challenges, opportunities and a framework for progress. *Hydrological Processes, 32*, 687–694. https://doi.org/10.1002/hyp.11434.

Pausas, J. G., & Keeley, J. E. (2019). Wildfires as an ecosystem service. *Frontiers in Ecology and the Environment, 17*(5), 289–295.

Paveglio, T., & Edgeley, C. (2017). Community diversity and hazard events: Understanding the evolution of local approaches to wildfire. *Natural Hazards, 87*(2), 1083–1108.

Paveglio, T., Norton, T., & Carroll, M. S. (2011). Fanning the flames? Media coverage during wildfire events and its relation to broader societal understandings of the hazard. *Human Ecology Review, 1*, 41–52.

Paveglio, T., Prato, T., Dalenberg, D., & Venn, T. (2014). Understanding evacuation preferences and wildfire mitigations among Northwest Montana residents. *International Journal of Wildland Fire, 23*(3), 435–444.
Paveglio, T. B., Moseley, C., Carroll, M. S., Williams, D. R., Davis, E. J., & Fischer, A. P. (2015). Categorizing the social context of the wildland urban interface: Adaptive capacity for wildfire and community "archetypes". *Forest Science, 61*(2), 298–310.
Paveglio, T. B., Carroll, M. S., Stasiewicz, A. M., Williams, D. R., & Becker, D. R. (2018). Incorporating social diversity into wildfire management: Proposing "pathways" for fire adaptation. *Forest Science, 64*(5), 515–532.
Peterson, J., Lahm, P., Fitch, M., George, M., Haddow, D., Melvin, M., Hyde, J., & Eberhardt, E. (Eds.). (2018) *NWCG smoke management guide for prescribed fire* (PMS 420-2. NFES 001279). Boise, ID: National Wildfire Coordinating Group.
Prior, T., & Eriksen, C. (2013). Wildfire preparedness, community cohesion and social–ecological systems. *Global Environmental Change, 23*(6), 1575–1586.
Quintiere, J. G. (2016). *Principles of fire behavior*. Boca Raton: CRC Press.
Rasker, R. (2015). Resolving the increasing risk from wildfires in the American West. *Solutions, 6* (2), 55–62.
Rego, F. C., & Colaço, M. C. (2013). Wildfire risk analysis. In A. H. El-Shaarawi & W. P. Piegorsch (Eds.), *Encyclopedia of environmetrics* (2nd ed.). Chichester: Wiley.
Rideout, D. B., Loomis, J., Ziesler, P. S., & Wei, Y. (2012). Comparing fire protection and improvement values at four major US National Parks and assessing the potential for generalized value categories. *International Journal of Safety and Security Engineering, 2*(1), 1–12.
Rideout, D. B., Ziesler, P. S., & Kernohan, N. J. (2014). Valuing fire planning alternatives in forest restoration: Using derived demand to integrate economics with ecological restoration. *Journal of Environmental Management, 141*, 190–200.
Rideout, D. B., Wei, Y., Kirsch, A., & Kernohan, N. (2017). STAR fire: Strategic budgeting and planning for wildland fire management. *Park Science, 32*(3), 34–41.
Schoennagel, T., Balch, J. K., Brenkert-Smith, H., Dennison, P. E., Harvey, B. J., Krawchuk, M. A., Mietkiewicz, N., Morgan, P., Moritz, M. A., Rasker, R., & Turner, M. G. (2017). Adapt to more wildfire in western North American forests as climate changes. *PNAS, 114*(18), 4582–4590.
Schultz, C., Hubers-Stearns, H., McCaffrey, S., Quirke, D., Ricco, G., & Moseley, C. (2018) *Prescribed fire policy barriers and opportunities: A diversity of challenges and strategies across the West.* Public Lands Policy Group Practitioner Paper No 2 and Ecosystem Workforce Program Working Paper No 86. University of Oregon, Corvallis, OR Retrieved April 20, 2019, from https://scholarsbank.uoregon.edu/xmlui/bitstream/handle/1794/23861/WP_86.pdf?sequence=1.
Schultz, C., Moseley, C., & Hubers-Stearns, H. (2019). *Planned burns can reduce wildfire risks, but expanding use of 'good fire' isn't easy.* Retrieved April 25, 2019, from https://theconversation.com/planned-burns-can-reduce-wildfire-risks-but-expanding-use-of-good-fire-isnt-easy-100806.
Sil, A., Azevedo, J. C., Fernandes, P. M., Regos, A., Vaz, A. S., & Honrado, J. P. (2019). (Wild)fire is not an ecosystem service. *Frontiers in Ecology and the Environment, 17*, 429–430.
Smith, A. M. S., Kolden, C. A., Paveglio, T. B., Cochrane, M. A., Bowman, D. M. J. S., Moritz, M. A., Kliskey, A. D., Alessa, L., Hudak, A. T., Hoffman, C. M., Lutz, J. A., Queen, L. P., Goetz, S. J., Higuera, P. E., Boschetti, L., Flannigan, M., Yedinak, K. M., Watts, A. C., Strand, E. K., Van Wagtendonk, J. M., Anderson, J. W., Stocks, B. J., & Abatzoglou, J. T. (2016). The science of firescapes: Achieving fire-resilient communities. *Bioscience, 66*(2), 130–146.
Smith, A. M., Kolden, C. A., & Bowman, D. M. (2018). Biomimicry can help humans to coexist sustainably with fire. *Nature Ecology & Evolution, 2*(12), 1827.
Sparhawk, W. N. (1925). The use of liability ratings in planning forest fire protection. *Journal of Agricultural Research, 30*(8), 693–792.
Steelman, T. (2016). US wildfire governance as a social-ecological problem. *Ecology and Society, 21*(4), 3. https://doi.org/10.5751/ES-08681-210403.

Stephens, S. L., Martin, R. E., & Clinton, N. E. (2007). Prehistoric fire area and emissions from California's forests, woodlands, shrublands, and grasslands. *Forest Ecology and Management, 251*, 205–216.

Stoll, A. M., & Green, L. C. (1958). *The production of burns by thermal radiation of medium intensity* (Paper Number 58-A-219). New York: American Society of Mechanical Engineers.

Stoll, A. M., & Green, L. C. (1959). Relationship between pain and tissue damage due to thermal radiation. *Journal of Applied Physiology, 14*, 373–382.

Termorshuizen, J. W., & Opdam, P. (2009). Landscape services as a bridge between landscape ecology and sustainable development. *Landscape Ecology, 24*, 1037–1052.

Theobald, D. M., & Romme, W. H. (2007). Expansion of the US wildland–urban interface. *Landscape and Urban Planning, 83*(4), 340–354.

Thomas, D., Butry, D., Gilbert, S., Webb, D., & Fung, J. (2017). *The costs and losses of wildfires a literature review*. National Institute of Standards and Technology Special Publication 1215:72

United Nations (UN). (2018). *Revision of the world urbanization prospects*. Retrieved June 20, 2019, from https://www.un.org/development/desa/publications/2018-revision-of-world-urbanization-prospects.html.

United Nations International Strategy for Disaster Reduction (UNISDR). (2009). *Terminology on disaster risk reduction*. Geneva: United Nations International Strategy for Disaster Reduction. Retrieved September 8, 2020, from https://www.unisdr.org/files/7817_UNISDRTerminologyEnglish.pdf.

US Department of Agriculture (USDA), & US Department of Interior (USDI). (n.d.). *National cohesive wildland fire management strategy*. Retrieved May 13, 2018, from https://www.forestsandrangelands.gov/strategy/.

Venn, T. J., & Calkin, D. E. (2007). Challenges of accommodating non-market values in evaluation of wildfire suppression in the United States. In *Proceedings of the American Agricultural Economics Association Annual Meeting*, Portland, OR.

Wieczorek, C. J., & Dembsey, N. A. (2016). Effects of thermal radiation on people: Predicting 1st and 2nd degree skin burns. In M. J. Hurley et al. (Eds.), *SFPE handbook of fire protection engineering*. New York: Springer.

Zárate, L., Arnaldos, J., & Casal, J. (2008). Establishing safety distances for wildland fires. *Fire Safety Journal, 43*(8), 565–575.

Part III
Managing Fuels, Fires, and Landscapes

Photo by Kari Greer/USFS

In this part of our book, *Fire science from chemistry to landscape management,* we address fuel dynamics and management first. Then readers learn about fire regimes (Fig. III.1) and landscape-scale management for fires. Fires can greatly alter landscapes, while the topography, vegetation, and environmental conditions influence fire behavior and effects for current and future fires (Fig. III.2). Then, we discuss integrated fire management, drawing upon the best ideas and practices worldwide. We include eight case studies of successful integrated fire management, each written by those engaged with fostering and applying innovative thinking. We know that readers will find much to stimulate their thinking. We hope readers will envision what innovative, effective fire management will be in their landscape.

Effective, integrated fire management results in more positive effects and fewer negative consequences in both the short- and long-term. Our key themes are strategies for effective fire management to benefit people and ecosystems while managing costs. Strategic fire management decisions during fires will shape landscapes for the future. We focus on integrated fire management but also address community-based management (FAO 2015). Community-based fire management is useful for local people working together (FAO 2011). Such approaches have developed through "grass-roots" efforts that are often aided by The Nature Conservancy, World Wildlife Fund for Nature, or other non-governmental organizations. TNC (2017) provided a framework for such efforts (Fig. III.3). All around the globe, people depend on forests and other lands for their livelihoods, and many have developed effective ways of managing the resources effectively. More importantly,

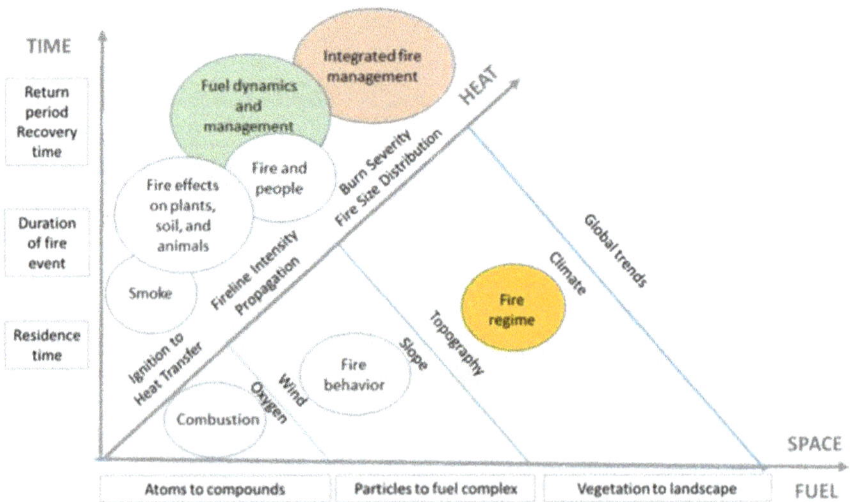

Fig. III.1 Fire regime is the third level of analysis in both temporal and spatial scales. The drivers are climate, topography, and vegetation. Fire regimes are related to the landscape scale and the fire return and vegetation recovery periods. Burn severity and fire size distribution are essential characteristics of the fire regime, which is dependent upon fuel dynamics and management. Integrated fire management is the global concept to have a coherent intervention in managing landscapes and the fire regimes, in their mutual relation as pattern and process

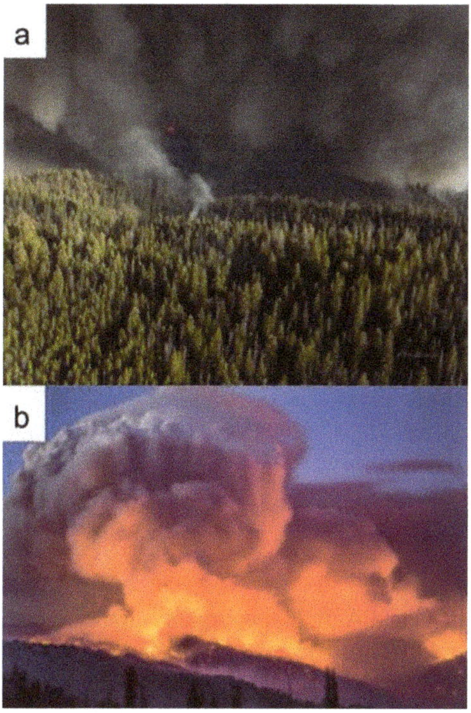

Fig. III.2 Pioneer Fire, Idaho 2016. (**a**) How this fire burned initially, and (**b**) what happened next reflects the history of fires and fire suppression here, as well as land use, topography, vegetation, and changing climate. Idaho has always had large fires, yet changing fuel conditions and warming climate have contributed to many huge fires in recent decades. The Pioneer fire was ignited by lightning in multiple locations. Even with active fire suppression and other fire management to limit fire spread into communities and other values at risk, the Pioneer Fire burned >76,000 ha with a fire management cost of > $100 million. (Photos by Kari Greer/USFS)

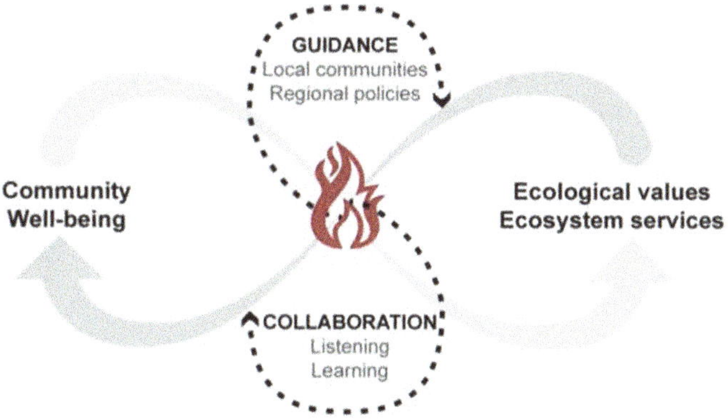

Fig. III.3 Fire management can help local communities achieve their goals while addressing regional policies and directives. Effective fire management can provide for both community well-being and the ecosystem services people value. People listening, learning, and collaborating before fires occur can plan for effective action during fires and thoughtful recovery after fires. Proactive, integrated fire management can achieve outcomes for human well-being and ecological values. (Redrawn from TNC 2017)

local people can foster local jobs and a sense of control over their future when they can manage surrounding landscapes themselves or in shared stewardship with other land managers (TNC 2017). Despite development pressures, giving voice to locals informs their choices and fosters action, and is an effective way for regional and national strategies to be implemented in many locations.

Fire management is proactive and must be more so in the future. Effective, integrated fire management capitalizes and builds upon prior fires (Fig. III.3). Decisions of using or not using fire or of suppressing or not suppressing fire have consequences. Fire managers are rewarded for the quality of their decisions. Those decisions must reflect risk analyses and proactive planning (Thompson et al. 2018).

In our final chapter, we discuss the implications of ongoing trends for the future. Thus, we end this part with a look into the future of fire science and fire management. Smoke may cloud the future, but we know that for people to live with fire well will require the work and creative ideas of many people, not just fire professionals.

References

Food and Agriculture Organization of the United Nations (FAO). (2011). *Community-based fire management: A review.* FAO Forestry Paper 166, Rome. Retrieved March 11, 2020, from http://www.fao.org/3/i2495e/i2495e.pdf.

Food and Agriculture Organization (FAO). (2015). *Community-based forestry.* Retrieved March 30, 2020, from http://www.fao.org/forestry/participatory/90729/en/.

The Nature Conservancy (TNC). (2017). *Strong voices, active choices: TNC's practitioner framework to strengthen outcomes for people and nature.* Arlington: The Nature Conservancy. Retrieved March 29, 2020, from https://www.nature.org/en-us/what-we-do/our-insights/perspectives/strong-voices-active-choices/.

Thompson, M. P., MacGregor, D. G., Dunn, C. J., Calkin, D. E., & Phipps, J. (2018). Rethinking the wildland fire management system. *Journal of Forest, 116*(4), 382–390.

Chapter 11
Fuel Dynamics and Management

Learning Outcomes

Through this chapter, we expect you as a reader to be able to

1. Identify the motivations for fuels treatments,
2. Describe the factors that influence live and dead fuel moisture,
3. Schoennagel et al. (2017) and Rhodes and Baker (2008) argued against investing in fuels treatments except near homes in the wildland urban interface because so few fuels treatments were challenged by fires within 10 years after treatment. In contrast, Hudak et al. (2011) and many others highlighted the efficacy of fuels treatments in wildfires. Briefly summarize the points for and against fuels treatments and make a science-based argument in support of your opinion,
4. Explain why mastication can alter fire intensity without removing fuels, and
5. Evaluate Keane's (2015) statement that fuels link fire behavior and effects. Do you agree? Why or why not? In your answer, include the implications for fuels management.
6. Use the interactive spreadsheets to challenge and defend your ideas about fuels and the effectiveness of fuel treatments in altering potential crown fire behavior.

Supplementary Information The online version of this chapter (https://doi.org/10.1007/978-3-030-69815-7_11) contains supplementary material, which is available to authorized users.

F. Castro Rego et al., *Fire Science*, Springer Textbooks in Earth Sciences, Geography and Environment, https://doi.org/10.1007/978-3-030-69815-7_11

11.1 Introduction

Fuels are broadly defined as any combustible material (NWCG 2006). For vegetation fires, fuels largely come from vegetation biomass as it grows and dies. Vegetation fuels are described within a hierarchical framework, from fuel particles to fuelbeds (Fig. 6.1).

The physical characteristics and distribution of vegetation fuels are highly variable over space and time due to many interacting ecological processes and human actions. Fuel dynamics have roughly two dimensions. *First,* fuels change physically as individual plants grow, die, and decompose, with consequences for the amount, structure, and composition of burnable biomass. These can be related to time-dependent flammability of disturbance-prone plants, senescence, and adaptation to fires (Rundel and Parsons 1979). *Second*, fuel moisture determines the extent to which fuels are available for combustion and, therefore, the rate and quantity of heat release. Fuel moisture varies widely and changes differently depending on fuel condition (dead or alive), size of fuel pieces, and other physical attributes, as well as environmental conditions.

Wildland fuels are often considered the most important factor influencing fire management, in part because fuels influence fire ignition, spread, and intensity. Fuels are the only part of the fire behavior triangle that can be manipulated, unlike weather and topography. Formulation of fire management strategies should begin by defining the desired fire regime, which shapes and is shaped by fuel dynamics in predictable ways (see Sect. 12.5). Fuels mediate human influences on fire behavior and effects. The ecology of fuels, understood as the tight connection between fuels, fire behavior, and fire effects, determines vegetation response and dynamics through complex feedbacks (Mitchell et al. 2009; Keane 2015). The concept of fuel ecology also implies envisioning fuels as ecosystem components with various functions rather than just fire-related biomass. For instance, standing dead trees are important nesting and perch sites, and once fallen, they are important habitats for small mammals, ants, and other insects, as well as bacteria and fungi as the trees slowly decompose into the soil. Litter accumulated on the soil surface is a source of nutrients as it decomposes and can protect the soil surface from raindrop impact and thus limit soil erosion potential. Organic matter on and in the soil holds soil particles together in aggregates, holds and releases soil nutrients and water, and are critical to nutrient cycling and soil productivity. Organic matter comes from surface litter as it breaks down physically and chemically and from fine plant roots that are constantly growing and dying. For more about these ecological considerations for fires, see Chaps. 9 and 12.

11.1.1 Dynamics of Fuel Load and Structure

Drivers of Temporal Changes

Biological mechanisms predominantly govern the character, magnitude, and organization of fuels over time, so there is an analogy with plant succession (Pyne et al.

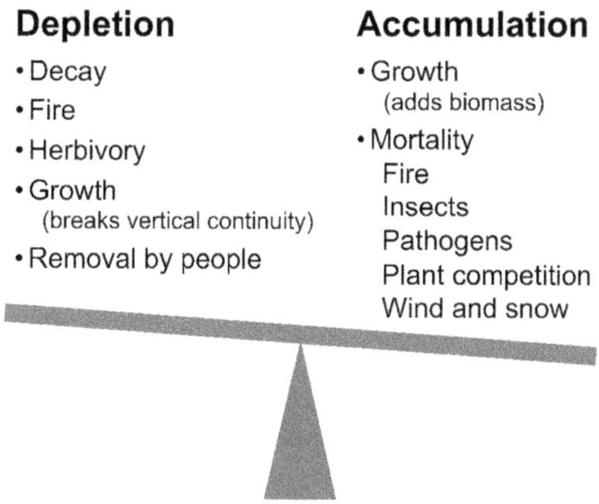

Fig. 11.1 Fuel load changes over time as a consequence of interacting processes (Adapted from Anderson and Brown 1988). People can be an important driver of fuel accumulation through land use and land or fire management practices and policies, e.g., afforestation or fire exclusion

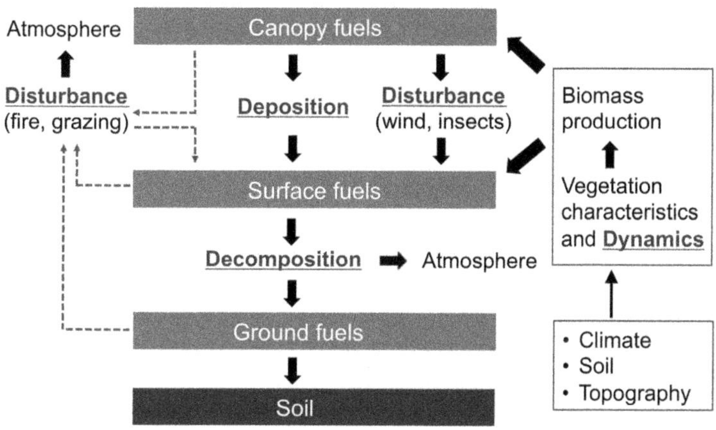

Fig. 11.2 Fuel dynamics are driven by Deposition, Decomposition, Disturbance, and vegetation Dynamics (the four Ds) and their interactions. (Adapted from Keane 2015)

1996). Thus, fuel succession expresses multi-year changes revealed through changes in fuel load and fuel structure, and like succession, the trajectories are not simple.

Anderson and Brown (1988) presented the temporal changes in fuel load as an outcome of the interplay between processes that either remove or add fuel (Fig. 11.1). Decomposition and plant growth drive the former and the latter, respectively, but disturbances play an important role. In particular, fire is both an agent of fuel depletion, through combustion, and fuel creation, through plant growth and mortality.

Keane (2015) proposed the four D's framework, where Deposition, Decomposition, Disturbance, and vegetation Dynamics drive fuel dynamics (Fig. 11.2). Overall, fuel dynamics reflect not just time and the legacy of past disturbances, including past fires and ongoing human actions, but also the constraints imposed by the physical environment (climate, topography, and soils).

Fuel deposition, also called fuel accretion and litterfall, is the outcome of leaves, twigs, branches, bark, and other plant parts falling and becoming dead surface fuels. Stems fall too, though sometimes not until long after trees die. Typically, deposition increases fuels below and near the source plants. Although people often alter how much and what fuels accumulate on the surface through deposition, fuels also naturally accumulate as plants and plant parts grow and die.

Decomposition (also known as decay) results in the breakdown of organic material into smaller pieces and simpler compounds. Insects, animals, and fire can speed physical fragmentation that in turn often favors decomposition. Decomposition can be quite slow, and fuels accumulate when and where biomass accumulation rates exceed decomposition rates. In places that are dry or cold or both, microbial activity is limited by moisture and temperature. Decomposition, like combustion, is a chemical reaction that releases carbon dioxide from the respiration of soil organisms. Decomposition, like combustion, is seldom complete, as lignin and other complex organic compounds that decompose slowly often accumulate in litter and duff. Because of decomposition and organisms that mix in mineral soil from below (Keane 2008), the mineral content of organic material on the soil surface may be relatively high (See Sect. 3.4). When it burns, surface organic matter is a source of heat. Unburned litter and duff can insulate the soil from heat and erosion while greatly influencing vegetation productivity through post-fire decomposition and release of nutrients (See Sect. 9.5.2).

Disturbances are ubiquitous within ecosystems. Disturbances shape ecosystem structure, function, and biodiversity. Following Pickett and White (1985), we define a disturbance as any biotic or abiotic event, force, or agent that alters ecosystem structure and function by causing mortality or damage. Disturbances have pronounced short-term effects on plant and animal populations and communities. Yet disturbances are critical to the long-term function and character of many ecosystems, especially ones where plants regeneration is disturbance-dependent. Individual disturbance events and their occurrence within the larger context of a disturbance regime, i.e., the cumulative effects of multiple disturbances over time, have complex effects on fuels. Fuels reflect the wide range of disturbance types and their magnitude (intensity and severity) and the spatial and temporal scales over which they occur. The intensity of a disturbance is an expression for the disturbance itself, for example, the heat released during a fire. In contrast, disturbance severity is a measure of its effect on organisms, communities, and ecosystems (See Sect. 12.3.3). For example, a low severity fire may result in the death of a few trees, while a high severity fire may kill all trees within an area. Disturbance intensity and severity are often positively linked (Heward et al. 2013), but the nature of the linkage can vary among disturbance types and ecosystems.

Vegetation dynamics are important, for as vegetation grows, biomass is added. Almost everywhere, vegetation is recovering from fire, wind, insects, pathogens, and human actions. Succession, the process of vegetation change through time, can follow multiple pathways, resulting in multiple stable states (Noble and Slatyer 1980). Bond and Keeley (2005) likened fire to a global herbivore because both fire and herbivory result in biomass that is typically far less than expected based on climate and soils. As fire and herbivory are prevalent in nearly all ecosystems, vegetation is always in some stage of succession. Plant invasion is a particular aspect of vegetation dynamics, and while invasive plants can either enhance or suppress fires, increases in fuel load and continuity after invasion by grasses, or by trees into meadows or other grasslands, is an increasingly important management concern (Brooks et al. 2004) (See Case Study 12.1).

As disturbances leave considerable residual standing and fallen vegetation, and many plants readily regrow or otherwise establish following disturbances, what we find at any location reflects the legacy of prior disturbances and prior vegetation structure and composition. Consequently, the fuels complex's long-term spatial and temporal characteristics reflect interactions among multiple disturbances and the social and biophysical factors influencing vegetation dynamics, decomposition, and deposition.

11.1.2 Disturbances, Fuels, and Fire

Fuel dynamics regulate the likelihood of disturbance by fire. In turn, disturbances can directly affect almost all attributes of the fuels complex. Still, the amount and distribution of fuels should be the focus as these are directly linked to potential fire behavior and effects. Interactions among disturbances (See Chap. 12) occur when the post-disturbance legacies influence the likelihood, type, and magnitude of subsequent disturbances or an ecosystem's ability to recover following disturbance (Buma 2015). Pyric herbivory, whereby fire shapes grazing by modifying animal behavior in terms of their feeding choices in space and time (Fuhlendorf et al. 2009), is a manifest example of two interacting disturbances with implications for fuel and fire dynamics. Disturbances impact fuels by reducing the existing biomass, converting live fuels to dead fuels, and combining these two processes.

Herbivory reduces fuel load. The effect depends upon the characteristics of the disturbance agent (e.g., anatomical differences such as mouth size, nutritional requirements, and forage preferences) and the ecosystem (e.g., plant composition, and plant physiology, nutritional status of plants) as well as the magnitude, season, and duration of the herbivory.

Lower fire spread rates in grazed grasslands compared to undisturbed grasslands is well documented (Cheney et al. 1993, 1998). Extensive grazing in southern European mountains (Fig. 11.3) works in tandem with fine-scale pastoral burning to create fine-grained fuel mosaics that inhibit the growth of large fires, even under extreme weather conditions (Fernandes et al. 2016). The demise of grazing in southern Russia's arid grasslands in the early 1990s made subsequent large fires possible (Dubinin et al. 2011). Bernardi et al. (2019) found that a higher density of domestic livestock across tropical regions is concomitant with lower fire frequency and higher cover of woody vegetation, implying that grass consumption decreases fire activity, allowing for woody plants to establish and grow. These examples attest to the ability of herbivores to influence fire through biomass consumption. However,

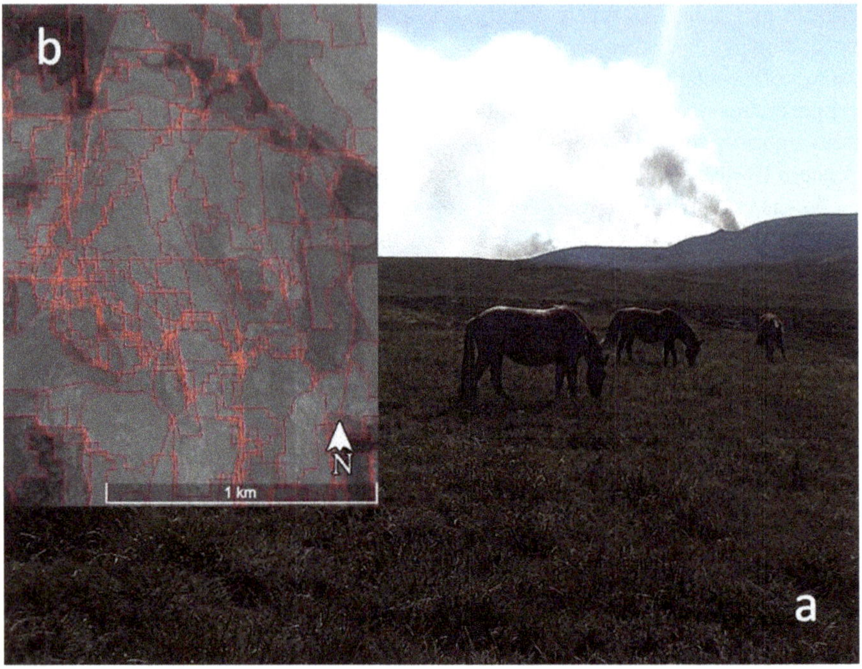

Fig. 11.3 (**a**) Shrub-dominated mountain pasture grazed by horses and cattle is maintained by frequent burning from autumn to spring in Castro Laboreiro in northwestern Portugal (Photo by P. Fernandes). (**b**) The map of fire perimeters (1975–2019, red lines) results interpreted from remote sensing does not fully reflect patch-mosaic granularity due to variable (in terms of size) omission of small fires over time; the whiter patches indicate more frequent fires. (Made with data from ICNF, n.d.)

as ecosystem engineers, wild and domestic animals affect fuels in a variety of ecosystems in ways other than through grazing and browsing. However, the effects are typically observed on finer spatial scales (Foster et al. 2020). Fuels may be compacted, especially when larger animals are involved, namely savanna megaherbivores. Fuels are made discontinuous by small and large animal trails, foraging, burrowing, and creating mounds such as those associated with the nests of birds or colonies of termites.

Disturbances, such as insects, pathogens, wind, and snow, convert living plant biomass to dead fuels through mortality. In addition to changing the abundance of dead fuels, disturbances often decrease canopy biomass and increase surface fuel loads as deposition occurs. While the latter follows the former in the event of biotic-related mortality, they are simultaneous in weather-related disturbances. The deposition of dead tree canopy fuels progresses in a stepwise fashion over time, from foliage to increasingly larger size classes of branches and ultimately ending with the tree bole. The rate of deposition is determined by the mortality agent, the tree species, the size of killed vegetation, and environmental factors such as soil, windiness, moisture, and the presence of decay fungi (Passovoy and Fulé 2006; Angers et al. 2011; Hoffman et al. 2012). Multiple interacting disturbances such as bark beetles, wind, and fire may produce novel conditions and long-term changes in landscape structure and function.

Finally, fire and some herbivorous insects influence the fuels complex through a combination of reducing fuels and converting live to dead fuels. Fuel consumption by fire occurs across the ground, surface, and canopy fuel layers, with the amount of reduction positively related to the disturbance magnitude. For example, crown fires under extreme environmental conditions can result in near-complete combustion of fine fuels on the ground surface and in tree and shrub crowns, and partial combustion of large-diameter dead down woody fuels (Call and Albini 1997; Stocks et al. 2004). However, wildfires under less extreme conditions tend to produce more heterogeneous fuel consumption, thus resulting in a much more heterogeneous post-fire fuels complex, e.g., Hudec and Peterson (2012). Extreme crown fires in conifer forests often result in live-to-dead conversion of stems and relatively large branches, whereas non-lethal surface fires primarily create litter from scorched foliage. See our discussion about fire and carbon in Chap. 9 as many ecosystem process modelers overestimate the carbon loss when forests burn when they assume that all above-ground biomass is consumed by fires (Stenzel et al. 2019).

The combination of drought and favorable host conditions across western North America has resulted in widespread tree mortality due to bark beetles (*Scolytinae* insects), such as the mountain pine beetle (*Dendroctonus ponderosae*). Increased extent and severity of future wildfires may result (Jenkins et al. 2014). The effects of bark beetle-caused tree mortality on fuels and potential fire behavior have been described using three broad temporal phases (Fig. 11.4).

The initial "red phase" occurs immediately after trees die and is characterized by the live-to-dead conversion of canopy fuels relative to the "green phase" that existed before the insect-induced tree mortality (Fig. 11.4). In the "red phase", lower canopy fuel moisture and alterations to foliar chemistry reduce the amount of heat energy

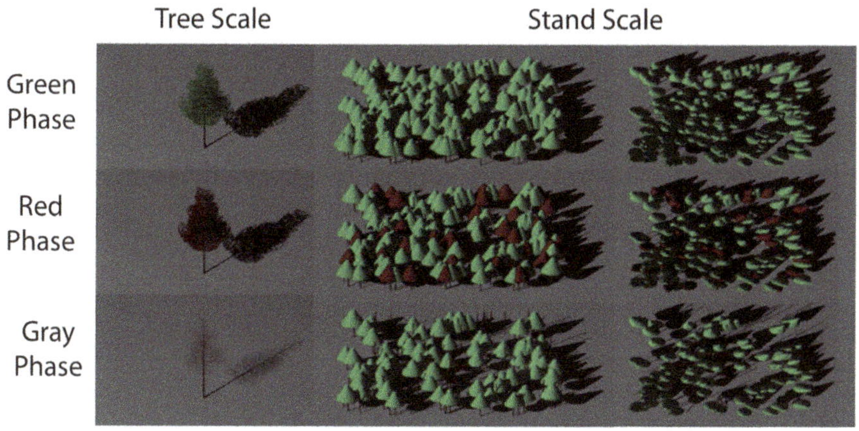

Fig. 11.4 Temporal phases after bark beetle-caused conifer mortality at the individual tree and stand scales. (From Hoffman et al. 2013)

required for crown ignition, which in turn increases the rate of spread and intensity of crown fires and burn severity (Jenkins et al. 2014; Hicke et al. 2012; Perrakis et al. 2014; Hoffman et al. 2015). Although most studies have suggested that bark beetles and fire behavior are positively linked during the "red phase", the strength of this linkage depends upon the level of tree mortality, pre-outbreak surface fuels, and burning conditions (Hoffman et al. 2012; Sieg et al. 2017). Within 1–3 years following tree mortality, the needles and small branches from killed trees begin to fall to the forest floor, reducing the canopy fuel load and increasing the surface fuel load. This time is referred to as the "gray phase" and is characterized by lower crown fire potential. However, the increased surface fuel loading and stronger winds associated with the loss of canopy biomass can magnify surface fire behavior and result in some passive crown fire. With time, fuel dynamics will be dominated by the continued deposition of large-diameter branch material and tree boles and the development of understory vegetation and regeneration. This "old phase" is characterized by increasing surface and canopy fuel load and decreasing tree crown base heights. Many people assume that these changes increase the potential for crown fire activity (Hicke et al. 2012; Stephens et al. 2018), but the degree to which this is true depends on the fuel amount and arrangement.

Changes in fuels and fire behavior after biotic-induced tree mortality are not restricted to bark beetle outbreaks. In Canada, multi-year defoliation by spruce budworm (*Choristoneura fumiferana*) kills balsam fir (*Abies balsamea*) and white spruce (*Picea glauca*) in boreal mixed conifer-deciduous stands. Fire potential then increases up to 5–8 years after tree mortality as crowns break and surface fuels accumulate (Stocks 1987).

Wind damage is another common disturbance that can significantly alter fuel conditions and fire behavior. For example, experimental burning in South Carolina forests dominated by either loblolly (*Pinus taeda*) or longleaf pine (*P. palustris*) in the wake of Hurricane Hugo, which on average decreased tree basal area by 35%, showed 87% and 7-fold increases in fire spread rate and flame length, respectively, due to fuel deposition (Wade et al. 1993, Fig. 11.5). Additionally, disturbance by wind decreases fuel moisture content within canopy gaps and favors an increase in the abundance of flammable grasses. However, wind can reduce litterfall, increase fuel patchiness, and promote succession to lower-flammability communities (Cannon et al. 2017). The abnormal fuel conditions created by high-magnitude hurricanes in the southeastern USA supports the idea of subsequent severe fires, as reviewed by Myers and Van Lear (1998) and confirmed by pollen and charcoal data analysis (Liu et al. 2008).

11.1.3 Modeling Fuel Accumulation

Olson (1963) proposed a simple asymptotical model for litter accumulation that balances fuel deposition and fuel decay:

Fig. 11.5 Fuel complex resulting from hurricane Hugo on the Francis Marion National Forest, South Carolina, USA. Total surface fuel load (up to 7.5 cm diameter), including duff, is 72.4 t ha^{-1}, of which 24% are coarse (>6 mm in diameter) dead woody fuels from pine trees. (Photograph from Wade et al. 1993)

$$w_L = w_{LS}\left(1 - e^{-bt}\right) \tag{11.1}$$

where w_L is the fuel load at moment t, w_{LS} is the maximum (or steady-state) fuel load, b is the decomposition rate, and t is time in years. The value of w_{LS} is given by litterfall divided by b, and hence it can be determined either experimentally or statistically by fitting Eq. (11.1) to data obtained across a sequence of times since fire; 3/b gives the time at which w_L reaches 95% of w_{LS}. The model assumes that $w_L = 0$ when $t = 0$. Still, the model can accommodate the decomposition of an initial fuel load (w_{L0}), e.g., the fuel remaining after a fire, by adding the decaying term $w_{L0}\,e^{-bt}$.

The Olson model assumes constant rates of fuel deposition and decay. However, seasonal variation occurs, as litterfall and b should respectively peak in summer and in winter in an evergreen forest under a temperate climate. Climate influences aside, variation on longer time scales is also expected, as litter production depends on the amount of canopy foliage, and the decomposition rate is influenced by vegetation type and structure and by fuel structure. To account for stand-development related effects, Fernandes et al. (2002) made litter load in maritime pine (*Pinus pinaster*) stands in Portugal also an empirical function of stand basal area (BA):

$$w_L = 2.025\left(BA^{0.677}\right)\left(1 - e^{0.276b}\right) \quad (11.2)$$

where w_L, BA, and t are in units of t ha^{-1}, m^2 ha^{-1}, and years, respectively.

Higher fuel accumulation rates allow for more frequent fires, which maintain lower fuel loads and lower fire intensity. Despite its shortcomings, the Olson curve is often used to describe fuel accumulation for fire management applications, namely to determine the ideal return interval of prescribed burning. It is commonly extended to other fuel layers (e.g., understory vegetation) and components (e.g., live fuels) (Fig. 11.6). Distinct fuel accumulation patterns are manifest, depending on the combination between w_{LS} and the rate at which fuels accumulate.

Fuel load dynamics can be exceedingly more nuanced and complex than portrayed by Olson's model. For example, the accumulation of downed woody debris and duff is initially low after forest stand-replacement wildfire, peaks on the short- to mid-term as fire-killed biomass accumulates on the forest floor, subsequently decreases through decomposition, and then increases as the trees regenerate and the forest reestablishes (Fig. 11.7). But post-fire fuel dynamics can be extremely

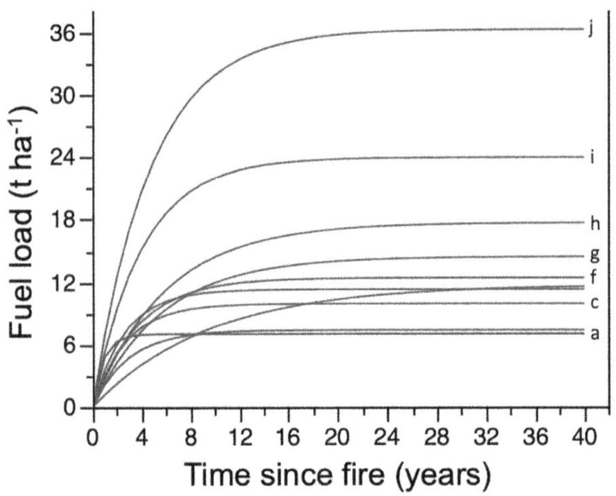

Fig. 11.6 Fuel accumulation described with Olson model: (a) Rainforest in southeastern Australia (Thomas et al. 2014), (b) Banksia woodland in southwestern Australia (Burrows and McCaw 1990), (c) Evergreen oak woodland in northeastern Spain (Ferran and Vallejo 1992), (d) Deciduous oak woodland in Ohio (Stambaugh et al. 2006), (e) Buttongrass moorland, Tasmania (Marsden-Smedley and Catchpole 1995), (f) Dry eucalypt forest in southeastern Australia (Thomas et al. 2014), (g) Dry heathland in Portugal (Fernandes and Rego 1998), (h) Dry eucalypt forest in southwestern Australia (Gould et al. 2011), (i) Pine forest in Florida (Sah et al. 2006), and (j) wet eucalypt forest in southwestern Australia (McCaw et al. 1996). The curves are for litter (a, c, d, f), litter and near-surface fuels (h), litter and elevated dead fuels (j), or total fuel load (b, e, g, i)

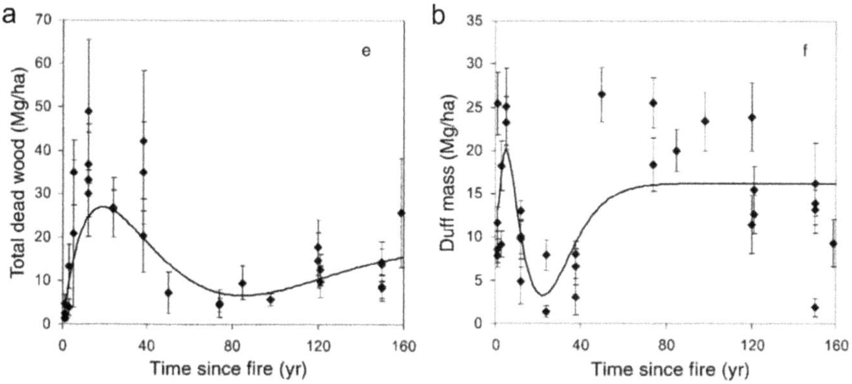

Fig. 11.7 Observed and modeled (curves) temporal patterns of (**a**) downed dead woody fuel and (**b**) duff along a 160-year chronosequence in the ponderosa pine forests of the Colorado Front Range, USA (Hall et al. 2006)

variable, depending on fire frequency, burn severity, site conditions, and the fuel component under consideration, as shown by a large study based on 182 sites sampled 1–24 years after ten large wildfires in central Idaho (Stevens-Rumann et al. 2020). Fuels increased post-fire, but less so when the site had been burned a few years earlier.

11.1.4 Fuel Dynamics and Plant Life Cycle

Depending on vegetation type, fuel dynamics can comprise important changes in properties other than fuel load or related metrics such as fuel depth or fuel cover. This is particularly noticeable when live fuels are a relevant component, as recognized early and modeled for grassland (McArthur 1966), shrubland (Rothermel and Philpot 1973), and woody understory (Hough and Albini 1978). In shrublands, dead fuel fraction increases with time, especially when the dominant species retain dead fuel in the canopy, and changes in bulk density and fuel partition by size class also occur. In northern Portugal's dry heathland, these dynamics (curve g in Figs. 11.6 and 11.8) concur to steady-state (asymptotic) fire behavior at ~15 years since fire, which matches well the region's median fire return interval (Fernandes et al. 2012a).

Grasslands go through seasonal growth cycles, with annual and perennial grasses differing in their seasonal growth and post-fire growth rates. Live biomass in senescing grasslands is gradually converted into dead fuels. This process is referred to as curing, and so the mixture of live and dead fuels changes throughout the growing season and increases the dead fuel fraction (Cheney and Sullivan 2008). Fire propagation in grassland requires a minimum curing level of ~20%, with fire spread rate rapidly increasing with increased curing (Cruz et al. 2015). This is because throughout the period of curing the mean fuel moisture content can vary

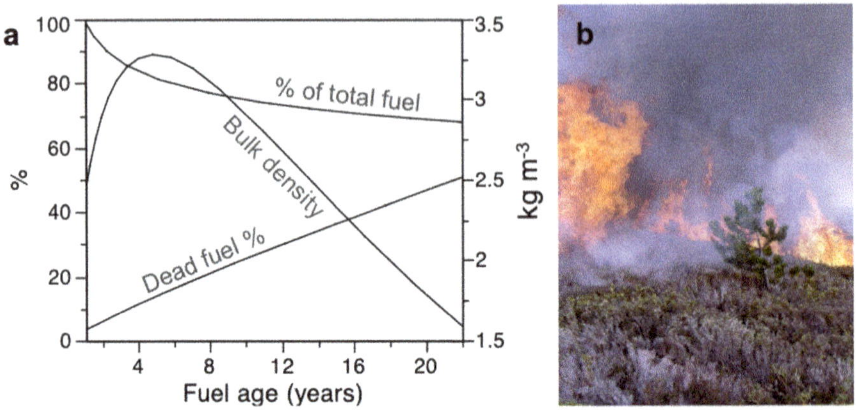

Fig. 11.8 (**a**) Structural dynamics of the fine fuels (diameter < 2.5 mm) in northern Portugal dry heathland of *Pterospartium tridentatum—Erica umbellata.* (Redrawn from Fernandes and Rego 1998). (**b**) Fire behavior under moderate fire weather conditions in a 21-year old stand, with senescent shrubs evident in the foreground. (Photo by Paulo Fernandes)

from above 300% to less than 10% (Cruz et al. 2015). Many studies have suggested that the effect of curing on fire spread is sigmoidal in nature (e.g., Cruz et al. 2015): there is little influence of live fuels on damping rate of spread at high levels of curing and a fairly linear relationship at moderate to low levels of curing.

The grass family also includes perennial evergreen species. Among these, bamboos display unique fuel dynamics on a time scale completely different from grasslands and savannahs. The flowering and fruiting of bamboo species are synchronous. It is followed by synchronous die-off that creates very high loads of fine, flammable fuels that can increase the likelihood of lightning-caused fires and facilitate crowning (Keeley and Bond 1999). *Chusquea culeou* is a prominent bamboo in southwestern South America, growing up to 6–8 m tall in the understory of dense deciduous *Nothofagus* forests and temperate rainforests (Fig. 11.9). These are not typically fire-friendly environments owing to high fuel moisture content (Kitzberger et al. 2016). However, a massive fuel hazard that persists for 4–5 years develops over large areas whenever *Chusquea* flowers, typically on 60–-70 year cycles. When combined with drought this enables large and severe fires that otherwise are not likely to occur (Armesto et al. 2009; Veblen et al. 2003).

11.2 Fuel Moisture Dynamics

Fuel moisture content (M) is by far the most temporally dynamic fuel property. As shown in previous chapters, M determines whether or not ignition and fire spread are possible. Moister fuels take longer to ignite and use more heat in the process. The burning rate decreases, less fuel is consumed, and so the flaming combustion of

Fig. 11.9 (**a**) Dead *Chusquea coleou* bamboo in the understory of rainforest dominated by the conifer *Fitzroya cupressoides* and the evergreen broadleaved *Nothofagus dombeyi* growing in Los Alerces National Park, Argentina. (**b**) Heavy litter load and dense clumps of culms over 2-m tall are evident. (Photos by Paulo Fernandes)

individual fuel particles takes longer (Nelson 2001). Consequently, directly or indirectly, fuel moisture content is a fundamental variable in fire danger rating and fire spread and fuel consumption models.

Fuel moisture dynamics differ between dead and live fuels. The moisture of the former reflects a passive (hygroscopic) response to the surrounding environment, whereas live fuels have physiological control over their moisture. Both live and dead fuel moisture reflect recent and long-term weather, but dead fuels respond more quickly to changing environmental conditions.

The water content of the live and dead vegetation involved in combustion plays a key role in determining fire spread and intensity. Fuel moisture varies at different time scales and changes differently between dead and live fuels. Fires spreading in live and dead fuels have different behavior as fires in live fuels can spread even when fuel moistures are above 100% (Weise et al. 2005).

Temporal variability in dead fuel moisture depends on the size of fuel particles. Compared to live fuel moisture, dead fuel moisture changes more rapidly in response to changes in temperature, humidity, and incoming solar radiation, which themselves depend upon the time of day, season, topography, and the vegetation structure.

The temporal variability of live fuel moisture is different from that of dead fuels. Unlike dead fuel moisture, which is primarily controlled through the loss or gain of water mass, live fuel moisture can be modified due to either a change in the actual mass of water present or through changes in the dry mass due to changes in plant phenology (Jolly et al. 2016). For example, jack pine (*Pinus banksiana*) and red pine (*P. resinosa*) dominated forests across much of North America experience a phenomenon known as the 'spring dip' in foliar moisture content just prior to new needle emergence (Van Wagner 1967; Jolly et al. 2016). The increased potential for crown fire during this period is often explained as a function of decreased moisture. However, several studies have indicated that the decline in foliar moisture content is driven by an increase in the dry mass content of the foliage, not a decline in the actual amount of water present in the foliage. This period is also associated with increased probability of crown fire behavior, as simulation results from Jolly et al. (2016) found that the increased amount of mass associated with the spring dip resulted in a shift from a surface fire to crown fire and an increase in the fire rate of spread and fireline intensity. Because of the different behavior between live and dead fuel moisture and resulting fire spread, the change from live to dead fuels is important to understand.

11.2.1 Dead Fuel Moisture

Dead fuels increase their moisture content through adsorption of water vapor, condensation, or precipitation, and decrease it through desorption and evaporation (Viney 1991). Dead fuels can hold increasingly more water within their cell walls until reaching the M fiber saturation point, usually 30–35%. Higher M values are possible depending on the amount of precipitated or condensed water at the surface of fuel particles and in their interstices and its absorption into cell cavities. Different mechanisms govern fuel moisture exchanges below and above cell saturation. Water vapor diffusion and permeability to water both vary with fuel properties at the particle and fuelbed levels, namely surface area-to-volume ratio and packing ratio (Nelson 2001). Fine fuels arranged in porous fuelbeds will lose or gain moisture quickly.

The temporal dynamics of dead fuel moisture content are mostly a function of variation in atmospheric conditions and precipitation patterns. However, different fuels (as defined by characteristics such as particle thickness, fuel layer depth and compactness, and position in the fuel profile) respond differently to those influences. Two related concepts are important to understand the dynamics of dead fuel moisture (M): equilibrium moisture content (EMC) and response time (Simard 1968; Byram and Nelson 2015). EMC is the eventual moisture content of dead fuels when exposed to constant relative humidity and ambient temperature. EMC is reached when there is no gain or loss of water between fuels and the adjoining air. Thus, current M lags behind EMC, even for rapidly responding extremely fine grass and moss fuels, and M at any given moment reflects the recent past conditions. For any given combination of relative humidity and air temperature, EMC is higher when fuels are losing (desorption) than when they are gaining (adsorption) water.

EMC, as well as the difference between desorption and adsorption curves, is observed in the laboratory but is seldom arrived at under natural conditions. This happens because air temperature and relative humidity vary continuously and because M is affected by additional variables, namely solar radiation and wind speed. Solar radiation warms the environment surrounding the fuel, and while wind cools fuels exposed to the sun, it also increases the evaporation rate. The rate at which a given fuel approaches EMC can be expressed by the fuel response time, or time lag constant, that follows an exponential curve and is defined as the time required for fuel to attain 63.2% of the change between the initial and the final M (Byram and Nelson 2015).

The time lag concept has been adopted by the US National Fire Danger Rating System (NFDRS, see Chap. 8) to assess M and its effect (Deeming et al. 1977). It is used to categorize dead fuels and partition their load in fuel inventories (Brown 1974) and fire behavior prediction models (Rothermel 1972). Three classes are considered, with time lags of 1, 10, and 100 h, respectively, described as fine, medium, and large fuels and corresponding to fuel particle diameters or thicknesses of <0.6, 0.6–2.5, and 2.5–7.5 cm. Those time lag classes can also be assumed as roughly and respectively representing the moisture contents of dead surface fuels directly exposed to weather influences (up to a 0.6-cm depth in the forest floor), the litter from just below the surface up to a 2.5-cm depth, and the rest of the forest floor up to a 10 cm depth (Deeming et al. 1977). The NFDRS also considers 1000-h fuels to account for the burn availability of larger (7.5–20 cm) downed wood and deeper (10–30 cm) layers of duff. Note that these response times are nominal and thus simplify natural variability. For example, Anderson (1990) found that the actual time lag of non-weathered fine fuels varied from 0.2 to 37 h as a function of the surface area-to-volume ratio of fuel particles and the packing ratio and depth of the fuelbed.

The Canadian Forest Fire Weather Index System (CFFWIS, See Chap. 8) includes three codes for the moisture status of three forest floor layers (Van Wagner 1987). The Fine Fuel Moisture Code (FFMC) represents fuels thinner than 1 cm in the top litter layer. The Duff Moisture Code (DMC) is indicative of the decomposing forest floor. The Drought Code (DC) represents deep and compact layers of mostly decomposed organic matter. The FFMC, DMC, and DC have nominal fuel depths of respectively 1.2, 7, and 18 cm and time lags of 16, 288, and 1248 h and so track dead fuel moisture content for fire danger rating purposes at daily to seasonal scales. While the Canadian and US methods are not strictly comparable, rough equivalents can be established between the FFMC and a composite of 1- and 10-h fuels, the DMC and 100-h fuels, and the DC and 1000-h fuels (Van Nest and Alexander 1999).

Forest floors waterlogged by prolonged rainfall or snowmelt have the highest and most uniform moisture contents, up to 400%. As shown by controlled experiments in

Fig. 11.10 Indoors drying curves for 7.6-cm thick *Pinus ponderosa* duff after simulated rainfall at a rate of 27 mm h^{-1}. Sections of the forest floor were cut, taken to the laboratory, wetted and allowed to dry under constant ambient conditions of temperature and relative humidity. (Redrawn from Stocks 1970)

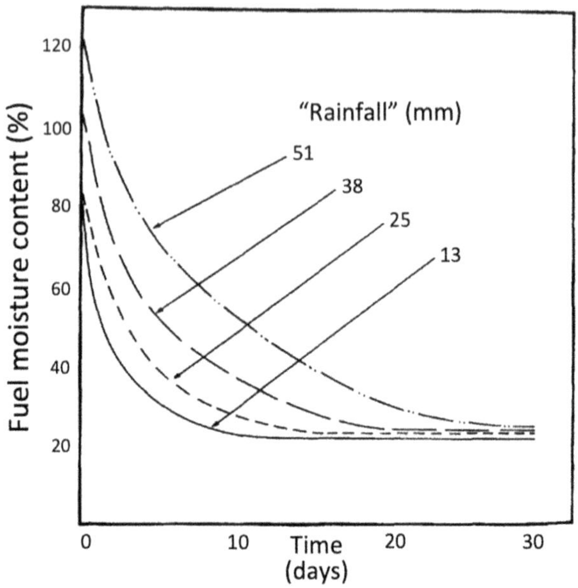

the laboratory (Stocks 1970, Fig. 11.10), the duff M immediately after a rain event is dependent on the amount of precipitation. The subsequent drying follows an exponential decay and converges to a final minimum M value. The influence of ambient weather on drying decreases with depth in the forest floor owing to increased shielding from surface conditions and, typically, higher compactness. Consequently, duff at increasingly deeper locations will dry at a slower rate. Marked inversions are possible in the forest floor's M profile, namely when the first rainfall event after a dry period is insufficient to wet the duff layer fully. Post-rainfall drying patterns are faster in more open vegetation types, as found by various studies cited by Matthews (2014).

The CFFWIS moisture codes can be converted to actual M (Van Wagner 1987), allowing inspection of the temporal dynamics of M variation among and between fuel layers. For example, M saturation after rainfall followed by a 4-month rainless period from late spring to the end of the summer, which is common in Mediterranean-type climates, is shown in Fig. 11.11. Under the air temperature and relative humidity conditions observed, the deep humus layer and fallen logs represented by the DC maintained M values above 100% for almost 3 months. Note that there are limitations in this usage of the DC, given the inherent differences between boreal (deeper) and Mediterranean (shallower) forest floors. Nevertheless, the overlying decomposing duff (characterized by the DMC) required just 3 weeks to dry to less than 100% M and in 2 months attained the steady-state M of 20%. In contrast, the precipitation influence on the M of surface fine dead fuels in the outermost litter layer, represented by the FFMC, vanishes in 2–3 days. Subsequent M fluctuation is solely due to variation in temperature, relative humidity, and wind speed.

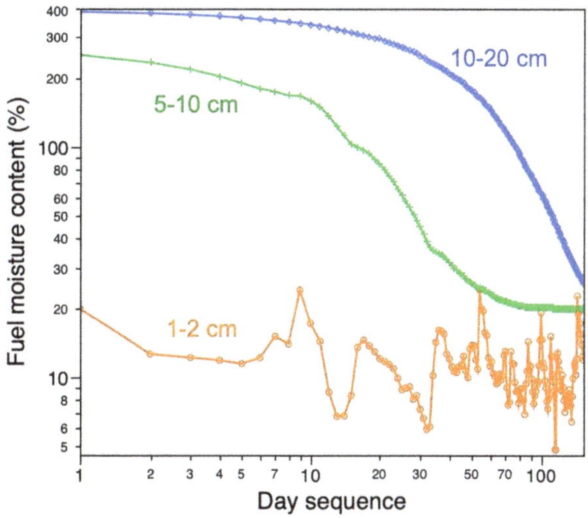

Fig. 11.11 Forest floor moisture contents at different depths converted from the Canadian FWI System moisture codes, respectively FFMC (1–2 cm), DMC (5–10 cm), and DC (10–20 cm). The estimates are based on observed data (May 1 to September 30, 2019) at the University of Trás-os-Montes and Alto Douro weather station (Vila Real, Portugal) but assuming moisture saturation at the onset of the time period and no rainfall until the end of it

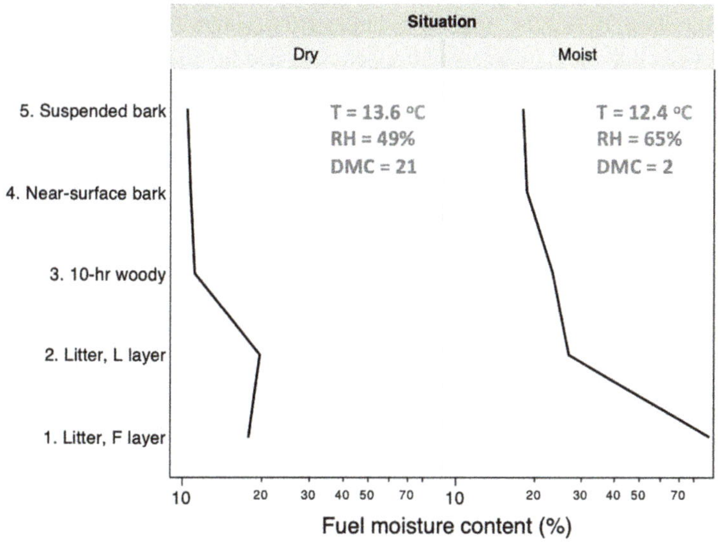

Fig. 11.12 Early afternoon vertical profile of dead fuel moisture content observed in a blue gum (*Eucalyptus globulus*) plantation in southern Portugal in 2 winter days, respectively dry and moist as determined by atmospheric conditions and recent rainfall. T, RH, and DMC are, respectively, in-stand (2-m height) ambient temperature and relative humidity and the Duff Moisture Code (DMC) of the Canadian FWI system from the nearest weather station. (Drawn from data on file, Pinto et al. 2014)

The fuel moisture in forests reflects weather, species, and position (Fig. 11.12). Comparing sampled fuel moisture contents in a *Eucalyptus globulus* plantation in southern Portugal between 2 winter days, respectively termed "dry" and "moist" illustrates the relevance of fuel position on a vertical axis by showing the entire profile of M variation for surface fuels. Stands of eucalypt species with smooth decorticating bark such as *E. globulus* have semi-detached bark streamers along the trunk and accumulate it around the tree base, posing spotting problems (Chap. 8). Compared with the moist (post rainfall) situation, the dry situation reflects three rainless weeks and warmer and drier atmospheric conditions. A pronounced difference in the M of the decomposing layer between the 2 days is manifest. However, M decreased in general with height, as suspended and elevated fuels are more exposed to weather influences. While on the "moist" day, the contrast is mostly between the F-layer litter and the other components, with poor distinction among the latter, the "dry" day features homogeneous M in the litter but at a substantially higher level (18–20%) than the overlaying fuels (~12%). Similar vertical gradients have been observed between L-layer litter and elevated dead fuels in understory shrubs in pine stands (Fernandes et al. 2009). The "moist" situation would likely produce a very low-intensity fire with partial removal of the litter and insignificant smoldering, but the "dry" situation would result in a more intense fire with homogeneously high fuel consumption, smoldering, and combustion of elevated bark.

Thus, both short and long-term dead fuel moisture differ between fuel layers. By monitoring those dynamics, directly or indirectly (through fire danger indexes), fire managers can link them to potential fire behavior and fire effects as part of planning for both the control and the use of fire.

The moisture content of fine dead fuels plays a critical role in fire behavior. Small decreases in M at the low end of its range (say 2–8%) correspond with disproportionately greater increases in the fire-spread rate (Chaps. 7 and 8). M can be determined directly by oven drying fuel samples, semi-directly through electrical resistance measurement, or using fuel moisture sticks as proxies. But these methods require equipment and, in the case of oven drying, time for processing, and they cannot be used for prediction in an operational context. It comes as no surprise, then, that huge efforts have been undertaken over the years to develop sound and reliable models of M for fire management purposes (Viney 1991; Matthews 2014).

The existing models range from simple empirical equations to process-based models based on energy and water balance conservation equations. Precipitation and condensation are difficult to tackle, and their influences are minor or absent during the more fire-prone seasons, days, and hours of the day. Many models therefore only consider vapor exchange processes and rely on air temperature and relative humidity to estimate either the EMC or actual M. For many practical purposes, the EMC can be considered an acceptable estimate of fine fuel M, as the lag of actual M in relation to EMC can be less than 1 h (Viney and Hatton 1989; Anderson and Anderson 2009). Models based primarily on vapor exchange and estimates of the antecedent and instantaneous air temperature, relative humidity, and precipitation have been and continue to be the more common approaches used by fire managers. More recently, the NFDRS has adopted the model of Nelson (2000), which also integrates the effects of evaporation, dew formation, and solar radiation. Predictions from simplified forms of "complete" process-based models are now available for some Australian fuels, e.g., Matthews (2014). Two examples of the

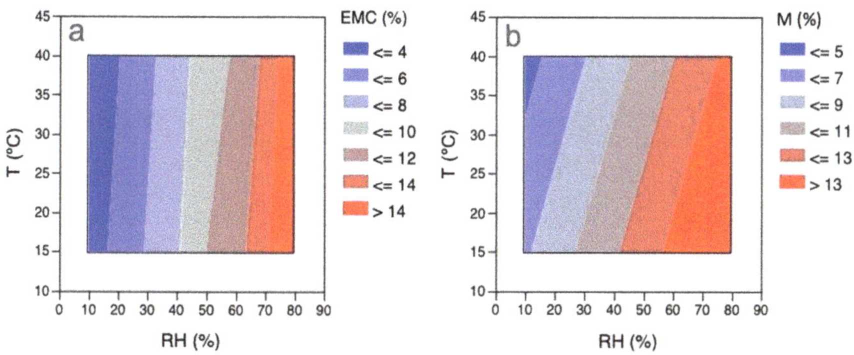

Fig. 11.13 Examples of (**a**) equilibrium (EMC) or (**b**) actual (M) moisture content of fine dead fuels predicted from air relative humidity (RH) and temperature (T). Estimates are from an empirical model (**a**, Simard 1968) and a process-based model calibrated for shrubland and assuming solar radiation above 500 W m^{-2} (clear sky in the early afternoon) (**b**, Anderson et al. 2015)

predicted EMC or M for fine dead fuels as a function of relative humidity and air temperature are shown in Fig. 11.13. For a given relative humidity value, the moisture content will decrease with increasing temperatures.

The moisture content of surface fine dead fuels experiences pronounced daily variation that can be described as a 24-h sinusoidal cycle with a minimum in the mid-afternoon and a maximum before sunrise (Viney 1992; Catchpole et al. 2001). This additional feature of fuel moisture dynamics is a combined outcome of variation in ambient weather and solar radiation throughout the day and consequently is affected by aspect and slope. For example, daytime variation in M (Fig. 11.14) can be estimated using the model of Rothermel et al. (1986), which essentially extends the Canadian FFMC code to integrate solar radiation. The minimum M contents were attained during the morning (9–12 AM), reflecting not just the weather conditions observed locally on a specific day (Fig. 11.14), but also the topographic context: a steep slope facing east, which is heated by the sun early in the morning.

Differences in M between stands (Fig. 11.14) can be significant, especially when they occur at the low end of the M range and consequently exacerbate differences in potential fire behavior between the three forest types (Pinto and Fernandes 2014). Note that the weather data collected inside stands indicate the combined effect of micrometeorology and solar radiation (as determined by stand structure). M was highest in the deciduous *Betula* stand, intermediate in the dense *Chamaecyparis* plantation, and lowest in the comparatively open *Pinus* stand.

The many variables that affect M (weather, topography, and vegetation) and their corresponding interactions in time and space make predicting M challenging. However, relevant progress has been made in mapping modeled M (Holden and Jolly 2011; Sullivan and Matthews 2013).

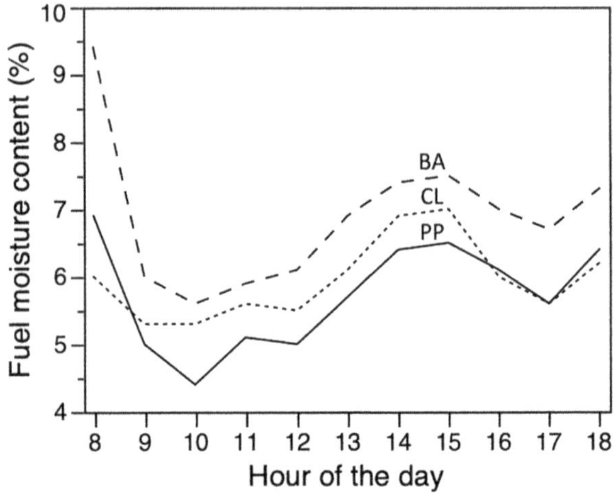

Fig. 11.14 Hourly daytime (8 AM to 6 PM) estimates of fine dead fuel moisture content in three forest stands in northern Portugal, respectively *Betula alba* (BA), *Chamaecyparis lawsoniana* (CL) and *Pinus pinaster* (PP), during one summer day. The stands are adjacent to one another and located at an elevation of 1100 m on an east-facing 25° slope. The estimates (Pinto and Fernandes 2014) were obtained with the M model of Rothermel et al. (1986) using within-stand measured weather and stand structure data

11.2.2 Live Fuel Moisture

Live fuels are an important or dominant component of the fuel complex in many vegetation types worldwide, including grasslands, shrublands, woodlands, and open forests. Live fuels are the vector of crown fire spread in conifer forests. A balance between two physiological processes governs the moisture content of live foliage: water uptake through the roots and water loss by transpiration. As these processes are related to water availability, the climate, environment, phenology, and species adaptations are essential factors. These processes also vary with the age of the leaves, resulting in significant differences between deciduous and evergreen species. Because of these relationships, the moisture content of leaves varies with the type of species and environment but also seasonally and diurnally. Van Wagner (1977) indicated that while deciduous broadleaves maintain FMC values from about 140 to 200% after the foliage-flushing period is over, the conifer forests of Canada most prone to crowning have values of foliar moisture content (FMC) from about 70 to 130% and eucalypts and chaparral are often at values of 100% or less. In the next sections, we will exemplify these relationships.

The Conifer Forests of North America

Most of the research on temporal variation in leaf moisture has been conducted in North America's crown fire-prone conifer forests. The seasonal trends in live moisture content of conifer needles were studied for jack pine (*Pinus banksiana*) and red pine (*P. resinosa*) by Van Wagner (1967) in the Petawawa Research Station in eastern Canada. Others continued similar studies, such as Jolly et al. (2016), who have carried out comparable work in Wisconsin (Fig. 11.15).

Similar trends were observed by Van Wagner (1967) for other North American conifer species, including white pine (*Pinus strobus*), balsam fir (*Abies balsamea*), and white spruce (*Picea glauca*). All conifer species show stable values throughout the year with a minimum moisture content of old leaves at spring (known as the spring dip) simultaneous with the flux of new leaves.

Temperate Deciduous Broad Leaves

Different authors in different parts of the world have studied the seasonal variation of leaf moisture of temperate deciduous broadleaves. Van Wagner (1967) addressed two important broadleaf species in eastern Canada: sugar maple (*Acer saccharum*) and trembling aspen (*Populus tremuloides*). Similar studies were conducted in France, where Leroy (1968) and Le Tacon and Toutain (1973) focused on two very important broadleaved species in temperate Europe: the European oak (*Quercus robur*) and the European beech (*Fagus sylvatica*) (Fig. 11.16).

In temperate conditions in North America and Europe, all the deciduous broadleaf species showed similar trends. Leaf moisture is very high (more than 200%) at the

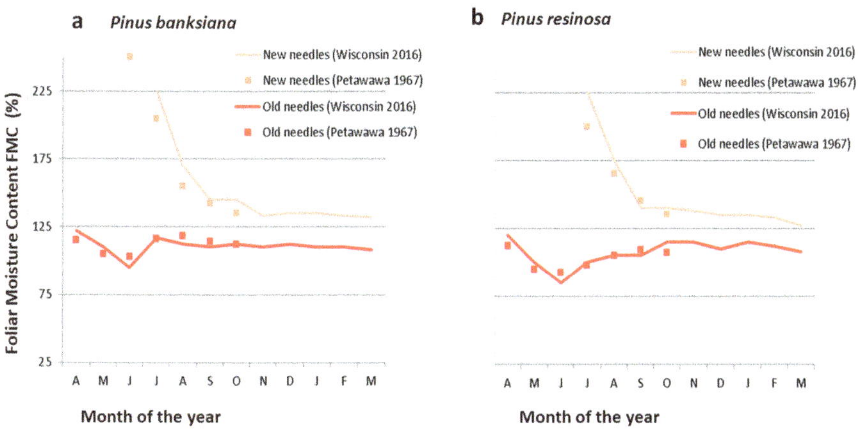

Fig. 11.15 Seasonal variation of foliar moisture content (FMC) of old and new needles of (**a**) jack pine (*Pinus banksiana*) and (**b**) red pine (*P. resinosa*) The graphs show remarkable agreement between results of the pioneering work of Van Wagner (1967) in eastern Canada (squares) with those obtained 50 years later by Jolly et al. (2016) in central Wisconsin (solid lines)

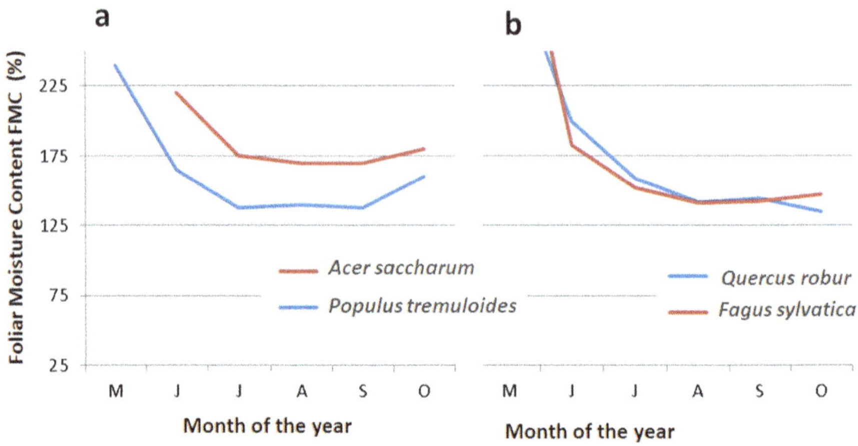

Fig. 11.16 Seasonal variation (May to October) of foliar moisture content for **a** sugar maple (*Acer saccharum*) and trembling aspen (*Populus tremuloides*) in eastern Canada (adapted from Van Wagner 1967) and **b** for the European oak (*Quercus robur*) and the European beech (*Fagus sylvatica*) in France. (Adapted from Leroy 1968 and Le Tacon and Toutain 1973)

beginning of the growing period (typically May). Leaf moisture subsequently decreased during summer, but was always relatively high (between 125 and 175%). These moisture values are beyond the thresholds for burning, justifying the inclusion of these species in studies related to crown fires only for comparison as "in Canada at least, only conifer stands will support crown fires" (Van Wagner 1967).

Evergreen Trees and Shrubs in Mediterranean-Type Climates

Different evergreen tree and shrub species show different adaptations to water stress under the same Mediterranean climate, exemplified by the Algarve region in southern Portugal (Fig. 11.17). Some tree species, such as pines (*Pinus pinaster* and *P. pinea*) and eucalypts (*Eucalyptus globulus*), keep a relatively constant foliar moisture content (around 125%) throughout the year. Other species like the strawberry tree (*Arbutus unedo*) show large variations around the average of FMC 125%, with a maximum in May and a minimum in September and October. Less pronounced but similar seasonal variation occurs for cork oak (*Quercus suber*) leaves with lower FMC reaching 75% in the fall.

In Mediterranean-type climates, live fuel moisture correlates well with moisture availability in the soil, as shown by Olsen (1960) for three chaparral shrub species, including chamise (*Adenostoma fasciculatum*), hoaryleaf ceanothus (*Ceanothus crassifolius*), and black sage (*Salvia mellifera*) in California. In the Mediterranean-type climate, soil moisture is low throughout summer and autumn. All chaparral species show low live moisture contents, indicating that they can burn readily after

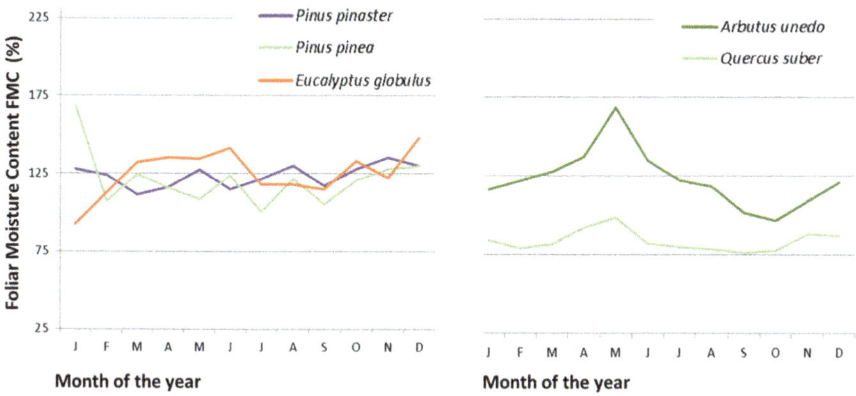

Fig. 11.17 Seasonal variation of foliar moisture content for five tree species in the Algarve region in southern Portugal. (Unpublished data from the authors)

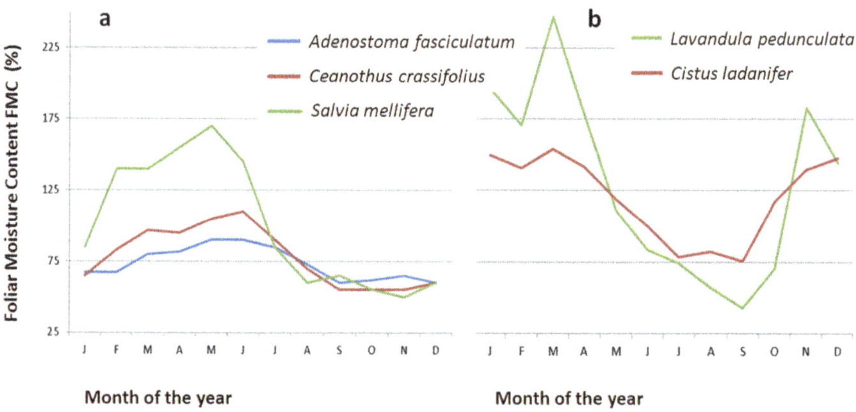

Fig. 11.18 Seasonal variation of (**a**) three chaparral species, chamise (*Adenostoma fasciculatum*), hoaryleaf ceanothus (*Ceanothus crassifolius*), and black sage (*Salvia mellifera*) in southern California, adapted from Olsen (1960), and (**b**) of two Mediterranean shrubs, French lavender (*Lavandula pedunculata*) and gum rockrose (*Cistus ladanifer*) in Algarve, Portugal. (Unpublished data from the authors)

July. Similar trends were observed for two Mediterranean shrubs, French lavender (*Lavandula pedunculata*) and gum rockrose (*Cistus ladanifer*) in the Algarve region in Portugal (Fig. 11.18).

Forests with Understory Shrubs

In general, forests support many different understory plants that occupy different vertical niches and distinct seasonal patterns in foliar moisture. Seasonal variations

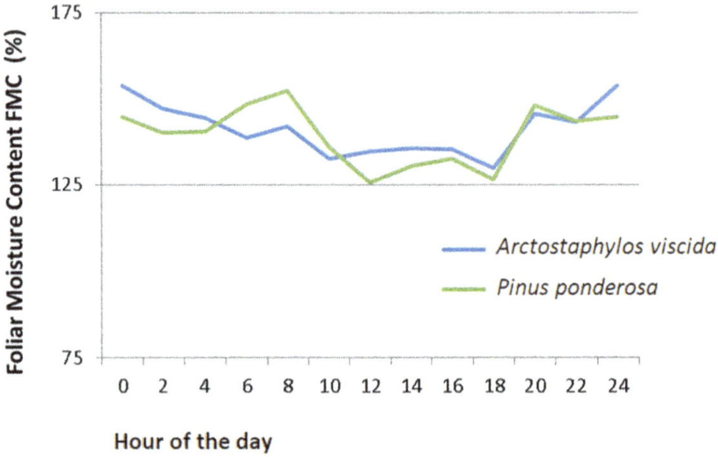

Fig. 11.19 Daily fluctuation of foliar moisture (FMC) of whiteleaf manzanita (*Arctostaphylos viscida*) and ponderosa pine (*Pinus ponderosa*). (Adapted from Philpot 1965)

of foliar moisture of pines and associated shrub species have been documented. Qi et al. (2016) compared the foliar moisture of lodgepole pine (*Pinus contorta*) with that of big sagebrush (*Artemisia tridentata*) in Montana. The foliar moisture content of the shrubs (manzanita and sagebrush) show a marked summer decline in response to soil moisture. This decline was especially sharp in big sagebrush as FMC decreased very rapidly from more than 175% in July to about 75% in September. The moisture content of the needles of the two pines (ponderosa and lodgepole) was relatively constant through time. The foliar moisture of old needles of the two pine species varied between 100 and 125%.

Live foliage moisture varies diurnally. Philpot (1963) studied ponderosa pine (*P. ponderosa*) and whiteleaf manzanita (*Arctostaphylos viscida*) in California (Fig. 11.19). The moisture is highest at night and lowest in the afternoon. Philpot (1965) demonstrated significant within-day differences in foliar moisture both for ponderosa pines (3–10 m tall) and shrubs (1–1.5 m tall) of whiteleaf manzanita during summer in California. These results agree with others obtained for *Pinus edulis* and *Ilex glabra* and summarized by Chandler et al. (1983), which suggest that the amount of moisture change throughout the day is closely correlated with temperature changes. As soil moisture in the rooting zone of woody species is relatively constant throughout the day, the main process driving the diurnal variation is transpiration. Leaf stomata are the main avenue for water loss from transpiration, and stomata usually close at night and open in the day in response to solar radiation, ambient temperature, air relative humidity, and wind. Philpot (1965) suggested that the diurnal fluctuation in both ponderosa pine and manzanita leaves' moisture content partly explains differences in fire behavior between night and midday.

Various environmental and physiological factors govern the moisture content of live fuels, making it more difficult to predict than dead fuel moisture. Further, most live fuel complexes include a mixture of species quite variable in foliage moisture content. Consequently, current tools and approaches to estimate FMC for fire management purposes are limited, including remote sensing (Yebra et al. 2013) and the establishment of relationships with drought indices or the moisture of the slowest-drying dead fuels (Burgan 1979; Pellizzaro et al. 2007). For this reason, fire managers often rely on FMC monitoring programs based on the destructive sampling of indicator species to provide estimates of live fuel moisture (Weise et al. 1998).

11.3 Fuels Management

Fuels treatments modify the amount, composition, and structure of the fuel complex to alter fire behavior or to minimize the negative impacts of future wildfires on ecosystem goods and services, cultural resources, and human communities (Hoffman et al. 2018). The limited scope of ignition control programs and the insufficiency of firefighting technology under elevated fire danger conditions, which account for most of the burned area, led Countryman (1974) to argue for a central role for proactive fuels management in risk reduction. Paradoxically, the need for fuels management is especially evident when high investment and organization levels result in prompt fire detection and suppression (Finney and Cohen 2003). However, fires surviving initial attack can easily turn into large and severe fires when unfavorable weather combines with high fuel hazard. Allocating much of the fire management budget to fire suppression-related activities, instead of to fuels treatments, can postpone and potentially magnify the impacts of undesired fire because it facilitates fuel buildup, in what is known as the "fire paradox" (Arno and Brown 1991).

Fuels treatments have become a valuable management tool, e.g., in dry forests in the western USA ecosystems where fire suppression and timber harvesting have led to increases in surface and canopy fuels within and around the wildland urban interface (WUI) (Graham et al. 2004; Hudak et al. 2011; Covington and Moore 1994; Stephens and Fulé 2005; Hessburg et al. 2005). Nonetheless, fuels management as a fundamental, broad-scale, and persistent component of fire management is scarce worldwide (see Chap. 13 for examples).

11.3.1 Fuels Management Strategies

Fuels management comprises three basic strategies: fuels reduction, fuels isolation, and fuel type conversion (Pyne et al. 1996). Although the goals of fuels reduction and conversion are to modify fire behavior, fuel isolation breaks up fuel continuity in the landscape to hinder fire spread. The techniques involved in the three basic strategies are similar; however, fuel isolation is implemented in the form of relatively linear fuelbreaks rather than across an area.

Fuel reduction decreases the quantity of fuels available for combustion, to which fire intensity responds linearly. Consequently, a fuel-reduced area slows down a wildfire, lowers flame size and heat release, and reduces spotting, the likelihood of plume-driven fire development, and smoldering combustion. Fuel reduction facilitates fire suppression operations directly, by modifying fire behavior, and indirectly by improving access, visibility, rate of containment-line construction, anchor points, safety, and optimal allocation of fire suppression resources (Fernandes 2015). Attacking the head of a typical wildfire is often impossible or unsuccessful. However, fire behavior varies markedly around the fire perimeter (Catchpole et al. 1993). Fuel reduction increases the extent of the fireline that can be tackled by direct attack and the associated spatiotemporal windows of opportunity. By allowing safer and more effective work on the flanks of a wildfire, fuel reduction decreases the potential for rapid fire growth when sudden shifts in wind direction and speed occur. Overall, fuel reduction increases fire control options and the corresponding effectiveness of fire suppression.

Although less often appreciated, fuel reduction also mitigates fire impacts (Chap. 9), such as soil heating, smoke production, carbon emissions, and plant injury and mortality, with potentially faster and more thorough post-fire recovery. By decreasing both flaming and non-flaming combustion, fuel reduction diminishes fire risk and the costs of wildfire suppression and post-fire rehabilitation (Fig. 11.20).

The fuel isolation strategy reduces the continuity of flammable vegetation by establishing narrow fuelbreaks of variable width, with residual trees (shaded fuelbreaks) or without trees, and within which fuels are reduced to confine wildfires. Ideally, fuelbreaks should be used as a basis for the gradual expansion of fuel reduction (Weatherspoon and Skinner 1996; Agee et al. 2000). Fuel isolation ranges

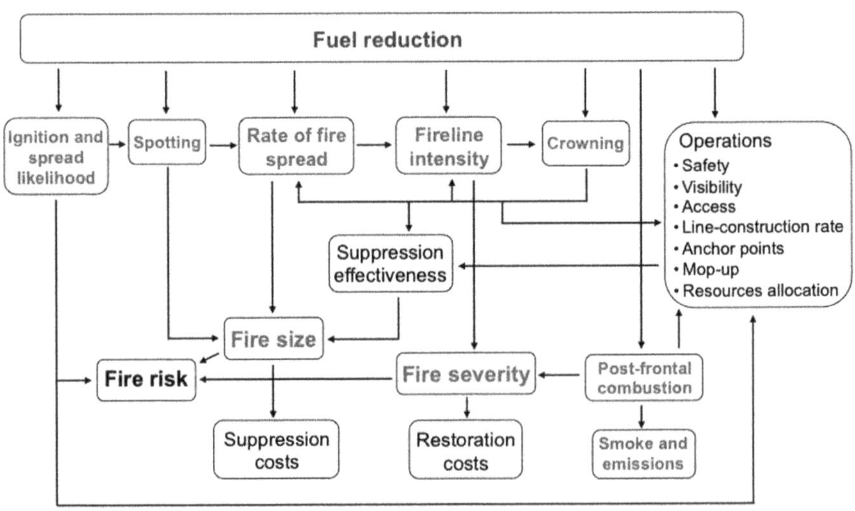

Fig. 11.20 Effects of fuel reduction treatments on fire behavior and effects, including the implications for fire suppression operations and costs

Fig. 11.21 Low-flammability environments can be achieved through forest type conversion, namely to deciduous broadleaves. (**a**) Patchy, low-intensity burning and self-extinction of a wildfire in *Betula pubescens* forest. (**b**) The green area denotes unscorched or unburned mixed broadleaved forest (mostly *Quercus robur* and *Castanea sativa*). (Photographs by Paulo Fernandes in northern Portugal)

from bare and narrow strips, typical of plantation forestry, to wide (>100 m) infrastructured fuelbreaks, i.e., including access routes and water points to support fire suppression. Fuel breaks can create conditions that expand the fire suppression capacity of ground resources and the effectiveness of aircraft drops (Weatherspoon and Skinner 1996). The design of a fuelbreak network can integrate and expand the diversity of existing land uses, and take advantage of topography and existing vegetation. Fuel isolation can take the form of "green" fuelbreaks (greenbelts) composed of low flammability species.

Similar to area-wide fuel reduction treatments, fuel conversion is expected to moderate the spread and effects of fire on the landscape, but by replacing vegetation, the effect may last longer (Fig. 11.21). The effectiveness in altering fire behavior depends on the overall fire environment in terms of physical fuel properties, fuel moisture, and wind speed (Pinto and Fernandes 2014). The conversion strategy is constrained by the options available and the resulting ecological changes. Depending on the context, it can be achieved by allowing plant succession to proceed, e.g., towards mesic or moister forest types in general, namely deciduous hardwoods or mixed deciduous-conifer stands.

The spatial layout of fuel reduction and conversion units in the landscape should be guided by factors such as the fire regime, fire management objectives, topography, site productivity, and the spatial pattern of values at risk (Ager et al. 2013). Fuelbreaks to facilitate fire suppression can be located to protect localized assets, e.g., the WUI, or to contain large fires at strategic locations. Area-wide treatments serve purposes of broad landscape protection or burn severity reduction; both purposes entail decreased fire behavior but, only the former actually implies a reduction in burn probability.

11.3.2 Fuel Reduction Principles and Techniques

Fuels treatments seek to modify fire behavior and/or effects, but a diversity of methods can be used depending on the specific management objective. For example, treatments within and around the WUI are often developed to reduce fire rate of spread, flame length, and intensity with the primary objective of facilitating fire suppression and protecting human life and property. Treatments away from the WUI may emphasize reducing fire intensity or the potential for crown fire with the primary goal of reducing burn severity so that fires can occur without negative impacts on ecological function (Reinhardt et al. 2008). Fuel treatments could also be designed to support the use of fire and to manage fires to burn through landscapes without loss of valued assets.

The choice of methods should be informed by an understanding of the role of fuel characteristics on fire behavior. Different fuel layers have different influences on fire behavior and affect different fire characteristics (Cheney 1990; Peterson et al. 2005) (Fig. 11.22). Compactness typically decreases from the bottom of the forest floor to the top of the understory. The finer fuels in litter and in low grassy or woody vegetation (plus moss and lichen in boreal forests) contribute to the leading edge of the flame front and drive surface fire spread. Coarse woody fuels and ground fuels such as duff do not add significantly to the heat flux at the fire front but are important contributors to the burnout time and total heat released during a fire. Compact fuelbeds, such as deep duff on the soil surface, do not support flaming combustion. However, the ascending heat from all fuels combined, plus flame contact from the combustion of tall shrubs and ladder fuels, can enable a crown fire, whose spread and intensity are influenced by foliar density and moisture in the canopy (Chap. 8).

The technical specifications of fuels treatments depend on factors such as vegetation type, the vertical distribution of fuel, and environmental impacts of the operations (Peterson et al. 2003). Fuel reduction in open vegetation is simply the removal or structural modification of the grass, shrub, or slash layer. Fire managers can design treatments to meet various goals. In conifer forests, fuels treatments are often designed to reduce the potential for crown fires because crown fires are associated with high rates of spread and fireline intensities, are more difficult to

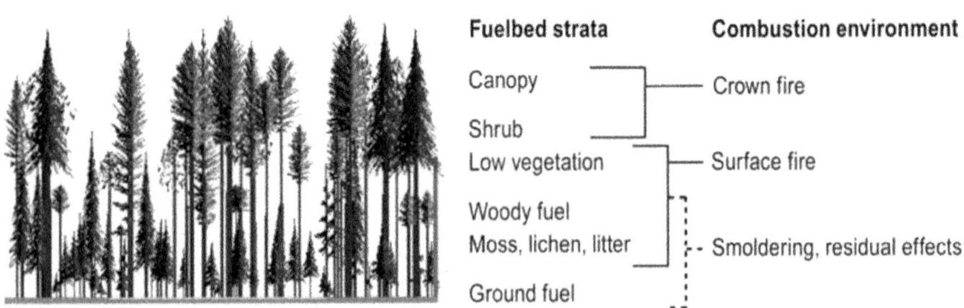

Fig. 11.22 Targeting different fuel strata for treatment impacts fire behavior differently. (From Peterson et al. 2005)

control, and pose a significant risk to life and property (Scott and Reinhardt 2001; Hoffman et al. 2018), as discussed in Chap. 8. Furthermore, crown fires are increasingly common in forests around the world. Fuels management to reduce the potential for crown fire ignition and spread is based on our understanding of the relationship between fuels and fire behavior. Four principles guide treatment design and define a hierarchy of treatment priorities (Graham et al. 2004; Agee and Skinner 2005):

1. Reduce surface fuels to decrease potential for high fire spread rate and intensity,
2. Break vertical continuity and minimize the likelihood of crown fire initiation by pruning trees to increase canopy base height and removing ladder fuels such as tall shrubs and small trees,
3. Thin the overstory to reduce the concentration of foliar biomass and reduce the possibility of tree-to-tree fire spread in an active crown fire, and
4. Remove smaller individuals and species with little resistance to fire to lessen tree mortality.

Surface Fuels Treatments

Various alternatives exist to reduce fuels underneath forest canopies and in open vegetation. Two general types of treatments can be distinguished: those that reduce fuels through consumption (e.g., prescribed burning and grazing, Fig. 11.23a, d) and those that rearrange fuels (e.g., mastication and other mechanical treatments, Fig. 11.23b, c). The latter make fuels less available for combustion, but often require supplementary treatment if fuels are to be removed completely.

Prescribed burning is particularly suited to accomplish fuel management on a significant spatial scale. Prescribed burning should conform to a predefined meteorological window (Fig. 11.24), as narrow as the specificity of treatment objectives dictates but wide enough to maximize the opportunities for success. The prescriptions are carefully chosen to result in fire behavior to accomplish the desired fuel consumption and fire effects. Although the fuel-reduction impact depends essentially on the moisture content gradient in surface and ground fuels, it is typically only the finer and more aerated components of the fuel complex that are substantially reduced with prescribed burning. However, prescribed burning can also consume or scorch ladder fuels in the lower canopy and kill dominated trees, hence increasing canopy base height and reducing canopy fuel load. In some locations, crown fires are prescribed. Planned fire or managed (under prescription) wildfire are the options of choice to simultaneously decrease fuel hazard and maintain or restore fire-adapted or fire-dependent ecosystems, such as the dry conifer forests of the western USA (Keane 2015). Often, prescribed burning fulfills other goals in addition to fuel reduction. Worldwide examples of prescribed burning programs as part of integrated fire management are presented in eight case studies in Chap. 13 and Case study 12.2.

Prescribed burning is less favored in other circumstances, such as those that involve risks to valued resources, e.g., to people especially at or near the WUI, or to plantation forestry of thin-barked trees. Several alternatives to prescribed burning

Fig. 11.23 Examples of common fuel treatments. (**a**) prescribed burning in southwestern Australia eucalypt woodland. (**b**) Mastication in western USA conifer forest. (**c**) Mechanical understory shredding in Portuguese pine forest. (**d**) Goat grazing maintaining a fuelbreak in Portugal. (Photographs by Paulo Fernandes, except (**b**) taken by Mike Battaglia)

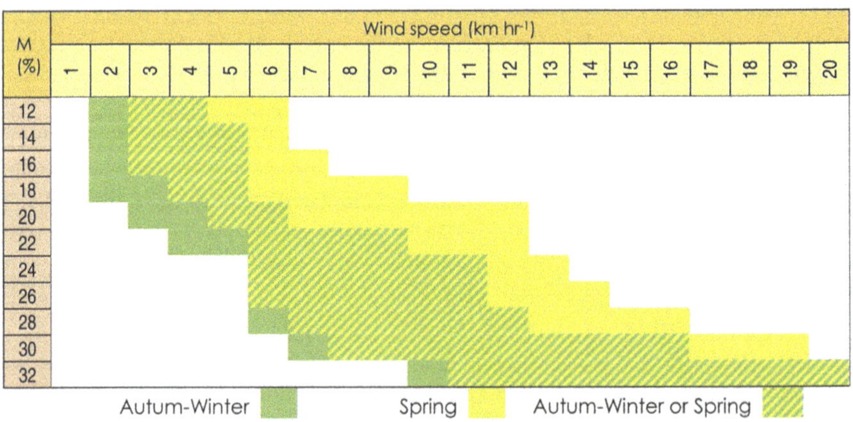

Fig. 11.24 Optimum burning window to reduce fuels in low (<1 m tall) dry heathland in Portugal dominated by the shrubs *Pterospartium tridentatum* and *Erica umbellata* as a function of elevated dead fuel moisture (M) content and 2-m wind speed in the open. Seasonal differences reflect differences in live fuel moisture content. (From Fernandes and Loureiro 2010)

exist, although they are often less cost-effective and have less impact on fuels. Motorized shrub cutting by hand crews only decreases fuel height. Mechanical treatments are constrained by accessibility, e.g., due to slope, and many require subsequent removal or on-site processing of the residual fuels to be effective. However, sufficiently compacted fuels can result from tractor-pulled mechanized equipment driving over the understory vegetation to crush and slash it, with or without incorporation in the forest floor. Chemical treatments with phytocides are efficient in controlling the woody understory, but temporarily increase fuel hazard due to conversion of live into dead fuel (Brose and Wade 2002, Mirra et al. 2017). The impact of livestock grazing (see Sect. 11.1.2) is selective and dispersed, as it depends on animal stocking rates and feeding preferences.

Impacts of surface fuels treatment in the medium to long run are strongly contingent on vegetation type and local soil and climate conditions. This dependence on local conditions hinders the formulation of generalized recommendations for fuel control, including the type and frequency of treatments. Operational sequences combining two or more techniques can offer the best results, as shown for fuelbreak maintenance in southern France (Rigolot and Etienne 1998).

Canopy Fuels Treatments: Thinning and Pruning

Silvicultural treatments to thin and prune forest stands are accomplished primarily through mechanical or manual treatments. Prescribed burning can result in a comparable effect, depending on tree crown base heights, fire intensity, and tree resistance to fire. Results are conditional on the structural impact achieved, i.e., the type and intensity of thinning and the subsequent development of vegetation (Graham et al. 2004). Thinning from below (or low thinning) (Fig. 11.25) is the most effective type of thinning for increasing the canopy base height, especially when codominant and dominant trees are also removed. When used in combination with other forest

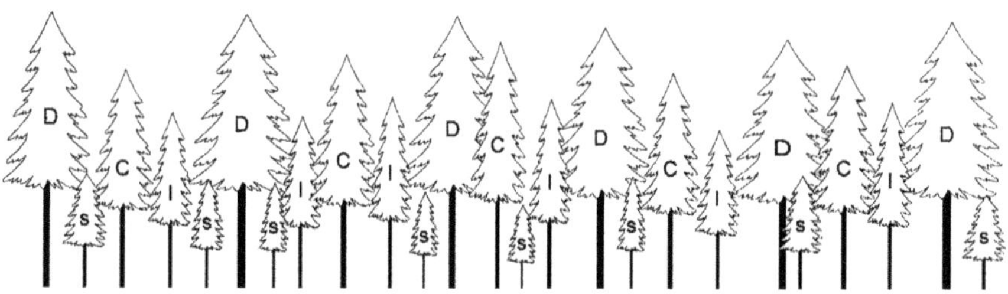

Fig. 11.25 A conifer forest with a mixture of dominant (D), codominant (C), intermediate (I), and suppressed (S) trees. The intensity of low thinning ranges from light to moderate to heavy, respectively, by removing only the suppressed, to also removing intermediate and codominant trees. Thinning can be spatially variable to further enhance the variation in forest structure spatially; this is sometimes done to enhance the wildlife habitat or aesthetics. In that case, dense clumps of trees with interconnected tree crowns may be left in the forest as long as the clumps are separated from one another. (From Graham et al. 1999)

treatments, thinning can produce interesting and heterogeneous stand structures (Peterson et al. 2003). Thinning from below is often used to transform dense stands of small trees into shaded fuelbreaks dominated by larger, more fire-resistant trees (Weatherspoon and Skinner 1996). Reducing the likelihood of active crown fire to a minimum requires decreasing tree canopy bulk density to 0.05–0.10 kg m^{-3} (Agee 1996; Van Wagner 1977). This level of thinning implies below the ideal density for maximizing tree growth for many species, and so there can be a trade-off between maximizing timber yield and minimizing crown fire hazard in forests managed for wood resources (Keyes and O'Hara 2002; Gomez-Vasquez et al. 2014).

Canopy interventions can simultaneously decrease and increase fire behavior potential (Agee et al. 2000; Graham et al. 2004). Relocating canopy fuels to the surface generates an extremely flammable fuel complex that will persist for a long time, especially in climates or sites that do not favor decomposition. Consequently, supplementary operations are advised, e.g., removal, pile and burn, broadcast burn, or mastication. However, the need for supplementary surface fuel treatments is not a general rule. For example, properly timed silvicultural treatments in *Pinus radiata* plantations do not require surface fuels treatments to be of value for fire suppression within a reasonable range of fire weather conditions (Cruz et al. 2017). Similarly, forest structure modification in dry conifer forests in the western USA can effectively reduce crown fire potential without subsequent surface fuel reductions (Fulé et al. 2012; Ziegler et al. 2017).

The reduction in canopy biomass associated with thinning reduces the amount of drag affecting the wind flow and increases the within and below canopy wind speeds. Furthermore, increased solar radiation associated with less canopy biomass influences fuel temperature and moisture and enhances understory vegetation development, especially in more productive sites. This last effect is nonetheless highly variable (Castedo-Dorado et al. 2012) and can be mitigated by the treatment of surface and ladder fuels (Weatherspoon 1996).

Mastication is a fuel treatment where machines are used to chip or mulch both living and dead trees and shrubs. Mastication is increasingly used as an alternative to prescribed burning or piling to reduce fire hazard in forests and shrublands. The practice has been recently studied as the shredded, irregular fuel particles in compact fuelbeds that result from mastication don't fit the assumptions of many fire behavior models. See Case Study 11.1.

Case Study 11.1 Mastication as a Fuels Treatment

Penelope Morgan, email: pmorgan@uidaho.edu
Department of Forest, Rangeland, and Fire Sciences, University of Idaho, Moscow

In mastication, trees and shrubs are chipped or mulched with a machine. Mastication is increasingly used as an alternative to prescribed burning or piling to reduce fire hazard in forests and shrublands. In this process, fuels are

(continued)

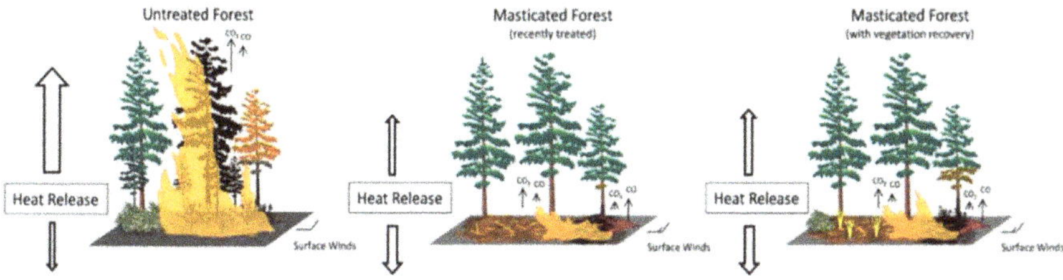

Fig. 11.26 Mastication treatments redistribute the fuels from tree and shrub to forest floor. The fuels are not removed from the site. If the masticated fuels don't decompose soon, they can add to the amount of fuel as vegetation recovers. (From Kreye et al. 2014)

Fig. 11.27 Fuel sampling pre-burn, fire behavior during burning, and fuel consumption evident after prescribed fire experiments conducted by Sparks et al. (2017) and Lyon et al. (2018). These photos are arranged with increasing fire intensity, as indicated by Fire Radiative Energy Density (FRED). (Photos from Sparks et al. 2017)

Case Study 11.1 (continued)
redistributed from crowns to dense, compact fuel layers on the surface (Fig. 11.26). Masticated fuels often burn with shorter flames and lower

(continued)

Case Study 11.1 (continued)
intensity than similar untreated fuels (Kreye et al. 2014). However, the potential for long-duration smoldering with related soil heating is high. Masticated fuelbeds retain fuel moisture and thus are more likely to smolder than similar amounts of fuels that are in less dense fuelbeds. The fuels are often shredded, resulting in irregular shapes (Keane et al. 2017). Between the shape of the pieces and the compact fuelbeds, masticated fuels don't fit the assumptions of many fire behavior models that fuels are of uniform and cylindrical shape. Masticated fuels burn less readily when aged (Kreye et al. 2014; Heinsch et al. 2018).

Costs of mastication treatments vary with the machine used and the material being masticated. Lyon et al. (2018) found that coarse mastication was faster and therefore 15% less expensive than fine mastication, yet the fire behavior was similar under their low intensity prescribed fire experiments. In fine mastication, there were few large pieces because the machine operator masticated each piece thoroughly, and this required more time to reposition the machine to process every stem. In contrast, in coarse mastication, large pieces of tree stems were left untreated.

During subsequent burning (Fig. 11.27), Lyon et al. (2018) found that the consumption of finely chipped, wet fuels was higher than for coarse wet fuels. However, when the fuels were relatively dry, coarse fuels had higher consumption than either fine, dry, or untreated fuels. The fuelbed characteristics (depth, piece size, and shape, decomposition rates, bulk density) vary with the machinery used in mastication, with the material that is masticated, how much biomass is masticated, and the time since mastication (Keane et al. 2017).

The extended smoldering combustion of masticated fuels (Heinsch et al. 2018; Lyon et al. 2018) suggest that fires burning in masticated fuels may result in more particulates in smoke near the ground. Masticated fuels burning in high wind conditions can produce embers, and the fuelbeds may ignite readily from embers (Kreye et al. 2014).

The ecological effects of mastication are poorly understood. The extended smoldering combustion likely results in soil heating, but only if the masticated fuelbeds burn. Unburned organic materials can insulate the soil. Unburned masticated layers on the soil surface likely limit evaporation from surface soil layers and thus act as a mulch that holds soil moisture into dry seasons. However, the mulch may also act as a physical barrier to seeds that germinate more successfully on bare mineral soil or for resprouting plants that are more likely to be stimulated when surface soils are less insulated. Further, it is possible that the presence of surface organic layers will alter the soil temperature and moisture and therefore the nutrient dynamics. As the layers of masticated fuels decompose, they will slowly release nutrients, but they may also limit the availability of nitrogen or other nutrients if the added carbon

(continued)

Case Study 11.1 (continued)
alters the carbon:nitrogen ratios enough that microbes absorb the available nitrogen leaving little for the plants. The ecological effects over time depend on the rate at which the accumulated biomass decomposes, and how these layers influence the soil temperature, moisture, and nutrient availability.

11.3.3 Fuels Treatment Effectiveness

Expectations Versus Reality

The assessment of fuel treatment effectiveness can be based on different criteria and is context dependent, e.g., one can either value the effect on wildfire extent or on burn severity. Either way, fuel treatment effectiveness depends on the influence of fuel structure and load on fire behavior versus the influence of weather and drought. The effect of treatments on wildfire spread can be barely noticeable when strong winds and dry fuels combine, especially in large fires (Banks and Little 1964; Keeley et al. 1999; McCarthy and Tolhurst 2001; Pye et al. 2003). It can be quickly suggested or concluded that treating fuels is futile in the face of severe fire weather or future climate change, but this point of view underestimates the impact on fire behavior and mistakenly assumes that fuel treatments are primarily intended to stop fires. Such expectation is excessive, as it implies unrealistic levels of success, consequently compromising an objective analysis (Finney and Cohen 2003). Further, stopping all fires can be counterproductive if fuels accumulate and then burn in a subsequent wildfire (Reinhardt et al. 2008).

What benefits should then be expected from fuel management? Suppose the goal is to decrease wildfire size. In that case, the assumption is that a properly treated area will expand fire management options and increase their effectiveness (Omi and Martinson 2002; Finney and Cohen 2003), an outcome of reducing fireline intensity to levels within fire control capacity (See Chap. 8). Several wildfire case studies have illustrated the value of fuel reduction to fire suppression strategies and outcomes (Cheney 2010; Tolhurst and McCarthy 2016). Treatment longevity can be seen as the length of time required for fire behavior to return to pre-treatment levels. However, from the practical perspective of fire control operations, the effective longevity of fuels treatments will be increasingly shorter as fire weather conditions worsen. While fuel reduction may not impact wildfire growth and burned areas under extreme fire weather, the decrease in energy release and thus burn severity will mitigate the environmental and socioeconomic impacts of wildfire (Reinhardt et al. 2008).

The previous paragraph considerations do not apply to the isolation strategy, whose success is measured only by the degree to which fire growth is curtailed. Reality defeats this expectation all too often, even when fuelbreaks are wide, as the likelihood of fire containment is influenced by factors such as fire fighter access and staffing, fuelbreak maintenance, fire weather, fire size and orientation, and spotting (Rigolot 2002; Weatherspoon and Skinner 1996; Syphard et al. 2011).

Fuel management is not equally effective in all vegetation types. Fuel dynamics and treatment longevity are vegetation type-specific. For example, savanna burning in northern Australia will only hinder fire spread before grass regrows and dries (Price et al. 2012), whereas treatments in mixed conifer forests in California can limit the potential for burning for up to 9 years (Collins et al. 2009). Different fuel components follow different recovery trajectories after treatment, which has implications for fire behavior over time, e.g., in dry eucalypt forests, the rate of build-up is litter>shrubs>bark (Gould et al. 2011). Different fire regimes require different fire and fuel management strategies (Gutsell et al. 2001), and treatment outcomes are more uncertain in crown-fire systems (Omi and Martinson 2002). In western USA forests, lesser relevance of fuel reduction is expected as the fire regime moves from frequent low-severity burning to infrequent stand-replacing fire (Schoennagel et al. 2004).

Assessments of Fuels Treatment Effectiveness

There are multiple approaches to evaluating fuels treatment (Fernandes and Botelho 2003; Fernandes 2015). They include:

1. Assessing the immediate physical impacts of treatments on the fuel complex and how fuels subsequently recover,
2. Expert opinion, which is conditioned by experience,
3. Simulations of fire behavior, affected by model capabilities and the assumptions adopted,
4. Documentation of the behavior or effects of wildfires in treated areas in comparison with adjacent untreated areas, as in Fig. 11.28, limited by data quality and quantity,
5. Observation of fire behavior and effects in experimental fires, the most authoritative method if high- to extreme-intensity fires are available, and.
6. Analysis of the fire regime, where the effects of fuel management can be confounded with other fire management activities.

The effectiveness of fuel treatments has been assessed at the stand scale (i.e., areas with a size from 10's to 100's of ha) using a combination of approaches, including assessing the immediate impacts on the fuels complex, computer simulations as well as post-fire case studies (Collins et al. 2007; Fulé et al. 2012; Hudak et al. 2011; Kalies and Kent 2016; Kennedy and Johnson 2014; Parsons et al. 2017; Safford et al. 2012; Stevens-Rumann et al. 2013; Ziegler et al. 2017). These studies have generally found that fuel treatments that use prescribed or managed fire alone, thinning alone, or a combination of the two can be successful at reducing the potential for high-severity crown fires compared to untreated areas. Still, there are differences among the treatment types (Fulé et al. 2012; Stephens et al. 2012). Prescribed fire alone can effectively reduce surface and canopy fuel load, raise the

Fig. 11.28 Examples of fuel treatments challenged by wildfire. (**a**) Treated (commercial harvest and prescribed fire) and (**b**) untreated *Pinus ponderosa* stand in the Santa Fé National Forest, New Mexico, USA with canopy bulk densities respectively of 0.021 and 0.118 kg m^{-3}, 1 year after wildfire (From Cram et al. 2006). (**c**) Wildfires stopped in a fuelbreak and **d** restrained by mosaic burning and grazing in northern Portugal shrubland. (Photographs by Paulo Fernandes)

crown base height, and reduce burn severity (Espinosa et al. 2019; Fulé et al. 2012; Knapp et al. 2005; Pollet and Omi 2002; Vaillant et al. 2009). Thinning can be effective at reducing burn severity. However, thinning treatments that do not limit post-treatment slash through whole-tree harvesting, piling and burning, or broadcast burning may not be effective at reducing the potential for high severity fires (Schmidt et al. 2008; Stephens et al. 2009). Similarly, treatments that utilize mastication may not be effective due to the increased surface fuel load (Battaglia et al. 2010; Jain et al. 2012; Kane et al. 2009; Kreye et al. 2014). Treatments that apply a combination of prescribed or managed fire along with mechanical methods are often the most effective at reducing the potential for crown fire initiation and spread (Prichard et al. 2010; Safford et al. 2012; Schwilk et al. 2009).

Following fuel treatments, overstory fuel load changes in response to the growth of existing trees, which can take advantage of the newly available growing space. The growth of understory and midstory vegetation also results in a decrease in the canopy base height over time. Because deposition often exceeds decomposition, the surface fuels increase as leaves, needles, cones, and branches accumulate. Several previous studies in the western USA have suggested that treatment longevity ranges from 10 to 20 year (Battaglia et al. 2008; Collins et al. 2011; Tinkham et al. 2016).

However, various factors may influence longevity including the type and intensity of treatment, site productivity, climate, and vegetation responses. Ultimately, fuel managers need to balance the maintenance of fuels treatments with the implementation of new fuels treatments across landscapes. North et al. (2012) suggested that managers increase both prescribed fire and managed wildfire to maintain current fuel treatments and expand the area treated. However, this approach will not be possible in all areas due to constraints related to air quality, wildlife habitat, and a growing wildland-urban interface.

Although treatment design and assessment usually occur at the stand scale, the size and severity of large wildfires indicate that treatments need to extend beyond individual stands to landscapes (Finney and Cohen 2003). Several field-based studies have shown reductions in burn severity at the landscape scale (Prichard and Kennedy 2014; Lydersen et al. 2017). Whether a consequence of fuel management or of prior wildfire (see Chap. 12), fuel age mosaics can be effective at controlling wildfire size and growth. This has been shown by aboriginal burning in arid grasslands in Australia (Bliege Bird et al. 2012), Baja California chaparral (Minnich and Chou 1997), pastoral burning in Portuguese shrublands (Fernandes et al. 2016), natural fire regimes in the western USA (Collins et al. 2009; Parks et al. 2015), and prescribed burning for hazard reduction in the eucalypt forests of southwestern Australia (Boer et al. 2009, See Case Study 13.1). The concept of burn leverage, i.e., the decrease in wildfire extent per unit of fuel-reduced area (Loehle 2004), can be used to assess the effectiveness of fuels treatments. Burn leverage has been quantified for prescribed burning and is modest, depending on the likelihood of wildfire-treatment encounters and the outcome of the encounter. One unit of prescribed fire replaces one unit of unplanned fire at best, but the typical ratio is about 3:1 (Price et al. 2015), and a 5–10% annual rate of landscape treatment is recommended (Fernandes 2015). As the required treatment effort is difficult to attain due to insufficient commitment, resources, or opportunities, high-value assets like wildland-urban interfaces are often prioritized for protection. However, such a strategy has a number of insufficiencies described by Case Study 13.1 and generally provides less net benefits in the long term than landscape-level treatments (Florec et al. 2020). Regardless of treatment strategy and objective, the location of treatments should be planned to achieve the maximum effect.

11.3.4 Decision Support and Optimization

Objective and quantitative criteria should guide the fuel treatment decision-making process. Fuel treatment recommendations and rules of thumb, e.g., for the distance between individual tree crowns or the width of fuelbreaks, have often lacked scientific and empirical support and require critical analysis (Alexander 2003). Likewise, erroneous ideas persist about the relationship between stand density and fire hazard, and objective specifications for silvicultural interventions to mitigate crown fire activity are recent.

Fire behavior is the link between formulating the desired fire-mitigation level and specifying the treatment characteristics. Fire behavior simulation can help justify the treatments, evaluate and compare alternative treatments, and anticipate how land use options and changes will change fire hazard (Roussopoulos and Johnson 1975). Thus, developing a prescription to change fuel quantity and structure such that the treatment is effective over a range of fire weather conditions necessarily involves fire behavior evaluation (Peterson et al. 2005). Through crown fire modeling (Chap. 8), quantitative objectives for treatments impact on stand structure can be defined to develop guidelines to reduce crowning potential in conifer forests (Alexander 1988; Graham et al. 1999; Reyes and O'Hara 2002). Compared to untreated conditions, fuels treatments effectively reduced the simulated rate of spread (Fig. 11.29) after understory elimination, and after alternative thinning intensities in a *Pinus ponderosa* / *Pseudotsuga menziesii* forest in Montana (Scott 2006, Fig. 11.29). The simulations integrated the effects of treatments on fuel structure and load and on the meteorological fire environment.

The main practical difficulty in developing silvicultural prescriptions based on fire behavior is to have information to quantify canopy fuel characteristics, namely foliar biomass equations, based on stand metrics that are both familiar to managers and easy to measure. Crown bulk density and crown base height are difficult to estimate in multilayered forests (Perry et al. 2004), and it would be convenient to have variables alternative to bulk density, preferably related to common structural descriptors of a forest stand (Reyes and O'Hara 2002), such as tree density and spacing, which are particularly interesting as an element of a thinning prescription (Peterson et al. 2005). Analysis of fire behavior simulations can be used to derive rules of thumb and other guidelines for end-users, e.g., Botequim et al. (2017), including density management diagrams indicating how fire hazard relates to stand density metrics throughout the development stages of an even-aged stand (Gomez-Vasquez et al. 2014).

Fuel management planning should also take place at broader spatial scales beyond the stand-scale analysis of fuel treatment alternatives and prescription development. The interconnection of plans designed for different scales requires an integrated approach that maximizes efficiency at each scale, which presupposes using tools appropriate to each scale, as in the demonstration project described by Long et al. (2003). Spatially explicit decision-support systems allow identifying areas where fires are most likely and how they can spread, and then decide on levels of hazard reduction and how to achieve them. Software that simulates fire growth in the landscape or estimates burn probability has been developed in the USA (Finney 1998, 2006), Canada (Parisien et al. 2005; Tymstra et al. 2010), and Australia (Tolhurst et al. 2008). The software equips the decision process with substantial analytical capacity, allowing comparisons between treatments and their spatial patterns.

Several operational, economic, social, and policy factors restrict the amount of land where fuels can be treated. Thus managers try to optimize treatment location and arrangement to attain the highest effectiveness possible. Treatment preferences can be defined on the basis of identifying land units corresponding to different

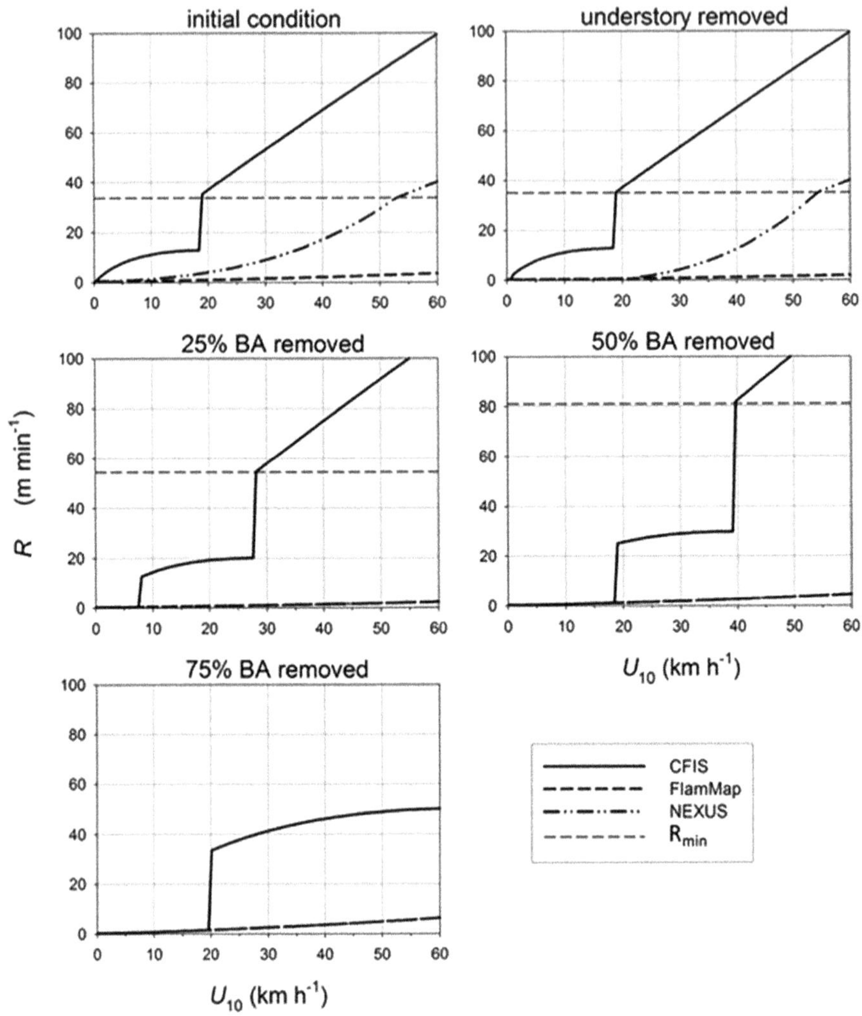

Fig. 11.29 Predicted fire spread rate in a *Pinus ponderosa / Pseudotsuga menziesii* stand as a function of wind speed in the open at a 10-m height (U_{10}) for the initial (untreated) condition, after understory removal, and following three levels of stand basal area (BA) reduction. Simulations are produced by two variants of linking Rothermel (1972), Van Wagner (1977), and Rothermel (1991) models, respectively FlamMap and NEXUS; and by CFIS, which combines Cruz et al. (2004, 2005) equations. R_{min} is the minimum rate of spread to maintain an active crown fire. Decreasing BA decreases canopy cover and canopy bulk density, and raises crown base height, all of which will reduce crown fire spread. These are especially important with higher wind speed and lower dead fine fuel moisture content. (Adapted from Scott 2006)

biophysical settings (Mislivets and Long 2003) or through the assessment of fireshed (akin to watershed) areas (Bahro et al. 2007). A fire risk framework is inherent to these approaches, where fire risk is understood as the combined outcome of likelihood (of ignition or burning), fire intensity, and fire effects (Miller and Ager 2013). In western Australia, targets for fuel-reduction burning are defined for combinations of fire management areas, distinguished by management objective, and fuel types (see Case Study 13.1). The fire risk framework is often preferred in the decision-making process, given the need to protect threatened values and because fuel treatments are costly.

The design of landscape fuel treatments requires managers to not only consider the factors that impact stand-scale effectiveness but how a treatment will impact fire spread and burn severity across a landscape including the proportion of the landscape treated and the placement of the treatment. Wildfires commonly grow larger than the individual fuel treatment units in their path. Significant slowdown of wildfire progression across landscapes requires a reasonable degree of overlap between treatment blocks in the direction of wildfire propagation (Fig. 11.30), which is unlikely when individual treatments are small and are dispersed according to random patterns (Finney 2001).

The cumulative effect of more and larger partially overlapped patches is that head fire spread is fragmented and a higher proportion of the fire front will spread by flanking, shifting the distribution of fire spread rates to lower values (Finney 2001). Consequently, the spatial organization of fuel treatments in terms of size, shape, orientation, and density is crucial regarding their ability and relevance to delay wildfire growth (Fig. 11.30). The spatial configuration of fuel treatments can be designed using the equations in Finney (2001) (Fig. 11.31) to arrive at strategically

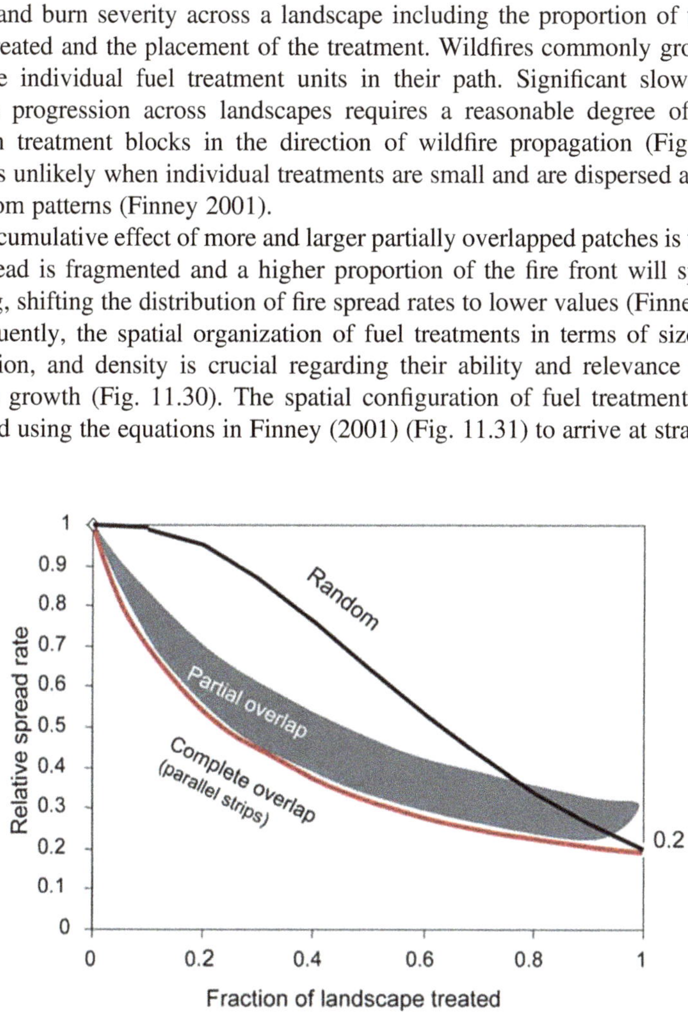

Fig. 11.30 Relative fire spread rate as a function of treated landscape fraction for distinct spatial patterns of treatment units. Compared to partial or total overlap amidst treatments, the random pattern requires the treatment of relatively large fractions of the landscape to result in a substantial reduction in the fire-spread rate. This analysis assumes uniform fuels, either untreated or treated. (From Finney 2004)

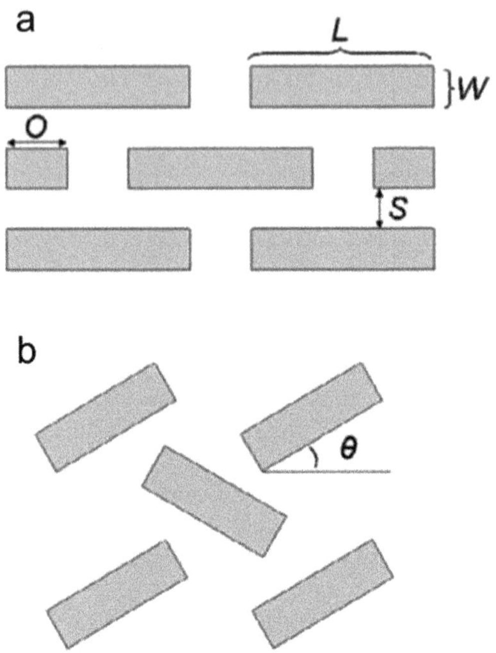

Fig. 11.31 Two spatial patterns of strategic fuel treatments based on partially overlapped treatments, whereby the inclination angle of the treatment unit (θ) is (**a**) constant or (**b**) variable. In the former, effectiveness is maximized when fire spreads at right angles to the treatments, while in the latter, fire growth is blocked regardless of fire spread direction. W, L, O, and S are respectively the treatment width, length, overlap, and separation distance. (From Parisien et al. 2006 based on Finney 2001)

placed areas of treatments (SPLATs). Landscape fire modeling suggests that 10–30% of the landscape needs to be strategically treated to reduce fire spread and intensity across that scale (Finney 2007; Schmidt et al. 2008). In Fig. 11.30, treating 20% of the landscape decreases the modeled fire spread rate by 5–60%, the low and high extremes corresponding respectively to random and to complete overlap patterns. It would be necessary to triple the treated area to obtain the same effect with random patterns, similar to the results from percolation models (Bevers et al. 2004; Loehle 2004). Fuel management planning should be a trade-off between minimizing treated area and creating spatial patterns that hinder wildfire expansion and alter fire behavior if the goal is to limit fire spread (Finney and Cohen 2003), but will be more successful if focused on fire management in support of land management goals. Further, landscape-scale fuel treatment planning can take advantage of landforms and vegetation patterns that are fuelbreaks. Fuel treatment regimes, with strategic fuels treatments implemented through time, can ensure ecosystem sustainability (Reinhardt et al. 2008). Still, a number of constraints may limit the ability to optimize landscape-scale fuel treatment, such as conservation status, land ownership, or access (Graham et al. 2004).

The design of fuelbreak networks should also be governed by strategic principles (Graham et al. 2004). As an alternative to a more exhaustive treatment of the landscape, and notwithstanding the limitations of the isolation strategy mentioned earlier, a spatially optimized network of fuelbreaks is preferable to random treatments over an equal proportion of the landscape (Loehle 2004). Strategically placed fuelbreaks could also support the use of prescribed fire and wildfires managed for

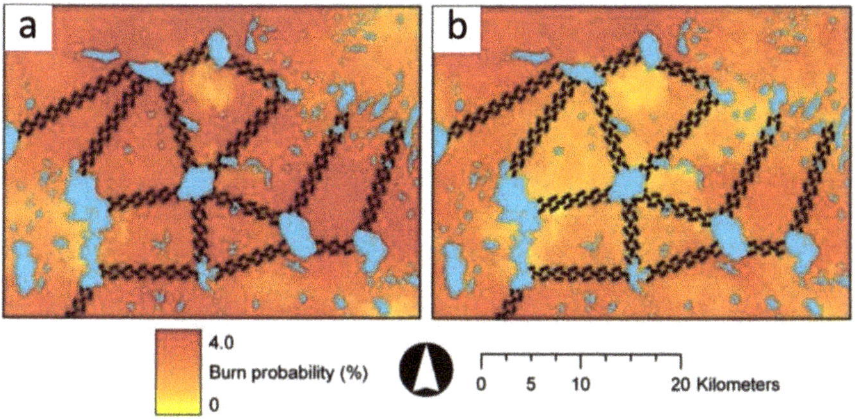

Fig. 11.32 Burn probability maps produced by the Burn-P3 model for (**a**) an untreated landscape in a boreal forest landscape (Prince Albert National Park in western Canada), and (**b**) for the same landscape after treatment according to a scenario linking lakes (in blue) where vegetation is converted to deciduous hardwoods. Treatment design used the equations in Finney (2001). Treatment units are 300-m wide and 900-m long (27 ha), are angled at 20° from the horizontal, and are organized in three rows separated by 200 m (see Fig. 11.31b). (From Parisien et al. 2006)

resource benefits. The fuelbreaks help managers move fires through the landscape, whether those fires are ignited purposefully or are unplanned fires that are delayed, herded, or otherwise managed to advance natural resource objectives while also protecting people, their property, and other values.

Methodologies have been developed to optimize fuel management in the landscape based on fire spread and growth simulation, e.g., by analyzing burn probability for different fuel treatment scenarios, such as in the example of Fig. 11.32. FlamMap (See Chap. 7) calculates fire behavior characteristics for every cell in the landscape under constant weather, corresponding to the conditions assumed for fuel treatment performance and enabling analysis of the effects of spatial fuel and topography patterns under those conditions. Each cell comprises rates of spread in all directions such that fire growth can be calculated for a given wind direction and ignition location. FlamMap integrates the Minimum Travel Time (MTT) algorithm (Finney 2002) that calculates fire travel time between landscape corners and computes fire growth by finding the paths with the minimum fire travel time. These calculations produce an arrival time grid that can be converted to fire progression maps but can also produce MTT paths (Fig. 11.33). This FlamMap feature is enabled by the Treatment Optimization Model (Finney 2007) and allows identifying and mapping the optimal fuels treatment locations that may disrupt the preferred pathways and slow down fire growth.

Multiple systems are available to support planning and analysis for fuel treatment planning. Several of them link together the spatial data, fire behavior prediction tools with their required inputs and outputs, and tools for visualizing alternative strategies. For instance, the Interagency Fuels Treatment Decision Support System (IFTDSS,

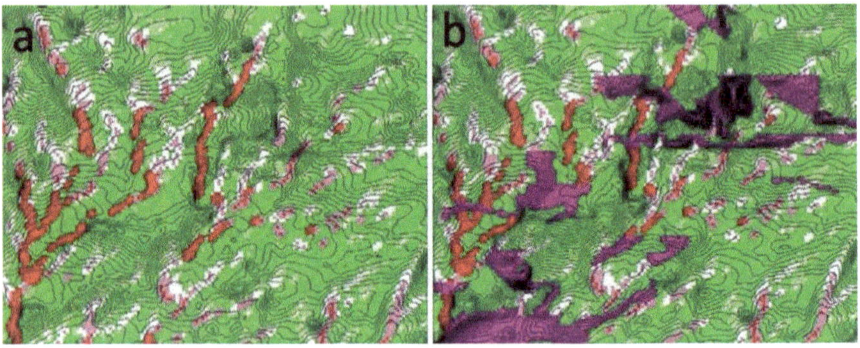

Fig. 11.33 (**a**) FlamMap fastest travel paths of fire across a landscape (red color) that account for most of the area burned are identified through the Minimum Travel Time (MTT) algorithm and (**b**) the locations and sizes of fuel treatments (15% of the landscape in this example, fuchsia color) that block those routes are then optimized with the Treatment Optimization Model. (From Finney 2004)

https://iftdss.firenet.gov/landing_page/) is a web-based framework for integrating software and data. Other fire management decision support systems exist, e.g., ARCFuels (Vaillant et al. 2013), and they are particularly useful for landscape-level planning. Many are modular enough that users can incorporate the tools they need or those for which they have the data. Some systems can be used to document what decisions were made when and on what data. Based on interviews with users, Noble and Paveglio (2020) highly recommended that end users be involved when planning the development of such tools and the training for using them.

11.4 Implications

Fuels provide a link between fire behavior and fire effects (Keane 2015). Fuel moisture is critically important to fire behavior and effects. Fuel moisture is dynamic, as it reflects changing environmental conditions. In warm, dry conditions, more fuels in more areas become available to burn if ignited, and the fuels load and vegetation structure and composition are conducive to fire spread.

The ecology of fuels, understood as the tight connection between fuels, fire behavior, and fire effects, determines vegetation response and dynamics through complex feedbacks (Mitchell et al. 2009; Keane 2015). The concept of fuel ecology also implies envisioning fuels as ecosystem components with various functions rather than just fire-related biomass.

Fuel management strategies have emerged as a critical tool in part because fuels are the only component of the fire behavior triangle (i.e., fuels, weather, and topography) that can be directly manipulated through management actions. Fuels management can alter burn severity even when fires burn under relatively extreme environmental conditions, but especially under moderate environmental conditions. Where and when fuels treatments alter fire behavior and effects, they can increase the resilience of ecosystems to fires and reduce smoke production. Then subsequent wildfires and prescribed fires may be more feasible and socially acceptable (Reinhardt et al. 2008).

Three basic strategies exist for fuel management: fuel reduction, fuel isolation, and fuel type conversion. Managers use various approaches to carry out fuel treatments, including mechanical, prescribed burning, grazing, and others. In all cases, the goal of fuels treatment is to modify the fuels complex to alter fire behavior, which can increase fire suppression effectiveness, reduce overall fire extent, decrease the exposure of localized assets to fire, and reduce burn severity. While multiple approaches can be used to evaluate fuels treatment, most studies have indicated that treatments are effective relative to their goals and objectives. However, it is important to recognize that the effectiveness of a fuels treatment depends upon the resulting fuels complex and the burning conditions. Fuels treatments are less effective when fires burn under extreme environmental conditions, particularly if spotting and extreme fire behavior occur. Furthermore, treatment effectiveness declines over time as vegetation regrows and thus will require follow-up treatments. Although fuel treatments are commonly implemented with the expressed purpose of assisting fire suppression, the treatments can also be part of integrated land management that seeks to increase ecosystem resilience and permits managers to reintroduce fire across landscapes (Stephens et al. 2020). Ideally, fuels treatments will contribute to long-term management objectives and draw on all we know about vegetation management. With fuels treatments, managers can proactively manage for future fires as they have more options when fires occur. See Chap. 12 for landscape management and restoration.

Decision support systems are increasingly used by managers in fuel management planning. The decision support systems are frameworks in which multiple models are linked together. Often the components are individual models for predicting fire behavior and effects, with the framework easing the task of formatting the required inputs for each model. Some decision support systems are quite useful not only in planning but also in documenting the basis of decisions.

Despite a long history of fuels management science, there is still considerable ongoing research related to wildland fuel dynamics and the links between fuels, fire behavior, and fire effects. Managers are rapidly learning by doing. Prescribed fire science and application are advancing. The social, economic, and political challenges are many (See Chap. 10). Addressing them takes collaboration across landscapes (See Chap. 12) as part of innovative integrated fire management (See examples in Chap. 13) is making opportunities out of challenges.

11.5 Interactive Spreadsheets: FUEL_DYNAMICS and CROWNFIRE_MITIGATION

Use the interactive spreadsheets, FUEL_DYNAMICS_v2.0 and CROWNFIRE_MITIGATION_v2.0 (Figs. 11.34 and 11.35) to interpret how predicted outputs are influenced by changing the inputs. Explore with FUEL DYNAMICS how fuel loads change in response to fuel structure descriptors and how they reaccumulate after prescribed fire depending on stand structure. Then, examine how the Canadian FWI codes for dead fuel moisture content can be used to design burn prescriptions to attain a given level of fuel consumption.

Using our CROWNFIRE_MITIGATION interactive spreadsheet, we encourage readers to deepen their understanding of the implications of fuel treatments for wildfire behavior. For a given combination of wind speed, dead fuel moisture content and slope, users can assess the effects on fire behavior of reducing surface fuel depth and load and modifying canopy structure through pruning and thinning.

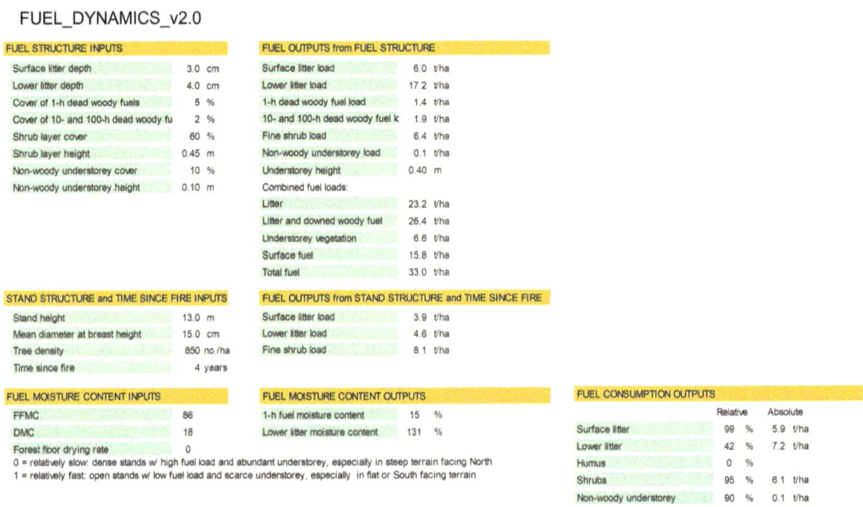

Fig. 11.34 Our interactive spreadsheet, FUEL_DYNAMICS_v2.0, will help you visualize how changing fuel-related inputs (vegetation structure, time since fire, fuel moisture) alters the predictions of different (and combined) categories of fuel load and fuel consumption. FUEL DYNAMICS was developed for maritime pine (*Pinus pinaster*) forests in Portugal by adapting components of the PiroPinus spreadsheet tool (Fernandes et al. 2012b) used to plan and evaluate prescribed fire operations

STAND STRUCTURE INPUTS		
Stand height	10.0	m
Height to live crown base	2.0	m
Mean diameter at breast height	20.0	cm
Tree density	200	nº/ha
Canopy cover	20	%
Canopy foliage load	4.1	t/ha
Canopy bulk density	0.05	kg/m3

INPUTS for SURFACE FIRE BEHAVIOR		
Fuel depth	1.50	m
Fuel load	5.0	t/ha
1-hr fuel moisture content	20.0	%
10-m open wind speed	20	km/h
Slope	10	º

SURFACE FIRE BEHAVIOR OUTPUTS			Fire models used
Forward rate of spread	8.2	m/min	Fernandes et al. (2009), Fernandes (2014)
Forward Byram's fireline intensity	1230	kW/m	

CROWN FIRE TRANSITION			
Threshold surface fireline intensity for transi	479	kW/m	Van Wagner (1977)
Crown fire transition	Yes		

CROWN FIRE BEHAVIOR in case of crowning			
Threshold rate of spread for active crowninç	59	m/min	Van Wagner (1977)
Crown fire type	Passive		
Rate of spread	3	m/min	Cruz et al. (2005)
Fireline intensity	800	kW/m	

Fig. 11.35 The CROWNFIRE_MITIGATION allows the user to define fuel and stand structure characteristics that are expected to minimize crown fire development and behavior. The spreadsheet combines compatible fire behavior models and criteria

References

Agee, J. K. (1996). The influence of forest structure on fire behavior. In: Proceedings of 17th Forest Vegetation Management Conference. Redding, CA, pp. 52–68

Agee, J. K., & Skinner, C. N. (2005). Basic principles of forest fuel reduction treatments. *Forest Ecology and Management, 211*(1–2), 83–96.

Agee, J. K., Bahro, B., Finney, M. A., Omi, P. N., Sapsis, D. B., Skinner, C. N., Van Wagtendonk, J. W., & Weatherspoon, C. P. (2000). The use of shaded fuelbreaks in landscape fire management. *Forest Ecology and Management, 127*(1-3), 55–66.

Ager, A. A., Vaillant, N. M., & McMahan, A. (2013). Restoration of fire in managed forests: A model to prioritize landscapes and analyze tradeoffs. *Ecosphere, 4*, art29.

Alexander, M. (1988). Help with making crown fire hazard assessments. In W Fischer, S Arno (comps), *Symposium & Workshop on Protecting People and Homes from Wildfire in the Interior West.* Gen Tech Rep INT-251. Ogden, UT: USDA Forest Service Intermountain Forest and Range Experiment Station, pp. 147–156.

Alexander, M. (2003). Understanding fire behaviour – The key to effective fuels management. In: *FERIC Fuels Management Workshop.* Hinton, AB: Hinton Training Centre.

Anderson, H. E. (1990). Moisture diffusivity and response time in fine forest fuels. *Canadian Journal of Forest Research, 20*, 315–325.

Anderson, S. A., & Anderson, W. R. (2009). Predicting the elevated dead fine fuel moisture content in gorse (*Ulex europaeus L.*) shrub fuels. *Canadian Journal of Forest Research, 39*, 2355–2368.

Anderson, H. E., & Brown, J. K. (1988). Fuel characteristics and fire behavior consideration in the wildlands. In WC Fisher, SF Arno (comp), *Protecting People and Homes from Wildfire in the Interior West.* Gen Tech Rep INT-251. Ogden, UT: USDA Forest Service Intermountain Research Station, pp. 124–130.

Anderson, W. R., Cruz, M. G., Fernandes, P. M., McCaw, L., Vega, J. A., Bradstock, R., Fogarty, L., Gould, J., McCarthy, G., Marsden-Smedley, J. B., Matthews, S., Mattingley, G., Pearce, G., & van Wilgen, B. (2015). A generic, empirical-based model for predicting rate of fire spread in shrublands. *International Journal of Wildland Fire, 24*, 443–460.

Angers, V. A., Gauthier, S., Drapeau, P., Jayen, K., & Bergeron, Y. (2011). Tree mortality and snag dynamics in North American boreal tree species after a wildfire: A long-term study. *International Journal of Wildland Fire, 20*, 751–763.

Armesto, J. J., Bustamante-Sanchez, M. E., Díaz, M. F., Gonzales, M. E., Holz, A., Nunez-Avila, M. C., & Smith-Ramírez, C. (2009). Fire disturbance regimes, ecosystem recovery and restoration strategies in Mediterranean and temperate regions of Chile. In A. Cerdá & P. R. Robichaud (Eds.), *Fire Effects on soils and restoration strategies* (pp. 537–567). Enfield: Science Publishers.

Arno, S. F., & Brown, J. K. (1991). *Overcoming the paradox in managing wildland fire in western wildlands* (pp. 40–46). Missoula: University of Montana, Montana Forest and Conservation Experiment Station.

Bahro, B., Barber, K. H., Sherlock, J. W., & Yasuda, D. A. (2007). Stewardship and fireshed assessment: A process for designing a landscape fuel treatment strategy. In: *Restoring Fire-Adapted Ecosystems: Proceedings of the 2005 National Silviculture Workshop*. Gen Tech Rep PSW-GTR-203. Berkeley, CA: USDA Forest Service Pacific Southwest Research Station, pp. 41–54.

Banks, W., & Little, S. (1964). The forest fires of April 1963 in New Jersey point the way to better protection and management. *Fire Control Notes, 25*, 3–6.

Battaglia, M. A., Smith, F. W., & Shepperd, W. D. (2008). Can prescribed fire be used to maintain fuel treatment effectiveness over time in Black Hills ponderosa pine forests? *Forest Ecology and Management, 256*, 2029–2038.

Battaglia, M. A., Rocca, M. E., Rhoades, C. C., & Ryan, M. G. (2010). Surface fuel loadings within mulching treatments in Colorado coniferous forests. *Forest Ecology and Management, 260*, 1557–1566.

Bernardi, R. E., Staal, A., Xu, C., Scheffer, M., & Holmgren, M. (2019). Livestock herbivory shapes fire regimes and vegetation structure across the global tropics. *Ecosystems, 22*, 1457–1465.

Bevers, M., Omi, P., & Hof, J. (2004). Random location of fuel treatments in wildland community interfaces: A percolation approach. *Canadian Journal of Forest Research, 34*, 164–173.

Bliege Bird, R. B., Codding, B. F., Kauhanen, P. G., & Bird, D. W. (2012). Aboriginal hunting buffers climate-driven fire-size variability in Australia's spinifex grasslands. *Proceedings of the National Academy of Sciences of the United States of America, 109*(26), 10287–10292.

Boer, M. M., Sadler, R. J., Wittkuhn, R., McCaw, L., & Grierson, P. F. (2009). Long- term impacts of prescribed burning on regional extent and incidence of wildfires – Evidence from fifty years of active fire management in SW Australian forests. *Forest Ecology and Management, 259*, 132–142.

Bond, W. J., & Keeley, J. E. (2005). Fire as a global 'herbivore': The ecology and evolution of flammable ecosystems. *Trends in Ecology & Evolution, 20*, 387–394.

Botequim, B., Fernandes, P. M., Garcia-Gonzalo, J., Silva, A., & Borges, J. G. (2017). Coupling fire behaviour modelling and stand characteristics to assess and mitigate fire hazard in a maritime pine landscape in Portugal. *European Journal of Forest Research, 136*, 527–542.

Brooks, M. L., D'Antonio, C. M., Richardson, D. M., Grace, J. B., Keeley, J. E., DiTomaso, J. M., Hobbs, R. J., Pellant, M., & Pyke, D. (2004). Effects of invasive alien plants on fire regimes. *BioSciences, 54*, 677–688.

Brose, P., & Wade, D. (2002). Potential fire behavior in pine flatwood forests following three different fuel reduction techniques. *Forest Ecology and Management, 163*(1–3), 71–84.

Brown, J. K. (1974). *Handbook for inventorying downed woody material*. Gen Tech Rep INT-16. Ogden, UT: USDA Forest Service Intermountain Forest and Range Experiment Station.

Buma, B. (2015). Disturbance interactions: Characterization, prediction, and the potential for cascading effects. *Ecosphere, 6*, 1–15.

Burgan, R. E. (1979). *Estimating live fuel moisture for the 1978 national fire danger rating system* (Vol. 226). Ogden: USDA Forest Service Intermountain Forest and Range Experiment Station.

Burrows, N. D., & McCaw, W. L. (1990). Fuel characteristics and bushfire control in banksia low woodlands in Western Australia. *Journal of Environmental Management, 31*, 229–236.

Byram, G. M., & Nelson, R. M. (2015). *An analysis of the drying process in forest fuel material*. e-Gen. Tech Rep SRS-200. Asheville, NC: USDA Forest Service Southern Research Station.

Call, P. T., & Albini, F. A. (1997). Aerial and surface fuel consumption in crown fires. *International Journal of Wildland Fire, 7*, 259–264.

Cannon, J. B., Peterson, C. J., O'Brien, J. J., & Brewer, J. S. (2017). A review and classification of interactions between forest disturbance from wind and fire. *Forest Ecology and Management, 406*, 381–390.

Castedo-Dorado, F., Gomez-Vazquez, I., Fernandes, P. M., & Crecente-Campo, F. (2012). Shrub fuel characteristics estimated from overstory variables in NW Spain pine stands. *Forest Ecology and Management, 275*, 130–141.

Catchpole, E. A., Alexander, M. E., & Gill, A. M. (1993). Elliptical-fire perimeter- and area-intensity distributions. *Canadian Journal of Forest Research, 23*, 1244–1124.

Catchpole, E. A., Catchpole, W. R., Viney, N. R., McCaw, W. L., & Marsden-Smedley, J. B. (2001). Estimating fuel response time and predicting fuel moisture content from field data. *International Journal of Wildland Fire, 10*, 215–222.

Chandler, C., Cheney, P., Thomas, P., Trabaud, L., & Williams, D. (1983). *Fire in forestry* (Vol. I: Forest fire behavior and effects). New York: Wiley.

Cheney, N. P. (1990). Quantifying bushfires. *Mathematical and Computer Modelling, 13*(12), 9–15.

Cheney, N. P. (2010). Fire behaviour during the Pickering Brook wildfire, January 2005 (Perth Hills Fires 71–80). *Conservation Science West Australia, 7*, 451–468.

Cheney, P., & Sullivan, A. (2008). *Grassfires: Fuel, weather and fire behaviour*. Clayton: CSIRO Publishing.

Cheney, N. P., Gould, J. S., & Catchpole, W. R. (1993). The influence of fuel, weather and fire shape variables on fire-spread in grasslands. *International Journal of Wildland Fire, 3*, 31–44.

Cheney, N. P., Gould, J. S., & Catchpole, W. R. (1998). Prediction of fire spread in grasslands. *International Journal of Wildland Fire, 8*, 1–13.

Collins, B. M., Moghaddas, J. J., & Stephens, S. L. (2007). Initial changes in forest structure and understory plant communities following fuel reduction activities in a Sierra Nevada mixed conifer forest. *Forest Ecology and Management, 239*, 102–111.

Collins, B. M., Miller, J. D., Thode, A. E., Kelly, M., van Wagtendonk, J. W., & Stephens, S. L. (2009). Interactions among wildland fires in a long- established Sierra Nevada natural fire area. *Ecosystems, 12*, 114–128.

Collins, B. M., Stephens, S. L., Roller, G. B., & Battles, J. J. (2011). Simulating fire and forest dynamics for a landscape fuel treatment project in the Sierra Nevada. *Forest Science, 57*, 77–88.

Countryman, C. (1974). *Can southern California wildland conflagrations be stopped?* Gen Tech Rep PSW-7. Berkeley, CA: USDA Forest Service Pacific Southwest Forest and Range Experiment Station.

Covington, W. W., & Moore, M. M. (1994). Southwestern ponderosa forest structure: Changes since Euro-American settlement. *Journal of Forestry, 92*, 39–47.

Cram, D. S., Baker, T. T., & Boren, J. (2006). *Wildland fire effects in silviculturally treated vs. untreated stands of New Mexico and Arizona*. Res Pap RMRS-RP- 55. Fort Collins, CO: USDA Forest Service Rocky Mountain Research Station.

Cruz, M. G., Alexander, M. E., & Plucinski, M. P. (2017). The effect of silvicultural treatments on fire behaviour potential in radiata pine plantations of South Australia. *Forest Ecology and Management, 397*, 27–38.

Cruz, M. G., Alexander, M. E., & Wakimoto, R. H. (2004). Modeling the likelihood of crown fire occurrence in conifer forest stands. *Forest Science, 50*(5), 640–658.

Cruz, M. G., Alexander, M. E., & Wakimoto, R. H. (2005). Development and testing of models for predicting crown fire rate of spread in conifer forest stands. *Canadian Journal of Forest Research, 35*(7), 1626–1639.

Cruz, M. G., Gould, J. S., Kidnie, S., Bessell, R., Nichols, D., & Slijepcevic, A. (2015). Effects of curing on grassfires: II. Effect of grass senescence on the rate of fire spread. *International Journal of Wildland Fire, 24*, 838–848.

Deeming, J. E., Burgan, R. E., & Cohen, J. D. (1977). *The National Fire-danger rating system–1978*. Gen Tech Rep INT-39. Ogden, UT: USDA Forest Service Intermountain Forest and Range Experiment Station.

Dubinin, M., Luschekina, A., & Radeloff, V. C. (2011). Climate, livestock, and vegetation: What drives fire increase in the arid ecosystems of southern Russia? *Ecosystems, 14*, 547–562.
Espinosa, J., Palheiro, P., Loureiro, C., Ascoli, D., Esposito, A., & Fernandes, P. M. (2019). Fire-severity mitigation by prescribed burning assessed from fire-treatment encounters in maritime pine stands. *Canadian Journal of Forest Research, 49*, 205–211.
Fernandes, P. M. (2015). Empirical support for the use of prescribed burning as a fuel treatment. *Current Forestry Reports, 1*, 118–127.
Fernandes, P. M., & Loureiro, C. (2010). *Handbook to plan and use prescribed burning in Europe*. Fire Paradox project (FP6-018505EC). Vila Real, Portugal: Universidade de Trás-os-Montes e Alto Douro.
Fernandes, P., & Rego, F. (1998). Changes in fuel structure and fire behaviour with heathland aging in Northern Portugal. In *Proceedings 13th Conference on Fire and Forest Meteorology*. Lorne: International Association of Wildland Fire.
Fernandes, P., Loureiro, C., Botelho, H., Ferreira, A., & Fernandes, M. (2002). Avaliação indirecta da carga de combustível em pinhal bravo. *Silva Lusitana, 10*, 73–90.
Fernandes, P., & Botelho, H. (2003). A review of prescribed burning effectiveness in fire hazard reduction. *International Journal of Wildland Fire, 12*, 117–128.
Fernandes, P. M., Botelho, H. S., Rego, F. C., & Loureiro, C. (2009). Empirical modelling of surface fire behaviour in maritime pine stands. *International Journal of Wildland Fire, 18*, 698–710.
Fernandes, P. M., Loureiro, C., Magalhães, M., Ferreira, P., & Fernandes, M. (2012a). Fuel age, weather and burn probability in Portugal. *International Journal of Wildland Fire, 21*, 380–384.
Fernandes, P. M., Loureiro, C., & Botelho, H. (2012b). PiroPinus: A spreadsheet application to guide prescribed burning operations in maritime pine forest. *Computers and Electronics in Agriculture, 81*, 58–61.
Fernandes, P. M., Monteiro-Henriques, T., Guiomar, N., Loureiro, C., & Barros, A. (2016). Bottom-up variables govern large-fire size in Portugal. *Ecosystems, 19*, 1362–1375.
Ferran, A., & Vallejo, V. R. (1992). Litter dynamics in post-fire successional forests of Quercus ilex. *Vegetatio, 99*, 239–246.
Finney, M. A. (1998). *FARSITE: Fire Area Simulator-model development and evaluation*. Res Pap RMRS-RP-4, revised 2004. Ogden, UT: USDA Forest Service Rocky Mountain Research Station.
Finney, M. A. (2001). Design of regular landscape fuel treatment patterns for modifying fire growth and behaviour. *Forest Science, 47*, 219–228.
Finney, M. A. (2002). Fire growth using minimum travel time methods. *Canadian Journal of Forest Research, 32*, 1420–1424.
Finney, M. A. (2004). Landscape fire simulation and fuel treatment optimization. In *Methods for integrating modeling of landscape change: Interior Northwest Landscape Analysis System. Gen Tech Rep PNW-GTR-610* (pp. 117–131). Portland: USDA Forest Service Pacific Northwest Research Station.
Finney, M. A. (2006). An overview of FlamMap fire modeling capabilities. In PL Andrews, BW Butler (comps), *Fuels Management - How to Measure Success: Conference Proceedings*, Portland, 28–30 March 2006. Proc RMRS-P-41. Fort Collins, CO: USDA Forest Service Rocky Mountain Research Station, pp. 213–220.
Finney, M. A. (2007). A computational method for optimizing fuel treatment locations. *International Journal of Wildland Fire, 16*, 702–711.
Finney, M., & Cohen, J. (2003). Expectation and evaluation of fuel management objectives. In P. Omi & L. Joyce (Eds.), *Fire, fuel treatments, and ecological restoration. Proc RMRS-P-29* (pp. 353–366). Ogden: USDA Forest Service Rocky Mountain Research Station.
Florec, V., Burton, M., Pannell, D., Kelso, J., & Milne, G. (2020). Where to prescribe burn: The costs and benefits of prescribed burning close to houses. *International Journal of Wildland Fire, 29*, 440–458.

Foster, C. N., Banks, S. C., Cary, G. J., Johnson, C. N., Lindenmayer, D. B., & Valentine, L. E. (2020). Animals as agents in fire regimes. *Trends in Ecology & Evolution, 35*, 346–356.
Fuhlendorf, S. D., Engle, D. M., Kerby, J., & Hamilton, R. (2009). Pyric herbivory: Rewilding landscapes through the recoupling of fire and grazing. *Conservation Biology, 23*, 588–598.
Fulé, P. Z., Crouse, J. E., Roccaforte, J. P., & Kalies, E. L. (2012). Do thinning and/or burning treatments in western USA ponderosa or Jeffrey pine-dominated forests help restore natural fire behavior. *Forest Ecology and Management, 269*, 68–81.
Gomez-Vasquez, I., Fernandes, P. M., Arias-Rodil, M., Barrio-Anta, M., & Castedo-Dorado, F. (2014). Using density management diagrams to assess crown fire potential in *Pinus pinaster* Ait. stands. *Annals of Forest Science, 71*, 473–484.
Gould, J. S., McCaw, L., & Cheney, P. N. (2011). Quantifying fine fuel dynamics and structure in dry eucalypt forest (Eucalyptus marginata) in Western Australia for fire management. *Forest Ecology and Management, 262*, 531–546.
Graham, R., Harvey, A., Jain, T., & Tonn, J. (1999). *The effects of thinning and similar stand treatments on fire behaviour in western forests. Gen Tech Rep PNW-463*. Portland: USDA Forest Service Pacifc Northwest Research Station.
Graham, R. T., McCaffrey, S., & Jain, T. B. (Tech Eds.) (2004). *Science basis for changing forest structure to modify wildfire behavior and severity*. Gen Tech Rep RMRS-GTR-120. Fort Collins, CO: USDA Forest Service Rocky Mountain Research Station.
Gutsell, S., Johnson, E., Miyanishi, K., Keeley, J., Dickinson, M., & Bridge, S. (2001). Varied ecosystems need different fire protection. *Nature, 409*, 977.
Hall, S. A., Burke, I. C., & Hobbs, N. T. (2006). Litter and dead wood dynamics in ponderosa pine forests along a 160-year chronosequence. *Ecological Applications, 16*, 2344–2355.
Heinsch, F. A., Sikkink, P. G., Smith, H. Y., & Retzlaff, M. L. (2018). *Characterizing fire behavior from laboratory burns of multi-aged, mixed-conifer masticated fuels in the western United States. RMRS-RP-107*. Fort Collins: USDA Forest Service Rocky Mountain Research Station.
Hessburg, P. F., Agee, J. K., & Franklin, J. F. (2005). Dry forests and wildland fires of the inland Northwest USA: Contrasting the landscape ecology of the pre-settlement and modern eras. *Forest Ecology and Management, 211*, 117–139.
Heward, H., Smith, A. M., Roy, D. P., Tinkham, W. T., Hoffman, C. M., Morgan, P., & Lannom, K. O. (2013). Is burn severity related to fire intensity? Observations from landscape scale remote sensing. *International Journal of Wildland Fire, 22*, 910–918.
Hicke, J. A., Johnson, M. C., Hayes, J. L., & Preisler, H. K. (2012). Effects of bark beetle-caused treex mortality on wildfire. *Forest Ecology and Management, 271*, 81–90.
Hoffman, C. M., Collins, B., & Battaglia, M. (2018). Wildland fuel treatments. In: Manzello, SL, ed. Encyclopedia of Wildfires and Wildland-Urban Interface (WUI) Fires. Springer, Cham. https://doi.org/10.1007/978-3-319-51727-8_83-1.
Hoffman, C. M., Sieg, C. H., McMillin, J. D., & Fulé, P. Z. (2012). Fuel loadings 5 years after a bark beetle outbreak in south-western USA ponderosa pine forests. *International Journal of Wildland Fire, 21*, 306–312.
Hoffman, C. M., Sieg, C. H., Morgan, P., Mell, W., Linn, R., Stevens-Rumann, C., McMillin, J., Parsons, R., & Maffei, H. (2013). *Progress in understanding bark beetle effects on fire behavior using physics-based models. Tech Brief CFRI-TB-1301*. Fort Collins: Colorado Forest Restoration Institute, Colorado State University.
Hoffman, C. M., Linn, R., Parsons, R., Sieg, C., & Winterkamp, J. (2015). Modeling spatial and temporal dynamics of wind flow and potential fire behavior following a mountain pine beetle outbreak in a lodgepole pine forest. *Agricultural and Forest Meteorology, 204*, 79–93.
Holden, Z. A., & Jolly, W. M. (2011). Modeling topographic influences on fuel moisture and fire danger in complex terrain to improve wildland fire management decision support. *Forest Ecology and Management, 262*, 2133–2141.
Hough, W. A., & Albini, F. A. (1978). *Predicting fire behavior in palmetto-gallberry fuel complexes. Res Pap SE-RP-174*. Asheville: USDA Forest Service Southeastern Forest Experiment Station.

Hudak, A. T., Rickert, I., Morgan, P., Strand, E., Lewis, S. A., Robichaud, P., Hoffman, C. M., & Holden, Z. A. (2011). *Review of fuel treatment effectiveness in forests and rangelands and a x from the 2007 megafires in central Idaho USA, Gen. Tech. Rep. RMRS-GTR-252*. Fort Collins: USDA Forest Service Rocky Mountain Research Station.

Hudec, J. L., & Peterson, D. L. (2012). Fuel variability following wildfire in forests with mixed severity fire regimes, Cascade Range, USA. *Forest Ecology and Management, 277*, 11–24.

Instituto da Conservação da Natureza e das Florestas (ICNF). (n.d.). *Portuguese Fire Atlas*. Retrieved July 15, 2020, from http://www2.icnf.pt/portal/florestas/dfci.

Jain, T. B., Battaglia, M. A., Han, H. S., Graham, R. T., Keyes, C. R., Fried, J. S., & Sandquist, J. E. (2012). *A comprehensive guide to fuel management practices for dry mixed conifer forests in the northwestern United States, Gen. Tech. Rep. RMRS-GTR-292*. Fort Collins: USDA Forest Service, Rocky Mountain Research Station.

Jenkins, M. J., Runyon, J. B., Fettig, C. J., Page, W. G., & Bentz, B. J. (2014). Interactions among the mountain pine beetle, fires, and fuels. *Forest Science, 60*, 489–501.

Jolly, W. M., Hintz, J., Linn, R., Kropp, R. C., Conrad, E. T., Parsons, R. A., & Winterkamp, J. (2016). Seasonal variation in red pine (Pinus resinosa) and jack pine (Pinus banksiana) foliar physio-chemistry and their potential influence on stand-scale wildland fire behavior. *Forest Ecology and Management, 373*, 167–178.

Kalies, E. L., & Kent, L. L. Y. (2016). Tamm review: Are fuel treatments effective at achieving ecological and social objectives? A systematic review. *Forest Ecology and Management, 375*, 84–95.

Kane, J. M., Varner, J. M., & Knapp, E. E. (2009). Novel fuelbed characteristics associated with mechanical mastication treatments in northern California and south-western Oregon, USA. *International Journal of Wildland Fire, 18*, 686–697.

Keane, R. E. (2008). Biophysical controls on surface fuel litterfall and decomposition in the northern Rocky Mountains, USA. *Canadian Journal of Forest Research, 38*, 1431–1445.

Keane, R. E. (2015). *Wildland fuel fundamentals and application*. New York: Springer.

Keane, R. E., Sikkink, P. G., & Jain, T. B. (2017). *Physical and chemical characteristics of surface fuels in masticated mixed-conifer stands of the US Rocky Mountains. Gen Tech Rep RMRS-GTR-370*. Fort Collins: USDA Forest Service Rocky Mountain Research Station.

Keeley, J. E., & Bond, W. J. (1999). Mast flowering and semelparity in bamboos: The bamboo fire cycle hypothesis. *The American Naturalist, 154*, 383–391.

Keeley, J., Fotheringham, C., & Morais, M. (1999). Reexamining fire suppression impacts on brushland fire regimes. *Science, 284*, 1829–1832.

Kennedy, M. C., & Johnson, M. C. (2014). Fuel treatment prescriptions alter spatial patterns of fire severity around the wildland–urban interface during the Wallow Fire, Arizona, USA. *Forest Ecology and Management, 318*, 122–132.

Keyes, C. R., & O'Hara, K. L. (2002). Quantifying stand targets for silvicultural prevention of crown fires. *Western Journal of Applied Forestry, 17*(2), 101–109.

Kitzberger, T., Perry, G. L. W., Paritsis, J., Gowda, J. H., Tepley, A. J., Holz, A., & Veblen, T. T. (2016). Fire-vegetation feedbacks and alternative states: Common mechanisms of temperate forest vulnerability to fire in southern South America and New Zealand. *New Zealand Journal of Botany, 54*, 247–272.

Knapp, E. E., Keeley, J. E., Ballenge, E. A., & Brennan, T. J. (2005). Fuel reduction and coarse woody debris dynamics with early season and late season prescribed fire in a Sierra Nevada mixed conifer forest. *Forest Ecology and Management, 208*, 383–397.

Kreye, J. K., Brewer, N. W., Morgan, P., Varner, J. M., Smith, A. M. S., Hoffman, C. M., & Ottmar, R. D. (2014). Fire behavior in masticated fuels: A review. *Forest Ecology and Management, 314*, 193–207.

Le Tacon, F., & Toutain, F. (1973). Variations saisonnières et stationnelles de la teneur en éléments minéraux des feuilles de hêtre (Fagus sylvatica) dans l'est de la France. *Annales des Sciences Forestières, 30*, 1–29.

Leroy, P. (1968). Variations saisonnières des teneurs en eau et éléments minéraux des feuilles de chêne (*Quercus pedunculata*). *Annales des Sciences Forestières, 25*, 83–117.

Liu, K. B., Lu, H., & Shen, C. (2008). A 1200-year proxy record of hurricanes and fires from the Gulf of Mexico coast: Testing the hypothesis of hurricane–fire interactions. *Quaternary Research, 69*, 29–41.

Loehle, C. (2004). Applying landscape principles to fire hazard reduction. *Forest Ecology and Management, 198*, 261–267.

Long, D., Ryan, K., Stratton, R., Mathews, E., Scott, J., Mislivet, M., Miller, M., & Hood, S. (2003). Modeling the effects of fuel treatments for the southern Utah fuel management demonstration project. In P. Omi & L. Joyce (Eds.), *Fire, fuel treatments, and ecological restoration. Proc RMRS-P-29* (pp. 387–395). Fort Collins: USDA Forest Service Rocky Mountain Research Station.

Lydersen, J. M., Collins, B. M., Brooks, M. L., Matchett, J. R., Shive, K. L., Povak, N. A., Kane, V. R., & Smith, D. F. (2017). Evidence of fuels management and fire weather influencing fire severity in an extreme fire event. *Ecological Applications, 27*, 2013–2030.

Lyon, Z., Morgan, P., Sparks, A., Stevens-Rumann, C., Keefe, R., & Smith, A. M. S. (2018). Fire behavior in masticated forest fuels: Lab and prescribed burn experiments. *International Journal of Wildland Fire, 27*, 280–292.

Marsden-Smedley, J. B., & Catchpole, W. R. (1995). Fire behaviour modelling in Tasmanian buttongrass moorlands I. Fuel characteristics. *International Journal of Wildland Fire, 5*, 203–214.

Matthews, S. (2014). Dead fuel moisture research: 1991–2012. *International Journal of Wildland Fire, 23*, 78–92.

McArthur, A. G. (1966). *Weather and grassland fire behaviour. Leaflet 100.* Canberra: Commonwealth of Australia, Forestry and Timber Bureau.

McCarthy, G., & Tolhurst, K. (2001). *Effectiveness of broadscale fuel reduction burning in assisting with wildfire control in parks and forests in Victoria. Fire Management Res Rep No 51.* Melbourne: Natural Resources and Environment.

McCaw, W. L., Neal, J. E., & Smith, R. H. (1996). Fuel accumulation following prescribed burning in young even-aged stands of karri (*Eucalyptus diversicolor*). *Australian Forestry, 59*, 171–177.

Miller, C., & Ager, A. A. (2013). A review of recent advances in risk analysis for wildfire management. *International Journal of Wildland Fire, 22*, 1–14.

Minnich, R. A., & Chou, Y. H. (1997). Wildland fire patch dynamics in the chaparral of southern California and northern Baja California. *International Journal of Wildland Fire, 7*, 221–248.

Mirra, I. M., Oliveira, T. M., Barros, A. M., & Fernandes, P. M. (2017). Fuel dynamics following fire hazard reduction treatments in blue gum (Eucalyptus globulus) plantations in Portugal. *Forest Ecology and Management, 398*, 185–195.

Mislivets, M., & Long, D. (2003). Prioritizing fuel management activities using watersheds and terrain units. In *5th Symposium Fire and Forest Meteorology & 2nd International Wildland Fire Ecology and Fire Management Congress* (pp. 1–7). Orlando: American Meteorological Society.

Mitchell, R. J., Hiers, J. K., O'Brien, J., & Starr, G. (2009). Ecological forestry in the Southeast: Understanding the ecology of fuels. *Journal of Forestry, 107*, 391–397.

Myers, R. K., & van Lear, D. H. (1998). Hurricane-fire interactions in coastal forests of the south: A review and hypothesis. *Forest Ecology and Management, 103*, 265–276.

Nelson, R. M. (2000). Prediction of diurnal change in 10-h fuel stick moisture content. *Canadian Journal of Forest Research, 30*, 1071–1087.

Nelson, R. M. (2001). Water relations of forest fuels. In E. A. Johnson & K. Miyanishi (Eds.), *Forest fires behavior and ecological effects* (pp. 79–149). San Diego: Academic Press.

Noble, P., & Paveglio, T. B. (2020). Exploring adoption of the wildland fire decision support system: End user perspectives. *Journal of Forestry, 118*, 154–171.

Noble, I. R., & Slatyer, R. O. (1980). The use of vital attributes to predict successional changes in plant communities subject to recurrent disturbances. *Vegetatio, 43*, 5–21.

North, M., Collins, B. M., & Stephens, S. (2012). Using fire to increase the scale, benefits, and future maintenance of fuels treatments. *Journal of Forestry, 110*, 392–401.

NWCG. (2006). *Glossary of wildland fire terminology. Publication PM205*. Boise: National Wildfire Coordinating Group (NWCG), National Interagency Fire Center.

Olsen, J. M. (1960). *Green-fuel moisture and soil moisture trends in southern California. Res Note 161*. Berkeley: USDA Forest Service Pacific Southwest Forest and Range Experiment Station.

Olson, J. S. (1963). Energy storage and the balance of producers and decomposers in ecological systems. *Ecology, 44*, 322–331.

Omi, P., & Martinson, E. (2002). Effectiveness of thinning and prescribed fire in reducing wildfire severity. In: DD Murphy, PA Stine (Eds.), *Proceedings of Sierra Nevada Science Symposium*, North Lake Tahoe, CA, October 7–9, 2002. Gen Tech Rep PSW-GTR-193 (pp. 87–92). Berkeley, CA: USDA Forest Service Pacific Southwest Research Station.

Parisien, M., Kafka, V., Hirsch, K. G., Todd, J. B., Lavoie, S. G., & Maczek, P. D. (2005). *Mapping wildfire susceptibility with the BURN-P3 simulation model. Inf Rep NOR-X-405*. Edmonton: Natural Resource Canada, Canadian Forest Service, Northern Forestry Centre.

Parisien, M. A., Junior, D. R., & Kafka, V. G. (2006). Using landscape-based decision rules to prioritize locations of fuel treatments in the boreal mixedwood of western canada. In PL Andrews, BW Butler (comps), *Fuels Management-How to Measure Success: Conference Proceeding*, Portland, 28–30 March 2006. Proc RMRS-P-41 (pp. 221–236). Fort Collins, CO: USDA Forest Service Rocky Mountain Research Station.

Parks, S. A., Holsinger, L. M., Miller, C., & Nelson, C. R. (2015). Wildland fire as a self-regulating mechanism: The role of previous burns and weather in limiting fire progression. *Ecological Applications, 25*, 1478–1492.

Parsons, R. A., Linn, R. R., Pimon, H. C., Sauer, J., Winterkamp, J., Sieg, C. H., & Jolly, M. (2017). Numerical investigation of aggregated fuel spatial pattern impacts on fire behavior. *Land, 6*, 43.

Passovoy, M., & Fulé, P. Z. (2006). Snag and woody debris dynamics following severe wildfires in northern Arizona ponderosa pine forests. *Forest Ecology and Management, 223*, 237–246.

Pellizzaro, G., Cesaraccio, C., Duce, P., Ventura, A., & Zara, P. (2007). Relationships between seasonal patterns of live fuel moisture and meteorological drought indices for Mediterranean shrubland species. *International Journal of Wildland Fire, 16*, 232–241.

Perrakis, D. D., Lanoville, R. A., Taylor, S. W., & Hicks, D. (2014). Modeling wildfire spread in mountain pine beetle-affected forest stands, British Columbia. *Fire Ecology, 10*, 10–35.

Perry, D., Jing, H., Youngblood, A., & Oetters, D. (2004). Forest structure and fire susceptibility in volcanic landscapes of the Eastern High Cascades, Oregon. *Conservation Biology, 18*, 913–926.

Peterson, D. L., Johnson, M. C., Agee, J. K., Jain, T. B., McKenzie, D., & Reinhardt, E. D. (2003). Fuels planning: managing forest structure to reduce fire hazard. In: Second International Wildland Fire Ecology and Fire Management Congress and Fifth Symposium on Fire and Forest Meteorology; 2003 November 16-20; Orlando, FL, USA Poster 3D. 5. Boston, MA: American Meteorological Society. Online: https://ams.confex.com/ams/FIRE2003/webprogram/Paper74459.html.

Peterson, D. L., Johnson, M. C., Agee, J. K., Jain, T. B., McKenzie, D., & Reinhardt, E. D. (2005). *Forest structure and fire hazard in dry forests of the western United States. PNW- GTR-268*. Portland: USDA Forest Service Pacific Northwest Research Station.

Philpot, C. W. (1963). *The moisture content of ponderosa pine and whiteleaf manzanita foliage in the Central Sierra Nevada. Res Note PSW-39*. Berkeley: USDA For Serv Pacific Southwest Research Station.

Philpot, C. W. (1965). *Diurnal fluctuation in moisture content of ponderosa pine and whiteleaf manzanita leaves. Res Note PSW-67*. Berkeley: USDA Forest Service Pacific Southwest Research Station.

Pickett, S. T. A., & White, P. S. (1985). *The ecology of natural disturbance and patch dynamics*. San Diego: Academic Press.

Pinto, A., & Fernandes, P. M. (2014). Microclimate and modelled fire behaviour differ between adjacent forest types in northern Portugal. *Forests, 5*, 2490–2504.

Pinto, A., Espinosa-Prieto, J., Rossa, C., Matthews, S., Loureiro, C., & Fernandes, P. (2014). Modelling fine fuel moisture content and the likelihood of fire spread in blue gum (Eucalyptus globulus) litter. In D. X. Viegas (Ed.), *Advances in Forest Fire Research* (pp. 353–359). Coimbra: Imprensa da Universidade de Coimbra.

Pollet, J., & Omi, P. N. (2002). Effect of thinning and prescribed burning on crown fire severity in ponderosa pine forests. *International Journal of Wildland Fire, 11*, 1–10.

Price, O. F., Russell-Smith, J., & Watt, F. (2012). The influence of prescribed fire on the extent of wildfire in savanna landscapes of western Arnhem Land, Australia. *International Journal of Wildland Fire, 21*, 297–305.

Price, O. F., Pausas, J. G., Govender, N., Flannigan, M., Fernandes, P. M., Brooks, M. L., & Bird, R. B. (2015). Global patterns in fire leverage: The response of annual area burnt to previous fire. *International Journal of Wildland Fire, 24*, 297–306.

Prichard, S. J., & Kennedy, M. C. (2014). Fuel treatments and landform modify landscape patterns of burn severity in an extreme fire event. *Ecological Applications, 24*, 571–590.

Prichard, S. J., Peterson, D. L., & Jacobson, K. (2010). Fuel treatments reduce the severity of wildfire effects in dry mixed conifer forest, Washington, USA. *Canadian Journal of Forest Research, 40*, 1615–1626.

Pye, J., Prestemon, J., Butry, D., & Abt, K. (2003). Prescribed burning and wildfire risk in the 1998 fire season in Florida. In P. Omi & L. Joyce (Eds.), *Fire, fuel treatments, and ecological restoration. Proc RMRS-P-29* (pp. 15–26). Ogden: USDA Forest Service Rocky Mountain Research Station.

Pyne, S., Andrews, P., & Laven, R. (1996). *Introduction to wildland fire* (2nd ed.). New York: Wiley.

Qi, Y., Jolly, W. M., Dennison, P. E., & Kropp, R. C. (2016). Seasonal relationships between foliar moisture content, heat content and biochemistry of lodgepole pine and big sagebrush foliage. *International Journal of Wildland Fire, 25*, 574–578.

Reinhardt, E. D., Keane, R. E., Calkin, D. E., & Cohen, J. D. (2008). Objectives and considerations for wildland fuel treatment in forested ecosystems of the interior western United States. *Forest Ecology and Management, 256*, 1997–2006.

Reyes, C., & O'Hara, K. (2002). Quantifying stand targets for silvicultural prevention of crown fires. *Western Journal of Applied Forestry, 17*, 101–109.

Rhodes, J. J., & Baker, W. L. (2008). Fire probability, fuel treatment effectiveness and ecological tradeoffs in western US public forests. *The Open Forest Science Journal, 14*, 1.

Rigolot, E. (2002). Fuel-break assessment with an expert appraisement approach. In D. X. Viegas (Ed.), *Forest fire research & wildland fire safety*. Rotherdam: Millpress.

Rigolot, E. & Etienne, M. (1998). Impact of fuel control techniques on Cistus monspeliensis dynamics. In Proceedings of the 13th Conference on Fire and Forest Meteorology, Ed. R. Weber, pp. 467–471. Fairfield: International Association of Wildland Fire.

Rothermel, R. C. (1972). *A mathematical model for predicting fire spread in wildland fuels. Res Pap INT-115*. Ogden: USDA Forest Service Intermountain Forest and Range Experiment Station.

Rothermel, R.C. (1991). Predicting behavior and size of crown fires in the northern Rocky Mountains. USDA Forest Service Research Paper INT-438.

Rothermel, R. C., & Philpot, C. W. (1973). Predicting changes in chaparral flammability. *Journal of Forestry, 71*, 640–643.

Rothermel, R., Wilson, R. A., Morris, G. A., & Sackett, S. S. (1986). *Modelling moisture content of fine dead wildland fuels. Research Paper INT-359*. Ogden: USDA Forest Service Intermountain Research Station.

Roussopoulos, P., & Johnson, V. (1975). *Help in making fuel management decisions. Res Pap NC-112*. St Paul: USDA Forest Service, North Central Forest Experiment Station.

Rundel, P. W., & Parsons, D. J. (1979). Structural changes in chamise (Adenostoma fasciculatum) along a fire-induced age gradient. *Journal of Range Management, 32*, 462–466.

Safford, H. D., Stevens, J. T., Merriam, K., Meyer, M. D., & Latimer, A. M. (2012). Fuel treatment effectiveness in California yellow pine and mixed conifer forests. *Forest Ecology and Management, 274*, 17–28.

Sah, J. P., Ross, M. S., Snyder, J. R., Koptur, S., & Cooley, H. C. (2006). Fuel loads, fire regimes, and post-fire fuel dynamics in Florida Keys pine forests. *International Journal of Wildland Fire, 15*, 463–478.

Schmidt, D. A., Taylor, A. H., & Skinner, C. N. (2008). The influence of fuels treatment and landscape arrangement on simulated fire behavior, Southern Cascade range, California. *Forest Ecology and Management, 255*, 3170–3184.

Schoennagel, T., Veblen, T., & Romme, W. (2004). The interaction of fire, fuels, and climate across Rocky Mountain forests. *BioScience, 54*, 661–676.

Schoennagel, T., Balch, J. K., Brenkert-Smith, H., Dennison, P. E., Harvey, B. J., Krawchuk, M. A., Mietkiewicz, N., Morgan, P., Moritz, M. A., Rasker, R., & Turner, M. G. (2017). Adapt to more wildfire in western North American forests as climate changes. *Proceedings of National Academy Science of the United States of America, 114*, 4582–4590.

Schwilk, D. W., Keeley, J. E., Knapp, E. E., McIver, J., Bailey, J. D., Fettig, C. J., Fiedler, C. E., Harrod, R. J., Moghaddas, J. J., Outcalt, K. W., & Skinner, C. N. (2009). The national fire and fire surrogate study: Effects of fuel reduction methods on forest vegetation structure and fuels. *Ecological Applications, 19*, 285–304.

Scott, J. H. (2006). *Comparison of crown fire modeling systems used in three fire management applications. Res Pap RMRS-RP-58*. Fort Collins: USDA Forest Service Rocky Mountain Research Station.

Scott, J. H., & Reinhardt, E. D. (2001). Assessing crown fire potential by linking models of surface and crown fire behavior. USDA Forest Service Research Paper RMRS-RP-29.

Sieg, C. H., Linn, R. R., Pimont, F., Hoffman, C. M., McMillin, J. D., Winterkamp, J., & Baggett, L. S. (2017). Fires following bark beetles: Factors controlling severity and disturbance interactions in ponderosa pine. *Fire Ecology, 13*, 1–23.

Simard, A. J. (1968). *The moisture content of forest fuels. I. A review of basic concepts. Information Rep FF-X-14*. Ottawa: Forest and Fire Res Institute, Forestry Branch, Department of Forestry and Rural Development.

Sparks, A. M., Smith, A. M. S., Talhelm, A. F., Kolden, C. A., Yedinak, K. M., & Johnson, D. M. (2017). Impacts of fire radiative flux on mature Pinus ponderosa growth and vulnerability to secondary mortality agents. *International Journal of Wildland Fire, 26*, 95–106.

Stambaugh, M. C., Guyette, R. P., Grabner, K. W., & Kolaks, J. (2006). Understanding Ozark forest litter variability through a synthesis of accumulation rates and fire events. In PL Andrews, BW Butler (comps), *Fuels Management-How to Measure Success: Conference Proceedings*, Portland, 28–30 March 2006. Proc RMRS-P-41 (pp. 321–332). Fort Collins, CO: USDA Forest Service Rocky Mountain Research Station.

Stenzel, J. E., Bartowitz, K. J., Hartman, M. D., Lutz, J. A., Kolden, C. A., Smith, A. M., Law, B. E., Swanson, M. E., Larson, A. J., Parton, W. J., & Hudiburg, T. W. (2019). Fixing a snag in carbon emissions estimates from wildfires. *Global Change Biology, 25*, 3985–3994.

Stephens, S. L., & Fulé, P. Z. (2005). Western pine forests with continuing frequent fire regimes: Possible reference sites for management. *Journal of Forestry, 103*, 357–362.

Stephens, S. L., Moghaddas, J. J., Edminster, C., Fiedler, C. E., Haase, S., Harrington, M., Keeley, J. E., Knapp, E. E., McIver, J. D., Metlen, K., & Skinner, C. N. (2009). Fire treatment effects on vegetation structure, fuels, and potential fire severity in western US forests. *Ecological Applications, 19*, 305–320.

Stephens, S. L., Collins, B. M., & Roller, G. B. (2012). Fuel treatment longevity in a Sierra Nevada mixed conifer forest. *Forest Ecology and Management, 285*, 204–212.

Stephens, S. L., Collins, B. M., Fettig, C. J., Finney, M. A., Hoffman, C. M., Knapp, E., North, M. P., Safford, H., & Wayman, R. B. (2018). Drought, tree mortality, and wildfire in forests adapted to frequent fire. *BioScience, 68*, 77–88.

Stephens, S. L., Battaglia, M. A., Churchill, D. J., Collins, B. M., Coppoletta, M., Hoffman, C. M., Lydersen, J. M., North, M. P., Parsons, R. A., Ritter, S. M., & Stevens, J. T. (2020). Forest restoration and fuels reduction: Convergent or divergent? *BioSciences, 71*(1), 85–101.

Stevens-Rumann, C., Shive, K., Fulé, P. Z., & Sieg, C. H. (2013). Pre-wildfire fuel reduction treatments result in more resilient forest structure a decade after wildfire. *International Journal of Wildland Fire, 22*, 1108–1117.

Stevens-Rumann, C. S., Hudak, A. T., Morgan, P., Arnold, A., & Strand, E. K. (2020). Fuel dynamics following wildfire in US Northern Rockies forests. *Frontiers in Forests and Global Change, 3*, 51.

Stocks, B. J. (1970). *Moisture in the forest floor - its distribution and movement. Publication no. 1271*. Ottawa: Department of Fisheries and Forestry, Canadian Forestry Service.

Stocks, B. (1987). Fire potential in the spruce budworm-damaged forests of Ontario. *The Forestry Chronicle, 63*, 8–14.

Stocks, B. J., Alexander, M. E., Wotton, B. M., Stefner, C. N., Flannigan, M. D., Taylor, S. W., Lavoie, N., Mason, J. A., Hartley, G. R., Maffey, M. E., & Dalrymple, G. N. (2004). Crown fire behaviour in a northern jack pine black spruce forest. *Canadian Journal of Forest Research, 34*, 1548–1560.

Sullivan, A. L., & Matthews, S. (2013). Determining landscape fine fuel moisture content of the Kilmore East 'Black Saturday' wildfire using spatially-extended point-based models. *Environmental Modelling and Software, 40*, 98–108.

Syphard, A. D., Keeley, J. E., & Brennan, T. J. (2011). Comparing the role of fuel breaks across southern California national forests. *Forest Ecology and Management, 261*, 2038–2048.

Thomas, P. B., Watson, P. J., Bradstock, R. A., Penman, T. D., & Price, O. F. (2014). Modelling surface fine fuel dynamics across climate gradients in eucalypt forests of south-eastern Australia. *Ecography, 37*, 827–837.

Tinkham, W. T., Hoffman, C. M., Ex, S. A., Battaglia, M. A., & Saralecos, J. D. (2016). Ponderosa pine forest restoration treatment longevity: Implications of regeneration on fire hazard. *Forests, 7*, 137.

Tolhurst, K. G., & McCarthy, G. (2016). Effect of prescribed burning on wildfire severity: A landscape-scale case study from the 2003 fires in Victoria. *Australian Forestry, 79*, 1–14.

Tolhurst, K., Shields, B., & Chong, D. (2008). Phoenix: Development and application of a bushfire risk management tool. *Australian Journal of Emergency Management, 23*, 47.

Tymstra, C. R., Bryce, W., Wotton, B. M., Taylor, S. W., & Armitage, O. B. (2010). *Development and structure of Prometheus: The Canadian wildland fire growth simulation model. Rep NOR-X-417*. Edmonton: Natuaral Resource Canada Canadian Forest Service, Northern Forestry Centre.

Vaillant, N. M., Fites-Kaufman, J. A., & Stephens, S. L. (2009). Effectiveness of prescribed fire as a fuel treatment in Californian coniferous forests. *International Journal of Wildland Fire, 18*, 165–175.

Vaillant, N. M., Ager, A. A., & Anderson, J. (2013). *ArcFuels10 system overview. Gen Tech Rep PNW-GTR-875*. Portland: USDA Forest Service Pacific Northwest Research Station.

Van Nest, T. A., & Alexander, M. E. (1999). *Systems for rating fire danger and predicting fire behavior used in Canada*. Paper presented at the National Interagency Fire Behavior Workshop, Phoenix, March 1–5, 1999.

Van Wagner, C. E. (1967). *Seasonal variation in moisture content of Eastern Canadian tree foliage and the possible effects on crown fires. Publ No 1024*. Chalk River: Department of Forestry and Rural Development, Petawawa Forest Experiment Station.

Van Wagner, C. E. (1977). Conditions for the start and spread of crown fire. *Canadian Journal of Forest Research, 7*, 23–34.

Van Wagner, C. E. (1987). *The development and structure of the Canadian Forest Fire Weather Index System, Tech Rep 35*. Ottawa, ON: Canadian Forestry Service.

Veblen, T. T., Kitzberger, T., Raffaele, E., & Lorenz, D. C. (2003). Fire history and vegetation changes in northern Patagonia, Argentina. In T. T. Veblen, W. L. Baker, G. Montenegro, &

T. W. Swetnam (Eds.), *Fire and climatic change in temperate ecosystems of the western Americas* (pp. 265–295). New York: Springer.

Viney, N. R. (1991). A review of fine fuel moisture modelling. *International Journal of Wildland Fire, 1*, 215–234.

Viney, N. R. (1992). Moisture diffusivity in forest fuels. *International Journal of Wildland Fire, 2*, 161–168.

Viney, N. R., & Hatton, T. J. (1989). Assessment of existing fine fuel moisture models applied to Eucalyptus litter. *Australian Forestry, 52*, 82–93.

Wade, D. D., Forbus, J. K., & Saveland, J. M. (1993). *Photo series for estimating post-hurricane residues and fire behavior in southern pine. Gen Tech Rep SE-82*. Asheville: USDA Forest Service Southeastern Forest Experiment Station.

Weatherspoon, C. (1996). Fire-silviculture relationships in Sierra forests. In Sierra Nevada Ecosystem Project: Final Report to Congress, vol. II, Assessments and Scientific Basis for Management Options, pp. 1167–1176. Davis: University of California.

Weatherspoon, C., & Skinner, C. (1996). Landscape-level strategies for forest fuel management. In *Sierra Nevada Ecosystem project: Final report to congress* (Vol. II: Assessments and Scientific Basis for Management Options, pp. 1471–1492). Davis: University of California Davis.

Weise, D. R., Hartford, R. A., & Mahaffey, L. (1998). Assessing live fuel moisture for fire management applications. In T. L. Pruden & A. B. Leonard (Eds.), *Assessing live fuel moisture for fire management applications, Misc Pub* (pp. 49–55). Berkeley: USDA Forest Service Pacific Southwest Experiment Station.

Weise, D. R., Zhou, X., Sun, L., & Mahalingam, S. (2005). Fire spread in chaparral—'go or no-go?'. *International Journal of Wildland Fire, 14*(1), 99–106.

Yebra, M., Dennison, P. E., Chuvieco, E., Riano, D., Zylstra, P., Hunt, E. R., Jr., Danson, F. M., Qi, Y., & Jurdao, S. (2013). A global review of remote sensing of live fuel moisture content for fire danger assessment: Moving towards operational products. *Remote Sensing of Environment, 136*, 455–468.

Ziegler, J. P., Hoffman, C. M., Battaglia, M., & Mell, W. (2017). Spatially explicit measurements of forest structure and fire behavior following restoration treatments in dry forests. *Forest Ecology and Management, 386*, 1–12.

Chapter 12
Fire Regimes, Landscape Dynamics, and Landscape Management

Learning Outcomes
Through this chapter, we expect readers to

1. Understand fire occurrence and effects, starting with temporal dynamics at points, and then expanding to landscapes including both space and time,
2. Describe the potential, uncertainty, and limitations for the different data sources used for describing recurring fires as fire regimes,
3. Critique the statement that "the ways in which climate, vegetation, and people drove historical fire regimes may hold important lessons for understanding what makes ecosystems resilient and managing them to adapt for the future",
4. Link fire effects on landscapes to landscape management, and
5. Describe one key theme common to two or three of the illustrative examples from around the globe used in this chapter.

12.1 Introduction

In this chapter, we address the spatial and temporal interactions between multiple fires across landscapes and discuss their implications for landscape dynamics and management. We build upon what you learned in previous chapters about the behavior and effects of individual fires and plants and fuel dynamics. Landscapes are heterogeneous areas of land made up of smaller patches of relatively homogenous vegetation. Landscape mosaics often contain habitats for plants and animals, which often include more than one patch. Fires are more likely in some habitats within landscape than in others, just as animals and plants find some locations more suitable than others, often termed selectivity (Rego et al. 2019).

F. Castro Rego et al., *Fire Science*, Springer Textbooks in Earth Sciences, Geography and Environment, https://doi.org/10.1007/978-3-030-69815-7_12

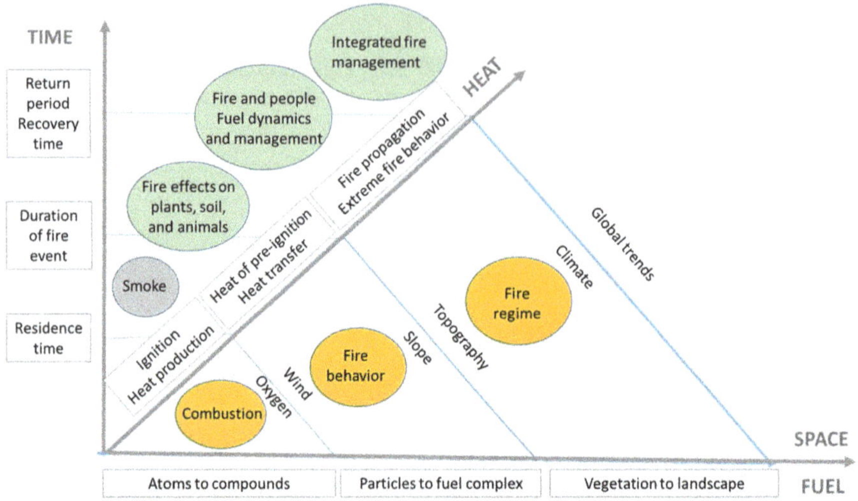

Fig. 12.1 The fire regime triangles are used here to illustrate similar controls on flames, fires (many flames), and fire regimes (many fires) over space and time. The triangles are nested. People alter many of the controls, especially fuels, at multiple scales

We begin this chapter by discussing what fire regimes are and why we seek to describe them. The pattern of recurring fires over time and space is the fire regime (Morgan et al. 2001; Krebs et al. 2010; Whitlock et al. 2010; Moritz et al. 2011) (Fig. 12.1). Some fire regime descriptors are temporal (frequency and seasonality), while others are spatial (area burned, fire size, and patch size distribution and pattern), and others reflect magnitude (severity and intensity). The temporal and spatial scales over which fire regimes are described must be identified as the environmental conditions and human activities that influence fire regimes have always changed and are now changing in novel ways (Krebs et al. 2010). We emphasize that variability in the effects of repeated fires is especially significant ecologically. Next, we describe the different data sources used to characterize past, current, and future fire regimes, using examples from the world's ecosystems. Then, we address changing fire regimes, and in so doing, discuss the interacting effects of people, climate, and fuels. The historical range of variability is often used as a reference for long-term resilience and a guide for management, yet there are limitations. Landscape dynamics are shaped by fire, other disturbances, and succession. We end with some case studies and other examples of the implications of alternative fire management strategies for landscape dynamics and management. The implications for integrated fire management are covered in Chap. 13.

Why do scientists and managers characterize and map fire regimes? Understanding how climate, topography, vegetation, and prior fires interact to shape recurring fires can inform land managers and scientists in forecasting fire occurrence and fire effects for the future, even as environmental conditions are changing. In particular, understanding of past and current fire regimes can inform management goals that encompass the restoration of long-term health and resilience. Restoration is "the process of assisting the recovery of an ecosystem that has been degraded, damaged, or destroyed" (Society for Ecological Restoration 2004). Resilience is the capacity of a system to absorb disturbance and reorganize while changing to retain essentially still the same function, structure, identity, and feedbacks (Society for Ecological Restoration, www.ser.org, Accessed 3 Dec 2019). Vulnerability and resilience are addressed in Chap. 10.

Fire regime information can be used to assess how likely fires are and how much variability there is in fire occurrence, especially if we use the historical patterns to understand what drives when and where fires are likely to occur. This understanding can inform choice among alternative management scenarios now and in the future, even as climate and other conditions change. Most importantly, fire regimes and landscape dynamics are critical to understanding. They can be very useful in communicating with others how dynamic landscapes are, why, and how some fires contribute to ecosystems that are less resilient to future disturbances, while other fires enhance resilience.

Regional fire years are those years when large areas burn with many fires burning in many different places at the same time. Conditions of weather and climate often drive such years, and forecasts can help us prepare for such years in advance. Whether fuels, climate, or both are most important in influencing fire occurrence and effects has tremendous implications for management in our rapidly changing world.

By the end of this chapter, you will be able to think critically about the implications of changing fire regimes, including current debates such as how natural are high severity fires, and how much high severity fire is too much. Answers to these questions have tremendous implications for landscape dynamics and management. You will be able to address fire and landscape issues such as the relative importance of climate and fuel in influencing past and future fire regimes, and how, when, and where prior fires will affect future fires depending on the time since the previous fire, weather, topography, and other environmental conditions.

12.2 Fire Regime Descriptors

The descriptors of recurring fires are drawn from those for repeated disturbances. Thus we can think of them in terms of time (e.g., frequency and season), space (e.g., area burned, fire size, and patch size distribution), and magnitude (e.g., severity and intensity) (Morgan et al. 2001; Keane et al. 2015). Frequency and severity are commonly used together to describe fire regimes. While this is convenient, other fire regime characteristics, as well as spatial and temporal variability, can be critically important to understanding the effects of repeated fires on ecosystems, for the variation is often more ecologically meaningful than the average. As some metrics vary with scale, it is important to identify the temporal and spatial domain for which fire regimes are described.

12.2.1 Temporal Fire Regime Descriptors and Metrics

Frequency is defined as the number of times a given point or area has burned within a given time. Frequency is often described by the interval between fires (called the fire return interval). Understanding why and how often fires occur is central to addressing questions of how often a plant or an organism will experience a fire or is likely to reproduce without experiencing a fire (see Sect. 12.5).

Frequency is most often applied to points or small areas (sometimes called point frequency) but can also be quantified for large areas (sometimes called area frequency) through fire rotation and fire cycle. Frequency is very scale-dependent. For example, fires may often burn somewhere in a large area but only rarely at a given point within the area. The median and range of fire return intervals indicate the number of years between successive fires. Often, fire frequency from individual trees in a small area is aggregated into a composite fire interval. This is most accurate when there are many fire intervals in the calculations. Bias can occur if not all the fire dates in an area are detected. Because the intervals between fires are highly skewed with a few long and many shorter intervals, the median is a more useful description than is the mean. Many scientists characterize fire frequency by fitting a using the Weibull distribution to the intervals and describing median and variability in fire intervals as moments of that distribution (Fig. 12.2), and many present box plots to illustrate the distributions.

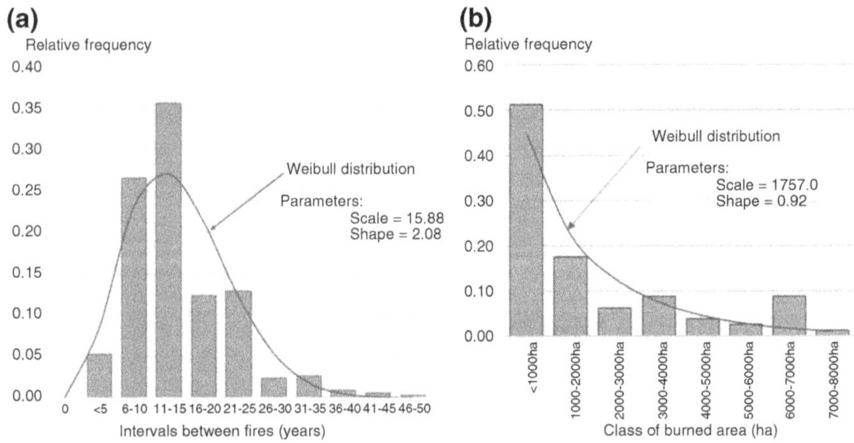

Fig. 12.2 The Weibull distribution is useful for characterizing both the central tendency (median) and the variability (as indicated by the scale and shape parameters of the distribution, respectively) of (**a**) The intervals between fires, 1687–1900, reconstructed from cross dated fire-scars in trees from systematically located sample points throughout the Dugout Creek (DCR) watershed in the Blue Mountains of Oregon, USA, and (**b**) fire sizes for the same area. Note that the data on fire size are truncated as those fires that burned more area than this watershed could not be quantified. See more on this study and methods later in this chapter (See Sect. 12.4.1) (Data from Heyerdahl et al. (2001, 2002); data are archived with the International Multiproxy Paleofire Database, NCEI 2020)

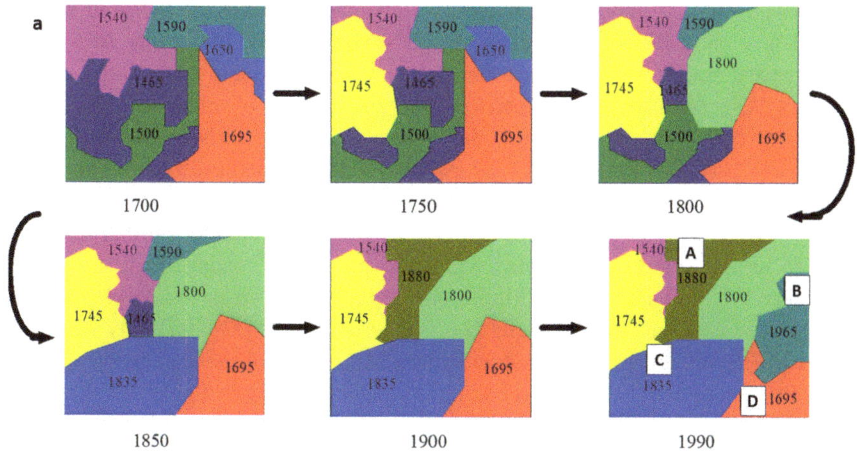

Fig. 12.3 Comparison of different measures of fire frequency for an area of 6730 ha. (**a**) The six maps are of the same 13,000 ha area every 50 years from 1700 to 1990. For each area of the forest, the date of the previous fires was reconstructed based upon the age of the trees. (**b**) Time since a fire in 1990 and the area for each polygon (redrawn from Fall 1999). (**c**) Fire frequency can be calculated from the fire intervals at each of four sample points labeled A, B, C, and D. Data are adapted from Heinselman (1973) based on fire history reconstructed from stand ages he mapped in the Boundary Waters Canoe Area in Minnesota in the USA

Annual burn probability, calculated as the number of times an area burned divided by the number of years, is another commonly used measure of fire frequency. Fernandes et al. (2012) used a fire atlas to calculate burn probability for large areas in Portugal. There, the burn probability increased linearly with time since fire, reflecting the influence of biomass accumulation and the potential utility of fuels treatments. Finney et al. (2008) used simulation models to assess the probability of fire across the USA (see Sect. 12.3.5 for description).

Measures of fire frequency can be calculated from maps of fire perimeters or from the size and ages of patches (Fig. 12.3). Sometimes the dates and areas burned in past fires are inferred from the age of trees, but this depends on being able to interpret the age structure and on the assumption that trees only establish after fires.

Fire rotation (FR), another measure of fire frequency, is defined as the length of time in years it takes for an area equal in size to the study area to burn (Heinselman 1973):

$$FR = \frac{t}{\sum_{i=1}^{n} \frac{a_i}{A_{land}}} \tag{12.1}$$

where t is the time in years, A_{land} is the area (size) of the landscape, a_i is the size of a single fire, and n is the total number of fires. FR has the advantage of incorporating both how large fires are and how often they occur; it is relatively simple to estimate. However, the same FR value can result from a few large or many small fires. The FR incorporates spatial and temporal variability, including parts of the area seldom burned and parts burned frequently, but a single number represents both. FR's primary assumption is that all past fires have been detected and accurately mapped, yet this is very difficult to achieve. The fire extent of older fires is likely underestimated as their evidence on the landscape could have been erased by more recent fires. A major limitation of FR is that variability measures cannot be calculated except by comparing different areas or different time periods.

Heinselman (1973) was one of the first scientists to describe a fire regime and FR. He argued for using the fire regime to guide fire management across the Boundary Waters Canoe Area, Minnesota, in the north-central USA. Heinselman (1973) estimated that 748,766 ha had burned across the 404,686 ha study area from 1542 to 1972, resulting in an FR of 243 years. Most (83% of the land area excluding lakes and streams) of the area burned in this time occurred in just nine individual fire years. Even in the moist coniferous forests in the area, fires were frequent enough that most forest areas in the entire landscape were recovering from fires. He argued that in this area fires should be managed to encourage burning in natural patterns and frequency.

The fire cycle is the median age of a forest, best calculated when the fires are stand-replacing and trees establish soon after each fire. The fire cycle is similar to FR in that it is based on the age of patches, and the median age depends on how often fires occur and how much area is burned in the fires. It is also dependent on the accurate aging of trees, which depends on dendrochronology (see some discussion of this below). Similar to FR, the fire cycle is a measure of fire frequency for an area.

The fire frequency measured at points and the fire frequency calculated for an area are related but seldom equal. If fires occur randomly (they don't, as some points on a landscape are more likely to burn than others, Fig. 12.3), the annual probability of fire is equivalent to the median fire interval or the inverse of fire rotation. When fires are large relative to the landscape area, the concepts of fire rotation and fire return interval are equivalent. Fire rotation is not very meaningful if fires are much more likely in some parts of the landscapes than in others. In that case, understanding fire return intervals at different locations is more useful.

The accuracy of all fire frequency measures depends on whether the evidence of all fires can be detected. If some fire dates are missed, or small or rare fires are not detected, fires will appear less frequent than they actually were. Conversely, if the dates of the same fire are misjudged as occurring in different years, fires could be judged to have occurred more frequently than actually occurred.

Temporal and spatial variability in fire frequency is often more critical ecologically than the mean or median fire frequency. Thus, it is important when describing fire regimes to include not only estimates of central tendency such as median, but also a measure of variability. Trees, such as ponderosa pine (*Pinus ponderosa*), longleaf pine (*P. palustris*), and maritime pine (*P. pinaster*), thrived in areas with fires historically burning every 2–25 years as large trees have thick bark and high, open crowns, yet young, small trees are readily killed by fire. Either occasional long fire-free intervals or spatial variability resulting in some areas within fires that were unburned or burned with low severity allowed some young trees to survive for long enough that they grew enough to survive one or more subsequent fires, i.e., they grew tall enough to have crowns above the heat from fires with bark thick enough to protect cambium (see Sect. 9.3 for fire effects on trees). Thus, the age structure and character of these forests historically and now, and many other forests, depended on variability in time and space around the median fire interval. Forests with multiple age classes and many clumps and openings often result in "old growth" forests resulting from frequent fires. Usually, these include old and large trees, a few young trees, and many snags and downed logs (Abella et al. 2007; Holden et al. 2007). The variable structure can be very important to biodiversity as multiple plants and animals depend on different structures in close proximity, such as old living and dead trees near meadows. In the absence of such variability, forests become more homogeneous. The spatial and temporal heterogeneity of fire regimes is not well captured in fire frequency unless we explicitly focus on it by calculating the variability. Fire occurrence can vary over even small areas and through time, and that can affect and be driven by the fuels and vegetation patchiness that develops as a result (more on such landscape dynamics later in this chapter).

The variability in the interval between recurring fires in a particular area interacts with species characteristics to influence the subsequent dynamics of plant species and communities. For example, the vital attributes developed in Australia by Noble and Slatyer (1980). Vital attributes of a species are those which are essential to its role in a vegetation replacement sequence. The first vital attribute (m) is the time required for a plant of a given species to reach reproductive maturity after disturbance, the second is its longevity (l), and the third is the longevity of the pool of propagules (e) of a plant of that species. Comparing the vital attributes of a species with the variability of an area's fire return can be useful in understanding the role of that species in community succession. Noble and Slatyer (1980) illustrate that relationship for communities with common tree species in Australia and in North America (Fig. 12.4).

Synchrony occurs when many different fires burn in different places at the same time. Heyerdahl et al. (2008b) and Morgan et al. (2008) found that the regional fire years (top 10% of all years) had significantly warmer springs followed by warmer, dryer summers compared to all the other years with less annual area burned. This was so whether they identified regional fire years based upon the proportion of all sites recording fire for a given year from 1650 to 1900, or based on annual area burned recorded in fire atlases for 1900–2003. Fire weather conditions often drive regional fire years, and climate forecasts can help managers prepare for such years in advance because in regional fire years, with many large fires burning simultaneously in many different locations, costs for fire suppression are high, and fire personnel can be overwhelmed as many fires threaten people and property at once. Synchrony of fires is one indication of top-down control by climate on fire regimes (see related discussion in Sect. 12.5.1).

Seasonality is also an important characteristic of the fire regime. The most common metric for the season of fires is the proportion of all fires or proportion of area burned in spring, summer, fall, or winter. Another aspect of seasonality is the length of the fire season. Westerling (2006) calculated the length of the fire season as

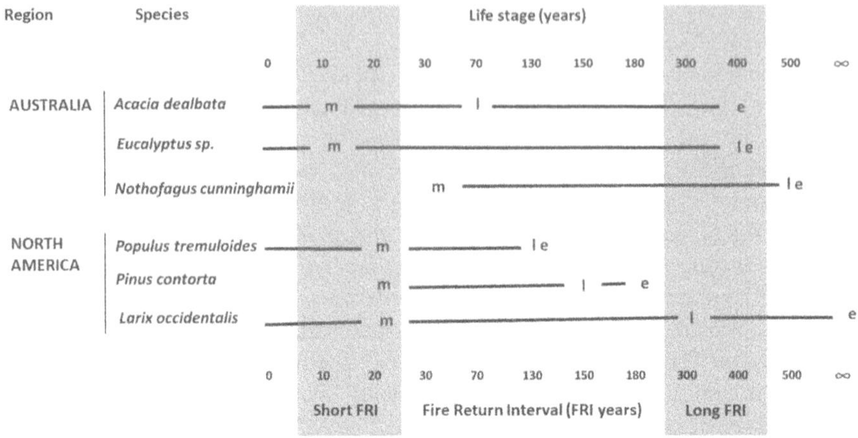

Fig. 12.4 Illustration of the use of species vital attributes in relation to fire return intervals FRI) for two situations, one for Australia with *Acacia dealbata*, *Eucalyptus* spp., and *Nothofagus cunninghamii*, and the other for North America with *Populus tremuloides*, *Pinus contorta*, and *Larix occidentalis.* Vital attributes include time to maturity (m), the longevity of the plant (l), and the longevity of the seed pool (e) (Noble and Slatyer 1980). All of the species can survive and thrive at the intermediate FRI. If fires become too frequent (FRI <10 years) *N. cunninghamii* in Australia and *P. contorta* in North America will be locally extinct if they do not survive the fires as they do not sprout and do not mature in time to produce seed before the next fire. If the interval between fires is too long (FRI >300 years) *P. contorta* and *P. tremuloides* would become locally extinct (until recolonized with new seedlings), and with FRI >450 years, only *N. cunninghamii* would be present. Here we assumed that adult *L. occidentalis* trees survive, while *P. tremuloides, A. dealbata*, and *Eucalyptus* spp. resprout and so they survive fires as buds that resprout after fire. Only a few very short or very long intervals in a local area can change species composition; those fire intervals could result from people igniting or suppressing fires or unusual fire weather conditions

the time between the earliest detected fire until the last detected fire. The season in which fires occur can influence their effects depending on the phenology of plants and animals. Seasonality of fires is important as many plants and animals adapted to survive and thrive in burned areas are quite vulnerable to fire effects when fires burn outside a given part of the year. Often, when people ignite fires, the fire season is longer than lightning-ignited fires (Balch et al. 2017).

Changing fire seasonality is often beneficial to many management objectives. Fire managers purposely change the seasonality of fires by prescribed burning in the autumn to spring and thus avoiding damaging summer wildfires (Fig. 12.5, See Case Studies 13.1 and 13.6 for examples from Australia). Prescribed burns may also be scheduled to avoid burning when animals have just had their young. All plants are more vulnerable to fire and slower to recover when they are burned when they are actively growing; plants that were dormant when fires occurred are more likely to survive and regrow following fires. The season of fire can also affect habitat for insects and other food sources for animals and thus indirectly affect population viability. With climate change already altering the timing of plant growth and flowering in many locations, many plants and animals may be “out of synchrony”

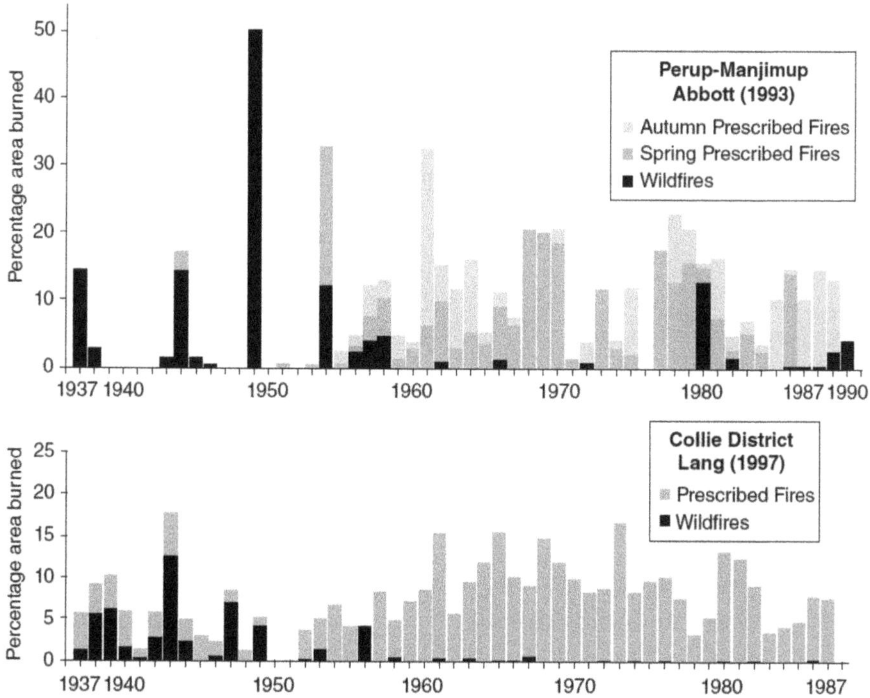

Fig. 12.5 The area burned by wildfires (black) greatly decreased where 9–15% of each of these two areas was burned each year with prescribed fires (gray). See related Case Studies 13.1 and 13.6 for integrated fire management success stories using prescribed burning in Australia. From Rego et al. (2019) based upon data from Abbott et al. (1993) and Lang (1997)

with the conditions they need for survival. Of concern to many is how life cycles of plants and animals will be affected when fires burn in quite different seasons than they did in the past.

12.2.2 *Spatial Fire Regime Descriptors and Metrics*

The area burned can be quantified as the mean or median size of individual fires, or as the total area burned or proportion of an area burned in a specified time such as 1 year. The area burned is often calculated based on mapped fire perimeters. The area burned has often been correlated with weather or climate, which scientists use to project the implications of changing climate for area burned into the future. See our discussion of the role of climate and fuels in fire regimes in Sect. 12.4.

Fire sizes are often highly skewed with many small fires and very few large fires (Fig. 12.2b), though the few largest fires may burn more total area than many small fires. Fire suppression efforts often alter fire size distributions. Thus, analysis of fire sizes can inform fire suppression capabilities, success, and strategies.

Patch size distributions of burned areas (e.g., Morgan et al. 2017) or vegetation are ecologically important. Patches, whether they are fires or patches of fire effects within larger fire perimeters, are variable in size and character, some of which are related to topography and the effects of prior disturbances, such as previous fires. Patches are defined as relatively uniform vegetation conditions different from the surrounding area. The size and shape of burned patches are critical to the ways ecosystems recover (Kemp et al. 2016, See Case Study 12.3), so quantifying the size and shape of patches through time can help understand the implications of change for vegetation composition and landscape dynamics. Spatial variability within patches can include the individuals, clumps, and openings that often result from previous disturbances (Churchill et al. 2013). This spatial variability in vegetation structure is positively linked to variability in fire behavior and severity through fine-scale pattern-process linkages. However, other factors, such as topography and extreme burning conditions, can alter this relationship. Hessburg et al. (2000, 2007) assessed patch size distributions from the oldest available (often 1930s) and more recent aerial photographs for watersheds. They found that patch size distribution of forests had been greatly affected by roads, logging, and fires, with smaller patches and less variable patch sizes in areas that had been logged, resulting in fragmentation of old forests in many areas. Thus, patch sizes reflect past disturbances and vegetation dynamics since disturbance. Haire and McGarigal (2009), Haire et al. (2013) found different patch size distributions within and beyond wilderness areas in central Idaho; they attributed differences to fire management and vegetation that results in variability in fuels. Archibald et al. (2010) found that in southern Africa, large fires seldom occur where human population density is high, and that humans greatly influence burned area extent, so much so that the effect of fire weather in driving variation between years is subdued where land use by people is intensive. In Portugal, the strength of the effect of fire weather as a driver of fire size follows biophysical and human population density gradients that determine the continuity and amount of fuels in the landscape (Fernandes 2019).

All fires are patchy. Landscapes consisting of favorable and unfavorable pixels distributed randomly, termed as neutral landscape models, have been proposed by Gardner et al. (1987), as a useful reference when evaluating the spread of disturbance and other landscape processes in real landscapes. Neutral landscapes have been used by conservation biologists to analyze landscape connectivity and distribution of populations (With 1997; With et al. 1997). Using the same principles and percolation theory, it is possible to predict the spread of disturbances as fire across landscapes consisting of pixels that are either favorable or not favorable for spread (Turner et al. 1989). An example of fire percolation simulations in neutral landscapes with different proportions of pixels favorable for fire spread is presented in Fig. 12.6, showing the patchiness resulting from the random component of the propagation process.

Patchiness of burned areas can be seen as patches of burned and unburned areas, as in the unburned islands (refugia) apparent in Fig. 12.6d, or in patches of different burn severity. Less than one-third of the area burned in large fires had burned with

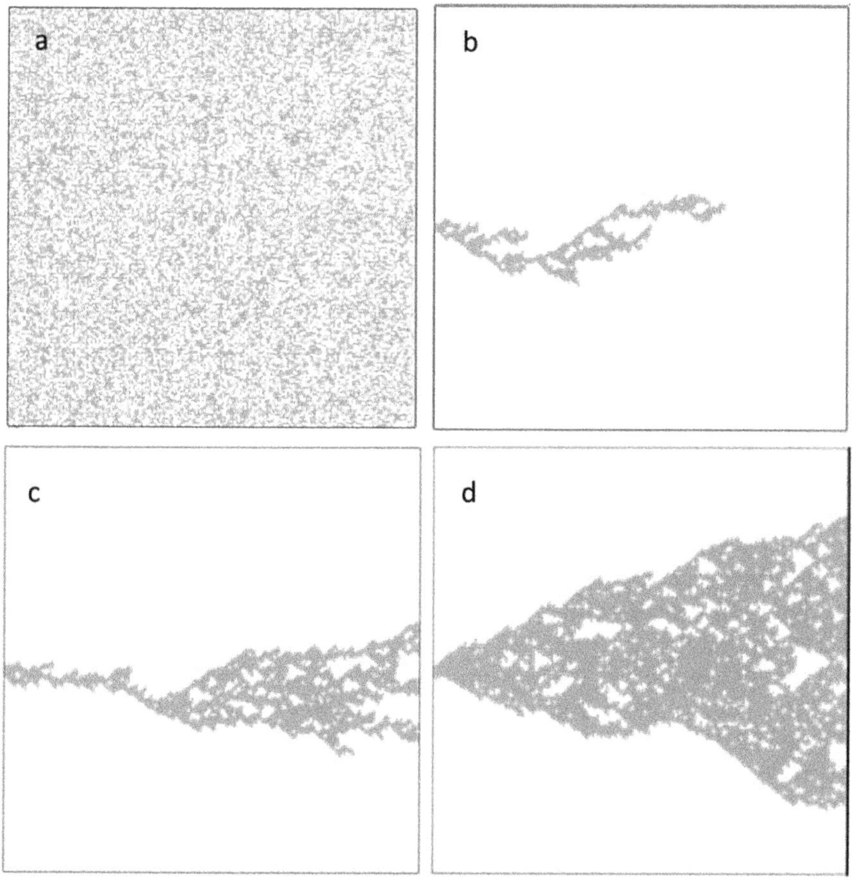

Fig. 12.6 Results from simulations of the percolation of a fire (i.e., ease of fire spread) from left to right, in neutral landscapes with randomly located pixels in different proportions. Pixels favorable for fire spread (gray) are intermixed with other pixels that cannot burn (white). (**a**) Neutral landscape with 50% of pixels favorable for fire spread. (**b**) Percolation of a fire in a landscape with 52% favorable pixels. (**c**) Percolation of fire with 53% favorable pixels. (**d**) Percolation with 57% favorable pixels. Simulations show that small changes in landscape composition might have very significant effects on fire propagation. Patchiness of fire is apparent in all simulations, even without the influence of topography, vegetation, wind patterns, land use, and other variables influencing landscape heterogeneity before fires occur, all of which could also influence patchiness

high severity in the USA since 1984 (Finco et al. 2012; Picotte et al. 2016). Birch et al. (2015) found that on 42 large forest fires, the proportion burned with high severity was poorly correlated with the area burned in a day. Within most areas burned in a single day, less than 13% of the area burned with high severity except under the most extreme conditions when 49% of the area burned in a single day burned with high severity. Spatial variability in fire behavior and vegetation is positively correlated with spatial variability in fire effects, critical for understanding the long-term ecological benefits or detriments from fire (see discussion in Chap. 9).

The pattern and spatial and temporal variability of fire regimes are seldom quantified yet ecologically important. Recognition of the importance of this variability across multiple temporal and spatial scales is fairly recent. It has been made possible by satellite imagery and other remote sensing techniques, geographic information systems, and the capacity for storing and analyzing large data sets.

12.2.3 Magnitude

Various measures can be used to estimate the magnitude of fires, including the energy released during combustion (e.g., fireline intensity and the total energy released; see Chap. 6) and the effects of fire on ecosystem structure and function. The fireline intensity for a given fire is often inferred from the flame length. Intensity is often unknown for fires that burned long ago, so it is not commonly included in historical fire regime descriptions. Alternatively, the predominant fire type (ground, surface, crown) within burned areas can be used to describe the magnitude of a fire, as it related to differences in fireline intensity and heat release rate. For fire ecology and land management, it often is more important to assess the magnitude in terms of the direct and indirect effects of fires on ecosystem structure and function than the fireline intensity.

Burn severity is the most commonly used measure of the magnitude of fires. Burn severity refers to the degree of ecological change resulting from wildland fire (Keeley 2009; Lentile et al. 2007; Morgan et al. 2001, 2014). Severity encompasses immediate post-fire effects (sometimes called fire severity). Burn severity includes short and long-term fire effects (Morgan et al. 2001, 2014; Keeley 2009). The differences in vegetation response to burn severity may be subtle and short-lived, or they may be pronounced and long-lasting. Whether the ecological consequences of burn severity are positive or negative, it is important to understand them, for they may shape how the next fire will burn and many ecosystem processes in the meantime.

Commonly, burn severity is classified based on percent overstory plant mortality from low (less than 20%) to moderate (20–70%) and high or stand-replacing (>70%) (Agee 1993). This is better referred to as vegetation burn severity to differentiate it from soil burn severity (Parson et al. 2010; Morgan et al. 2014). Burn severity is measured and mapped post-fire (often comparing pre to post-fire as described below) in support of post-fire management. See Chap. 9 for how burn severity is assessed on the ground and from satellite imagery.

When large individual patches or extensive area burned with high severity, policymakers, land managers, and scientists often become concerned because ecosystems in these patches are, by definition, highly altered. Where fires burn with high severity, much of the above-ground biomass of plants can be killed or partially consumed. In areas burned with high severity, post-fire tree regeneration may be delayed, where seed sources are far away (Stevens-Rumann et al. 2018; Stevens-Rumann and Morgan 2019, see Case Study 12.3), soil heating may be great, soil erosion potential may be high (Parson et al. 2010; Robichaud et al. 2013), and wildlife habitat is significantly altered (Turner et al. 1997; Keeley et al. 2008; Lentile et al. 2007; Romme et al. 2011). Areas burned with high severity will have different vegetation trajectories than areas burned with low severity (Morgan et al. 2015; Lewis et al. 2017). Soil burn severity is of particular concern because it can influence soil erosion potential. Calling these "severe" can be misleading because it implies that the fire effects were catastrophic. However, many plants and animals survive and thrive in areas burned with high severity (DellaSala and Hanson 2015; Hutto et al. 2016). Ecological diversity can benefit from pyrodiversity when fires burn with a mix of low, moderate, and high severity and unburned islands (see Chap. 9). Landscapes encompassing burns of different ages and severity can provide a mix of habitats (Morgan et al. 2015; Lewis et al. 2017).

Fires are almost all of mixed and variable severity when enough fires are included (e.g., over enough area or enough time). Most individual fires are of mixed severity when viewed at broad scales. All fires encompass some areas of no to very low mortality and some areas of high to complete mortality, whether viewed in terms of area burned in a day (Birch et al. 2014, 2015) or full fire perimeters. Most fire regimes are a mix of severity classes, and it is important to characterize them accordingly, perhaps as mixtures of probability distributions.

Burn severity is influenced by multiple environmental factors, including fuels and vegetation, and time since previous fire or other disturbances (Parks et al. 2014b, 2018; Stevens-Rumann et al. 2016). Fire weather influences burn severity (Birch et al. 2015; Dillon et al. 2011; Keyser and Westerling 2017), and it will likely be increasingly influential as climate changes (Parks et al. 2018). Burn severity is increasingly analyzed relative to vegetation, topography, fire weather, and people to understand how managers can shape fire effects. See related discussion of the soil heating in Sect. 9.3, assessing burn severity in Sect. 9.7, and the relative importance of climate and fuels for fires in Sect. 12.4.

Accurate, consistent, and timely burn severity maps are used in managing and rehabilitating wildfires. In the USA, Burned Area Emergency Response (BAER) teams often map soil burn severity using both field observations and remotely-sensed imagery to assess where treatments may reduce the potential of soil erosion to affect values at risk (Parson et al. 2010). The process for creating burn severity maps from a combination of pre-fire and post-fire satellite imagery is well documented (Key and Benson 2006; Miller and Thode 2007; Morgan et al. 2014; Parks et al. 2018). Since burn severity from one fire can alter fire behavior, and therefore severity, of a subsequent fire (Parks et al. 2014b, Prichard et al. 2017; Stevens-Rumann et al. 2016), satellite-derived post-fire burn severity maps can be very useful in planning for and managing subsequent fires in the same location.

12.2.4 Perspective on Fire Regimes

The different aspects of fire regimes are related to one another (Fig. 12.7). Fire regimes are driven by environmental conditions (ignitions, vegetation, climate, and topography) that influence both fire behavior and effects. Fire behavior and the management of fuels and fires influence fuels. After discussing various data sources used to describe fire regimes, we discuss how fire regimes change in response to climate, fuels, people, and other influences in later sections.

Our fire history methods are more useful for characterizing fire frequency and size than for describing other aspects of fire regimes (Morgan et al. 2001). Therefore, we know more about fire frequency and size than fire severity, rotation, variability, and other fire regime characteristics. Accuracy and precision of fire history data are seldom quantified, and variability of fire regime characteristics over time and space is seldom evaluated.

Frequency and severity are often used together to describe fire regime classes (Fig. 12.8) (Morgan et al. 2001). However, describing fire regimes with frequency and severity doesn't include the all-important idea of spatial complexity. A combination of descriptors for fire regimes will likely be useful. We suggest using one metric for frequency (e.g., fire return interval), one spatial metric (e.g., fire size distribution or patch size distribution), and one metric of magnitude (e.g., burn severity) together with measures of variability. Ideally, seasonality will also be included. The size of the area and the time considered should be indicated in any descriptor of the fire regime. Likely, the size distribution of fires or burned patches or similar measures has not been included because it is difficult to reconstruct these for

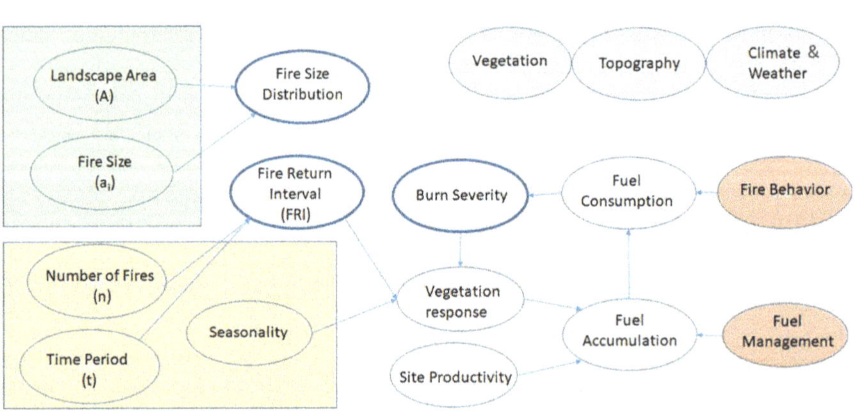

Fig. 12.7 The fire regime for a landscape can be characterized with fire return interval (FRI, a measure of fire frequency that is influenced by the number of fires and the time period assessed), the size distribution of fires or burned patches (influenced by fire size and landscape area), seasonality, and burn severity (influenced by fuel consumption that is in turn influenced by fuel accumulation since the last fire which depends upon site productivity and vegetation response). Thus fire regimes represent spatial (*green*), temporal (*yellow*), and human (*orange*) influences. Drivers of fire regimes (*gray*) include vegetation, climate (and weather), and topography (*orange*)

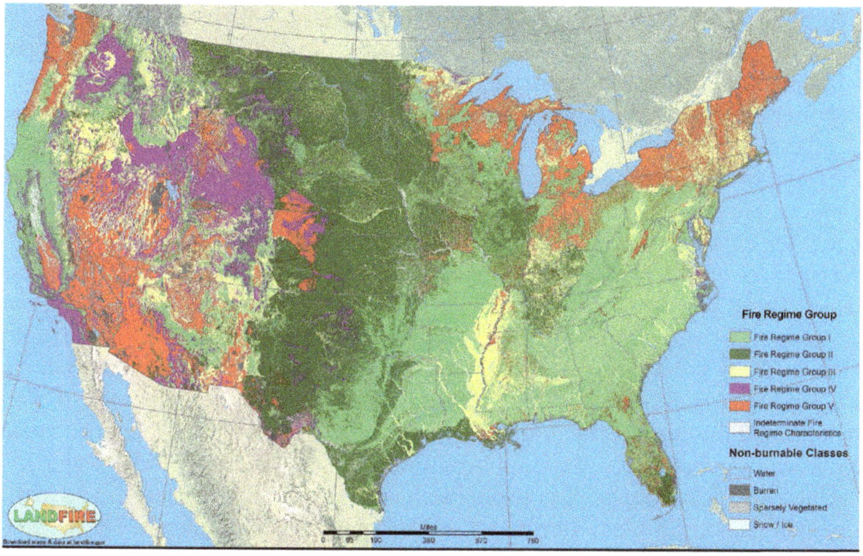

Fig. 12.8 Every place has a fire history. Historical fire regimes for the conterminous USA were mapped by LANDFIRE based upon vegetation, available data, expert opinion, and models. The mapped fire regimes are Fire Regime I: 0 to 35-year frequency, low to mixed burn severity; Fire Regime II: 0 to 35-year frequency, stand-replacement burn severity; Fire Regime III: 35 to 200-year frequency, low to mixed burn severity; Fire Regime IV: 35 to 200-year frequency, stand-replacement burn severity; and Fire Regime V: 200+ year frequency, and stand-replacement burn severity (Landfire n.d.)

historical fire regimes. However, as we described above, remote sensing, historical documents, simulation models, and other sources can suggest patch size distributions, whether patches are defined by burn severity or fire size. Further, many remotely sensed data can help us think about fire across the same continuous gradients that determine species distributions and vegetation types.

It is all too easy to think of a vegetation type as having a particular historical and current fire regime. In reality, historical fire regimes varied as environmental conditions varied over space and time. Also, how likely a specific place is to burn is influenced by the context. Two trees in the same vegetation types with the same soils and topographic setting may have different fire histories if one is surrounded by areas that frequently burn, in sharp contrast to one surrounded by very moist vegetation that doesn't burn often. Further, appreciating and embracing the complexity inherent in landscapes and fires helps us to be aware of the beauty and ecological surprises that ecosystem services sometimes depend upon.

Scientists have learned much about fire regimes, advancing our knowledge at different scales, yet challenges remain. Linking information obtained from various methods for a comprehensive understanding of fire regimes is challenging but needed (Morgan et al. 2001; McLaughlan et al. 2020). Integrating across scales is needed to further inform our understanding of fire ecology if we are to forecast the implications of changing climate and other conditions for fires and their effects in the future (McLaughlan et al. 2020). Unfortunately, mixed-severity fire regimes are widespread yet difficult to characterize. We know more about the fire regimes in dry forests that historically burned mostly with frequent, low-severity fires that scarred trees we can now date, or in cold forests that burned with stand-replacing fires that left relatively distinct age cohorts that can be used to map the extent of previous burned patches (see limitations in Sect. 12.3). Compared to forests, we know less about the historical fire regimes in woodlands, shrublands, and grasslands, yet many of the landscapes we care about and seek to manage include a mixture of these vegetation types.

Landscape heterogeneity influences both how fires burn and how landscapes recover from fires. Fires and vegetation in many areas have been affected by urbanization, roads, the legacy of prior logging and disturbances, farming, grazing, mining, and fire suppression. Variations of these factors, along with topography and environmental conditions, have resulted in varying sizes and character of vegetation patches within landscapes. This heterogeneity can alter fire return intervals and patch size distribution. In many forests, vegetation composition and landscape patterns have been altered by land use, including but not limited to fire suppression (Hessburg et al. 2000, 2007, 2015). Prior fires can limit the extent and severity of subsequent fires, but this varies with vegetation, weather, time since prior fire, ignitions, and fire suppression. Many landscapes were and are a mosaic of patches recovering from prior disturbances. Subsequent fires interact with this heterogeneity. Often, heterogeneity begets heterogeneity in positive feedback. Where human actions have homogenized landscapes, e.g., by suppressing small fires burning under mild environmental conditions and therefore allowing the successional advance to similar vegetation compositions and structure over large areas, fires have been burning with larger and more homogeneous effects.

Fires burning under less extreme conditions could enhance spatial heterogeneity, species composition, and future fire resilience of vegetation. Even if they don't burn much land area, these fires can alter the effects of future fires and their patch size distribution. However, fires burning under less extreme conditions are often the ones we can readily suppress. The fires burning under extreme conditions are the ones most likely to further alter patch size distributions. Even a small change in the number of very large patches will greatly alter how future landscapes respond to future fires (Hessburg et al. 2007).

Describing fire regimes is challenging for several reasons. *First,* it is difficult to characterize a fire regime with a single metric, and we must consider variability. Variability is often more important than mean or median. For instance, occasional, relatively long intervals between fires may allow trees or other plants to establish and grow large enough to survive the next fire. Unusually short intervals between fires may kill vegetation that established after the first fire while also killing enough of the seed sources that vegetation trajectories are changed. Spatial variability in fire effects is crucial to the landscape heterogeneity that develops; it reflects the interaction between the fire, vegetation, and environmental conditions as they vary across topography and through time. Fire effects are heterogeneous at fine to broad scales, and they reflect fire behavior (flaming and smoldering combustion heat and heat transfer), and the interaction between fires, fuels, and vegetation. *Second,* some metrics, especially frequency, are influenced by the scale at which they are measured. *Third,* many aspects of fire regimes are difficult to measure. For these reasons, broad classes of fire frequency and burn severity are often used to describe historical fire regimes (Fig. 12.8). *Fourth,* little empirical data addresses both long time periods and broad spatial scales, and multiple approaches can be used together. *Fifth,* our information is incomplete. We want the information for whole landscapes and for many landscapes for which we lack fire history data. Careful observation of fire effects, combining methods, and modeling are all promising in overcoming our challenges. Heward et al. (2013) demonstrated an important opportunity for involving fire personnel in research to accomplish research objectives, but this must be done without jeopardizing primary work roles (Lentile et al. 2000). Others are embedding research teams into fire management teams. These and other approaches are needed to obtain data on many large fires. Advancing our understanding of how the conditions before, during, and after fires are related depends on spatially coincident observations on actively burning wildfires (Kremens et al. 2010). It is easier to make measurements on actively burning prescribed fires than on intensely burning wildfires. Thus, we have more observations in some ecosystems and geographic areas (such as Australian grasslands, southeastern pine forests, some boreal forests) but few in others. Likewise, we have more observations from prescribed burns than from intensely burning wildfires.

Fire regimes are changing, often in response to interactions among climate, land use, and vegetation. Climate acts on both vegetation and fire, further complicating interactions. Fire regimes have been mapped globally. Archibald et al. (2013) described 'pyromes' based on their statistical analysis of fire frequencies, intensities, burned areas, and fire season length as interpreted from satellite imagery available in recent decades (Fig. 12.9). They concluded it will be difficult to predict how global

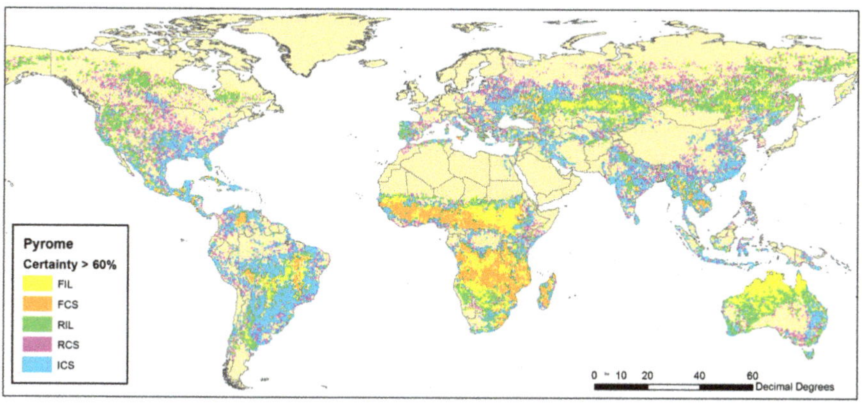

Fig. 12.9 Current global fire regimes were mapped after being classified into five pyromes based upon fire frequencies, intensities, burned areas, and fire season length (From Archibald et al. 2013)

fire regimes will be influenced by global change because the same pyromes are found in different vegetation types and in different climatic conditions. Further, there are relatively few places globally where we have a sense of what fire regimes were historically to give us perspective on how they are changing. As a result, we rely heavily upon models and upon inferences from comparing across space.

12.3 Data Sources for Describing Fire Regimes

We use proxies for inferring historical and modern fire regimes. Here we describe the available data sources and their strengths and limitations. The available data sources vary in temporal and spatial resolution (Fig. 12.10). Most cannot reach far enough back to a point when humans did not influence fire regimes, and they often, therefore, reflect the long history of people using fire. The decay of data back through time influences the temporal and spatial scale of the inferences that can be made (Swetnam et al. 1999). Many researchers draw on multiple lines of evidence to strengthen their conclusions.

Archived fire history data are useful for describing broad patterns. Data inferred from natural proxies, such as dated fire scars from tree rings and charcoal from lakes and bogs globally, are archived in the International Multiproxy Paleofire Database (NCEI 2020), a public archive maintained by the USA National Oceanic and Atmospheric Administration in Boulder, Colorado. Such data have been used to assess climate drivers of fire occurrence from long times and large areas. For instance, Marlon et al. (2012) described a "late-twentieth century fire deficit" relative to biomass burning over the prior 1500 years in the western USA (Fig. 12.11). This deficit suggests that we are accumulating biomass on the landscape that could fuel future fires, for one of the paradoxes of successfully suppressing fires is that the next fire may be more intense. Similarly, Marlon et al. (2008) used a network of archived data to describe long-term global patterns of fire with implications for atmospheric carbon (Fig. 12.12).

Recognizing the influence of global sea surface temperature anomalies on fires has been made possible by large fire history data from many places around the globe. These patterns are often not recognized without long-term data (e.g., Swetnam and Betancourt 1990; Marlon et al. 2008, 2012). They are the result of sea surface temperatures, such as the El Nino Southern Oscillation or the Pacific Decadal Oscillation, affecting the global circulation of winds and climate that alters the timing of spring and drought. Understanding these influences is crucial to forecasting the implications of changing global circulation for future fire regimes.

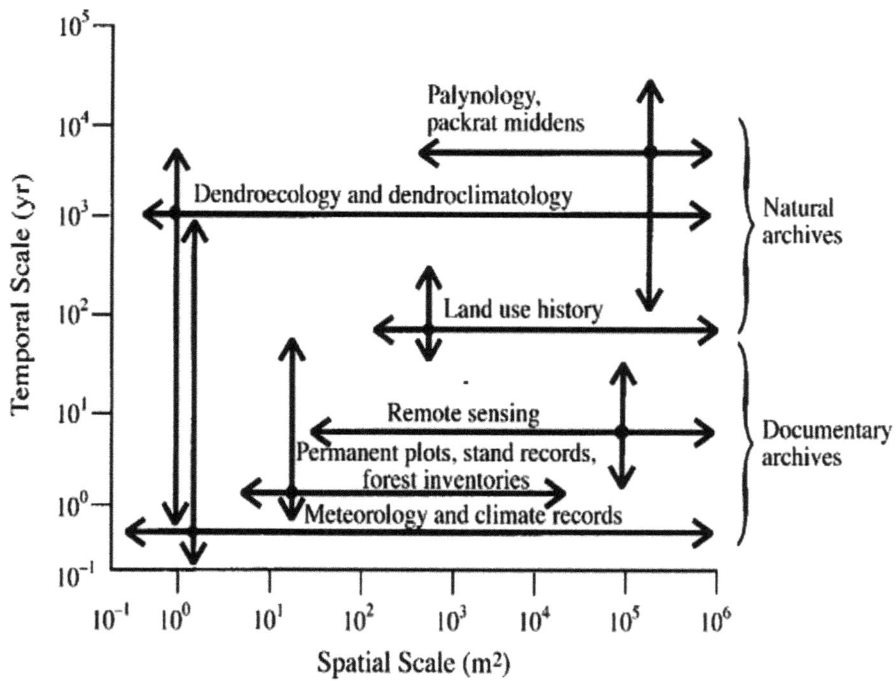

Fig. 12.10 Sources of data for describing historical and modern fire regimes vary in their temporal and spatial scales (Swetnam et al. 1999)

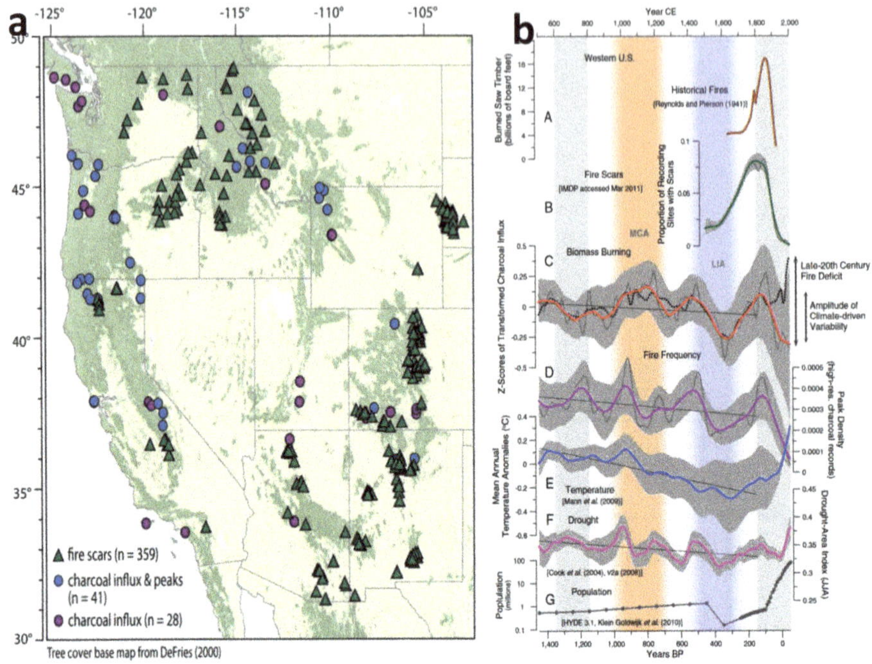

Fig. 12.11 In this 1500-year perspective for fires in the western USA, Marlon et al. (2012) drew upon charcoal and pollen studies at many sites across the western USA (**a**) and interpreted (**b**) a "late-twentieth century fire deficit" relative to historical biomass burning (Line C), due to less frequent fires (Line D) even as temperatures, drought, and precipitation have increased (Lines E, F, and G) in recent decades beyond background variability (shown in gray around the mean). Both (**a**) and (**b**) are from Marlon et al. (2012)

12.3.1 Tree Rings

Where trees form annual rings, they can be used to assess tree ages or fire scars from which past fire occurrence, size, and effects can be inferred. Tree-ring data are often quite precise spatially and temporally for recent centuries (usually up to 400 years), but only if the tree rings can be accurately dated. Dendrochronology is the science of tree rings in which crossdating the varying patterns of wide and narrow rings is critical to assigning the exact calendar year a given tree ring formed; dendroecology uses those approaches to study ecological processes (Harley et al. 2018). This temporal accuracy is required for accurate estimates of fire frequency and developing correlations with climate data. Crossdating helps correct for missing and false rings, and can be used to date both living trees and long-dead logs, stumps, or other wood pieces. With sanding and magnification, fire scars can often be assigned to early, middle, or late in the annual tree ring to assess season of fire. Sanding and magnification also make it possible to count the density of resin ducts (Sparks et al. 2017; Hood et al. 2015) to assess defense against bark beetles, to assess past and current

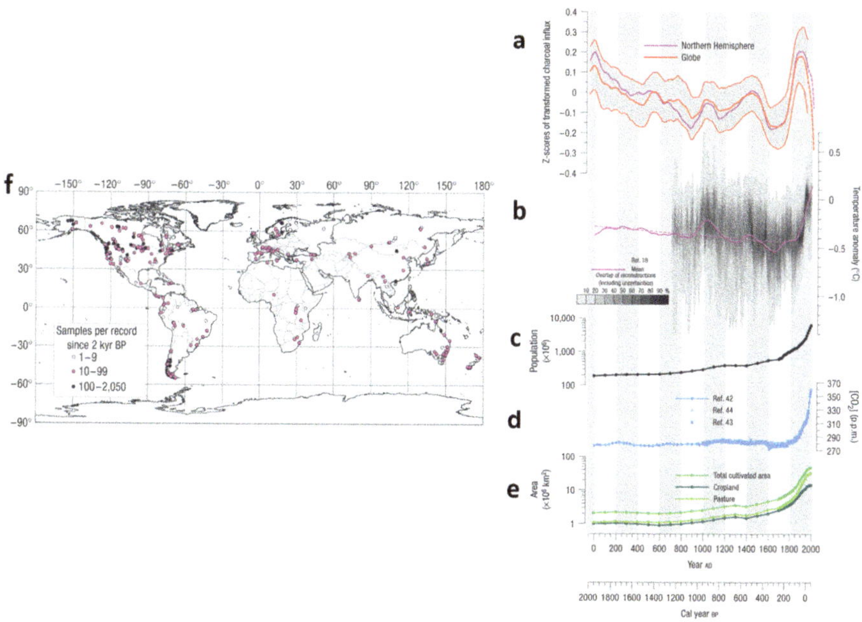

Fig. 12.12 Both people and climate influenced global fires over the last 2000 years. (a) Global trends in charcoal in sediments. (b) Temperature relative to the long-term average, (c) human population, (d) atmospheric CO_2, and (e) area of different land uses. (f) Data on charcoal in sediments was from these locations around the world. (From Marlon et al. 2008)

forest age structure, to compare host and non-host trees to assess past crown defoliation by insects, to inform models of ecosystem processes, and in many other applications of tree growth and survival information (Harley et al. 2018).

When trees are injured by but survive fires, those events can be recorded in the annual growth rings (Fig. 12.13a). When injured, ponderosa pine and some other trees produce resin. The resin-infused wood around scars decays very slowly, which helps preserve it. Once a fire scars a tree, it is more likely to scar again in future fires. The triangular "cat face" typically extends from the ground up with relatively smooth margins that differentiate it from injury due to mechanical damage from other trees falling or animals. Not all trees record fire. The fires must be sufficiently severe to kill cambium partially, but not the whole tree. Sampling old logs and stumps can provide long-term records (Fig. 12.13), but there is some urgency of collecting such old pieces before decay, or subsequent fires make it impossible for us to extract the stories we can infer from them.

Sampling strategies vary. In their study of fire and climate across the northern Rocky Mountains, Heyerdahl et al. (2008a, b) targeted particular locations and trees in those locations that had many old fire scars. They specifically sought trees with many well-preserved scars to get the most scars over time. Such targeted sampling is useful for obtaining a long record to examine the correlations of fire occurrence with climate over long time periods. Systematically sampling across landscapes can better

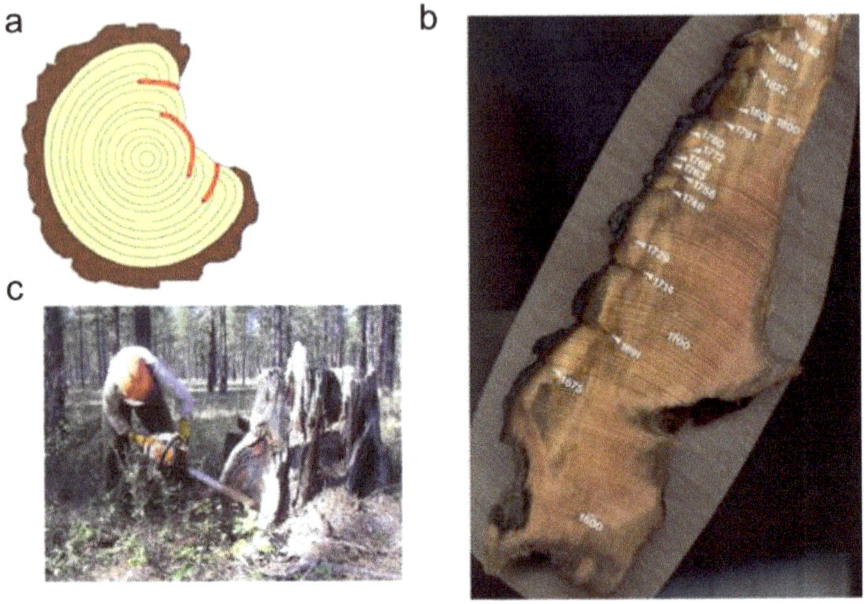

Fig. 12.13 (**a**) Many trees produce annual growth rings with alternating light-colored rings of the larger cells of the earlywood and the darker colored ring of the smaller and more dense latewood. When the cambium of a tree is injured by fires but the tree survives, we can date the resulting scars visible in the tree rings (here highlighted in red). (**b**) Often, long-dead stumps and logs hold a long history of fires from when they were living trees. (**c**) Partial cross sections can be removed from living or dead trees, then sanded so that individual scars can be cross dated to the exact calendar year using dendrochronology. The scars can be dated and analyzed relative to climate to inform our forecasting of fires for the future. Photographs by James Riser II and Emily Heyerdahl from Heyerdahl et al. (2008a)

represent the fire history of the whole surrounding landscapes (Falk et al. 2011). Heyerdahl et al. (2001) were some of the first to use fire scars and tree ages collected at systematically located sample points across watersheds to examine the degree to which fire frequency varied with topography and climate (Fig. 12.14). They found that climate was important but that fire frequency varied with local topography. This approach is now widely used. For example, Merschel et al. (2018) sampled systematically to understand the spatial variability of fire frequency and size. They found that historical fires were driven by top-down, coarse-scale factors including climate (fires were both widespread and synchronous during extended dry seasons), by bottom-up, fine-scale topography including elevation but not slope or aspect, and mesoscale landscape context and edaphic factors influencing fire spread (Fig. 12.15). Thus, data from grid points can be used to evaluate patterns across scales.

Often fire scars are combined with data on tree age at systematically located plots. Combining data is useful to investigate mixed-severity fire regimes. Interpreting such tree age data can be challenging. As only some trees establish soon after fires while others establish through time, fire dates based solely on tree ages may be

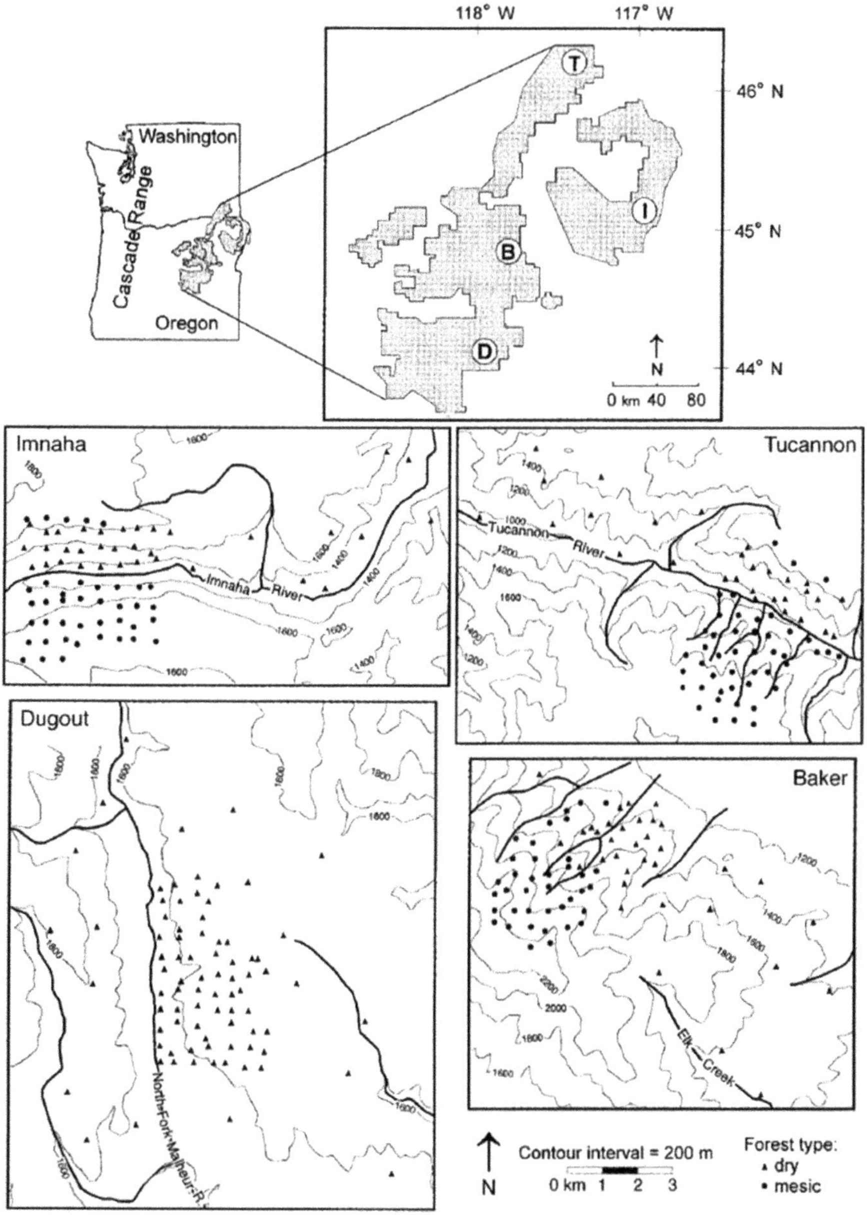

Fig. 12.14 Heyerdahl et al. (2001) used fire scars and tree ages collected at regular intervals across four watersheds to examine the degree to which fire frequency varied with topography. They found that fire frequency responded to both "top-down" controls (broad climatic patterns over space and time) and "bottom-up" controls (local topography, including aspect, and landscape context) (Heyerdahl et al. 2001)

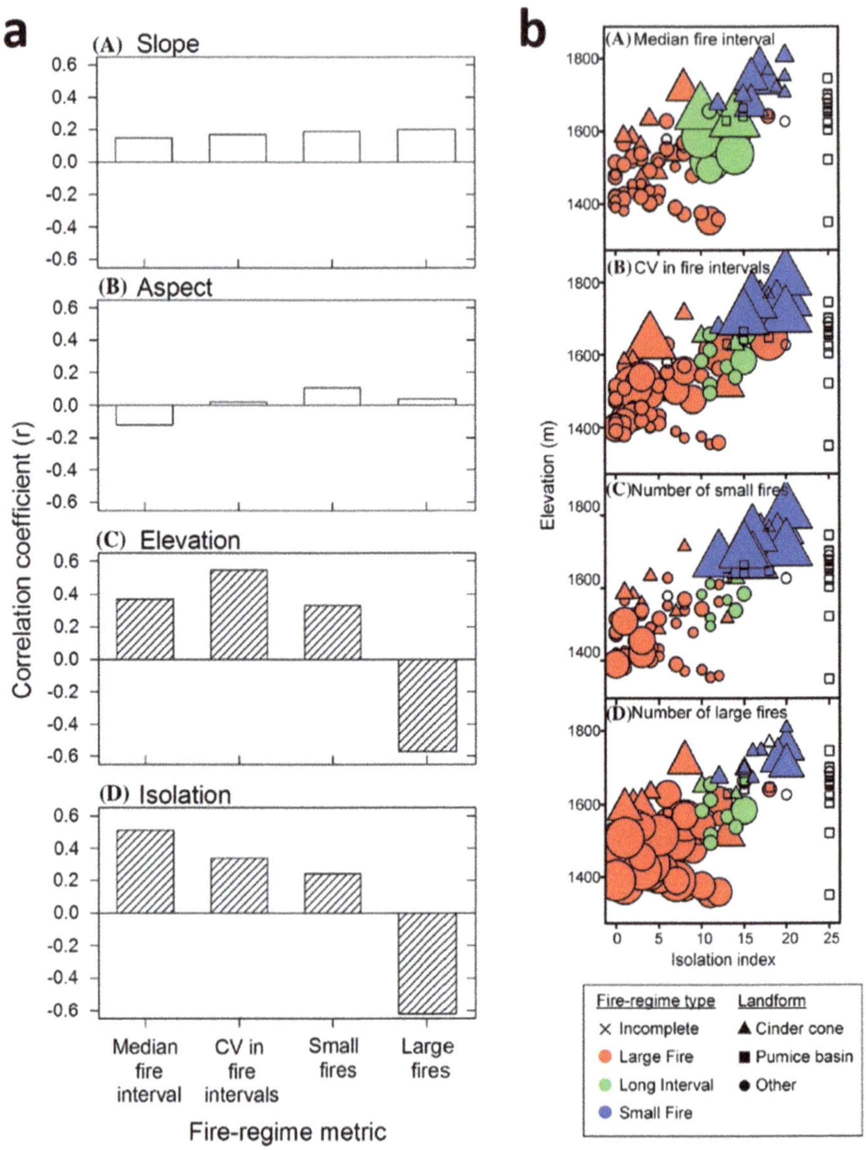

Fig. 12.15 Systematically sampling tree ages, dated fires, and fire sizes allowed Merschel et al. (2018) to understand how (**a**) the median interval between fires (an indicator of fire frequency), the coefficient of variation between intervals (a measure of variability), and the number of small and large fires did not vary with slope or aspect but differed significantly (shaded bars) with elevation, and (**b**) the degree to which parts of the landscape were isolated from fire spread. These were mixed conifer forests in the dry interior forests of western Oregon (From Merschel et al. 2018)

approximate. Further, because it is challenging to assess tree age at the point of the root-shoot interface without cutting each tree down and dissecting it, most age data are imprecise as trees take a variable number of years to reach sample height. Also, evidence of prior fires may burn up and be erased in subsequent fires.

Some have drawn upon forest inventory data and early surveying records. For instance, Hagmann et al. (2013) effectively used tree ages recorded in detailed forest inventory to characterize past forest composition, structure, and fire history in central Oregon. However, while both Forest Inventory and Analysis data (USDA n.d.-b) and General Land Office survey data (BLM, n.d.) are systematic samples of trees across all forest in the USA, the data are poorly suited for assessing forest age structure (Stevens et al. 2016) and burn severity because relatively few trees were sampled at each point (Fulé et al. 2014).

Multiple researchers have used tree ages to make time-since-fire maps, which are useful for calculating area frequency. Because this approach relies upon accurately aging cohorts of trees that established post-fire, it is more reliable where most of the trees established soon after a fire that killed all or most of the trees present before the fire. Most of the calculations based on these data rely on the assumption that the frequency of disturbance is a spatially homogeneous, stationary Poisson process resulting in patch age-class distribution that is relatively stable over time (steady-state shifting mosaic). If small, old stands are not detected or not aged accurately, the disturbance frequency will be overestimated (Finney 1995). Further, it can be challenging to infer past fire dates when trees don't establish in abundance soon after fires, or when fires are frequent and of mixed severity, so age structure patches are not distinct.

12.3.2 Charcoal and Pollen from Sediments

Charcoal records from sediments in bogs and lakes, debris flows, or landscape depressions can be used to reconstruct multi-century fire histories. The presence and density of charred particles of organic matter accumulated in sediments where they were deposited by wind and water during and after fires (Figs. 12.16 and 12.17) can be used to infer past fires and vegetation. Larger particles are assumed to be from closer sources, and peak accumulations must be differentiated from background levels. The carbon in charred pieces can be dated using radiocarbon and other techniques. Peaks in charcoal density in sediment cores are used to date fires. Because transportation and deposition of secondary charcoal continue for some years after a fire, fires' exact age is seldom known unless combined with tree ages. Even when the charcoal particles are directly dated by radiocarbon, dating precision is within a 5% error, and the charcoal could be from one fire or fires in multiple years. Pierce et al. (2004) dated charcoal in debris flows in Yellowstone and Idaho to assess long-term patterns in severe fires. The pollen within the sediments can also be analyzed and is useful in inferring the long-term interaction between fire, vegetation, and climate.

Fig. 12.16 Dates of past fires can be inferred from charcoal accumulated in the sediments of some small lakes, bogs, or depressions where charred organic matter is carried by wind or surface water (NCEI n.d)

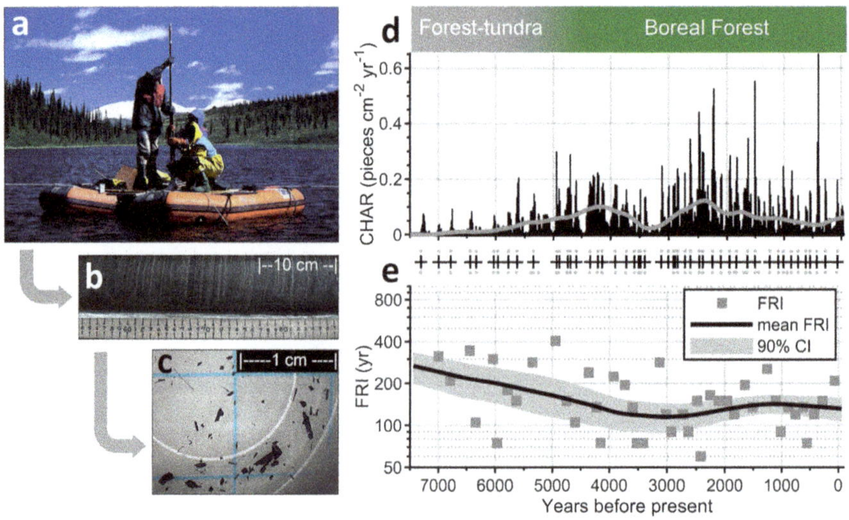

Fig. 12.17 A fire history record from the Alaskan boreal forest reconstructed from lake sediments. (**a**) Lake sediments are collected from a raft using a manually operated piston corer. (**b**) A portion of a lake sediment core with laminations, indicating little sediment mixing and a high-resolution record; more recent layers are on the left and deeper, older layers are on the right. (**c**) Macroscopic charcoal pieces from one 0.25-cm slice of a lake sediment core, representing approximately 10 years of time. (**d**) Continuous samples of charcoal across several meters of sediment cores from one lake, creating a time series of charcoal accumulation rates (CHAR). Peaks in CHAR are identified statistically when they exceed a threshold above background CHAR (grey line). Peaks are interpreted as fire events, indicated by the "+" symbols, with interpretations from alternative thresholds indicated with the grey "." symbols. (**e**) Individual fire events are analyzed as fire return intervals (FRI) with individual FRIs indicated by grey squares and the 1000-year mean FRI indicated by the black line. When the modern boreal forest developed in interior Alaska about 5000 years ago, fire frequency and the biomass burned per fire increased significantly, indicated here by greater CHAR and shorter FRIs. Data are from Code Lake, in the south-central Brooks Range, Alaska, as published in Higuera et al. (2009) (Photographs by Philip Higuera) (**a**) Footprint Lake, Brooks Range; (**b**) laminated sediments from Little Isac Lake, Noatak National Preserve; and (**c**) macroscopic charcoal from Last Chance Lake, Brooks Range)

There are many advantages to fire histories inferred from charcoal and pollen in sediments. Typically, these paleoecological data have great temporal depth at a few points, but not an annual resolution as age is interpolated from lead and radiocarbon dating. Although these paleoecological data are less temporally or spatially precise than those based on tree rings, they can be invaluable for providing long-term perspectives (Whitlock et al. 2010). When data from extensive networks of such sites are analyzed, researchers can infer the long-term dynamics of fire, vegetation, and climate (Figs. 12.11, 12.12, and 12.17). Such data become very useful for informing models that can then be used to forecast the potential effects of different climate conditions for the future on the fire-vegetation-climate system (e.g., Hu et al. 2015).

12.3.3 Historical Documents

People have archived maps, databases, and historical accounts about fires, all of which can be used to infer aspects of past and current fire regimes. Maps of daily or hourly perimeters are often made during the management of large fires, usually from aerial surveys or inferred from satellite imagery. Fire perimeter maps, often called fire atlases, provide spatially explicit fire perimeters that can be used to characterize past fire extent, area burned, and fire probability. Rollins et al. (2001, 2002) used fire perimeters mapped and interpreted from vegetation and old aerial photographs in two very different large wilderness areas in the Rocky Mountains USA to explore the influence of topography, climate, and fuels on fire probability (Fig. 12.18). Areas burned repeatedly were non-random as they were strongly affected by topography, vegetation, and climate. It is strong evidence of these variables' influence that they are important despite great differences in topography, vegetation, and climate in the two areas they studied.

Long-term trends in burn severity are seldom characterized due to a lack of data. Morgan et al. (2017) used 30 years of Landsat satellite imagery with historical aerial photographs to describe trends in burn severity for 133 years over 346,265 ha of complex terrain. They found that burn severity has not increased in recent decades when viewed over this long time and large area.

In Portugal, Fernandes et al. (2014) analyzed Forest Service records to identify fire regime shifts, highlighting how land use and fuel dynamics (through afforestation and changes in biomass removal by people and livestock) modulate the influence of fire weather on area burned. Sequeira et al. (2019) used historical documents from local councils and other sources to build databases of past fire size, locations, and causes for parts of Spain and Portugal for multiple centuries. They defined fire types based on land use, fires, and the local human population to identify 'pyrotransitions'. Both landscape and social changes explained the changes in fire regimes through time and the differences between similar landscapes in Spain and Portugal.

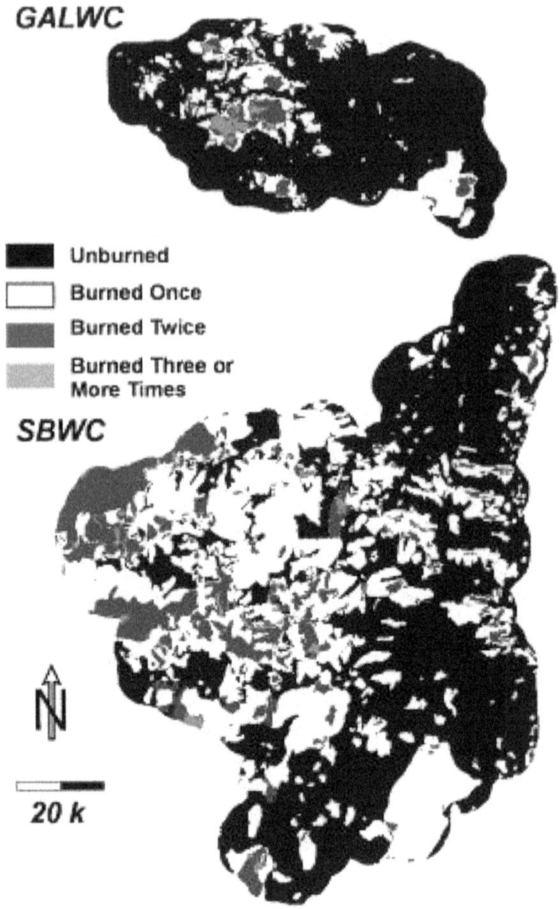

Fig. 12.18 Historical maps of past fires can be digitized and overlaid to create maps of burn probability. In the 486,673-ha Gila-Aldo Leopold Wilderness Complex in New Mexico (GALWC, 1909–1993) and 785,090-ha Selway-Bitterroot Wilderness Complex in Idaho and Montana (SBWC, 1880–1996), Rollins et al. (2002) found that very few areas had burned more than once despite the long records of fire perimeters, frequent lightning ignitions, and limited fire suppression in the rugged, remote terrain. In recent decades in both areas, many fires have been managed to play their natural ecological role rather than being aggressively suppressed. From Rollins et al. (2002)

Data on individual fire locations, their size, and whether they were caused by lightning or by humans is often available from provincial, state, or national sources. In the USA, the Fire Program Analysis fire occurrence database (Short 2017) is a spatial database of wildfires that occurred from 1992 to 2015. It includes 1.88 million geo-referenced wildfire records, representing 56.7 million ha burned from 1992 to 2015. Abatzoglou et al. (2016) and Balch et al. (2017) analyzed these data for the western USA. They found that fires ignited by lightning accounted for most of the area burned in 1992–2013. However, humans caused 84% of all fires, and human-ignited fires contributed substantially to fire threats to people and ecosystems, in part because people ignited fires in more places and over longer seasons than did lightning.

Limitations exist. Historical data are only available for some locations. Lack of fires in the record could be because of missing data or other errors rather than a lack of fires (Morgan et al. 2014). Interpretation requires skill, and results can be qualitative or only semi-quantitative. The minimum fire size recorded varies across time due to changes in criteria and in technology. Like other human archives, missing data or uneven accuracy through time are inherent. Errors in spatial location and fire occurrence date are difficult to assess and likely vary through time (Morgan et al. 2014). Further, fire perimeter maps often include but don't delineate the many included areas that were not burned or burned with low severity (Krawchuk et al. 2016; Meddens et al. 2016). See our discussion of the ecological importance of these "fire refugia" in Sect. 9.4.

12.3.4 Remote Sensing

Fire occurrence is often mapped from remotely sensed data from satellites, airplanes, unmanned aircraft systems (UAS), fire towers, or other ground-based systems. Remotely sensed data can often be consistently gathered over large areas from satellites or airplanes (Fig. 12.19), including areas where it is dangerous or difficult for people to access safely and quickly. Actively burning fires can be mapped based on the difference in reflection from surrounding unburned areas. However, to be detected, fires need to either be large relative to the sensor's pixel size or, if smaller, fires must be burning intensely enough to saturate the sensor detectors (Roy et al. 2013). For some sensors on some platforms, including many satellites, clouds and smoke may block some of the visible wavelengths used in assessing fires and vegetation. Further, we can often map patches or other continuous indicators of spatial variability when we use remotely sensed data. The data can be used to map fires and fire regimes across large areas, as was done for China by Chen et al. (2017), and compare them to other locations. However, accurate interpretation of remotely-sensed data requires field data to provide "ground-truth". Comparisons to recorded

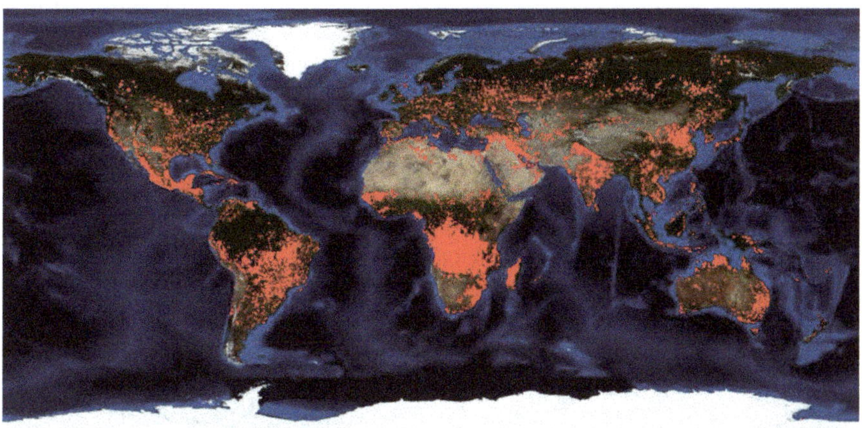

Fig. 12.19 Actively burning fires as mapped from MODIS for June 6–12, 2019 (NASA 2019)

fire data and maps are useful for assessing accuracy, but not all fires are recorded and mapped accurately in any database.

Fireline intensity and fire perimeters are often inferred from Moderate Resolution Imaging Spectroradiometer (MODIS) using the Fire Radiative Power (FRP, units are Watts). Total biomass consumption can be inferred using Fire Radiative Energy (FRE, units are Joules). Thus MODIS is used for global analyses of fire regimes (Roy et al. 2010, 2013). Launched in 1999 and 2002, MODIS was designed for detecting fires. Heward et al. (2013) used observations from fire fighters observing fire behavior on active wildfires as part of their jobs. They found that the fire intensity inferred from MODIS FRP was reasonably accurate. Further, burn severity interpreted from Landsat imagery using dNBR agreed with the 90th percentile of MODIS FRP, suggesting that ecological effects could also be interpreted at least for the highest FRP values. However, that is more likely true for overstory trees killed by intense fires than it is for soils affected by smoldering fires. MODIS is used for global detection of active fires and burned area mapping (Roy et al. 2008). Many other satellite sensors are used, often with change detection algorithms (Roy et al. 2008, 2013). The use of remotely-sensed data has greatly increased in recent decades, and this trend will continue for both science and management (See Chap. 14).

Satellite sensors provide the only way to monitor fire occurrence and characteristics globally, including smoke plumes (Roy et al. 2013). Hantson et al. (2015) mapped the fire size distribution worldwide and found that fire sizes varied greatly with climate and human activity. Chen et al. (2017) described six fire regimes in China by combining the burned area inferred from MODIS with local records of active fires (Fig. 12.20). They found that more than 78% of the land area in China was affected by fire from 2001 to 2016 and that people ignited almost all the fires. Chen et al. (2017) suggested that the relatively large, infrequent fires that burned within a short fire season and high inter-annual variability in area burned were likely accidental fires. In contrast, they described the small, frequent forest fires in southern China burning during a long fire season with low inter-annual variability as fires being used in vegetation management. Cropland and grassland fires had different traits and different potential drivers, including both climate and human activity as found by Hantson et al. (2015). Chen et al. (2017) forecasted that fire density and fire danger will increase with increased temperature and drought in the future, but the spatial pattern of fires will depend on the spatial pattern of humans using the land and either purposefully or accidentally lighting fires, suppressing them and altering the fuel for vegetation fires. Likely, these latter ideas apply globally.

Remotely-sensed data have limitations, some of which can be overcome when combining data from multiple sensors, ground-based field observations, and other data. Satellite data are typically only available since the 1970s and 1980s. This more than 30 years of data is remarkably useful, particularly for providing information on spatial variability continuously across fires, but the total time span limits their use for describing long-term fire regimes, and we know this is much shorter than the life span of many plants. Repeated data from the same site may be many days apart from the same satellite, making it impossible to map fire behavior and effects more

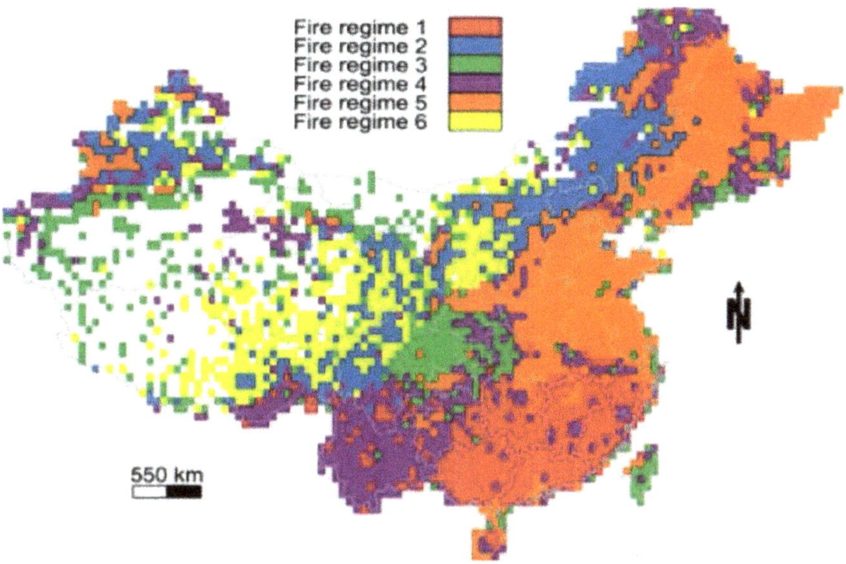

Fig. 12.20 Six different fire regimes were identified for China based on MODIS satellite imagery and active fire data. Chen et al. (2017) used cluster analysis on these data to interpret frequency, size, season, and interannual variability. From Chen et al. (2017)

frequently. However, data from some sensors can provide more frequent images and images at fine spatial resolution (Roy et al. 2013). Mapping global fire regimes is challenging for many reasons, including the very large datasets required, and likely will be more successful for burned areas than for active fires (Roy et al. 2013).

Burn severity is commonly assessed using pre- and post-fire satellite imagery (Key and Benson 2006; Miller and Thode 2007; Morgan et al. 2014; Parks et al. 2014b). As burn severity from one fire can alter the extent and burn severity of a subsequent fire (Parks et al. 2014b, Stevens-Rumann et al. 2016; Prichard et al. 2017), satellite-derived post-fire burn severity maps can be very useful in planning for and managing subsequent fires in the same location. See Sects. 9.7 and 12.3.6 for more discussion and examples useful for inferring burn severity from remotely-sensed data.

A variety of metrics can be used with remotely-sensed imagery to infer fire extent, burn severity, and patches burned or not (Fig. 12.21). The Normalized Burn Ratio (NBR), differenced Normalized Burn Ratio (dNBR), Relativized differenced Normalized Burn Ratio (RdNBR) (Dillon et al. 2011; Key and Benson 2006; Miller and Thode 2007; Dillon et al. 2020) and Relativized Burn Ratio (Parks et al. 2014a) are all calculated using the near- (NIR) and mid-infrared (MIR) bands from Landsat satellite imagery.

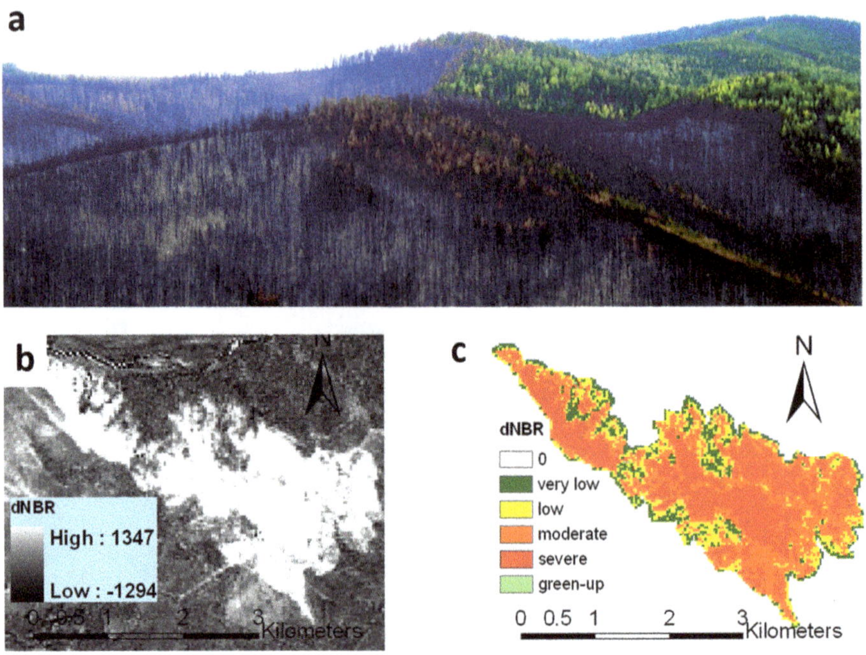

Fig. 12.21 Burn severity for the Cooney Ridge fire, Montana. (**a**) Tree crowns are black and the ground surface is ash-covered in areas burned with high severity, while tree crowns are green in areas unburned or burned with low severity, and brown tree crowns indicate conifer trees are dead but fine fuels in crowns were not consumed, an indication that the fire burned with moderate severity. Photo by AT Hudak. (**b**) Continuous dNBR interpreted from satellite imagery. (**c**) Classified burn severity based upon the dNBR values (From Hudak et al. 2018)

For example:

$$NBR = \frac{(NIR - MIR)}{(NIR + MIR)} \tag{12.2}$$

where NIR is band 4 and MIR is band 7 for Landsat 5 and 7. Then, dNBR is estimated as:

$$dNBR = \left(\left(NBR_{prefire} - NBR_{postfire}\right) \times 1000\right) - dNBR_{offset} \tag{12.3}$$

where dNBRoffset is the average dNBR value from pixels in relatively homogeneous and unchanged areas outside of the burn perimeter. The offset is included to adjust for differences between the pre- and post-fire images due to other factors than fires, such as different phenology or environmental conditions (Miller et al. 2009; Parks et al. 2014a). RdNBR is calculated as:

$$RdNBR = \frac{dNBR}{\left|NBR_{prefire}\right|^{0.5}} \tag{12.4}$$

The prefire and postfire images are often selected just before the fire and 1 year later to minimize differences in plant phenology, called extended assessments, as the full fire effects may not be immediately apparent (Eidenshink et al. 2007). However, in some ecosystems with very rapid plant recovery, such as grasslands, the pre- and post-fire images are selected from the same growing season (just a few weeks or months post fire (Eidenshink et al. 2007). Making the burn severity metric relative to the amount of biomass present pre-fire helps make more consistent assessments over landscapes characterized by widely different plant communities (Miller and Thode 2007).

The Relativized Burn Ratio (RBR) was proposed and evaluated by Parks et al. (2014a) as a robust modification of dNBR. They reported that RBR was better correlated with field measures of burn severity (Composite Burn Index, see Chap. 9) than either dNBR or RdNBR for 18 large fires in the western USA. RBR includes a minor adjustment to dNBR so that the denominator will never be less than zero. RBR is calculated as:

$$RBR = \frac{dNBR}{NBR_{prefire} + 1.001} \tag{12.5}$$

The Monitoring Trends in Burn Severity project (www.mtbs.gov) is an effort to map both the perimeters and burn severity within large fires (>400 ha in the western USA, >200 ha in the eastern USA) inferred from Landsat Thematic Mapper satellite imagery (available since 1984) across the United States. MTBS analysts choose the threshold index values between burn severity classes. Using different thresholds for different fires makes it very difficult to compare fire to fire and across years effectively. Hence, many people recommend using the continuous data and carefully making choices about thresholds, only varying from standard ones when needed for your particular application (Morgan et al. 2014). See also Roy et al. (2013), Dillon et al. (2020), and references therein.

Careful selection of satellite imagery, indices, and timing is critical for effectively mapping burn severity. For example, in the southeastern USA, Picotte and Robertson (2011) found burn severity needed to be assessed at a different time and using different thresholds than one might use to differentiate high and low severity burns in the western USA. The choice of an index (RdNBR, dNBR, NBR) and the methods for determining thresholds for severity classes must be context-specific (i.e., forest vs. non-forest, initial vs. extended assessments) (Dillon et al. 2020). The choice of index and data sources will depend on objectives. Morgan et al. (2014) argued for being explicit about what is being measured, how, and why.

12.3.5 Simulating Fire Regimes

Simulation models are very useful for projecting current and future fire regimes (e.g., for areas without detailed fire regime reconstruction) and across space (e.g., landscapes) (Keane et al. 2003). Because simulation models help us explore the relative importance and interaction of input variables, they can help us learn. Simulation models of fire regimes can inform our scientific understanding and sometimes inform management. For instance, simulation models have helped us understand how "top-down" influences, such as climate, can vary with meso- and fine-scale "bottom-up" environmental conditions. Together, these influence landscape fire regimes. Landscape models can incorporate the effects of the legacy of past fires and other disturbances. Simulation models can help us understand the roles and interactions between climate and fuels and how these are changing from place to place and through time. Often the understanding that comes from developing simulation models is more important than specific predictions. Simulation models can be used to explore forecasts for future fires over both a long time and large areas. However, simulation models depend upon some historical and modern records for calibration which may not capture the effect of long-term variability in climate, land use, and ecosystem processes (Keane et al. 2015).

Finney et al. (2008) simulated burn probability for the conterminous USA, demonstrating that simulation models were useful for broad-scale fire risk assessments. Fire probability of occurrence and fire size distributions approximated historical observations. Most of the area burned by large fires burned under relatively extreme weather conditions and active fire suppression efforts and were thus less influenced by local topography and fuels that would be important under more moderate conditions. The many fires that were suppressed when they were small contributed little to the actual or simulated fire probabilities. Surely the burn probability would be different if some of the 98% of all fires that were suppressed when they were small grew to their full potential. The simulation results of Finney et al. (2008) indicated that fire size was influenced by fuels, ignition location, and weather. To the degree that fuels are important, landscape-level fuel management could alter fire spread and affect burn probabilities even when fires burned under relatively extreme weather conditions. See Sect. 12.4.1 for discussion of the relative influence of climate, fuels, people, and other factors on fire regimes.

12.3.6 Combining Methods to Characterize Past, Present, and Possible Future Fire Regimes

Combining methods is often both necessary and useful. Heyerdahl et al. (2014) combined tree age structure, dated fire scars, and simulation modeling with FlamMap to assess past and future fire effects and fire behavior in lodgepole pine

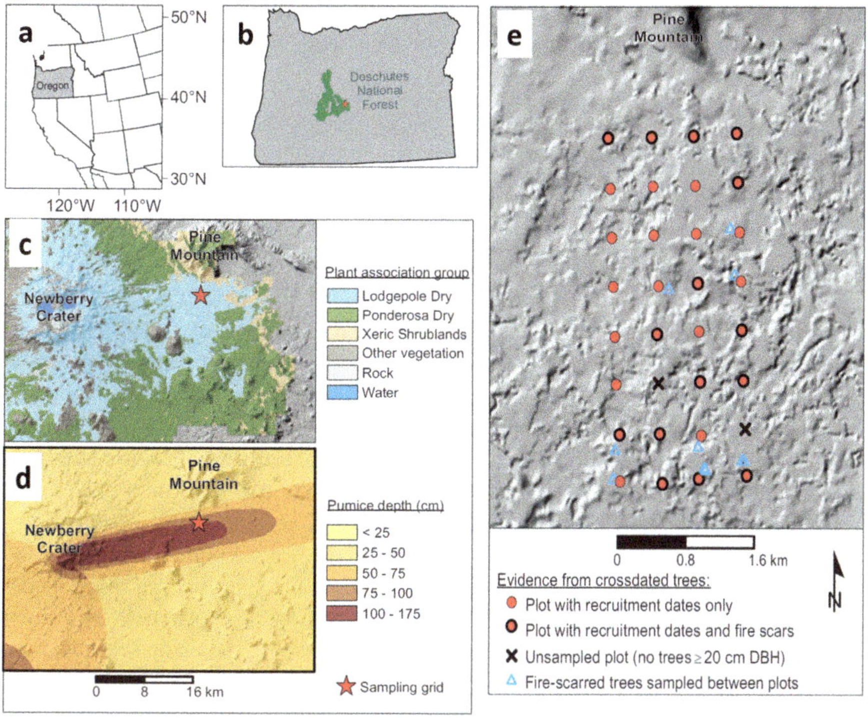

Fig. 12.22 Mixed severity fire regimes were assessed for (**a**) Oregon's (**b**) Pumice Plateau (**c**) dominated by lodgepole pine. Heyerdahl et al. (2014) used (**d**) fire behavior predictions made using the FlamMap model to complement tree age data from crossdated increment cores on systematically located plots (From Heyerdahl et al. 2014)

forests (Fig. 12.22). They found that surface fires were dominant historically and are most likely to occur in the future.

Combining multiple approaches could help overcome challenges associated with limited fire history data in many grasslands, shrublands, and woodlands. Fire frequency can be inferred from fire scars on trees growing on adjacent forested sites, area burned can be assessed with historical records, and simulation models can be used to evaluate the variability in fire regime characteristics based on the data available. Alternatively, people used field observations and plant ecology with experimental fires burned at different intervals to see what resulted in desirable fire effects, e.g., to examine how frequent burning could restore tallgrass prairie on the Curtis Prairie in Wisconsin. Similarly, understanding mixed fire regimes will require a mix of methods.

Using multiple lines of evidence has potential benefits. When results are in agreement, conclusions are more robust. New and important scientific insights are both possible and likely when scientists employ multiple approaches to analysis. This is especially so in mixed-severity fire regimes and across broad spatial scales.

Landscape heterogeneity is challenging, but combinations of dendrochronology, historical aerial photographs, land survey data, and simulation modeling show promise.

12.4 Changing Fire Regimes Through Time and over Space

Globally, fires have burned an average of 450 M ha annually. The area burned inferred from satellite imagery has declined by almost 25% between 1998 and 2015 as both the number and mean size of fires declined (Andela et al. 2017). Decreases were most pronounced in savannas in South America and Africa and in the grasslands of Asia in response to climate and changing land uses as indicated by cropland area, livestock density, and human population density (Andela et al. 2017). In many temperate zones, much area has burned in large fires in recent years, and more large fires are projected for the future for many areas of the globe (Bowman et al. 2017; Abatzoglou and Williams 2016; Dennison et al. 2014, and many others). Fires increasingly threaten people and their property as the human population grows and expands into wildlands. In particular, just a few fires, those that burn under extremely hot, dry, and windy conditions when fuel load and connectivity are not a limitation, account for most of the area burned. In contrast, the many fires that burn under relatively mild environmental conditions have more positive and fewer negative ecological and social effects. As people wrestle with the challenges associated with large severe fires, understanding the dominant drivers can help inform strategies. Where fuels are a primary driver, fuels management could be effective. Still, where the underlying driver is primarily climate, then many experts would argue that fuels management is not likely to be as effective in altering fire behavior. Where both fuels and climate are influencing fires and their effects (this applies in most places), solutions will need to be creative, strategic, and adaptive. No single solution will always work everywhere.

Globally, ecosystem services have been affected as fire regimes have changed in almost all terrestrial ecosystems from historical to modern times (Fig. 12.23). Ecoregions with degraded and very degraded fire regimes cover 53% and 8%, respectively, of the Earth's land area (Shlisky et al. 2007). Of the Earth's ecoregions, more than half are fire-dependent, and another quarter are fire-sensitive. Only 25% of the terrestrial world assessed exhibits an intact fire regime. Biodiversity is lost where fire regimes are degraded (Shlisky et al. 2007), especially where key plant and animal species are dependent on fire for long-term population viability. Causes of degraded fire regimes globally include urban development, livestock farming, ranching and agriculture, fire use and fire suppression, resource extraction (including energy production, mining, and logging), and climate change (Shlisky et al. 2007).

Past fire regimes have shaped plants, animals, and ecosystem processes. When fire regimes experience rapid alterations, species once well adapted to fires may no longer be resilient to future fires. Johnstone et al. (2016) call this "resilience debt" when ecosystems are less able to recover between disturbances or because post-fire environmental conditions have changed. We must expect resilience debt due to

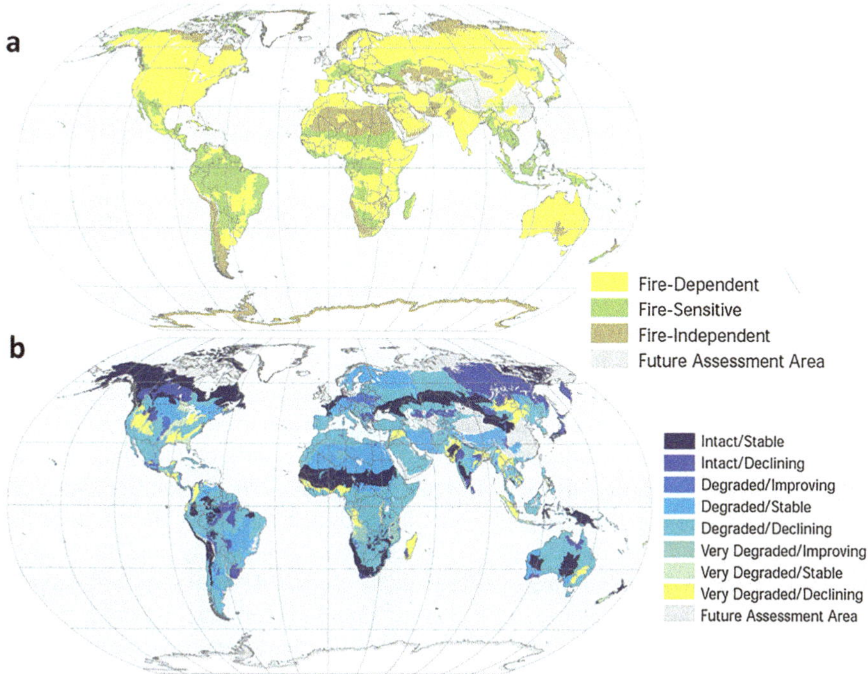

Fig. 12.23 Fire regimes have changed globally according to the Global Fire Assessment. (**a**) More than half of all terrestrial ecoregions are fire-dependent (i.e., they need fire to function well as ecosystems) and another quarter are fire-sensitive (i.e., fires are destructive to key ecosystem elements and functions). (**b**) Most fire regimes have degraded across the Globe's ecoregions. These assessments were developed through expert opinions comparing historical to current fire regimes (From Shlisky et al. 2007)

global change which is bringing change in vegetation with invasive species, change in human activities that alter land use, and often novel climate with different temperature and precipitation and variability in all of these. Some have used the historical range of variability in fire regimes as a clue to what species and ecosystems have tolerated in the past, reasoning that mimicking those past conditions could favor ecosystem components and functions we don't fully understand. Remember, however, that we should be thinking about resilience for future conditions and that the historical range of variability is best used as a guide, not a goal (See Sect. 12.4.2).

12.4.1 Climate, Fuels, and People How and Where Fire Regimes Change

Climate, fuels, and people are all implicated in changes in areas burned and burn severity, and the many extreme fires of recent decades. What has been and will be the relative importance of climate, fuels, and human actions in our rapidly changing world, and how will these vary from place to place? Knowing this can help us advance scientific understanding (how does the relative influence of drivers and their interactions vary and why?) and shape effective management (how can we get more positive and fewer negative impacts of fire?). People can and often do alter fuels, but when and where is this more or less effective in altering fire behavior, and how will that change as climate changes? Let's address these questions with examples. These examples illustrate the changing influences of fuels, climate, and people, and how they interact to shape fire regimes in different ways in different places.

The ponderosa pine forests of the Jemez Mountains in New Mexico in the southwestern USA are one of the few places with a long, detailed fire history from tree rings and historical information to disentangle the interactions among people, fire, vegetation, and climate (Swetnam et al. 2016). For many centuries, fires in these forest landscapes were frequent, with mostly surface fires burning fine fuels. Regionally, interannual climate variations drive fire occurrence through their influence on fuel abundance (as prior rains can increase fine fuels) and availability (low fuel moistures develop during droughts). Land uses by the many people living in the Jemez mountains 1300–1680 limited widespread fires as people ignited many small fires. In the eighteenth and nineteenth centuries, climate drove fire occurrence and extent, and then late in the nineteenth century, intensive domestic livestock grazing limited the amount and continuity of the grass, and Native Americans were removed from large areas, with fire suppression, settlement, and roads further limiting fire occurrence during most of the twentieth century (Swetnam et al. 2016). Fuels accumulated, and forests became denser as trees regenerated in abundance in meadows and forests. Recent very large fires in the southwestern USA reflect changing forest fuels, warmer droughts, and land use that has increased the area and continuity of dense, multi-layered forest canopies, particularly in warm, dry forests (Swetnam et al. 2016).

Fires were also historically important to the ecosystems of cold forests in the northern Rocky Mountains of the USA (Brunelle et al. 2005; Whitlock et al. 2008; Murray et al. 1998, 2000). However, fires were historically less frequent than in nearby dry mixed-conifer forests. Hessburg et al. (2000, 2007) concluded that the landscape-scale heterogeneity of cold forests has changed through human action and advancing succession, though not as much as in dry mixed-conifer forests at lower elevations. All forests have been affected by roads and suppression of fires burning under less extreme conditions. Murray et al. (1998, 2000) documented landscape trends (1753–1993) of whitebark pine forests in the mountains along Idaho and Montana's border. Morgan et al. (2014) found very little difference in the fire extent in dry and cold forests during 1900–2006. They interpreted this as reflecting the influence of twentieth-century fire suppression and other land uses, especially in dry forests, but also in cold forests.

Several concepts are key to understanding changes in fire regimes. *First*, it is important to be clear what aspects of fire regimes are the focus of analysis, for fuels and topography can affect burn severity more than area burned, but less so when fires burn under extreme weather and climate conditions (See Chaps. 8 and 9). Much of

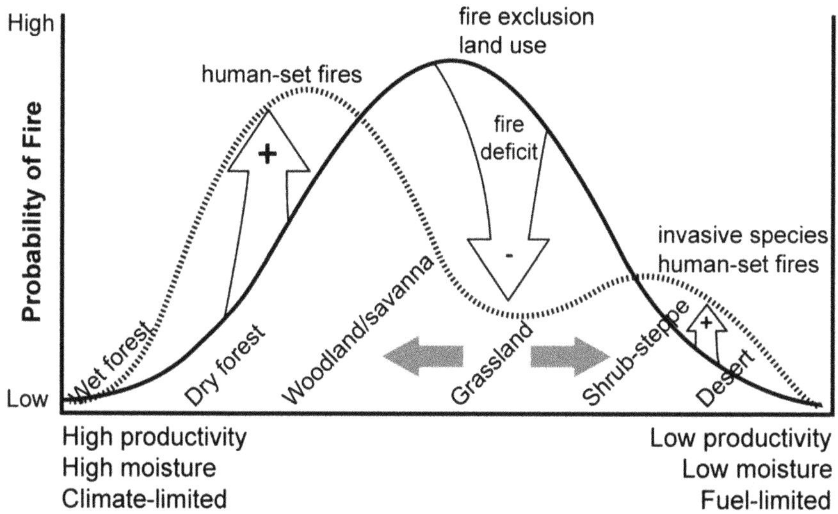

Fig. 12.24 Historically (solid curve), fires were most likely on sites of intermediate biomass productivity. In many ecosystems, the probability of fire is different now (dashed line) than in the past. In wet forests, biomass was historically and is currently often abundant, but only available to burn when long-term warm, dry conditions dried the fuels ("climate-limited"). In areas with low biomass productivity, such as shrub-steppe and deserts, fires were historically uncommon ("fuel-limited"), yet currently (dashed curve) fires are much more common where invasive grasses provide fuel and people ignite fires purposely or accidentally. People have often altered the probability of fires burning, creating a fire "deficit" (large arrow pointing down with minus sign) in many areas of moderate productivity, such as grasslands and savannas, while people have made a fire "surplus" (large arrows pointing up with plus sign) in many dry forests and elsewhere. Note that in many terrestrial ecosystems, biomass production exceeds decomposition, resulting in accumulating biomass that may burn if ignited under hot, dry, windy conditions (Redrawn from McWethy et al. 2013)

the discussion about spatial and temporal controls of fire regimes, including in this section, focus on fire frequency, some on burn severity, and almost none on seasonality or spatial variability. See Chap. 9 for related discussion. *Second*, people, climate, and vegetation influence fire regimes as almost all terrestrial landscapes may burn when environmental conditions align. Places with very high biomass productivity often have much standing biomass that is not likely to burn except when enabled by extended drought and wind (Fig. 12.24). Fires here are climate-limited (also termed flammability-limited) (Krawchuk and Moritz 2011; Prichard et al. 2017). With the increasing probability of extended and warmer droughts, such areas are more likely to burn in the future. Where biomass productivity is low, as in many semi-arid ecosystems and deserts, fires are currently much more common than they were historically, especially where fuels are now more continuous and abundant from invasive grasses. These ecosystems were historically fuel-limited (Krawchuk and Moritz 2011; Pausas and Paula 2012; Prichard et al. 2017), but the added fuels enable fires to burn and spread. More fires occur, and more area is burned at intermediate productivity (Bowman et al. 2011). The coincidence of sufficient fuel production and long dry periods with ignitions means that much of the world's land surface will support fires in parts of most years and especially so in some years. Only the most arid (with low productivity) and the wettest (with limited flammability) will not burn. At intermediate biomass productivity and moisture (most temperate lands), people have significantly altered current fire probability relative to historical conditions through igniting fires, suppressing fires, and excluding fires with land uses that limit fire spread. Understanding where fires are burning in ways that are quite different from those to which plants and animals are adapted is key to mitigating changing fire regimes. *Third*, of the many influences on fire regimes, some are local "bottom-up", such as topography, fuels, and ignitions, while other influences are "top-down" such as climate and regional or national policies because they act across large areas. Thus, while climate and weather influence both fire and vegetation "top-down" leading to regional and global patterns, local topography and fuels provide "bottom-up" controls. *Fourth*, people have influenced fire regimes everywhere with their land uses and fire suppression, and invasive species. In some places there is a fire deficit, elsewhere we have a fire surplus. Change has and will happen, but not in the same way or to the same degree everywhere. What changes is partially up to people. Here we'll try to put these influences and their interactions in perspective.

Fuels are central to understanding changing fire regimes. Without fuels to burn, fires are unlikely to burn and spread, and this will change many of the ecosystem services we as people care deeply about depending upon vegetation which is the fuel. Fuels are the link between fire behavior and fire effects (Keane 2015). Managing fuels (Chap. 11) can greatly alter fires, especially the ecological effects of fires. In "climate-limited" ecosystems, fires only burn under extreme conditions when the abundant fuels become dry enough for long enough. The ponderosa pine forests of the Jemez (see previous examples, this section) were historically fuel-limited but now have an abundance of fuel continuous vertically (vertical fuel ladders) and horizontally (landscapes with large patches of fuels) that can burn intensely when fires ignite under very hot, dry and windy conditions. Similarly, the quantity and availability of fuels and their moisture have contributed to the current flammability of many landscapes, increasing the likelihood that fires will ignite readily, with the potential to burn large areas with high intensity in the USA and around the globe. These concepts are quite useful in understanding the degree to which climate or fuel or both will drive future fire regimes, and therefore the degree to which climate alone is a useful predictor of area burned and fire effects (Pausas and Paula 2012). Many areas have a "fire deficit" while others have a "fire surplus" (Fig. 12.24). The term "fire deficit" is usually applied to current conditions relative to historical fire frequency documented from fire-scarred trees or other historical data (Marlon et al. 2012, Fig. 12.11).

Bradstock (2010) described four different "switches" for fires. All of which must be "on" for fires to spread and each has different influences and scales. Fires don't spread without ignition from lightning or people, but ignition alone is insufficient unless biomass is present in sufficient quantity and continuity, dry enough to burn,

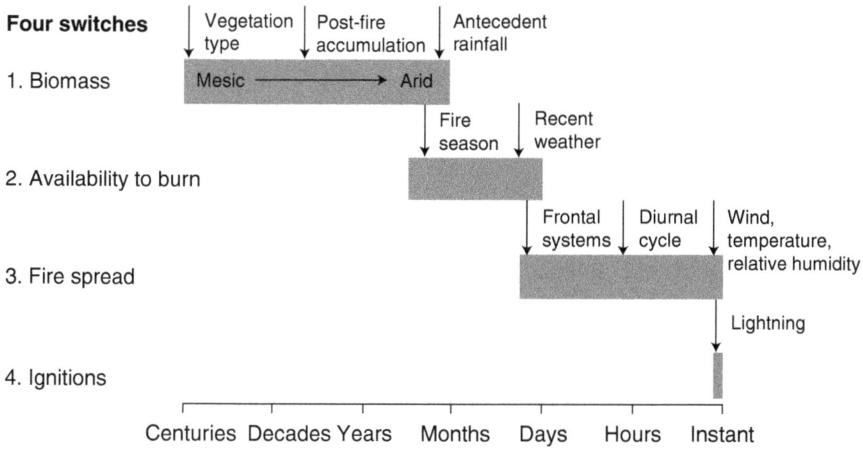

Fig. 12.25 For fires to spread, all four different switches need to be "on", and each of these switches reflects different time scales (From Bradstock 2010 and Murphy et al. 2011)

and ambient weather conditions are conducive. Murphy et al. (2011) related these factors to time (Fig. 12.25). These proximal factors are ultimately influenced by climate, soils, vegetation, people, and time (Bradstock 2010). One important implication of the four-switch model is that fire regimes are complex, reflecting the interactions of different factors, with potentially different primary controls in different locations.

People have had an overriding influence on fire probability across the entire productivity gradient because they alter fuels directly through fuels treatments (see Chap. 11), logging and grazing, other forms of biomass removal, changes in land use policies, and the suppression and ignition of fires. People are also agents of global change that includes altered climate and the proliferation of invasive species. By igniting and suppressing fires, people often find paradoxical results. The more they suppress fires, the more biomass accumulates, which can support larger, more severe fires, especially when those fires ignite during extreme fire weather conditions that will be more prevalent in many years and in many areas with changing climate.

Climate is an important "top-down" influence. Climate is an important driver of the annual area burned (Higuera et al. 2015; Morgan et al. 2008; Whitlock et al. 2008). Fire weather influences fire behavior and resultant burn severity (see discussion in Sect. 10.7) (Birch et al. 2015; Dillon et al. 2011; Keyser and Westerling 2017). Although fire weather has less influence on burn severity than on area burned, fire weather will likely be increasingly important as climate changes (Parks et al. 2016). However, this depends to some degree on how vegetation biomass is affected by climate, and also how available those fuels are to burn (a function of fuel moisture, see Chap. 11). If the area burned increases in the future as many have predicted, and fires burn soon after prior fires, then there may be insufficient or at least greatly altered fuels for subsequent fires.

Fuels, vegetation, topography, and site-specific fuel moisture are all "bottom-up" influences on fire intensity and burn severity. Elevation and topographic position influence site productivity. Broad-scale climatic patterns are strongly influenced by elevational gradients. Thus, fine- and mesoscale environmental conditions can alter how "top-down" climate and weather are "felt" through the site-specific temperature, precipitation, and solar radiation conditions. Analysis of fire regimes based on climate alone misses these effects. Climate data alone typically don't explain changes in vegetation composition and landscape patterns, just as those factors alone are not enough to forecast implications of fires without also considering the climate. Likewise, land use and other human actions are pervasive influences.

Compared to dry sites with sparse vegetation, fires are less likely to occur on moist, productive sites with more fuels available, although fires there typically burn with higher severity. In contrast, in the eastern USA, fuel moisture influenced burn severity in both forest and non-forest settings (overall and in 6 of 8 regions for forests and 5 of 8 regions for non-forest) compared to green biomass (indicated by NDVI) and elevation (Dillon et al. 2020). Dillon et al. (2020) suggested that these findings reflect a generally climate-limited scenario for wildfires in the eastern USA. In other words, the amount of available fuel is more consistent across the eastern USA, and less driven by topographic gradients. In many locations worldwide, humans have increased fire frequency by igniting fires or introducing grasses or other plants that burn readily. There is positive feedback with grasses in many locations, especially introduced grasses that fuel fires that perpetuate grasses in the grass-fire cycle (D'Antonio and Vitousek 1992). See Case Study 12.1.

Both fuels and climate are important, but few analyses combine fuels and climate. Both fuel and climate influence both the area burned and the burn severity, but in different ways in different places. The large fires of recent decades globally are linked to both changing fuel availability and warm and persistent droughts. Simultaneously, land use and fire suppression have fostered increased homogeneity of vegetation structure and fuel conditions across parts of many landscapes, furthering the potential for large patches to burn with high severity. The latter reflect the legacies of prior fires and other disturbances, varying rates of succession, and land use. In Portugal, while large fires only develop when fire weather is conducive, fire size and effects are primarily a function of landscape-level fuel continuity and variability in fuel accumulation as determined by previous fires (Fernandes et al. 2016a, b). Parks et al. (2018) analyzed the relative influence of fuels, topography, climate, and fire weather on burn severity. They argued that in forests of the western USA, live fuels (i.e., green vegetation) most influence the burn severity followed in importance by fire weather tied to the specific day of burning, then long-term climate, and finally, topography. Their results contrast with previous studies that found topography to be one of the more important drivers of burn severity (e.g., Birch et al. 2015; Dillon et al. 2011; Estes et al. 2017). They pointed out, as did Dillon et al. (2011), that topography is an indirect measure of vegetation and fuel distribution and can appear more important when specific variables representing vegetation are not included in analyses. Parks et al. (2018) concluded that the amount of pre-fire vegetation is more important than any inherent topographic factor.

With that said, some combination of topographic variables did still add value to their statistical models in every scenario, even after accounting for vegetation and fuel moisture. These results suggest that manipulating the fuels, e.g., through fuels treatments or through fuels accumulation in the absence of other disturbances, influences the ecological effects of fires, especially in some topographic settings. More analyses of the multi-scale interactions among fuel properties, fire weather, topography, and climatic predictors of fire extent and burn severity are needed to determine how they vary along environmental gradients and at what scale (McLaughlan et al. 2020). Forkel et al. (2017) evaluated the performance of fire-enabled dynamic global vegetation models against 20 years of area burned interpreted from MODIS satellite data. Such models are widely used to predict the implications of changing climate for ecosystem processes. Forkel et al. (2017) found that burned area variability differed with vegetation types. They concluded that vegetation effects on fires, such as fuel structure and fuel moisture, as well as fire ignition and spread, need to be included in dynamic global vegetation models if they are to better simulate year to year variation in the burned area across the globe. In other words, both vegetation and climate matter to fire occurrence, fire extent, and burn severity.

Prior fires and other disturbances greatly influence burn probability and burn severity. Fires are somewhat "self-regulating", as previous fires can limit the size and severity of subsequent fires. Because fuels are consumed and reduced for a time after fires, there is less available fuel to burn in subsequent wildfires. Time and productivity influence how long prior fires will limit the probability and burn severity of reburn and therefore serve as fuels treatments (Moritz et al. 2011; Parks et al. 2015a, Fernandes et al. 2012, 2016a, b). The effect depends on vegetation development as grasses may rapidly grow after fires to soon provide fuel for future fires, but trees and shrubs grow slower and may shade surface fuels and shelter them from the wind. Thus, previously burned landscapes can reburn in later fires (Prichard et al. 2017; Stevens-Rumann et al. 2016; Parks et al. 2015b). Fuels, time, biomass accumulation rates (See Chap. 11) are all important, but so are wind and other weather conditions (and therefore climate) and topography (Prichard et al. 2017). Past fires limit fire spread most effectively soon after fires. Parks et al. (2014b) found that there was a smaller effect of prior fires on fire size during extreme weather, as indicated by large Energy Release Component, which is an indicator of long-term drought (Fig. 12.26). They also found that prior fires were less effective the longer since previous fires. The effect was shorter (6 years) under warm, dry climate than under cool, wet climate (14–18 years).

Hurteau et al. (2019) projected that area burned in California forests would increase significantly in the next few decades, driven by the changing climate. The area burned increased at a slightly slower rate when fuels and prior fires were considered along with climate in their simulation model (Fig. 12.27).

One of the most prevalent results of changed fire regimes is that many landscapes are missing the small- and medium-sized fires that burned under mild to moderate weather and fuels (Haugo et al. 2019). These are the fires most commonly suppressed by people. Though the cumulative area burned is low, by altering

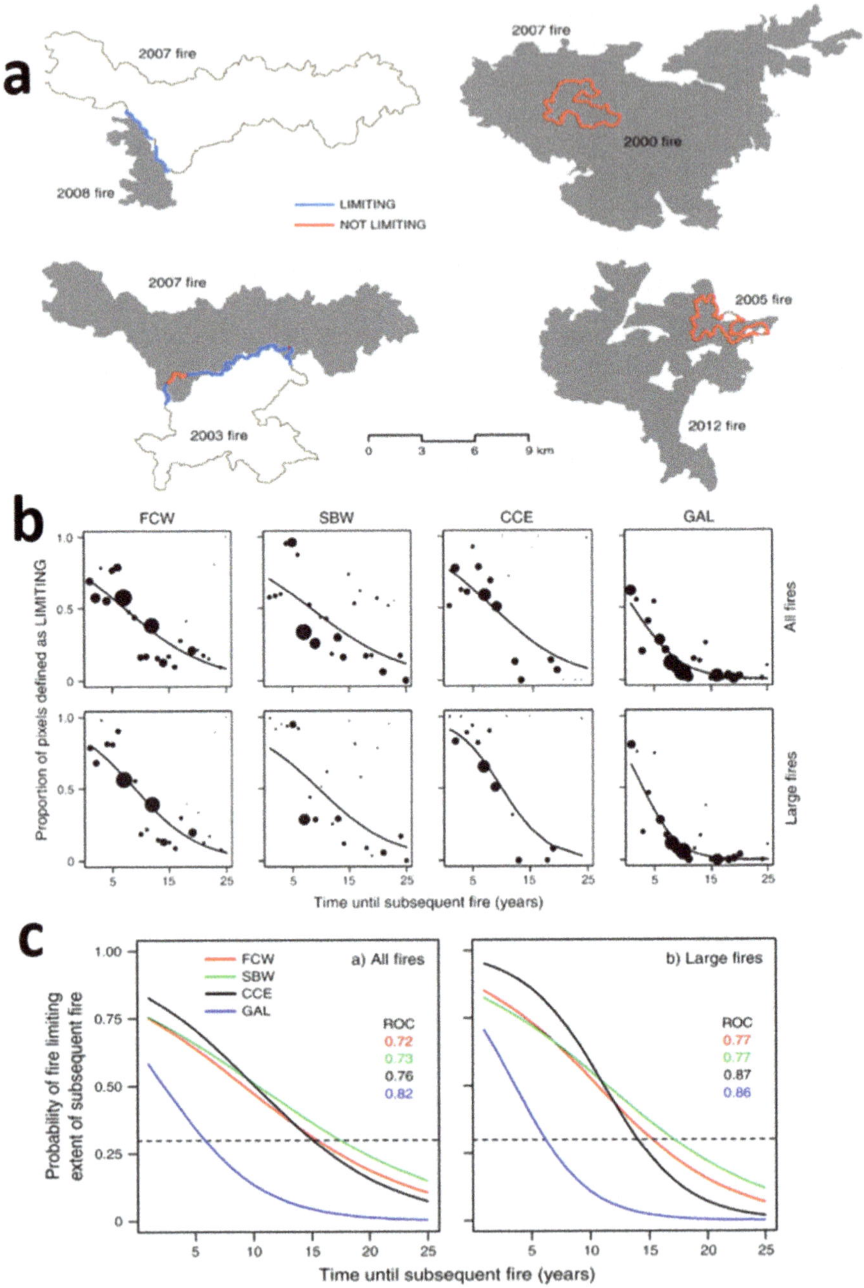

Fig. 12.26 Prior fires limited the spread of subsequent fires but less so in extreme conditions and with increasing time since the prior fire. These analyses were in four large wilderness areas in the western USA where fires have burned with limited suppression given the remote locations, rugged terrain, and few threats to human life, property, and other values, and consistent with managing for resource benefits and as a natural process. (**a**) Illustrations of prior fires (*red perimeter*) limiting or

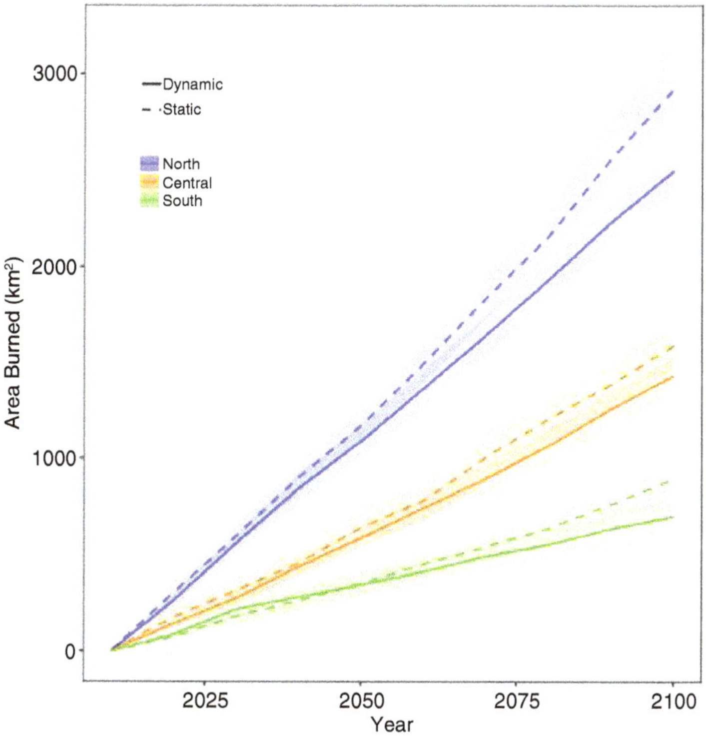

Fig. 12.27 Simulated area burned for forests at three different latitudes in California. The dynamic simulations take into account prior fires and climate, while the static simulations are based solely on climate. As climate changes, the area burned in large fires is projected to increase in many areas whether or not fuel dynamics from prior fires are considered. Simulations were made using the LANDIS-II ecosystem process model (From Hurteau et al. 2019)

vegetation over many small areas they shape landscape heterogeneity that influences how readily subsequent fires will grow and how vegetation will respond to those subsequent fires. Even small changes in the proportion of large patches can greatly alter how vegetation responds to fires, and thus subsequent landscape patterns.

People have influenced climate with implications for fires. Human-caused climate change contributed half of the area burned in recent decades in the western USA

Fig. 12.26 (continued) not limiting the extent of later fires (*gray*). (**b**) The effect declines with time for all fires and for large fires, but more quickly in the warmer and dryer Gila-Aldo Leopold wilderness in New Mexico (GAL) than in the Frank Church (FCW) in Idaho, the Selway-Bitterroot (SBW) in Idaho and Montana, and the Crown of the Continent (CCE) in Montana, USA. (**c**) Subsequent fires were less likely to be limited by prior fires when later fires burned under more extreme climatic conditions; colored lines from black to red represent increasing Energy Release Component (ERC) values (From Parks et al. 2015a)

(Abatzoglou and Williams 2016). People have also suppressed fires and altered vegetation through land uses. Haugo et al. (2019) found that relative to the historical fire frequency, current fires were much less frequent and many were burning with much higher burn severity. They attributed this to fire suppression efforts plus logging, roads, and other land uses in forests across Oregon and Washington, USA. They quantified the degree of change and used this to map the need for restoration to increase forest resilience. See Sect. 10.2.1 about resilience and fires. Parks et al. (2015b) developed predictive models of area burned based on climate (1984–2012 based on geographic areas with limited land use and fire suppression and then projected their model results for the western USA. They found that many forested areas had less area burned than expected (presumably due to fire suppression, land use, as well as roads and other barriers to fire spread). In comparison, some extensive shrubland and grassland areas had much more area burned than expected based on climate alone, presumably due to an abundance of non-native annual grasses such as cheatgrass (*Bromus tectorum*). People have altered fire occurrence, vegetation composition, and landscape homogeneity over extensive areas.

Invasive species have changed current and future fire regimes in many ecosystems. Examples are juniper trees (they are native but invasive, see Case Study 12.1) and annual grasses that have in many locations fueled very frequent fires in what has been called the grass-fire cycle (Case Study 12.1). We note that invasive species sometimes reduce fire frequency and severity, especially when the invasive species facilitate moisture retention on-site or otherwise are of low flammability.

We know from earlier chapters that weather and fuel moisture affect fire behavior, soil heating, and burn severity. Energy Release Component and similar indices that reflect the combined effect of temperature and drought on the moisture levels in fuel can be useful for forecasting fire size and intensity (Jolly and Freeborn 2017). Similarly, the fuel moisture of 1000-h time lag fuels integrates the effect of temperature, relative humidity, and precipitation amount. When these large fuels and duff burn, the long duration burning that often results can contribute to severe fire effects (Morgan et al. 2014). However, topography and vegetation moderate the influences of climate.

As climate changes, the climate will be increasingly important in shaping fires, fuels, and vegetation response. Fuels management will still be needed, though an ecologically and socially appropriate mix of fuel management tools and practices is needed (Moritz et al. 2018). Fuels management is more broadly accepted when it occurs near homes. There is more scientific debate and less general acceptance by people when fuels far from homes are treated, particularly if the justification is focused on protecting those homes.

Will climate or fuel or both, or other factors determine future fire regimes? Pausas and Paula (2012) postulated that in dry Mediterranean ecosystems, fuel abundance is quite important in controlling fires, and therefore that climate alone won't be enough to forecast future fire regimes. They argue that in dry ecosystems, it is the vegetation structure that influences the availability of fuel and thus is most limiting as every year the potential fire season is long. Over large areas of the Earth, there are seasons for vegetation production followed by prolonged dry seasons that make some or all of those fuels available for burning. The coincidence of sufficient fuel production and long dry periods with ignitions means that much of the world's land surface will support fires in parts of most years and especially so in some years. Only the most arid (with low productivity) and the wettest (with limited flammability) will not burn. Fire regimes reflect interactions of climate, weather, topography, fuels, and vegetation, and these reflect legacies of prior disturbance and management.

12.4.2 *Historical Range of Variability (HRV), Future Range of Variability (FRV), and Resilience*

Understanding how ecosystems functioned and fire interacted in the past can help people anticipate the future. Historical Range of Variability (sometimes called Natural Range of Variability) is the minimum to the maximum value of selected ecosystem characteristics through time (Fig. 12.28). Fire frequency is commonly

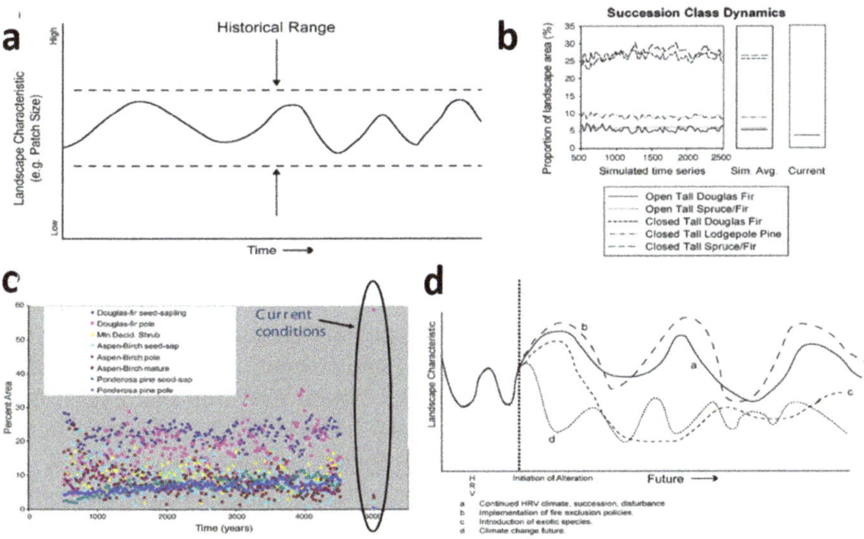

Fig. 12.28 (**a**) The historical range of variability (HRV) is characterized based on the minimum to maximum values of a landscape characteristic, such as median patch size or frequency of fires, over a past time period. (**b**, **c**) One can contrast the current landscape to historical conditions as done in this case for the proportion of the landscape in different forest composition and structure classes. Here, current conditions are "out of whack" as they differ so greatly from historical abundance with likely implications for habitat for many birds, fire hazard, and ecosystem functions. Observations may be reconstructed from field observations or models or both. (**d**) The legacy of past conditions can shape the future, but so do climate, succession, disturbances, and land uses. Where the ecosystems depart greatly from HRV, the forest ecosystems may be less sustainable (From Keane et al. 2009)

used, as is the proportion of land area occupied by a particular vegetation class, such as old-growth ponderosa pine.

Historical ecological understanding and simulation modeling based upon it can help people understand the effects of land use, climate, and fires. HRV is a useful, albeit flawed, guide for achieving management objectives based upon long-term health and sustainability, but it should not be the goal (Landres et al. 1999; Swetnam et al. 1999; Keane et al. 2009; Keeley et al. 2009).

HRV is useful for communicating and understanding change (Landres et al. 1999) and fostering ecosystem resilience to future changes. Historical ecology informs models that can then be used to test hypotheses about the implications of global change. HRV is less useful as a guide when conditions are expected to be very novel (e.g., through climate change, great abundance of invasive species, high density of human population, etc.), and where management objectives are very focused on individual species of concern (Keane et al. 2009; Landres et al. 1999). As Harari (2017) states, the value of history is not to predict the future but to enhance understanding of the past to envision alternative futures. HRV is less useful when it is interpreted for a relatively short time or under the assumption that background climate is stationary through time, for climate has varied through time, as have vegetation and disturbance (Whitlock et al. 2010). HRV can be oversimplified if applied with the assumption that one fire history characterizes all places with a given vegetation type. HRV is scale-dependent.

Hessburg et al. (2015) recommended using both the Historical Range of Variability (HRV) and Future Range of Variability (FRV) as guides in restoring the resilience of landscapes to future fires. They define FRV as a reference representing predicted future characteristics of a landscape; FRV can be generated from simulation models or approximated from other parts of a landscape representing those future conditions (Keane et al. 2009; Hessburg et al. 2015).

Haugo et al. (2015) used a comparison of the historical and current abundance of forest ecosystem structure classes to assess the "need" for restoration to improve forest health and ecosystem services and, therefore, future resilience to fires across Oregon and Washington. They argued that forest composition structure needed to change on 40% of the area of coniferous forested lands. They called for a substantial increase in the pace and scale of thinning, and low severity fire, especially in dry forests, and managing for succession to older forest structures in many forests.

Resilience for the future will require ongoing learning from history, current conditions, and simulation models (Schoennagel et al. 2017). HRV is useful even in a rapidly changing world. It is also useful when the focus is on resilience to future disturbances, not reconstructing the past. That depends on using HRV to build our understanding of how ecosystems "work", and then using that understanding to guide our management for the future.

Case Study 12.1 The Grass-Fire Cycle is Fueled by Invasive Species and Positive Feedback with Fire

Penelope Morgan, email: pmorgan@uidaho.edu
Department of Forest, Rangeland, and Fire Sciences, University of Idaho, Moscow, ID, USA

Alien plant invasion is one of the widespread global changes with broad implications for the native biodiversity and ecosystem services we depend upon. Globally, invasive species have altered ecosystems in many places. Invasive species are especially problematic when they alter fire frequency, size, severity, and seasonality, often with great consequences for native plants and landscape dynamics (Brooks et al. 2004). The "grass-fire cycle" (Fig. 12.29) is one such common and widespread phenomenon. In a positive feedback loop, often triggered by land clearing and introduction of non-native grasses, the grasses fuel fires and then increase following fires and then rapidly accumulate to fuel future fires that are both large and frequent (D'Antonio and Vitousek 1992; Brooks et al. 2004).

Some invasive species compete quite successfully with native plants, and often thereby alter ecosystem services as well as fire regimes. Such species are often, but not always, non-native. They may be called alien, introduced, exotic, or weeds, but not all such plants become invasive. If invasive species thrive, they can alter the amount, timing, and spatial continuity of fuel. Introduced grasses may not always lead to positive feedback with the area burned. For instance, McGranahan et al. (2013) found that an invasive grass

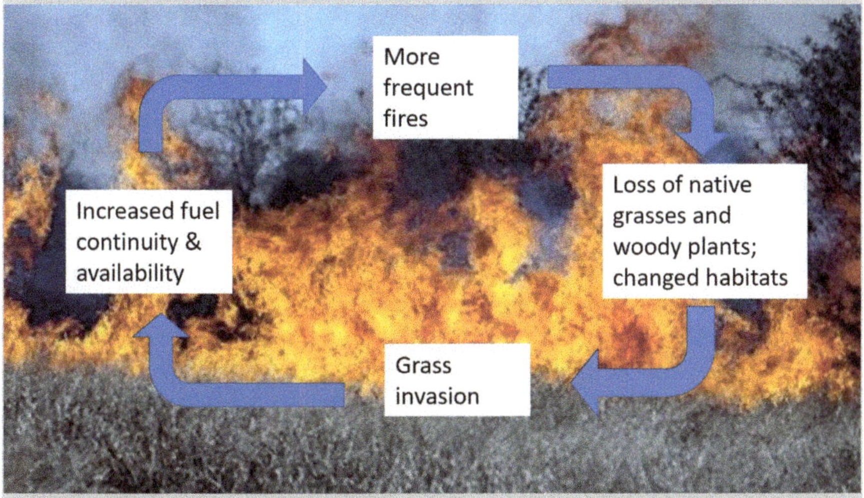

Fig. 12.29 Land use and introduction of invasive grasses can alter fire regimes, often resulting in the conversion of forests or shrublands to dominance by grass in savanna, shrublands, or grasslands. The positive feedback makes it difficult to interrupt the grass-fire cycle

(continued)

Case Study 12.1 (continued)

increased live fuel proportion and reduced the potential fire spread during the late summer season in a tallgrass prairie.

Greater sage-grouse (*Centrocercus urophasianus*) and many other species that need sagebrush in a fine-scale mosaic with bunchgrasses and forbs have declined so much that many are threatened with extinction. Their habitat has become limited as very frequent fires fueled by introduced annual grasses, including cheatgrass (*Bromus tectorum*), have burned large patches within the Great Basin area of the western USA (Coates et al. 2016). Now, few islands of Wyoming big sagebrush (*Artemisia tridentata* ssp. *wyomingensis*) remain in what was once the "sagebrush sea". The annual grasses establish in abundance before and after wildfires, increasing fuel continuity, and they readily burn within 1 or 2 years after a prior fire. Fires readily kill sagebrush plants that can take years to establish and grow long enough to produce seed. The introduced annual grasses are well adapted to the semi-arid habitats, especially where they can establish in fall or early spring, survive winter as seedlings and then out-compete native bunchgrasses for the soil moisture from winter snowpack. In contrast, at higher elevations, where annual grasses are much less abundant, fire suppression has been so effective that western juniper (*Juniperus occidentalis*) and Rocky Mountain Douglas-fir (*Pseudotsuga menziesii* var. *glauca*) have invaded areas once dominated by mountain big sagebrush (*Artemisia tridentata* ssp. *vaseyana*). Fire once limited these native tree species to rocky areas, but now sagebrush and grasses are eventually replaced as tree canopy cover increases. Similarly, western juniper and other tree species are invading areas dominated by mountain big sagebrush (*Artemisia tridentata* ssp. *vaseyana*) where fires are now less frequent than historically (see Case Study 12.2 for landscape implications). The area of Wyoming and mountain big sagebrush have been greatly reduced by the change in fire regimes—too much fire and too little fire, respectively. This "Goldilocks" problem can only be addressed if we don't assume there is a single one-size-fits-all solution, we are proactive and innovative, and we learn and adapt landscape management. Already there have been many changes in land and vegetation management on both private and public (both state and federal) lands under the conservation plan that was developed with unprecedented, collaborative problem-solving efforts across the range of the greater sage-grouse. The Sage Grouse Initiative (2017) is a partnership-based, science-driven, nationally-funded effort in which voluntary incentives are designed to foster proactive conservation. More than 1500 ranchers are partners along with government agencies in 11 states since 2010, conserving 2.2 million ha of sage-grouse and sagebrush habitat. Tools include prescribed burning, grazing with domestic livestock, post-fire management, and fire suppression, all to keep the sage-grouse from being further endangered and keeping ranchers

(continued)

Case Study 12.1 (continued)
ranching. However, the challenges are many for maintaining populations of greater sage-grouse now and in the future (Figs. 12.30 and 12.31).

In the grass-fire cycle described by D'Antonio and Vitousek (1992) with examples from all over the globe, invasive grasses increase fuel load or flammability or both, which leads to more fires that are more frequent, larger, or burn with higher intensity (or all three) (Fig. 12.32). Together these and competition lead to mortality of large trees and lack of tree regeneration and, in turn, an increase in the invasive grass in a positive feedback cycle. Grasses are common invasive species world-wide and given their biomass, grasses can readily fuel fires when it is dry. As many grasses resprout or readily establish in disturbed areas, grasses commonly increase post-fire to favor more frequent, intense, and large fires. Bowman et al. (2014) evaluated claims that a tropical savanna tree in northern Australia was being eliminated by high-intensity fires

Fig 12.30 The greater sage-grouse is one of the many sagebrush obligate birds whose population has greatly declined as habitat changed through multiple effects, including habitat change due to changing fire regimes (Pacific Southwest Region U.S. Fish and Wildlife Service 2006)

(continued)

Case Study 12.1 (continued)

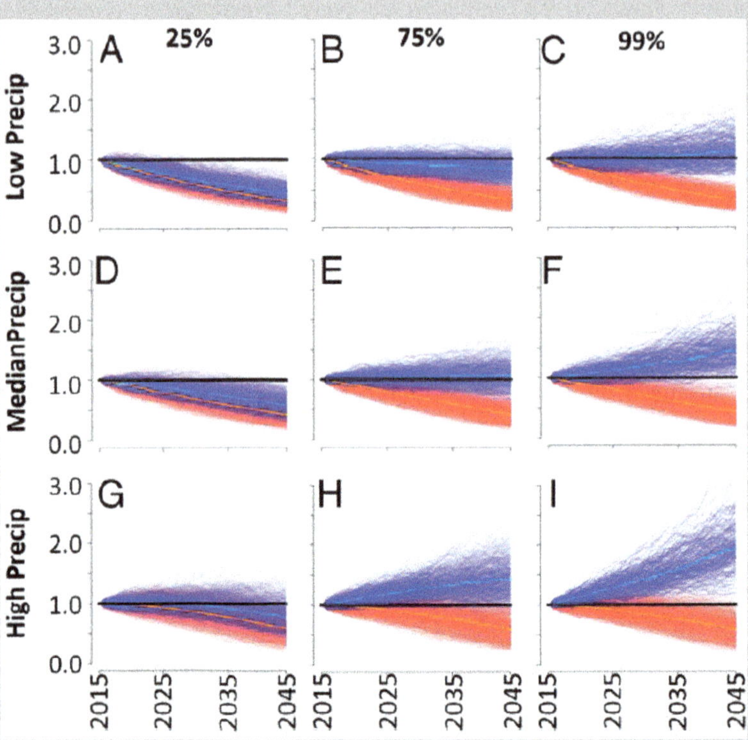

Fig. 12.31 Cumulative area burned would have to be reduced by 75–99% in the next 30 years within 5 km of the breeding areas for greater sage-grouse populations to increase (blue), and only if precipitation is above the median according to this simulation model. Here, the black line is a stable population (From Coates et al. 2016)

fueled by a non-native grass. They burned plots experimentally to document the higher intensity fires with the grasses present and found no tree seedlings established. Rossiter-Rachor et al. (2008) found that as gamba grass (*Andropogon gayanus*) introduced from Africa increased in abundance, fires burned with such high intensity that nitrogen was lost. Thus, in the grass-fire cycle, ecosystem processes are affected as well as vegetation composition and structure.

Breaking the grass-fire cycle is challenging (Brooks et al. 2004, Fig. 12.32). In the earliest stages (green and yellow in the figure), managers can seek to control or eradicate the species they have identified as potentially problematic. They can seek to maintain the historical fire regime. This proactive management is less intensive and more likely to be successful than trying to control or

(continued)

Case Study 12.1 (continued)

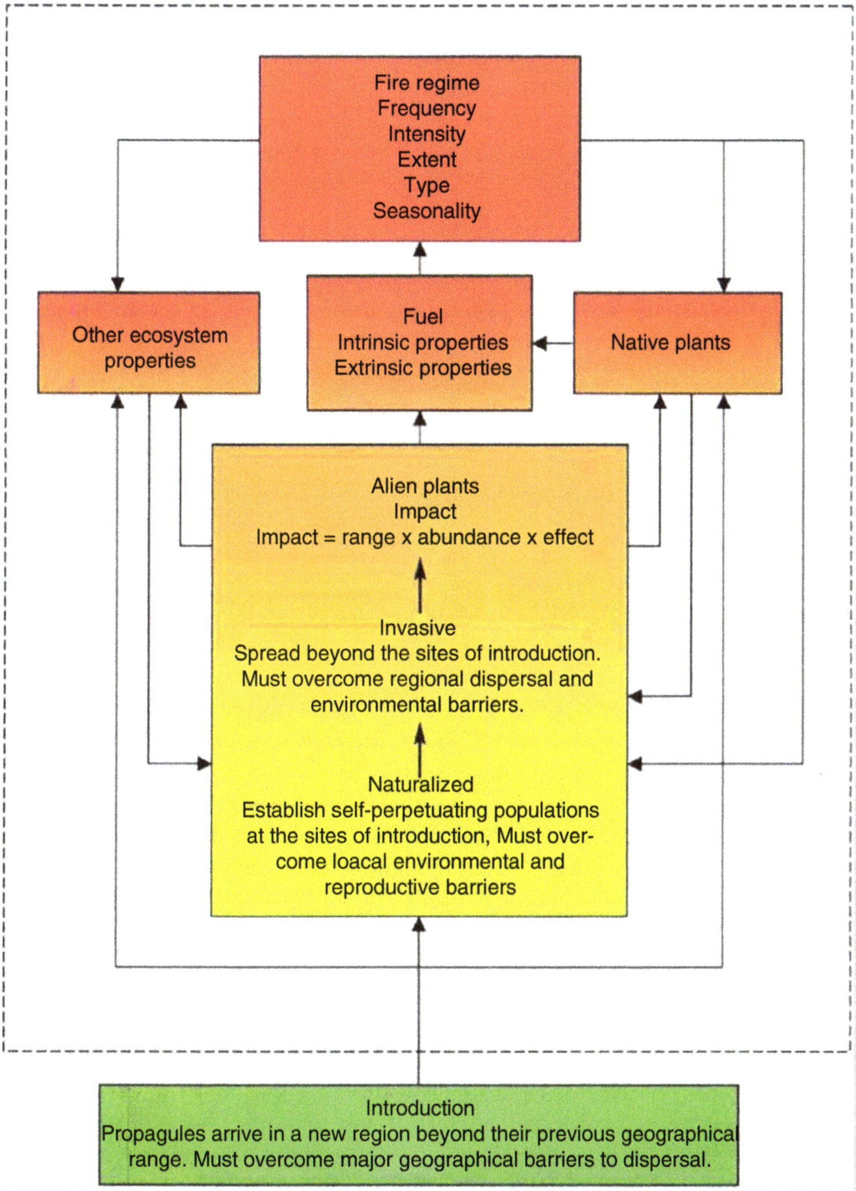

Fig. 12.32 Invasive species can alter fire regimes. For instance, introduced annual or perennial grasses often increase the fuel continuity and fuel repeated fires that are too frequent for the native flora to survive even if they were adapted to less frequent fires. Alternatively, invasive plants can alter fuels so that fires are less likely or fires burn in a different season than the one to which native plants were adapted (From Brooks et al. 2004)

(continued)

Case Study 12.1 (continued)
eliminate the invasive species and restore the historical fire regime once the fire regimes have been altered (red in Fig. 12.32). Brooks et al. (2004) recommended managing ignition sources and fuels where invasive plants have promoted larger and more frequent fires. Such strategies could include creating fuel breaks, planting or seeding with fire-resistant plants, and fire prevention. Where invasive species result in smaller and less frequent fires or alter the seasonality and patch sizes, igniting prescribed fires once fuels have been enhanced is potentially useful, particularly if coupled with managing land use to favor native species. In both cases, working where the fire regimes are just beginning to change and where ecosystems are fairly healthy will likely be more successful in preventing or reversing degradation in vegetation composition and fire regimes.

12.5 Landscape Dynamics and Landscape Management

Once we consider fire history and fire effects over space and time, we immediately recognize that fires have implications for landscape dynamics. Indeed, there is a rich interaction between fires and landscapes depending on climate, past fires, vegetation, topography, and other factors. Disturbances are the norm. Vegetation is almost always recovering from disturbances. Fires and other disturbances are a critical part of ecosystems.

Disturbance interactions are an important part of understanding the dynamics of landscapes. The combination and sequence of repeated disturbances influence the severity of subsequent fires and forest recovery. For instance, Stevens-Rumann and Morgan (2016) found that repeated fires affected the density of tree seedlings that subsequently established with very few tree seedlings found when either the earlier or later fires burned with high severity. Similarly, tree seedling density was less in the areas burned with either prior bark beetle or prior wildfire. Stevens-Rumann et al. (2016) found that prior wildfires influenced the severity of subsequent fires, similar to the findings of Parks et al. (2014b). Teske et al. (2012) and Parks et al. (2015a, b, 2016) found that prior fires limited the extent of subsequent fires, though this varied with weather and topography (Holsinger et al. 2016). Flower et al. (2014) found that defoliation by spruce budworm did not increase the probability that fires occurred in grand fir forests, but forest composition and structure, including fire exclusion, can influence the severity of defoliation which in turn can affect how subsequent fires burn. Hood et al. (2015) concluded that low-severity fire increased ponderosa pine tree defense against bark beetle attacks. Thus, prior fires or other disturbances may increase or decrease the likelihood, spatial extent, or magnitude of subsequent disturbances. The combined effect can be more pronounced than either disturbance alone, especially if either or both of the disturbances are severe, and the subsequent disturbance occurs so soon after the first one that the vegetation has not recovered.

Where prior disturbances increase the size of patches burned with high severity in subsequent fires, vegetation recovery rate and composition could be altered by the limited availability of seed sources from outside those patches.

Past landscape changes influence fuels, and therefore, burned area and burn severity and vegetation structure and composition. Patch size distribution will be especially important in influencing landscape dynamics. When a landscape burns with very large patches, especially if those large patches burn with high severity, there are many consequences for ecosystem function. The seed sources for post-fire regeneration of trees and other plants may be far away (see Case Study 12.3), habitat for animals changes, especially those that need hiding and thermal cover near more open areas, soil erosion potential is often higher and less carbon may be sequestered (see Sect. 9.5.3). Managing for future resilience to future fires in a rapidly changing world will require applying what we learn from history to some future range of variability, where fires burn and ecosystems respond in both similar and different ways (Moritz et al. 2018). Further, fuels management alone is not restoration (Stephens et al. 2020). Some fuels management is often crucial to successful ecosystem restoration. However, not all fuels treatments further restoration goals (Stephens et al. 2020). Restoration of ecological processes usually needs fire, though some combination of mechanical treatments and fire are often effective.

Where fire suppression has been effective, most fires are suppressed when they are small, and fires are more easily suppressed when fuel moisture and wind are mild. As a result, the numerous small- and medium-sized fires that burned under less-than-extreme conditions of weather and fuels are missing (Haugo et al. 2019). Even though they likely didn't burn much land area individually, such fires cumulatively shaped landscape patch patterns. They often fostered landscape heterogeneity as their effects often limited the spread or the severity of subsequent fires and limited the creation of large, homogeneous patches in subsequent fires. Importantly, the variety of previously burned and recovering patches in the landscape influenced how subsequent fires burned and vegetation responded to subsequent fires that burned under more extreme weather and fuel conditions. Hessburg et al. (2007) showed with simulation modeling that even relatively small changes in the proportion of large patches alter the resilience of landscapes to future fires as indicated by their patch size distribution.

Succession, the process of vegetation response to disturbances, follows multiple pathways reflecting both the structure and composition of vegetation when burned and burn severity. Tepley et al. (2013) give one example (Fig. 12.33). Succession, including multiple stable states, can be characterized using relatively simple state and transition models or more complex ecosystem process models (Keane et al. 2015). Sometimes there is a shifting mosaic that results in a steady state of the proportion of different successional stages, even though the spatial arrangement of those stages may change over time (typically characterized by dominant plants and structure). However, the steady-state mosaic only occurs if disturbances are much smaller than the landscape area and vegetation recovery is rapid relative to the frequency (Turner et al. 1993). Thus landscape equilibrium is scale-dependent, and other landscape dynamics are possible (Fig. 12.34).

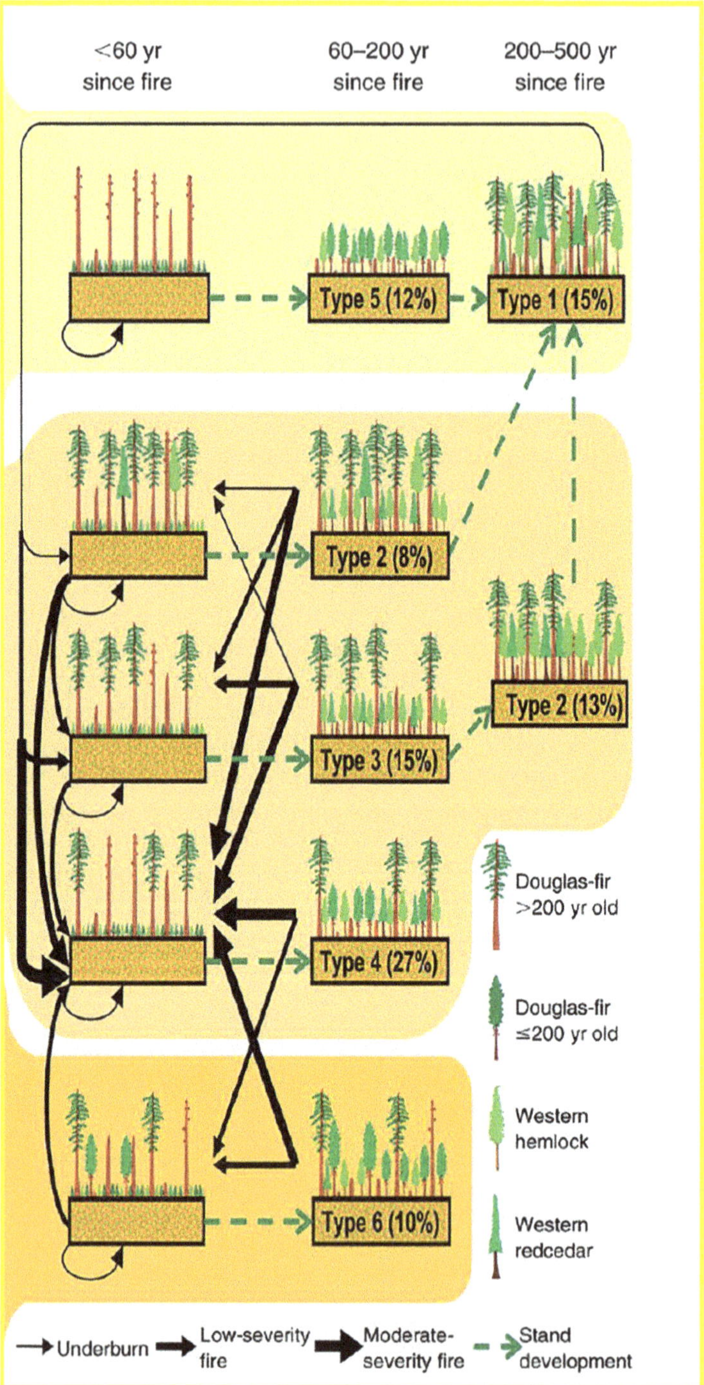

Fig. 12.33 Multiple pathways of succession in moist, productive Douglas-fir and western hemlock forests in the Cascades mountains of Oregon, USA. In the absence of fire (dashed lines) succession

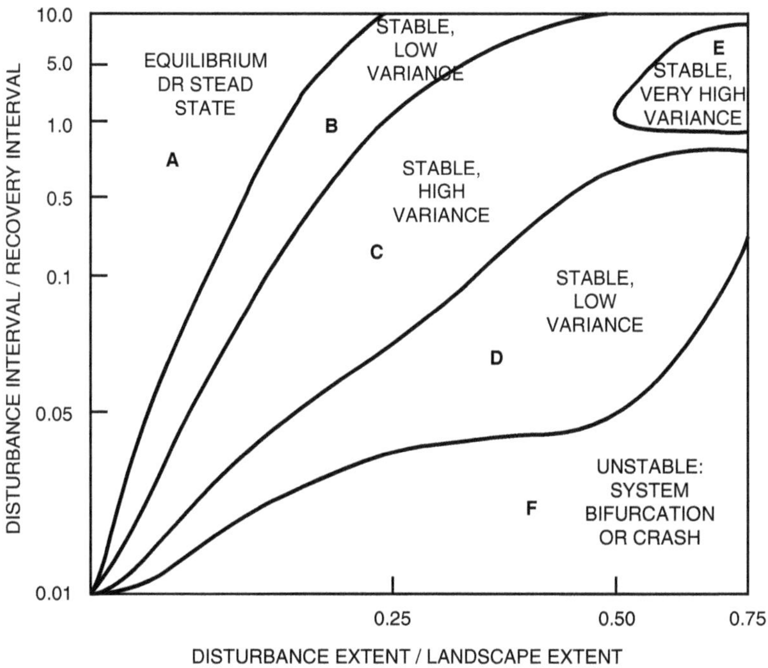

Fig. 12.34 Landscape dynamics vary with scale. Equilibrium or steady-state conditions are only likely when the extent of disturbances, such as stand-replacing fires, are much smaller than the size of the landscape assessed, and when vegetation recovers quickly relative to time between disturbances. Variability over time and space similarly varies with scale (From Turner et al. 1993)

12.5.1 Modeling Landscape Dynamics to Inform Landscape Management

Models are useful to explore the implications of alternative land management and disturbance scenarios. Though sometimes limited by their assumptions, models enable exploring the implications of landscape change, some of which may be unexpected. Here we share several examples.

Fig. 12.33 (continued) proceeds from young to mature and old-growth forests (the three columns, respectively). The pathway of succession varies with burn severity (the darker the arrow, the more severe the fire, except that stand-replacing fires are not shown as they are assumed to all lead to the upper left corner) and the vegetation condition when fires occur. Burn severity and stand structure influence the resulting forest structure, and that in turn influences succession. Infrequent stand-replacing fires with single-cohort eventually develop the complex vertical structure of old-growth. Episodic fires of moderate burn severity result in many trees surviving in multiple cohorts of mostly shade-tolerant species. Frequent fires of low burn severity can favor a mix of trees of different sizes in complex stand structures. (From Tepley et al. 2013)

Bunting et al. (2007) used state and transition models to compare the landscape composition under alternative fire management scenarios, including wildfire only and with different percentages of the landscapes burned with prescribed burning. In Case Study 12.2 they illustrate their findings for one large watershed in the Owyhee uplands in southwestern Idaho, USA. Their results have implications for sage grouse and other sagebrush-obligate birds, and fire behavior and biodiversity when the landscape patch pattern changes through time. They emphasized the value of prescribed burning and fire management that increased the area burned over time and space.

Case Study 12.2 Landscape Dynamics and Management: The Western Juniper Woodland Story

Stephen C. Bunting, email: sbunting@uidaho.edu; *Eva K. Strand*, email: evas@uidaho.edu

College of Natural Resources, University of Idaho, Moscow, ID, USA

Juniper and pinyon/juniper woodland encroachment has been a well-documented problem on rangelands in the USA, including the Great Basin (Miller et al. 2005, 2008; Sankey and Germino 2008), Colorado Plateau (Miller and Tausch 2001), and Great Plains (Engle and Kulbeth 1992) for decades. These semi-arid landscapes were historically mosaics of woodland, shrub steppe, and other vegetation types. During the recent 150–200 years, they have become increasingly dominated by juniper and pinyon/juniper woodlands. The encroachment usually results in the structural simplification of the landscape as it becomes increasingly dominated by woodland with little sagebrush and herbaceous vegetation as these are shaded out by trees. Changes with this encroachment have often been reduced community and landscape diversity, reduced herbaceous biomass production, changes in the fuel complex, altered watershed characteristics, and loss of habitat for many plants and animals, particularly those associated with shrub steppe (Miller et al. 2005). The causes of woodland encroachment into the shrub steppe include (1) climate change and climate variation, (2) historical livestock grazing which reduced the herbaceous biomass and herbaceous competition, (3) passive fire suppression resulting from livestock grazing, roads, and other agricultural development, and (4) active fire suppression (Miller et al. 2005). The goal of our research was to estimate how much area within a landscape needed to burn with either wildfire or prescribed fire in order to maintain sagebrush steppe vegetation on the landscape. Landscape composition is influenced by the interaction between successional processes, natural disturbance regimes, and management. This example illustrates how landscape composition, diversity, and pattern can be influenced by fire management in steppe and woodland vegetation in Idaho.

(continued)

Fig. 12.35 Typical sagebrush steppe/western juniper woodland mosaic. Many plants and animals depend on the open areas dominated by grass and shrubs. In the absence of fires, trees become increasingly abundant (Photograph by Stephen C. Bunting)

Case Study 12.2 (continued)

We studied the encroachment of western juniper (*Juniperus occidentalis*) into sagebrush shrub steppe in the Owyhee Mountains in southwestern Idaho, USA. The sagebrush steppe consists primarily of two types dominated by mountain big sagebrush (*Artemisia tridentata* ssp. *vaseyana*), and little sagebrush (*A. arbuscula*). On some transitional sites, a fine-scale mixture of the two species occurs (Fig. 12.35).

Methods

We selected three watersheds (each ~6000 ha) for study. The current vegetation for two Potential Vegetation Types (PVT) of interest was classified into one of six successional stages representing the successional continuum (Fig. 12.36). The three successional stages constituting the early stages of woodland development are referred to as Phase 1, 2, and 3 as developed by Miller et al. (2005). Vegetation in each watershed was mapped into polygons of similar vegetation reflectance with a supervised classification using >740 ground control points within a Landsat 7 ETM satellite image from 2 August 2002 (Fig. 12.37). The polygon groups were assigned a successional stage based on the field classification of the ground control points and this was mapped (Fig 12.37). Successional models for the two PVTs were developed using tree age estimates at the ground control points and the Vegetation Dynamics Development Tool (VDDT). VDDT provides a state-and-transition modeling framework for analyzing landscape succession while accounting for

(continued)

Fig. 12.36 Six juniper woodland successional vegetation stages. Following a fire, a successional stage dominated by herbaceous vegetation develops (upper left). In the next successional stage, sagebrush has re-colonized the site (upper center). In Phase 1, juniper seedlings and small trees begin to occupy the site (upper right). As the trees grow and become co-dominant with the steppe vegetation Phase 2 develops (lower left). In Phase 3, the woodland vegetation is dominating ecological processes (lower center), and eventually, mature woodlands develop that can last on the site for more than 1000 years in the absence of major disturbance (lower right). Fire in early and mid successional stages (Phase 2 and earlier) generally results in a return to herbaceous grasslands while vegetation resulting from a fire in Phase 3 and mature juniper is less predictable (Photograph in the lower right by Eva Strand, all others by Stephen Bunting)

Case Study 12.2 (continued)

disturbances and management actions (Fig. 12.38). Data related to the historical occurrence of wildfire for the Owyhee Mountains was derived from the US Bureau of Land Management historical fire perimeters data. The data regarding pre-European fire history was taken from published fire history studies of the Owyhee Mountains conducted by Burkhardt and Tisdale (1976). VDDT is not spatially explicit. Therefore, the VDDT models were implemented in the Tool for Exploratory Landscape Scenario Analyses

(continued)

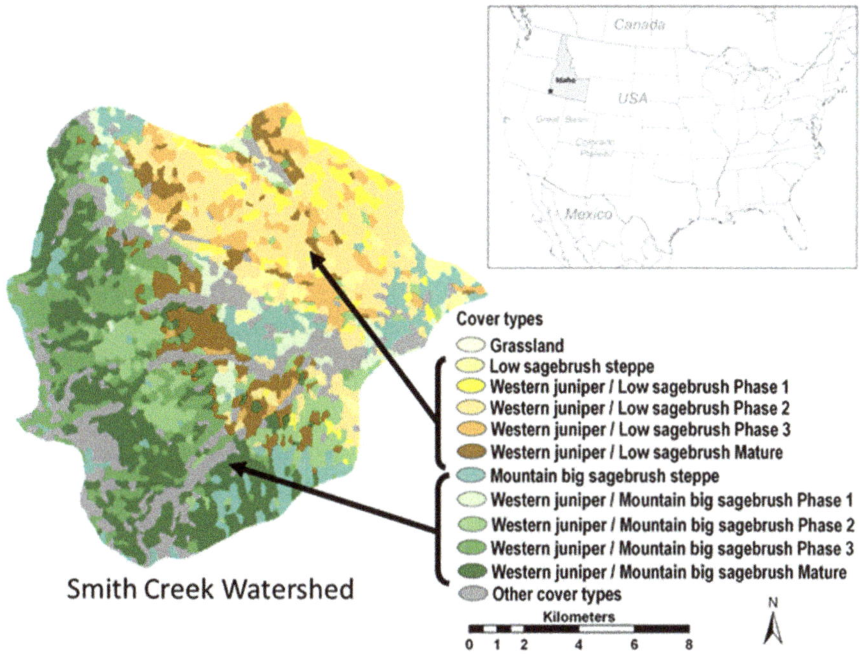

Fig. 12.37 Current vegetation cover type map based on a classified Landsat image of the Smith Creek watershed

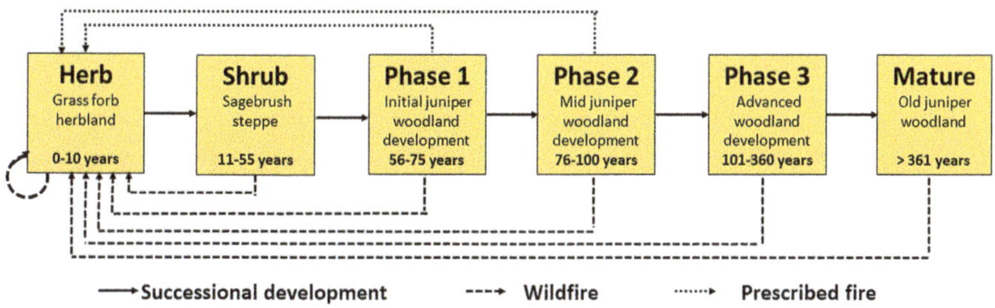

Fig. 12.38 Successional model developed using the Vegetation Dynamics Development Tool. Solid lines indicate transitions between phases in the absence of disturbance. Dashed lines indicate vegetation change following wildfire and prescribed fire. Phases shown in Fig. 12.36. (Diagram by S. C. Bunting and E. K. Strand)

Case Study 12.2 (continued)

(TELSA Version 3.3), a spatially-explicit, deterministic succession modeling tool with stochastic properties. We used TELSA to predict future landscape compositions of the three watersheds that would result from a variety of fire management strategies. Within the model, random prescribed fires were "ignited" in the sagebrush steppe and Phase 1 and Phase 2 juniper successional stages. Particular attention was given to these two Phases because they have the attributes that are most dynamic and subject to successional change. Because the establishment of juniper often occurs relatively rapidly and these two phases retain species found in sagebrush steppe, transitions between these phases occur readily. In addition, the loss of sagebrush-dominated vegetation significantly affects the abundance and quality of habitat for many plant and animal species. We compared results for several scenarios of differing levels of prescribed burning (2, 5, 7, 10% of the watershed/decade) and a scenario reflecting the current wildfire management (fire suppression). The models were run ten times for each scenario in each watershed to characterize stochastic variability. The average and variation for each scenario for each watershed were then compiled and statistically analyzed. The historical watershed composition was estimated from tree ages by essentially running the TELSA model backward in time. This approach was limited to a maximum period of 150–200 years because it is based on the assumption that the probabilities of plant establishment and growth, and of fire occurrence historically were similar to the current probabilities.

Results

Since approximately 1850, western juniper woodland area increased 8- to 10-fold on the three watersheds with very large patches of continuous trees where historically the vegetation had many smaller patches of trees interspersed with patches of shrub steppe. Most of the tree encroachment occurred into what had previously been sagebrush shrub steppe but encroachment also affected other vegetation types such as aspen woodland and dry meadows. The resulting encroachment woodlands primarily consisted of the open young woodlands (Phase 2) and young multi-story woodlands (Phase 3).

Under the current wildfire management strategy of rapid and aggressive wildfire suppression, the area of sagebrush steppe and Phase 1 juniper woodlands will continue to decline into the future (Fig. 12.39). As a consequence, landscape diversity and evenness will decline, as would the number of habitat boundaries between patch types. Patch size increased as patches of differing patch type coalesced into the same patch type of later stages (Phase 3 and mature woodlands). Distances between patches of the earlier successional stages (herbaceous, sagebrush steppe, and Phase 1 woodland) increased.

Prescribed burning 2, 5, and 7% of the landscapes per decade, focusing on the burning of Phase 1 and Phase 2 woodlands, would increasingly slow the

(continued)

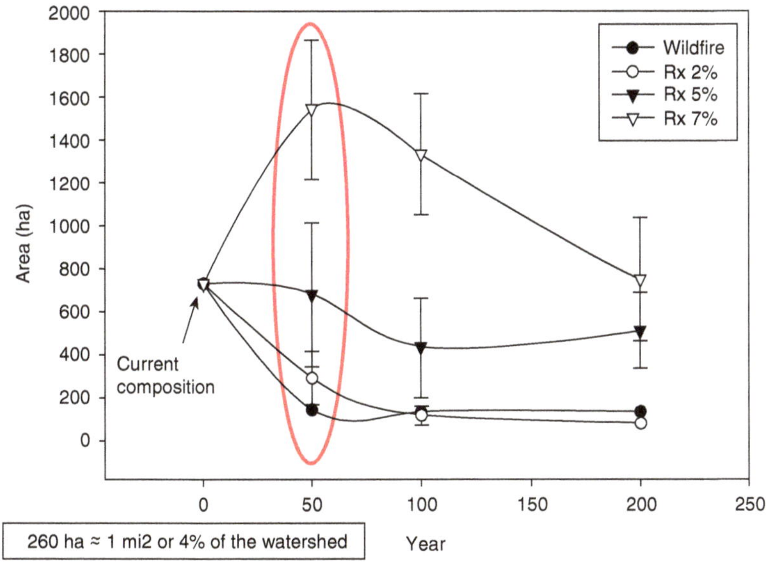

Fig. 12.39 Predicted change in the total area of mountain big and little sagebrush steppe within Smith Creek watershed under different fire scenarios. Note that 345 ha is about 5% of the watershed (graphics by the case study coauthors, Steve Bunting and Eva Strand, used with their permission)

Case Study 12.2 (continued)

decline of sagebrush steppe vegetation (Fig. 12.39). However, sagebrush steppe-dominated vegetation continued to become less abundant under all scenarios. Given the weather variation between different years and, consequently, the ability to implement prescribed burning, burned area is expressed as a percentage per decade. The random application of prescribed fire allowed some Phase 1 and 2 patches to not be burned and transition into Phase 3 woodlands where they were immune to prescribed fire treatments. Prescribed burning more than 10% of the landscapes per decade also resulted in the loss of sagebrush habitat. Areas were being reburned so quickly (less than 20 years between fires) that the sagebrush shrubs did not have time to develop and the burned areas remained dominated by herbaceous vegetation. In order to maintain the present levels of sagebrush-dominated vegetation habitat, some percentage of Phase 3 woodland must be treated to bring it back into the early successional loop. However, since this successional stage is less resilient with respect to sagebrush steppe species establishment, recovery of sagebrush steppe habitat on these sites required longer time periods than for the Phase 1 and 2 vegetation, as can be seen in Fig. 12.39.

(continued)

Case Study 12.2 (continued)

The oldest western juniper woodland stands commonly contained trees that were 800–1000 years old or more. These stands were frequently found on rocky ridges and sites where they were unlikely to burn under most wildfire conditions. However, old tree stands also occurred on more fire-prone sites with deeper soil and more level topography. We concluded that if a stand of trees could serendipitously avoid being burned for 2 or 3 fire events, the trees could grow to a size that would enable them to likely survive subsequent fires. Fire-scarred trees were present but not common in the watersheds. Very old juniper added greatly to the diversity of the landscapes and provided habitat for many species, including many birds that are unlikely to live elsewhere. The old juniper stands provide unique habitat characteristics such as nest cavities and logs. The old stands are also more likely to have an uneven-aged structure than the younger woodlands. As a consequence, many plant and animal species are only found in the older juniper woodlands. Because old juniper woodland stands may require 500–800 years to develop, care should be taken to not unnecessarily treat the areas. Also, past wildfires have burned through very old juniper woodland stands. Thus, some Phase 3 woodlands should be retained in appropriate locations on the landscape to allow for the development of future old mature stands to replace those that may be lost to wildfire or inappropriate management activities. Efforts should be taken to preserve these mature stands when using prescribed burning and managing fire and fuels in these landscapes.

We did not account for livestock grazing in our modeling. However, it is well known that the previous removal of herbaceous biomass by livestock can influence prescribed burning success and resulting pattern and affect subsequent post-burn recovery.

Summary

Landscape models that include successional and disturbance influences were extremely useful in assessing the long-term effects of management actions in the sagebrush steppe and western juniper-dominated landscape. There are ecological concerns regarding the downward population trends of many sagebrush-obligate species, such as the greater sage-grouse (*Centrocercus urophasianus*), pigmy rabbit (*Brachylagus idahoensis*), sage thrasher (*Oreoscoptes montanus*) and sage sparrow (*Artemisiospiza nevadensis*). Both spatially explicit and non-spatial models are useful, but some landscape characteristics such as mean patch size, the pattern of patches, the mean distance between areas of the same patch type, and boundary length can only be assessed with spatially-explicit models. Other characteristics such as landscape diversity and patch type evenness and dominance can be assessed by either type of model. We emphasize that the maps of patch types on the landscape when the processes occur at random are only representations of one

(continued)

Case Study 12.2 (continued)
of an infinite number of outcomes and are not necessarily "real" maps of the future landscape.

The models were run for 200 years into the future. While 200 years is well beyond the typical management framework, the results of these longer runs were important in our analysis because some trends in landscape change were not apparent with shorter runs of 25–50 years. For example, the loss to wildfire and development of old mature juniper woodlands operates on an entirely different time scale than our typical management planning.

Wildfire management has changed here, partially as a result of this landscape-level research. Local managers are using prescribed fire and other vegetation treatments in expanding juniper woodlands. As scientists, we worked closely with local and regional managers. We worked as partners to co-produce the knowledge to inform and evaluate management. The simulation model results were instrumental in helping both scientists and managers to explore and understand the implications of different management scenarios. In the last 10–15 years, large wildfires have also become more common, likely due to the warming climate that has led to a longer, drier summer season allowing wildfires to ignite and spread. As a result of concern for sagebrush-obligate species and the susceptibility of sagebrush to fire, mechanical treatments on juniper are now being implemented more often and over larger areas. Succession, disturbance, and management will continue to affect the area and pattern of sagebrush steppe and juniper woodlands into the future.

Loehman et al. (2018) used simulation models to forecast fire frequency, area burned, and area burned with high severity for two different landscapes in the southwestern USA. The forests there are dominated by ponderosa pine and other species. They used two different landscape models, FireBGCv2 (Keane et al. 2011) and LANDIS-II (Scheller et al. 2007), to project forests and wildfire under contemporary and both Warm-Dry and Hot-Arid scenarios with four different management scenarios. They found that regardless of management technique, the ecosystems reorganized under the effects of changing climate (Fig. 12.38). Intensive management somewhat reduced the severity and extent of fires but did not prevent forest ecosystem reorganization, including the loss of key species, altered nutrient cycles, and carbon sequestering. Thus, they concluded that we will have to change our concept of the desired future conditions or use novel approaches to landscape management.

Managing ecosystems under a changing climate requires being flexible and adapting approaches. Reestablishing fire regimes is a key principle of forest ecosystem and landscape restoration (e.g., Hessburg et al. 2015) under current climate scenarios. Multiple treatments, used alone or in concert, may be needed to restore ecosystems for many areas (Reinhardt et al. 2008; Stephens et al. 2020). We recommend that while restoration can be informed by historical conditions, the focus should be on resilience for the future that includes rapidly changing climate. When Flatley and Fulé (2016) used a combination of simulation models to explore the consequences of climate change and fire regimes for dry and mesic mixed conifer forests in the southwestern USA, they concluded that longer (compared to historical conditions) intervals between frequent fires would be needed in the dry forests coupled with protection of mesic forests from fire if they were to be maintained. Similarly, Loehman et al. (2018) concluded that intensive management with thinning and prescribed burning was not enough to forestall ecosystem change under future hot and dry climate scenarios (Fig. 12.40).

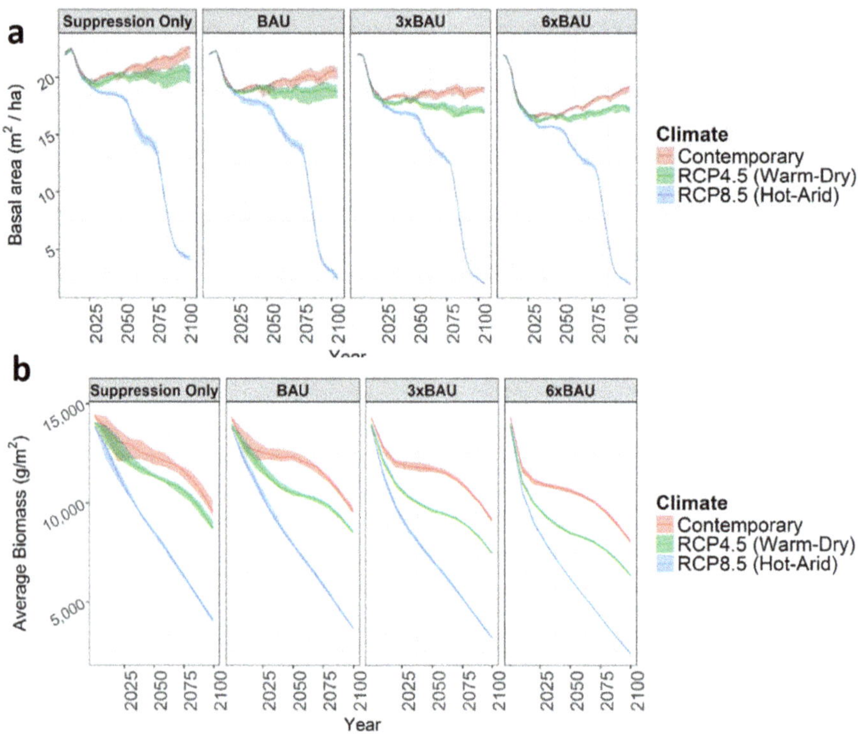

Fig. 12.40 Simulated future forests under projected climate change scenarios with fire suppression only (90% of ignitions suppressed successfully), and then with suppression combined in each of three scenarios with an increased burned area representing 76-, 22- and 11-year rotation of thinning and prescribed fire (here referred to as BAU, 3×BAU, and 6×BAU where BAU means Business As Usual). Simulations were made with FireBGCv2 for (**a**) the Jemez Mountains in New Mexico and (**b**) the Kaibab Plateau in Arizona using LANDIS-II with other models. (From Loehman et al. 2018)

Table 12.1 Landscape management can be guided by these key ideas, adapted from seven principles of forest restoration described by Hessburg et al. (2015)

Principles for restoring resilience of forest landscapes to future fires and changing climate
Manage forest landscapes as patchworks across multiple scales to foster landscape processes
Use topography to guide restoration of vegetation to ensure heterogeneity and habitat
Enable and encourage disturbance and succession consistent with natural regimes
Manage patch size distributions and then allow them to adapt to topography, climate, and disturbance
Consider patches within patches to foster heterogeneity over space and through time
Retain and protect the large and old trees, as well as large snags and logs, as these are both the legacy and backbone of many forest ecosystems
Collaborate to coordinate management across boundaries to accommodate diverse land ownership and management objectives and to coordinate access, ecosystem services

12.5.2 Landscape Restoration, Resilience to Future Fires, and Changing Climate

In many ecosystems around the world, there is great emphasis on increasing the pace and scale of restoration of resilience to future fires, usually through reducing fuel hazard, improving wildlife habitats, and otherwise increasing the probability that future fires will have more positive and fewer negative effects. Dustin Doane, a fire manager on the Payette National Forest in Idaho, said, "More prescribed fire is one of the most critical change agents in these efforts. Therefore, the demand for responsible, efficient, and effective applications of prescribed fire is paramount in today's environment. Successful efforts will require good plans, experienced practitioners, community support, and partner participation." The safety of fire fighters and other people is important, as is the degree of exposure of people to smoke, the costs of action or inaction, and the scales and locations where treatments can be effective must all be considered. See the related Case Study 13.4, and other success stories of Integrated Fire Management in Chap. 13. Resilience is defined and discussed in Chap. 10.

Landscape management can be guided by seven principles (Table 12.1). Heterogeneity at multiple scales is critical to ecological restoration and other landscape management goals (Hessburg et al. 2015). Use topography and other guides, and realize that not every patch needs to be treated. If we are to manage landscapes for resilience to future fires in the changing climate, we likely cannot accomplish meaningful change without active management, such as prescribed burning and vegetation manipulation, including tree cutting, and strategically managing wildfires. Hessburg et al. (2015) also recommended increasing heterogeneity within stands and across landscapes in keeping with topography and disturbance by fire, insects, and pathogens that would have maintained and enhanced clumps and openings at multiple spatial scales. They suggested adapting vegetation management to leave such diversity in forest structure and composition through time and across space as recommended by North et al. (2019), Churchill et al. (2013), and others.

Hessburg et al. (2000) compared current to historical aerial photographs for several watersheds across the forests of the northwestern USA. They found that the patch sizes and forest structure have changed in all forests, including not just the drier forests but also moist mixed-conifer and cold, subalpine forests. Haugo et al. (2015) showed that current conditions have significantly departed from the past, reflecting increased forest area, increased tree density, more canopy layering, and increased continuity of these conditions across forest landscapes. In such landscapes, fires can more readily spread and grow because early seral forests are underrepresented, and continuous mid- and late seral conditions are over-represented. These forests are more prone to large patches burned with high severity than they were historically (Hessburg et al. 2016). Though there are many plants and animals, including (e.g., woodpeckers), that depend on severely burned forests for their habitat (DellaSala and Hanson 2015; Hutto et al. 2016), it is unlikely that there is too little severely burned area. When more area and more large patches burn with high severity, this will affect many ecosystem functions, including soil erosion potential, nutrient cycling, primary production and decomposition.

High-severity fire can be a catalyst for ecosystem change in the face of changing climate (Stevens-Rumann et al. 2015, 2018; Walker et al. 2018; Stevens-Rumann and Morgan 2019), particularly where there are large patches burned with high severity in forests. Then, post-fire recovery can be slow or follow a very different trajectory if trees don't regenerate and forests become non-forest. Post-fire management can sometimes address this. Such management must be strategically focused on locations where post-fire tree planting will result in forest establishment, or where prior management has increased the potential for seed source trees to survive when fires occur (Stevens-Rumann and Morgan 2019). Managers can focus on multiple strategies to increase survival of the trees that will be the seed source for future forest regeneration. They can also focus on sites where post-fire planting is likely to be successful even as climate changes and reburning from repeated fires is less likely. See Case Study 12.3.

Grazing and burning can be useful together in managing native prairie vegetation for the benefit of herbivores and other animals and native plants. The idea is to burn parts of large pastures through time. As herbivores move about (many will prefer to graze in recently burned areas), they will subject the land to a spatially varying mix of intensive to low herbivory on vegetation that is unburned in places, burned recently in others, and burned long ago in still other parts of the area. Given the high diversity of plant species in prairies, some will be favored immediately, some will be eaten soon after fires, and others only long after burning. The many native grasses and forbs regrow or otherwise recover following fire, and thrive when the accumulated dead grass is consumed by fire, thus opening space. The herbivores, including ungulates, insects, and rodents, while free to graze throughout the pasture, preferentially consume some but not all plants and will concentrate in some but not all areas. The combined effect is high diversity with patches of different conditions over space and through time. Developed by Fuhlendorf and Engle (2001), the 'patch burning' approach has been adapted and applied by private landowners raising livestock on native prairie vegetation. In addition to favoring landscape heterogeneity with multiple ecological benefits, the landowners can sometimes effectively incorporate this as part of their management system.

Case Study 12.3 Post-Fire Tree Regeneration in a Changing Climate

Camille Stevens-Rumann, email: C.Stevens-Rumann@colostate.edu
Forest and Rangeland Stewardship, Colorado State University, Fort Collins, CO
Penelope Morgan, email: pmorgan@uidaho.edu
Department of Forest, Rangeland, and Fire Sciences, University of Idaho, Moscow, ID, USA

Fires will be an agent of ecosystem change as climate changes (Stevens-Rumann et al. 2018; Walker et al. 2018; Davis et al. 2019). Consider recent studies of post-fire tree regeneration or lack thereof. In many locations across the western USA (see review by Stevens-Rumann and Morgan 2019), the Mediterranean region (Pausas et al. 2008), and elsewhere, researchers and managers are concerned. Even 7–25 years after large fires, many formerly forested sites have so few tree seedlings that forests will be replaced by persistent shrublands, grasslands, or woodlands, especially for warm, dry sites burned in recent decades. Why?

Stevens-Rumann et al. (2018) found that one of the causes of low tree seedling densities in burned areas after 62 large fires in the US Rocky Mountains was the warmer and dryer conditions post-fire. These warm and dry conditions were particularly prevalent in fires that occurred after 2000 compared to wildfires in the 1980s and 1990s (Fig. 12.41). They compared the climate conditions on each site to the 30-year averages on each site. Once the distance to seed source was accounted for, the burn severity class identified by the proportion of overstory trees surviving fires or satellite-derived burn severity was not a significant influence on post-fire tree regeneration (Kemp et al. 2016; Stevens-Rumann et al. 2018). Similar trends and causes have occurred in numerous wildfires across the western USA (Stevens-Rumann and Morgan 2019). Davis et al. (2019) aged tree seedlings on multiple sites across the western USA to reach a similar conclusion. They found that the warm, dry conditions of recent years have exceeded the thresholds for the successful post-fire establishment of ponderosa pine and Douglas-fir tree seedlings on warm, dry sites (Fig. 12.41). The adult trees on many sites globally are experiencing high levels of tree mortality, usually in response to drought and bark beetles (van Mantgem et al. 2009, 2013; Allen et al. 2010; Hicke et al. 2015) even in the absence of fire. High rates of tree mortality are predicted due to the increasing vulnerability of forests to drought-induced tree mortality as climate changes (Allen et al. 2010; Hicke et al. 2012; Anderegg et al. 2015). In

(continued)

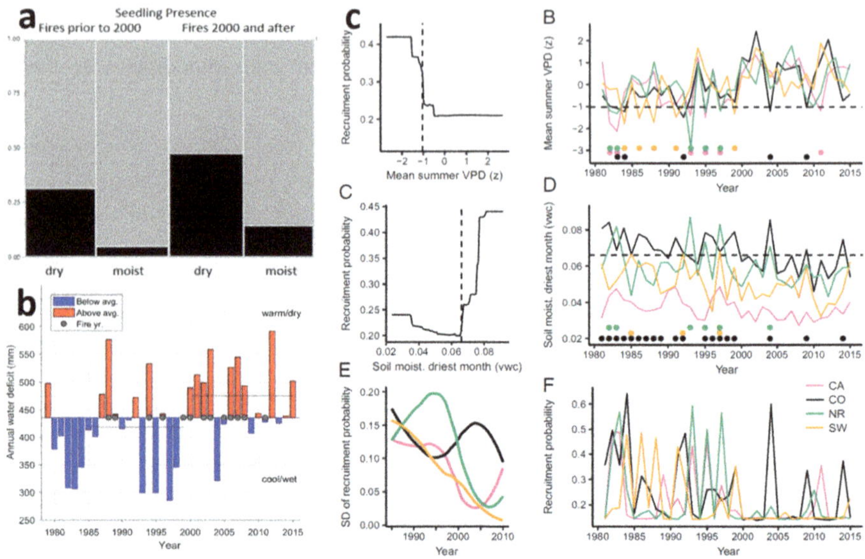

Fig. 12.41 (**a**) Many of the sites at low elevations within large fires sampled in California (CA), Colorado (CO), the US northern Rockies (NR) and the southwestern USA (SW) have poor regeneration. (**b**, **c**) Drought (summer vapor pressure deficit VPD, z score) and soil moisture during the driest month (volumetric water content, vwc) have exceeded thresholds for successful recruitment, in many years. This is increasingly so in recent years. As a result, recruitment probability of tree seedlings is highly variable year to year, and very low in recent years. Analyses accounted for the effect of burn severity and distance from the seed source. (**a** and **b** from Stevens-Rumann and Morgan 2016; graphs in **c** are from Davis et al. 2019)

Case Study 12.3 (continued)

the Sierra Nevada Mountains of southern California, more than 100 million trees have died in recent years in response to acute drought, bark beetles, and high forest density where fires in recent decades have been far less frequent than historically (Stephens et al. 2018). Other disturbances could induce additional tree mortality, thus further limiting tree seed sources for forest regeneration.

What are the implications for management? What proactive strategies show promise for managing before and during fires, and how can post-fire management be more effective and strategic?

With wildfires as a potential catalyst for abrupt ecosystem change, especially in the face of climate change, there are concerns about the resilience of many forests (Fig. 12.41). If and where grasslands or shrublands replace forests post fire, there may be less carbon stored (Liang et al. 2018), changed habitat for many wildlife species, and other ecosystem services society values.

(continued)

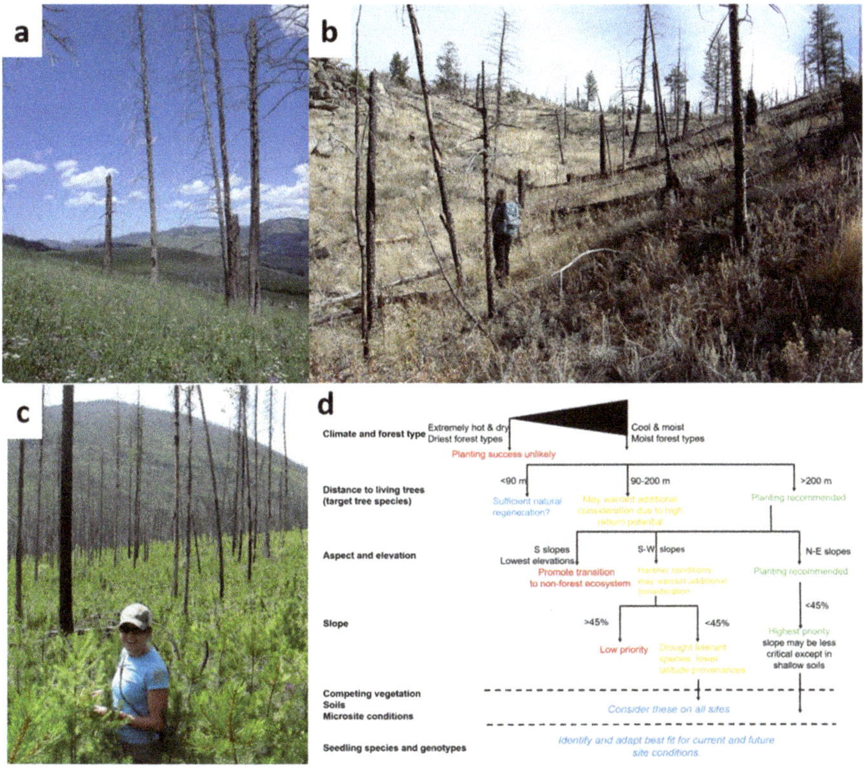

Fig. 12.42 (**a**) On many warm, dry sites, no or very few tree seedlings have regenerated following large fires (Photograph by Daniel Donato). (**b**) This is sometimes related to distance to the seed source, but trees are less likely to successfully establish if the climate is too warm and dry (Photograph by Monica Rother). (**c**) For lodgepole pine with serotinous cones and for many tree species, seedlings establish in abundance post fire (Photo used with permission from Kerry Kemp). (**d**) This decision tree, developed by scientists and managers working together, suggests that on the warmest and driest sites, post-fire tree regeneration is unlikely whether trees are planted or not. Also, where the climate is very favorable, planting may not be needed either. On intermediate sites, managers could strategically target planting tree seedlings in areas beyond the reach of seed sources, and also beyond the edge where reburns are more likely to occur (From Stevens-Rumann et al. 2019)

Case Study 12.3 (continued)

With the changing climate, we can expect the loss of forests on relatively warm and dry sites, especially at the lower treeline (Fig. 12.42).

Strategic management decisions can use this knowledge on seedling establishment limitations to promote the resilience of forests to future disturbances (North et al. 2019). Here are four key factors for consideration, as outlined in multiple publications and amongst managers. *First*, projections of areas

(continued)

Case Study 12.3 (continued)
susceptible to high severity fire could be combined with future climate projections to assess potential ecosystem shifts due to changing climate. These projections coul help identify where prescribed burning or other treatments could help facilitate the adaptation of ecosystems to changing climate. *Second*, managers of forests and fires could foster the survival of the trees that will provide seed sources for future regeneration now and in the future. Surviving trees provide locally adapted seed, which can aid in successful regeneration through time if the conditions are suitable for tree seedling establishment and growth. This strategy includes managing so that more areas burn under less extreme conditions and with more smaller patches, prescribed burning to limit the accumulation of fuels to the point where trees are likely to die in subsequent wildfires, and thinning to foster the development of larger trees more likely to survive fires. Perhaps you can think of this as ensuring "mama" trees can survive to provide seed rain year after year near where baby trees can grow. Walker et al. (2018) found there were more tree seedlings in the parts of the 2011 Los Conchas Fire that burned with low severity and where prior prescribed burns or wildfires had burned recently. *Third,* Stevens-Rumann and Morgan (2019) developed a decision tree to assist managers in their strategic decisions about where and what to plant post fire to foster forests. They suggest that managers avoid planting trees on sites that are so warm and dry that planted trees are unlikely to survive, and also on cooler and wetter sites where trees will naturally regenerate post fire so no planting is needed. Action must be tempered by local knowledge and experience. Further, they suggested not planting near edges of burned areas as they are both within "reach" of a seed source and more likely to burn if the surrounding area burns again. North et al. (2019) suggest planting clumps of "founder trees", especially in relatively inaccessible areas where planting is challenging. As different tree species respond differently to warm and dry conditions (Davis et al. 2018, 2019), planting multiple species and genomes and fire-resistant trees likely to survive future fires is advisable. *Fourth*, tree seedlings may not establish and grow following fires on some sites. There will be areas where forests will not regenerate, and will instead be replaced by shrublands and grasslands, regardless of intervention. Managing for a resilient community of understory plants will help the land during global change.

12.6 Landscape Management Perspectives

The ways in which climate, vegetation, and people influenced historical fire regimes might hold important lessons for understanding what makes ecosystems resilient and managing them to adapt for the future (Morgan et al. 2020). Past disturbances affected the vegetation with which fires interact, as does the weather at the time of the fire, climate, and topography. Past landscape change, including vegetation structure and composition, and therefore fuels, influenced both area burned and burn severity. Landscapes thus have inertia, as landscapes reflect legacies of past disturbance. Once the patch size distribution changes, subsequent fire effects will often reinforce it (under less extreme conditions) or overcome it (under more extreme conditions).

Fires currently burn more areas than we affect with fuels and other land management in many areas worldwide. Fires must be considered and planned for in landscape management, for they will occur. We can manage fires and their effects to help achieve landscape management goals. Although fires can be a blunt tool, we can often harness fires' power to change landscapes. A landscape framework is useful. The Washington Department of Natural Resources (2020) has a scientifically sound, strategic management plan in place to address landscape-scale forest health across mixed public and privately owned forest lands. With five interlocking goals guiding landscape-scale assessment, action, and monitoring over 0.5 million ha, the plan boldly seeks to reduce the risk of uncharacteristic fires and other disturbances to protect people and ecosystems, enhance economic well-being of rural communities, manage for future resilience of forests and watersheds (Washington DNR 2020). There and elsewhere, experience with recent large fires has driven innovation in policy to guide strategic decisions before, during, and after fires. Cross-boundary, collaborative, landscape-scale work with fire can benefit multiple parts of social-ecological systems (see Chaps. 13 and 14).

With large fires affecting landscapes now, strategic thinking is needed for burned areas. Three strategies build on assessing fire effects within the burns to understand where and managing burned areas and the surrounding landscape to build toward landscape management goals. *First,* areas where fires did good work toward landscape management goals can be appreciated. Active management may be needed there in the future, but doing nothing there may be the best strategy until then. *Second,* where fires did some good work within individual burned areas, some additional treatments may be needed to meet long-term goals including resilience to future fires. This could include prescribed burning or other fuels treatment, rehabilitation to reduce soil erosion potential, salvage logging, planting, and other treatments to support local economies while furthering long-term landscape management goals. *Third,* some parts of some burned areas may require accepting transition to quite different conditions and different trajectories than those that existed before the fire. In some burned areas, especially those burned with high tree mortality and near lower timberline, forests may not reestablish even with intensive planting efforts. There, accepting and managing transition to different vegetation may be the best strategy (Stevens-Rumann and Morgan 2019, See Case Study 12.3). Planting and seeding could focus on shifting plant composition to more drought-tolerant species.

People will continue to debate the best management to achieve resilient social-ecological systems. Many managers seek to balance the benefits of fire with negative impacts. All management decisions, including no action, have consequences for future fires and vegetation. Evaluating best management requires projecting what fire effects, short-term and long-term, will occur and balancing the ecosystem products and services people want with what is ecologically appropriate, socially acceptable, and economically feasible. To respond to rapidly changing landscape and social conditions, managers will need to be highly adaptive and adept at learning from what works well and what doesn't. Collaborative partnerships, including scientists, managers, and other citizens, as appropriate, are useful for monitoring, learning from the results of management decisions, and communicating effectively.

Managing lands and fires can be challenging. Currently, the area of fuels treatments, including thinning, prescribed burning, grazing, and other actions, are dwarfed by the need and by the area burned (Schoennagel et al. 2017). To influence the spread and severity of future fires, treating large areas and managing large fires in strategic ways will be needed. Yet, such efforts are often controversial, especially when there is distrust, risk, and uncertainty. Trust grows from engaged partners collaborating together to build local understanding about how and why landscapes have changed and what those landscapes may be in the future under different management scenarios. People can build trust by working in areas of common ground (Fig. 12.43) where there is agreement, such as prescribed burning and fuels treatments in dry forests and near towns or other values at risk. People, from land managers to society at large, must wrestle with and decide what future proportion and pattern of area burned and burn severity might be desirable and possible in each locality. Management decisions ideally reflect local and broader objectives and costs, fire fighter safety, public safety, smoke, and other local concerns and regional or national priorities. If the goal is to manage landscape change and fire regimes, then any management actions will need to be strategically located and large enough to alter landscape dynamics. Both opportunities and constraints vary geographically,

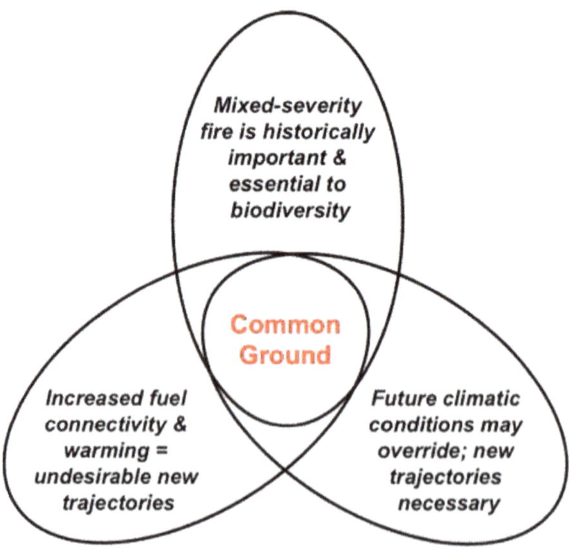

Fig. 12.43 There is often much common ground where different objectives overlap. Working from the "zones of agreement" can help gain experience and trust needed to work through decisions elsewhere (From Moritz et al. 2018)

but planning across landscapes can enable meeting shared goals while accommodating different specific objectives and constraints in different parts of the landscapes. Decisions can be paralyzed by a lack of information and agreement about the interpretation of scientific findings and social will. Yet, a decision of no action has consequences that must be considered. Engaging stakeholders in discussing alternative management choices has built trust and consensus, as has working in those areas where there is greater agreement. Further, we must appreciate how important it is for people to feel safe, so coupling landscape management with helping communities feel less vulnerable to fires is important.

Many people support aggressive fire suppression and fuels treatments within and immediately adjacent to the WUI. However, vegetation management beyond the WUI influences forest resilience, smoke, water quantity and quality, and many other ecosystem services people value. Appropriate strategies for managing fire far from the WUI differ from those within and adjacent to the WUI. There is broad agreement amongst fire scientists that fire is a fundamental ecological process (McLaughlan et al. 2020). Further, combinations of low-, moderate-, and high-severity fire are essential to ecological processes, including landscape resilience to future fires in forests and grasslands (Moritz et al. 2018). Managers are often open to using fire, depending on fuel conditions and the land management objectives, but they will need many tools. Fire alone may not suffice.

Climate and fuels will both be increasingly important as we go forward. Where climate change results in extended droughts and extreme fire weather, more large fires are likely, but in some areas, actively managing fuels can reduce the proportion of the area burned with high severity depending on the topography, vegetation and fuels. Uncertainties when projecting historical trends into the future abound. Many such uncertainties reflect the interactions, environmental gradients, invasive species, land uses, and social trends, including fire policy and how we value and manage ecosystem services. Nonetheless, we can say with certainty that landscapes will change and that those changes will often be the result of fires. Don't let fear of fire drive us to attempt exclusion of fire. In many landscapes, more fires is needed, not less, for resilience to future fires and to meet management goals. Prioritize prescribed burning and managing wildfires where we can.

With the pressure of global change, strategic choices are urgent. As Aplet and Cole (2010) suggest, we can resist change (perhaps guided by historical conditions), accept change, or guide change. Guiding change could result in transformation to more novel ecosystem conditions. Consider using vegetation management, including prescribed fires to aid wildfire by preparing anchor points, fuel breaks around valued assets, including peoples' homes. Fuels treatments and fire management can increase the potential for success in achieving long-term sustainability of social-ecological systems.

Loud voices and competing science are challenging. Moritz et al. (2018) highlighted the differing science perspectives that could inform best management practices. These include what proportions of low-, moderate-, and high-severity fire are natural and desirable, and how and where landscapes and fire regimes have changed since the nineteenth century. Moritz et al. (2018) highlighted common ground amongst competing objectives and scientifically-based viewpoints about fires (Fig. 12.43). Often, there is more common ground than we think, particularly when we think long-term (about future generations) and build trust. With so many scientists offering their perspectives, the sometimes fractious debates will continue about what to do where and which science is sounder. Trust becomes all the more important. Clear communication, including listening, is important to building trust.

Actively using fire can help increase the resilience of our landscapes to future fires. Our communities and local economies depend on healthy ecosystems and healthy watersheds. A little smoke now from prescribed burns helps avoid a lot of smoke in the future. Prescribed fires are often a key tool in restoration. Prescribed fires can promote healthy ecosystems, reduce the likelihood that future fires will burn severely, and often reduce the smoke impacts on public health. Fire is part of our land's personality. Fires were part of healthy forests for millennia, and many plants and animals depend on fires. However, severe fires over large areas, especially ones with large patches severely burned, can harm plants and animals. In fire-adapted ecosystems, "no fire" is not one of our choices. Prescribed burns are done carefully with a lot of planning based on ecological understanding, building on decades of local, on-the-ground knowledge, and cutting edge fire and smoke modeling tools to inform choice of when, where and how to conduct prescribed fires.

Proactive, strategic management is useful, especially where future fires are highly likely. Thompson et al. (2016) recommended a framework called Potential Operational Delineations implemented before fires occur to support risk-informed management during fires. In this and similar approaches, maps and historical data are used to assess where fires are most likely to occur and where they are most likely to spread, and then the conditions under which fire effects would be a net positive or net negative outcome for the local management goals. They form the basis of strategic fire response zones for each geographical location, so fire management can effectively advance resource and land management. Similarly, Hessburg et al. (2015) recommend using topography as a template for applying their seven principles of landscape management.

References

Abatzoglou, J. T., & Williams, A. P. (2016). Impact of anthropogenic climate change on wildfire across western US forests. *PNAS, 113*, 11770–11775. https://doi.org/10.1073/pnas.1607171113.

Abatzoglou, J. T., Kolden, C. A., Balch, J. K., & Bradley, B. A. (2016). Controls on interannual variability in lightning-caused fire activity in the western US. *Environmental Research Letters, 11*(4), 045005. https://doi.org/10.1088/1748-9326/11/4/045005/meta.

Abbott, I., VanHeurck, P., & Burbidge, T. (1993). Ecology of the pest ins jarrah leaf miner (Lepidoptera) in relation to fire and timber harvesting in Jarrah Forest in western Australia. *Australian Forestry, 56*, 264–275.

Abella, S. R., Covington, W. W., Fulé, P. Z., Lentile, L. B., Sanchez Meador, A. J., & Morgan, P. (2007). Past, present, and future old growth in frequent-fire conifer forests of the western United States. *Ecology and Society, 12*(2), 1.

Agee, J. K. (1993). *Fire ecology of Pacific Northwest forests*. Washington, DC: Island Press.

Allen, C. D., Macalady, A. K., Chenchouni, H., Bachelet, D., McDowell, N., Vennetier, M., Kitzberger, T., Rigling, A., Breshears, D. D., Hogg, E. T., & Gonzalez, P. (2010). A global overview of drought and heat-induced tree mortality reveals emerging climate change risks for forests. *Forest Ecology and Management, 259*, 660–684. https://doi.org/10.1016/j.foreco.2009.09.001.

Andela, N., Morton, D. C., Giglio, L., Chen, Y., Van Der Werf, G. R., Kasibhatla, P. S., DeFries, R. S., Collatz, G. J., Hantson, S., Kloster, S., & Bachelet, D. (2017). A human-driven decline in global burned area. *Science, 356*(6345), 1356–1362.

Anderegg, W. R. L., Hicke, J. A., Fisher, R. A., Allen, C. D., Aukema, J., Bentz, B., Hood, S., Lichstein, J. W., Macalady, A. K., McDowell, N., & Pan, Y. (2015). Tree mortality from drought, insects, and their interactions in a changing climate. *New Phytologist, 208*, 674–683. https://doi.org/10.1111/nph.13477.

Aplet, G. H., & Cole, D. N. (2010). *The trouble with naturalness: Rethinking park and wilderness goals*. Beyond naturalness: rethinking park and wilderness stewardship in an era of rapid change, 12, 21–22. Island Press, Washington, DC

Archibald, S., Nickless, A., Govender, N., Scholes, R. J., & Lehsten, V. (2010). Climate and the inter-annual variability of fire in southern Africa: A meta-analysis using long-term field data and satellite-derived burnt area data. *Global Ecology and Biogeography, 19*(6), 794–809.

Archibald, S., Lehmann, C. E., Gómez-Dans, J. L., & Bradstock, R. A. (2013). Defining pyromes and global syndromes of fire regimes. *Proceedings of the Natinal Academy of Sciences, 110*(16), 6442–6447.

Balch, J. K., Bradley, B. A., Abatzoglou, J. T., Nagy, R. C., Fusco, E. J., & Mahood, A. L. (2017). Human-started wildfires expand the fire niche across the United States. *PNAS, 114*(11), 2946–2951. https://doi.org/10.1073/pnas.1617394114.

Birch, D. S., Morgan, P., Kolden, C. A., Hudak, A. T., & Smith, A. M. S. (2014). Is proportion burned severely related to daily area burned. *Environmental Research Letters, 9*, 064011. https://doi.org/10.1088/1748-9326/9/6/064011\.

Birch, D. S., Morgan, P., Kolden, C. A., Abatzoglou, J. T., Dillon, G. K., Hudak, A. T., & Smith, A. M. S. (2015). Daily weather and other factors influencing burn severity in central Idaho and western Montana, 2005-2007 and 2011. *Ecosphere, 6*(1), 17. https://doi.org/10.1890/ES14-00213.1.

Bowman, D. M. J. S., Balch, J., Artaxo, P., Bond, W. J., Cochrane, M. A., D'Antonio, C. M., DeFries, R., Johnston, F. H., Keeley, J. E., Krawchuk, M. A., & Kull, C. A. (2011). The human dimension of fire regimes on Earth. *Journal of Biogeography, 38*(12), 2223–2236. https://doi.org/10.1111/j.1365-2699.2011.02595.x.

Bowman, D. M. J. S., MacDermott, H. J., Nichols, S. C., & Murphy, B. P. (2014). A grass–fire cycle eliminates an obligate-seeding tree in a tropical savanna. *Ecology and Evolution, 4*(21), 4185–4194.

Bowman, D. M. J. S., Williamson, G., Kolden, C. A., Abatzoglou, J. T., Cochrane, M. A., & Smith, A. M. S. (2017). Human exposure and sensitivity to globally extreme wildfire events. *Nature Ecology & Evolution, 1*(3), 1–6. https://doi.org/10.1038/s41559-016-0058.

Bradstock, R. A. (2010). A biogeographic model of fire regimes in Australia: Current and future implications. *Global Ecology and Biogeography, 19*(2), 145–158.

Brooks, M. L., D'Antonio, C. M., Richardson, D. M., Grace, J. B., Keeley, J. E., DiTomaso, J. M., Hobbs, R. J., Pellant, M., & Pyke, D. (2004). Effects of invasive alien plants on fire regimes. *BioScience, 54*(7), 677–688.

Brunelle, A., Whitlock, C., Bartlein, P., & Kipfmueller, K. (2005). Holocene fire and vegetation along environmental gradients in the Northern Rocky Mountains. *Quaternary Science Reviews, 24*(20), 2281–2300.

Bunting, S. C., Strand, E. K., & Kingery, J. L. (2007). Landscape characteristics of sagebrush-steppe/juniper woodland mosaics under various modeled prescribed fire regimes. *Proceedings of Timbers Fire Ecology Conference, 23*, 50–57.

Bureau of Land Management (BLM). (n.d.) *General Land Office records.* Retrieved March 25, 2019, from https://glorecords.blm.gov/search/default.aspx?searchTabIndex=0&searchByTypeIndex=1.

Burkhardt, J. W., & Tisdale, E. W. (1976). Causes of juniper invasion in southwestern Idaho. *Ecology, 57*, 472–484.

Chen, D., Pereira, J. M., Masiero, A., & Pirotti, F. (2017). Mapping fire regimes in China using MODIS active fire and burned area data. *Applied Geography, 85*, 14–26.

Churchill, D. J., Larson, A. J., Dahlgreen, M. C., Franklin, J. F., Hessburg, P. F., & Lutz, J. A. (2013). Restoring forest resilience: From reference spatial patterns to silvicultural prescriptions and monitoring. *Forest Ecology and Management, 291*, 442–457.

Coates, P. S., Ricca, M. A., Prochazka, B. G., Brooks, M. L., Doherty, K. E., Kroger, T., Blomberg, E. J., Hagen, C. A., & Casazza, M. L. (2016). Wildfire, climate, and invasive grass interactions negatively impact an indicator species by reshaping sagebrush ecosystems. *PNAS, 113*(45), 12745–12750.

D'Antonio, C. M., & Vitousek, P. M. (1992). Biological invasions by exotic grasses, the grass/fire cycle, and global change. *Annual Review of Ecological Systems, 23*(1), 63–87.

Davis, K. T., Higuera, P. E., & Sala, A. (2018). Anticipating fire-mediated impacts of climate change using a demographic framework. *Functional Ecology, 32*(7), 1729–1745.

Davis, K. T., Dobrowski, S. Z., Higuera, P. E., Holden, Z. A., Veblen, T. T., Rother, M. T., Parks, S. A., Sala, A., & Maneta, M. P. (2019). Wildfires and climate change push low-elevation forests across a critical climate threshold for tree regeneration. *PNAS, 116*(13), 6193–6198.

DellaSala, D. A., & Hanson, C. T. (2015). Ecological and biodiversity benefits of megafires. In D. A. DellaSala & C. T. Hanson (Eds.), *The ecological importance of mixed-severity fires* (pp. 23–54). Amsterdam: Elsevier.

Dennison, P. E., Brewer, S. C., Arnold, J. D., & Moritz, M. A. (2014). Large wildfire trends in the western United States, 1984–2011. *Geophysical Research Letters, 41*(8), 2928–2933.

Dillon, G. K., Holden, Z. A., Morgan, P., Crimmins, M. A., Heyerdahl, E. K., & Luce, C. H. (2011). Both topography and climate affected forest and woodland burn severity in two regions of the western US, 1984 to 2006. *Ecosphere, 2*(12), 130. https://doi.org/10.1890/ES11-00271.1.

Dillon, G. K., Panunto, M. H., Davis, B., Morgan, P., Birch, D., & Jolly, W. M. (2020). *Development of a severe fire potential map for the contiguous United States. RMRS-GTR-415.* Fort Collins: USDA Forest Service Rocky Mountain Research Station.

Eidenshink, J., Schwind, B., Brewer, K., Zhu, Z. L., Quayle, B., & Howard, S. (2007). A project for monitoring trends in burn severity. *Fire Ecology, 3*(1), 3–21.

Engle, K. M., & Kulbeth, J. D. (1992). Growth dynamics of crowns of eastern red cedar at 3 locations in Oklahoma. *Journal of Range Management, 45*, 301–305.

Estes, B. L., Knapp, E. E., Skinner, C. N., Miller, J. D., & Preisler, H. K. (2017). Factors influencing fire severity under moderate burning conditions in the Klamath Mountains, northern California, USA. *Ecosphere, 8*(5), e01794.

Falk, D. A., Heyerdahl, E. K., Brown, P. M., Farris, C., Fulé, P. Z., McKenzie, D., Swetnam, T. W., Taylor, A. H., & Van Horne, M. L. (2011). Multi-scale controls of historical forest-fire regimes: New insights from fire-scar networks. *Frontiers in Ecology and the Environment, 9*(8), 446–454.

Fall, J. (1999). *An introductory tutorial on common methods for determining fire frequency.* Burnaby: Simon Fraser University. Retrieved May 19, 2020, from http://www.rem.sfu.ca/forestry/download/download.htm.

Fernandes, P. M. (2019). Variation in the Canadian Fire Weather Index thresholds for increasingly larger fires in Portugal. *Forests, 10*, 838. https://doi.org/10.3390/f10100838.

Fernandes, P., Loureiro, C., Magalhães, M., & Fernandes, M. (2012). Fuel age, weather and burn probability in Portugal. *International Journal of Wildland Fire, 21*, 380–384.

Fernandes, P. M., Loureiro, C., Guiomar, N., Pezzatti, G. B., Manso, F., & Lopes. (2014). The dynamics and drivers of fuel and fire in the Portuguese public forest. *Journal of Environmental Management, 146*, 373–382. https://doi.org/10.1016/j.jenvman.2014.07.049.

Fernandes, P. M., Monteiro-Henriques, T., Guiomar, N., Loureiro, C., & Barros, A. (2016a). Bottom-up variables govern large-fire size in Portugal. *Ecosystems, 19*, 1362–1375. https://doi.org/10.1007/s10021-016-0010-2.

Fernandes, P. M., Pacheco, A. P., Almeida, R., & Claro, J. (2016b). The role of fire suppression force in limiting the spread of extremely large forest fires in Portugal. *European Journal of Forest Research, 135*, 253–262.

Finco, M., Quayle, B., Zhang, Y., Lecker, J., Megown, K. A., & Brewer, C. K. (2012). Monitoring trends and burn severity (MTBS): Monitoring wildfire activity for the past quarter century using Landsat data. In: RS Morin, GC Liknes (comps.), *Moving from status to trends: Forest Inventory and Analysis (FIA) symposium 2012*. Gen Tech Rep NRS-P-105 (pp. 222–228). Newton Square: USDA Forest Service Northern Research Station.

Finney, M. A. (1995). The missing tail and other considerations for the use of fire history models. *International Journal of Wildland Fire, 5*(4), 197–202.

Finney, M. A., Seli, R. C., McHugh, C. W., Ager, A. A., Bahro, B., & Agee, J. K. (2008). Simulation of long-term landscape-level fuel treatment effects on large wildfires. *International Journal of Wildland Fire, 16*(6), 712–727.

Flatley, W. T., & Fulé, P. Z. (2016). Are historical fire regimes compatible with future climate? Implications for forest restoration. *Ecosphere, 7*(10), e01471.

Flower, A., Gavin, D. G., Heyerdahl, E. K., Parsons, R. A., & Cohn, G. M. (2014). Western spruce budworm outbreaks did not increase fire risk over the last three centuries: A dendrochronological analysis of inter-disturbance synergism. *PLoS One, 9*(12), e114282.

Forkel, M., Andela, N., Harrison, S. P., Lasslop, G., van Marle, M., Chuvieco, E., Dorigo, W., Forrest, M., Hantson, S., Heil, A., & Li, F. (2017). Emergent relationships on burned area in global satellite observations and fire-enabled vegetation models. *Biogeosciences, 16*, 57–76.

Fuhlendorf, S. D., & Engle, D. M. (2001). Restoring heterogeneity on rangelands: Ecosystem management based on evolutionary grazing patterns: We propose a paradigm that enhances heterogeneity instead of homogeneity to promote biological diversity and wildlife habitat on rangelands grazed by livestock. *BioScience, 51*(8), 625–632.

Fulé, P. Z., Swetnam, T. W., Brown, P. M., Falk, D. A., Peterson, D. L., Allen, C. D., Aplet, G. H., Battaglia, M. A., Binkley, D., Farris, C., & Keane, R. E. (2014). Unsupported inferences of high-severity fire in historical dry forests of the western United States: Response to Williams and Baker. *Global Ecology and Biogeography, 23*(7), 825–830.

Gardner, R. H., Milne, B. T., Turner, M. G., & O'Neill, R. V. (1987). Neutral models for the analysis of broad-scale landscape pattern. *Landscape Ecology, 1*, 19–28.

Hagmann, R. K., Franklin, J. F., & Johnson, K. N. (2013). Historical structure and composition of ponderosa pine and mixed-conifer forests in south-central Oregon. *Forest Ecology and Management, 304*, 492–504.

Haire, S. L., & McGarigal, K. (2009). Changes in fire severity across gradients of climate, fire size, and topography: A landscape ecological perspective. *Fire Ecology, 5*, 86–103.

Haire, S. L., McGarigal, K., & Miller, C. (2013). Wilderness shapes contemporary fire size distributions across landscapes of the western United States. *Ecosphere, 4*, art15.

Hantson, S., Pueyo, S., & Chuvieco, E. (2015). Global fire size distribution is driven by human impact and climate. *Global Ecology and Biogeography, 24*(1), 77–86.

Harari, Y. N. (2017). *Homo Deus: A brief history of tomorrow*. New York: Random House.

Harley, G., Baisan, C., Brown, P., Falk, D., Flatley, W., Grissino-Mayer, H., Hessl, A., Heyerdahl, E., Kaye, M., Lafon, C., & Margolis, E. (2018). Advancing dendrochronological studies of fire in the United States. *Fire, 1*(1), 11. https://doi.org/10.3390/fire1010011.

Haugo, R., Zanger, C., DeMeo, T., Ringo, C., Shlisky, A., Blankenship, K., Simpson, M., Mellen-McLean, K., Kertis, J., & Stern, M. (2015). A new approach to evaluate forest structure restoration needs across Oregon and Washington, USA. *Forest Ecology and Management, 335*, 37–50.

Haugo, R. D., Kellogg, B. S., Cansler, C. A., Kolden, C. A., Kemp, K. B., Robertson, J. C., Metlen, K. L., Vaillant, N. M., & Restaino, C. M. (2019). The missing fire: Quantifying human exclusion of wildfire in Pacific Northwest forests, USA. *Ecosphere, 10*(4), e02702.

Heinselman, M. L. (1973). Fire in the virgin forests of the Boundary Waters Canoe Area, Minnesota. *Quaternary Research, 3*(3), 329–382.

Hessburg, P. F., Smith, B. G., Salter, R. B., Ottmar, R. D., & Alvarado, E. (2000). Recent changes (1930s-1990s) in spatial patterns of interior northwest forests, USA. *Forest Ecology and Management, 136*, 53–83.

Hessburg, P., Salter, R., & James, K. (2007). Re-examining fire severity relations in pre-management era mixed conifer forests: Inferences from landscape patterns of forest structure. *Landscape Ecology, 22*, 5–24.

Hessburg, P. F., Larson, A. J., Churchill, D. J., Haugo, R. D., Miller, C., Spies, T. A., North, M. P., Povak, N. A., Belote, R. T., Singleton, P. A., Gaines, W. L., Keane, R. E., Aplet, G. H., Stephens, S. L., Morgan, P., Bisson, P. A., Rieman, B. E., Salter, R. B., & Reeves, G. H. (2015). Restoring fire-prone forest landscapes: Seven core principles. *Landscape Ecology, 30*(10), 1805–1835. https://doi.org/10.1007/s10980-015-0218-0.

Hessburg, P. F., Spies, T. A., Perry, D. A., Skinner, C. N., Taylor, A. H., Brown, P. M., Stephens, S. L., Larson, A. J., Churchill, D. J., Povak, N. A., Singleton, P. H., McComb, B., Zielinski, W. J., Collins, B. M., Salter, R. B., Keane, J. J., Franklin, J. F., & Riegel, G. (2016). Tamm Review: Management of mixed-severity fire regime forests in Oregon, Washington, and Northern California. *Forest Ecology and Management, 366*, 221–250.

Heward, H., Smith, A. M., Roy, D. P., Tinkham, W. T., Hoffman, C. M., Morgan, P., & Lannom, K. O. (2013). Is burn severity related to fire intensity? Observations from landscape scale remote sensing. *International Journal of Wildland Fire, 22*(7), 910–918.

Heyerdahl, E. K., Brubaker, L. B., & Agee, J. K. (2001). Spatial controls of historical fire regimes: A multiscale example from the interior west, USA. *Ecology, 82*(3), 660–678.

Heyerdahl, E. K., Brubaker, L. B., & Agee, J. K. (2002). Annual and decadal climate forcing of historical fire regimes in the interior Pacific Northwest, USA. *Holocene, 12*(5), 597–604.

Heyerdahl, E. K., Morgan, P., & Riser II, J. P. (2008a). *Crossdated fire histories (1650 to 1900) from ponderosa pine-dominated forests of Idaho and western Montana*. Gen Tech Rep RMRS-GTR-214WWW.Fort Collins: USDA Forest Service Rocky Mountain Research Station.

Heyerdahl, E. K., Morgan, P., & Riser, J. P., II. (2008b). Multi-season climate synchronized widespread historical fires in dry forests (1650-1900), Northern Rockies, USA. *Ecology, 89* (3), 705–716.

Heyerdahl, E. K., Loehman, R. A., & Falk, D. A. (2014). Mixed-severity fire in lodgepole pine dominated forests: Are historical regimes sustainable on Oregon's Pumice Plateau, USA. *Canadian Journal of Forest Research, 44*(6), 593–603.

Hicke, J. A., Johnson, M. C., Hayes, J. L., & Preisler, H. K. (2012). Effects of bark beetle-caused tree mortality on wildfire. *Forest Ecology and Management, 271*, 81–90.

Hicke, J. A., Meddens, A. J. H., & Kolden, C. A. (2015). Recent tree mortality in the western United States from bark beetles and forest fires. *Forest Science, 62*(2), 141–153. https://doi.org/10.5849/forsci.15-086.

Higuera, P. E., Brubaker, L. B., Anderson, P. M., Hu, F. S., & Brown, T. A. (2009). Vegetation mediated the impacts of postglacial climate change on fire regimes in the south-central Brooks Range, Alaska. *Ecological Monographs, 79*(2), 201–219.

Higuera, P. E., Abatzoglou, J. T., Littell, J. S., & Morgan, P. (2015). The changing strength and nature of fire-climate relationships in the northern Rocky Mountains, USA, 1902-2008. *PLoS One, 10*(6), e0127563. https://doi.org/10.1371/journal.pone.127563.

Holden, Z. A., Morgan, P., Rollins, M. G., & Kavanagh, K. L. (2007). Effects of multiple fires on stand structure in two southwestern wilderness areas, USA. *Fire Ecology, 3*(2), 18–24.

Holsinger, L., Parks, S. A., & Miller, C. (2016). Weather, fuels, and topography impede wildland fire spread in western US landscapes. *Forest Ecology and Management, 380*, 59–69.

Hood, S., Sala, A., Heyerdahl, E. K., & Boutin, M. (2015). Low-severity fire increases tree defense against bark beetle attacks. *Ecology, 96*(7), 1846–1855.

Hu, F. S., Higuera, P. E., Duffy, P., Chipman, M. L., Rocha, A. V., Young, A. M., Kelly, R., & Dietze, M. C. (2015). Arctic tundra fires: Natural variability and responses to climate change. *Frontiers in Ecology and the Environment, 13*(7), 369–377.

Hudak, A. T., Freeborn, P. H., Lewis, S. A., Hood, S. M., Smith, H. Y., Hardy, C. C., Kremens, R. J., Butler, B. W., Teske, C., Tissell, R. G., Queen, L. P., Nordgren, B. L., Bright, B. C., Morgan, P., Riggan, P. J., Macholz, L., Lentile, L. P., Riddering, J. P., & Mathews, E. E. (2018). The Cooney Ridge Fire Experiment: An early operation to relate pre-, active, and post-fire field and remotely sensed measurements. *Fire, 1*(1), 10. https://doi.org/10.3390/fire1010010.

Hurteau, M. D., Liang, S., Westerling, A. L., & Wiedinmyer, C. (2019). Vegetation-fire feedback reduces projected area burned under climate change. *Scientific Reports, 9*(1), 1–6.

Hutto, R. L., Keane, R. E., Sherriff, R. L., Rota, C. T., Eby, L. A., & Saab, V. A. (2016). Toward a more ecologically informed view of severe forest fires. *Ecosphere, 7*(2), e01255.

Johnstone, J. F., Allen, C. D., Franklin, J. F., Frelich, L. E., Harvey, B. J., Higuera, P. E., Mack, M. C., Meentemeyer, R. K., Metz, M. R., Perry, G. L. W., Schoennagel, T., & Turner, M. G. (2016). Changing disturbance regimes, ecological memory, and forest resilience. *Frontiers in Ecology and the Environment, 14*, 369–378. https://doi.org/10.1002/fee.1311.

Jolly, W. M., & Freeborn, P. H. (2017). Towards improving wildland firefighter situational awareness through daily fire behaviour risk assessments in the US Northern Rockies and Northern Great Basin. *International Journal of Wildland Fire, 26*(7), 574–586.

Keane, R. E. (2015). *Wildland fuel fundamentals and applications*. New York: Springer.

Keane, R. E., Cary, G. J., & Parsons, R. (2003). Using simulation to map fire regimes: An evaluation of approaches, strategies, and limitations. *International Journal of Wildland Fire, 12*(4), 309–322.

Keane, R. E., Hessburg, P. F., Landres, P. B., & Swanson, F. J. (2009). The use of historical range and variability (HRV) in landscape management. *Forest Ecology and Management, 258*(7), 1025–1037.

Keane, R. E., Loehman, R. A., & Holsinger, L. M. (2011). *The FireBGCv2 landscape fire and succession model: A research simulation platform for exploring fire and vegetation dynamics*. Gen Tech Rep RMRS-GTR-255. Fort Collins: USDA Forest Service Rocky Mountain Research Station.

Keane, R. E., McKenzie, D., Falk, D. A., Smithwick, E. A., Miller, C., & Kellogg, L. K. (2015). Representing climate, disturbance, and vegetation interactions in landscape models. *Ecological Modelling, 309*, 33–47.

Keeley, J. E. (2009). Fire intensity, fire severity and burn severity: A brief review and suggested usage. *International Journal of Wildland Fire, 18*(1), 116–126.

Keeley, J. E., Brennan, T., & Pfaff, A. H. (2008). Fire severity and ecosystem responses following crown fires in California shrublands. *Ecological Applications, 18*(6), 1530–1546. https://doi.org/10.1890/07-0836.1.

Keeley, J. E., Aplet, G. H., Christensen, N. L., Conard, S. G., Johnson, E. A., Omi, P. N., Peterson, D. L., & Swetnam, T. W. (2009). *Ecological foundations for fire management in North American forest and shrubland ecosystems*. Gen Tech Rep PNW-GTR-779. Portland: USDA Forest Service Pacific Northwest Research Station.

Kemp, K. B., Higuera, P. E., & Morgan, P. (2016). Fire legacies impact conifer regeneration across environmental gradients in the US northern Rockies. *Landscape Ecology, 31*(3), 619–636.

Key, C. H., & Benson, N. C. (2006). Landscape assessment: Sampling and analysis methods. In: DC Lutes, RE Keane, JF Caratti, CH Key, NC Benson, S Sutherland, LJ Gangi (Eds.), *FIREMON: Fire effects monitoring and inventory system*. Gen Tech Rep RMRS-GTR-164-CD. Fort Collins: USDA Forest Service Rocky Mountain Research Station.

Keyser, A., & Westerling, A. L. (2017). Climate drives inter-annual variability in probability of high severity fire occurrence in the western United States. *Environmental Research Letters, 12*(6), 065003.

Krawchuk, M. A., & Moritz, M. A. (2011). Constraints on global fire activity vary across a resource gradient. *Ecology, 92*, 121–132.

Krawchuk, M. A., Haire, S. L., Coop, J., Parisien, M. A., Whitman, E., Chong, G., & Miller, C. (2016). Topographic and fire weather controls of fire refugia in forested ecosystems of northwestern North America. *Ecosphere, 7*(12), 1–18.

Krebs, P., Pezzatti, G. B., Mazzoleni, S., Talbot, L. M., & Conedera, M. (2010). Fire regime: History and definition of a key concept in disturbance ecology. *Theory in Biosciences, 129*(1), 53–69.

Kremens, R. L., Smith, A. M., & Dickinson, M. B. (2010). Fire metrology: Current and future directions in physics-based measurements. *Fire Ecology, 6*(1), 13–35.

Landfire. (n.d.) *Landfire. Fire Regime Maps*. Retrieved May 30, 2019, from https://www.landfire.gov/geoareasmaps/2012/CONUS_FRG_c12.jpg.

Landres, P. B., Morgan, P., & Swanson, F. J. (1999). Evaluating the utility of natural variability concepts in managing ecological systems. *Ecological Applications, 9*(4), 1179–1188.

Lang, S. (1997). *Burning in the bush: A spatio-temporal analysis of Jarrah forest fire regimes*. Honours Thesis. Geography, Australian National University, Canberra.

Lentile, L. B., Morgan, P., Hardy, C., Hudak, A., Means, R., Ottmar, R., Robichaud, P., Sutherland, E., Szymoniak, J., Way, F., Fites-Kaufman, J., Lewis, S., Mathews, E., Shovic, H., & Ryan, K. (2000). *Value and challenges of conducting rapid response research on wildland fires*. Gen Tech Rep RMRS-GTR-193. Fort Collins: USDA Forest Service Rocky Mountain Research Station.

Lentile, L. B., Morgan, P., Hudak, A. T., Bobbitt, M. J., Lewis, S. A., Smith, A. M., & Robichaud, P. R. (2007). Post-fire burn severity and vegetation response following eight large wildfires across the western United States. *Fire Ecology, 3*(1), 91–108.

Lewis, S. A., Hudak, A. T., Robichaud, P. R., Morgan, P., Satterberg, K. L., Strand, E. K., Smith, A. M., Zamudio, J. A., & Lentile, L. B. (2017). Indicators of burn severity at extended temporal scales: A decade of ecosystem response in mixed-conifer forests of western Montana. *International Journal of Wildland Fire, 26*(9), 755–771.

Liang, S., Hurteau, M. D., & Westerling, A. L. (2018). Large-scale restoration increases carbon stability under projected climate and wildfire regimes. *Frontiers in Ecology and the Environment, 16*(4), 207–212.

Loehman, R., Flatley, W., Holsinger, L., & Thode, A. (2018). Can land management buffer impacts of climate changes and altered fire regimes on ecosystems of the southwestern United States? *Forests, 9*(4), 192. https://doi.org/10.3390/f9040192.

van Mantgem, P. J., Stephenson, N. L., Byrne, J. C., Daniels, L. D., Franklin, J. F., Fulé, P. Z., Harmon, M. E., Larson, A. J., Smith, J. M., Taylor, A. H., & Veblen, T. T. (2009). Widespread increase of tree mortality rates in the western United States. *Sciences, 323*(5913), 521–524.

van Mantgem, P. J., Nesmith, J. C., Keifer, M., Knapp, E. E., Flint, A., & Flint, L. (2013). Climatic stress increases forest fire severity across the western United States. *Ecology Letters, 16*(9), 1151–1156.

Marlon, J. R., Bartlein, P. J., Carcaillet, C., Gavin, D. G., Harrison, S. P., Higuera, P. E., Joos, F., Power, M. J., & Prentice, I. C. (2008). Climate and human influences on global biomass burning over the past two millennia. *Nature Geoscience, 1*(10), 697. https://doi.org/10.1038/ngeo313.

Marlon, J. R., Bartlein, P. J., Gavin, D. G., Long, C. J., Anderson, R. S., Briles, C. E., Brown, K. J., Colombaroli, D., Hallett, D. J., Power, M. J., & Scharf, E. A. (2012). Long-term perspective on wildfires in the western USA. *PNAS, 109*(9), E535–E543.

McGranahan, D. A., Engle, D. M., Miller, J. R., & Debinski, D. M. (2013). An invasive grass increases live fuel proportion and reduces fire spread in a simulated grassland. *Ecosystems, 16*(1), 158–169.

McLaughlan, K. K., Higuera, P. E., Miesel, J., Rogers, B. M., Schweitzer, J., Shuman, J. K., Tepley, A. J., Varner, J. M., Veblen, T. T., Adalsteinsson, S. A., & Balch, J. K. (2020). Fire as a fundamental ecological process: Research advances and frontiers. *Journal of Ecology*. https://doi.org/10.1111/1365-2745.13403.

McWethy, D. B., Higuera, P. E., Whitlock, C., Veblen, T. T., Bowman, D. M. J. S., Cary, G. J., Haberle, S. G., Keane, R. E., Maxwell, B. D., McGlone, M. S., & Perry, G. L. W. (2013). A conceptual framework for predicting temperate ecosystem sensitivity to human impacts on fire regimes. *Global Ecology and Biogeography, 22*(8), 900–912.
Meddens, A. J. H., Kolden, C. A., & Lutz, J. A. (2016). Detecting unburned areas within wildfire perimeters using Landsat and ancillary data across the northwestern United States. *Remote Sensing of Environment, 186*, 275–285.
Merschel, A. G., Heyerdahl, E. K., Spies, T. A., & Loehman, R. A. (2018). Influence of landscape structure, topography, and forest type on spatial variation in historical fire regimes, Central Oregon, USA. *Landscape Ecology, 33*(7), 1195–1209.
Miller, R. F., & Tausch, R. J. (2001). The role of fire in pinyon and juniper woodlands: A descriptive analysis. In KEM Galley, TP Wilson (Eds.), *Proceedings of the Invasive Species Workshop: The Role of Fire in the Control and Spread of Invasive Species. Fire Conference 2000: The First National Congress on Fire Ecology, Prevention, and Management* (pp. 15–30). Misc Pub No. 11, Tall Timbers Research Station, Tallahassee.
Miller, J. D., & Thode, A. E. (2007). Quantifying burn severity in a heterogeneous landscape with a relative version of the delta Normalized Burn Ratio (dNBR). *Remote Sensing of Environment, 109*(1), 66–80.
Miller, R. F., Bates, J. D., Svejcar, T. J., Pierson, F. B., & Eddleman, L. E. (2005). *Biology, ecology, and management of western juniper*. Tech Bull 152, Corvallis: Oregon State University Agricultural Experiment Station.
Miller, R. F., Tausch, R. J., McArthur, E. D., Johnson, D. D., & Sanderson, S. C. (2008). *Age structure and expansion of piñon-juniper woodlands: A regional perspective in the Intermountain West*. Res Pap RMRS-RP-69. Fort Collins: USDA Forest Service Rocky Mountain Research Station.
Miller, J. D., Safford, H. D., Crimmins, M., & Thode, A. E. (2009). Quantitative evidence for increasing forest fire severity in the Sierra Nevada and Southern Cascade mountains, California and Nevada, USA. *Ecosystems, 12*, 16–32.
Morgan, P., Hardy, C. C., Swetnam, T. W., Rollins, M. G., & Long, D. G. (2001). Mapping fire regimes across time and space: Understanding coarse and fine-scale fire patterns. *International Journal of Wildland Fire, 10*, 349–342.
Morgan, P., Heyerdahl, E. K., & Gibson, C. E. (2008). Multi-season climate synchronized widespread forest fires throughout the 20th-century, Northern Rocky Mountains, USA. *Ecology, 89* (3), 717–728.
Morgan, P., Heyerdahl, E. K., Miller, C., & Wilson, A. M. (2014). Northern Rockies pyrogeography: An example of fire atlas utility. *Fire Ecology, 10*(1), 14–30. https://doi.org/10.4996/fireecology.1001014.
Morgan, P., Moy, M., Droske, C. A., Lewis, S. A., Lentile, L. B., Robichaud, P. R., Hudak, A. T., & Williams, C. J. (2015). Vegetation response to burn severity, native grass seeding, and salvage logging. *Fire Ecology, 11*(2), 31–58.
Morgan, P., Hudak, A. T., Wells, A., Baggett, L. S., Parks, S. A., Bright, B. C., & Green, P. (2017). Multidecadal trends in area burned at high severity in the Selway-Bitterroot Wilderness Area 1880-2012. *International Journal of Wildland Fire, 26*(11), 930–943.
Morgan, P., Heyerdahl, E. K., Strand, E. K., Bunting, S. C., Riser II, J. P., Abatzoglou, J. T., Nielsen-Pincus, M., & Johnson, M. (2020). Fire and land cover change in the Palouse Prairie–forest ecotone, Washington and Idaho, USA. *Fire Ecology, 16*(1), 1–17.
Moritz, M. A., Hessburg, P. F., & Povak, N. A. (2011). Native fire regimes and landscape resilience. In D. McKenzie, C. Miller, & D. A. Falk (Eds.), *The landscape ecology of fire* (pp. 51–86). Dordrecht: Springer.
Moritz, M. A., Topik, C., Allen, C. D., Hessburg, P. F., Morgan, P., Odion, D. C., Veblen, T. T., & McCullough, I. M. (2018). *A statement of common ground regarding the role of wildfire in forested landscapes of the western United States*. Fire Research Consensus Working Group

Final Report. Retrieved January 21, 2019, from https://live-ncea-ucsb-edu-v01.pantheonsite.io/sites/default/files/2020-02/WildfireCommonGround.pdf.

Murphy, B., Williamson, G. J., & Bowman, D. M. (2011). Fire regimes: Moving from a fuzzy concept to geographic entity. *The New Phytologist, 192*, 316–318.

Murray, M. P., Bunting, S. C., & Morgan, P. (1998). Fire history of an isolated subalpine mountain range of the Intermountain Region, United States. *Journal of Biogeography, 25*, 1071–1080.

Murray, M. P., Bunting, S. C., & Morgan, P. (2000). Landscape trends (1753-1993) of whitebark pine (*Pinus albicaulis*) forests in the West Big Hole Range of Idaho/Montana, USA. *Arctic, Antarctic, and Alpine Research, 32*(4), 412–418.

National Aeronautics and Space Administration (NASA). (2019). *Fire Information for Resource Management Systems (FIRMS)*. Retrieved June 12, 2019, from https://firms.modaps.eosdis.nasa.gov/map/#t:adv;d:2019-06-05..2019-06-12;l:viirs,modis_a,modis_t;@0.0,0.0,2z.

National Centers for Environmental Information (NCEI). (2020). *International Multiproxy Paleofire Database, Asheville*. Retrieved August 23, 2020, from https://www.ncdc.noaa.gov/data-access/paleoclimatology-data/datasets/fire-history.

National Centers for Environmental Information (NCEI). (n.d.) *Fire history*. Retrieved November 2019, from https://www.ncdc.noaa.gov/data-access/paleoclimatology-data/datasets/fire-history.

Noble, I. R., & Slatyer, R. O. (1980). The use of vital attributes to predict successional changes in plant communities subject to recurrent disturbances. *Vegetation, 43*(1–2), 5–21.

North, M. P., Stevens, J. T., Greene, D. F., Coppoletta, M., Knapp, E. E., Latimer, A. M., Restaino, C. M., Tompkins, R. E., Welch, K. R., York, R. A., & Young, D. J. (2019). Tamm Review: Reforestation for resilience in dry western US forests. *Forest Ecology and Management, 15* (432), 209–224.

Pacific Southwest Region U.S. Fish and Wildlife Service. (2006). *Centrocercus urophasianus*. Wikimedia Commons. Retrieved March 20, 2020, from https://commons.wikimedia.org/w/index.php?curid=12016910.

Parks, S. A., Dillon, G. K., & Miller, C. (2014a). A new metric for quantifying burn severity: The relativized burn ratio. *Remote Sensing, 6*(3), 1827–1844.

Parks, S. A., Miller, C., Nelson, C. R., & Holden, Z. A. (2014b). Previous fires moderate burn severity of subsequent wildland fires in two large western US wilderness areas. *Ecosystems, 17*, 29–42.

Parks, S. A., Holsinger, L. M., Miller, C., & Nelson, C. R. (2015a). Wildland fire as a self-regulating mechanism: The role of previous burns and weather in limiting fire progression. *Ecological Applications, 25*(6), 1478–1492.

Parks, S. A., Miller, C., Parisien, M.-A., Holsinger, L. M., Dobrowski, S. Z., & Abatzoglou, J. (2015b). Wildland fire deficit and surplus in the western United States, 1984–2012. *Ecosphere, 6*(12), 275. https://doi.org/10.1890/ES15-00294.1.

Parks, S. A., Miller, C., Holsinger, L. M., Baggett, L. S., & Bird, B. J. (2016). Wildland fire limits subsequent fire occurrence. *International Journal of Wildland Fire, 25*, 182–190. https://doi.org/10.1071/WF15107.

Parks, S. A., Holsinger, L. M., Panunto, M. H., Jolly, W. M., Dobrowski, S. Z., & Dillon, G. K. (2018). High-severity fire: Evaluating its key drivers and mapping its probability across western US forests. *Environmental Research Letters, 13*(4), 044037.

Parson, A., Robichaud, P. R., Lewis, S. A., Napper, C., & Clark, J. T. (2010). *Field guide for mapping post-fire soil burn severity*. Gen Tech Rep RMRS-GTR-243. Fort Collins: USDA Forest Service Rocky Mountain Research Station.

Pausas, J. G., & Paula, S. (2012). Fuel shapes the fire–climate relationship: Evidence from Mediterranean ecosystems. *Global Ecology and Biogeography, 21*(11), 1074–1082.

Pausas, J. G., Llovet, J., Rodrigo, A., & Vallejo, R. (2008). Are wildfires a disaster in the Mediterranean basin? – A review. *International Journal of Wildland Fire, 17*, 713–723. https://doi.org/10.1071/WF07151.

Picotte, J. J., & Robertson, K. (2011). Timing constraints on remote sensing of wildland fire burned area in the southeastern US. *Remote Sensing, 3*(8), 1680–1690.

Picotte, J. J., Peterson, B., Meier, G., & Howard, S. M. (2016). 1984–2010 trends in fire burn severity and area for the conterminous US. *International Journal of Wildland Fire, 25*(4), 413–420. https://doi.org/10.1071/WF15039.
Pierce, J. L., Meyer, G. A., & Jull, A. T. (2004). Fire-induced erosion and millennial-scale climate change in northern ponderosa pine forests. *Nature, 432*(7013), 87–90.
Prichard, S. J., Stevens-Rumann, C. S., & Hessburg, P. F. (2017). Tamm review: Shifting global fire regimes: Lessons from reburns and research needs. *Forest Ecology and Management, 396*, 217–233.
Rego, F. C., Bunting, S. C., Strand, E. K., & Godinho-Ferreira, P. (2019). *Applied landscape ecology*. Hoboken: Wiley.
Reinhardt, E. D., Keane, R. E., Calkin, D. E., & Cohen, J. D. (2008). Objectives and considerations for wildland fuel treatment in forested ecosystems of the interior western United States. *Forest Ecology and Management 256*(12), 1997–2006.
Robichaud, P. R., Wagenbrenner, J. W., Lewis, S. A., Ashmun, L. E., Brown, R. E., & Wohlgemuth, P. M. (2013). Post-fire mulching for runoff and erosion mitigation; Part II: Effectiveness in reducing runoff and sediment yields from small catchments. *Catena, 105*, 93–111.
Rollins, M. G., Swetnam, T. W., & Morgan, P. (2001). Evaluating a century of fire patterns in two Rocky Mountain Wilderness areas using digital fire atlases. *Canadian Journal of Forest Research, 31*(12), 2107–2123.
Rollins, M. G., Morgan, P., & Swetnam, T. (2002). Landscape scale controls over 20th century fire occurrence in two large Rocky Mountain (USA) wilderness areas. *Landscape Ecology, 17*, 539–557.
Romme, W. H., Boyce, M. S., Gresswell, R., Merrill, E. H., Minshall, G. W., Whitlock, C., & Turner, M. G. (2011). Twenty years after the 1988 Yellowstone fires: Lessons about disturbance and ecosystems. *Ecosystems, 14*(7), 1196–1215.
Rossiter-Rachor, N. A., Setterfield, S. A., Douglas, M. M., Hutley, L. B., & Cook, G. D. (2008). Andropogon gayanus (gamba grass) invasion increases fire-mediated nitrogen losses in the tropical savannas of northern Australia. *Ecosystems, 11*(1), 77–88.
Roy, D. P., Boschetti, L., Justice, C. O., & Ju, J. (2008). The collection 5 MODIS burned area product—Global evaluation by comparison with the MODIS active fire product. *Remote Sensing of Environment, 112*(9), 3690–3707.
Roy, D. P., Boschetti, L., & Giglio, L. (2010). Remote sensing of global savanna fire occurrence, extent and properties. In M. J. Hill & N. P. Hanan (Eds.), *Ecosystem function in global savannas: Measurement and modeling at landscape to global scales* (pp. 239–255). Boca Raton: CRC Press, Taylor and Francis.
Roy, D. P., Boschetti, L., & Smith, A. M. (2013). Satellite remote sensing of fires. In C. M. Belcher (Ed.), *Fire phenomena and the Earth system: An interdisciplinary guide to fire science* (pp. 77–90). London: Wiley.
Sage Grouse Initiative. (2017). *Grazing management in perspective: A compatible tool for Sage Grouse Conservation*. Retrieved July 17, 2019, from https://www.sagegrouseinitiative.com/grazing-management-perspective-compatible-tool-sage-grouse-conservation/.
Sankey, T. T., & Germino, M. J. (2008). Assessment of juniper encroachment with the use of satellite imagery and geospatial data. *Rangeland Ecology & Management, 61*, 412–418.
Scheller, R. M., Domingo, J. B., Sturtevant, B. R., Williams, J. S., Rudy, A., Gustafson, E. J., & Mladenoff, D. (2007). Design, development, and application of LANDIS-II, a spatial landscape simulation model with flexible temporal and spatial resolution. *Ecological Modelling, 201*, 409–419.
Schoennagel, T., Balch, J. K., Brenkert-Smith, H., Dennison, P. E., Harvey, B. J., Krawchuk, M. A., Mietkiewicz, N., Morgan, P., Moritz, M. A., Rasker, R., & Turner, M. G. (2017). Adapt to more wildfire in western North American forests as climate changes. *PNAS, 114*(18), 4582–4590.

Sequeira, C. R., Montiel-Molina, C., & Rego, F. C. (2019). Landscape-based fire scenarios and fire types in the Ayllón massif (Central Mountain Range, Spain), 19th and 20th centuries. *Cuadernos de Investigación Geográfica,* 2020(46). https://doi.org/10.18172/cig.3796.

Shlisky, A., Waugh, J., Gonzalez, P., González, M., Manta, M., Santoso, H., Rodrıguez-Trejo, D., Swaty, R., Schmidt, D., Kaufmann, M., Myers, R., Alencar, A., Kearns, F., Johnson, D., Smith, J. L., Zollner, D., Fulks, W., & Nuruddin, A. (2007). *Fire, ecosystems and people: Threats and strategies for global biodiversity conservation.* Global Fire Initiative Tech Rep 2. Arlington: The Nature Conservancy. Retrieved May 30, 2019, from http://www.tncfire.org/documents/fire_ecosystems_and_people.pdf.

Short, K. C. (2017). *Spatial wildfire occurrence data for the United States, 1992-2015* [FPA_FOD_20170508] (4th edn). Fort Collins: USDA Forest Service Res Data Archive. https://doi.org/10.2737/RDS-2013-0009.4. Retrieved June 2, 2019.

Society for Ecological Restoration International Science & Policy Working Group (SER). (2004). *Primer on ecological restoration*. Retrieved July 19, 2019, from https://www.ctahr.hawaii.edu/littonc/PDFs/682_SERPrimer.pdf.

Sparks, A. M., Smith, A. M., Talhelm, A. F., Kolden, C. A., Yedinak, K. M., & Johnson, D. M. (2017). Impacts of fire radiative flux on mature Pinus ponderosa growth and vulnerability to secondary mortality agents. *International Journal of Wildland Fire, 26*(1), 95–106.

Stephens, S. L., Collins, B. M., Fettig, C. J., Finney, M. A., Hoffman, C. M., Knapp, E. E., North, M. P., Safford, H., & Wayman, R. B. (2018). Drought, tree mortality, and wildfire in forests adapted to frequent fire. *BioScience, 68*(2), 77–88.

Stephens, S. L., Battaglia, M. A., Churchill, D. J., Collins, B. M., Coppoletta, M., Hoffman, C. M., Lydersen, J. M., North, M. P., Parsons, R. A., Ritter, S. M., & Stevens, J. T. (2020). Forest restoration and fuels reduction: Convergent or divergent? *BioSciences, 71*(1), 85–101.

Stevens, J. T., Safford, H. D., North, M. P., Fried, J. S., Gray, A. N., Brown, P. M., Dolanc, C. R., Dobrowski, S. Z., Falk, D. A., Farris, C. A., & Franklin, J. F. (2016). Average stand age from forest inventory plots does not describe historical fire regimes in ponderosa pine and mixed-conifer forests of western North America. *PLoS One, 11*(5), e0147688.

Stevens-Rumann, C., & Morgan, P. (2016). Repeated wildfires alter forest recovery of mixed-conifer ecosystems. *Ecological Applications, 26*(6), 1842–1853. https://doi.org/10.1890/15-1521.1.

Stevens-Rumann, C. S., & Morgan, P. (2019). Tree regeneration following wildfires in the western US: A review. *Fire Ecology, 15*(1), 15.

Stevens-Rumann, C., Morgan, P., & Hoffman, C. (2015). Bark beetles and wildfires: How does forest recovery change with repeated disturbance in mixed conifer forests? *Ecosphere, 6*(6), 1–7. https://doi.org/10.1890/ES14-00443.1.

Stevens-Rumann, C., Prichard, S. J., Strand, E. K., & Morgan, P. (2016). Prior wildfires influence burn severity of subsequent fires. *Canadian Journal of Forest Research, 46*(11), 1375–1385.

Stevens-Rumann, C. S., Kemp, K. B., Higuera, P. E., Harvey, B. J., Rother, M. T., Donato, D. C., Morgan, P., & Veblen, T. T. (2018). Evidence for declining forest resilience to wildfires under climate change. *Ecology Letters, 21*(2), 243–252.

Stevens-Rumann, C., Morgan, P., Davis, K., Kemp, K., & Blades, J. (2019). *Post-fire tree regeneration (or lack thereof) can change ecosystems. Science Review No 5.* Kalispell: Northern Rockies Fire Science Network. Retrieved March 2020, from https://www.nrfirescience.org/sites/default/files/TreeRegenerationReviewFinal_compressed_0.pdf.

Swetnam, T. W., & Betancourt, J. L. (1990). Fire-southern oscillation relations in the southwestern United States. *Sciences, 249*(4972), 1017–1020.

Swetnam, T. W., Allen, C. D., & Betancourt, J. L. (1999). Applied historical ecology: Using the past to manage for the future. *Ecological Applications, 9*(4), 1189–1206.

Swetnam, T. W., Farella, J., Roos, C. I., Liebmann, M. J., Falk, D. A., & s. (2016). Multiscale perspectives of fire, climate and humans in western North America and the Jemez Mountains, USA. *Philosophical Transactions of Royal Society of London B: Biological Science, 371*(1696), 20150168.

Tepley, A. J., Swanson, F. J., & Spies, T. A. (2013). Fire-mediated pathways of stand development in Douglas-fir/western hemlock forests of the Pacific Northwest, USA. *Ecology, 94*(8), 1729–1743.

Teske, C. C., Seielstad, C. A., & Queen, L. P. (2012). Characterizing fire-on-fire interactions in three large wilderness areas. *Fire Ecology, 8*, 82–106.

Thompson, M. P., Bowden, P., Brough, A., Scott, J. H., Gilbertson-Day, J., Taylor, A., Anderson, J., & Haas, J. R. (2016). Application of wildfire risk assessment results to wildfire response planning in the southern Sierra Nevada, California, USA. *Forests, 7*(3), 64.

Turner, M. G., Gardner, R. H., Dale, V. H., & O'Neill, R. V. (1989). Predicting the spread of disturbance across heterogeneous landscapes. *Oikos, 55*, 121–129.

Turner, M. G., Romme, W. H., Gardner, R. H., O'Neill, R. V., & Kratz, T. K. (1993). A revised concept of landscape equilibrium: Disturbance and stability on scaled landscapes. *Landscape Ecology, 8*, 213–227.

Turner, M. G., Romme, W. H., Gardner, R. H., & Hargrove, W. W. (1997). Effects of fire size and pattern on early succession in Yellowstone National Park. *Ecological Monographs, 67*(4), 411–433.

USDA Forest Service. (n.d.-a) *FlamMap*. Retrieved December 18, 2019, from https://www.firelab.org/project/flammap.

USDA Forest Service. (n.d.-b) *Forest Inventory and analysis*. Retrieved March 25, 2019, from https://www.fia.fs.fed.us.

Walker, R. B., Coop, J. D., Parks, S. A., & Trader, L. (2018). Fire regimes approaching historic norms reduce wildfire-facilitated conversion from forest to non-forest. *Ecosphere, 9*(4), e02182.

Washington Department of Natural Resources. (2020) 20-year forest health strategic plan Eastern Washington. Retrieved November 20, 2020, https://www.dnr.wa.gov/publications/rp_forest_health_20_year_strategic_plan.pdf

Westerling, A. L., Hidalgo, H. G., Cayan, D. R., & Swetnam, T. W. (2006). Warming and earlier spring increase western US forest wildfire activity. *Science, 313*(5789), 940–943.

Whitlock, C., Marlon, J., Briles, C., Brunelle, A., Long, C., & Bartlein, P. (2008). Long-term relations among fire, fuel, and climate in the north-western US based on lake-sediment studies. *International Journal of Wildland Fire, 17*(1), 72–83.

Whitlock, C., Higuera, P. E., McWethy, D. B., & Briles, C. E. (2010). Paleoecological perspectives on fire ecology: Revisiting the fire-regime concept. *Open Ecology Journal, 5*(3), 1.

With, K. A. (1997). The Application of neutral landscape models in conservation biology. *Conservation Biology, 11*(5), 1069–1080.

With, K. A., Gardner, R. H., & Turner, M. G. (1997). Landscape connectivity and population distributions in heterogeneous environments. *Oikos, 1*, 151–169.

Chapter 13
Integrated Fire Management

Learning Outcomes
At the conclusion of this chapter, we expect you will be able to

1. Identify the characteristics of successful integrated fire management,
2. Discuss the challenges for effective integrated fire management for a particular landscape of interest to you,
3. Explain what you could learn from the individual case studies that would be particularly relevant for a particular landscape of interest to you, and
4. Explain how integrated fire management includes and goes well beyond fire suppression.

13.1 What Is Integrated Fire Management and Why Do We Need It?

Fires so greatly influence many landscapes that effectively managing fires is integral to successful natural resource management. Attaining the objectives of fire management is conditional on a number of factors, including the resources available, but starts by adequately understanding fire-related processes and our impact on them (Martell 2001). A diverse array of activities falls under the fire management umbrella, but achieving control (be it of fuels, ignitions, or fire spread) is the common denominator.

Integrated Fire Management engages with all fire-related information and activities consistent with and furthering vegetation and land management goals. Rather than focusing on minimizing the area burned, integrated fire management seeks to maximize the net benefits of fires, including both using fires and suppressing fires effectively to further the strategic goals. Many have called for integrated fire

F. Castro Rego et al., *Fire Science*, Springer Textbooks in Earth Sciences, Geography and Environment, https://doi.org/10.1007/978-3-030-69815-7_13

management (Table 13.1). All over the world, scientists, managers, and policymakers have called for fire management that is more holistic than what has often been a primary focus on fire control, suppression, and exclusion (Table 13.1). Integrated Fire Management evolved from but is significantly different than an approach based on fire suppression alone (Table 13.1). The latter comes from a command and control perspective. Integrated fire management can only be successful if we think of fire as more than something to be suppressed immediately. Instead, we must plan for the long-term effects of fire as part of sustainable management.

Integrated Fire Management has a long history, and yet it is relatively new. Many Indigenous cultures practiced informal Integrated Fire Management. They had few other tools for manipulating vegetation for utilitarian purposes, especially across large areas, and they found many cultural and spiritual uses of fire (Huffman 2013; Bowman et al. 2009, 2011). Today's fire managers often apply lessons learned from Indigenous and other people who have lived for generations in places where they found ways for living with and using fire. Fire suppression was the focus of early professional foresters, mostly influenced by forestry education grounded in central Europe. Western science sought solutions to the "fire problem" by understanding fire behavior and fire danger and means for preventing, detecting, and suppressing fires. The science of fire ecology has become increasingly popular since the middle 1900s (Komarek 1976), and now informs fire science and management broadly. The ecological role of fire is increasingly well understood and valued, though it can be difficult to balance protecting people and property from fires with the ecological imperative for ecosystems to burn. As a result, the use of fire and multiple fire management strategies are both parts of integrated fire management today.

After this brief introduction, we have eight success stories of integrated fire management from all over the world. Together they illustrate how integrated fire management is adapted and applied through the knowledge of a place and people even while policies adapt too. In the last section, we discuss the implications and recommendations.

13.2 Global Success Stories

Here we highlight global success stories. Each is written by local experts who have been centrally involved with these successful implementations of integrated fire management. In each, the authors address the history and context and then describe how programs are implemented and challenges addressed. All of the global success stories include extensive use of prescribed fires. All have overcome and still face challenges, but each has lessons to be learned that could be applied elsewhere. We close the chapter with broader insights drawn from these success stories and published literature.

Table 13.1 Integrated fire management contrasts with suppression-centered fire management. Both are designed to protect people and property

	Suppression-centered fire management	Integrated fire management
Objectives and overall approach	Minimize area burned. Large fires equate to higher losses	Minimize fire-induced damage and the difference between the negative and positive impacts of fire. Larger fires do not necessarily equate to higher impacts. Holistic
Society & policy	Wildfire perceived solely as an emergency to be addressed by civil protection. Separation between fire and forest management. Unbalanced allocation of resources to fire suppression	Living with fire. Integration of fire and forest management. More equitable allocation of resources between suppression and fire mitigation/forest and land management
Fire suppression	Rigid, full force, regardless of resources at risk, burn conditions, and costs. Focus on civil protection decreases effectiveness	Flexible. Variable in effort and timing. Deliberate planned response, weighing consequences, and including monitoring and limited or no-suppression options. Rationalized costs. Increased effectiveness
Environmental/ecological issues	Fire is solely a damaging disturbance.	Ecosystems require fire regimes that are consistent with current and future ecological and social goals
Socioeconomic issues	Traditional burning subject to social coercion. No consideration for Traditional Ecological Knowledge	Fire provides ecosystem services. Traditional Ecological Knowledge is considered. Involvement with local communities and reinstatement or regulation of their burning practices
Fuels management	Absent or restricted to an isolation strategy (fire- and fuel-breaks)	Extended fuel reduction/modification programs, including preventive silviculture and area-wide/mosaic treatments, often through prescribed burning
Planning and decision support	One-size fits all	Guided by forests and land management goals and policies. Hierarchical, from the global objectives of resources management to the specific aims of fire management, to the formulation of strategies, tactics, and actions. Fire management zoning. Accounting for cross-sectional dynamics and as a social-ecological problem. Consistent and compatible across agencies, clear and comprehensive, and spatially and temporally scalable. Risk-based
Monitoring	Cost, area burned	Cost, area burned, effectiveness in meeting strategic objectives emphasizing outcomes, area meeting desired natural resource management objectives

(continued)

Table 13.1 (continued)

	Suppression-centered fire management	Integrated fire management
Governance	Command and control	Integrative, comprising cooperative planning and deliberative processes and facilitated by inclusive, participatory, and reflexive practices and mechanisms, enabling adaptive strategies. Regular policy assessment

Adapted from Fernandes (2020) and Fernandes et al. (2020), which was synthesized from these sources: Birot (2009), Calkin et al. (2011, 2015), Chandler et al. (1983), Egging and Barney (1979), Fernandes et al. (2013), Fischer (1980), IUFRO (2018), Lotan (1979), Meyer et al. (2015), Minor and Boyce (2018), Moritz et al. (2014), Myers (2006), North et al. (2015), O'Laughlin (2005), Pacheco et al. (2015), Ruane (2018), Russell-Smith et al. (2013), Silva et al. (2010), Steelman (2016), TNC (2017), Twidwell et al. (2019)

13.2.1 Prescribed Fires Alter Wildfires

In southwestern Australia (Case Study 13.1), Neil Burrows describes how prescribed burning has altered the pattern of extensive wildfires. When 8% of the landscape is burned each year, wildfires burn less total area and fire fighters find that suppression is eased where prior fires have consumed fuels.

Case Study 13.1 Managing with Fire in Forests of Southwestern Australia

Neil D. Burrows, email: neil.burrows@dbca.wa.gov.au
Department of Biodiversity, Conservation and Attractions, Perth, Western Australia, Australia

Introduction

Fires have shaped the biodiversity of southwestern Australian eucalypt forests for thousands of years (Hassell and Dodson 2003). For millennia, fires were important for the physical and spiritual well-being of Noongar Aboriginal people, who used fire skillfully and frequently for a myriad of reasons (Hallam 1975), creating a quasi-stable fire-induced mosaic of vegetation at different growth stages. As a result, large, damaging wildfires were rare (Burrows et al. 1995). European settlement some 200 hundred years ago saw the demise of traditional Aboriginal burning. Initially, the fire-phobic European-trained foresters adopted a 'fire suppression' policy, believing all fires to be harmful to the forest. However, this failed to prevent large, damaging bushfires and in the 1950s, the policy was changed to include prescribed burning to reduce the fuel hazard. Initially, the impetus for regular prescribed burning across broad areas arose from the need to reduce the impact of large, high-intensity fires on human life, property, and forest values. However,

(continued)

Case Study 13.1 (continued)

because these ecosystems are fire-maintained, prescribed fire is increasingly used to meet multiple land use objectives including the protection of conservation, cultural and environmental values such as wildlife, water and soil, and for maintaining healthy ecosystems (Burrows 2008; McCaw 2012).

The Fire Environment

Southwestern Australia has a Mediterranean-type climate with cool, moist winters and warm, dry summers. Mean annual rainfall in the jarrah forest varies from about 600 mm in the eastern part of its range to about 1200 mm in western and southern parts of its range, while the mean annual rainfall in karri forest is 1200–1300 mm. Typical of Mediterranean-type climates, rainfall is strongly seasonal with about 80% of annual rainfall falling during the six consecutive wettest months (May–October). Consequently, fuels are dry enough to burn for 6–8 months of the year. Maximum summer temperatures regularly exceed 35 °C while winter maxima are usually 15–20 °C. Since the 1970s, there has been a steady decrease in rainfall across most of the region, prolonging the fire season by up to several months in some years (McCaw 2012). Weather factors influencing the fire environment include coastal sea breezes, strong easterly winds from the warm, dry interior, abrupt wind, temperature, and relative humidity changes associated with prefrontal troughs and occasional incursions of tropical cyclones. Both human (deliberate and accidental) and lightning-caused fires are common during the dry summer months.

Remnant native eucalypt forests and associated ecosystems, including shrublands and wetlands, extend over an area of about 2.5 million ha of predominantly public land managed for multiple purposes including conservation, timber production, water catchment protection, and recreation. Forest ecosystems occur primarily on undulating land surfaces and nutrient-poor soils derived from Precambrian granite and gneiss substrates that have undergone prolonged leaching, erosion, and deposition. Dry sclerophyll forests are dominated by jarrah (*Eucalyptus marginata*) and marri (*Corymbia calophylla*) averaging around 20–30 m in height. The structure of the species-rich understorey vegetation varies according to soil and climate butthe veg is mostly <2 m high with a canopy cover of 35–70%. Wet sclerophyll (karri) forests are dominated by karri (*E. diversicolor*) and can reach heights of up to 85 m, with dense understoreys up to 10 m high (Fig. 13.1). The dominant fine fuels in all these forests are dead leaves, twigs, and bark that accumulate on the forest floor (surface fuel), suspended dead material (near-surface fuel), shrubs (elevated fuel), and bark on standing trees. In long unburnt forests, total fuel loads can reach 15–20 and 50–60 tonnes ha^{-1} in jarrah and karri forests respectively (McCaw et al. 2002; Gould et al. 2011).

Prescribed Burning for Fuels Management

(continued)

Fig. 13.1 (**a**) Dry sclerophyll jarrah (*Eucalyptus marginata*) forest. (**b**) Aerial view of prescribed burning in a mosaic of jarrah forest and shrubland. (**c**) Wet sclerophyll karri (*E. diversicolor*) forest. To the left of the road is 15 years since prescribed fire, and to the right of the road is 3 years since prescribed fire. (Photographs by Neil Burrows)

Case Study 13.1 (continued)

The combination of hot, dry, windy weather and accumulations of flammable fuels can give rise to large, intense bushfires that threaten communities, critical infrastructure, conservation values, and other assets. Forest fires generate intensity from the amount of fuel that burns and the rate at which it burns, which is primarily a function of fuel structure, fuel moisture content, wind, and other weather conditions, and topography. Science, history and fire fighter experience have shown that managing the build-up of flammable fuels is the most effective way to reduce the potential fire intensity, thereby reducing intensity, size, damage potential and suppression difficulty (Fernandes and Botelho 2003; McCaw et al. 2008; Boer et al. 2009; McCaw 2012). Prescribed burning is a cost-effective means of achieving this in these fire-adapted forests without causing long-term harm (Wittkuhn et al. 2011; Burrows et al. 2019). Landscape-scale prescribed fire has been used extensively to manage fuels in southwestern Australian forests since the mid-1950s, although the extent of burning was limited prior to the destructive bushfires of 1961 (Fig. 13.2). Reducing fuel load and altering fuel structure can mitigate key aspects of

(continued)

Case Study 13.1 (continued)
wildfire behavior including the rate of spread, flame dimensions, spotting, and fireline intensity (McCaw et al. 2008). The contribution of prescribed burning to mitigating the effects of wildfires has been quantified in various ways including using basic fire behavior science, well-documented case studies, and analysis of historical fire statistics (Underwood et al. 1985; Fernandes and Botelho 2003; Boer et al. 2009).

Fuel reduction improves the safety, efficiency, and effectiveness of fire suppression, especially when fuels are less than about 6 years old (McCaw 2012). There is a strong inverse relationship between the areal extent of prescribed burning and of wildfire (Sneeuwjagt 2008; Boer et al. 2009). Landscape-scale prescribed burning of southwestern forests since the 1960s has reduced the area burnt by wildfire, hence, the extent of wildfire loss and damage. The inverse relationship between the extent of prescribed burning and wildfire is evident in the historical data for the entire 2.5 million ha of the southwest forest region over six decades (Figs. 13.2 and 13.3). Reasons for the gradual reduction in prescribed burning since the mid-1990s and the associated increase in the area burnt by wildfire (Fig. 13.2) include reduced windows of opportunity for prescribed burning due to changing climate, changing land uses, population growth, industrial legacies (fire-sensitive mining rehabilitation and regrowth forests), a reduced capacity to conduct burns, and smoke and air quality issues. In order to maintain the long-term average annual area burned by wildfire to <1% of the forest region, about 8% of the region needs to

(continued)

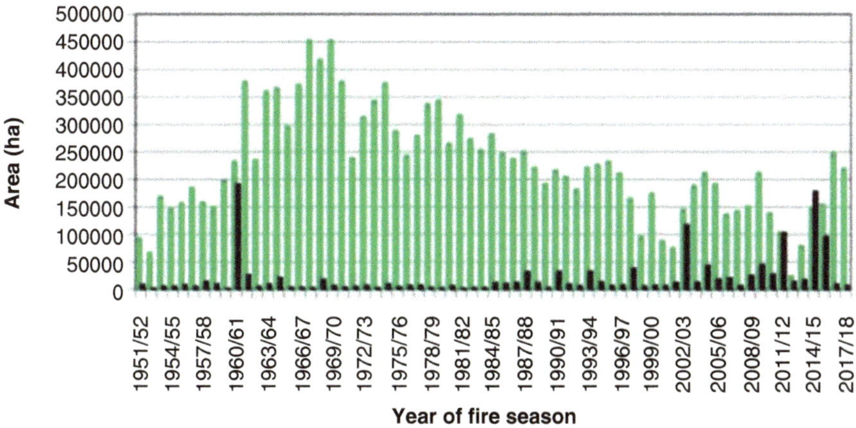

Fig. 13.2 Annual area of the 2.5 million ha of southwest Australia forests burned by prescribed fire (black bars) and wildfire (green bars) from 1951/1952 to 2017/2018. (Data from annual reports by the Western Australian Department of Biodiversity, Conservation and Attractions, and its predecessors)

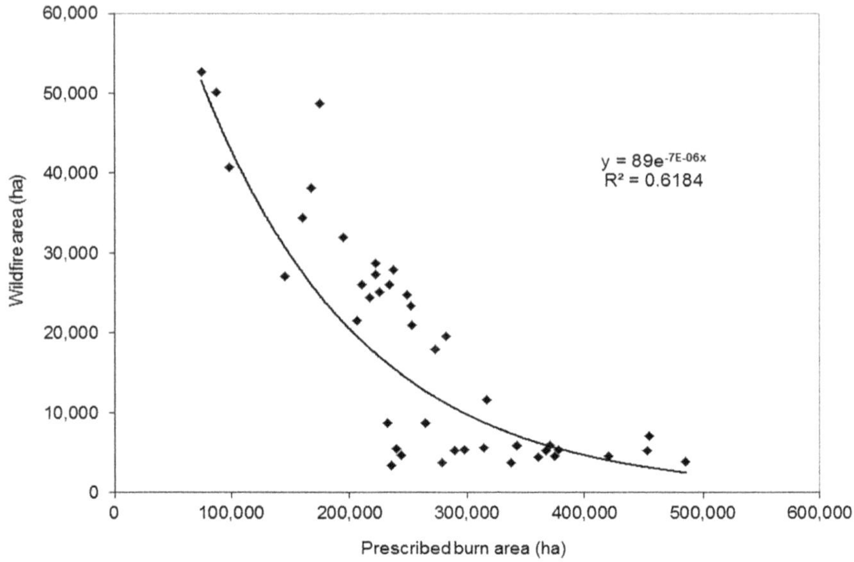

Fig. 13.3 Area of southwest forest area burned by prescribed fire (mean of 4 years) with the area burnt by wildfire (mean of 4 years). (Adapted from Sneeuwjagt 2008)

Case Study 13.1 (continued)

be prescribed burnt each year, equating to about 200,000 ha (Fig. 13.3). Based on historical data, this level of landscape burning reduced the number, extent, and frequency-size distribution of wildfires. Furthermore, the length of time that sites remained unburnt by wildfire doubled to about 9 years (Boer et al. 2009). Fuel reduction had a detectable effect on the incidence and extent of wildfires for up to 6 years after prescribed burning, consistent with the scientific knowledge of fuel dynamics and field observations of the important contribution of fuel-reduced areas to wildfire suppression (Gould et al. 2011).

For the curvilinear inverse relationship between area burnt by prescribed fire and by wildfire (Fig. 13.3), the x and y coordinates for the inflection point are ~200,000 ha (prescribed burn area) and ~20,000 ha (wildfire area). For this reason, and based on experience, the current fire management policy has a prescribed burning target of 200,000 ha per year representing about 8% of the public forest estate. Based on the historical data (Figs. 13.2 and 13.3), this will result in about 0.8% of the forest estate being burnt by wildfire per annum, and about 40% of the forest region carrying fuels ≤4 years old, providing opportunities for safer and effective fire suppression. Having significant areas of low fuel in which fires will either go out or exhibit mild behavior allows fire

(continued)

Case Study 13.1 (continued)
fighters greater flexibility in prioritizing the allocation of fire fighting resources, especially in the event of multiple fire outbreaks.

Six decades of widespread prescribed burning has significantly reduced the extent and severity of wildfires, saving human lives and substantially reducing damage to infrastructure, economic losses, and disruption and trauma to communities who live in and around the forests. Although not well documented, it is evident that prescribed burning has also reduced the harmful effects of large and intense wildfires on other values including wildlife, water catchments, and soils.

It has been claimed that, based on computer simulations, it is only necessary to reduce fuel hazard in the immediate vicinity, 100–500 m or so, of the peri-urban interface and around assets, and that burning beyond this is ineffective (e.g., Furlaud et al. 2018). Aside from the limitations of computer simulations to adequately represent the complexities and nuances of landscape-scale fire behavior, prescribed burning, and fire suppression, neglecting to treat fuels in the broader landscape will result in a cycle of large and damaging bushfires. Prior to adopting a landscape-scale prescribed burning policy in the mid-1950s, southwestern forest fire managers relied on a system of five chain-wide (~100 m) fuel reduced buffers, or 'green belts', around communities and other assets including forest regeneration, and an assumed ability to suppress wildfires. This system failed under severe fire weather conditions because the heavy fuel build-up in the surrounding forest resulted in very high-intensity wildfires which could not be suppressed at their peak, and which, via spotting and ember storms, breached the narrow fuel buffers and then impacted settlements. These fires also caused extensive damage to timber, water catchments, and other forest values. Decades of experience and knowledge of severe fire behavior dictates that unless the buffers are several kilometers wide, a large well-developed fire burning in heavy, long unburnt forest fuels under severe fire weather conditions will generate fireballs of hot gas and dense smoke, and also a blizzard of embers that will breach narrow buffers, besieging urban areas. If the bushfire approaching the interface buffer is moderate to low intensity as a result of low fuel loads and/or low fire danger rating, a narrower buffer may be effective.

Another shortcoming of only treating fuels at the interface is that it places everything outside the treated buffer at risk to wildfire damage. This includes the critical infrastructure of state-level significance such as major transport corridors, infrastructure associated with power and energy generation and distribution, water supply catchments, pipelines and pumping stations, major cables and towers, and major wastewater treatment sites. Wildfires in the broader landscape may have a significant impact on the livelihood of

(continued)

Case Study 13.1 (continued)
individuals or community economic sustainability, such as infrastructure of local or regional significance, agricultural land, major industries such as mines, refineries, manufacturing plants, and native and plantation timber industries. Such fires also threaten other significant built, natural or cultural assets, such as areas of transient population density and low resilience to bushfire, including holiday homes, hobby farms and recreation and camping sites, fire-vulnerable Aboriginal or European heritage sites, significant ecological communities or species habitats, and natural areas with specific fire regime requirements. Also, fire fighters will be expected to fight fires beyond the interface. Fires burning in long unburnt, heavy forest fuels will be dangerous and difficult to control, even under moderate fire weather conditions, and impossible to control under more severe weather conditions. For narrow buffers around communities to effectively stop a high-intensity forest fire burning under severe weather conditions (although too narrow to capture embers), they would need to be burnt, or otherwise treated, every 2–3 years. Given that there are thousands of kilometers of the convoluted urban interface areas, much of which is private property, it is not feasible to install and maintain a system of 100–500 m fuel reduced buffers to a standard that they will stop a running fire. To protect communities and other values, it is necessary to both reduce flammable fuels from the interface out as far as is practical and to manage fuels in the broader landscape.

A risk-based framework is being used to manage bushfire risk through fuel management on public lands in Western Australia, including the southwestern forest region. This establishes risk criteria, including indicators of acceptable bushfire risk. The indicators establish targets for fuel management that are applicable statewide but can be customized to meet local circumstances. The framework provides the context for fuel management in Western Australia, including descriptions of the fire (climate, weather, fuels) and social (land use, cultural considerations) environments that contribute to risk. Areas of relatively homogenous risk context are described as Bushfire Risk Management Zones, of which eight are identified. Each Bushfire Risk Management Zone is further divided into Fire Management Areas, based on the management intent. These are areas where fuels will be managed primarily to (1) buffer settlements, (2) buffer critical infrastructure, (3) manage fuels in the landscape to prevent or buffer large bushfires, or (4) for other land management outcomes. The location, extent, and indicators of acceptable bushfire risk of each Fire Management Area are modified according to the nature and distribution of assets and potential fire behavior in the landscape. The risk criteria established in the framework are converted to spatially-represented targets for fuel management in each Bushfire Risk Management Zone. These underpin fuel management planning on public lands.

(continued)

Case Study 13.1 (continued)

Prescribed Burning for Biodiversity Conservation

While the use of prescribed burning to mitigate wildfire risk focuses on managing the accumulation and structure of fuel, long-term studies show that prescribed burning does not pose a risk to biodiversity (e.g., Wittkuhn et al. 2011; Burrows et al. 2019). However, there are situations where the application of fire for specific biodiversity conservation outcomes focuses on managing components of the fire regime considered important for the maintenance of habitats for selected species. These components include the interval between fires, fire seasonality, intensity, scale, and patchiness of burning (Burrows 2008). Fire management for biodiversity outcomes is guided by biodiversity conservation objectives operating at a range of spatial and temporal scales. These objectives have a foundation in ecological theory and are based on knowledge gained through experiments, retrospective studies, and monitoring. Responses to fire have been documented for many species of flora and fauna, including threatened taxa, in forests and associated ecosystems (Wittkuhn et al. 2011; Pekin et al. 2012; Burrows et al. 2019).

The primary objectives of fire management for conserving biodiversity at the landscape scale are (1) to maintain a diverse representation of ecosystem seral stages and habitat conditions and, (2) to protect fire-sensitive and fire-independent ecosystems and niches, including riparian zones, aquatic ecosystems, rock outcrops and peat wetlands (Burrows 2008). Strategies to achieve these objectives include maintaining a mosaic of fire management units within the landscape at different times since fire, including recently burnt and long unburnt, and units burnt in different seasons. Ideally, the mosaic will include three biologically important fire regime components: (1) time since last fire, (2) fire frequency, and (3) fire season. In the southwestern forest region, fire-sensitive and fire-independent ecosystems embedded in the forest matrix are usually less flammable than the surrounding ecosystems because they remain damp for extended periods or because fuels are naturally discontinuous (Burrows 2008; Shedley et al. 2018). Low intensity prescribed fires in spring or late autumn will burn the more flammable ecosystems in the landscape but not the less flammable ones. Making the landscape less vulnerable to large, intense wildfires reduces the risk of fire-sensitive ecosystems being damaged by severe wildfires. For example, large granite outcrops that occur throughout the forests provide habitat for both fire-sensitive species, such as obligate seeding plants with long juvenile periods (e.g., *Calothamnus* and *Acacia* spp.), and fire-independent species (e.g., *Borya* spp. and cryptogams). These outcrop ecosystems require relatively long intervals between fires and can be damaged by intense wildfires that can develop in the surrounding forests if forest fuels have not been regularly prescribed burnt (Burrows 2013). Because vegetation (fuel) on rock outcrops is discontinuous, the surrounding forests,

(continued)

Case Study 13.1 (continued)
which contain continuous fuels, can be burnt under mild weather conditions in spring or autumn without burning the rock outcrop habitats.

For almost 60 years, prescribed burning has been used in southwestern Australia forests to achieve multiple objectives, supported by applied research into fire behavior and fire ecology, and resulting in a significant reduction in the areal extent of bushfires and bushfire losses without loss of biodiversity or environmental damage. This has been achieved by understanding the role of fire in these environments, by understanding fire behavior and by understanding the importance of continuing with wise anthropogenic planned burning, which commenced with the arrival of Noongar people thousands of years ago. In recent times, prescribed burning has been achieved by good planning, a skilled and adequately resourced workforce, political and community support, and ongoing investment in applied fire research. Climate variability, population growth, air quality concerns, and land-use and use legacies and land use changes present ongoing challenges for forest fire managers to maintain an effective prescribed burning program that protects communities and the environment from damaging wildfires into the future.

13.2.2 Conserving Biodiversity Using Integrated Fire Management

Fire is widely used to conserve biological diversity. In the eastern Pyrenees (Case Study 13.2), Eric Rigolot and Bernard Lambert explain how prescribed fires have been instrumental in managing vegetation and fuels with broad implications for biodiversity and social values. In Kruger National Park in South Africa, fires are managed for habitat for the large charismatic animals, the species, and the landscape diversity of the vegetation (Case Study 13.3). There, the lessons learned from long-term experiments using fire at different intervals and seasons have informed integrated fire management. In both areas, scientists have been engaged in assessing and informing the burning program. The landscape scale of these case studies is impressive and necessary for accomplishing long-term ecological, fire management, and social goals.

Case Study 13.2 Prescribed Burning: An Integrated Management Tool Meeting Many Needs in the Pyrénées-Orientales Region in France
Eric Rigolot, email: eric.rigolot@inra.fr
UR629, INRAE, URFM, 84914, Avignon, France
and Bernard Lambert, email: bernard.lambert66@orange.fr
Société d'Elevage des Pyrénées-Orientales, Prades, France

(continued)

Case Study 13.2 (continued)

The Practice of Pastoral Fires in the Pyrenees

For thousands of years across the entire Pyrenees mountain range, humankind's relationship with the environment has been closely associated with fire. According to the Greek historian Diodorus Silucus, the very name of Pyrenees would come from the Greek Pyros, which means fire. Cutting and then burning forests for agriculture and grazing has been carried out for at least several 1000 years, including regular pastoral fires on rangelands encroached with trees and shrubs, burning wood in forges and charcoal kilns, wildfires of various origins, often of low intensity until the middle of the twentieth century but then more devastating in the modern era (Métailié 1981). The use of fire for agricultural, pastoral, and industrial purposes involved highly elaborate methods integrated into a coherent agri-silvi-pastoral system.

Since its creation in the seventeenth century, the French forestry commission has sought to control pastoral fires in order to conserve timber and forests. However, along the last century, the socio-economic decline in the mountainous area and the resulting shrub encroachment led to a significant change in fire-related practices. At the same time, the administrative framework, initially supportive of these agri-pastoral fires, gradually turned against pastoral burning.

Consequently, pastoral society was confronted with a dual dilemma. First, once-grassy areas used for grazing were gradually becoming overgrown by shrubs and trees due to the decline of pastoralism and farming and many areas were purposefully reforested by the national Forest Service. The *breeder-gatherers* needed to reopen their pastoral areas threatened with closure when shrublands grew. The second dilemma was to find a solution to uncontrolled pastoral burns that regularly escalated into wildfires.

Unable to suppress the fires and limit the ecological, economic, and social impacts of changing lands, farmers and herders wished to reintroduce burning practices. This involved setting up an institutional structure to help farmers take back ownership of fire so they could use fire while carefully managing its ecological and social effects. It was in this context that the first institutional European experiment supporting the use of fire by rural communities was carried out in the easternmost part of the French Pyrenees.

The Pyrénées-Orientales Region in France

The Pyrénées-Orientales region (Fig. 13.4) comprises vast wildlands formed at the end of the nineteenth century as land uses changed and many people moved away from rural lands, which allowed a gradual invasion of shrubs and trees (Fig. 13.5). Roura (2002) found that area in grasslands and prairies declined by 70% from 1953 to 2000 as they were replaced by forests and heathland dominated by *Cytisus oromediterraneus*. Today, the biodiver-

(continued)

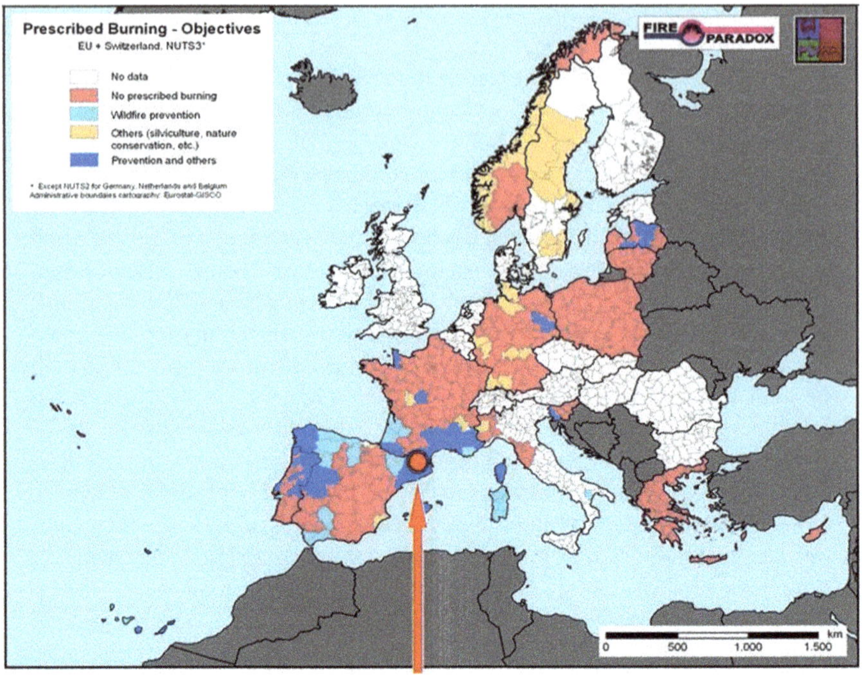

Fig. 13.4 Location of the Pyrénées-Orientales region at the eastern extremity of the Pyrenees mountain range (*arrow and red dot*). This map shows the type of prescribed burning objectives in Europe following the Nomenclature of Territorial Units for Statistics 3 scale. (Adapted from Lazaro and Montiel 2010)

Case Study 13.2 (continued)

sity of the open grasslands of these Mediterranean mountains is threatened and shrub encroachment constitutes a formidable source of fuel favouring large fires.

Agri-silvi-pastoral activities have always been the matrix of biological diversity in open areas, as well as contributing to the diversity of the landscapes in this French region, which extends over a wide altitudinal gradient from sea level to 2921 m, thus presenting a wide range of environments of major interest for their cultural and biological heritage. This region is home to 12 natural habitats of European interest including three priority ones, and a quarter of the heritage flora of the Catalan Pyrenees. Moreover, among the heritage vertebrate fauna, 40% of nesting birds use these open and semi-open habitats to breed. Regulatory measures aimed at protecting the environment are now gaining increased public support locally and nationally.

(continued)

Fig. 13.5 In many areas, grasslands have been invaded by shrubs and trees such as broom (*Cytisus oromediterraneus*) and mugho pine (*Pinus mugo*). (Photograph by B Lambert)

Case Study 13.2 (continued)

The invasion by shrubs and trees coupled with the decline in traditional uses for pastoral, agricultural, forest, and hunting purposes, these vast spaces have also seen their social functions diversify with the explosion of nature-related tourist activities. At the crossroads of uses and expectations, many farmers, shepherds, foresters, fire fighting services, and "environmentalists" have joined forces to invent a common use of fire in order to satisfy new social demands in these culturally important and biodiverse environments.

Creation and Purpose of the Prescribed Burning Unit

To this end, an action research initiative was set up between 1984 and 1987 to pool the efforts of the scientific community and the agricultural profession to "relearn" fire control and the response of ecosystems. This experimental pilot programme totalizing on the whole 150 ha, comprised ten plots of land reflecting the diversity of the heathland and rangelands in the fire-sensitive

(continued)

Case Study 13.2 (continued)

area. Given the positive results of the initial trials and the increasing demand for pastoral fires by farmers, in autumn 1986, the agricultural profession and the local authorities decided to support the creation of the first prescribed burning unit in France to be project managed by the agricultural profession (Lambert 2010). The unit is a departmental entity run by a pastoral agency, with active participation of national forest service personnel over the first 10 years, then replaced by local fire fighters, with support from time to time by national military units of civil protection. In this part of the Pyrenees, local beneficiaries do not participate directly in prescribed burns, but can achieve safety fuel-breaks beforehand.

The aims of the Pyrénées-Orientales' prescribed burning unit were to (1) train and raise the awareness of farmers on the use of fire, (2) possibly replace them to perform pastoral burnings that were difficult to control, (3) carry out, as a priority, burnings in areas susceptible to major fires, (4) provide training to departmental fire fighters, as well as national reinforcements, (5) provide a training school for future bosses of prescribed burning units, (6) add burning to the range of tools used to manage heritage species, including grey partridge (*Perdrix perdrix*) and Pyrenean chamois (*Rupicapra rupicapra*), and natural habitats of European interest, (7) ensure technical exchanges with Spanish fire fighters in the framework of the European programme and, in particular, the Generalitat of Catalonia, and Vila Réal University in Portugal, and, since 2006, (8) train local fire fighters in the use of backing fire techniques for wildfire suppression.

Scientific Support to Meet Multi-Functional Objectives

The prescribed burning unit came about in close collaboration with the research community. Since the winter of 1984/1985, the pastoral improvement programme on the use of fire was created with the Inra Ecodevelopment Research group of Avignon, the first experimental burning operations were supervised by the Cemagref of Aix en Provence (now INRAE) following a visit to North America, and monitoring vegetation as temperature measurements thanks to the Cnrs-Cefe experiment carried out in the Montpellier heathland. Later, information collected from each site was compiled and organised by Inra (now INRAE) Ecology of Mediterranean Forest research group (URFM) as part of the European FireTorch programme (Botelho et al. 2002).

This long-term collaboration made it possible to demonstrate the effects of repeated burnings on heathland dynamics in order to determine the most appropriate management scenarios and the most relevant development indicators (Rigolot et al. 1998) and to redefine the use of fire in technical sequences to manage pastoral environments in the Mediterranean mountains. The impacts of prescribed burning on fauna were analysed, as well as the benefits

(continued)

Case Study 13.2 (continued)
of using fire to conserve the habitats of birds (Pons et al. 2003) and the management of game birds. The relationships between institutional uses (prescribed burning) and traditional fires (pastoral fires) were analysed (Fernandes et al. 2013).

After 30 years of operation of the Pyrénées-Orientales' prescribed burning unit, an audit was conducted by an independent research group following a request from the administration to identify the main points of success from an organisational point of view and to report any threats to the system's sustainability (Métailié et al. 2017).

Strong Points of the Prescribed Burning Unit

The Pyrénées-Orientales' prescribed burning unit plays a key role in maintaining rangeland resources and open landscapes. For 32 years, the unit annually treated between 600 and 1200 ha (40–80 plots of land) totaling 26,000 ha and participating in the maintenance of 14% of the 120,000 ha of areas used for livestock grazing (Fig. 13.6). The size of the plots of land varies

(continued)

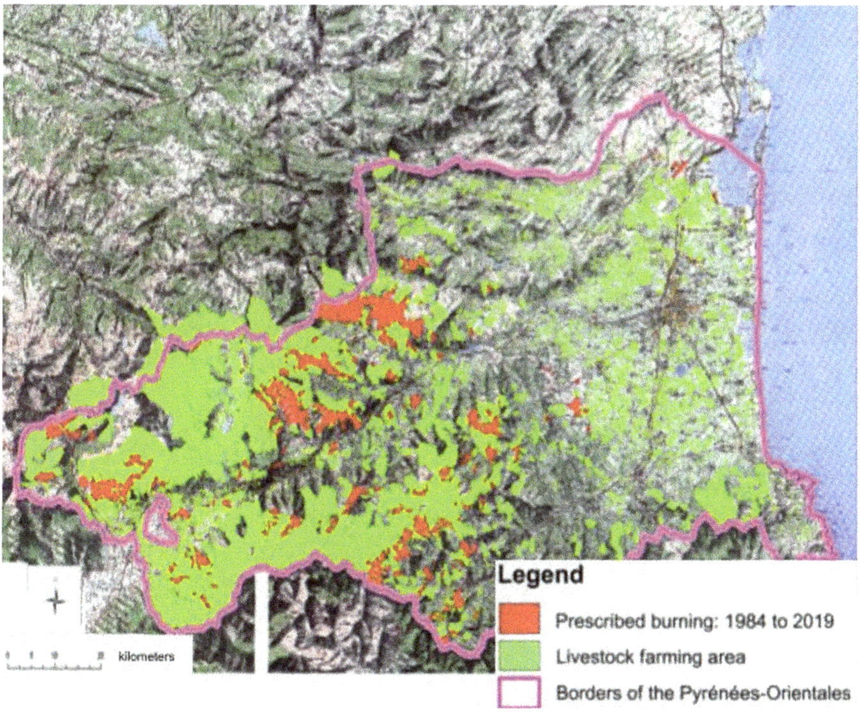

Fig. 13.6 Map of the livestock farming area in the Pyrénées-Orientales (120,000 ha) and zones treated using prescribed burning between 1984 and 2019. (Map made by B Lambert)

Case Study 13.2 (continued)
between 0.5 and 179 ha with an average size of 9 ha. The prescribed burning projects are split evenly between the fire-sensitive Mediterranean area and the montane and subalpine zones.

Prescribed burning is now recognised as a tool that promotes biodiversity. Managers frequently combine prescribed burning with mechanical or other techniques to achieve vegetation management and social goals. Increasingly, fire is perceived by scientists and managers of natural areas as an integral part of the evolutionary cycle of certain ecosystems. Managers use fire in technical sequences and specify the appropriate regime for each context. Mosaic burning comprising burned patches of a few acres to one hectare, carried out in several phases over one or more seasons, has become standard practice in the upper heathlands of the Pyrénées-Orientales (Rigolot et al. 1998).

The extensive practice of prescribed burning in the Pyrénées-Orientales has also been a success in terms of suppressing wildland fires. Thanks to prescribed burning practices, we have observed a strong decline in fires of a pastoral origin since the 1990s. In addition to contributing to fuel management, prescribed burning also acts on the causes of forest fires. Prescribed burning now receives institutional recognition via a body of laws and regulations, as well as being accompanied by a well-established training programme. Fire fighters implementing prescribed burning follow a highly effective training programme, which they put into practice fighting forest fires and, in particular, by familiarising themselves with suppression fire practices. Fire fighters working in the Pyrénées-Orientales, thanks to their high levels of involvement and regular practice of burning in the winter, since 2004 have been systematically using backing fires during the local summer wildfire suppression campaigns, as well as supporting their colleagues elsewhere in France and Catalonia, and even Sweden in the summer of 2018.

The systematic recording of costs on prescribed burning intervention worksheets shows that these costs are acceptable. The least expensive operations are those carried out in pre-mountainous areas (i.e., in non-wooded heathland) where the estimated costs are between €50 and €100/ha. Forest fire prevention operations carried out in wooded areas are generally more expensive costing between €100 and €600/ha compared to €800 to €2400/ha for mechanical clearing and €2000–€4000 for manual clearing. Therefore, prescribed burning, as carried out by the Pyrénées-Orientales prescribed burning unit, is economically viable.

The prescribed burning unit is led by the Pyrénées-Orientales' livestock breeding society, a professional organisation that ensures that fires are properly used. Their work includes consultations with a large number of local, regional, and even cross-border partners to develop recommendations and guidelines suitable for all stakeholders. With the aim of greater transparency

(continued)

Case Study 13.2 (continued)
and to reach a wider audience, a real-time information website has been developed (www.risque-incendie.com).

Due to its seniority, the Pyrénées-Orientales' prescribed burning unit is a reference not only in the pastoral network of the Pyrenees mountain range but also at the national level where it has acted as a facilitator of the national prescribed burning network for 15 years (Lambert 2010).

Weaknesses and Threats

Despite this success, the programme faces climatic and social pressures (Rigolot and Lambert 2017). Increasingly variable weather conditions require greater responsiveness and high levels of professionalism: after years of logistics implementation (1987–1989) followed by a long period of consolidation (1990–2002), recent trends show a marked shortening in the periods suitable for prescribed burning due to climate change. With a late autumn season that is too dry and too hot, the most favourable winter period is between the end of January and the beginning of March (the campaign period is now under 30 days compared to over 70 days in the 1990s). Faced with increasingly volatile and unpredictable weather patterns, it is necessary to demonstrate increasing levels of professionalism and responsiveness to take advantage of these unpredictable time windows. This means adopting a strategy that is both flexible in terms of its implementation and its ability to carry out several simultaneous operations during these increasingly short time windows.

The demand for more consultations has led to a reduction in the size of the worksites. Burning prescriptions are now the domain of a more complex and larger group of stakeholders compared to the erstwhile forester/farmer. This means lengthier consultations, increasingly complex specifications, more limited burning practices, and increased higher costs.

Social acceptance of prescribed burning has always been challenging, even more so today as opposition to this practice by urban groups is growing. Excellent results in terms of fire prevention have led to a loss of awareness on the risks of these new urban groups. These groups increasingly dominate local commissions and municipal councils, and pressure elected officials to reject practices they consider unsightly, a source of dissatisfaction for holidaymakers who get dirty, and which some see as being dangerous, inefficient, responsible for erosion, deforestation and, more recently, air pollution. This last point is of particular sensitivity during high-pressure conditions in and around populated areas on the plain and the coast, including ski resorts. The sharp drop in the number of wildfire areas in recent years is reducing the motivation of elected officials to support prescribed burning. While objectively the risk of wildfire remains, elected officials, faced with significantly improved annual reports, are tempted to support less prescribed burning thus reducing public funding. Public funding currently provides more than 75% of

(continued)

Case Study 13.2 (continued)
the operating costs of the unit, which is around €100,000 per annum and per 1000 ha treated.

Conclusion

Prescribed burning has reached a crossroads in the Pyrénées-Orientales. Over 30 years of development supported by scientific research have resulted in an effective technical and organisation model. It has achieved the dual objectives of controlling wildfires due to poor pastoral burning practices and meeting farmers' needs in terms of maintaining and renewing the pastoral resources. But is it really efficient? The per-hectare financial and human resources required for operations are high. However, the strictly agronomic and forest fire prevention points of view are only a partial vision of the benefits of integrated fire management in this area. Maintaining wildlife habitats, strengthening fire fighters' skills through prescribed burning practice, maintaining mountain landscapes and supporting the remaining farmers in these rural communities are positive outcomes that should be better evaluated from an environmental, social and economic point of view in order to give a full picture of the efficiency of prescribed burning in the Pyrénées-Orientales region. While biotechnological research is still required in some areas, notably into combustion management to ensure better smoke control, the social science and humanities research is needed to continue improving integrated fire management this Mediterranean mountain region. Ultimately, political science must inform public decision-making in terms of the inevitable trade-offs that exist between risk management and the multiple uses of mountainous areas and nature conservation.

Case Study 13.3 Integrated Fire Management in Kruger National Park

Navashni Govender, email: navashni.govender@sanparks.org
Conservation Management, South African National Parks, Conservation Management, Kruger National Park, Skukuza, South Africa

African savannas are driven by fluctuations in rainfall, herbivory, nutrients and fires (Sankaran et al. 2005). In Africa and the Kruger National Park (KNP), fires are ignited by people, whether on purpose or accidentally, and by lightning (Archibald et al. 2009). Lightning fires are less common and usually do not burn large expanses of natural veld (van Wilgen 2009). The fauna and flora in these fire-prone ecosystems have co-evolved with fire, resulting in a resilient fire-adapted system with many fire-adapted plant species (Bond and Keeley 2005). In southern Africa, people have been using and controlling fires to manipulate their environment for thousands of years.

(continued)

Case Study 13.3 (continued)
Humans used fires to manage their environments for agricultural purposes, to cycle soil nutrients and to control herbivore movements to facilitate easy hunting of wildlife, therefore the KNP acknowledges humans as a "natural" ignition source. The main objectives for fire management in KNP is to mimic the role that fire plays in maintaining African savannas by improving biodiversity, wildlife habitats and ecosystem services, whilst specifically considering fire-herbivory interactions, by evaluating and responding appropriately to fire threats facing infrastructure and human lives.

Established in 1926, KNP, is one of the largest proclaimed and officially protected natural areas in the world. The park covers approximately 2 million ha, occupying almost 2.5% of the total land surface area in South Africa (Fig. 13.7). It is situated in the north-eastern region of South Africa and is separated from adjoining Moçambique by the Lebombo mountain range in the east and from Zimbabwe by the Limpopo valley in the north. It is elongated with a total length of approximately 320 km and a mean width of 65 km (Govender et al. 2012). The park's mean annual rainfall is approximately 500 mm, with a rainfall gradient from around 350 mm in the north to around 750 mm in the south. The climate of KNP is well suited to support regular veld fires, with the wet season from October to March that results in fuel accumulation and the dry season, which extends from April to September, which encourages veld fires. The park is distinctively divided in two by its geology, with granitic sandy soils occurring on the western half of the park and basaltic clay soils on the eastern half. There are two major river systems (the Nkomati system in the south and Limpopo system in the north). The vegetation of the park is characterized as an open-wooded savanna, dominated by trees in the genera *Acacia* (*Senegalia* and *Vachellia*), *Combretum*, *Sclerocarya*, and *Colophospermum* (Fig. 13.8b). The flora of the park comprises ±2000 taxa, including over 400 tree and shrub species and over 220 grass species. The fauna of the park includes 148 mammals, 53 fish, 35 amphibians, 118 reptiles and ±500 bird species.

The Evolution of Fire Management in Kruger National Park (KNP)

The first park Warden, Colonel James Stevenson-Hamilton, enforced a general ban on any deliberate burning because he believed that fires had a detrimental impact on vegetation and wildlife. In 1935, the National Parks Board decided that fires should not be explicitly banned but rather controlled, and in 1937 Stevenson-Hamilton suggested the bush be burnt every second year to avoid the accumulation of moribund material. In 1947, Stevenson-Hamilton suggested that half the available fuel in KNP should be burnt every year between February and April in order to promote low-intensity fires burning while vegetation is still green. Due to the limited resources available,

(continued)

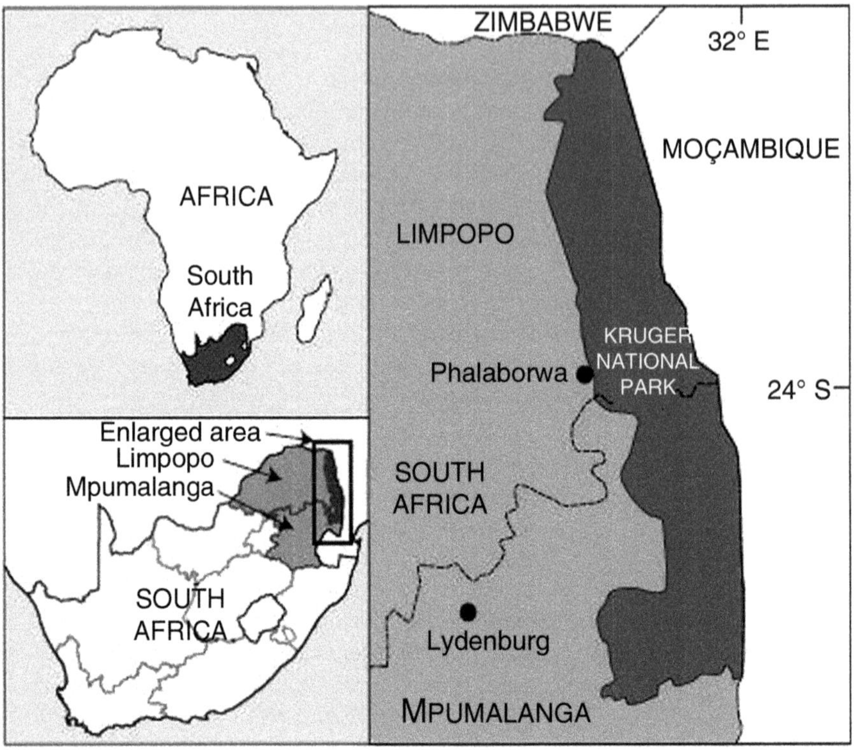

Fig. 13.7 The location of Kruger National Park within South Africa and southern Africa (Govender et al. 2012)

Case Study 13.3 (continued)
most of these policies were nearly impossible to implement, and thus between 1926 and 1947 the KNP fire policy was essentially "laissez-faire".

In the mid-1940s, prescribed burning was prohibited to limit land degradation and soil erosion. The 1948–1956 period was known as the Fire Suppression or Protection Era (van Wilgen 2009).

Between 1957 and 1980, KNP managers implemented a Fixed Prescribed Burning strategy by establishing a graded firebreak network with more than 400 burn blocks ranging in size between 50 and 24,000 ha (van Wilgen 2009). Fires were applied every 3 years in Spring (after the first rains) (Govender et al. 2012). In 1981, this rigid burning program was declared unsuitable and adapted to allow for seasonal variation in the timing of prescribed burns, whilst retaining the 3-year rotation (van Wilgen et al. 2014). This Flexible Prescribed Burning strategy lasted until 1991 when KNP managers shifted their fire strategy towards a "Natural" Fire Policy in which lightning fires were

(continued)

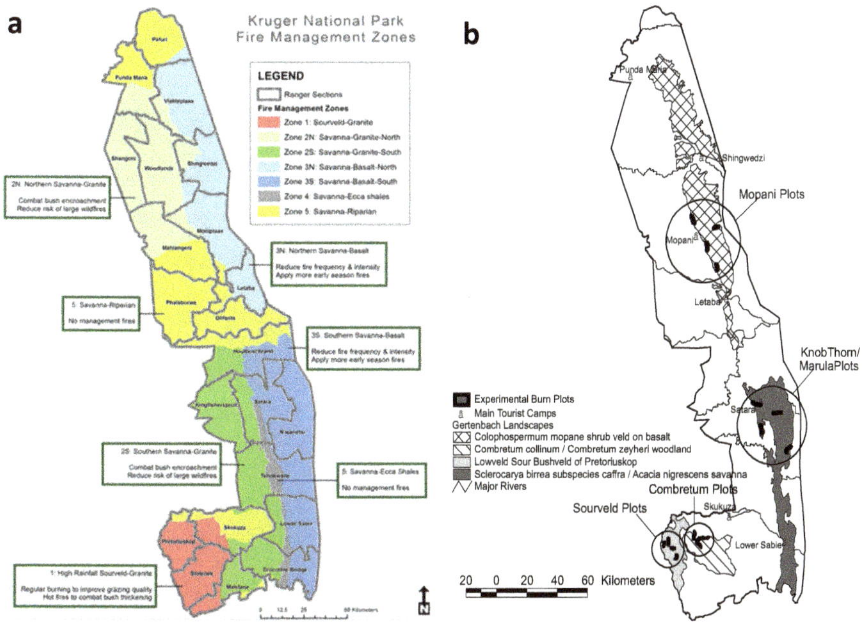

Fig. 13.8 Maps of Kruger National Park: (**a**) Distribution of the Fire Management Zones (FMZ) (van Wilgen et al. 2014). (**b**) The location of the Long-Term Experimental Burn Plots (Biggs et al. 2003)

Case Study 13.3 (continued)

allowed to burn to their fullest extent and were no longer confined by burn blocks (van Wilgen et al. 2004). In recognition of the substantial extent of area burnt per year by people, the fire management strategy was adapted once again in 2001 when management officially recognized the role of people in the landscape (van Wilgen et al. 2004; Govender et al. 2012).

Our Integrated Fire Management Strategy, implemented since 2001, allows for multiple ignition sources such as lightning, game rangers and migrants traversing the park from Mozambique into South Africa (Govender et al. 2012). The amount to burn would be calculated based on the preceding 2 years' rainfall and fuel load accumulation (van Wilgen et al. 2014). This strategy aimed at promoting variability by influencing fire intensities and spatial patterns whilst allowing for lighting-ignited fires and acknowledging the occurrence of inevitable wildfires. In 2012, the Integrated Fire Management Strategy was updated to include Fire Management Zones delineated based on the underlying geology, fire return period, and mean annual rainfall (Fig. 13.8a, Smit et al. 2013). In each zone, different fire strategies are implemented to achieve specific ecological objectives.

(continued)

Fig. 13.9 Impact of the same fire on two different *T. sericea* trees. (**a**) The fire burns into the cambium of the tree with high elephant utilization, which can eventually lead to the death of the tree. (**b**) Here, the fire burned on the surface of the tree only, as a result of an intact bark due to no prior elephant use of the tree. (Photographs by Tercia Strydom)

Case Study 13.3 (continued)

Fire and Herbivory Interactions

Globally, fire and large mammal herbivores (grazers and browsers) are two of the key consumers of above-ground plant biomass; they affect vegetation structure and shape the landscape. Positive feedbacks between fire and herbivory (grazers and browsers) can maintain ecosystems in alternate states between a savanna-grassland and savanna-forest mosaic. The co-evolvement of fire and herbivory adapted plant communities in African savannas and the KNP has allowed managers to use these processes as management tools to achieve various ecological objectives. For example, elephants debark trees, exposing vulnerable cambium tissue to excessive heat by fires; such trees eventually succumb to successive fires (Fig. 13.9). The removal or exclusion of either browsing by elephants (Fig. 13.10) or fire (Fig. 13.11) can increase tree woody biomass, cover, and structure resulting in decreased grass cover which reduces the spread and occurrence of fires in areas, thereby promoting the recruitment and establishment of mature trees (Hempson et al. 2019).

(continued)

Fig. 13.10 Long-term fixed-point photographs ((**a**) 1985, (**b**) 1995, (**c**) 2005 and (**d**) 2015) at the Nwashitsumbe roan enclosure where elephants have been excluded since its establishment in 1969 illustrating the increase in woody vegetation biomass, cover, and structure. (Photographs by Danny Govender)

Case Study 13.3 (continued)

Hempson et al. (2019) provided the elegant conceptual diagram (Fig. 13.12) of alternate fire-grass and fire-herbivore (grazing lawn) stable states along a productivity gradient. The current fire policy in KNP provides for fire-dominated grasslands (Zone 1, see Fig. 13.8a) by increasing fire intensity and frequency in these landscapes and herbivore and grazer dominated grasslands (Zone 2S & 2N, see Fig. 13.8a) by reducing fire frequency and intensity thereby increasing grazing lawns and more palatable grass species.

Legislation

The National Veld and Forest Fire Act (No. 101 of 1998) promotes the use of prescribed fire and intends to prevent and combat wildfires in mountainous, veld, and forested areas throughout the country, thereby resulting in greater attention to fire safety and protection. The KNP is a member of the Greater Kruger Fire Protection Association (GKFPA) that covers the entire KNP and some of the nearby private nature reserves (Klaserie, Timbavati, Balule, and Umbabat private nature reserves; and Sabi Sand Wildtuin (SSW); and MalaMala Game Reserve).

(continued)

Fig. 13.11 Long-term experimental burning plots in the central area of the Kruger National Park. The two plots are separated by a strip of cleared land. Fire has been excluded from the plot on the right for 65 years, resulting in dense and closed woodland, whereas the plot on the left has been subjected to fire every 2 years, resulting in a more open landscape. (Photograph by Navashni Govender)

Case Study 13.3 (continued)

Members strategically use fire within the GKFPA for ecological maintenance of biodiversity and for managing the fire risks in the area. Monitoring is required, as is reviewing and updating the rules and regulations that govern resource sharing, training standards, required equipment, firebreaks, and communication). Through the GKFPA, landowners cooperate for better and safer veld fire management practices.

Fire Research: The Long-Term Fire Experimental Burn Plots

Our understanding of the effects of fire on the ecosystem is often supported by research based on the experimental application of selected fire regimes on fixed areas. In the early 1950s, fire research began formally in the KNP with the establishment in 1954 of a long-term fire experiment (Biggs et al. 2003). The main aim of the experiment was to study the effects of fire frequency and season on the vegetation of the KNP under the grazing pressure of indigenous herbivores.

(continued)

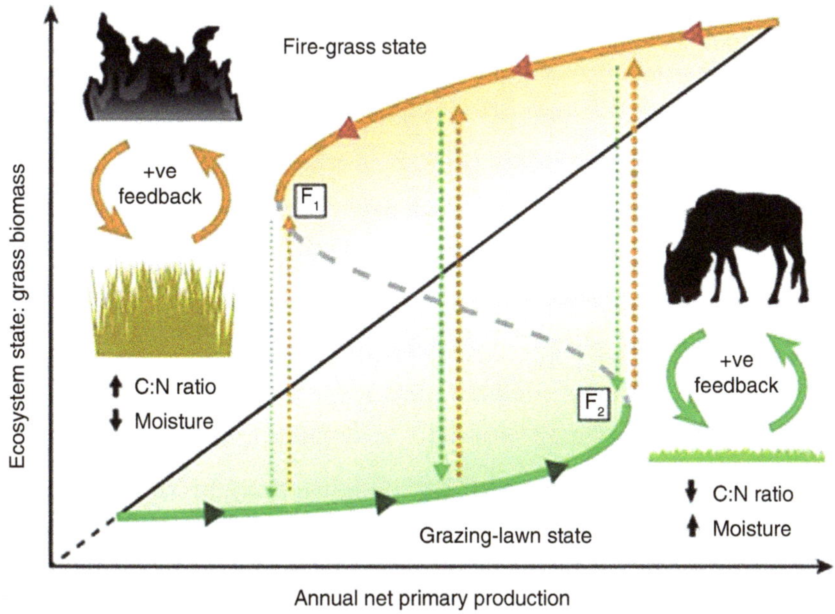

Fig. 13.12 Conceptual diagram illustrating the Alternate Grassland States fire-dominated *(orange)* and grazer-dominated (*green*) that can exist in the landscape. Within each state, there are stabilized positive feedbacks with fire (*solid orange arrows*) and grazers (*solid green arrows*). Transitions between states (*broken lines*) result in changes of available resources, such as rainfall, nutrients, light and C:N ratio, along the black diagonal line which represents an increase in grass biomass with high annual net primary production. (Adapted from Hempson et al. 2019)

Case Study 13.3 (continued)

Fire frequency treatments applied are annual (in August only), biennial and triennial intervals, and fire season treatment are in February (late summer), April (autumn), August (winter), October (after the first spring rains) and December (early summer)). After 1974, quadrennial and sexennial burns in October were added to the experiment. Each plot is approximately 7 ha, and treatments are replicated four times in each of the four major landscapes in the park (Fig. 13.8b) (Biggs et al. 2003).

Future Developments

All fires in KNP are mapped for ecological and safety reasons (Govender et al. 2012). The ecological requirements are defined by the KNP Fire Threshold of Potential Concerns (TPCs) (van Wilgen et al. 2014) and the safety criteria is that of reduction of fire risk and time since last fire. With the use of a computerized Decision Support System (DSS) the current fire TPC's (fire intensity and extent for each FMZ) are evaluated on an annual basis (Attorre et al. 2015). The DSS allows for the collection, organization, calculation,

(continued)

Case Study 13.3 (continued)

analysis, and visualization of important fire-related data, including precipitation, vegetation, and fuel biomass, area burnt, fire frequency, and intensity.

Since 2004, KNP has been using remote sensing techniques for fire mapping and monitoring purposes (Govender et al. 2012). Moderate Resolution Imaging Spectroradiometer (MODIS) instruments onboard the EOS-AM (Terra) and EOS-PM (Aqua) satellites, which were launched in 2001, are used to detect fires and create KNP's fire scar maps. With the launch of Sentinel 2 and Sentinel 3 satellites in 2015 by the European Space Agency (ESA) KNP will now migrate from MODIS to Sentinel-2 with higher resolution satellite imagery. This will improve the detection of smaller and low-intensity fires, minimize false detections, allow for more accurate fire scar perimeter mapping, and help quantify the effective temperature and fire radiative power (FRP).

Burn quotas were determined based on results from the Veld Condition Assessments (particularly biomass) and fuel load estimates from a model developed using 2 years' preceding rainfall totals. Since 2012, fuel loads available to burn were predicted using the preceding 2 years' rainfall model (van Wilgen et al. 2004; Archibald et al. 2009). However, a major shortfall herewith is the low density and wide spatial extent of the rainfall stations used to collect rainfall data. A total of only 23 rainfall stations (point data) scattered across 2 million hectares and data interpolation is bound to result in less accurate fuel load estimates further away from where rainfall is actually measured. Currently, we are investigating the possibility of using high-resolution satellite imagery to estimate fuel loads and fire probability. To measure vegetation characteristics, we're exploring the use of Normalized Difference Vegetation Index (NDVI) and/or Enhanced Vegetation Index (EVI) data derived from Sentinel-2 imagery. Improved fuel load and vegetation greenness estimates combined with time since last fire, from higher spatial and temporal resolution imagery will go a long way in providing more accurate burn quotas for the KNP fire managers.

Poaching has become a major problem facing rhino populations here and around the world and in KNP rhino poaching has increased substantially in recent years. KNP rangers now use fire to track poachers in the veld. Rangers burn blocks where poaching activity is high. After the fire, poachers are more readily tracked in the ash and bare soil and are more visible until the thick grass cover recovers. Prescribed fires are burnt to encourage rhinos (improved grazing from the new green flush) to move into areas that are deemed safer and away from parks perimeter.

Conclusions

Fire is a key process and management tool within our African savanna ecosystems. Fire management in KNP has changed at least seven times since

(continued)

Case Study 13.3 (continued)
the park's proclamation in 1926. Strategic Adaptive Management is core to KNP's management that is consistently informed by the best available information, implemented, and monitored. KNP managers use fire for promoting and enhancing biodiversity and ecosystem processes for the benefits of wildlife and people, protecting human lives and property from fire threats.

13.2.3 Working with Partners Through Shared Stewardship and Cooperatives

Landscape-scale integrated fire management must include working effectively with many people, as fire and smoke cross boundaries between lands. In central Idaho, efforts to greatly increase the scale and pace of forest restoration requires strategically using prescribed and wildfires for landscape-scale burning to accomplish objectives (Case Study 13.4). There, managers of public lands work with people in local communities to hear and address concerns about fire and smoke, and with managers of adjacent public and private lands to meet shared objectives. As Doane and coauthors explain, they use innovative practices for planning and implementing integrated fire management. Ongoing monitoring and adaptation are key to achieving their long-term goals (Case Study 13.4). Changing policies and laws aid them. Collaborative decision-making has brought people together to address broad land and fire management challenges.

In the prairie grasslands and woodlands of the Great Plains in central North America, almost all of the land is privately owned, yet for ecological and economic reasons, many land managers there use and manage fires. Bauman et al. (Case Study 13.5) communicate the imperative for burning, as well as the innovative ways that people have overcome the challenges. The Nature Conservancy, a non-governmental organization, has been especially effective in envisioning a future of burning where neighbors are helping neighbors achieve their own objectives while providing for vibrant grasslands, ranches, and people. Here, many landowners have formed prescribed burning cooperatives to share knowledge, training, equipment, and labor.

Case Study 13.4 Integrated Fire Management: Landscape Fire on the Payette National Forest in Idaho, USA

Dustin Doane, email: dustin.doane@usda.gov; *Phil Graeve*, email: phillip.graeve@usda.gov; *Patrick Schon*, email: patrick.schon@usda.gov; *Erin Phelps*, email: erin.phelps@usda.gov
Payette National Forest, New Meadows, Idaho, USA

(continued)

Case Study 13.4 (continued)

Why We Are Using Fire

Historically, fires were common across the 1 million ha landscape of the Payette National Forest in central Idaho, with roughly 28,000 ha burned on average each year (Table 13.2). Fire is not only natural; fire is critical for the health of our ecosystems. We use fire on the Payette National Forest because we understand that reestablishing a healthy relationship with fire is the only way to be successful land managers.

Aggressive fire suppression and other land uses since the early 1900s led to an accumulation of ground, surface, and canopy fuels, an increase in overall tree densities, and a shift in forest composition toward less fire-resilient species. This departure from historical conditions across central Idaho is directly correlated to the extent to which key ecosystem components have been altered, thereby reducing the current resiliency of forest landscapes to disturbances. The primary disturbance risks are now uncharacteristic wildland fire, native insect outbreaks, and introduced insects and diseases, all of which kill many trees. The same conditions that lead to undesirable fire effects also favor the proliferation of insects and diseases beyond endemic levels that can lead to high rates of tree mortality as stressed trees are less able to recover from disease or insect outbreaks than their healthy counterparts. Uncharacteristic disturbances are those differing in fire size, spatial patterns, severities, and frequencies from historical such that large changes in ecosystems result (Singleton et al. 2019).

Uncharacteristic wildfires threaten a variety of ecosystem functions and values. Threatened and sensitive wildlife species such as northern Idaho ground squirrel and great grey owl have evolved with the historical type and frequency of fire (Table 13.2), so the changes in tree density and forest encroachment into meadows since 1900 have likely affected these and many other species negatively. In addition, some of the very reasons the National Forests exist and are valued can be greatly impacted by uncharacteristic wildfires, such as recreation access, grazing forage, and sustainable timber production. Uncharacteristic fire behavior can also significantly increase the risk to people in communities, including emergency responders. These undesirable consequences are driving the need to increase the proactive management and restoration of fire on the landscape, and motivating all area land managers and community members to take action.

Use of Fire

Fire plays a key role in restoring landscape conditions and increasing the ability to protect values at risk, from homes and infrastructure to wildlife habitat and recreational opportunities. This is recognized nationally and locally, within the USFS, cooperating agencies, state governments, and by many people in the surrounding communities (USDA 2018).

(continued)

Table 13.2 Types of fire historically on the Payette National Forest in Central Idaho

Approximate percentage of area burned annually	Forest type	Burn severity	Historical fire return interval[a]
50	Diverse understory of grasses, forbs, and low shrubs with a large-diameter fire-resilient overstory trees, primarily ponderosa pine (*Pinus ponderosa*) and Douglas-fir (*Pseudotsuga menziesii*)	Low	Frequent: 5–30 years between fires
30	Stands dominated by Douglas-fir, western larch (*Larix occidentalis*), and grand fir (*Abies grandis*), and open forests of whitebark pine (*Pinus albicaulis*)	Mixed	Less frequent: 5–100 years between fires
1	Lodgepole pine (*Pinus contorta*), Englemann spruce (*Picea engelmannii*), subalpine fir (*Abies lasiocarpa*), and Douglas-fir	High	Low frequency: 25–300 years between fires
20	Non-forested grass and shrublands	Variable	Low to high frequency, depending on adjacent forest types

From Payette National Forest public documents
[a]Historical refers to 1300–1900 AD

Case Study 13.4 (continued)

In the last 10 years, the use of prescribed fire has become more common on the Payette National Forest, and through extensive public outreach and education efforts, the relationship between prescribed fire and the local communities has become increasingly positive. The short-term goal for the Payette National Forest is to burn 10,000 ha per year with prescribed fire, which is considerably higher than the 1000 ha accomplished in 2010 (Fig. 13.13).

The ability to manage a naturally-ignited wildfire for resource objectives is also an important tool on the Payette National Forest. The Payette National Forest administers approximately 310,800 ha of the Frank Church-River of No Return Wilderness Area (FCRONR). Fires have generally been managed to play their natural role since the establishment of this Wilderness Area in 1980. There are some exceptions when it comes to the immediate risk to private inholdings or other significant values requiring protection, but in general, the management policy for the FCRONR acknowledges that wildfires are natural processes and that efforts should be made to minimize effects from suppression activities on wilderness characteristics. In other words, the emphasis is on protecting wilderness characteristics from suppression actions (e.g., constructing containment lines, use of chainsaws or helicopters, etc.) rather than protecting them from wildfire itself.

(continued)

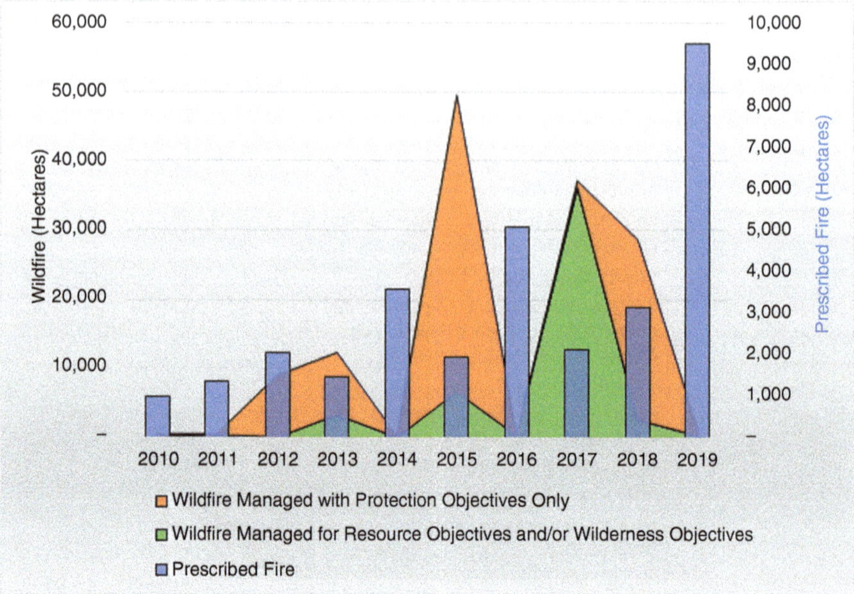

Fig. 13.13 Wildfire and prescribed fire (ha) across the Payette National Forest (fiscal years 2010–2019, Payette National Forest 2019). Burning of piles made by hand or machine is not included. (Compiled from Payette National Forest data)

Case Study 13.4 (continued)

Managing lightning-ignited fires to meet resource objectives outside of the FCRONR is slowly expanding since it became an option with the revision of the Payette National Forest Land and Resource Management Plan (Forest Plan) in 2003, with approximately six fires per year outside wilderness managed for resource objectives. The Forest Plan categorized the National Forest into three areas of management options (Fig. 13.14): areas where only protection objectives are available (red), areas where wilderness and protection objectives areas are available (green), and areas where resource and protection objectives are available (blue). Managing lightning-ignited fires for resource objectives is not currently allowed across approximately 100,000 ha (~11% of the National Forest area, red in Fig. 13.14); wildfire events in those areas must be managed with a focus on protecting human and other resource values. Therefore, prescribed fire is the only means to restore fire as a process, and it is often used along with other fuels treatments near communities.

Because the risk of smoke impacts and the consequences of prescribed fire escaping containment are heightened near communities, thinning and other mechanical activities are often relied upon to facilitate the application of

(continued)

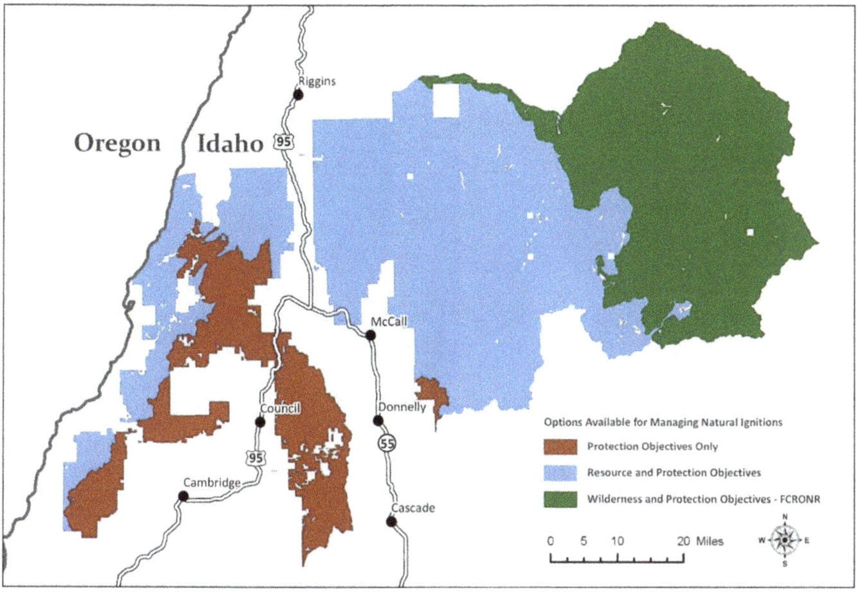

Fig. 13.14 Fire management options across the Payette National Forest. US 95 is the main north-south highway in Idaho. McCall and Cascade are the primary communities, but there is extensive residential development on private lands, especially around the towns and adjacent to the National Forest System lands. (Map courtesy of Payette National Forest GIS Specialist)

Case Study 13.4 (continued)

prescribed fire as well as mimic fire effects when the application of fire poses too great a risk to people and property. This may include thinning of understory trees, removal of fire-intolerant tree species, increasing the base heights of the canopy, and piling of recent or existing dead and down woody material for burning at a later date when managers can ignite piles safely with minimal smoke impacts to people.

Prescribed Fire Planning

To meet the overall demand for restoring fire to the landscape, planning must be robust and bold. Forest personnel have found the following to be effective and efficient for planning for use of prescribed fire. Project areas need to be large (20,000 to >100,000 ha (50,000 to >250,000 ac) and include lands managed by adjacent National Forests, the State of Idaho, the Bureau of Land Management, local governments, and/or private lands) because fires, insects, and disease do not recognize jurisdictional boundaries. Use roads and other significant existing barriers to fire spread as project boundaries. Treatments within stands of historically frequent fire need to be continuous across the landscape for greatest effectiveness (Menakis et al. 2016). Forests treated with

(continued)

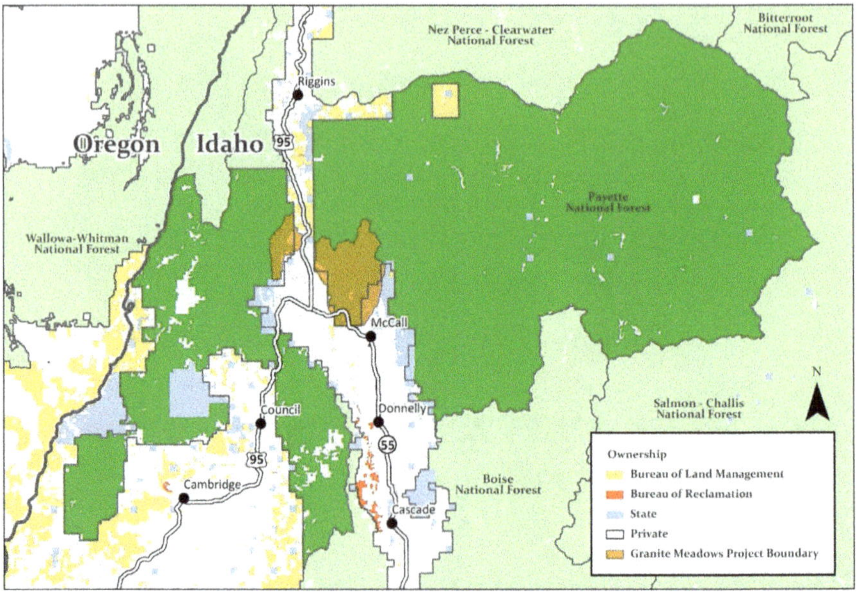

Fig. 13.15 Land managers adjacent to the Payette National Forest; Granite Meadows Project in *orange*. Map courtesy of Payette National Forest GIS Specialist

Case Study 13.4 (continued)
isolated, light, and patchy treatments have limited resilience to large wildfires. Objectives need to be general so that desired conditions can be met among the various weather and environmental conditions. Planning decisions should allow for 20 or more years of burning and include maintenance burning and other treatments. This provides clarity to the temporal scale of implementation and allows implementers the ability to continue to make efforts to maintain or improve conditions. Continued growth in the application and management of fire is dependent on social acceptance among agency personnel, partners, and the community. This can be difficult as many stakeholders derive much of their experiences with fire from very large, damaging, and costly wildfires. They don't have another lens from which to view the positive aspects of fire on the landscape. Therefore, a first step is to understand the resource values most salient to these critical stakeholders, be it recreation, timber, grazing, water, etc., then connect the maintenance of those values with the role of fire.

Example of Landscape-Scale, Collaborative Planning

The Granite Meadows project is a large, cross-boundary look at the health of our collective landscape and an example of how changes cannot be made if we don't share stewardship responsibilities across jurisdictional lines (Fig. 13.15). The project encompasses approximately 28,000 ha of Payette

(continued)

Case Study 13.4 (continued)
National Forest lands and also includes over 6500 ha managed by the State of Idaho, the US Bureau of Land Management, and private owners. Upon completion of the environmental analysis process, the Payette National Forest plans to enter into agreements with these area land managers and owners to treat the landscape as a whole by way of the Wyden Amendment. The Wyden Amendment (US Public Law 105-277, Section 323 as amended by Public Law 111-11 Section 3001 or 4001) allows the Forest Service to enter into cooperative agreements with willing Federal, Tribal, State, and local governments, private and nonprofit entities, and private landowners to benefit resources within watersheds on National Forest lands. Twenty years of planned activities include approximately 10,000 ha of commercial tree harvest treatments, 19,400 ha of noncommercial vegetative treatments, 31,600 ha of prescribed fire, and a variety of watershed health activities and road and recreational improvements. This project is based in part on recommendations provided by the Payette Forest Coalition (PFC). The PFC consists of stakeholders from a broad range of interests including conservation groups, timber industry, recreational groups, and state and county government.

Prescribed Fire Strategies

Strategies for applying fire across the Payette National Forest change with landscape conditions. Remote and roadless areas, including but not limited to the designated wilderness, are priority areas for the use of natural ignitions, and prescribed fire is an excellent tool to facilitate greater use of those natural ignitions, especially as a means to protect the human values that do exist in these remote areas. When used within the footprints of the large wildfires of the recent past that exhibit very homogenous and extensive coarse woody debris loadings, prescribed fire can provide a network of habitats and connective corridors for wildlife in addition to adding diversity to age class and structure of the previously burned areas and surrounding landscape.

Many efficiencies are gained by burning across large areas, especially when burning large areas within a network of roads. The fire and engineering managers work together to reduce ladder fuels and tree densities along roadways. This improves roadway safety (e.g., travel conditions and sight distance) and drainage and eases the application and management of prescribed fire. Recent maintenance burns in these landscapes have also shown that tree mortality, escapes, and smoke impacts are far less than with the initial prescribed burns. Costs of the maintenance burns are generally 30–50% of the cost of initial applications of prescribed fire.

When applying fire near significant values, such as homes and timber, on nearby private lands, the amount of burn preparation increases (e.g., prior thinning to reduce ladder fuels, notifications, coordination with partnering agencies, etc.), and the resources allocated typically increase. Smoke

(continued)

Case Study 13.4 (continued)
management becomes more labor-intensive and limits the timing and size of burns in and near WUI areas. The cost per hectare increases in the WUI, but it also drops significantly after the first prescribed burn. Our fire management strategy within the WUI is to prepare many sites through mechanical means then follow with prescribed fire. Maintenance using prescribed fire is planned at intervals frequent enough to maintain low surface fuel conditions and thus limit the threat of wildfire to values. People in the communities around the Payette National Forest, including numerous Home Owners Associations and individual landowners, are currently increasing their own efforts to improve forest resilience on their lands, which includes preparing for and using prescribed fire.

Regardless of geographic location, when conditions permit, the Payette National Forest personnel work cohesively in applying fire to the landscape. Spring burning begins below the snow line among the lower elevation ponderosa pine communities (900–1200 m elevation) in March and April (Fig. 13.16) and lasts into June among the mid-elevation Douglas-fir and

(continued)

Fig. 13.16 Early spring burning in stands dominated by ponderosa pine and Douglas-fir; fire backing downslope with 0.3 m flame lengths. (Photograph by Dustin Doane)

Fig. 13.17 Late spring burning in stands dominated by Douglas-fir and grand fir. (Photograph by Patrick Schon)

Case Study 13.4 (continued)

grand fir (1500–1800 m elevation) forests (Fig. 13.17). All elevations are typically available to burn in the fall (900–2400 m elevation). Burning can start in late August and continue into October. On the Payette National Forest, we seek to use all seasons to treat a single large block (2000–6000 ha in size). This often involves burning drier forest types on the south- and west-facing aspects in early spring (Figs. 13.16 and 13.17), then burning east-facing aspects in late spring, then burning north-facing aspects and other relatively wetter forest types within the block during the fall (Figs. 13.18 and 13.19). Implementing fires in two or three large blocks concurrently across the forest will compound the efficiencies gained (sharing aircraft, personnel, etc.) by treating large areas rather than individual stands. We expect and manage for unburned patches within the burn perimeter in order to create a mosaic of burned and unburned across the landscape.

Challenges and Solutions

Poor smoke dispersion is the most limiting factor in the ability to apply fire across landscapes. It is common to have 8 h of acceptable weather and fuel

(continued)

Fig. 13.18 Landscape-scale fall burn in grasslands and adjacent stands dominated by ponderosa pine and Douglas-fir. (Photograph by Dustin Doane)

Case Study 13.4 (continued)
conditions to meet objectives and conditions desirable for fire containment, but ignitions are generally limited to two to 4 h in order to allow for adequate smoke dispersal late in the day. On the Payette National Forest we have found the following to be very effective in overcoming these challenges:

- Having multiple burn blocks of various sizes prepared to receive fire across various watersheds and fuel conditions. This affords burners options to continue applying fire daily while allowing air sheds of recently burned blocks time for "scrubbing" (i.e., allowing time for the smoke to clear out of the airshed).
- Having more blocks prepared to receive fire than what may be feasible to accomplish in a given burn season is vital. Extensive burn preparation is accomplished throughout the snow-free months by personnel from all resource areas and managers of adjacent National Forests and other lands. Implementing multiple large burn blocks allows burners to take advantage of treating more of the landscape on those days with excellent smoke dispersion.

(continued)

Fig. 13.19 Early fall burning among high elevation forests of whitebark pine and subalpine fir among remote and rugged terrain; ignited by helitorch. (Photograph by Dustin Doane)

Case Study 13.4 (continued)

- Expanding the number of months available to the application of fire has increased opportunities to burn as well as days of "scrubbing".
- Increasing the use of fire and concurrently, seizing every opportunity to talk with the public and partners about the "why" as well as the "what" and the "how". The growing understanding of fire's multiple benefits has greatly improved the acceptance of low levels of residual smoke as continued applications of fire have increased community acceptance.

Another challenge is the tendency for some agency specialists to focus almost exclusively on one potential negative outcome from the application of fire (e.g., the mortality of legacy trees or smoke impacting recreational use for one afternoon), and focusing on the greatest potential consequence without giving much weight to the probability of that event occurring. One "no" can override a thousand "yeses". Differing perceptions of risk and tolerance for these risks are a challenge to communicate within the organization. This has limited the use of fire (prescribed or wildfire) or various mechanical means to achieve desired conditions.

(continued)

Case Study 13.4 (continued)

To address these challenges, we have found that bringing specialists together to delve into the following items has been effective in broadening the lens from which specialists view fire and thus building support for the application of both prescribed and wildfires with resource objectives. Conducting these discussions in the field and for specific projects has been effective.

- The historical role of fire on the various fire regimes of the forest and specifically, the area of concern
- Long-term social and ecological risk of not restoring fire, including the consequences of not burning before wildfires occur
- Short- and long-term risk to various resources of concern within the area
- Locations where the application of fire is not a concern
- Fire effects that are acceptable within areas of concern

Additionally, the Payette National Forest has benefitted from the national Prescribed Fire Training Center (https://www.fws.gov/fire/pftc/index.shtml). Having agency administrators, fire, and other resource specialists participate in the training and outreach has helped all to understand what other agencies and units are doing to overcome challenges and be successful. We share experiences, try out new ideas, and build a network for support at various levels of the organization, all of which are invaluable in the pursuit of improving the effectiveness of not only our use of fire but also our relationships with people inside and outside the Payette National Forest.

Keys to Success

There is no template to restoring fire to this 1 million-hectare landscape. It is being constructed now. There will always be a reason not to move forward, there will always be hurdles to overcome, and there will always be a risk of failure. Success requires collaborating effectively with partners and stakeholders to make the most informed decisions with an eye toward the greatest good for the greatest number for the long-term, hiring people with drive and grit, and facilitating an environment where planners and implementers are not afraid to fail.

> **It is not the critic who counts**; not the man who points out how the strong man stumbles, or where the doer of deeds could have done them better. **The credit belongs to the man who is actually in the arena**, whose face is marred by dust and sweat and blood; who strives valiantly; who errs, who comes short again and again, because there is no effort without error and shortcoming; but who does actually strive to do the deeds; who knows great enthusiasms, the great devotions; who spends himself in a worthy cause; who at the best knows in the end the triumph of high achievement, and who at the worst, **if he fails, at least fails while daring greatly**, so that his place shall never be with those cold and timid souls who neither know victory nor defeat. Theodore Roosevelt (April 23, 1910)

Case Study 13.5 From Normal to Scary to Necessary: Innovations in Great Plains Fire Use

Pete Bauman, email: pete.bauman@sdstate.edu
South Dakota State University Extension, Watertown, SD, USA
Joe Blastick, email: jblastick@tnc.org
The Nature Conservancy, Clear Lake, SD, USA
Sean Kelly, email: Sean.Kelly@sdstate.edu
South Dakota State University Extension, Winner, SD, USA

North America's Great Plains had historically immense grasslands in the heart of the continent. Complex interactions of climate, grazing, and fire yielded a diversity of life that rivals any of Earth's most charismatic landscapes. The vast majority of fires were ignited by people to manipulate vegetation with multiple goals, including attracting grazing animals for food, clothing, and other needs. The interaction of fire and grazing resulted in a shifting mosaic of vegetation age and structure that harbored a sea of life. Thus, grazing and fire were intertwined into the very fabric of the indigenous culture and ecology of the Great Plains.

Today, we have a much different Great Plains. Why? Exposure to foreign diseases decimated North America's indigenous people, and as they were reduced, removed, or relocated, their once widespread use of fire as a tool was essentially eliminated. These changes altered plant community composition and allowed for the expansion of woody species once limited by frequent fire. European settlement quickly reduced the vast herds of grazers while also suppressing fire for the protection of personal and public property. Wild ungulates gave way to domestic livestock and large open landscapes were replaced by fenced pastures. Other developments, such as roads, railroads, enhanced watercourses, and crop fields, limited fire spread and aided in fire suppression. The resulting patchwork of vegetation and fuels sharply contrasts with the historical spatial or temporal patterns associated with prior centuries of fire, grazing, and land use interactions (Twidwell et al. 2013).

Consequently, grasslands have been dominated by grazing alone or complete rest from either fire or grazing. Today, we are witnessing the return of 'necessary' fire by innovative persons and groups who are motivated to use fire to conserve grassland-dependent species and ecological functionality ***OR*** by those who are motivated to maintain healthy grassland communities for profitable livestock operations.

Understanding Fire Impacts: Yesterday, Today, and Tomorrow

Fire impacts vegetation. Understanding fire effects requires knowledge of native, non-native, and invasive plant species. Non-native species are now common and invasive species (some native) are often considered in the goals and success of prescribed fires. Today's fires may either stimulate healthy native plant communities through biomass production, seed production, and

(continued)

Fig. 13.20 Grasslands with burned, unburned, grazed, and ungrazed patches often have high species diversity as with The Nature Conservancy's (TNC) Chippewa Prairie in full bloom following a spring prescribed burn. (Photograph by Joe Blastick, The Nature Conservancy)

Case Study 13.5 (continued)
control of invasive species, or they may degrade those same communities by favoring invasion by undesirable species, so effective fire execution is important. Vegetation response is ultimately based on the season of burn, intensity, frequency, land use history, and the relative health of the plant community (Fig. 13.20).

Historically, fire and grazing interaction would have likely resulted in shifts of a post-fire plant community from early successional broadleaf plants to more dominant grasses over the course of several years. Today, however, land managers using patch-burn grazing (see patch-burn grazing section) or other methods must consider how all species will be affected by fires and grazing, including stimulation of undesirable species that may require additional tools or a shift in fire and grazing strategies. Without this awareness, a fire event might just be 'burning for the sake of burning' which in today's Great Plains might not be ecologically, socially, or economically justified.

Why Burn in the Great Plains?

Fire is useful for maintaining and enhancing healthy native grassland communities…*for the most part* (Fig. 13.21). Well-timed, appropriately applied fire requires art and science. Few tools can rival the effects of a

(continued)

Fig. 13.21 Male and female Dakota Skippers butterflies (*Hesperia dacotae*) rest on native prairie coneflower (*Echinacea purpurea*) in full July bloom after a mid-May fire at TNC's Hole in the Mountain Prairie. (Photograph by Joe Blastick, The Nature Conservancy)

Case Study 13.5 (continued)

well-planned and executed fire. The effects of fire can be partially mimicked with other tools. However, when done well, fire provides the best potential of any tool to achieve certain desired results quickly and efficiently, especially over large areas. This is not to imply that the use of fire is easy and risk-free. The art of fire is to find balance in stimulating the greatest good while mitigating risks associated with a potential fire escape.

A good example is using fire to manage big bluestem (*Andropogon gerardii*) and smooth bromegrass (*Bromus inermis*). Big bluestem is a native, warm-season perennial grass that is desirable for wildlife and livestock forage. Smooth bromegrass, on the other hand, is an exotic, cool-season, perennial grass that was introduced into the Great Plains and now occupies extensive areas and is generally considered undesirable for wildlife and livestock and is an indicator of unhealthy grassland, as it can displace more resilient and nutritious native grasses if not kept at bay. As a warm-season grass, big bluestem grows in the heat of the growing season, whereas smooth bromegrass grows in the cooler spring and fall months. If burned too early in the spring

(continued)

Case Study 13.5 (continued)

while both species are still dormant, one risks stimulating the growth of the smooth bromegrass due to the early warming of the soil, causing it to sprout and flourish. Burning later in the spring when the smooth bromegrass is green, lush, and growing will significantly harm the smooth bromegrass while stimulating the native big bluestem, which will then itself flourish by producing more forage, cover, and viable seed. Burning in mid-summer while the big bluestem is green and lush, one may harm big bluestem and stimulate the smooth bromegrass to flourish with regrowth in the cool fall. Now consider that many grassland communities have over 200 species of native grasses, forbs, and shrubs, along with diverse insects, animals, and other life below and above ground. Finally, consider invasive species concerns, profitability, and politics and it is easy to understand that successful prescribed fire requires both art and science!

Challenges and Innovations in Integration of Agency- and Private Landowner-Led Prescribed Fire

Over decades, federal, state, and non-government organizations (here collectively referred to as 'agencies') have developed most of our modern tools and techniques for wildfire suppression. Many agencies have also developed or adapted wildfire suppression programs to prescribed burning on lands they own and manage, leading to methods and innovations that allow them to use fires to accomplish vegetation and habitat goals such as restoration or recovery. Generally, today's agency-led prescribed burning is accomplished under the adoption of a wildfire 'system' known the National Wildfire Coordinating Group (NWCG); initially developed in 1976 by the Departments of Agriculture and Interior and which now includes several other non-federal affiliates. This system has allowed personnel across various agencies a means of coordination related to communication, training, qualifications, human resources, mechanical resources, health and safety, and fire planning complexity analysis (NWCG 2019).

Most agencies who practice both wildfire suppression and prescribed fire implementation now adhere to the NWCG standards across all aspects of their fire management programs. However, because the NWCG system was developed for wildfire response and suppression, partnering NWCG-compliant agencies with non-NWCG prescribed fire practitioners, such as landowner burn cooperatives, small or local conservation NGOs, local volunteer fire departments, or prescribed fire contractors can complicate prescribed fire operations. This is especially true in the Great Plains states with predominantly privately-owned lands where wildfire response is mostly by rural volunteer fire departments (VFDs).

Under NWCG policies, all personnel involved in a fire event must meet stringent basic training and physical fitness requirements, which can be

(continued)

Case Study 13.5 (continued)

challenging for non-agency staff such as landowners, VFDs, or private fire service providers. Without the basic qualifications being met by all persons, agency personnel often cannot assist in training or fire events that include landowners or VFDs in live-fire scenarios which of course provide the best opportunities for learning and skills building. Therefore, agency personnel often must avoid participation in live-fire events, including live-fire training or burning on private lands, which can diminish the mutual transfer of knowledge and skills.

This is not to suggest that NWCG affiliation completely hinders all prescribed fire cooperation or training among these groups. Agencies conducting prescribed fire in the Great Plains recognize the enormous need for fire application across this vast landscape, and they understand that the inclusion of private landowners is essential. Private landowners also recognize this need, and while there are private landowner groups who are very experienced, others who are less experienced desperately desire appropriate training and guidance in their prescribed fire efforts.

Mutual desire to advance fire has yielded creative options including mutual training opportunities, University Extension outreach (some including live-fire workshops), memorandums of understanding for mutual acceptance of training standards, and various other efforts that improve fire communications, training, and application. In South Dakota (SD), for example, partner agencies, including The Nature Conservancy (TNC), The US Fish and Wildlife Service, SD Department of Game, Fish, and Parks, SD State University Extension, the Natural Resources Conservation Service, Pheasants Forever, and others formed the Prairie Coteau Habitat Partnership. This partnership cooperatively supports private lands burning and has developed classroom-based landowner training addressing fire planning, burn unit preparation, weather, safety, ignition techniques, water handling, tools, equipment, and communications.

Perhaps one of the best examples of persistent innovation by an agency can be found in TNC's fire program in the Great Plains. Challenges related to basic NWCG requirements for landowners already discussed still persist, but from the Sheyenne Delta region of North Dakota through the Loess Hills of Iowa down to the Cross Timbers region of Oklahoma and into Texas, TNC fire program managers are finding innovative ways to cooperate with agencies, landowners, and VFDs to advance fire training and objectives. Part of TNC's ongoing commitment to ensuring prescribed fire remains a viable option in the Great Plains and other areas is the coordination of the popular, multi-partner Prescribed Fire Training Exchanges (TREX) programs. The TREX training events are geared toward fire personnel in landscapes or regions who share similar needs for training and skills in their fuel type or terrain. The TREX program embraces diversification and encourages participation from a great

(continued)

Fig. 13.22 Multi-agency personnel gather with private landowners to conduct sand table fire scenario training supported by the Great Plains Fire Science Exchange. (Photograph by Pete Bauman)

Case Study 13.5 (continued)
variety of fire personnel and skill levels, including landowners in some cases. Further, in partnership with NGO, Federal, and State Agencies, TNC supports outreach to private landowners through training and coaching, even if cooperative live-fire training is not always possible (Fig. 13.22).

In keeping with its commitment to innovation in prescribed fire, TNC now supports live-fire training for individuals who have no prior classroom or field experience and who do not meet any of the minimum NWCG requirements through a new Supervised Participant designation. TNC can allow a limited number of Supervised Participants on the fire line as long as they are under the direct supervision of a fire-qualified coach. This experience allows individuals to be exposed to the process of live prescribed fire in a safe and controlled learning environment. Beyond TNC's efforts, other live-fire training workshops open to landowners and VFD personnel still occur in some areas where a fire culture has persisted over time, such as through university Extension services in Texas, Oklahoma, and Kansas, but they are currently uncommon in the northern Great Plains.

(continued)

Case Study 13.5 (continued)

Managing Prescribed Fire Liability and Risk

Liability and risk are two terms often associated with prescribed burning. Weir et al. (2020) described liability as the 'legal responsibility for one's acts or omissions', whereas risk is defined as 'the likelihood of liability for or loss from exposure to a potentially harmful event'. In the Great Plains, liability and risk generally fall on the person or entity who owns the land where the fire is being conducted. State or federal agencies or non-governmental organizations like The Nature Conservancy that conduct prescribed fire are insured and address liability and risk through a variety of options including memorandums of understanding, liability waivers, and other such tools where appropriate. Private landowners who choose to use fire are often covered under their own insurance policies under individual state law. In general terms, criminal liability and negligence are generally of limited concern if the private landowner has done due diligence in prescribed fire planning and is adhering to state statutes (Weir et al. 2017, 2020). In instances where a service provider is conducting the prescribed fire on behalf of the landowner, liability, risk, and responsibility are often shared as per the agreement between the parties involved, subject to existing law.

The Rise of Landowner Prescribed Burn Associations

For private landowners, the promise of improved ecology or diversity is often not enough incentive to adopt such a complex tool as burning. While there is an economic justification for fire through increased post-burn grassland forage production, the economic payoff alone rarely provides enough of a catalyst to convince a non-burning landowner to begin using fire. However, when the economics and ecology of fire align, we see private landowners earnestly embrace fire as a tool. Such is the case with the woody species encroachment that is threatening ranchlands throughout the Great Plains, largely due to the lack of fire over time. Ironically, landowners are discovering that the only reasonable tool left at their disposal for woody species control is the one tool that was put away decades ago: fire (Twidwell et al. 2013).

For example, eastern redcedar (*Juniperus virginia*) and Rocky Mountain juniper (*J. scopolurum*), both native species, are dramatically changing the vegetation dynamics of the central Great Plains. Expansion of these species is now one of the leading threats to grasslands. Woody species expansion represents a degradation of the grasslands similar to physical land conversion, but it differs in that it is a slow degradation from within that often goes unnoticed until drastic measures are needed. Cedar encroachment from Texas through the central Dakotas is appropriately dubbed the 'green glacier'.

Landowners and agencies have taken note, and there are active campaigns against woody encroachment throughout the Great Plains. While mechanical means are useful, fire is necessary to accomplish control at a meaningful scale

(continued)

Case Study 13.5 (continued)
and at a reasonable cost, and landowners have responded with the formation of landowner-led and managed Prescribed Burn Associations (PBAs). For example, Nebraska's Loess Canyon Rangeland Alliance has burned nearly 60,000 acres of private rangeland since 2002 (Loess Canyons Rangeland Alliance n. d.), while South Dakota's Mid-Missouri River Prescribed Burn Association, one of the newest in the country, has treated approximately 1000 acres in its first 2 years (Kelly 2018). For perspective, over the last 10 years, federal agency-led prescribed burns have treated 1.6 million acres annually across all 50 states, meanwhile private landowners in Oklahoma and Kansas burned just over three million acres annually over the same period. Ultimately, success stems from the desire of private individuals to overcome obstacles in order to accomplish their goals (Fig. 13.23).

Private burn cooperatives have proven to be resilient and harbor certain traits that help them achieve success. Burn cooperatives are motivated to accomplish specific goals and are formed by dedicated individuals who truly have something personal to lose if they fail. Their motivation to use fire safely and efficiently is born of great need. Many of these cooperatives have formal structures with directors and advisors, elected leadership, bylaws, and rules for membership, service, and participation. Over 50 local Prescribed Burn Associations have been initiated in the USA since the mid-1990s (Fig. 13.24).

Landowner cooperatives become highly skilled in their home landscapes and are serious about safety. Written burn plans, equipment and resource coordination, communication, and fire escape response plans are common among the best landowner cooperatives. Negative events, such as burned fences, a fire that escapes control lines and burns neighboring property, or negligence are extremely rare. Landowner cooperatives have a track record comparable to professional fire agencies (Weir et al. 2015) (Fig. 13.25).

Successful landowner cooperatives are flexible. They have proven that agency protocols are not always necessary for successful fire implementation in the Great Plains. They have sifted through the fundamentals of sound prescribed burn protocols without becoming weighed down in unnecessary policies or procedures. Communication, planning, maps, reliable resources, and basic training are part of successful private burn cooperatives, as each cooperative sorts out what works best.

The future of burning in the Great Plains will continue to be defined by innovation from persons unwilling to allow our grassland landscapes to deteriorate. Agencies and NGOs will continue to be challenged with balancing protocols and procedures with basic fire needs and limited financial resources. Landowners will be challenged to educate their neighbors and communities on the need for fires while proving they can handle the fires necessary to sustain healthy grasslands in the Great Plains. Landowners will continue to be served

(continued)

Fig. 13.23 Landowners participate in a cooperative burn on the Edwards Plateau of Texas. "This PBA traces its roots back to Dr. Butch Taylor, who hosted a landowner field day at the Sonora, TX research station and promoted the concept of less reliance on agencies and more reliance on 'neighbor helping neighbor' to accomplish private land burn objectives". (Photograph and quote by Ray Hinnant)

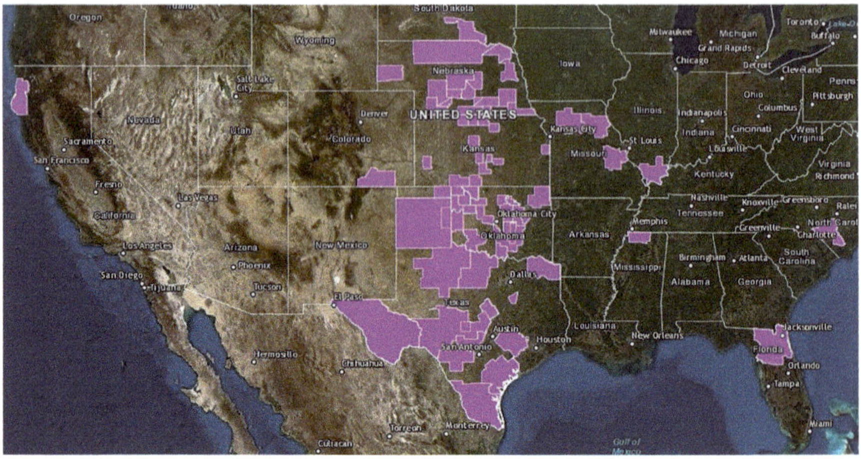

Fig. 13.24 Distribution of landowner-led Prescribed Burn Cooperatives (*in purple*) in the USA (www.gpfirescience.org, Accessed 6 Sep 2019)

Fig. 13.25 Private landowners gather for prayer prior to implementing a cooperative prescribed burn on the Edwards Plateau of Texas. (Photograph by Ray Hinnant)

Case Study 13.5 (continued)
through groups dedicated to disseminating information, such as the Great Plains Fire Science Exchange, various university extension services, TNC, and many agencies.

13.2.4 Addressing Contemporary Challenges by Adapting Traditional Burning Practices

In the savanna landscapes of northern Australia, Indigenous and other people are burning, using practices informed by and adapted from traditional knowledge. Yet the goal is a contemporary one: reducing carbon emissions from burning landscapes, as described by Russell-Smith and Murphy in Case Study 13.6. Burning in the early dry season rather than in the late dry season can alter emissions because it alters how fires burn. In the process, the landscape changes, often favoring wildlife and plants and local culture.

In the forests of northern California, prescribed fires have cultural values as the primary objective. As Huffman and others describe in Case Study 13.7, the traditional use of fire is being revived. While supporting Indigenous culture, the burning is often done in partnership with The Nature Conservancy and agency partners whose people gain valuable experience and training in prescribed burning and fire management while also learning about the local fire ecology important to both people and plants. This is a great example of the successful efforts of partnering with people in local communities to further effective fire management. The Nature Conservancy has been partnering with Indigenous people on fire and land management, fostering local leadership worldwide (TNC 2017). TNC also leads the Fire Adapted Communities Learning Network with funding from the US government.

Case Study 13.6 Contemporary Fire Management in Australia's Fire-Prone Northern Savannas

Jeremy Russell-Smith, email: Jeremy.Russell-Smith@cdu.edu.au; *Brett P. Murphy*, email: Brett.P.Murphy@cdu.edu.au
Research Institute for the Environment and Livelihoods, Charles Darwin University, Darwin, Northern Territory, Australia

Tropical savanna landscapes constitute the most fire-prone of the Earth's biomes and produce most greenhouse gas emissions from biomass burning annually (van der Werf et al. 2017). The savannas of northern Australia (Fig. 13.26), especially in the northernmost 1.2 million km^2 region receiving >600 mm annual rainfall, are no exception. Today, fire regimes in this region are characterized by very frequent fires (typically recurring every 1–3 years),

(continued)

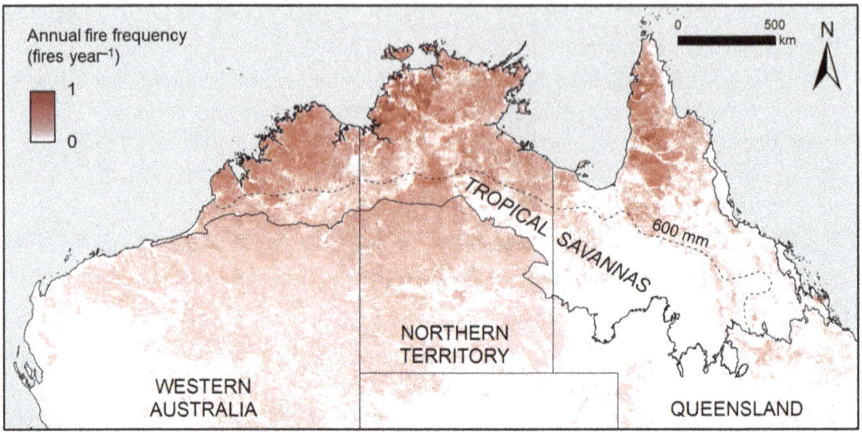

Fig. 13.26 The extent of the tropical savannas in northern Australia, across three states, with annual fire frequency, based on the MODIS satellite record (2000–2017), ranging from 0 (not burnt) to 1 (burnt 17 times in 17 years). The 600 mm mean annual rainfall isohyet is indicated by the dashed line. (Map made by Brett P. Murphy)

Case Study 13.6 (continued)

mostly as extensive wildfires (typically >100 km^2) in the latter part of the 7–8 month dry season (typically April–November) under relatively severe fire-weather conditions (windy, high temperature, and low humidity) (Yates et al. 2008; Edwards et al. 2015). Fires in Australia's tropical savannas produce about 65% of the continent's greenhouse gas emissions from biomass burning (Murphy et al. 2019), despite representing only about 26% of the land area.

The current seasonal fire pattern (i.e., most fires burning in the late dry season (LDS August–November) is the result of the breakdown of traditional Indigenous (Aboriginal) modes of fire and resource management, commencing in the late nineteenth Century associated with the advent of European pastoralism and disruption of relatively fine-scale (multi-hectare scale) burning practices undertaken throughout the year, but particularly prior to the LDS (Russell-Smith et al. 2003). The ecological impacts of such contemporary LDS fire regimes in northern Australia are increasingly well documented and understood, including significant deleterious impacts on animals (including invertebrates and vertebrates), especially those with restricted home ranges and specialized habitat requirements (Lawes et al. 2015); fire-vulnerable vegetation types, especially those supporting fire interval-sensitive taxa (i.e., obligate seeders) (Bowman and Panton 1993); and soil erosion and sedimentation processes, especially on slopes ≥5° (Russell-Smith et al. 2006).

Landscape-scale fire management across this relatively high-rainfall, fire-prone savanna region has proven particularly problematic. The rural

(continued)

Case Study 13.6 (continued)
population is very sparse (averaging <0.2 persons km^{-2}), with few associated infrastructure and management resources. The terrain is generally flat to undulating with limited natural (e.g., watercourses) and built (e.g., roads and tracks) barriers to fire spread, especially under the relatively severe fire-weather conditions typical of the LDS. Despite being within a rich nation, Australia's tropical savannas have suffered from relatively meager resourcing of fire management by Australian Commonwealth, State and Territory governments in the late twentieth Century, mainly because of the region's remoteness: the vast majority of the Australian population lives in southern Australia, and the Northern Territory has a very small population (about 245,000 people spread over 1.42 million km^2).

Development of Commercial Savanna Burning Projects

The signing of the Kyoto Protocol in 1997 opened the door to the development of a *savanna burning* greenhouse gas (GHG) emissions abatement methodology. This is based essentially on emulating traditional savanna fire management practices (focused on the undertaking of strategic landscape-scale burning under relatively mild early dry season (EDS, April–July) fire-weather conditions), in order to reduce fuel loads over extensive areas especially by reinforcing existing barriers to fire spread (e.g., watercourses and previously burnt areas). Such management reduces the risk of extensive emissions-intensive LDS wildfires. Implementation of this initial trial program was undertaken on 28,000 km^2 of Indigenous-owned lands adjoining the World-Heritage Kakadu National Park, in the West Arnhem Land region of the Northern Territory (Fig. 13.27). This region is renowned as a globally significant center of plant and animal endemism, as well as a stronghold for Indigenous culture. Indigenous Traditional Owners own outright or share title (Native Title) across the vast majority of northern Australia's tropical savannas; both tenure types give Indigenous landowners (represented by various statutory land councils or Aboriginal Corporations) the right to manage these lands, and exclude people and activities such as mining and pastoralism.

As described in detail elsewhere (e.g., Russell-Smith et al. 2009, 2013), core elements of the West Arnhem Land Fire Abatement (WALFA) program have involved three elements. *First,* building the capacity of Indigenous landowners to manage fuels and fires over a vast landscape that contained few access tracks and is mostly topographically rugged. *Second,* developing a nationally and internationally credible and accredited GHG emissions accounting methodology, where GHG emissions from EDS fires broadly contribute half those produced in the LDS (per unit of burnt area). *Third,* and perhaps the most challenging of all, developing the governance capacity of local Indigenous institutions to administer the complex implementation,

(continued)

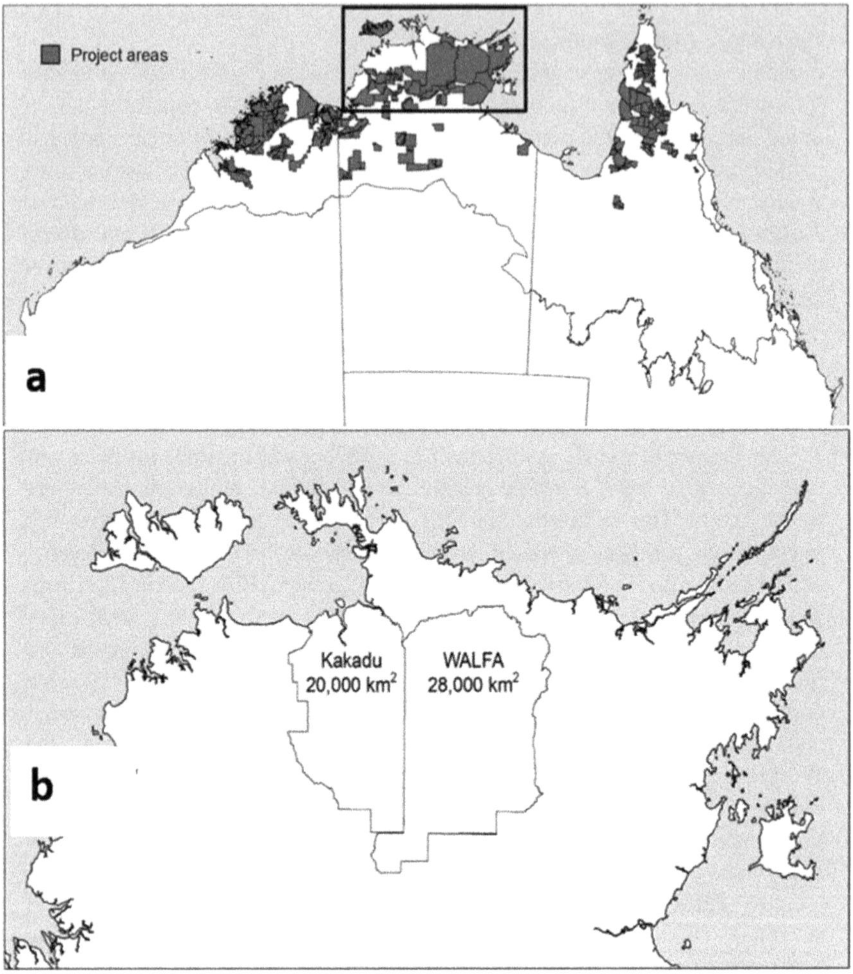

Fig. 13.27 (**a**) Registered savanna burning projects in the tropical savannas of northern Australia. (**b**) The savanna burning accounting methodology that underpins these projects was first developed for the West Arnhem Land Fire Abatement (WALFA) program, adjacent to World Heritage Kakadu National Park. (Map made by Brett P. Murphy)

Case Study 13.6 (continued)
accounting, monitoring, and reporting systems required for the undertaking of a sophisticated commercial operation.

By 2004, the GHG emissions abatement methodology was considered sufficiently robust for the Australian Government's then Australian Greenhouse Office to endorse the accounting processes. In 2006, West Arnhem

(continued)

Case Study 13.6 (continued)
Indigenous landowners entered into a 17-year contract with a multinational energy corporation to annually offset 100,000 tons CO_2 equivalent (t CO_2-e, a basic unit of GHG) of GHG emissions through the implementation of the WALFA program (Fig. 13.27). As of 2019, the WALFA program has continued to deliver the contracted emissions, essentially through changing the pre-project LDS-dominated fire regime to one where most burning is now undertaken under relatively mild fire-weather conditions in the EDS (see example given in Russell-Smith et al. 2013).

In 2011, the Australian Government introduced its Carbon Farming Initiative (CFI) emissions-trading legislation, and in 2013 the *Savanna burning* accounting methodology was formally legislated as the first approved national methodology determination under the CFI Act. Since that time, there have been subsequent amendments and revisions both to the CFI and to the *Savanna burning* methodology itself, notably with additional accounting of dead organic matter biomass sequestration (previously the methodology only accounted for the avoidance of methane and nitrous oxide emissions). Further updates and revision to the current *Savanna burning* methodology are anticipated to occur over the next year or 2, particularly with respect to accounting for the effects of severe fires on live tree biomass (via reductions in growth and survival), fuel loads, and related measurement processes. Importantly, the original CFI emissions trading scheme, and its subsequent replacement with the current taxpayer-funded Emissions Reduction Fund (ERF), have both provided a stable national market platform for the trading of carbon credits (where one Australian Carbon Credit Unit (ACCU) = 1 t CO_2-e), and where currently 1 ACCU ~ US$10.

Current Fire Management Patterns across the Northern Savannas

Incentivized savanna fire management programs are helping transform fire regimes in many fire-prone northern Australian regions. For example, since 2013, ~25% of the entire 1.2 million km^2 northern Australian region is now under a formally registered ERF project (Fig. 13.27). This has resulted in significant achievements in GHG abatement at project scales and has facilitated the implementation of more conservative prescribed EDS fire regimes over extensive regions (Fig. 13.28). It is likely that this shift in fire regimes has benefited a range of plants, animals, and ecological communities, especially those that are sensitive to frequent high-intensity fires.

This management paradigm shift has not necessarily fully addressed complementary ecological challenges, especially the requirement to deliver relatively small patchy fires at hectare and multiple-hectare scales (e.g., Woinarski et al. 2005; Yates et al. 2008). For example, a recent assessment of the fire size distribution in the WALFA project area illustrated that fire patch sizes have decreased significantly with EDS prescribed burning, however, the total area

(continued)

Fig. 13.28 Warddeken Ranger, Greg Lippo conducting early dry season burning in the Warddeken Indigenous Protected Area, Arnhem Land. (Photograph by Rowand Taylor)

Case Study 13.6 (continued)

burnt is still substantially dominated by large fires, exceeding 10 km^2, both in EDS and LDS periods (Evans and Russell-Smith 2019). Such large fires (Fig. 13.29) are thought to have negative impacts on biodiversity, especially animal species with small home range sizes, such as small mammals (Woinarski and Winderlich 2014). Furthermore, it is important to acknowledge that amongst Australian ecologists there are legitimate concerns about the potential for perverse biodiversity outcomes resulting from the roll-out of savanna burning projects across northern Australia. The primary concern is that extensive EDS prescribed burning, without reducing the total area burnt, may be detrimental to certain fire-sensitive species. However, it is well established that low-intensity EDS fires are significantly more patchy (i.e., leaving more internally unburnt sites) than typically very extensive and more intense LDS fires (Price et al. 2003). At this stage, there is no direct evidence of perverse biodiversity outcomes—but, as the savanna burning industry develops, there is a clear need for more research to understand both the short- and long-term effects of this style of fire management on northern Australia biodiversity and ecological functions.

(continued)

Fig. 13.29 Approaching late dry season firestorm in Eucalyptus savanna open woodland. (Photograph by Jeremy Russell-Smith)

Case Study 13.6 (continued)

Irrespective of the biodiversity benefits of enhanced resourcing of fire management in northern Australian savannas, there are enormous social benefits. The tropical savannas are home to a significant proportion of Australia's Indigenous population—a highly economically marginalized segment of Australian society. The vast majority of the region is comprised of Indigenous-owned lands, where economic opportunities are severely lacking. The savanna burning industry has brought much needed economic activity to Indigenous lands, providing culturally-appropriate '*on country*' employment opportunities for Indigenous people. In addition to the direct economic benefits to Indigenous people, the savanna burning industry has contributed resources to enable Indigenous people to re-engage with important cultural aspects of fire and land management.

The inclusion of biomass sequestration components in Australia's updated *savanna burning* GHG accounting methodology also presents additional project implementation challenges given that, unlike the accounting of abatement of GHG emissions through EDS prescribed burning as an annual management activity, sequestered carbon is considered as a property right with long-term permanency obligations (either 25 or 100 years) under Australian law (Dore

(continued)

Case Study 13.6 (continued)
et al. 2014). These complexities are compounded both by different legislative and regulatory carbon rights frameworks operating in respective northern Australian jurisdictions (Queensland, Northern Territory, Western Australia), as well as under different tenure (e.g., freehold, pastoral leasehold, and co-existing Indigenous (Native Title)) arrangements (Dore et al. 2014).

Despite these considerable policy challenges, the implementation of commercial savanna burning opportunities to date has radically transformed what has been an intractable landscape-scale fire management problem in fire-prone northern Australia savannas to one which is providing demonstrable cultural, ecological, and economic benefits. This style of fire management is already being advocated for other tropical savanna regions beyond Australia (Lipsett-Moore et al. 2018).

Case Study 13.7 Indigenous Cultural Burning and Fire Stewardship

Frank K. Lake, email: frank.lake@usda.gov
USDA Forest Service, Pacific Southwest Research Station Fire and Fuels Program, 1700 Bayview Dr. Arcata, CA, USA
Mary R. Huffman, email: mhuffman@TNC.ORG
Indigenous Peoples Burning Network, The Nature Conservancy, Lyons, CO, USA
Don Hankins, email: dhankins@csuchico.edu
California State University-Chico, Department of Geography and Planning, Chico, CA, USA

Introduction

Since time immemorial, Indigenous peoples of fire-prone regions have evolved and adapted with fire. Many Indigenous cultures became fire-dependent as they modified lightning-driven fire regimes. Indigenous peoples created cultural fire regimes that shaped many ecosystems, created unique habitats, and contributed to population dynamics. The interdependent nature of Indigenous fire systems makes these some of the most integrated fire management systems in the world.

Creation accounts and other traditional stories are integrated with traditional law for Indigenous peoples (Black 2011; Eriksen and Hankins 2014; Hankins 2018). They also help Indigenous learners to understand a range of physical, social, ecological and metaphysical factors related to fire. The stewardship of fire across diverse ecosystems is derived from the complex interaction of story, intergenerational and cross-gender teachings, ceremonial observances, and subsistence practices.

(continued)

Fig. 13.30 Many indigenous peoples have traditional stories about fire that help foster their wise stewardship of fire and other resources. In this book, based on a traditional story from the Karuk tribe, coyote brings fire to the people. (Book cover art by Sylvia Long, printed by Chronicle Books)

Case Study 13.7 (continued)

Many of these stories narrate struggles by ancestral beings with vastly destructive fires. Following these tribulations, people were gifted with the ability to use fire for various purposes. For each Indigenous group, the central figures of the story are different and appropriate to the given landscape, but they reveal practical lessons (Fig. 13.30). In this Indigenous world view, fire is *spirit and relation*, which engenders respect and reciprocity. Fire is a sacred gift, and since humans are one of the very few species that can capture fire (from lightning ignitions), store it (as coals in fire hearths), and create it (using fire drills, flints, matches, torches), Indigenous peoples take very seriously the ethical applications of fire and responsibility for fire stewardship.

(continued)

Fig. 13.31 Smoke from a prescribed burn for cultural objectives along the Klamath River in northern California near the creation place of the Karuk People. Photo of the Mid-Klamath Watershed Council. (Photograph by Will Harling)

Case Study 13.7 (continued)

Throughout climatic cycles over millennia, many Indigenous peoples have tended their world with fire (Fig. 13.31), based upon place-specific and intraspecific knowledge. Since ancestral times, fire and other practices have been passed down as an obligation to past, present, and unborn generations to maintain resilient and sustainable ecosystems (Hankins 2018). Traditional cultural fire practices diversified fire's effects spatially and temporally in consideration of desired outcomes for habitat and species in the post-fire environment. In this way, fire became fundamental to Indigenous economies and quality of life. It is this ecocultural relationship with fire and the knowledge of fire effects that inform Indigenous fire stewardship (Lake and Christianson 2019).

Across the USA, Canada, Australia, and other countries, Indigenous fire knowledge and practices were disrupted by European colonization and western societal commitments to fire suppression and control. Genocide, population relocation, assimilation, centralization of fire governance, incarceration,

(continued)

Case Study 13.7 (continued)

and fines for fire use severely constrained the perpetuation of Indigenous fire knowledge and practices. In turn, the erosion of cultural fire regimes had broad implications for many fire-prone ecosystems. Despite this ecocultural devastation, Indigenous peoples around the world are revitalizing their traditional cultural fire systems in today's context.

Indigenous Fire Systems Are Sophisticated, Viable and Ongoing

Indigenous cultural burning practices are distinguished from other fire management (e.g., prescribed burning or fire suppression) by traditional law, stewardship obligations and responsibilities, purposes, post-fire outcomes, and the right to burn (Eriksen and Hankins 2014). Indigenous fire systems integrate dozens of physical, ecological, and social factors (Huffman 2013). Further, Indigenous communities from fire-prone ecosystems have developed fire-dependent cultures in which mental, physical, emotional, and spiritual health are connected with traditional fire knowledge and practices, and with the health of the Earth (Lake and Christianson 2019).

The protocols for fire use and stewardship comes from the teachings of responsibility, which can vary with local environments. Generally, fire is not to be used or controlled for one's own greed or benefit. Fire use should be considerate of the impacts and potential effects it has on humans and for relations in nature (e.g., other species). The responsibilities of those guiding fire conduct and applications differ by cultural roles, leadership, and position in one's tribe/band, village/clan, family/house, gender, and age. As each is taught at a young age about fire, trained through life for fire use and entrusted with varying levels of responsibility for burning, the complexity of the rules of engagement change. In many Indigenous cultures, children are taught about fire, then allowed to use fire with supervision. Depending on their level of maturity, responsibility, and knowledge, older children expand and increase the complexity and scope of their use of fire through adulthood. Skillful burning is acknowledged by the resulting quality of species used for foods, regalia, basketry, tools, and materials that support traditional economies, as well as the patterning of fire to create landscape heterogeneity. The misuse of fire (e.g., setting fire at inappropriate times and locations) is also addressed by the traditional laws of many Indigenous peoples, sometimes with severe penalties as a consequence.

Integration of Western Science and Traditional Knowledge

Contemporary Indigenous fire practices often integrate or recognize the support of scientific knowledge. Planning and implementing prescribed burns may combine uses of traditional knowledge of weather systems, plant phenology, or other indicators to determine appropriate times for burning, but those may also be validated with data from meteorological forecasts, or technical tools and meters for measuring fire conditions in the field. Geographic

(continued)

Case Study 13.7 (continued)

information systems are used for planning and analysis, and global positioning systems help to map fire perimeters or resource patches of interest to the community. Many contemporary Indigenous fire systems practices integrate traditional and western scientific knowledge and understanding. Wildland fire has become both a cultural and an academic, professional pursuit among younger Indigenous community members who are seeking to uphold traditional responsibilities or interests while preparing for mainstream employment in wildland fire research and management. Whereas western science knowledge has strength in reductionist reasoning, traditional knowledge has strength in seeing phenomena as holistically interconnected (Lake et al. 2017).

Challenges to the Continuation of Indigenous Fire Stewardship

Cultural fire practitioners face significant challenges in upholding the responsibilities of Indigenous fire stewardship. Specific challenges include dichotomies between Indigenous and contemporary laws, application of fire suppression training standards to cultural burning, failure to recognize spiritual relationships and time-sensitive ceremonial obligations, and perceptions of Indigenous fire stewardship as a historical process without relevance to current or future conditions. In contrast to the dominant fire fighting systems, which are built with mobile systems that transfer people and equipment to wherever fires are the worst, Indigenous fire stewardship is place-based. Traditional knowledge is passed from generation to generation by elders and community members through stories, direct teachings, and practices. Though different in character and purpose, training for cultural burning is no less rigorous. In some cultures, responsibilities for fire stewardship are inherited or carried only by designated subsets of a community (Stewart et al. 2002; White 2004; Eriksen and Hankins 2014). The video "Revitalizing our Relationship with Fire" (Klamathmedia 2018) illustrates an Indigenous community's perspective on how these two fire cultures differ.

In the United States, we know of no cases in which traditional Indigenous fire training, leadership, and decision-making authority stands on equal footing with the dominant western fire fighting system. There is currently no pathway for practicing traditional cultural burning apart from the training and permitting required by the National Wildfire Coordinating Group (NWCG) Group (https://www.nwcg.gov), state, or local government fire institutions. In addition, Indigenous communities face generation gaps in which young people pursue mainstream lifestyles and move away to pursue career opportunities. Practicing traditional lifeways during intermittent visits home is often insufficient to achieve the requisite knowledge and skill required of Indigenous cultural fire practitioners.

Blended approaches are possible, however. In northern California, Indigenous communities in the ancestral territories of the Yurok, Hupa, and Karuk

(continued)

Fig. 13.32 Young person from the Yurok Tribe gathering acorns using a traditional burden basket. Acorns are a nutritious, fire-dependent, and culturally important Native food. The basket is woven from hazel stems sprouted after a cultural burn. (Photograph by Margo Robbins)

Case Study 13.7 (continued)

Tribes are building professional fire programs in keeping with the standards of the NWCG as a means of providing career opportunities. At the same time, the non-profit Yurok Cultural Fire Management Council is providing training for traditional family-led burns in the area. Local school curricula in the ancestral territories of the three tribes include cultural burning and special events that celebrate fire-dependent native foods and materials for basketry and ceremonial regalia (Fig. 13.32).

In some Indigenous communities, youth are being taught about cultural fire stewardship responsibilities, then encouraged to seek and attain western academic fire degrees and professional qualifications. This integration of Indigenous and Western fire knowledge and management practices are adapting to contemporary socio-economic, cultural, and governance systems. It is the emergence of this new era of Indigenous western-trained researchers, managers, and practitioners that link historical relationships of people and landscapes with modern societal challenges of living with fire. In this way,

(continued)

Case Study 13.7 (continued)

policies, management, and research can be aligned to support effective fire stewardship in support of Indigenous values (Lake et al. 2017).

Building Cross-Cultural Fire Partnerships

Where Indigenous leaders find it acceptable, their communities are working across cultures to accelerate the revitalization of cultural burning practices. For example, in 2013, Miwko? fire practitioners developed and conducted a training burn with professional fire fighters from multiple agencies in central California (including other tribal organizations). The purpose was to restore fire to a culturally important area while also building awareness of cultural burning needs and practices. Similar partnerships between individual tribes and non-profit organizations, universities, for-profit fire companies, or government agencies are becoming more common.

In the USA, Indigenous peoples are also participating in the Fire Learning Network (TNC 2018), which is a suite of national networks focused on the restoration of fire regimes, fire-adapted communities, Prescribed Fire Training Exchanges (TREX), cultural burning, and more. TREX prescribed fire training exchanges are known for their effectiveness in building fire partnerships; among these, the Yurok TREX is known specifically for its emphasis on cultural burning (TNC 2019; Azzuz 2017).

While Memoranda of Understanding (MOUs) are generally not legally binding, they can serve as a formal expression by the signatories of their intent to work together toward shared goals. On October 4, 2019, the Leech Lake Band of Ojibwe of the Minnesota Chippewa Tribe and the Chippewa National Forest signed an MOU that outlined a process and a framework for managing a 750,000-acre area where the Leech Lake Reservation and the national forest overlap (USDA Forest Service and Leech Lake Band of Ojibwe 2019). Following a century of conflict stemming from a legal violation by the US Forest Service (USFS), this MOU records the intent of the USFS to manage the vegetation of the overlap area according to cultural values and pre-defined standards articulated by the tribe with tribal involvement in decision-making.

The Contribution of Biological Stations and University Reserves

Biological research stations and ecological reserves, such as those associated with universities, offer important opportunities to integrate Indigenous fire systems into learning landscapes. Since 2007, the Big Chico Creek Ecological Reserve of California State University—Chico has burned 200–300 acres each year in diverse ecosystems for multiple objectives. The implementation of these burns is guided by a Miwko? traditional cultural practitioner and professor, reserve staff and volunteers, and local Mechoopda tribal practitioners and others. These burns provide students and future fire practitioners an appropriate place to practice prescribed burning with

(continued)

Fig. 13.33 Mechoopda woman learning to burn with guidance to enhance a patch of *Juncus sp.* at the Big Chico Creek Ecological Reserve near Chico, California, USA. (Photograph by Eli Goodsell)

Case Study 13.7 (continued)
integrated perspectives and knowledge transfer among participants (Figs. 13.33 and 13.34).

Revitalizing Cultural Burning, Reconciliation, and Adaptation to Climate Change

Among the many important outcomes of Canada's reconciliation process with Indigenous peoples are substantial financial investments in the wellbeing of Indigenous communities. This includes funding to prepare for the anticipated impacts of climate change, including increased wildfires. For example, the Xwisten First Nation has a long tradition of fire management for cultural purposes. With funds from the First Nations Adapt Program of the Department of Indigenous Services Canada, the Xwisten Nation is working with the First Nations Emergency Services Society of British Columbia to integrate Indigenous cultural values into climate change adaptation planning (FNESS 2019). Revitalizing cultural burning practices and using Indigenous knowledge are key components of their adaptation approach (Fig. 13.35).

(continued)

Fig. 13.34 A vigorous blue oak (*Quercus douglasiana*) (*center of photo*) stands in a spring-fed meadow stewarded with Indigenous fire at the Big Chico Creek Ecological Reserve near Chico, California, USA. (Photograph by Don Hankins)

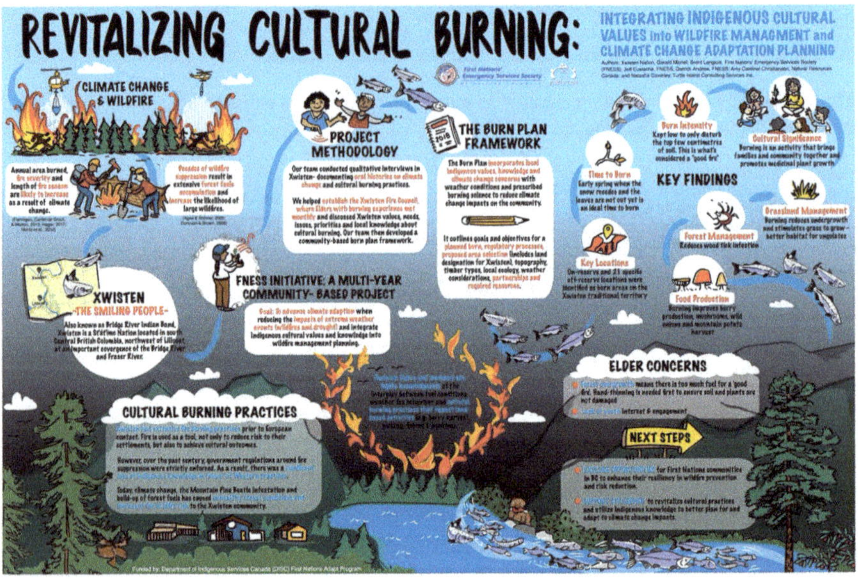

Fig. 13.35 Conceptual diagram of cultural burning (Xwisten Nation et al. 2018)

Case Study 13.7 (continued)

Networking for Greater Impact: The Indigenous Peoples Burning Network and Other Networks

The Indigenous Peoples Burning Network (IPBN) (TNC 2019) is a support network among Native American communities that are revitalizing their traditional fire cultures in a contemporary context. It is one of a group of Fire Learning Networks administered by The Nature Conservancy (TNC) and funded primarily by the USDA Forest Service and the US Department of Interior Office of Wildland Fire. Though currently focused on revitalizing fire culture across the USA, these characteristics reflect worldwide standards articulated in the United Nations Declaration on the Rights of Indigenous Peoples (UNDRIP) (United Nations 2007). The principles of self-determination; free, prior and informed consent; and protection of intellectual property articulated in UNDRIP are central to the operation of the IPBN.

In addition to the IPBN, fire-related networks of multiple Indigenous communities are also growing in Australia and Latin America. Started as a project in 2015, the Participatory and Intercultural Fire Management Network brought Indigenous communities, academic institutions, and government agencies together with a goal to include Indigenous fire practices in government policy in Venezuela, Brazil, and Guyana (Bilbao et al. 2019; COBRA 2019). The National Indigenous Fire Network (n.d.) initiated its knowledge-sharing in 2016 focused on restoring Aboriginal fire, cultural knowledge, and practice in Australia, where this network emphasized Indigenous peoples representing and controlling their knowledge for all Australians. The Firesticks Alliance Indigenous Corporation in Australia is another network, founded in 2018. It supports cultural burning for healthy communities and healthy landscapes with key considerations for respect, responsibility, and "embedding cultural connection within contemporary natural resource management practices" (Firesticks Alliance Indigenous Corporation 2019).

What Can Non-Indigenous Partners Do to Help?

Non-Indigenous fire researchers, land managers, resource specialists, and policy-makers can support Indigenous peoples' efforts toward fire stewardship and sovereignty. If you are not a member of an Indigenous community, you can still make a commitment in your own work and take action to advance Indigenous fire stewardship. Approaches include learning about the history of the Indigenous peoples in the place where you are working, contacting traditional cultural practitioners whose ancestral lands you are studying or managing, even if they currently live far away (see the interactive map at https://native-land.ca/), and walking part of that ancestral territory together with associated Indigenous peoples to learn their fire-related concerns and aspirations. You can invite cultural practitioners to guide the development or revision of management plans and practices you and your organization are

(continued)

Case Study 13.7 (continued)
using. Connect with other groups, agencies, and organizations that are supporting equitable partnerships with Indigenous peoples. Integrate Indigenous peoples and traditional knowledge into fire research to support diversity and inclusivity of thought and practices (Lake et al. 2017). Finally, advocate for including Indigenous peoples through community participatory frameworks in forestry, fuels and wildland fire planning, research/monitoring, new science to advance understandings around Indigenous fire and identifying research needs, and adaptive management.

13.2.5 Burning in Highly Urbanized Landscapes

Across Florida, there is a strong fire culture that expects and supports prescribed burning along with the smoke produced (Case Study 13.8). Despite many people and many cities, fire is used to manage vegetation and wildlife habitat in many different ecosystems throughout Florida and the surrounding southern region of the USA. Both law and practice support the right to burn and state certification of people that supports training and effective practice. Just as the idea of Prescribed Fire Councils has spread from Florida to other locations in the USA, many of the laws, policies and practices have been and could be more widely adopted to support integrated fire management that combines fire suppression, education, and prescribed burning.

Case Study 13.8 Pioneering, Progressive, and Persistent: Florida's Fire Management Is Fire Use

Leda Kobziar, email: lkobziar@uidaho.edu
University of Idaho, Moscow, Idaho, USA
J. Morgan Varner, email: mvarner@talltimbers.org
Tall Timbers Research Station, Tallahassee, FL, USA

"What are we going to burn today?" is the question Steve Miller, former Chief of the St. John's River Water Management District in central Florida, would ask his team every morning. Not "if" or "whether", but "*what*". If they didn't answer, his team would have to defend the decision *not* to burn—a turn of the tables compared to what most public lands managers experience in the USA. The sentiment reflects a uniquely committed, progressive approach to prescribed burning for public lands management in a state with more than 20 million residents live and where over 19% of the landscape is considered wildland-urban interface (Fig. 13.36). Over 909,000 ha on average are burned annually in Florida, equivalent to 7% of the state's total land area, or nearly 21% of the land managed by local, state, and federal agencies (Nowell et al. 2018; Florida Natural Areas Inventory ND). That Florida's prescribed fire

(continued)

Fig. 13.36 Prescribed fire is used in the WUI in central Florida to reduce fuel loads, maintain ecosystems, and promote ecosystem functions. (Photograph by Peter Henn)

Case Study 13.8 (continued)
accomplishments have been sustained in light of 26% population growth since 2000, a large percentage of its lands in private ownership, and continuing natural-to-urban land cover conversions, is remarkable.

Fire use is successful in Florida because its many values have withstood the tests of time. Prescribed fire use is generally aimed at one of two often overlapping goals: fuels reduction and ecological benefits. Burn objectives also differ by ecosystem and landowner type. Most state and federal fire use practitioners identify fuels reduction and promotion of wildlife habitat as primary objectives, while private landowners employ fire for both fuels reduction and for silvicultural purposes of reducing competition for target timber species (Kobziar et al. 2015). Across the southeastern USA, fire managers report a significant but relatively short-term wildfire risk reduction benefit from prescribed burning and therefore apply frequent fire return intervals whenever appropriate. Although limited budgets, staffing, and challenges associated with the wildland-urban interface have been identified as impediments to increased prescribed fire use, Florida implements prescribed burns on

(continued)

Case Study 13.8 (continued)
nearly twelve times the land area burned in wildfires state-wide on average (Nowell et al. 2018). How Florida became a global leader in prescribed fire use is a multifaceted story that is the product of intersecting ecological, political, and social forces under the influence of pioneering scientists, managers, and legislators.

How Policy Requires and Protects the Use of Prescribed Fire in Florida

European Settlers adopted Native American fire use practices, in addition to bringing with them traditions of fire use from western Europe (Pyne 2017). Although fire suppression gained some ground in the South in the 1930s, pioneering defenders of fire's natural role in Florida's ecosystems made scientific as well as cultural arguments for its continued use. Herbert Stoddard, a self-trained naturalist and co-founder of the Tall Timbers Research Station in Tallahassee, Florida, proved unequivocally that prescribed burning the understory of pine forests promoted northern bobwhite quail (*Colinus virginianus*) habitat in his 1931 publication, "The Bobwhite Quail: Its Habits, Preservation, and Increase". Importantly, prescribed burning was shown not only to promote wildlife and reduce hazardous fuels but to be the foundation for a lucrative and influential hunting industry. Such complementarity of benefits would continue to be an influential driver of Florida's fire management culture.

Fifty years later, as Florida initiated its long-standing trajectory of population and development growth, prescribed fire use declined. With increasing wildfires, fire professionals convinced the state legislature that vegetation management was needed near overgrown, absentee homeowner properties. In response, Florida passed the Hawkins Act of 1977 (FS 590.125(4)) which gave the state forestry agency permission to reduce wildfire hazards on private lands, since it was in the public interest. As the least expensive and most effective tool for doing so, the agency used prescribed fire to achieve wildfire hazard mitigation objectives, with complementary benefits to conserving and restoring fire-dependent ecosystems.

Yet the increasing population of retired people, coupled with a widely publicized Florida Supreme Court decision to assign liability for a prescribed fire-related smoke fatality to a landowner, exacerbated fear of personal, economic losses if prescribed fires did not go as planned. In response, a blue-ribbon panel was formed to identify solutions to fire use declines, resulting in yet another pioneering act of legislation. The Florida Prescribed Burning Act of 1990 (now FL State Statute 590.125 (3)) was the first of its kind in the USA and has since inspired similar legislation across numerous southern states. The legislation defines and prescribes how burns are authorized, planned, and conducted, and for properly trained landowners and managers reduces the risk of liability in the unlikely event that fire escapes or smoke cause injuries. The Act included special protection for individuals who earned a Prescribed

(continued)

Case Study 13.8 (continued)

Burn Manager Certification and set standards for the training and continuing education of certified fire use practitioners. Twenty-three states now have prescribed burn certification programs modeled after the Florida program. The Florida Prescribed Burning Act is based on these and other declarations:

(a) *The application of prescribed burning is a land management tool that benefits the safety of the public, the environment, and the economy of the state. The Legislature finds that:*

1. *Prescribed burning reduces vegetative fuels within wildland areas. Reduction of the fuel load reduces the risk and severity of wildfire, thereby reducing the threat of loss of life and property, particularly in urban areas.*
2. *Most of Florida's natural communities require periodic fire for maintenance of their ecological integrity. Prescribed burning is essential to the perpetuation, restoration, and management of many plant and animal communities. Significant loss of the state's biological diversity will occur if fire is excluded from fire-dependent systems.*

The state forestry agency was urged to "maximize the opportunities for prescribed burning" through its burn authorization ("permit") program. Importantly, the habitat conservation *and economic* benefits of fire in Florida's ecosystems was formally recognized by the State. This helps public land managers comply with federal laws (e.g., Endangered Species Act) while incorporating prescribed fire into their management plans. It also allows these managers to argue that funding for training, personnel, and equipment should be allocated to prescribed fire programs for maintaining biological diversity; this is a value to all of Florida's citizens.

The Ecological Imperative

The Florida Natural Areas Inventory lists 23 terrestrial and 19 wetland communities in Florida: 16 of these are considered fire-maintained. The drivers of prescribed fire use in Florida and across the region are reduction in fire hazard, for restoration or maintenance of ecosystems and their numerous imperiled species, or for manipulation of game habitat. Arguably, all three are ecological, for fuels recover rapidly in the climate and setting in Florida (Fig. 13.37), the high plant and animal diversity is a function of the setting and frequent fires, as are the primary game species.

To some degree, all of Florida's fire-prone ecosystems illustrate rapid post-fire vegetation recovery. Resprouting is a prevailing trait in widespread and dominant grasslands (e.g., in wetlands) and shrublands that drive wildfire hazards across the state. The palms, most notably the shrub *Serenoa repens* (Fig. 13.37), resprout from above-ground fire-resistant rhizomes. Heaths and

(continued)

Fig. 13.37 In Florida flatwoods pine ecosystems, a high rate of vegetation recovery coupled with high flammability even when the fuel moisture content is high makes frequent fire use imperative for reducing wildfire hazards. (Photographs by Jesse Kreye)

Case Study 13.8 (continued)
other woody shrubs (*Ilex glabra* and *I. coriacea* among others) resprout basally in rapid sequence.

From a fire hazard perspective, rapid recovery translates into the need to apply fire or other complementary treatments at high frequencies (2–5 or more fires per decade in many Florida ecosystems) in order to diminish hazard and maintain ecosystem assemblages. The fire-prone flora reflects a wide suite of traits to facilitate the spread of fire in an otherwise humid subtropical climate. Frequent fires are ignited in upland pine ecosystems, in the north dominated by *Pinus palustris*, *P. elliottii*, *P. taeda*, *P. echinata,* and *P. serotina*, and in the southern Florida peninsula forests dominated by *P. densa*. These pines cast flammable litter that sustains frequent fires that burn with an intensity sufficient to injure or kill competitors and maintain characteristic open woodlands favored by many co-occurring rare plants and animals. Less frequent, though of much greater in intensity are the fires ignited in *P. clausa* stands in the central and southern peninsula "scrub". These ecosystems similarly harbor an abundance of imperiled species, particularly herbaceous endemics. Highly

(continued)

Fig. 13.38 Continuum of fire across three of Florida's fire-maintained ecosystems. Although fire return intervals (FRI) vary from one to 50 years, each of these ecosystems is fire-dependent. (Photographs by Leda Kobziar and Steve Miller)

Case Study 13.8 (continued)
flammable humid grasslands and shrublands are interspersed with these upland dominants and also burn frequently. Even wetland systems burn frequently, such as the famous Everglades "river of grass", where wildfires are predominantly ignited by late-summer lightning storms and often burn during rainfall (Fig. 13.38).

Prescribed fire's use for upland animals, primarily two native bird species, is a major impetus across land ownerships. In addition to the northern bobwhite quail mentioned above, the red-cockaded woodpecker ("RCW"; *Leuconotopicus borealis*) is an imperiled native bird across the broader southeastern USA which depends on frequently burned pine forest habitat. Especially on federal lands, RCW habitat restoration and maintenance using prescribed fire is a major objective of overall forest management and results in hundreds of thousands of acres burned each year. Both public land managers and private landowners benefit from the complementarity of habitat conservation, wildfire risk reduction, and even timber proceeds associated with habitat-benefitting thinning operations (Stephens et al. 2019).

The availability of prescription windows, with weather conditions suitable for applying fires that are neither too intense nor too marginal to meet

(continued)

Case Study 13.8 (continued)

objectives, is critical to Florida's fire use success. In Florida as in many landscapes, smoke management is a major impediment to the application of fire. The focus of smoke management efforts is to evacuate smoke into the atmospheric mixing layers and allow transport winds to disperse smoke away from developed areas and roadways. Chiodi et al. analyzed 30 years of weather data for daily optimal mixing and transport conditions that overlapped with conditions both conducive to safe burning as well as meeting ecological and fuels reduction objectives. Florida's winter-spring seasons (January to April across most of the state) have frequent and somewhat consistent (year-to-year) availability for burning. Nowell et al. (2018) showed that private and public land managers occupied these windows with between 900 and 1900 prescribed burns per month conducted during the months of January–March, compared with fewer than 500 on average during other months.

Sustaining Success: Training, Collaborations, and Continuing Education

The backbone of the Florida Prescribed Burning Act's success is its science-based, continually updated definition of "accepted forestry practices", coupled with the training and education required for Florida prescribed burn managers. As accepted forestry practices have changed, so too has the Act, with updates in 1999 and 2013 to broaden allowances for smoldering combustion and to increase the length of the burn day for certified individuals. In the most recent version of the Act, aspiring Florida certified prescribed burn managers must complete classroom and experiential training with currently certified burn managers, have their plans and management of an actual prescribed burn evaluated by the state forestry District Manager, and maintain their credentials through successful certified burns and continuing education. Rules designate procedures by which certification can be revoked by the state at any time. The program is designed to ensure that the public understands that although prescribed burning is a landowner's right, the state must authorize all burns, and provides the greatest liability protection by ensuring substantial training and mastery of the latest fire science and tools.

One critical mechanism for knowledge exchange and continuing fire science education in Florida are the state's three Prescribed Fire Councils (PFCs). Membership in councils has no barriers. Public and private land managers, members of the public, land management agency personnel, and regulators are brought together biannually to engage in knowledge exchange and networking. Scientists and managers from across the region present findings with clear implications for prescribed fire use, including fuels management results, modeling tools, meteorological analyses, fire's ecological effects, social and political topics, and guidance for complying with state and federal regulations.

(continued)

Case Study 13.8 (continued)

The three PFCs in Florida were the first-ever established and have served as a model for the growth of an international coalition now comprised of PFCs in 31 states (Melvin 2018).

Continuing education through the PFCs and other knowledge exchange networks (e.g., the Southern Fire Exchange) must then be applied to the landscape, and Florida's fire professionals have developed unique, lasting partnerships for sustained use of prescribed fire. Private-public partnerships such as the Collaborative Forest Restoration Landscape Cooperative Program, the Tall Timbers Research Station, the USDA Forest Service, the US Fish and Wildlife Service, the Southeast Fire Ecology Partnership, the Longleaf Alliance, The Nature Conservancy, the Prescribed Fire Training Center, and others increase prescribed fire application through shared knowledge, planning, and resources. Private-public partnerships have been identified as critical not only for prescribed fire use but also for wildfire suppression as shared planning can promote both and thereby enhance public safety. For example, during Florida's largest wildfire on record, the Bugaboo Scrub Fire of 2007, combined public-private acres treated with prescribed fire near and in the Osceola National Forest were credited with preventing the wildfire from reaching the town of Lake City (pers. comm., Jim Karels, Director, Florida Forest Service). Prescribed fire numbers are closely aligned with wildfire numbers on a state level, where higher numbers of prescribed fires burned prior to wildfires lead to fewer wildfires.

Keeping Fire on our Side: Sharing the Benefits of Prescribed Burning

Florida and neighboring states have been intentional in presenting a consistent and unified approach to public outreach regarding the use and effectiveness of prescribed fire. For example, in 2011 State Fire Chiefs from Florida, Georgia, and South Carolina partnered with the US Forest Service and the Tall Timbers Research Station to analyze and respond to public sentiments regarding prescribed fires. The not-surprising finding that the public didn't enjoy seeing images of flames prompted a new approach to fire management media and outreach. The public information campaigns that resulted, VisitMyForest.org and GoodFires.org, provide concise and definitive descriptions of the benefits and use of prescribed fire for maintaining healthy ecosystems and reducing wildfire risk, and has been adopted by 13 additional state agencies in the South (Fig. 13.39).

Part of Florida's success is that its prescribed fire program is fully accessible to and integrated with the public, with demonstrations and other community events dedicated to fire use education throughout the year and highlighted during Florida's legislated *Prescribed Fire Awareness Week.* Public school teachers in the state learn about fire ecology and fire use in forestry tours sponsored by the Florida Forest Service. These tours highlight

(continued)

Fig. 13.39 The Good Fires message spread to neighboring states throughout the USA, including this media outreach from the Ashland Forest Resiliency Stewardship Project

Fig. 13.40 Images from a coloring book included in K-12 teacher materials for Fire in Florida's Forests educational program. (Florida Forest Service)

Case Study 13.8 (continued)

the use of prescribed fire for ecosystem maintenance and reduction of wildfire hazards, as well as provide educational resources for lessons about prescribed burning (Fig. 13.40).

The state's public information officers are engaged with local communities and regularly provide updates on prescribed burning activities through social and traditional media. As a result, from these and other public outreach campaigns, even in the case of a rare escaped prescribed burn, most Florida citizens express support for the continued use of prescribed fire.

The Florida model for prescribed fire holds tremendous promise in the USA and perhaps more broadly. Within the USA, regional adoption of prescribed

(continued)

Case Study 13.8 (continued)
fire councils, legislation protecting fire managers, and public education campaigns have supported prescribed fire application in spite of population growth and increased scrutiny regarding regional air quality (Melvin 2018). The export of the Florida model to fire-prone western states and more broadly across the USA has so far resulted in some successes and identification of other hurdles. California and the Pacific West have rapidly developed prescribed fire councils, undergone recent changes to state laws, and have observed subtle increases in prescribed burn extent. Many western states are dominated by federal land ownership that contrasts with the Florida private-public mixture and represents different impediments to implementation. Other areas of the USA have seen increased interest in prescribed fire. Major gains could be realized if western states followed the Florida model (Stephens et al. 2019). To date, the southern states represent nearly 80% of all prescribed burn extent annually across the USA (Melvin 2018). Following the Florida model that incorporates policy changes and implementation, a greater understanding of the ecological role of fire in native ecosystems, how the use of fire minimizes economic challenges, and reuniting humans with fire offers tremendous promise for other fire-prone regions of the USA and perhaps globally.

13.3 Applying Integrated Fire Management Effectively

Clearly, effective integrated fire management requires the innovative, widespread use of fire to complement fire suppression and vegetation management objectives. Integrated fire management includes practices informed by both science and local knowledge, all while adapting and learning from monitoring for effectiveness. The best fire practitioners work with partners to engage them in strategic, integrated fire management. The best fire practitioners integrate the "big picture" while taking the time to manage fire to fit people and place, and thus address local needs while furthering healthy ecosystems for people and nature.

As we described in the introduction to this chapter, integrated fire management includes the perspective of individual wildfires as damaging events. Thus, integrated fire management includes the prevention, preparation, response, recovery, and mitigation steps that are focused on managing individual fires as disasters (Fig. 13.41). However, integrated fire management *also* encompasses planning for and responding to many fires over time and space, with a particular focus on increasing both the benefits and lessening the negative impacts of fires. Fire use, including prescribed burning, along with mechanical treatments, are part of integrated fire management. Many prescribed fire programs have been justified because of the way fires can be used to manage fuels and ecological benefits. Managers must use many different tools to accomplish integrated fire management, despite the

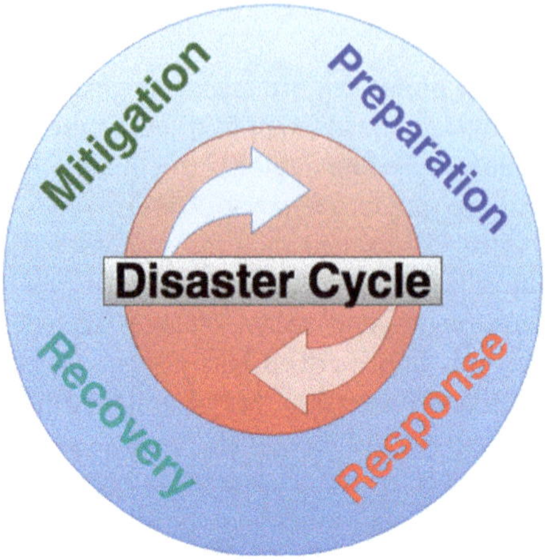

Fig. 13.41 Integrated Fire Management includes and goes beyond the multiple aspects of prevention of fires, preparation for fires, and both early detection and rapid response to fires, as well as recovery from and mitigation of fire effects when those fires are considered wildfire disasters. The same aspects are sometimes termed reduction, readiness, response, and recovery (Kyle Schwartz, https://commons.wikimedia.org/w/index.php?search=disaster+management+cycle&title=Special%3ASearch&go=Go&ns0=1&ns6=1&ns12=1&ns14=1&ns100=1&ns106=1#/media/File:Disaster_Cycle.png)

challenges of prescribed burning, including smoke management, risk management, and costs. Strategically managing wildfires, mechanical fuel treatments, and other approaches will all be part of effective integrated fire management. Sustainable management depends on finding approaches that are socially acceptable, ecologically appropriate, and economically feasible.

Ideally, integrated fire management incorporates the many different dimensions of fire. Thus, fires are not necessarily always a disaster or a damaging event with negative social and ecological consequences. Instead, fires can also have positive consequences if kept in a spatial and temporal balance with the landscape. In this perspective, fire use is central to integrated fire Management from the traditional use by shepherds and farmers to the modern use of fire in fuels and habitat management or in wildfire suppression by trained professionals (Fig. 13.42). Ecosystem services can be delivered through the use of fire, often informed by traditional knowledge and practice. Often, traditional knowledge and local practices can be adapted to address the implications of novel challenges, including but not limited to climate change (Fernandes et al. 2020; Fernandes 2020). Changing from a focus on limiting area burned and controlling fire is needed if we are to overcome what some have called the "fire fighting trap" (Collins et al. 2013) or the "fire paradox" (Arno and Brown 1991, Silva et al. 2010). The paradoxical trap occurs where very efficient and effective suppression of all fires, especially those burning under mild environmental conditions, have decreased the patchiness of vegetation at the landscape scale, and accumulated fuels have contributed to increased fire hazard so that the few fires that

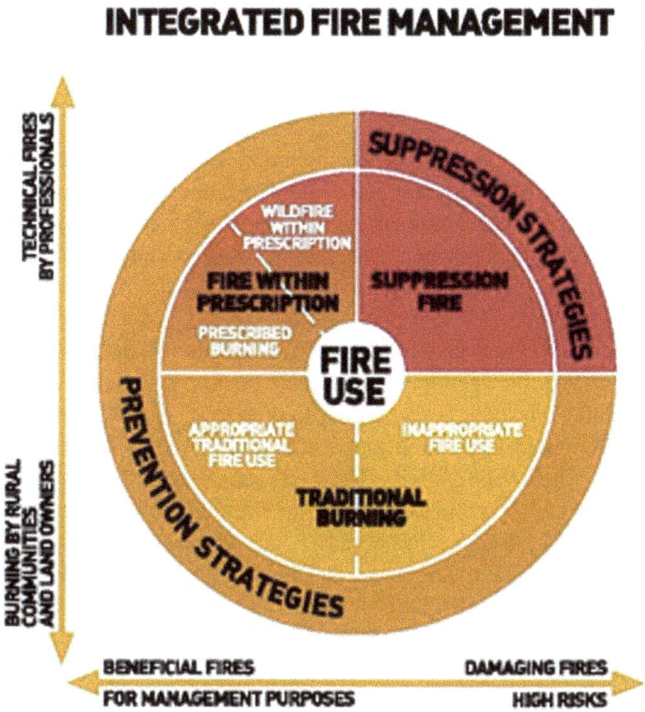

Fig. 13.42 Planning for and using fire is central to integrated fire management. Key is making fire management socially and ecologically appropriate to people and place while also being fiscally responsible and providing the well-being of people and ecosystems (Silva et al. 2010)

escape fire suppression burn with high intensity, and a high proportion of the total area burned burns under extreme environmental conditions (Fernandes et al. 2020; Moreira et al. 2020). One indication of the fire paradox is an increase in variability in annual area burned where fires are actively suppressed (Fernandes et al. 2020). The variability results because fire suppression is more effective in years of mild conditions, and then in years of more extreme conditions, many areas burn even with aggressive prevention, preparedness, and suppression. In a changing world, one with many warmer droughts and more extreme weather in many areas as global climate changes, paradoxical results are likely without a paradigm shift (Moreira et al. 2020). It is often more popular with the public and politicians to respond in years of many large fires and extensive areas burned by building up suppression capacity. Such years can result in high costs of suppression and high societal impacts, yet such years could also trigger changes in fire policies and strategies, just as they have in the past (Fernandes et al. 2020, Moreira et al. 2020). Minimizing the area burned is much easier to communicate and understand as a goal. However, in recent decades we have learned that a sole focus on reduced area burned as the metric of success will inevitably fail because extreme fire events are inevitable if suppression is not combined with strategic and proactive fuels management (Moreira et al. 2020). Integrated fire management is complex, but crucial, especially if we are to coexist with fires over the long-term as the world changes. See Chaps. 10–12 for managing fire effects, fuels, and landscape dynamics.

Effective fire management requires collaboration with others, as fires often burn across diverse landscapes. Working across boundaries, whether those boundaries are between lands managed by different people, or those boundaries are within society, will be needed. Schultz et al. (2018) highlighted how such coordination can address the barriers to prescribed burning and turn them into opportunities. Similarly, effective collaboration can build capacities to transform fire management (Schultz and Moseley 2019).

All over the world, local people are working together to accomplish integrated fire management that fits their place and their people. Often, the most effective approaches are the ones that come from local people. For example, Indigenous people and local communities in Brazil consider themselves the guardians of the forest (If Not Us Then Who? 2018). They use and manage fires in ways that are culturally meaningful and useful to them. In the Republic of Santa Cruz in Bolivia, a trusted older woman observes the weather and assesses fire danger to inform prescribed burning by village farmers (FAN 2013). Thus integrated fire management is not just done by professionals. Indeed, if fire management is solely directed by and done by professionals, it may fail, as local engagement is necessary.

Thompson et al. (2018) argued for changing fire management to emphasize being proactive, focused on long-term effectiveness in meeting strategic objectives. They and others urge managers to take a proactive response to unplanned wildfire ignitions. Responses can be pre-planned based on topography, management goals, and where fires are likely to spread and where that spread could be redirected, encouraged, or stopped. Fire practitioners may herd, delay, encourage, or suppress fires. The strategic management goal is for net positive effects in both the short- and the long-term while thinking strategically (beyond tactics) using risk-based decision making. Proactive assessment and planning drive decisions during fires. Post-fire assessments of decisions focus on their quality and intent (Thompson et al. 2018) while people and organizations constantly learn to be highly effective.

Altering fuels through using fire and managing fire can reduce the burn severity and therefore alter the ecological effects so that we can reap the benefits and manage the impacts for ecosystems and society as large areas burn. Integrated fire management for the future will include adapting existing tools, including mechanical fuel treatments, prescribed fires, and prevention, and suppression of some but not all fires (often this will mean different strategies at different times and on different parts of a single fire), and managing wildfires. No single tool is enough. Managers can potentially choose from their whole "toolbox" to achieve the most sustainable, long-term fire and natural resource management. No single solution, such as logging or limiting all logging, will accomplish desired objectives in all forests. Both less-aggressive fire suppression and expanded use of managed wildfire under relatively moderate weather conditions will often be useful to reduce costs, ensure fire fighter safety, and foster social and ecological resilience. People will adopt new technologies and ways of working together strategically to adapt and mitigate climate and other global changes. People must be clear about both uncertainties, and about goals. Identifying appropriate goals and measures of successful progress is key; the metrics need to be measurable, meaningful, and useful. Based on the case studies of successful application of integrated management in this chapter, those metrics could include social and ecological indicators such as carbon or other greenhouse gas emissions, biodiversity, economic losses and gains, soil erosion potential, human lives affected by fire and smoke, etc.

Even if it is already proactive and strategic, fire management must become more so in light of changing climate, policies, and human goals. Managers may find that both less-aggressive fire suppression and expanded use of managed wildfire under relatively moderate weather conditions can aid them when and where reducing the vulnerability of people and natural resources to fires is the objective. Managing wildfires may be one important way to achieve relatively widespread fuels and vegetation change at the spatial scales and in the short timeframe needed. Monitoring is crucial. All management, including doing nothing, has consequences. Monitoring is not just about whether what was planned was accomplished; it must also be about whether we had the desired effect, and about learning when we try new fire management strategies over large areas and long times. Often, monitoring by multiple people with the mix of backgrounds is needed to measure diverse project effects. Especially if the findings are timely and transparent to all, collective learning can build common understanding and trust (National Forest Foundation n.d.).

There are barriers to integrated fire management. New paradigms can be difficult to accept and implement. Fire suppression is highly visible in the media while the longer-term effects of fuels management and proactive strategies are less simple stories to tell well. When societal costs are high, money and personnel that were budgeted for proactive work often get reassigned to suppression. For many people, the media focus on fires as disasters with images of peoples' homes threatened and burned wildlife can increase fear of fire and then support for suppression. Focusing on whole landscapes, not just the area immediately adjacent to peoples' homes in the wildland-urban interface, can be controversial, while it is also more costly as fire management near homes takes more time and care.

Managers may find that both less-aggressive fire suppression and expanded use of managed wildfire under relatively moderate weather conditions can aid them where reducing the vulnerability of people and natural resources to fires is the objective. Managing wildfires may be one important way to achieve relatively widespread vegetation change at the spatial scales and in the short timeframe needed. Likely this means nimble adjustment of both strategies and tactics. For instance, not all parts of a given fire will necessarily be managed under the same strategy and the strategy can change through time on the same fire. The perspective needs to infuse the full fire cycle, for actions during fire influence fire effects, and post-fire management is also preparation for the next fire.

Landscapes are particularly important, for it is at the landscape scale that the broader regional and national policies meet the local, site-specific needs and opportunities. It is at the landscape scale that management and planning for diverse objectives and vegetation and sites are often possible so that more people meet their management goals more of the time. Thinking and managing across landscapes may result in different prescriptions than what might be considered optimal for individual stands of vegetation if each stand had been managed in isolation. Cooperation on road networks, communication, and shared equipment often result, as does the effectiveness in addressing cross-boundary issues such as wildlife populations and yield of high-quality water. It is not easy to manage fire across all lands, engaging all the different agencies and other land managers with their variety of objectives, yet fire doesn't respect boundaries. At the landscape scale, we can begin to address fire for "All hands, All lands" as called for in the National Cohesive Wildland Fire Management Strategy for the USA and similar national policies elsewhere. At the landscape scale, community-based fire management can empower local communities to promote human well-being and conservation.

The pathway to resilience takes recognizing risk, then adapting to and mitigating risks (Calkin et al. 2011, 2015; Smith et al. 2016; Thompson et al. 2018; Fernandes et al. 2020). Often this requires embracing diverse perspectives. No single "one-size-fits-all" approach will work.

Successful Integrated Fire Management requires finding local solutions to broad challenges. This is clearly illustrated in the case studies. Fire and the social-ecological environment are too complex to represent easily, and so it takes creativity, conversation, trying and learning. Shared stewardship is essential, and that means sharing leadership, innovation, and success.

References

Archibald, S., Roy, D. P., Van Wilgen, B. W., & Scholes, R. J. (2009). What limits fire? An examination of drivers of burnt area in Southern Africa. *Global Change Biology, 15*, 613–630. https://doi.org/10.1111/j.1365-2486.2008.01754.x.

Arno, S. F., & Brown, J. K. (1991). Overcoming the paradox in managing wildland fire. *Western Wildlands, 17*(1), 40–6.

Attorre, F., Govender, N., Hausmann, A., Farcomeni, A., Guillet, A., Scepi, E., Smit, I. P., & Vitale, M. (2015). Assessing the effect of management changes and environmental features on the spatio-temporal pattern of fire in an African Savanna: Fire spatio-temporal pattern. *Journal for Nature Conservation, 28*, 1–10.

Azzuz, E. (2017). *Elizabeth Azzuz talks about cultural fire*. Retrieved November 20, 2019, from https://www.youtube.com/watch?v=aV5ZaIEdukI&t=4s.

Biggs, R., Biggs, H. C., Dunne, T. T., Govender, N., & Potgieter, A. L. F. (2003). Experimental burn plot trial in the Kruger National Park: History, experimental design and suggestions for data analysis. *Koedoe, 46*, 1–15.

Bilbao, B., Mistry, J., Millán, A., & Berardi, A. (2019). Sharing multiple perspectives on burning: Towards a participatory and intercultural fire management policy in Venezuela, Brazil and Guyana. *Fire, 2*(3), 39. https://doi.org/10.3390/fire2030039.

Birot, Y. (2009). *Living with wildfires: What science can tell us - a contribution to the science-policy dialogue*. Joensuu: European Forest Institute.

Black, C. F. (2011). *The land is the source of the law: A dialogic encounter with Indigenous jurisprudence*. London: Routledge.

Boer, M. M., Sadler, R. J., Wittkuhn, R. S., McCaw, L., & Grierson, P. F. (2009). Long-term impacts of prescribed burning on regional extent and incidence of wildfires – Evidence from 50 years of active management in SW Australian forests. *Forest Ecology and Management, 259*, 132–142.

Bond, W. J., & Keeley, J. E. (2005). Fire as a global 'herbivore': The ecology and evolution of flammable ecosystems. *Trends in Ecology & Evolution, 20*, 387–394.

Botelho, H., Fernandes, P., Rigolot, E., Rego, F., Guarnieri, F., Bingelli, F., Vega, J. A., Prodon, R, Molina, D., Gouma, V., & Leone, V. (2002). Main outcomes of the Fire Torch project: A management approach to prescribed burning in Mediterranean Europe. In D.X. Viegas (Ed.), *Proceedings of IV ICFFR,* Luso, Portugal, November 2002, pp 18–23.

Bowman, D. M. J. S., & Panton, W. J. (1993). Decline of Callitris intratropica in the Northern Territory: Implications for pre- and post-colonisation fire regimes. *Journal of Biogeography, 20*, 373–381.

Bowman, D. M., Balch, J. K., Artaxo, P., Bond, W. J., Carlson, J. M., Cochrane, M. A., D'Antonio, C. M., DeFries, R. S., Doyle, J. C., Harrison, S. P., & Johnston, F. H. (2009). Fire in the Earth system. *Science, 324*(5926), 481–484.

Bowman, D. M., Balch, J., Artaxo, P., Bond, W. J., Cochrane, M. A., D'Antonio, C. M., DeFries, R., Johnston, F. H., Keeley, J. E., Krawchuk, M. A., & Kull, C. A. (2011). The human dimension of fire regimes on Earth. *Journal of Biogeography, 38*(12), 2223–2236.

Burrows, N. (2008). Linking fire ecology and fire management in south-west Australian forest landscapes. *Forest Ecology and Management, 255*, 2394–2406.

Burrows, N. (2013). Fire dependency of a rock outcrop plant Calothamnus rupestris (Myrtaceae) and implications for managing fire in south-western Australian forests. *Australian Journal of Botany, 61*, 81–88.

Burrows, N., Ward, B., & Robinson, A. (1995). Jarrah forest fire history from stem analysis and anthropological evidence. *Australian Forestry, 58*, 7–16.

Burrows, N., Ward, B., Wills, A., Williams, M., & Cranfield, R. (2019). Fine-scale change in jarrah forest understorey vegetation assemblages over time is independent of fire regime. *Fire Ecology, 15*, 10. https://doi.org/10.1186/s42408-019-0025-0.

Calkin, D. E., Finney, M. A., Ager, A. A., Thompson, M. P., & Gebert, K. M. (2011). Progress towards and barriers to implementation of a risk framework for US federal wildland fire policy and decision making. *Forest Policy and Economics, 13*, 378–389.

Calkin, D. E., Thompson, M. P., & Finney, M. A. (2015). Negative consequences of positive feedbacks in US wildfire management. *Forest Ecosystems, 2*, 9.

Chandler, C., Cheney, P., Thomas, P., Trabaud, L., & Williams, D. (1983). *Fire in forestry.* New York: Wiley.

COBRA. (2019). *Participatory and Intercultural Fire Management Network.* Retrieved November 11, 2019, from http://projectcobra.org/participatory-and-intercultural-fire-management-network/.

Collins, R. D., de Neufville, R., Claro, J., Oliveira, T., & Pacheco, A. P. (2013). Forest fire management to avoid unintended consequences: A case study of Portugal using system dynamics. *Journal of Environmental Management, 130*, 1–9.

Dore, J., Michael, C., Russell-Smith, J., Tehan, M., & Caripes, M. (2014). Carbon projects and Indigenous land tenure in northern Australia. *Rangeland Journal, 36*, 389–402.

Edwards, A. C., Russell-Smith, J., & Meyer, C. P. (2015). Contemporary fire regime risks to key ecological assets and processes in north Australian savannas. *International Journal of Wildland Fire, 24*, 857–870.

Egging, L. T., & Barney, R. J. (1979). Fire management: A component of land management planning. *Journal of Environmental Management, 3*, 15–20.

Eriksen, C., & Hankins, D. L. (2014). The retention, revival, and subjugation of Indigenous fire knowledge through agency fire fighting in Eastern Australia and California. *Society and Natural Resources, 27*(12), 1288–1303. https://doi.org/10.1080/08941920.2014.918226.

Evans, J., & Russell-Smith, J. (2019). Delivering effective savanna fire management for defined biodiversity conservation outcomes: An Arnhem Land case study. *International Journal of Wildland Fire*. https://doi.org/10.1071/WF18126.

Fernandes, P. M. (2020). Sustainable fire management. In W. L. Filho et al. (Eds.), *Life on land, encyclopedia of the UN sustainable development goals*. Switzerland: Springer. https://doi.org/10.1007/978-3-319-71065-5_119-1.

Fernandes, P. M., & Botelho, H. S. (2003). A review of prescribed burning effectiveness in fire hazard reduction. *International Journal of Wildland Fire, 12*, 117–128.

Fernandes, P. M., Davies, G. M., Ascoli, D., Fernández, C., Moreira, F., Rigolot, E., Stoof, C. R., Vega, J. A., & Molina, D. (2013). Prescribed burning in southern Europe: Developing fire management in a dynamic landscape. *Frontiers in Ecology and the Environment, 11*, e4–e14.

Fernandes, P. M., Delogu, G. M., Leone, V., & Ascoli, D. (2020). Wildfire policies contribution to foster extreme wildfires. In F. Tedim, V. Leone, & S. McGee (Eds.), *Extreme wildfire events and disasters* (pp. 187–200). Amsterdam: Elsevier. https://doi.org/10.1016/B978-0-12-815721-3.00010-2.

Florida Natural Areas Inventory (ND). Florida Resources and Environmental Analysis Center, Florida State University, Tallahassee FL. Retrieved August 24, 2019, from https://www.fnai.org/.

Firesticks Alliance Indigenous Corporation. (2019). Retrieved November 11, 2019, from https://www.firesticks.org.au/.

First Nations' Emergency Services Society of British Columbia (FNESS). (2019). *First nations adapt program*. Retrieved November 11, 2019, from https://www.fness.bc.ca/core-programs/forest-fuel-management/first-nations-adapt-program.

Fischer, W. (1980). Fire management techniques for the 1980's. *Ames Forester, 67*, 23–28.

Fundación Amigos de la Naturaleza (FAN). (2013). *Community-based fire management in Bolivia*. Retrieved November 20, 2019, from https://www.youtube.com/watch?v=MQrt_FHFLYk&t=6s.

Furlaud, J. M., Williamson, G. J., & Bowman, D. M. J. S. (2018). Simulating the effectiveness of prescribed burning at altering wildfire behaviour in Tasmania, Australia. *International Journal of Wildland Fire, 27*(1), 15–28. https://doi.org/10.1071/wfi706.

Gould, J. S., McCaw, W. L., & Cheney, N. P. (2011). Quantifying fine fuel dynamics in dry eucalypt forest (Eucalyptus marginata) in Western Australia. *Forest Ecology and Management, 252*, 531–546.

Govender, N., Mutanga, O., & Ntsala, D. (2012). Veld fire reporting and mapping techniques in the Kruger National Park, South Africa, from 1941 to 2011. *African Journal of Range and Forage Science, 29*, 63–73. https://doi.org/10.2989/10220119.2012.697918.

Hallam, S. J. (1975). *Fire and hearth: A study of Aboriginal usage and European usurpation in south western Australia*. Canberra: Australian Institute of Aboriginal Studies.

Hankins, D. L. (2018). Ecocultural equality in the Miwko? Waali?. *San Francisco Estuary and Watershed Science, 16*(3), 1–11. https://doi.org/10.15447/sfews.2018v16iss3art1.

Hassell, C. W., & Dodson, J. R. (2003). The fire history of south-west Western Australia prior to European settlement in 1826-1829. In I. Abbott & N. Burrows (Eds.), *Fire in ecosystems of south-west Western Australia: Impacts and management* (pp. 71–86). Leiden: Backhuys Publishers.

Hempson, G. P., Archibald, S., Donaldson, J. E., & Lehmann, C. E. R. (2019). Alternate grassy ecosystem states are determined by palatability–flammability trade-offs. *Trends in Ecology & Evolution, 34*, 286–290.

Huffman, M. R. (2013). The many elements of traditional fire knowledge: Synthesis, classification, and aids to cross-cultural problem solving in fire-dependent systems around the world. *Ecology and Society, 18*(4), 3. https://doi.org/10.5751/ES-05843-180,403.

If Not Us Then Who? (2018). *Respect indigenous & local knowledge: Community fire management*. Retrieved November 20, 2019, from https://www.youtube.com/watch?v=niq5ywQzmZo&t=14s.

International Union of Forest Research Organizations (IUFRO). (2018). Global fire challenges in a warming world: Summary note of a global expert workshop on fire and climate change. In F.-N. Robinne, J. Burns, P. Kant, M. D. Flannigan, M. Kleine, B. de Groot, & D. M. Wotton (Eds.), *Occasional Paper 32*. Vienna: IUFRO. Retrieved July 19, 2019, from https://www.iufro.org/publications/article/2019/01/23/occasional-paper-32-global-fire-challenges-in-a-warming-world/.

Kelly, S., (Ed.). (2018). *Mid-Missouri River prescribed burn association news*. Retrieved September 6, 2019, from www.midmissouririverpba.com.

Karuk Tribe (2018). *Revitalizing our relationship with fire*. Klamathmedia.org. Retrieved November 20, 2019, from https://www.youtube.com/watch?v=SF3MNpuqzSg.

Kobziar, L. N., Godwin, D., Taylor, L., & Watts, A. C. (2015). Perspectives on trends, effectiveness, and impediments to prescribed burning in the southern US. *Forests, 6*(3), 561–580.

Komarek, E. V. (1976). Fire ecology review. *Proceedings Tall Timbers Fire Ecology Conference, 14*, 201–216.

Lake, F. K., & Christianson, A. C. (2019). Indigenous fire stewardship. In S. L. Manzello (Ed.), *Encyclopedia of wildfires and wildland-urban interface (WUI) fires*. Cham: Springer. https://doi.org/10.1007/978-3-319-51727-8_225-1.

Lake, F. K., Wright, V., Morgan, P., McFadzen, M., McWethy, D., & Stevens-Rumann, C. (2017). Returning fire to the land: Celebrating traditional knowledge and fire. *Journal of Forestry, 115* (5), 343–353. https://doi.org/10.5849/jof.2016-043R2.

Lambert, B. (2010). The French prescribed burning network and its professional team in Pyrénées-Orientales: Lessons drawn from 20 years of experience. In C Montiel, D Kraus (Eds.), *Best practices of fire use – Prescribed burning and suppression fire programmes in selected case-study Regions in Europe*. EFI Research Report 24, pp. 89–106.

Lawes, M. J., Murphy, B. P., Fisher, A., Woinarski, J. C. Z., Edwards, A. C., & Russell-Smith, J. (2015). Fire size and small mammal declines in northern Australia: evidence from long-term monitoring in Kakadu National Park. *International Journal of Wildland Fire, 24*, 712–722.

Lazaro A, Montiel C (2010) Overview of prescribed burning policies and practices in Europe and other countries. In: Sande Silva J, Rego F, Fernandes P, Rigolot E (eds) Towards integrated fire management - Outcomes of the European Project Fire Paradox. EFI Research Report 23, European Forest Institute, Joensuu, p 137–150.

Lipsett-Moore, G. J., Wolff, N. H., & Game, E. T. (2018). Emissions mitigation opportunities for savanna countries from early dry season fire management. *Nature Communications, 9*, 2247.

Loess Canyons Rangeland Alliance. (n.d.) Retrieved September 4, 2019, from https://www.loesscanyonsburngroup.com.

Lotan, J. E. (1979). Integrating fire management into land-use planning: A multiple-use management research, development, and applications program. *Environmental Management, 3*, 7–14.

Martell, D. L. (2001). Forest fire management. In E. A. Johnson & Miyanishi (Eds.), *Forest fires* (pp. 527–583). San Diego: Academic Press.

McCaw, W. L. (2012). Managing forest fuels using prescribed fire – A perspective from southern Australia. *Forest Ecology and Management, 15*(294), 217–224. https://doi.org/10.1016/j.foreco.2012.09.012.

McCaw, W. L., Neal, J. E., & Smith, R. H. (2002). Stand characteristics and fuel accumulation in a sequence of even-aged karri (*Eucalyptus diversicolor*) stands in south-west Western Australia. *Forest Ecology and Management, 158*, 263–271.

McCaw, W. L., Gould, J. S., & Cheney, N. P. (2008). Quantifying the effectiveness of fuel management in modifying fire behaviour. In: *Fire, environment and society – from research to practice. 15th Conference of the Australasian Fire and Emergency Services Authorities Council*, Adelaide.

Melvin, M. A. (2018). *National Prescribed Fire Use Survey Report*. Coalition of Prescribed Fire Councils, Newton, GA. Retrieved April 2, 2019, from http://www.prescribedfire.net/resources-links.

Menakis, J., Romero, F., Vaillant, N., & Johnson, M. (2016). *Fuels treatment effectiveness review of the Canyon Creek Complex on the Malheur National Forest*. USDA Forest Service. Retrieved October 21, 2019, from http://forestrestorationworkshop.org/wp-content/uploads/FTE-Review-Canyon-Creek-Complex2016_0623.pdf.

Métailié, J. P. (1981). *Le feu pastoral dans les Pyrénées centrales* (p. 293). Barousse, Oueil, Larboust: Editions du CNRS.

Métailié, J. P., Daupras, F., Faerber, J., Lerigoleur, E., Maire, E., & de Munnik, N. (2017). *Evaluation des pratiques de brûlage dirigé dans les Pyrénées-Orientales - Synthèse des perceptions et besoins de différentes catégories d'acteurs impliqués dans les actions de la cellule de brûlage dirigé des Pyrénées-Orientales de 1984 à 2015* (Evaluation of prescribed burning practices in the Pyrénées-Orientales - Synthesis of the perceptions and needs of different categories of actors involved in the actions of the prescribed burning cell of the Pyrénées-Orientales from 1984 to 2015). Evaluation of Rapport d'audit 2016 et 2017, p 200, annexes.

Meyer, M. D., Roberts, S. L., Wills, R., Brooks, M., & Winford, E. M. (2015). Principles of effective USA federal fire management plans. *Fire Ecology, 11*, 59–83.

Minor, J., & Boyce, G. A. (2018). Smokey Bear and the pyropolitics of United States forest governance. *Political Geography, 62*, 79–93.

Moreira, F., Ascoli, D., Safford, H., Adams, M. A., Moreno, J. M., Pereira, J. M. C., Catry, F. X., Armesto, J., Bond, W., González, M. E., Curt, T., Koutsias, N., McCaw, L., Price, O., Pausas, J. G., Rigolot, E., Stephens, S., Tavsanoglu, C., Vallejo, V. R., Van Wilgen, B. W., Xanthopoulos, G., & Fernandes, P. M. (2020). Wildfire management in Mediterranean-type regions: Paradigm change needed. *Environmental Research Letter, 15*, 011001. https://doi.org/10.1088/1748-9326/ab541e.

Moritz, M. A., Batllori, E., Bradstock, R. A., Gill, A. M., Handmer, J., Hessburg, P. F., Leonard, J., McCaffrey, S., Odion, D. C., Schoennagel, T., & Syphard, A. D. (2014). Learning to coexist with wildfire. *Nature, 515*, 58–66.

Murphy, B. P., Prior, L. D., Cochrane, M. A., Williamson, G. J., & Bowman, D. M. J. S. (2019). Biomass consumption by surface fires across Earth's most fire prone continent. *Global Change Biology, 25*, 254–268.

Myers, R. L. (2006). *Living with fire - Sustaining ecosystems & livelihoods through integrated fire management*. The Nature Conservancy. Retrieved November 21, 2019, from http://www.conservationgateway.org/Files/Pages/living-fire.aspx

National Forest Foundation (n.d.). Multiparty Monitoring. National Forest Foundation, Missoula, MT, USA. Retrieved November 20, 2020, from https://www.nationalforests.org/collaboration-resources/

National Indigenous Fire Network. (n.d.). Retrieved November 11, 2019, from https://www.facebook.com/pg/indigenousfire/about/?ref=page_internal.

National Wildfire Coordinating Group (NWCG). (2019). Retrieved September 6, 2019, from www.nwcg.gov.

North, M. P., Stephens, S. L., Collins, B. M., Agee, J. K., Aplet, G., Franklin, J. F., & Fulé, P. Z. (2015). Reform forest fire management. *Science, 349*, 1280–1281.

Nowell, H. K., Holmes, C. D., Robertson, K., Teske, C., & Hiers, J. K. (2018). A new picture of fire extent, variability, and drought interaction in prescribed fire landscapes: insights from Florida government records. *Geophysical Research Letters, 45*(15), 7874–7884.

O'Laughlin, J. (2005). Policy issues relevant to risk assessments, balancing risks, and the national fire plan: Needs and opportunities. *Forest Ecology and Management, 211*, 3–14.

Pacheco, A. P., Claro, J., Fernandes, P. M., de Neufville, R., Borges, J. G., Oliveira, T., & Rodrigues, J. C. (2015). Cohesive fire management within an uncertain environment: A review of risk handling and decision support systems. *Forest Ecology and Management, 347*, 1–17.

Payette National Forest. (2019). *Payette National Forest annual fire report dataset*. McCall: USDA Forest Service, Payette National Forest.

Pekin, B. K., Wittkuhn, R. S., Boer, M. M., Macfarlane, C., & Grierson, P. F. (2012). Response of plant species and life forms diversity to variable fire histories and biomass in jarrah forests of southwest Australia. *Austral Ecology, 37*, 330–338.

Pons, P., Lambert, B., Rigolot, E., & Prodon, E. (2003). The effects of grassland management using fire on habitat occupancy and conservation of birds at a mosaic landscape. *Biodiversity and Conservation, 12*, 1843–1860.

Price, O., Russell-Smith, J., & Edwards, A. (2003). Fine-scale patchiness of different fire intensities in sandstone heath vegetation in northern Australia. *International Journal of Wildland Fire, 12*, 227–236.

Pyne, S. J. (2017). *Fire in America: a cultural history of wildland and rural fire*. University of Washington Press, Seattle WA.

Rigolot, E., & Lambert, B. (2017). *Prospect on prescribed burning development in France*. Paper presented at International Congress on Prescribed Fires, Barcelona, 1–3 February 2017.

Rigolot, E., Etienne, M., & Lambert, B. (1998). Different fire regime effects on a Cytisus purgans community. In L. Trabaud (Ed.), *Fire management and landscape ecology* (pp. 137–145). Missoula: International Association of Wildland Fire.

Roura, N. (2002). *Evolució de la vegetació en un paisatge rural de muntanya: sud del massís de Madres i Mont Coronat, Pirineus orientals (Evolution of vegetation in a rural mountain landscape: South of the Madres massif and Mont Coronat, Eastern Pyrenees) (1953-2000)*. Treball de recerca, Universitat de Girona i Réserve Naturelle de Nohèdes.

Ruane, S. (2018). Using a worldview lens to examine complex policy issues: a historical review of bushfire management in the South West of Australia. *Local Environment, 23*, 777–795.

Russell-Smith, J., Yates, C. P., Edwards, A., Allan, G. E., Cook, G. D., Cooke, P., Craig, R., Heath, B., & Smith, R. (2003). Contemporary fire regimes of northern Australia: Change since Aboriginal occupancy, challenges for sustainable management. *International Journal of Wildland Fire, 12*, 283–297.

Russell-Smith, J., Yates, C. P., & Lynch, B. (2006). Fire regimes and soil erosion in north Australian hilly savannas. *International Journal of Wildland Fire, 15*, 551–556.

Russell-Smith, J., Whitehead, P. J., & Cooke, P. M. (Eds.). (2009). *Culture, ecology and economy of savanna fire management in northern Australia: Rekindling the Wurrk tradition*. Melbourne: CSIRO Publications.

Russell-Smith, J., Cook, G. D., Cooke, P. M., Edwards, A. C., Lendrum, M., Meyer, C. P., & Whitehead, P. J. (2013). Managing fire regimes in north Australian savannas: Applying customary Aboriginal approaches to contemporary global problems. *Frontiers in Ecology and the Environment, 11*, e55–e63. https://doi.org/10.1890/120251.

Sankaran, M., Hanan, N. P., Scholes, R. J., Ratnam, J., Augustine, D. J., Cade, B. S., Gignoux, J., Higgins, S. I., Le Roux, X., Ludwig, F., Ardo, J., Banyikwa, F., Bronn, A., Bucini, G., Caylor, K. K., Coughenour, M. B., Diouf, A., Ekaya, W., Feral, C. J., February, E. C., Frost, P. G. H., Hiernaux, P., Hrabar, H., Metzger, K. L., Prins, H. H. T., Ringrose, S., Sea, W., Tews, J., Worden, J., & Zambatis, N. (2005). Determinants of woody cover in African savannas. *Nature, 438*, 846–849. https://doi.org/10.1038/nature04070.

Schultz, C. A., & Moseley, C. (2019). Collaborations and capacities to transform fire management. *Science, 366*(6461), 38–40.

Schultz, C. A., Huber-Stearns, H., McCaffrey, S., Quirke, D., Ricco, G., & Moseley, C. (2018). *Prescribed fire policy barriers and opportunities: a diversity of challenges and strategies across the West*. Ecosystem Workforce Program Working Paper 86, Oregon State University, Corvallis and Public Lands Policy Group Practitioner Paper 2. Fort Collins: Colorado State University. Retrieved May 18, 2020, from https://scholarsbank.uoregon.edu/xmlui/bitstream/handle/1794/23861/WP_86.pdf?sequence=1.

Shedley, E., Burrows, N., Coates, D. J., & Yates, C. J. (2018). Using bioregional variation in fire history and fire response attributes as a basis for managing threatened flora in a fire-prone Mediterranean climate biodiversity hotspot. *Australian Journal of Botany, 66*, 134–143.

Silva, J. S., Rego, F., Fernandes, P., & Rigolot, E. (2010). Introducing the fire paradox. In J. S. Silva, F. Rego, P. Fernandes, & E. Rigolot (Eds.), *Towards integrated fire management – Outcomes of the European Project Fire Paradox. EFI Res Rep 23* (pp. 3–6). Joensuu: European Forest Institute.

Singleton, M. P., Thode, A. E., Sánchez Meador, A. J., & Iniguez, J. M. (2019). Increasing trends in high-severity fire in the southwestern USA from 1984 to 2015. *Forest Ecology and Management, 433*, 709–719.
Smit, I. P. J., Smit, C. F., Govender, N., Linde, M., & van der MacFadyen, S. (2013). Rainfall, geology and landscape position generate large-scale spatiotemporal fire pattern heterogeneity in an African savanna. *Ecography, 36*, 447–459. https://doi.org/10.1111/j.1600-0587.2012.07555.x.
Smith, A. M., Kolden, C. A., Paveglio, T. B., Cochrane, M. A., Bowman, D. M., Moritz, M. A., Kliskey, A. D., Alessa, L., Hudak, A. T., Hoffman, C. M., & Lutz, J. A. (2016). The science of firescapes: Achieving fire-resilient communities. *Bioscience, 66*(2), 130–146.
Sneeuwjagt, R. J. (2008). Prescribed burning: How effective is it in the control of large bushfires? *In Fire, Environment and Society – from research to practice. 15th Conference of the Australasian Fire and Emergency Services Authorities Council*, Adelaide, pp. 419–435.
Steelman, T. (2016). U.S. wildfire governance as social-ecological problem. *Ecological Society, 21* (4), –3. https://doi.org/10.5751/ES-08681-210403.
Stephens, S. L., Kobziar, L. N., Collins, B. M., Davis, R., Fulé, P. Z., Gaines, W., Ganey, J., Guldin, J. M., Hessburg, P. F., Hiers, K., & Hoagland, S. (2019). Is fire "for the birds"? How two rare species influence fire management across the US. *Frontiers in Ecology and the Environment, 17* (7), 391–399.
Stewart, O. C., Lewis, H. T., & Anderson, M. K. (2002). *Forgotten fires: Native Americans and the transient wilderness*. Norman: University of Oklahoma Press.
The Nature Conservancy. (2018). *Fire learning network*. Retrieved November 11, 2019, from https://www.conservationgateway.org/ConservationPractices/FireLandscapes/FireLearningNetwork/Pages/fire-learning-network.aspx.
The Nature Conservancy. (2019). *Indigenous peoples burning network*. Retrieved November 11, 2019, from http://www.conservationgateway.org/ConservationPractices/FireLandscapes/Pages/IPBN.aspx.
The Nature Conservancy (TNC). (2017). *Strong voices, active choices: TNC's practitioner framework to strengthen outcomes for people and nature*. Arlington: The Nature Conservancy.
The Nature Conservancy (TNC). (2019). *Prescribed fire training exchanges*. Retrieved September 4, 2019, from http://www.conservationgateway.org/ConservationPractices/FireLandscapes/HabitatProtectionandRestoration/Training/TrainingExchanges/Pages/fire-training-exchanges.aspx.
Thompson, M. P., MacGregor, D. G., Dunn, C. J., Calkin, D. E., & Phipps, J. (2018). Rethinking the wildland fire management system. *Journal of Forestry, 116*(4), 382–390.
Twidwell, D., Rogers, W. E., Fuhlendorf, S. D., Wonkka, C. L., Engle, D. M., Weir, J. R., Kreuter, U. P., & Taylor, C. A. (2013). The rising Great Plains fire campaign: Citizens' response to woody plant encroachment. *Frontiers in Ecology and the Environment, 11*, e64–e71. https://doi.org/10.1890/130015.
Twidwell, D., Wonkka, C. L., Wang, H.-H., Grant, W. E., Allen, C. R., Fuhlendorf, S. D., Garmestani, A. S., Angeler, D. G., Taylor, C. A., Kreuter, U. P., & Rogers, W. E. (2019). Coerced resilience in fire management. *Journal of Environmental Management, 240*, 368–337.
Underwood, R. J., Sneeuwjagt, R. J., & Styles, H. G. (1985). The contribution of prescribed fire to forest fire control in Western Australia: Case studies. In JR Ford (Ed.), *Symposium on fire ecology and management in Western Australian ecosystems*. WAIT Environmental Studies Group, Report No 14.
United Nations. (2007). *United Nations Declaration on the Rights of Indigenous Peoples. Document 61/295*. Retrieved November 3, 2019, from https://www.un.org/development/desa/indigenouspeoples/wp-content/uploads/sites/19/2018/11/UNDRIP_E_web.pdf.
US Department of Agriculture (USDA). (2018). *Toward shared stewardship across landscapes: An outcome-based investment strategy*. Washington, DC: USDA Forest Service. Retrieved October 21, 2019, from https://www.fs.fed.us/sites/default/files/toward-shared-stewardship.pdf.

USDA Forest Service, Leech Lake Band of Ojibwe. (2019). *Memorandum of understanding between the USDA Forest Service Chippewa National Forest and the Leech Lake Band of Ojibwe of the Minnesota Chippewa Tribe*. Retrieved November 2, 2019, from https://www.fs.usda.gov/Internet/FSE_DOCUMENTS/fseprd672397.pdf.

Weir, J., Twidwell, D., & Wonkka, C. L. (2015). *Prescribed burn association activity, needs, and safety record: A survey of the Great Plains*. Retrieved September 6, 2019, from https://www.gpfirescience.org/uploads/annountspublis/pdfs/PrescribedBurnAssocSurvey.pdf.

Weir, J., Coffey, R. S., Russell, M. L., Baldwin, C. E., Twidwell, D., Cram, D., Bauman, P., & Fawcett, J. (2017). *Prescribed burning: Spotfires and escapes*. NREM-2903. Stillwater: Oklahoma Cooperative Extension Service. Retrieved September 6, 2019, from http://pods.dasnr.okstate.edu/docushare/dsweb/Get/Document-10793/NREM-2903web.pdf.

Weir, J., Bauman, P., Cram, D., Kreye, J. K., Baldwin, C., Fawcett, J., Treadwell, M., Scasta, J. D., & Twidwell, D. (2020). *Prescribed fire: Understanding liability, laws, and risk*. NREM-2905. Stillwater: Oklahoma Cooperative Extension Service. http://pods.dasnr.okstate.edu/docushare/dsweb/Get/Document-11743/NREM-2905web.pdf.

van der Werf, G. R., Randerson, J. T., Giglio, L., Thijs, T., van Leeuwen, T. T., Chen, Y., Rogers, B. M., Mu, M., van Marle, M. J. E., Morton, D. C., Collatz, G. J., Yokelson, R. J., & Kasibhatla, P. S. (2017). Global fire emissions estimates during 1997-2016. *Earth System Science Data, 9*, 697–720.

White, G. (2004). Restoring the cultural landscape. In *American perspectives on the wildland-urban interface* (pp. 25–29). Quincy: National Wildland-Urban Interface Fire Program, Firewise Communities.

van Wilgen, B. W. (2009). The evolution of fire management practices in savanna protected areas in South Africa. *South African Journal of Science, 105*, 343–349.

van Wilgen, B. W., Govender, N., Biggs, H. C., Ntsala, D., & Funda, X. N. (2004). Response of Savanna fire regimes to changing fire-management policies in a large African National Park: Fire regimes in an African Park. *Conservation Biology, 18*, 1533–1540. https://doi.org/10.1111/j.1523-1739.2004.00362.x.

van Wilgen, B. W., Govender, N., Smit, I. P. J., & MacFadyen, S. (2014). The ongoing development of a pragmatic and adaptive fire management policy in a large African savanna protected area. *Journal of Environmental Management, 132*, 358–368. https://doi.org/10.1016/j.jenvman.2013.11.003.

Wittkuhn, R. S., McCaw, L., Wills, A. J., Robinson, R., Andersen, A. N., Van Heurck, P., Farr, J., Liddelow, G., & Cranfield. (2011). Variation in fire interval sequences has minimal effects on species richness and composition in fire-prone landscapes of south-west Western Australia. *Forest Ecology and Management, 261*, 965–978.

Woinarski, J. C. Z., & Winderlich, S. (2014). *A strategy for the conservation of threatened species and threatened ecological communities in Kakadu National Park 2014-2024*. Darwin: NESP Northern Australia Environmental Resources Hub. Retrieved October 27, 2019, from https://www.nespnorthern.edu.au/wp-content/uploads/2015/10/kakadu_strategy_-_31-10-14_0.pdf.

Woinarski, J. C. Z., Williams, R. J., Price, O., & Rankmore, B. (2005). Landscapes without boundaries: Wildlife and their environments in northern Australia. *Wildlife Research, 32*, 377–388.

Xwisten Nation, Michel, G., Langlois, B., Eustache, J., Andrew, D., Cardinal, C. A., & Caverley, N. (2018). *Revitalizing traditional burning: Integrating indigenous cultural values into wildfire management and climate change adaptation planning*. Kamloops: First Nations' Emergency Services Society (FNESS).

Yates, C. P., Edwards, A. C., & Russell-Smith, J. (2008). Big fires and their ecological impacts in Australian savannas: Size and frequency matters. *International Journal of Wildland Fire, 17*, 768–781.

Chapter 14
Futuring: Trends in Fire Science and Management

Learning Outcomes
After reading and thinking about the material in this chapter, you will be able to:

1. Discuss and give examples of the implications of ongoing and future trends in fire science and management,
2. Synthesize the ideas of integrated fire science with those from the previous chapter on integrated fire management, and
3. Identify trends and challenges for fire science and management that apply in specific cases, and suggest some proactive solutions.

14.1 Introduction

Fires have shaped the evolution of plants and animals over millennia and humans have shaped fire regimes for a long time in the different regions of the world. Even if there is not a general appreciation of the many ecosystem services that fires influence, humans have relied and continue to rely upon many ecosystem services from fires. The social perspectives we have about fire have shaped ecological effects and will shape future fires greatly.

Fires can provision, regulate ecosystem processes, or otherwise provide culturally important ecosystem services. By creating open spaces, fires were an evolutionary force for many of the plants and animals upon which people depend (Pausas and Keeley 2019, Fig. 14.1). But fires can also decrease provisioning and regulation of ecosystem services such as wood production and erosion control, and produce ecosystem disservices, namely material and health disservices like infrastructure damage and air pollution (Sil et al. 2019).

F. Castro Rego et al., *Fire Science*, Springer Textbooks in Earth Sciences, Geography and Environment, https://doi.org/10.1007/978-3-030-69815-7_14

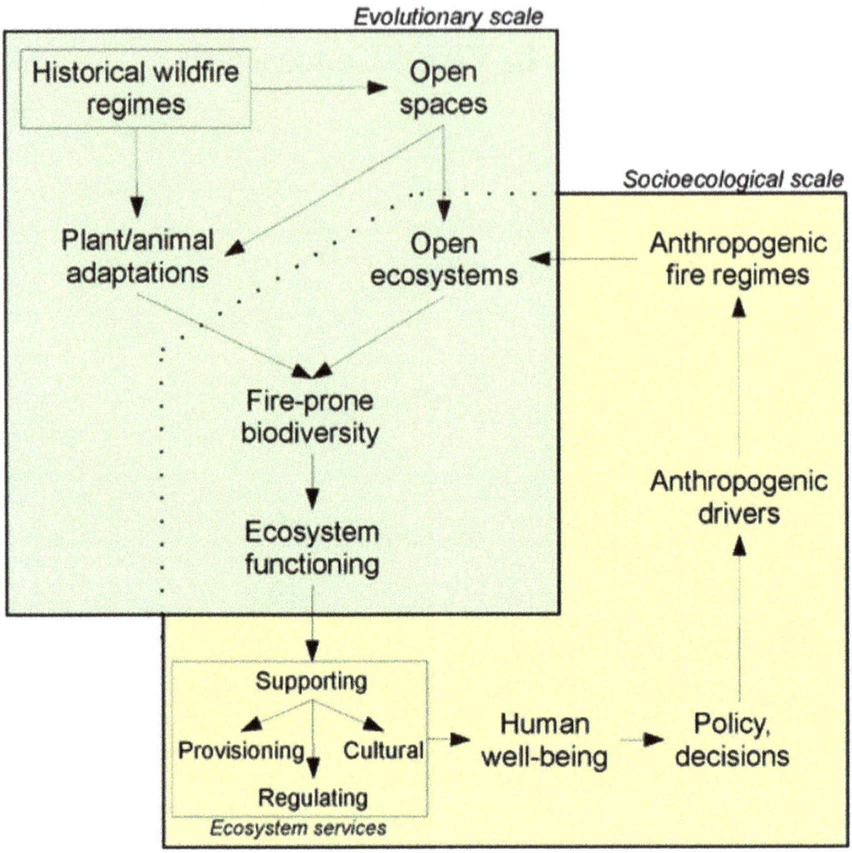

Fig. 14.1 Fires have shaped the evolution of plants and animals over millennia, and humans have shaped fire regimes. (From Pausas and Keeley 2019)

Humans will likely continue to change the land uses and climate and both extreme and other fires will continue to bring smoke, policy issues, costs, and societal discussions. Fires will continue to be important to society with their social and economic impacts, for fires shape ecosystems, affect and respond to climate, and fires are essential to ecosystem health, ecosystem services, and water and carbon cycles. Globally, nearly 450 Mha have burned annually (Andela et al. 2017). Although the global area burned has declined by almost 25% in recent decades (Andela et al. 2017), many scientists predict that the area burned by extreme fires will increase. In this chapter, we highlight ongoing trends that will shape the future of fire science and management.

Changing social-ecological systems and climate are two aspects of global change that are occurring widely but with uncertain consequences for fires, ecosystems, and people. Providing the range of ecosystem services people value while protecting people, property, and economies from the adverse effects of fire and smoke given global change has greatly increased the complexity of fire science and management (Fig. 14.1) as described in Chap. 13 on Integrated Fire Management. To address these challenges, fire science is increasingly interdisciplinary as scientists address ecological and social aspects of fires while recognizing the complexities of integrating across local to global spatial scales, and from immediate to long-term temporal scales (McLaughlan et al. 2020).

Access to new technology, big data, and data analytics are transforming fire science and management. In addition, there is increasing emphasis on collaborations among disciplines, and between scientists and managers. As a result, there is also an increasing trend of more education and training. Ideally, these trends will make our communities more fire-adapted, our ecosystems and landscapes more resilient to future fires, and help guide safe and effective fire response. These trends are already apparent in some national fire management strategies such as the USA's National Cohesive Wildland Fire Management Strategy (Fig. 14.2).

Fig. 14.2 The National Cohesive Wildland Fire Management Strategy was developed for the USA through collaboration among many people from federal, state and local government agencies, multiple non-governmental organizations at these levels, and the public. The strategy integrates people and places for resilience to fires. It is centered on what we know and what we will continue to learn from science and experience

Fire is part of human history, present, and future. Comprehending why is fundamental to understanding the processes, changes, and consequences at local, regional, and national scales. The development of regional fire scenarios for Spain (Montiel et al. 2019) or the comparison of different areas in Spain and Portugal (Sequeira et al. 2019) based on historical fire research are examples of the importance of such studies. However, these analyses focus on particular areas or regions. For global analyses, we need to focus on global changes, including the drivers that operate globally that include climate change and social trends.

14.2 Global Changes Already Influence Fires and Fire Effects

Global changes, including climate change and human population change, are already influencing the occurrence, size, and ecological and social effects of fires. Climate change has already contributed to an increase in the occurrence of extreme and catastrophic fires, longer fire seasons (>18.5% longer worldwide, Jolly et al. 2015, Fig. 14.3) and an increase in the annual area burned in many areas (Williams and Abatzoglou 2016, and others) even as the area burned globally has decreased (Andela et al. 2017). Many large fires around the world have been costly to suppress and have resulted in considerable losses of human life and property (Bowman et al. 2011; Lannom et al. 2014; Doerr and Santín 2016). These trends, driven by global warming and a history of land management practices will be part of the Anthropocene, this epoch when people strongly influence Earth processes. See Chap. 8 for discussion of extreme fires.

As with climate change, demographic changes are occurring worldwide. Globally, human populations are changing their geographic distribution and their social, political, and economic relationships to natural resources and to fire. While many rural areas, especially in areas of low productivity, are depopulating, the global population is increasing with more people living in urban areas. In some regions, many wildland-urban areas are extensive and growing rapidly. As a result, more fires are damaging and judged as being extreme. All of these trends and others mean that people and the ecosystem services we value are increasingly vulnerable to fire and smoke in many places around the globe. Society must find ways to live with fire and to foster the good work fires can do in landscapes while reducing ecosystem vulnerability and negative consequences for people. See Chap. 10 for our discussion of vulnerability and resilience. See Chap. 12 for how climate, fuels, and prior fires are affecting how fires burn.

Landscapes reflect and influence changes. Social changes have altered the fuels that burn when fires ignite, and therefore the size and intensity and severity of fires. We might expect more extreme fires in the future, particularly if most of the smaller fires that are burning under relatively mild wind and fuel dryness continue to be suppressed in the future. Landscapes have changed greatly through land use, so

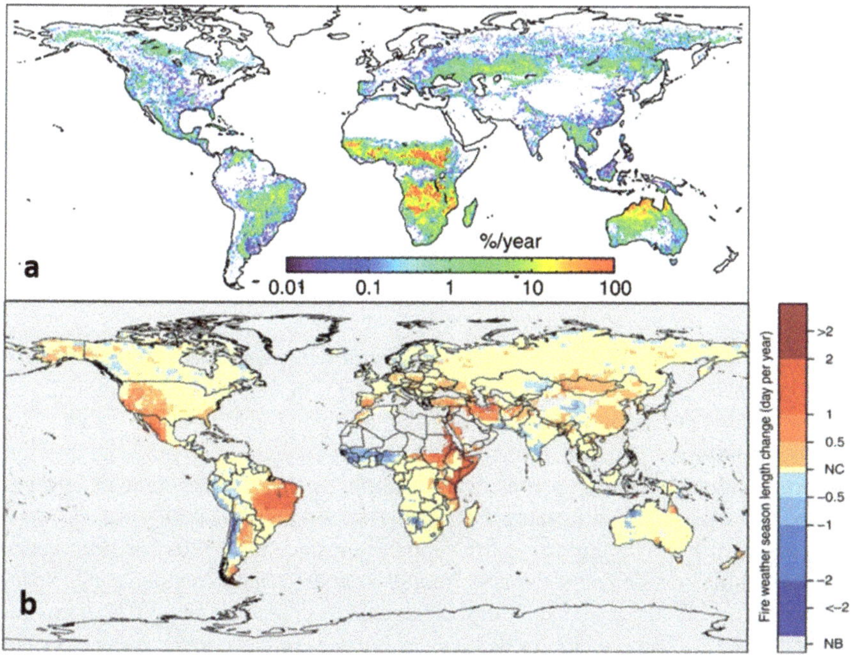

Fig. 14.3 (**a**) Extensive area burned in recent decades (1997–2013) and more is expected in the future (Giglio et al. 2013). (**b**) Globally, long fire weather seasons were more frequent in most places but not everywhere. On average, global fire seasons are 18.5% longer globally in the time period considered (1979–2013), and this trend is likely to continue into the future (Jolly et al. 2015)

much so that the vegetation trajectories are often novel, especially under the influence of changing climate. For some ecosystems, future trajectories may be quite different from the historical range of variability (HRV, See Chap. 12), especially with invasive species.

Future conditions will be increasingly novel. Uncertainty is certain. Current trends for the relationship between fires and people can be determined, and their legacy will shape future ecosystem responses to fires. We know that fire regimes and vegetation response to fires will change with climate and social trends. In the future, fires will likely occur in places and burn in ways that are unknown to the plants and animals that often depend on fires to maintain their habitat and unfamiliar to those who study and manage them. This uncertainty also arises from many other unexpected sources. Fire management organizations have been involved in the responses to floods, earthquakes, and other hazards. Fire managers are effective leaders, but these assignments compound the stress of longer fire seasons, financial oversight on decisions in managing large fires, and the complexities for managing fires burning across boundaries with multiple different objectives. Currently, the viral COVID-19 disease poses great challenges for society. The leaders in fire organizations wrestle with the novel requirements of maintaining physical distance when fighting fires as they worry about how smoke exposure will interact with COVID-19 exposure for fire personnel and the public (Rover 2020). Increasingly, fire scientists and managers are learning to expect the unexpected.

14.2.1 Climate Change: More Extreme Wildfires with More Severe Impacts

Changing climate is already influencing fires worldwide, and it will become increasingly important as climate change trends continue. The impacts vary regionally. In general, more extreme wildfires are occurring as a combination of the weather and drought conditions, the fire proneness of the landscapes, and more people and property in the path of large fires. The example of the "Black Summer" fires in Australia in 2019–2020 illustrates these changes (See Sect. 14.2.3).

In southern Europe, aggressive fire suppression since the 1990s has been generally successful in decreasing the area burned in many countries despite increasing trends in fire danger and landscape flammability (Turco et al. 2016; Curt and Frejaville 2018). Some of the most tragic wildfire events occurred in Spain (1994, 2006 and 2017), Portugal (2003, 2005 and 2017) and Greece (2000, 2007, and 2018) suggesting a new wildfire context in Europe defined by extreme surges in fire growth and heat release (Rego et al. 2018).

The effect of global warming on the area burned is clear. In California, Williams et al. (2019) attributed much of the five-fold increase in areas burned in recent decades to anthropogenic climate warming. Increased temperature of 1.4 °C since 1970 has contributed to more summer and fall fires by increasing evaporation from soils and vegetation and drying fuels. Williams and Abatzoglou (2016) similarly attributed more than half of the increase in area burned in recent decades across the conterminous United States to anthropogenic climate change. Williams et al. (2019) also highlighted the challenges of continuing warming for increased area burned in the future with impacts varying from place to place as they are altered by fire and land management, ignitions by people and lightning, vegetation types, and their interactions.

The effect of changing precipitation with global warming is also clear. Warmer droughts foster vegetation stress and mortality and favor fires, though these effects vary from place to place and it is more difficult to predict changes in precipitation than changes in temperature. Holden et al. (2018) found that the annual area of forest burned was greater when low precipitation occurred during the fire season (summer and fall) in the western USA. They found that the influence of the number of rainy days on area burned was more than 2.5 times greater than the net effect of short-term drought as indicated by vapor pressure deficit, and both were substantially more important than winter snowpack. If these relationships hold into the future, the combination of warmer temperatures and more frequent droughts, especially during the fire season, will have many and far-reaching ecological and socioeconomic implications in addition to fires themselves. If there is less water in streams in late summer because streamflow peaked earlier, and less moisture in the soils to support plant growth and establishment, this could result in tree and shrub crowns dry enough to fuel intensely burning fires and alter ecosystem recovery from fires. Already, Davis et al. (2019), Stevens-Rumann and Morgan (2019) and Stevens-Rumann et al. (2018) found that many warm, dry sites now forested may have crossed a threshold for successful tree establishment following large forest fires in the western USA. If so, then some forests could be replaced by shrublands or other vegetation, especially at lower timberline. Similarly, trees are failing to regenerate on many sites in the Mediterranean basin following more severe or more frequent fires, namely in evergreen oak woodland (Acácio et al. 2009; Guiomar et al. 2015) and mountain pine forests (Martín-Alcón and Coll 2016). See Chap. 9 and Case Study 12.3 for more discussion on post-fire vegetation recovery changing with changing climate.

Changing climate has influenced the area burned directly and indirectly through interaction with fuels. In the western USA, less snowpack in the spring due to warmer springs, warmer summer temperatures leading to lower fuel moisture, and decreased summer precipitation are all implicated, yet few analyses include all three or their interactions. Further, ongoing changes in vegetation interact with climate changes to influence future fires, yet few studies have investigated the effects of interactions among changing climate, fuel complexes, fires and other disturbances on the future area burned. Hurteau et al. (2019) found that relative to considering climate only, including fuels as affected by previous fires reduced estimates of future area burned by 14% while emissions of carbon and particulates were reduced by 12% and 13%, respectively when fuels and climate were simulated together for forests of the Sierra Nevada mountains of California. Most importantly, the vegetation-fuels-fire feedbacks were more pronounced for the largest fires. The effect of altered fuels is short-lived and depends on repeated fires, including prescribed burns that could be used to help manage forests at low and middle elevations (Hurteau et al. 2019). Wet periods that promote grass followed by dry periods resulting in low fuel moisture can be especially important in open "fuel-limited" systems where fine fuels that accumulate with moisture and then dry are important for fueling fire spread (Williams et al. 2019). Climate influences vegetation directly and indirectly through fires, while burn severity and consequent vegetation recovery are also influenced by other factors such as topography that also interact in multiple ways to complicate the interplay between climate, fuels, and fire.

On a global scale, fires influence the carbon cycle. Carbon, both terrestrial and atmospheric, is affected by fire regimes, but the process is not simple. Even when fires burn severely, much carbon remains in burned trees and logs, as well as in many unburned areas, and this is not reflected in many of the simulation models used to forecast the implications of fires for carbon emissions from burned forests (Stenzel et al. 2019). Forests stored less carbon and had lower carbon uptake where fires burned with high severity (Hurteau et al. 2019; Stenzel et al. 2019).

In summary, fires burn large areas annually across Earth's land area (Fig. 14.3), and fire seasons are getting longer all around the globe (Jolly et al. 2015). Likely this reflects earlier springs, later falls and warmer droughts, all of which will influence fires directly and also indirectly through effects on vegetation and people, and these will, in turn, affect the carbon sequestered (or not) in ecosystems. See our discussion of burn severity in Sect. 9.6 and 12.2, carbon in Sect. 9.5, and changing fire regimes in Sect. 12.5.

14.2.2 Social Changes: New Challenges and Opportunities

Fire is increasingly recognized as a social-ecological system. Though fire is a biophysical process, fire science, management, and policy are social, political, and economic, and these all reflect peoples' perceptions about fire and fire risk. Fires have always and will increasingly reflect social, political, and economic forces. Worldwide, humans ignite many more fires than lightning does (e.g., Balch et al. 2017 for the western USA). Human values shape land use, fire response, and the policies that shape both fire response and land use. Perceptions of fire will ultimately shape the size, intensity, and effects of future fires. This will be increasingly true as human influence expands around the globe. Fires made us human, and people are reshaping the role of fire on Earth (Bowman et al. 2011; Pyne 2015).

Fire science and management are increasingly welcoming and learning from diverse perspectives and social science is fundamental. In many traditional communities, shamans and wise women and wise men taught others based on what they observed and tried. They shared traditional knowledge through stories and examples (Huffman 2014). This is the earliest fire knowledge, yet these diverse perspectives have seldom been welcomed by western science until quite recently.

The role of women in fire science and management has been often overlooked. It is true that much of the initial work on western fire science has been associated with men, as pictured in the first chapters of this book. This was caused by the historical societal biases for funding, social norms, and related opportunities. However, these historical biases have fortunately changed to a much more balanced situation in the past decades. Smith and Strand (2018) highlighted 146 women leaders in fire science. This and a similar earlier article (Smith et al. 2018a, b) have fostered many conversations about how we can all work to promote diversity in our discipline. Increasingly, women and others are contributing diverse perspectives to enrich fire science and management.

In spite of progress, discrimination is still occurring globally. McDonald (2012) highlighted how prevalent sexual harassment is. Gender discrimination and sexual harassment are widely experienced by women in wildland fire management (Fig. 14.4, AFE 2016a). This issue must be addressed if fire science and management are to benefit from the many different perspectives a diverse workforce brings. We expect that more women will work in fire science and management as the breadth of opportunities and needs become clear, and we hope that they will be increasingly represented in fire leadership roles. We believe that the groups who are generally

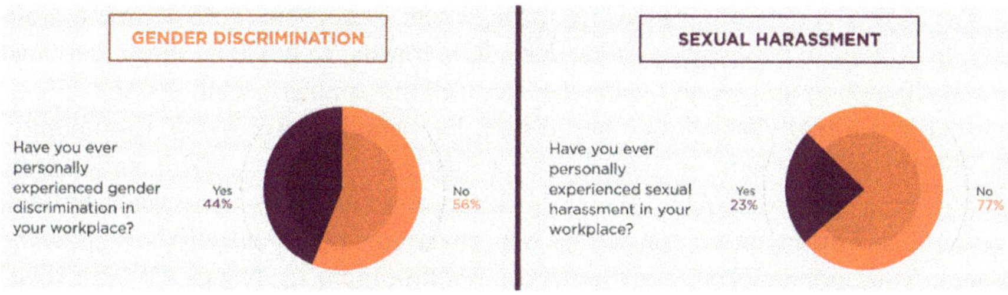

Fig. 14.4 Many of the 342 male and female respondents to an international survey of fire scientists and managers said they had experienced gender discrimination or sexual harassment. (From AFE 2016b)

under-represented, including women, have unique talents and perspectives, and that they can play a critical role in advancing problem-solving in both fire science and fire management. We need more opportunities "where women and men can discuss and understand current issues and work together to build a more inclusive, supportive culture in fire" such as the Women Training Exchange (Lenya Quinn-Davidson, personal communication; Stamper 2017). We believe that the fresh approaches and insights that come with gender, racial, and disciplinary diversity will help address the increasing complexity of the fire challenges for society.

A major trend in global social changes is that human populations are increasing in many wildland-urban areas where fires are likely to threaten people and their property when surrounding vegetation burns. Many rural areas are declining in population as urban areas grow. These trends influence peoples' familiarity with fire as well as the social and political acceptance (or more often fear) of fires and smoke. While human well-being is closely linked to fires and their consequences (Huffman 2014), the often strong emotional reaction to fires reflects both fascination and fear. Social beliefs about fire vary with traditional and local knowledge, gender, social classes, and ethnicities. These beliefs influence fire management strategies around the world. Some strategies will build from embracing anecdotal, qualitative, and experience-based learning more typical of traditional knowledge and integrating that with the ideas from western science. Other strategies come with a mindful focus on social justice, including valuing ecosystems and their services. Community-based fire management strategies that focus on the challenges and knowledge of local ecosystems and people while responding to regional and national priorities will become increasingly common.

Globally, fire management is increasingly complex and challenging. There is widespread public attention, in part because fire is compelling enough that many people have an opinion. Further, global change will force attention to linkages between fire ignition, behavior, and effects, forcing us to explore where and how we can sequester carbon in fire-prone environments. See Chap. 9 for our discussion of fire and carbon in ecosystems. See Case Study 13.6 to learn how carbon sequestration can increase and cultural values increase through altering the fire and changing the season of fire use.

We expect community-based fire management to become more common globally, as billions of people worldwide depend on forests, woodlands, shrublands, and grasslands for food, grazing, watershed protection, or other social, economic, cultural, and spiritual values important to rural livelihoods (FAO 2011). Community-based fire management is useful, for it fits fire to places and people while empowering people (FAO 2011). Such approaches have developed through "grass-roots" efforts often assisted by non-governmental organizations such as the Nature Conservancy (TNC) or the World Wildlife Fund for Nature (WWF). TNC (2017) provided a framework for such efforts (Fig. 14.5). This is especially important in fire-adapted ecosystems where conservation of biodiversity and ecosystem services are objectives. As Indigenous people manage or have tenure rights to over 25% of the world's land, and their territories include much of the global biodiversity and forest carbon, their fire and vegetation management actions matter globally. Local people can foster local jobs and a sense of control over their future when they can manage surrounding landscapes themselves or in shared stewardship with other land managers (TNC 2017). Despite development pressures, giving voice to locals that informs their choices and fosters action is critical to sustaining efforts for conservation and thriving communities (Fig. 14.5).

Fires are increasingly managed across boundaries (Schultz and Moseley 2019). Those boundaries are often geographical, as fires spread from land managed by one entity to adjacent land managed by another entity. Fires also move across social boundaries as different groups of people affected by a single fire may have very different perceptions and experiences with fire. It is not easy to manage fire across all lands, engaging all the different agencies and other land managers with their variety of objectives, yet fires do not respect boundaries, and effective response depends on

Fig. 14.5 Fire can be part of the community-based management that is part of thriving communities filled with people whose voices are heard in making collaborative choices about actions that further community goals. This encompasses fire and broader social-ecological system goals. (Adapted from TNC 2017)

changing policies and practices at multiple scales (Schultz and Moseley 2019). At the landscape scale, we can begin to address fire for "all lands, all hands" as called for in the National Cohesive Wildland Fire Management Strategy in the USA and similar policies in other countries. At the landscape scale, the different expectations and objectives can often be met in different, complementary locations to accomplish effective fire management across boundaries.

If we are to live with and benefit from fires, we need fire-adapted homes and communities in fire-resilient landscapes. Policies and programs are responding to fires, but we hope and expect that fire response will be increasingly proactive and based on understanding. To engage effectively with fire, people will have to accept and manage risk and communicate that effectively, collaborate with partners who may have different values, objectives, and experiences than their own, and build trust and credibility around local solutions to regional and global challenges (Enquist et al. 2017) including fire.

14.2.3 Global Change and the Australian "Black Summer" Fires

Vegetation fires are an intrinsic element of terrestrial ecosystems under seasonally dry climates. Fires affect an annual average of about 4.5 million km^2 of the Earth's surface. However, until recently the global relevance of fire was hardly acknowledged because most of the burned area coincided with sparsely populated regions, such as tropical and temperate savannas, grasslands, and boreal forests. Wildfires have become more prominent in recent years, a consequence of their heightened impacts as measured by the loss of human life and assets. The tragic fires of 2017, 2018, 2019, and 2020 in Portugal, Chile, Greece, the USA, and Australia are vivid in the collective memory. Tragic fires have spurred policy review and changes in the past, just as they are doing now in Australia (Morgan et al. 2020).

Wildfires in the Brazilian Amazon and in southeastern Australia were the highlights of 2019 and early 2020. The fires in the Amazon were mainly a collateral consequence of the loss or degradation of natural forest cover, rather than its cause. They reflect slash and agricultural burning in recently deforested areas, as the moist environment of evergreen tropical forests typically inhibits fire spread. In contrast, the Australian fires have been influenced by climate change, which induces more severe and lengthy fire seasons, social change that has more people and cities in the path of the fires and smoke, and changing fuel conditions as a result of fewer low-intensity fires in the recent compared to the historical past. What then are the implications? What are the lessons to be learned from the Australian "Black Summer" fires? Climate change, fuels, and social change have all contributed. Are the very large fires that burned in Australia in 2019–2020 harbingers of the future?

Southeastern Australia is no stranger to devastating fires, well documented in the region since the nineteenth century. Nonetheless, the recent wildfires are a new

Fig. 14.6 Wildfires burning on 31 Dec 2019 near Bateman's Bay in southeastern Australia. Flames are readily apparent as are the clouds of smoke and the top of the dense pyrocumulus cloud evident in the lower right (Contains modified Copernicus Sentinel data [2019]/Sentinel Hub/Processed by Pierre Markuse)

phenomenon, given their overall extent. In 2019–2020, many individual fires burned more than 100,000 hectares (Boer et al. 2020), and the fires and their smoke were readily visible from space (Fig. 14.6). Boer et al. (2020) analyzed satellite data worldwide for the past 20 years and found that the 2019–2020 Australian fires burned an unprecedented 21% of the area of Australian temperate broadleaf forest, a far higher proportion than other biomes with <5% burned for most and 8–9% for Asian and African tropical and subtropical dry broadleaf forest biomes. As contributing factors, Boer et al. (2020) and others cited deep, extended drought and extreme heat associated with sea surface temperature anomalies in the Indian Ocean, as well as wind and many ignitions. Worldwide, this is the first time that fires of this extent have burned in forest-dominated landscapes adjoined by areas densely populated by people. In Australia, the recent fires burned in many forests, including some without prior historical records of fire, and some adapted to very infrequent fires (Gill 1975); this could signal a tipping point that will result in changed vegetation types. Between June 2019 and March 2020, the fires burned 18.6 million ha, likely killed more than one billion animals, burned almost 5900 buildings, and killed 34 people while displacing and inconveniencing thousands of people. Smoke from fires exposed people to harmful air quality in many Australian cities even when those were far from the flames, resulting in 417 estimated excess deaths, 3151 hospitalizations, and 1305 asthma-related emergency presentations (Borchers Arriagada et al. 2020). In just 2 months, the fires released more than 350 million tonnes of CO_2 into the atmosphere (Sanderson and Fisher 2020). The cost of A$4.4 billion will likely exceed the cost of the Black Saturday fires that burned in 2009, with additional financial impacts on businesses and local communities.

What contributed to these fires, and what are the long-term implications? Eucalypt forests in the region form very large and continuous patches, and the patches have become more continuous as many of the historically frequent fires were suppressed or limited by land use. Human activities such as prescribed fires and vegetation management have been constrained in many of the parks and public forests, and fire hazard reduction has been mostly limited to the immediate vicinity of urban areas. This is a combined outcome of increasingly passive forest management and a focus on emergency response to wildfire rather than on mitigation. Vegetation moisture was at critically low levels following extended drought with multiple years with below-normal precipitation. Maximum daily temperatures in excess of 40 °C combined with very dry and strong winds and unstable atmosphere favored fast and intense fire spread. Finally, the region was swept by successive waves of dry lightning that ignited most of the fires. The relationship between fire and climate is complex, however, and vegetation conditions have contributed as well in Australia and globally (Forkel et al. 2019).

Fighting fires will be increasingly unsuccessful in preventing large fires and their threats to people and property unless we also address the social and political conditions (Morgan et al. 2020), including public attitudes that abdicate responsibility for learning to live with and protect homes from fires. Fires here and elsewhere around the globe may well be the agent of climate change, but they are also partially the result of policy and land use. Whether the likelihood of extreme fire weather conditions will increase in the future will depend on the long-term and uncertain results of the individual and societal policies addressing climate change. Society will have to learn to live with and adapt to this new fire environment, by enabling both long-known and new fire risk reduction strategies conducive to fire-resilient communities and ecosystems. Thus, although fires are a biophysical force, fires also reflect social and political attitudes.

Currently, Australian authorities are again discussing ways to increase prescribed burning, including cultural burning (Morgan et al. 2020), for reducing fire hazard and for biodiversity conservation in Australian landscapes. Likely the burned areas and fuel treatments will need to extend well beyond the areas immediately adjacent to or within the wildland-urban interface. Another potentially useful strategy is managing the burned areas as the vegetation recovers, for burning, thinning, or other treatments could to help foster the desired vegetation conditions into the future. Perhaps Australia will change their fire staffing and equipment which currently relies very heavily upon local volunteer fire fighters. Some have suggested that such "firies" be paid or have other financial incentives for the extensive time they spent on fires this year and may spend on fires in the future.

Worldwide, with attention heightened by the fires, people are demanding Australia and other nations to address climate change. Some scientists, citizens, and policymakers have viewed these fires as a sign of changing climate. That the large fires and smoke have affected most Australian citizens directly or indirectly could foster conversations about changing climate and societal response. The social impacts that are highly visible on social media, the millions of wild and domestic animals injured or killed, and the immense cost of fire fighting even when many of the fire fighters are volunteers, have all led to public and private anger that could fuel societal efforts to address climate change. Cultural burning and other uses of fires offer viable alternative fire management. Certainly, time will tell if the fires of 2019–2020 are recognized as agents of climate change. We encourage people to adapt and mitigate future climate change effects.

Turning the complex challenges of fire into opportunities in Australia and elsewhere will require innovative and integrated fire management (Chap. 13, Morgan et al. 2020). In Australia and across the globe, people must rethink our approaches to fire. We can recognize and fear fire as a destructive force even as we use and celebrate fires as forces of renewal and tools for managing healthy social-ecological systems. Using fire, including prescribed burning for cultural values and for managing fire hazard, and in sustainable vegetation management, is part of embracing fire as a means for caring for our planet.

For more on extreme fires see Chap. 8. For more on the relative influence of climate, fuels, and people on fires see Sect. 12.5. See Case Study 13.1 for the success of strategically burning at the landscape scale and treating more than the area immediately adjacent to the wildland-urban interface in southwestern Australia. See Case Study 13.6 where prescribed burning in during the early dry season in northern Australia has reduced carbon emissions, provided cultural values, and increased biodiversity.

14.3 Developing Technology and Bigger Data

Although technology has long been useful in advancing fire science and management, recent advances in computing and communication technologies have led to the emergence of next-generation data collection techniques, data analytics, and advanced modeling and simulation capabilities. These technologies are currently capable of generating terabytes, even petabytes, of data, challenging the way we think about data management and analysis. However, these tools are allowing us to address increasingly complex questions at finer spatial and temporal resolutions and across increasingly broad areas.

Although there are a number of emerging technologies that have the potential to impact the future of fire science and management we focus on five major advancements in data collection, data analysis, and simulation including:

1. Increasing resolution of spatial, spectral, and temporal data from satellite imagery
2. Light Detection and Ranging (LiDAR)
3. Digital aerial photogrammetry and Unmanned Aircraft Systems (UAVs)
4. Wireless sensor networks
5. “Big data” and simulation

14.3.1 Increasing Resolution of Spatial, Spectral, and Temporal Data from Satellite Imagery

Satellite-based remote sensing has become an important cost-effective data source for mapping fuels, detecting fires, assessing fire behavior and effects, and for planning fuel treatments and post-fire vegetation response. Current satellite- and airborne-based platforms include a variety of sensor types (e.g., optical, thermal, hyperspectral, LiDAR, active and passive microwave) and cover a wide range of spatial, spectral, and temporal resolutions and extents. Although the spatial coverage provided by satellite-based remote sensing is an important tool for assessing vegetation, fuels, and fires across large areas, satellite systems with global coverage often do not contain sufficient spatial or temporal resolution to provide the detailed characterization of fuels complexes and fire behavior often required for local management decisions. Continued developments in sensor design and improved affordability of sensor platforms are providing new opportunities for satellite-based remote sensing to provide data at increasingly fine spatial, temporal, and spectral resolutions. For example, DigitalGlobe's WorldView-3 and 4 satellites are capable of providing panchromatic imagery at a resolution of 31 cm, 8-band multispectral imagery with a resolution of 1.24 m, and shortwave infrared imagery at a resolution of 3.7 m and clouds, aerosols, vapors, ice and snow data at a resolution of 30 m at a specific location every 24 h. Next-generation satellite sensors are currently being evaluated as tools to produce high-resolution maps of wildland fuels for planning, including fuel treatments, restoration, assessing burn severity and monitoring long-term effects of fire on vegetation (Warner et al. 2017). In addition to increased spatial resolution, future investments in the development of multi- and hyperspectral sensors are likely to play an important role in advancing satellite-based remote sensing capabilities (NAS 2018). In addition to advancements in sensor capabilities, there is an increase in the use of data collected by microsatellites. Microsatellites are relatively low cost, small satellites that can be deployed in relatively large numbers relative to traditional satellites (Butler 2014). The use of a relatively large number of satellites increases their overall temporal resolution and ground coverage of the data (Butler 2014). Ultimately, large networks of microsatellites could provide nearly real-time capabilities to fire scientists and managers to detect and monitor fires even in remote areas.

14.3.2 Light Detection and Ranging (LiDAR)

Light Detection and Ranging (LiDAR) technology is another increasingly important data collection tool in fire science and management. LiDAR works by rapidly emitting light from a laser and measuring the time it takes for each emitted light particle to travel to an object and back, enabling users to precisely calculate the location and spatial configuration of objects. LiDAR data can be collected from a number of platforms including airborne, terrestrial, and satellite-based systems. While airborne-based LiDAR data are commonly used to quantify forest structure for several decades (Lefsky et al. 1999; Lim et al. 2003; Hall et al. 2005; Roberts et al. 2005; González-Olabarria et al. 2012), recent developments in LiDAR sensor design have reduced acquisition costs. For example, several countries (e.g., Finland, Poland, England, Sweden, USA, and Spain) currently have or are pursuing national airborne based LiDAR datasets to assist in forest and fuels inventories.

In addition to advancements in airborne LiDAR, there have also been a number of technological breakthroughs in the development and use of satellite-based LiDAR platforms. One example of such technology is NASA's Global Ecosystem Dynamics Investigation LiDAR (GEDI) mission which deployed a high-resolution LIDAR on the International Space Station. One advantage of satellite-based LiDAR sensors is that they have the potential to provide global data.

While airborne and satellite-based platforms are allowing for LiDAR data to be collected across greater extents and at lower costs (Wulder et al. 2008), there are still challenges with using such technologies to quantify surface and ladder fuels that are not directly visible to the sensor due to the foliage and branches above them (Lovell et al. 2003, Andersen et al. 2005). As a complement to airborne-based lidar platforms, a number of researchers are investigating the potential use of terrestrial-based LiDAR platforms (Newnham et al. 2015; Loudermilk et al. 2009). Although terrestrial LiDAR is not commonplace in fire science and management it has shown considerable promise for characterizing surface and canopy fuels at fine scales and in three dimensions (Rowell and Seielstad 2012; Rowell et al. 2016) and in a supporting role along with airborne based LIDAR in broad-scale fuels inventory that can aid in planning for fire and vegetation management.

14.3.3 Digital Aerial Photogrammetry and Unmanned Aircraft Systems (UAVs)

Low weight, low-cost unmanned aircraft (UAV) are another tool that is revolutionizing data collection in fire science and management. UAVs can come in a range of sizes, have various flight times that can be scheduled to accomplish desired tasks, and be equipped with a variety of sensors that allow them to quantify the fuel complex, locate and map fire perimeters, estimate the rate of spread and fireline intensity, identify spot fires, quantify current meteorological conditions across a fire area, provide data on air quality, and other data before, during, and after fires (Casbeer et al. 2005, Merino et al. 2006, 2012; Chisholm et al. 2013, Shin et al. 2018, Lin et al. 2018, Moran et al. 2019). While it is possible for UAV platforms to provide 3D characterizations of the fuels complex using LiDAR, recent developments in Structure for Motion (SfM) and multi-view stereo algorithms allow for three-dimensional (3D) information to be characterized using sequences of overlapping two-dimensional (2D) images. The combination of relatively low-cost

UAV platforms, cameras, and SfM approaches to produce 3D characterizations of vegetation structure similar to airborne LiDAR systems has led to a rapid increase in UAV-based SfM approaches in fire science and management (Leberi et al. 2010; Iglhaut et al. 2019). UAVs can also be equipped with communications technology allowing them to improve communications among the many different people involved during fire incidents (Merwaday and Guvenc 2015). Because of the relatively low-cost, high resolution and flexibility of UAVs to attach various types of sensors (e.g., multi- and hyperspectral, LiDAR), UAV platforms may be useful in assessing burn severity with an accuracy on par with or above those of the satellite-based sensors that are currently used (Fernández-Guisuraga et al. 2018, Samiappan et al. 2019, McKenna et al. 2017). Ultimately UAVs provide a major step forward in ensuring that fire scientists and managers can collect appropriate data at spatial and temporal scales in a cost-effective manner.

14.3.4 Wireless Sensor Networks

Wireless sensor networks are another emerging technology being used in fire science and management. Wireless sensor networks expand the current sampling capabilities by enabling data collection across large areas with high temporal frequency. Wireless sensors can collect a variety of physical parameters (e.g., temperature, humidity, wind speed), chemical data (e.g., carbon dioxide, volatile organic compounds and particulate matter) or images (e.g., infrared detectors). The data collected by the sensor network is then transmitted via cyberinfrastructure to the end-user for analysis and interpretation. Data processing can be embedded within wireless networks so that information can be used to assess data quality and update sampling protocols in real-time (e.g., increasing sampling rates in response to a perturbation detected in the data). Wireless network sensors are currently being used to improve fire detection (Hefeeda and Bagheri 2009; Lloret et al. 2009; Aslan et al. 2012; Bouabdellah et al. 2013), conduct real-time monitoring of fire weather and behavior (Hartung and Han 2006; Gao et al. 2014), and to predict fire behavior and assess risk (Son 2006; García et al. 2008). Although current technologies can only support relatively small sensor networks, further advancements in power generation technology such as solar power and bio-batteries along with low power sensors along with advancements in computing technologies, cyberinfrastructure, and software (Porter et al. 2005; Allen et al. 2018) will continue to increase the size, coverage and sampling frequency of wireless sensor networks in fire science and management.

14.3.5 "Big Data" and Simulation

The advancements in data collection capabilities along with the availability and access to open data (i.e., data that anyone is free to use, reuse and redistribute, Culina et al. 2018) have resulted in the rapid ability to collect and access large volumes of data. The volume of data worldwide increased by over 800% over the last 5 years and is expected to continue to double every 2 years (Gantz and Reinsel 2011; Chen et al. 2014). While the volume of data collected is a key aspect of "big data", the variety of data (e.g., tabular, image, and text) and the velocity or speed at which data are collected, and the reliability (often referred to as veracity) are also important aspects of working with and using "big data" (LaDeau et al. 2017; Farley et al. 2018). We expect that "big data" will support major breakthroughs in science, enough so that some scientists suggest that this represents a distinct fourth scientific paradigm complementing empirical descriptions of natural phenomena, theoretical modeling, and generalization, and simulation approaches (Hey 2009). In addition to paving the way to new scientific discovery, "big data" is expected to transform the way we prepare for, respond to, manage, and recover from fires. However, for fire scientists and managers to take advantage of "big data", they need continued development of:

1. Cyberinfrastructure that allows for a wide variety of data to be integrated and made available,
2. Statistical approaches that can integrate a wide variety of data types across spatial and temporal scales,
3. Computing infrastructure that can effectively deal with the volume and velocity of data being collected,
4. Training and education that includes data science skills

Evolving cyberinfrastructure will support the storage, management, integration, and sharing of various sources of data, and allow data visualization and analysis. The continued development of these technologies will be particularly important as big-data solutions become increasingly used in real-time decision making during fires. For example, next-generation cyberinfrastructures will need to access and process a variety of data sources (e.g., satellite, UAV, and networks of sensors) to make real-time predictions that then allow managers to make predictions of fire spread and intensity that can inform fire management actions. Future advances in cyberinfrastructure will continue to increase data transfer speeds, improve data storage and management efficiencies, and connect and link data sets from around the world.

The large sample sizes and high dimensionality of "big data" presents a number of statistical challenges, including spurious correlations among explanatory variables, increased risk of type-two error, nonnormality, and spatial/temporal autocorrelation. All of these limit the usefulness of many classical statistical approaches (Dray et al. 2012; Fan et al. 2014; Durden et al. 2017). Two approaches that are being increasingly used to overcome this challenge are Bayesian statistics and machine learning. Bayesian statistics are highly flexible. They can deal with multiple data types that span a range of spatial scales, and they represent the uncertainty present in the data (McCarthy 2007; Cressie et al. 2009). However, there are challenges with scaling Bayesian approaches to "big data", leading to increasing computational needs. Machine learning techniques are another increasingly common and flexible approach for working with "big data". Similar to Bayesian methods, machine learning techniques are highly flexible in that they can deal with multiple types of data that are highly correlated and nonlinear. Machine learning methods are a relatively broad class of approaches that are classified depending upon the desired outcome (Olden et al. 2008). Supervised learning approaches, including classification and regression trees and artificial neural networks, build mathematical models from data that contain both the dependent and independent data similar to many of the traditional statistical methods. On the other hand, unsupervised machine learning methods use only input data and are thus useful for identifying clusters or other patterns similar to classical clustering methods. Although the results of machine learning approaches can often be difficult to interpret, they are powerful tools for making use of "big data" in science and management.

Analyses of "big data" have already allowed significant advances in characterizing fire activity and understanding of fire regimes and their drivers and the role of fire in the Earth system at regional to continental and global scales. For these purposes, worldwide databases of climate variables, lightning activity, fire weather, plant productivity, land use and land cover, and human population density and footprint come together with remotely-sensed fire detections, burned areas, and fire characteristics. Global examples include the modeling of fire incidence metrics, e.g., burned area fraction, from environmental and human-related variables (Krawchuk et al. 2009; Bistinas et al. 2014; Knorr et al. 2014; Kelley et al. 2019), quantification and modeling of fire emissions (Van der Werf et al. 2010; Andela et al. 2016), identification of global "pyromes" (i.e., multi-faceted fire regime classes, Archibald et al. 2013), analysis of fire size variation (Hantson et al. 2015), and establishment of fire-climate relationships (Abatzoglou et al. 2018).

Process-based simulation modeling, sometimes called physics-based or mechanistic modeling, has also emerged as a powerful tool in fire science and management (Hoffman et al. 2018; Loehman et al. 2020; McLaughlan et al. 2020). Process-based models attempt to explicitly represent the relevant components, processes, and interactions that drive system behavior. These models can be viewed as a virtual world that acts as a new kind of experimental system (Winsberg 2001; Winsberg 2003; Peck 2004) that allow researchers and managers to conduct experiments that would be impossible, too risky, or costly in the real world, or to investigate novel ecosystems for which there is no historical analog (Cuddington et al. 2013; Gustafson 2013). For example, experiments which would potentially result in the ignition and spread of crown fires, such as studying the effect of bark beetles or of various fuels treatments on fire behavior, are often considered too risky, costly and difficult to conduct safely, and have therefore been studied using process-based models instead (Hoffman et al. 2012; Ziegler et al. 2017; Parsons et al. 2017; Sieg et al. 2017). Process-based models are also increasingly being used to understand the impacts of management decisions under global change (He et al. 2002; Borys et al. 2016; Keane et al. 2019). Not only can simulations foster numerical experimentation they can also complement traditional experimentation by suggesting new hypotheses that can be tested, informing sampling strategies and assisting in the interpretation of empirical data (Lenhard 2007; Hoffman et al. 2018). Such approaches will likely be used to explore alternative scenarios that could then be implemented on the ground. Nevertheless, it is important to remember that models inherently oversimplify their representation of some phenomena and necessarily ignore others, and therefore are not a complete representation of the true system being modeled. Given the inevitable limitations and uncertainties associated with models, it is critical that they are continuously evaluated through verification, validation, and uncertainty quantification. As suggested by Box (1979) "all models are wrong, but some are useful."

14.4 Integrating Fire Science and Management

One thing that seems clear is that the scale and complexity of challenges faced by wildland fire scientists and managers are increasing. While future fire scientists and managers will have a vast array of methods and tools to help them measure, monitor and make predictions about wildland fires, they will also increasingly engage in interdisciplinary, transdisciplinary and translational collaborative research to address these challenges (Gibbons et al. 1994, Brandt et al. 2013, Enquist et al. 2017, Smith et al. 2018, Knapp et al. 2019). As such, wildland fire science in the future will bridge the disciplinary silos that have been historically characteristic. This approach will not only include collaboration among various disciplines involved in wildland fire science (e.g., natural sciences, social sciences, and engineering) but also engage the end-users of research including land managers, policymakers, the public, and private institutions in the co-production of knowledge. We believe that this trend will mean that wildland fire sciences are directly motivated by the problems and challenges it addresses rather than the disciplinary concepts, methods, and approaches used. By engaging participants with different backgrounds, perspectives, and cultures, our fundamental understanding and applicability of wildland fire science will be enhanced. This requires shared language and strategies to integrate methods from different disciplines (Lawrence and Despres 2004; Brandt et al. 2013). Increasingly integrated fire science relies on the use of the internet and new communication tools to bring together collaborators who are geographically, temporally, and culturally separated. Integrated wildland fire science not only integrates scientists, stakeholders, and decision-makers but develops trust and a shared understanding and frequent and ongoing engagement among the participants, thus ultimately allowing for the translation of science into management strategies and tools that are applied (Kemp et al. 2015; Scholz and Steiner 2015; Blades et al. 2016).

Collaborative efforts to integrate fire science and management have also been developing worldwide. One such effort was made in the framework of the project Fire Paradox (2006–2010), funded by the European Commission, that brought together 36 partners from 16 countries, from Argentina to South Africa and Mongolia, including experts from the USA, Canada and Australia (Fig. 14.7). The project objective was to create a scientific and technical basis for new practices and integrated fire management policies. Proposals for policy change in Europe through a Fire Framework Directive towards Integrated Fire Management were suggested

Fig. 14.7 Field discussions between fire scientists and managers during the plenary meeting of the Fire Paradox project in 2006 in Las Palmas in the Canary Islands. (Photograph by Paulo Fernandes, co-author)

(Rego et al. 2010) and a collection of best practices of fire use, including prescribed burning and suppression fire was produced (Montiel and Kraus 2010). This included the innovative development of fire professional groups for fire use and analysis (GAUF) in Portugal, which were very active in using suppression fire (Salgueiro 2010). Fire Paradox was a good example of the integration between fire science and management that has advanced both.

14.5 Advancing Education and Training

Education and Training are two main ways to integrate fire science in practice. Over the last several decades, wildland fire has increasingly become a critical aspect of land management, through fuels management, ecosystem restoration, and continued protection of human life and property. Although natural resource education and training programs have often included classes on wildland fire science as an elective, there is a trend to require all students in disciplines which support land and fire management (e.g., forestry, natural resources, ecology, civil service) to learn about both fire management and ecology. In addition to increased recognition of wildland fire as an essential topic in natural resources, there is also a trend for developing specialized educational programs including minors, concentrations, and even entire majors about fires at universities. Such programs often recognize the need for fire fighters, fire scientists, and fire managers to have knowledge in multiple disciplines, including physical sciences, ecology, and social sciences, while also being adept at communicating clearly, anticipating and resolving conflicts, and facilitating discussions (Schwartz et al. 2017). The curricula integrate perspectives from multiple disciplines. In Europe, the PyroLife project (Pyrolife 2019) is training 15 doctoral

Fig. 14.8 Fire professionals learn through experience, education, and training. Effective preparation for the future will require more education, and the ability to effectively use technology while making decisions under uncertainty. (From Wells 2011)

students on integrated fire management, targeting fire risk (quantification, reduction, and communication) under the sign of diversity (interdisciplinarity, intersectionality, geography, and gender).

Furthermore, we applaud the increased recognition that fire professionals of the future will gain knowledge throughout their careers through a combination of experience, education, and training (Fig. 14.8) (Kobziar et al. 2009, Wells 2011, Spencer et al. 2015). This recognition is leading to the development of new models of wildfire training and education which integrate each of the three aspects. Recently the Association for Fire Ecology has developed both an individual and academic certification program which emphasizes the importance of linking education, training, and experiences for the development of fire professionals (AFE 2020). Training programs such as the Prescribed Fire Training Exchanges (TREX) established by The Nature Conservancy (TNC 2018) in the USA, or FlameWork in Portugal (Seamon 2019, Fig. 14.9), seek to increase local fire management capacity by creating collaborative learning opportunities which integrate experience, education, and training. Such integrated training programs also foster opportunities for fire professionals across a range of experiences, backgrounds, locations, and cultures to learn from one another while meeting land management objectives. Often, those objectives are increasingly ecological in addition to reducing fire hazard. Soft skills are included, such as communicating with the public directly and through media. See Case Study 13.5 for more on TREX.

Although prescribed fire has long been accepted as an important tool in fire management, there is a trend to increase prescribed fire science (Hiers et al. 2020) and to develop a dedicated prescribed fire workforce. While it has historically been assumed that the knowledge gained from studying wildfires and tools used to suppress wildfires are appropriate for planning and conducting prescribed fire, there are a number of unique properties of prescribed fires (e.g., the ability to

Fig. 14.9 FlameWork international prescribed burning exchanges held in Portugal in 2019 were very successful (Photograph by Carlos Trindade)

manipulate fire behavior and effects through time and space through altering ignition patterns) that differentiate them from wildfires (Hiers et al. 2020). We urge emphasizing the ongoing trend to increase prescribed fire research that spans all aspects of wildland fire science (e.g., fuels, fire behavior, fire effects, and ecological impacts, and social sciences) and use of the advancements in technology mentioned in Sect. 14.3. At the same time that prescribed fire science is increasing, there is also a trend to develop new prescribed fire training programs within a number of countries, states, and provinces. While the standards for such training programs can vary widely, they typically include a combination of practical experiences and training and education that covers a diverse set of topics including the law, public relations, fire behavior and meteorology, fire ecology, and smoke management. Thus, they support both planning for and implementing prescribed fires in comprehensive programs.

14.6 The Future of Fire

The trends identified in this chapter will be critical in addressing ongoing and future challenges. To prepare for future opportunities we need to address these questions and others we have not even thought to ask. What comes after people recognize fire as both an effect of and an agent of global change and especially of climate change? What is next once people accept fires as an essential and pervasive influence in forests, woodlands, shrublands, and grasslands? How might we envision managing to enable fires to move through landscapes, and where and when is that possible? What if we understand that fire can be transformed from a threat to medicine for land and a culturally important component? If we as a society are able to respond to these challenges, we can then more often celebrate some fires, use more fires in some locations, and be less threatened by wildfires. This is a fire paradox. We can respect and use the power of fire to change landscapes. Then we will use the positive feedback cycle between changing fire regimes and the landscapes that can result in more balanced landscapes with adequate fire regimes. This has substantial implications for people and nature.

The current global changes and the expected future trends call for focusing less exclusively on fire suppression and more on fire use and preparedness. What if many people become simultaneously fire fighters and fire lighters, or what if we have as many fire lighters as we have fire fighters? What if a proportion of funds now used to fight fires were instead targeted toward planning and using fires to accomplish landscape management goals, both social and biophysical? What if we had a cohesive strategy that fosters fire-adapted communities in resilient landscapes with effective use of fires and response to fires? Once we have a more nuanced and realistic view of fires, how will our perceptions and language support for innovative fire management change?

Science can inform societal reaction to the challenging complexities of fire-related issues today, including costs, threats to people and property, ecological values, and impacts of fires. Collaboration and effective multi-way communication can build trust. Proactive and strategic fire management is needed, as are innovative technologies and ways of working together strategically to adapt and mitigate climate changes and other global changes to local ecosystems and local people while responding to regional and national priorities. We must focus on clear, strategic goals. We must be clear about uncertainties but we must not let uncertainty keep us from moving forward and learning as we go.

We authors dream that people will use fires as part of effective efforts to adapt and shape future fires and smoke. Time will tell if we achieve fire-adapted homes and communities in fire-resilient landscapes in ways that are socially just and sustainable. We hope and work to shape proactive approaches to fires that are good for our planet and people. Such efforts will be place-based, and filled with people learning from each other. We need innovative approaches that provide for the essential role of fire while reducing societal and ecosystem vulnerability to fire. Then, preparing for, enduring, and recovering from fires could include celebrating and using fire.

Ultimately, people must learn to balance realities. Wildfires and smoke will occur, some of those fires will be large and smoke will affect many people. Yet fires are part of the personality of forests, woodlands, shrublands, and grasslands, and without fires, these systems change. Those fires provide many of the ecosystem services people value, so let's learn from the many successful cases how to protect people, property, and economies from the adverse effects of fire and smoke. Both can be accomplished in landscapes where fires burn with an ecologically appropriate mix of low, moderate, and high severity, and with patch sizes and spatial patterns (Moritz et al. 2018). This requires engaging with fire and with people to find ways to sustainably use landscapes in ways that are ecologically appropriate, financially feasible, and socially acceptable.

As we move forward in what some have called the "Era of Megafires" (Hessburg 2017) or the "Pyrocene" (Pyne 2018), wildfires will continue to influence vegetation change, and therefore the goods and services people receive from ecosystems. We will keep learning from fires through rapidly changing science. We can choose how to manage fires to help shape how those wildfires affect future fires, land, and people for both the short- and long-term. Indeed, managing vegetation and communities so that they are resilient to fires is a worthy goal, and wildfires can help us achieve that resilience. If we do not engage with fires, using them to help us adapt and accomplish our land and resource management goals, then there will likely be widespread vegetation change at multiple spatial scales. Increasingly, society's environmental goals will include carbon sequestration, resilience, and adaptation to global change, all while effectively managing fires and their attendant smoke to increase positive impacts and lessen negative impacts on people and ecosystem services.

We must. We can. We will. We hope that our book is a contribution in that direction.

References

Abatzoglou, J. T., Williams, A. P., Boschetti, L., Zubkova, M., & Kolden, C. A. (2018). Global patterns of interannual climate–fire relationships. *Global Change Biology, 24*(11), 5164–5175.

Acácio, V., Holmgren, M., Rego, F., Moreira, F., & Mohren, G. M. (2009). Are drought and wildfires turning Mediterranean cork oak forests into persistent shrublands? *Agroforestry Systems, 76*(2), 389–400.

Allen, B. M., Nimmo, D. G., Ierodiaconou, D., VanDerWal, J., Koh, L. P., & Ritchie, E. G. (2018). Futurecasting ecological research: The rise of technology. *Ecosphere, 9*(5), e02163. https://doi.org/10.1002/ecs2.2163.

Andela, N., Van Der Werf, G. R., Kaiser, J. W., Van Leeuwen, T. T., Wooster, M. J., & Lehmann, C. E. (2016). Biomass burning fuel consumption dynamics in the tropics and subtropics assessed from satellite. *Biogeosciences, 13*(12), 3717–3734.

Andela, N., Morton, D. C., Giglio, L., Chen, Y., Van Der Werf, G. R., Kasibhatla, P. S., DeFries, R. S., Collatz, G. J., Hantson, S., Kloster, S., & Bachelet, D. (2017). A human-driven decline in global burned area. *Sci, 356*(6345), 1356–1362. https://doi.org/10.1126/science.aal4108.

Andersen, H. E., McGaughey, R. J., & Reutebuch, S. E. (2005). Estimating forest canopy fuel parameters using LIDAR data. *Remote Sensing of Environment, 94*(4), 441–449.

Archibald, S., Lehmann, C. E., Gómez-Dans, J. L., & Bradstock, R. A. (2013). Defining pyromes and global syndromes of fire regimes. *Proceedings of National Academy of Sciences, 110*(16), 6442–6447.

Aslan, Y. E., Korpeoglu, I., & Ulusoy, Ö. (2012). A framework for use of wireless sensor networks in forest fire detection and monitoring. *Computers, Environment and Urban Systems, 36*(6), 614–625. https://doi.org/10.1016/j.compenvurbsys.2012.03.002.

Association for Fire Ecology (AFE). (2016a). *Sexual harassment and gender discrimination in wildland fire management must be addressed.* Position Paper: Sexual harrasment and gender discrimination. Eugene: Association for Fire Ecology. Retrieved May 30, 2020, from https://static1.squarespace.com/static/5ea4a2778a22135afc733499/t/5eadf7c2c40da37246e2cd68/1588459459469/AFE+2016+position+paper+on+discrimination+final+11-25.pdf.

Association for Fire Ecology (AFE). (2016b). *Sexual harassment and gender discrimination in wildland fire management must be addressed.* Position Paper: Sexual harrasment and gender discrimination. Eugene: Association for Fire Ecology. Retrieved May 30, 2020, from https://fireecology.org/sexual-harassment-position-paper.

Association for Fire Ecology (AFE). (2020). *Wildland Fire Professional Certification Program.* Association for Fire Ecology. Retrieved June 19, 2020, from https://fireecology.org/professional-certification.

Balch, J. K., Bradley, B. A., Abatzoglou, J. T., Nagy, R. C., Fusco, E. J., & Mahood, A. L. (2017). Human-started wildfires expand the fire niche across the United States. *Proceedings of the National Academy of Sciences, 114*(11), 2946–2951. https://doi.org/10.1073/pnas.1617394114.

Bistinas, I., Harrison, S. P., Prentice, I. C., & Pereira, J. M. C. (2014). Causal relationships vs. emergent patterns in the global controls of fire frequency. *Biogeosciences, 11*, 5087–5101.

Blades, J. J., Klos, P. Z., Kemp, K. B., Hall, T. E., Force, J. E., Morgan, P., & Tinkham, W. T. (2016). Forest managers' response to climate change science: Evaluating the constructs of boundary objects and organizations. *Forest Ecology and Management, 15*(360), 376–387.

Boer, M. M., de Dios, V. R., & Bradstock, R. A. (2020). Unprecedented burn area of Australian mega forest fires. *Nature Climate Change, 10*(3), 171–172. https://doi.org/10.1038/s41558-020-0716-1.

Borchers Arriagada, N., Palmer, A. J., Bowman, D. M., Morgan, G. G., Jalaludin, B. B., & Johnston, F. H. (2020). Unprecedented smoke-related health burden associated with the 2019–20 bushfires in eastern Australia. *The Medical Journal of Australia.* https://doi.org/10.5694/mja2.50545.

Borys, A., Suckow, F., Reyer, C., Gutsch, M., & Lasch-Born, P. (2016). The impact of climate change under different thinning regimes on carbon sequestration in a German forest district. *Mitigation and Adaptation Strategies for Global Change, 21*(6), 861–881.

Bouabdellah, K., Noureddine, H., & Larbi, S. (2013). Using wireless sensor networks for reliable forest fires detection. *Procedia Computer Science, 19*, 794–801.

Bowman, D. M., Balch, J., Artaxo, P., Bond, W. J., Cochrane, M. A., D'antonio, C. M., DeFries, R., Johnston, F. H., Keeley, J. E., Krawchuk, M. A., & Kull, C. A. (2011). The human dimension of fire regimes on Earth. *Journal of Biogeography, 38*(12), 2223–2236. https://doi.org/10.1111/j.1365-2699.2011.02595.x.

Box, G. E. (1979). Robustness in the strategy of scientific model building. In R. L. Launer & G. N. Wilkinson (Eds.), *Robustness in statistics* (pp. 201–236). Cambridge: Academic Press.

Brandt, P., Ernst, A., Gralla, F., Luederitz, C., Lang, D. J., Newig, J., Reinert, F., Abson, D. J., & von Wehrden, H. (2013). A review of transdisciplinary research in sustainability science. *Ecological Economics, 92*, 1–15.

Butler, D. (2014). Many eyes on Earth. *Nature, 5050*, 143–144.

Casbeer, D. W., Beard, R. W., McLain, T. W., Li, S. M., & Mehra, R. K. (2005). Forest fire monitoring with multiple small Unmanned Air Vehicles (UAVs). In *IEEEProceedings of the American Control Conference*, Portland, 8–10 June 2005, pp. 3530–3535.

Chen, M., Mao, S., & Liu, Y. (2014). Big data: A survey. *Mobile Networks and Applications, 19*(2), 171–209.

Chisholm, R. A., Cui, J., Lum, S. K., & Chen, B. M. (2013). UAV LiDAR for below-canopy forest surveys. *Journal of Unmanned Vehicle Systems, 1*(1), 61–68.

Cressie, N., Calder, C. A., Clark, J. S., Hoef, J. M. V., & Wikle, C. K. (2009). Accounting for uncertainty in ecological analysis: The strengths and limitations of hierarchical statistical modeling. *Ecological Applications, 19*(3), 553–570.

Cuddington, K., Fortin, M. J., Gerber, L. R., Hastings, A., Liebhold, A., O'Connor, M., & Ray, C. (2013). Process-based models are required to manage ecological systems in a changing world. *Ecosphere, 4*(2), 1–12.

Culina, A., Baglioni, M., Crowther, T. W., Visser, M. E., Woutersen-Windhouwer, S., & Manghi, P. (2018). Navigating the unfolding open data landscape in ecology and evolution. *Nature Ecology and Evolution, 2*(3), 420–426.

Curt, T., & Frejaville, T. (2018). Wildfire policy in Mediterranean France: How far is it efficient and sustainable? *Risk Analysis, 38*(3), 472–488. https://doi.org/10.1111/risa.12855.

Davis, K. T., Dobrowski, S. Z., Higuera, P. E., Holden, Z. A., Veblen, T. T., Rother, M. T., Parks, S. A., Sala, A., & Maneta, M. P. (2019). Wildfires and climate change push low-elevation forests across a critical climate threshold for tree regeneration. *PNAS, 116*(13), 6193–6198.

Dray, S., Pélissier, R., Couteron, P., Fortin, M. J., Legendre, P., Peres-Neto, P. R., Bellier, E., Bivand, R., Blanchet, F. G., De Cáceres, M., & Dufour, A. B. (2012). Community ecology in the age of multivariate multiscale spatial analysis. *Ecological Monographs, 82*, 257–275. https://doi.org/10.1890/11-1183.1.

Doerr, S. H., & Santín, C. (2016). Global trends in wildfire and its impacts: Perceptions versus realities in a changing world. *Philosophical Transactions of Royal Society B, 371*(1696), 20150345. https://doi.org/10.1098/rstb.2015.0345.

Durden, J. M., Luo, J. Y., Alexander, H., Flanagan, A. M., & Grossmann, L. (2017). Integrating "big data" into aquatic ecology: Challenges and opportunities. *Limnology and Oceanography Bulletin, 26*(4), 101–108.

Enquist, C. A., Jackson, S. T., Garfin, G. M., Davis, F. W., Gerber, L. R., Littell, J. A., & Hiers, J. K. (2017). Foundations of translational ecology. *Frontiers in Ecology and the Environment, 15* (10), 541–550.

Fan, J., Fang, H., & Liu, H. (2014). Challenges of big data analysis. *National Science Review, 1*, 293–314. https://doi.org/10.1093/nsr/nwt032.

Farley, S. S., Dawson, A., Goring, S. J., & Williams, J. W. (2018). Situating ecology as a big-data science: Current advances, challenges, and solutions. *Biosciences, 68*(8), 563–576. https://doi.org/10.1093/biosci/biy068.

Fernández-Guisuraga, J., Sanz-Ablanedo, E., Suárez-Seoane, S., & Calvo, L. (2018). Using unmanned aerial vehicles in postfire vegetation survey campaigns through large and heterogeneous areas: Opportunities and challenges. *Sensors, 18*(2), 586.

Food and Agriculture Organization of the United Nations (FAO). (2011). *Community-based fire management: A review*. FAO Forestry Paper 166, Rome. Retrieved March 11, 2020, from http://www.fao.org/3/i2495e/i2495e.pdf.

Forkel, M., Andela, N., Harrison, S. P., Lasslop, G., Van Marle, M., Chuvieco, E., Dorigo, W., Forrest, M., Hantson, S., Heil, A., & Li, F. (2019). Emergent relationships with respect to burned area in global satellite observations and fire-enabled vegetation models. *Biogeosciences, 16*, 57–76. https://doi.org/10.5194/bg-16-57-2019.

Gantz, J., & Reinsel, D. (2011). Extracting value from chaos. *IDC iView*, pp. 1–12.

Gao, L., Bruenig, M., & Hunter, J. (2014). Estimating fire weather indices via semantic reasoning over wireless sensor network data streams. *International Journal of Web and Semantic Technology, 5*(4), 1–20.

García, E. M., Serna, M. Á., Bermúdez, A., & Casado, R. (2008). Simulating a WSN-based wildfire fighting support system. In: *Proceedings of 14th IEEE International Workshop on Parallel and Distributed Processing with Applications*, Melbourne, 8–10 December 2008, pp. 896–902.

Gibbons, M., Limoges, C., Nowotny, H., Schwartzman, S., Scott, P., & Throw, M. (1994). *The new production of knowledge: The dynamics of science and research in contemporary societies*. London: Sage.
Giglio, L., Randerson, J. T., & van der Werf, G. R. (2013). Analysis of daily, monthly, and annual burned area using the fourth-generation global fire emissions database (GFED4). *Journal of Geophysical Research – Biogeosciences, 118*(1), 317–328.
Gill, A. M. (1975). Fire and the Australian flora: A review. *Australian Forestry, 38*(1), 4–25. https://doi.org/10.1080/00049158.1975.10675618.
González-Olabarria, J. R., Rodríguez, F., Fernández-Landa, A., & Mola-Yudego, B. (2012). Mapping fire risk in the Model Forest of Urbión (Spain) based on airborne LiDAR measurements. *Forest Ecology and Management, 282*, 149–156.
Guiomar, N., Godinho, S., Fernandes, P. M., Machado, R., Neves, N., & Fernandes, J. P. (2015). Wildfire patterns and landscape changes in Mediterranean oak woodlands. *Science of Total Environment, 536*, 338–352.
Gustafson, E. J. (2013). When relationships estimated in the past cannot be used to predict the future: Using mechanistic models to predict landscape ecological dynamics in a changing world. *Landscape Ecology, 28*(8), 1429–1437.
Hall, S. A., Burke, I. C., Box, D. O., Kaufmann, M. R., & Stroker, J. M. (2005). Estimating stand structure using discrete-return LiDAR: An example from low density, fire prone ponderosa pine forests. *Forest Ecology and Management, 208*, 189–209.
Hantson, S., Pueyo, S., & Chuvieco, E. (2015). Global fire size distribution is driven by human impact and climate. *Global Ecology and Biogeography, 24*(1), 77–86.
Hartung, C., & Han, R. (2006). FireWxNet: A multi-tiered portable wireless system for monitoring weather conditions in wildland fire environments. In: *Proceedings of 4th International Conference on Mobile Systems, Applications and Services*, Uppsala, 19–22 June 2006. New York: Association for Computing Machinery, pp. 28–41.
He, H. S., Mladenoff, D. J., & Gustafson, E. J. (2002). Study of landscape change under forest harvesting and climate warming-induced fire disturbance. *Forest Ecology and Management, 155*(1–3), 257–270.
Hefeeda, M., & Bagheri, M. (2009). Forest fire modeling and early detection using wireless sensor networks. *Ad Hoc and Sensor Wireless Networks, 7*, 169–224.
Hessburg, P. (2017, April 24). *Era of mega-fires: How do you want your fire? How do you want your smoke?* Multi-media event presentation sponsored by NRFIRESCIENCE.ORG. Missoula: University of Montana.
Hey, A. J. (Ed.). (2009). *The fourth paradigm: Data-intensive scientific discovery* (Vol. 1). Redmond WA: Microsoft Research.
Hiers, J. K., O'Brien, J. J., Varner, J. M., Butler, B. W., Dickinson, M., Furman, J., Gallagher, M., Godwin, D., Goodrick, S. L., Hood, S. M., Hudak, A., Kobziar, L. N., Linn, R., Loudermilk, E. L., McCaffrey, S., Robertson, K., Rowell, E. M., Skowronski, N., Watts, A. C., & Yedinak, K. M. (2020). Prescribed fire science: The case for a refined research agenda. *Fire Ecology, 16* (1), 1–15.
Hoffman, C., Morgan, P., Mell, W., Parsons, R., Strand, E. K., & Cook, S. (2012). Numerical simulation of crown fire hazard immediately after bark beetle-caused mortality in lodgepole pine forests. *Forest Science, 58*(2), 178–188.
Hoffman, C. M., Sieg, C. H., Linn, R. R., Mell, W., Parsons, R. A., Ziegler, J. P., & Hiers, J. K. (2018). Advancing the science of wildland fire dynamics using process-based models. *Fire, 1* (2), 32.
Holden, Z. A., Swanson, A., Luce, C. H., Jolly, W. M., Maneta, M., Oyler, J. W., Warren, D. A., Parsons, R., & Affleck, D. (2018). Decreasing fire season precipitation increased recent western US forest wildfire activity. *Proceedings of the National Academy of Sciences, 115*(36), e8349–e8357. https://doi.org/10.1073/pnas.1802316115.
Huffman, M. (2014). Making a world of difference in fire and climate change. *Fire Ecology, 10*(3), 90–101.

Hurteau, M. D., Liang, S., Westerling, A. L., & Wiedinmyer, C. (2019). Vegetation-fire feedback reduces projected area burned under climate change. *Scientific Reports, 9*(1), 2838. https://doi.org/10.1038/s41598-019-39,284-1.

Iglhaut, J., Cabo, C., Puliti, S., Piermattei, L., O'Connor, J., & Rosette, J. (2019). Structure from motion photogrammetry in forestry: A review. *Current Forestry Reports, 5*(3), 155–168.

Jolly, W. M., Cochrane, M. A., Freeborn, P. H., Holden, Z. A., Brown, T. J., Williamson, G. J., & Bowman, D. M. (2015). Climate-induced variations in global wildfire danger from 1979 to 2013. *Nature Communications, 6*, 7536. https://doi.org/10.1038/ncomms8537.

Keane, R. E., Gray, K., Davis, B., Holsinger, L. M., & Loehman, R. (2019). Evaluating ecological resilience across wildfire suppression levels under climate and fuel treatment scenarios using landscape simulation modeling. *International Journal of Wildland Fire, 28*(7), 533–549.

Kelley, D. I., Bistinas, I., Whitley, R., Burton, C., Marthews, T. R., & Dong, N. (2019). How contemporary bioclimatic and human controls change global fire regimes. *Nature Climate Change, 9*(9), 690–696.

Kemp, K. B., Blades, J. J., Klos, P. Z., Hall, T. E., Force, J. E., Morgan, P., & Tinkham, W. T. (2015). Managing for climate change on federal lands of the western United States: Perceived usefulness of climate science, effectiveness of adaptation strategies, and barriers to implementation. *Ecology and Society, 20*(2). https://doi.org/10.5751/ES-07522-200,217.

Kobziar, L. N., Rocca, M. E., Dicus, C. A., Hoffman, C., Sugihara, N., Thode, A. E., Varner, J. M., & Morgan, P. (2009). Challenges to educating the next generation of wildland fire professionals in the United States. *Journal of Forestry, 107*(7), 339–345.

Knapp, C. N., Reid, R. S., Fernández-Giménez, M. E., Klein, J. A., & Galvin, K. A. (2019). Placing transdisciplinarity in context: A review of approaches to connect scholars, society and action. *Sustainability, 11*(18), 4899.

Knorr, W., Kaminski, T., Arneth, A., & Weber, U. (2014). Impact of human population density on fire frequency at the global scale. *Biogeosciences, 11*(4), 1085–1102.

Krawchuk, M. A., Moritz, M. A., Parisien, M. A., Van Dorn, J., & Hayhoe, K. (2009). Global pyrogeography: The current and future distribution of wildfire. *PLoS One, 4*(4), e5102. https://doi.org/10.1371/journal.pone.0005102.

LaDeau, S. L., Han, B. A., Rosi-Marshall, E. J., & Weathers, K. C. (2017). The next decade of big data in ecosystem science. *Ecosystems, 20*(2), 274–283.

Lannom, K. O., Tinkham, W. T., Smith, A. M., Abatzoglou, J., Newingham, B. A., Hall, T. E., Morgan, P., Strand, E. K., Paveglio, T. B., Anderson, J. W., & Sparks, A. M. (2014). Defining extreme wildland fires using geospatial and ancillary metrics. *International Journal of Wildland Fire, 23*(3), 322–337.

Lawrence, R. J., & Despres, C. (2004). Futures of transdisciplinarity. *Futures, 36*, 397–405.

Leberi, F., Irschara, A., Pock, T., Meixner, P., Gruber, M., Scholz, S., & Wiechert, A. (2010). Point clouds: LiDAR versus three-dimensional vision. *Photogrammetric Engineering & Remote Sensing, 76*, 1123–1134.

Lefsky, M. A., Cohen, W. B., Acker, S. A., Parker, G. G., Spies, T. A., & Harding, D. (1999). Lidar remote sensing of the canopy structure and biophysical properties of Douglas-fir western hemlock forests. *Remote Sensing of Environment, 70*(3), 339–361.

Lenhard, J. (2007). Computer simulation: The cooperation between experimenting and modeling. *Philosophy in Science, 74*, 176–194.

Lim, K., Treitz, P., Wulder, M., St-Onge, B., & Flood, M. (2003). LiDAR remote sensing of forest structure. *Progress in Physical Geography, 27*(1), 88–106.

Lin, Z., Liu, H. T., & Wotton, M. (2018). Kalman filter-based large-scale wildfire monitoring with a system of UAVs. *IEEE Transactions on Industrial Electronics, 66*(1), 606–615.

Lloret, J., Garcia, M., Bri, D., & Sendra, S. (2009). A wireless sensor network deployment for rural and forest fire detection and verification. *Sensor Nodes, 9*(11), 8722–8747.

Loehman, R. A., Keane, R. E., & Holsinger, L. M. (2020). Simulation modeling of complex climate, wildfire, and vegetation dynamics to address wicked problems in land management. *Frontiers in Forest and Global Change*. https://doi.org/10.3389/ffgc.2020.00003.

Loudermilk, E. L., Hiers, J. K., O'Brien, J. J., Mitchell, R. J., Singhania, A., Fernandez, J. C., Cropper, W. P., & Slatton, K. C. (2009). Ground-based LIDAR: A novel approach to quantify fine-scale fuelbed characteristics. *International Journal of Wildland Fire, 18*(6), 6.

Lovell, J. L., Jupp, D. L., Culvenor, D. S., & Coops, N. C. (2003). Using airborne and ground-based ranging lidar to measure canopy structure in Australian forests. *Canadian Journal of Remote Sensing, 29*(5), 607–622.

Martín-Alcón, S., & Coll, L. (2016). Unraveling the relative importance of factors driving post-fire regeneration trajectories in non-serotinous Pinus nigra forests. *Forest Ecology and Management, 361*, 13–22.

McCarthy, M. A. (2007). *Bayesian methods for ecology*. Cambridge: Cambridge University Press.

McDonald, P. (2012). Workplace sexual harassment 30 years on: A review of the literature. *International Journal of Management Reviews, 14*, 1. https://doi.org/10.1111/j.1468-2370.2011.00300.x.

McKenna, P., Erskine, P. D., Lechner, A. M., & Phinn, S. (2017). Measuring fire severity using UAV imagery in semi-arid central Queensland, Australia. *International Journal of Remote Sensing, 38*(14), 4244–4264.

McLaughlan, K. K., Higuera, P. E., Miesel, J., Rogers, B. M., Schweitzer, J., Shuman, J. K., Tepley, A. J., Varner, J. M., Veblen, T. T., Adalsteinsson, S. A., & Balch, J. K. (2020). Fire as a fundamental ecological process: Research advances and frontiers. *Journal of Ecology*. https://doi.org/10.1111/1365-2745.13403.

Merino, L., Caballero, F., Martínez-de Dios, J. R., Ferruz, J., & Ollero, A. (2006). A cooperative perception system for multiple UAVs: Application to automatic detection of forest fires. *Journal of Field Robotics, 23*(3–4), 165–184.

Merino, L., Caballero, F., Martínez-de-Dios, J. R., Maza, I., & Ollero, A. (2012). An unmanned aircraft system for automatic forest fire monitoring and measurement. *Journal of Intelligent & Robotic Systems, 65*(1), 533–548.

Merwaday, A., & Guvenc, I. (2015). UAV assisted heterogeneous networks for public safety communications. In: *Proceedings of 2015 IEEE Wireless Communications and Networking Conference Workshops (WCNCW)*, pp. 329–334.

Montiel C, Kraus D (Eds). (2010). *Best practices of fire use – Prescribed burning and suppression fire programmes in selected case-study regions in Europe*. European Forest Institute EFI Research Report 24, Joensuu.

Montiel, C., Karlsson, O., & Galiana, L. (2019). Regional fire scenarios in Spain: Linking landscape dynamics and fire regime for wildfire risk management. *Journal of Environmental Management, 233*, 427–439.

Moran, C. J., Seielstad, C. A., Cunningham, M. R., Hoff, V., Parsons, R. A., Queen, L., Sauerbrey, K., & Wallace, T. (2019). Deriving fire behavior metrics from UAS imagery. *Fire, 2*(2), 36.

Morgan, G. W., Tolhurst, K. G., Poynter, M. W., Cooper, N., McGuffog, T., Ryan, R., Wouters, M. A., Stephens, N., Black, P., Sheehan, D., & Leeson, P. (2020). Prescribed burning in south-eastern Australia: History and future directions. *Australian Forestry, 83*(1), 4–28. https://doi.org/10.1080/00049158.2020.1739883.

Moritz, M. A., Topik, C., Allen, C. D., Hessburg, P. F., Morgan, P., Odion, D. C., Veblen, T. T., & McCullough, I. M. (2018) *A statement of common ground regarding the role of wildfire in forested landscapes of the western United States*. Fire Research Consensus Working Group Final Report. SNAPP and NCEAS. Retrieved January 15, 2020, from https://www.nceas.ucsb.edu/files/research/projects/WildfireCommonGround.pdf.

National Academies of Sciences, Engineering, and Medicine (NAS). (2018). *Thriving on our changing planet: A decadal strategy for Earth observation from space*. Washington, DC: National Academies Press. https://doi.org/10.17226/24938.

The Nature Conservancy (TNC). (2017). *Strong voices, active choices: TNC's practitioner framework to strengthen outcomes for people and nature*. Arlington: The Nature Conservancy. Retrieved March 29, 2020, from https://www.nature.org/en-us/what-we-do/our-insights/perspectives/strong-voices-active-choices/.

The Nature Conservancy (TNC). (2018). *Prescribed fire training exchanges*. Retrieved April 11, 2020, from https://www.conservationgateway.org/ConservationPractices/FireLandscapes/HabitatProtectionandRestoration/Training/TrainingExchanges/Pages/fire-training-exchanges.aspx.

Newnham, G. J., Armston, J. D., Calders, K., Disney, M. I., Lovell, J. L., Schaaf, C. B., Strahler, A. H., & Danson, F. M. (2015). Terrestrial laser scanning for plot-scale forest measurement. *Current Forestry Report, 1*(4), 239–251.

Olden, J. D., Lawler, J. J., & Poff, N. L. (2008). Machine learning methods without tears: A primer for ecologists. *The Quarterly Review of Biology, 83*(2), 171–193.

Parsons, R., Linn, R., Pimont, F., Hoffman, C., Sauer, J., Winterkamp, J., & Jolly, W. (2017). Numerical investigation of aggregated fuel spatial pattern impacts on fire behavior. *Land, 6*(2), 43.

Pausas, J. G., & Keeley, J. E. (2019). Wildfires as an ecosystem service. *Frontiers in Ecology and the Environment, 17*(5), 289–295. https://doi.org/10.1002/fee.2044.

Peck, S. L. (2004). Simulation as experiment: A philosophical reassessment for biological modeling. *Trends in Ecology & Evolution, 19*, 530–534.

Porter, J., Arzberger, P., Braun, H.-W., Bryant, P., Gage, S., Hansen, T., Hanson, P., Lin, C.-C., Lin, F.-P., Kratz, T., Michener, W., Shapiro, S., & Williams, T. (2005). Wireless sensor networks for ecology. *BioSciences, 55*(7), 561–572. https://doi.org/10.1641/0006-3568(2005)055[0561:WSNFE]2.0.CO;2.

Pyne, S. J. (2015). How humans made fire, and fire made us human. *AEON*. Retrieved April 20, 2020, from https://aeon.co/essays/how-humans-made-fire-and-fire-made-us-human.

Pyne, S. J. (2018). Big fire; or introducing the pyrocene. *Fire, 1*(1), 1. https://doi.org/10.3390/fire1010001.

PyroLife Project. (2019). *PyroLife project*. Retrieved June 19, 2020, from https://pyrolife.lessonsonfire.eu/pyrolife-project/.

Rego, F., Rigolot, E., Fernandes, P., Montiel, C., & Sande Silva, J. (2010). *Towards integrated fire management*. European Forest Institute EFI Policy Brief 4, Joensuu.

Rego, F. C., Moreno, J. M., Vallejo, V. R., & Xanthopoulos. (2018). Forest fires. Sparking firesmart policies in the EU. In N. Faivre (Ed.), *Research & innovation projects for policy. Climate action and resource efficiency*. Brussels: European Commission.

Roberts, S. D., Dean, T. J., Evans, D. L., McCombs, J. W., Harrington, R. L., & Glass, P. A. (2005). Estimating individual tree leaf area in loblolly pine plantations using LiDAR-derived measurements of height and crown dimensions. *Forest Ecology and Management, 213*(1–3), 54–70.

Rowell, E., & Seielstad, C. (2012). Characterizing grass, litter, and shrub fuels in longleaf pine forest pre-and post-fire using terrestrial LiDAR. In: *Proceedings of 12th international SilviLaser*, Vancouver, 16–19 September, pp. 16–19.

Rowell, E., Loudermilk, E. L., Seielstad, C., & O'Brien, J. J. (2016). Using simulated 3D surface fuelbeds and terrestrial laser scan data to develop inputs to fire behavior models. *Canadian Journal of Remote Sensing, 42*(5), 443–459.

Rover, B. (2020). *Wildfires and the pandemic – What's ahead*. International Association of Wildland Fire. Retrieved June 5, 2020, from https://www.iawfonline.org/article/2020-04-wildfires-pandemic-whats-ahead-wfca/.

Salgueiro, A. (2010). The Portuguese National Programme on Suppression Fire: GAUF Team Actions. In: C. Montiel, D. Kraus (Eds.), *Best practices of fire use – Prescribed burning and suppression fire programmes in selected case-study regions in Europe*. European Forest Institute EFI Research Report 24, Joensuu, pp. 123–136.

Samiappan, S., Hathcock, L., Turnage, G., McCraine, C., Pitchford, J., & Moorhead, R. (2019). Remote sensing of wildfire using a small unmanned aerial system: Post-Fire mapping, vegetation recovery and damage analysis in Grand Bay, Mississippi/Alabama, USA. *Drones, 3*(2), 43.

Sanderson, B. M., & Fisher, R. A. (2020). A fiery wake-up call for climate science. *Nature Climate Change, 10*(3), 175–177. https://doi.org/10.1038/s41558-020-0707-2.

Seamon, G. (2019). FlameWorks. *Tall Timbers*. Retrieved June 19, 2020, from https://resilience-blog.com/wp-content/uploads/2019/11/FlameWork_eJournal_Fall2019_pp42-46.pdf.

Sequeira, C. R., Rego, F., Montiel-Molina, C., & Morgan, P. (2019). Half-century changes in LULC and fire in two Iberian inner mountain areas. *Fire, 2*(3), 45. https://doi.org/10.3390/fire2030044.

Scholz, R. W., & Steiner, G. (2015). The real type and ideal type of transdisciplinary processes: Part I—theoretical foundations. *Sustainability Science, 10*, 527–544.

Schwartz, M. W., Hiers, J. K., Davis, F. W., Garfin, G. M., Jackson, S. T., Terando, A. J., Woodhouse, C. A., Morelli, T. L., Williamson, M. A., & Brunson, M. W. (2017). Developing a translational ecology workforce. *Frontiers in Ecology and the Environment, 15*(10), 587–596.

Schultz, C. A., & Moseley, C. (2019). Collaborations and capacities to transform fire management. *Science, 366*(6461), 38–40.

Shin, P., Sankey, T., Moore, M., & Thode, A. (2018). Evaluating unmanned aerial vehicle images for estimating forest canopy fuels in a ponderosa pine stand. *Remote Sensing, 10*(8), 1266.

Sieg, C. H., Linn, R. R., Pimont, F., Hoffman, C. M., McMillin, J. D., Winterkamp, J., & Baggett, L. S. (2017). Fires following bark beetles: Factors controlling severity and disturbance interactions in ponderosa pine. *Fire Ecology, 13*(3), 1–23.

Sil, A., Azevedo, J., Fernandes, P. M., Regos, A., Vaz, A. S., & Honrado, J. (2019). (Wild)fire is not an ecosystem service. *Frontiers in Ecology and the Environment, 17*(8), 429–430.

Smith, A., Goldammer, J. G., & Bowman, D. M. (2018a). Introducing Fire: A transdisciplinary journal to advance understanding and management of landscape fires from local to global scales in the past, present, and future. *Fire, 1*(1), 2. https://doi.org/10.3390/fire1010002.

Smith, A. M. S., & Strand, E. K. (2018). Recognizing women leaders in fire science: Revisited. *Fire, 1*, 45. https://doi.org/10.3390/fire1030045.

Smith, A. M. S., Kolden, C. A., Prichard, S. J., Gray, R. W., Hessburg, P. F., & Balch, J. K. (2018b). Recognizing women leaders in fire science. *Fire, 1*, 30.

Son, B. (2006). A design and implementation of forest-fires surveillance system based on wireless sensor networks for South Korea mountains. *International Journal of Computer Science and Network Security, 6*(9B), 124–130.

Stamper, A. (2017). *Women on fire, lighting up a new path*. International Association of Wildland Fire. Retrieved April 20, 2020, from https://www.iawfonline.org/article/women-on-fire-lighting-up-a-new-path/.

Stenzel, J. E., Bartowitz, K. J., Hartman, M. D., Lutz, J. A., Kolden, C. A., Smith, A. M., Law, B. E., Swanson, M. E., Larson, A. J., Parton, W. J., & Hudiburg, T. W. (2019). Fixing a snag in carbon emissions estimates from wildfires. *Global Change Biology, 25*(11), 3985–3994.

Stevens-Rumann, C. S., & Morgan, P. (2019). Tree regeneration following wildfires in the western US: A review. *Fire Ecology, 15*(1), 15.

Stevens-Rumann, C. S., Kemp, K. B., Higuera, P. E., Harvey, B. J., Rother, M. T., Donato, D. C., Morgan, P., & Veblen, T. T. (2018). Evidence for declining forest resilience to wildfires under climate change. *Ecology Letters, 21*(2), 243–252.

Turco, M., Bedia, J., Liberto, F. D., Fiorucci, P., von Hardenberg, J., Koutsias, N., Llasat, M.-C., Xystrakis, F., & Provenzale, A. (2016). Decreasing fires in Mediterranean Europe. *PLoS One, 11*(3), e0150663. https://doi.org/10.1371/journal.pone.0150663.

Van der Werf, G. R., Randerson, J. T., Giglio, L., Collatz, G. J., Mu, M., Kasibhatla, P. S., Morton, D. C., DeFries, R. S., Jin, Y., & van Leeuwen, T. T. (2010). Global fire emissions and the contribution of deforestation, savanna, forest, agricultural, and peat fires (1997-2009). *Atmospheric Chemistry and Physics, 10*(23), 11707–11735.

Warner, T. A., Skowronski, N. S., & Gallagher, M. R. (2017). High spatial resolution burn severity mapping of the New Jersey Pine Barrens with WorldView-3 near-infrared and shortwave infrared imagery. *International Journal of Remote Sensing, 38*(2), 598–616. https://doi.org/10.1080/01431161.2016.1268739.

Wells, G. (2011). Preparing tomorrow's fire professionals: Integration of education, training, and experience through science-management partnerships. *Fire Science Digest, 9*. USDI and USDO

Joint Fire Science Program, Boise. Retrieved December 13, 2019, from https://www.firescience.gov/Digest/FSdigest9.pdf.

Williams, A. P., & Abatzoglou, J. T. (2016). Recent advances and remaining uncertainties in resolving past and future climate effects on global fire activity. *Current Climatic Change Reports, 2*(1), 1–14. https://doi.org/10.1007/s40641-016-0031-0.

Williams, A. P., Abatzoglou, J. T., Gershunov, A., Guzman-Morales, J., Bishop, D. A., Balch, J. K., & Lettenmaier, D. P. (2019). Observed impacts of anthropogenic climate change on wildfire in California. *Earth's Future, 7*(8), 892–910. https://doi.org/10.1029/2019EF001210.

Winsberg, E. (2001). Simulations, models, and theories: Complex physical systems and their representations. *Philosophy in Science, 68*, S442–S454.

Winsberg, E. (2003). Simulated experiments: Methodology for a virtual world. *Philosophy in Science, 70*, 105–125.

Wulder, M. A., Bater, C. W., Coops, N. C., Hilker, T., & White, J. C. (2008). The role of LiDAR in sustainable forest management. *The Forestry Chronicle, 84*(6), 807–826.

Ziegler, J. P., Hoffman, C., Battaglia, M., & Mell, W. (2017). Spatially explicit measurements of forest structure and fire behavior following restoration treatments in dry forests. *Forest Ecology and Management, 386*, 1–12.

Index

F. Castro Rego et al., *Fire Science*, Springer Textbooks in Earth Sciences, Geography and Environment, https://doi.org/10.1007/978-3-030-69815-7

D

E

GPSR Compliance
The European Union's (EU) General Product Safety Regulation (GPSR) is a set of rules that requires consumer products to be safe and our obligations to ensure this.

If you have any concerns about our products, you can contact us on

ProductSafety@springernature.com

In case Publisher is established outside the EU, the EU authorized representative is:

Springer Nature Customer Service Center GmbH
Europaplatz 3
69115 Heidelberg, Germany

www.ingramcontent.com/pod-product-compliance
Ingram Content Group UK Ltd.
Pitfield, Milton Keynes, MK11 3LW, UK
UKHW020713071026
15590UKWH00010B/483
9783030698171

Systems Biology

Shigetada Nakanishi • Ryoichiro Kageyama
Dai Watanabe
Editors

Systems Biology

The Challenge of Complexity

Springer

Editors
Shigetada Nakanishi, M.D., Ph.D.
Director
Osaka Bioscience Institute
Osaka 565-0874
Japan

Ryoichiro Kageyama, M.D., Ph.D.
Director, Professor
Institute for Virus Research
Kyoto University
Kyoto 606-8507
Japan

Dai Watanabe, M.D., Ph.D.
Professor
Department of Biological Sciences
Faculty of Medicine, and Department
of Molecular and System Biology
Graduate School of Biostudies
Kyoto University
Kyoto 606-8501
Japan

ISBN: 978-4-431-87703-5
e-ISBN: 978-4-431-87704-2
DOI: 10.1007/978-4-431-87704-2

Springer Tokyo Berlin Heidelberg New York

Library of Congress Control Number: 2009922675

Printed in Japan

Printed on acid-free paper

Springer is a part of Springer Science+Business Media
springer.com

Preface

Biological signaling pathways dynamically interact with one another to form complex information networks intracellularly, intercellularly, and eventually at the level of the organism. Biology and medicine have conventionally focused on identification and characterization of functional elements in biological signaling pathways. Recently, research in this field has been pursuing a new approach, systems biology, to understand the dynamics, complexity, and physiological functions of the biological signaling networks. Instead of reductionistic analyses or large-scale studies of biomolecules piece by piece, systems biology emphasizes the need for interdisciplinary methods and analysis of the regulation and operation of information networks at the systems level.

The Uehara Memorial Symposium 2008, entitled "Systems Biology: The Challenge of Complexity," was convened in Tokyo from June 30 to July 2, 2008. The aim of the symposium was to bring together leading scientists in the field of systems biology to discuss the latest research on function and dysfunction of biological networks. The following are the main topics addressed in the symposium:

1. Dynamics and complexity of intercellular networks
2. Integrative mechanism of intracellular information networks
3. Dysfunction of information networks and related diseases

An explicit aim of the symposium was to contribute to an understanding of the complexity and dynamics of biological systems and to direct our efforts toward the development of novel strategies for disease control.

We are very pleased to be able to publish the proceedings of this exciting symposium.

Osaka, Japan

Shigetada Nakanishi, Editor

Contents

Contributors

Kentaro Abe
Department of Biological Sciences, Faculty of Medicine, and Department of Molecular and System Biology, Graduate School of Biostudies, Kyoto University, Yoshida, Sakyo-ku, Kyoto, Japan

George J. Augustine
Department of Neurobiology, Duke Medical Center, Durham, NC, USA

Eshel Ben-Jacob
School of Physics and Astronomy, Raymond and Beverly Sackler Faculty of Exact Sciences, Tel-Aviv University, Tel-Aviv, Israel

Upinder S. Bhalla
National Centre for Biological Sciences, TIFR, Bellary Road, Bangalore, India

Toshiya Endo
Department of Chemistry, Graduate School of Science, Nagoya University, Nagoya, Japan

Hisataka Fujimoto
Department of Biological Sciences, Faculty of Medicine, and Department of Molecular and System Biology, Graduate School of Biostudies, Kyoto University, Yoshida, Sakyo-ku, Kyoto, Japan

Susumu Goyama
Department of Hematology and Oncology, Graduate School of Medicine, University of Tokyo, The University of Tokyo Hospital, Bunkyo-ku, Tokyo, Japan

Sten Grillner
Nobel Institute for Neurophysiology, Department of Neuroscience, Karolinska Insititutet, Stockholm, Sweden

Taku Hasegawa
Department of Biological Sciences, Faculty of Medicine, and Department of Molecular and System Biology, Graduate School of Biostudies, Kyoto University, Yoshida, Sakyo-ku, Kyoto, Japan

Chiharu Higashida
Department of Pharmacology, Kyoto University Faculty of Medicine, Sakyo-ku, Kyoto, Japan

Russell Hill
Computational Biology and Neurocomputing, School of Computer Science and Communication, Royal Institute of Technology, Stockholm, Sweden

Motoshi Ichikawa
Department of Hematology and Oncology, Graduate School of Medicine, University of Tokyo, The University of Tokyo Hospital, Bunkyo-ku, Tokyo, Japan

Kentaro Ito
Division of Oncology, The Institute of Medical Science, The University of Tokyo, Minato-ku, Tokyo, Japan

Hideo Iwasaki
Department of Electrical Engineering & Bioscience, Waseda University, and PRESTO, Japan Science & Technology Agency (JST), Shinjuku-ku, Tokyo, Japan

Kevin A. Janes
Koch Institute for Integrative Cancer Research, Center for Cell Decision Processes, Department of Biology, and Department of Biological Engineering, Massachusetts Institute of Technology, Cambridge, MA, USA; Department of Cell Biology, Harvard Medical School, Boston, MA, USA

Ryoichiro Kageyama
Institute for Virus Research, Kyoto University, Kyoto, Japan, and Japan Science and Technology Agency, CREST, Kyoto, Japan

Hideki Katagiri
Division of Advanced Therapeutics for Metabolic Diseases, Center for Translational and Advanced Animal Research, Tohoku University Graduate School of Medicine, Aoba-ku, Sendai, Japan

Mitsuo Kawato
ATR Computational Neuroscience Laboratories, Seika, Kyoto, Japan

Shigeru Kondo
Division of Biological Science, Graduate School of Science, Nagoya University, Chikusa-ku, Nagoya, Aichi, Japan

Ryo Kunimoto
Department of PharmacoInformatics, Graduate School of Pharmaceutical Sciences, Kyoto University, Kyoto, Japan

Shinya Kuroda
Department of Biophysics and Biochemistry, Graduate School of Science, The University of Tokyo, and CREST, Japan Science and Technology Agency, Bunkyo-ku, Tokyo, Japan

Mineo Kurokawa
Department of Hematology and Oncology, Graduate School of Medicine, University of Tokyo, The University of Tokyo Hospital, Bunkyo-ku, Tokyo, Japan

Jens Looser
University of Wuerzburg, Wuerzburg, Germany

Ryosuke Matsui
Department of Biological Sciences, Faculty of Medicine, and Department of Molecular and System Biology, Graduate School of Biostudies, Kyoto University, Sakyo-ku, Kyoto, Japan

Masahiro Morita
Division of Oncology, The Institute of Medical Science, The University of Tokyo, Minato-ku, Tokyo, Japan

Georg Nagel
University of Wuerzburg, Wuerzburg, Germany

Masahiro Nakagawa
Department of Hematology and Oncology, Graduate School of Medicine, University of Tokyo, The University of Tokyo Hospital, Bunkyo-ku, Tokyo, Japan

Shigetada Nakanishi
Osaka Bioscience Institute, Osaka, Japan

Atsushi Nambu
Division of System Neurophysiology, National Institute for Physiological Sciences, and Department of Physiological Sciences, Graduate University for Advanced Studies, Myodaiji, Okazaki, Aichi, Japan

Takeshi Nitta
Division of Experimental Immunology, Institute for Genome Research, University of Tokushima, Tokushima, Japan

Yasutaka Niwa
Institute for Virus Research, Kyoto University, Kyoto, Japan and Japan Science and Technology Agency, CREST, Kyoto, Japan

Hideaki Ogasawara
National Institute of Information and Communications Technology, Seika, Kyoto, Japan, and ATR Computational Neuroscience Laboratories, Seika, Kyoto, Japan

Yasushi Okuno
Department of PharmacoInformatics, Graduate School of Pharmaceutical Sciences, Kyoto University, Kyoto, Japan

Yu-ichi Ozaki
Department of Biophysics and Biochemistry, Graduate School of Science, University of Tokyo, and CREST, Japan Science and Technology Agency, Bunkyo-ku, Tokyo, Japan

H. Christian Reinhardt
Koch Institute for Integrative Cancer Research, Center for Cell Decision Processes, Department of Biology, and Department of Biological Engineering, Massachusetts Institute of Technology, Cambridge, MA, USA

Kazuki Saito
Chiba University, Graduate School of Pharmaceutical Sciences, Inage-ku, Chiba, Japan, and RIKEN Plant Science Center, Tsurumi-ku, Yokohama, Japan

Fumiaki Sato
Department of Nanobio Drug Discovery, Graduate School of Pharmaceutical Sciences, Kyoto University, Kyoto, Japan

Kazuharu Shimizu
Department of Nanobio Drug Discovery, Graduate School of Pharmaceutical Sciences, Kyoto University, Kyoto, Japan

Hiromi Shimojo
Institute for Virus Research, Kyoto University, Kyoto, Japan, and Japan Science and Technology Agency, CREST, Kyoto, Japan

Toru Suzuki
Division of Oncology, The Institute of Medical Science, The University of Tokyo, Minato-ku, Tokyo, Japan

Yousuke Takahama
Division of Experimental Immunology, Institute for Genome Research, University of Tokushima, Tokushima, Japan

Keiko Tanaka
Department of Biophysics and Biochemistry, Graduate School of Science, University of Tokyo, Tokyo, Japan, and PRESTO, Japan Science and Technology Agency, Tokyo, Japan

Kazuya Terasawa
Department of Genomic Drug Discovery Science, Kyoto University, Sakyo-ku, Kyoto, Japan

Soken Tsuchiya
Department of Nanobio Drug Discovery, Graduate School of Pharmaceutical Sciences, Kyoto University, Kyoto, Japan

Gozoh Tsujimoto
Department of Genomic Drug Discovery Science, Kyoto University, Sakyo-ku, Kyoto, Japan

Shinsuke Uda
Department of Biophysics and Biochemistry, Graduate School of Science, University of Tokyo, CREST, Japan Science and Technology Agency, Bunkyo-ku, Tokyo, Japan

Hiroki R. Ueda
Laboratory for Systems Biology and Functional Genomics Unit, Center for Developmental Biology, RIKEN, Chuo-ku, Kobe, Hyogo, Japan

Peter Wallén
Computational Biology and Neurocomputing, School of Computer Science and Communication, Royal Institute of Technology, Stockholm, Sweden

Dai Watanabe
Department of Biological Sciences, Faculty of Medicine, and Department of Molecular and System Biology, Graduate School of Biostudies, Kyoto University, Sakyo-ku, Kyoto, Japan

Naoki Watanabe
Department of Pharmacology, Kyoto University Faculty of Medicine, Sakyo-ku, Kyoto, Japan

Cornelis J. Weijer
Division of Cell and Developmental Biology, School of Life Sciences, University of Dundee, Wellcome Trust Biocentre, Dundee, UK

Michael B. Yaffe
Koch Institute for Integrative Cancer Research, Center for Cell Decision Processes, Department of Biology, and Department of Biological Engineering, Massachusetts Institute of Technology, Cambridge, MA, USA

Tadashi Yamamoto
Division of Oncology, The Institute of Medical Science, The University of Tokyo, Minato-ku, Tokyo, Japan

Shinya Yamanaka
Center for iPS Cell Research & Application (CiRA) / Institute for Frontier Medical Sciences, Kyoto University, Kyoto, Japan

Part I
Dynamics and Complexity of Intercellular Networks

Integrative Synaptic Mechanisms of the Neural Network

Shigetada Nakanishi

Introduction

The cerebellum is a key neural substrate for motor coordination and motor learning [1,2]. The main neural circuitry of the cerebellum consists of two excitatory glutamatergic pathways: the pons–mossy fiber–granule cell–parallel fibers and the inferior olive–climbing fibers [1]. The inputs of mossy fibers and climbing fibers converge on Purkinje cells in the cerebellar cortex and deep cerebellar nuclei [1]. Besides this main circuitry, Golgi cells receive excitatory parallel fiber input and suppress granule cell transmission via the inhibitory gamma-aminobutyric acid (GABA) transmitter. The cerebellar network has been well characterized by electrophysiology and morphology, and its input/output relationship is also relatively well quantitatively assessed by both electrophysiological and behavioral analysis. Our investigations have been concerned with the regulatory mechanisms of integrative synaptic transmission, which is responsible for motor learning and motor coordination in the cerebellar circuitry.

Neural Circuits for Conditioned Eyeblink Motor Learning

The cerebellum is critically implicated in motor learning [2,3]. The conditioned eyeblink response is a typical cerebellum-dependent motor learning event [2–5]. When a conditioned sensory stimulus such as a tone is paired with an unconditioned periorbital electrical stimulus, the trained animal shows a tone-dependent eyeblink response in the absence of unconditioned stimulation (Fig. 1). Information about the conditioned stimulus is transmitted through the pons, mossy fibers, granule cells, and parallel fibers, whereas information about the unconditioned

---

S. Nakanishi
Osaka Bioscience Institute, 6-2-4 Furuedai, Suita, Osaka 565-0874, Japan

S. Nakanishi et al. (eds.), *Systems Biology*,
DOI:10.1007/978-4-431-87704-2_1, © Springer 2009

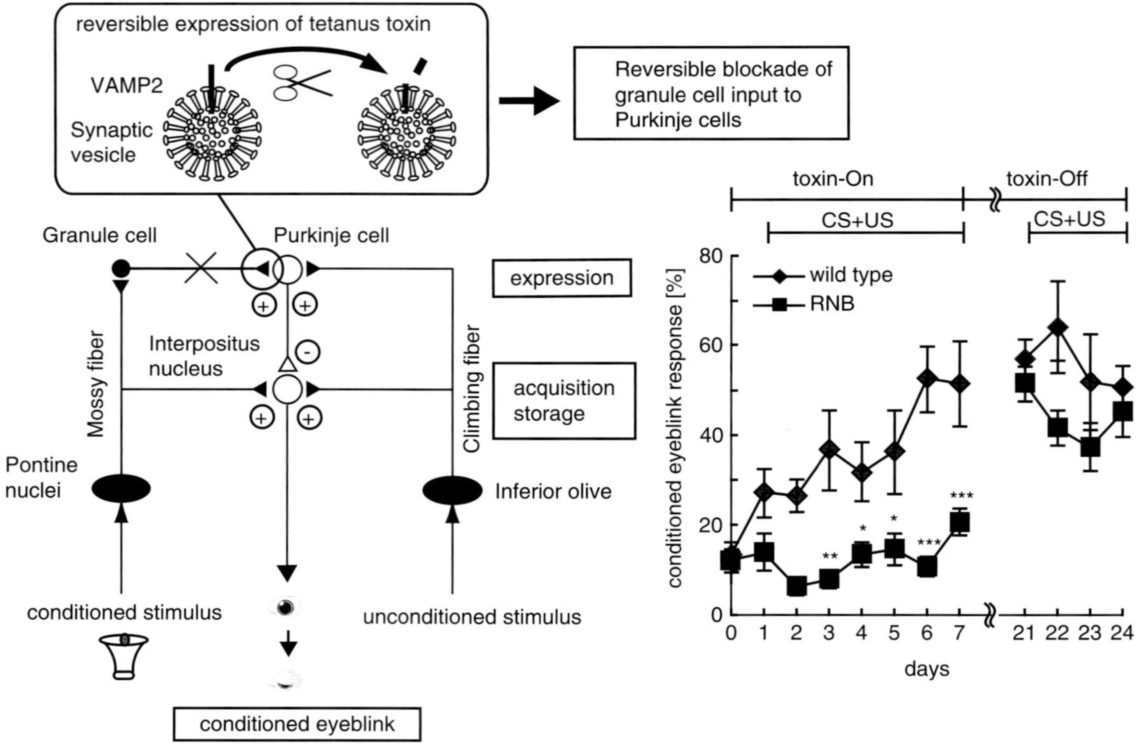

Fig. 1 Mechanistic analysis of conditioned eyeblink motor learning by reversible neurotransmission blocking techniques. *Left:* The neural circuitry responsible for conditioned eyeblink motor learning. In this circuitry, granule cell input to Purkinje cells was selectively and reversibly blocked by the tetracycline derivative DOX-dependent expression of tetanus toxin in granule cells. *Right:* Conditioned eyeblink responses in wild-type mice and reversible neurotransmission blocking (RNB) mice. In RNB mice, granule cell transmission was blocked and reversibly recovered by DOX-dependent expression of tetanus toxin. *CS*, conditioned stimulus; *US*, unconditioned stimulus

stimulus is transmitted through the inferior olive and climbing fibers (Fig. 1). Importantly, this information converges on both Purkinje cells and deep cerebellar nuclei, namely, the interpositus nucleus. However, the relative importance and the underlying mechanism of cerebellum-dependent eyeblink motor learning have remained largely unclear.

To address the neural circuits and mechanisms underlying the conditioned eyeblink response, we developed a novel technique termed reversible neurotransmission blocking (RNB) [6] (Fig. 1). With the RNB technology, we generated two lines of transgenic mice. One line of transgenic mice selectively expressed the tetracycline-activating transcription factor in granule cells under the control of the $GABA_A$ $\alpha6$ promoter. In the second line of transgenic mice, the fusion protein of tetanus toxin and green fluorescent protein (GFP) was controlled by this transcription factor. Tetanus toxin is a bacterial toxin that specifically cleaves the synaptic vesicle VAMP2 and in turn blocks transmitter release from the synaptic vesicles [7]. When these two lines of transgenic mice were mated, the tetracycline derivative doxycycline (DOX)-dependent expression of tetanus toxin was specifically observed in granule cells. Consequently, VAMP2 was selectively and reversibly cleaved in granule cells, depending on the administration and omission of DOX [6]. The RNB mice were used to explore the mechanism underlying conditioned eyeblink motor learning with the following rationale (Fig. 1). When granule cell input to Purkinje cells is selectively blocked, the conditioned signal is not transmitted to the Purkinje cells, but this information is still conveyed to the interpositus nucleus. Therefore, the reversible blockade of granule cell transmission can delineate the role of the two information pathways in the conditioned eyeblink response.

We performed electrophysiology and behavioral experiments on conditioned eyeblink motor learning [8]. We confirmed the reversibility of blockade of granule cell transmission to Purkinje cells by extracellular recording of Purkinje cells in awake animals. This blockade, however, had no influence on climbing fiber transmission. We therefore succeeded in manipulating the reversible blocking of granule cell transmission to Purkinje cells without any impairment of response to the climbing fiber input. We then tested the conditioned eyeblink response in this model animal (Fig. 1). In RNB mice, the conditioned eyeblink response disappeared during the administration of DOX and was recovered by the omission of DOX. Importantly, when granule cell transmission was recovered by the withdrawal of DOX, the normal conditioned response was immediately induced at the beginning of the second conditioning session of the pretrained RNB mice (Fig. 1). This finding explicitly demonstrates that although the conditioned response is not expressed during DOX treatment, this memory is acquired and stored in RNB mice during DOX treatment. We further confirmed that a bilateral lesion of the interpositus nucleus completely abolished memory acquisition and storage in DOX-withdrawn RNB mice [8].

These results demonstrate that information of the conditioned signal to Purkinje cells is necessary for the expression of the conditioned response, but the memory is acquired and stored despite the absence of the conditioned signal to Purkinje cells

(Fig. 1). The most plausible interpretation of our study is that the blockade of granule cell transmission relieves tonic Purkinje cell inhibition [9,10], and that the interpositus nucleus induces latent neural plasticity in response to paired conditioned and unconditioned signals. This neural plasticity should be critical for memory acquisition and storage and would allow prompt expression of the conditioned response once the expression process is restored by the recovery of granule cell transmission [8]. We are currently investigating whether and how neural plasticity is induced at the interpositus nucleus during DOX treatment in RNB mice.

The Regulatory Role of Golgi Cells

In the cerebellar circuitry, Golgi cells also receive excitatory parallel fiber input and in turn suppress granule cell excitation through the inhibitory GABA transmitter [11,12] (Fig. 2). Golgi cell-mediated feedback inhibition is thought to be important for filtering mossy fiber input before distributing this input to the Purkinje cells.

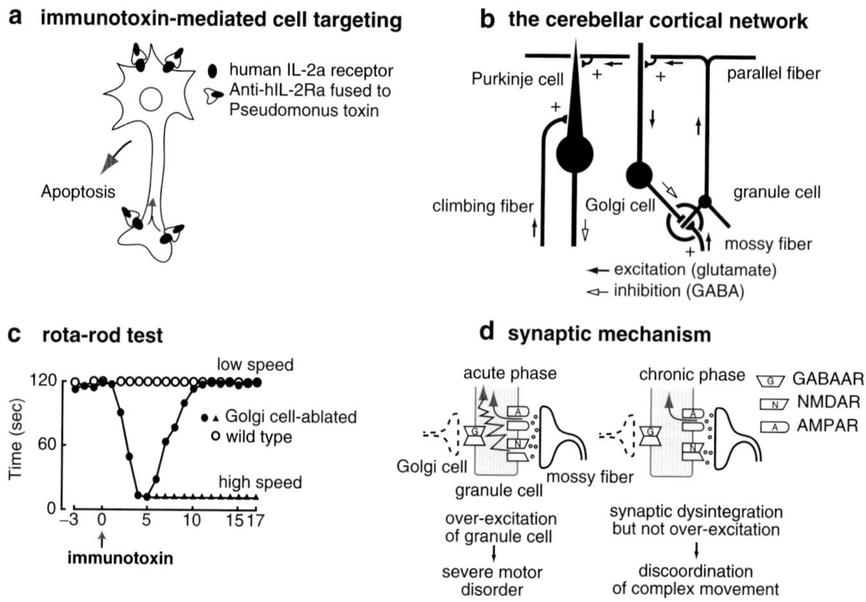

Fig. 2 Integrative mechanisms of neurotransmission at the Golgi cell–mossy fiber–granule cell synapse. **a** Principle of the immunotoxin-mediated cell targeting technology. **b** Neural circuitry of the cerebellar cortex. **c** Ability of wild-type and Golgi cell-eliminated transgenic mice to stay on a slowly rotating and a rapidly rotating rota-rod. **d** A model for synaptic mechanisms underlying motor disorders in the acute phase and the functional compensation of motor disorders in the chronic phase after Golgi cell elimination

However, Golgi cells represent less than 1% of the cerebellar cell population, and the functional role of Golgi cells in the cerebellar circuitry remains to be determined.

To address the functional role of Golgi cells in the cerebellar circuitry, we developed another technique termed immunotoxin-mediated cell targeting (IMCT) as a collaborative work with Ira Pastan in the National Institutes of Health (NIH) [13] (Fig. 2). In this technology, we generated transgenic mice that expressed the human interleukin (IL)-2 receptor under the control of a neuron-specific promoter. We then injected the recombinant immunotoxin that is composed of the human IL-2 receptor monoclonal antibody fused to *Pseudomonas* exotoxin. The immunotoxin binds to the membrane human IL-2 receptor, and this complex is internalized and then kills target cells, thereby leading to elimination of specific neuronal cells.

We constructed a transgene that contained the metabotropic glutamate receptor 2 (mGluR2) promoter, followed by the human IL-2 receptor fused to GFP. In the cerebellum of transgenic mice, the fluorescence of the GFP–IL-2 receptor complex was specifically observed in mGluR2-immunoreactive Golgi cells. Intrathecal injection of immunotoxin eliminated Golgi cells in transgenic mice but not in wild-type mice. Importantly, granule cells, Purkinje cells, and basket and stellate cells all remained intact after immunotoxin injection.

Immunotoxin-injected transgenic mice showed severe acute ataxia. When the rota-rod test was conducted at a low speed, Golgi cell-eliminated transgenic mice failed this task acutely after immunotoxin injection, but the motor disorders gradually recovered thereafter (Fig. 2). However, these mice still failed to stay on the more rapidly rotating rod. The movement disorder was never observed in immunotoxin-treated wild-type mice.

We addressed how synaptic transmission is severely perturbed in the acute phase of Golgi cell elimination and how this perturbation is at least partly compensated in the chronic phase. We performed a series of electrophysiological and optical recordings of cerebellar slices of wild-type mice and Golgi cell-eliminated transgenic mice [13]. These experiments indicated a novel mechanism for synaptic integration at mossy fiber–granule cell–Golgi cell synapses [13,14] (Fig. 2). Under normal circumstances, GABA inhibition and *N*-methyl-D-aspartic acid (NMDA) receptors play an important role in synaptic integration responsible for complex motor coordination. When Golgi cells are eliminated, GABA is depleted, and this depletion results in overexcitation of NMDA receptors and causes severe acute motor disorders. In the chronic phase, NMDA receptors are adaptively downregulated, thus relieving overexcitation of granule cells. Importantly, AMPA receptors remain unchanged before and after Golgi cell elimination, and these AMPA receptors allow animals to manage simple motor movement. However, NMDA receptors are attenuated, thus causing functional deficits in directing more complex motor coordination. Our studies demonstrate that the cooperative actions of excitatory and inhibitory receptors are important not only for integrative brain functioning but also for the compensation of brain dysfunction.

Regulatory Mechanism of Golgi Cell Inhibition

Golgi cell-mediated feedback inhibition is also known to be so strong that well-defined control of this feedback inhibition is necessary [15,16]. Otherwise, the strong inhibition would block information transmission of mossy fiber input to Purkinje cells. For example, strong facial stimulation has been shown to evoke spike discharges and then induce long silencing of putative Golgi cells in rats [17]. This silencing of Golgi cells is assumed to be important for information transmission of the strong tactile stimuli to Purkinje cells. To address how Golgi cell inhibition is properly controlled, we focused on abundantly expressing, postsynaptic mGluR2 in Golgi cells [18], because mGluR2 is often coupled to the inhibitory signaling cascade [19,20].

We performed a series of whole-cell recordings of GFP-positive Golgi cells after electrical stimulation of parallel fibers in cerebellar slices of wild-type and mGluR2 knockout mice [18] (Fig. 3). These experiments indicated that mGluR2 activates the G protein-coupled inwardly rectifying K^+ channel (GIRK) and suppresses Golgi cell excitability. Furthermore, immunostaining characterization revealed that Kir 3.2 of the GIRK K^+ channel is involved in mGluR2-mediated Golgi cell suppression.

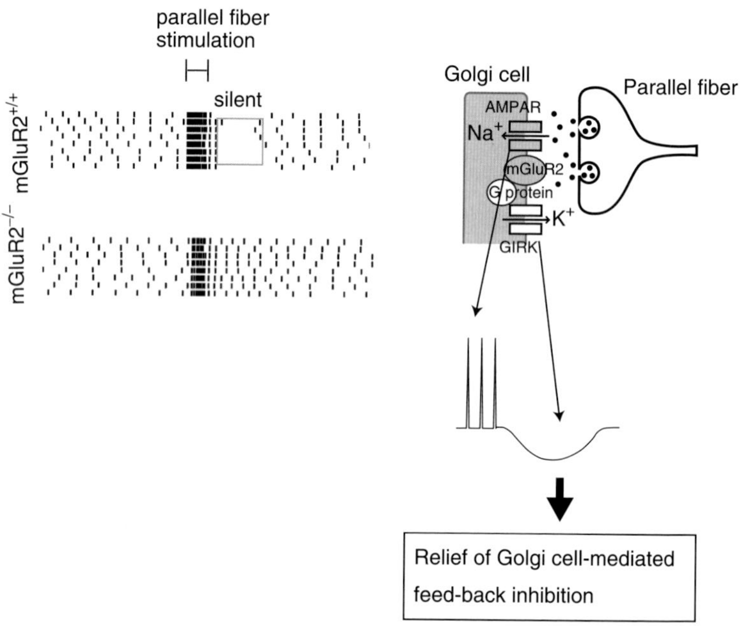

Fig. 3 Input-dependent, metabotropic glutamate receptor 2 (*mGluR2*)-mediated suppression of Golgi cell inhibition. *Left:* Spike raster plots of mGluR2$^{+/+}$ and mGluR2$^{-/-}$ Golgi cells recorded in cerebellar slices before and after strong parallel fiber stimulation. Strong parallel fiber stimulation induced spike discharges in both mGluR2$^{+/+}$ and mGluR2$^{-/-}$, and a long-lasting silent phase followed in mGluR2$^{+/+}$ but not in mGluR2$^{-/-}$. *Right:* A model of mGluR2-mediated suppression following AMPA receptor-mediated excitation of the Golgi cell

We then investigated how the postsynaptic mGluR2 senses different extents of granule cell input at Golgi cells [18]. These results explicitly indicated that when Golgi cells receive glutamatergic input from parallel fibers, AMPA receptors first excite Golgi cells and mGluR2 then suppresses Golgi cell excitability in a stimulus strength-dependent manner (Fig. 3). Therefore, the postsynaptic mGluR2 is capable of sensing the strength of presynaptic granule cell input and relieving Golgi cell-mediated feedback inhibition in a stimulus strength-dependent manner [18]. Consequently, this mechanism enables transmission of strong mossy fiber input to Purkinje cells. Furthermore, because Golgi cells project divergent axon terminals on multiple granule cells, the control by postsynaptic mGluR2 is capable of influencing a regional population of granule cells and plays an important role in spatiotemporal regulation of the cerebellar circuitry.

Concluding Remarks

The cerebellar network is organized in an orderly manner, and synaptic transmission in the cerebellar circuitry is regulated dynamically by specialized mechanisms. Our investigation of eyeblink conditioning has revealed that different local circuits in the cerebellar network participate in distinct processes of cerebellum-dependent eyeblink motor learning. In the mechanism of Golgi cell-mediated feedback inhibition, GABA inhibition and glutamate excitation by different types of glutamate receptors cooperatively and differentially regulate synaptic transmission in mossy fiber–granule cell synapses and play an important role in integrative and adaptive mechanisms for motor coordination. Moreover, the dual regulation by ionotropic and metabotropic glutamate receptors is crucial in sensing and properly transmitting different strengths of glutamate-mediated neural information in the cerebellar circuitry. Neural information is thus hierarchically processed and integrated at different levels of the neural network.

References

1. Llinas RR, Walton KD. Cerebellum. In: Shepherd GM (ed) The Synaptic Organization of the Brain. Oxford: Oxford University Press, 1998:255–288.
2. Ito M. Cerebellar long-term depression: characterization, signal transduction, and functional roles. Physiol Rev 2001;81(3):1143–95.
3. Christian KM, Thompson RF. Neural substrates of eyeblink conditioning: acquisition and retention. Learn Mem 2003;10(6):427–55.
4. Mauk MD. Roles of cerebellar cortex and nuclei in motor learning: contradictions or clues? Neuron 1997;18(3):343–6.
5. De Zeeuw CI, Yeo CH. Time and tide in cerebellar memory formation. Curr Opin Neurobiol 2005;15(6):667–74.
6. Yamamoto M, Wada N, Kitabatake Y, Watanabe D, Anzai M, Yokoyama M, Teranishi Y, Nakanishi S. Reversible suppression of glutamatergic neurotransmission of cerebellar granule cells in vivo by genetically manipulated expression of tetanus neurotoxin light chain. J Neurosci 2003;23(17):6759–67.

7. Schiavo G, Benfenati F, Poulain B, Rossetto O, Polverino de Laureto P, DasGupta BR, Montecucco C. Tetanus and botulinum-B neurotoxins block neurotransmitter release by proteolytic cleavage of synaptobrevin. Nature (Lond) 1992;359(6398):832–5.
8. Wada N, Kishimoto Y, Watanabe D, Kano M, Hirano T, Funabiki K, Nakanishi S. Conditioned eyeblink learning is formed and stored without cerebellar granule cell transmission. Proc Natl Acad Sci U S A 2007;104(42):16690–5.
9. Ito M. Long-term depression. Annu Rev Neurosci 1989;12:85–102.
10. Linden DJ. Neuroscience. From molecules to memory in the cerebellum. Science 2003;301(5640):1682–5.
11. Gabbiani F, Midtgaard J, Knopfel T. Synaptic integration in a model of cerebellar granule cells. J Neurophysiol 1994;72(2):999–1009.
12. Singer W. Neurophysiology: the changing face of inhibition. Curr Biol 1996;6(4):395–7.
13. Watanabe D, Inokawa H, Hashimoto K, Suzuki N, Kano M, Shigemoto R, Hirano T, Toyama K, Kaneko S, Yokoi M, Moriyoshi K, Suzuki M, Kobayashi K, Nagatsu T, Kreitman RJ, Pastan I, Nakanishi S. Ablation of cerebellar Golgi cells disrupts synaptic integration involving GABA inhibition and NMDA receptor activation in motor coordination. Cell 1998;95(1):17–27.
14. Nakanishi S. Synaptic mechanisms of the cerebellar cortical network. Trends Neurosci 2005;28(2):93–100.
15. Brickley SG, Cull-Candy SG, Farrant M. Development of a tonic form of synaptic inhibition in rat cerebellar granule cells resulting from persistent activation of GABAA receptors. J Physiol 1996;497(pt 3):753–9.
16. Wall MJ, Usowicz MM. Development of action potential-dependent and independent spontaneous GABAA receptor-mediated currents in granule cells of postnatal rat cerebellum. Eur J Neurosci 1997;9(3):533–48.
17. De Schutter E, Bjaalie JG. Coding in the granular layer of the cerebellum. Prog Brain Res 2001;130:279–96.
18. Watanabe D, Nakanishi S. mGluR2 postsynaptically senses granule cell inputs at Golgi cell synapses. Neuron 2003;39(5):821–9.
19. Nakanishi S. Molecular diversity of glutamate receptors and implications for brain function. Science 1992;258(5082):597–603.
20. Yokoi M, Kobayashi K, Manabe T, Takahashi T, Sakaguchi I, Katsuura G, Shigemoto R, Ohishi H, Nomura S, Nakamura K, Nakao K, Katsuki M, Nakanishi S. Impairment of hippocampal mossy fiber LTD in mice lacking mGluR2. Science 1996;273(5275):645–7.

The Molecular and Cellular Design of Networks in Motion

Sten Grillner, Peter Wallén, and Russell Hill

Introduction

In this chapter, we review the extensive knowledge gained on the lamprey nervous system through an interactive process between experiments and modeling [1–3]. The organization of the locomotor system is to a large extent conserved through vertebrate phylogeny, and it is therefore also pertinent to explore to what extent this knowledge can be applied to the more complex mammalian nervous system. The overall aim is to account for this complex set of behaviors, based on an understanding of the intrinsic cellular mechanisms determining the operation of the different neuronal networks.

The different subsystems involved in the control of locomotion can be represented as follows (Fig. 1).

- A neural system responsible for selection of the appropriate behavior, in this case locomotion. Striatum, the input layer of the basal ganglia, has an important role in this context. Striatum receives phasic input from pallium (cortex) and thalamus and a modulatory input from the dopamine system. The GABAergic striatal neurons have a high threshold for activation. When activated they can indirectly release specific motor programs by inhibiting the GABAergic output neurons of the basal ganglia (pallidum) that at rest provide tonic inhibition of the different motor programs [4] (Grillner 2006).
- A command system that, when released from basal ganglia inhibition, can elicit locomotion by activating the pattern generating circuits in the spinal cord. Two command systems for locomotion, the mesencephalic locomotor region (MLR)

S. Grillner
Nobel Institute for Neurophysiology, Department of Neuroscience, Karolinska Insititutet, Retzius väg 8, SE-171 77 Stockholm, Sweden

S. Nakanishi et al. (eds.), *Systems Biology*,
DOI:10.1007/978-4-431-87704-2_2, © Springer 2009

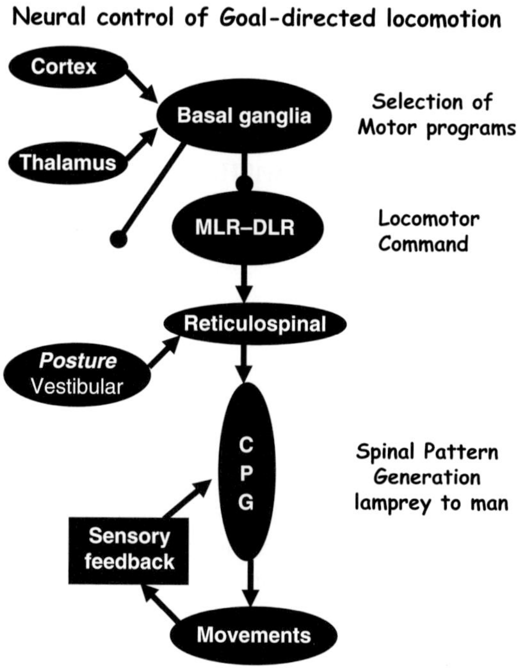

Fig. 1 Subsystems involved in the control of motor behavior, such as locomotion. Selection of a certain motor program is performed in the basal ganglia, which receive inputs from cortex (pallium) and thalamus. The basal ganglia activate indirectly command centers in the diencephalic locomotor region (*DLR*) and the mesencephalic locomotor region (*MLR*) through disinhibition. They activate the spinal central pattern generators (*CPG*) for locomotion via the reticulospinal (RS) neurons. In the brainstem, information is further integrated based on visual, sensory, and vestibular inputs to control both steering and posture. As in higher vertebrates, the spinal cord CPG neurons are modulated by local sensory feedback

and the diencephalic locomotor region (DLR), have been defined. They act via a symmetrical activation of reticulospinal neurons that turn on the spinal circuits.

- Segmental and intersegmental networks (central pattern generators, CPGs) located at the spinal level. The CPGs contain the necessary timing information to activate the different motoneurons in the appropriate sequence to produce the propulsive movements.
- The segmental burst-generating network in the lamprey contains excitatory interneurons that within the pool of interneurons provide excitation. The alternating pattern between the left and right side is provided by reciprocal inhibitory connections between pools of excitatory interneurons.
- A sensory control system, sensing the locomotor movements, helps to compenate for external perturbations by a feedback action on the spinal CPGs.

- A control system for steering the body toward different goals. The steering commands are superimposed on the basic locomotor activity and will bias the control signals so as to steer the movements to the left or right side or in other orientations [5].
- During locomotion the body moves with the dorsal side up, regardless of perturbations. This "postural" control system for orientation of the body in the gravity field depends on bilateral vestibular input that detects any deviation from the appropriate orientation of the head, whether tilt to the left or right or changes in pitch angle during locomotion. These vestibular effects are mediated via brainstem interneurons to reticulospinal neurons on the left or right side, respectively, that can elicit compensatory movements that restore the body orientation [6].

Basic Mechanisms of Burst Generation

One major problem when studying vertebrate pattern generation has been the intrinsic function of the networks controlling behaviors such as respiration and locomotion. In the case of lower vertebrates such as the lamprey and the frog embryo, the locomotor networks are comparatively well understood. At the segmental level, recurring locomotor bursts can be generated even in a hemisegment, provided that the excitability is high enough [7,8]. The burst generation is primarily dependent on a pool of excitatory interneurons (EINs). Within this pool EINs excite other EINs (AMPA and NMDA receptors), although mutual excitation between individual EINs has not been observed (Figs. 2, 3) [9]. These pools of EINs form burst-generating kernels within the CPG. The excitatory drive to the EIN kernel is provided by the locomotor command regions via glutamate reticulospinal neurons, whereas the burst generation primarily results from the intrinsic synaptic and membrane properties of the EINs. Although there are inhibitory premotor interneurons [10] that can be active during locomotion, they are not essential for burst generation, because they can be blocked without any effects on burst frequency [7].

The EINs have overshooting action potentials and a fast and a slow AHP similar to that of motoneurons [11,12]. Although they, for reason of their small size, have been studied in lesser detail than the larger spinal neurons, their properties appear very similar.

Each hemisegment is estimated to have around 30 EINs. In addition to the segmental EIN interaction, EINs have brief ascending (three segments) and longer descending (six to eight segments) axonal branches [11,13]. Each EIN will thus receive excitation not only from segmental EINs but also from EINs located both rostrally and caudally to itself. The connectivity ratio within the pool of EINs is estimated to be approximately 10%.

If a pool of model or real EINs is activated from the brainstem or experimentally by a bath applying glutamate agonists, burst activity will be produced, the rate of which depends on the excitatory drive. The initial depolarization of EINs will make the EINs with the lowest threshold start firing, and they will in turn activate other

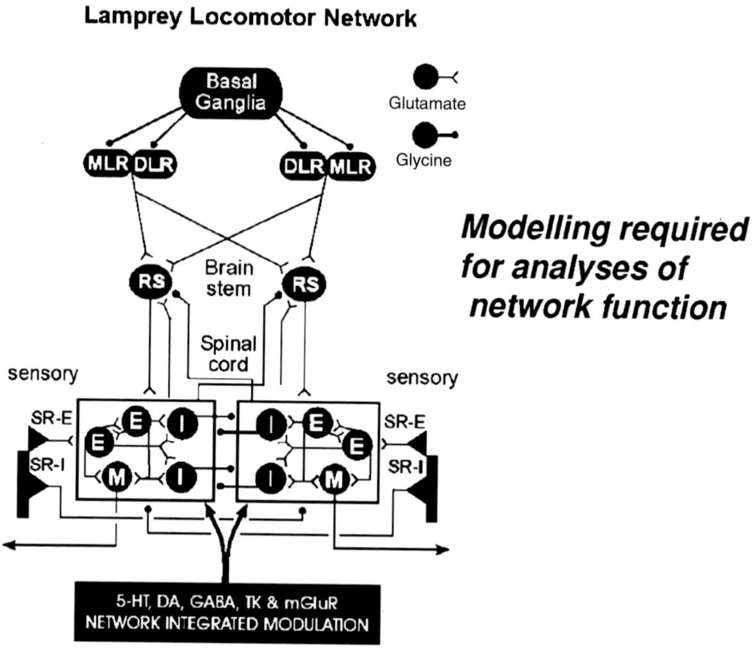

Fig. 2 Locomotor network of the lamprey. Schematic representation of the forebrain, brainstem, and spinal components of the neural circuitry that generates rhythmic locomotor activity. All neuron symbols denote populations rather than single cells. The reticulospinal (*RS*), glutamatergic neurons excite all classes of spinal interneurons and motoneurons. The excitatory interneurons (*E*) excite all types of spinal neurons, i.e., the inhibitory glycinergic interneurons (*I*) that cross the midline to inhibit all neuron types on the contralateral side and motoneurons (*M*). The stretch receptor neurons are of two types: one excitatory (*SR-E*), which excites ipsilateral neurons, and one inhibitory (*SR-I*), which crosses the midline to inhibit contralateral neurons. RS neurons receive excitatory synaptic input from the diencephalic and the mesencephalic locomotor regions (*DLR* and *MLR*), which in turn receive input from the basal ganglia as well as visual and olfactory input. In addition, metabotropic receptors are also activated during locomotion and are an integral part of the network (*5-HT*, 5-hydroxytryptamine; *DA*, dopamine; *GABA*, gamma-aminobutyric acid; *TK*, tachykinin; and *mGluR*, metabotropic glutamate receptor)

EINs synaptically. When the EINs are depolarized, the membrane depolarization can be boosted by the opening of both voltage-dependent NMDA receptors and low-voltage-activated Ca^{2+} channels. This process will be responsible for the initiation of the burst (Fig. 4). During the ongoing burst activity, Ca^{2+} will enter through several different channels including both NMDA channels and a variety of voltage-dependent Ca^{2+} channels. The Ca^{2+} levels will in turn activate Ca^{2+}-dependent K^+ channels (K_{Ca}) that will pull the membrane potential downward and close voltage-dependent channels (e.g., NMDA), and thereby terminate the burst. In addition to K_{Ca} channels, K_{Na} channels also contribute to the hyperpolarization [14,15]. Following the termination of the burst the Ca^{2+} levels will be enhanced, which will lead to a hyperpolarization caused by K_{Ca} activation that will gradually decline as

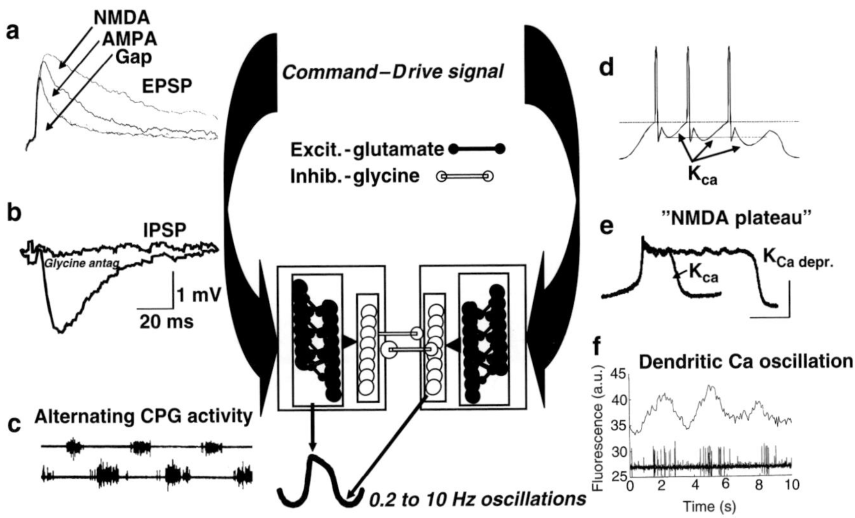

Fig. 3 Schematic representation of the spinal networks for locomotion. The spinal networks that coordinate locomotion are located at the segmental level and consist of a pool of excitatory glutamatergic interneurons, which provide excitation within the pool of interneurons (*filled symbols* = glutamatergic). They are activated from the locomotor command centers in mesencephalon and diencephalon via reticulospinal glutamatergic axons. The descending axons provide an excitatory postsynaptic potential (*EPSP*), which has three components: AMPA, NMDA, and gap junctions (**a**). The glutamatergic interneurons in turn activate glycinergic interneurons (*open symbols*) with axons that inhibit neurons at the contralateral side (**b**). (*IPSP*, inhibitory postsynaptic potential.) Excitatory interneurons project to ipsilateral interneurons that receive out-of-phase inhibition from the contralateral side; this results in alternating efferent activity between the left and right side (**c**). One feature of major importance for terminating burst activity is a calcium (*Ca*) entry in soma and dendrites (**f**) during activity, which results in the activation of calcium-dependent potassium channels (K_{Ca}). The action potential is followed by a slow after-hyperpolarization (sAHP) caused by K_{Ca}. The summation of the sAHPs lead to frequency adaptation (**d**), and the K_{Ca} also contributes to the termination of the plateau potentials (**e**). Time and voltage calibrations as indicated in the figures except for **e**

the cells reduce cytoplasmic Ca^{2+} levels by uptake. When the EIN population has recovered from the K_{Ca} activation, a new burst will be initiated by the background excitatory drive.

Generation of Alternating Activity

The generation of alternating activity between two pairs of antagonists, such as the left and right side of a segment, is a standard element in many networks. The burst-generating kernels on each side in the lamprey segment can produce recurring burst activity by themselves. By connecting them through reciprocal inhibition, alternating

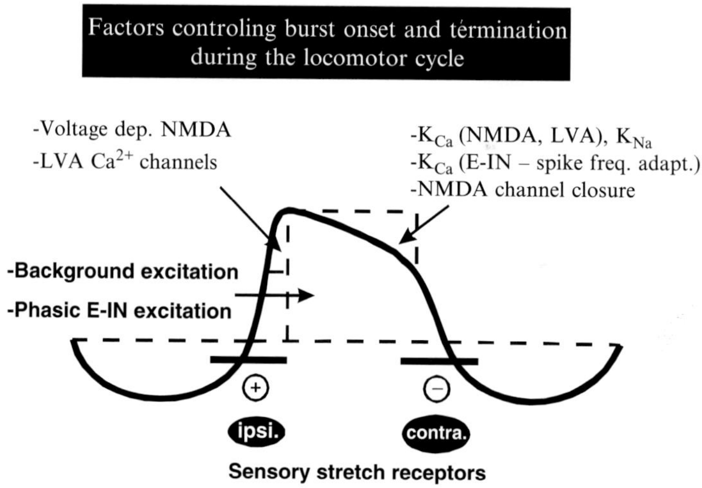

Fig. 4 Factors controlling burst onset and termination. Several different factors contribute to the initiation of the depolarizing phase, its maintenance, and its termination. In addition to conventional synaptic receptor activation, voltage-dependent NMDA receptors, low-voltage activated Ca^{2+} channels (LVA-CA), and Na^{+} channels may be activated. Ca^{2+} and Na^{+} will enter the cell through these channels, cause activation of K_{Ca}, and thereby a progressive hyperpolarization leading to closure of the NMDA channels. The initiation of the depolarizing phase is facilitated by activation of ipsilateral (*ipsi.*) excitatory stretch receptor neurons, whereas the termination of the depolarized phase is partially a result of activation of contralateral (*contra.*) inhibitory stretch receptor neurons. E-IN, excitatory interneurone

activity will occur (Figs. 2, 3). The EINs drive, in addition to ipsilateral motoneurons, also inhibitory commissural interneurons that provide inhibition of the different neurons on the contralateral side [12,16,17]. Thus, when one side is bursting the contralateral side will be inhibited, and as the burst is terminated the ipsilateral side will initiate an excitatory burst from the background excitatory drive [18]. This side will then in turn inhibit the first side and so forth.

Intersegmental Network

The lamprey swims normally by producing an undulatory wave propagated from head to tail (Fig. 5). The faster the wave is propagated backward along the body, the faster the animal will move forward through the water [19–21]. The delay between the activation of each segment is around 1% of the cycle duration regardless of whether the cycle duration is 2 s or 0.10 s. As the lamprey has about 100 segments, this means that the delay from head to tail fin will be constant around one full cycle whether the lamprey swims fast or slowly. This rostrocaudal propagation

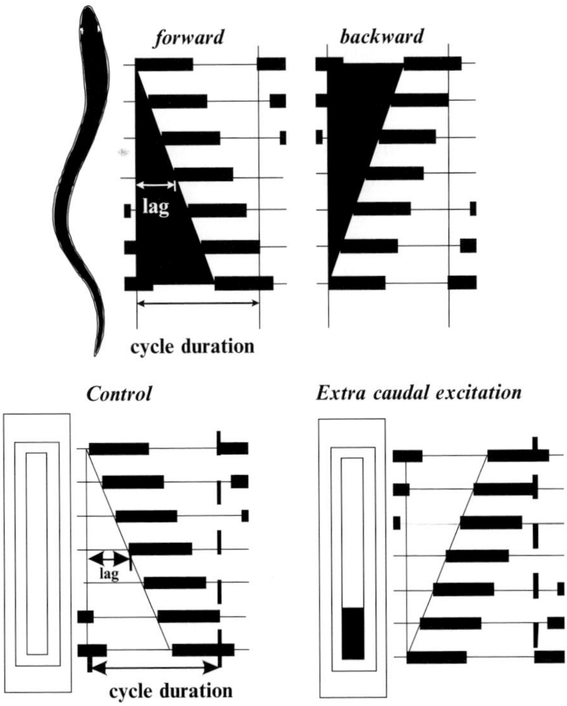

Fig. 5 Lamprey spinal cord intersegmental coordination. During swimming, a mechanical wave is transmitted along the body by sequential activation of the muscle segments. When the animal moves forward, there is a lag between consecutive segments in the spinal cord. This lag is always a certain proportion of the cycle duration (i.e., a constant phase lag). It can be reversed into a wave that is propagated from tail to head, as during backward swimming. In the isolated spinal cord preparation, a rostral to caudal lag is also seen. This pattern can be reversed if, for example, extra excitation is added to the caudal spinal cord. The caudal segments then receive a higher inherent frequency and can then entrain the more rostral segmental networks

of the wave also occurs during locomotor activity in the isolated spinal cord (Fig. 5) and is thus basically part of the standard spinal pattern generation.

If caught in a corner, the lamprey can also swim backward [22] by reversing the direction of the mechanical wave (Fig. 5). A reversed-phase coupling can be produced in the isolated spinal cord if the caudal part of the spinal cord has higher excitability than more rostral segments [23]. Essentially, the segments with the highest excitability will lead the activity and those with lower excitability will follow with a delay.

The EINs extend over an area of around 3 segments rostral to the cell body and 8 segments caudally [13]. This pattern applies to all EINs along the spinal cord, and they thus form a continuous network that is not separated by segmental boundaries. Such a hemicord network, only including the EIN connectivity, can generate not only bursting but also an intersegmental rostrocaudal lag. The rostrocaudal phase lag can be generated in a simulation with 100 segments and 3000 EINs. This number of neurons and connectivity approaches that which is found biologically. If the relative excitability along the cord is modified so that the rostral part has lower excitability then the caudal one, the phase lag is reversed, as found experimentally. The phase lag remains constant along the spinal cord with both a forward and a backward coordination. This finding thus shows that a segmental coupling of exclusively excitatory interneurons with appropriate cellular and intersegmental connectivity can produce an appropriate phase coupling for forward and backward swimming.

The hemicord coordination is of interest from the analytical point of view, but physiologically there is always a reciprocal activity at the left and the right side, resulting in the propulsive locomotor movements. Segmentally, the EINs also activate commissural inhibitory interneurons that ensure that the activity is alternating in each segment. The intersegmental coordination has also been modeled with a more theoretical approach [24].

Steering

What we have considered so far is the neural basis of symmetrical locomotor movements. To make the movements behaviorally meaningful, we need in addition to steer them. Steering movements to the left or right side are achieved through an asymmetrical activation of reticulospinal neurons on the left and right side, particularly involving the middle and posterior rhombencephalic reticular nuclei [5,25,26]. This activation results in longer and more intense bursts on the side toward which a turning response occurs (Fig. 6B). Reproducible steering responses to the left or right side can be elicited in a reduced brainstem–spinal cord in vitro preparation by trigeminal stimuli that, via lower brainstem interneurons, impinge on the reticulospinal neurons [27]. The basic brainstem circuitry for turning is thus comparatively simple (Fig. 6A). The reticulospinal neurons impinge on both the EINs and the commissural inhibitory interneurons and will thereby excite to provide longer bursts on the same side and concurrently also a longer inhibition of the contralateral side. These results have been analyzed both experimentally and through modeling [28].

The foregoing results illustrate the basic neural machinery for turning, but this does not unravel the mechanisms by which the animal itself will elicit goal-directed steering toward a particular object or prey. Tectum (superior colliculus) is an important structure in this context. It receives input from the eye that is organized in a retinotopic fashion. It provides thus a sensory map that is aligned to a motor map, which can elicit eye movements to a target and also orientation movements of the body and finally locomotion [29], seemingly toward an object that gave rise to activation of the retina (Fig. 6C). The tectal circuitry in interaction with the basal ganglia presumably is responsible for this action; however, this is outside the scope of this brief review.

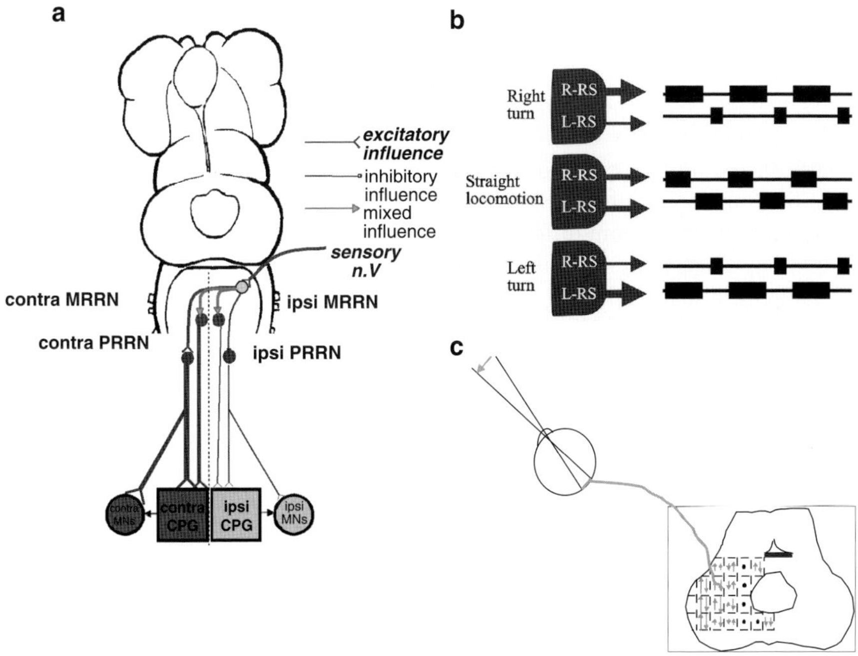

Fig. 6 Exploration of the mechanisms controlling steering in the lamprey. **a** During experimental conditions, steering signals can be produced by trigeminal sensory inputs (*sensory n. V*) that via interneurons activate preferentially contralateral medial and posterior rhombencephalic reticulospinal neurons (*MRRN* and *PRRN*) that project to the spinal cord [26]. *Black*, excitation; *gray*, inhibition. The descending signals affect both the CPG and the motoneurons (*MNs*) directly. **b** An asymmetrical activation of *RS* neurons results in a steering movement toward the more strongly activated side. Behaviorally, this will allow the animal to escape, or swim away from, the original stimulus (compare **a**). **c** In **a** and **b,** the turning machinery itself is illustrated: goal-directed steering toward a certain object requires sensory processing. The tectum receives input from the eye in a retinotopic fashion, thus forming a motor map, which can elicit both appropriate eye and orientation movements toward a given target [29], which will result in steering if the animal is swimming

Network Properties Can Be Radically Modified via G Protein-Coupled Receptors

We have discussed the network operation in terms of fast synaptic interaction on the millisecond level acting via ionotropic glutamate, glycine, and GABA receptors ($GABA_A$). At all levels of the central nervous system (CNS) there are also slower forms of synaptic transmission mediated by metabotropic receptors acting at second, minute, or hour time scales. These transmitter systems not only influence mood and attention but also fine tune a variety of sensory and motor functions. These systems include peptidergic, aminergic (5-HT, noradrenaline, dopamine), endocannabinoid, nitric oxide, metabotropic glutamate ($mGluR_{1-7}$), and GABA receptors ($GABA_B$). Each of these receptors may change the properties of a network by modifying the action of specific cellular and synaptic targets, for instance, by phosphorylating a specific ion channel subtype. Practically all networks analyzed, whether vertebrate or invertebrate, are subject to many forms of modulation that can fundamentally modify cellular and network properties. The functional significance of this type of modulation is presumably to be able to adapt and optimize the function of a network to varying behavioral demands. A wealth of information has emerged from different vertebrate and invertebrate model systems, but for reasons of space we illustrate here general principles using three examples from the lamprey system (Table 1).

Table 1 Metabotropic amino acid, aminergic, and peptidergic G protein-mediated modulation of ion channel, synaptic, cellular, and network activity in the lamprey spinal cord

Modulation: ion channels → network activity

	Presyn. gating	$Ca_v1.3$ (LVA)	$Ca_v2.1$-2.2 (HVANP/Q)	IP3	SK1-K_{Ca}	K_{Na}	K^+	NMDA	Network (frequency)
$GABA_B$	*Inh*	↘	↘		(↘)	*0*			↘
$mGluR_1$	*0*	*0*	*0*			*0*	↘	↗	↗
$mGluR_5$	*0*	*0*	*0*	*Ca^{2+} osc* ↗		*0*			↘
$mGluR_{II\text{-}III}$	*Inh*		*0*			*0*			↘
5-HT_{1A}	*Inh*	↘	↘		(↘)	*0*			↘
D2	*Inh*	↘			(↘)	*0*			↘
TK	*Facil.*				↘		↘	↗	↗
NPY	*Inh*								*0*
SOM		*0*	*0*		*0*		↗		↘
NT		*0*	*0*		*0*				↗

The table summarizes the results of a number of studies (see text). The effects of different transmitters and receptors on different targets are listed in the columns on the right. Gray backgrounds indicate receptors known to be activated endogenously during fictive locomotion. The presynaptic actions can be targeted to sensory afferents, excitatory or inhibitory interneurons and descending reticulospinal axons. Different transmitters have selective actions on different cellular targets (Inh indicates presynaptic inhibition and Facil facilitation). The locomotor network phasically modulates, in each cycle, the synaptic transmission from sensory afferents and interneurons. The modulation of HVA_{Ca}, LVA_{Ca}, K_{Ca} and K^+ and NMDA channels is indicated with a downward arrow for depression and an upward arrow for facilitation. Again, the effects may be specific to particular cell types. Finally, the effects on the network level have been studied on the background of locomotor activity (arrows relate to locomotion burst frequency), and in related modeling experiments. 5-HT, 5-hydroxytryptamine (serotonin) receptor; D_2, type 2 dopamine receptor; HVA, high voltage activated; mGluR, metabotropic glutamate receptor; NPY, neuropeptide Y; SOM, somatostatin; NT, neurotensin; TK, tachykinin

The lamprey CPG is embedded in a number of modulatory systems [3,30], some of which are activated as soon as the network starts to operate. Neurons of the spinal 5-HT system are located in the midline where they form a dense plexus, from which 5-HT is released in a paracrine fashion onto the dendrites of network interneurons [31]. 5-HT reduces the current carried through N-type Ca^{2+} channels [32], which in turn results in a reduction of the amplitude of the slow K_{ca}-dependent after hyperpolarization (sAHP) [33,34]. In addition, 5-HT has presynaptic inhibitory effects on excitatory synapses [32]. The 5-HT-mediated reduction of the K_{Ca} current results not only in a reduction of the sAHPs but also in a higher frequency of action potentials and a reduced spike frequency adaptation. The net effect at the network level is a reduced burst frequency, with longer and more intense bursts and a more regular burst pattern. The 5-HT system thus contributes to the operation of the CPG by modifying the properties of the network interneurons and the transmitter they release at their synapses. 5-HT appears to be involved in the control of most vertebrate locomotor networks. Indeed, in mammals as in the lamprey, the raphe-spinal and parapyramidal 5-HT neurons are turned on as locomotion begins [35,36].

Metabotropic glutamate receptors [37] are activated as soon as glutamatergic neurons become active during locomotion. One subtype, $mGluR_1$ [38], is activated during locomotor activity and acts to speed up locomotor activity through several distinct cellular mechanisms, mediated in part by phospholipase C–protein kinase C and IP_3 pathways. $MGluR_1$ activation enhances the NMDA receptor current, reduces a leak conductance, and triggers the release of endocannabinoids, which in turn reduces glycinergic synaptic current from contralateral neurons via a presynaptic action. These three cellular mechanisms are convergent and produce net excitatory and stabilizing effects on the network.

Tachykinins (including substance P) are released during activity in the locomotor network and exert an excitatory effect to induce a more regular burst pattern [39,40]. When the action of tachykinins is blocked by receptor-specific antagonists, burst frequency is reduced. Similarly to $mGluR_1$, the tachykinins act via a protein kinase C-mediated facilitation of NMDA currents. The tachykinins also modify the activity-dependent synaptic plasticity in crossed inhibitory synapses; although the first IPSP in a spike train may be unchanged, the subsequent IPSPs decline in amplitude much faster than under control conditions, thus enhancing burst frequency [41]. If tachykinins are applied in high concentrations, their action may last over 24 h.

Although fast synaptic actions determine motor pattern in a given instant, the response pattern of the neurons and the efficacy of their synapses are, to a large extent, determined by the actions of these modulatory systems, which help the network adapt to different behavioral challenges. In the lamprey (Table 1), the molecular targets of several modulators have been identified. Because intrinsic network function and behavior is also well understood, it is possible to estimate the effect of modulation of a target gene on the actual motor behavior and thus bridge the gap from gene to behavior [3].

Concluding Remarks

The aim of this brief review is to serve as an introduction to the lamprey model system in the study of the cellular and molecular bases of vertebrate motor behavior and in particular the intrinsic function of these neuronal networks.

Acknowledgments We hereby acknowledge the support of the European Commission (Neurobotics, SG, Spinal Cord Repair, Lampetra) and The Swedish Research Council (S.G., J.).

References

1. Grillner, S. Neuronal networks in motion from ion channels to behaviour. An R Acad Nac Med (Madr) 2006;123(2):297–8.
2. Grillner, S., et al. Neural bases of goal-directed locomotion in vertebrates: an overview. Brain Res Rev 2008;57(1):2–12.
3. Grillner, S. The motor infrastructure: from ion channels to neuronal networks. Nat Rev Neurosci 2003;4(7):573–86.
4. Grillner, S. Biological pattern generation: the cellular and computational logic of networks in motion. Neuron 2006;52(5): 751–66.
5. Wannier, T., et al. Differential effects of the reticulospinal system on locomotion in lamprey. J Neurophysiol 1998;80(1):103–12.
6. Deliagina, T.G., et al. Vestibular control of swimming in lamprey. III. Activity of vestibular afferents: convergence of vestibular inputs on reticulospinal neurons. Exp Brain Res 1992;90(3):499–507.
7. Cangiano, L. and S. Grillner. Mechanisms of rhythm generation in a spinal locomotor network deprived of crossed connections: the lamprey hemicord. J Neurosci 2005;25(4):923–35.
8. Cangiano, L. and S. Grillner. Fast and slow locomotor burst generation in the hemispinal cord of the lamprey. J Neurophysiol 2003;89(6):2931–42.
9. Parker, D. and S. Grillner. The activity-dependent plasticity of segmental and intersegmental synaptic connections in the lamprey spinal cord. Eur J Neurosci 2000;12(6):2135–46.
10. Buchanan, J.T. and S. Grillner. A new class of small inhibitory interneurones in the lamprey spinal cord. Brain Res 1988;438(1-2):404–7.
11. Buchanan, J.T., et al. Identification of excitatory interneurons contributing to generation of locomotion in lamprey: structure, pharmacology, and function. J Neurophysiol 1989;62(1):59–69.
12. Biro, Z., R.H. Hill, and S. Grillner. 5-HT modulation of identified segmental premotor interneurons in the lamprey spinal cord. J Neurophysiol 2006;96:931–5.
13. Dale, N. Excitatory synaptic drive for swimming mediated by amino acid receptors in the lamprey. J Neurosci 1986;6(9):2662–75.
14. Hill, R., et al. Apamin blocks the slow AHP in lamprey and delays termination of locomotor bursts. Neuroreport 1992;3(10):943–5.
15. Wallen, P., et al. Sodium-dependent potassium channels of a Slack-like subtype contribute to the slow afterhyperpolarization in lamprey spinal neurons. J Physiol 2007;585(Pt 1):75–90.
16. Buchanan, J.T. Identification of interneurons with contralateral, caudal axons in the lamprey spinal cord: synaptic interactions and morphology. J Neurophysiol 1982;47(5):961–75.
17. Ohta, Y. and S. Grillner. Monosynaptic excitatory amino acid transmission from the posterior rhombencephalic reticular nucleus to spinal neurons involved in the control of locomotion in lamprey. J Neurophysiol 1989;62(5):1079–89.
18. Hellgren, J., S. Grillner, and A. Lansner. Computer simulation of the segmental neural network generating locomotion in lamprey by using populations of network interneurons. Biol Cybern 1992;68(1):1–13.

19. Wallen, P. and T.L. Williams. Fictive locomotion in the lamprey spinal cord in vitro compared with swimming in the intact and spinal animal. J Physiol (Lond) 1984;347:225–39.
20. Williams, T., et al. Locomotion in lamprey and trout: the relative timing of activation and movement. J Exp Biol 1989;143:559–66.
21. Grillner, S. On the generation of locomotion in the spinal dogfish. Exp Brain Res 1974;20(5):459–70.
22. Islam, S.S., et al. Pattern of motor coordination underlying backward swimming in the lamprey. J Neurophysiol 2006;96(1):451–60.
23. Matsushima, T. and S. Grillner. Neural mechanisms of intersegmental coordination in lamprey: local excitability changes modify the phase coupling along the spinal cord. J Neurophysiol 1992;67(2):373–88.
24. Cohen, A.H., et al. Modelling of intersegmental coordination in the lamprey central pattern generator for locomotion. Trends Neurosci 1992;15(11): 434–8.
25. Deliagina, T.G., P.V. Zelenin, and G.N. Orlovsky. Encoding and decoding of reticulospinal commands. Brain Res Brain Res Rev 2002;40(1-3):166–77.
26. Deliagina, T.G., et al. Activity of reticulospinal neurons during locomotion in the freely behaving lamprey. J Neurophysiol 2000;83(2):853–63.
27. Fagerstedt, P., et al. Lateral turns in the lamprey. II. Activity of reticulospinal neurons during the generation of fictive turns. J Neurophysiol 2001;86(5):2257–65.
28. Kozlov, A.K., et al. Modeling postural control in the lamprey. Biol Cybern 2001;84(5):323–30.
29. Saitoh, K., A. Ménard, and S. Grillner. Coordination of eye and locomotor movement elicited from the lamprey tectum. In: SFN Annual Meeting 2004, San Diego.
30. El Manira, A. and P. Wallen. Mechanisms of modulation of a neural network. News Physiol Sci 2000;15:186–191.
31. Christenson, J.P., et al. 5-Hydroxytryptamine depresses reticulospinal excitatory postsynaptic potentials in motoneurones of the lamprey. In: Fuxe L.F.A.K. (ed) Volume Transmission in the Brain: Novel Mechanisms for Neural Transmission. Raven Press, New York: 1991:159–70.
32. Hill, R.H., et al. 5-HT inhibits N-type but not L-type $Ca(2^+)$ channels via 5-HT1A receptors in lamprey spinal neurons. Eur J Neurosci 2003;18(11):2919–24.
33. Wallen, P., et al. Effects of 5-hydroxytryptamine on the afterhyperpolarization, spike frequency regulation, and oscillatory membrane properties in lamprey spinal cord neurons. J Neurophysiol 1989;61(4):759–68.
34. Biro, Z., R.H. Hill, and S. Grillner. 5-HT modulation of identified segmental premotor interneurons in the lamprey spinal cord. J Neurophysiol 2006;96(2):931–5.
35. Jacobs, B.L. and C.A. Fornal. Serotonin and motor activity. Curr Opin Neurobiol 1997;7(6):820–5.
36. Liu, J. and L.M. Jordan. Stimulation of the parapyramidal region of the neonatal rat brain stem produces locomotor-like activity involving spinal 5-HT7 and 5-HT2A receptors. J Neurophysiol 2005;94(2):1392–404.
37. El Manira, A., et al. Metabotropic glutamate receptors provide intrinsic modulation of the lamprey locomotor network. Brain Res Brain Res Rev 2002;40(1-3):9–18.
38. Kettunen, P., et al. Neuromodulation via conditional release of endocannabinoids in the spinal locomotor network. Neuron 2005;45(1):95–104.
39. Parker, D., W. Zhang, and S. Grillner. Substance P modulates NMDA responses and causes long-term protein synthesis-dependent modulation of the lamprey locomotor network. J Neurosci 1998;18(12):4800–13.
40. Perez, C.T., R.H. Hill, and S. Grillner. Endogenous tachykinin release contributes to the locomotor activity in lamprey. J Neurophysiol 2007;97(5):3331–9.
41. Parker, D. and S. Grillner. Activity-dependent metaplasticity of inhibitory and excitatory synaptic transmission in the lamprey spinal cord locomotor network. J Neurosci 1999;19(5):1647–56.

Bacterial Complexity: More Is Different on All Levels

Eshel Ben-Jacob

Introduction

This chapter presents new clues, drawn from the morphogenesis of bacterial colonies, for the challenges of complexity posed in the integrative study of biological systems (systemic or systems biology). Some exciting observations of complex cooperative behavior of bacteria in colonies are presented, guided by the assumption that they might shed new light on the foundations and evolution of biocomplexity in general. The chapter is aimed at researchers of different disciplines—microbiology, biology, chemistry, physics, mathematics, and computer science—so, to make it comprehensible to all, I have avoided the use of specialized terminology and limited the presented details.

Bacteria are not the simple, solitary creatures of limited capabilities they were long believed to be. The impression that bacteria act as unsophisticated, solitary creatures stems from years of laboratory experiments in which they were grown under artificial conditions. However, under the harsh conditions in the wild, or in the laboratory under growth conditions that mimic natural environments, these versatile organisms work as a team and employ chemical communication to form highly complex colonies of 10^9–10^{12} bacteria each [1–11]. Such colonies behave much like a multicellular organism with cell differentiation, distribution of tasks, and, in some cases, even modules that act like reproductive organs.

One aspect of these behaviors has to do with collective engineering of complex spatial organization of the colony, with the bacteria forming different patterns to better cope with the environment [8–11]. To form such multicellular superorganisms, new functional features appear at every level from the internal cellular gel and genome states to the growth of the colony as a whole, thus facilitating a high level of functional complexity. Bacteria cannot store genetically all the relevant information required for creating colonial patterns. In a new scenario that I propose they need not do so, because the respective units (the individual bacteria) assume newly co-generated

E. Ben-Jacob
School of Physics and Astronomy, Raymond and Beverly Sackler Faculty of Exact Sciences, Tel-Aviv University, 69978 Tel-Aviv, Israel

S. Nakanishi et al. (eds.), *Systems Biology*,
DOI:10.1007/978-4-431-87704-2_3, © Springer 2009

traits and faculties that are not explicitly stored in the genetic information of the individuals [8–13]. The required contextual information is cooperatively generated by using internally stored information, as well as new information extracted from the environment. Thus, these bacteria only require genetically stored information on how to produce perceptive faculties and how these capabilities, along with the guidelines for using them, may be employed to generate new knowledge as required [8–13].

Engineered Complexity of the *Paenibacillus dendritiformis* Bacteria

To study the mechanisms and principles by which the bacteria regulate (engineer) the complexity of colonial patterns, we developed a new strain of lubricating bacteria, *Paenibacillus dendritiformis* [8]. The patterns shown in Fig. 1 are generated during colony development of these bacteria in response to growth on nutrient-poor and hard surfaces. To cope with this situation, the bacteria collectively produce a lubricating layer of fluid that allows them to swim on the hard surface. The engineering capabilities of the bacteria are reflected by their ability to perform collective sensing of the level of nutrients and the hardness of the surface. Consequently, they regulate the genome level to adjust the secretion of lubrication and its viscosity to generate colonies whose structure matches the growth conditions.

Chemotactic Signaling

To achieve even greater engineering efficiency, bacteria employ the mechanisms of chemotactic signaling. Chemotaxis is cell movement in response to gradients in the concentration of a chemical agent [8–11]. The movement can be biased either toward higher concentrations (attractive) or away from high concentrations (repulsive). Bacteria are too short to detect chemical gradients, yet swimming bacteria found a smart solution to detect gradients and bias their movement accordingly. For attractive (repulsive) bias, they detect the concentration as they swim, and if the concentration increases (decreases) they delay their tumbling. We emphasize that this process involves sensing (of the chemical level), memory (of the measured concentration), and information processing—evaluation of the change in concentration between measurements.

The most familiar case of chemotaxis is attraction to an external chemical such as a nutrient. There is evidence that such chemotaxis occurs in these colonies and is responsible for an increased expansion rate and colony "bushiness" at intermediate values of the nutrient concentration (Fig. 1e). At relatively higher nutrient levels, additional bacterial density variations are visible within the branches (Fig. 1d). Model simulations can be used to test the idea that the patterns result from bacterial

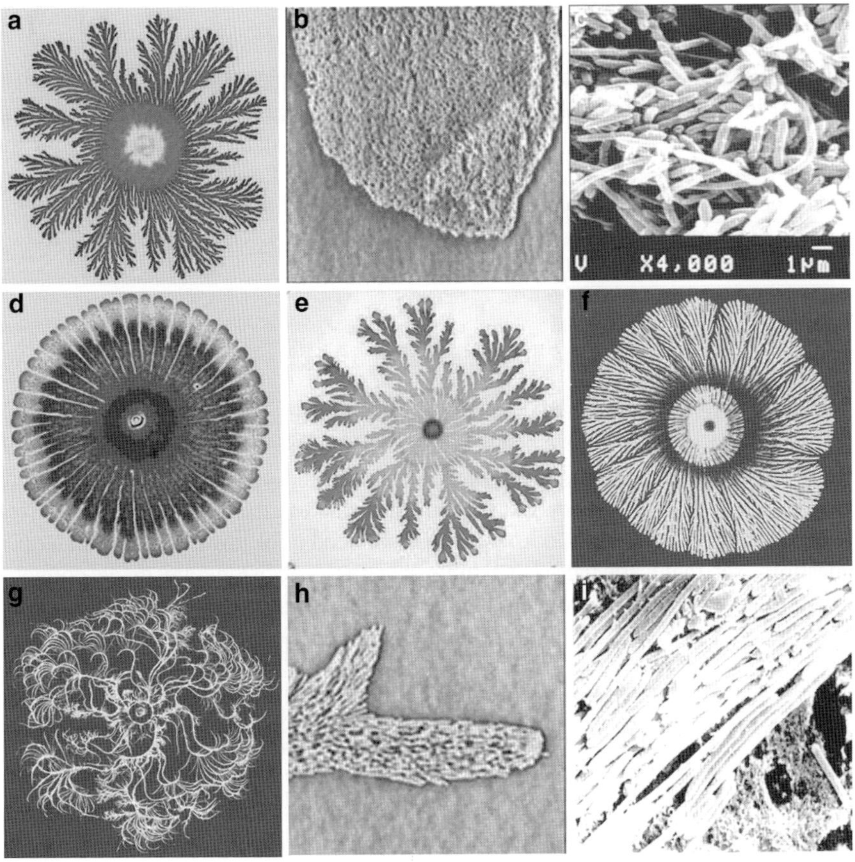

Fig. 1 Colonial morphogenesis of the *Paenibacillus dendritiformis* bacteria when grown on hard and food-depleted substrate. The colony starts from a droplet (5 μl) inoculation at the center of the Petri dish (*darker dot*). After inoculation, the bacteria go through an "embryonic" colonial stage for several hours (presumably of assessment of the environmental conditions), and only then does the colony start to expand outward on the surface [8–11]. **a** It takes the colony about 2 days to reach the observed size of about 5 cm in diameter shown here. **b** Snapshot from a video clip taken through an optical microscope with 500× magnification. The video clip observations reveal bacteria swimming: segments of straight motion for about 1–3 s at a speed of about 1 μm/s, interrupted by bacterial tumbling terminating in a random new direction; that is why the bacterial movement is usually modeled as random walk. **c** A scanning electron micrograph. Note the variability in the individual bacteria. It is now understood that phenotypic diversity is not arbitrary but collectively regulated to afford the colony elevated group flexibility [48]. **d** Typical pattern developed at higher food levels when attractive chemotactic signaling is activated (as is concluded by comparison with model simulations). **e** Typical pattern when food chemotaxis dominates the growth at intermediate levels of food. **f** Colonial pattern at a low level of food when repulsive chemotactic signaling is assumed to be intensified. The colors are added to indicate the bacterial density. As explained in the text, the *P. dendritiformis* bacteria can perform morphotype transition from branching to a chiral pattern exhibiting the formation of curly branches (**g**). **h** Snapshot from a video clip taken through an optical microscope at 500×. The video clip observations reveal that upon elongation the cells alter their collective movement from the typical run-and-tumble of the swimming cells (**b**) to a coordinated forward-backward movement, which yields, as shown by model simulations, an organized twist of the branches with a specified handedness. To bring this about, the bacteria suppress cell division and become elongated, as observed via scanning electron microscope (**i**). Transmission electron microscope observations reveal that the cells contain several chromosomes

utilizing attractive chemotactic signaling. Very different patterns form at low nutrient levels (Fig. 1f). To explain the mechanism, we recall that part of the branch-making dynamics relies on the cells going into a nonmotile state further back from the colony front, where the nutrient level is extremely low. Based on model simulation, it was proposed that cells emit a repellent chemical as they are entering this state, leading to the observed colony pattern in this case.

Open Question

So, the picture that emerges is that the basic branching pattern is sculpted by the combined action of a variety of chemotactic strategies. As these different influences sort themselves out, changing conditions and changing bacterial strains always lead to new structures. Exactly how information from the outside is utilized to help decide which, if any, of these processes need to be turned on is still an open question. Yet there are increasing hints that these decisions are made cooperatively, much like what is known to be true regarding the collective decision making of bacteria to sporulate or the decision to share genetic information.

Reversible and Inheritable Identity Switching

A remarkable trait of the *P. dendritiformis* bacteria is their identity-switching ability; cooperatively, they can make drastic alterations of their internal genomic state, effectively transforming themselves into practically different cells [5,8]. Under conditions somewhat more favorable to motion, such as growth on a softer substrate, the bacteria engineer radically different classes of chiral colony patterns in which the branches are thinner and curl in the same direction (Fig. 1g). Accompanying the colonial structure is a designed genome change: the bacteria are now programmed to become manyfold longer and have multiple chromosomes. The morphotype transitions are both inheritable: the identity is maintained during Luria-Bertani (LB) growth and even throughout sporulation/germination. Yet the transitions are reversible; for example, the reverse transitions from chiral to ordinary branching occur on harder substrates (when higher bacteria densities are required to produce sufficient amounts of lubrication). Optical microscope observations during colony development reveal the following: upon elongation, the cells alter their collective motion from the typical run-and-tumble to a coordinated forward-backward motion with limited tumbling [*].

---

[*] Bacterial images and video clips are available from PhysicaPlus, the online magazine of the Israel Physical Society http://physicaplus.org.il and at Ben Jacob's home page http://star.tau.ac.il/~eshel/.

Complexity of the *Paenibacillus vortex* Bacteria

Some bacterial strains organize their colonies by generating modules, each consisting of many bacteria, which are used as building blocks for the colony as a whole. This behavior is observed, for example, in the lubricating bacteria *Paenibacillus vortex* shown in Fig. 2 [5,8–11]. The modules of this lubricating strain are groups of bacteria that move around a common center. Model simulation suggests that the vortices are generated by the action of attractive chemotaxis and possibly physical links between bacteria [14]. It is also suggested that the vortices are "pushed out" in response to a repulsive chemotactic agent secreted by bacteria at the center of the colony. The bacteria in each vortex have high length variability, with some of the cells being tens of micrometers long with several chromosomes. Genome sequencing results indicate that this strain (and the *P. dendritiformis*) has genes for both flagella and pili motility [15].

The Complexity Challenge of Modular Organization

Maintaining the integrity of the individual vortices while they serve as higher-order building blocks of the colony requires advanced communication: each cell has to follow far more complex dynamics, being part of both a specific vortex and the whole colony, so that it can adjust its activities accordingly. A greater challenge is

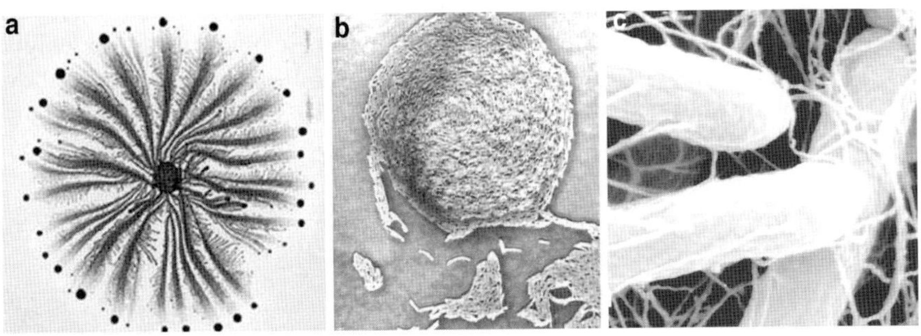

Fig. 2 Modular colonial organization of the *Paenibacillus vortex* bacteria. In (**a**) we see the colony organization. The diameter of the colony shown is about 5 cm. Each *dot* is a module or cluster of bacteria that move; hence, the term vortex. **b** Each vortex (500× picture taken from a video clip) is composed of many cells that swarm collectively around their common center at about 10 µm/s. The vortices vary in size from tens to millions of bacteria, according to their location in the colony. The cells in the vortex replicate, and the vortex expands in size and moves outward as a unit, leaving behind a trail of motile but usually nonreplicating cells, the vortex branch. The dynamics of the vortices is quite complicated and includes attraction, repulsion, merging, and splitting of vortices. Yet, from this complex, seemingly chaotic dynamics, a colony with complex but nonarbitrary organization emerges [*]. **c** Electron microscope view of the individual bacteria showing flagella and pili (courtesy of I. Colin [14])

posed by the formation of new vortices that emerge in the trail behind a vortex. Following initiation signals, the nonmotile cells in the trail start to secrete lubricating fluid and begin to move quite rapidly as a turbulent "biofluid," until an eddy forms and turns into a new vortex as the cells in the eddy generate strong coupling. The entire process appears to be under advanced communication-based cooperative control that goes hand in hand with genome-wide changes in the individual cells. For the integrity of the vortices to be maintained (as the cells replicate), the changes at the gene network level have to be inheritable. The results support the idea of "more is different on all levels" and that it afforded by epigenetic memory.

Cell Differentiation

The idea of epigenetic memory that goes together with cell differentiation is more directly supported by the observations shown in Fig. 3 even after several generations of growth, colonies started from inocula taken from the center of a colony significantly differ in pattern from those started from inocula taken from the leading vortices.

Complexity Transfer

We developed a simple experimental procedure to further test the idea of cell differentiation and task distribution. The procedure involves extension of the idea of replica plating. A large-diameter (several centimeters) cylinder made of Plexiglas covered with a thin sterile paper is used as a stamp. It is (gently) placed above a mother colony and then on another plate with fresh substrate. Figure 4 shows how the pattern of the mother colony is thus copied. In principle, the cells that are

Fig. 3 Precursors of cell differentiations. **a** The mother colony from which cells were taken (by picking). The growth patterns of colonies taken from the center in **a**, the *darker area* from which the mother colony started, are shown in **b**. **c** The growth patterns of colonies started from cells taken from large vortices at the front of the mother colony

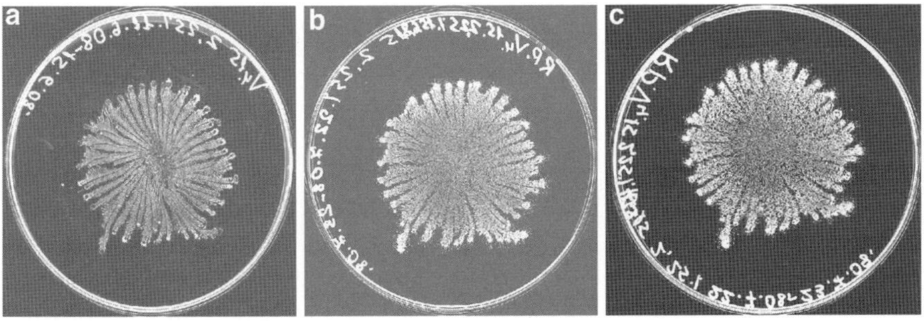

Fig. 4 Replicating colony pattern. **a** The mother colony. Negative colors are used; *white* indicates high bacteria density. **b** A colony grown from replica plating of the mother colony. **c** A colony grown from a successive stamping

transferred to the fresh dish could grow and expand in all directions. Instead, they follow quite accurately the complex pattern of the mother colony, as each cell keeps its functional role in the mother colony.

Functional Complexity and Learning from Experience

In natural habitats, bacteria are regularly exposed to nonlethal (subinhibitory) levels of antibiotic material. Previously [8], I proposed the idea of "functional complexity": the complexity of the colony organization has an important functional role as it affords the bacteria with higher adaptability to cope with environmental stresses. For example, upon encountering antibiotic stress, the bacteria re-shape the colony pattern [8,9,16], as is shown in Fig. 4. Recently it has been suggested that bacteria also possess (epigenetic) memory of the past [9,13], which enables them to keep track of how they handled previous encounters with antibiotics (Fig. 1c). Bacteria learn from past experience and can cope better upon a second encounter with the same antibiotic, as is reflected by the fact that the colony expands faster and has a more complex pattern (Fig. 5). This effect can be erased by growth in neutral conditions. One possibility is that this effect is connected with a genetic shift in the population. Another possibility involves heritable (epigenetic) effects at the genome level.

Information Processing on All Levels

As was pointed out [8–13], bacteria sense the environment and perform internal information processing (according to the internally stored information) to extract latent information embedded in the complexity of their environment. The latent

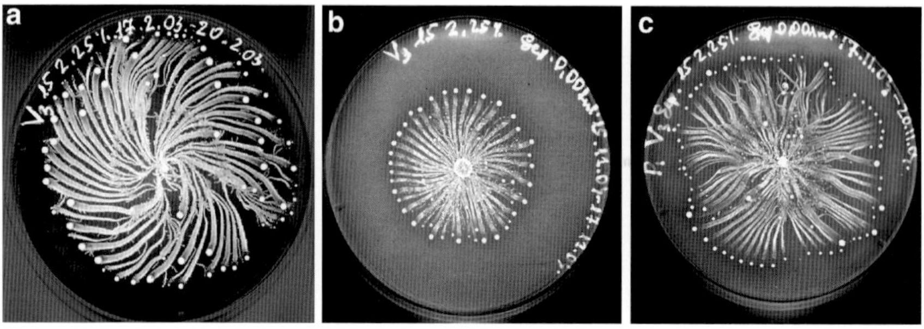

Fig. 5 Learning from experience. The observations show the response of the *Paenibacillus vortex* bacteria to nonlethal levels of Septrin. **a** Normal growth pattern in the absence of antibiotic. The effect of first exposure of the bacteria to the antibiotic is shown in **b**; the response in a second encounter is shown in **c**. Learning from experience (**c**) is manifested by the fact that upon second encounter with the antibiotic the colony expands faster and the pattern has higher complexity [9]

information is then converted into usable or "active" information that affects the bacterium activity as well as intracellular changes [12,13,17].

Although an individual bacterium can sense only a limited area between replications, a colony that is composed of billions of bacteria can sense a large area and over long time periods. Collectively, bacteria can glean relevant latent information embedded in the complex environment over larger distances and longer periods of time. To coordinate such cooperative ventures, bacteria have developed various methods of biochemical communication such as quorum sensing by using a variety of mediators, which range from simple molecules to polymers, peptides, complex proteins, genetic material, and even "cassettes of genetic information" such as plasmids and viruses [8,9,12,13,18]. The intercellular communication is afforded by the evolution of highly complex and intricate intracellular signaling mechanisms involving signal transduction networks [19] and genetic language to turn genes off and on [20]. These mechanisms exhibit functional complexity, which generates intrinsic meaning for contextual interpretations of chemical messages and for formulating appropriate complex responses.

The observations shown in Fig. 6 illustrate that bacteria exchange information extracted from the environment, process the information, and act accordingly. The *P. vortex* bacteria, while moving on M-H agar, form snakelike swarms hundreds of bacteria wide [14]. The swarms can expand very efficiently toward several nutrient sources placed on the surface. To coordinate such search strategy, the individual bacteria must respond not only to local gradients in food concentration but also to differences in concentration between remote locations. Hence, they must exchange the information they detect and compare local conditions to those detected at different locations.

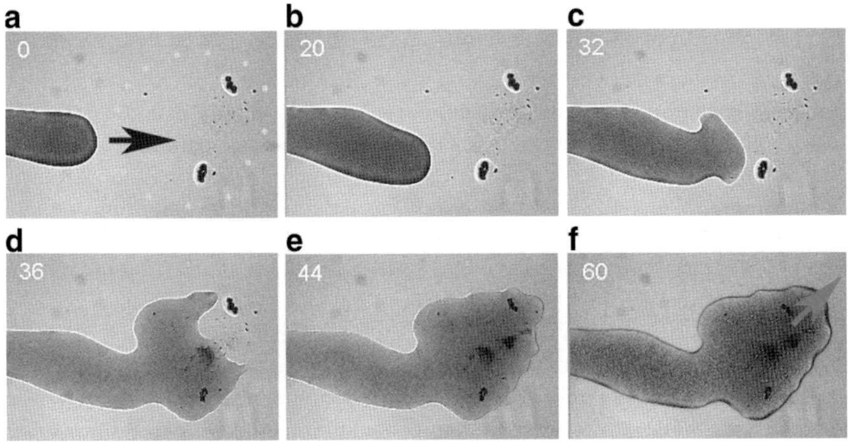

Fig. 6 Swarming intelligence of the *Paenibacillus vortex* bacteria. The pictures (×50) show a snapshot from a video clip of a branch of a colony of *P. vortex* bacteria moving on 0.3% w/v Mueller–Hinton agar (M-H agar). The branch is extending into an area where extracellular material derived from washes of swarming cells was delivered by toothpick and allowed to soak into the agar. Time of frame capture is noted in seconds. **a** Starting ($t = 0$): Area of extract outlined in *blue dots* with direction of cell mass elongation shown by the *black arrow*. **b** *Dark marks* inside the area of the extract are disturbances caused by the toothpick contacting the agar ($t == 20$). **c** Stage $t = 32$: Cell mass starts to disperse as it contacts the area of the extract. **d** Stage $t = 36$. **e** Stage $t = 44$: Cell mass has dispersed into area of extract. **f** Stage $t = 60$: Additional cells are moving into this area from further back in the colony; the cell mass is growing in volume at the tip, and extends in the direction of the *red arrow*. (Video courtesy of I. Colin [14])

Concluding Remarks and Looking Ahead

I have shown how complex colonial forms (patterns) emerge through the communication-based singular interplay between individual bacteria (the micro-level) and the colony (the macro-level). Each bacterium is, by itself, a biotic autonomous system with its own internal cellular gel that possesses informatics capabilities (storage, processing, and interpretation of information). These capabilities afford the cell a certain freedom to select its response to biochemical messages it receives, including self-alteration and broadcasting messages to initiate alterations in other bacteria. Such self-plasticity and decision-making capabilities elevate the level of bacterial cooperation during colonial self-organization.

As the individuals in a growing colony begin to respond to the colony itself (i.e., information flow from the colony to the individual), these individuals respond by regulating their movements, growth rates, the various tasks they perform, the chemical signals they send to other bacteria, and even their gene network state (phenotypic state) according to the received signals.

The essential new lesson learned from bacteria is that colonial high complexity provides the degree of plasticity and flexibility required for better durability and adaptability of the whole colony to a dynamic environment. According to this picture, new features collectively emerge during biotic self-organization on every level, from the membranes and cytoplasm to the whole colony. The cells thus assume newly co-generated traits and abilities that are not explicitly stored in the genetic information of the individuals. For example, bacteria cannot genetically store all the information required for creating the colonial patterns. In the new picture, they do not need to do so because the required information is cooperatively generated as self-organization proceeds by bacterial communication, informatics, and self-plasticity capabilities. Thus, the bacteria need only have genetically stored the guidelines for producing these capabilities and use them to generate new information as required.

It seems that bacteria have some sort of collective memory by which they keep track of how they handled their previous encounters with antibiotics. They know how to collectively glean information from the environment, "talk" with each other, distribute tasks, generate collective memory, and turn their colony into a *cybernetic system*—a massive "brain" that can perform distributed information processing, learn from past experience, and possibly alter the genome organization or even create new genes to better cope with novel challenges [12,13,17].

Acknowledgments This chapter is a concise summary of a lecture presented in the Uehara meeting on "Systems Biology: The Challenge of Complexity" in Tokyo, 2008. I am most thankful to the Uehara Foundation and the organizers of the meeting for hospitality and support of my stay in Tokyo. I am thankful to Inna Brainis for technical assistance and to Ingham Colin for his collaboration and for providing the pictures shown in Fig. 3 and the clips shown in Fig. 6. The research presented here has been supported in part by the Maguy-Glass Chair in Physics of Complex Systems and the Tauber Fund.

References

1. Matsushita, M. and Fujikawa, H. (1990) Diffusion-limited growth in bacterial colony formation. Physica A 168:498–506
2. Ben-Jacob, E., et al. (1994) Generic modeling of cooperative growth patterns in bacterial colonies. Nature (Lond) 368:46–49
3. Cohen, I., Czirok, A. and Ben-Jacob, E. (1996) Chemotactic based adaptive self-organization during colonial development. Physica A 233:678–698.
4. Shapiro, J.A. and Dworkin, M. (1997) Bacteria as Multicellular Organisms. Oxford University Press, Oxford
5. Ben-Jacob, E. et al. (1998) Cooperative organization of bacterial colonies: from genotype to morphotype. Annu Rev Microbiol 52:779–806
6. Shapiro, J.A. (1998) Thinking about bacterial populations as multicellular organisms. Annu Rev Microbiol 52:81–104
7. Ben Jacob, E., Cohen, I. and Levine, H. (2000) Cooperative self-organization of microorganism. Adv Phys 49:395–554
8. Ben-Jacob, E. (2003) Bacterial self-organization: co-enhancement of complexification and adaptability in a dynamic environment. Philos Trans R Soc Lond A361:1283–1312
9. Ben Jacob, E. et al. (2004) Bacterial linguistic communication and social intelligence. Trends Microbiol 12(8):366–372
10. Ben Jacob, E., Aharonov, Y. and Shapira, Y. (2005) Bacteria harnessing complexity. Biofilms 1:239–263

11. Ben Jacob, E. and Levine, H. (2006) Self engineering capabilities of bacteria. J R Soc Interface 3:197–214
12. Ben-Jacob, E. and Shapira, Y. (2005) Meaning-based natural intelligence vs. information-based artificial intelligence. In: Ben-Noon C. (ed) Cradles of Creativity. Sharey Mishpat, Jerusalem
13. Ben-Jacob, E., Shapira, Y. and Tauber, A.I. (2006) Seeking the foundations of cognition in bacteria: from Schrödinger's negative entropy to latent information. Physica A 359:495–524
14. Ingham, J.C. and Ben-Jacob, E. (2008) Swarming and complex pattern formation in *Paenibacillus* vortex studied by imagine and tracking cells. BMC Microbiol 8:1–36
15. We have completed sequencing the genome of the two strains *P. dendritiformis* and *P. vortex*. The data will soon be published and posted: see http://star.tau.ac.il/~eshel/
16. Ben Jacob, E. et al. (2000) Bacterial cooperative organization under antibiotic stress, Physica A 282:247–282
17. Ben-Jacob, E. (1998) Bacterial wisdom, Gödel's theorem and creative genomic webs. Physica A 248:57–76
18. Bassler, L.B. and Losicck, R. (2006) Bacterially speaking. Cell 125:237–246
19. Ptashne, M. and Gann, A. (2002) Genes and Signals. Cold Spring Harbor Press, Cold Spring Harbor, NY
20. Searls, D.B. (2002) The language of genes. Nature (Lond) 420:211–217

How Animals Get Their Skin Patterns: Fish Pigment Pattern as a Live Turing Wave

Shigeru Kondo

Reaction-Diffusion (RD) Model

There are several theoretical mechanisms that are able to generate spatial patterns autonomously without any pre-pattern.[1,2] Among them, the most plausible in the biological system is the reaction-diffusion (RD) mechanism, which was first presented by A. Turing in 1952[3,4] and mathematically refined by mathematical biologists.[5,6] In the model, the spatial pattern is made as stationary waves generated by the interactive RD of putative chemical substances. According to the mathematical study using computer simulation, an RD system is able to generate stable and evenly spaced patterns when the whole network satisfies a condition: "local activation and long-range inhibition."[6] The spatial patterns made by the system (e.g., "RD pattern" or "Turing pattern") do not need any pre-pattern and autonomously regenerates when artificially disturbed.

However, in spite of its theoretical importance, until very recently Turing's theory was not widely accepted by experimental biologists, for two major reasons. The first is that the main concept of the theory, "the pattern is made by wave," is quite unfamiliar to many experimental biologists. The second is the difficulty of showing the existence of "wave." To prove that the RD mechanism functions in a morphogenetic event, we need to show that the pattern possesses the dynamic nature of the RD wave.

Biological Phenomena in Which the Involvement of the RD Mechanism Is Suspected

The first clear experimental support for this mechanism came in chemical experiments in the early 1990s. Castets et al. and Ouyang and Swinney succeeded in making the Turing pattern using complex chemical reactions.[7,8] Their work encouraged biologists

S. Kondo
Division of Biological Science, Graduate School of Science, Nagoya University, Furo-cho, Chikusa-ku, Nagoya City, Aichi Prefecture 464-8602, Japan

S. Nakanishi et al. (eds.), *Systems Biology*,
DOI:10.1007/978-4-431-87704-2_4, © Springer 2009

to study the theoretical mechanism, and Kondo and Asai presented the first reliable data of the existence of the Turing pattern in biological systems in 1995.[9] They recorded the pattern change of the pigment pattern of a marine angelfish, *Pomacanthus imperator*, and suggested that the time course of the pattern change is identical to that predicted by Turing's theory. The involvement of the Turing mechanism is now seriously investigated experimentally in many morphological events: hair pattern of mammals,[10–12] feather pattern of birds,[13–15] regeneration of *Hydra*,[16–18] and right–left determination in the vertebrates.[19–21] In these cases, a candidate of the core functional molecules of pattern formation was proposed. For example, in the case of mouse hair pattern, Sick et al. suggested that the WNT and DKK proteins play the roles of the putative activator and inhibitor in the reaction-diffusion system.[12] Moreover, by artificially changing the parameters of the interactions, it is possible to induce the pattern change that is predicted from the simulation, suggesting that the RD mechanism underlies the determination of the hair distribution. However, most of these morphogenetic events are irreversible, and the patterns that we can observe are completed and fixed ones. Therefore, it is usually impossible to directly detect the existence of the "wave" in the course of the pattern-forming event.

Observation of the Characteristic Movement of the RD Wave in Animal Skin

To date, pigmentation patterns in animal skins,[1,22] feathers of birds,[13,14] and shells of snails[23] are the only examples in which we can detect the dynamic nature of Turing waves as a time course of the pattern change. Especially, the two-dimensional (2-D) skin pattern of fish is quite convenient to study because their waves are sometimes alive even when the fish has grown up into an adult.

For example, when a striped angelfish (*Pomacanthus imperator*) grows, the branching points of the stripes slide horizontally as the zip opens and add a number of stripes; eventually the spacing between the stripes remains stable (Fig. 1a).[9] In the case of the spotted catfish (*Plecostoms*), both division of the spots and insertion of the new spots occur to retain the density and size of the spots (Fig. 1b).[24] Both stripes and the spots are the most typical 2-D patterns generated by the RD mechanism, and the time course of the pattern change possesses the characteristics of the dynamics of RD waves, strongly suggesting that the RD mechanism underlies the process of pigment pattern formation of fish.

Pigment Pattern of Zebrafish Retains the Dynamic Nature of the RD Wave During the Postembryonic Stage

About two decades ago, the zebrafish, a small fish species with very clear stripes in the trunk and fins, was selected as the model animal for biological study and for the use of the genomic information and molecular genetic technologies that are available to

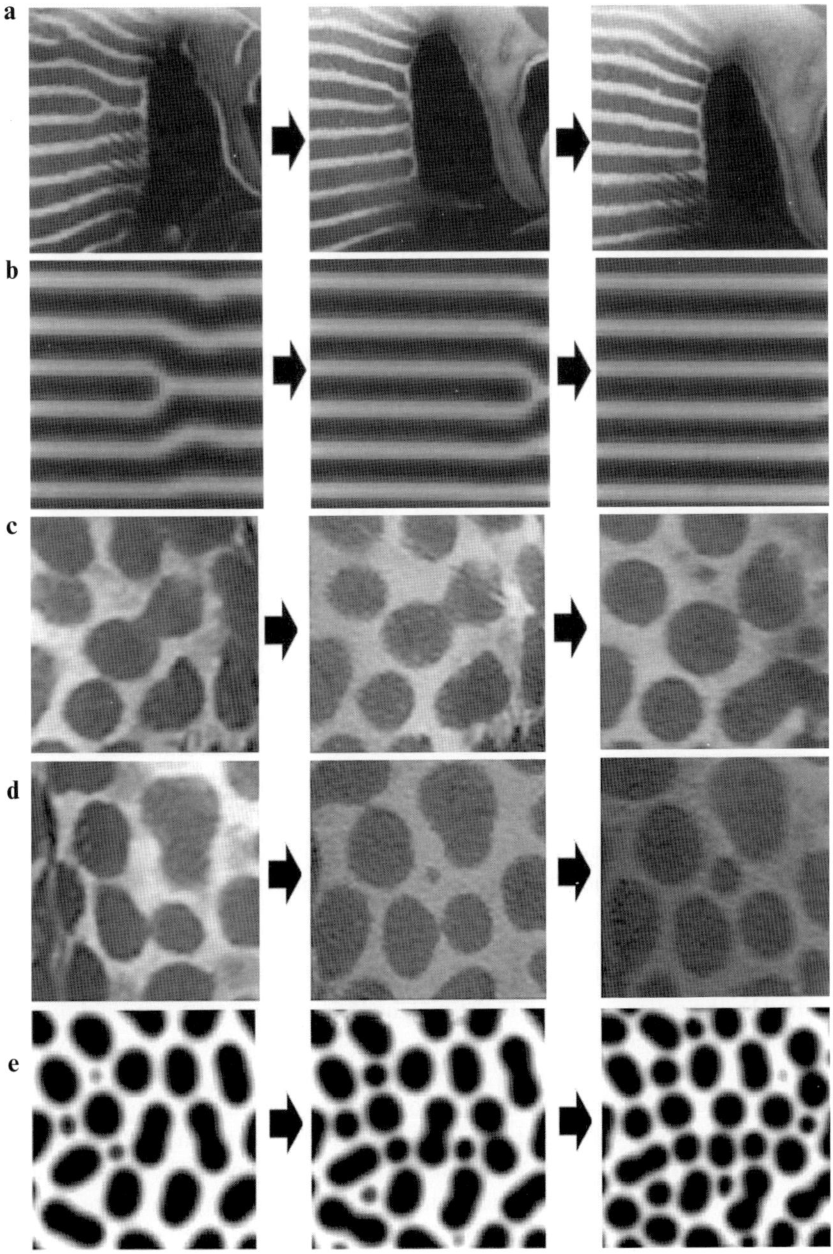

Fig. 1 Dynamic pattern change that occurred in the fish and simulation. **a** Change of stripe pattern occurred in the skin of *Pomacanthus imperator* during 90 days; the prediction made by simulation of the reaction diffusion mechanism is shown in **b**. Sliding of the branch point was observed in both systems. **c, d** Change of spot pattern as occurred in the skin of a catfish, *Plecostoms*, during 14 days; the simulation is shown in **e**. Division (**c**) and insertion (**d**) of the spot are observed in both real fish and the simulation based on the reaction-diffusion (RD) model. (From [9] and [24])

identify the molecular mechanism of Turing formation. The pattern of skin pigmentation in zebrafish is composed of three types of pigment cells distributed in the hypodermis: melanophores, the main component of dark stripes; xanthophores, the main component of light stripes; and iridophores.[25,26] Although differing from the stripes of *Pomacanthus imperator* in that the stripes of zebrafish do not become rearranged during normal growth, artificial disturbance of the pattern can induce the characteristic pattern change that is specific to the RD mechanism. By using a laser to kill the pigment cells in specific regions, it is possible to induce a dynamic movement of the stripes that is characteristic of the stationary wave of the RD mechanism. Yamaguchi et al.[27] continuously killed the melanophores in the dorsal two black stripes and observed that the ventral black stripe bent and moved dorsally to fill the vacant space (Fig. 2). This movement of the stripes clearly shows that the pigment pattern of the zebrafish is made and maintained dynamically by the RD, or a very similar, mechanism.

The fact that killing of the pigment cells induced the pattern change in the experiments suggests that the pigment cells do not simply make the hidden pattern visible but are the major players in the pattern formation.[27] Therefore, to uncover the mechanism, the most critical aspect should be the identification of the interactions between the two types of pigment cells (melanophores and xanthophores).

Mutants That Affect Both Development of Pigment Cells and the Resulting Pattern

The zebrafish mutants those have abnormal skin patterns are classified into two categories: those with a defect in the development of pigment cells; and those with normal development of pigment cells at the embryonic stage, but which have a disrupted pattern in the adult fish. Molecular genetic studies of the mutants in the first category have shown that *nacre*,[28] *sparse*,[29] and *rose*[30] are required for the development of melanophores, and that *panther*[31] is required for the development of xanthophores. The *nacre* gene codes a transcription factor, Mitf1, that is required by the cells to develop to the melanophore precursor; *sparse* and *panther* code the class II receptor tyrosine kinases, c-*kit* and c-*fms*, respectively; *rose* codes the endothelin receptor b1; and *sparse* fish lack the embryonic melanophores, but the adult melanophores are normally integrated in the stripes (Fig. 3). The number of embryonic melanophores in *rose* fish is normal, but a reduced number of adult melanophores compose the disturbed stripe.[32] The double mutant of these genes loses almost all the melanophores, suggesting that there are two different populations in the melanophores, and only the adult type is required for the stripe pattern formation.

Interestingly, when one of the pigment cell types fails to develop, other types of pigment cells fail to localize to their normal positions.[28–31,33,34] *nacre* fish lack all melanophores and *panther* fish lack xanthophores. In both cases, the remaining pigment cells cannot form clear stripes but disperse randomly or form aggregates of uncertain shape. However, when a pigment cell type that is lost in the mutant is introduced, clear stripes

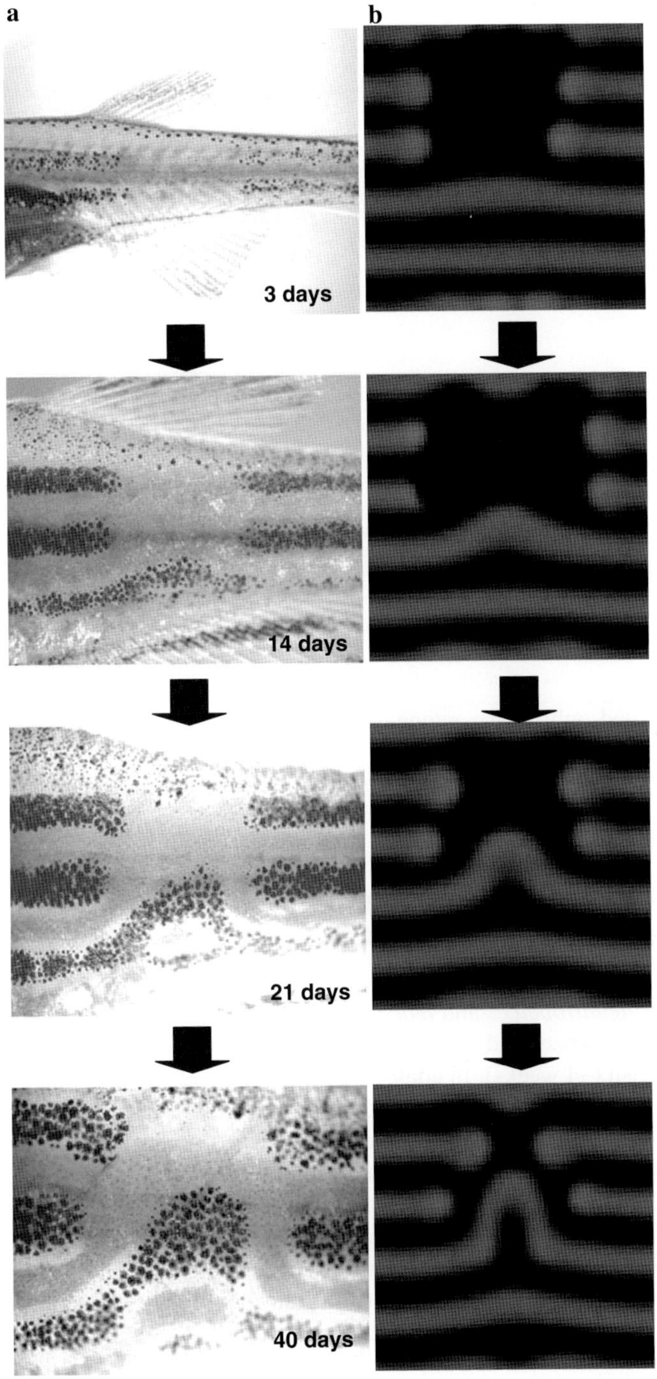

Fig. 2 Dynamic change of the zebrafish pigment pattern induced by laser experiment. **a** Time course of the pattern change. Melanophores in the upper two black stripes are killed by laser. **b** Time-lapse captured images of the regeneration process produced by simulation of the reaction-diffusion mechanism. (From [27])

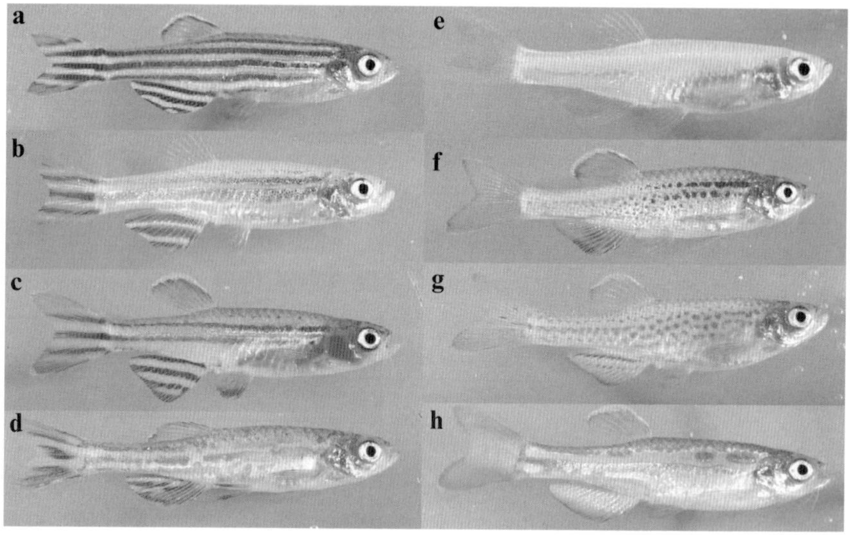

Fig. 3 Pigment patterns of zebrafish mutants: wild type (**a**), *sparse* (**b**), (c)*nacre* (**c**), *puma* (**d**), *panther* (**e**), *jaguar* (*obelix*) (**f**), and *leopard* (**g**)

are regenerated only in the region where both types of pigment cells exist. This experimental result shows that mutual interaction between melanophores and xanthophores plays a critical role in the generation of skin pigmentation pattern.[25,35–37]

Mutants That Affect the Pattern But Do Not Affect the Development of Pigment Cells

Mutants of *jaguar/obelix*[37] and *leopard*[37,38] belong to this second category. In these mutants, development and distribution of the pigment cells is normal in embryos and young fish (~4 weeks). In adult fish, although development of pigment cells is normal, their spatial arrangement (pattern) is changed, suggesting that these genes are specifically required for pattern formation. The jaguar mutant fish have wider stripes and the leopard fish have spotted patterns. The chimera experiment done by Maderspacher with these mutants suggested that *jaguar* (*obelix*) is required in melanophores to promote their aggregation and to control boundary integrity, and *leopard* regulates homotypic interaction within both melanophores and xanthophores, and interaction between the two, thus controlling boundary shape.[37] The positional cloning of the *jaguar* mutant revealed that this gene codes a inwardly rectifying potassium channel, Kir7.1.[39] kir channels are a group of ion channels that transfer potassium ion unidirectionally from the outside to the inside of the cell.[40–44] It is shown in the study of cultured cells that this class of channels is responsible for the stability of the membrane potential and sensitivity to the external signals.[45–47] Recently, Jantzi et al. reported that the kir channel (Kir2.2) facilitates cell-to-cell communication that propagates the contraction signal in the hamster retractor muscle feed artery.[48] In zebrafish, the Kir7.1 gene is expressed in the melanophores,[39] suggesting that the channel controls the interaction of the melanophores. The responsible gene for the *leopard* mutant is cloned and identified to be a component of gap junction, connexin41.8.[49] Connexin41.8 is expressed in many kinds of skin cells including melanophores and xanthophores. As the expression level of the connexin41.8 gene is very low in any cell of the zebrafish skin, very little is yet known about the function and the role of the molecule in the pigment pattern formation. Johnson et al. reported recently that the other class of gap junction gene, connexin43, codes the *shortfin (sof)* gene, which controls the length of the fins of the zebrafish.[50] This fact is quite interesting because it implies that a similar mechanism could control the size of limbs and the 2-D patterns in the skin.

Future Directions

Although molecular genetic studies have been identifying the genes and molecules involved in pigment pattern formation of fish, the questions as to how they are organized and how they generate the RD wave remain largely unanswered. To understand pigment pattern formation, it is necessary to integrate all the molecular reactions and deduce how such a complex system behaves. For this purpose, the framework of RD mechanism is required. In most of the mathematical models of the RD systems, the putative molecular network is composed of chemicals that control the synthesis of the molecules and diffuse them in the field. However, it is

Table 1

Mutant	Gene	Pattern	Melanophore	Xanthophore	Iridophore
sparse	kit/receptor tyrosine kinase	Stripe, boundary is slightly ambiguous	EM absent	Present	Present
rose	ednrb1/G protein-coupled receptor	Fewer stripes, but normal in fins	LM absent	Present	Absent
puma	Not cloned	Purely formed	LM absent	Present	Present
nacre	mitf1/transcription factor	Clusters of xanthophores, but purely striped	Absent	Present	Present
panther	fms /receptor tyrosine kinase	Scattered melanophores	Fewer	Absent	Present
jaguar (*obelix*)	kir7.1/potassium channel	Wider stripes	Present	Present	Present
leopard	Connexin41.8/ gap junction	Spots	Present	Present	Present

Source: Modified from [25]

also possible to compose an equivalent network with the interaction of two types of cell (melanophores and xanthophores). Therefore, there is no theoretical difficulty in applying the model to the zebrafish pigment pattern formation.

Recent mathematical studies on the RD and the related autonomous pattern-forming system revealed that the necessary conditions for the spatial pattern (spots and stripes) formation is the combination of "local activation" and "long-range inhibition."[51] Therefore, to find the core molecules of the RD mechanism, measuring the effective distance for each signaling molecule is essential. Use of advanced live-imaging techniques will help to assess whether the functioning molecules behave as the model predicts and will also help to develop the correct molecular-level network. With the cooperation of advanced molecular genetics and mathematical modeling, the long-unsettled question of how animal skin patterns are generated will be solved in the near future.

References

1. Murray, J. D. & Oster, G. F. Generation of biological pattern and form. IMA J Math Appl Med Biol **1:**51–75 (1984).
2. Meinhardt, H. Models of Biological Pattern Formation. Academic Press, London (1982).
3. Turing, A. The chemical basis of morphogenesis. Philos Trans R Soc Lond B **237:**37–72 (1952).
4. Gierer, A. & Meinhardt, H. A theory of biological pattern formation. Kybernetik **12:**30–39 (1972).
5. Murray, J. Mathematical Biology. Springer, Berlin (2003).
6. Meinhardt, H. & Gierer, A. Pattern formation by local self-activation and lateral inhibition. Bioessays **22:**753–760 (2000).
7. Castets, V., Dulos, E., Boissonade, J. & De Kepper, P. Experimental evidence of a sustained standing Turing-type nonequilibrium chemical pattern. Phys Rev Lett **64:**2953–2956 (1990).
8. Ouyang, Q. & Swinney, H. Transition from a uniform state to hexagonal and striped Turing patterns. Nature (Lond) **352:**610–612 (1991).
9. Kondo, S. & Asai, R. A reaction-diffusion wave on the skin of the marine angelfish *Pomacanthus*. Nature (Lond) **376:**765–768 (1995).
10. Nagorcka, B. N. Evidence for a reaction-diffusion system as a mechanism controlling mammalian hair growth. Biosystems **16:**323–332 (1983).
11. Jung, H. S. et al. Local inhibitory action of BMPs and their relationships with activators in feather formation: implications for periodic patterning. Dev Biol **196:**11–23 (1998).
12. Sick, S., Reinker, S., Timmer, J. & Schlake, T. WNT and DKK determine hair follicle spacing through a reaction-diffusion mechanism. Science **314:**1447–1450 (2006).
13. Harris, M. P., Williamson, S., Fallon, J. F., Meinhardt, H. & Prum, R. O. Molecular evidence for an activator-inhibitor mechanism in development of embryonic feather branching. Proc Natl Acad Sci U S A **102:**11734–11739 (2005).
14. Prum, R. O. & Williamson, S. Reaction-diffusion models of within-feather pigmentation patterning. Proc Biol Sci **269:**781–792 (2002).
15. Jiang, T. X. et al. Integument pattern formation involves genetic and epigenetic controls: feather arrays simulated by digital hormone models. Int J Dev Biol **48:**117–135 (2004).
16. Gierer, A. et al. Regeneration of hydra from reaggregated cells. Nat New Biol **239:**98–101 (1972).
17. Bode, H. R. Head regeneration in *Hydra*. Dev Dyn **226:**225–236 (2003).

18. Technau, U. et al. Parameters of self-organization in *Hydra* aggregates. Proc Natl Acad Sci U S A **97:**12127–12131 (2000).
19. Nakamura, T. et al. Generation of robust left-right asymmetry in the mouse embryo requires a self-enhancement and lateral-inhibition system. Dev Cell **11:**495–504 (2006).
20. Hamada, H. et al. Role of asymmetric signals in left-right patterning in the mouse. Am J Med Genet **101:**324–327 (2001).
21. Hamada, H., Meno, C., Watanabe, D. & Saijoh, Y. Establishment of vertebrate left-right asymmetry. Nat Rev Genet **3:**103–113 (2002).
22. Murray, J. D., Deeming, D. C. & Ferguson, M. W. Size-dependent pigmentation-pattern formation in embryos of *Alligator mississippiensis*: time of initiation of pattern generation mechanism. Proc R Soc Lond B Biol Sci **239:**279–293 (1990).
23. Meinhardt, H. The Algorithmic Beauty of Sea Shells. Springer, Berlin (2003).
24. Asai, R., Taguchi, E., Kume, Y., Saito, M. & Kondo, S. Zebrafish leopard gene as a component of the putative reaction-diffusion system. Mech Dev **89:**87–92 (1999).
25. Kelsh, R. N. Genetics and evolution of pigment patterns in fish. Pigment Cell Res **17:**326–336 (2004).
26. Hirata, M., Nakamura, K., Kanemaru, T., Shibata, Y. & Kondo, S. Pigment cell organization in the hypodermis of zebrafish. Dev Dyn **227:**497–503 (2003).
27. Yamaguchi, M., Yoshimoto, E. & Kondo, S. Pattern regulation in the stripe of zebrafish suggests an underlying dynamic and autonomous mechanism. Proc Natl Acad Sci U S A **104:**4790–4793 (2007).
28. Lister, J. A., Robertson, C. P., Lepage, T., Johnson, S. L. & Raible, D. W. nacre encodes a zebrafish microphthalmia-related protein that regulates neural-crest-derived pigment cell fate. Development (Camb) **126**:3757–3767 (1999).
29. Parichy, D. M., Rawls, J. F., Pratt, S. J., Whitfield, T. T. & Johnson, S. L. Zebrafish sparse corresponds to an orthologue of c-kit and is required for the morphogenesis of a subpopulation of melanocytes, but is not essential for hematopoiesis or primordial germ cell development. Development (Camb) **126:**3425–3436 (1999).
30. Parichy, D. M. et al. Mutational analysis of endothelin receptor b1 (rose) during neural crest and pigment pattern development in the zebrafish *Danio rerio*. Dev Biol **227:**294–306 (2000).
31. Parichy, D. M., Ransom, D. G., Paw, B., Zon, L. I. & Johnson, S. L. An orthologue of the kit-related gene fms is required for development of neural crest-derived xanthophores and a subpopulation of adult melanocytes in the zebrafish, *Danio rerio*. Development (Camb) **127:**3031–3044 (2000).
32. Johnson, S. L., Africa, D., Walker, C. & Weston, J. A. Genetic control of adult pigment stripe development in zebrafish. Dev Biol **167:**27–33 (1995).
33. Rawls, J., EM., M. & SL., J. How the zebrafish gets its stripes. Dev Biol **240:**301–314 (2001).
34. Johnson, S. L., Africa, D., Horne, S. & Postlethwaite, J. H. Half-tetrad analysis in zebrafish: mapping the ros mutation and the centromere of linkage group I. Genetics **139:**1727–1735 (1995).
35. Parichy, D. M. & Turner, J. M. Temporal and cellular requirements for Fms signaling during zebrafish adult pigment pattern development. Development (Camb) **130:**817–833 (2003).
36. Parichy, D. M., Turner, J. M. & Parker, N. B. Essential role for puma in development of postembryonic neural crest-derived cell lineages in zebrafish. Dev Biol **256:**221–241 (2003).
37. Maderspacher, F. & Nusslein-Volhard, C. Formation of the adult pigment pattern in zebrafish requires leopard and obelix dependent cell interactions. Development (Camb) **130:**3447–3457 (2003).
38. Kirschbaum, F. Untersuchungen ueber das Farbmuster der Zebrabarbe Brachydanio rerio (Cyprinidae, Teleostei). Wilhelm Roux's Arch **177:**129–152 (1975).
39. Iwashita, M. et al. Pigment pattern in jaguar/obelix zebrafish is caused by a Kir7.1 mutation: implications for the regulation of melanosome movement. PLoS Genet **2:**e197 (2006).

40. Doring, F. et al. The epithelial inward rectifier channel Kir7.1 displays unusual K^+ permeation properties. J Neurosci **18:**8625–8636 (1998).
41. Nakamura, N., Suzuki, Y., Ikeda, Y., Notoya, M. & Hirose, S. Complex structure and regulation of expression of the rat gene for inward rectifier potassium channel Kir7.1. J Biol Chem **275:**28276–28284 (2000).
42. Kim, S. J. et al. Inwardly rectifying K^+ channels in the basolateral membrane of rat pancreatic acini. Pflugers Arch **441:**331–340 (2000).
43. Kusaka, S. et al. Functional Kir7.1 channels localized at the root of apical processes in rat retinal pigment epithelium. J Physiol **531:**27–36 (2001).
44. Shimura, M. et al. Expression and permeation properties of the K(+) channel Kir7.1 in the retinal pigment epithelium. J Physiol **531:**329–346 (2001).
45. Wischmeyer, E., Doring, F. & Karschin, A. Stable cation coordination at a single outer pore residue defines permeation properties in Kir channels. FEBS Lett **466:**115–120 (2000).
46. Suzuki, Y. et al. Expression of the K^+ channel Kir7.1 in the developing rat kidney: role in K^+ excretion. Kidney Int **63:**969–675 (2003).
47. Yasuda, K. et al. Expression and functional properties of unique inward rectifier K^+ channel Kir7.1 in the porcine iris and retinal pigment epithelium. Curr Eye Res **27:**279–287 (2003).
48. Jantzi, M. C. et al. Inward rectifying potassium channels facilitate cell-to-cell communication in hamster retractor muscle feed arteries. Am J Physiol Heart Circ Physiol **291:**H1319–H1328 (2006).
49. Watanabe, M. et al. Spot pattern of leopard *Danio* is caused by mutation in the zebrafish connexin41.8 gene. EMBO Rep **7:**893–897 (2006).
50. Iovine, M. K., Higgins, E. P., Hindes, A., Coblitz, B. & Johnson, S. L. Mutations in connexin43 (GJA1) perturb bone growth in zebrafish fins. Dev Biol **278:**208–219 (2005).
51. Meinhardt, H. & Gierer, A. Pattern formation by local self-activation and lateral inhibition [see comment]. Bioessays **22:**753–760 (2000).

Characterization and Application of Natural Light-Sensitive Proteins

Jens Looser and Georg Nagel

Introduction

Photoreceptors

Light is essential to life as the primary source of energy, but it is also sensed by many organisms to gain information about the environment. For this second purpose, several photoreceptors evolved in nature, among them the rhodopsins and the blue-light sensors. Rhodopsins are found not only in the animal kingdom but also in prokaryotes, Archaea, and green algae—so far, however, not in higher plants. The UV/blue-light-sensing photoreceptors for phototropism and photomorphogenesis in plants were found to be flavoproteins, of which homologues were found in all kingdoms. Although the signaling chains initiated by these photoreceptors in their natural cellular environment still have to be worked out, especially in plants, some of these photoreceptors could be successfully transplanted to other organisms to act as light-regulated molecular tools. Here we describe briefly the properties of and applications for type I rhodopsins and a photoactivated adenylyl cyclase (PAC), a flavoprotein from *Euglena gracilis*.

Type I Rhodopsins

Rhodopsins are proteins with seven transmembrane helices (7TM) that are light sensitive via the chromophor retinal (vitamin A). Several rhodopsins regulate, via a signaling chain, the activity of ion channels. These ion channels are sometimes called "light sensitive" although their light sensitivity is completely indirect. Animal rhodopsins belong to the large family of G protein-coupled receptors. All rhodopsins consist of a

J. Looser and G. Nagel
University Wuerzburg, JvSachs-Platz 2, 97082 Wuerzburg, Germany

S. Nakanishi et al. (eds.), *Systems Biology*,
DOI:10.1007/978-4-431-87704-2_5, © Springer 2009

protein moiety, the opsin, and a covalently linked (as Schiff base to a lysine in TM7) chromophor. The chromophor is 11-*cis* retinal in the ground state of animal rhodopsins (type II rhodopsins). Archaea, bacteria, fungi, and green algae contain similar (still 7TM) but different rhodopsins, which, however, have other functions. These type I, or microbial type, rhodopsins contain all-*trans* retinal in the ground state and are not G protein-coupled receptors but light-activated ion pumps, providing cells with a mechanism to convert light energy into chemical energy, or photoreceptors, enabling phototaxis of microbes. Bacteriorhodopsin (BR) and halorhodopsin (HR) use photons for active transport of H^+ to the outside and Cl^- to the cytoplasm, respectively, whereas the sensory rhodopsins I and II couple to their specific transducer (Schmies et al., 2001) to trigger a signaling chain to the flagellar motor.

Blue-Light Sensors

Photoreceptors for blue light (and UV-A) were first hypothesized to exist in plants where they are involved in phototropism (growth to or from the light source) and photomorphogenesis (light-dependent development of the plant). From their action spectrum it could be speculated that they are flavoproteins, i.e., that they contain a UV-A/blue-light-absorbing flavin derivative as chromophor. These derivatives are most commonly flavin mononucleotide (FMN) and flavin adenine dinucleotide (FAD), derivatives of riboflavin (vitamin B_2). Currently known light-sensing flavoproteins are divided in three groups: the FMN-using phototropins, and the FAD-using cryptochromes and BLUF proteins (BLUF, blue light using FAD, a domain on these photoreceptors). Their signaling pathways are different; however, they often involve kinase activity and autophosphorylation. PAC, a BLUF protein from *E. gracilis*, is the first and so far only known light-activated adenylyl cyclase.

Results and Discussion

Bacteriorhodopsin and Halorhodopsin

BR and HR were purified from their natural source and rigorously studied with respect to their photocycle and three-dimensional structure. For both light-driven ion pumps the structure, even of photocycle intermediates, is now known to be about 2 Å, and transport of H^+ (BR) or Cl^- (HR) through the protein is well understood. Although biochemical assays could show potential dependence of light-activated H^+ transport in BR (Michel and Oesterhelt, 1976), it was not possible to gain exact data about the voltage dependence of ion pumping, neither from the tiny Archaebacteria nor from purified (and reconstituted) protein. An alternative is heterologously expressed type I rhodopsins. The first functional heterologous expression of bacteriorhodopsin was demonstrated by Büldt and coworkers in *Schizosaccharomyces pombe* (Hildebrandt et al., 1989, 1993). As these fission

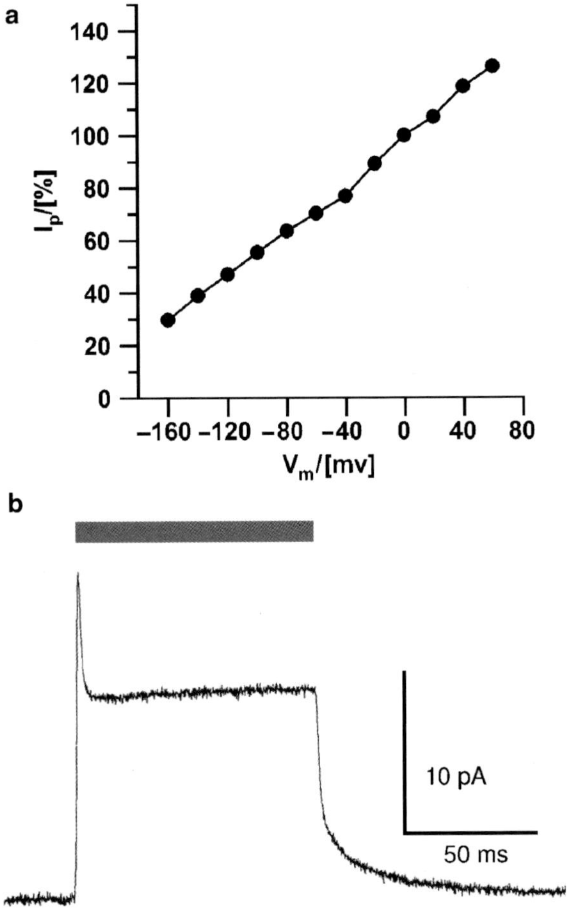

Fig. 1 Photocurrent caused by light-activated H^+ pumping of bacteriorhodopsin (BR), expressed in oocytes of *Xenopus laevis*. **a** Current–voltage relationship of stationary BR photocurrent in whole oocytes. Photocurrent is plotted as percentage of photocurrent at 0 mV. (Adapted from Nagel et al., 1998, with permission.) **b** BR photocurrent at 0 mV in giant excised inside-out membrane patch. Duration of light flash is indicated by *grey bar*

yeast cells are too small for electrophysiological analysis, we expressed BR in oocytes of *Xenopus laevis*, large cells (>1 mm diameter) that allow easy access for two-electrode voltage-clamp analysis (Nagel et al., 1995). Light-induced currents were strongly voltage dependent but always outward directed in the measured voltage range (from −160 mV to +60 mV), as expected for a H^+-exporting ion pump (Fig. 1a). The action spectrum of these photocurrents showed a maximum at 570 nm, well in accordance with the absorption maximum of 568 nm for BR (Nagel et al., 1995). For a better time resolution, we recorded from giant (20–30 μm diameter) excised inside-out oocyte membrane patches (Hilgemann, 1989; Nagel et al., 1992), expressing BR. Upon illumination an outward current increased rapidly and desensitized to a stationary photocurrent (~70% at 0 mV) during continuous illumination, before returning to the dark current with two time constants of ~1 ms and ~15 ms (Fig. 1b). Obviously this light-activated current will hyperpolarize a non-voltage-clamped cell quickly and reversibly.

We then expressed the light-driven chloride pump halorhodopsin from *Natronomonas pharaonis* (NpHR) or from *Halobacterium salinarum* (HsHR) in oocytes. Photocurrents were again outward directed in the whole voltage range and strongly dependent on extracellular chloride (Fig. 2a). For NpHR, we deter-

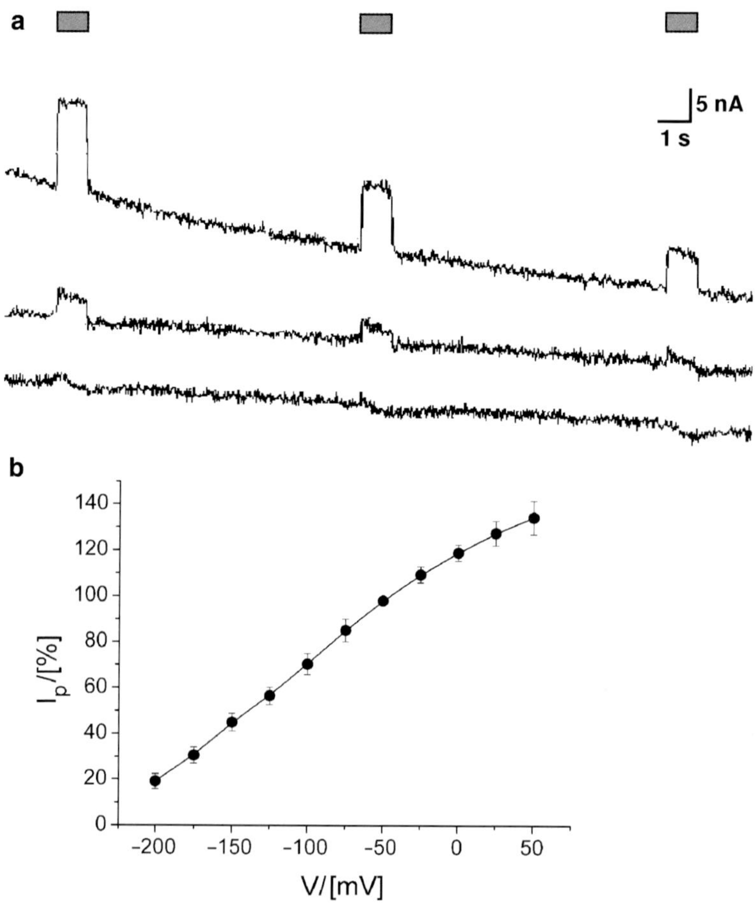

Fig. 2 Light-activated Cl^- pumping by halorhodopsin (NpHR). **a** Continuous recording from a whole oocyte during exchange of 121 mM Cl^- against D,L-aspartate (from *top to bottom*). Light flashes are indicated by *grey bars*. (From Zhang et al., 2007, with permission.) **b** Current–voltage relationship of stationary NpHR photocurrent in whole oocytes. Photocurrent is plotted as percentage of photocurrent at –50 mV. (From Zhang et al., 2007, with permission)

mined a $K_{0.5}$ of 16 mM Cl^-; chloride affinity was lower for HsHR ($K_{0.5}$ = 32 mM Cl^-) (Zhang et al., 2007). The current–voltage relationship of light-driven chloride pumping is quite similar to the I-V of BR pump currents (Fig. 2b), again allowing us to hyperpolarize a HR-expressing cell, simply by illumination. The action spectrum maximum for light-activated, hyperpolarizing current we determined to be 580 nm (Zhang et al., 2007). Our colleagues at Stanford University (Deisseroth and coworkers) and University of Frankfurt (Gottschalk and coworkers) then showed that NpHR is ideally suited to suppress action potentials in rat neurons or muscle contractions in live *Caenorhabditis elegans*, respectively, by illumination with amber light (Zhang et al., 2007).

Channelrhodopsins

Type I rhodopsins were suggested to trigger phototaxis in green algae (Foster et al., 1984) where photoreceptor currents occur with a delay of less than 30 μs after flash stimulation (Sineshchekov et al., 1990; Harz & Hegemann, 1991). This finding suggested an intimate link between photoreceptor and electrical conductance. It was not possible to clone these opsins, but recently overlapping partial cDNA sequences appeared in an EST database (Asamizu et al., 2000) of the green alga *Chlamydomonas reinhardtii* that encodes opsin-related proteins. The sequences were named channelopsins (Chop1 and Chop2) by us (Nagel et al., 2002). The core regions (about 300 N-terminal amino acid residues of more than 700) comprise seven hypothetical transmembrane segments with sequence similarity (15%–20%) to the archaeal sensory rhodopsins (SR), the ion transporters bacteriorhodopsin (BR) and halorhodopsin (HR) (Nagel et al., 2002; Sineshchekov et al., 2002; Suzuki et al., 2003). Sequence analysis suggests that K296 in Chop1 and K257 in Chop2 is the retinal-binding amino acid; 60% of the 22 amino acids that are in direct contact with the retinal in BR are identical (9 amino acids, aa) or conservatively exchanged (4 aa) in Chop1. Photocurrent measurements on *C. reinhardtii* cell populations, in which the ratio of Chop1 and its homologue Chop2 has been changed by an antisense approach, demonstrated that both functional channelrhodopsins, ChR1 and ChR2, contribute to the photoreceptor currents, but the mechanism remained obscure (Sineshchekov et al., 2002). Antibodies against Chop1 detected expression in the eye spot (Suzuki et al., 2003) where previously the tightly coupled photoreceptor–ion conductance complex was localized (Harz and Hegemann, 1991).

We expressed channelrhodopsin-1 and channelrhodopsin-2 in oocytes, either as full-length proteins of more than 700 amino acids or as the N-terminal half of 346 (ChR1-346) and 315 (ChR2-315) amino acids, comprising essentially the seven transmembrane, rhodopsin-typical helices (Nagel et al., 2002, 2003) (Fig. 3a). No differences in light-induced electrical currents were observed between full-length and shortened versions. Light-induced currents of channelrhodopsins were, however, quite different from light-induced currents of BR or HR as they showed a clear reversal potential (Fig. 3b,c). Therefore it could be concluded

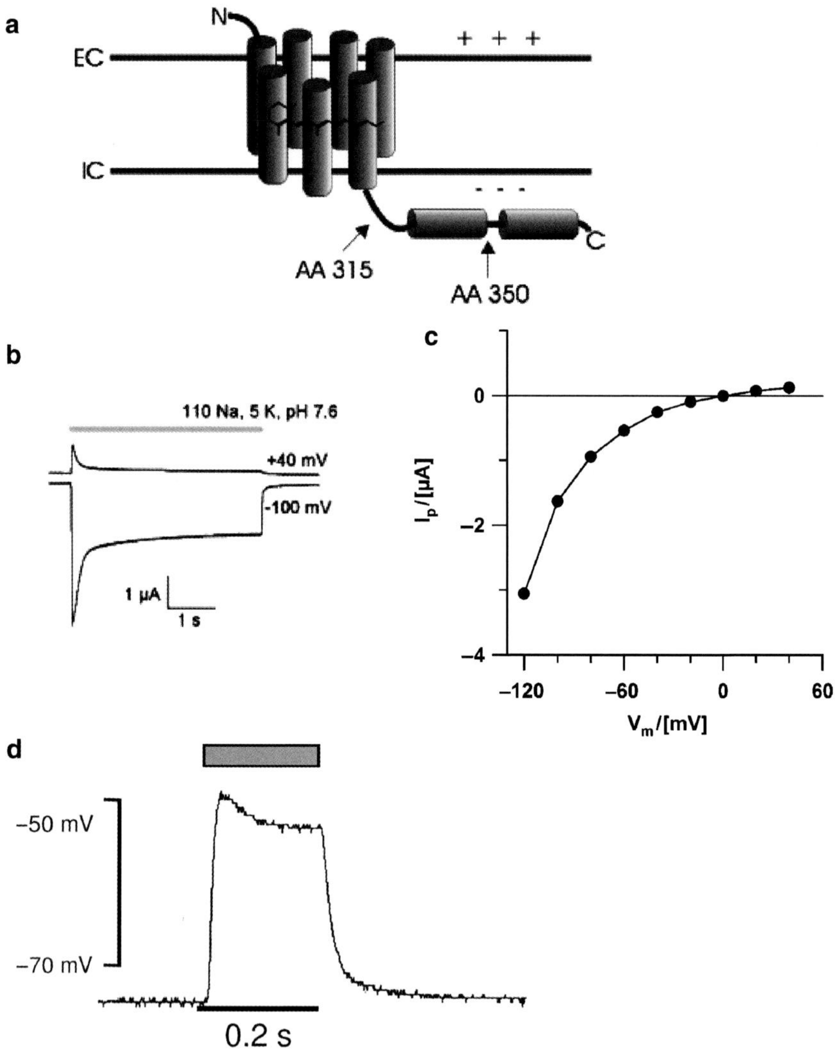

Fig. 3 Structure and function of channelrhodopsin-2 (ChR2). **a** Cartoon of the ChR2-structure comprising a 7TM segment, a rab-like domain (320–350), and a Gly-rich domain (350–737). *EC*, extracellular side; *IC*, intracellular side; *AA*, amino acids. (From Nagel et al., 2003, with permission.) **b** Light-activated current in whole oocyte, expressing ChR2-315 (first 315 amino acids of ChR2), at +40 mV and at −100 mV. Duration of light flash is indicated by *grey bar*. (From Nagel et al., 2003, with permission.) **c** Current–voltage relationship of stationary ChR2 photocurrent in whole oocytes in Na^+-containing Ringer's solution, as in **b** (pH 07.6). (Adapted from Nagel et al., 2003, with permission.) **d** Ability of ChR2 to depolarize cells by illumination with blue light. Current clamp measurement of an illuminated (*grey bar*) oocyte. (From Nagel et al., 2003, with permission)

that channelrhodopsins are directly light-gated ion channels. Further experiments with exchange of extracellular ions led to the conclusion that channelrhodopsin-1 (ChR1) is a light-gated proton channel (Nagel et al., 2002) and that channelrhodopsin-2 (ChR2) is a light-gated cation channel with permeabilities for protons > Li^+ > Na^+ > K^+ > Ba^{2+} ≈ Ca^{2+} but no permeability to Mg^{2+} or to anions (Nagel et al., 2003). As expected from a proton channel, at neutral pH, light-activated currents of ChR1 were much smaller than of ChR2, although we recently noticed also some small cation permeability of ChR1 at pH 9, in accordance with a recent publication by Berthold et al. (2008). The I-V relation of light-induced ChR2 currents is inward rectifying with a reversal potential near 0 mV (Fig. 3c), clearly indicating that activation of ChR2 will lead to a collapse of membrane potential. Indeed, light activation of ChR2 in oocytes or mammalian cells leads to a drastic depolarization (Fig. 3d). The action spectrum of ChR2 photocurrents showed a maximum between 460 and 480 nm (Nagel et al., 2003), conveniently blue-shifted by more than 100 nm with respect to the maximum of the hyperpolarizing NpHR.

Expression of ChR2 in neurons made these light sensitive, allowing light flashes to induce action potentials at high frequency and with great reliability (Boyden et al., 2005). Expression of ChR2 in nerve or muscle cells of *C. elegans* provided a convenient way to depolarize these cells and achieve dramatic behavioral effects in live animals (Nagel et al., 2005).

Photoactivated Adenylyl Cyclases (PACα and PACβ)

Recently the photoreceptor for photophobic response of *Euglena gracilis* was shown to be a tetramer of two highly homologous BLUF flavoproteins, PACα and PACβ (Iseki et al., 2002). Watanabe and coworkers showed that the purified photoreceptor of ~400 kDa had some adenylyl cyclase activity in the dark, which was about 100 fold stimulated by illumination with UV-A or blue light. We expressed these proteins together and separately in oocytes and tested for cAMP generation with an immunoassay on oocyte extracts or by measuring electrical conductance of whole oocytes coexpressing PAC and cAMP-sensitive ion channels (Schröder-Lang et al., 2007). Surprisingly, PACα alone showed adenylyl cyclase activity, which was strongly increased by short flashes of blue light (Fig. 4a). Coexpression of PACα and PACβ made no difference to expression of PACα alone, but expression of PACβ alone showed much weaker activity, comparable to the activity of 100-fold less PACα. For very sensitive detection of cAMP we used CFTR, an anion channel that is activated by (oocyte-endogenous) cAMP-dependent protein kinase (PKA, $K_{0.5}^{cAMP} \approx 100$ nM). For fast detection of cAMP, however, we coexpressed PACα with the directly cAMP-gated cation channel CNGA2-C460W/E583M ($K_{0.5}^{cAMP} \approx 1$ μM) (Fig. 4b). As can be seen from the rapid change of electrical current, PACα responds very rapidly to light with cAMP production ($\tau < \approx 10$ ms) upon switching on the light and also must decrease its activity very rapidly in the dark, as deduced from experiments with light flashes of different duration (Schröder-Lang et al., 2007). PACα and PACβ are composed of two BLUF domains (F1, F2), binding FAD, and two cyclase domains (C1, C2), which are known to dimerize in homologous cyclases. By mutation of conserved residues in C1 or C2, we were able to completely knock down the activity of PACα. When we then coexpressed

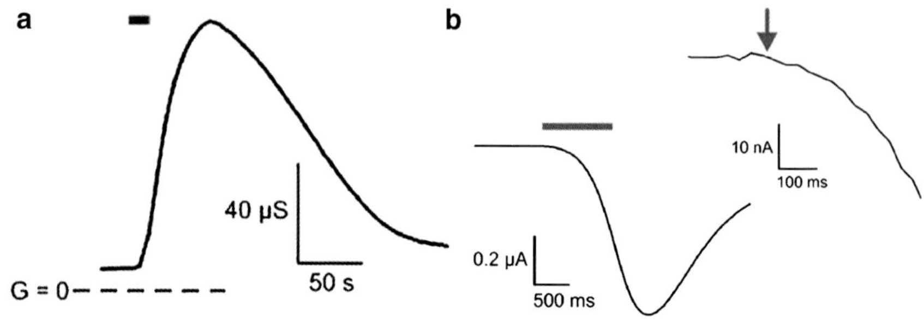

Fig. 4 Fast and reversible light-induced [cAMP] increase in oocytes via PACα, measured by cAMP-sensitive ion channels. **a** Light-induced increase of CFTR-mediated conductance by 20 s illumination (*grey bar*) of coexpressed PACα. (Adapted from Schröder-Lang et al., 2007, with permission.) **b** Light-induced (1 s; *grey bar*) current increase of CNGA2 mutant channel, coexpressed with PACα. Onset is very fast, as shown in *inset* (*arrow* indicates start of illumination). (Adapted from Schröder-Lang et al., 2007, with permission)

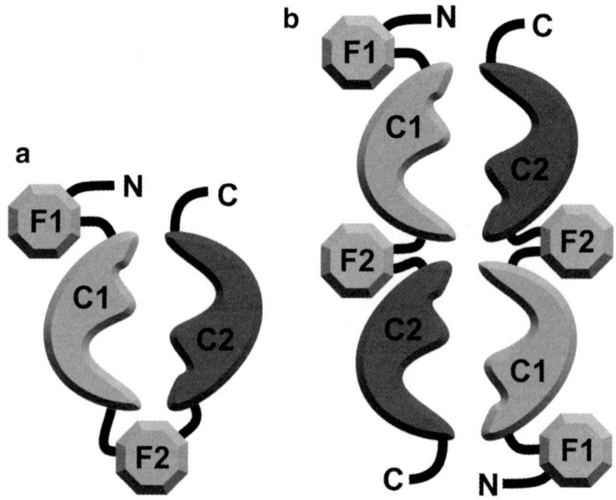

Fig. 5 Models of domain arrangement in PACα. **a** This model of a PACα monomer could explain construction of a cyclase from two C domains. **b** Dimer model of PACα, as suggested by a mixture of lethal C1 and C2 mutants that showed restored cyclase activity, whereas each mutant alone had no activity

nonfunctional C1 and C2 mutants of PACα with CFTR, we partially restored dark- and light-induced activity of PACα, which strongly suggests that PACα dimerizes to build functional intermolecular C1–C2 domains (Fig. 5).

Acknowledgments This work was supported by the German Research Association (DFG) to G.N. (NA407/6).

References

Asamizu E, et al. (2000) Generation of expressed sequence tags from low-CO_2 and high-CO_2 adapted cells of *Chlamydomonas reinhardtii*. DNA Res **7:**305–7.

Berthold P, Tsunoda SP, Ernst OP, Mages W, Gradmann D, Hegemann P (2008) Channelrhodopsin-1 initiates phototaxis and photophobic responses in *Chlamydomonas* by immediate light-induced depolarization. Plant Cell **20:**1665–77.

Boyden ES, Zhang F, Bamberg E, Nagel G, Deisseroth K (2005) Millisecond-timescale, genetically targeted optical control of neural activity. Nat Neurosci **8(9):**1263–8.

Foster KW, Saranak J, Patel N, Zarilli G, Okabe M, Kline T, Nakanishi K (1984) A rhodopsin is the functional photoreceptor for phototaxis in the unicellular eukaryote *Chlamydomonas*. Nature (Lond) **311:**756–9.

Harz H, Hegemann P (1991) Rhodopsin-regulated calcium currents in *Chlamydomonas*. Nature (Lond) **351:**489–91.

Hildebrandt V, Ramezani-Rad M, Swida U, Wrede P, Grzesiek S, Primke M, Büldt G (1989) Genetic transfer of the pigment bacteriorhodopsin into the eukaryote *Schizosaccharomyces pombe*. FEBS Lett **243:**137–40.

Hildebrandt V, Fendler K, Heberle J, Hoffmann A, Bamberg E, Büldt G (1993) Bacteriorhodopsin expressed in *Schizosaccharomyces pombe* pumps protons through the plasma membrane. Proc Natl Acad Sci U S A **90:**3578–82.

Hilgemann DW (1989) Giant excised cardiac sarcolemmal membrane patches: sodium and sodium-calcium exchange currents. Pflügers Arch. 415(2):247-9.

Holland EM, Braun FJ, Nonnengässer C, Harz H, Hegemann P (1996) The nature of rhodopsin-triggered photocurrents in *Chlamydomonas*. I. Kinetics and influence of divalent ions. Biophys J **70:**924–31.

Iseki M, Matsunaga S, Murakami A, Ohno K, Shiga K, Yoshida K, Sugai M, Takahashi T, Hori T, Watanabe M (2002) A blue-light-activated adenylyl cyclase mediates photoavoidance in *Euglena gracilis*. Nature (Lond) **415(6875):**1047–51.

Michel H, Oesterhelt D (1976) Light-induced changes of the pH gradient and the membrane potential in *H. halobium*. FEBS Lett **65:**175–8.

Nagel G, Hwang T-C, Nastiuk KL, Nairn AC, Gadsby DC (1992) The protein kinase A-regulated cardiac Cl^- channel resembles the cystic fibrosis transmembrane conductance regulator. Nature (Lond) **360:**81–4.

Nagel G, Möckel B, Büldt G, Bamberg E (1995) Functional expression of bacteriorhodopsin in oocytes allows direct measurement of voltage dependence of light induced H^+ pumping. FEBS Lett **377:**263–6.

Nagel G, Kelety B, Möckel B, Büldt G, Bamberg E (1998) Voltage dependence of proton pumping by bacteriorhodopsin is regulated by the voltage-sensitive ratio of M1 to M2. Biophys J **74:**403–12.

Nagel G, Ollig D, Fuhrmann M, Kateriya S, Musti AM, Bamberg E, Hegemann P (2002) Channelrhodopsin-1: a light-gated proton channel in green algae. Science **296:**2395–8.

Nagel G, Szellas T, Huhn W, Kateriya S, Adeishvili N, Berthold P, Ollig D, Hegemann P, Bamberg E (2003) Channelrhodopsin-2, a directly light-gated cation-selective membrane channel. Proc Natl Acad Sci U S A **100:**13940–5.

Nagel G, Brauner M, Liewald JF, Adeishvili N, Bamberg E, Gottschalk A (2005) Light-activation of Channelrhodopsin-2 in excitable cells of *Caenorhabditis elegans* triggers rapid behavioral responses. Curr Biol **15(24):**2279–84.

Schmies G, Engelhard M, Wood PG, Nagel G, Bamberg E (2001) Electrophysiological characterization of specific interactions between bacterial sensory rhodopsins and their transducers. Proc Natl Acad Sci U S A **98:**1555–9.

Schröder-Lang S, Schwärzel M, Seifert R, Strünker T, Kateriya S, Looser J, Watanabe M, Kaupp UB, Hegemann P, Nagel G (2007) Fast manipulation of cellular cAMP level by light in vivo. Nat Methods **4:**39–42.

Sineshchekov OA, Litvin FF, Keszthelyi L (1990) Two components of photoreceptor potential in phototaxis of the flagellated green alga *Haematococcus pluvialis*. Biophys J **57:**33–9.

Sineshchekov OA, Jung K-H, Spudich JL (2002) From the cover: Two rhodopsins mediate phototaxis to low- and high-intensity light in *Chlamydomonas reinhardtii*. Proc Natl Acad Sci U S A **99:**8689–4.

Suzuki T, Yamasaki K, Fujita S, Oda K, Iseki M, Yoshida K, Watanabe M, Daiyasu H, Toh H, Asamizu E, Tabata S, Miura K, Fukuzawa H, Nakamura S, Takahashi T (2003) Archaeal-type rhodopsins in *Chlamydomonas*: model structure and intracellular localization. Biochem Biophys Res Commun **301:**711–7.

Zhang F, Wang L-P, Brauner M, Liewald JF, Kay K, Watzke N, Wood PG, Bamberg E, Nagel G, Gottschalk A, Deisseroth K (2007) Multimodal fast optical interrogation of neural circuits. Nature (Lond) **446:**633–9.

Systems Biology of Mammalian Circadian Clocks

Hiroki R. Ueda

Introduction: Systems Biology as "Biology After Identification"

Recent large-scale efforts in genome sequencing, expression profiling, and functional screening have produced an embarrassment of riches for life sciences researchers, and biological data can now be accessed in quantities that are orders of magnitude greater than were available even a few years ago. The growing need for interpretation of data sets, as well as the accelerating demand for their integration to a higher-level understanding of life, has set the stage for the advent of systems biology [1,2], in which biological processes and phenomena are approached as complex and dynamic systems. Systems biology is a natural extension of molecular biology and can be defined as "biology after identification of key gene(s)." We see systems biological research as a multistage process, beginning with the comprehensive identification and quantitative analysis of individual system components and their networked interactions, and leading to the ability to control existing systems toward the desired state and design new ones based on an understanding of structure and underlying dynamical principles (Fig. 1).

The Mammalian Circadian Clock as a Model System

We have taken the mammalian circadian clock as an initial model system that exhibits system-level dynamical and structural properties to develop research strategies and technologies for the studies of complex and dynamic biological systems. The mammalian circadian clock consists of complexly integrated feed-back and feed-forward loops [3] and also exhibits well-defined dynamical properties [4],

H.R. Ueda
Laboratory for Systems Biology and Functional Genomics Unit, Center for Developmental Biology, RIKEN, 2-2-3 Minatojima-minamimachi, Chuo-ku, Kobe, Hyogo 650-0047, Japan

S. Nakanishi et al. (eds.), *Systems Biology*,
DOI:10.1007/978-4-431-87704-2_6,

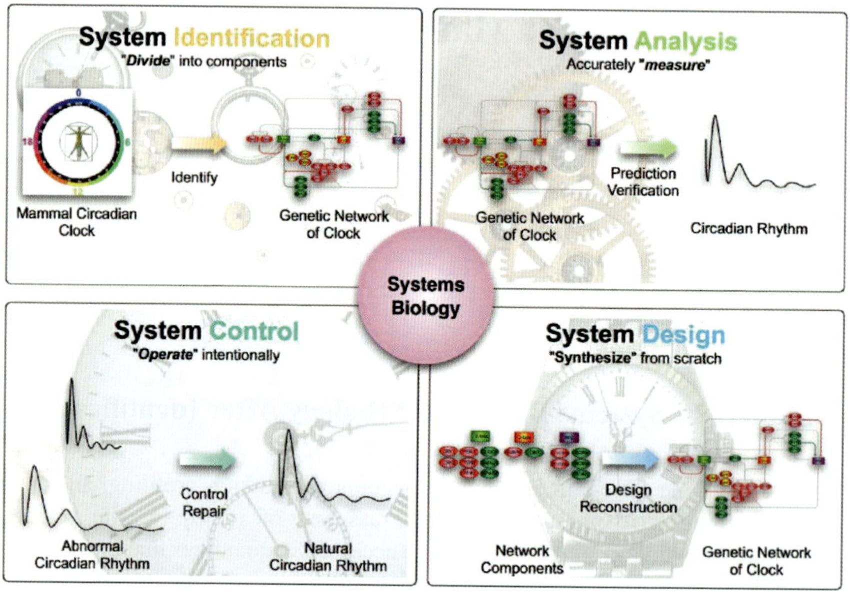

Fig. 1 Systems biology. Systems biological research starts with comprehensive identification (*upper left panel*). In this step, individual system components and their networked interactions are comprehensively identified. In the second step, to derive the design principle of a target system, the behavior of the system is predicted and validated through an accurate measurement with perturbations (*upper right panel*). An understanding of the design principle of the system is essential to derive the method of controlling the system toward the desired state (*lower left panel*). Finally, the level of understanding is confirmed by reconstruction of the system (*lower right panel*)

including (1) endogenous oscillations of an approximately 24-h period; (2) entrainment to external environmental changes (temperature and light cycle); (3) temperature compensation over a wide range of temperature; and (4) synchronization of multiple cellular clocks despite the inevitable molecular noise. All these dynamic properties would be difficult to elucidate without utilizing such system-level approaches. In addition to its advantages as a basic model system for systems biological research, the function of the circadian clock is intimately involved in the control of metabolic and physiological processes [3,5], and its dysregulation is associated with the onset and development of numerous human diseases, including sleep disorders, depression, and dementia (Fig. 2). An improved understanding at the system level promises to provide biomedical and clinical investigators with a powerful new arsenal for attacking these conditions. In the following sections, we describe in detail systems biology strategies and technologies (especially, system identification and system design) in addition to their application to the specific question of mammalian circadian clocks.

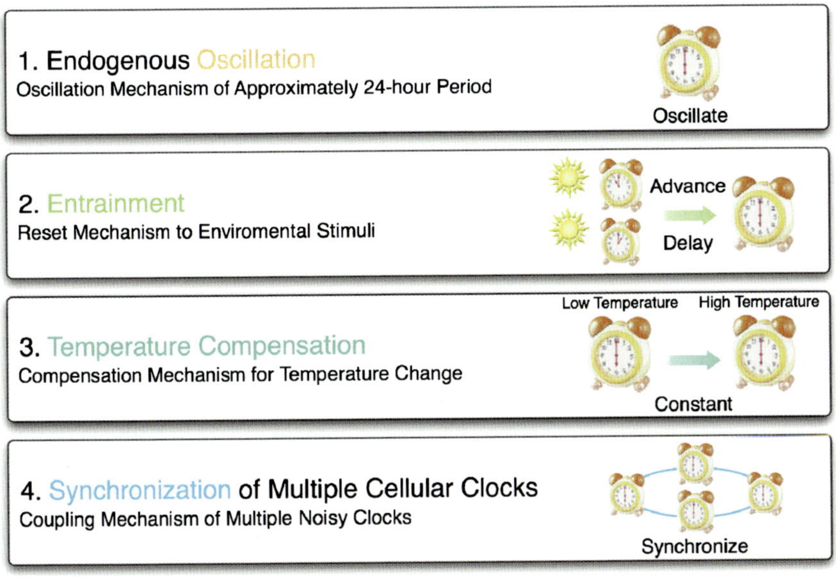

Fig. 2 Dynamic properties of mammalian circadian clocks. Although circadian clocks appear to be cell autonomous oscillators (**1**), the clock can synchronize with the environment (**2**). Period length is kept constant irrespective of temperature (**3**). In a multicell tissue such as the suprachiasmatic nucleus (SCN), the clocks synchronize with each other (**4**)

System Identification of Clocks

Following the completion of genome projects for species such as mouse and human, genome-wide resources such as siRNA or cDNA libraries have undergone considerable expansion. Development of high-throughput technologies also assists in the efficient use of these resources. These genome-wide resources and technologies as well as genome-associated information currently allow us to comprehensively identify system components of interest (system identification).

Circadian clocks of multicellular organisms consist of complexly integrated regulatory loops with positive or negative regulators known as clock genes [6–19]. The transcriptional regulation network of these genes forms a circadian clock oscillator, which is known to control output genes and affect physiological and metabolic processes [3,5]. Although some transcriptional regulations of identified clock genes have been the subject of previous studies, a system-level understanding of circadian clocks remains to be elucidated. In this section on identification of clocks, we provide results of our system identification of mammalian circadian clocks [20,21].

The mammalian circadian master clock is primarily located in the suprachiasmatic nucleus (SCN) [3]. Transcript analyses have indicated that circadian clocks are not restricted to the SCN but are found in several tissues [22], including liver and cultured fibroblast cells such as Rat-1 [23,24] or NIH3T3 [25] cells. The mechanisms underlying circadian rhythms are also known to be conserved across species [4]. At the basic core of the clock lies a transcriptional/translational feedback loop [26–28], whose primary components are known as "clock genes" [6–19]. For example, in the mouse system, transcription factors CLOCK and BMAL1 proteins dimerize and directly and indirectly activate transcription of the *Per* and *Cry* genes through E-box elements (5'-CACGTG-3') [26,27]. The PER and CRY proteins accumulate in the cytosol, and are then translocated following phosphorylation into the nucleus, where they inhibit the activity of CLOCK and BMAL1 [3]. The turnover of the inhibitory PER and CRY proteins leads to a new cycle of activation by CLOCK and BMAL1 via E-box elements. Despite the reporting of many transcriptional regulations of each gene, however, an overview of circadian clock core network remains to be put forward.

Complicated networks cannot be elucidated without access to both (1) comprehensive identification of network circuits and (2) accurate measurement of system dynamics. In a previous attempt to comprehensively identify the circadian clock core network, we first quantitatively and comprehensively measured genome-wide gene expression using a high-density oligonucleotide probe array [29] and identified genes showing circadian oscillation with characteristic expression patterns through biostatistics (Fig. 3). The second step involved comprehensively determining the transcription start sites [30] (TSSs) and conserved noncoding regions to construct the genome-wide promoter/enhancer database. Using these data, we predicted that there was a relationship between expression patterns of identified genes and DNA regulatory elements on their promoter/enhancer regions. We found clock-controlled elements (CCEs), E-boxes (5'-CACGTG-3') [31]/E'-boxes (5'-CACGTT-3') [21,32], D-boxes (5'-TTATG[C/T]AA-3') [33], or RREs (5'-[A/T]A[A/T]NT[A/G]GGTCA-3') [34] are distributed throughout the oscillatory genes.

To determine the role of these elements in the circadian clock, we utilized an cell culture system with which we can monitor circadian rhythms in transcriptional dynamics using a destabilized luciferase (d*Luc*) reporter driven by clock-controlled

Fig. 3 (continued) peak time. Colors in descending order from *red* to *black* to *green* represent the normalized data. From the obtained data, we identified a set of genes rhythmically expressed under both LD and DD. We classified 101 genes in the SCN and 393 genes in the liver as "significantly rhythmic under both LD and DD." **c** Temporal expression profiles of transcription factors in the SCN (*upper panel*) and liver (*lower panel*) under constant darkness (DD) conditions. Relative mRNA levels under DD condition of the indicated genes were measured with a quantitative polymerase chain reaction (QPCR) assay, in which *GAPDH* expression was used as an internal control. Data were normalized so that the average copy number (QPCR) over a 12-point time course is 1.0. Circadian expression of transcription factors having functional and evolutionary conserved E-boxes (*Dbp*, *Dec1*, and *Dec2*), both of E-boxes/E'-boxes and D-boxes (*Per1*, *Per2*, *RevErbAα*, and *RevErbAβ*), D-boxes (*Per3*, *Rorα*, and *Rorβ*), RREs (*Bmal1*, *Clock*, *Npas2*, and *E4bp4*), both of E-boxes/E'-boxes and RREs (*Cry1* and *Rorγ*), on their noncoding regions. *Clock* and *Rorγ* were constitutively expressed in the SCN

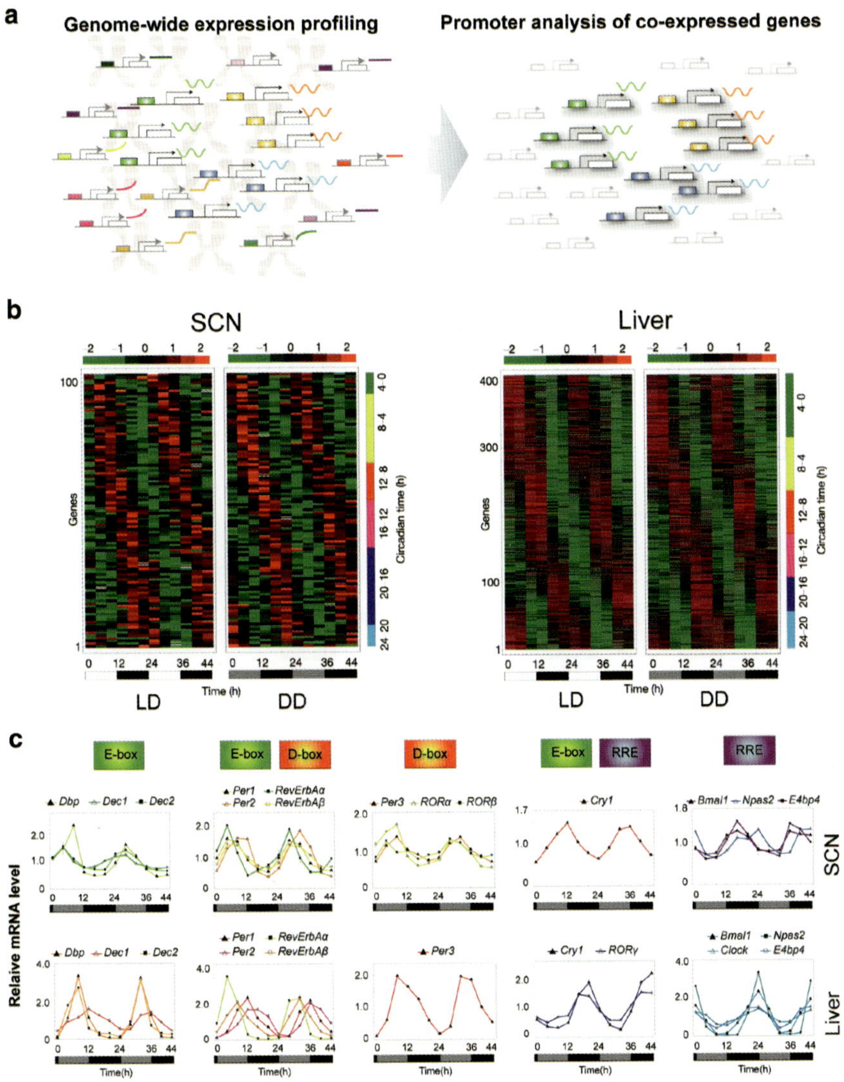

Fig. 3 **a** Strategy for identification of clock-controlled elements (CCEs). Gene expression information was obtained by performing comprehensive expression profiling (*left panel*). Through statistical analysis, genes with a special characteristic pattern of expression (circadian oscillation) were selected (*right panel*; oscillatory genes are indicated by their *color*). DNA regulatory elements for specific issue (i.e., expression timing) were predicted by combining expression pattern information and transcriptional regulatory elements information from promoter regions. **b** Genome-wide expression profiles in mouse central (*SCN; left panel*) and peripheral (*liver; right panel)* clocks. Total RNA was extracted every 4 h during light/dark cycles (LD) or constant darkness (DD) over 2 days and used to determine genome-wide gene expression profiles with the Affymetrix mouse high-density oligonucleotide probe array (GeneChip). Data were normalized so that the average signal intensity and standard deviation over 12-point time courses were 0.0 and 1.0, respectively. *Columns* represent time points, and *rows* represent genes that were organized by

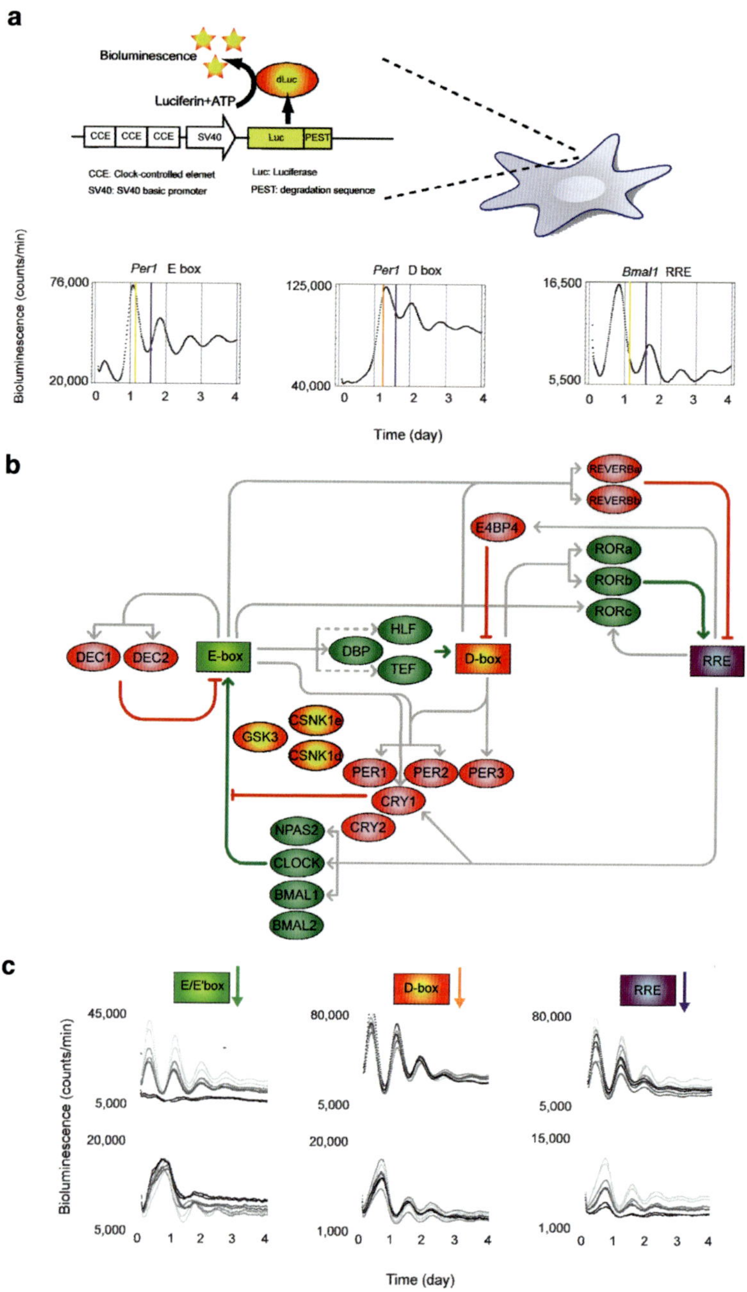

Fig. 4 **a** Schematic overview of experiment. Cultured mammalian cell (Rat-1) was transfected with d*Luc* under the regulation of CCE and SV40 basic promoter. The circadian change of the bioluminescence was monitored by a PMT detector over several days (*upper panel*). Representative circadian rhythms of bioluminescence from wild-type CCE fused to the SV40 basic promoter driving a d*Luc* reporter. The circadian bioluminescence phase from the *Per2* promoter and that of

promoters (Fig. 4A). In this cell culture system, named the "*in cellulo* cycling assay," we transiently transfected reporter constructs into cultured Rat-1 cells, stimulated with dexamethasone, and measured their bioluminescence. Dexamethasone was administrated to induce macroscopic circadian oscillations in the cultured cells. Through the genome-wide searching described above, we found CCEs on 16 clock/clock-controlled genes promoter/enhancers. Then, using the *in cellulo* cycling assay system, we were able to reveal that functionally and evolutionary conserved E/E'-boxes are located on noncoding regions of nine genes (*Per1*, *Per2*, *Cry1*, *Dbp*, *Ror*γ, *RevErbA*α/*Nr1d1*, *RevErbA*β/*Nr1d2*, *Dec1*/*Bhlhb2*, and *Dec2*/*Bhlhb3*), D-boxes on those of seven genes (*Per1*, *Per2*, *Per3*, *RevErbA*α, *RevErbA*β, *Ror*α, and *Ror*β), and RREs on those of six genes (*Bmal1*/*Arntl*, *Clock*, *Npas2*, *Cry1*, *E4bp4*/*Nfil3*, and *Ror*γ}. Based on this functional and conserved transcriptional regulatory mechanism, we succeeded in drawing transcriptional circuits underlying mammalian circadian rhythms (Fig. 4B) [21].

Our analysis further suggested that regulation of E/E'-boxes is the topological vulnerability point in mammalian circadian clocks. We functionally verified this concept using *in cellulo* cycling assay systems (Fig. 4C). Overexpression of repressors of E/E'-box regulation (CRY1 [27]), RRE regulation (REVERBAα [14,20]), or D-box regulation (E4BP4 [16]) affected circadian rhythmicity in *Per2* or *BMAL1* promoter activity. The effects were different, however, between each repressor, and the severest effect was observed when the E/E'-box was attacked. Such different modes of effect cannot be explained by mere quantitative differences in the strength of these three repressors, indicating that there is some qualitative difference between E/E'-box, D-box, and RRE regulation in circadian rhythmicity [21].

Design of Clocks

We herein describe in detail the final step of systems biology, system design, reconstruction of the system using a hypothesized design principle that has been revealed through the efforts of a combination strategy of system identification, system analysis, and system control for validating the sufficiency of the system components. Three specific features of circadian oscillatory wave are the period length, the amplitude, and the peak phase. In this section, we introduce our recent study on the design principles on phase determination using an in vitro cell culture system as the "physical simulator" [35].

Fig. 4 (continued) the *Bmal1* promoter are indicated by *yellow* and *purple* lines, respectively (*bottom panels*). **b** Schematic representation of transcriptional network of mammalian circadian clock. Genes and CCEs are depicted as *ellipsoids* and *rectangles*, respectively. Transcriptional/translational activation and repression are depicted as *gray*, *green*, and *red* lines, respectively. **c** Effect of repression on each CCE. The *E/E'-boxes*, *D box*, and *RRE* are repressed by overproduction of CRY1, E4BP4, and REVERBAα, respectively. The consequences of those repressions were monitored by *Per2*-dLuc (*upper panel*) and *Bmal1*-dLuc (*lower panel*). d*Luc*, destabilized luciferase; CCE, clock-controlled elements; PMT, photomultiplier tube

Design of Circadian Circuits: Phase Determination

Although the network structure composed by clock genes and clock-controlled DNA elements (CCEs; Fig. 4) has been described comprehensively, the dynamic principles governing this transcriptional circuit remain elusive. One of the key issues concerning the logic of the mammalian circadian clock is how the circadian oscillating genes expression peaks ("phases") are determined. In attempting to elucidate this issue, we have started a project for understanding the molecular logic of phase-controlling system(s). For this purpose, we have extended our in vitro cycling assay system to the "physical simulator," with which we can implement artificial circuits of interest. We used this system to prove the sufficiency of the components we predict in natural circadian phase-controlling mechanisms.

Mammalian circadian clocks are composed of complexly integrated transcriptional regulatory loops. To reach a system-level understanding of the circadian clock controlling system, we focused on the three main transcriptional regulation elements (CCEs), E/E'-box, D-box, and RRE (Fig. 4). We first focused on the daytime transcriptional regulation mediated via the D-box. The transcriptional activator DBP activates genes expression via the D-box, whereas the transcriptional repressor E4BP4 represses expression (Fig. 5A, upper panel). *Dbp* is regulated by the E-box,

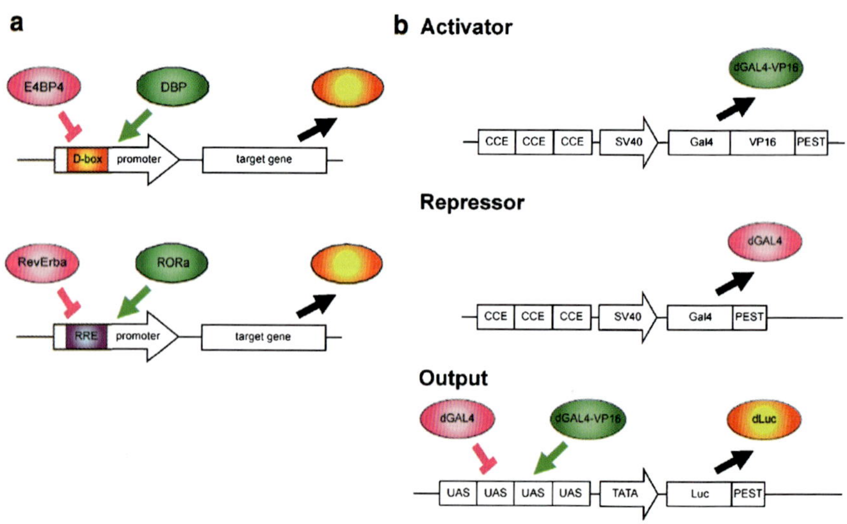

Fig. 5 Synthetic transcriptional circuit of mammalian circadian clocks. **a** Hypothetical models of daytime and nighttime output regulation. In natural circuits, *Dbp* and *E4bp4* control the expression via the D-box. Similarly, *RevErbA*α and *Ror*α participate in the regulation via the RRE. **b** The artificial transcriptional system. dGAL4-VP16 and dGAL4 were used as the activator and repressor. These transcriptional factors were expressed under the control of three tandem repeats of a CCE (E'-box of *Per2*, D-box of *Per3*, or RRE of *Bmal1*). The artificial activator and repressor competitively bind the four tandem repeats of UAS in the artificial promoter to regulate the output reporter gene *dLuc*. For observation of the dynamic behavior of this artificial circuit, NIH3T3 cells were transiently transfected with plasmids harboring the activator, repressor, or reporter, and the reporter activity was monitored under a real-time bioluminescence measuring system

the morning control element, while *E4bp4* is under the regulation of the RRE, the nighttime control element. On the basis of this transcriptional circuit information, we hypothesized that a morning activator and a nighttime repressor can determine the daytime transcriptional output mediated through the D-box. Similarly, RRE activators (e.g., *Rorα*) are expressed during the daytime under the control of the D-box, and the RRE repressors (e.g., *RevErbAα*) seem strongly influenced by the morning element (E'-box; Fig. 5A, lower panel). We have therefore also been able to make the simple hypothesis that a daytime activator and a morning repressor can specify the nighttime transcriptional output mediated through the RRE. To test these hypotheses, we adopted a synthetic approach to physically simulate an identified structure and observed the resulting dynamics using artificial transcriptional circuits.

To design and implement artificial transcriptional circuits in mouse NIH3T3 cells, which have a self-oscillating circadian clock, we developed an in vitro cycling assay system composed by the following three components: an artificial activator (a destabilized GAL4-VP16 fusion protein: dGAL4-VP16), an artificial repressor (a destabilized GAL4 protein: dGAL4), and an output reporter [a destabilized *luciferase* (*dLuc*) gene driven by a minimal TATA box fused with four tandem repeats of UAS, the GAL4-binding sequence] (Fig. 5B). In this system, the input from artificial activator and repressor competitively regulates the output and reporter gene expression through the UAS (Fig. 5B, lower panel), and the dynamics of the output could be monitored through the light intensity of the expressed dLuc protein under a real-time bioluminescence measuring system. As the artificial activator and repressor are driven by the SV40 basic promoter fused with three tandem repeats of CCEs, we could control the expression timing of these artificial regulators via the morning (E'-box), daytime (D-box), or nighttime (RRE) elements. We reasoned that if (and only if) the phases of the transcriptional activator(s) and repressor(s) are acceptable determinants of the phase of the downstream transcriptional output, then it should be possible to generate the natural phases using synthetic transcriptional regulators and promoters. If this were proved to be so, the findings would prove to be very significant, as transcription factors are regulated by various posttranscriptional mechanisms, including translation, phosphorylation, ubiquitination, sumoylation, and nuclear transportation, which are thought to contribute, at least in part, to the phase decision of the downstream transcriptional outputs.

Using the in vitro cycling assay system as a physical simulator, we first tested our hypothesis for daytime output by examining the dynamic behavior of the transcriptional output generated from the competition of an artificial morning activator controlled via E'-box and a nighttime repressor controlled via RRE. The phases of the activator and repressor in this experiment were detected at circadian time (CT) 4.0 $\pm$ 0.29 ($n = 2$), and CT17.1 $\pm$ 0.37 ($n = 2$). The transcriptional output driven by these regulators exhibited circadian oscillation with a phase at CT7.7 $\pm$ 0.85 ($n = 2$), which is very close ($\leq$1.0 h) to the corresponding natural daytime (D-box) phase, CT8.7 $\pm$ 1.13 (Fig. 6A). These results suggest that morning activation and nighttime repression are sufficient to determine the daytime transcriptional output. Importantly, we could not observe high-amplitude circadian oscillation by expressing the morning activator or nighttime repressor alone. From these, we concluded that both morning activation and nighttime repression are necessary for daytime output.

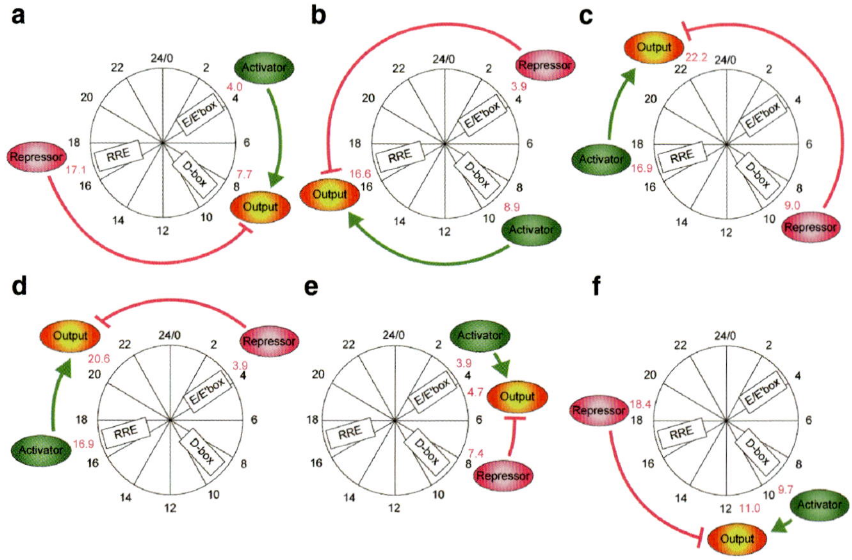

Fig. 6 Promoter activities of an activator, repressor, and output in different artificial transcriptional circuits: morning activator under E'-box control and nighttime repressor under RRE control (**a**); daytime activator under D-box control and morning repressor under E'-box control (**b**); nighttime activator under RRE control and daytime repressor under D-box control (**c**); nighttime activator and morning repressor (**d**); morning activator and daytime repressor (**e**); and daytime activator and nighttime repressor (**f**). The schemes summarize the representative promoter activities of each artificial circuit monitored by bioluminescence from NIH3T3 cells, where an activator, repressor, and output phases are indicated with their peak time indicated by circadian time (*red numbers*)

We next tested the hypothesis for nighttime output by examining another artificial circuit consisting of an artificial daytime activator under D-box control and a morning repressor under E'-box control. The phases of a transcriptional activator and repressor in this experiment were at CT8.9 ± 0.28 (n = 2) and CT3.9 ± 0.13 (n = 2). The transcriptional output driven by the regulators exhibited circadian oscillation with its phase at CT16.6 ± 1.04 (n = 2), which is very close (≤1.0 h) to the corresponding natural nighttime phase (CT17.0 ± 0.81) (Fig. 6B). These results also suggest that daytime activation and morning repression are necessary and sufficient to generate the circadian nighttime output, implying that the input phases of transcriptional regulators can determine the phases of transcriptional output. These findings led us to reason that various combinations of transcriptional regulators with CCEs for the three basic circadian phases (morning, daytime, and nighttime) might not only generate the basic phases but also other phases. Indeed, we succeeded in generating various phases though simple combinations of the transcriptional regulators of the three basic circadian phases (Fig. 6C–F).

Importantly, even if the phase of the activator was not changed, the advanced or delayed phase of the repressor led to a corresponding change in the phase of transcriptional output (Fig. 6). Similarly, a change in the activator phase also led to a corresponding alteration in the output phase (Fig. 6). We can deduce from these findings that mammalian circadian clocks can start from three basic phases to the final, practically continuous, phases through transcriptional cascades, which have been observed in vivo in the central and peripheral clock tissues (e.g., SCN and liver). Moreover, this concept is supported by our recent observations that many transcription factors are rhythmically expressed at various phases in the liver (Fig. 3) and that some seem to be under the direct control of the circadian clock via CCEs. In addition, our in vitro findings lead to the concept that the natural transcription activator(s) and repressor(s) that bind the similar DNA elements are expected to generate additional output phases because they are expressed in different in vivo phases.

In these experiments, we showed that the transcriptional regulation of upstream transcription factors can determine the phase of the downstream output. This study presents a synthetic approach to the "proof-by-synthesis" of transcriptional logic, which provides us with a new strategy, not only to investigate the requirements for identified components and/or their interactions but also to reveal as-yet unidentified components or interactions. These results reveal design principles for elucidating continuous transcriptional outputs observed in vivo and for the logical design of artificial promoters working at novel phases. Logical synthesis of artificial circuits with defined structure and observation of their dynamics provide an alternative strategy applicable for the investigation of complex and dynamic biological systems such as mammalian circadian clocks.

Conclusion

After identification of key clock genes, there are increasing demands for higher-order understanding of design principles in mammalian circadian clocks. In this chapter, we introduced systems biological approaches, beginning with comprehensive identification (System Identification) and quantitative analysis (System Analysis) of individual clock components and their networked interactions, and leading to the ability to control existing systems toward the desired state (System Control) and to design new ones based on an understanding of structure and underlying dynamical principles (System Design). We also described the detailed examples in the application of Systems Identification and System Design for the mammalian circadian clocks. We strongly believe that it is now high time to fully integrate these systems biology approaches for the solution of various system-level questions in mammalian circadian clocks.

Acknowledgments We thank our collaborators. System Identification of clocks is a collaboration with Drs. W. Chen, A. Adachi, H. Wakamatsu, S. Hayashi, T. Takasugi, M. Nagano, K. Nakahama, Y. Suzuki, S. Sugano, M. Iino, Y. Shigeyoshi, S. Hashimoto, M. Sano, and M. Machida. System Analysis of clocks is a collaboration with Drs. T.K. Sato, R.G.. Yamada, H. Ukai, J.E. Baggs, L.J. Miraglia, T.J. Kobayashi, D.K. Welsh, S.A. Kay, and J.B. Hogenesch. System Control of clocks is the collaboration with Drs. H. Ukai, T.J. Kobayashi M. Nagano,

K. Masumoto, M. Sujino, T. Kondo, K. Yagita, and Y. Shigeyoshi. System Design of Clocks is a collaboration with Drs. M. Nakajima, M. Ishida, M. Ukai-Tadenuma, and T. Kasukawa. The molecular timetable method is a collaborative work with Drs. W. Chen, Y. Minami, S. Honma, K. Honma, M. Iino, and S. Hashimoto. The theoretical study on Synchronization of Clocks is a collaboration with K. Hirose and M. Iino. We also thank Drs. I. Nikaido, H. Ukai, R.G. Yamada, T.J. Kobayashi, M. Nakajima, Y. Minami, and A. Wada for creation of the figures and critical reading of the manuscript. This research was supported by an intramural Grant-in-Aid from CDB, Director's Fund from CDB, President's Fund from RIKEN, Grant-in-Aid for Genome-Network Project, and Scientific Research on Priority Areas "Systems Genomics" from MEXT, Japan, and a Grant-in-Aid for NEDO project from METI, Japan.

References

1. Kitano H (2002) Systems biology: a brief overview. Science 295:1662–1664
2. Kitano H (2002) Computational systems biology. Nature (Lond) 420:206–210
3. Reppert SM, Weaver DR (2002) Coordination of circadian timing in mammals. Nature (Lond) 418:935–941
4. Dunlap JC, Loros JJ, DeCoursey PJ (eds) (2004) Chronobiology: Biological Timekeeping. Sinauer Associates, Sunderland, MA
5. Panda S, Hogenesch JB, Kay SA (2002) Circadian rhythms from flies to human. Nature (Lond) 417:329–335
6. King DP, Zhao Y, Sangoram AM et al (1997) Positional cloning of the mouse circadian clock gene. Cell 89:641–653
7. Bunger MK, Wilsbacher LD, Moran SM et al (2000) Mop3 is an essential component of the master circadian pacemaker in mammals. Cell 103:1009–1017
8. Bae K, Jin X, Maywood ES et al (2001) Differential functions of mPer1, mPer2, and mPer3 in the SCN circadian clock. Neuron 30:525–536
9. Zheng B, Albrecht U, Kaasik K et al (2001) Nonredundant roles of the mPer1 and mPer2 genes in the mammalian circadian clock. Cell 105:683–694
10. van der Horst GT, Muijtjens M, Kobayashi K et al (1999) Mammalian Cry1 and Cry2 are essential for maintenance of circadian rhythms. Nature (Lond) 398:627–630
11. Vitaterna MH, Selby CP, Todo T et al (1999) Differential regulation of mammalian period genes and circadian rhythmicity by cryptochromes 1 and 2. Proc Natl Acad Sci U S A 96:12114–12119
12. Lowrey PL, Shimomura K, Antoch MP et al (2000) Positional syntenic cloning and functional characterization of the mammalian circadian mutation tau. Science 288:483–492
13. Sato TK, Panda S, Miraglia LJ et al (2004) A functional genomics strategy reveals Rora as a component of the mammalian circadian clock. Neuron 43:527–537
14. Preitner N, Damiola F, Lopez-Molina L et al (2002) The orphan nuclear receptor REV-ERB-alpha controls circadian transcription within the positive limb of the mammalian circadian oscillator. Cell 110:251–260
15. Wuarin J, Schibler U (1990) Expression of the liver-enriched transcriptional activator protein DBP follows a stringent circadian rhythm. Cell 63:1257–1266
16. Mitsui S, Yamaguchi S, Matsuo T et al (2001) Antagonistic role of E4BP4 and PAR proteins in the circadian oscillatory mechanism. Genes Dev 15:995–1006
17. Honma S, Kawamoto T, Takagi Y et al (2002) Dec1 and Dec2 are regulators of the mammalian molecular clock. Nature (Lond) 419:841–844
18. Zylka MJ, Shearman LP, Weaver DR et al (1998) Three period homologs in mammals: differential light responses in the suprachiasmatic circadian clock and oscillating transcripts outside of brain. Neuron 20:1103–1110

19. Reick M, Garcia JA, Dudley C et al (2001) NPAS2: an analog of clock operative in the mammalian forebrain. Science 293:506–509
20. Ueda HR, Chen W, Adachi A et al (2002) A transcription factor response element for gene expression during circadian night. Nature (Lond) 418:534–539
21. Ueda HR, Hayashi S, Chen W et al (2005) System-level identification of transcriptional circuits underlying mammalian circadian clocks. Nat Genet 37:187–192
22. Yamazaki S, Numano R, Abe M et al (2000) Resetting central and peripheral circadian oscillators in transgenic rats. Science 288:682–685
23. Balsalobre A, Damiola F, Schibler U (1998) A serum shock induces circadian gene expression in mammalian tissue culture cells. Cell 93:929–937
24. Yagita K, Tamanini F, van Der Horst GT et al (2001) Molecular mechanisms of the biological clock in cultured fibroblasts. Science 292:278–281
25. Akashi M, Nishida E (2000) Involvement of the MAP kinase cascade in resetting of the mammalian circadian clock. Genes Dev 14:645–649
26. Gekakis N, Staknis D, Nguyen HB et al (1998) Role of the CLOCK protein in the mammalian circadian mechanism. Science 280:1564–1569.
27. Kume K, Zylka MJ, Sriram S et al (1999) mCRY1 and mCRY2 are essential components of the negative limb of the circadian clock feedback loop. Cell 98:193–205.
28. Shearman LP, Sriram S, Weaver DR et al (2000) Interacting molecular loops in the mammalian circadian clock. Science 288:1013–1019
29. Lipshutz RJ, Fodor SP, Gingeras TR et al (1999) High density synthetic oligonucleotide arrays. Nat Genet 21:20–24
30. Suzuki Y, Taira H, Tsunoda T et al (2001) Diverse transcriptional initiation revealed by fine, large-scale mapping of mRNA start sites. EMBO Rep 2:388–393
31. Hogenesch JB, Gu YZ, Jain S et al (1998) The basic-helix-loop-helix-PAS orphan MOP3 forms transcriptionally active complexes with circadian and hypoxia factors. Proc Natl Acad Sci U S A 95:5474–5479
32. Yoo SH, Ko CH, Lowrey PL et al (2005) A noncanonical E-box enhancer drives mouse Period 2 circadian oscillations in vivo. Proc Natl Acad Sci U S A 102:2608–2613
33. Falvey E, Marcacci L, Schibler U (1996) DNA-binding specificity of PAR and C/EBP leucine zipper proteins: a single amino acid substitution in the C/EBP DNA-binding domain confers PAR-like specificity to C/EBP. Biol Chem 377:797–809
34. Harding HP, Lazar MA (1993) The orphan receptor Rev-ErbA alpha activates transcription via a novel response element. Mol Cell Biol 13:3113–3121
35. Ukai-Tadenuma M, Kasukawa T, Ueda HR (2008) Proof-by-synthesis of the transcriptional logic of mammalian circadian clocks. Nat Cell Biol 10:1154–1163

Toward Understanding How the Immune System Establishes a Diverse Yet Self-Tolerant T-Cell Repertoire: Stepwise Roles of Thymic Microenvironments

Takeshi Nitta and Yousuke Takahama

Introduction

The thymus is an organ that supports the development and repertoire formation of T lymphocytes (1). Thymic parenchyma consists of leukocytic cells called thymocytes, the majority of which belong to the T-lymphoid lineage, and various stromal cells including thymic epithelial cells (TEC) (2). Thymic stromal cells provide multiple signals to support manifold processes of thymocyte development that are essential for the supply of circulating T lymphocytes (3). In response to these signals, developing thymocytes undergo proliferation, differentiation, and relocation to generate mature T lymphocytes that carry a diverse yet self-tolerant repertoire of T-cell antigen receptors (TCR) (4). These steps of T-lymphocyte development take place in anatomically discrete regions of the thymus where a variety of specialized stromal cells are localized (5).

T lymphocytes arise from hematopoietic stem cell-derived T-lymphoid progenitor cells that migrate to the thymus (6). Most immature hematopoietic cells that have just entered the thymus lack the expression of CD4 and CD8 and therefore belong to CD4/CD8 double-negative (DN) thymocytes (7, 8). The development of DN thymocytes is associated with the dynamic relocation of the cells in thymic parenchyma; T-lymphoid progenitor cells in adult mouse thymus are mostly localized in the corticomedullary junction, the area between deep cortex and medulla (9), whereas thymocytes migrate toward the capsular region of the thymus during differentiation and develop into CD4/CD8 double-positive (DP) thymocytes (10). DP thymocytes expressing TCR on the cell surface are localized in the cortex. DP thymocytes move actively within the cortical microenvironment (11, 12), probably seeking TCR interaction with major histocompatibility complex (MHC)-encoded molecules that are associated with self-peptides. Cortical DP thymocytes that interact via their TCR with the self-peptide–MHC complex are selected for survival or death depending on the avidity of the interaction (13, 14). DP thymocytes that receive TCR signals with ligand interactions of weak avidity and nonextensive aggregation are induced to survive

T. Nitta and Y. Takahama
Division of Experimental Immunology, Institute for Genome Research,
University of Tokushima, Tokushima 770-8503, Japan

S. Nakanishi et al. (eds.), *Systems Biology*,
DOI:10.1007/978-4-431-87704-2_7, © Springer 2009

and differentiate into mature thymocytes, the process referred to as positive selection (15, 16). By contrast, DP thymocytes that receive TCR signals with ligand interactions of strong avidity and extensive aggregation are destined to die (17, 18), a process referred to as negative selection. During positive selection, the differential kinetics of TCR–ligand interactions determines cell lineage to become either $CD4^{+}$ $CD8^{-}$ or $CD4^{-}$ $CD8^{+}$ single-positive (SP) thymocytes (19). Positively selected thymocytes relocate to thymic medulla, where they further interact with self-peptides displayed in the medullary microenvironment (20). Medullary TEC (mTEC) express a diverse set of genes representing peripheral tissues (21), thereby contributing to the establishment of self-tolerance in thymic medulla. A nuclear factor called autoimmune regulator (AIRE) participates in this promiscuous gene expression in mTEC (22). Consequently, a diverse yet self-tolerant TCR repertoire is formed in the thymus, and mature T lymphocytes with such a TCR repertoire are released to the circulation. Thus, T-cell repertoire formation consists of stepwise fate determinations of thymocyte development in different thymic microenvironments. The dynamic relocation of developing thymocytes within thymic microenvironments is crucial for T-cell repertoire selection.

The aforementioned control of T-lymphocyte development in the thymus is supported in multiple thymic microenvironments that are formed by different sets of thymic stromal cells. Thymic stromal cells are composed of TEC and other stromal cells, including mesenchymal cells, endothelial cells, and hematopoietic cells, such as dendritic cells (DC) and macrophages (23). TEC consist of at least two major populations, cortical TEC (cTEC) and mTEC, which play crucial roles in the development and repertoire formation of T lymphocytes. cTEC and mTEC are derived from common progenitor cells that are generated from the endoderm of the third pharyngeal pouch (24–26). The development of TEC is dependent on many transcription factors, including Tbx1, Hoxa3, Pax1, and Foxn1 (27–30). Thymic mesenchymal cells are also crucial for supporting thymus generation and T-lymphocyte development (31, 32). It is also fascinating to note that thymic stromal cells that support T-lymphocyte development are supported, in turn, by developing thymocytes. The bidirectional signal exchanges between thymocytes and thymic stromal cells are appreciated as "thymic crosstalk" (33–35).

To understand how T lymphocytes are selected to establish a diverse yet self-tolerant repertoire in the thymus, we performed genome-wide screening for genes that are expressed in the thymus and that affect thymus development. We find that β5t-containing thymoproteasomes specifically expressed by cTEC play a pivotal role in the positive selection of CD8 SP thymocytes. We also find that positively selected thymocytes begin expressing CCR7, a chemokine receptor, and thereby relocate to thymic medulla where mTEC produce CCR7 ligands. Furthermore, we find that tumor necrosis factor (TNF) superfamily (TNFSF) ligands, including RANKL and CD40L, are produced by positively selected thymocytes and pivotally regulate mTEC development and thymic medulla formation. Thus, our study shows that sequential encounter of developing thymocytes with a unique set of self-molecules in the cortical microenvironment and another set of self-molecules in the medullary microenvironment is essential for the generation of a diverse yet self-tolerant repertoire of T lymphocytes.

Positive Selection in Thymic Cortex and β5t-Containing Thymoproteasomes

A number of studies support the notion that low-affinity self-peptides present in thymic cortex are responsible for the positive selection of developing thymocytes. However, the nature of positively selecting peptides and positively selecting antigen-presenting cells remains elusive (36). A single MHC–peptide complex expressed by cTEC can produce a diverse repertoire of T lymphocytes (37), suggesting that any peptide that causes low-avidity TCR engagement might be capable of triggering positive selection of thymocytes. It was also suggested that, rather than initial positive selection, subsequent negative selection establishes repertoire formation of T lymphocytes (38). Furthermore, it was shown that the experimental capability of inducing positive selection is not limited to cTEC but can be detected in fibroblasts and DC (39–41). It was also shown that developing thymocytes and thymus-reentering T lymphocytes can induce positive selection of thymocytes (42–45). Thus, the possibility that cTEC, or any other cells in the thymic cortex, have any specialized capability to induce positive selection seemed unlikely until recently.

Our recent discovery of a novel subunit of the 20S proteasome, β5t, has revealed a unique capability of cTEC to support the positive selection of thymocytes (46). Proteasomes are multicatalytic protease complexes that are responsible for regulated proteolysis in eukaryotic cells and for the generation of antigenic peptides presented by class I MHC molecules (47, 48). The 20S proteasome is responsible for the proteolytic activity of the proteasome and is composed of 28 subunits (two α-rings with α1 to α7 subunits and two β-rings with β1 to β7 subunits). Among these subunits, β1, β2, and β5 are responsible for caspase-like, trypsin-like, and chymotrypsin-like catalytic activities, respectively (49, 50). Interferon-γ induces the production of a new set of catalytic subunits, β1i, β2i, and β5i, to replace their constitutive counterparts, β1, β2, and β5, thereby forming immunoproteasomes, a proteasome complex that possesses increased chymotrypsin-like activity and participates in efficient antigen presentation and immune responses (51). On the other hand, the newly identified catalytic subunit β5t is incorporated into the 20S proteasome instead of β5 or β5i, together with β1i and β2i (46). Because this novel proteasome containing β5t is specifically expressed in the thymus and exclusively in cTEC, it is termed a thymoproteasome.

In comparison to β5-containing standard proteasomes and β5i-containing immunoproteasomes, β5t-containing thymoproteasomes exhibit reduced chymotrypsin-like activity but normal caspase-like activity and normal trypsin-like activity (46). Proteasomes are responsible for the production of MHC class I-binding peptides and are the sole enzymes that determine the C-termini of the peptides (52, 53). Hydrophobic C-terminal anchor residues of the peptides are essential for high-affinity peptide binding into the clefts of MHC class I complexes (54). Chymotrypsin-like activity carried by β5 and β5i is important for the production of high-affinity MHC class I ligands (55). Thus, it is possible that cTEC generate a unique set of MHC class I-associated peptides that are different from those present in any other cells.

The generation of $CD4^- CD8^+$ (CD8SP) thymocytes that express high levels of TCR is severely reduced in β5t-deficient mice. The selective reduction of CD8SP

T lymphocytes is also observed in the spleen of these mice. On the other hand, DP and CD4SP thymocytes as well as peripheral CD4 T lymphocytes are unaffected. The absence of β5t does not affect cortical or medullary architecture or overall thymus size, indicating that β5t is essential for neither the development of cTEC nor the generation of normal thymic architecture. These results demonstrate that β5t is required for the development of CD8SP T lymphocytes in the thymus and suggest the possibility that β5t is associated with positive selection of CD8SP T lymphocytes (46).

It is possible that thymoproteasomes in cTEC may be somehow involved in providing costimulatory signals that are specifically required for the generation of CD8SP T lymphocytes rather than CD4SP T lymphocytes. However, since the generation of CD8SP T lymphocytes is specifically affected in β5t-deficient mice, and as the surface expression of MHC class I molecules on cTEC of β5t-deficient mice is comparable to that of normal mice, β5t-containing thymoproteasomes in cTEC are likely involved in producing class I MHC-loaded peptides that provide TCR signals required for positive selection of class I MHC-restricted CD8SP T lymphocytes (46). β5t specifically limits chymotryptic activity that cleaves peptide bonds after hydrophobic amino acid residues in cTEC, and therefore, thymoproteasomes predominantly produce low-affinity class I MHC ligands specifically in cTEC. These low-affinity class I MHC ligands may limit the duration and/or avidity of interaction with TCR and contribute to inducing positive selection of the majority of CD8SP T lymphocytes (46, 56, 57).

The results from β5t-deficient mice reveal that cTEC possess unique protein degradation activity that might lead to the production of a unique set of class I MHC-associated peptides necessary for the generation of CD8 T lymphocytes. This unique protein degradation activity of cTEC might not be limited to class I MHC-associated peptides but might also occur in class II MHC-associated peptides, because cathepsin L, a lysosomal protease that is highly expressed by cTEC and not mTEC (58), is required for the optimal generation of CD4SP T lymphocytes (59, 60). A unique lysosomal degradation activity might be functional in cTEC, which is mediated by cathepsin L in a manner analogous to thymoproteasomes. Thus, the unique character of cTEC protein degradation and self-peptide presentation may be pivotal for the positive selection of thymocytes in both CD4SP and CD8SP lineages (56, 57). These findings of cTEC would not only highlight the significant roles of thymic cortex in T-lymphocyte development but also further our understanding of the molecular mechanisms of T-lymphocyte repertoire selection.

CCR7 Regulates Relocation of Positively Selected Thymocytes to Medulla

Newly generated immature DP thymocytes move randomly in the thymic cortex, as shown by in vitro real-time imaging of intact thymus lobes (11, 61). This “random walk motility” is also observed by noninvasive intravital imaging of thymocytes in the fish thymus (12). On the other hand, immature thymocytes at early ontogeny before CD4 and CD8 expression appear dormant in the thymus (12). Thus, the differentiation into DP thymocytes possibly coincides with the acquisition of cellular motility, which represents the activity of cells that dynamically seek TCR interaction with the MHC–peptide complex in the cortical microenvironment. DP thymocytes that actively move within the cortex interact with cTEC and pause their motility upon TCR interaction with MHC–peptide ligands (62).

Positively selected thymocytes are induced to survive and to relocate from the cortex to the medulla. Positive selection coincides with the appearance of a thymocyte population that displays rapid and directed migration toward the medulla (11). Upon positive selection, DP thymocytes elevate CCR7 expression on the cell surface (63–65). CCR7 ligands, CCL19 and CCL21, in the thymus are predominantly expressed by mTEC and mostly localized in the medulla (65). Consequently, positively selected thymocytes are attracted to the medulla through CCR7-mediated chemotaxis. We found that positively selected thymocytes in mice deficient for CCR7 or CCR7 ligands are defective in accumulation in the medulla and are localized in the cortex (65, 66). Mice deficient for CCR7 signals exhibit autoimmunity to peripheral tissues (66–68). Thymocytes generated without CCR7 ligands are potent in inducing autoimmune exocrinopathy in mice, and thus are defective in establishing central tolerance (66). Therefore, the migration of positively selected thymocytes to the medulla is essential for the establishment of central tolerance, and the proper relocation of developing thymocytes within multiple thymic microenvironments is necessary for repertoire formation of T lymphocytes (20, 66).

TNFSF Regulation of Medulla Formation

The microenvironment of thymic medulla is mainly composed of mTEC and hematopoietic cells, including mature thymocytes and DC. Similar to cTEC, mTEC are derived from endodermal precursor cells that are generated at the third pharyngeal pouch (25, 26, 69). Thymic medulla is a specialized microenvironment where developing thymocytes establish tolerance to systemic self-antigens, including peripheral tissue-specific antigens. Indeed, T-lymphocyte development within the thymus that is defective in thymic medulla formation leads to failure in establishing self-tolerance, resulting in autoimmune disorders (70–72). In the thymic medulla, mTEC express a variety of "tissue-specific" genes, for example, the gene encoding insulin whose expression is restricted to β-islet cells in the pancreas (73–76). This "ectopic" expression of tissue-specific genes by mTEC is called "promiscuous gene expression" and is responsible for the establishment of self-tolerance through the presentation of tissue-specific antigens to developing thymocytes (21, 77, 78). AIRE, a nuclear protein that is predominantly expressed by mTEC, is associated with the promiscuous gene expression of mTEC (22, 79–82).

The formation of thymic medulla is associated with the development of mature thymocytes in the thymus. An early study showed that medulla formation is defective in *scid* mice, in which thymocyte development is arrested at the DN stage and is restored by reconstitution with wild-type hematopoietic cells (33). Studies of mice deficient for positive selection, including TCRα-deficient mice, ZAP70-deficient mice, and MHC class I and class II double-deficient mice, confirmed that the formation of thymic medulla is dependent on the generation of positively selected mature thymocytes (83–86). In mice deficient for CCR7 or CCR7 ligands, medulla formation is mildly defective (65, 66), suggesting that optimum development of thymic medulla requires the relocation of positively selected thymocytes. Thus, signals produced by positively selected thymocytes crucially regulate mTEC development and medulla formation.

We recently identified that positive selection promotes mTEC proliferation and thereby nurtures the formation of thymic medulla (87). We found that positively selected thymocytes express RANKL, a TNFSF member ligand (87). The number of mTEC is reduced in mice deficient for RANKL, and the enforced expression of RANKL in mice deficient for positive selection restores mTEC cellularity and medulla formation, indicating that RANKL mediates thymic crosstalk signals for the optimal formation of thymic medulla (87). Osteoprotegerin, a naturally occurring soluble RANKL inhibitor that binds to RANKL and inhibits RANKL binding to its signaling receptor RANK, is expressed by mTEC and regulates mTEC development and medulla formation, because mice deficient for osteoprotegerin exhibit hypercellularity of mTEC and an enlarged thymic medulla (87). The RANKL receptor, RANK, is more strongly expressed in mTEC than in cTEC and is required for the development of mTEC expressing AIRE (87, 88). RANKL is also expressed by $CD4^{+}$ $CD3^{-}$ lymphoid tissue inducer (LTi) cells and is involved in the differentiation of mTEC during embryogenesis (88–90).

Positively selected thymocytes also produce CD40L, another TNFSF member ligand (87, 91). In contrast to RANKL, CD40L is not expressed in thymic LTi cells (87). The enforced expression of CD40L in vivo produces an enlarged thymic medulla, suggesting that CD40L signals can promote thymic medulla formation (92, 93). Analysis using fetal thymus organ culture shows that CD40L as well as RANKL facilitates mTEC development through classical and nonclassical NF-κB pathways (94). Mice deficient for both CD40 and RANKL exhibit severe defect in mTEC development and thymic medulla formation, whereas the single deficiency of CD40 causes only a mild defect in thymic medulla (94). Thus, it is likely that RANKL and CD40 cooperate with each other to promote mTEC development. It is also possible that RANKL and CD40 may sequentially regulate mTEC development; RANKL produced by LTi cells plays a role in mTEC development during embryogenesis, whereas RANKL and CD40L produced by positively selected mature thymocytes essentially promote postnatal increase of mTEC cellularity (87, 94).

Concluding Remarks

The fact that the thymus contains at least two distinct microenvironments, the cortex and the medulla, has been known for more than 100 years (95–98). Studies conducted over a span of 40 years since the discovery of the immunological function of the thymus have remarkably advanced our understanding of lymphocyte biology in relationship to T-lymphocyte development and selection in the thymus. However, the developmental and molecular biology of thymic stromal cells has remained quite vague until recently. Recent achievements have enabled us to study the biology of thymic microenvironments using several molecules that represent key functions of thymic stromal cells. Those molecules include Foxn1, Delta-like ligands, AIRE, and β5t. Several molecules that mediate the development of thymic stromal cells, such as RANKL and CD40L, have also been identified. These outcomes should provide a solid foundation for further studies of thymic microenvironments to better understand the molecular mechanism of the development and repertoire formation of T cells and the therapeutic reconstitution of functional thymus for various clinical situations (99, 100). Unveiling of the molecular identities of thymic microenvironments has just begun.

Acknowledgments The support by Uehara Memorial Foundation and MEXT was acknowledged. T.N. was a JSPS Research Fellow.

References

1. Miller, J. F. A. P. (1961) Immunological function of the thymus. Lancet 2:748–749.
2. Sainte-Marie, G. and Leblond, C. P. (1964) Cytologic features and cellular migration in the cortex and medulla of thymus in the young adult rat. Blood 23:275–299.
3. Kingston, R., Jenkinson, E. J., and Owen, J. J. (1985) A single stem cell can recolonize an embryonic thymus, producing phenotypically distinct T-cell populations. Nature (Lond) 317:811–813.
4. von Boehmer, H. (1988) The developmental biology of T lymphocytes. Annu. Rev. Immunol. 6:309–326
5. Jenkinson, E. J., Owen, J. J., and Aspinall, R. (1980). Lymphocyte differentiation and major histocompatibility complex antigen expression in the embryonic thymus. Nature (Lond) 284:177–179.
6. Le Douarin, N. M. and Jotereau, F. V. (1975) Tracing of cells of the avian thymus through embryonic life in interspecific chimeras. J. Exp. Med. 142:17–40.
7. Bhandoola, A., Sambandam, A., Allman, D., Meraz, A., and Schwarz, B. (2003) Early T lineage progenitors: new insights, but old questions remain. J. Immunol. 171:5653–5658.
8. Scollay, R., Wilson, A., D'Amico, A., Kelly, K., Egerton, M., Pearse, M., Wu, L., and Shortman, K. (1988) Developmental status and reconstitution potential of subpopulations of murine thymocytes. Immunol. Rev. 104:81–120.
9. Lind, E. F., Prockop, S. E., Porritt, H. E., and Petrie, H. T. (2001). Mapping precursor movement through the postnatal thymus reveals specific microenvironments supporting defined stages of early lymphoid development. J. Exp. Med. 194, 127-134.
10. Wilson, A., Petrie, H. T., Scollay, R. and Shortman, K. (1989). The acquisition of CD4 and CD8 during the differentiation of early thymocytes in short-term culture. Int. Immunol. 1:605–612.
11. Witt, C. M., Raychaudhuri, S., Schaefer, B., Chakraborty, A. K., and Robey, E. A. (2005) Directed migration of positively selected thymocytes visualized in real time. PLoS Biol. 3:e160.
12. Li, J., Iwanami, N., Hoa, V. Q., Furutani-Seiki, M., and Takahama, Y. (2007) Noninvasive intravital imaging of thymocyte dynamics in medaka. J. Immunol. 179:1605–1615.
13. von Boehmer, H. (1994) Positive selection of lymphocytes. Cell 76:219–228.

14. Allen, P. M. (1994) Peptides in positive and negative selection: a delicate balance. Cell 76:593–596.
15. Ashton-Rickardt, P. G., Van Kaer, L., Schumacher, T. N., Ploegh, H. L., and Tonegawa, S. (1993) Peptide contributes to the specificity of positive selection of CD8+ T cells in the thymus. Cell 73:1041–1049.
16. Takahama, Y., Suzuki, H., Katz, K. S., Grusby, M.J., and Singer, A. (1994) Positive selection of CD4+ T cells by TCR ligation without aggregation even in the absence of MHC. Nature (Lond) 371:67–70.
17. Ashton-Rickardt, P. G. and Tonegawa, S. (1994) A differential-avidity model for T-cell selection. Immunol. Today 15:362–366.
18. Sebzda, E., Wallace, V. A., Mayer, J., Yeung, R.S., Mak, T. W., and Ohashi, P. S. (1994) Positive and negative thymocyte selection induced by different concentrations of a single peptide. Science 263:1615–1618.
19. Singer, A. (2002) New perspectives on a developmental dilemma: the kinetic signaling model and the importance of signal duration for the CD4/CD8 lineage decision. Curr. Opin. Immunol. 14:207–215.
20. Takahama, Y. (2006) Journey through the thymus: stromal guides for T-cell development and selection. Nat. Rev. Immunol. 6:127–135.
21. Klein, L. and Kyewski, B. (2000) "Promiscuous" expression of tissue antigens in the thymus: a key to T-cell tolerance and autoimmunity? J. Mol. Med. 78:483–494.
22. Anderson, M. S., Venanzi, E. S., Klein, L., Chen, Z., Berzins, S. P., Turley, S. J., von Boehmer, H., Bronson, R., Dierich, A., Benoist, C., and Mathis, D. (2002) Projection of an immunological self shadow within the thymus by the aire protein. Science 298:1395–1401.
23. Boyd, R. L., Tucek, C. L., Godfrey, D. I., Izon, D. J., Wilson, T. J., Davidson, N. J., Bean, A. G., Ladyman, H. M., Ritter, M. A., and Hugo, P. (1993) The thymic microenvironment. Immunol. Today. 14:445–459.
24. Manley, N. R. and Blackburn, C. C. (2003) A developmental look at thymus organogenesis: where do the non-hematopoietic cells in the thymus come from? Curr. Opin. Immunol. 15:225–232.
25. Rossi, S. W., Jenkinson, W. E., Anderson, G., and Jenkinson, E. J. (2006) Clonal analysis reveals a common progenitor for thymic cortical and medullary epithelium. Nature (Lond) 441:988–991.
26. Bleul, C. C., Corbeaux, T., Reuter, A., Fisch, P., Monting, J. S., and Boehm, T. (2006) Formation of a functional thymus initiated by a postnatal epithelial progenitor cell. Nature (Lond) 441:992–996.
27. Lindsay, E. A., Vitelli, F., Su, H., Morishima, M., Huynh, T., Pramparo, T., Jurecic, V., Ogunrinu, G., Sutherland, H. F., Scambler, P. J., Bradley, A., and Baldini, A. (2001) Tbx1 haploinsufficiency in the DiGeorge syndrome region causes aortic arch defects in mice. Nature (Lond) 410:97–101.
28. Manley, N. R. and Capecchi, M. R. (1998) Hox group 3 paralogs regulate the development and migration of the thymus, thyroid, and parathyroid glands. Dev. Biol. 195:1–15.
29. Su, D. M. and Manley, N. R. (2000) Hoxa3 and pax1 transcription factors regulate the ability of fetal thymic epithelial cells to promote thymocyte development. J. Immunol. 164:5753–5760.
30. Nehls, M., Pfeifer, D., Schorpp, M., Hedrich, H., and Boehm, T. (1994) New member of the winged-helix protein family disrupted in mouse and rat nude mutations. Nature (Lond) 372:103–107.
31. Anderson, G., Jenkinson, E. J., Moore N. C., and Owen, J. J. T. (1993) MHC class II-positive epithelium and mesenchyme cells are both required for T-cell development in the thymus. Nature (Lond) 362:70–73.
32. Jenkinson, W. E., Jenkinson, E. J., and Anderson, G. (2003) Differential requirement for mesenchyme in the proliferation and maturation of thymic epithelial progenitors. J. Exp. Med. 198:325–332.
33. Shores, E.W., Van Ewijk, W., and Singer, A. (1991) Disorganization and restoration of thymic medullary epithelial cells in T cell receptor-negative scid mice: evidence that receptor-bearing lymphocytes influence maturation of the thymic microenvironment. Eur. J. Immunol. 21:1657–1661.

34. Ritter, M. A. and Boyd, R. L. (1993) Development in the thymus: it takes two to tango. Immunol. Today 14:462–469.
35. van Ewijk, W., Shores, E. W., and Singer, A. (1994) Crosstalk in the mouse thymus. Immunol. Today 15:214–217.
36. Starr, T. K., Jameson, S. C., and Hogquist, K. A. (2003) Positive and negative selection of T cells. Annu. Rev. Immunol. 21:139–176.
37. Ignatowicz, L., Kappler, J., and Marrack, P. (1996) The repertoire of T cells shaped by a single MHC/peptide ligand. Cell 84:521–529.
38. Huseby, E. S., White, J., Crawford, F., Vass, T., Becker, D., Pinilla, C., Marrack, P., and Kappler, J. W. (2005) How the T cell repertoire becomes peptide and MHC specific. Cell 122:247–260.
39. Pawlowski, T., Elliott, J. D., Loh, D. Y., and Staerz, U. D. (1993) Positive selection of T lymphocytes on fibroblasts. Nature (Lond) 364:642–645.
40. Hugo, P., Kappler, J. W., McCormack, J. E., and Marrack, P. (1993) Fibroblasts can induce thymocyte positive selection in vivo. Proc. Natl. Acad. Sci. U. S. A. 90:10335–10339.
41. Yasutomo, K., Lucas, B., and Germain, R. N. (2000) TCR signaling for initiation and completion of thymocyte positive selection has distinct requirements for ligand quality and presenting cell type. J. Immunol. 165:3015–3022.
42. Li, W., Kim, M. G., Gourley, T. S., McCarthy, B.P., Sant'Angelo, D. B., and Chang, C. H. (2005) An alternate pathway for CD4 T cell development: thymocyte-expressed MHC class II selects a distinct T cell population. Immunity 23:375–386.
43. Choi, E. Y., Jung, K. C., Park, H. J., Chung, D. H., Song, J. S., Yang, S. D., Simpson, E., and Park, S. H. (2005) Thymocyte–thymocyte interaction for efficient positive selection and maturation of CD4 T cells. Immunity 23:387–396.
44. Horai, R., Mueller, K. L., Handon, R. A., Cannons, J. L., Anderson, S. M., Kirby, M. R., and Schwartzberg, P. L. (2007) Requirements for selection of conventional and innate T lymphocyte lineages. Immunity 27:775–785.
45. Kirberg, J., Bosco, N., Deloulme, J. C., Ceredig, R., and Agenes, F. (2008) Peripheral T lymphocytes recirculating back into the thymus can mediate thymocyte positive selection. J. Immunol. 181:1207–1214.
46. Murata, S., Sasaki, K., Kishimoto, T., Niwa, S., Hayashi, H., Takahama, Y., and Tanaka, K. (2007) Regulation of $CD8^+$ T cell development by thymus-specific proteasomes. Science 316:1349–1353.
47. Rock, K. L. and Goldberg, A. L. (1999) Degradation of cell proteins and the generation of MHC class I-presented peptides. Annu. Rev. Immunol. 17:739–779.
48. Kloetzel, P. M. (2001) Antigen processing by the proteasome. Nat. Rev. Mol. Cell. Biol. 2:179–187.
49. Coux, O., Tanaka, K., and Goldberg, A. L. (1996) Structure and functions of the 20S and 26S proteasomes. Annu. Rev. Biochem. 65:801–847.
50. Baumeister, W., Walz, J., Zühl, F., and Seemüller, E. (1998) The proteasome: paradigm of a self-compartmentalizing protease. Cell 92:367–380.
51. Tanaka, K. and Kasahara, M. (1998) The MHC class I ligand-generating system: roles of immunoproteasomes and the interferon-gamma-inducible proteasome activator PA28. Immunol. Rev. 163:161–176.
52. Cascio, P., Hilton, C., Kisselev, A. F., Rock, K. L., and Goldberg, A.L. (2001) 26S proteasomes and immunoproteasomes produce mainly N-extended versions of an antigenic peptide. EMBO J. 20:2357–2366.
53. Rock, K. L., York, I. A., and Goldberg, A. L. (2004) Post-proteasomal antigen processing for major histocompatibility complex class I presentation. Nat. Immunol. 5:670–677.
54. Young, A. C., Nathenson, S. G., and Sacchettini, J. C. (1995) Structural studies of class I major histocompatibility complex proteins: insights into antigen presentation. FASEB J. 9:26–36.
55. Fehling, H. J., Swat, W., Laplace, C., Kühn, R., Rajewsky, K., Müller, U., and von Boehmer, H. (1994) MHC class I expression in mice lacking the proteasome subunit LMP-7. Science 265:1234–1237.

56. Murata, S., Takahama, Y., and Tanaka, K. (2008) Thymoproteasome: probable role in generating positively selecting peptides. Curr. Opin. Immunol. 20(2):192–196.
57. Takahama, Y., Tanaka, K., and Murata, S. (2008) Modest cortex and promiscuous medulla for thymic repertoire formation. Trends Immunol. 29:251–255.
58. Honey, K. and Rudensky, A. Y. (2003) Lysosomal cysteine proteases regulate antigen presentation. Nat. Rev. Immunol. 3:472–482.
59. Nakagawa, T., Roth, W., Wong, P., Nelson, A., Farr, A., Deussing, J., Villadangos, J. A., Ploegh, H., Peters, C., and Rudensky, A. Y. (1998) Cathepsin L: critical role in Ii degradation and CD4 T cell selection in the thymus. Science 280:450–453.
60. Honey, K., Nakagawa, T., Peters, C., and Rudensky, A. (2002) Cathepsin L regulates $CD4^+$ T cell selection independently of its effect on invariant chain: a role in the generation of positively selecting peptide ligands. J. Exp. Med. 195:1349–1358.
61. Bousso, P., Bhakta, N. R., Lewis, R. S., and Robey, E. (2002) Dynamics of thymocyte-stromal cell interactions visualized by two-photon microscopy. Science 296:1876–1880.
62. Bhakta, N. R., Oh, D. Y., and Lewis, R. S. (2005) Calcium oscillations regulate thymocyte motility during positive selection in the three-dimensional thymic environment. Nat. Immunol. 6:143–151.
63. Ngo, V. N., Tang, H. L., and Cyster, J. G. (1998) Epstein-Barr virus-induced molecule 1 ligand chemokine is expressed by dendritic cells in lymphoid tissues and strongly attracts naive T cells and activated B cells. J. Exp. Med. 188:181–191.
64. Campbell, J. J., Pan, J., and Butcher, E. C. (1999) Developmental switches in chemokine responses during T cell maturation. J. Immunol. 163:2353–2357.
65. Ueno, T., Saito, F., Gray, D. H., Kuse, S., Hieshima, K., Nakano, H., Kakiuchi, T., Lipp, M., Boyd, R. L., and Takahama, Y. (2004) CCR7 signals are essential for cortex-medulla migration of developing thymocytes. J. Exp. Med. 200:493–505.
66. Kurobe, H., Liu, C., Ueno, T., Saito, F., Ohigashi, I., Seach, N., Arakaki, R., Hayashi, Y., Kitagawa, T., Lipp, M., Boyd, R. L., and Takahama, Y. (2006) CCR7-dependent cortex-to-medulla migration of positively selected thymocytes is essential for establishing central tolerance. Immunity 24:165–177.
67. Höpken, U. E., Wengner, A. M., Loddenkemper, C., Stein, H., Heimesaat, M. M., Rehm, A., and Lipp, M. (2007) CCR7 deficiency causes ectopic lymphoid neogenesis and disturbed mucosal tissue integrity. Blood 109:886–895.
68. Davalos-Misslitz, A. C. M, Rieckenberg, J., Willenzon, S., Worbs, T., Kremmer, E., Bernhardt, G., and Förster R. (2007) Generalized multi-organ autoimmunity in CCR7-deficient mice. Eur. J. Immunol. 37:613–622.
69. Rodewald, H. R., Paul, S., Haller, C., Bluethmann, H., and Blum, C. (2001) Thymus medulla consisting of epithelial islets each derived from a single progenitor. Nature (Lond) 414:763–768.
70. Weih, F., Carrasco, D., Durham, S. K., Barton, D. S., Rizzo, C. A., Ryseck, R. P., Lira, S. A., and Bravo, R. (1995) Multiorgan inflammation and hematopoietic abnormalities in mice with a targeted disruption of RelB, a member of the NF-kB/Rel family. Cell 80:331–340.
71. DeKoning, J., DiMolfetto, L., Reilly, C., Wei, Q., Havran, W., and Lo, D. (1997) Thymic cortical epithelium is sufficient for the development of mature T cells in relB-deficient mice. J. Immunol. 158:2558–2566.
72. Boehm, T., Scheu, S., Pfeffer, K., and Bleul, C. C. (2003) Thymic medullary epithelial cell differentiation, thymocyte emigration, and the control of autoimmunity require lympho-epithelial cross talk via LTbR. J. Exp. Med. 198:757–769.
73. Heath, V. L., Moore, N. C., Parnell, S. M., and Mason, D.W. (1998) Intrathymic expression of genes involved in organ specific autoimmune disease. J. Autoimmun. 11:309–318.
74. Klein, L., Klein, T., Rüther, U., and Kyewski, B. (1998) CD4 T cell tolerance to human C-reactive protein, an inducible serum protein, is mediated by medullary thymic epithelium. J. Exp. Med. 188:5–16.
75. Sospedra, M., Ferrer-Francesch, X., Domínguez, O., Juan, M., Foz-Sala, M., and Pujol-Borrell, R. (1998) Transcription of a broad range of self-antigens in human thymus suggests a role for central mechanisms in tolerance toward peripheral antigens. J. Immunol. 161:5918–5929.

76. Werdelin, O., Cordes, U., and Jensen, T. (1998) Aberrant expression of tissue-specific proteins in the thymus: a hypothesis for the development of central tolerance. Scand. J. Immunol. 47:95–100.
77. Klein, L., Klugmann, M., Nave, K. A., Tuohy, V.K., and Kyewski, B. (2000) Shaping of the autoreactive T-cell repertoire by a splice variant of self protein expressed in thymic epithelial cells. Nat. Med. 6:56–61.
78. Derbinski, J., Schulte, A., Kyewski, B., and Klein, L. (2001) Promiscuous gene expression in medullary thymic epithelial cells mirrors the peripheral self. Nat. Immunol. 2:1032–1039.
79. Heino, M., Peterson, P., Kudoh, J., Nagamine, K., Lagerstedt, A., Ovod, V., Ranki, A., Rantala, I., Nieminen, M., Tuukkanen, J., Scott, H. S., Antonarakis, S. E., Shimizu, N., and Krohn, K. (1999) Autoimmune regulator is expressed in the cells regulating immune tolerance in thymus medulla. Biochem. Biophys. Res. Commun. 257:821–825.
80. Heino, M., Peterson, P., Sillanpää, N., Guérin, S., Wu, L., Anderson, G., Scott, H. S., Antonarakis, S. E., Kudoh, J., Shimizu, N., Jenkinson, E. J., Naquet, P., and Krohn, K. J. (2000) RNA and protein expression of the murine autoimmune regulator gene (Aire) in normal, RelB-deficient and in NOD mouse. Eur. J. Immunol. 30:1884–1893.
81. Zuklys, S., Balciunaite, G., Agarwal, A., Fasler-Kan, E., Palmer, E., and Holländer, G. A. (2000) Normal thymic architecture and negative selection are associated with Aire expression, the gene defective in the autoimmune-polyendocrinopathy-candidiasis-ectodermal dystrophy (APECED) J. Immunol. 165:1976–1983.
82. Derbinski, J., Gäbler, J., Brors, B., Tierling, S., Jonnakuty, S., Hergenhahn, M., Peltonen, L., Walter, J., and Kyewski, B. (2005) Promiscuous gene expression in thymic epithelial cells is regulated at multiple levels. J. Exp. Med. 202:33–45.
83. Philpott, K. L., Viney, J. L., Kay, G., Rastan, S., Gardiner, E. M., Chae, S., Hayday, A. C., and Owen, M. J. (1992) Lymphoid development in mice congenitally lacking T cell receptor $\alpha\beta$-expressing cells. Science 256:1448–1452.
84. Surh, C. D., Ernst, B., and Sprent, J. (1992) Growth of epithelial cells in the thymic medulla is under the control of mature T cells. J. Exp. Med. 176:611–616.
85. Negishi, I., Motoyama, N., Nakayama, K., Nakayama, K., Senju, S., Hatakeyama, S., Zhang, Q., Chan, A. C., and Loh, D. Y. (1995) Essential role for ZAP-70 in both positive and negative selection of thymocytes. Nature (Lond) 376:435–438.
86. Nasreen, M., Ueno, T., Saito, F., and Takahama, Y. (2003) In vivo treatment of class II MHC-deficient mice with anti-TCR antibody restores the generation of circulating CD4 T cells and optimal architecture of thymic medulla. J. Immunol. 171:3394–3400.
87. Hikosaka, Y., Nitta, T., Ohigashi, I., Yano, K., Ishimaru, N., Hayashi, Y., Matsumoto, M., Matsuo, K., Penninger, J. M., Takayanagi, H., Yokota, Y., Yamada, H., Yoshikai, Y., Inoue, J., Akiyama, T., and Takahama, Y. (2008) The cytokine RANKL produced by positively selected thymocytes fosters medullary thymic epithelial cells that express autoimmune regulator. Immunity 29(3):438–450.
88. Rossi, S. W., Kim, M. Y., Leibbrandt, A., Parnell, S. M., Jenkinson, W. E., Glanville, S.H., McConnell, F. M., Scott, H. S., Penninger, J. M., Jenkinson, E. J., Lane, P. J., and Anderson, G. (2007) RANK signals from $CD4^{+}3^{-}$ inducer cells regulate development of Aire-expressing epithelial cells in the thymic medulla. J. Exp. Med. 204:1267–1272.
89. Anderson, G., Lane, P. J., and Jenkinson, E. J. (2007) Generating intrathymic microenvironments to establish T-cell tolerance. Nat. Rev. Immunol. 7:954–963.
90. White, A. J., Withers, D. R., Parnell, S. M., Scott, H. S., Finke, D., Lane, P. J., Jenkinson, E. J., and Anderson, G. (2008) Sequential phases in the development of Aire-expressing medullary thymic epithelial cells involve distinct cellular input. Eur. J. Immunol. 38:942–947.
91. Fuleihan, R., Ahern, D., and Geha, R. S. (1995) CD40 ligand expression is developmentally regulated in human thymocytes. Clin. Immunol. Immunopathol. 76:52–58.
92. Dunn, R. J., Luedecker, C. J., Haugen, H. S., Clegg, C. H., and Farr, A. G. (1997) Thymic overexpression of CD40 ligand disrupts normal thymic epithelial organization. J. Histochem. Cytochem. 45:129–141.
93. Clegg, C. H., Rulffes, J. T., Haugen, H. S., Hoggatt, I. H., Aruffo, A., Durham, S. K., Farr, A. G., and Hollenbaugh, D. (1997) Thymus dysfunction and chronic inflammatory disease in gp39 transgenic mice. Int. Immunol. 9:1111–1122.

94. Akiyama, T., Shimo, Y., Yanai, H., Qin, K., Ohshima, D., Maruyama, Y., Asaumi, Y., Kitazawa, J., Takayanagi, T., Penninger, J. M., Matsumoto, M., Nitta, T., Takahama, Y., and Inoue, J. (2008) The tumor necrosis factor family receptors RANK and CD40 cooperatively establish the thymic medullary microenvironment and self-tolerance. Immunity 29:423–437.
95. Hassall, A. H. (1849) The Microscopic Anatomy of the Human Body, in Health and Disease. Highley, London.
96. Symington, J. (1898) The thymus gland in the marsupialia. J. Anat. Physiol. 32(Pt 2):278–291.
97. Lewis, T. (1904) Observations upon the distribution and structure of haemolymph glands in mammalia and aves, including a preliminary note on the thymus. J. Anat. Physiol. 38(Pt 3):312–324.
98. Goodall, A. (1905) The post-natal changes in the thymus of guinea-pigs, and the effect of castration on thymus structure. J. Physiol. 32:191–198.
99. van den Brink, M. R., Alpdogan, O., Boyd, R. L. (2004) Strategies to enhance T-cell reconstitution in immunocompromised patients. Nat. Rev. Immunol. 4:856–867.
100. Gray, D. H., Ueno, T., Chidgey, A. P., Malin, M., Goldberg, G. L., Takahama, Y., and Boyd, R. L. (2005) Controlling the thymic microenvironment. Curr. Opin. Immunol. 17:137–143.

The Mechanisms of Motor Programming for Learned Vocalization in Songbirds

Hisataka Fujimoto, Taku Hasegawa, Ryosuke Matsui, Kentaro Abe, and Dai Watanabe

Introduction

Vocal signals, the fundamental interface for communication, convey information in their acoustic and temporal structures. Neural systems must extract the information from complex sequences of sound elements and, in turn, generate the motor commands for vocal sequences. Because humans acquire vocal communication after birth, the neural systems must establish a precise correspondence between sensory and motor representations of the vocal signals and control vocal behavior under the guidance of auditory information [1]. Recent advances in noninvasive technologies, such as functional magnetic resonance imaging (MRI) and positron emission tomography, have revealed brain areas involved in human vocal communication and imaged the metabolic signals reflecting their auditory–vocal association [2]. However, little is known about the neural processing that underlies sensory–motor integration for the learned vocal signals.

Songbirds, a diverse group of avian species classified as the order Passeriformes, provide an attractive model system. These avian species, much like humans, learn to imitate complex vocal signals of their elders during early life and maintain their spectral and temporal structures throughout their whole life [3–6]. Toward the understanding of the brain function to enable learned vocal communication, recent studies investigating the neural basis of songbird vocal behavior have been addressing a key issue: How do individual neurons establish sensory–motor correspondence for learned communication signals?

H. Fujimoto, T.Hasegawa, R.Matsui, K.Abe, and D.Watanabe
Department of Biological Sciences, Faculty of Medicine, and Department of Molecular and System Biology, Graduate School of Biostudies, Kyoto University, Yoshida, Sakyo-ku, Kyoto 606-8501, Japan

S. Nakanishi et al. (eds.), *Systems Biology*,
DOI:10.1007/978-4-431-87704-2_8, © Springer 2009

Neural Substrates for Learned Vocal Communication in Songbirds

In the songbird brain circuits, two subsystems play significant roles in the perception and expression of the learned vocal signals [7]. One subsystem is the telencephalic auditory system that comprises the primary and secondary auditory areas [8] (Fig. 1). Another subsystem is the song control system, a motor system specialized for learned vocalization [9], including the motor control pathway critical for vocal patterning [9,10] and the basal ganglia pathway necessary for audition-dependent vocal plasticity [11–13] (Fig. 1). The nucleus HVC (used as a proper name; formerly called the high vocal center), the hierarchically highest telencephalic motor area in the song control system, links the motor control pathway with the basal ganglia pathway (Fig. 1). Importantly, HVC receives input indirectly from the primary and secondary auditory areas [8,14] (Fig. 1). HVC could, therefore, serve essential functions as a sensorimotor integration site for vocal communication and as a gateway for auditory information flow into the song control system.

BOS Selectivity and Auditory Gating in the Vocal Control System

Because auditory feedback is required both before and after establishment of song vocalization [3,13,15–17], the neurons sensitive to self-generated vocalization could play an important role in learning and maintenance of the vocal signals [18].

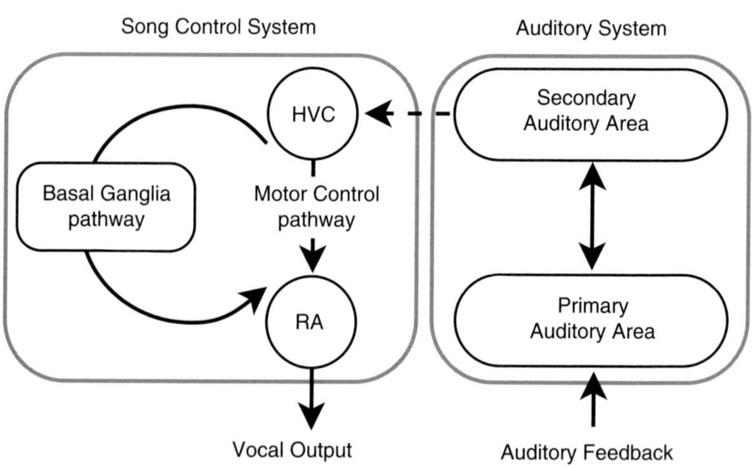

Fig. 1 Simplified diagram of the avian auditory system and song control system (*RA*, the robust nucleus of arcopallium). Not all nuclei of the song control system are shown

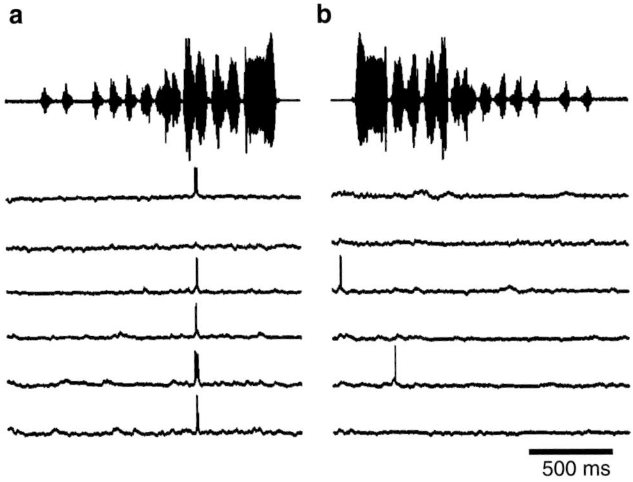

Fig. 2 Individual bird's own song (BOS)-selective response in the song control system. Auditory responses of a HVC neuron were recorded intracellularly under anesthetized states. Stereotyped time-locked spike bursts were evoked in the presence of BOS (**a**), whereas not in the presence of reversed song (**b**). *Upper traces* and the *lower six traces* represent auditory stimuli and membrane potentials, respectively

Neural recordings from an anesthetized songbird have revealed that neurons in the song control system display highly selective auditory response to the individual bird's own song (BOS) [19–24] (Fig. 2). Although neurons in the primary and secondary auditory areas also show activity in response to BOS playback [23,25], they preferentially respond to the conspecific song sequences rather than to the individual BOS [25,26]. At present, the precise origin of BOS selectivity remains unclear, but a distinct enhancement of BOS selectivity begins to emerge at the level of HVC in the song control system [23,27] (Fig. 2). The presence of the auditory selectivity to the self-generated vocal signals in HVC is consistent with the idea that HVC plays a critical part in the modification of motor commands for vocal signals under the guidance of auditory feedback.

Generally, single-neuron sensory responses can change depending on the states of the neural networks, reflecting the behavioral states of the animals [28]. It is, therefore, important to investigate the effects of behavioral state changes on auditory responses in the song control system. A remarkable state-dependency of the auditory response in the song control system was first demonstrated in HVC [29]; auditory responses of HVC neurons observed in anesthetized or sleep states are greatly suppressed to the same stimuli under awake unrestrained conditions [29]. Although mechanisms and functions of the state-dependent gating of auditory responses remain unknown, the absence of state-dependent modulation in the primary or secondary auditory areas indicates that auditory signaling is elaborately regulated in HVC [29], acting as the junction point between the auditory and song control system (Fig. 1). In addition, these findings emphasize the importance of studying properties of high-order integrative neural systems, such as HVC, carefully under natural conditions [29–35].

Recording Neural Activity in Songbirds Under Natural Conditions

Characterizing the activities of individual neurons involved in songbird vocal communication poses some technical challenges. Song vocalization is a natural social behavior, and songbirds will sing only under unrestrained conditions. The device for measuring neural activities should be compact and lightweight so as not to disturb their communication behavior. Moreover, the positioning of electrodes should be remotely and flexibly manipulated to isolate single-neuron signals.

To overcome the difficulties in recording, various recording devices have been developed [30,32,35,36]. We have devised a miniature motorized microdrive system, whose basic design was previously described [36]. In brief, the microdrive system, which weighs less than 1.5 g, includes three motorized positioners and headstage preamplifiers. The individual electrodes are held by shuttles that travel along motor shafts. Each motor is equipped with a planetary gear system, whose reduction ratio is 337 to 1. As the result, the output torque is sufficient to drive shuttles, and the motion of each electrode is controlled with a positional resolution of less than 0.2 μm.

The Auditory–Vocal Association in HVC Neurons

On the basis of anatomical studies, HVC neurons are classified into three types: projection neurons that innervate telencephalic motor area, the robust nucleus of arcopallium (RA) (HVC_{RA} neurons) [9], projection neurons that extend their axons into basal ganglia area X (HVC_X neurons) [9], and interneurons.

Both HVC_{RA} neurons and HVC_X neurons are almost silent in awake nonsinging state and generate brief bursts of firing sparsely during vocalization [32,35,37]. The bursts of these projection neurons are highly stereotyped, tightly time-locked to the song sequences [32,37], suggesting that the robustness of the song vocalization is most likely to reflect this remarkable feature of the neural ensemble in HVC. Although HVC_{RA} neurons display no response to the auditory stimuli [35,37], a substantial proportion of HVC_X neurons respond selectively to the individual BOS playback [35]. Furthermore, the firing patterns of BOS-selective response observed in these HVC_X neurons are nearly identical to the firing patterns when the birds vocalize the same song sequences [35]. This observation raises the possibility that activity during singing is the result of auditory feedback. However, disruption of auditory feedback does not alter the pattern of this singing-related activity [35]. This evidence indicates that the activity in BOS-selective HVC_X neurons during vocalization carries not auditory feedback but motor-related information. Because HVC_X neurons project to the basal ganglia (Fig. 1), as is necessary for audition-dependent vocal plasticity, the motor-related information represented in these neurons might be compared to the auditory feedback to improve vocal performance.

Both HVC_{RA} and HVC_{X} neurons are reciprocally interconnected with interneurons [38], which extend their axons locally within HVC. Because the activity and plasticity of the HVC_{RA} and HVC_{X} neurons are strictly regulated by GABA-mediated synaptic actions of these interneurons [38,39], it is of interest whether and how auditory information modulates GABA-mediated inhibition of the HVC interneurons. To address this question, we studied the interneuronal activity in HVC under awake unrestrained conditions using the microdrive system described above (Fujimoto, unpublished data).

HVC interneurons can be identified by their spontaneous activity in the awake nonsinging state [32,35,37]. A subclass of HVC interneurons displays tonic bursts of spiking in response to the BOS playback. The background spontaneous activities of these interneurons are greatly suppressed by auditory stimuli other than the individual BOS playback. Thus, the auditory responses observed in these HVC interneurons are BOS selective. In contrast to BOS-selective HVC_{X} neurons, the firing properties of these interneurons in singing behavior are quite different from those in sensory state. The BOS-selective interneurons become active before the onset of the song vocalization, and the burst firings continue until vocalization is terminated. BOS-selective HVC interneurons may, therefore, suppress the auditory information before onset of singing and act as a switch from the auditory state to the auditory-insensitive motor state.

Conclusions

Recent studies of songbird vocal communication have revealed that two types of neurons in the HVC, the telencephalic motor area for vocal behavior, display selective response to self-generated learned vocal signals. One is a subtype of projection neurons that transmit similar codes to the basal ganglia both in experiencing and in executing the learned vocal signals [35] (Fig. 1). The other is a subtype of interneurons that regulate local circuits differently between sensory phase and motor phase. Studies focusing on the connectivity, activity, and plasticity of these neuronal cell types will be helpful for the further understanding of the neural basis not only of songbird vocal behavior but also of human vocal communication.

Acknowledgments This work was supported by a grant from the Uehara Memorial Foundation.

References

1. Kuhl PK (1993) Developmental speech perception: implications for models of language impairment. Ann N Y Acad Sci 682:248–263.
2. Rizzolatti G, Craighero L (2004) The mirror-neuron system. Annu Rev Neurosci 27:169–192.
3. Konishi M (1965) The role of auditory feedback in the control of vocalization in the white-crowned sparrow. Z Tierpsychol 22:770–783.

4. Immelmann K (1969) Song development in the zebra finch and other estrildid finches. In: Hinde RA (ed) Bird Vocalizations. Cambridge University Press, Cambridge.
5. Doupe AJ, Kuhl PK (1999) Birdsong and human speech: common themes and mechanisms. Annu Rev Neurosci 22:567–631.
6. Tchernichovski O, Mitra PP, Lints T, Nottebohm F (2001) Dynamics of the vocal imitation process: how a zebra finch learns its song. Science 291:2564–2569.
7. Jarvis ED, Güntürkün O, Bruce L, Csillag A, Karten H, Kuenzel W, Medina L, Paxinos G, Perkel DJ, Shimizu T, Striedter G, Wild JM, Ball GF, Dugas-Ford J, Durand SE, Hough GE, Husband S, Kubikova L, Lee DW, Mello CV, Powers A, Siang C, Smulders TV, Wada K, White SA, Yamamoto K, Yu J, Reiner A, Butler AB, Avian Brain Nomenclature Consortium (2005) Avian brains and a new understanding of vertebrate brain evolution. Nat Rev Neurosci 6:151–159.
8. Reiner A, Perkel DJ, Bruce LL, Butler AB, Csillag A, Kuenzel W, Medina L, Paxinos G, Shimizu T, Striedter G, Wild M, Ball GF, Durand S, Güntürkün O, Lee DW, Mello CV, Powers A, White SA, Hough G, Kubikova L, Smulders TV, Wada K, Dugas-Ford J, Husband S, Yamamoto K, Yu J, Siang C, Jarvis ED, Gütürkün O, Avian Brain Nomenclature Forum (2004) Revised nomenclature for avian telencephalon and some related brainstem nuclei. J Comp Neurol 473:377–414.
9. Nottebohm F, Stokes TM, Leonard CM (1976) Central control of song in the canary, *Serinus canarius*. J Comp Neurol 165:457–486.
10. Vicario DS (1991) Organization of the zebra finch song control system: II. Functional organization of outputs from nucleus *Robustus archistriatalis*. J Comp Neurol 309:486–494.
11. Bottjer SW, Halsema KA, Brown SA, Miesner EA (1989) Axonal connections of a forebrain nucleus involved with vocal learning in zebra finches. J Comp Neurol 279:312–326.
12. Scharff C, Nottebohm F (1991) A comparative study of the behavioral deficits following lesions of various parts of the zebra finch song system: implications for vocal learning. J Neurosci 11:2896–2913.
13. Brainard MS, Doupe AJ (2000) Interruption of a basal ganglia-forebrain circuit prevents plasticity of learned vocalizations. Nature (Lond) 404:762–766.
14. Vates GE, Broome BM, Mello CV, Nottebohm F (1996) Auditory pathways of caudal telencephalon and their relation to the song system of adult male zebra finches. J Comp Neurol 366:613–642.
15. Leonardo A, Konishi M (1999) Decrystallization of adult birdsong by perturbation of auditory feedback. Nature (Lond) 399:466–470.
16. Sakata JT, Brainard MS (2006) Real-time contributions of auditory feedback to avian vocal motor control. J Neurosci 26:9619–9628.
17. Tumer EC, Brainard MS (2007) Performance variability enables adaptive plasticity of 'crystallized' adult birdsong. Nature (Lond) 450:1240–1244.
18. Konishi M (2004) The role of auditory feedback in birdsong. Ann N Y Acad Sci 1016:463–475.
19. McCasland JS, Konishi M (1981) Interaction between auditory and motor activities in an avian song control nucleus. Proc Natl Acad Sci U S A 78:7815–7819.
20. Doupe AJ, Konishi M (1991) Song-selective auditory circuits in the vocal control system of the zebra finch. Proc Natl Acad Sci U S A 88:11339–11343.
21. Margoliash D, Fortune ES (1992) Temporal and harmonic combination-sensitive neurons in the zebra finch's HVc. J Neurosci 12:4309–4326.
22. Vicario DS, Yohay KH (1993) Song-selective auditory input to a forebrain vocal control nucleus in the zebra finch. J Neurobiol 24:488–505.
23. Lewicki MS, Arthur BJ (1996) Hierarchical organization of auditory temporal context sensitivity. J Neurosci 16:6987–6998.
24. Dave AS, Yu AC, Margoliash D (1998) Behavioral state modulation of auditory activity in a vocal motor system. Science 282:2250–2254.
25. Grace JA, Amin N, Singh NC, Theunissen FE (2003) Selectivity for conspecific song in the zebra finch auditory forebrain. J Neurophysiol 89:472–487.

26. Theunissen FE, Amin N, Shaevitz SS, Woolley SM, Fremouw T, Hauber ME (2004) Song selectivity in the song system and in the auditory forebrain. Ann N Y Acad Sci 1016:222–245.
27. Janata P, Margoliash D (1999) Gradual emergence of song selectivity in sensorimotor structures of the male zebra finch song system. J Neurosci 19:5108–5118.
28. Fontanini A, Katz DB (2008) Behavioral states, network states, and sensory response variability. J Neurophysiol 100:1160–1168.
29. Schmidt MF, Konishi M (1998) Gating of auditory responses in the vocal control system of awake songbirds. Nat Neurosci 1:513–518.
30. Hessler NA, Doupe AJ (1999) Singing-related neural activity in a dorsal forebrain-basal ganglia circuit of adult zebra finches. J Neurosci 19:10461–10481.
31. Luo M, Fee MS, Katz LC (2003) Encoding pheromonal signals in the accessory olfactory bulb of behaving mice. Science 299:1196–1201.
32. Hahnloser RH, Kozhevnikov AA, Fee MS (2002) An ultra-sparse code underlies the generation of neural sequences in a songbird. Nature (Lond) 419:65–70.
33. Leonardo A (2004) Experimental test of the birdsong error-correction model. Proc Natl Acad Sci U S A 101:16935–16940.
34. Leonardo A, Fee MS (2005) Ensemble coding of vocal control in birdsong. J Neurosci 25:652–661.
35. Prather JF, Peters S, Nowicki S, Mooney R (2008) Precise auditory-vocal mirroring in neurons for learned vocal communication. Nature (Lond) 451:305–310.
36. Fee MS, Leonardo A (2001) Miniature motorized microdrive and commutator system for chronic neural recording in small animals. J Neurosci Methods 112:83–94.
37. Kozhevnikov AA, Fee MS (2007) Singing-related activity of identified HVC neurons in the zebra finch. J Neurophysiol 97:4271–4283.
38. Mooney R, Prather JF (2005) The HVC microcircuit: the synaptic basis for interactions between song motor and vocal plasticity pathways. J Neurosci 25:1952–1964.
39. Rosen MJ, Mooney R (2003) Inhibitory and excitatory mechanisms underlying auditory responses to learned vocalizations in the songbird nucleus HVC. Neuron 39:177–194.

Dynamic Model of the Basal Ganglia Functions and Movement Disorders

Atsushi Nambu

Introduction

For control of voluntary movements, interactions between the cerebral cortex and the basal ganglia through the corticobasal ganglia loop are essential. The basal ganglia receive cortical inputs, process the information, and send it back to the original cerebral cortex via the thalamus to assist cortical activity (Fig. 1). The internal segment of the globus pallidus (GPi) and the substantia nigra pars reticulata (SNr) are the output nuclei of the basal ganglia. On the other hand, the striatum is the input station and receives direct excitatory cortical inputs. In the current model of basal ganglia organization, the striatum projects to the output nuclei, via two major projection systems, the *direct* and *indirect* pathways [2]. The direct pathway arises from GABAergic striatal neurons containing substance P and projects monosynaptically to the GPi/SNr. The indirect pathway arises from GABAergic striatal neurons containing enkephalin and projects polysynaptically to the GPi/SNr by way of a sequence of connections involving the external segment of the globus pallidus (GPe) and subthalamic nucleus (STN).

Cortico-STN-GPi/SNr *Hyperdirect* Pathway

The cerebral cortex projects topographically to the STN, as well as to the striatum [8,17]. Thus, recent studies propose the idea that the STN is another input station of the basal ganglia, forming the cortico-STN-GPi/SNr *hyperdirect* pathway [19], whose conduction velocity is faster than those of the direct and indirect pathways (Fig. 1). Indeed, stimulation of the motor cortices induces an early, short-latency excitation, followed by an inhibition and a late excitation in monkey GPe, GPi, and

A. Nambu
Division of System Neurophysiology, National Institute for Physiological Sciences, and Department of Physiological Sciences, Graduate University for Advanced Studies (SOKENDAI), 38 Nishigonaka, Myodaiji, Okazaki 444-8585, Japan

S. Nakanishi et al. (eds.), *Systems Biology*,
DOI:10.1007/978-4-431-87704-2_9,

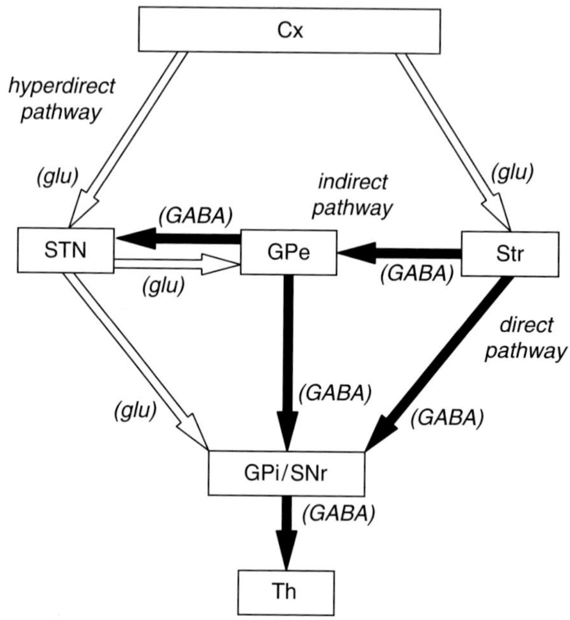

Fig. 1 Basic circuitry of the basal ganglia, comprising the Cx-STN-GPi/SNr *hyperdirect*, Cx-Str-GPi/SNr *direct*, and Cx-Str-GPe-STN-GPi/SNr *indirect* pathways. *Open* and *filled arrows* represent excitatory glutamatergic (*glu*) and inhibitory GABAergic (*GABA*) projections, respectively. *Cx*, cerebral cortex; *GPe*, external segment of the globus pallidus; *GPi*, internal segment of the globus pallidus; *SNr*, substantia nigra pars reticulata; *STN*, subthalamic nucleus; *Str*, striatum; *Th*, thalamus. (Modified from [19])

SNr neurons [12,18,21,23]. The early excitation in the GPi is derived from the cortico-STN-GPi/SNr hyperdirect pathway, whereas the inhibition and late excitation are mediated by the cortico-striato-GPi/SNr direct and cortico-striato-GPe-STN-GPi/SNr indirect pathways, respectively.

Activity of GPi Neurons

The GPi receives topographical inputs from the cerebral cortex through the hyperdirect, direct, and indirect pathways. GPi neurons in the forelimb region of the motor territory (ventral two-thirds of caudal GPi) change their activity, either inhibitory or excitatory, relating to forelimb movements [6,7,20,22]. These changes are presumably the result of the combination of inhibitory GABAergic inputs from the striatum, and excitatory glutamatergic inputs from the STN, which reflects the flow of information through the direct, indirect, and hyperdirect pathways. We do not know much about the extent of the contribution of each input to GPi activity during movements. However, an interesting finding is that GPi activity during voluntary limb movements always displays an increase rather than a decrease in discharge [6,7,20,22], suggesting a large contribution of inputs from the STN.

Dynamic Model of the Basal Ganglia Functions

GPi/SNr neurons are GABAergic and fire at high frequency; thus, neurons in their target structures, such as the thalamus, are inhibited continuously (Fig. 2). When striatal neurons are activated by cortical inputs, the striatal neurons inhibit GPi/SNr activity through the striato-GPi/SNr direct pathway. The continuous inhibition from the output nuclei to the target structures is transiently removed (disinhibition), and thalamic neurons are activated. The increased activity of the thalamus is transmitted to the motor cortex, and finally evokes a selected motor program. On the other hand, signals through the hyperdirect and indirect pathways have excitatory effects on the GPi/SNr and thus have inhibitory effects on thalamic neurons. Considering the respective axonal conduction velocities, signals through the hyperdirect pathway first actively inhibit thalamic neurons, then those through the direct pathway disinhibit them, and finally those through the indirect pathway inhibit them again. Thus, signals through the hyperdirect and indirect pathways make clear initiation and termination of the selected motor program [19].

In addition to such a temporal aspect, the enhancement by differential inputs through the hyperdirect, direct, and indirect pathways may work in a spatial domain as well. The *center-surround model* of basal ganglia functions proposes focused selection of an appropriate motor program and inhibition of competing motor

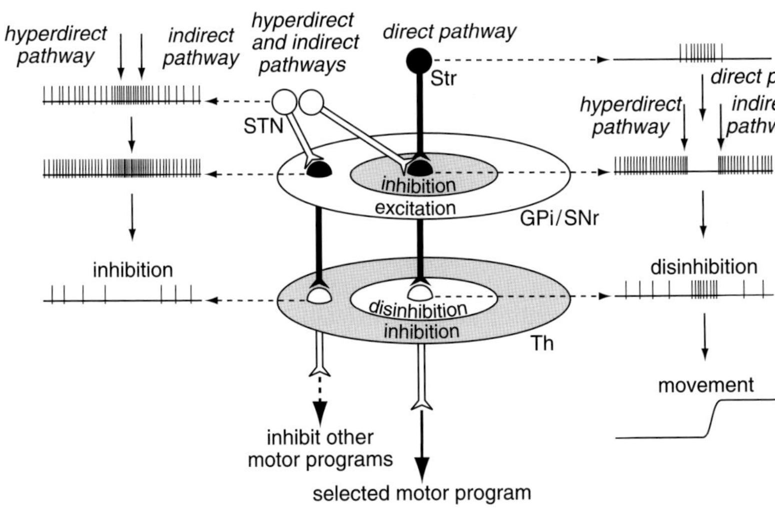

Fig. 2 Spatial and temporal distributions of basal ganglia activity during voluntary movements based on the dynamic model. Signals through the *direct pathway* inhibit GPi neurons in the center area, activate thalamic neurons by disinhibition, and finally release a selected motor program. On the other hand, signals through the *hyperdirect* and *indirect pathways* have broad excitatory effects on *GPi* neurons in temporal and spatial domains, make clear initiation and termination of the selected motor program, and inhibit other unnecessary competing motor programs. *Open* and *filled* neurons represent excitatory glutamatergic and inhibitory GABAergic neurons, repectively. *STN*, subthalamic nucleus; *Str*, striatum. (Modified from [16])

programs [10,14,15,19] (Fig. 2). Thalamic neurons are usually inhibited by continuous GABAergic inputs from the GPi. Signals through the direct pathway inhibit a specific group of GPi neurons in the center area, and thalamic and cortical neurons are released from inhibition and become active, resulting in the release of the selected motor program. On the other hand, signals through the hyperdirect and indirect pathways excite other groups of GPi neurons in the surrounding area and increase inhibition on thalamic neurons that mediate other competing motor programs. Inhibition in the center area of the GPi by striatal GABAergic inputs and excitation in the surrounding area by glutamatergic STN inputs coordinate the execution of the selected motor program and the inhibition of other competing motor programs.

Based on the *temporal* and *spatial* inputs to the target structures through the hyperdirect, direct, and indirect pathways, only the selected motor program is executed at the selected timing, and other competing motor programs are canceled.

Pathophysiology of Movement Disorders

The neural mechanisms underlying hypokinetic and hyperkinetic movement disorders are currently explained as changes in the *static* state of the basal ganglia by the standard model. The impaired balance of activity between the direct and indirect pathways causes an increase or decrease in the mean firing rate of GPi/SNr neurons [1,5]. For example, in Parkinson's disease, activity through the indirect pathway is increased and activity through the direct pathway is reduced. Both changes enhance the mean firing rate of GPi/SNr neurons, suppress the activity of thalamic and cortical neurons, and finally cause akinesia. On the other hand, in dystonia, activity through both direct and indirect pathways is increased, reduces the mean firing rate of GPi/SNr neurons, excites the activity of thalamic and cortical neurons by disinhibitory mechanism, and finally causes involuntary movements.

The dynamic model can better explain the pathophysiology of hypokinetic and hyperkinetic movement disorders by introducing the temporal and spatial aspects of the basal ganglia functions (Fig. 3). In normal monkeys, cortical stimulation evokes a triphasic response composed of an early excitation, an inhibition, and a late excitation in GPi neurons. In monkeys rendered parkinsonian by infusion of 1-methyl-4-phenyl-1,2,3,6-tetrahydropyridine (MPTP), the duration of the inhibition in GPi neurons evoked by cortical stimulation is reduced, and the amplitude of early and late excitations is increased. The intravenous injection of L-DOPA transiently improves parkinsonian akinesia, and the cortically evoked response pattern in the GPi is restored to the normal triphasic pattern. These observations suggest that in parkinsonian state, when a voluntary movement is about to be initiated by cortical mechanisms, signals through the hyperdirect and indirect pathways expand and suppress larger areas of the thalamus and cortex than in the normal state, and a signal through the direct pathway is reduced (Fig. 3, left). Thus, smaller areas of the thalamus and cortex are disinhibited for a shorter period of time than in the normal state, and not only the competing motor program, but also the selected motor program, cannot be released, resulting in the akinesia of Parkinson's disease.

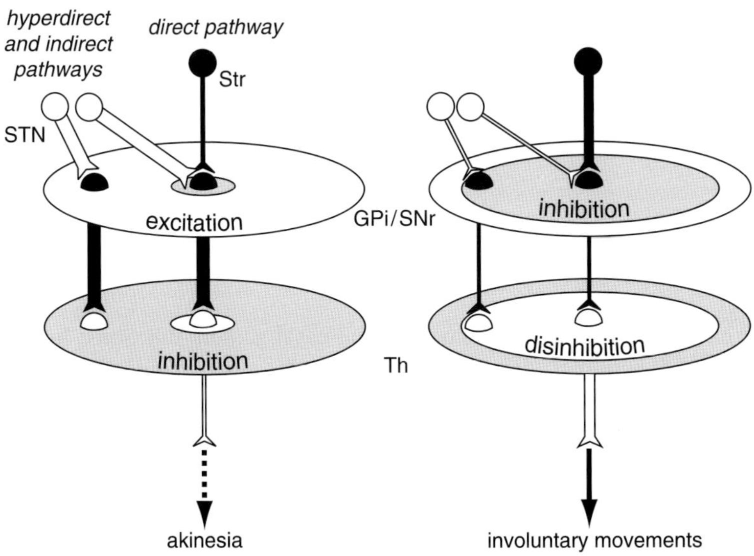

Fig. 3 Schematic diagram showing the activity changes in the basal ganglia during movement disorders. Increased signals through the *hyperdirect and indirect pathways* and a reduced signal through the *direct pathway* result in reduced disinhibition in the Th/Cx, leading to hypokinetic disorders (*left*). An increased signal through the direct pathway and reduced signals through the hyperdirect and indirect pathways result in increased disinhibition in the Th/Cx, leading to the involuntary movements observed in hyperkinetic disorders (*right*)

On the other hand, cortical stimulation produces long inhibitions in the GPe/GPi of monkeys during hemiballism, which is induced by STN blockade. In the transgenic mouse model of dystonia, cortical stimulation also induces a long-lasting inhibition in the GPe/GPi [4]. Thus, cortically evoked a long-lasting inhibition is considered to be a common feature of hyperkinetic disorders. Even a tiny neuronal activity originating in the cortex is transmitted through the basal ganglia circuitry and finally induces a strong and long-lasting inhibition in the wide areas of the GPi (Fig. 3, right). Therefore, the wide areas of the thalamus and cortex that mediate both the selected and competing motor programs are activated in an uncontrollable manner, resulting in the involuntary movements observed in hyperkinetic disorders.

Mechanism of the Effectiveness of Stereotaxic Surgery

Recent advances in stereotaxic surgery have highlighted the STN as a target structure. In monkeys and humans, a lesion or high-frequency stimulation (HFS) of the STN can ameliorate parkinsonian motor symptoms [3,9,11,13]. In our experiments, blockade of STN activity by the local injection of $GABA_A$ agonist muscimol also improves the motor activity of parkinsonian monkeys. The blockade of the STN diminishes the early and late excitations evoked by cortical stimulation and prolongs the duration of the inhibition in the GPi. According to the dynamic model, the main cause of akinesia is a reduced disinhibition in the thalamus and cortex during movements. The blockade of the STN restores the disinhibition in the cortex and thalamus during movement, and releases the selected motor program. On the other hand, the mechanism of STN-HFS is not clear, and whether HFS excites or inhibits STN neuronal activity is open to argument [13].

Concluding Remarks

I have introduced the dynamic model of the basal ganglia functions: Information through the hyperdirect, direct and indirect pathways dynamically controls the activity of the thalamus and cortex, and releases only the selected motor program at the selected timing. The dynamic model can also explain the pathophysiology of movement disorders and the mechanism of the effectiveness of stereotaxic surgery. The dynamic model will lead us not only to better understanding of the normal functions of the basal ganglia, but also to better treatment of movement disorders.

Acknowledgments This study was supported by the Uehara Memorial Foundation and Grants-in-Aid for Scientific Research (B) from the Ministry of Education, Culture, Sports, Science and Technology of Japan.

References

1. Albin RL, Young A B, Penney, JB (1989) The functional anatomy of basal ganglia disorders. Trends Neurosci 12:366–375
2. Alexander GE, Crutcher MD (1990) Functional architecture of basal ganglia circuits: neural substrates of parallel processing. Trends Neurosci 13:266–271
3. Bergman H, Wichmann T, DeLong MR (1990) Reversal of experimental parkinsonism by lesions of the subthalamic nucleus. Science 249:1436–1438
4. Chiken S, Shashidharan P, Nambu A (2008) Cortically evoked long-lasting inhibition of pallidal neurons in a transgenic mouse model of dystonia. J Neurosci 28:13967–13977
5. DeLong MR (1990) Primate models of movement disorders of basal ganglia origin. Trends Neurosci 13:281–285
6. Georgopoulos AP, DeLong MR, Crutcher MD (1983) Relations between parameters of step-tracking movements and single cell discharge in the globus pallidus and subthalamic nucleus of the behaving monkey. J Neurosci 3:1586–1598
7. Hamada I, DeLong MR, Mano N (1990) Activity of identified wrist-related pallidal neurons during step and ramp wrist movements in the monkey. J Neurophysiol 64:1892–1906
8. Hartmann-von Monakow K, Akert K, Künzle H (1978) Projections of the precentral motor cortex and other cortical areas of the frontal lobe to the subthalamic nucleus in the monkey. Exp Brain Res 33:395–403
9. Hashimoto T, Elder CM, Okun MS, Patrick SK, Vitek JL (2003) Stimulation of the subthalamic nucleus changes the firing pattern of pallidal neurons. J Neurosci 23:1916–1923
10. Hikosaka O, Takikawa Y, Kawagoe R (2000) Role of the basal ganglia in the control of purposive saccadic eye movements. Physiol Rev 80:953–978

11. Jahanshahi M, Ardouin CMA, Brown RG, Rothwell JC, Obeso J, Albanese A, Rodriguez-Oroz MC, Moro E, Benabid AL, Pollak P, Limousin-Dowsey P (2000) The impact of deep brain stimulation on executive function in Parkinson's disease. Brain 123:1142–1154
12. Kita H, Nambu A, Kaneda K, Tachibana Y, Takada M (2004) Role of ionotropic glutamatergic and GABAergic inputs on the firing activity of neurons in the external pallidum in awake-monkeys. J Neurophysiol 92:3069–3084
13. Lozano AM, Dostrovsky J, Chen R, Ashby P (2002) Deep brain stimulation for Parkinson's disease: disrupting the disruption. Lancet Neurol 1:225–231
14. Mink JW (1996) The basal ganglia: focused selection and inhibition of competing motor programs. Prog Neurobiol 50:381–425
15. Mink JW, Thach WT (1993) Basal ganglia intrinsic circuits and their role in behavior. Curr Opin Neurobiol 3:950–957
16. Nambu A (2007) Globus pallidus internal segment. In: Tepper JM, Abercrombie ED, Bolam JP (eds) GABA and the Basal Ganglia: From Molecules to Systems. Progress in Brain Research, vol 160. Elsevier, Amsterdam, pp 135–150
17. Nambu A, Takada M, Inase M, Tokuno H (1996) Dual somatotopical representations in the primate subthalamic nucleus: evidence for ordered but reversed body-map transformations from the primary motor cortex and the supplementary motor area. J Neurosci 16:2671–2683
18. Nambu A, Tokuno H, Hamada I, Kita H, Imanishi M, Akazawa T, Ikeuchi Y, Hasegawa N (2000) Excitatory cortical inputs to pallidal neurons via the subthalamic nucleus in the monkey. J Neurophysiol 84:289–300
19. Nambu A, Tokuno H, Takada M, (2002) Functional significance of the cortico-subthalamo-pallidal "hyperdirect" pathway. Neurosci Res 43:111–117
20. Nambu A, Yoshida S, Jinnai K (1990) Discharge patterns of pallidal neurons with input from various cortical areas during movement in the monkey. Brain Res 519:183–191
21. Tachibana Y, Kita H, Chiken S, Takada M, Nambu A (2008) Motor cortical control of internal pallidal activity through glutamatergic and GABAergic inputs in awake monkeys. Eur J Neurosci 27:238–253
22. Turner RS, Anderson ME (1997) Pallidal discharge related to the kinematics of reaching movements in two dimensions. J Neurophysiol 77:1051–1074
23. Yoshida S, Nambu A, Jinnai K (1993) The distribution of the globus pallidus neurons with input from various cortical areas in the monkey. Brain Res 611:170–174

Part II
Integrative Mechanism of Intracellular Information Networks

A Posttranslational Chemical Circadian Oscillator in Cyanobacteria

Hideo Iwasaki

Introduction

To understand dynamic biological behaviors at the system level in the "post-genome" era, it is important to elucidate the complicated molecular network generating biological information in addition to classical analysis on a single gene–function relationship. Periodic temporal (rhythmic) and spatial (morphological) pattern formations are both typical examples of dynamic and complex phenomena widely observed in organisms from bacteria to plants and humans. To elucidate such spatiotemporal dynamics at the molecular systems level, integration of molecular genetics, molecular biology, biochemistry, comprehensive genomic and post-genomic approaches, quantitative analyses, and mathematical considerations is necessary. In addition, synthetic biological approaches to reconstitute such phenomena in vitro and/or in vivo are also helpful to validate reality and find problems in the proposed molecules or models. For this type of analysis, we have analyzed cyanobacteria as one of the simplest organisms to exhibit circadian rhythms and regular pattern formations with cell differentiation. Here, I summarize a brief history and the current state of molecular, biochemical, and synthetic analysis on the cyanobacterial circadian clock and refer to some prospects on extensive researches on morphological development in filamentous cyanobacteria.

Circadian Rhythms

Life is full of a variety of rhythms, such as brain waves, calcium waves, metabolic oscillations, respiration rhythms, heart rate, the sleep/wake cycle, and annual cycles of flowering and reproduction. This variety is most likely because rhythmicity can

---

H. Iwasaki
Department of Electrical Engineering & Bioscience, Waseda University, and PRESTO, Japan Science & Technology Agency (JST), 2-2 Wakamatsu-cho, Shinjuku, Tokyo 162-8480, Japan

S. Nakanishi et al. (eds.), *Systems Biology*,
DOI:10.1007/978-4-431-87704-2_10,

be the origin of a variety of biological functions and information. First, autonomous reproduction or propagation per se is essentially periodic. It is inevitable because it is a replicating process based on rebuilding something that takes some time (or time delay). Second, rhythmicity enables biological processes to occur at their proper timing, thereby prohibiting mix-ups in complicated reaction networks. Rhythms may be important as a basis for frequency responses, even in biological informatics. Recent studies suggest that frequency responses are observed not only in higher-order brain informative processes but also in cellular signal transduction (see chapter by Dr. Shinya Kuroda). Third, rhythms are recursive and, therefore, predictable. Thus, they are useful tools to predict and adapt to periodic environmental changes such as day/night alternation and seasonal cycles. Circadian, circatidal, and circannual rhythms are typical examples of such biological oscillations.

Circadian rhythm, an endogenous oscillation with a period of ~24 h, is one of the most intensively analyzed biological oscillations. It is observed in cellular, physiological, behavioral, and population parameters, which are ubiquitously observed in many organisms. Circadian rhythms are characterized by three features. First, the rhythms persist even in the absence of external cues with a period length (free-running period) that is close to, but not always exactly equal to, the 24-h period of the earth's rotation. Second, circadian rhythms can be entrained to the external earth's cycles, such as the predictable alternation of light and dark. Third, the period length is relatively stable at different constant temperatures: that is, the free-running period is temperature compensated. All these features support the idea that circadian clocks are physiologically significant for organisms as adaptive systems to daily changes in environmental conditions on the earth, such as light, temperature, and humidity. Furthermore, the circadian clock is suggested as a fundamental basis of photoperiodic behavior in plants, insects and mammals). In humans, circadian clocks are involved in many physiological regulations, such as the wake/sleep cycle, hormone secretions, and body temperature. Such a human circadian control has recently become a medical consequence, because a loss of clock function causes depression, jet lag, or insomnia.

Schematically, the circadian organization can be considered to be organized by three elements: the circadian oscillator that generates basic oscillation, input pathways that transfer external environmental information to the oscillator, and output pathways from the oscillator which regulate many temporal events. Among them, here I focus on the molecular mechanism of the circadian oscillator in cyanobacteria.

Molecular Model of the Circadian Oscillator: A Transcription-Translation-Based Oscillator (TTO) Model

Circadian rhythms are observed in unicellular organisms and even in an isolated nerve cell. Thus, the circadian clock consists of an intracellular molecular network. Molecular genetics of circadian clocks started from isolation of three *Drosophila* eclosion rhythm mutants and mapping them to a single genomic locus named *period* (*per*) in 1971 (Konopka and Benzer, 1971). It took 19 years, however, to develop the experimental suggestion of the first molecular model for the *per* gene function in circadian rhythm generation (Hardin et al., 1990). This model is based on a rhythmic expression of a clock gene itself via an autofeedback control in which the clock gene expression is negatively inhibited by its own translational product (clock protein) with a time delay (transcription-translation feedback oscillator model; TTO model). Retrospectively, this type of transcription rhythms was theoretically proposed for the first time in 1965 by Goodwin (Goodwin, 1965), which required 25 years to obtain experimental validation at the level of molecular biology. After the first *per* model, many more factors and reactions have been found and therefore the models have become much more complicated (e.g., Ueda et al., 2005; see also chapter by Dr. Hiroki R. Ueda). For example, to ensure such a long-period rhythmicity, it was suggested that a time lag was important between cytosolic accumulation of PER protein and suppression of its own transcription in the nucleus. Accordingly, the stability of PER protein is likely regulated by multiple pathways including complex formation, phosphorylation, and ubiquitination.

Interestingly, amino acid sequences of clock proteins are not conserved among animals, plants, fungi, and bacteria. Nevertheless, the TTO model has become a milestone in not only circadian clock research in any organisms (reviewed by Dunlap, 2004) but also in developmental clock in somitogenesis (Hirata et al., 2002). However, we have found an alternative mechanism to generate circadian oscillation in cyanobacteria, which persisted even after complete elimination of transcription and translation reactions (Tomita et al., 2005).

Dilemma in Diazotrophic Cyanobacteria and Adaptive Periodic Spatiotemporal Pattern Formations

Among many model organisms, cyanobacteria, the simplest organisms to exhibit circadian rhythms, have become intensively analyzed to address fundamental questions in circadian biology. How was such a prokaryotic circadian control discovered in nature?

Several cyanobacterial strains are able to fix molecular nitrogen, as are some other prokaryotic organisms. A key enzyme for this reaction, nitrogenase, is extremely sensitive to oxygen that is inevitably generated by photosynthesis. Therefore, nitrogen-fixing (diazotrophic) Cyanobacteria have developed two different strategies to divide the contradictory metabolic pathways. One is a spatial separation of two metabolisms specifically adopted in filamentous (multicellular) diazotrophic cyanobacteria that differentiate a cell specialized for nitrogen fixation, named the heterocyst. The heterocyst shuts off photosystem II (PSII) activity and induces the nitrogenase protein, in contrast to the vegetative cells producing oxygen through full-chain photosynthesis. Interestingly, the spatial pattern of heterocyst development varies among cyanobacterial stains. Most noticeably, *Anabaena* (*Nostoc*) sp. PCC

7120, a genetically tractable filamentous strain, develops the heterocyst at every 8–12 cells in dividing filaments. Thus, heterocyst formation can be an interesting model system to analyze spatial pattern formation with bona fide cell differentiation.

However, this design cannot explain the ability of fixing nitrogen and oxygenic photosynthesis in unicellular or nonheterocystous diazotrophic cyanobacteria. An alternative strategy, "temporal separation," has been proposed for these cyanobacteria. In 1985, nitrogenase activity in the nonheterocystous, filamentous cyanobacterium *Oscillatoria* sp. was shown to exhibit daily rhythms peaking at night that persisted even after transfer to continuous light (LL) conditions (Stal and Krumbein, 1985). Thereafter, two other groups also reported robust daily oscillation of nitrogenase activity in unicellular cyanobacteria under constant conditions (Mitsui et al., 1986; Grobbelaar et al., 1986). Especially, Huang and colleagues were the first researchers who realized the cause was endogenous circadian oscillation, as reported for eukaryotes (Grobbelaar et al., 1986).

Cyanobacterial Circadian Clock Mechanism: Ticking Time Beyond TTO

Among the many cyanobacterial strains, *Synechococcus elongatus* PCC 7942 is most intensively analyzed in circadian biology because this strain is genetically tractable and exhibits highly reproducible high-amplitude gene expression rhythms monitored by a bioluminescence reporter (Fig. 1; Kondo et al., 1993; Iwasaki and Dunlap, 2000; Iwasaki and Kondo, 2000; Iwasaki and Kondo, 2004). Note that this strain is not a diazotrophic strain and cannot fix nitrogen. Interestingly, by promoter-trap analysis, this strain shows circadian rhythms in essentially all gene promoter activities (Liu et al., 1995). Cyanobacterial transcription rhythms require each of three clock

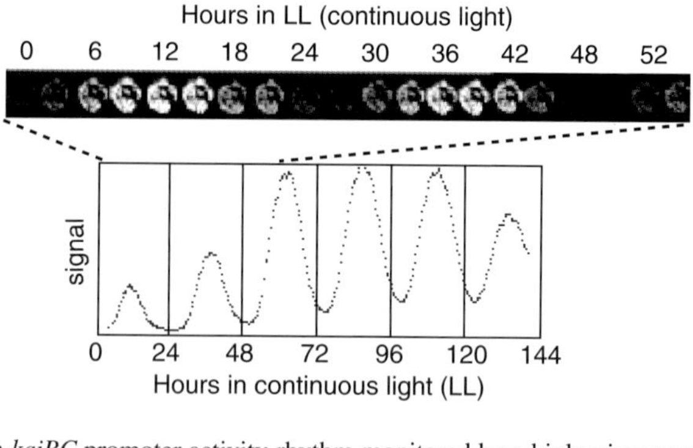

Fig. 1 Circadian *kaiBC* promoter activity rhythm monitored by a bioluminescence reporter under continuous light (*LL*) conditions. The *kaiBC* promoter was fused to a promoter-less bacterial luciferase operon, *luxAB*, from *Vibrio harveyi*, and introduced into *Synechococcus elongatus*. The transformed cells were grown on solid media under the light, transferred to the dark for 12 h to reset the clock, returned to continuous light (*LL*) conditions, and then subjected to bioluminescence monitoring using a cooled charge-coupled device (-CCD) camera or photomultiplier tubes

genes, *kaiA*, *kaiB*, and *kaiC*, which form a gene cluster (Ishiura et al., 1998). *kai* means "cycle" in Japanese, and many clock mutations with a period ranging from 14 to 80h, and even arrhythmia, have been mapped to the *kai* genes. *kaiB* and *kaiC* form an operon to be coexpressed in a circadian fashion with very high amplitude. Note that, differing from eukaryotic clock proteins such as CLOCK and BMAL1 (CYCLE), all Kai proteins lack DNA-binding motifs. However, overexpression of *kaiC* suppresses its own transcription to be arrhythmic. Disruption and overexpression of *kaiA* suppresses and enhances *kaiBC* expression level, respectively. Based on these genetic experiments, Ishiura et al. (1998) proposed that circadian *kaiBC* expression is based on an autoregulatory feedback mechanism in which KaiC negatively regulates its own transcription, whereas KaiA activates it. This scheme is essentially the same as the classical TTO mechanism proposed for eukaryotic clocks.

Nevertheless, we found that circadian rhythm in the phosphorylation state of KaiC protein persisted after transfer to continuous darkness (DD), under which *kai* gene transcription was completely suppressed even in the presence of excess amounts of transcription/translation inhibitors (Fig. 2; Tomita et al., 2005). Importantly, the

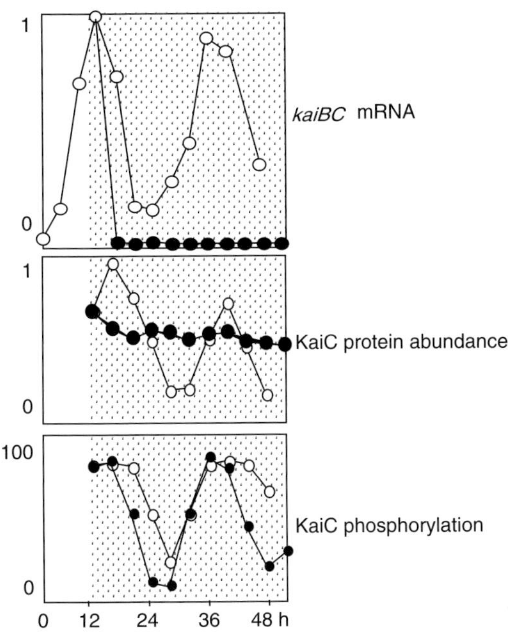

Fig. 2 KaiC phosphorylation cycle in the absence of transcription/translation feedback. Under continuous light (LL) conditions (*open circles*), mRNA from the *kaiBC* operon (*top*) exhibits a high-amplitude circadian rhythm, and the level of KaiC protein accumulation (*middle*) and the magnitude of KaiC phosphorylation (*bottom*) also cycle in a circadian fashion with a delay by several hours. When cells are transferred from hour 12 in the light to continuous darkness (DD) (*filled circles*), most of the mRNAs including *kaiA* and *kaiBC* transcripts were dramatically and immediately nullified to an undetectable level. Interestingly, the KaiA, KaiB, and KaiC proteins become stabilized in the dark to be constantly (arrhythmically) present without de novo protein synthesis. Nevertheless, the KaiC phosphorylation cycle persisted normally. Note that this KaiC phosphorylation rhythm in DD is not affected at all in the presence of an excess amount of transcription/translation inhibitors

period length of the circadian KaiC phosphorylation rhythm under DD was temperature compensated and affected by known clock mutations mapped to the *kai* genes, as was observed under continuous light (LL) conditions. This finding was the first experimental evidence of circadian control that is not dependent on the TTO scheme.

More dramatically, Nakajima et al. (2005) succeeded in reconstitution of the KaiC phosphorylation cycle in vitro by simply incubating KaiA, KaiB, and KaiC recombinant proteins at an appropriate molecular ratio in the presence of ATP. Again, the period of the in vitro reconstituted KaiC phosphorylation rhythm is temperature compensated, and the period in bioluminescence in particular *kai* mutants is consistent with the period in the in vitro phosphorylation cycle using the same mutant variants of KaiC (Nakajima et al. 2005). Therefore, it was concluded that this posttranslational oscillation is the core to the temporal integration of the cyanobacterial circadian system. This observation is particularly important because it was the first experiment to determine a "sufficient condition (factors)" for driving circadian rhythms. Conversely, molecular genetic approaches have been powerful tools to identify molecules essential for rhythm generation, i.e., to search for the "necessary condition." It was surprising that such a robust and long period oscillation as circadian rhythms was based on a chemical network composed of only three kinds of proteins. Although autocatalytic chemical oscillations such as Belousov–Zhabotinsky (BZ) reactions have been well known, the period length of these is much shorter than that of circadian rhythms: incorporating a time delay by separating molecules with semipermeable membranes was necessary for known chemical oscillations with a period of more than a couple of hours. Temperature compensation of chemical oscillations in a homogeneous aqueous solution has been rarely known.

How is the KaiC phosphorylation cycle generated? Interestingly, KaiC itself harbors both autophosphorylating and autodephosphorylating activities for Ser 431 and Thr 432 residues. KaiA activates the KaiC autophosphorylation, while KaiB antagonizes the KaiA function both in vitro and in vivo (Nishiwaki et al., 2000; Iwasaki et al., 2002; Kitayama et al., 2003; Williams et al., 2002). To generate enzymatic oscillation, the ratio of the Kai proteins and sequential dynamic complex formations among Kai proteins are important (Kageyama et al., 2006). Consistent with this, the importance of such cyclic protein complex formation has been also suggested in vivo (Iwasaki et al., 1999; Iwasaki et al., 2000; Iwasaki et al., 2002; Kageyama et al., 2003). KaiC has two ATP-binding motifs and exhibits ATP-dependent hexamer formation (Nishiwaki et al., 2000; Mori et al., 2002). Terauchi et al. (2007) recently found a weak ATPase activity of KaiC, and revealed that 1 molecule of KaiC catalyzed only 15 ATP molecules per day. This enzymatic rate is much slower than that of known slow enzymes such as AAA-type ATPases (less than 1/1000). They also found that KaiC ATPase activity showed a circadian cycle in a reaction mixture exhibiting KaiC phosphorylation rhythm. Importantly, the ATPase activity of KaiC is temperature compensated regardless of the presence or absence of KaiA or KaiB protein. This nature was also observed even two phosphorylation residues were substituted with alanines, indicating that this is not dependent on the phosphorylation state of KaiC (Terauchi et al., 2007). Generally, in an oscillating system, total energy in the system is in proportion to its frequency. Consistent with this idea, Terauchi et al. (2007) examined ATPase activities of five KaiC mutant proteins that gave rise to a different period length in vivo and in vitro. Strikingly, the result demonstrated that the ATPase activity even in the absence of KaiA and KaiB proteins was proportional to the frequency of resulting rhythmicity in the presence of KaiA and KaiB in vitro. This finding strongly supports that energy from ATP hydrolysis by KaiC is the key factor in determining the period length (Terauchi et al., 2007).

More recently, Kitayama et al. (2008) demonstrated that even in the absence of the rhythmic KaiC phosphorylation cycle some *kai* mutant strains in vivo exhibited temperature-compensated transcription/translation rhythms under LL conditions. In a *kaiA*-overexpressor mutant strain, the level of KaiC phosphorylation was constitutively elevated, whereas circadian rhythm in the *kaiBC* promoter activity persisted. Moreover, the rhythms in *kaiBC* expression and KaiB and KaiC accumulation were observed even after two phosphorylation residues of KaiC were replaced with Glu to mimic a constitutively phosphorylated form. Again, the KaiC phosphorylation cycle per se is not likely to be an essential process for Kai-dependent temperature-compensated transcriptional rhythms (Kitayama et al., 2008). Retrospectively, this situation is reminiscent of the residual low-amplitude, unstable transcription cycle in a *sasA*-null mutant strain, in which extremely downregulated KaiC protein was constitutively phosphorylated (Iwasaki et al., 2000; Takai et al., 2006). SasA is a KaiC-interacting sensory histidine kinase necessary for robust circadian oscillations (Iwasaki et al., 2000). Because the KaiC ATPase activity is the simplest reaction that is directly associated with basic circadian properties, it may independently regulate both the KaiC phosphorylation cycle and the transcriptional cycle. Thus, there is much accumulated information on the coexistence of at least four types of feedback loops in the circadian system. The detailed molecular mechanism of all these circuits and the relationships among them remain to be resolved.

Toward Understanding Spatial Pattern Formation in Multicellular (Filamentous) Cyanobacteria

Spatial extension of dynamic networks generating rhythms can be used also for a mechanism generating spatial pattern formations (morphology), represented by the clock-and-wavefront model for somitogenesis and the Turing model. Thus, methods and models used for circadian biology may be applied and extended to studies on periodic pattern formations. As described here, filamentous (multicellular) diazotrophic cyanobacteria often develop heterocysts (Fig. 3A). Particularly, *Anabaena* sp. PCC 7120 differentiates heterocysts at about every ten cells. Many mutations affecting heterocyst formations have been isolated, while a basis for such regular pattern formation still remains to be solved (reviewed by Golden and Yoon, 2003; Zhang et al., 2006).

Thus, we are interested in four questions to be addressed in *Anabaena*. First, what is the mechanism for the one-dimensional periodic pattern formation? Second,

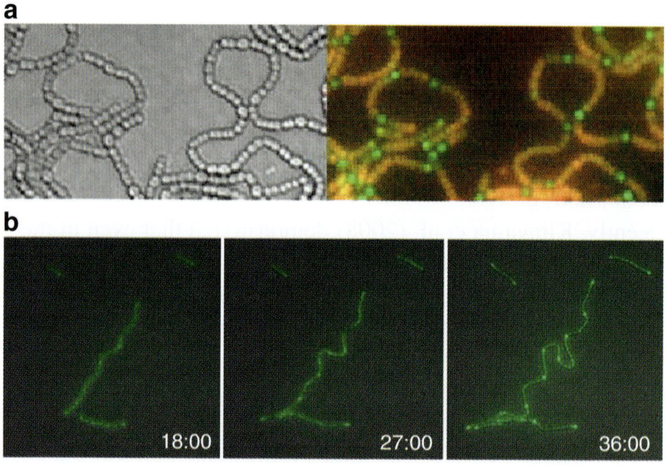

Fig. 3 Heterocyst formation in *Anabaena*. **a** *Anabaena* cells harboring a *gfp* reporter gene fused to the promoter of a gene named *hetR*, essential for heterocyst formation, were grown in the absence of nitrogen in solid media to develop heterocysts, which are larger than vegetative cells (*left*, bright field) and accumulate *hetR::gfp* signals (*green* signals in the *right panel*). *Red* signals in the *right panel* indicate autofluorescence from chlorophyll in vegetative cells. Note that the red signal correlating to the photosynthetic activity is nullified in the heterocysts. **b** Example of time-lapse fluorescence observation during heterocyst formation in the *hetR::gfp* strain. *Numbers* indicate hours after nitrogen depletion

are neighboring clock cells coupled to each other? Third, are clock systems equivalent between normal (vegetative) cells and heterocysts? Finally, is differentiation gated by the circadian clock? We initially set up an automated monitoring system to grow bacteria and to observe morphogenesis and gene expression profiles under the microscope (Fig. 3B). Using this system, we succeeded in obtaining the first "bacterial cell lineage" (Iwasaki, unpublished results). Moreover, we observed low-amplitude circadian rhythms in expression of genes involved in morphogenesis.

For experimentally validating models of circadian rhythms, pulsed (transient) administration of a genetic inducer or reagents is quite effective to perturb the oscillatory network that gives rise to phase-shift of the rhythm, for example. Such a perturbation experiment for spatial patterns similarly requires local administration of a reagent, metabolite, or inducer. Thus, application of microelectrical mechanical systems (MEMS) and microfluidic techniques seems very useful for administrating reagents to specific cells in an *Anabaena* filament to test which type of perturbation would be observed following morphogenesis. Such studies are now underway.

Acknowledgments I thank Dr. Takao Kondo and his laboratory members for the fruitful long-term collaboration in the *Synechococcus* circadian clock project. I also thank our laboratory members, especially Hironori Asai and Shunsuke Iwamori, for the *Anabaena* pattern formation project. This work was supported in part by the Uehara Memorial Foundation, grants-in-aid from the Japanese Society for Promotion of Science (No. 19657019 and 20370072), and the ERATO Aihara complexity modeling project (Japan Science and Technology Agency).

References

Dunlap JC (2004) The molecular mechanism of circadian oscillators. In: Chronobiology: Biological Timekeeping. Sinauer Associates, Sunderland, MA, pp 212–253

Golden JW, Yoon HS (2003) Heterocyst development in *Anabaena*. Curr Opin Microbiol 6:557–563

Goodwin BC (1965) Oscillatory behavior in enzymatic control processes. Adv Enzyme Regul 3:425–438

Grobbelaar N, Huang T-C, Lin H-Y, Chow T-J (1986) Dinitrogen-fixing endogenous rhythm in *Synechococcus* RF-1. FEMS Microbiol Lett 37:173–177

Hardin PE, Hall JC, Rosbash M (1990) Feedback of the *Drosophila period* gene product on circadian cycling of its messenger RNA levels. Nature (Lond) 343:536–540

Hirata H, Yoshiura S, Ohtsuka T, Bessho Y, Harada T, Yoshikawa K, Kageyama R (2002) Oscillatory expression of the bHLH factor Hes1 regulated by a negative feedback loop. Science 298:840–843

Ishiura M, Kutsuna S, Aoki S, Iwasaki H, Andersson CR, Tanabe A, Golden SS, Johnson CH, Kondo T (1998) Expression of a gene cluster *kaiABC* as a circadian feedback process in cyanobacteria. Science 281:1519–1523

Iwasaki H, Dunlap JC (2000) Microbial circadian oscillatory systems in *Neurospora* and *Synechococcus*: models for cellular clocks. Curr Opin Microbiol 3:189–196

Iwasaki H, Kondo T (2000) The current state and problems of circadian clock studies in cyanobacteria. Plant Cell Physiol 41:1013–1020

Iwasaki H, Kondo T (2004) Circadian timing mechanism in the prokaryotic clock system of cyanobacteria. J Biol Rhythms 19:436-444

Iwasaki H, Taniguchi Y, Ishiura M, Kondo T (1999) Physical interactions among circadian clock proteins, KaiA, KaiB and KaiC, in cyanobacteria. EMBO J 18:1137–1145

Iwasaki H, Williams SB, Kitayama Y, Ishiura M, Golden SS, Kondo T (2000) A KaiC-interacting sensory histidine kinase, SasA, necessary to sustain robust circadian oscillation in cyanobacteria. Cell 101:223–233

Iwasaki H, Nishiwaki T, Kitayama Y, Nakajima M, Kondo T (2002) KaiA-stimulated KaiC phosphorylation in circadian timing loops in cyanobacteria. Proc Natl Acad Sci U S A 99:15788–15793

Kageyama H, Kondo T, Iwasaki H (2003) Circadian formation of clock protein complexes by KaiA, KaiB, KaiC, and SasA in cyanobacteria. J Biol Chem 278:2388–2395

Kageyama H, Nishiwaki T, Nakajima M, Iwasaki H, Oyama T, Kondo T (2006) Cyanobacterial circadian pacemaker: Kai protein complex dynamics in the KaiC phosphorylation cycle in vitro. Mol Cell 23:161–171

Kitayama Y, Iwasaki H, Nishiwaki T, Kondo T (2003) KaiB functions as an attenuator of KaiC phosphorylation in the cyanobacterial clock system. EMBO J 22:2127–2134

Kitayama Y, Nishiwaki T, Terauchi K, Kondo T (2008) Dual KaiC-based oscillations constitute the circadian system of cyanobacteria. Genes Dev 22:1513–1521

Kondo T, Strayer CA, Kulkarni RD, Taylor W, Ishiura M, Golden SS, Johnson CH (1993) Circadian rhythms in prokaryotes: luciferase as a reporter of circadian gene expression in cyanobacteria. Proc Natl Acad Sci U S A 90:5672–5676

Konopka RJ, Benzer S (1971) Clock mutants of *Drosophila melanogaster*. Proc Natl Acad Sci U S A 68:2112–2116

Liu Y, Tsinoremas NF, Johnson CH, Lebedeva NV, Golden SS, Ishiura M, Kondo T (1995) Circadian orchestration of gene expression in cyanobacteria. Genes Dev 9:1469–1478

Mitsui A, Kumazawa S, Takahashi A, Ikemoto H, Cao S, Arai T (1986) Strategy by which nitrogen-fixing unicellular cyanobacteria grow photoautotrophically. Nature (Lond) 323:720–722.

Mori T, Saveliev SV, Xu Y, Stafford WF, Cox MM, Inman RB, Johnson CH (2002) Circadian clock protein KaiC forms ATP-dependent hexameric rings and binds DNA. Proc Natl Acad Sci U S A 99:17203–7208

Nakajima M, Imai K, Ito H, Nishiwaki T, Murayama Y, Iwasaki H, Oyama T, Kondo T (2005) Reconstitution of circadian oscillation of cyanobacterial KaiC phosphorylation in vitro. Science 308:414–415

Nishiwaki T, Iwasaki H, Ishiura M, Kondo T (2000) Nucleotide binding and autophosphorylation of clock protein KaiC as a circadian timing process of cyanobacteria. Proc Natl Acad Sci U S A 97:495–499

Stal LJ, Krumbein WE (1985) Nitrogenase activity in the non-heterocystous cyanobacterium *Oscillatoria* sp. grown under alternating light-dark cycles. Arch Microbiol 143:67–71

Takai N, Nakajima M, Oyama T, Kito R, Sugita C, Sugita M, Kondo T, Iwasaki H (2006) A KaiC-associating SasA–RpaA two-component regulatory system as a major circadian timing mediator in cyanobacteria. Proc Natl Acad Sci U S A 103:12109–12114

Terauchi K, Kitayama Y, Nishiwaki T, Miwa K, Murayama Y, Oyama T, Kondo T (2007) ATPase activity of KaiC determines the basic timing for circadian clock of cyanobacteria. Proc Natl Acad Sci U S A 104:16377–16381

Tomita J, Nakajima M, Kondo T, Iwasaki H (2005) No transcription–translation feedback in circadian rhythm of KaiC phosphorylation. Science 307:251–254

Ueda HR, Hayashi S, Chen W, Sano M, Machida M, Shigeyoshi Y, Iino M, Hashimoto S (2005) System-level identification of transcriptional circuits underlying mammalian circadian clocks. Nat Genet 37:187–192

Williams SB, Vakonakis I, Golden SS, LiWang AC (2002) Structure and function from the circadian clock protein KaiA of *Synechococcus elongatus*: a potential clock input mechanism. Proc Natl Acad Sci U S A 99:15357–15362

Zhang CC, Laurent S, Sakr S, Peng L, Bédu S (2006) Heterocyst differentiation and pattern formation in cyanobacteria: a chorus of signals. Mol Microbiol 59:367–375

The Apoptotic Signaling Network Dynamically Interprets the Outputs of Individual Signaling Pathways in an Early Analog and Late Digital Manner

Kevin A. Janes, H. Christian Reinhardt, and Michael B. Yaffe

Introduction

Changes in cell behavior are determined by a network of proteins that actively transmits signaling information (Irish et al., 2004; Jordan et al., 2000; Pawson, 2004). Modulation of total levels of key proteins, as well as their posttranslational states and enzymatic activities, could all potentially act as "molecular signals" that are monitored and interpreted. To understand how complex signal-transduction events control cell responses, these molecular signals must be measured dynamically and then mapped to downstream changes in cell outcome (Gaudet et al., 2005; Janes et al., 2005). Such approaches can accurately predict diverse cellular responses and suggest how molecular information is organized and propagated within the cell (Miller-Jensen et al., 2007).

One crucial cell decision is whether to die or to survive (Green and Evan, 2002; Kyriakis, 2001; Tran et al., 2004). Apoptosis is a common mechanism of cell death in response to many intrinsic and extrinsic cytotoxic cues such as cytokines and DNA damage (Strasser et al., 2000; Zhou and Elledge, 2000). Neighboring cells can also directly induce apoptosis in a target cell by secreting prodeath factors. Although cell death is important for normal development and tissue homeostasis, it is irreversible, requiring apoptosis to be tightly controlled. Deregulated apoptosis is known to be involved in many human diseases such as cancer (Rudin and Thompson, 1997; Vaux, 1993).

Tumor necrosis factor (TNF) is the founding member of a family of death ligands that plays critical roles in the regulation of apoptosis and inflammation (Aggarwal, 2003; Baud and Karin, 2001). In response to TNF, both pro-apoptotic and pro-survival

K.A. Janes, H.C. Reinhardt, and M.B. Yaffe
Koch Institute for Integrative Cancer Research, Center for Cell Decision Processes, Department of Biology, and Department of Biological Engineering, Massachusetts Institute of Technology, Cambridge, MA, USA

K.A. Janes
Department of Cell Biology, Harvard Medical School, Boston, MA 02115, USA

S. Nakanishi et al. (eds.), *Systems Biology*,
DOI:10.1007/978-4-431-87704-2_11, © Springer 2009

pathways are activated via caspases and IKK–NF-κB (Karin and Ben-Neriah, 2000; Riedl and Shi, 2004). TNF also activates stress pathways, including JNK and p38–MK2, which play pro- or anti-apoptotic roles depending upon the timing and duration of signaling and the activation state of other pathways (Deng et al., 2003; Janes et al., 2005; Lamb et al., 2003; Roulston et al., 1998; Ventura et al., 2004). Thus, predicting the apoptotic response of cells to TNF is complicated because the pathways and molecular signals involved are dynamic, context dependent, and directly antagonistic (Micheau and Tschopp, 2003; Miller-Jensen et al., 2007; Wajant et al., 2003).

In addition to direct signaling through the TNF receptor, TNF drives secretion of the cytokines transforming growth factor-α (TGF-α), interleukin-1α (IL-1α), and interleukin-1 receptor antagonist (IL-1ra) to initiate a cascade of autocrine signaling events through the epidermal growth factor (EGF) and IL-1 receptors (Janes et al., 2006). These induced autocrine ligands collaborate with circulating growth factors, such as EGF and insulin, to influence cell sensitivity to TNF (Garcia-Lloret et al., 1996; Remacle-Bonnet et al., 2000). How the collective information from pro-death, pro-survival, and autocrine ligands is processed through the same network of molecular signals to control apoptosis is unclear.

We have recently formalized a data-driven computational approach for investigating the TNF apoptotic-signaling network together with the pro-survival networks of EGF and insulin (Gaudet et al., 2005; Janes et al., 2006; Janes et al., 2004). The state of the network was estimated by measuring signaling proteins within various pathways that were directly implicated in the mechanisms of cytokine signaling, apoptosis, or both. We collected large-scale dynamic measurements of molecular signals that included the phosphorylation of critical receptors, adaptors, and transcription factors, the catalytic activity of protein kinases, and the cleavage state of caspases. In parallel, several apoptotic readouts induced by ten different cytokine–stimulus combinations were measured in human HT-29 colon adenocarcinoma cells. The complete set of 7980 molecular signals and 1440 apoptotic responses allowed us to model network function from primary data without the need to agglomerate kinetic parameters, intracellular concentrations, and protein–protein interactions from the literature.

We modeled how molecular signals may work together to control apoptosis using partial least squares regression (Janes and Yaffe, 2006). In a partial least squares model, signaling events ("variables") conveying similar information are combined into supervariables called principal components that are trained to predict outcomes such as cell death. Applying this approach to our dataset revealed many statistically coupled molecular signals whose dynamics were linked to apoptosis. The resulting model accurately captured the apoptotic responses induced by various cytokine combinations and correctly predicted the effect of perturbations, which were outside the initial model training (Janes et al., 2005). The success of the model also led to mechanistic hypotheses about apoptotic regulation that were subsequently verified by follow-up experiments, validating the usefulness of the data-driven modeling approach (Janes et al., 2005).

Counterintuitively, one can learn as much from failures of a model as from its successes. In engineering sciences, where modeling is a much more established technique, "failure analysis" is routinely used to define the normal operating range of complex systems, such as engines, bridges, and circuit boards (for recent examples, see Chansawat et al., 2006; Khatir et al., 2007; and Truman and Booker, 2007). Failure-directed approaches have not yet been applied to models of biological systems, but failed models can identify gaps in existing knowledge (Price et al., 2003) and reveal new mechanisms that were not captured by the original model (Hua et al., 2005; Miller-Jensen et al., 2007).

Here, we investigated whether the failure regimes of data-driven models could be used to identify fundamental properties of signal transduction in biological networks—namely, whether integrated cellular response networks could interpret the state of distinct signal transduction pathways in a digital (i.e., binary '1' or '0') or analog manner. To accomplish this, we reduced signaling dynamic range by grouping similar levels of signaling into a finite number of "bins" and then assessed the ability of the binned signals to predict apoptosis. We found for the TNF-induced autocrine circuits (Janes et al., 2006) that the signals induced by autocrine IL-1α can be selectively perceived by the network as digital information when dynamic range is compressed into a single steplike threshold. The response to autocrine TGF-α, conversely, requires analog signaling, with dynamic range spread evenly across signal strength.

Results

To investigate the relative importance of signal strength and dynamic range in a biologically relevant network, it was necessary to start with both a high-quality signaling dataset and a validated data-driven model. Our signaling dataset comprised 19 molecular signals measured at 13 time points from 0 to 24 h after stimulation with one of ten different combinations of TNF, EGF, and insulin (see Experimental Procedures for details). We analyzed molecular signals that were mechanistically connected to pathways downstream of the TNF, EGF, or insulin receptors and upstream of apoptosis. Measurements were collected from three or six biological replicates for a total of 7980 individual protein measurements.

Signaling dynamics are critical for many cell responses (Marshall, 1995; Pawson, 2004), but there is only a limited understanding of how dynamic signals are interpreted (Murphy and Blenis, 2006). Therefore, using the time course data, we extracted 30–40 time-dependent descriptors ("signaling metrics") that summarize the duration, activation, and deactivation of each molecular signal (Janes et al., 2005). Together, these metrics capture the information in a time course that is conceivably available to mediate cell outcomes, such as apoptosis. For each stimulus combination, four different cell-death markers were measured in triplicate by flow cytometry at three time points from 12 to 48 h to quantify the percentage of cells undergoing apoptosis. Together, the fused signaling-apoptosis dataset that served as the basis for model training consisted of 6600 signaling metrics (660 metrics per treatment × 10 treatments) and 120 apoptotic readouts (4 readouts × 3 time points × 10 treatments).

Our goal was to predict apoptosis (Y) from the measured signaling network state (X). The simplest relationship between these two datasets is a linear one: $Y = X\beta$, where Y is a matrix of 10 rows of treatments and 12 columns of apoptotic readouts at different time points, X is a matrix of 10 rows of treatments and 660 columns of signaling metrics, and β is a matrix of regression coefficients that describes how each metric contributes to each apoptotic readout (660 rows × 12 columns). There are not enough treatment conditions to calculate unique solutions for the complete matrix of regression coefficients (β), in the way that is commonly done for linear regression, curve fitting, etc. However, one can break down the matrix of signaling metrics into a series of vector products, which multiply a "scores" vector (s) and a "loadings" vector (Equation 1):

$$X \approx \begin{bmatrix} s_{1,1} \\ s_{2,1} \\ \vdots \\ s_{10,1} \end{bmatrix} [l_{1,1}\ l_{1,2} \cdots l_{1,660}] + \begin{bmatrix} s_{1,2} \\ s_{2,2} \\ \vdots \\ s_{10,2} \end{bmatrix} [l_{2,1}\ l_{2,2} \cdots l_{2,660}] + \ldots = s_1 l_1^T + S_2 l_2^T + \ldots$$

where s is a vector of 10 rows (1 for each observation), l is a vector of 660 rows (1 for each signaling metric), and ^{T} is the vector transpose (a function that swaps row and column elements). Each score-loading product ($s_1 l_1^T$, $s_2 l_2^T$, etc.) represents a "principal component" of the dataset. Often, data matrices can be approximated with only a few principal components (our model requires three, for example).

This approximation simplifies the dataset by reducing its dimensionality. Dimensionality reduction is advantageous in many other biological applications: colorimetric assays (using a one-dimensional calibration line that relates the two dimensions of absorbance and concentration), multiple sequence alignments (reducing many sequences to a few phylogenies), etc. Here, the reduced number of scores vectors is used to predict Y more efficiently: $Y = TB$, where T is a matrix of concatenated scores vectors (10 rows of treatments × 3 columns of principal components), and B is a matrix of regression coefficients, describing how each element in a score vector contributes to each apoptotic readout (3 rows × 12 columns). (Note that the number of elements in B that need to be estimated from the data is far smaller than the number of elements in β.) Numerical algorithms are available to extract the optimal principal components from X for predicting the data in Y (Geladi and Kowalski, 1986). The result is a "partial least squares" model linking X and Y through a small set of principal components derived from the data (Janes and Yaffe, 2006). By using molecular signals in X that are mechanistically connected to apoptosis (Y, see above), these principal components can give insight into the events directly responsible for the apoptotic phenotype.

Applying the partial least squares algorithms to our dataset yielded a three-component model that could fit the apoptotic readouts to within 99% (Janes et al., 2005). To assess the stability of the model, we cross-validated the predictions by leaving out one of the cytokine treatments and training a model that contained only nine treatments. This "leave-one-out" model was then used to predict the withheld treatment condition, and the cross-validation procedure was iterated for all treatments in the dataset. Apoptotic readouts induced by stimulus combinations withheld from the training dataset were captured to within 94% of their measured values (Janes et al., 2005), indicating that the full model is stable and quantitatively predictive.

Importantly, the success of these predictions is not simply a consequence of starting with a high number of signaling metrics, a subset of which might always predict any Y matrix. Using the same metrics and treatment conditions in a three-component model, we were unable to predict the starting TNF-EGF-insulin concentrations (36% accuracy) or the receptor occupancy for the stimulating cytokines (0% accuracy, data not shown). These Y matrices have fourfold fewer elements (three cytokine levels or receptor occupancies, compared to 12 apoptotic readouts) and yet cannot be predicted. We conclude that the apoptotic model is not based on false discovery and can be used as the foundation for exploring signaling dynamic range.

Autocrine-Feedback Circuits Are Selectively Perceived as Analog or Digital Signals

Signal-transduction and gene-regulatory networks have a dense feedback connectivity that resembles the layout of electrical circuits (Hasty et al., 2002; McAdams and Shapiro, 1995). This observation raises the question of whether natural and engineered circuits transmit the same types of information with similar requirements for signal strength and dynamic range. The discrete, four-base code of the genome and the binary nature of many cell fate responses suggests that the intervening signaling networks might operate digitally (Hood and Galas, 2003). If true, then dynamic range would be of limited importance, because all the information would be encoded in the abrupt transitions from one digital state to the next. However, mathematical models based on digital regulation have often required additional analog pathways to capture experimental observations, suggesting that dynamic range could be essential for many pathways (Altan-Bonnet and Germain, 2005; McAdams and Shapiro, 1995; McAdams and Shapiro, 2003; Ruf et al., 2006). It is not known whether network-level data quantitatively support digital or analog signal processing as a mechanism for controlling cellular outcomes.

The original data-driven apoptosis model is based on analog signal processing: across the full range of observations for each metric, the model assumes a smooth, mathematically continuous relationship between measured values ("signal activation") and how that value is interpreted as a contributor to apoptosis ("signal output") (Janes et al., 2005; Janes and Yaffe, 2006) (Fig. 1A: "Analog response"). To test whether this analog relationship was required, we explored the predictive power of discretized models that incorporate digital signaling elements by grouping similar observations into discrete "bins" (Fig. 1A). Each bin for a signaling metric reflects an interval of activation states that has lost its dynamic range and thus relays the same output. Using many bins is only slightly different from the starting dataset. As the number of bins decreases, however, more dynamic range is lost and the observations become coarse grained, with two bins representing an entirely on–off digital signal.

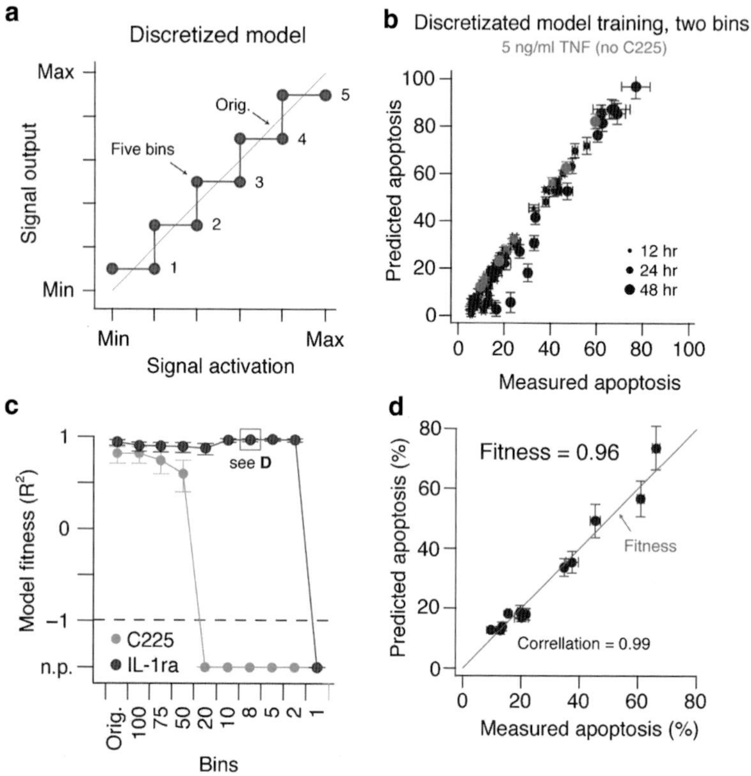

Fig. 1 Selective failure of progressively discretized models linking signal activation to signal output for predicting apoptosis. **a** Discretization of the experimentally measured signal activation state (abscissa) into with five bins of signal output (*ordinate*) used for model construction. Signal activation and signal output were uniformly discretized into a fixed number of bins (here, five bins) based on the minimum (*Min*) and maximum (*Max*) observed values for each signaling metric. Points at the discontinuities (*vertical lines*) were assigned to the upper bin values. The original continuous "analog" model was constructed using the measured signal activation-modeled signal output response shown by the *diagonal line*. **b** Training of the apoptosis prediction model using discretized signal outputs from ten different conditions of tumor necrosis factor (TNF), insulin, and epidermal growth factor (EGF) stimulation. In model training, data from one of the ten conditions were removed, the model was trained on the remaining nine datasets, and then used to predict four different readouts of apoptosis at 12, 24, and 48 h after the omitted treatment using the discretized signaling metrics from that treatment. This approach was then iteratively repeated for all ten conditions. Results of a two-bin model are shown. Sample data from one specific condition (5 ng/ml TNF) are highlighted in *blue*. **c** Predictions of apoptosis after perturbation of autocrine feedback in models with increasing discretization. Model fitness was calculated for predictions of the C225 (*green*) and interleukin (IL)-1ra (*blue*) perturbations after discretizing the signaling metrics as described in **a**. The specific data point for the model fitness shown in **d** is *boxed in purple*. **d** An example of good model fitness for modeling apoptosis after disruption of the IL-1 feedback loop using an eight-bin discretized model. The Pearson correlation coefficient is indicated. Perfect model fitness (fitness = 1) is represented by the *purple line*

After each discretization, we trained a new "discretized model" using the binned signaling data to capture the original measurements of apoptosis induced by TNF, EGF, and insulin. This retraining is necessary, because manipulating the dataset without updating the model will mathematically violate the underlying equations of the partial least squares approach (scores and loadings vectors, for example, will no longer approximate X; see above). To restrict discretization to the data itself, all models were fit with continuous regression coefficients. We found that the ability of the discretized models to predict the observed cytokine-induced apoptotic responses was indistinguishable from the original analog model based on predictions of cytokine treatments that were withheld from the training data (Fig. 1B). Therefore, discretization did not prevent the construction of a stable data-driven model that predicts apoptosis induced by TNF, EGF, and insulin.

To gauge the predictive power of the discretized models in a manner independent of the training data, we used perturbations of two TNF-induced autocrine circuits that were not present in the training set: (1) 5 ng/ml TNF in the presence of C225 (Fig. 1C, "C225"), a monoclonal antibody that blocks binding of autocrine TGF-α to its cognate receptor; and (2) 100 ng/ml TNF in the presence of exogenous IL-1ra (Fig. 1C, "IL-1ra"), a receptor antagonist that prevents signaling from autocrine IL-1α. The original analog apoptosis model could accurately predict the extent of apoptosis resulting from both of these perturbations (Janes et al., 2005) (Fig. 1C, "Analog"), thus providing a validated test set to compare the quality of the discretized model variants.

For each variant, we compared the prediction of four apoptotic readouts—phosphatidyl serine exposure, membrane permeability, nuclear fragmentation, and cleaved caspase-3/cleaved cytokeratin-18—against the values measured experimentally at 12, 24, and 48 h (Janes et al., 2005). Failed models often lose the quantitative accuracy of their predictions but still retain a qualitative ability to discriminate high versus low levels of cell responses (K.A.J. and M.B.Y., unpublished observations; see below). We therefore quantified the accuracy of predictions by using a formula for model "fitness" (Gaudet et al., 2005) that postulates a one-to-one match between apoptosis predicted by the model and that measured by experiment (Fig. 1D, "Perfect fitness"). This definition of fitness is more stringent because it penalizes model variants that are correlative qualitatively but not predictive quantitatively (Fig. 1C, "n.p.").

When the signaling data were progressively discretized from 100 bins down to 1 bin, we observed two distinct patterns of model failure (Fig. 1C). For the C225-antibody blockade of autocrine TGF-α, the extent of TNF-induced apoptosis could not be captured accurately by models when fewer than 50 bins were used to discretize the signaling metrics. This finding indicates that fine gradations in signaling are critical for the network to interpret TGF-α autocrine signals and suggests that the intracellular network interprets TGF-α feedback in an analog manner. By contrast, IL-1ra perturbation of TNF-induced apoptosis appeared largely insensitive to the level of discretization. High-quality predictions could be achieved by a model derived from as few as two signaling bins (i.e., "on" or "off"), raising the possibility that autocrine IL-1α signaling is perceived in a digital manner in these cells. Thus, although TNF activates both the TGF-α and IL-1α autocrine circuits (Janes et al., 2006), our analysis raises the possibility that the molecular signals activated by these extracellular feedbacks are transmitted or interpreted differently within the intracellular network.

Discussion

The selective failure of the autocrine perturbations during discretization matches the apparent function of the autocrine ligands that were blocked. For example, C225 disrupts binding of autocrine TGF-α, which is constitutively released in HT-29 cells but upregulated by around twofold within minutes after TNF addition (Janes et al., 2006). This autocrine tuning of a preexisting stimulus suggests that TGF-α signaling is interpreted as an analog input, which would explain the rapid failure of the C225 prediction after discretization. Conversely, IL-1ra prevents binding of autocrine IL-1α, which is barely detectable in the medium of unstimulated cells but is induced dramatically between 4 and 8 h after TNF is added (Janes et al., 2006). The extracellular "state" of IL-1α is thus essentially digital (present or absent). Accordingly, predictions of apoptosis following IL-1α blockade with IL-1ra show no sensitivity to discretization with as few as two bins.

In addition, the sequential release of TGF-α (early) and IL-1α (late) parallels the transition from analog decision making at early times to digital cell-death execution at late times. Autocrine TGF-α collaborates with direct signaling from the TNF receptor to interpret the cellular microenvironment and decide whether to undergo apoptosis, whereas IL-1α further amplifies the extent of caspase activation and drives a larger percentage of cells to death (Janes et al., 2006). This observation indicates that an analog-to-digital conversion of signaling is taking place at some point between cytokine receptors and caspase-effector proteins. Future network breakpoint analyses using perturbations of signaling proteins will allow us to determine the analog and digital nodes of the intracellular network, as we have done here for the TNF-induced autocrine cascade.

References

Aggarwal, B. B. (2003). Signalling pathways of the TNF superfamily: a double-edged sword. Nat Rev Immunol 3:745–756.

Alon, U., Camarena, L., Surette, M. G., Aguera y Arcas, B., Liu, Y., Leibler, S., and Stock, J. B. (1998). Response regulator output in bacterial chemotaxis. EMBO J 17:4238–4248.

Altan-Bonnet, G., and Germain, R. N. (2005). Modeling T cell antigen discrimination based on feedback control of digital ERK responses. PLoS Biol 3:e356.

Baud, V., and Karin, M. (2001). Signal transduction by tumor necrosis factor and its relatives. Trends Cell Biol 11:372–377.

Chansawat, K., Yim, S. C. S., and Miller, T. H. (2006). Nonlinear finite element analysis of a FRP-strengthened reinforced concrete bridge. J Bridge Eng 11:21–32.

Deng, Y., Ren, X., Yang, L., Lin, Y., and Wu, X. (2003). A JNK-dependent pathway is required for TNF-alpha-induced apoptosis. Cell 115:61–70.

Garcia-Lloret, M. I., Yui, J., Winkler-Lowen, B., and Guilbert, L. J. (1996). Epidermal growth factor inhibits cytokine-induced apoptosis of primary human trophoblasts. J Cell Physiol 167:324–332.

Gaudet, S., Janes, K. A., Albeck, J. G., Pace, E. A., Lauffenburger, D. A., and Sorger, P. K. (2005). A compendium of signals and responses triggered by prodeath and prosurvival cytokines. Mol Cell Proteomics 4:1569–1590.

Geladi, P., and Kowalski, B. R. (1986). Partial least-squares regression: a tutorial. Anal Chim Acta 185:1–17.

Green, D. R., and Evan, G. I. (2002). A matter of life and death. Cancer Cell 1:19–30.

Hasty, J., McMillen, D., and Collins, J. J. (2002). Engineered gene circuits. Nature (Lond) 420:224–230.

Hood, L., and Galas, D. (2003). The digital code of DNA. Nature (Lond) 421:444–448.

Hua, F., Cornejo, M. G., Cardone, M. H., Stokes, C. L., and Lauffenburger, D. A. (2005). Effects of bcl-2 levels on fas signaling-induced caspase-3 activation: molecular genetic tests of computational model predictions. J Immunol 175:985–995.

Irish, J. M., Hovland, R., Krutzik, P. O., Perez, O. D., Bruserud, O., Gjertsen, B. T., and Nolan, G. P. (2004). Single cell profiling of potentiated phospho-protein networks in cancer cells. Cell 118:217–228.

Janes, K. A., and Yaffe, M. B. (2006). Data-driven modelling of signal-transduction networks. Nat Rev Mol Cell Biol 7:820–828.

Janes, K. A., Kelly, J. R., Gaudet, S., Albeck, J. G., Sorger, P. K., and Lauffenburger, D. A. (2004). Cue-signal-response analysis of TNF-induced apoptosis by partial least squares regression of dynamic multivariate data. J Comput Biol 11:544–561.

Janes, K. A., Albeck, J. G., Gaudet, S., Sorger, P. K., Lauffenburger, D. A., and Yaffe, M. B. (2005). A systems model of signaling identifies a molecular basis set for cytokine-induced apoptosis. Science 310:1646–1653.

Janes, K. A., Gaudet, S., Albeck, J. G., Nielsen, U. B., Lauffenburger, D. A., and Sorger, P. K. (2006). The response of human epithelial cells to TNF involves an inducible autocrine cascade. Cell 124:1225–1239.

Jordan, J. D., Landau, E. M., and Iyengar, R. (2000). Signaling networks: the origins of cellular multitasking. Cell 103:193–200.

Karin, M., and Ben-Neriah, Y. (2000). Phosphorylation meets ubiquitination: the control of NF-[kappa]B activity. Annu Rev Immunol 18:621–663.

Khatir, Z., Lefebvre, S., and Saint-Eve, F. (2007). Experimental and numerical investigations on delayed short-circuit failure mode of single chip IGBT devices. Microelectron Reliab 47:422–428.

Kyriakis, J. M. (2001). Life-or-death decisions. Nature (Lond) 414:265–266.

Lamb, J. A., Ventura, J. J., Hess, P., Flavell, R. A., and Davis, R. J. (2003). JunD mediates survival signaling by the JNK signal transduction pathway. Mol Cell 11:1479–1489.

Marshall, C. J. (1995). Specificity of receptor tyrosine kinase signaling: transient versus sustained extracellular signal-regulated kinase activation. Cell 80:179–185.

McAdams, H. H., and Shapiro, L. (1995). Circuit simulation of genetic networks. Science 269:650–656.

McAdams, H. H., and Shapiro, L. (2003). A bacterial cell-cycle regulatory network operating in time and space. Science 301:1874–1877.

Micheau, O., and Tschopp, J. (2003). Induction of TNF receptor I-mediated apoptosis via two sequential signaling complexes. Cell 114:181–190.

Miller-Jensen, K., Janes, K. A., Brugge, J. S., and Lauffenburger, D. A. (2007). Common effector processing mediates cell-specific responses to stimuli. Nature (Lond) 448:604–608.

Murphy, L. O., and Blenis, J. (2006). MAPK signal specificity: the right place at the right time. Trends Biochem Sci 31:268–275.

Pawson, T. (2004). Specificity in signal transduction: from phosphotyrosine-SH2 domain interactions to complex cellular systems. Cell 116:191–203.

Price, N. D., Papin, J. A., Schilling, C. H., and Palsson, B. O. (2003). Genome-scale microbial in silico models: the constraints-based approach. Trends Biotechnol 21:162–169.

Remacle-Bonnet, M. M., Garrouste, F. L., Heller, S., Andre, F., Marvaldi, J. L., and Pommier, G. J. (2000). Insulin-like growth factor-I protects colon cancer cells from death factor-induced apoptosis by potentiating tumor necrosis factor alpha-induced mitogen-activated protein kinase and nuclear factor kappaB signaling pathways. Cancer Res 60:2007–2017.

Riedl, S. J., and Shi, Y. (2004). Molecular mechanisms of caspase regulation during apoptosis. Nat Rev Mol Cell Biol 5:897–907.

Roulston, A., Reinhard, C., Amiri, P., and Williams, L. T. (1998). Early activation of c-Jun N-terminal kinase and p38 kinase regulate cell survival in response to tumor necrosis factor alpha. J Biol Chem 273:10232–10239.

Rudin, C. M., and Thompson, C. B. (1997). Apoptosis and disease: regulation and clinical relevance of programmed cell death. Annu Rev Med 48:267–281.

Ruf, F., Park, M. J., Hayot, F., Lin, G., Roysam, B., Ge, Y., and Sealfon, S. C. (2006). Mixed analog/digital gonadotrope biosynthetic response to gonadotropin-releasing hormone. J Biol Chem 281:30967–30978.

Strasser, A., O'Connor, L., and Dixit, V. M. (2000). Apoptosis signaling. Annu Rev Biochem 69:217–245.

Tran, S. E., Meinander, A., and Eriksson, J. E. (2004). Instant decisions: transcription-independent control of death-receptor-mediated apoptosis. Trends Biochem Sci 29:601–608.

Truman, C. E., and Booker, J. D. (2007). Analysis of a shrink-fit failure on a gear hub/shaft assembly. Eng Fail Anal 14:557–572.

Vaux, D. L. (1993). Toward an understanding of the molecular mechanisms of physiological cell death. Proc Natl Acad Sci U S A 90:786–789.

Ventura, J. J., Cogswell, P., Flavell, R. A., Baldwin, A. S., Jr., and Davis, R. J. (2004). JNK potentiates TNF-stimulated necrosis by increasing the production of cytotoxic reactive oxygen species. Genes Dev 18:2905–2915.

Wajant, H., Pfizenmaier, K., and Scheurich, P. (2003). Tumor necrosis factor signaling. Cell Death Differ 10:45–65.

Zhou, B. B., and Elledge, S. J. (2000). The DNA damage response: putting checkpoints in perspective. Nature (Lond) 408:433–439.

Specific Features of Transient Ras and Sustained Rap1 Activation

Yu-ichi Ozaki, Shinsuke Uda, and Shinya Kuroda

Epidermal growth factor (EGF) and nerve growth factor (NGF) induce transient and sustained extracellular signal-regulated kinase (ERK) activation, respectively, and regulate cell growth and differentiation in PC12 cells (Fig. 1a) (Gotoh et al. 1990; Qiu and Green 1992; Traverse et al. 1992; Vaudry et al. 2002). Transient and sustained ERK activation in PC12 cells depends on transient Ras activation and sustained Rap1 activation, respectively (Fig. 1b) (Vaudry et al. 2002; York et al. 1998). We have previously modeled EGF- and NGF-dependent ERK activation in PC12 cells, and found that transient Ras and consequent ERK activation depends on rapidly increasing rates of EGF and NGF, whereas sustained Rap1 and consequent ERK activation depends on the final concentration of NGF (Sasagawa et al. 2005). We here construct simple Ras and Rap1 models whose inactivation is dependent on and independent of stimulation, respectively (Fig. 1c), and summarize how the modes of inactivation of Ras and Rap1 makes the features distinct.

We described the Ras model in dimensionless form as follows:

$$\frac{d}{dt}GEF = R - (1+R)\cdot GEF \tag{1}$$

$$\frac{d}{dt}GAP = q\left(p\cdot R - \left(1+p\cdot R\right)\cdot GAP\right) \tag{2}$$

$$\frac{d}{dt}Ras = s\left(k\cdot GEF - \left(k\cdot GEF + GAP\right)\cdot Ras\right) \tag{3}$$

where *R*, *GEF*, *GAP*, and *Ras* correspond, respectively, to phosphorylated receptors, activated SOS, activated RasGAP, and activated Ras. Parameters *p*, *q*, *k*, and *s*, denote, respectively, the association constant of *GAP* activation, the relative rate constant of *GAP* activation compared to *GEF* activation, the equilibrium constant of *Ras*, and the rate constant of *Ras* activation (see Appendix). Total amounts of

Y. Ozaki, S. Uda, and S. Kuroda
Department of Biophysics and Biochemistry, Graduate School of Science,
University of Tokyo, CREST, Japan Science and Technology Agency, Hongo 7-3-1,
Bunkyo-ku, Tokyo 113-0033, Japan

S. Nakanishi et al. (eds.), *Systems Biology*,
DOI:10.1007/978-4-431-87704-2_12, © Springer 2009

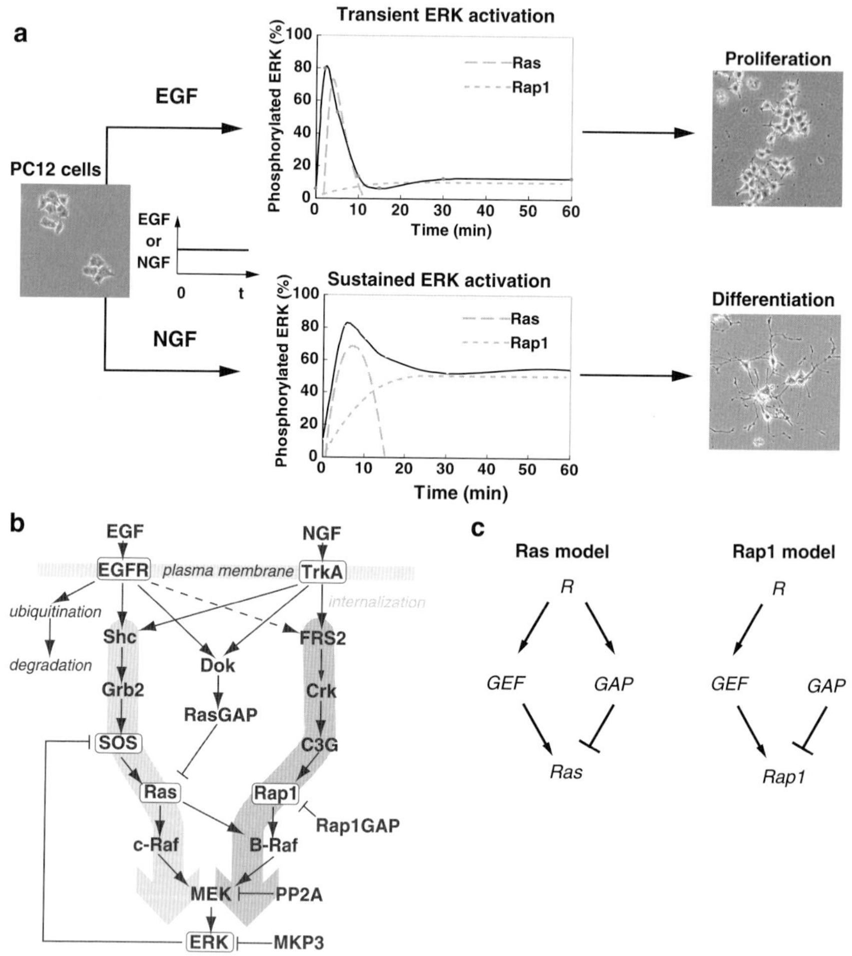

Fig. 1 Transient and sustained extracellular signal-regulated kinase (ERK) activation in PC12 cells. **a** Constant epidermal growth factor (EGF) and nerve growth factor (NGF) stimulation induces transient, and transient and sustained ERK activation, resulting in cell proliferation and differentiation, respectively. **b** EGF- and NGF-dependent ERK signaling networks in PC12 cells. *Thick arrows* denote Ras- and Rap1-dependent networks, which are responsible for transient and sustained ERK activation, respectively. **c** Overview of Ras and Rap1 models. *Arrows and bars* indicate activation and inactivation, respectively. *R* corresponds to phosphorylated receptors in **b**; *GEF* in the Ras and Rap1 models correspond to SOS- and C3G-dependent networks in **b**, respectively; *GAP* in the Ras and Rap1 models correspond to the RasGAP-dependent network and Rap1GAP in **b**, respectively. GEF, guanine nucleotide exchange factor; GAP, GTPase-activating protein

GEF, *GAP*, and *Ras* are set at 1, and the dissociation constant of *GEF* is set at 1. The rate constant of *GEF* activation becomes 1 when *R* is small.

Similarly, we described the Rap1 model in dimensionless form as follows:

$$\frac{d}{dt}GEF = R - (1 + R) \cdot GEF \tag{4}$$

$$GAP = p \tag{5}$$

$$\frac{d}{dt}Rap1 = s\left(k \cdot GEF - \left(k \cdot GEF + GAP\right) \cdot Rap1\right) \tag{6}$$

where the parameters and assumptions are the same as those in the Ras model except that *GAP* is a constant *p* regardless of stimulation.

From Equations 1–3, we derived a steady-state solution of the Ras model as follows:

$$GEF = \frac{R}{1+R} \tag{7}$$

$$GAP = \frac{p \cdot R}{1 + p \cdot R} \tag{8}$$

$$Ras = \frac{k}{k + p \cdot \dfrac{1+R}{1+p \cdot R}} \tag{9}$$

When $p > 1$ and $R < 1/p$, *GEF* and *GAP* are asymptotically proportional to *R* (Fig. 2a, upper panel), and *Ras* is constant regardless of *R* (Fig. 2a, lower panel). When $1/p < R < 1$, *GAP* becomes saturated while *GEF* is still proportional to *R*, and *Ras* is increased with *R*. When $R > 1$, both *GEF* and *GAP* are saturated and *Ras* becomes constant.

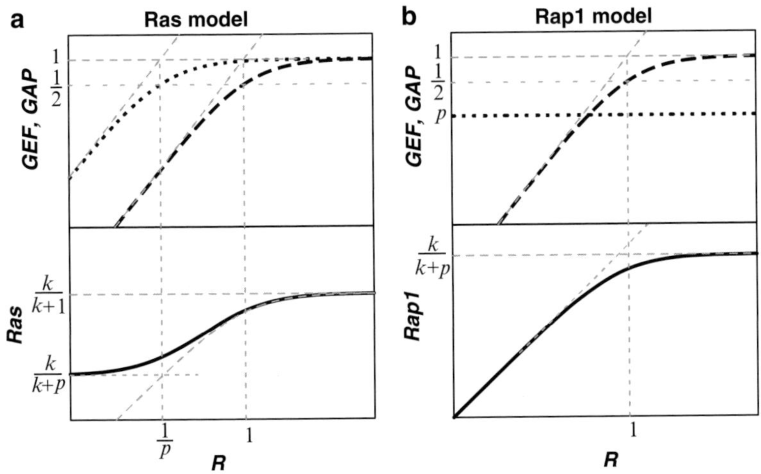

Fig. 2 Steady-state responses of Ras and Rap1 models: Ras model (**a**); Rap1 model (**b**). *Upper panels* indicate steady-state responses of *GEF* and *GAP* as indicated by *dashed and dotted lines*, respectively. K_d values, which give 50% of the maximum values, of *GEF* and *GAP* are 1 and 1/p, respectively. *Lower panels* indicate steady-state response of *Ras* and *Rap1*. Note that the log-log plot is used for all panels

From Equations 4–6, we also derived a steady-state solution of *GEF* and *Rap1* in response to *R*:

$$GEF = \frac{R}{1+R} \tag{10}$$

$$Rap1 = \frac{k}{k + p\left(1 + \frac{1}{R}\right)} \tag{11}$$

Note that *GAP* is constant in the Rap1 model. When $R < 1$, *GEF* is proportional to *R* (Fig. 2b, upper panel), and *Rap1* is also proportional to *R* (Fig. 2b, lower panel). When $R > 1$, *GEF* becomes saturated, and *Rap1* becomes constant.

In particular, when *R* is small, *Ras* becomes constant and independent from *R*, whereas *Rap1* is proportional to *R*. This result highlights one of the unique features of Rap1 systems: Rap1 captures amplitude of stimulation. Because *GEF*, *GAP*, *Ras*, and *Rap1* are not saturated when *R* is small, we can further simplify both Ras and Rap1 models:

$$\frac{d}{dt}GEF = R - GEF \tag{1'}$$

$$\frac{d}{dt}GAP = q\left(p \cdot R - GAP\right) \tag{2'}$$

$$\frac{d}{dt}Ras = s\left(k \cdot GEF - GAP \cdot Ras\right) \tag{3'}$$

$$\frac{d}{dt}GEF = R - GEF \tag{4'}$$

$$GAP = p \tag{5'}$$

$$\frac{d}{dt}Rap1 = s\left(k \cdot GEF - GAP \cdot Rap1\right) \tag{6'}$$

Initial conditions of *GEF*, *GAP*, *Ras*, and *Rap1* are set at those of the steady-state solution with $R = 1$, $p = 1$, and $k = 1$ without loss of generality. Hereafter, we used these simplified Ras and Rap1 models for further analyses. Note that constant *R* induces transient *Ras* activation when $0 < q < 1$.

Step stimulation of *R* induces transient increase of *Ras* (Fig. 3a–c). When $t < 1/q$, *GEF* and *GAP* are similarly increased (Fig. 3b), and *Ras* is also increased (Fig. 3c). When $t > 1/q$, *GEF* becomes constant while *GAP* is increased, and *Ras* is decreased; this makes a transient peak of *Ras*. From Equation 3′, *Ras* becomes closer to the *GEF*/*GAP* ratio as *s* becomes larger (Fig. 3c).

We did ramp stimulation of *R* by modified ramp function $R = (1 - e^{-r\,t}) + 1$, where *r* denotes increasing rate of *R*. Therefore, the ramp stimulation remains a

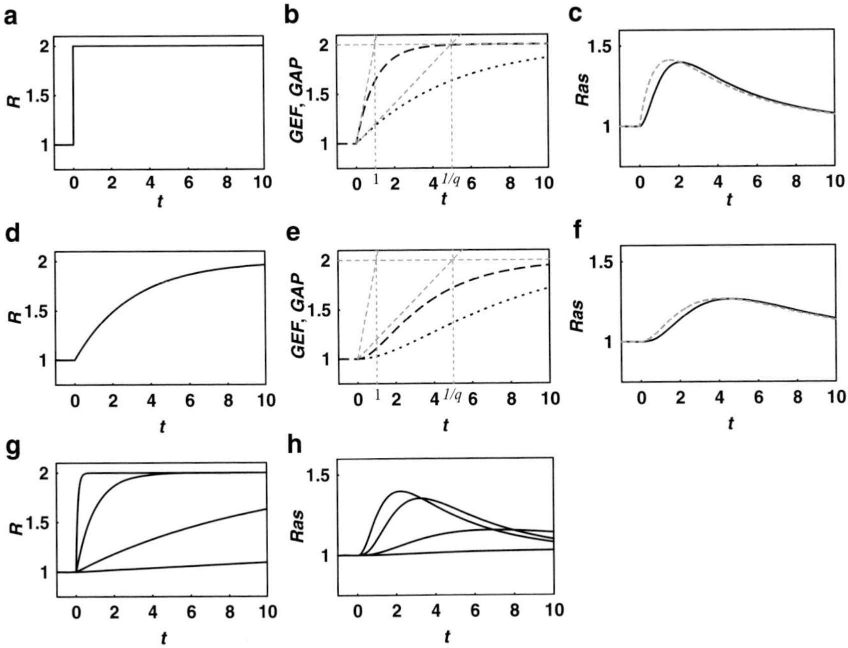

Fig. 3 Transient *Ras* response to step and ramp stimulation. **a–c** Responses to step stimulation. **d–f** Responses to ramp stimulation. **a**, **d** The step and ramp stimulation of *R*. Ramp stimulation of *R* is given by $R = (1 - e^{-rt}) + 1$ with $r = 0.333$. **b**, **e** *GEF* (*dashed line*) and *GAP* (*dotted line*) responses with $q = 0.2$. The time constant of *GEF* and *GAP* are 1 and $1/q$ as indicated, respectively. **c**, **f** *Ras* response with $s = 2$. *Gray dashed lines* in **c** and **f** indicate the *GEF*/*GAP* ratio. **h** *Ras* responses to various ramp stimulations of *R* with $r = 0.01$, 0.1, 1, and 10 (**g**)

finite dose at $t \rightarrow \infty$. Ramp stimulation of *R* induces slower *GEF* and *GAP* increase (Fig. 3d, e), resulting in a smaller *GEF*/*GAP* ratio (Fig. 3f); this delays and decreases the transient peak of *Ras* (Fig. 3f). Importantly, the transient peak of *Ras* is delayed and decreased as *r* decreases (Fig. 3g,h). Note that the final amplitude of *R* is the same regardless of the value of *r*, indicating that the Ras system captures the increasing rate of stimulation, rather than the final amplitude of stimulation. Thus, the Ras and Rap1 systems capture both the increasing rate and final amplitude of stimulation, respectively.

In PC12 cells, Ras activation occurs more rapidly than Rap1 activation (Sasagawa et al. 2005; York et al. 1998), suggesting faster and slower GEF activation for Ras and Rap1, respectively. Therefore, we analyzed the case where *GEF* activation for *Ras* is faster than that for *Rap1* (Fig. 4). Step stimulation of *R* induces transient *Ras* and sustained *Rap1* increase, respectively (Fig. 4a–c). The impulse stimulation of *R*, approximated by a pulse, sufficiently induces transient *Ras* increase, but not sustained *Rap1* increase (Fig. 4d–f). In contrast, the ramp stimulation of *R* sufficiently induces sustained *Rap1* increase but not transient *Ras* increase

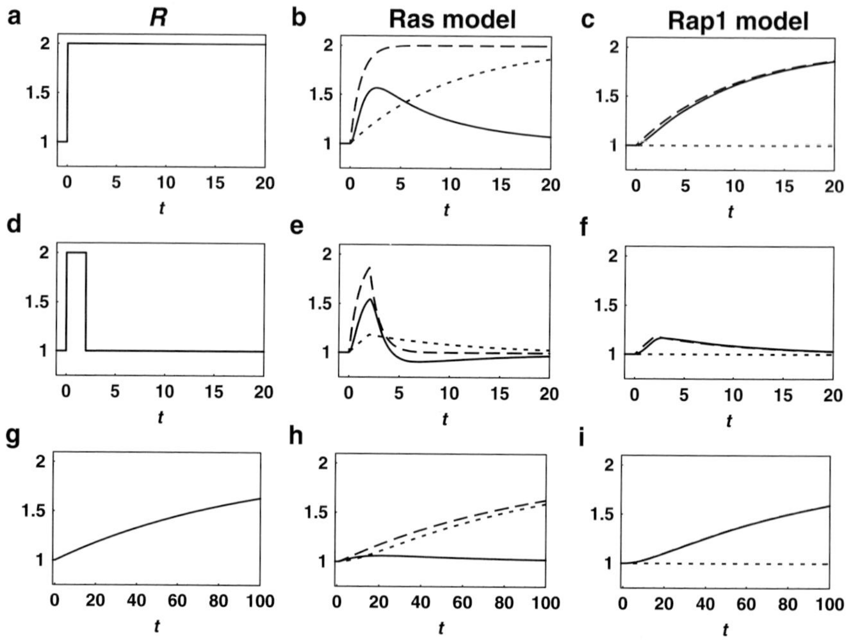

Fig. 4 Distinct *Ras* and *Rap1* responses to step, impulse, and ramp stimulation. **a–c** Responses to step stimulation. **d–f** Responses to impulse stimulation. The pulse width is set at 2. **g–i** Responses to ramp stimulation with $r = 0.01$. **a**, **d**, **g** Stimulation of R; **b**, **e**, **h** responses in *Ras* model; **c**, **f**, **i** responses in the *Rap1* model. The time constant of *GEF* activation for the Rap1 model is set at 10. All responses are calculated with $q = 0.1$ and $s = 2$. Notation for line styles is the same as in Fig. 3

(Fig. 4g–i). These results indicate that the impulse and ramp stimulation of R can specifically induce *Ras* and *Rap1* increase, respectively. This finding highlights the distinct features of the Ras and Rap1 systems. The Ras system specifically captures a rapid temporal pattern of stimulation rather than a slow temporal pattern, whereas the Rap1 system specifically captures a slow temporal pattern of stimulation rather than a rapid temporal pattern.

Appendix

We derive Equations 1–6 from Fig. 1c as follows. Molecule A has two states, inactive (A_i) and active (A^*). Conversion of A between the two states is catalyzed by effector molecules E_1^* and E_2^* as indicated in the following reaction:

$$A_i \underset{r_2 \cdot E_2^*}{\overset{r_1 \cdot E_1^*}{\rightleftarrows}} A^*$$

Forward reaction rate from A_i to A^* is assumed to be proportional to the product of A_i and E_1^* with rate constant r_1, and backward reaction rate is similarly assumed. We obtain the equation of molecule A as follows:

$$\frac{d}{dt}A^* = r_1 \cdot E_1^* \cdot A_i - r_2 \cdot E_2^* \cdot A^*.$$

A_i can be eliminated when an amount of A is conserved throughout the reaction. Similarly, we obtain Equations 1–6. Similar to biochemical reactions, we define dissociation and association constants by r_2/r_1 and r_1/r_2, respectively.

Acknowledgments We thank laboratory members for critically reading this manuscript, and the Uehara Memorial Foundation for the grant of systems biology.

References

Gotoh Y, Nishida E, Yamashita T et al (1990) Microtubule-associated-protein (MAP) kinase activated by nerve growth factor and epidermal growth factor in PC12 cells. Identity with the mitogen-activated MAP kinase of fibroblastic cells. Eur J Biochem 193:661–669.

Qiu MS, Green SH (1992) PC12 cell neuronal differentiation is associated with prolonged p21ras activity and consequent prolonged ERK activity. Neuron 9:705–717.

Sasagawa S, Ozaki Y, Fujita K et al (2005) Prediction and validation of the distinct dynamics of transient and sustained ERK activation. Nat Cell Biol 7:365–373.

Traverse S, Gomez N, Paterson H et al (1992) Sustained activation of the mitogen-activated protein (MAP) kinase cascade may be required for differentiation of PC12 cells. Comparison of the effects of nerve growth factor and epidermal growth factor. Biochem J 288(pt 2):351–355.

Vaudry D, Stork PJ, Lazarovici P et al (2002) Signaling pathways for PC12 cell differentiation: making the right connections. Science 296:1648–1649.

York RD, Yao H, Dillon T et al (1998) Rap1 mediates sustained MAP kinase activation induced by nerve growth factor. Nature (Lond) 392:622–666.

Control and Alteration of Protein Traffic in the Cell

Toshiya Endo

Mitochondrial Protein Traffic

Normal eukaryotic cell functions rely on cellular systems of protein trafficking control that ensure correct targeting and assembly of specific sets of proteins for membrane-bounded compartments or organelles. The known principles of protein traffic are that destination signals are encoded in the transported proteins themselves and that they are decoded by receptors in the target membranes, for example [1].

Mitochondria are two-membrane-bounded organelles consisting of 1000–2000 different proteins, more than 99% of which are synthesized in the cytosol as precursor proteins and subsequently imported into mitochondria. Mitochondrial protein import and sorting to one of the four subcompartments, the outer membrane, intermembrane space, inner membrane, and matrix, are mediated by proteinaceous machineries called translocators in the outer and inner mitochondrial membranes [2–5] (Fig. 1). The translocators cooperate with each other to achieve precise as well as efficient protein delivery to their intramitochondrial destinations. A translocator is a protein complex that performs multiple functions. It functions as a receptor for recognition of the targeting as well as sorting signals, and it provides a protein-conducting channel through which precursor proteins go across the membrane in an unfolded state. It also offers a driving force to achieve a vectorial movement of the translocating polypeptide chain.

Recognition of Mitochondrial Targeting Signals

The first step of mitochondrial protein traffic is the entry into the mitochondria at the outer membrane. There are multiple gates to enter mitochondria: Tom20, Tom22, and Tom70 [6]. Those receptors are the subunits of the outer membrane translocator, the TOM40 complex, and have receptor domains facing the cytosol.

T. Endo
Department of Chemistry, Graduate School of Science, Nagoya University, Nagoya, Japan

S. Nakanishi et al. (eds.), *Systems Biology*,
DOI:10.1007/978-4-431-87704-2_13, © Springer 2009

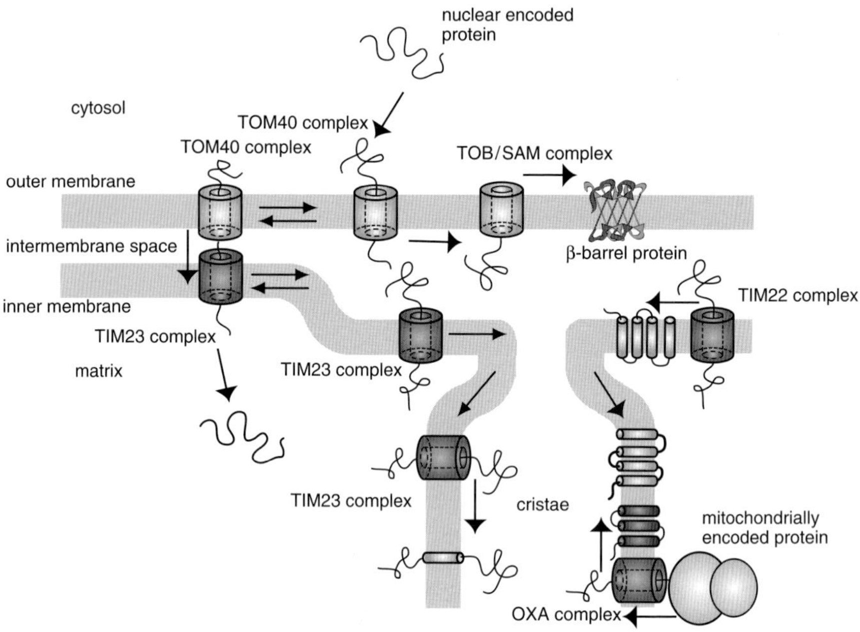

Fig. 1 Mitochondrial protein traffic. The translocators, the *TOM40* and *TOB/SAM* complexes in the outer membrane, and the *TIM23*, *TIM22*, and *OXA* complexes in the inner membrane control traffic of 1000–2000 different mitochondrial proteins

Mitochondrial precursor proteins can be classified into two groups, which are recognized by different receptors. Most mitochondrial proteins destined for the matrix or inner membrane are synthesized as precursor proteins with an N-terminal cleavable presequence. The presequence contains a mitochondrial targeting signal, which is mainly recognized by Tom20 and Tom22. On the other hand, many polytopic inner membrane proteins including the carrier protein family lack presequences. They have instead internal mitochondrial targeting signals in the mature part, which are recognized by Tom70.

We previously revealed the mechanism of discrimination of the mitochondrial targeting signal in the presequence from secretory signals in the signal sequence of secretory proteins by the receptor Tom20 [6,7]. Mitochondrial presequences do not share distinct consensus sequences, but instead display common physicochemical properties. Presequences are rich in positively charged residues and have the potential to form an amphiphilic helical structure. On the other hand, signal sequences of secretory proteins that are recognized by the cytosolic receptor SRP (signal recognition particle) contain a cluster of hydrophobic amino acids with two or more positive charges at the N-terminal region. We determined the nuclear magnetic resonance (NMR) structure of the receptor domain of rat Tom20 in a complex with a presequence peptide [7]. The presequence peptide of rat aldehyde dehydrogenase does not take an ordered structure in solution, but when bound to the hydrophobic groove of Tom20, it takes a helical structure. Because the hydrophobic groove of Tom20 is shallow, the amphiphilic helical structure of the presequence nicely fits the hydrophobic groove. The side chains of the hydrophobic residues are aligned on one side of the amphiphilic helix and are oriented toward the hydrophobic binding groove. On the other hand, hydrophilic and/or positively charged residues are located on the other side of the helix, thereby exposing their side chains to the aqueous solution.

In contrast to Tom20, because the binding groove of SRP is much deeper than that of Tom20 [8], the helical structure of the secretory signal peptide is fully accommodated into the binding groove of SRP, so that hydrophobic residues on both sides of the helix can interact with hydrophobic residues of the groove. Therefore, the difference in the depths of the hydrophobic grooves between Tom20 and SRP could allow precise recognition of the amphiphilic helix and the hydrophobic helix by their cognate receptors. This concept explains well the fact that mitochondrial targeting signals lack a strict consensus sequence motif and mitochondrial receptors have to recognize rather a broad range of sequences of mitochondrial targeting signals. Nevertheless, recognition by Tom20 may not be sufficient to explain the highly precise targeting of mitochondrial precursor proteins.

Our current understanding is that mitochondrial presequences are recognized by an array or chain of presequence-binding receptor proteins [2–5]. For example, after crossing the outer membrane through the TOM40 channel, the presequence is recognized by another receptor site (*trans* site) of the TOM40 complex on the intermembrane-space side of the outer membrane. Then, proteins destined for the matrix have to cross the inner membrane, and therefore they are recognized by Tim50, another receptor of the TIM23 complex, which is a translocator in the inner membrane for protein translocation across the inner membrane [9,10]. The presequence is transferred from the outer membrane receptor Tom20 and the *trans* site of the TOM40 complex to the inner membrane receptor Tim50, most likely following the affinity gradient between Tom20 and Tim50.

What receives the presequence next? It is Tim23 of the TIM23 complex that receives the presequence from Tim50. This step now requires input of energy, i.e., the membrane potential across the inner membrane. The energy is likely required for clearance of the presequence from the high-affinity presequence binding site. The presequence is then transferred to the matrix side through the import channel of the TIM23 complex, to be bound by mitochondrial Hsp70 in the matrix. Mitochondrial Hsp70 pulls in the presequence to achieve vectorial movement of the precursor protein.

Alteration of the Targeting Signal–Receptor Pair

Because we now understand the details of the recognition mechanism of mitochondrial targeting signals by receptor components along the import pathway, it is time for us to attempt to artificially alter the mitochondrial protein traffic by changing the import receptors. The basic idea is that mitochondrial proteins have a mitochondrial targeting signal, which is recognized by mitochondrial import receptors such as Tom20. Then, we change the signal recognition site of the receptor proteins, so that standard mitochondrial targeting signals cannot be recognized by the altered receptor. Then we screen for new targeting signals that are recognized by the altered receptor. Combination of the altered receptor and new targeting signals could be used to change the bulk of resident proteins in mitochondria, resulting in formation of mitochondria with a new function.

Generation of Mutant Tom20 with Altered Substrate Specificity

Although we determined the NMR structure of rat Tom20 in a complex with a presequence, the three-dimensional structure of yeast Tom20 or its complex with a presequence is not available. Therefore, we estimated the structure of yeast Tom20 by homology modeling using the software DS Modeling (Accelrys, San Diego, CA, USA). Then we put a presequence of yeast cytochrome oxidase subunit IV onto the yeast Tom20 structure. As we know that Tom20 has a hydrophobic groove that can accommodate the hydrophobic side of the amphiphilic helix of the presequence, we replaced several hydrophobic residues in the groove with hydrophilic and positively charged lysine. When we replaced wild-type Tom20 with those mutant Tom20s in yeast cells, the mutant cells cannot grow rapidly, as we expected for mutant Tom20 with altered substrate specificity.

To choose a Tom20 mutant that fits our purpose best, we analyzed protein levels of the Tom20 mutants. Some mutant Tom20s showed lowered protein levels, suggesting that they are not stable in cells. We thus chose Tom20 mutants including V97K-I133K, which exhibits a sufficiently high expression level.

Next, we checked that mutant Tom20s do not affect the assembly structures of the TOM40 complex. For this purpose, we performed blue-native polyacrylamide gel electrophoresis (BN-PAGE) analyses of the solubilized mitochondria with mutant Tom20s. The intact TOM40 complex with wild-type Tom20 showed a band of a 450-kDa complex on a BN-PAGE gel. Some mutant Tom20s caused breakdown of the TOM40 complex, but the others did not. Of note, mutant Tom20 V97K-I133K showed that it does not affect the 450-kDa complex of the TOM40 complex.

Screening System to Select Mutant Presequences Recognized by Mutant Tom20

Next, we set up a system to screen for mutant presequences that are recognized by mutant Tom20 (V97K-I133K) but not by wild-type Tom20 (Fig. 2). Yeast (*Saccharomyces cerevisiae*) can produce energy by both alcohol fermentation and respiration. Respiration requires the electron transport system on the inner membrane, including cytochrome oxidase. We thus first disrupted the *COX IV* gene for

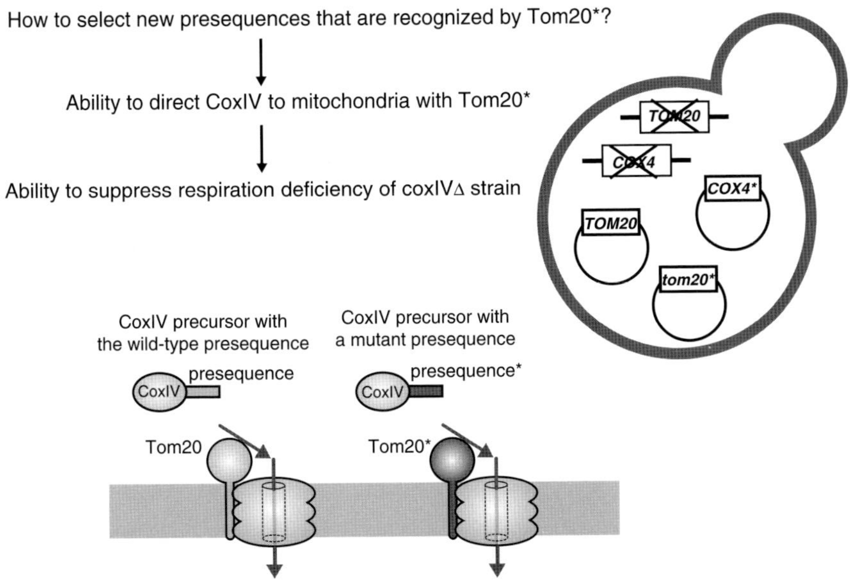

Fig. 2 Strategy to screen for mutant presequences to be recognized by altered Tom20

the cytochrome oxidase subunit IV precursor. This strain can grow on a fermentable medium such as glucose but not on a nonfermentable medium such as glycerol. However, when the gene for Cox IV precursor was inserted into a plasmid to be introduced in this cell, the expressed Cox IV was transported to mitochondria to make the mitochondria respiration competent.

Now, if expression of wild-type Tom20 is shut off, and instead mutant Tom20 is expressed, the CoxIV precursor cannot enter the mitochondria because its presequence is not recognized by mutant Tom20. Now we can use this system to screen for mutant presequences that can be recognized by mutant Tom20, because such a presequence can direct CoxIV to make the mitochondria respiration competent. We are at the moment screening for such presequences. Once we obtain altered presequences that can be recognized by altered Tom20 (V97K-I113K), but not by the original wild-type Tom20, we could introduce a set of mitochondria with a new presequence and their cognate Tom20 mutant into yeast cells to change the bulk of mitochondrial proteins.

Future Perspective

Even if we obtain a set of altered Tom20-preseuqence pairs, we need to overcome several practical problems to reach our goal. For example, mutant presequences for altered Tom20 should be adjusted for best recognition by a chain of downstream receptors along the import pathway, otherwise the mutant presequences cannot guide passenger proteins to the desired intramitochondrial compartments in new mitochondria with altered Tom20. Another problem is the mitochondrial fusion. Because mitochondria frequently undergo fusion and fission that would cause mixing of mitochondrial contents [11], we need to prevent such mixing to keep distinct sets of mitochondria intact after their delivery to new mitochondria with altered Tom20. After successful resolution of those problems, we hope that we will be able to make new mitochondria with a new function by altering mitochondrial targeting signals and their cognate receptor protein.

Acknowledgments I thank K. Mizutani, T. Sato, and T. Yoshihisa for their collaboration in this work.

References

1. Schatz, G. and Dobberstein, B. (1996). Common principles of protein translocation across membranes. Science 271:1519–1526.
2. Endo, T., Yamamoto, H., and Esaki, M. (2003). Functional cooperation and separation of translocators in protein import into mitochondria, the double-membrane bounded organelles. J. Cell Sci. 116:3259–3267.
3. Wiedemann, N., Frazier, A.E., and Pfanner, N. (2004). The protein import machinery of mitochondria. J. Biol. Chem. 279:14473–14476.
4. Neupert, W. and Herrmann, J.M. (2007). Translocation of proteins into mitochondria. Annu. Rev. Biochem. 76:723–749.
5. Koehler, C.M. (2004). New developments in mitochondrial assembly. Annu. Rev. Cell Dev. Biol. 20:309–335.
6. Endo, T. and Kohda, D. (2002). Functions of outer membrane receptors in mitochondrial protein import. Biochim. Biophys. Acta 1592:3–14.
7. Abe, Y., Shodai, T., Muto, T., Mihara, K., Torii, H., Nishikawa, S., Endo, T., and Kohda, D. (2000). Structural basis of presequence recognition by the mitochondrial protein import receptor Tom20. Cell 100:551–560.
8. Keenan, R.J., Freyman, D.M., Walter, P., and Stroud, R.M. (1998). Crystal structure of the signal sequence binding subunit of the signal recognition particle. Cell 94:181–191.
9. Geissler, A., Chacinska, A., Truscott, K.N., Wiedemann, N., Brandner, K., Sickmann, A., Meyer, H.E., Meisinger, C., Pfanner, N., and Rehling, P. (2002). The mitochondrial presequence translocase: an essential role of Tim50 in directing preproteins to the import channel. Cell 111:507–518.
10. Yamamoto, H., Esaki, M., Kanamori, T., Tamura, Y., Nishikawa, S., and Endo, T. (2002.) Tim50 is a subunit of the TIM23 complex that links protein translocation across the outer and inner mitochondrial membranes. Cell 111:519–528.
11. Okamoto, K. and Shaw, J.M. (2005). Mitochondrial morpology and dynamics in yeast and multicellular eukaryotes. Annu. Rev. Genet. 39:503–536.

Plant Functional Genomics Based on Integration of Metabolomics and Transcriptomics: Toward Plant Systems Biology

Kazuki Saito

Introduction

By taking the advantages of development of sequencing technology, the genome sequences of several plant species have been revealed: these include *Arabidopsis thaliana* [1], *Oryza sativa* (rice) [2], *Populus trichocarpa* (poplar) [3], *Vitis vinifera* (grapevine) [4], and *Sorghum bicolor* (sorghum) (http://www.phytozome.net/sorghum). However, even in *Arabidopsis*, which is most extensively investigated in terms of gene function, only half the genes have been functionally annotated based on sequence similarity to known genes, and the function of only about 11% has been confirmed with evidence [5]. Therefore, the identification of the function of unknown genes is currently a major subject in plant genomics. Analyses of the changes in plants in which specific genes are either overexpressed (*gain-of-function* experiment) or knocked out (*loss-of-function* experiment) are generally used for decoding the functions of genes. Systematic analyses of the transcriptome and metabolome, in particular, correlating the expression pattern of genes with the accumulation pattern of metabolites, could be an excellent way for deducing the functions of genes, even if these engineered plants do not show apparent phenotypic alternation [6–9].

In plants, a large number of genes involved in primary and secondary metabolism are present to form multigene families, for example, in *Arabidopsis*, 30 terpene synthase genes, 272 cytochrome P450 genes, 107 glycosyltransferase genes, and 130 ABC protein genes [10]. These genes are believed to be involved in the synthesis, modification, degradation, and/or transport of particular metabolites in plants. Compared with the model plant *Arabidopsis*, the situation is more complicated in the case of other plants, even if their genome sequences are available, because of the lack of feasible genetic resources for functional investigation such as large mutant panels and full-length cDNA collections. The integration of metabolic profiles with gene expression profiles can provide hints for the identification of functions

K. Saito
Chiba University, Graduate School of Pharmaceutical Sciences, Inage-ku, Chiba 263-8522, Japan, and RIKEN Plant Science Center, Tsurumi-ku, Yokohama 230-0045, Japan

S. Nakanishi et al. (eds.), *Systems Biology*,
DOI:10.1007/978-4-431-87704-2_14, © Springer 2009

of unknown metabolic genes, regardless of model or nonmodel plants. With the recent advances of sophisticated bioinformatics tools and analytical technology, the systems biology approach becomes more realistic to solve biological problems. In this chapter, the functional genomics study of combining transcriptome and metabolome is discussed, leading to the development of plant systems biology.

Functional Identification of Genes Involved in Metabolism by Integration of Transcriptomics and Metabolomics

A set of genes involved in a certain biological process are generally coregulated and thus coexpressed under the control of a shared regulatory system. Therefore, if an unknown gene is coexpressed with known genes of a particular biological process, this unknown gene is likely to be involved in this process. This co-occurrence principle can be extended to metabolite co-accumulation relationships with the expression pattern of genes in the particular pathway in which the metabolite is involved. Given that a pathway is modulated by mutation or stress, this modulation can be characterized by metabolic profiling. By using this combined strategy of gene coexpression and metabolic profiling under given conditions, a comprehensive prediction of genes involved in a particular process can be achieved. Then, once the genes of interest are prioritized by gene-to-metabolite connection, reverse genetics and biochemistry of recombinant proteins are conducted by taking the advantages of the publicly available rich bioresources of *A. thaliana*. Finally, comprehensive gene identification and discovery of networks of genes and metabolites can be elucidated (Fig. 1) [9]. Several studies that successfully applied this principle are discussed hereafter.

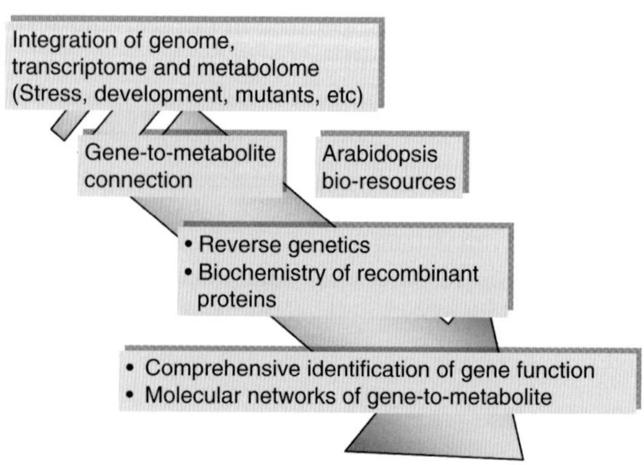

Fig. 1 Concept for integrated post-genomics study

Flavonoid-Related Genes

A comprehensive analysis of the metabolome and transcriptome of *A. thaliana* overexpressing the *PAP1* gene encoding an MYB transcription factor was carried out for the identification of novel gene functions involved in flavonoid biosynthesis [11]. For metabolome analysis, two metabolic profiling methods were applied: flavonoid-targeted analysis by high-performance liquid chromatography-mass spectrometry (LC-MS) and nontargeted analysis by Fourier transform ion-cyclotron mass spectrometry (FTICR-MS). The advantages of FTICR-MS are its ultrahigh-resolution capacity and high m/z accuracy. This combined analysis revealed the specific accumulation of cyanidin and quercetin derivatives and identified 8 novel anthocyanins from an array of putative 1,800 metabolites in *PAP1* overexpressing plants. The transcriptome analysis of more than 20,000 genes on a DNA microarray revealed the induction of 38 genes by ectopic overexpression of the *PAP1* transcription factor. Several genes with unidentified functions, encoding biosynthetic enzymes, transporters, and transcription factors, were induced by *PAP1*. Two putative glycosyltransferase genes (At5g17050 and At4g14090) induced by *PAP1* expression were confirmed to encode flavonoid 3-*O*-glucosytransferase and anthocyanin 5-*O*-glucosyltransferase, respectively, from the enzymatic activity of their recombinant proteins and results of the analysis of anthocyanins in the respective gene knockout mutants. Furthermore, the functions of a novel acyltransferase [12] and a glutathione *S*-transferase [13] were confirmed by the same approach. This innovative means of predicting novel gene functions has been further extended by using publicly available datasets of the *Arabidopsis* transcriptome.

Glycosylation plays a major role in the remarkable chemical diversity of flavonoids in plants, including *A. thaliana*. However, the wide diversity of proteins encoded by the large gene family of type 1 glycosyltransferases (UGT) hampers the determination of the biochemical function of each gene solely from its primary sequence. Transcriptome coexpression analysis combined with a reverse genetics approach led to the identification of a gene dominantly determining the flavonoid patterns in *Arabidopsis* [14]. This investigation efficiently applied transcriptome coexpression analysis with the ATTED-II database [15] and its mirror site Correlated Gene Search at PRIMe (http://prime.psc.riken.jp/), which are the secondary databases derived from a public *Arabidopsis* transcriptome dataset with Affymetrix ATH1 GeneChip [16]. The expression pattern of a UGT gene denoted as UGT1 in Fig. 2 was found to be highly correlated with known flavonoid biosynthetic genes. No C-7 rhamnosylated flavonols were detected in two knockout mutants by insertion of T-DNA into this *UGT89C1* gene. This specific metabolite deficiency in the mutants was restored by stable transformation with the genomic fragment containing intact *UGT89C1*. Biochemical tests using recombinant proteins also indicated that *UGT89C1* functions as flavonol 7-*O*-rhamnosyltransferase. As an another example, the metabolic profiling of the mutant lacking a methyltransferase gene *AtOMT1* (At5g54160) narrowed down by the coexpression analysis showed that gene is involved not only in the production of lignins and sinapoyl

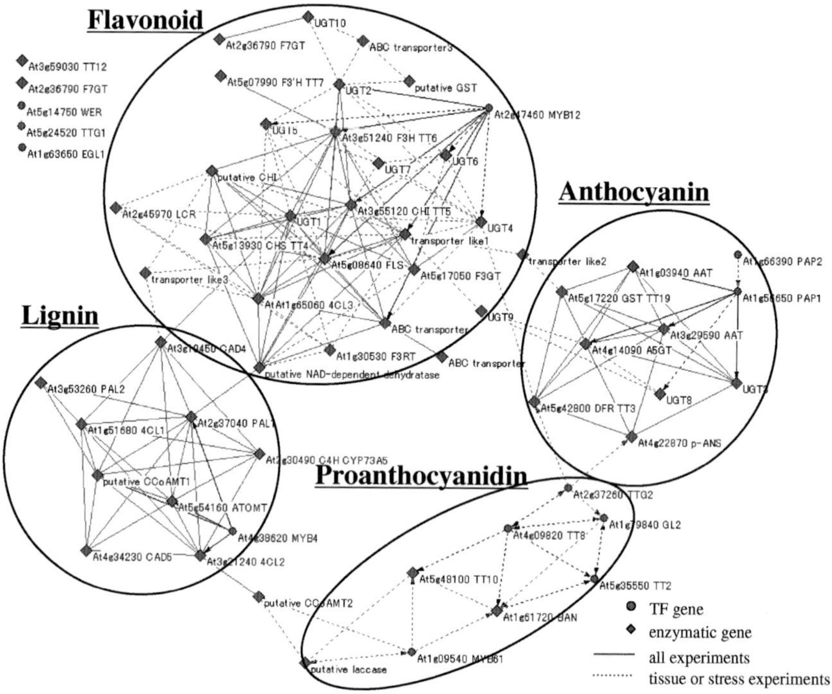

Fig. 2 Coexpression graph of *Arabidopsis* genes related to phenylpropanoid and flavonoids (courtesy of T. Tohge and K. Yonekura-Sakakibara)

esters but also in the methylation of flavonols forming isorhamnetin [17]. These study demonstrates that the integration of transcriptome coexpression analysis with a reverse genetic approach is a versatile tool for elucidation of the multigene family of metabolic pathways in *Arabidopsis*.

Sulfur Metabolism-Related Genes

The feasibility of the batch-learning self-organizing map (BL-SOM) for high-throughput analysis of post-genomic data on plant biology has been illustrated [18]. BL-SOM is a modification of the conventional self-organizing map (SOM), which provides colored feature maps independent of the order of data input. The nutritional stress experiments on *A. thaliana* were analyzed by computation with BL-SOM for the data sets of metabolome and transcriptome obtained by FTICR-MS and DNA array. BL-SOM analyses of nontargeted metabolome data provided metabolic fingerprinting of global responses to nutritional stress and development. The excellent clustering performance of BL-SOM led to metabolite annotation from accurate m/z values, suggesting a metabolic mechanism responding to sulfur deficiency. Transcriptome data were also analyzed by BL-SOM, and genes responding to sulfur deficiency were classified according to their expression patterns. The results showed that functionally related genes were nicely clustered in the same or neighbor lattice points on BL-SOM feature maps.

This principle of coexpression and co-accumulation by BL-SOM analysis was further applied to metabolomic and transcriptomic data of *Arabidopsis* grown under sulfur deficiency to decode novel gene function [19, 20]. The metabolism of glucosinolates was shown to be coordinately regulated under sulfur starvation. Three uncharacterized putative sulfotransferase genes clustering together with known glucosinolate biosynthesis genes were candidates for involvement in biosynthesis. In vitro enzymatic assays of the recombinant gene products prioritized by the integrated analysis confirmed their functions as desulfoglucosinolate sulfotransferases. Several genes involved in sulfur assimilation clustered with *O*-acetylserine, which is considered a positive regulator of these genes. The genes involved in anthocyanin biosynthesis clustered with the gene encoding a transcriptional factor that upregulates specifically anthocyanin biosynthesis genes.

The discovery of two R2R3-MYB transcription factors that positively control the biosynthesis of glucosinolates in *Arabidopsis* was achieved by an integrated omics approach [21]. Combined transcriptome coexpression analysis of publicly available condition-independent data and the condition-specific (i.e., sulfur deficiency in this case) data identified *MYB28* and *MYB29* as candidate transcription factor genes specifically involved in the regulation of aliphatic glucosinolate production. Analysis of a knockout mutant and ectopic expression of the gene demonstrated that *MYB28* is a positive regulator for basal-level production of aliphatic glucosinolates. In contrast, *MYB29* presumably plays an accessory function for methyl jasmonate-mediated induction of a set of aliphatic glucosinolate biosynthetic genes. Overexpression of *MYB28* in *Arabidopsis* cultured suspension cells resulted in the production of large amounts of glucosinolates. Because an *Arabidopsis* cell suspension produces no glucosinolates, this study indicates the possibility of efficient industrial production of glucosinolates by manipulation of these transcription factors. A working model for regulation of glucosinolate production involving these genes was postulated.

Systems Analysis of Genotype-Dependent Metabolic Regulation by Metabolomic Approach

Metabolites are not only the catalytic products of enzymatic reactions but also the active regulators or the ultimate phenotypic representatives of homeostasis in highly complex cellular processes. Such a mode of regulation can be revealed by biochemical networks. The metabolic network between wild-type and two mutant [*methionine-over accumulation 1* (*mto1*) and *transparent testa4* (*tt4*)] plants was investigated regarding the alteration of metabolite accumulation in *A. thaliana* [22]. In the GC-TOF/MS analysis, quantitative information was acquired regarding more than 170 metabolites. In addition to the conventional analysis regarding the levels of metabolites, metabolite-to-metabolite correlations across individual plants caused by tiny environmental fluctuations was analyzed by a novel score, z-score of index metabolite, describing the correlation of a given metabolite to others. Although the two mutants showed no apparent morphological abnormalities, the overall correlation values in *mto1* were much lower than those of the wild-type and *tt4* plants. These findings suggested the loss of overall network stability caused by the uncontrolled accumulation of methionine. In the *tt4* mutant, a new correlation between malate and sinapate was observed, although the levels of malate, sinapate, and sinapoylmalate remain unchanged, suggesting an adaptive reconfiguration of the network module by this mutation. Gene expression correlations presumably responsible for these metabolic networks were determined using the metabolite correlations as clues. As the conclusions, two Arabidopsis mutants, *mto1* and *tt4*, exhibited the following changes in the entire metabolome network: the overall loss of metabolic stability (*mto1*) or the generation of a metabolic network of a backup pathway for the lost physiological functions (*tt4*). The expansion of metabolite correlation to gene expression correlation provides detailed insights into the systemic understanding of the plant cellular process regarding metabolomes and transcriptomes.

Conclusions and Future Prospects

Recent advances of sophisticated bioinformatics tools and analytical technology allowed the systems biological approach to be more realistic for solving biological problems that have not been well addressed by simple reductionism. We foresee the development of systems biology driven by large datasets of transcriptome, metabolome, proteome, and fluxome. The recently developed ultrahigh-speed sequencing technology is likely to be exploited for meeting the challenge of sequencing a whole genome or large cDNA collections of a variety of plant species, including crops and pharmaceutical plants. The systems biology approach based on construction of mathematical models using large omics datasets will open a new avenue for our better understanding of hidden networks of molecular elements (gene, transcripts, proteins, and metabolites) in plant life. This information will be eventually applicable to biotechnology in the new era where plant biotechnology could play a more indispensable role in our lives.

Acknowledgments I thank my colleagues who have participated in this project. Part of this study has been supported by Uehara Memorial Foundation, Japan.

References

1. The Arabidopsis Genome Initiative (2000) Analysis of the genome sequence of the flowering plant *Arabidopsis thaliana*. Nature (Lond) 408:796–815.
2. International Rice Genome Sequencing Project (2005) The map-based sequence of the rice genome. Nature (Lond) 436:793–800.

3. Tuskan, G. A., Difazio, S., Jansson, S., Bohlmann, J., Grigoriev, I., Hellsten, U., Putnam, N., Ralph, S., Rombauts, S., Salamov, A., Schein, J., Sterck, L., Aerts, A., Bhalerao, R. R., Bhalerao, R. P., Blaudez, D., Boerjan, W., Brun, A., Brunner, A., Busov, V., Campbell, M., Carlson, J., Chalot, M., Chapman, J., Chen, G. L., Cooper, D., Coutinho, P. M., Couturier, J., Covert, S., Cronk, Q., Cunningham, R., Davis, J., Degroeve, S., Dejardin, A., Depamphilis, C., Detter, J., Dirks, B., Dubchak, I., Duplessis, S., Ehlting, J., Ellis, B., Gendler, K., Goodstein, D., Gribskov, M., Grimwood, J., Groover, A., Gunter, L., Hamberger, B., Heinze, B., Helariutta, Y., Henrissat, B., Holligan, D., Holt, R., Huang, W., Islam-Faridi, N., Jones, S., Jones-Rhoades, M., Jorgensen, R., Joshi, C., Kangasjarvi, J., Karlsson, J., Kelleher, C., Kirkpatrick, R., Kirst, M., Kohler, A., Kalluri, U., Larimer, F., Leebens-Mack, J., Leple, J. C., Locascio, P., Lou, Y., Lucas, S., Martin, F., Montanini, B., Napoli, C., Nelson, D. R., Nelson, C., Nieminen, K., Nilsson, O., Pereda, V., Peter, G., Philippe, R., Pilate, G., Poliakov, A., Razumovskaya, J., Richardson, P., Rinaldi, C., Ritland, K., Rouze, P., Ryaboy, D., Schmutz, J., Schrader, J., Segerman, B., Shin, H., Siddiqui, A., Sterky, F., Terry, A., Tsai, C. J., Uberbacher, E., Unneberg, P., et al. (2006) The genome of black cottonwood, *Populus trichocarpa* (Torr. & Gray). Science 313:1596–604.
4. Jaillon, O., Aury, J. M., Noel, B., Policriti, A., Clepet, C., Casagrande, A., Choisne, N., Aubourg, S., Vitulo, N., Jubin, C., Vezzi, A., Legeai, F., Hugueney, P., Dasilva, C., Horner, D., Mica, E., Jublot, D., Poulain, J., Bruyere, C., Billault, A., Segurens, B., Gouyvenoux, M., Ugarte, E., Cattonaro, F., Anthouard, V., Vico, V., Del Fabbro, C., Alaux, M., Di Gaspero, G., Dumas, V., Felice, N., Paillard, S., Juman, I., Moroldo, M., Scalabrin, S., Canaguier, A., Le Clainche, I., Malacrida, G., Durand, E., Pesole, G., Laucou, V., Chatelet, P., Merdinoglu, D., Delledonne, M., Pezzotti, M., Lecharny, A., Scarpelli, C., Artiguenave, F., Pe, M. E., Valle, G., Morgante, M., Caboche, M., Adam-Blondon, A. F., Weissenbach, J., Quetier, F. & Wincker, P. (2007) The grapevine genome sequence suggests ancestral hexaploidization in major angiosperm phyla. Nature (Lond) 449:463–7.
5. The Multinational Arabidopsis Steering Commitee (2007) The Multinational Coordinated *Arabidopsis thaliana* Functional Genomics Project. Annual Report 2007.
6. Bino, R. J., Hall, R. D., Fiehn, O., Kopka, J., Saito, K., Draper, J., Nikolau, B. J., Mendes, P., Roessner-Tunali, U., Beale, M. H., Trethewey, R. N., Lange, B. M., Wurtele, E. S. & Sumner, L. W. (2004) Potential of metabolomics as a functional genomics tool. Trends Plant Sci 9:418–25.
7. Oksman-Caldentey, K. M. & Saito, K. (2005) Integrating genomics and metabolomics for engineering plant metabolic pathways. Curr Opin Biotechnol 16:174–9.
8. Schauer, N. & Fernie, A. R. (2006) Plant metabolomics: towards biological function and mechanism. Trends Plant Sci 11:508–16.
9. Saito, K., Hirai, M. & Yonekura-Sakakibara, K. (2008) Decoding genes by coexpression network and metabolomics: 'majority report by precogs.' Trends Plant Sci 13:36–43.
10. D'Auria, J. C. & Gershenzon, J. (2005) The secondary metabolism of *Arabidopsis thaliana*: growing like a weed. Curr Opin Plant Biol 8:308–16.
11. Tohge, T., Nishiyama, Y., Hirai, M. Y., Yano, M., Nakajima, J., Awazuhara, M., Inoue, E., Takahashi, H., Goodenowe, D. B., Kitayama, M., Noji, M., Yamazaki, M. & Saito, K. (2005) Functional genomics by integrated analysis of metabolome and transcriptome of *Arabidopsis* plants over-expressing an MYB transcription factor. Plant J 42:218–35.
12. Luo, J., Nishiyama, Y., Fuell, C., Taguchi, G., Elliott, K., Hill, L., Tanaka, Y., Kitayama, M., Yamazaki, M., Bailey, P., Parr, A., Michael, A. J., Saito, K. & Martin, C. (2007) Convergent evolution in the BAHD family of acyl transferases: identification and characterization of anthocyanin acyl transferases from *Arabidopsis thaliana*. Plant J 50:678–95.
13. Wangwattana, B., Koyama, Y., Nishiyama, Y., Kitayama, M., Yamazaki, M. & Saito, K. (2008) Characterization of PAP1-upregulated glutathione S-transferase genes in *Arabidopsis thaliana*. Plant Biotechnol 25:191–6.
14. Yonekura-Sakakibara, K., Tohge, T., Niida, R. & Saito, K. (2007) Identification of a flavonol 7-*O*-rhamnosyltransferase gene determining flavonoid pattern in *Arabidopsis* by transcriptome coexpression analysis and reverse genetics. J Biol Chem 282:14932–41.

15. Obayashi, T., Kinoshita, K., Nakai, K., Shibaoka, M., Hayashi, S., Saeki, M., Shibata, D., Saito, K. & Ohta, H. (2007) ATTED-II: a database of co-expressed genes and cis elements for identifying co-regulated gene groups in *Arabidopsis*. Nucleic Acids Res 35:D863–9.
16. Goda, H., Sasaki, E., Akiyama, K., Maruyama-Nakashita, A., Nakabayashi, K. et al. (2008) The AtGenExpress hormone- and chemical-treatment data set: Experimental design, data evaluation, model data analysis, and data access. Plant J 55:526–42.
17. Tohge, T., Yonekura-Sakakibara, K., Niida, R., Watanabe-Takahashi, A. & Saito, K. (2007) Phytochemical genomics in *Arabidopsis thaliana*: a case study for functional identification of flavonoid biosynthesis genes. Pure Appl Chem 79:811–23.
18. Yano, M., Kanaya, S., Altaf-UI-Amin, M., Kurokawa, K., Hirai, M. Y. & Saito, K. (2006) Integrated data mining of transcriptome and metabolome based on BL-SOM. J Comput Aid Chem 7:125–36.
19. Hirai, M. Y., Yano, M., Goodenowe, D. B., Kanaya, S., Kimura, T., Awazuhara, M., Arita, M., Fujiwara, T. & Saito, K. (2004) Integration of transcriptomics and metabolomics for understanding of global responses to nutritional stresses in *Arabidopsis thaliana*. Proc Natl Acad Sci U S A 101:10205–10.
20. Hirai, M. Y., Klein, M., Fujikawa, Y., Yano, M., Goodenowe, D. B., Yamazaki, Y., Kanaya, S., Nakamura, Y., Kitayama, M., Suzuki, H., Sakurai, N., Shibata, D., Tokuhisa, J., Reichelt, M., Gershenzon, J., Papenbrock, J. & Saito, K. (2005) Elucidation of gene-to-gene and metabolite-to-gene networks in *Arabidopsis* by integration of metabolomics and transcriptomics. J Biol Chem 280:25590–5.
21. Hirai, M. Y., Sugiyama, K., Sawada, Y., Tohge, T., Obayashi, T., Suzuki, A., Araki, R., Sakurai, N., Suzuki, H., Aoki, K., Goda, H., Nishizawa, O. I., Shibata, D. & Saito, K. (2007) Omics-based identification of *Arabidopsis* Myb transcription factors regulating aliphatic glucosinolate biosynthesis. Proc Natl Acad Sci U S A 104:6478–83.
22. Kusano, M., Fukushima, A., Arita, M., Jonsson, P., Moritz, T., Kobayashi, M., Hayashi, N., Tohge, T. & Saito, K. (2007) Unbiased characterization of genotype-dependent metabolic regulations by metabolomic approach in *Arabidopsis thaliana*. BMC Syst Biol 1:53.

A Possible Role of Homeostasis Between Monomeric and Filamentous Actin in Filament Nucleation Revealed by Pharmacokinetic Modeling

Naoki Watanabe and Chiharu Higashida

Introduction

The currently prevailing method to elucidate molecular functions in vivo is to knock out or knock down gene expression and to observe the resultant phenotypic changes. For example, fibroblasts extend pseudopods called lamellipodia toward the source of the platelet-derived growth factor (PDGF) gradient. We now know tens of molecules, both cell signaling intermediates and cytoskeleton regulators, involved in such cell polarization and migration. They interact with each other in a complicated manner both spatially and temporally. Our laboratory has recently carried out a screening to examine the effect of knockdown of Rho family GTPases on PDGF-induced chemotaxis. Many of these molecules have been implicated in the regulation of the cytoskeleton. Depletion of several Rho family GTPases had profound and distinct effects on cell morphology during migrating (Monypenny, J. et al., submitted). But given the complexity of the molecular interplay, it often remains unclear whether the induced phenotype is a direct or an indirect consequence of individual treatment. Even with the data showing clear morphological phenotypes, the question as to whether individual molecules function locally or globally, for example, is not easy to answer.

How, then, can we investigate such fast and complex mechanisms in detail?

Currently, we are applying several approaches to solve this question. One is directly capturing biochemical reactions quantitatively in living cells using "single-molecule observation of cytoskeletal reorganization." The second is "fast perturbation of the system using small compounds." Drug treatment provides useful information in combination with time-lapse imaging as it enables us to capture initial effects of perturbation. Then, we apply "systematic configuration of biochemistry" to discover hidden connections between biochemical functions in the system. The actin system is one of the best characterized systems biochemically in

N. Watanabe and C. Higashida
Department of Pharmacology, Kyoto University Faculty of Medicine, Yoshida Konoe-cho, Sakyo-ku, Kyoto 606-8501, Japan

S. Nakanishi et al. (eds.), *Systems Biology*,
DOI:10.1007/978-4-431-87704-2_15, © Springer 2009

vitro, and much information is available. Here we describe a specific topic on our recent discovery of a role of homeostasis between monomeric and filamentous actin in filament nucleation. In this example, systematic integration of biochemical parameters by the use of pharmacokinetic simulation gave us an insight into an unexpected outcome induced by drug manipulation of cellular actin systems.

Turnover Between Filamentous and Globular Actin in Cells

In cells, actin exists in two forms, monomeric G-actin and filamentous F-actin. Current modern microscopy allows single-molecule observation of fluorescently labeled actin at the filamentous state in living cells. Cytoskeletal proteins stay attached to cellular structures on the order of seconds and stop random diffusion temporarily. This property makes it feasible to visualize single molecules of cell structure-associated probes when introduced at a very low density. In this way, we can visualize and measure spatiotemporal dynamics of assembly and disassembly of cytoskeletons in living cells [1,2].

As observed in our single-molecule imaging using GFP-actin as a probe [1], F-actin turnover is very dynamic. In typical cultured cells, the conversion between G-actin and F-actin occurs in every one-half minute to several minutes [1,3–5]. Another property commonly found in cultured cells is that G-actin and F-actin exist at roughly a 1:1 ratio. The cellular concentration of G-actin may exceed 100 μM. Therefore, similarly abundant G-actin-sequestering proteins are required to bind and prevent G-actin from polymerization (Fig. 1).

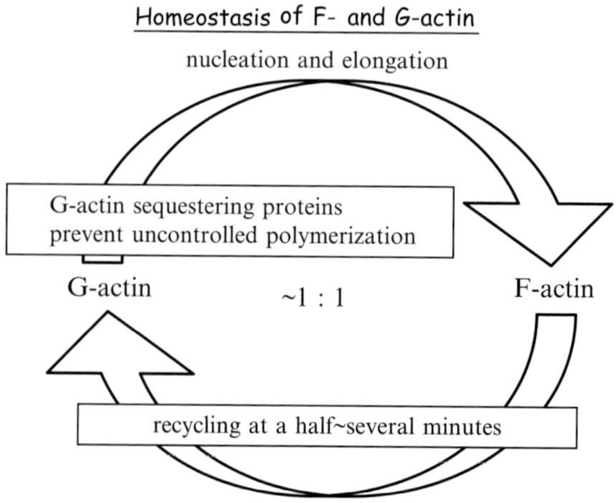

Fig. 1 Actin filament turnover in cells. In typical cultured cells, conversion between monomeric G-actin and filamentous F-actin occurs in every one-half minute to a few minutes. Another common property is that G-actin and F-actin exist roughly at a 1:1 ratio. The concentration of G-actin in cells may exceed 100 μM. G-actin-sequestering proteins bind G-actin and prevent it from polymerization

To form de novo filaments from G-actin, actin assembly undergoes two steps, nucleation and elongation [6]. In the nucleation step, the formation of actin dimers and trimers occurs frequently in solution. However, dissociation of subunits is also very fast, and this reaction barely proceeds forward. Once filament nuclei consisting of four actin subunits are formed, they act as a filament, and actin elongation proceeds steadily. In cells, elongation is driven largely by the profilin–actin complex. The rate of actin nucleation depends on free G-actin squared to cubed concentration. Filament nucleation occurs spontaneously in physiological salt condition, but this step requires a long time in typical in vitro experiments.

In cells, to nucleate filaments efficiently, actin nucleators are required. One of the major actin nucleators is Arp2/3 complex [7,8]. Arp2/3 complex nucleates the filaments off the side of preexisting filaments. This branched network is abundantly found at the leading edge of migrating cells. On the other hand, the Formin family (Formins) is responsible for the formation of a different set of actin filaments such as actin stress fibers and cytokinetic contractile rings. These Formin-dependent actin structures are characterized as being composed of long actin filaments.

There is another difference between two actin nucleators. The Arp2/3 complex binds the slow-growing end of the new filament and allows filament growth at the fast-growing end. By contrast, after nucleating new actin filament, Formins stay associated with its fast-growing end during filament growth. This unique property is useful to assemble long actin filaments because it can prevent the association of Capping protein. Capping protein inhibits elongation of actin. The anti-Capping protein activity of Formins thus explains why Formin-dependent filaments are long.

Low-Dose Actin Monomer Inhibitors Promote Actin Nucleation by mDia1 in Cells But Not In Vitro

Previously, we identified a member of Formins called mDia1 as an effector of the small GTPase Rho [9]. Rho is a molecular switch that regulates cytokinesis, smooth muscle contraction, and actin stress fiber formation in cells. Formins share conserved FH1 and FH2 regions in their C-terminal halves and also play a critical role in cytokinesis and cell polarity of various species. The GTP-bound form of Rho binds the N-terminal region of mDia1. This binding opens autoinhibited mDia1 to release the activity of its FH1–FH2 unit [10]. The FH1–FH2 region of mDia1 has a strong promoting effect on actin stress formation in cells [10]. With the FH2 domain, Formins nucleate actin filaments and bind the growing ends of filaments [11,12]. Remarkably, our single-molecule observation of the FH1–FH2 domain mutant of mDia1 has revealed fast long-range directional movement of molecules [13]. By marked contrast, however, we barely observed processive movement of full-length mDia1 in cells.

Recently, we sought cellular conditions where directional movement of native full-length mDia1 is induced frequently [14]. Interestingly, we came across an unexpected drug, latrunculin B, as a strong inducer of processive movement of mDia1. LatB binds G-actin near its ATP-binding site. LatB prevents actin monomer from assembling into filaments. LatB is a drug widely used to probe actin-dependent cellular processes. When administrated at a high concentration, over 1 μM, LatB stops the movement of mDia1 because of a loss in the polymerization-competent G-actin. However, we observed that at a low dose, LatB dramatically increases the frequency of emergence of processively moving full-length mDia1. Processive mDia1Full was frequently observed as early as 10 s after the perfusion of 100 nM LatB. In addition, the speed of mDia1 rapidly decreased to less than 40%. This finding provided important information in our simulation, which is described next.

After a series of experiments, we found that the FH1–FH2 or FH2 region of mDia1 alone can respond to low-dose LatB and exhibit increased frequency of processive movement. We also confirmed that Rho is required for full-length mDia1 being in an open conformation but is not a direct mediator of LatB-increased actin nucleation by mDia1 in cells. We also tested if LatB directly promotes mDia1-catalyzed actin assembly in vitro. However, under various conditions tested, LatB only had a negative effect on mDia1FH2-mediated actin filament assembly in vitro.

Paradoxical Increase in Free G-Actin Revealed by Pharmacokinetic Simulation of Latrunculin B Treatment

Given the distinct effects of LatB on mDia1-catalyzed actin nucleation between in cells and in vitro, we reevaluated the pharmacological effect of LatB in cellular actin systems. We built a pharmacokinetic simulation to estimate changes in the cytoplasmic concentration of G-actin, either free or bound to thymosin-β4 (Tβ4), profilin, and latrunclin B (LatB) in our experiments. In the model, two major G-actin-sequestering proteins are considered. Profilin binds G-actin noncompetitively with latrunculin. On the other hand, the binding of latrunculin to G-actin substantially reduces the affinity between Tβ4 and G-actin [15]. Given the large amounts of the two proteins, we expect that the effect of other minor G-actin-binding proteins can largely be represented by these two proteins with different competitiveness against LatB.

The simulation scheme is illustrated in Fig. 2. The scheme can be described by simultaneous differential equations. Involvement of actin filament turnover was also considered based on the observed decreased speed of mDia1. Unexpectedly, simulation analysis predicted a rapid several-fold increase in free G-actin concentration that occurs within 1 min. Although some parameters such as the ratio between G-actin and Tβ4 have not been precisely defined yet, we consistently observed several-fold increases in free G-actin within a reasonable range of parameter values in calculation. The estimated increases were in good agreement with the time course of the increase in the number of processively moving mDia1Full molecules. It is of note that the increase in free G-actin occurs despite LatB accumulating in the cytoplasm at the several micromolar level.

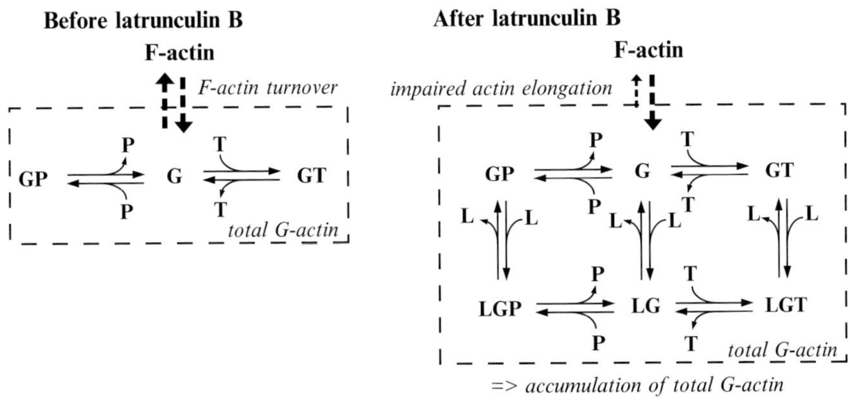

Fig. 2 Pharmacokinetic modeling of latrunculin B treatment. *G*, *L*, *T*, and *P* represent G-actin, LatB, thymosin-β4 (Tβ4), and profilin, respectively. The scheme was described by differential equations that estimated time-dependent concentration changes of each G-actin species either free or bound to LatB, Tβ4, and/or profilin. LatB is a widely used research tool to probe actin-dependent cellular processes. Intuitively, LatB is conceived of decreasing polymerization-competent G-actin (*downward on the right scheme*). However, our simulation revealed the more complex behavior of the system. LatB binds G-actin noncompetitively with profilin, whereas LatB-bound G-actin binds Tβ4 with a substantially reduced affinity [15]. Might this property cause a paradoxical increase in drug-free G-actin? Just the addition of low-dose LatB in the circuit (*squares surrounded by dashed lines*) affected the free G-actin concentration only marginally. Instead, a change in the balance between F-actin and total G-actin had a profound effect on the free G-actin concentration. In experiments, we observed a decrease in actin elongation speed by mDia1 and an ~30% F-actin decrease in cells treated with low-dose LatB. Incorporation of the observed shift from F- to G-actin into the model led us to find a paradoxical effect of LatB, which is to increase free G-actin several fold [14]

F-actin turnover and its modulation by LatB are the key for the induction of the paradoxical G-actin increase in the simulation. If we turned off the term for mimicking the imbalance between actin assembly and disassembly in the calculation, no obvious change in the free G-actin concentration was observed over time. This finding indicates that an increase in total G-actin plays a pivotal role. Our model predicts that a ~30% increase in the total G-actin results in the several-fold increase in free G-actin regardless of various changes in other parameters. Hence, although the primary pharmacological effect of LatB is inactivation of G-actin, LatB treatment also induces a 1.2- to 1.3-fold increase in total G-actin, which may lead to a several-fold free G-actin increase accompanied by saturation of the buffering capacity of G-actin-sequestering proteins. Our finding highlights an unprecedented example of pharmacological effects in which an antagonist increases the concentration of drug-free targets in a complex biological system [14].

Figure 3 summarizes our current model for dual regulation of mDia1-catalyzed actin assembly. We predict that Rho signaling determines the ratio of the opened and closed forms of mDia1, whereas an increase in free G-actin efficiently converts

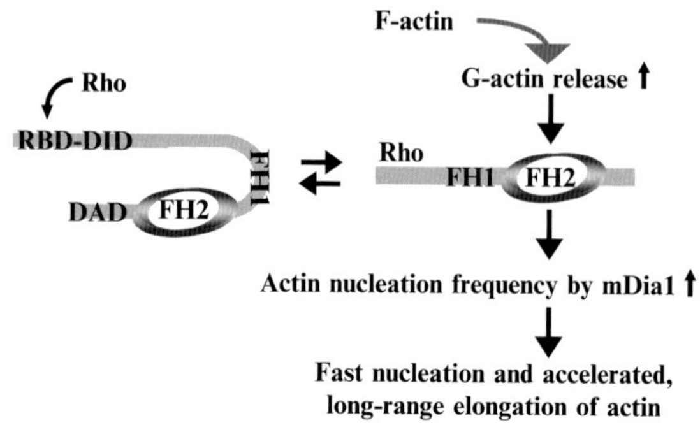

Fig. 3 The feedback actin polymer restoration mechanism involving *mDia1*. Rho signaling determines the ratio between the closed (*left*) and the opened (*right*) form of mDia1, whereas an increase in free G-actin efficiently converts "opened" mDia1 to the fully active actin nucleating state, leading to fast long-range actin filament assembly in cells

"opened" mDia1 to the fully active actin-nucleating and processive actin-capping state, leading to fast, long-range actin filament assembly.

Concluding Remarks

Currently, we are investigating the physiological relevance of G-actin regulation of actin nucleation. Our single-molecule imaging results suggest that the G-actin regulation mechanism may apply not only for mDia1 but also for some other actin nucleators. We also observed coupling of subcellular localization between mDia1-associated actin nucleation and actin-depolymerizing molecules [14]. We therefore predict that concentration fluctuation of G-actin and its partners may play a role in various cell morphological regulations.

Modeling and computational analysis would be vital to our understanding of complex biological systems in the near future. Yet, we need well-defined parameters and better understanding of individual interconnections between biological functions. Our current analysis highlights that we still do not realize "systems" as simple as this case. We anticipate step-by-step revelation of hidden secrets in biology by the joint approach between various research fields, including systems biology.

References

1. Watanabe, N., and Mitchison, T. J. (2002). Single-molecule speckle analysis of actin filament turnover in Lamellipodia. Science 295:1083–1086.
2. Miyoshi, T., Tsuji, T., Higashida, C., Hertzog, M., Fujita, A., Narumiya, S., Scita, G., and Watanabe, N. (2006). Actin turnover-dependent fast dissociation of capping protein in the

dendritic nucleation actin network: evidence of frequent filament severing. J Cell Biol 175:947–955.

3. Theriot, J. A., and Mitchison, T. J. (1991). Actin microfilament dynamics in locomoting cells. Nature (Lond) 352:126–131.
4. McGrath, J. L., Tardy, Y., Dewey, C. F., Jr., Meister, J. J., and Hartwig, J. H. (1998). Simultaneous measurements of actin filament turnover, filament fraction, and monomer diffusion in endothelial cells. Biophys J 75:2070–2078.
5. Mallavarapu, A., and Mitchison, T. (1999). Regulated actin cytoskeleton assembly at filopodium tips controls their extension and retraction. J Cell Biol 146:1097–1106.
6. Pollard, T. D., Blanchoin, L., and Mullins, R. D. (2000). Molecular mechanisms controlling actin filament dynamics in nonmuscle cells. Annu Rev Biophys Biomol Struct 29:545–576.
7. Welch, M. D., Rosenblatt, J., Skoble, J., Portnoy, D. A., and Mitchison, T. J. (1998). Interaction of human Arp2/3 complex and the Listeria monocytogenes ActA protein in actin filament nucleation. Science 281:105–108.
8. Mullins, R. D., Heuser, J. A., and Pollard, T. D. (1998). The interaction of Arp2/3 complex with actin: nucleation, high affinity pointed end capping, and formation of branching networks of filaments. Proc Natl Acad Sci U S A 95:6181–6186.
9. Watanabe, N., Madaule, P., Reid, T., Ishizaki, T., Watanabe, G., Kakizuka, A., Saito, Y., Nakao, K., Jockusch, B. M., and Narumiya, S. (1997). p140mDia, a mammalian homolog of *Drosophila diaphanous*, is a target protein for Rho small GTPase and is a ligand for profilin. EMBO J 16:3044–3056.
10. Watanabe, N., Kato, T., Fujita, A., Ishizaki, T., and Narumiya, S. (1999). Cooperation between mDia1 and ROCK in Rho-induced actin reorganization. Nat Cell Biol 1:136–143.
11. Evangelista, M., Zigmond, S., and Boone, C. (2003). Formins: signaling effectors for assembly and polarization of actin filaments. J Cell Sci 116:2603–2611.
12. Watanabe, N., and Higashida, C. (2004). Formins: processive cappers of growing actin filaments. Exp Cell Res 301:16–22.
13. Higashida, C., Miyoshi, T., Fujita, A., Oceguera-Yanez, F., Monypenny, J., Andou, Y., Narumiya, S., and Watanabe, N. (2004). Actin polymerization-driven molecular movement of mDia1 in living cells. Science 303:2007–2010.
14. Higashida, C., Suetsugu, S., Tsuji, T., Monypenny, J., Narumiya, S. and Watanabe, N. (2008). G-actin regulates rapid induction of actin nucleation by mDia1 to restore cellular actin polymers. J Cell Sci 121:3403–3412.
15. Yarmola, E. G., Somasundaram, T., Boring, T. A., Spector, I., and Bubb, M. R. (2000). Actin-latrunculin A structure and function. Differential modulation of actin-binding protein function by latrunculin A. J Biol Chem 275:28120–28127.

Molecules, Networks, and Memory

Upinder S. Bhalla

Introduction

A profound evolution of modeling scope and scale has occurred in the field as we have sought to understand how memory works at the level of molecular networks. We have moved from an initial concept of a small number of relatively simple synaptic functions to the current appreciation of the complexity of function and cellular mechanisms that support these functions.

One of the initial triumphs of the field was to find a molecular correlate of the crucial property of Hebbian associativity [1]. A list of further functional modules of the synapse might include the detection and selection of memory-triggering activity patterns; chemical circuits that store information through feedback and switching; presynaptic release control; movement of key molecules into and out of the synapse; and activity-triggered protein synthesis (Fig. 1). More recent data suggest further modules for the formation and maintenance of synaptic contacts and pre- and postsynaptic structure. This long list of synaptic functions may suggest a pessimistic view of our ability to either complete the list or understand the signaling mechanisms. I propose, instead, that the list is a roadmap, and hence a clear statement of where modeling studies need to go. The system is undoubtedly complex, and therein lies the challenge for the field. Typical models for these functional modules contain tens to hundreds of molecular species and reactions. In keeping with the theme of this symposium and book, I suggest that the challenge of biological complexity is best met squarely, by developing models that embrace the complexity but are not overwhelmed by it. This systematic process of model building, one synaptic function at a time, is the basis for an eventual quantitative understanding of the synapse and its role in memory.

We owe our initial view of synaptic function in memory to two streams of neuroscience: anatomy, from Cajal, and psychology, from Donald Hebb. At the risk of

U.S. Bhalla
National Centre for Biological Sciences, TIFR, Bellary Road, Bangalore 560065, India

S. Nakanishi et al. (eds.), *Systems Biology*,
DOI:10.1007/978-4-431-87704-2_16, © Springer 2009

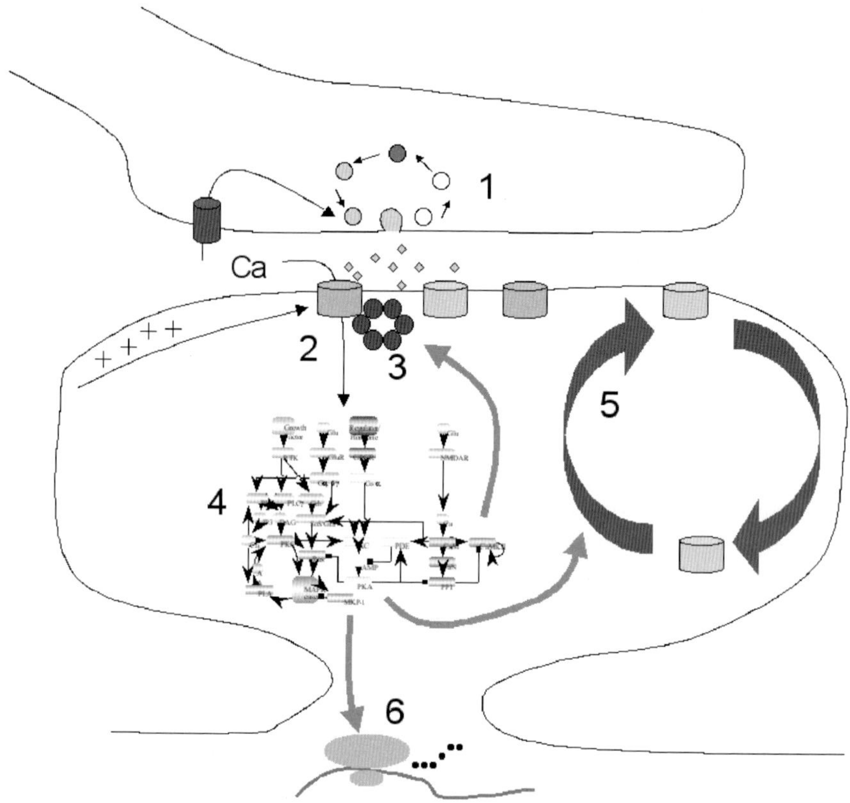

Fig. 1 Modules in synaptic function. *1* Presynaptic vesicle release control. *2* Associativity. The *N*-methyl D-asparate (NMDA) channel opens only when the input neurotransmitter is associated with postsynaptic depolarization. *3* Bistability. The hexagonal set of molecules represents calcium calmodulin-dependent type II kinase (CaMKII), which may act as a bistable switch. *4* Pattern selectivity. The complex of signaling pathways leads to different outcomes depending on input patterns in time and on cellular context. *5* Molecular movement. Here the movement of the alpha-amino-3-hydroxy-5-methyl-4-isoxazolepropionic acid (AMPA) receptor is shown, which is regulated by activity. *6* Dendritic protein synthesis

doing many contemporary scientists an injustice, one could attribute the ideas of connectivity to Cajal and the ideas of associativity to Hebb. The key breakthrough that completed this early picture of associative memories came from the work of Ascher and Nowak in the 1980s. They showed that the *N*-methyl D-asparate (NMDA) receptor could implement the crucial function of input-output associativity at the synapse [2]. This function turned out to be a relatively simple process of magnesium ion blockage of the receptor until the channel was depolarized. Thus, ion flow through the NMDA receptor could only occur if input activity (glutamate neurotransmitter) were associated with output activity (depolarization of the post-synaptic cell). This finding was the first indication of how synaptic logic could implement crucial memory functions. Thus, at this stage the field had a satisfyingly consistent view of memory in terms of connectivity and the rules that led to change in connectivity.

The next level of inquiry led into the mechanistic details of synaptic function and into intracellular signaling networks. To recapitulate, some of the key functional modules are chemical circuits that store information; presynaptic release control; selection of memory triggering activity patterns; movement of synaptic molecules; and activity-triggered protein synthesis. We have pursued development of detailed signaling models of these functional modules as a roadmap to understanding synaptic function.

Information Storage

After associativity, there followed one of the central questions of synaptic function: How do synapses store information? By the time this question was addressed at the molecular level, the electrical properties of synaptic modification were well established [1]. It was known that synaptic conductances increased and stayed elevated for long periods following even a brief calcium influx. This sustained change suggested that some structural or chemical activity was causing the synapse to switch to a new state of higher conductance. This was remarkable enough, but the duration of the change posed a special problem, because individual molecules turn over in days or less [3]. Taking a cue from computer memories, the concept of a bistable chemical "switch" became a central theme of these studies. Such switches typically have a feedback process such that they can either remain in a quiescent state or a self-sustaining state of high activity (Fig. 2i–iii). One candidate switching molecule was calcium calmodulin-dependent type II kinase (CaMKII). Many theoretical models [4,5] have suggested that CaMKII may act as a bistable switch. However, experimental studies have revealed a great deal of complexity in this putative switch, and the details are still being worked out. For example, CaMKII translocates between a soluble fraction in the synapse and a particulate fraction at the postsynaptic density (PSD) [6]. We have shown in our models that this translocation process may be a critical contributor to long-lasting bistability [7]. In the soluble form, CaMKII can be rapidly dephosphorylated by protein phosphatase 2A (PP2A). However, at the PSD the access to the kinase is restricted to protein phosphatase 1 (PP1), which is saturable. The balance of CaMKII autophosphorylation and PP1 dephosphorylation at the PSD may therefore be in the range where bistability is possible (Fig. 2iv).

Other pathways have also been proposed for synaptic switching, and several of these now have experimental support. We have proposed that the mitogen-activated protein kinase (MAPK) pathway forms a positive feedback loop with phospholipase A2 and protein kinase C [8] (Fig. 2v). We have subsequently provided experimental evidence that this feedback loop does indeed function in fibroblasts [9].

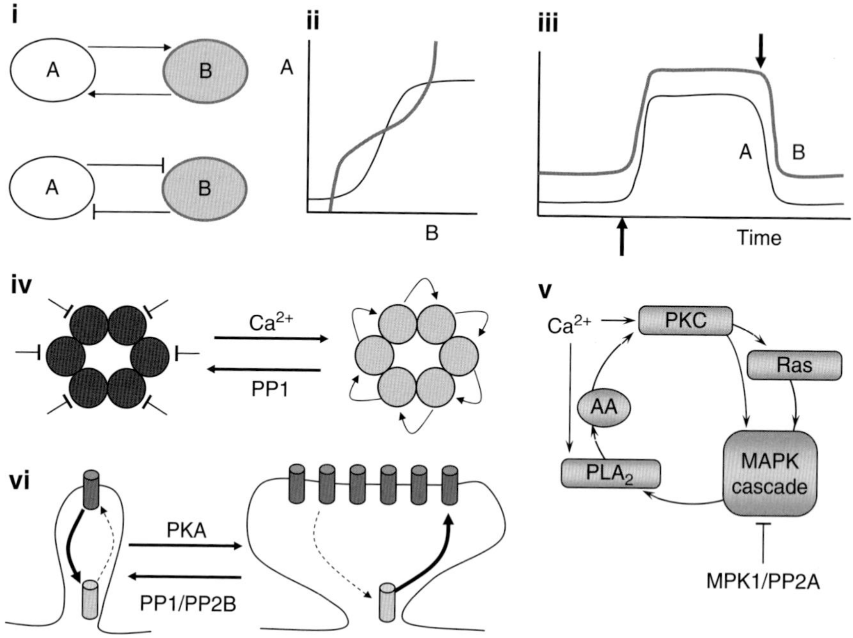

Fig. 2 Bistability in synaptic plasticity. **i** Two possible feedback circuits that may give bistability. The *top* one is mutually excitatory; the *bottom* one is mutually inhibitory. The *net sign* around the loop is positive in both cases. **ii** Bistability analysis using superimposed dose–response curves for *A* and *B* with respect to each other. The presence of three intersection points indicates bistability. **iii** Functioning of a bistable switch. The system remains stable in a low state of activity until a brief stimulus arrives (*lower arrow*). It then switches to a state of high activity, thus "remembering" the first stimulus. It remains active until an inhibitory stimulus arrives (*upper arrow*), at which point it switches back to the low state. **iv** CaMKII bistability. The CaMKII subunits phosphorylate and therefore activate each other. The initial trigger is Ca^{2+} influx. To turn the system off, protein phosphatase 1 (PP1) must dephosphorylate enough subunits to arrest the feedback cycle. **v** Mitogen-activated protein kinase (MAPK) bistability through a feedback loop involving phospholipase A2 (PLA2) and protein kinase C (PKC). Ca^{2+} is an activating stimulus, and the phosphatases MKP1 and PP2A can turn off the activity. **vi** AMPA receptor cycling and bistability. In the low state, only a few receptors are present at the synaptic junction. In the high state, many receptors are present, and this saturates the dephosphorylation process required to remove them

Other groups have investigated its role in cerebellar long-term depression (LTD) [10], and very recent experimental evidence for this is presented in this volume. We have reported a further, unexpected switch that emerged from an analysis of receptor trafficking (Fig. 2vi) [7]. This switch behaves somewhat differently from the feedback loops of the CaMKII and MAPK switches. Instead of self-activating feedback, the bistability arises from saturation of transport to and from the synapse. Interestingly, one of the saturated molecules is the same phosphatase (PP1) that is involved in CaMKII regulation.

Overall, there are now several candidate mechanisms for synaptic information storage. It is now considered likely that multiple biochemical switches may act in concert to pass synaptic "memory" through multiple phases of formation and consolidation [1,11]. Our current models of such switches incorporate tens to hundreds of molecular species and many hundreds of reaction steps [7,8,12]. This unresolved complexity has come as a challenge to the field, specially given the simplicity of the initial synaptic operation of associativity.

Presynaptic Release Control

The presynaptic terminal has been somewhat neglected from a modeling point of view, despite its crucial role in synaptic transmission and considerable experimental and theoretical evidence for its role in information storage. This is partly accounted for by its relative experimental inaccessibility, which deprives us of the wealth of signaling and physiological data that have contributed so much to understanding postsynaptic signaling. Recent experimental and modeling studies have begun to make inroads into this system [13]. Short-term plasticity effects such as paired-pulse facilitation or depression have been modeled in terms of the calcium dynamics at the presynaptic terminal. One of the key unknowns is the mechanisms by which presynaptic signaling contributes to changes in the total synaptic efficacy. This mechanism presumably involves some form of retrograde signal from the postsynaptic side to account for associativity with postsynaptic activity. Clearly, any complete picture of synaptic function must account for the details of these presynaptic events.

Pattern Selectivity

Synapses are highly selective about circumstances that may lead to plasticity. These circumstances may include strong activity, specific temporal patterns of activity, and context in the form of other synaptic inputs to the cell as well as hormonal and broadcast neurotransmitter input. The synapse must continually monitor all these inputs to decide whether to maintain its synaptic weight at a steady level or to increase or decrease it. Such decisions tie into plasticity switches as well as to structural change and control of protein synthesis.

We began analyzing this issue with the observation that different signaling pathways exhibit a very wide range of time courses. Using this as a starting point, we tested our models to see if a set of four major kinase pathways at the synapse might exhibit selectivity for different activity patterns [14,15]. This turned out to be the case, and the models also predicted mechanisms by which these tuning properties might be modulated by neurotrophins or hormones. We followed this initial observation with a systematic analysis of patterns used to elicit long-term potentiation (LTP), and found that the MAPK pathway was tuned to spaced synaptic input in a manner reminiscent of one of the strongest LTP-inducing patterns

[16]. In contrast, CaMKII was best activated for brief, strong input. We experimentally tested these predictions and confirmed that the MAPK cascade, but not the CaMKII pathway, exhibited tuned activity tuning to spaced input in a manner very similar to our initial predictions [16]. In ongoing work, we find that the protein synthesis pathways may also be capable of some degree of pattern selectivity.

These studies converge to indicate that synapses implement a very effective system of pattern selectivity through their signaling networks.

Movement of Synaptic Molecules

Synapses are highly dynamic structures. They undergo changes in their morphology as a function of time, activity, and system cues such as stress and neurotrophins. As a first step to characterizing the complex details of synaptic remodeling, researchers have begun to look at how molecules move around in the synapse. Such studies have been carried out at exquisite, single-molecule detail to look at how synaptic ultrastructure affects molecular diffusion [17]. We have been examining less spatially detailed, but chemically complex, active movement of molecules at the synapse and dendrite.

One of the simple mechanisms for such movement harnesses diffusion by the regulation of binding sites at different target locations. We examined control of two important synaptic molecules by this mechanism: CaMKII and the alpha-amino-3-hydroxy-5-methyl-4-isoxazolepropionic acid (AMPA) receptor. CaMKII moves into the PSD following strong calcium influx, whereas the AMPA receptor translocation is governed by its phosphorylation state. Although our model was designed primarily to monitor receptor movement, this system turned out to have the unexpected emergent property of bistability [7]. The "down" state of the system had very few AMPA receptors at the PSD and hence corresponded to a weak or silent synapse. In this state, the basal levels of PP1 at the PSD were sufficient to rapidly dephosphorylate the receptor and thus mark it for movement back into the reserve pools (Fig. 2vi). The "up" state had ~150 receptors at the PSD, a level comparable to active synapses. In this condition, the PP1 became saturated, and therefore the removal of AMPA receptors was balanced by their insertion into the PSD. We predicted that the saturation of PP1 could also be a mechanism for coupling with the movement of CaMKII to the PSD. One of the particularly striking features of this coupled bistable system was that it was robust with respect to stochasticity. Chemical noise compromises the stability of bistable chemical switches, leading to spontaneous state changes, which would be disastrous for a memory switch. In the AMPAR-CaMKII switch, we found that the spontaneous switching time was well over a year, making it an attractive candidate for long-term memory storage [7].

Although these initial models are interesting, they are clearly at a very early stage for characterizing the complex structural and molecular movement events that are involved in synapse formation and memory. It seems likely that more complete models will require the integration of three-dimensional reaction-diffusion calculations [17] along with mechanical models of structural components of the synapse.

Activity-Triggered Protein Turnover

In concert with regulating the movement of its constituent molecules, the synapse must also regulate their supply. This control involves a delicate balance between molecular import and export to the soma and local synthesis and degradation. We have very recently developed models for one aspect of this process: activity-dependent protein at dendritic sites adjacent to synapses. This site turns out to be a convergence point for multiple signals, including neurotrophins, at least two calcium-responsive pathways, and cellular metabolic inputs. Our simulations show that the pathways are capable of differential activation to different activity patterns involved in LTP and LTD, potentially leading to selective expression of different proteins depending on context. We had anticipated that activity-dependent protein synthesis might be a locus for yet another synaptic bistable switch, because it has numerous potential positive feedback loops. For example, several components of the protein synthesis machinery are themselves synthesized locally. Furthermore, brain-derived neurotrophic factor (BDNF), one of the factors that activates the pathway, is also synthesized. Nevertheless, our quantitative analysis suggests that none of the known feedback mechanisms can give rise to bistability in this system. We speculate that this is an illustration of functional modularity in the synapse, in that the switching properties are separate from the protein synthesis control mechanisms.

Discussion

Associativity, information storage, presynaptic release, pattern selectivity, molecular movement, and protein synthesis: this weighty list accounts for many of the known functions the synapse must support to store memories. It is an intimidating list of functions, encompassing as it does much of cell biology and physiology. Nevertheless, it also serves as a roadmap for experiments and, increasingly, for models that can encapsulate the complexity and define the essential steps.

Over the years, my group and others have built a parts list of these key processes that underlie memory, in the form of increasingly detailed biochemical and biophysical computer models. Although these modules already are complex, we have been impelled by our own and other experiments to introduce still more detail. These elaborations include stochastic “noisy” chemical steps, diffusion, and interactions with ion channels and cellular biophysics. I propose that one of the key lessons of these modeling studies is that synaptic complexity is neither an accident, nor something that can easily be abstracted away. Instead, it is an essential part of how the cell constructs a reliable submicrometer machine that performs complex decisions and stores information for a lifetime.

References

1. Bliss TV, Collingridge GL (1993) A synaptic model of memory: long-term potentiation in the hippocampus. Nature (Lond) 361:31–39.
2. Ascher P, Nowak L (1988) The role of divalent cations in the *N*-methyl-D-aspartate responses of mouse central neurones in culture. J Physiol 399:247–266.
3. Ehlers MD (2003) Activity level controls postsynaptic composition and signaling via the ubiquitin-proteasome system. Nat Neurosci 6:231–242.
4. Miller P, Zhabotinsky AM, Lisman JE, Wang XJ (2005) The stability of a stochastic CaMKII switch: dependence on the number of enzyme molecules and protein turnover. PLoS Biol 3:e107.
5. Zhabotinsky AM (2000) Bistability in the Ca(2+)/calmodulin-dependent protein kinase-phosphatase system. Biophys J 79:2211–2221.
6. Shen K, Teruel MN, Connor JH, Shenolikar S, Meyer T (2000) Molecular memory by reversible translocation of calcium/calmodulin-dependent protein kinase II. Nat Neurosci 3:881–886.
7. Hayer A, Bhalla US (2005) Molecular switches at the synapse emerge from receptor and kinase traffic. PLoS Comput Biol 1:137–154.
8. Bhalla US, Iyengar R (1999) Emergent properties of networks of biological signaling pathways. Science 283:381–387.
9. Bhalla US, Ram PT, Iyengar R (2002) MAP kinase phosphatase as a locus of flexibility in a mitogen-activated protein kinase signaling network. Science 297:1018–1023.
10. Kuroda S, Schweighofer N, Kawato M (2001) Exploration of signal transduction pathways in cerebellar long-term depression by kinetic simulation. J Neurosci 21:5693–5702.
11. Kennedy MB, Beale HC, Carlisle HJ, Washburn LR (2005) Integration of biochemical signalling in spines. Nat Rev Neurosci 6:423–434.
12. Bhalla US (2004) Signaling in small subcellular volumes. II. Stochastic and diffusion effects on synaptic network properties. Biophys J 87:745–753.
13. Hosoi N, Sakaba T, Neher E (2007) Quantitative analysis of calcium-dependent vesicle recruitment and its functional role at the calyx of Held synapse. J Neurosci 27:14286–14298.
14. Bhalla US (2002) Biochemical signaling networks decode temporal patterns of synaptic input. J Comput Neurosci 13:49–62.
15. Bhalla US (2002) Mechanisms for temporal tuning and filtering by postsynaptic signaling pathways. Biophys J 83:740–752.
16. Ajay SM, Bhalla US (2004) A role for ERKII in synaptic pattern selectivity on the time-scale of minutes. Eur J Neurosci 20:2671–2680.
17. Coggan JS, Bartol TM, Esquenazi E, Stiles JR, Lamont S, et al. (2005) Evidence for ectopic neurotransmission at a neuronal synapse. Science 309:446–451.

Systems Biology Meets Single-Cell Physiology: Role of a Positive-Feedback Signal Transduction Network in Cerebellar Long-Term Synaptic Depression

Keiko Tanaka and George J. Augustine

Introduction

It is widely assumed that persistent changes in the brain, such as those occurring during learning, memory, development, and various pathological conditions, are mediated by long-lasting, plastic changes in synaptic efficacy [1]. Long-term depression (LTD) of excitatory transmission at cerebellar Purkinje cells is an attractive system for studies of such long-lasting synaptic plasticity [2]. LTD arises from interaction of the two excitatory synaptic inputs that innervate Purkinje cells: climbing fibers (CF) and parallel fibers (PF). Simultaneous activation of these inputs causes a LTD of the efficacy of PF synapses [2].

The signals generated by the activity of these synapses are well understood. CF activation depolarizes the Purkinje cell and produces a rise in the intracellular Ca^{2+} This rise in $[Ca^{2+}]_i$ is necessary and sufficient to account for the CF contribution to LTD [3–5]. PF synapses produce several signals that are important for LTD. PF activity produces two messengers within Purkinje cells: diacylglycerol (DAG) and inositol trisphosphate (IP3). DAG is important for LTD induction, via its well-known ability to activate protein kinase C (PKC) [6–8], while IP3 contributes to LTD signaling by releasing Ca^{2+} from intracellular stores and thereby elevating $[Ca^{2+}]_i$ [9–12]. PF activity also appears to activate nitric oxide synthase in the presynaptic terminals of PFs, generating the membrane-permeant gas, nitric oxide (NO), that is also involved in LTD [13–17].

As its name implies, LTD is a slow process that gradually develops over 10–30 min and lasts for hours and perhaps for much longer (Fig. 1A). An initial phase of LTD results from a reduction in the number of functional alpha-amino-3-hydroxy-5-methyl-4-isoxazolepropionic acid (AMPA)-type glutamate receptors

K. Tanaka
Department of Biophysics and Biochemistry, Graduate School of Science, University of Tokyo, Tokyo, Japan, and PRESTO, Japan Science and Technology Agency, Tokyo, Japan

G.J. Augustine
Department of Neurobiology, Duke Medical Center, Box 3209, Durham, NC 27710, USA

S. Nakanishi et al. (eds.), *Systems Biology*,
DOI:10.1007/978-4-431-87704-2_17, © Springer 2009

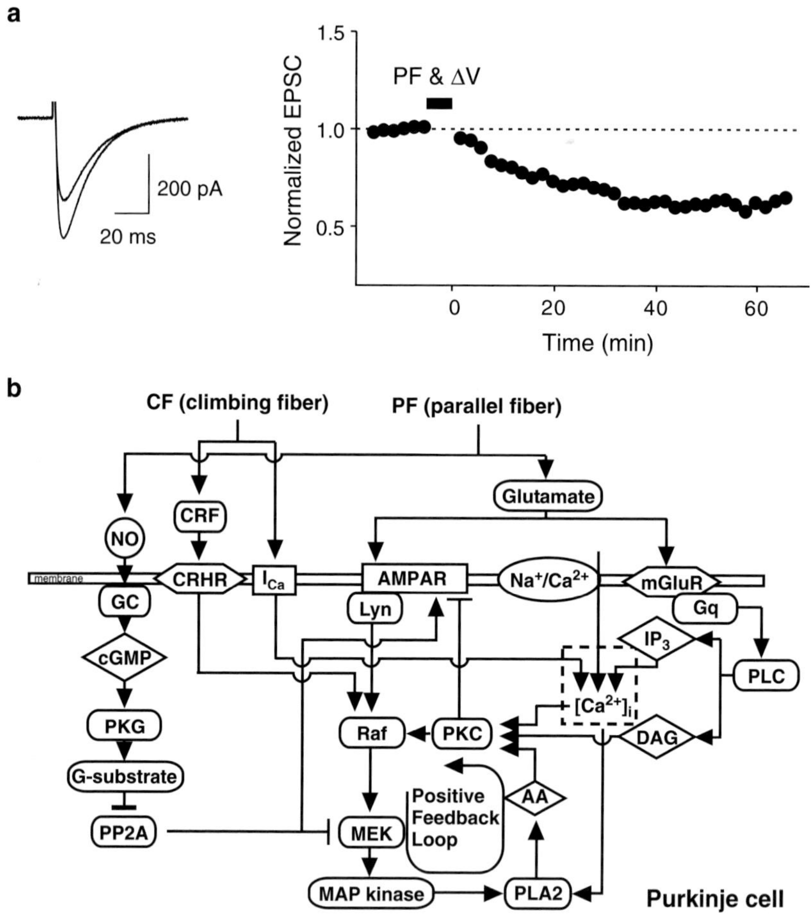

Fig. 1 **a** Long-term depression (LTD) can be elicited by pairing parallel fiber (*PF*) synaptic activity with depolarization of the Purkinje cell membrane potential, which causes a time-dependent (*right*) reduction in the amplitude of excitatory postsynaptic currents (*EPSCs*) elicited by stimulating PFs (*left*). **b** Diagram of the signaling components incorporated in the computational model of Kuroda et al. [24]

that mediate PF synaptic transmission [18]. This reduction is caused by internalization of these receptors via a clathrin-dependent endocytosis process [19,20] that is triggered by PKC phosphorylation of the GluR2 and GluR3 subunits of the AMPA receptors. A second component of LTD occurs more than an hour after LTD induction (at room temperature). This component is caused by changes in gene transcription and translation [21–23], although the affected genes have not yet been identified.

It is important to understand how the second messenger signals produced by synaptic activity, which typically disappear within a few seconds or less after synaptic activity ends [3,11], are converted into the slow reactions that cause LTD of PF synaptic transmission. A comprehensive "systems biology" style computational model was (Fig. 1B) [24]. This computational model accounts for the component of LTD that is caused by changes in AMPA receptor trafficking and uses AMPA receptor phosphorylation as a readout for LTD. The model is capable of predicting many of the features of LTD, including the stimulus requirements for LTD induction and the response of LTD to several experimental manipulations. The central feature of this model is a Ca^{2+}-triggered positive feedback loop involving many signal transduction proteins, including protein kinase C (PKC), the mitogen-activated protein kinase (MAPK), and phospholipase A2 (PLA2) [25]. This feedback loop is postulated to create a bistable system that allows PKC to be activated in a regenerative way once the positive feedback loop is activated by the signals generated by synaptic activity. Thus, the model predicts that the positive feedback loop translates the brief signals produced by synaptic activity into enduring biochemical events, specifically a sustained activation of PKC, that yield the slow and gradual development of LTD after synaptic activity ends. The goal of the work described in this chapter was to provide experimental tests of this computational model.

Calcium Requirements of LTD

Our first experimental indication of the positive feedback loop in LTD was observed when we investigated the relationship between $[Ca^{2+}]_i$ and the amount of LTD [20]. In these studies, we photolysed a "caged" Ca^{2+} compound, DMNPE-4, to increase $[Ca^{2+}]_i$ directly in the postsynaptic Purkinje cells. In this way, we could also control the length of time that $[Ca^{2+}]_i$ was elevated. As seen in Fig. 2A, for a given duration of $[Ca^{2+}]_i$ elevation, the relationship between $[Ca^{2+}]_i$ and the amount of LTD was sigmoidal, with no LTD produced at lower $[Ca^{2+}]_i$ and full LTD elicited at higher $[Ca^{2+}]_i$. These relationships could be described by the Hill equation, with a Hill coefficient of approximately 5 indicating cooperative triggering of LTD by Ca^{2+}. Prolonging the duration of $[Ca^{2+}]_i$ elevation caused a leftward shift in this relationship (closed triangles and diamonds in Fig. 2A), while decreasing the duration of $[Ca^{2+}]_i$ elevation caused a rightward shift (closed circles and squares in Fig. 2A). Thus, the duration of $[Ca^{2+}]_i$ elevation influences the Ca^{2+} requirements for LTD induction. This dynamic property indicates that LTD is induced by a "leaky integrator" process [20], similar to the "leaky integrate and fire" concept often used in models of neuronal electrical signaling [26,27].

These features of the triggering of LTD by Ca^{2+} suggest that LTD results from Ca^{2+} activating signaling processes that have regenerative properties but behave as a leaky integrator. Because such properties could arise from a positive feedback process, our collaborators T. Doi, H. Ogasawara, and M. Kawato determined whether the computational model of Kuroda et al. [24] could explain our results. Even though this model (Fig. 1B) was not originally designed to simulate the results of our experiments that manipulated $[Ca^{2+}]_i$, in fact only a few minor

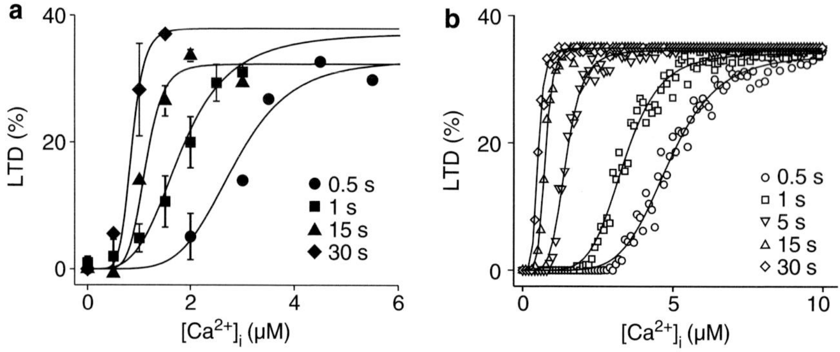

Fig. 2 Relationship between peak *[Ca²⁺]ᵢ* and *LTD* for different durations of $[Ca^{2+}]_i$ elevation, obtained from experimental measurements (**a**) and predicted by computational simulations (**b**). (From Tanaka et al. [20])

modifications were needed for this model to reproduce all major features of our experimental measurements of the Ca^{2+} dependence of LTD (Fig. 2B).

The computational model predicted that disruption of the positive feedback loop, by preventing activation of PKC via arachidonic acid (AA), should reduce the Ca^{2+} sensitivity of LTD induction. This prediction was tested experimentally by examining the effect of a PLA2 inhibitor on the $[Ca^{2+}]_i$ dependence of LTD. We found a close correspondence between the experimental results and the predictions of the computational model [20], supporting the idea that both the cooperative activation and the leaky integrator properties of LTD triggering by Ca^{2+} arise, at least in part, from the positive feedback loop working downstream of Ca^{2+}.

Reciprocal Activation of PKC and MAPK Required for LTD

Although the experimental measurements of the relationship between $[Ca^{2+}]_i$ and LTD were consistent with the computational model, we sought more direct tests of the positive feedback loop model. We next tested the essence of this loop, the predicted mutual activation of PKC and MAPK by each other, by determining experimentally (1) whether MAPK is downstream of PKC, and (2) whether PKC is also downstream of MAPK [28].

To address the first question, we first measured activation of MAPK by biochemical and immunohistochemical analysis using an antibody that selectively recognizes the phosphorylated, activated form of MAPK [29,30]. LTD induction caused a pronounced increase in MAPK phosphorylation in Purkinje cells, and this phosphorylation was blocked by a PKC inhibitor, bisindolylmaleimide I (BIM), indicating that activation of MAPK during LTD induction requires an upstream action of PKC. We then tested whether this interaction between PKC and MAPK is required for LTD. For this purpose, a PKC activator, 12-*O*-tetradecanoyl

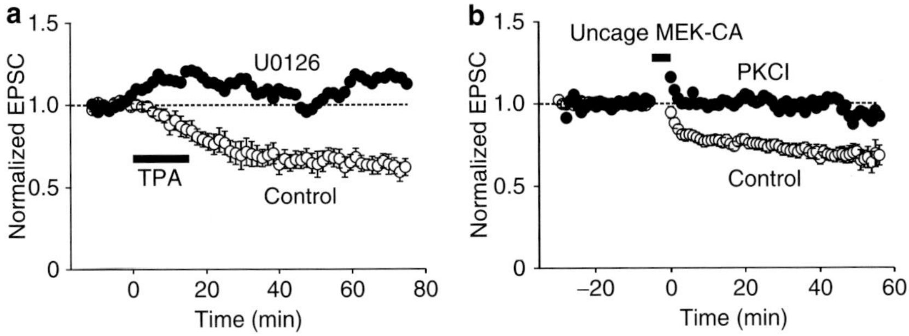

Fig. 3 **a** LTD induced by activating protein kinase C (PKC) by 12-*O*-tetradecanoyl phorbol-13-acetate (TPA) was blocked by U0126, indicating that PKC works by activating mitogen-activated protein kinase (MAPK). **b** LTD induced by activating MAPK by uncaged MEK-CA was blocked by PKCI, indicating that MAPK works by activating PKC. (From Tanaka and Augustine [28])

phorbol-13-acetate (TPA), was used to induce LTD; as previously reported [31–33], TPA treatment caused LTD (open circles in Fig. 3A). This LTD resulting from direct activation of PKC was blocked by an inhibitor that blocks the MAPK pathway, U0126 (closed circles in Fig. 3A), indicating that MAPK works downstream of PKC. Thus, these results demonstrate that PKC activates MAPK during LTD and that this interaction is required for LTD induction.

A similar strategy was used to determine whether MAPK also acts upstream of PKC. Prolonged activation of PKC, detected by translocation of PKC to the plasma membrane [34], was observed after LTD induction and was blocked by U0126. This finding indicates that PKC activation is sustained via an upstream action of MAPK. To determine whether this interaction is required for LTD, we asked whether a PKC inhibitor could prevent LTD induced by direct activation of MAPK. Toward this goal, we covalently attached a NVOC-Cl caging group [35] onto a pseudophosphorylated, constitutively active form of MEK (MEK-CA), a kinase that phosphorylates/activates MAPK [30]. In vitro analysis demonstrated that applying UV light to the caged MEK-CA permits temporally precise activation of MAPK. We then introduced the caged MEK-CA into Purkinje cells through the patch pipette. Photolysis of caged MEK-CA induced LTD of PF synaptic transmission (open circles in Fig. 3B). This LTD induced by direct activation of MAPK was blocked by a PKC inhibitor, the PKC inhibitory peptide (PKCI; closed circles in Fig. 3B). Taken together with the results showing MAPK-dependent PKC activation during LTD, these results indicate that MAPK also activates PKC.

In summary, our results demonstrate that the reciprocal activation of PKC and MAPK is important for LTD, as predicted by the computational model. We thus conclude that the positive feedback loop is active during LTD and is required for LTD.

Positive Feedback Loop Prolongs Duration of PKC Activation

The computational model [24] predicts that the main purpose of the positive feedback loop is to prolong the duration of PKC activity in time. We tested this prediction via two series of experiments. The first examined the time window of PKC activity by applying a PKC inhibitor, BIM, at several time points after LTD-inducing stimuli. In the absence of BIM treatment, LTD was produced by pairing of PF stimulation with depolarization of the Purkinje cell membrane potential (PF&ΔV) [3,36] (open circles of Fig. 4A). However, when BIM was applied as late as 20 min after the PF&ΔV stimulus, LTD was reduced or eliminated. The closed circles of Fig. 4A show the results of an experiment where BIM was applied 10 min after the stimulus: while the stimulus caused excitatory postsynaptic current (EPSC) amplitude to depress initially, this depression reversed after BIM treatment. In contrast, applying BIM 30 min or later after the stimulus did not affect LTD. The relationship between the time of BIM application and the degree of LTD impairment indicated that PKC remains active for at least 20–30 min after the LTD-inducing stimulus ends (Fig. 4B). Thus, activation of PKC is prolonged, as predicted by the model.

The second series of experiments determined whether this prolonged activation of PKC is the result of the positive feedback loop. For this purpose, we applied a PLA2 inhibitor (OBAA) at different times after the stimulus, just as already described for BIM. As was the case for BIM, OBAA was effective in reducing LTD even when applied 20–30 min after the end of the stimulus. The relationship between the time of OBAA application and the degree of LTD impairment closely paralleled the relationship for BIM application. We therefore conclude that the positive feedback loop is responsible for sustaining activation of PKC for at least the first 20–30 min after synaptic activity.

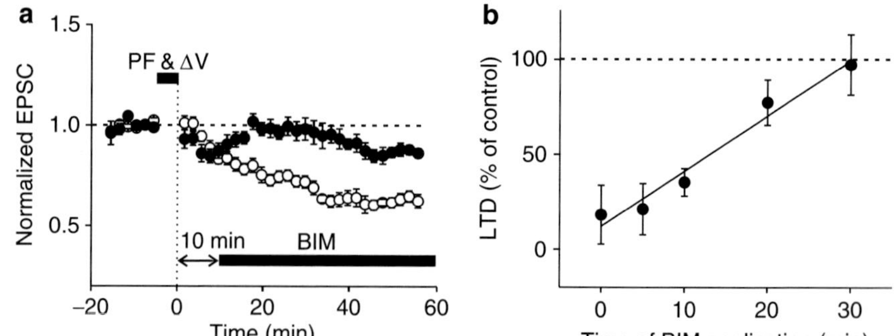

Fig. 4 **a** Application of bisindolylmaleimide I (*BIM*) 10 min after pairing of PF stimulation with depolarization of the Purkinje cell membrane potential (*PF&ΔV*) caused LTD. **b** Relationship between the time of BIM application and the degree of LTD impairment. (From Tanaka and Augustine [28])

Summary

Our experimental results provide several lines of evidence consistent with the hypothesis that the positive feedback loop first proposed on the basis of "systems biology" computational models is important for LTD:

1. Elevating postsynaptic $[Ca^{2+}]_i$, by photolysis of a caged Ca^{2+} compound, indicated a sigmoidal relationship between $[Ca^{2+}]_i$ and LTD. This cooperative behavior was predicted by the computational model.
2. The duration of the rise in $[Ca^{2+}]_i$ influenced the apparent Ca^{2+} affinity of LTD, being described by a leaky integrator process. The computational model was also capable of simulating these properties of Ca^{2+}-triggered LTD.
3. Disrupting this cycle experimentally also produced changes in the Ca^{2+} dependence of LTD that were predicted by the computational model.
4. Histochemical "epistasis" experiments demonstrate the reciprocal activation of PKC and MAPK during LTD.
5. Physiological experiments demonstrate that such reciprocal activation of PKC and MAPK is required for LTD.
6. Timed application of enzyme inhibitors reveals that this positive feedback loop causes PKC to be active for more than 20 min, allowing sufficient time for LTD expression.

Based on these results, we conclude that the positive feedback loop is an essential component of the molecular signaling cascade responsible for LTD induction (Fig. 5). Our work therefore provides experimental support for the proposal that the PKC/MAPK positive feedback loop serves as a general mechanism for sustaining intracellular signaling in time [25].

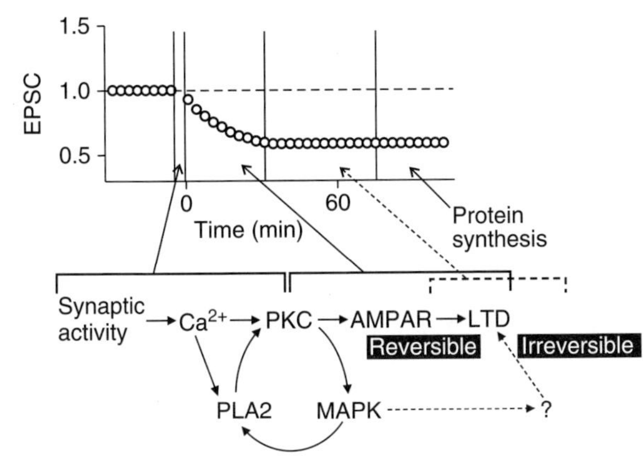

Fig. 5 Model for time course of signal transduction events involved in LTD. (From Tanaka and Augustine [28])

It appears that the increases in $[Ca^{2+}]_i$ associated with synaptic activity disappear within 1 s after synaptic activity ceases. However, the resulting activation of PKC turns on the positive feedback loop, which sustains PKC activity for at least 20–30 min and leads to the gradual expression of LTD as AMPA receptors are phosphorylated [37,38] and internalized [19,20]. Because application of either BIM or OBAA caused LTD to be reversed, it appears that initially the internalization of AMPA receptors can be reversed if the positive feedback loop is not maintained. Application of these inhibitors at later times does not reverse LTD, indicating that internalization becomes stabilized by this time point (Fig. 5). At times later than 60–90 min after stimulation, synthesis of new proteins maintains LTD [22]. Whether this phase of LTD relies on the positive feedback loop or any of its components remains to be determined.

This positive feedback model of LTD is reminiscent of the regenerative molecular mechanisms that have been implicated in other forms of long-lasting synaptic plasticity, such as autophosphorylation of the Ca/CaM-dependent protein kinase II that is important for hippocampal long-term synaptic potentiation [39] and activation of the prion-like protein, CPEB, which is important for long-term synaptic plasticity in *Aplysia* [40]. Thus, it is possible that positive feedback loops are a general mechanism for translating transient signals generated by brief synaptic activity into sustained biochemical activity during long-term synaptic plasticity.

References

1. Pittenger C, and Kandel ER (2003) In search of general mechanisms for long-lasting plasticity: *Aplysia* and the hippocampus. Philos Trans R Soc Lond B Biol Sci 358:757–763
2. Ito M (2001) Cerebellar long-term depression: characterization, signal transduction, and functional roles. Physiol Rev 81:1143–1195
3. Konnerth A, Dreessen J, and Augustine GJ (1992) Brief dendritic calcium signals initiate long-lasting synaptic depression in cerebellar Purkinje cells. Proc Natl Acad Sci U S A 89:7051–7055
4. Sakurai M (1990) Calcium is an intracellular mediator of the climbing fiber in induction of cerebellar long-term depression. Proc Natl Acad Sci U S A 87:3383–3385
5. Linden DJ, Dickinson MH, Smeyne M, et al (1991) A long-term depression of AMPA currents in cultured cerebellar Purkinje neurons. Neuron 7:81–89
6. Crépel F, and Krupa M (1988) Activation of protein kinase C induces a long-term depression of glutamate sensitivity of cerebellar Purkinje cells. An in vitro study. Brain Res 458:397–401
7. Linden DJ (1994) Input-specific induction of cerebellar long-term depression does not require presynaptic alteration. Learn Mem 1:121–128
8. Matsuda S, Launey T, Mikawa S, et al (2000) Disruption of AMPA receptor GluR2 clusters following long-term depression induction in cerebellar Purkinje neurons. EMBO J 19:2765–2774
9. Khodakhah K, and Armstrong CM (1997) Induction of long-term depression and rebound potentiation by inositol trisphosphate in cerebellar Purkinje neurons. Proc Natl Acad Sci U S A 94:14009–14014
10. Inoue T, Kato K, Kohda K, et al (1998) Type 1 inositol 1,4,5-trisphosphate receptor is required for induction of long-term depression in cerebellar Purkinje neurons. J Neurosci 18:5366–5373

11. Finch EA, and Augustine GJ (1998) Local calcium signalling by inositol-1,4,5-trisphosphate in Purkinje cell dendrites. Nature (Lond) 396:753–756
12. Miyata M, Finch EA, Khiroug L, et al (2000) Local calcium release in dendritic spines required for long-term synaptic depression. Neuron 28:233–244
13. Ito M, and Karachot L (1990) Messengers mediating long-term desensitization in cerebellar Purkinje cells. Neuroreport 1:129–132
14. Daniel H, Hemart N, Jaillard D, et al (1993) Long-term depression requires nitric oxide and guanosine 3′:5′ cyclic monophosphate production in rat cerebellar Purkinje cells. Eur J Neurosci 5:1079–1082
15. Boxall AR, and Garthwaite J (1996) Long-term depression in rat cerebellum requires both NO synthase and NO-sensitive guanylyl cyclase. Eur J Neurosci 8:2209–2212
16. Lev-Ram V, Jiang T, Wood J, et al (1997) Synergies and coincidence requirements between NO, cGMP, and Ca^{2+} in the induction of cerebellar long-term depression. Neuron 18:1025–1038
17. Reynolds T, and Hartell NA (2001) Roles for nitric oxide and arachidonic acid in the induction of heterosynaptic cerebellar LTD. Neuroreport 12:133–136
18. Linden DJ (2001) The expression of cerebellar LTD in culture is not associated with changes in AMPA-receptor kinetics, agonist affinity, or unitary conductance. Proc Natl Acad Sci U S A 98:14066–14071
19. Wang YT, and Linden DJ (2000) Expression of cerebellar long-term depression requires postsynaptic clathrin-mediated endocytosis. Neuron 25:635–647
20. Tanaka K, Khiroug L, Santamaria F, et al (2007) Ca^{2+} requirements for cerebellar long-term synaptic depression: role for a postsynaptic leaky integrator. Neuron 54:787–800
21. Nakazawa K, Karachot L, Nakabeppu Y, et al (1993) The conjunctive stimuli that cause long-term desensitization also predominantly induce c-Fos and Jun-B in cerebellar Purkinje cells. Neuroreport 4:1275–1278
22. Linden DJ (1996) A protein synthesis-dependent late phase of cerebellar long-term depression. Neuron 17:483–490
23. Karachot L, Shirai Y, Vigot R, et al (2001) Induction of long-term depression in cerebellar Purkinje cells requires a rapidly turned over protein. J Neurophysiol 86:280–289
24. Kuroda S, Schweighofer N, and Kawato M (2001) Exploration of signal transduction pathways in cerebellar long-term depression by kinetic simulation. J Neurosci 21:5693–5702
25. Bhalla US, and Iyengar R (1999) Emergent properties of networks of biological signaling pathways. Science 283:381–387
26. Knight BW (1972) Dynamics of encoding in a population of neurons. J Gen Physiol 59:734–766
27. Fohlmeister JF, Poppele RE, and Purple RL (1977) Repetitive firing: a quantitative study of feedback in model encoders. J Gen Physiol 69:815–848
28. Tanaka K, and Augustine GJ (2008) A positive feedback signal transduction loop determines timing of cerebellar long-term depression. Neuron 59:608–620
29. Nishida E, and Gotoh Y (1993) The MAP kinase cascade is essential for diverse signal transduction pathways. Trends Biochem Sci 18:128–131
30. Cobb MH, and Goldsmith EJ (1995) How MAP kinases are regulated. J Biol Chem 270:14843–14846
31. Linden DJ, and Connor JA (1991) Participation of postsynaptic PKC in cerebellar long-term depression in culture. Science 254:1656–1659
32. Endo S, and Launey T (2003) ERKs regulate PKC-dependent synaptic depression and declustering of glutamate receptors in cerebellar Purkinje cells. Neuropharmacology 45:863–872
33. Kondo T, Kakegawa W, and Yuzaki M (2005) Induction of long-term depression and phosphorylation of the δ2 glutamate receptor by protein kinase C in cerebellar slices. Eur J Neurosci 22:1817–1820
34. Newton AC (2001) Protein kinase C: structural and spatial regulation by phosphorylation, cofactors, and macromolecular interactions. Chem Rev 101:2353–2364
35. Marriott G, Roy P, and Jacobson K (2003) Preparation and light-directed activation of caged proteins. Methods Enzymol 360:274–288

36. Crépel F, and Jaillard D (1991) Pairing of pre- and postsynaptic activities in cerebellar Purkinje cells induces long-term changes in synaptic efficacy *in vitro*. J Physiol 432:123–141
37. Chung HJ, Steinberg JP, Huganir RL, et al (2003) Requirement of AMPA receptor GluR2 phosphorylation for cerebellar long-term depression. Science 300:1751–1755
38. Steinberg JP, Takamiya K, Shen Y, et al (2006) Targeted in vivo mutations of the AMPA receptor subunit GluR2 and its interacting protein PICK1 eliminate cerebellar long-term depression. Neuron 49:845–860
39. Sanhueza M, McIntyre CC, and Lisman JE (2007) Reversal of synaptic memory by Ca^{2+}/calmodulin-dependent protein kinase II inhibitor. J Neurosci 27:5190–5199
40. Bailey CH, Kandel ER, and Si K (2004) The persistence of long-term memory: a molecular approach to self-sustaining changes in learning-induced synaptic growth. Neuron 44:49–57

Computational Models of Cerebellar Long-Term Memory

Hideaki Ogasawara and Mitsuo Kawato

Introduction

Our brain is capable of learning new things while maintaining old memory. Retention of information requires stability, and new learning requires plasticity. As a memory device, neurons have to meet these contradictory requirements (the "stability versus plasticity dilemma" [1]), but stochastic noise makes this duty still more difficult. The dendritic spine, the key unit of neuronal information processing, is very small (~1 μm or less in diameter) and contains only a limited number of each molecular species. For instance, the number of α-amino-3-hydroxy-5-methylisoxazole-4-propionic acid (AMPA)-type glutamate receptors (AMPARs) in a parallel fiber (PF)–Purkinje cell (PC) synapse is as small as 4 to 73 [2]. In such a minute environment, stochastic fluctuations come into play and, affect the signaling pathways underlying memory formation and maintenance. How do neurons handle the stability versus plasticity dilemma without being overwhelmed by the noise? In this chapter, we address this issue by reviewing several theoretical studies of cerebellar long-term depression (LTD) and simulating simple models.

Cerebellar LTD

The main neurons and wirings in the cerebellar cortex include PCs, PFs, and climbing fibers (CFs). PCs provide the sole output from the cortex, and each PC receives two types of excitatory inputs: one from hundreds of thousands of PFs and the other from a single CF. Marr-Albus-Ito theory [3–5] states that their neuronal circuit underlies

H. Ogasawara
National Institute of Information and Communications Technology. 2-2-2, Hikaridai, Seika, Kyoto, 619-0288, Japan

H. Ogasawara and M. Kawato
ATR Computational Neuroscience Laboratories, 2-2-2 Hikaridai, Seika, Kyoto 619-0288, Japan

S. Nakanishi et al. (eds.), *Systems Biology*,
DOI:10.1007/978-4-431-87704-2_18, © Springer 2009

supervised learning of the cerebellum. In this theory, PFs provide a sensorimotor context to PCs, while CFs carry teaching signals that modify PF-PC synapses in an associative manner.

In fact, the transmission efficacy of a PF–PC synapse is reduced when the CF and PF are repetitively and synchronously activated (cerebellar LTD). The molecular mechanisms are comprehensively reviewed elsewhere [6–11]. Briefly, PF firing induces production of inositol 1,4,5-trisphosphate (IP_3) through the mGluR1 metabotropic glutamate receptor pathway. CF firing depolarizes the PC and induces Ca^{2+} influx through voltage-gated calcium channels (Fig. 1a). Therefore, [IP_3] and [Ca^{2+}] represent PF and CF activities, respectively (In this chapter, [substance] stands for the concentration of the substance.) Synergistic increase of [IP_3] and [Ca^{2+}] activates IP_3 receptors (IP_3Rs) located in a calcium store of the dendritic spine and results in an enormous Ca^{2+} release. Ca^{2+} activates protein kinase C (PKC), which in turn phosphorylates the GluR2 subunit of AMPARs. Phosphorylated receptors are removed from the postsynaptic membrane through endocytosis.

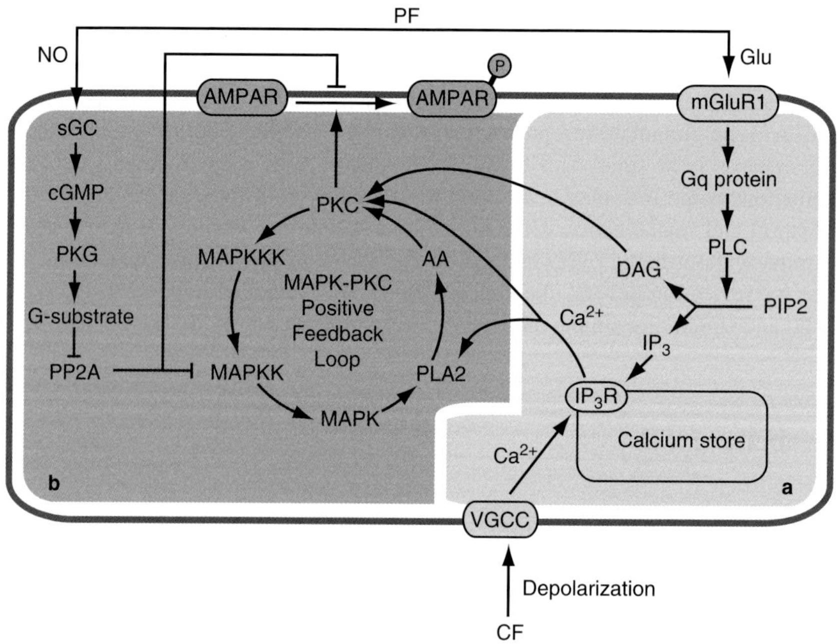

Fig. 1 Schematic view of signaling cascades in cerebellar long-term depression (LTD). Reactions surrounded by the *bold gray line* take place inside the dendritic spine. **a** Parallel fiber–climbing fiber (PF-CF) coincidence detection mechanisms (*light gray area; right side*) [14]. *Glu*, glutamate; *PKG*, cGMP-dependent protein kinase; *PIP2*, phosphatidylinositol bisphosphate. **b** The mitogen-activated protein kinase (MAPK)–PKC positive feedback loop [13] and its peripherals (*dark gray area; left side*). *AA*, arachidonic acid; *AMPA*, α-amino-3-hydroxy-5-methylisoxazole-4-propionic acid; *AMPAR*, AMPA-type glutamate receptor; *DAG*, diacyl glycerol. (From [15], with permission of S. Karger AG, Basel)

Essential molecules for LTD include mitogen-activated protein kinase (MAPK), MAPK kinase (MAPKK), MAPKK kinase (MAPKKK), PKC, phospholipases A2 (PLA2) and C (PLC), arachidonic acid (AA), IP_3, and many others [6–11].

Models to Explain Cerebellar LTD

In numerous situations such as development and cell-cycle progression, cells take discrete states and jump from one to another, avoiding intermediate states. For example, frog eggs are either mature or immature, and there is no half-mature egg [12]. This kind of switch-like behavior is termed "bistability." Bistability is particularly important in neurobiology because it is implicated in the storage of cellular information. Previous studies have shown that Ca^{2+}-induced Ca^{2+} release (CICR) and the MAPK-PKC positive feedback loop in the PC are among many bistable dynamics of the cell (Fig. 1) [13,14] (reviewed in [15]).

Calcium Dynamics Model

Calcium release via IP_3Rs is a supralinear function of calcium influx; the greater an input stimulus (calcium influx) is, the much more greater the output signal (calcium release) is. Doi et al. [14] simulated the molecular mechanism and demonstrated that a positive feedback loop $Ca^{2+} \rightarrow IP_3Rs \rightarrow Ca^{2+}$ is responsible for the supralinearity of calcium release. They also showed that the optimal PF-CF delay (~100 ms) for calcium release and LTD is attributable to the slow mGluR1 pathway that takes time to biochemically produce IP_3, rather than to calcium kinetics and the binding properties of the receptor [16] (see also comments in [17]).

MAPK-PKC Positive Feedback Loop

Kuroda et al. [13] modeled the intracellular signaling cascade of cerebellar LTD (Fig. 1b). Their model predicted that MAPK and PKC form a bistable positive feedback loop, which mediates the intermediate phase (<~40 min) of LTD. Subsequent neurobiological experiments strongly supported their predictions [18,19].

Cascade of Excitable Dynamics

How do neurons resolve the stability versus plasticity dilemma [1] in stochastic noise? Taking into account the fact that bistability is ubiquitous in signaling pathways [20], one of us has proposed a cascade of excitable dynamics as a likely

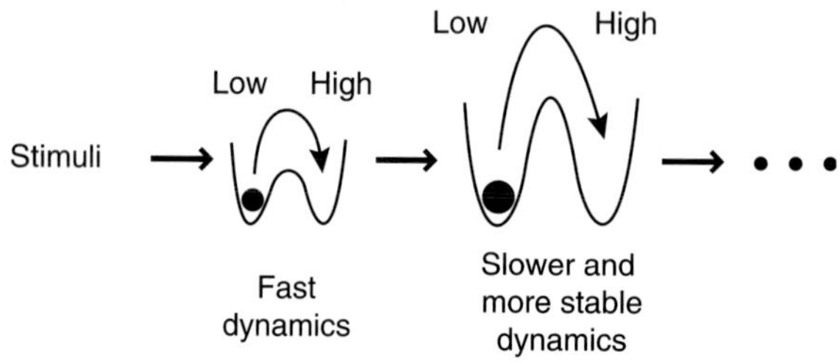

Fig. 2 Threshold cascade model (see text for explanation). (From [15], with permission of S. Karger AG, Basel)

mechanism (Fig. 2) [21]. In the model (detailed in [15]), multiple bistable dynamics with various time constants are connected in tandem, in the order of fast to slow. Each dynamics has an energy barrier that correlates with its time constant (the faster a switch is, the lower its barrier is), and an active dynamics stimulates the next one that is more stable and requires larger inputs for activation. In this way, when external stimuli activate the fastest system, the activity is transmitted from one to another slower one, and finally to the slowest one. Such a cascade of multiple bistable dynamics will form long-term memory that is stable, still open to modifications, and robust to noise. The model is applicable to at least the first two steps of cerebellar LTD, that is, CICR and the MAPK-PKC positive feedback loop (Fig. 1).

Protein Kinase Mζ Model

Based on these studies and an abundance of biological literature, we have proposed a conceptual model [15] in which cerebellar LTD is induced and maintained by successive activation of three bistable dynamics with different time scales: CICR, the MAPK-PKC positive feedback loop, and protein kinase Mζ (PKMζ)-induced PKMζ synthesis.

PKMζ is a persistently active enzyme, consisting of the catalytic domain of an atypical PKC isoform, PKCζ. It is expressed in a brain-specific manner. A series of experiments performed mainly by Sacktor's group demonstrates accumulating evidence that PKMζ plays a pivotal role in long-term maintenance of memory (exhaustively reviewed in [22]). Briefly, a procedure to induce hippocampal LTP (LTD) upregulates (downregulates) PKMζ. Injection of a PKMζ inhibitor, ZIP, to the hippocampus reverses LTP in vivo and produces persistent loss of 1-day-old spatial information. More surprisingly, an injection of ZIP to the rat neocortex, which is regarded as the long-term repository of memory, erases associative memories as old as 1 month [23]. These findings strongly suggest that PKMζ is the (semi-)permanent form of memory trace. PKMζ is also expressed in the cerebellar cortex [24,25]. Because cerebellar plasticity is regarded as a mirror image of hippocampal plasticity and they share much of their signaling networks [6,7,9,10,26], PKMζ activity is likely to be central to cerebellar LTD, as it is to hippocampal LTP.

Sometimes hippocampal studies advance far ahead of its cerebellar counterparts. For instance, calmodulin-dependent protein kinase II was discovered to be essential for hippocampal plasticity and learning in 1992 [27], but it was not until 2006 that the kinase was shown to be also necessary for cerebellar memory [28]. One of the main purposes of this model is to shorten the interval by calling for experiments to characterize the role of PKMζ in the cerebellum.

Pathways

PKMζ induces PKMζ expression, and this enzyme can act in two steps; either directly on its expression or on actin polymerization, which facilitates protein synthesis (Fig. 3a). Thus, we postulate that PKMζ forms a bistable switch by a combination of multi-step ultrasensitivity and positive feedback [29,30]. Central to the model are three positive feedback loops, i.e., CICR (Fig. 3b, circle 1), the MAPK-PKC positive feedback loop (Fig. 3b, circle 2), and PKMζ-induced PKMζ expression (Fig. 3b, circle 3). Their decay time constants are less than 1 s, tens of minutes, and weeks or longer, respectively. Importantly, these excitable dynamics of different time scales are connected in tandem. When stimuli activate the quickest dynamics (Fig. 3b, loop 1) repetitively, activity is transmitted from the quickest, to the intermediate (Fig. 3b, loop 2), and thereafter to the slowest and most stable dynamics (Fig. 3b, loop 3). As a consequence, long-term stability and plasticity of memory are established. In this sense, this model is a more concrete representation of the cascade model of excitable dynamics [21] (Figs. 2, 3c).

Various States of Memory

This model schematizes various experimental observations of cerebellar synaptic plasticity. In the basal state, PP2A is active and inhibits the MAPK-PKC positive feedback loop (Fig. 3d).

LTP Induction

Postsynaptic LTP is induced in an NO-dependent manner at low [Ca^{2+}] [31,32]. NO, whose synthesis is triggered by PF spikes, facilitates exocytosis of AMPARs in a *N*-ethylmaleimide-sensitive factor (NSF)-dependent manner [33] (Fig. 3e).

Induction and Maintenance of LTD

When [Ca^{2+}] and [NO] are increased by conjunctive PF–CF activity, NO releases the MAPK-PKC positive feedback loop from PP2A inhibition, and Ca^{2+} activates PKC, which then phosphorylates and internalizes AMPARs (Fig. 3f). In the intermediate phase (Fig. 3g), the activated MAPK-PKC positive feedback loop maintains PKC activity and receptor endocytosis. During, and subsequent to, the late phase (Fig. 3h),

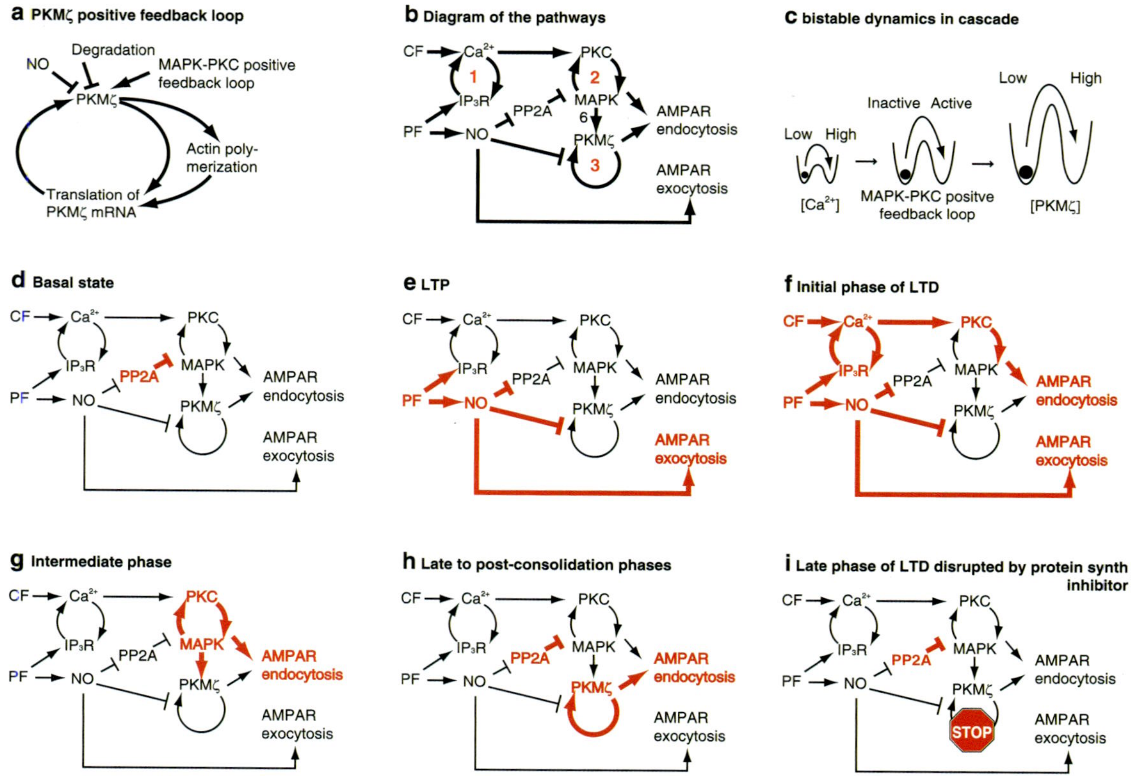

Fig. 3 Conceptual model of cerebellar long-term memory (see text for explanation). **a** Schematic diagram of long-term memory. *Sharp and blunt arrows* indicate excitatory and inhibitory pathways, respectively. **b** The protein kinase Mζ (PKMζ) positive feedback loop. **c** The cascade model [21] representation of the cerebellar long-term memory model. **d–i** Various states of memory. *Bold lines* and *thin lines* indicate activated and inactivated pathways, respectively; *bold text* and *thin text* indicate active or increased molecules and inactive or decreased molecules, respectively. **d** The basal state. **e** LTP. **f** The initial, intermediate (**g**), and late to post-consolidation (**g**) phases of LTD. i The late phase of LTD disrupted by application of a protein synthesis inhibitor. (From [15], with permission of S. Karger AG, Basel)

newly synthesized PKMζ maintains AMPAR phosphorylation and internalization for a long period.

Disruption of Late-Phase LTD by Application of a Protein Synthesis Inhibitor

The model predicts that protein synthesis inhibitor disrupts late-phase LTD by preventing PKMζ expression (Fig. 3i), whereas it does not affect the initial and intermediate phases. Once PKMζ is sufficiently expressed, only minimal synthesis of PKMζ is required to compensate for its degradation and maintain LTD. During this period, transient application of a protein synthesis inhibitor will fail to switch off the PKMζ positive feedback loop, so long as the level of remaining PKMζ is above the threshold for maintaining loop activity.

Simulation Study of Single or Interlinked Bistable Switches

We have proposed that a cascade of bistable dynamics is a possible mechanism that resolves the stability versus plasticity dilemma in stochastic noise (Figs. 2, 3) [15,21]. To test this hypothesis computationally, we build simple simulation models. One model contains a single loop (Fig. 4a), and the other contains two interlinked

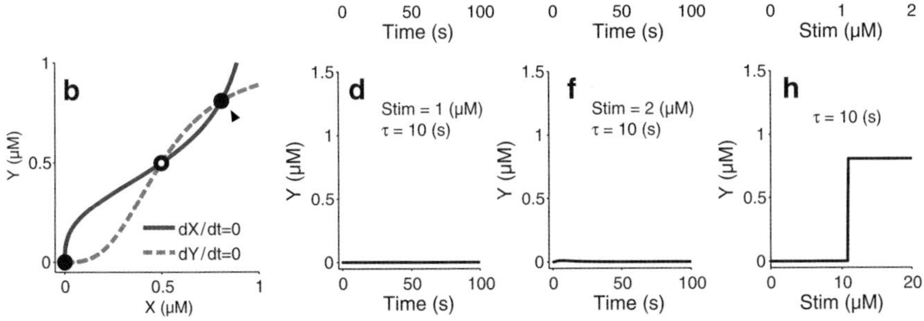

Fig. 4 Deterministic runs of the single-loop model. **a** Schematic diagram of the model. *X* and *Y* activate each other, and a stimulus (*stim*) is given to *X*. **b** Nullclines and stable (*filled circles*) and unstable (*open circle*) steady states. In the stable steady states, the system is either active (*arrowhead*) or inactive. Stim = 0 throughout. **c–f** Time courses of *Y* during and after stimulation. Stim is either 1 μM (**c, d**) or 2 μM (**e, f**) and given for 1 s in the beginning of each run. **g, h** Dose–response relationship. The loop with the smaller τ is more sensitive to stimulation. τ is 1 s (**c, e, g**) or 10 s (**d, f, h**)

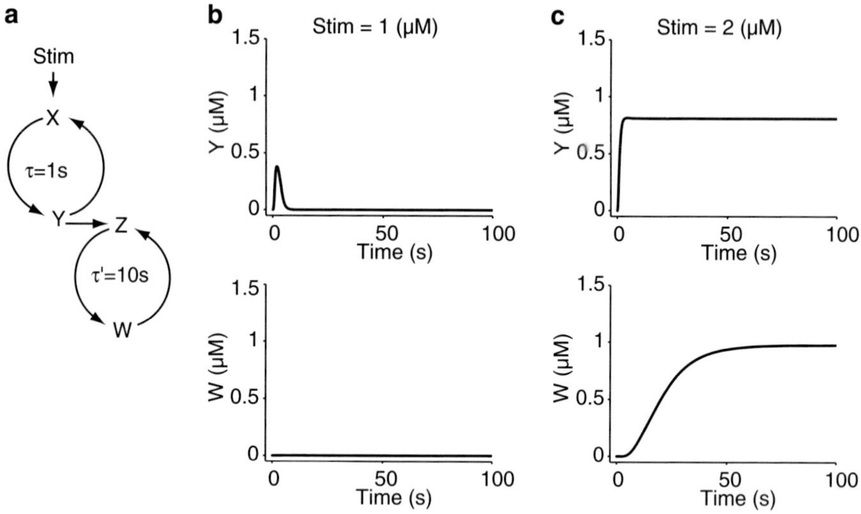

Fig. 5 Deterministic runs of the double-loop model. **a** Schematic diagram of the model. A stimulus (Stim) activates *X* of the first loop, and *X* and *Y* activate each other. *Y* also activates *Z* of the second loop, and *Z* and *W* activate each other. The first loop is faster than the second loop. **b, c** Time courses of *Y* and *W* during and after stimulation. The second loop has the same time constant as the single loop in Fig. 4d,f,h, but is more sensitive to stimuli because the first loop boosts the inputs. τ and τ' are 1 s and 10 s, respectively. Stim is either 1 μM (**b**) or 2 μM (**c**) and given for 1 s in the beginning of each run

loops (Fig. 5a). Note that our aim here is to capture the basic properties of isolated or coupled switches by using least complicated models, and not to realistically represent biological signaling pathways. Each loop consists of a couple of mutually activating molecules (X and Y as well as Z and W in Figs. 4a, 5a). The ordinary differential equations (ODEs) for the single loop model and the first loop of the double loop model are as follows:

$$\tau \frac{dX}{dt} = \frac{Y^n}{Y^n + K^n} X_{inact} - X + Stim(t)$$

$$\tau \frac{dY}{dt} = \frac{X^n}{X^n + K^n} Y_{inact} - Y$$

where $n = 3$ and $K = 0.5\,\mu\text{M}$. The inactive forms of X and Y, X_{inact} and Y_{inact}, are constant at 1 μM and do not change in concentrations even when they are consumed (buffered molecules)[1]. *Stim* is a time-dependent stimulus that induces *X*. Similarly, the ODEs for the second loop of the double loop model are as follows:

[1] The buffered species (e.g. X_{inact}) may appear to violate mass conservation, but actually do not; we assume a situation in which these molecules are abundant in the cell and freely diffuse into or out of the system, keeping their concentrations constant.

$$\tau' \frac{dZ}{dt} = \frac{W^n}{W^n + K^n} Z_{inact} - Z + Y$$

$$\tau' \frac{dW}{dt} = \frac{Z^n}{Z^n + K^n} W_{inact} - W$$

The inactive forms of Z and W, Z_{inact} and W_{inact}, are constant at 1 μM. The time constants τ and τ' determine the rates of reactions; for instance, the bigger (smaller) τ is, the more slowly (faster) X and Y change. Numerical integration of the ODEs yields deterministic solutions, and a Monte-Carlo method [34] serves for stochastic simulations.

Although the interlinked loops superficially resemble the model of Brandman et al. [35], there are crucial differences between these dynamic systems. First, the stimulus acts on the first loop in our model (Fig. 5a), whereas it acts directly on both loops in theirs. Second, in contrast to our model, the bistable property of the positive feedback loops in [35] is not inherent but depends on the level of the stimulus.

Single-Loop Model

First, we need to characterize the single-loop model (Fig. 4a) by performing deterministic simulations. Figure 4b indicates steady-state solutions given that one of the two variables (X or Y) is kept constant (nullclines); the points of intersection are equilibrium points of the system. The figure indicates that the loop is a bistable system, having two stable steady states and an unstable saddle point. In one stable steady state, X and Y are active (~0.8 μM; arrowhead), and in the other stable steady state they are inactive. Figure 4c–f shows the time course of Y during and after transient stimulation. When fast kinetics is assumed ($\tau = 1$ s), the loop responds only transiently to a weak stimulus (Fig. 4c), whereas it is fully and permanently activated by a strong stimulus (Fig. 4e). At equilibrium, the loop is totally inactive or stereotypically active depending on the strength of the stimulus (Fig. 4g), which reflects the all-or-none nature of the bistable switch. A loop of slower kinetics ($\tau = 10$ s) is less sensitive to stimuli (Fig. 4d,f,h) because the stimuli are too short in duration (for 1 s) and end before the state of the loop crosses the threshold.

Next, we consider stochasticity. In deterministic simulations, X and Y do not change once they reach the active state (Fig. 4b; arrowhead), but it does not mean that no reaction is taking place there. Rather, each molecule is constantly being activated and inactivated at the same rate, appearing to be static. However, in a very small system such as a dendritic spine where the number of molecules is severely limited, stochastic fluctuations are no longer negligible. Figure 6 plots several stochastic runs of a minute system (0.05 μm^3) corresponding to a tiny spine [36]. In this system, a concentration of 1 μM is equivalent to 30 molecules. Starting from the active state, the system is shaken by stochastic fluctuations and eventually pushed off to the inactive state (Fig. 6a–c); it would permanently

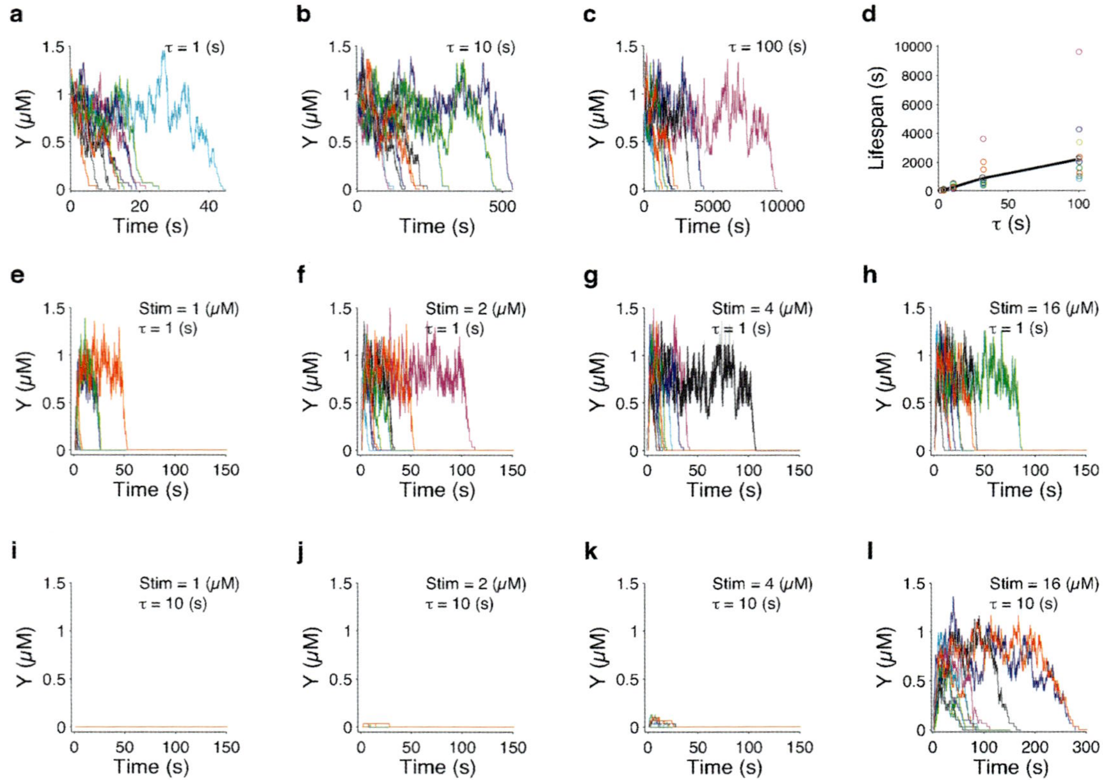

Fig. 6 Stochastic runs of the single-loop model. **a–c** Time courses of *Y* starting from the active state. The larger τ is, the longer the lifespan of the active state tends to be. τ is either 1 s (**a**), 10 s (**b**), or 100 s (**c**). **d** Linear relationship between τ and average lifespan. *Circles* indicate lifespan observed in individual runs and the *bold line* indicates their averages. **f–l** Time courses of *Y* during and after stimulation. The loop with a small τ (1 s) is sensitive to stimulation but, it can remain at the active state only for a short period. In contrast, the loop with a larger τ (10 s) is less sensitive to stimuli, but once activated, it maintains activity for a long time. Note the different time scale of **l**. Stim is either 1 μM (**e, i**), 2 μM (**f, j**), 4 μM (**g, k**), or 16 μM (**h, l**) and is given for 1 s in the beginning of each run

remain at the active state in deterministic simulation. The lifespan of the active state is proportional to τ (Fig. 6d); the slower (faster) the loop is, the longer (shorter) period its activity persists for. It should be noted that no further reaction will take place once both X and Y reach zero, as the Monte Carlo method [34] assumes that the probability of any reaction to occur in the future is proportional to X and Y. Thus, in stochastic simulations the inactive state is practically the only stable steady state, and the active state is no longer permanent. When τ is small (1 s), the loop is activated for a while by transient stimuli that are sufficiently strong (Fig. 6e–h). In contrast, when τ is larger (10 s), the loop is less sensitive to stimulation (Fig. 6i–k); once activated by a very strong stimulus, however, the loop often maintains its activity for a long time (Fig. 6l). These findings indicate that in a stochastic situation there is a trade-off between stability and sensitivity.

Interlinked Loop Model

What if two loops are coupled? We expect that a cascade of bistable dynamics makes stable and plastic memory. In particular, we are interested in a slow loop interlinked with and activated by a faster loop (Fig. 5a) because it mimics some of the cerebellar LTD steps (Fig. 3c). Figure 5b,c shows deterministic runs of the model. The faster loop (X and Y) behaves similarly as the single-loop model with fast kinetics (Fig. 4c,e). It is activated only transiently by a weak stimulus (Fig. 5b) and permanently by a strong stimulus (Fig. 5c). The slower loop (Z and W), which would not be activated by the direct input of a strong stimulus (Fig. 4f), is gradually activated by the same size of stimulus (Fig. 5c), and eventually reaches the active state. This increase in sensitivity is caused by amplification of the signal by the faster loop; its time-integrated output ($\int Ydt$) is infinitely larger than the time integral of the transient stimulus ($\int Stim(t)dt$). The interlinked loops are concluded to be as sensitive as the single fast loop.

When stochasticity is considered, a more interesting feature emerges. As is seen in the stochastic runs of the single loop model (Fig. 6e–h), a strong stimulus brings the faster loop to the active state, where the loop hangs around for a while before it eventually returns to the inactive state (Fig. 7, upper panels). Although activity of the fast loop is only transient, in some cases it contributes sufficiently large input to the slower loop (Fig. 7, lower panels). Once activated by the first loop, the second loop more often than not maintains its activity for a considerably long time. If the system did not consist of two loops but rather a single slow loop, it would be much less sensitive to the stimuli, as is evident from Fig. 6i–k. The output of the system, W, indicates an important feature that the couple of fast and slow loops is more sensitive to stimuli than a single loop with slow kinetics, whereas its activity is more stable than that of a single loop with fast kinetics.

This simple model demonstrates that a cascade of bistable dynamics has a combination of highly desirable features for a memory device, that is, robustness to noise (stability) and sensitivity to input stimuli (plasticity). Such switch cascades

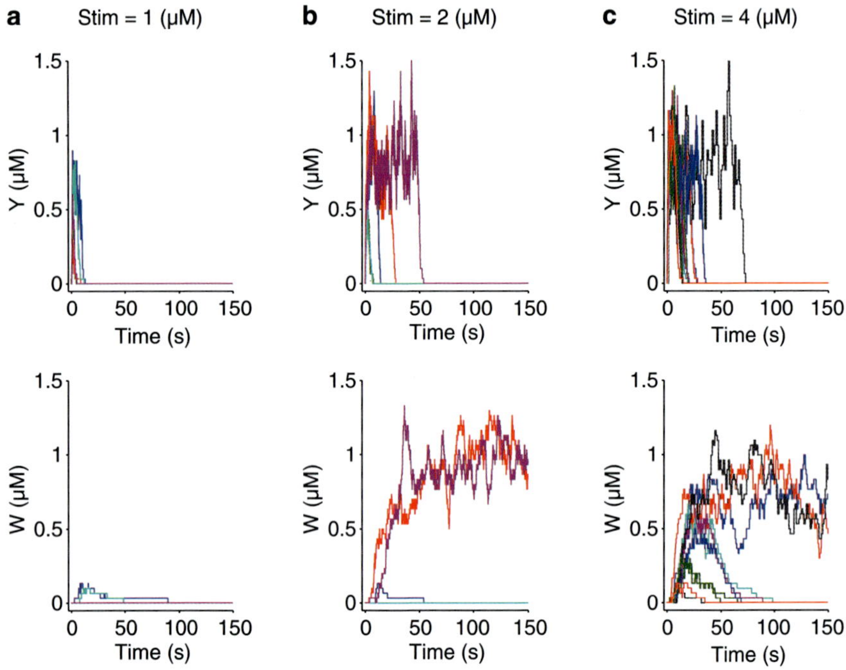

Fig. 7 Stochastic runs of the double-loop model. The *upper and lower panels* show time courses of the first loop and its downstream slower loop, respectively. The output of the system, *W*, shows that this double loop model is more sensitive to stimuli than a single loop with slow kinetics, whereas its activity is more stable than that of a single loop with fast kinetics. τ and τ' are 1 s and 10 s, respectively. Stim is either 1 μM (**a**), 2 μM (**b**), or 4 μM (**c**) and is given for 1 s in the beginning of each run

are biologically implementable, as is exemplified by the pairing of CICR and MAPK-PKC positive feedback loops [13,14].

Conclusions

In this chapter, we reviewed theoretical studies on cerebellar LTD. CICR and the MAPK-PKC positive feedback loop form bistable switches connected in tandem, and we hypothesize that PKMζ is another switch that follows. Simulations of single or interlinked bistable positive feedback loops supported our hypothesis that a cascade of excitable dynamics is a likely mechanism that resolves the stability versus plasticity dilemma.

Acknowledgments We thank Dr. Sungho Hong of Okinawa Institute of Science and Technology for his helpful comments. This work was supported by a grant from the Uehara Memorial Foundation.

References

1. Abraham WC, Robins A (2005) Memory retention: the synaptic stability versus plasticity dilemma. Trends Neurosci 28:73–78
2. Masugi-Tokita M, Tarusawa E, Watanabe M, Molnar E, Fujimoto K, Shigemoto R (2007) Number and density of AMPA receptors in individual synapses in the rat cerebellum as revealed by SDS-digested freeze-fracture replica labeling. J Neurosci 27:2135–2144
3. Marr D (1969) A theory of cerebellar cortex. J Physiol 202:437–470
4. Ito M (1970) Neurophysiological aspects of the cerebellar motor control system. Int J Neurol 7:162–176
5. Albus JS (1971) A theory of cerebellar function. Math Biosci 10:25–61
6. Ito M (2001) Cerebellar long-term depression: characterization, signal transduction, and functional roles. Physiol Rev 81:1143–1195
7. Ito M (2002) The molecular organization of cerebellar long-term depression. Nat Rev Neurosci 3:896–902
8. Hartell NA (2002) Parallel fiber plasticity. Cerebellum 1:3–18
9. Ito M (2006) Cerebellar circuitry as a neuronal machine. Prog Neurobiol 78:272–303
10. Jörntell H, Hansel C (2006) Synaptic memories upside down: bidirectional plasticity at cerebellar parallel fiber-Purkinje cell synapses. Neuron 52:227–238
11. Iino M (2006) Ca^{2+}-dependent inositol 1,4,5-trisphosphate and nitric oxide signaling in cerebellar neurons. J Pharmacol Sci 100:538–544
12. Ferrell JE Jr, Machleder EM (1998) The biochemical basis of an all-or-none cell fate switch in *Xenopus* oocytes. Science 280:895–898
13. Kuroda S, Schweighofer N, Kawato M (2001) Exploration of signal transduction pathways in cerebellar long-term depression by kinetic simulation. J Neurosci 21:5693–5702
14. Doi T, Kuroda S, Michikawa T, Kawato M (2005) Inositol 1,4,5-trisphosphate-dependent Ca^{2+} threshold dynamics detect spike timing in cerebellar Purkinje cells. J Neurosci 25:950–961
15. Ogasawara H, Doi T, Kawato M (2008) Systems biology perspectives on cerebellar long-term depression. NeuroSignals 16:300–317
16. Sarkisov DV, Wang SS (2008) Order-dependent coincidence detection in cerebellar Purkinje neurons at the inositol trisphosphate receptor. J Neurosci 28:133–142
17. Ogasawara H (2008) The calcium kinetics and inositol trisphosphate receptor properties shape the asymmetric timing window of coincidence detection. J. Neurosci 28:4293–4294
18. Tanaka K, Khiroug L, Santamaria F, Doi T, Ogasawara H, Ellis-Davies GC, Kawato M, Augustine GJ (2007) Ca^{2+} requirements for cerebellar long-term synaptic depression: role for a postsynaptic leaky integrator. Neuron 54:787–800
19. Tanaka K, Augustine GJ (2008) A positive feedback signal transduction loop determines timing of cerebellar long-term depression. Neuron 59:608–620
20. Ramakrishnan N, Bhalla US (2008) Memory switches in chemical reaction space. PLoS Comput Biol 4:e1000122
21. Kawato M (2008) From 'understanding the brain by creating the brain' towards manipulative neuroscience. Philos Trans R Soc Lond B Biol Sci 363:2201–2214
22. Sacktor TC (2008) PKMζ, LTP maintenance, and the dynamic molecular biology of memory storage. Prog Brain Res 169:27–40
23. Shema R, Sacktor TC, Dudai Y (2007) Rapid erasure of long-term memory associations in the cortex by an inhibitor of PKMζ. Science 317:951–953
24. Hernandez AI, Blace N, Crary JF, Serrano PA, Leitges M, Libien JM, Weinstein G, Tcherapanov A, Sacktor TC (2003) Protein kinase Mζ synthesis from a brain mRNA encoding an independent protein kinase Cζ catalytic domain. Implications for the molecular mechanism of memory. J Biol Chem 278:40305–40316
25. Oster H, Eichele G, Leitges M (2004) Differential expression of atypical PKCs in the adult mouse brain. Brain Res Mol Brain Res 127:79–88
26. Lynch MA (2004) Long-term potentiation and memory. Physiol Rev 84:87–136

27. Silva AJ, Stevens CF, Tonegawa S, Wang Y (1992) Deficient hippocampal long-term potentiation in alpha-calcium-calmodulin kinase II mutant mice. Science 257:201–206
28. Hansel C, de Jeu M, Belmeguenai A, Houtman SH, Buitendijk GH, Andreev D, De Zeeuw CI, Elgersma Y (2006) αCaMKII Is essential for cerebellar LTD and motor learning. Neuron 51:835–843
29. Goldbeter A, Koshland DE Jr (1981) An amplified sensitivity arising from covalent modification in biological systems. Proc Natl Acad Sci U S A 78:6840–6844
30. Ferrell JE Jr (1999) Building a cellular switch: more lessons from a good egg. Bioessays 21:866–870
31. Shibuki K, Kimura S (1997) Dynamic properties of nitric oxide release from parallel fibres in rat cerebellar slices. J Physiol 498(pt 2):443–452
32. Lev-Ram V, Wong ST, Storm DR, Tsien RY (2002) A new form of cerebellar long-term potentiation is postsynaptic and depends on nitric oxide but not cAMP. Proc Natl Acad Sci U S A 99:8389–8393
33. Kakegawa W, Yuzaki M (2005) A mechanism underlying AMPA receptor trafficking during cerebellar long-term potentiation. Proc Natl Acad Sci U S A 102:17846–17851
34. Gillespie DT (1977) Exact stochastic simulation of coupled chemical reactions. J Phys Chem 81:2340–2361
35. Brandman O, Ferrell JE Jr, Li R, Meyer T (2005) Interlinked fast and slow positive feedback loops drive reliable cell decisions. Science 310:496–498
36. Nimchinsky EA, Sabatini BL, Svoboda K (2002) Structure and function of dendritic spines. Annu Rev Physiol 64:313–353

Biogenesis and Function Mechanisms of Micro-RNAs and Their Role as Oncogenes and Tumor Suppressors

Soken Tsuchiya, Kazuya Terasawa, Ryo Kunimoto, Yasushi Okuno, Fumiaki Sato, Kazuharu Shimizu, and Gozoh Tsujimoto

Introduction

Micro-RNAs (miRNAs) are evolutionarily conserved small noncoding RNAs (20–23 nucleotides). MiRNAs regulate various physiological pathways such as differentiation, proliferation, and apoptosis by negative regulation of the gene expressions at the posttranscriptional level [1–3]. Currently, more than 800 human miRNAs have been identified and registered in the miRNA database miRBase [4]. Strikingly, 30% of protein-coding transcripts in humans is predicted to be regulated by miRNAs [5,6]. Recently, miRNAs have been reported to work as oncogenes or tumor suppressor genes and be directly involved in the initiation, progression, and metastasis of various cancers [7–9]. Therefore, we focus on the role that miRNAs play in cancer and the use of miRNAs in drug discovery. Collection of evidence suggests that miRNAs can be potentially useful for understanding tumorigenesis and finding novel strategies for cancer diagnosis and therapy.

Biogenesis of miRNAs

MiRNAs are generated in multiple steps (Fig. 1). Initially, miRNAs are transcribed by RNA polymerase II as long RNA precursors (pri-miRNAs) [10–12]. Pri-miRNAs are usually several kilobases in length, and contain a 7-methyl

S. Tsuchiya, F. Sato, K. Shimizu
Department of Nanobio Drug Discovery, Graduate School of Pharmaceutical Sciences, Kyoto University, Kyoto, Japan

K. Terasawa and G. Tsujimoto
Department of Genomic Drug Discovery Science, Kyoto University, 46-29 Yoshida Shimoadachi, Sakyo-ku, Kyoto 606-8501, Japan

R. Kunimoto and Y. Okuno
Department of PharmacoInformatics, Graduate School of Pharmaceutical Sciences, Kyoto University, Kyoto, Japan

S. Nakanishi et al. (eds.), *Systems Biology*,
DOI:10.1007/978-4-431-87704-2_19, © Springer 2009

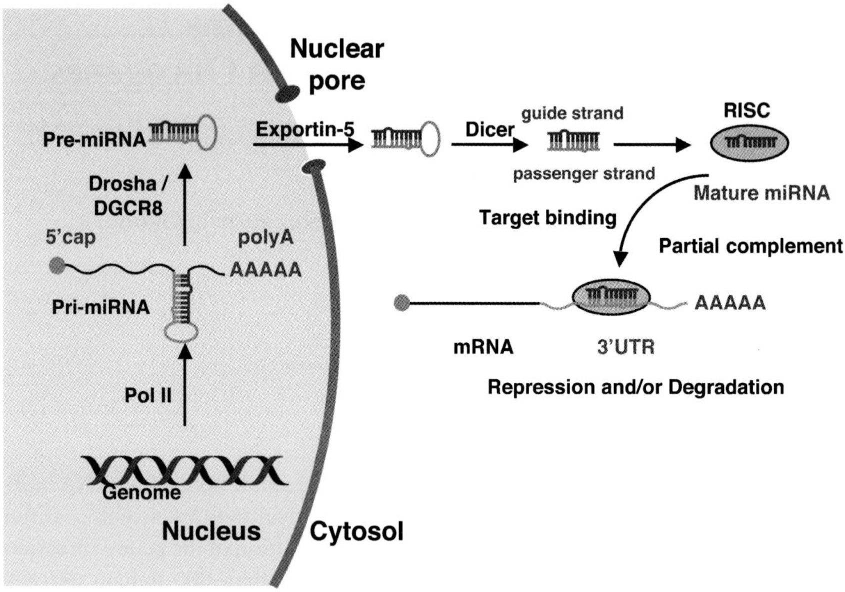

Fig. 1 Schematic diagram of biogenesis and function in micro-RNAs (miRNAs). MiRNAs are transcribed by RNA polymerase II and sequentially processed by Drosha/DGCR8 and Dicer. MiRNA-loaded RNA-induced silencing complex (*RISC*) causes the cleavage or translational silencing of target mRNAs

guanosine cap structure and a poly(A) tail similar to protein-coding mRNAs. The pri-miRNAs are processed into precursors of approximately 70 nucleotides (pre-miRNAs) with a stem-loop structure and a two-nucleotide 3′-overhang by the RNase III enzyme, Drosha, and a double-stranded RNA-binding protein, DGCR8/Pasha [13–15]. Then, pre-miRNAs are transported to the cytoplasm by a member of the Ran transport receptor family, exportin-5 [16,17]. Pre-miRNAs exported in the cytoplasm are further processed by another RNase III enzyme, Dicer, and unwound by a helicase [18]. Finally, only one mature miRNA strand (guide strand) is incorporated into a RNA-induced silencing complex (RISC) that mediates cleavage or translational inhibition of target mRNAs, while the other strand (passenger strand) is quickly degraded [19–21]. The stability of the base pairs at the 5′-end of the duplex determines which strand is incorporated in RISC [22,23]. RISC is composed of Dicer, Argonaute2 (Ago2), and the double-strand RNA-binding protein, TRBP [21,24], and cleaves target mRNAs more efficiently by using pre-miRNAs rather than the duplex RNAs that do not have the stem-loop structure, suggesting that processing by Dicer may be coupled with assembly of the mature miRNA into RISC [21]. The incorporated miRNA guides the RISC to the complementary sequence in the 3′-untranslated region (UTR) of target mRNAs. MiRNAs base-pair to the 3′-UTR of target mRNA with perfect or near-perfect complementarity, leading to the target mRNA degradation by Ago2, a component of RISC [25]. In contrast, partial base-pairing between a miRNA and a target mRNA leads to translational silencing of a target mRNA without RNA degradation [26]. In partial base-pairing, the binding of some nucleotides in the 5′-region of miRNAs has been indicated to be functionally important by systematic mutation experiments [27,28].

The Role of miRNAs in Cancer

Recently, it has been reported that the expression of several miRNAs is altered in a variety of human cancers, suggesting potential roles of miRNAs in tumorigenesis [29]. Calin et al. [30] showed that more than 50% of miRNAs were located in cancer-associated genomic regions or in fragile sites. In fact, miR-15a and miR-16 genes exist as a cistronic cluster at 13q14, which is deleted or downregulated in most cases (~68%) of B-cell chronic lymphocytic leukemias [31]. Cimmino et al. [32] found that both these miRNAs negatively regulate the expression of B cell lymphoma 2 (BCL2), which inhibits apoptosis and is present in many types of cancer including leukemias. In fact, overexpression of miR-15 and miR-16 in the MEG-01 cell line induces apoptotic cell death.

Alterations in gene copy number of miRNAs are detected in a variety of human cancers [33–35]. Zhang et al. [33] showed that miRNAs exhibited high-frequency genomic alterations in human ovarian cancer, breast cancer, and melanoma using high-resolution array-based comparative genomic hybridization. Hayashita et al. [34] found that the expression and gene copy number of the miR-17–92 cluster composed of seven miRNAs is increased in lung cancer cell lines, especially with small-cell lung cancer histology. Enforced expression of miRNAs included in this polycistronic cluster enhances cell proliferation in a lung cancer cell line. The increase in expression and gene copy number of miR-17–92 cluster was also found in B-cell lymphomas [35]. The expression of miRNAs in this cluster is upregulated by c-Myc, whose expression and/or function is one of the most common abnormalities in human cancers, and miR-17-5p and miR-20a in this miR-17–92 cluster negatively regulate the expression of transcriptional factor E2F1 [36]. Furthermore, it was indicated that the miR-17–19b cluster included in miR-17–92 cluster inhibited apoptotic cell death and accelerated c-Myc-induced lymphomagenesis in mice reconstituted with miR-17–19b cluster-overexpressed hematopoietic stem cells [35]. Furthermore, the miR-17–92 cluster has been reported to augment angiogenesis in vivo by downregulation of antiangiogenic thrombospondin-1 and connective tissue growth factor in Ras-transformed colonocytes [37].

Global expression profiling analysis of protein-coding genes is known to be useful for cancer diagnoses and prognosis predictions [38]. Recently, Lu et al. [29] indicated that miRNA expression profiles can successfully classify poorly differentiated cancers that cannot be classified by mRNA expression profiles. Accordingly, miRNA expression profiles are more accurately correlated with clinical severity of cancer malignancy than protein-coding gene expression profiles. This result indicates the potential of miRNA expression profiles in cancer classification and prognosis prediction [29].

Because miRNAs act as oncogenes or tumor suppressor genes, miRNAs are potential targets of therapeutic strategies. Recently, Krutzfeldt et al. [39] indicated that chemically engineered oligonucleotides, termed antagomirs, efficiently inhibited miRNAs in vivo. Additionally, it is reported that introduction of 2′-*O*-methoxyethyl phosphorothioate antisense oligonucleotide of miR-122 (abundant in the liver; regulates cholesterol and fatty acid metabolism) decreases plasma cholesterol levels and improves liver steatosis in mice with diet-induced obesity [40]. These findings indicate that antisense oligonucleotides are also potential targets for drug discovery, suggesting the possibility that intractable cancers may become curable by overexpression and/or inhibition of miRNAs. However, for miRNAs to be used in gene therapy, further improvements to make miRNAs more effective and less toxic than other cancer therapy are required.

Prediction of Target mRNAs

Identification of the miRNA-targeted genes provides deep biological insights to understand the new mechanisms of the physiopathological phenomena. Therefore, development of a comprehensive assay to rapidly identify target mRNAs would greatly assist understanding of miRNAs and lead to novel therapeutic approaches against cancer. Although there are many bioinformatics approaches for target prediction, which have predicted that a single miRNA may have hundreds of target genes [5,6,41–45], they have room for many improvements in accuracy and completeness of prediction, respectively. Thus, we have developed a new computational approach to extract regulatory networks between miRNAs and their target mRNAs (Fig. 2). We assumed that the expression pattern of miRNA would present inverse correlation with that of its corresponding target mRNA. Hence, selecting the inverse correlative pairs among the miRNA–target pairs predicted by the sequence-based algorism (ex. miRanda) from expression profiles of mRNAs and miRNAs, we have successfully reduced the false-positive pairs of target prediction results. Moreover, overlapping target information of the predicted pairs onto the coexpressed gene networks, we have also generated the gene networks coregulated by the common miRNA. We hope the inferred networks can be helpful to understand the biological systems of the miRNAs world.

It has been established that the abnormalities of specific miRNA expression contribute to tumorigenesis. Therefore, miRNAs are expected to be powerful tools for cancer classification, diagnosis, and prognosis prediction as well as potential targets of cancer therapy. Furthermore, identification of target mRNAs regulated by miRNAs, elucidation of the oncogenic or tumor suppressive molecular mechanisms by miRNAs and identification of genetic variation in miRNAs and miRNA–target sites of mRNAs may lead to the discovery of new molecular targets related to oncogenesis.

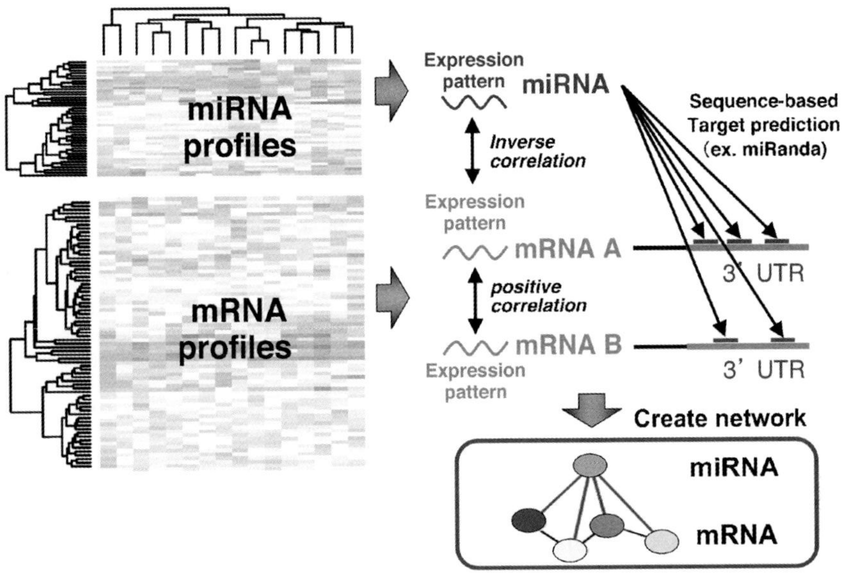

Fig. 2 Scheme of formulation of networks between miRNAs and their target mRNAs

References

1. Bartel DP (2004) MicroRNAs: genomics, biogenesis, mechanism, and function. Cell 116:281–297
2. Tsuchiya S, Okuno Y, Tsujimoto G (2006) MicroRNA: biogenetic and functional mechanisms and involvements in cell differentiation and cancer. J Pharmacol Sci 101:267–270
3. Pasquinelli AE, Reinhart BJ, Slack F et al (2000) Conservation of the sequence and temporal expression of let-7 heterochronic regulatory RNA. Nature (Lond) 408:86–89
4. Griffiths-Jones S, Grocock RJ, van Dongen S et al (2006) miRBase: microRNA sequences, targets and gene nomenclature. Nucleic Acids Res 34:D140–D144
5. Lewis BP, Burge CB, Bartel DP (2005) Conserved seed pairing, often flanked by adenosines, indicates that thousands of human genes are microRNA targets. Cell 120:15–20
6. Xie X, Lu J, Kulbokas EJ et al (2005) Systematic discovery of regulatory motifs in human promoters and 3′ UTRs by comparison of several mammals. Nature (Lond) 434:338–345
7. Esquela-Kerscher A, Slack FJ (2006) Oncomirs: microRNAs with a role in cancer. Nat Rev Cancer 6:259–269
8. Calin GA, Croce CM (2006) MicroRNA–cancer connection: the beginning of a new tale. Cancer Res 66:7390–7394
9. Calin GA, Croce CM (2006) MicroRNA signatures in human cancers. Nat Rev Cancer 6:857–866
10. Lee Y, Jeon K, Lee JT et al (2002) MicroRNA maturation: stepwise processing and subcellular localization. EMBO J 21:4663–4670
11. Cai X, Hagedorn CH, Cullen BR (2004) Human microRNAs are processed from capped, polyadenylated transcripts that can also function as mRNAs. RNA 10:1957–1966
12. Lee Y, Kim M, Han J et al (2004) MicroRNA genes are transcribed by RNA polymerase II. EMBO J 23:4051–4060
13. Lee Y, Ahn C, Han J et al (2003) The nuclear RNase III Drosha initiates microRNA processing. Nature (Lond) 425:415–419

14. Gregory RI, Yan KP, Amuthan G et al (2004) The Microprocessor complex mediates the genesis of microRNAs. Nature (Lond) 432:235–40
15. Han J, Lee Y, Yeom KH et al (2006) Molecular basis for the recognition of primary microRNAs by the Drosha-DGCR8 complex. Cell 125:887–901
16. Yi R, Qin Y, Macara IG et al (2003) Exportin-5 mediates the nuclear export of pre-microRNAs and short hairpin RNAs. Genes Dev 17:3011–3016
17. Lund E, Guttinger S, Calado A et al (2004) Nuclear export of microRNA precursors. Science 303:95–98
18. Hutvagner G, McLachlan J, Pasquinelli AE et al (2001) A cellular function for the RNA-interference enzyme Dicer in the maturation of the let-7 small temporal RNA. Science 293:834–838
19. Matranga C, Tomari Y, Shin C et al (2005) Passenger-strand cleavage facilitates assembly of siRNA into Ago2-containing RNAi enzyme complexes. Cell 123:607–620
20. Rand TA, Petersen S, Du F et al (2005) Argonaute2 cleaves the anti-guide strand of siRNA during RISC activation. Cell 123:621–629
21. Gregory RI, Chendrimada TP, Cooch N et al (2005) Human RISC couples microRNA biogenesis and posttranscriptional gene silencing. Cell 123:631–640
22. Khvorova A, Reynolds A, Jayasena SD (2003) Functional siRNAs and miRNAs exhibit strand bias. Cell 115:209–216
23. Schwarz DS, Hutvágner G, Du T et al (2003) Asymmetry in the assembly of the RNAi enzyme complex. Cell 115:199–208
24. Chendrimada TP, Gregory RI, Kumaraswamy E et al (2005) TRBP recruits the Dicer complex to Ago2 for microRNA processing and gene silencing. Nature (Lond) 436:740–744
25. Meister G, Landthaler M, Patkaniowska A et al (2004) Human Argonaute2 mediates RNA cleavage targeted by miRNAs and siRNAs. Mol Cell 15:185–197
26. Hutvágner G, Zamore PD (2002) A microRNA in a multiple-turnover RNAi enzyme complex. Science 297:2056–2060
27. Doench JG, Sharp PA (2004) Specificity of microRNA target selection in translational repression. Genes Dev 18:504–511
28. Kiriakidou M, Nelson PT, Kouranov A et al (2004) A combined computational-experimental approach predicts human microRNA targets. Genes Dev 18:1165–1178
29. Lu J, Getz G, Miska EA et al (2005) MicroRNA expression profiles classify human cancers. Nature (Lond) 435:834–838
30. Calin GA, Sevignani C, Dumitru CD et al (2004) Human microRNA genes are frequently located at fragile sites and genomic regions involved in cancers. Proc Natl Acad Sci U S A 101:2999–3004
31. Calin GA, Dumitru CD, Shimizu M et al (2002) Frequent deletions and down-regulation of micro-RNA genes miR15 and miR16 at 13q14 in chronic lymphocytic leukemia. Proc Natl Acad Sci U S A 99:15524–15529
32. Cimmino A, Calin GA, Fabbri M et al (2005) miR-15 and miR-16 induce apoptosis by targeting BCL2. Proc Natl Acad Sci U S A 102:13944–13949
33. Zhang L, Huang J, Yang N et al (2006) microRNAs exhibit high frequency genomic alterations in human cancer. Proc Natl Acad Sci U S A 103:9136–9141
34. Hayashita Y, Osada H, Tatematsu Y et al (2005) A polycistronic microRNA cluster, miR-17–92, is overexpressed in human lung cancers and enhances cell proliferation. Cancer Res 65:9628–9632
35. He L, Thomson JM, Hemann MT et al (2005) A microRNA polycistron as a potential human oncogene. Nature (Lond) 435:828–833
36. O'Donnell KA, Wentzel EA, Zeller KI ey al (2005) c-Myc-regulated microRNAs modulate E2F1 expression. Nature (Lond) 435:839–843
37. Dews M, Homayouni A, Yu D et al (2006) Augmentation of tumor angiogenesis by a Myc-activated microRNA cluster. Nat Genet 38:1060–1065
38. Ramaswamy S, Tamayo P, Rifkin R et al (2001) Multiclass cancer diagnosis using tumor gene expression signatures. Proc Natl Acad Sci U S A 98:15149–15154

39. Krutzfeldt J, Rajewsky N, Braich R et al (2005) Silencing of microRNAs in vivo with 'antagomirs'. Nature (Lond) 438:685–689
40. Esau C, Davis S, Murray SF et al (2006) miR-122 regulation of lipid metabolism revealed by in vivo antisense targeting. Cell Metab 3:87–98
41. John B, Enright AJ, Aravin A et al (2004) Human MicroRNA targets. PLoS Biol 2:e363
42. Kiriakidou M, Nelson PT, Kouranov A et al (2004) A combined computational-experimental approach predicts human microRNA targets. Genes Dev 18:1165–1178
43. Lewis BP, Shih IH, Jones-Rhoades MW et al (2003) Prediction of mammalian microRNA targets. Cell 115:787–798
44. Krek A, Grun D, Poy MN et al (2005) Combinatorial microRNA target predictions. Nat Genet 37:495–500
45. Miranda KC, Huynh T, Tay Y et al (2006) A pattern-based method for the identification of microRNA binding sites and their corresponding heteroduplexes. Cell 126:1203–1217

Part III
Dysfunction of Information Networks and Related Diseases

The Pathogenetic Significance of Deregulated Transcription Factors in Hematological Malignancies

Masahiro Nakagawa, Susumu Goyama, Motoshi Ichikawa, and Mineo Kurokawa

Introduction

Hematological malignancies have been treated with traditional chemotherapy, such as alkylating agents, anthracycline, or antimetabolites. These conventional chemotherapies are not specific to malignant cells, and severe side effects are unavoidable with intensive chemotherapy. Therefore, it might be difficult to improve the cure rate further with conventional chemotherapy. Hematopoietic stem cell transplantation (HSCT) is a powerful and curable treatment, but it is very invasive. Patients sometimes suffer from severe side effects, such as graft-versus-host disease (GVHD), and HSCT may also cause premature death. In addition, it is sometimes difficult to find appropriate donors. In recent years, with the great advances in molecular biology, genetics, biochemistry, and systems biology, we gradually come to understand the molecular mechanisms of pathogenesis in hematological malignancies. Molecular targeted therapy targets molecules or pathways that play critical roles in maintaining hematological malignancies. In contrast to conventional chemotherapy, molecular targeted therapy is highly specific to malignant cells and has minimal side effects. Currently, the most successful molecular targeted therapy is imatinib, a relatively specific inhibitor of the BCR-ABL tyrosine kinase in chronic myeloid leukemia (CML). Imatinib dramatically changed the standard therapy and indication of HSCT for CML [1,2]. However, in contrast to CML, which is relatively homogeneous, molecular pathogenesis and therapeutic targets are largely unknown in many hematological malignancies. Therefore, we definitely need to elucidate molecular mechanisms of pathogenesis in other types of hematopoietic malignancies. Classically, we analyzed the functions of molecules individually, which method was not necessarily efficient. To extensively clarify molecular pathogenesis in many hematological malignancies, we applied bioinformatics and systems biology. We made genome-wide analyses of intracellular

M. Nakagawa, S. Goyama, M. Ichikawa, and M. Kurokawa
Department of Hematology and Oncology, Graduate School of Medicine,
University of Tokyo, The University of Tokyo Hospital, 7-3-1 Hongo,
Bunkyo-ku, Tokyo, 113-8655, Japan

S. Nakanishi et al. (eds.), *Systems Biology*,
DOI:10.1007/978-4-431-87704-2_20, © Springer 2009

changes caused by leukemia-associated transcriptional factors, with the intention to comprehend total molecular pathogenesis and to identify therapeutic targets.

Deregulation of Transcription Factors in Hematological Malignancies

Hematopoiesis is a precisely regulated process that generates terminally differentiated cells in the peripheral blood from immature progenitors from bone marrow. The number of mature cells is tightly maintained in spite of the short life of these cells. For this strict regulation, hematopoietic progenitors must proliferate, differentiate, or undergo apoptosis, according to the individual circumstances. These processes are controlled by the combined network of multiple hematopoietic transcription factors [3]. Deregulation of these process caused by genetic mutations results in the development of hematological malignancies (Fig. 1). Specific chromosomal translocations are detected in nearly half of cases with acute leukemia. It was revealed that hematopoietic transcription factors are frequent targets for leukemia-associated gene rearrangements. This finding is consistent with the fact that hematopoiesis can be achieved through well-organized cooperation between hematopoietic transcription factors: among these are *Evi-1* (ectopic viral integration site 1) and *AML1* (acute myeloid leukemia 1).

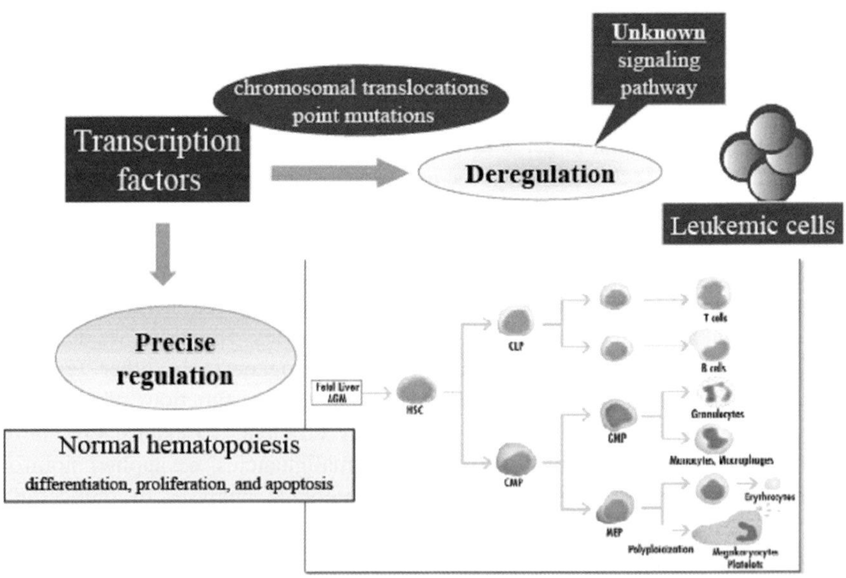

Fig. 1 Deregulation of transcription factors in leukemia. Transcription factors precisely regulate hematopoietic system for differentiation, proliferation, and apoptosis. Deregulation of transcription factors causes hematopoietic malignancies by unknown signaling pathways

AML1 in Normal and Malignant Hematopoiesis

AML1 is a transcription factor first isolated from t(8;21)(q21;q22) and later in t(3;21)(q26;q22), t(12;21)(p13;q22), and t(16;21)(q24;q21) translocations found in human leukemia in which the aberrant fusion genes AML1/ETO and AML1/Evi-1, TEL/AML1, and AML1/MTG16, respectively, were generated [4]. These fusion proteins have a common functional property whereby they act as a dominant negative inhibitor for the normal AML1 allele. For example, the AML1/ETO fusion protein retains the ability to bind to the *cis* elements of AML1 target genes but does not initiate gene transcription. This property enables the fusion protein to interfere with the transcriptional activity of native AML1. Similar in vitro data have been obtained for TEL/AML1 and AML1/Evi-1. In addition, AML1/ETO elicits a dominant inhibitory effect on AML1 function during hematopoietic development. AML1/ETO knock-in mice recapitulate an AML1-null phenotype; they die in utero, although they have one fully intact AML1 allele. It is shown that AML1 dimerizes with the common β-subunit, CBFβ, to bind to its target sequences known as PEBP2 sequences and regulates a variety of hematopoietic lineage-specific genes, including interleukin-3, granulocyte-macrophage colony-stimulating factor, macrophage colony-stimulating factor receptor, neutrophil elastase, granzyme B, myeloperoxidase, neutrophil defensin, and subunits of the T-cell and B-cell antigen receptor, in cooperation with other transcription factors [4]. AML1 is absolutely required for mouse embryogenesis and hematopoiesis, because AML1-null embryos die at midgestation (by E12.5) with massive hemorrhage in the central nervous system and complete effacement of definitive hematopoiesis in the fetal liver, although primitive erythropoiesis is preserved [5,6]. The analysis on AML1 function in adults has been hampered by premature lethality of AML1 knockout (KO) mice. Recently, we have generated AML1 conditional knockout (cKO) mice in which AML1 is disrupted specifically in hematopoietic compartments after birth, and found that AML1 negatively regulates quiescent hematopoietic stem cells (HSCs) in adult hematopoiesis, whereas it is required for megakaryocytic maturation and lymphocytic development [7,8].

Evi-1 in Normal and Malignant Hematopoiesis

Evi-1 was first identified as a common locus of retroviral integration in myeloid tumors in AKXD mice [9]. In humans, Evi-1 is located on chromosome 3q26, and rearrangements on chromosome 3q26 often activate Evi-1 expression in acute myeloid leukemia (AML) and myelodysplastic syndrome (MDS) [10,11]. Although these rearrangements are infrequent in AML, they are of remarkable prognostic value. Patients with these karyotypes are characterized by elevated platelet count and lack of response to antileukemic therapy [12]. Elevated Evi-1 expression occurs with high frequency in AML patients without 3q26 abnormalities and is associated with unfavorable outcomes [13,14]. Thus, Evi-1 is one of the key factors that predict poor survival in leukemia patients. Previous studies revealed that Evi-1 possesses diverse functions as an oncoprotein. Evi-1 antagonizes growth-inhibitory effects of transforming growth factor-beta (TGF-β) by interacting with Smad3 [15], protects cells from stress-induced cell death by inhibiting c-Jun N-terminal kinase (JNK) [16], increases the expression of endogenous c-Jun and c-fos, resulting in activation of AP-1 [17], and blocks granulocytic differentiation of myeloid cells [18]. Evi-1 knockout mice die in utero by E10.5 exhibiting widespread hypocellularity, hemorrhaging, and disruption in the development of paraxial mesenchyme [22]. To investigate the physiological role of Evi-1 in adult hematopoiesis, we have generated Evi-1 cKO mice in which Evi-1 is disrupted specifically in hematopoietic compartments after birth, and found that Evi-1 is essential for proliferation of both HSCs and leukemic cells [19].

The Signaling Pathway Regulated by AML1 or Evi-1

Deregulation of the transcriptional network caused by functional impairments of AML1 or Evi-1 is the key to comprehend molecular pathogenesis of hematopoietic malignancies. Therefore, we focused on the signaling pathway regulated by them. In normal conditions, precise regulation of AML1 or Evi-1 maintains hematopoiesis. Once their functions are impaired, the deregulated network causes hematological malignancies. First, we obtained transcriptome from the cells just after deleting AML1. We purified lineage-negative, Sca-1-positive, c-kit-positive (LSK) cells as HSCs from cKO mice and retrovirally transduced Cre-ER to them. To detect immediate targets of AML1, we collected the cells just after deleting AML1 by activation of Cre-ER using 4-OHT and analyzed them with cDNA microarray. Next, we combined transcriptome analysis and knowledge-based database for various information networks. We used gene set enrichment analysis (GSEA) [20,21], which is a computational method that determines whether an a priori defined set of genes shows statistically significant, concordant differences between two biological states, such as AML1 deficient or AMLI competent. We found several signal pathways regulating cell proliferation, inflammation, or apoptosis are activated in AML1-deficient cells. Next, we obtained transcriptome from the cells shortly after deleting Evi-1, by retrovirus containing Cre instead of

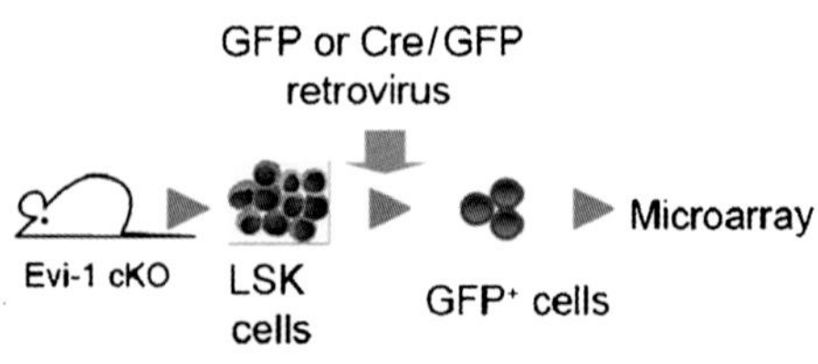

Fig. 2 Transcriptome from the cells just after deleting *Evi-1*. Lineage-negative, Sca-1-positive, c-kit-positive (*LSK*) cells were isolated as hematopoietic stem cells (HSCs) from *Evi-1* conditional knockout (*cKO*) mice. *GFP* or *Cre/GFP* retroviral infection was performed on these cells, and GFP-positive cells were collected shortly after deleting Evi-1 by Cre recombinase activity. Gene expression profiles were compared in cDNA microarray analysis between GFP and Cre/GFP cells, that is, Evi-1 wild-type and Evi-1 knockout cells

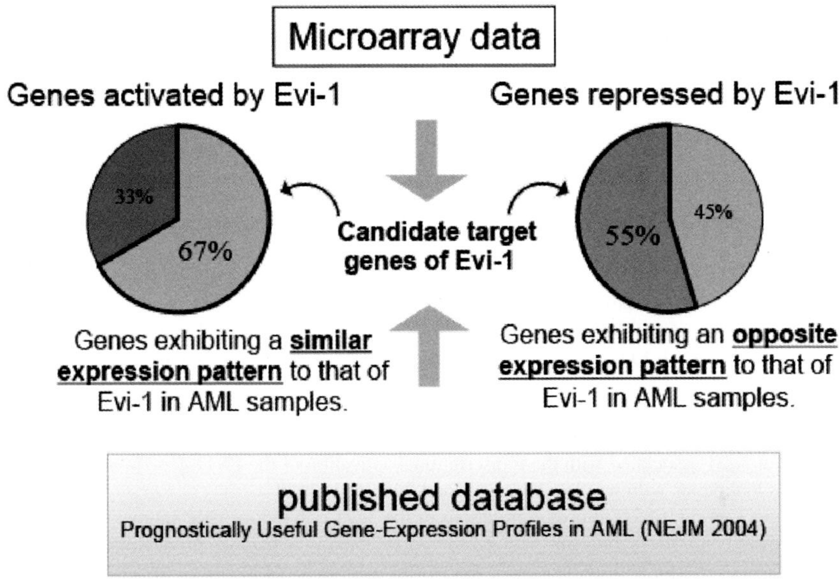

Fig. 3 Analysis of gene expression pattern using human acute myeloid leukemia (AML) samples. Gene expression pattern was compared with human AML samples. Candidate target genes of Evi-1 were extracted by investigating their expression pattern. Genes exhibiting similar expression pattern to that of Evi-1 are possibly activated by Evi-1. On the other hand, genes exhibiting the opposite expression pattern to that of Evi-1 are possibly repressed by Evi-1

Cre-ER (Fig. 2). This time, we compared gene expression patterns using human AML samples [14]. Candidate target genes of Evi-1 were extracted by investigating their expression pattern (Fig. 3). Genes exhibiting a similar expression pattern to that of Evi-1 are possibly activated by Evi-1. On the other hand, genes exhibiting the opposite expression pattern to that of Evi-1 may be repressed by Evi-1. From these two types of microarray analyses, we picked some candidates related to HSCs and platelet formation, because Evi-1 knockout mice exhibited decreased HSCs and platelet production. For example, GATA2 is an important gene that regulates HSCs, as previously reported that Evi-1 regulates fetal HSC proliferation through GATA2 expression [22]. It awaits further investigation whether these candidate target genes could be therapeutic targets in AML1 or Evi-1-related leukemia.

References

1. Druker BJ et al (2006) Five-year follow-up of patients receiving imatinib for chronic myeloid leukemia. N Engl J Med 35523:2408–2417
2. Hehlmann R et al (2007) Drug treatment is superior to allografting as first-line therapy in chronic myeloid leukemia. Blood 10911:4686–4692
3. Iwasaki H, Akashi K (2007) Hematopoietic developmental pathways: on cellular basis. Oncogene 2647:6687–6696

4. Lutterbach B, Hiebert SW (2000) Role of the transcription factor AML-1 in acute leukemia and hematopoietic differentiation. Gene (Amst) 2452:223–235
5. Okuda T et al (1996) AML1, the target of multiple chromosomal translocations in human leukemia, is essential for normal fetal liver hematopoiesis. Cell 842:321–330
6. Wang Q et al (1996) Disruption of the Cbfa2 gene causes necrosis and hemorrhaging in the central nervous system and blocks definitive hematopoiesis. Proc Natl Acad Sci U S A 938:3444–3449
7. Ichikawa M et al (2004) AML-1 is required for megakaryocytic maturation and lymphocytic differentiation, but not for maintenance of hematopoietic stem cells in adult hematopoiesis. Nat Med 103:299–304
8. Ichikawa M et al (2008) AML1/Runx1 negatively regulates quiescent hematopoietic stem cells in adult hematopoiesis. J Immunol 1807:4402–4408
9. Mucenski ML et al (1988) Identification of a common ecotropic viral integration site, Evi-1, in the DNA of AKXD murine myeloid tumors. Mol Cell Biol 81:301–308
10. Ogawa S et al (1996) Abnormal expression of Evi-1 gene in human leukemias. Hum Cell 94:323–332
11. Suzukawa K et al (1994) Identification of a breakpoint cluster region 3 of the ribophorin I gene at 3q21 associated with the transcriptional activation of the EVI1 gene in acute myelogenous leukemias with inv(3)(q21q26). Blood 848:2681–2688
12. Pintado T et al (1985) Clinical correlations of the 3q21;q26 cytogenetic anomaly. A leukemic or myelodysplastic syndrome with preserved or increased platelet production and lack of response to cytotoxic drug therapy. Cancer (Phila) 553:535–541
13. Barjesteh van Waalwijk van Doorn-Khosrovani S et al (2003) High EVI1 expression predicts poor survival in acute myeloid leukemia: a study of 319 de novo AML patients. Blood 1013:837–845
14. Valk PJ et al (2004) Prognostically useful gene-expression profiles in acute myeloid leukemia. N Engl J Med 35016:1617–1628
15. Kurokawa M et al (1998) The oncoprotein Evi-1 represses TGF-beta signalling by inhibiting Smad3. Nature (Lond) 3946688:92–96
16. Kurokawa M et al (2000) The evi-1 oncoprotein inhibits c-Jun N-terminal kinase and prevents stress-induced cell death. EMBO J 1912:2958–2968
17. Tanaka T et al (1994) Evi-1 raises AP-1 activity and stimulates c-fos promoter transactivation with dependence on the second zinc finger domain. J Biol Chem 26939:24020–24026
18. Morishita K et al (1992) Expression of the Evi-1 zinc finger gene in 32Dc13 myeloid cells blocks granulocytic differentiation in response to granulocyte colony-stimulating factor. Mol Cell Biol 121:183–189
19. Goyama S et al (in press) Cell Stem Cell
20. Mootha VK et al (2003) PGC-1alpha-responsive genes involved in oxidative phosphorylation are coordinately downregulated in human diabetes. Nat Genet 343:267–273
21. Subramanian A et al (2005) Gene set enrichment analysis: a knowledge-based approach for interpreting genome-wide expression profiles. Proc Natl Acad Sci U S A 10243: 15545–15550
22. Yuasa H et al (2005) Oncogenic transcription factor Evi1 regulates hematopoietic stem cell proliferation through GATA-2 expression. EMBO J 2411:1976–1987

Ultradian Oscillation Networks in Somite Segmentation and Other Biological Events

Yasutaka Niwa*, Hiromi Shimojo*, and Ryoichiro Kageyama

Introduction

When you observe the surrounding environment, you notice that it is composed of many events that are periodic, although they seem at first glance to be stable. Mesopotamian culture knew that the moon waxes and wanes in a monthly cycle and that the position of the sun changes with a period of a year. Our internal environment, which is kept relatively constant by forces that promote homeostasis, also consists of many cyclic events. We usually wake up in the morning and go to bed at night. When this rhythm is disturbed by jet lag, one feels uncomfortable for a few days. No one doubts that this circadian rhythm is one of the most famous and essential rhythms in our body. By contrast, developmental processes seem to be well organized and programmed in advance. However, from the perspective of a single cell, there is noise in both intracellular and intercellular environments, and there should be some mechanisms for being resistant to such noise.

Recently several groups showed that the expression of several genes is not just constant. Rather, expression oscillates, and does so robustly with a periodicity of less than 24 h. That is an ultradian rhythm, and it occurs in developmental tissues or cultured cells when you observe them not as a whole but at the single-cell level, or if you observe their synchronized behaviors [14,15,20,21,29,31]. Here we would like to review these ultradian rhythms and their biological meaning in a developmental and cellular context such as somitogenesis and neurogenesis in mice and several signaling pathways in cultured cells.

Y. Niwa, H. Shimojo, and R. Kageyama
Institute for Virus Research, Kyoto University, Kyoto, Japan and Japan Science and Technology Agency, CREST, Kyoto, Japan

*Equal contribution

S. Nakanishi et al. (eds.), *Systems Biology*,
DOI:10.1007/978-4-431-87704-2_21,

Somite Segmentation Clock

Somitogenesis

Considering the skeleton of the vertebrate body, one would conclude that our body consists of a beautiful segmental structure. Examples are the ribs and the vertebrae (Fig. 1a). These metameric structures are already defined at the developmental stage by their precursors, which are called somites (Fig. 1b). A somite is a transient structure and gives rise to skeletal muscle and dermis in addition to ribs and vertebrae. One new pair of somites is generated with a certain periodicity by being separated from the most anterior part of the presomitic mesoderm (PSM), which is located in the most caudal part of the embryo. This process is accompanied by morphological boundary formation, which is called segmentation. For mouse, chicken, and zebrafish, respectively, the segmentation period is 120, 90, and 30 min.

Discovery of Molecular Evidence

This extraordinarily periodical event has been the object of mathematical modeling for a long time [7]. However, its molecular mechanism has been examined for just a decade. The first molecular clue was the discovery of the oscillatory expression of *c-hairy1* in the chick PSM [25]. This gene expression pattern varied among littermates and moved from posterior to anterior with the same periodicity as the segmentation period [25]. This variation in gene expression pattern was thought to

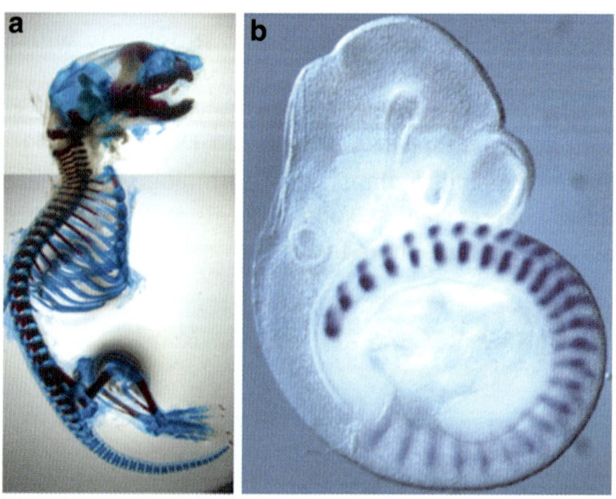

Fig. 1 Mouse skeleton and somites. **a** Skeletal staining of a mouse at P0. **b** Visualization of somites in a mouse embryo at E9.5 by in situ hybridization for *Uncx4.1*, which is a marker gene of the somite

be caused by the oscillatory changes in expression level in each cell. Subsequently, *c-hairy1* counterparts in mouse and zebrafish were found to show oscillatory expression patterns [16,26]. These molecules, or the system that generates these oscillatory expression patterns, were thought to account for the segmentation period, and they are now called a segmentation clock. Real-time imaging of the expression level of these genes recently showed that this oscillation continuously occurs at the single-cell level [21].

Molecular Mechanism of the Segmentation Clock

Negative Autoregulation of the Transcription Factor Hes7

In mouse somitogenesis, a gene for the basic helix-loop-helix (bHLH) transcription factor Hes7, a counterpart in mice of *c-hairy1*, has a critical role [5,6]. This gene is specifically expressed in the PSM in an oscillatory pattern [5]. This oscillatory expression is essential for proper somitogenesis because both persistent expression and loss of expression of Hes7 lead to irregular formation of ribs and vertebrae [5,13]. Furthermore, Hes7 represses its own promoter directly and is quickly removed by the ubiquitin-proteasome pathway [6]. This negative feedback loop targeting its own transcription seems to explain its oscillatory expression pattern in the PSM (Fig. 2).

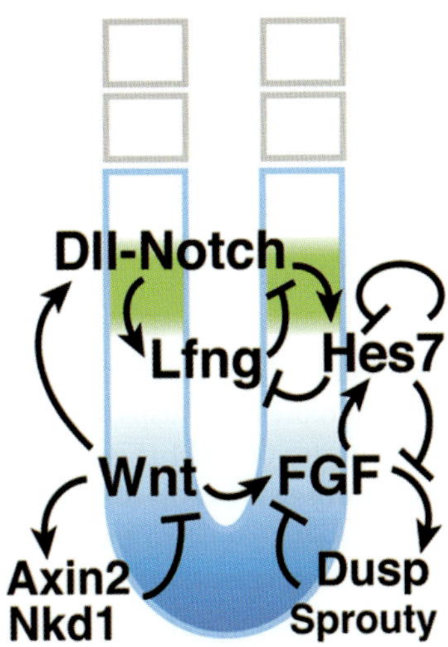

Fig. 2 Schematic representation of the mouse segmentation clock mechanism. Molecules in three major pathways and Hes7 interact with each other and form complex networks, which are thought to make the robustness of the mouse segmentation clock

Negative Feedback Loop in Notch Signaling

Another important molecule in mouse somitogenesis is Lunatic fringe (Lfng), which is a glycosyltransferase for Notch. As does Hes7, Lfng shows oscillatory expression in the PSM, and both persistent expression and loss of expression of Lfng result in the irregular formation of ribs and vertebrae [10,28,32]. Notch signaling induces Lfng expression, whereas Lfng inhibits Notch activity, forming another negative feedback loop (Fig. 2) [8]. This negative feedback loop generates oscillations in the activity of the Notch signaling pathway, and loss of Lfng leads to the failure of Notch signaling to oscillate [22], suggesting that Notch signaling plays an essential role in the segmentation clock. Furthermore, mutants with defects in somitogenesis in mice and zebrafish are mainly the result of the Notch pathway genes [16,26]. There is some controversy about the role of Notch signaling in the segmentation clock in mice, but there is no doubt that the Notch pathway is essential for proper somitogenesis.

Negative Feedback Loop in Wnt and FGF Signaling

In addition to Notch signaling, several molecules of Wnt and fibroblast growth factor (FGF) signaling also show oscillatory expression patterns in the PSM [9]. Among them, Axin2, which is an inhibitor of Wnt signaling, was first reported to oscillate out of phase from Lfng (Fig. 2) [3]. Recent genome-wide studies searching for oscillatory molecules in the PSM showed that many genes in Wnt and FGF signaling, as well as Notch signaling, show oscillatory expression in the PSM [9]. These results suggest that not only Hes7 but also Notch, Wnt, and FGF signaling have their own feedback loops, and that these feedback loops couple with each other and make the segmentation clock in the PSM robust.

Interaction Between Feedback Loops

Hes7 represses not only itself but also Lfng, which is thought to be an inhibitor of Notch signaling [5,6]. Hes7 also represses Dusp4, an inhibitor of FGF/MAPK signaling, which oscillates in the posterior PSM (Fig. 2) [23]. Periodic repression by Hes7 is essential for the oscillatory expression of Lfng and Dusp4 [23], which are thought to be important for their own negative feedback signaling loops. So, it is likely that Hes7 is a core of the segmentation clock and regulates oscillations in both Notch and FGF signaling pathways [23]. Actually, a recent report shows that Hes7 oscillation still remains even in the absence of Notch signaling [23]. On the other hand, the amplitude and anterior propagation of Hes7 oscillations are affected without Notch signaling [23]. Because dissociated, cultured PSM cells quickly lose their stable 2-h cycle oscillations, it is possible that, in mice, Notch signaling also regulates the segmentation clock via intercellular synchronization of the unstable oscillators in different cells, as was first proposed in zebrafish [17,18,21].

Oscillations in Neural Progenitors

Neural Development and Notch Signaling

Neural development proceeds by strictly regulated steps: from a cell proliferation phase, to a neurogenic phase, and then to a gliogenic phase (Fig. 3). During these processes, neural progenitors change their competency, giving rise to distinct types of cells [1,12]; this means that maintenance of neural progenitors until later stages of development is essential to generate cells both in correct number and with full cell-type diversity. Notch signaling plays an important role in maintenance of neural progenitors [2,11]. When Notch ligands such as Delta-like1 (Dll1) expressed in neighboring cells bind to Notch receptors, the intracellular domain of Notch (act-Notch) is cleaved and released from the membrane and goes to the nucleus (Fig. 4).

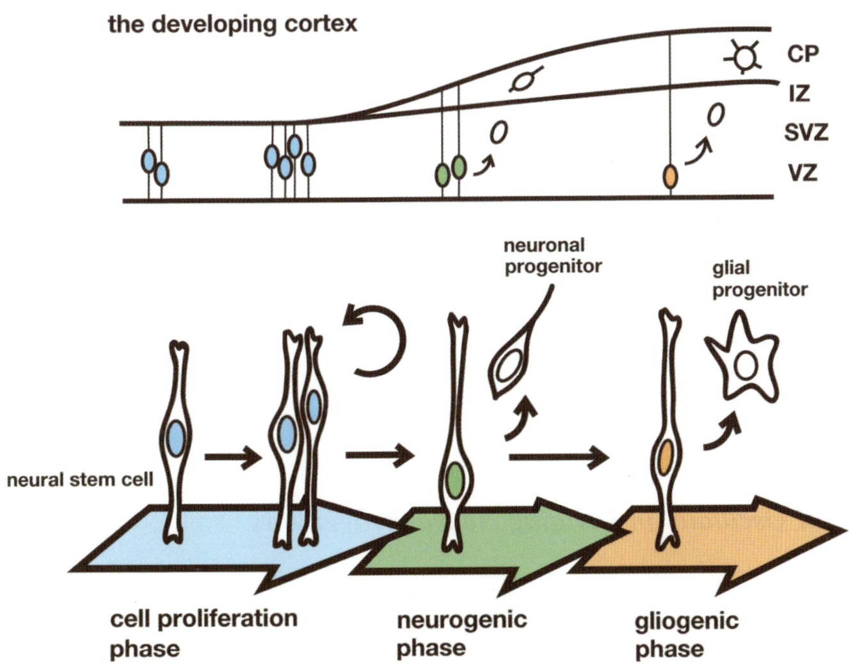

Fig. 3 Neural stem cells change their competency during neural development. Neural development proceeds according to the following steps: cell proliferation phase, neurogenic phase, and gliogenic phase. During these steps, neural stem cells change their competency and give rise to different cell types such as neuronal progenitors and glial progenitors. *VZ*, ventricular zone; *SVZ*, subventricular zone; *IZ*, intermediate zone; *CP*, cortical plate

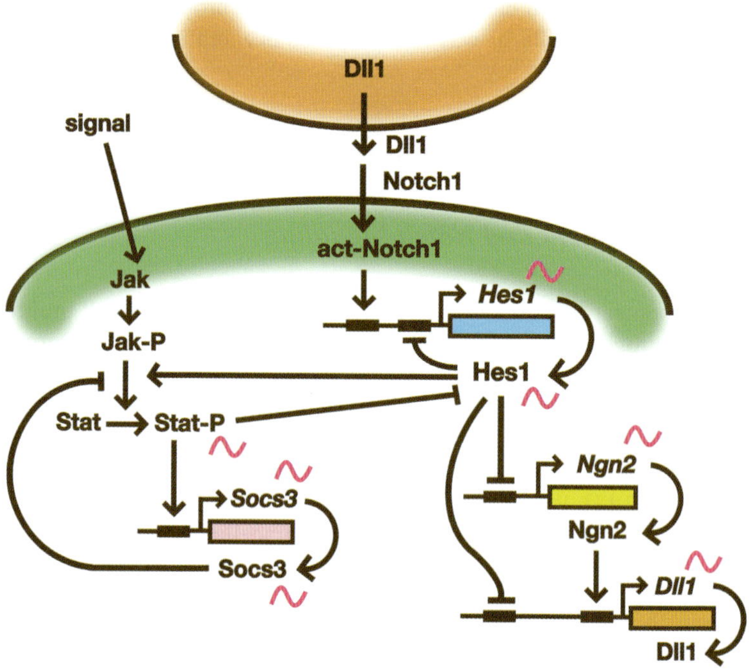

Fig. 4 Ultradian oscillation networks in neural progenitors. In neural progenitors, the expressions of Hes1, Ngn2, and Dll1 oscillate, thereby inducing mutual activation of Notch signaling in neighboring cells and maintaining Hes1 expression and the undifferentiated states. In addition to Notch signaling, Jak-Stat signaling molecules oscillate. Oscillations in Notch signaling and Jak-Stat signaling regulate each other

act-Notch forms a complex with RBP-J, and this complex activates expression of the bHLH transcriptional repressors Hes1 and Hes5 [24]. Hes genes downregulate the expression of proneural genes such as Neurogenin2 (Ngn2) and inhibit neuronal differentiation [4,27,19]. We now focus on oscillations in Notch signaling to better understand neural development.

Hes1 Oscillates in Dividing Neural Progenitors

In cultured cells, Hes1 expression oscillates with a period of about 2 h [14,21]. This oscillation is induced by serum stimulation or Notch activation, and is regulated by negative feedback: Hes1 can repress its own expression by directly binding to its own promoter [30], followed by rapid disappearance of both Hes1 mRNA and Hes1 protein, because they are extremely unstable, and then by the next round of expression. In this way, Hes1 autonomously displays an oscillatory expression mode.

In the developing central nervous system, Hes1 is expressed in the ventricular zone, a proliferative region, and the expression of Hes1 shows various levels in each dividing progenitor cell, suggesting the possibility that Hes1 expression oscillates in dividing neural progenitors. Using a real-time imaging method, Hes1 expression was found to be dynamically changing at the single-cell level (Fig. 4) [29].

Oscillations of Notch Signaling Regulate Maintenance of Neural Progenitors

The Hes1 target genes Ngn2 and Dll1 are expressed in dividing neural progenitors and show an inverse correlation with Hes1 expression, suggesting that these genes oscillate in neural progenitor cells. Interestingly, the expression of Ngn2 and Dll1 is found to be dynamically changing in dividing neural progenitors at the single-cell level (Fig. 4). Furthermore, inhibition of Notch signaling, a condition known to induce neuronal differentiation, leads to downregulation of Hes1 and sustained upregulation of Ngn2 and Dll1. These findings suggest that Hes1 oscillation leads to Ngn2 and Dll1 oscillations in dividing neural progenitors by periodically repressing their expression (Fig. 4). Oscillation of the expression of the Notch ligand Dll1 also activates Notch signaling in neighboring neural progenitors, thereby maintaining Hes1 oscillations and the undifferentiated state of these cells. As neural development proceeds, the expression of Hes1 is downregulated, whereas Ngn2 and Dll1 expressions are upregulated in a sustained manner, leading to neuronal differentiation. Ngn2 oscillation is apparently not sufficient, but sustained upregulation is required for neuronal differentiation. It is likely that Ngn2 oscillation induces expression of subsets of downstream genes such as Dll1 and is advantageous for maintenance of neural progenitors.

Jak-Stat Signaling Is Required for Hes1 Oscillation

In addition to Notch signaling, the expression of molecules of Jak-Stat and Smad signaling pathways also oscillates in response to serum in cultured fibroblast cells [31]. Furthermore, the stability of Hes1 protein is regulated by Stat signaling: Hes1 protein is stabilized in the absence of Stat signaling, whereas it is destabilized by constitutive activation of Stat signaling. Both sustained activation and inactivation of Stat signaling abolish Hes1 oscillation, suggesting that oscillations in Stat signaling regulate Hes1 oscillation by controlling the stability of Hes1 protein (Fig. 4). Moreover, loss of Hes1 oscillation leads to cell-cycle arrest, suggesting that the expression mode of signaling pathway molecules affect the downstream cellular event.

In the developing central nervous system, Hes1 oscillation disappears because of inhibition of Jak-Stat signaling, suggesting that Jak-Stat signaling also regulates Hes1 oscillation in neural progenitors.

Conclusions

A growing number of genes in various tissues and cells have been reported to be expressed cyclically with periods shorter than 24 h. These ultradian oscillation networks play an important role in biological events. During somitogenesis, the ultradian oscillation networks function as a segmentation clock, one of the biological clocks that regulate the timing of developmental events. Nevertheless, there are issues that remain to be determined. One of these questions is this: What is the master factor, acting as a pacemaker of the segmentation clock?

On the other hand, in non-PSM cells, there are few reports that discuss biological events, such as neural development, from the point of view of different gene expression modes. However, a growing number of oscillating genes have been reported. For instance, expression of NF-κB and p53 also oscillates [20,15]. These results suggest that oscillatory expression is more general and plays a more important role in cellular events than was previously thought, although the full significance of these systems remains unclear. The study of these ultradian oscillations in non-PSM cells is just getting started.

Many issues remain regarding ultradian rhythms, such as their molecular mechanisms, crosstalk of various signaling pathways, and downstream biological events. By elucidating the whole picture of oscillatory networks, we could characterize the sophisticated systems that regulate molecular events, cellular events, and biological events in more detail.

References

1. Alvarez-Buylla A, Garcia-Verdugo JM, Tramontin AD (2001) A unified hypothesis on the lineage of neural stem cells. Nat Rev Neurosci 2:287–293
2. Artavanis-Tsakonas S, Rand MD, Lake RJ (1999) Notch signaling: cell fate control and signal integration in development. Science 284:770–776
3. Aulehla A, Wehrle C, Brand-Saberi B, Kemler R, Gossler A, Kanzler B, Herrmann BG (2003) *Wnt3a* plays a major role in the segmentation clock controlling somitogenesis. Dev Cell 4:395–406
4. Bertrand N, Castro DS, Guillemot F (2002) Proneural genes and the specification of neural cell types. Nat Rev Neurosci 3:517–530
5. Bessho Y, Miyoshi G, Sakata R, Kageyama R (2001) Dynamic expression and essential functions of *Hes7* in somite segmentation. Genes Dev 15:2642–2647
6. Bessho Y, Hirata H, Masamizu Y, Kageyama R (2003) Periodic repression by the bHLH factor Hes7 is an essential mechanism for the somite segmentation clock. Genes Dev 17:1451–1456
7. Cooke J, Zeeman EC (1976) A clock and wavefront model for control of the number of repeated structures during animal morphogenesis. J Theor Biol 58:455–476
8. Dale JK, Maroto M, Dequeant ML, Malapert P, McGrew M, Pourquié O (2003) Periodic Notch inhibition by Lunatic Fringe underlies the chick segmentation clock. Nature (Lond) 421:275–278
9. Dequéant M-L, Glynn E, Gaudenz K, Wahl M, Chen J, Mushegian A, Pourquié O (2006) A complex oscillating network of signaling genes underlies the mouse segmentation clock. Science 314:1595–1598

10. Evrard YA, Lun Y, Aulehla A, Gan L, Johnson RL (1998) *lunatic fringe* is an essential mediator of somite segmentation and patterning. Nature (Lond) 394:377–381
11. Gaiano N, Fishell G (2002) The role of Notch in promoting glial and neural stem cell fates. Annu Rev Neurosci 25:471–490
12. Götz M, Huttner WB (2005) The cell biology of neurogenesis. Nat Rev Mol Cell Biol 6:777–788
13. Hirata H, Bessho Y, Kokubu H, Masamizu Y, Yamada S, Lewis J, Kageyama R (2004) Instability of Hes7 protein is crucial for the somite segmentation clock. Nat Genet 36:750–754
14. Hirata H, Yoshiura S, Ohtsuka T, Bessho Y, Harada T, Yoshikawa K, Kageyama R (2002) Oscillatory expression of the bHLH factor Hes1 regulated by a negative feedback loop. Science 298:840–843
15. Hoffmann A, Levchenko A, Scott ML, Baltimore D (2002) The IκB-NF-κB signaling module: temporal control and selective gene activation. Science 298:1241–1245
16. Holley SA (2007) The genetics and embryology of zebrafish metamerism. Dev Dyn 236:1422–1449
17. Horikawa K, Ishimatsu K, Yoshimoto E, Kondo S, Takeda H (2006) Noise-resistant and synchronized oscillation of the segmentation clock. Nature (Lond) 441:719–723
18. Jiang Y-J, Aerne BL, Smithers L, Haddon C, Ish-Horowicz D, Lewis J (2000) Notch signaling and the synchronization of the somite segmentation clock. Nature (Lond) 408:475–479
19. Kageyama R, Ohtsuka T, Kobayashi T (2007) The Hes gene family: repressors and oscillators that orchestrate embryogenesis. Development (Camb) 134:1243–1251
20. Lev Bar-Or R, Maya R, Segel LA, Alon U, Levine AJ, Oren M (2000) Generation of oscillations by the p53-Mdm2 feedback loop: a theoretical and experimental study. Proc Natl Acad Sci U S A 97:11250–11255
21. Masamizu Y, Ohtsuka T, Takashima Y, Nagahara H, Takenaka Y, Yoshikawa K, Okamura H, Kageyama R (2006) Real-time imaging of the somite segmentation clock: revelation of unstable oscillators in the individual presomitic mesoderm cells. Proc Natl Acad Sci U S A 103:1313–1318
22. Morimoto M, Takahashi Y, Endo M, Saga Y (2005) The Mesp2 transcription factor establishes segmental borders by suppressing Notch activity. Nature (Lond) 435:354–359
23. Niwa Y, Masamizu Y, Tianxiao L, Nakayama R, Deng Chu-Xia, Kageyama R (2007) The initiation and propagation of Hes7 oscillation are cooperatively regulated by Fgf and notch signaling in the somite segmentation clock. Dev Cell 13:298–304
24. Ohtsuka T, Ishibashi M, Gradwohl G, Nakanishi S, Guillemot F, Kageyama R (1999) *Hes1* and *Hes5* as Notch effectors in mammalian neuronal differentiation. EMBO J 18:2196–2207
25. Palmeirim I, Henrique D, Ish-Horowicz D, Pourquié O (1997) Avian hairy gene expression identifies a molecular clock linked to vertebrate segmentation and somitogenesis. Cell 91(5):639–648
26. Rida PC, Le Minh N, Jiang YJ (2004) A Notch feeling of somite segmentation and beyond. Dev Biol 265:2–22
27. Ross SE, Greenberg ME, Stiles CD (2003) Basic helix-loop-helix factors in cortical development. Neuron 39:13–25
28. Serth K, Schuster-Gossler K, Cordes R, Gossler A (2003) Transcriptional oscillation of lunatic fringe is essential for somitogenesis. Genes Dev 17:912–925
29. Shimojo H, Ohtsuka T, Kageyama R (2008) Oscillations in Notch signaling regulate maintenance of neural progenitors. Neuron 58:52–64
30. Takebayashi K, Sasai Y, Sakai Y, Watanabe T, Nakanishi S, Kageyama R (1994) Structure, chromosomal locus, and promoter analysis of the gene encoding the mouse helix-loop-helix factor HES-1, Negative autoregulation through the multiple N box elements. J Biol Chem 269:5150–5156
31. Yoshiura S, Ohtsuka T, Takenaka Y, Nagahara H, Yoshikawa K, Kageyama R (2007) Ultradian oscillations of Stat, Smad, and Hes1 expression in response to serum. Proc Natl Acad Sci U S A 104:11292–11297
32. Zhang N, Gridley T (1998) Defects in somite formation in lunatic fringe-deficient mice. Nature (Lond) 394:374–377

The Cellular Basis of *Dictyostelium* Morphogenesis

Cornelis J. Weijer

Introduction

One of the central aims of the study of development is to understand how distinct cellular behaviours, for example, division, differentiation, apoptosis, and movement, are coordinated in space and in time to result in reproducible pattern formation and morphogenesis. Coordination of these cellular behaviours requires extensive communication between cells of different types and between the cells and their environment. The social amoeba *Dictyostelium discoideum,* a simple genetically tractable organism situated at the threshold of single and multicellular organisms in the evolutionary tree of life, is well suited for the study of these interactions because its genome has been sequenced and it is amenable to experimental manipulation through targeted gene disruption and replacement (Kreppel, Fey et al. 2004; Eichinger, Pachebat et al. 2005). *Dictyostelium* cells normally live as single cells in the leaf litter of the soil where they feed on bacteria and divide by binary fission. Under starvation conditions, up to several hundred thousand cells aggregate chemotactically to form a multicellular structure the slug, that directed by light and temperature gradients migrates to the surface of the soil to form a fruiting body. The fruiting body is composed of a stalk supporting a mass of spores. The spores are a dormant stage which after dispersal may germinate to release amoebae, thus closing the life cycle (Fig. 1). We describe here key aspects of the signalling mechanisms coordinating cellular behaviours responsible for pattern formation and morphogenesis.

The Control of Cell Movement During Development

Because *Dictyostelium* development takes place in the absence of food under starvation conditions, only limited cell divisions occur during multicellular development. Morphogenesis therefore primarily results from the movement of

---

C.J. Weijer
Division of Cell and Developmental Biology, School of Life Sciences,
University of Dundee, Wellcome Trust Biocentre, Dundee DD1 5EH, UK

S. Nakanishi et al. (eds.), *Systems Biology*,
DOI:10.1007/978-4-431-87704-2_22, © Springer 2009

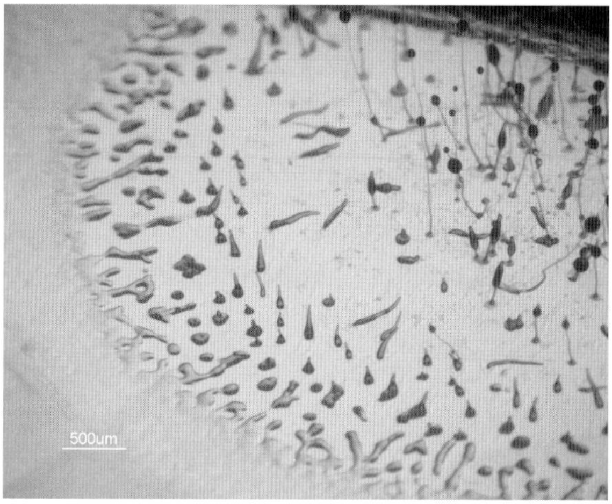

Fig. 1 Colonie of Dictyostelium cells on a bacterial lawn, showing a cross section through the life cycle of *Dictyostelium*. A single amoeba was inoculated in a lawn of bacteria. The amoebae eat the bacteria and form a plaque. At the edge of the plaque (*bottom left*), the amoebae are feeding; behind the feeding front the cells start to starve and form aggregation streams. Once the cells have aggregated, they form tipped mounds that transform into standing slugs. The slugs topple over and migrate away, and after a variable period of migration the slugs stop moving and transform into fruiting bodies. The fruiting body stands on the substrate with a basal disk that anchors a 2- to 3-mm-long stalk, which supports a spore head containing hundreds of thousands of spores; 80% of the population of cells that aggregate form spores. Under suitable conditions spores disperse and germinate to form new colonies, while the stalk cells die and do not propagate. The whole cycle from the initiation of starvation to fruiting body formation takes 24 h at ~20°C

individual differentiating cells into a relatively complex structure, the proportions of which are essentially independent of how many cells exactly aggregate. Key questions are which signals guide the movement behaviour of thousands of cells during development, which signals control differentiation, and how do movement and differentiation interact?

Aggregation

Starvation induces changes in the gene expression programme that results in the cells acquiring the ability to produce and secrete and degrade cyclic adenosine monophosphate (cAMP) (Saran, Meima et al. 2002; Iranfar, Fuller et al. 2003). Through the expression of the cAMP receptor, the cells also acquire the ability to respond chemotactically to cAMP gradients. It has emerged that chemotaxis results from the polarisation of the cytoskeletal dynamics persistently along the cAMP gradient. Unstimulated amoeboid cells are continuously changing shape by extension and retraction of pseudopods in all directions, resulting in a random walk (Soll, Wessels et al. 2003; Sasaki, Janetopoulos et al. 2007). In the presence of an external gradient of a chemoattractant such as cAMP the cells persistently extend successive pseudopods in the direction of rising cAMP concentration, while at the same time suppressing the extension of lateral pseudopods, which results in efficient movement up the gradient (Andrew and Insall 2007). Much current research is directed towards understanding how cells detect cAMP gradients, polarise their cytoskeleton, and move in response to these cAMP gradients (Chen, Iijima et al. 2007; Franca-Koh, Kamimura et al. 2007; Insall and Andrew 2007). Cells move by extending pseudopods at their leading edge, a process driven by actin polymerisation. For extension to occur, the assembly of the myosin thick filaments in the cortex at the site of extension needs to be inhibited, which involves phosphorylation of the tail of the myosin heavy chain on several threonine residues (Yumura, Yoshida et al. 2005; Bosgraaf and van Haastert 2006; Goldberg, Wolpin et al. 2006). Cells also need to pull up their back end and suppress the extension of lateral pseudopods, which involves, besides the well-characterised role of myosin II, also the action of members of the myosin I family and is dependent on internal cAMP levels (Falk, Wessels et al. 2003; Zhang, Heid et al. 2003). To move forward the cells must gain traction form the substrate on which they are moving, which involves the formation of multiple transient (10–20 s) actin contact sites that have been shown to transduce traction forces to the substrate (Bretschneider, Diez et al. 2004; Uchida and Yumura 2004). It appears that cells may undergo alternating phases of actin-driven extension at the front and myosin-driven contraction at the back (Iwadate and Yumura 2008). Much work is directed towards the investigation of the molecular mechanisms that result in signal detection, cell polarisation and its translation in directed movement. This work has been extensively reviewed elsewhere recently and is not covered here in detail (Willard and Devreotes 2006; Janetopoulos and Firtel 2008).

Aggregation is caused by periodic cAMP synthesis and secretion by cells in the aggregation centre. Binding of cAMP to the serpentine transmembrane cAMP receptor results in stimulation of the signal transduction cascade that leads to the activation of an adenylylcyclase (ACA), which within tens of seconds produces cAMP, part of which is secreted to the outside (Fig. 2). The secreted cAMP binds to the receptor and thus is part of an autocatalytic feedback loop, resulting in a rapid increase of cAMP production. However, stimulation of the receptor also activates an adaptation process that with a small time delay results in the inhibition of adenylyl cyclase activation and a cessation of cAMP production. As cAMP diffuses away into the extracellular medium and is also degraded by secreted cAMP phosphodiesterases, this results in a drop in cAMP levels, which in turn results in deadaptation of the cells (Fig. 2). Both excitation and adaptation depend on the receptor-dependent activation of a heterotrimeric G protein, resulting in activation of Ras, which in turn activates phosphatidyl inositide 3 kinase (PI3 kinase), which phosphorylates phosphatidylinositide(4,5)phosphate (PIP2) to phosphatidyl-(3,4,5)-phosphate (PIP3). PIP3 then activates many downstream pathways, leading to chemotaxis and adenylcyclase activation (Mahadeo and Parent 2006). The activation of adenylcyclase is critically dependent on the PIP3-dependent binding of CRAC (cytosolic regulator of adenylyl cyclase) to the membrane where it is

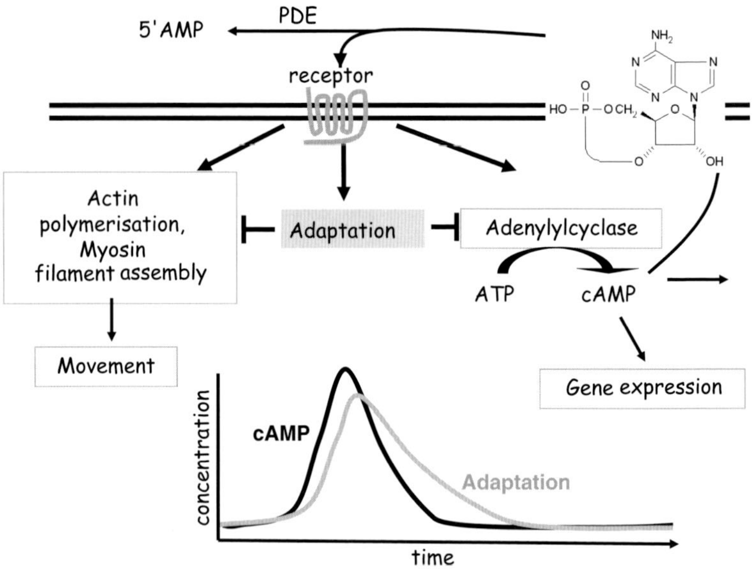

Fig. 2 Cyclic adenosine monophosphate (cAMP) signalling activates cAMP relay and chemotaxis. Extracellular cAMP binds to a transmembrane serpentine cAMP receptor and activates two distinct signal transduction pathways. One pathway leads to organisation of the actin myosin cytoskeleton and chemotaxis. The second pathway results in activation of the aggregation stage adenylylcyclase (ACA). Activation of ACA results in cAMP production, part of which is secreted and binds to the receptor to form a positive feedback loop, where a little extracellular cAMP results in the production of more. Binding of cAMP to the receptor also activates an adaptation process that inhibits both ACA activation and the chemotactic signal transduction pathway. Once adaptation reaches a full response, the production of cAMP ceases and cAMP secretion stops. As cAMP is continuously degraded by an extracellular cAMP phosphodiesterase, extracellular cAMP levels fall, which allows the cells to de-adapt and start a new cAMP response

activated, which is required for adenylyl cyclase activation and chemotaxis (Comer, Lippincott et al. 2005).

Detection and amplification of this signal by surrounding cells coupled with desensitisation of the cAMP-producing cells results in the propagation of waves of cAMP away from the aggregation centre (Fig. 3). Cells detect the rising phase of the wave and move in the direction of increasing cAMP concentration; once the waves passes, the cells are adapted and are insensitive to the falling phase of the wave and therefore do not turn around and chase the wave after it has passed. These cAMP waves therefore guide the cells towards the aggregation centre, where they accumulate into a three-dimensional aggregate, the mound (Dormann and Weijer 2001; Dormann and Weijer 2003). During the synchronised chemotactic movement phase cells elongate, while during the falling phase of the waves the cells are amoeboid in shape. The large-scale spatiotemporal patterns of cells behaviour can be visualised as changes in light scattering since moving elongated cells scatter more light (Fig. 3A) These waves can thus successfully be used to visualize and measure the spatiotemporal dynamics of wave propagation at all stages of development (Fig. 3B,C,D).

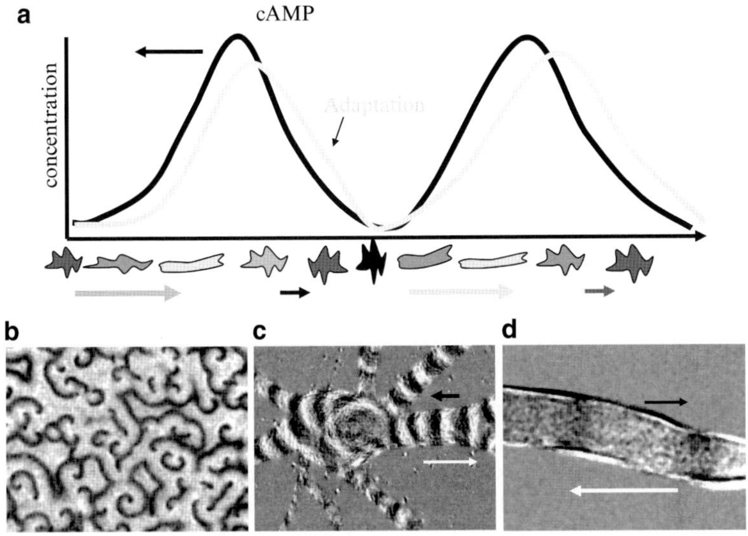

Fig. 3 Optical density waves reflect cAMP waves at different stages of development. **a** Aggregation centres initiate cAMP waves. These propagate from the centre outward (*black arrow*). Cells detect the rising phase of the cAMP wave and move in the direction of higher cAMP concentrations (*bottom figures and arrows*; colour of cells and arrows indicates degree of light scattering). During their chemotactic movement the cells elongate, and this changes their light-scattering properties. When during the rising phase of the wave many cells move in synchrony, this results in increased light-scattering waves, reflecting the rising phase of the cAMP signal. **b** Spiral optical density waves during the early aggregation phase, when the cells are still in a monolayer on agar. **c** Optical density waves in a streaming aggregate. In the body of the aggregate, multi-armed spiral waves rotate counterclockwise, throwing off individual wavefronts that propagate down the streams to the periphery of the aggregate. *White arrow* indicates the direction of wave propagation; *black arrow* indicates the direction of cell movement. **d** A slug migrating to the *right*, showing two dark optical density waves that travel from right to left (*white arrow*); cells move to the right following the tip (*black arrow*)

Recently, this method has been used to perform a high-throughput analysis of cAMP signalling mutants (Sawai, Guan et al. 2007). Initially the cells move towards the aggregation centre as individuals, but after 10–20 waves have passed they form bifurcating aggregation streams in which the cells make head-to-tail contacts via a calcium-independent adhesion molecule, contact site A, and side-to-side contacts via calcium-dependent cadherins (Wong, Yang et al. 2002; Harris, Ravandi et al. 2003). Stream formation is dependent on the localisation of ACA in the rear of the aggregating cells, resulting in polarised cAMP secretion from the back of the cells (Kriebel, Barr et al. 2003). cAMP wave propagation can be observed at the individual cell level by following the localised translocation of PIP3 at the leading edge of the cell (Dormann, Weijer et al. 2002; Dormann, Weijer et al. 2004). The number of cells in aggregation streams appears to be controlled by the local concentration of a secreted extracellular high molecular weight protein complex, a counting factor, which through modulation of movement and adhesion may control the numbers of cells that stably migrate in an aggregation stream (Jang and Gomer 2008)

Mound and Slug Formation

After the cells have aggregated, they form a hemispherical structure—the mound. Mounds are characterised by rotating waves of cAMP that direct the counterrotational periodic movement of the cells. Cells start to differentiate into prespore and prestalk cells during aggregation, based on physiological biases such as nutritional state and cell-cycle position at the time of starvation already present in the population before aggregation (Araki, Abe et al. 1997). As a result, there is little correlation between the time of arrival in the mound and differentiation fate. Therefore, initially the prestalk and prespore cell types display a salt-and-pepper distribution in the mound (see Fig. 5A, later in this chapter). A subpopulation of prestalk cells sort out to form the tip, and the slug tip guides the movement of all other cells, thus acting as an organiser (Weijer 2004). Use of a temperature-sensitive ACA mutant has shown that ACA activity is required for cells to be able to sort to the tip (Patel, Guo et al. 2000).The tip action as an organiser can be mimicked by the periodic injection of cAMP pulses of the right frequency and duration (Dormann and Weijer 2001), suggesting that the tip is a source of periodic cAMP waves, in agreement with the fact that prestalk cells express ACA and the extracellular cAMP phosphodiesterase pdeA (Verkerke-van Wijk, Fukuzawa et al. 2001; Weening, Wijk et al. 2003). It is not yet known which signals control tip cell fate (see below), but it is becoming clear that to proceed from the aggregate to the mound stage cell–cell adhesion and or contact start to play an important role. Mutants defective in the putative single-pass transmembrane contact molecules lagC, lagD cannot proceed beyond the aggregation stage and are defective in tip formation (Kibler, Svetz et al. 2003). There is evidence that *Dictyostelium* may possess several integrin-like adhesion molecules (Cornillon, Froquet et al. 2008), and it has been known that mutants in talin, paxillin, and a lim domain protein, which are thought to couple adhesion molecules to the actin cytoskeleton, are all defective in cell sorting (Chien, Chung et al. 2000; Tsujioka, Yoshida et al. 2004; Bukharova, Weijer et al. 2005). Sorting of prestalk cells towards the tip requires the invasive movement of prestalk cells through a tightly packed mass of other. Myosin II is absolutely required for progression beyond the mound stage, and it is known that prestalk cells express higher levels of myosin II (Eliott, Vardy et al. 1991; Springer, Patterson et al. 1994). For prestalk cells to sort, they need to assemble more myosin thick filaments than prespore cells (Dormann and Weijer, unpublished observations). Together all these observations suggest that tip formation is the result of cellular properties specific to tip-forming cells. Tip cells express high levels of ACA, resulting in increased signalling ability, therefore allowing them to set up a new independent signalling centre that can outcompete the aggregation centre organising the mound. They also assemble more myosin II thick filaments, critical to their ability to produce more force in response to a cAMP signal. This latter property allows them to push other cells aside and reaggregate within the mound to form the tip (Dormann and Weijer, in preparation).

In slugs, optical density waves can be seen to propagate from the middle of the prestalk zone to the back, reflecting the periodic movement of the cells forward

(Fig. 3). These optical waves are strictly dependent on the tip. Cells in the tip often rotate perpendicular to the direction of slug migration, especially when it is lifted from the substrate. In the back of the slug the cells move periodically forward, and all cells move on average with slug speed. It has been shown that the assumption of cAMP wave propagation and chemotaxis in response to these waves is, in principle, sufficient to explain morphogenesis from a single cell via aggregation and stream and mound formation to cell sorting and slug formation. The interactions between cell signalling and cell movement can be described by relatively simple mathematical models in a robust way, and it would appear that these process are sufficient to explain *Dictyostelium* morphogenesis (Fig. 4) (Vasiev and Weijer 2003; Umeda and Inouye 2004). However, the situation is almost certainly more

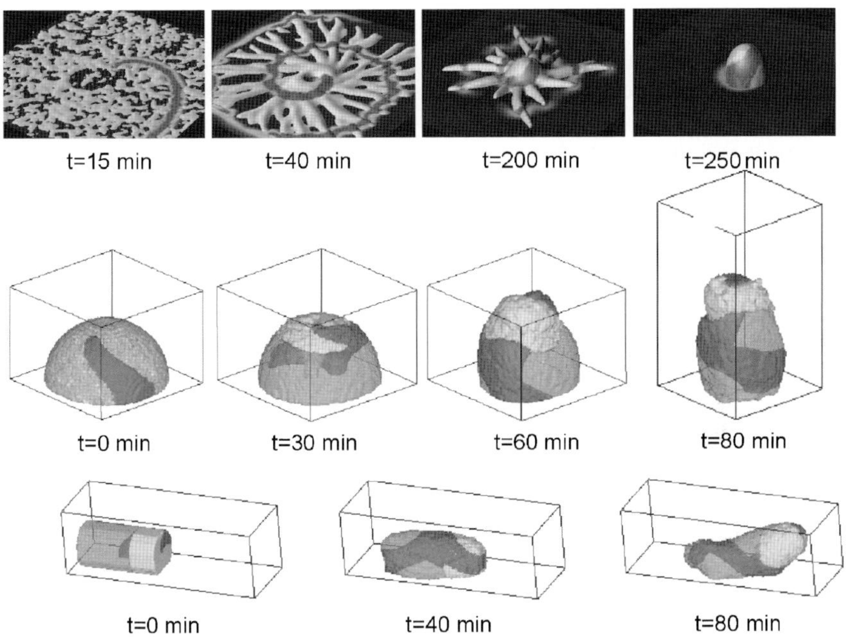

Fig. 4 Model calculation of wave propagation and cell movement from aggregation to slug migration using a hydrodynamic model. The *top row* depicts the aggregation up to the mound stage. The first image starts with the randomly distributed cells (*grey*), which are organised by a spiral wave of cAMP (*dark grey*). They form aggregation streams and finally a hemispherical mound (Vasiev, Siegert et al. 1997). The *middle row* shows cell sorting and the formation of a slug. The mound consists of two cells types: 20% light grey prestalk cells and 80% dark grey prespore cells. They are initially randomly mixed. The cAMP waves (*darkest grey*) organises the movement of the cells. In the model the assumption is that the prestalk cells are more excitable and develop more movement force in response to a cAMP wave. As a result of this, they move towards the centre of the mound and up to form the tip. The separation of the cells feeds back on the signal propagation, resulting in the formation of a twisted scroll wave; this leads to an intercalation of the cells and an upward extension of the slug (Vasiev and Weijer 1999). The *bottom row* shows that a slug organised by a scroll wave can move (Vasiev and Weijer 2003)

complex since strains lacking the aggregation-stage adenylylcyclase can still form slugs when they overexpress the catalytic subunit of protein kinase A, suggesting either that there either exists a ACA-independent mechanism to produce periodic cAMP signals, for instance, involving cAMP generation by other adenylylcyclases, ACB or ACG, and the recently discovered cAMP-stimulated cAMP phosphodiesterase (Meima, Weening et al. 2003), or that there exists altogether different mechanisms that can control cell movement such as contact following (Umeda and Inouye 2002). The latter mechanism does not, however, explain which signals direct the movement of the cells in the tip.

Differentiation

It is well established that *Dictyostelium* slugs can form as few as a few hundred cells and can contain up to several million cells, while the proportions between spore and stalk cells remain relatively constant (Rafols, Amagai et al. 2001; Maruo, Sakamoto et al. 2004). A major goal is to understand the relationship between cell movement and the signals that control differentiation. These signals must be able to maintain the correct proportioning of the prespore and prestalk cell types in an environment of extensive cell movement and changes in shape of the slug. In the slug the different cell types are arranged in a simple axial pattern: pstA cells in the tip, a band of pstO cells that form the upper and part of the lower cup of the fruiting body, prespore cells, precursors for spores, which are intermingled anterior-like cells, and rearguard cell precursors to the lower cup and basal disk in the back of the slug (Fig. 5A) (Williams 2006). It seems evident that this requires adaptive signalling dynamics, but the signals and the details of their regulation are not yet understood in detail (Fig. 5B). cAMP pulses control the expression of aggregation-stage genes necessary for cAMP relay and cell–cell contact, and cAMP is necessary for prespore gene expression in later development (Saran, Meima et al. 2002; Iranfar, Fuller et al. 2003). Prespore cells in turn produce DIF (differentiation-inducing factor, a small cholinated hexaphenone), which controls the differentiation of pstO cells (Kay and Thompson 2001; Maeda, Sakamoto et al. 2003; Thompson, Fu et al. 2004; Saito, Taylor et al. 2006). DIF spreads by simple diffusion from the prespore zone in adjacent regions where it controls the differentiation of prestalk O cells and possibly rear guard cells (Fukuzawa, Abe et al. 2003). Cells in the pstA zone express ACA, and studies investigating the cyclic AMP-dependent nuclear translocation of the transcription factor statA have shown that cAMP levels are high in the tip, while cAMP is lower elsewhere in the slug (Dormann, Abe et al. 2001; Verkerke-van Wijk, Fukuzawa et al. 2001), compatible with the idea that all cells in the tip relay the cAMP signal and only the anterior-like cells in the rest of the slug do so. However, the signals that control the expression of functionally important tip-enriched molecules such as ACA and myosin II still have to be identified. It also remains to be seen how accurate cell-type proportioning is achieved in slugs despite their vastly different sizes and dynamic changes in shape.

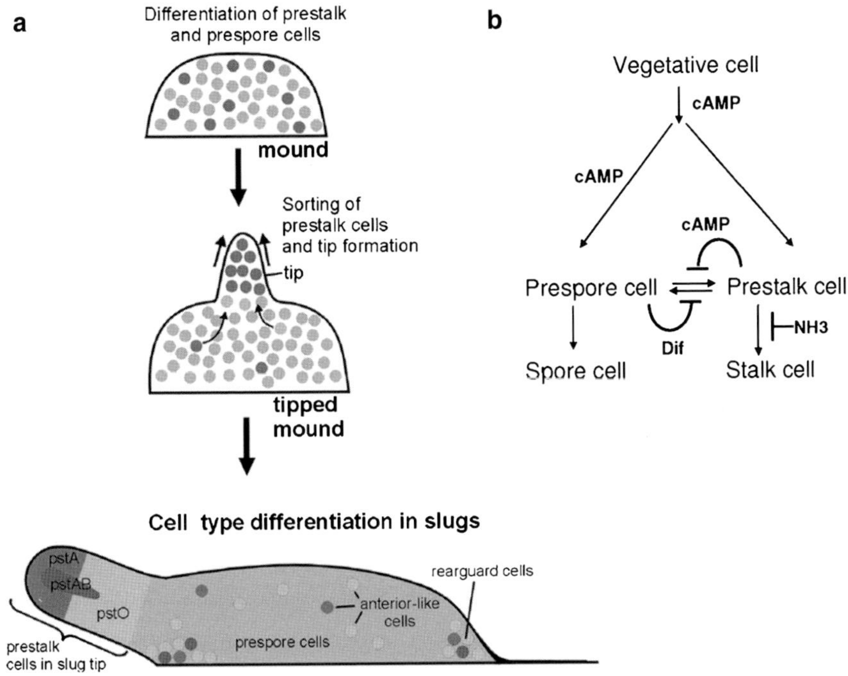

Fig. 5 Cell types and cell-type proportioning in *Dictyostelium*. **a** During aggregation, cells start to differentiate into prestalk cells (*dark grey*) and prespore cells (*light grey*). Because cells arrive in the mound in random order, they form a salt-and-pepper distribution of cell types. After a variable time of movement, the prestalk cells reaggregate to form the tip and the initial prestalk zone. The tip guides the movement of all the other cells, and the structure elongates to from a slug that topples over and migrates away. In the slugs, there are at least four cell types. The tip is made up of PstA (*dark grey*) cells, followed by a cohort of PstO (*light grey*) cells, which together form the prestalk zone. The prestalk zone is followed by the prespore zone in which mainly prespore cells are intermingled with anterior-like cells, cells of prestalk character that do not sort but express ACA and relay the cAMP signal. In the back of the slug, the rearguard cells are found. **b** Prespore and prestalk cells differentiate from vegetative-stage cells. The early differentiation in aggregation-stage cells requires cAMP pulses. The cells then differentiate into prestalk and prespore cells. Prestalk cells initiate cAMP waves, and extracellular cAMP is needed for prespore gene expression. Prespore cells make differentiation-inducing factor (*DIF*), which is necessary for PstO cell differentiation. Prestalk cells secrete DIF-ase that inactivates DIF. Prestalk cells differentiate into stalk cells, and prespore cells differentiate into spores. Stalk differentiation is inhibited by NH3. An open question is how cell-type proportioning works quantitatively, which is the subject of further modelling studies

The switch from migrating slugs to culmination appears to be controlled by a fall in ammonia concentration. The identification of a number of ammonia transporters, some of which are expressed in the very tip and when deleted show a slugger phenotype, supports the importance of ammonia as a morphogen (Kirsten, Xiong et al. 2005; Singleton, Kirsten et al. 2006). Ammonia most likely signals through the histidine kinase DhkC to the response regulator domain of the internal cAMP phosphodiesterase RegA, which is a major determinant in the control of intracellular cAMP levels (Singleton, Zinda et al. 1998; Saran, Meima et al. 2002). High ammonia is expected to result in activation of regA and low internal cAMP levels. A drop in ammonia is expected to result in a rise of intracellular cAMP and stalk cell differentiation.

In conclusion, *Dictyostelium* is, besides being a system of choice to investigate the molecular mechanisms underlying cell polarity and chemotaxis, also an excellent model system to investigate the cellular principles that underlie multicellular tissue formation and morphogenesis.

Acknowledgments This work was supported by the BBSRC and the Wellcome Trust.

References

Andrew, N. and R. H. Insall (2007). Chemotaxis in shallow gradients is mediated independently of PtdIns 3-kinase by biased choices between random protrusions. Nat Cell Biol 9(2):193–200.

Araki, T., T. Abe, et al. (1997). Symmetry breaking in *Dictyostelium* morphogenesis: evidence that a combination of cell cycle stage and positional information dictates cell fate. Dev Biol 192(2):645–648.

Bosgraaf, L. and P. J. van Haastert (2006). The regulation of myosin II in *Dictyostelium*. Eur J Cell Biol 85:969–979.

Bretschneider, T., S. Diez, et al. (2004). Dynamic actin patterns and Arp2/3 assembly at the substrate-attached surface of motile cells. Curr Biol 14(1):1–10.

Bukharova, T., G. Weijer, et al. (2005). Paxillin is required for cell-substrate adhesion, cell sorting and slug migration during *Dictyostelium* development. J Cell Sci 118(pt 18):4295–4310.

Chen, L., M. Iijima, et al. (2007). PLA2 and PI3K/PTEN pathways act in parallel to mediate chemotaxis. Dev Cell 12(4):603–614.

Chien, S., C. Y. Chung, et al. (2000). The *Dictyostelium* LIM domain-containing protein LIM2 is essential for proper chemotaxis and morphogenesis. Mol Biol Cell 11(4):1275–1291.

Comer, F. I., C. K. Lippincott, et al. (2005). The PI3K-mediated activation of CRAC independently regulates adenylyl cyclase activation and chemotaxis. Curr Biol 15(2):134–139.

Cornillon, S., R. Froquet, et al. (2008). Involvement of Sib proteins in the regulation of cellular adhesion in *Dictyostelium discoideum*. Eukaryot Cell 7(9):1600–1605.

Dormann, D., T. Abe, et al. (2001). Inducible nuclear translocation of a STAT protein in *Dictyostelium* prespore cells: implications for morphogenesis and cell-type regulation. Development (Camb) 128(7):1081–1088.

Dormann, D. and C. J. Weijer (2001). Propagating chemoattractant waves coordinate periodic cell movement in *Dictyostelium* slugs. Development (Camb) 128(22):4535–4543.

Dormann, D. and C. J. Weijer (2003). Chemotactic cell movement during development. Curr Opin Genet Dev 13(4):358–364.

Dormann, D., G. Weijer, et al. (2002). Visualizing PI3 kinase-mediated cell-cell signaling during *Dictyostelium* development. Curr Biol 12(14):1178–1188.

Dormann, D., G. Weijer, et al. (2004). In vivo analysis of 3-phosphoinositide dynamics during *Dictyostelium* phagocytosis and chemotaxis. J Cell Sci 117(pt 26):6497–6509.

Eichinger, L., J. A. Pachebat, et al. (2005). The genome of the social amoeba *Dictyostelium discoideum*. Nature (Lond) 435(7038):43–57.

Eliott, S., P. H. Vardy, et al. (1991). The distribution of myosin-II in *Dictyostelium discoideum* slug cells. J Cell Biol 115:1267–1274.

Falk, D. L., D. Wessels, et al. (2003). Shared, unique and redundant functions of three members of the class I myosins (MyoA, MyoB and MyoF) in motility and chemotaxis in *Dictyostelium*. J Cell Sci 116(pt 19):3985–3999.

Franca-Koh, J., Y. Kamimura, et al. (2007). Leading-edge research: PtdIns(3,4,5)P3 and directed migration. Nat Cell Biol 9(1):15–17.

Fukuzawa, M., T. Abe, et al. (2003). The Dictyostelium prestalk cell inducer DIF regulates nuclear accumulation of a STAT protein by controlling its rate of export from the nucleus. Development (Camb) 130(4):797–804.

Goldberg, J. M., E. S. Wolpin, et al. (2006). Myosin light chain kinase A is activated by cGMP-dependent and cGMP-independent pathways. FEBS Lett 580(8):2059–2064.

Harris, T. J., A. Ravandi, et al. (2003). Cytoskeleton interactions involved in the assembly and function of glycoprotein-80 adhesion complexes in *Dictyostelium*. J Biol Chem 278(4):2614–2623.

Insall, R. and N. Andrew (2007). Chemotaxis in *Dictyostelium*: how to walk straight using parallel pathways. Curr Opin Microbiol 10(6):578–581.

Iranfar, N., D. Fuller, et al. (2003). Genome-wide expression analyses of gene regulation during early development of *Dictyostelium discoideum*. Eukaryot Cell 2(4):664–670.

Iwadate, Y. and S. Yumura (2008). Actin-based propulsive forces and myosin-II-based contractile forces in migrating *Dictyostelium* cells. J Cell Sci 121(pt 8):1314–1324.

Janetopoulos, C. and R. A. Firtel (2008). Directional sensing during chemotaxis. FEBS Lett 582:2075–2085.

Jang, W. and R. H. Gomer (2008). Combining experiments and modelling to understand size regulation in *Dictyostelium discoideum*. J R Soc Interface 6(5 suppl 1):S49–S58.

Kay, R. R. and C. R. Thompson (2001). Cross-induction of cell types in *Dictyostelium*: evidence that DIF-1 is made by prespore cells. Development (Camb) 128(24):4959–4966.

Kibler, K., J. Svetz, et al. (2003). A cell-adhesion pathway regulates intercellular communication during *Dictyostelium* development. Dev Biol 264(2):506–521.

Kirsten, J. H., Y. Xiong, et al. (2005). Ammonium transporter C of *Dictyostelium discoideum* is required for correct prestalk gene expression and for regulating the choice between slug migration and culmination. Dev Biol 287(1):146–156.

Kreppel, L., P. Fey, et al. (2004). dictyBase: a new *Dictyostelium discoideum* genome database. Nucleic Acids Res 32(database issue):D332–D333.

Kriebel, P. W., V. A. Barr, et al. (2003). Adenylyl cyclase localization regulates streaming during chemotaxis. Cell 112(4):549–560.

Maeda, M., H. Sakamoto, et al. (2003). Changing patterns of gene expression in dictyostelium prestalk cell subtypes recognized by in situ hybridization with genes from microarray analyses. Eukaryot Cell 2(3):627–637.

Mahadeo, D. C. and C. A. Parent (2006). Signal relay during the life cycle of *Dictyostelium*. Curr Top Dev Biol 73:115–140.

Maruo, T., H. Sakamoto, et al. (2004). Control of cell type proportioning in *Dictyostelium discoideum* by differentiation-inducing factor as determined by in situ hybridization. Eukaryot Cell 3(5):1241–1248.

Meima, M. E., K. E. Weening, et al. (2003). Characterization of a cAMP-stimulated cAMP phosphodiesterase in *Dictyostelium discoideum*. J Biol Chem 278(16):14356–14362.

Patel, H., K. D. Guo, et al. (2000). A temperature-sensitive adenylyl cyclase mutant of *Dictyostelium*. EMBO J 19(10):2247–2256.

Rafols, I., A. Amagai, et al. (2001). Cell type proportioning in *Dictyostelium* slugs: lack of regulation within a 2.5-fold tolerance range. Differentiation 67(4-5):107–116.

Saito, T., G. W. Taylor, et al. (2006). Identification of new differentiation inducing factors from *Dictyostelium discoideum*. Biochim Biophys Acta 1760(5):754–761.

Saran, S., M. E. Meima, et al. (2002). cAMP signaling in *Dictyostelium*. Complexity of cAMP synthesis, degradation and detection. J Muscle Res Cell Motil 23(7-8):793–802.

Sasaki, A. T., C. Janetopoulos, et al. (2007). G protein-independent Ras/PI3K/F-actin circuit regulates basic cell motility. J Cell Biol 178(2):185–191.

Sawai, S., X. J. Guan, et al. (2007). High-throughput analysis of spatio-temporal dynamics in *Dictyostelium*. Genome Biol 8(7):R144.
Singleton, C. K., J. H. Kirsten, et al. (2006). Function of ammonium transporter A in the initiation of culmination of development in *Dictyostelium discoideum*. Eukaryot Cell 5(7):991–996.
Singleton, C. K., M. J. Zinda, et al. (1998). The histidine kinase dhkC regulates the choice between migrating slugs and terminal differentiation in *Dictyostelium discoideum*. Dev Biol 203(2):345–357.
Soll, D. R., D. Wessels, et al. (2003). Computer-assisted reconstruction and motion analysis of the three-dimensional cell. Sci World J 3:827–841.
Springer, M. L., B. Patterson, et al. (1994). Stage-specific requirement for myosin II during *Dictyostelium* development. Development (Camb) 120:2651–2660.
Thompson, C. R., Q. Fu, et al. (2004). A bZIP/bRLZ transcription factor required for DIF signaling in *Dictyostelium*. Development (Camb) 131(3):513–523.
Tsujioka, M., K. Yoshida, et al. (2004). Talin B is required for force transmission in morphogenesis of *Dictyostelium*. EMBO J 23(11):2216–2225.
Uchida, K. S. and S. Yumura (2004). Dynamics of novel feet of *Dictyostelium* cells during migration. J Cell Sci 117(pt 8):1443–1455.
Umeda, T. and K. Inouye (2002). Possible role of contact following in the generation of coherent motion of *Dictyostelium* cells. J Theor Biol 219(3):301–308.
Umeda, T. and K. Inouye (2004). Cell sorting by differential cell motility: a model for pattern formation in *Dictyostelium*. J Theor Biol 226(2):215–224.
Vasiev, B., F. Siegert, et al. (1997). A hydrodynamic model for *Dictyostelium discoideum* mound formation. J Theor Biol 184(4):441.
Vasiev, B. and C. J. Weijer (1999). Modeling chemotactic cell sorting during *Dictyostelium discoideum* mound formation. Biophys J 76(2):595–605.
Vasiev, B. and C. J. Weijer (2003). Modelling of *Dictyostelium discoideum* slug migration. J Theor Biol 223(3):347–359.
Verkerke-van Wijk, I., M. Fukuzawa, et al. (2001). Adenylyl cyclase A expression is tip-specific in *Dictyostelium* slugs and directs StatA nuclear translocation and CudA gene expression. Dev Biol 234(1):151–160.
Weening, K. E., I. V. Wijk, et al. (2003). Contrasting activities of the aggregative and late PDSA promoters in *Dictyostelium* development. Dev Biol 255(2):373–382.
Weijer, C. J. (2004). *Dictyostelium* morphogenesis. Curr Opin Genet Dev 14(4):392–398.
Willard, S. S. and P. N. Devreotes (2006). Signaling pathways mediating chemotaxis in the social amoeba, *Dictyostelium discoideum*. Eur J Cell Biol 85(9-10):897–904.
Williams, J. G. (2006). Transcriptional regulation of Dictyostelium pattern formation. EMBO Rep 7(7):694–698.
Wong, E., C. Yang, et al. (2002). Disruption of the gene encoding the cell adhesion molecule DdCAD-1 leads to aberrant cell sorting and cell-type proportioning during *Dictyostelium* development. Development (Camb) 129(16):3839–3850.
Yumura, S., M. Yoshida, et al. (2005). Multiple myosin II heavy chain kinases: roles in filament assembly control and proper cytokinesis in *Dictyostelium*. Mol Biol Cell 16(9):4256–4266.
Zhang, H., P. J. Heid, et al. (2003). Constitutively active protein kinase A disrupts motility and chemotaxis in *Dictyostelium discoideum*. Eukaryot Cell 2(1):62–75.

Metabolic Information Highway: Interorgan Metabolic Communication Via the Autonomic Nervous System

Hideki Katagiri

Introduction

Our research goals are the development of therapeutic strategies for diabetes mellitus.

The incidence of obesity is rising at an alarming rate in much of the world [1]. Obesity, especially visceral obesity, is prone to be associated with hypertension, glucose intolerance, and dyslipidemia, collectively termed the metabolic syndrome. Independently of hypercholesterolemia, especially increased levels of oxidized low density lipoproteins [2], the metabolic syndrome increases the risk for atherosclerosis and cardiovascular morbidities [1]. To overcome obesity-related diabetes and the metabolic syndrome, it seems to be necessary to treat obesity itself. Therefore, we have attempted to unravel and manipulate the regulatory systems governing body weight as well as energy metabolism.

Leptin and Leptin Resistance

What endogenous mechanisms do we have for energy homeostasis?

Leptin, one of the adipokines, is a major contributor to energy homeostasis. Leptin is secreted mainly from adipocytes into the circulating blood, in proportion to fat stores, and binds to its receptor in the hypothalamus, leading to decreased expression of orexic neuropeptide (NPY) and increased expression of anorexic neuropeptide (POMC), resulting in suppression of food intake. Therefore, when energy storage is increased, leptin secretion is increased, leading to suppression of food intake and thus weight reduction. Conversely, decreased energy storage in adipose tissue decreases leptin secretion, resulting in weight gain. Through this mechanism, the leptin system contributes to fixing body weight within a certain range [3].

H. Katagiri
Division of Advanced Therapeutics for Metabolic Diseases, Center for
Translational and Advanced Animal Research, Tohoku University Graduate School of Medicine,
2-1 Seiryo-machi, Aoba-ku, Sendai 980-8575, Japan

S. Nakanishi et al. (eds.), *Systems Biology*,
DOI:10.1007/978-4-431-87704-2_23, © Springer 2009

However, this system is impaired in the obese state. According to obesity development, food intake amounts are gradually increased [4]. Thus, obesity itself further enhances food intake, forming a vicious cycle for further worsening obesity. Such obesity-enhanced food intake is explained by decreased hypothalamic sensitivity to leptin. In the obese state, circulating leptin concentrations are high, but the responses of hypothalamic cells to leptin are markedly impaired. This impairment of leptin responses is called leptin resistance. Therefore, hypothalamic leptin resistance is an important mechanism maintaining obesity and a potential target for fighting obesity.

Interorgan Metabolic Communication

Then, what endogenous mechanisms other than the leptin system do we have for energy homeostasis, especially energy expenditure?

Metabolism does not go on independently in different tissues and organs, but rather in a coordinated and regulated manner throughout the body. This coordinated metabolic regulation involving several organs/tissues requires interorgan communication systems. Because the human body is a multiorgan entity, I believe that the metabolic communication network among organs is essential for metabolic homeostasis.

Humoral factors, such as insulin and adipokines, are known to play important roles in this communication. However, using tissue-specific transgenic and knockout mice, unexpected metabolic phenotypes are reported in remote tissues other than the gene-disrupted tissue, which means the presence of as yet unknown systems for metabolic communication.

Therefore, our goal is to identify ways of (1) improving leptin sensitivity, as well as (2) other endogenous mechanisms maintaining energy homeostasis, via interorgan metabolic communication.

I looked for a strategy that would give us some insight into the mechanisms underlying metabolic crosstalk among organs and tissues. I suspected that, if metabolism could be acutely altered in just one organ, it would be much easier to analyze acute effects on metabolism in other remote tissues. Intervening in this step would give us an understanding of the mechanisms. In addition, the metabolic effects might be compensatory and thereby beneficial for systemic metabolism, possibly making it a potential therapeutic target. Therefore, the interorgan communication system could become a therapeutic target for the metabolic syndrome.

Therefore, we selected an adenoviral gene transfer system to express proteins that alter metabolism in a single tissue/organ of mice that had become obese and diabetic [5]. Using this strategy, in addition to well-known humoral signals, we have recently shown the importance of afferent neuronal signals for interorgan metabolic communication.

Metabolic Information Highways

Afferent Signals for Improving Hypothalamic Leptin Resistance from Visceral Adipose Tissue

As I already described, obesity induces leptin resistance. Therefore, I wondered what happens to hypothalamic leptin sensitivity when adipose tissue is acutely diminished in obese mice. As uncoupling protein (UCP)-1 is known to enhance cellular metabolism by blocking ATP synthesis in mitochondria, we expressed UCP1 in intraabdominal fat tissues of obese mice.

Direct injection of UCP1 adenovirus into the epididymal fat of mice with high fat diet-induced obesity resulted in very limited but significant expression of UCP1 in adipocytes, with no detectable expression in other tissues. Limited expression of UCP1 in intraabdominal fat suppressed overeating in mice with diet-induced obesity, suggesting improvement of leptin resistance in the hypothalamus. To examine whether leptin sensitivity truly had improved, we performed leptin tolerance testing. Administration of leptin markedly decreased food intake as compared with the control mice. In addition, although daily leptin administration did not decrease the body weight of control mice, because of leptin resistance, the body weight of UCP-1 mice did decrease quite a lot. Thus, leptin sensitivity was clearly shown to be improved by UCP1 expression in intraabdominal adipose tissues. Furthermore, UCP1 expression in visceral adipose tissue actually altered hypothalamic neuropeptide expression: NPY expression was decreased and POMC expression was increased. These results demonstrate that enhanced metabolism in intraabdominal fat tissue does indeed improve leptin sensitivity in the hypothalamus. Then, to elucidate the underlying mechanism, we attempted to intervene in intertissue communication.

Except for markedly decreased leptin levels, no adipokines were altered. Therefore, we hypothesized that neuronal signals are involved in this effect. So, we dissected nerve bundles innervating bilateral epididymal fat tissues. Ten days after nerve dissection, LacZ or UCP1 adenovirus was injected into epididymal fat tissues. Nerve dissection blocked the suppression of overeating seen in sham-operated mice. These findings suggest that neuronal signals from intraabdominal fat tissue are involved in regulating food intake [6].

Thus, visceral adipose tissue secretes leptin into the circulating blood and also regulates its own sensitivity in the hypothalamus via a neuronal pathway (Fig. 1). Food intake is likely to be precisely regulated by these dual signals from visceral adipose tissue. Our findings may provide evidence of a third key role of adipose tissue. The first one is energy storage, and the second one is adipokine secretion. This third adipose tissue role is to transmit afferent signals about energy status to the brain [7].

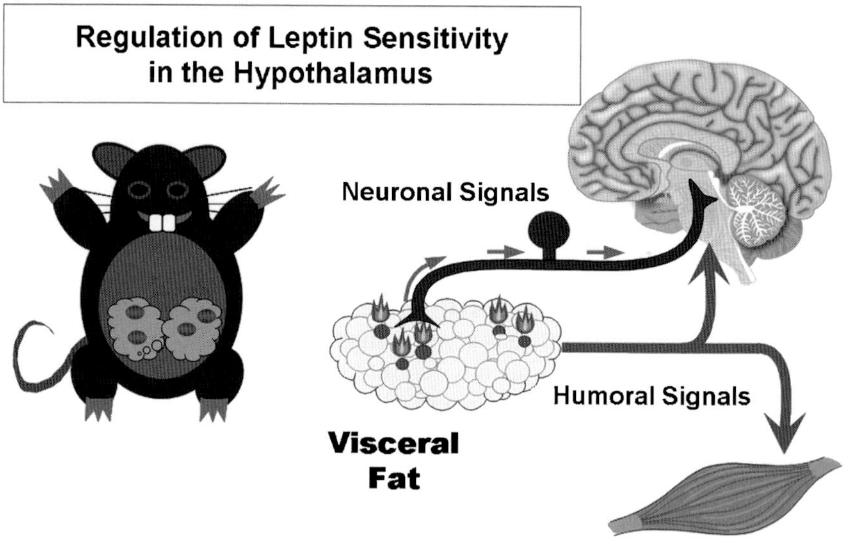

Fig. 1 Metabolic information highway from visceral adipose tissue. Neuronal signals from visceral adipose tissue regulate hypothalamic leptin sensitivity

Afferent Signals Enhancing Energy Expenditure from the Liver

We have another question: What happens when energy storage is increased in the liver?

To address this question, we attempted to express peroxisome proliferator-activated receptor-gamma (PPARγ) in the liver. PPARγ is a strong transcriptional factor that induces expression of genes involved in lipid synthesis and lipid uptake [8]. Although PPARγ expression in the liver is very low in the lean state, hepatic expression of PPARγ, especially PPARγ2, is functionally enhanced in a number of obesity models, including mice with genetically induced and high fat diet-induced obesity, as well as human obese subjects. Furthermore, liver-specific disruption of PPARγ in genetically obese mice reportedly prevents hepatic steatosis [9], suggesting PPARγ expression in the liver to play an important role in development of hepatic steatosis. Therefore, to enhance hepatic lipid accumulation, a recombinant adenovirus encoding PPARγ2 was intravenously administered. Systemic infusion of recombinant adenoviruses into mice through the tail vein resulted in PPARγ2 being expressed primarily in the liver, with no detectable expression in other tissues such as fat, muscle, or brain.

As expected, hepatic PPARγ2 expression induced severe steatosis in the liver. In contrast, intriguingly, in PPARγ2 mice, adipose tissues were remarkably diminished, by 50%, in just 1 week. Basal metabolic rates were increased, by

30%, in these mice, while food intakes were unchanged. As a result, high fat chow-induced weight gain was markedly suppressed. Surprisingly, despite marked hepatic steatosis, glucose tolerance and insulin sensitivity were markedly improved in PPARγ2 mice, as demonstrated by glucose tolerance and insulin tolerance testing.

The metabolic phenotypes associated with hepatic PPARγ2 expression are summarized here. In the liver, hepatic PPARγ2 expression induced abundant lipid accumulation. In contrast, in the periphery, fat accumulation was markedly decreased and insulin sensitivity was strikingly improved. In addition, systemic basal metabolic rates were significantly increased. As a result, systemic glucose tolerance was remarkably improved by hepatic PPARγ2 expression. These remote effects are very beneficial for fighting obesity and diabetes.

We obtained data suggesting the involvement of sympathetic nerve activation in increased lipolysis and energy expenditure. Therefore, the target of hepatic signals is not in the periphery, but rather the brain. At this time as well, we speculated about neuronal involvement. To determine whether the beneficial remote effects involve the vagal nerve, we dissected its hepatic branch.

Seven days after selective hepatic vagotomy, recombinant adenovirus was administered. In mice subjected to hepatic vagotomy, hepatic PPARγ2 expression increased liver weight and hepatic triglyceride content essentially as it had in sham-operated mice. Thus, the hepatic phenotypes are not affected by hepatic vagotomy.

In contrast, intriguingly, the decreases in brown adipocyte sizes and white adipose tissue weight as well as increases in serum free fatty acids (FFA) levels, which reflect increased lipolysis, and basal metabolic rates were completely blocked by selective hepatic vagotomy. In addition, pharmacological afferent blockade of the hepatic vagus similarly blocked these remote tissue effects.

Taken together, these observations show that hepatic PPARγ2 expression conveys metabolic information to the brain via the afferent vagus, resulting in efferent sympathetic activation, which enhances lipolysis and energy expenditure and improves obesity-related insulin resistance and diabetes. Hepatic PPARγ expression is physiologically induced when energy intake is excessive. Therefore, it may function as a feedback mechanism against excess energy intake, preventing the development of obesity by enhancing energy expenditure [10] (Fig. 2). This concept was introduced as the Metabolic Information Highway in "This Week" in *Science*, the issue in which our article was published. The autonomic nervous system is the pavement for this highway.

Let us compare this machinery with the leptin system. In the leptin system, adipose tissue recognizes excess energy accumulation and releases leptin to the brain via circulating blood. On the other hand, with this neuronal machinery, the liver senses energy storage and releases signals to the brain via the afferent vagus. While leptin mainly suppresses appetite, hepatic neuronal signals mainly enhance energy expenditure. Both inhibit weight gain. We are now working to identify which molecules activate the vagal nerve and what portions of the brain integrate peripheral metabolic signals.

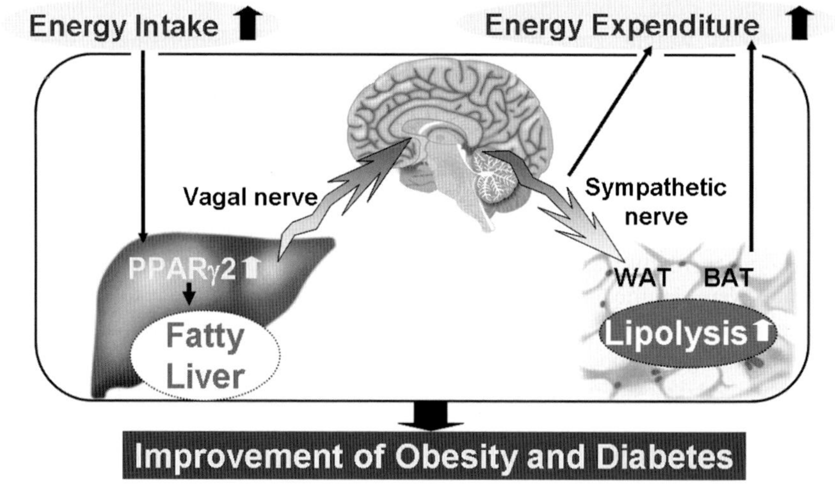

Fig. 2 Metabolic information highway from the liver. Neuronal signals from the liver regulate systemic energy metabolism to prevent the development of obesity by enhancing energy expenditure. *PPARγ*, peroxisome proliferator-activated receptor-gamma; *WAT*, white adipose tissue; *BAT*, brown adipose tissue

Conclusion

Two metabolic information highways via afferent neuronal pathways are presented: neuronal signals from adipose tissue affect hypothalamic leptin sensitization and neuronal signals from the liver regulate systemic energy expenditure. In addition to these two systems, we are now identifying another metabolic information highway from the liver to the pancreas [11]. Furthermore, the afferent signals originating in hepatic PPARγ2 expression appear to be involved in the development of obesity-related diseases (Uno K. and Katagiri H. et al., unpublished data). Thus, growing evidence has revealed the important roles of afferent neuronal signals in interorgan metabolic communication.

Collectively, the involvement of afferent neuronal signals highlights the importance of the central nervous system. The brain receives various forms of metabolic information from peripheral organs/tissues via two avenues, humoral factors and neuronal signals. These inputs from the periphery are probably integrated and processed in the brain, leading to the transmission of regulatory signals for appropriate systemic responses [1]. In addition, humoral and neuronal signals affect each other, as exemplified by the findings that adiponectin expressions are regulated by sympathetic activity [12]. Further elucidation of these regulatory systems may facilitate unraveling the mechanisms underlying metabolic homeostasis and development of the metabolic syndrome as a state of dysregulation. Moreover, targeting of the coordinated regulatory system is a potential therapeutic strategy for several types of diabetes as well as the metabolic syndrome [13].

Acknowledgments We thank Prof. Y. Oka and all other members of Division of Molecular Metabolism and Diabetes, Tohoku University Graduate School of Medicine. We also thank Ms. I. Sato and K. Kawamura for technical support. This work was supported by the Uehara Memorial Foundation.

References

1. Katagiri H, Yamada T, Oka Y. Adiposity and cardiovascular disorders: disturbance of the regulatory system consisting of humoral and neuronal signals. Circ Res 2007;101(1):27–39.
2. Ishigaki Y, Katagiri H, Gao J, Yamada T, Imai J, Uno K, Hasegawa Y, Kaneko K, Ogihara T, Ishihara H, Sato Y, Takikawa K, Nishimichi N, Matsuda H, Sawamura T, Oka Y. Impact of plasma oxidized low-density lipoprotein removal on atherosclerosis. Circulation 2008;118 (1):75–83.
3. Friedman JM, Halaas JL. Leptin and the regulation of body weight in mammals. Nature (Lond) 1998;395(6704):763–70.
4. Gao J, Katagiri H, Ishigaki Y, Yamada T, Ogihara T, Imai J, Uno K, Hasegawa Y, Kanzaki M, Yamamoto TT, Ishibashi S, Oka Y. Involvement of apolipoprotein E in excess fat accumulation and insulin resistance. Diabetes 2007;56(1):24–33.
5. Ishigaki Y, Katagiri H, Yamada T, Ogihara T, Imai J, Uno K, Hasegawa Y, Gao J, Ishihara H, Shimosegawa T, Sakoda H, Asano T, Oka Y. Dissipating excess energy stored in the liver is a potential treatment strategy for diabetes associated with obesity. Diabetes 2005;54(2):322–32.
6. Yamada T, Katagiri H, Ishigaki Y, Ogihara T, Imai J, Uno K, Hasegawa Y, Gao J, Ishihara H, Niijima A, Mano H, Aburatani H, Asano T, Oka Y. Signals from intra-abdominal fat modulate insulin and leptin sensitivity through different mechanisms: neuronal involvement in food-intake regulation. Cell Metab 2006;3(3):223–9.
7. Rosen ED, Spiegelman BM. Adipocytes as regulators of energy balance and glucose homeostasis. Nature (Lond) 2006;444(7121):847–53.
8. Desvergne B, Michalik L, Wahli W. Transcriptional regulation of metabolism. Physiol Rev 2006;86(2):465–514.
9. Matsusue K, Haluzik M, Lambert G, Yim SH, Gavrilova O, Ward JM, Brewer B Jr, Reitman ML, Gonzalez FJ. Liver-specific disruption of PPAR-gamma in leptin-deficient mice improves fatty liver but aggravates diabetic phenotypes. J Clin Invest 2003;111(5):737–47.
10. Uno K, Katagiri H, Yamada T, Ishigaki Y, Ogihara T, Imai J, Hasegawa Y, Gao J, Kaneko K, Iwasaki H, Ishihara H, Sasano H, Inukai K, Mizuguchi H, Asano T, Shiota M, Nakazato M, Oka Y. Neuronal pathway from the liver modulates energy expenditure and systemic insulin sensitivity. Science 2006;312(5780):1656–9.
11. Imai J, Katagiri H,Yamada Y, Ishigaki Y, Suzuki T, Kudo H, Uno K, Hasegawa Y, Gao J, Kaneko K, Ishihara H, Niijima A, Nakazato M, Asano T, Minokoshi Y, Oka Y. (2008) Regulation of Pancreatic β cell Mass by Neuronal Signals from the Liver. Science 2008;322 (5905): 1250–4.
12. Imai J, Katagiri H, Yamada T, Ishigaki Y, Ogihara T, Uno K, Hasegawa Y, Gao J, Ishihara H, Sasano H, Oka Y. Cold exposure suppresses serum adiponectin levels through sympathetic nerve activation in mice. Obesity (Silver Spring) 2006;14(7):1132–41.
13. Yamada T, Oka Y, Katagiri H. Inter-organ metabolic communication involved in energy homeostasis: potential therapeutic targets for obesity and metabolic syndrome. Pharmacol Ther 2008;117(1):188–98.

Involvement of the CCR4-NOT Deadenylase Complex in the Control of Cell Growth

Masahiro Morita, Kentaro Ito, Toru Suzuki, and Tadashi Yamamoto

Introduction

The Tob/BTG family comprises six proteins, Tob, Tob2, ANA, BTG1, BTG2/PC3/TIS21, and PC3B, which share a common amino-terminal domain (1). All these proteins, when overexpressed, suppress growth of NIH3T3 cells (2). There is evidence that the Tob/BTG family of proteins is involved in regulation not only of cell growth but also of differentiation and development. For instance, BTG1 is thought to be involved in myogenesis induced by triiodothyronine (3). Analysis of tob-deficient mice revealed that Tob is involved in bone development (4). Although they lack DNA-binding domains, the Tob/BTG proteins are generally viewed as transcriptional cofactors. For example, Tob interacts with Smad in BMP2 signaling (4) and in T-cell anergy (5). BTG2 enhances Hoxb9-mediated transcription (6). Both Tob and BTG2 reduce cyclin D1 expression (7,8), possibly by recruiting histone deacetylase to the cyclin D1 promoter (9), contributing to G_0/G_1 arrest. On the other hand, Tob/Btg family proteins are also implicated in translational regulation by regulating the deadenylase activity or by interacting with the polyA binding proteins (10).

The CCR4-NOT complex is a large (>1 MDa) multi-subunit protein complex and is conserved from yeast to humans (11). The mammalian CCR4-NOT complex consists of 10 Cnot proteins: Cnot1-Cnot4, Ccr4a/Cnot6, Ccr4b/Cnot6L, Caf1/Cnot7, Pop2/Cnot8, Caf40/Cnot9, and Caf130/Cnot10 (Table 1). Yeast CCR4-NOT has been considered to be a global transcription complex that regulates a variety of genes such as the nonfermentative gene either positively or negatively (11). Accumulating evidence also shows that the mammalian Cnot proteins interact with proteins in the transcription machinery. For example, Cnot1 and Cnot7 are reported to interact with estrogen receptor-α (12,13). Cnot7 is also shown to interact with retinoid X receptor-β (14). Another report showed that Cnot9 is associated with retinoic acid receptor-α (15). Therefore, the mammalian CCR4-NOT complex appears to be involved in various transcription events controlled by the nuclear

M. Morita, K. Ito, T. Suzuki, and T. Yamamoto
Division of Oncology, The Institute of Medical Science, The University of Tokyo, 4-6-1
Shirokane-dai, Minatok-ku, Tokyo 108-8639

S. Nakanishi et al. (eds.), *Systems Biology*,
DOI:10.1007/978-4-431-87704-2_24, © Springer 2009

Table 1 Function of core subunits of the CCR4-NOT complex

Human		Yeast
CNOT1	Scaffold protein, interacts with Era	NOT1
CNOT2	Repressor for transcription?	NOT2
CNOT3	Helicase?	NOT3
CNOT4	Ubiquitin ligase	NOT4
		NOT5
CNOT6	Deadenylase	CCR4
CNOT6L	Deadenylase	
CNOT7	Deadenylase, associates with RXRb	Caf1
CNOT8	Deadenylase	
CNOT9	Associates with RAR	Caf40
CNOT10	?	Caf130

receptors. In addition the CCR4-NOT complex associates with components of general transcription factor TFIID complex (11). On the other hand, evidence is accumulating that the Cnot6 and Cnot7 encode major deadenylases, suggesting that this complex also plays a role in RNA degradation. Therefore, it appears that the CCR4-NOT complex has two functions in RNA metabolism: cytoplasmic function to regulate mRNA turnover and nuclear function to regulate mRNA synthesis. The mechanism by which these dual activities are coordinated remains largely elusive. It is worthy to mention that the CCR4-NOT complex is vitally important for the cells, because disruption of the complex results in induction of apoptosis (Ito et al., unpublished data).

Association of Tob with CCR4-NOT

Intracellular signaling is mediated by protein–protein interaction as well as multiprotein complexes, and mass spectrometric analysis is commonly used to identify components of protein complexes. We used a proteomics approach to identify Tob-interacting proteins.

To obtain sufficient quantities of Tob-interacting proteins, we produced Tob-Flag expressing HeLa S3 clones. Cells that stably expressed Tob-Flag grew more slowly than the original HeLa S3 cells, as was expected from the antiproliferative activity of Tob. After Tob-Flag-expressing HeLa S3 cells were cultured, their lysates were prepared and subjected to tandem-affinity purification of Tob-Flag with anti-Flag and anti-Tob monoclonal antibodies. The purified proteins were separated by gel electrophoresis and subjected to MALDI-TOF/TOF mass spectrometry analysis, which identified components of the CCR4-NOT complex, RNA helicases, RNA-binding proteins, and tankyrase-binding protein TAB182 as Tob-interacting proteins (16).

Because the yeast Not1 protein serves as a scaffold protein of the yeast CCR4-NOT complex (11), we assumed that human Cnot1 might also act as a scaffold. To substantiate the interaction between Tob and the CCR4-NOT complex, we examined whether Cnot1 interacts with Tob. Co-immunoprecipitation assays of the lysates of COS7 cells transfected with plasmid carrying Flag-Cnot1 and Myc-Tob together showed that Myc-Tob interacted with Flag-Cnot1 either directly or indirectly. We further showed that other components of the complex, Cnot3 and Cnot6L, interact with Tob by using the lysates of the cells transfected with plasmids carrying Tob-myc and Flag-Cnot3 or Flag-Cnot6L (16).

Biological Function of Deadenylase Subunits of the CCR4-NOT Complex

In eukaryotes, the cap structure and the poly(A) tail protect functional mRNA from degradation by exonucleases. In the general turnover pathways, mRNA degradation is initiated by shortening of the poly(A) tail. Most often, deadenylation induces decapping of the target mRNA through a poorly known mechanism. This action triggers the rapid destruction of the mRNA in P-bodies by the 5'-to-3' exonuclease Xrn1. Alternatively, deadenylation is followed by the exosome-mediated degradation of the mRNA in the 3'-to-5' direction (17,18). Apparently, deadenylation is a critical step for mRNA turnover, as poly(A) shortening removes the targeted mRNA from the translatable pool and initiates the mRNA degradation process. However, little is known about how poly(A) tail shortening is initiated and regulated. Accumulating evidence shows that CCR4-NOT is a major deadenylase that triggers mRNA decay (18). In mammals, two orthologues of Ccr4, Cnot6 and Cnot6L, and two orthologues of Pop2, Cnot7 and Cnot8, have been identified. However, the respective contributions of these factors to mRNA deadenylation remain unclear. The mechanism by which the CCR4-NOT complex is recruited to the poly(A) sequence at the 3'-end of mRNA is not known, either.

Analysis of a Component of the CCR4-NOT Complex: Subcellular Localization and Activity of Cnot6L

Previous reports indicated that yeast Ccr4p is involved in two cellular functions: transcriptional regulation in the nucleus and deadenylation of mRNAs in the cytoplasm (11,19). To assess the possible function of the mammalian homologue of Ccr4, we examined the subcellular localization of Cnot6L protein in NIH 3T3 cells transfected with expression plasmid encoding GFP-Cnot6L. Immunofluorescence analysis revealed that GFP-Cnot6L was present predominantly in the cytoplasm. The results were confirmed by immunoblotting the biochemically prepared subcellular fraction with anti-Cnot6L antibodies. The majority of the Cnot7 proteins are present in the cytoplasm (20), although Cnot7 could also cooperate with nuclear transcription factors (14).

Yeast CCR4p is a catalytic component of the cytoplasmic deadenylase complex, and its C-terminal region is highly homologous to apurinic endonucleases (9). Putative catalytic residues present in the C-terminal region of Ccr4p were conserved in Cnot6 and Cnot6L. We examined whether Cnot6L was associated with deadenylase activity by an in vitro deadenylase assay using single-stranded poly(A) RNA as a substrate. The GST-tagged protein purified from GST-tagged Cnot6L-transfected HEK293T cells cleaved the poly(A) RNA substrate from the 3'-end (21). Introduction of mutation at the active sites of nuclease/phosphatase family proteins in Cnot6L, which is Glu195Ala (E195A), Glu239Ala (E239A), Asp410Ala (D410A), Asp489Ala (D489A), or His529Ala (H529A) substitution, destroyed the deadenylase activity (21).

Suppression of Cnot6L Impairs the Growth of NIH 3T3 Cells

To clarify the biological significance of Cnot6L, we generated a retrovirus-based vector expressing Cnot6L shRNAs under the control of the U6 promoter. The retroviral vector also expressed the puromycin phosphotransferase gene, which allowed selection of transduced cells with puromycin. By examining NIH 3T3 cells infected with the viruses, we found that *Cnot6L* mRNA expression was almost completely inhibited by Cnot6L-specific shRNA. We also found that the proliferation rate of Cnot6L-depleted cells was significantly lower than that of control shRNA-treated cells. Severe reduction of proliferation of Cnot6L-depleted cells was also observed in colony formation experiments. Flow cytometric analysis showed that the proportion of G_0/G_1 phase cells (83%) was elevated and that of S-phase cells (5%) was reduced by Cnot6L depletion in comparison with control cells (G_0/G_1, 57%; S, 16%) (Fig. 1a). These data suggested that Cnot6L depletion inhibited progression of the cell cycle from G_0/G_1 to S phase (21).

To examine the significance of the Cnot6L deadenylase activity in cell growth, we prepared retrovirus vectors expressing RNAi-refractory wild-type Cnot6L (rCnot6L-WT) and mutant Cnot6L (rCnot6L-E239A) that has no deadenylase activity in vitro. The expression levels of rCnot6L-WT and rCnot6L-E239A were similar in immunoblotting experiments. Ectopic expression of rCnot6L restored the proliferation of Cnot6L-depleted cells, confirming that the growth defect was not the result of off-target effects by Cnot6L shRNA. Importantly, ectopic expression of rCnot6L-E239A did not rescue the growth defect caused by Cnot6L depletion. The profile of expression of cell cycle-related genes (described below) in rCnot6L-expressing cells was similar to that in control cells. On the other hand, in the rCnot6L-E239A-expressing cells, $p27^{Kip1}$ and cyclin D1 expression and the phosphorylation state of the RB protein did not return to a level similar to that in control cells. These data suggested that the deadenylase activity of Cnot6L played an important part in cell growth regulation (21).

Table 2 Genes expressed differentially between control and Cnot6L RNAi cells

	Gene	Characteristics	Fold change
Increased in Cnot6L-depleted cells	*p27*^{Kip}	CDK inhibitor	2.1
Decreased in Cnot6L-depleted cells	Nr2fl	Nuclear factor	−3.2
	Ddit3	Transcription	−2.2
	F2f5	Transcription	−4.9
	Lig3	DNA replication	−2.0
	Pml	Transcription	−2.5

Alteration of p27^{Kip1} mRNA Expression in Cnot6L-Depleted Cells

To investigate the molecular mechanism that underlies G_0/G_1 arrest in Cnot6L-depleted cells, we examined the expression of cell cycle-related proteins, including cyclins, CDKs, and CDK inhibitors, by immunoblotting (Fig. 1b). We found that expression of p27^{Kip1} was increased significantly by Cnot6L depletion, whereas expression of p21^{Cip1}, CDK2, and CDK4 in the Cnot6L-depleted cells was similar to that in control cells. Levels of cyclin D1 and cyclin A were decreased in Cnot6L-depleted cells. Consistent with these results, the unphosphorylated form of RB became dominant in Cnot6L-depleted cells. Thus, increased expression of p27^{Kip1} in Cnot6L-depleted cells would contribute to G_0/G_1 arrest of the cell-cycle progression (21).

We next compared the expression profiles of the cellular genes between control and Cnot6L-depleted NIH 3T3 cells by DNA microarray analysis. Among 1508 cell cycle-related genes described in the analysis, the level of *p27*^{Kip1} mRNA was prominently increased in Cnot6L-depleted cells (Table 2). We further confirmed that the *p27*^{Kip1} mRNA level was increased in the absence of Cnot6L, whereas the *p21*^{Cip1} mRNA level was not affected. By examining the half-life of *p27*^{Kip1} mRNA using actinomycin D-treated cells, we found that the rate of decline in *p27*^{Kip1} transcript levels was lower in Cnot6L-depleted cells than in control cells. The half-lives of *p27*^{Kip1} mRNA were 1.8 h and 1.3 h in Cnot6L-depleted and control cells, respectively. The possibility that the *p27*^{Kip1} mRNA level was increased by transcriptional activation was ruled out by examining the promoter activity of the *p27*^{Kip1} gene in the absence or presence of Cnot6L.

We then performed PAT assays to show that the poly(A) tail of *p27*^{Kip1} mRNA, but not that of *p21*^{Cip1} mRNA, was significantly longer in Cnot6L-depleted cells than that in control cells. These data suggested that Cnot6L functions as a deadenylase for the poly(A) tail of *p27*^{Kip1} mRNA in vivo (21).

Mechanism of Cnot6L Action

To investigate the mechanism by which *p27*^{Kip1} mRNA stability is affected by Cnot6L deadenylase, we performed UTR (untranslated region) luciferase assays. The 5'-UTR and 3'-UTR of *p27*^{Kip1} mRNA were inserted upstream and downstream of the luciferase

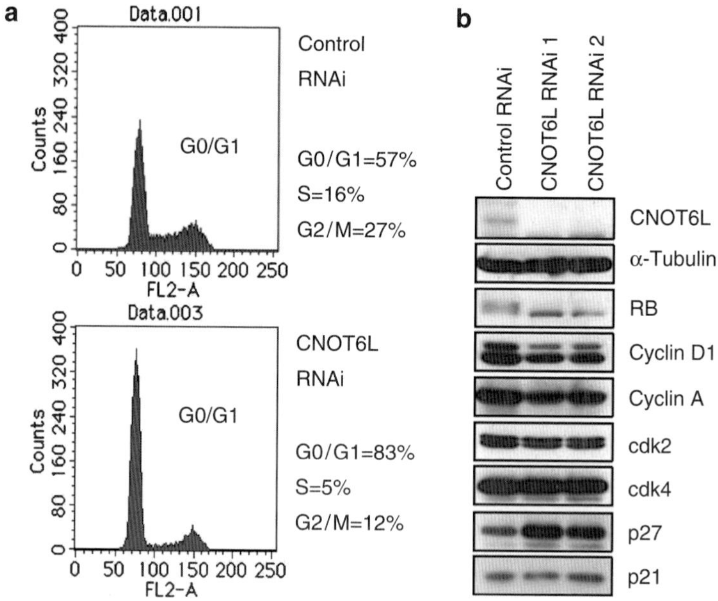

Fig. 1. Cnot6L RNAi inhibits the growth of NIH 3T3 cells. **a** The DNA contents of control and Cnot6L-depleted cells were determined by fluorescence-activated cell sorting. The percentages of cells in the G_0/G_1, S, and G_2/M phases of the cell cycle were quantified. **b** Expression levels of cell-cycle regulatory proteins were examined by Western blotting. The proteins were identified and quantified with anti-Cnot6L, anti-α-tubulin, anti-RB, anti-cyclin A, anti-cyclin D1, anti-cdk2, anti-cdk4, anti-$p27^{Kip1}$, and anti-$p21^{Cip1}$ antibodies. Detection of α-tubulin was performed as an internal control for protein loading

coding region, respectively. We transfected control or Cnot6L-depleted NIH 3T3 cells with luciferase constructs (control Luc, 5'-UTR plus Luc, and Luc plus 3'-UTR) and determined the luciferase activities in cell lysates. The results showed that the luciferase activity of 5'-UTR plus Luc was not significantly affected by the presence or absence of Cnot6L. In contrast, the luciferase activity of Luc plus 3'-UTR was decreased more than that of Luc only in the control cells. This result was most likely due to ARE (AU-rich element) sequences of $p27^{Kip1}$ mRNA, which would promote rapid deadenylation-dependent mRNA degradation (22). Importantly, in the Cnot6L-depleted cells, the luciferase activities of Luc plus 3'-UTR were not decreased. These data suggested that Cnot6L contributes to the deadenylation and degradation of $p27^{Kip1}$ mRNA through the 3'-UTR ARE-mediated mechanism in vivo.

Effects of Tob/Btg on Deadenylase Activity of CCR4-NOT

To address the biological significance of the interaction between Tob and the CCR4-NOT complex, we examined whether the deadenylase activity of the CCR4-NOT complex is affected by Tob using anti-Cnot6L immunoprecipitates. Cnot6L and associated proteins were purified from HEK293T cells transfected with GFP-Cnot6L expression plasmid with or without pME-Flag-Tob. The deadenylase assay with purified proteins showed that cleavage of the poly(A) RNA substrate occurred. Importantly, anti-GFP immunoprecipitates from the pME-Flag-Tob cotransfected cells contained the Flag-Tob protein and deadenylated the poly(A) substrates less efficiently. The data suggested that the in vitro deadenylase activity of the CCR4-NOT complex was negatively regulated by Tob (16). Consistently, Cnot7 deadenylase assays with wild-type BTG2 revealed the suppressive role of BTG2 in the regulation of Cnot7 deadenylase activity. X-ray crystallography-based structural analysis as well as co-immunoprecipitation analysis revealed that Cnot7 directly interact with Tob and BTG2. BTG2-mediated suppression was hampered by introducing mutation that disrupts the interaction with Cnot7 (Yang et al., unpublished data).

Discussion

We provide evidence suggesting that Tob cooperates with the CCR4-NOT complex to regulate the length of the poly(A) tail of mRNAs and thereby their translation. Each mRNA has a distinct degradation rate. Although the precise mechanism of deadenylation remains to be established, the deadenylation rate of each mRNA would influence the mRNA degradation rate. Sequence elements that are responsible, at least in part, for mRNA turnover by promoting deadenylation have been identified in the c-Fos, interleukin 2, and tumor necrosis factor-alpha genes. One of the best-studied and most prevalent elements is the ARE, which is found in the 3'-UTR of mRNAs (22). AREs in the interleukin 2 mRNA and tumor necrosis factor-alpha mRNA promote rapid deadenylation-dependent mRNA decay. We also show here that the ARE in the 3'-UTR of $p27^{Kip1}$ mRNA is involved in Cnot6L-mediated deadenylation. Probably Cnot6L and/or its associated molecules interact with the proteins binding to the ARE of $p27^{Kip1}$ mRNA. Because the 3'-poly(A) tail can enhance translation, deadenylation influences the efficiency of translation. In fact, we detected increment of $\mathrm{p27^{Kip1}}$ in the Cnot6L-depleted cells, which induced cell growth suppression. Thus, deadenylases are crucial for gene expression and therefore are likely to be important for many biological processes.

A recent report shows that GW182 protein that is required for P-body integrity interacts with the CCR4-NOT deadenylases (23). GW182 also interacts with the AGO (argonaut) family proteins, micro-RNA effectors (24). These observations raise the possibility that the CCR4-NOT deadenylase utilizes the micro-RNA system to recognize the target mRNAs. In addition, target specificity would be also determined by the interaction of the CCR4-NOT complex with various AU-rich binding proteins. We previously showed that Tob interacts with poly(A)-binding proteins PABP and iPABP and is involved in translational control (10). Ezzeddine et al. recently reported that Tob interacts with PABPC1 and Cnot6/Cnot7 simultaneously and regulates deadenylation positively (25). Tob may help recruit the CCR4-NOT complex to the vicinity of the poly(A) sequence of mRNAs by binding

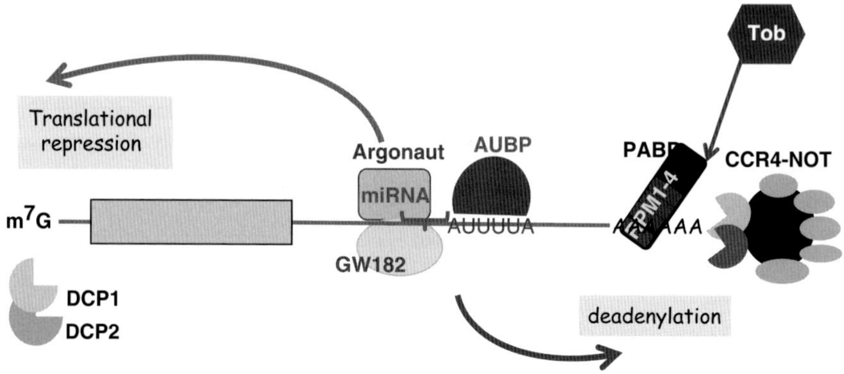

Fig. 2 Possible cooperation of micro-RNA and its partner Argonaut, ARE-binding protein, poly(A)-binding protrein, Tob, and the CCR4-NOT complex in the translational control. *miRNA*, micro-RNA; *AUBP*, AU-rich element-binding protein; *PABP*, poly(A)-binding protein

with the poly A-binding protein (26) (Fig. 2). The possible tripartite interaction of Tob, CCR4-NOT, and PABP/iPABP would destabilize target mRNAs, resulting in suppression of translation. However, the scenario could be more complicated because Tob is able to inhibit deadenylase activity in vitro (which leads to stabilization of mRNA and enhancement of translation). Further studies are needed to clarify how the deadenylase activity of the CCR4-NOT complex is regulated by Tob and/or PABP in vivo.

There are many unanswered questions. Is each deadenylase (Cnot6L, Cnot6, Cnot7, Cnot8) functionally redundant? Are the deadenylases cooperating to control mRNA stability? Does each deadenylase have preference in targeting mRNA? Do they have any activities other than the exonuclease activity? How does the CCR4-NOT complex differentially utilize the Tob family proteins? Further extensive studies are required to answer these interesting questions.

Acknowledgments Our work presented here were supported by grants from Uehara Memorial Foundation and the Ministry of Education, Science, Culture, Sports, and Technology, Japan.

References

1. Tirone F (2001) The gene *PC3* (*TIS21/BTG2*), prototype member of the *PC3/BTG TOB* family: regulator in control of cell growth, differentiation, and DNA repair? J Cell Physiol 187:155–165
2. Jia S, Meng A (2007) Tob genes in development and homeostasis. Dev Dyn 236:913–921
3. Rodier A et al. (1999) BTG1: a triiodothyronine target involved in the myogenic influence of the hormone. Exp Cell Res 249:337–348
4. Yoshida Y et al. (2000) Negative regulation of BMP/Smad signaling by Tob in osteoblasts. Cell 103:1085–1097

5. Tzachanis D et al. (2001) Tob is a negative regulator of activation that is expressed in anergic and quiescent T cells. Nat Immunol 2:1174–1182
6. Prevot D et al. (2000) The leukemia-associated protein Btg1 and the p53-regulated protein Btg2 interact with the homeoprotein Hoxb9 and enhance its transcriptional activation. J Biol Chem 275:147–153
7. Guardavaccaro D et al. (2000) Arrest of G(1)-S progression by the p53-inducible gene PC3 is Rb dependent and relies on the inhibition of cyclin D1 transcription. Mol Cell Biol 20:1797–1815
8. Suzuki T et al. (2002) Phosphorylation of three regulatory serines of Tob by Erk1 and Erk2 is required for Ras-mediated cell proliferation and transformation. Genes Dev 16:1356–1370
9. Yoshida Y et al. (2003) Mice lacking a transcriptional corepressor Tob are predisposed to cancer. Genes Dev 17:1201–1206
10. Okochi K et al. (2005) Interaction of anti-proliferative protein Tob with poly(A)-binding protein and inducible poly(A)-binding protein: implication of Tob in translational control. Genes Cells 10:151–163
11. Collart MA, Timmers HT (2004) The eukaryotic Ccr4-not complex: a regulatory platform integrating mRNA metabolism with cellular signaling pathways? Prog Nucleic Acid Res Mol Biol 77:289–322
12. Winkler GS et al. (2006) Human Ccr4-Not complex is a ligand-dependent repressor of nuclear receptor-mediated transcription. EMBO J 25:3089–3099
13. Prevot D et al. (2001) Relationships of the antiproliferative proteins BTG1 and BTG2 with CAF1, the human homolog of a component of the yeast CCR4 transcriptional complex: involvement in estrogen receptor alpha signaling pathway. J Biol Chem 276:9640–9648
14. Nakamura T et al. (2004) Oligo-astheno-teratozoospermia in mice lacking Cnot7, a regulator of retinoid X receptor β. Nat Genet 36:528–533
15. Hiroi N et al. (2002) Mammalian Rcd1 is a novel transcriptional cofactor that mediates retinoic acid-induced cell differentiation. EMBO J 21:5235–5244
16. Miyasaka et al. (2008) Interaction of antiproliferative protein Tob with the CCR4-NOT deadenylase complex. Cancer Sci 99:755–761
17. Meyer S et al. (2004) Messenger RNA turnover in eukaryotes: pathways and enzymes. Crit Rev Biochem Mol Biol 39:197–216
18. Garneau NL et al. (2007) The highways and byways of mRNA decay. Nat Rev Mol Cell Biol 8:113–126
19. Denis CL, Chen J (2003) The CCR4-NOT complex plays diverse roles in mRNA metabolism. Prog Nucleic Acids Res Mol Biol 73:221–250
20. Ikematsu N et al. (1999) Tob2, a novel anti-proliferative Tob/BTG1 family member, associates with a component of the CCR4 transcriptional regulatory complex capable of binding cyclin-dependent kinases. Oncogene 18:7432–7441
21. Morita M et al. (2007) Depletion of mammalian CCR4b deadenylasetriggers increment of the *p27*^{Kip1} mRNA level and impaires cell growth. Mol Cell Biol 27:4980–4990
22. Chen CYA, and Shyu AB (1995) AU-rich elements: characterization and importance in mRNA degradation. Trends Biochem Sci 20:465–470
23. Behm-Ansmant I et al. (2006) mRNA degradation by microRNAs and GW182 requires both CCR4:NOT deadenylase and DCP1:DCP2 decapping complexes. Genes Dev 20:1885–1898
24. Ding L, Han M (2007) GW182 family proteins are crucial for microRNA-mediated gene silencing. Trends Cell Biol 17:411–416
25. Ezzeddine N et al. (2007) Human TOB, an antiproliferative transcription factor, is a poly(A)-binding protein-dependent positive regulator of cytoplasmic mRNA deadenylation. Mol Cell Biol 27:7791–7801
26. Funakoshi Y et al. (2007) Mechanism of mRNA deadenylation: evidence for a molecular interplay between translation termination factor eRF3 and mRNA deadenylases. Genes Dev 23:3135–3148

Reconstruction of Nuclear Reprogramming by Defined Factors

Shinya Yamanaka

Introduction

Embryonic stem (ES) cells are pluripotent stem cells derived from the inner cell mass of blastocyst. The generation of mouse ES cells (Evans et al., 1981; Martin, 1981) resulted in the "knockout mouse" technology (Hooper et al., 1987; Doetschman et al., 1987). The generation of ES cell lines from human blastocyst-stage embryos (Thomson et al., 1998) has opened up the possibility of using these cells in cell transplantation therapies for juvenile diabetes, Parkinson's disease, heart failure, and spinal cord injury. The usage of human embryos and tissue rejection, however, remain a concern for human ES cells. One strategy to overcome such issues is to reprogram the nuclei of differentiated cells to an ES cell-like, pluripotent state.

Reprogramming by Fusion with ES Cells

Miller and Ruddle demonstrated that thymocytes acquired pluripotency upon fusion with embryonal carcinoma (EC) cells (Miller and Ruddle, 1976), and similar results were later obtained by electrofusion with embryonic germ (EG) cells (Tada et al., 1997) and mouse ES cells (Tada et al., 2001). Transplantation of these cells into nude mice resulted in the formation of teratomas consisting of various tissues from all three germ layers, confirming the pluripotency of these cells. Reprogramming by fusion with human ES cells was also reported in 2005 (Cowan et al., 2005; Yu et al., 2006). These results suggest that ES cells contain factors that induce pluripotency in somatic cells (PIF, pluripotency-inducing factor).

S. Yamanaka
Center for iPS Cell Research & Application (CiRA) / Institute for Frontier Medical Sciences, Kyoto University, Kyoto 606-8507, Japan

S. Nakanishi et al. (eds.), *Systems Biology*,
DOI:10.1007/978-4-431-87704-2_25,

Generation of Induced Pluripotent Stem (iPS) Cells

Fusion-induced reprogramming showed that ES cells contain PIF. These factors also play are likely play important roles in the maintenance of pluripotency in ES cells. Based on this hypothesis, we selected three groups of PIF candidates (Takahashi and Yamanaka, 2006). The first group consists of transcription factors specifically expressed in ES cells, including Nanog, Oct-3/4, Sox2, UTF1, Sall4, Sox15, and Rex1. The second group consists of growth- and tumor-related genes that play roles in ES cells, including c-Myc, Stat3, β-catenin, Grb2, KLF4, TCL1, and ERas. The third group consists of genes that are also specifically expressed in ES cells, but with less defined functions, including ECAT1, ESG1, Fbx15, DNMT3L, ECAT8, GDF3, ECAT15-1, ECAT15-2, Fthl17, and Stella.

To test these candidates, we developed a system in which the induction of pluripotency can be detected by marker gene expression. This system utilized Fbx15, which is specifically expressed in ES cells and early embryos but is dispensable for self-renewal of ES cells (Tokuzawa et al., 2003). The βgeo cassette (a fusion of β-galactocidase and the neomycin-resistant gene) was knocked into the mouse Fbx15 genes by homologous recombination. ES cells homozygous for βgeo knock-in ($Fbx15^{\beta geo/\beta geo}$) were resistant to a high concentration of G418 (up to 12 mg/ml), whereas somatic cells derived from $Fbx15^{\beta geo/\beta geo}$ mice were sensitive to the selection. Even a partial induction of pluripotency would make somatic cells resistant to a normal concentration of G418 (0.3 mg/ml).

We introduced the PIF candidate into $Fbx15^{\beta geo/\beta geo}$ mouse embryonic fibroblasts (MEF) by retrovirus-mediated transfection. No G418-resistance colonies were obtained with any single factor. However, by combining four factors (Oct3/4, Sox2, c-Myc, and Klf4), multiple G418-resistant colonies were obtained (Fig. 1). These cells showed morphology and proliferation similar to ES cells. Furthermore, when transplanted into nude mice, these ES-like cells produced teratomas containing various tissues of the three germ layers. We designated these cells iPS cells for induced pluripotent stem cells (Takahashi and Yamanaka, 2006).

Fbx15-selected iPS cells, however, were not able to contribute to adult chimeric mice when transplanted into blastocysts. DNA microarray analyses revealed significant difference in global gene expression patterns between Fbx15-selected iPS cells ES cells. In addition, the promoter regions of Oct-3/4 and Nanog remained methylated in Fbx15-selected iPS cells. These data demonstrated that reprogramming in Fbx15-selected iPS cells was not complete.

To obtain better iPS cells, three groups utilized Nanog as a selection maker (Wernig et al., 2007; Okita et al., 2007; Maherali et al., 2007). We expected that Nanog would serve as a better marker because it is more tightly associated with pluripotency than is Fbx15 (Pan and Thomson, 2007). All the three groups showed that Nanog-selected iPS cells are nearly indistinguishable from ES cells in gene

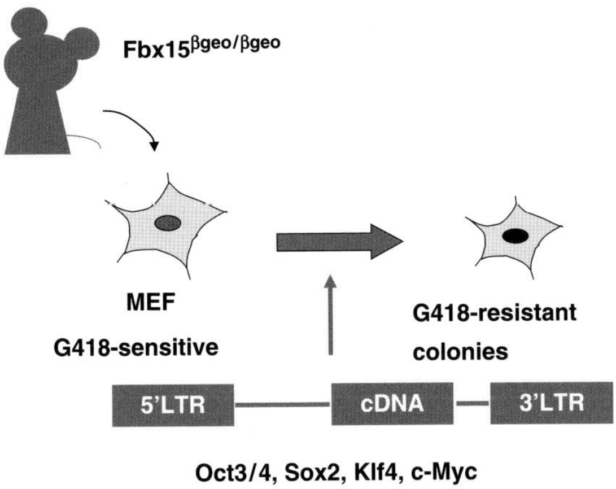

Fig. 1 The induction of iPS cells by Oct-3/4, Sox2, c-Myc, and KLF4. Retroviral transfection of *Oct-3/4, Sox2, c-Myc, and KLF4* into mouse fibroblasts results in the induction of pluripotent iPS cells

expression and epigenetic status. Furthermore, Nanog-selected iPS cells contributed to adult and germline-competent chimeras.

How Are iPS Cells Generated by the Four Factors?

ES cells are similar to tumor cells in that they are immortal and proliferate rapidly. Thus, generation of iPS cells from somatic cells should also require transformation, which might be achieved by the two tumor-related gene products, c-Myc and KLF4. The Myc protein induces transformation (Adhikary and Eilars, 2005). However, it also elicits p53-dependent apoptosis. KLF4 might be required to suppress p53- and c-Myc-induced apoptosis (Rowland and Peeper, 2006). KLF4, in contrast, activates p21 and suppresses proliferation. The Myc protein can alleviate this cytostatic effect of KLF4 by suppressing *p21*. The balance between c-Myc and KLF4 might be critical for transformation in iPS cells. In addition, Myc proteins probably loosen chromatins of somatic cells by binding to numerous sites throughout the genome and by recruiting multiple histone acetylases complexes (Knoepfler et al., 2006). Oct-3/4 and Sox2 are likely to direct cell fate from tumor cells to pluripotent cells during iPS cell generation. KLF4 may also function as a co-factor of Oct-3/4 and Sox2 (Nakatake et al., 2006).

The four factors, however, cannot explain every aspect of iPS cell induction. Among fibroblasts that express all the four transgenes, only a small portion (less than 1%) can become iPS cells. Future studies are required to determine the molecular mechanism for iPS cell generation by the four factors.

References

Adhikary S, Eilers M (2005) Transcriptional regulation and transformation by Myc proteins. Nat Rev Mol Cell Biol 6:635–645.

Cowan CA, Atienza J, Melton DA, Eggan K (2005) Nuclear reprogramming of somatic cells after fusion with human embryonic stem cells. Science 309:1369–1373.

Doetschman T, Gregg RG, Maeda N, Hooper ML, Melton DW, Thompson S, Smithies O (1987) Targetted correction of a mutant HPRT gene in mouse embryonic stem cells. Nature (Lond) 330:576–578.

Evans MJ, Kaufman MH (1981) Establishment in culture of pluripotential cells from mouse embryos. Nature (Lond) 292:154–156.

Hooper M, Hardy K, Handyside A, Hunter S, Monk M (1987) HPRT-deficient (Lesch-Nyhan) mouse embryos derived from germline colonization by cultured cells. Nature (Lond) 326:292–295.

Knoepfler PS, Zhang XY, Cheng PF, Gafken PR, McMahon SB, Eisenman RN (2006) Myc influences global chromatin structure. EMBO J 25:2723–2734.

Maherali N, Sridharan R, Xie W, Utikal J, Eminli S, Arnold K, Stadtfeld M, Yachechko R, Jaenisch JT, Plath K, Hochedlinger K (2007) Directly reprogrammed fibroblasts show global epigenetic remodelling and widespread tissue contribution. Cell Stem Cell 1:55–70.

Martin GR (1981) Isolation of a pluripotent cell line from early mouse embryos cultured in medium conditioned by teratocarcinoma stem cells. Proc Natl Acad Sci U S A 78: 7634–7638.

Miller RA, Ruddle FH (1976) Pluripotent teratocarcinoma-thymus somatic cell hybrids. Cell 9:45–55.

Nakatake Y, Fukui N, Iwamatsu Y, Masui S, Takahashi K, Yagi R, Yagi K, Miyazaki J, Matoba R, Ko MS, Niwa H (2006) Klf4 cooperates with Oct3/4 and Sox2 to activate the Lefty1 core promoter in embryonic stem cells. Mol Cell Biol 26:7772–7782.

Okita K, Ichisaka T, Yamanaka S (2007) Generation of germ-line competent induced pluripotent stem cells. Nature (Lond) 448:313–317.

Pan G, Thomson JA (2007) Nanog and transcriptional networks in embryonic stem cell pluripotency. Cell Res 17:42–49.

Rowland BD, Peeper DS (2006) KLF4, p21 and context-dependent opposing forces in cancer. Nat Rev Cancer 6:11–23.

Tada M, Tada T, Lefebvre L, Barton SC, Surani MA (1997) Embryonic germ cells induce epigenetic reprogramming of somatic nucleus in hybrid cells. EMBO J 16:6510–6520.

Tada M, Takahama Y, Abe K, Nakatsuji N, Tada T (2001) Nuclear reprogramming of somatic cells by in vitro hybridization with ES cells. Curr Biol 11:1553–1558.

Takahashi K, Yamanaka S (2006) Induction of pluripotent stem cells from mouse embryonic and adult fibroblast cultures by defined factors. Cell 126:663–676.

Thomson JA, Itskovitz-Eldor J, Shapiro SS, Waknitz MA, Swiergiel JJ, Marshall VS, Jones JM (1998) Embryonic stem cell lines derived from human blastocysts. Science 282:1145–1147.

Tokuzawa Y, Kaiho E, Maruyama M, Takahashi K, Mitsui K, Maeda M, Niwa H, Yamanaka S (2003) Fbx15 is a novel target of Oct3/4 but is dispensable for embryonic stem cell self-renewal and mouse development. Mol Cell Biol 23:2699–2708.

Wernig M, Meissner A, Foreman R, Brambrink T, Ku M, Hochedlinger K, Bernstein BE, Jaenisch R (2007) In vitro reprogramming of fibroblasts into a pluripotent ES cell-like state. Nature (Lond) 448:318–324.

Yu J, Vodyanik MA, He P, Slukvin, II, Thomson JA (2006) Human embryonic stem cells reprogram myeloid precursors following cell–cell fusion. Stem Cells 24:168–176.

Index